Sixteenth International Seaweed Symposium

Developments in Hydrobiology 137

Series editor
H. J. Dumont

Sixteenth International Seaweed Symposium

Proceedings of the Sixteenth International Seaweed Symposium,
held in Cebu City, Philippines, 12–17 April 1998

Edited by

Joanna M. Kain (Jones), Murray T. Brown and Marc Lahaye

*Reprinted from Hydrobiologia, volumes 398/399 (1999)
and the Journal of Applied Phycology, volume 11, no. 1 (1999)*

Kluwer Academic Publishers
Dordrecht / Boston / London

Library of Congress Cataloging-in-Publication Data

A C.I.P. Catalogue record for this book is available from the Library of Congress.

ISBN 0-7923-5932-I

Published by Kluwer Academic Publishers,
P.O. Box 17, 3300 AA Dordrecht, The Netherlands

Sold and distributed in North, Central and South America
by Kluwer Academic Publishers,
101 Philip Drive, Norwell, MA 02061, U.S.A.

In all other countries, sold and distributed
by Kluwer Academic Publishers Group,
P.O. Box 322, 3300 AH Dordrecht, The Netherlands

Printed on acid-free paper

All Rights reserved
© 1999 Kluwer Academic Publishers

No part of the material protected by this copyright notice may be reproduced
or utilized in any form or by any means, electronic or mechanical,
including photocopying, recording or by any information storage and
retrieval system, without written permission from the copyright owner.

Printed in the Netherlands

Dedication

(1919–1996)

This volume of the Proceedings of the XVIth International Seaweed Symposium is dedicated to Maxwell S. Doty in recognition of his outstanding contributions, both to the International Seaweed Association in furthering its objectives, and to the successful establishment of seaweed farming, particularly in the Far East. Max Doty was elected to the Executive of the ISA in 1971, at the time of the ISS in Sapporo, and he was the first person from the United States to hold a position on the Executive Council. Max served as President of the ISA for two terms and upon retirement he was offered Honorary Life Membership in the ISA. Max was an outstanding scientist, a gifted and visionary seaweed farmer and a sincere friend to all those who were privileged to know him. Max is truly missed by members of the International Seaweed Association and throughout the international community.

Appreciation

Maxwell S. Doty died on April 8, 1996, just hours prior to receiving a telephone call from China which was to inform him the last photo on Chinese *Porphyra* cultivation had been captured for his book on commercial marine algae. It is only appropriate that the Proceedings of the International Seaweed Symposium, held in Cebu, Philippines, be dedicated to his memory, energy and passion to his craft which he maintained for nearly sixty years until the day of his passing.

A review of Max Doty's life and professional career was eloquently penned by Gerry Kraft, Phycologia (1997) 36: 82–90. The larger question remains, how will history define Max Doty's contribution to science? His phycological legacy goes far beyond his significant research interests of algal taxonomy, productivity, distribution, intertidal stratification, phycocolloids and marine agronomy. What is the definition of a great teacher or scientist? Is it his publication record of nearly two hundred efforts? Is it the development of topic defining hypotheses, 'critical tide factors' or the development of 'clod cards'? Is it the effect the development of marine agronomy had on the general public? Just ask the farmers and fishermen of the Philippines, Indonesia and Pacific island nations. Or is it the time, energy and passion offered to the graduate students one leaves behind?

Selfishly, as Max Doty's final doctoral candidate, I offer this dedication on behalf of the 20 masters and 13 doctoral students he supervised. His students, practising on nearly all the earth's continents, understand the debt we all owe to Max. The world he occupied, 'the sixth floor' of the St. John's Botany building at the University of Hawaii at Manoa, was for a critical period of our lives, the centre of our universe.

'For the splendid record in Botany 684', a simple inscription placed into a book from his personal library and offered to one student each semester representing excellence during Doty's premier graduate candidate training device – Monday afternoon seminar. Regardless of the semester's workload these two credits dominated each student's efforts. Participants emerged from this pressure grinder, humbled but self-assured, armed with the ability to develop cogent thoughts and responses before crowds of 2 to 2000 with clarity and confidence.

Max Doty's office was a shrine to organized chaos. The most coveted position was the single chair in his inner office. Max Doty would spend hours with students in private discussions. He took delight in probing and challenging each student's research project and effort. His insight and enthusiasm was contagious. It was in these private sessions that students could be overwhelmed by his passion for the development of phycology and marine agronomy. As an 'old school' mentor, he forced his students to be both independent and creative in their scientific development. Doty demanded each student be the world's expert in their chosen topic prior to initiating their first experiment, often resulting in gargantuan literature reviews. Even his insistence of mastering two foreign languages, limited to those dialects in which the classic literature was published, forced his students into a greater understanding of their developing phycological heritage.

A mentor can only lead by example in which his work ethic was legendary. One was always amazed by his insistence on returning to the sixth floor or the field so quickly after an illness or injury. Doty was back in the lab within three weeks of open-heart surgery, this after conducting business from the ICU. His profession was an obsession, making phycology both his carrier as well as his hobby. Many of his students maintain his obsession in one form or another. For that the record is clear. Dr William Brostoff (Doty, Ph.D. 1988) liked to record phycological lineages. I know the students he left behind are proud of theirs and respect the debt we all owe to one man, Dr Maxwell Stanford Doty.

Ira A. Levine
PhycoGen, Inc.
Portland, Maine 04101

Contents

Dedication	V
Appreciation	VII
International Seaweed Association Executive Council	XV
National Organizing Committee and National Sponsors	XVI
Acknowledgments	XVII–XVIII
Reviewers	XIX
Marinalg International Awards	XXI–XXII
Student Paper Awards	XXII
Opening Ceremony and Plenary Session	XXIII
The seaweed industry and R & D: a solid partnership	
by W.G. Padolina	XXV–XXVI
List of Registrants	XXVII–L

1. Plenary Presentations

Diversity of the seaweed flora of the Philippines and its utilization
    by G.C. Trono, Jr. ... 1–6
Red algal polysaccharide industry: economics and research status at the turn of the century
    by D.F. Kapraun ... 7–14
A conceptual framework for marine agronomy
    by B. Santelices ... 15–23

2. Systematics and Floristics

Observations on the phylogenetic systematics and biogeography of the Solieriaceae (Gigartinales, Rhodophyta) inferred from *rbc*L sequences and morphological evidence
    by S. Fredericq, D.W. Freshwater & M.H. Hommersand ... 25–38
Development of the extant diversity in *Halimeda* is linked to vicariant events
    by W.H.C.F. Kooistra, M. Calderón & L.W. Hillis ... 39–45
Phylogeny of Alariaceae (Phaeophyta) with special reference to *Undaria* based on sequences of the RuBisCo spacer region
    by H.S. Yoon & S.M. Boo ... 47–55
A new agarophyte, *Curdiea balthazar* sp. nov. (Gracilariales, Rhodophyta), from the Three Kings Islands, northern New Zealand
    by W.A. Nelson, G.A. Knight & R. Falshaw ... 57–63
Marine benthic algae of North East Herald Cay, Coral Sea, South Pacific
    by A.J.K. Millar ... 65–74
Changes of the benthic algal flora of the Tremiti Islands (southern Adriatic) Italy
    by M. Cormaci & G. Furnari ... 75–79

3. Life History

Cell–cell recognition during fertilization in the red alga, *Aglaothamnion oosumiense* (Ceramiaceae, Rhodophyta)

    by S.-H. Kim & G.H. Kim 81–89

Morphology and reproduction of the adventive Mediterranean rhodophyte *Polysiphonia setacea*

    by F. Rindi, M.D. Guiry & F. Cinelli 91–100

Long-term and diurnal carpospore discharge patterns in the Ceramiaceae, Rhodomelaceae and Delesseriaceae (Rhodophyta)

    by J.A. West & D.L. McBride 101–113

Porphyra sp. (Bangiales, Rhodophyta): reproduction and life form

    by A. Candia, S. Lindstrom & E. Reyes 115–119

Life history, in culture, of the obligate epiphyte *Porphyra moriensis* (Bangiales, Rhodophyta)

    by M. Notoya & A. Miyashita 121–125

Culture studies of *Porphyra dentata* and *P. pseudolinearis* (Bangiales, Rhodophyta), two dioecious species from Korea

    by N.-G. Kim 127–135

4. Phenology and Population Studies

Biological basis for the management of *Gigartina skottsbergii* (Gigartinales, Rhodophyta) in southern Chile

    by R. Westermeier, A. Aguilar, J. Sigel, J. Quintanilla & J. Morales 137–147

Reproductive biology of *Gigartina skottsbergii* (Gigartinaceae, Rhodophyta) from Chile

    by M. Avila, A. Candia, M. Núñez & H. Romo 149–157

Population structure and reproduction of the carrageenophyte *Chondracanthus pectinatus* in the Gulf of California

    by I. Pacheco-Ruíz & J.A. Zertuche-González 159–165

Seasonal variations of growth and agar composition of *Gracilaria multipartita* harvested along the Atlantic coast of Morocco

    by T. Givernaud, A. El Gourji, A. Mouradi-Givernaud, Y. Lemoine & N. Chiadmi . 167–172

Biomass and agar assessment of three species of *Gracilaria* from Negros Island, central Philippines

    by H.P. Calumpong, A. Maypa, M. Magbanua & P. Suarez 173–182

The effects of a simulated harvest on *Porphyra* (Bangiales, Rhodophyta) in South Africa

    by N.J. Griffin, J.J. Bolton & R.J. Anderson 183–189

Effects of seasonal growth rate on morphological variation of *Undaria pinnatifida* (Alariaceae, Phaeophyceae)

    by M.D. Stuart, C.L. Hurd & M.T. Brown 191–199

Phenology of *Sargassum* spp. (Sargassaceae, Phaeophyta) from Reunion Rocks, KwaZulu-Natal, South Africa

    by R.D. Gillespie & A.T. Critchley 201–210

Population and alginate yield and quality assessment of four *Sargassum* species in Negros Island, central Philippines

    by H.P. Calumpong, A.P. Maypa, & M. Magbanua 211–215

The ecological effects of mining discharges on subtidal habitats dominated by macroalgae in northern Chile: population and community level studies
    by J.A. Vásquez, J.M.A. Vega, B. Matsuhiro & C. Urzúa 217–229

5. Ecology

Obtaining absorption spectra from individual macroalgal spores using microphotometry
    by M.H. Graham & B.G. Mitchell 231–239
Evaluating substances that facilitate algal spore adhesion
    by B. Santelices & D. Aedo 241–246
Preliminary observations on the development of kelp gametophytes endophytic in red algae
    by D.J. Garbary, K.Y. Kim, T. Klinger & D. Duggins 247–252
Using *in situ* substratum sterilization and fluorescence microscopy in studies of microscopic stages of marine macroalgae
    by M.S. Edwards 253–259
The sea star *Asterina pectinifera* causes deep-layer sloughing in *Lithophyllum yessoense* (Corallinales, Rhodophyta)
    by D. Fujita 261–266
Spatial variability in secondary metabolite production by the tropical red alga *Portieria hornemannii*
    by D.B. Matlock, D.W. Ginsburg & V.J. Paul 267–273
Influence of *Ecklonia radiata* kelp canopy on structure of macro-algal assemblages in Marmion Lagoon, Western Australia
    by G.A. Kendrick, P.S. Lavery & J.C. Phillips 275–283

6. Responses to Environmental Factors

Factors affecting sporulation of *Gracilaria cornea* (Gracilariales, Rhodophyta) carposporophytes from Yucatan, Mexico
    by A. Guzmán-Urióstegui & D. Robledo 285–290
Porphyra cultivation in Alaska: conchocelis growth of three indigenous species
    by M.S. Stekoll, R. Lin & S.C. Lindstrom 291–297
Studies on the growth, in culture, of two forms of *Porphyra lacerata* from Japan
    by M. Notoya & K. Nagaura 299–303
Effect of tissue nitrogen and phosphorus quota on growth of *Porphyra yezoensis* blades in suspension cultures.
    by J.T. Hafting 305–314
Effects of nitrogen source, N:P ratio and N-pulse concentration and frequency on the growth of *Gracilaria cornea* (Gracilariales, Rhodophyta) in culture
    by L. Navarro-Angulo & D. Robledo 315–320
Photosynthetic and respiratory responses of the agarophyte *Gelidiella acerosa* collected from tidepool, intertidal and subtidal habitats
    by E.T. Ganzon-Fortes 321–328
The effect of light on growth and agar content of *Gelidium pulchellum* (Gelidiaceae, Rhodophyta) in culture
    by I. Sousa-Pinto, E. Murano, S. Coelho, A. Felga & R. Pereira 329–338

Effects of environmental factors and plant growth regulators on growth of the red alga *Gracilaria vermiculophylla* from Shikoku Island, Japan
    by N.S. Yokoya, H. Kakita, H. Obika & T. Kitamura 339–347
Carbon acquisition strategies of the red alga *Eucheuma denticulatum*
    by M. Granbom & M. Pedersén 349–354
Relationship of CO_2 concentrations to photosynthesis of intertidal macroalgae during emersion
    by K. Gao, Y. Ji & Y. Aruga 355–359
A theoretical analysis and field evaluation of a light and temperature model of production by *Ecklonia cava*
    by M. Honda 361–374
Effects of copper pollution on the ultrastructure of *Lessonia* spp.
    by P.I. Leonardi & J.A. Vasquez 375–383

7. Chemistry

Further evaluation of the structure of the polysaccharide from *Plocamium costatum* with the use of set theory
    by I.J. Miller 385–389
Biology and agar composition of *Gelidium sesquipedale* harvested along the Atlantic coast of Morocco
    by A. Mouradi-Givernaud, L.A. Hassani, T. Givernaud, Y. Lemoine & O. Benharbet   391–395
A comparative analysis of agarans from commercial species of *Gracilaria* (Gracilariales, Rhodophyta) grown *in vitro*
    by J. Macchiavello, R. Saito, G. Garófalo & E.C. Oliveira 397–400
Pyruvated carrageenans from *Solieria robusta* and its adelphoparasite *Tikvahiella candida*
    by A. Chiovitti, A. Bacic, G.T. Kraft, D.J. Craik & M.-L. Liao 401–409
Monthly changes in the content of fucans, their constituent sugars and sulphate in cultured *Laminaria japonica*
    by M. Honya, H. Mori, M. Anzai, Y. Araki & K. Nisizawa 411–416

8. Cultivation

Open-water aquaculture of the red alga *Chondrus crispus* in Prince Edward Island, Canada
    by T. Chopin, G. Sharp, E. Belyea, R. Semple & D. Jones 417–425
Recent advances in the understanding of the biological basis for *Gigartina skottsbergii* (Rhodophyta) cultivation in Chile
    by A.H. Buschmann, J.A. Correa & R. Westermeier 427–434
Economic feasibility of *Sarcothalia* (Gigartinales, Rhodophyta) cultivation
    by M. Avila, E. Ask, B. Rudolph, M. Nuñez & R. Norambuena 435–442
Comparison of the performance of the agarophyte, *Gracilariopsis bailinae*, and the milkfish, *Chanos chanos*, in mono- and biculture
    by L.B. Alcantara, H.P. Calumpong, M.R. Martinez-Goss, E.G. Meñez & A. Israel . 443–453
Upwelling and fish-factory waste as nitrogen sources for suspended cultivation of *Gracilaria gracilis* in Saldanha Bay, South Africa
    by R.J. Anderson, A.J. Smit & G.J. Levitt 455–462

Outplanting of laboratory-generated carposporelings of *Gracilariopsis bailinae* off northern Philippines

    by S.F. Rabanal & R.V. Azanza 463–468

Review of genetic engineering of *Laminaria japonica* (Laminariales, Phaeophyta) in China

    by S. Qin, G.-Q. Sun, P. Jiang, L.-H. Zou, Y. Wu & C.-K. Tseng 469–472

A new method of *Laminaria japonica* strain selection and sporeling raising by the use of gametophyte clones

    by D. Li, Z.-G. Zhou, H. Liu & C. Wu 473–476

9. Exploitation

Development of commercial *Kappaphycus* production in the Line Islands, Central Pacific

    by D.M. Luxton & P.M. Luxton 477–486

The status of commercial algal utilization in New Zealand

    by W.L. Zemke-White, G. Bremner & C.L. Hurd 487–494

The optimal utilization of kelp resources in the southern Cape area of South Africa

    by K.W.G. Rotmann 495–499

Gelidium robustum agar: quality characteristics from exploited beds and seasonality from an unexploited bed at Southern Baja California, México

    by Y. Freile-Pelegrín, D. Robledo & E. Serviere-Zaragoza 501–507

Gracilariopsis lemaneiformis beds along the west coast of the Gulf of California, Mexico

    by I. Pacheco-Ruíz, J.A. Zertuche-González, F. Correa-Díaz, F. Arellano-Carbajal & A. Chee-Barragán 509–514

The following papers (pages 515–650) are reprinted from the *Journal of Applied Phycology*, volume 11, no. 1, pages 1–136 (1999).

Guest Editors: Murray T. Brown and Marc Lahaye

10. Chemistry and Chemical Composition

Chemical composition and ^{13}C NMR spectroscopic characterisation of ulvans from *Ulva* (Ulvales, Chlorophyta)

    by M. Lahaye, E. Alvarez-Cabal Cimadevilla, R. Kuhlenkamp, B. Quemener, V. Lognoné & P. Dion 515–521

Direct determination of alginate content in brown algae by near infra-red (NIR) spectroscopy

    by S.J. Horn, E. Moen & K. Østgaard 523–527

Oleic acid is the main fatty acid related with carotenogenesis in *Dunaliella salina*

    by H. Mendoza, A. Martel, M. Jiménez del Río & G. García Reina 529–533

Alginate stability during high salt preservation of *Ascophyllum nodosum*

    by E. Moen, B. Larsen, K. Østgaard & A. Jensen 535–539

Chemical characteristics and gelling properties of agar from two Philippine *Gracilaria* spp. (Gracilariales, Rhodophyta)

    by N.E. Montaño, R.D. Villanueva & J.B. Romero 541–548

Polysaccharides from the red seaweed *Bostrychia montagnei*: chemical characterization

    by M.D. Noseda, S. Tulio & M.E.R. Duarte 549–554

Chemical structure and gel properties of carrageenans from algae belonging to the Gigartinaceae and Tichocarpaceae, collected from the Russian Pacific coast
    by I.M. Yermak, Y.H. Kim, E.A. Titlynov, V.V. Isakov & T.F. Solov'eva 555–562

Isolation and characterisation of a fourth hemagglutinin from the red alga, *Gracilaria verrucosa*, from Japan
    by H. Kakita, S. Fukuoka, H. Obika & H. Kamishima 563–570

11. Pollution and Remediation

Copper, copper mine tailings and their effect on marine algae in Northern Chile
    by J.A. Correa, J.C. Castilla, M. Ramírez, M. Varas, N. Lagos, S. Vergara, A. Moenne, D. Román & M. T. Brown 571–581

Factors influencing seaweed responses to eutrophication: some results from EU-project EUMAC
    by W. Schramm 583–592

Modelling complexation and electrostatic attraction in heavy metal biosorption by *Sargassum biomass*
    by S. Schiewer 593–601

Ecological engineering in aquaculture: use of seaweeds for removing nutrients from intensive mariculture
    by M. Troell, P. Rönnbäck, C. Halling, N. Kautsky & A. Buschmann 603–611

12. Physiology, Cell and Molecular Biology

Occurrence of closely spaced genes in the nuclear genome of the agarophyte *Gracilaria gracilis*
    by A.O. Lluisma & M.A. Ragan 613–618

'Seed' production of *Porphyra* spp. by tissue culture
    by M. Notoya 619–624

Strain selection in *Kappaphycus alvarezii* var. *alvarezii* (Solieriaceae, Rhodophyta) using tetraspore progeny
    by E. José de Paula, R.T.L. Pereira & M. Ohno 625–635

Efficient utilisation of high photon irradiance for mass production of photoautotrophic micro-organisms
    by A. Richmond & N. Zou 637–641

Time-dependent attachment mechanism of bacterial pathogen during ice-ice infection in *Kappaphycus alvarezii* (Gigartinales, Rhodophyta)
    by D.B. Largo, K. Fukami & T. Nishijima 643–650

Subject Index 651–655

Chemicals Index 657–658

Taxonomic Index 659–666

Author Index 667–668

INTERNATIONAL SEAWEED ASSOCIATION

EXECUTIVE COUNCIL

Dimitri Stancioff
(United States)
President

Mark A. Ragan
(Canada)
Immediate Past President

Joe T. Baker (Australia)	**Masao Ohno** (Japan)
Peter Gacesa (United Kingdom)	**Marianne Pedersén** (Sweden)
Guillermo Garcia-Reina (Spain)	**Wayne J. Sander** (United States)
Klaus Lüning (Germany)	**Bernabé Santelices** (Chile)
Dennis McHugh (Australia)	**Adelaida Semesi** (Tanzania)
Eurico C. de Oliveira (Brazil) Secretary	**Parker S. Laite** (United States) Treasurer

Honorary Life Member

Jack L. McLachlan
(Canada)
Editor-in-Chief

NATIONAL ORGANIZING COMMITTEE

Gavino C. Trono, Jr Chairman
Filipina B. Sotto Vice Chairman
Rhodora V. Azanza Secretary
Enrique Inciong Treasurer

Flerida Arce	**Benson U. Dakay**	**Cesario Pagdilao**
Hilconida P. Calumpong	**Miquel D. Fortes**	**Maximo Ricohermoso**
Anastacio Cambonga	**Anicia Q. Hurtado-Ponce**	**Tita B. Tomayao**

NATIONAL SPONSORS

Department of Science & Technology, Philippines
FMC Marine Colloids, Philippines
Genu Philippines, Inc.
Japan Seaweed Association
Marinalg International
National Academy of Science and Technology
Shemberg Marketing Corporation
Southeast Asian Fisheries Development Center
Land Bank of the Philippines
MCPI Corporation
Westmont Bank of the Philippines

Acknowledgements

The International Seaweed Association greatly appreciates the generous financial contributions received from the following:

Marinalg International, Paris, France

Ceamsa – Ponte Vedra, Spain
China Seaweed Industrial Association – Beijing, China
Danisco Ingredients, Braband, Denmark
FMC Corp., Food Ingredients Division, Brussels, Belgium
Hercules Inc., Copenhagen Pectin Division, Lille Skensved, Denmark
Hispanagar/Sobigel, Burgos, Spain
Marokagar, Casablanca, Morocco
Monsanto PLC, Waterfield, UK
Pronova Biopolymer As, Drammen, Norway
Quest, Cork, Ireland
Setexam, Kenitra, Morocco
SKW Biosystems, France

Japan Seaweed Association
(c/o Dr Masao Ohno)

Chuo-Kasei Co. Ltd., Osaka
Fujicco Co. Ltd., Kobe
Hamada Honten Co. Ltd., Hyogo
Ina Food Co. Ltd., Nagano
Kadoya Co. Ltd., Kobe
Kaneryo Kaiso Co. Ltd., Kumamoto
Kimitsu Chemical Industries Co. Ltd., Tokyo
Marine Science Co. Ltd., Tokyo
Marutomo Co. Ltd.. Ehime
Matsuura Co. Ltd., Tokushima
Mukawa Laboratory, Shimizu
Navocul Cosmetic Co. Ltd., Tokyo
Philagar (Japan) Corp., Tokyo
Riken Shokuhin Co. Ltd., Miyagi
Seitaro Arai & Co. Ltd., Yokohama
Shirako Co. Ltd., Tokyo
Suzuki Consulting Engineer Office, Shizuoka
Taito Co. Ltd., Kobe
Taiyo Kagaku Co. Ltd., Mie
Yamachu Co. Ltd. Oita

Yamagataya Nori Co. Ltd., Tokyo

Seaweed Industry Association of the Philippines, Cebu City, Philippines

Seaweed processors

Geltech Hayco, Mabolo, Cebu
Genu Products (Phil.) Inc., Banilad, Cebu
King Agro Marine, San Juan, Metro Manila
LM Zamboanga United Trading, Zamboanga City
Marcel Carrageenan, Quezon City. Metro Manila
MCPI Corp., Tugbongan, Cebu
NATUM, Zamboanga City
Philippines-Bioindustries, Inc., Makati, Metro Manila
Quest International Philippines Corp., Mactan, Cebu
Shemberg Corporation, Mandaue, Cebu
Zamboanga Fish Trading, Zamboanga City

Others

Seaweed buyers and farmers as a group

Other Contributors

Individuals

Dr. Harris "Pete" Bixler
Former Senate President Neptali Gonzales

Corporations

FMC Marine Colloids Philippines Inc.
Genu Products (Phil.) Inc.
Kopenhagen Pektin Factory
Marcel Carrageenan
MCPI Corp.

Myeong Shin Chemicals
Shemberg Marketing Corp.

Banks

Land Bank of the Philippines
Westmont Bank of the Philippines

Government Agencies and NGOs

Bureau of Fisheries and Marine Resources of
   the Department of Agriculture
Department of Science and Technology
National Academy of Science and Technology
Southeast Asia Fisheries Development Center
   (SEAFDEC)

Reviewers

The Editors of the Proceedings of the XVIth International Seaweed Symposium are very grateful to the following people who reviewed manuscripts, in some cases more than one each (in brackets):

Mireille Amat
Robert Anderson (2)
Put Ang (2)
Monique Axelos
Rhodora Azanza (2)
Inka Bartsch
John Beardall
Sven Beer
Amha Belay
Dominique Bertrand
Carolyn Bird
Harris Bixler
John Bolton (2)
Peter Bond
Michael Borowitzka (2)
Catherine Boyen
Juliet Brodie (2)
Alejandro Buschmann (4)
Jim Callow
Maureen Callow
Arturo Candia (2)
Anthony Chapman (2)
Donald Cheney
Lionel Chevolot
Anong Chirapart (2)
Thierry Chopin
Marina Ciancia
Eric Coppejans (2)
Juan Correa (2)
James Craigie
Beatrice Darcy-Vrillon
Chris Dawes
Stylianos Delivopoulos
Robert DeWreede (2)
Maria Donkin
Ed Drew
Matthew Dring
Louis Druehl
Patrick Durand
Len Evans
Ruth Falshaw (2)
Xiugeng Fei
Marcia Figueiredo (2)

Felix Figueroa (2)
Robert Fletcher
Michael Foster (2)
Linda Franklin
Michael Friedlander (2)
Giovanni Furnari
Richard Furneaux
Paul Gabrielsen
David Garbary (3)
Valrie Gerard
Thierry Givernaud
Margaret Gordon (2)
Michael Gretz
Michael Guiry (4)
Dennis Hanisak (2)
Gavin Hardy
Savas Haritonidis
Jacqueline Hemingson
Terry Holt (4)
Max Hommersand
Kanji Hori
Arne Jensen (2)
Lena Kautsky
Nils Kautsky
Bertrand Keaffer
Derek Keats
Gary Kendrick
Nina Klochkova
Wendy Knoop
Svein Knutsen (2)
Jean Michel Kornprobst
Bjorn Larsen
Ming-Long Liao
Sandra Lindstrom (3)
Mark Littler
Vincent Lognoné
David Luxton (2)
Dennis McHugh (2)
Jack McLachlan (2)
Alan Millar
Ian Miller (2)
Aziza Mouradi-Givenaud
Ivka Munda

Erminio Murano (2)
Wendy Nelson (4)
Trevor Norton (2)
Miguel Noseda
Masahiro Notoya (3)
Eurico de Oliveira (2)
Bernard Onno
Marianne Pedersén
Siew Moi Phang (2)
Hans Porse
Philippe Potin
David Price
Mark Ragan
Hans Rakels
G Subba Rangaiah
John Raven (2)
Catherine Renard
Bruce Robertson (2)
David Rogers
Craig Sanderson
Rui Santos
Winfrid Schramm
Ester Serrao
Kjersti Sjøtun
Pauline Snoeijs
Isabel Sousa-Pinto
Robin South
Michael Stekoll (2)
Dagmar Stengel
Carlos Storz
Ian Tan (2)
Gavino Trono (2)
Anatoly Usov (2)
John van der Meer
Julio Vasquez (3)
Valerie Vreeland
Bob Waaland (2)
Di Walker
John West (2)
Christian Wiencke
Michael Wynne
Charles Yarish (2)
Jason Young

Marinalg International Awards

Mr Secretary, Mr Chairman, Ladies and Gentlemen,

The International Seaweed Symposium is really a great event and particularly the XVIth, because, every 3 years, it offers the unique opportunity for all the participants, scientists, university experts and also technicians and specialists of the industry and business managers to have contacts, dialogues and exchanges during these days in such a unique and nice place, and consequently beaming one from each other.

And the topics range is so large and so diverse from marine algal genetics, to morphology and taxonomy of micro to macroalgae's, physiology and biochemistry properties and structure, cultivation and ecology.

The Marinalg Awarding is one of the important and now traditional moments of the International Seaweed Symposium for 12 years. Indeed, I remember with emotion the scientist, the late L. Lamar Whitney who conceived the "Whitney Challenge Award" with the Marinalg responsible. Some of you have known this personality.

Marinalg International is a world-wide association founded in 1976 with members being on every continent including Northern and Southern America, Asia and Europe, in 12 countries. The members are manufacturing seaweed hydrocolloids: The main applications are food, pharmaceutical and cosmetic industries.

How does Marinalg determine the winners of the awards ?

The papers presented at the International Seaweed Symposium are always evaluated only after their publication.

Each company, member of Marinalg, assigns its scientists to read all of the papers and gives six months to choose their top five choices.

The results are forwarded to the Marinalg Headquarters in Paris (France). Mr. Couchoud, our Adviser, determines the formalization of the methodical and systematic ranking.

Thanks to this methodical approach, the researchers are assured that their work is being critically and world-wide read by their peers in industry. And through this way of selection, the Industry is fully aware of the latest developments in algal research, and, finally, those scientists who would like to see their research efforts applied in practice, can learn what the current thinking is in the economical world,

Today it gives me great pleasure on behalf of Marinalg to announce the winners of the Marinalg awards for papers presented at the XVth International Seaweed Symposium in 1995.

First prize: Rhodora Azanza-Corrales, Tehane T. Aliaza & Nemesio E. Montaño (Marine Science Institute, University of the Philippines) 'Recruitment of *Eucheuma* and *Kappaphycus* on a farm in Tawi-Tawi, Philippines'

Second prize: Ramiro Rojas H., Nelson León M. & Ramiro Rojas O. (Departamento de Cultivo de Algas, Chiloé, Chile) 'Practical and descriptive techniques for *Gelidium rex* (Gelidiales, Rhodophyta) culture'

Third prize: Suzanne Fredericq (Smithsonian Institution), Max H. Hommersand (University of North Carolina) & D. Wilson Freshwater (Center for Marine Science Research, Wilmington, 'The molecular systematics of some agar- and carrageenan-containing marine red algae based on *rbc*L sequence analysis'

Fourth prize: James S. Craigie (National Research Council of Canada) & Juan A. Correa (Pontificia Universidad Catolica de Chile), Etiology of infectious diseases in cultivated *Chondrus crispus* (Gigartinales, Rhodophyta)

[All papers were published in the Proceedings of the Fifteenth International Seaweed Symposium.]

Congratulations to all of you for your work!

I have only one final announcement to authors presenting papers here at the XVIth ISS. the Board of Directors of Marinalg has agreed once more to present you with the 'Whitney Challenge Marinalg Award', next time, in 2001. And the winners will be announced at the XVIIth ISS.

Thanks to all contributors and good luck to each of you!

JEAN-CLAUDE ATTALE
President, Marinalg International
85 Boulevard Haussmann
F-75008 Paris, France

Student Paper Awards

First Prize: Jeff T. Hafting, University of British Columbia, Canada, for 'Propagation of *Porphyra yezoensis* blades in suspension cultures via conchospores and monospores, and the effect of tissue nitrogen and phosphorus quota on blade growth'

Second prize: G. W. Maneveldt, University of the Western Cape, South Africa, for 'Explaining the dominance of the low shore intertidal zone by the coralline crust, *Spongites yendoi*'

Third prize: Annie George, University of Malaya, Malaysia, for 'Survey of potential disease causing organisms associated with *Gracilaria changii*' and Sung Ho Kim, Kongju National University, Korea, for 'Cell–cell recognition during the fertilization in a red alga, *Aglaothamnion oosumiense* (Ceramiaceae, Rhodophyta)'

Student Travel Awards

Fellowship grants from the XVIth ISS Organizing Committee were awarded to: Elena Varela-Alvarez (Ireland); Gan Sook Yee (Malaysia); Annie George (Malaysia); Russell Gillespie (South Africa); David Ginsburg (Guam, USA); Alexander Kamnev (Russia); Rufus Kitto (Yemen); Lydia Ladah (Mexico); Roberto Aurelio Lopez (Mexico); Elena Nevrova (Ukraine); Antonina Podkoritova (Russia); Lynda Poole (United Kingdom); S. P. Samarakoon (Sri Lanka); R. Selvaraj (India); Dagmar Stengel (Ireland); Anatoly Usov (Russia); Boya Volesky (Canada); Alexander Vershinin (Russia); Xing-Hong Yan (Japan); Irina Yermak (Russia); Lindsey Zemke-White (New Zealand); Zhi-Gang Zhou (China); Tatiana Zvyaginstseva (Russia)

Opening Ceremony and Plenary Session

Platform personnel at the opening ceremony, left to right: Dimitri Stancioff – President, International Seaweed Association; Edgardo Gomez – Director, Marine Science Institute, University of the Philippines, Diliman; Jean Claude Attale – President, Marinalg International; William Padolina – Secretary, Department of Science & Technology, Philippines; Rhodora Azanza – Secretary, Local Organizing Committee; Nemesio Montano – Marine Science Institute, University of the Philippines; Gavino Trono – Chairman, Local Organizing Committee; Maximo Ricohermoso – MCPI Corporation, Philippines

Visayas Ballroom during a plenary session

The seaweed industry and R&D: a solid partnership

It is my sincere privilege to keynote the Opening Ceremonies of the XVIth International Seaweed Symposium (ISS). Let me begin by taking this opportunity to welcome all our invited speakers and delegates who have taken the time and effort to come to Cebu City, the Philippines' Queen City of the South. Thanks to the Cebu City officials for their support.

We, as this year's host, are more than honoured to make your stay in our country a most productive and memorable experience. And as organizers, we look forward to your valuable inputs that will surely contribute to the success of this important gathering. Let me ask you to give a round of warm applause to the international and national organizing committees and the hosts – MSI, USC & SIAP.

Permit me to make a special welcome to Prof. Arne and Prof. Synnove Jensen, of Norway. Professor Jensen was my best professor during a brief post-doctoral stint at the Organic Chemistry Laboratories of the then Norwegian Institute of Technology in Trondheim. Congratulations to Rhodora Azanza and Dr Montaño for sharing the best paper.

Seaweeds are essentially photosynthetic organisms that act as primary producers of organic matter at the food chain. Seaweeds, like all other aquatic algae forms, also serve an important role as the providers of oxygen for other aquatic life.

Throughout history, seaweeds have been cultivated and harvested as food by the peoples of Asia, Europe, and Polynesia. The Japanese often sit down and partake of a meal of kombu, nori, and wakame dipped in soy sauce, while the Hawaiians have always included limu (seaweed) in their daily fare.

The quantum leaps achieved in the phycological sciences over the last 25 years have made this aquatic resource much more than a traditional food source. Global expertise in seaweed biology and chemistry has been advancing at a geometric clip, and as scientists you have a first-hand view of how such expertise is rapidly being translated into commercial success.

For this reason, the Department of Science and Technology (DOST) – in close collaboration with marine R&D institutes from both government and the private sector – is vigorously supporting the research and development initiatives of the seaweed sector by taking a client-oriented approach. Our multidisciplinary teams are mandated to identify natural marine products that have commercial potential as pharmaceuticals, agrichemicals, pharmacological research tools, health products, or fine chemicals.

Through the Philippine Council for Aquatic and Marine Research and Development (PCAMRD), DOST's institutional interventions are anchored on the clear goal of enhancing the country's competitive standing as a major player in the global seaweed market.

The seaweed industry is presently concentrated on a certain number of species. *Kappaphycus alvarezii* represents the bulk of aquatic production. On the other hand, *Eucheuma denticulatum* (carrageenophyte), *Gracilaria* spp. (agarophytes), *Caulerpa lentillifera* (sea grapes), *Gelidielia acerosa* (agarophyte) and *Sargassum* spp. (for animal feed/alginate) are also harvested and processed in our waters but in lesser quantities.

Global research and development, however, is expanding its scope to discover new and potentially useful extracts from hitherto overlooked species. For instance, researchers at Canada's Institute for Marine Biosciences (IMB) have recently developed a potential plant growth stimulant extracted from *Ascophyllum nodosum* or the common rockweed.

Philippine seaweed research, for its part, is paying closer attention to *Gracilaria*. Studies supported by PCAMRD have already led to the identification of a number of species that produce high-grade agar for colloidal purposes. Specifically, we have identified *G. firma* as an ideal stock for open reef farming, while *G. heteroclada* has proven to be more suited for pond culture. We are at present making available both methods for culture of interested seaweed farmers and entrepreneurs.

PCAMRD has also launched initial efforts to rehabilitate the *Sargassum* beds of Philippine seas. PCAMRD, with the support and co-operation of other government and non-government agencies, is presently compiling a national inventory and assessment of natural stocks of economically-viable species such as agarophytes and alginophytes. At the moment, natural stocks are the major source of biomass for these species.

Work has also been done on the management of these natural stocks for their conservation. The demand for dried *Kappaphycus alvarezii* for kappa carrageenan manufacture has increased tremendously during the last two years especially here in Cebu City, and farming areas are now undergoing expansion to supply this shortage.

Among the developing countries, it is important that they address valid farmer concerns such as plummeting stock productivity and the steady decline of colloid quality. I urge that serious and sustained research interventions must be initiated in order to accelerate the pace of genomics development, scientific collaboration to prevent protectionism groups from strangling free trade by setting up no tariff trade barriers.

In this age of information, the delegates of this symposium, especially those involved in large-scale DNA sequencing and sequence analysis, must undergo more intensive networking activities in order that information resources be utilized for the general uplifting of the seaweed processing sector. You must consciously strive to promote the development of new scientific and technological knowledge and to find ways that they are shared for the benefit and the advancement of all.

In conclusion, it is our common mission to spur development in our respective countries through the application of marine and phycological sciences. Aquaculture represents an important opportunity for sustainable economic growth in our coastal communities. As marine scientists, it is important that you continue working in close partnership with industry, academia, and other government agencies in order to:

- take a multidisciplinary approach to the research and commercial relevance of seaweed, algae, and marine plant aquaculture;
- help maintain the domestic and international reputation of our marine-based products;
- seek new biologically-active substances in extracts from marine organisms; and
- aggressively explore new business opportunities for marine-based technologies.

Let me end with the story of the great French military leader, General Lyautey, who once asked his gardener to plant a particular tree. The gardener protested that such a tree grew very slowly and would not reach maturity for a hundred years. The General replied, "in that case, we have no time to lose. Plant it this afternoon." We, too, have no time to lose. We must continue to build the research agenda that will shape the seaweed industry deep into the twenty-first century. Thank you and good day.

WILLIAM G. PADOLINA
Secretary of the Department of Science and Technology, Philippines

List of Registrants

ADNAN Hariadi
Copenhagen Pectin A/S
JL Pendidikan I/B-8
Perum. Graha Kert 1
80224 Denpasar
Indonesia

AMANO Hideomi
Laboratory of Marine Biochemistry
Faculty of Bioresources
Mie University
Tsu, Mei 514-8507
Japan
e-mail: amano@bio.mei-u.ac.jp

AMAT Mireille A.
SKW Biosystems
Usine de Baupte
F-50500 Carentan
France

ANDERSON Robert J.
Sea Fisheries Research Institute
Private Bag X2
Roggebaai 8012
Cape Town
South Africa
e-mail: anderson@botzoo.uct.ac.za

ANG Put O. Jr.
Department of Biology
The Chinese University of Hong Kong
Shatin,
NT Hong Kong
China
e-mail: put-ang@cuhk.edu.hk

ARMISEN Rafael
Hispanagar S. A.
Calle Lopez Bravo No. 1
Poligono de Villalonquesar
Burgos 09080
Spain
e-mail: hispanagar_research@cyl.com

ARUGA Yusho
Tokyo University of Fisheries
Konan- 4, Minato-ku
Tokyo 108-8477
Japan
e-mail: yaruga@tokyo-u-fish.ac.jp

ARULAMPALAM Premala
FAO SPADP Phase II
Cl-UNDP
Private Mail Bag
Suva
Fiji
e-mail: faoaqua@mailhost.sopac.org.fj

ASK Erick
FMC Marine Colloids Phil. Inc.
Mandaue City
Cebu
Philippines
e-mail: 1726.1626@compiserve.com

ATTALE Jean Claude
Marinalg
85 Boulevard Haussmann
F-75008 Paris
France

AVILA Marcela
Instituto de Fomento Pesquero
Division de Acuicultura
Casilla 665, Puerto Montt
Chile
e-mail: mavila@ifop.cl

AZANZA Rhodora V.
Marine Science Institute
University of the Philippines
Diliman,
Quezon City 1101
Philippines
e-mail: rhod@msi01.cs.upd.edu.ph

BAO Ying
 EMBLC-Institute of Oceanology
 Chinese Academy of Sciences
 7 Nanhai Road
 Qingdao 266071
 China
 email: feixg@ms.qdio.ac.cn

BARTSCH Inka
 Biologische Anstalt Helgoland
 Notkestrasse 31
 D-22607 Hamburg
 Germany

BEACH Pamela J.
 DANISCO Ingredients
 200 New Century Parkway
 New Century, KS 66031
 U.S.A.

BIXLER Harris "Pete" J.
 Shemberg USA Inc.,
 33 MT. Ephraim Road
 Searsport, ME 04974
 U.S.A.
 e-mail: pbixler@agate.net

BJORNSATER Bo
 Department of Botany
 Stockholm University
 S-106 91 Stockholm
 Sweden
 e-mail: bjornsat@botan.su.se

BOALCH Gerald T.
 Marine Biological Association of the U.K.
 The Laboratory, Citadel Hill
 Plymouth
 Devon PL1 2PB
 U.K.
 e-mail: gtb@wpo.nerc.ac.uk

BOKN Tor L.
 Norwegian Institute for Water Research
 P.O. Box 173
 Kjelsaas
 N-0875 Oslo
 Norway
 e-mail: tor.bokn@niva.no

BOLTON John J.
 Botany Department
 University of Cape Town
 Private Bag
 Rondebosch 7700
 South Africa
 e-mail: bolton@botzoo.uct.ac.za

BOO Sung Min
 Department of Biology
 Chungnam National University
 Daejon 305 764
 Korea
 e-mail: smboo@hanbat.chungnam.ac.kr

BOROWITZKA Michael A.
 School of Biological Sciences and Biotechnology
 Murdoch University
 Perth, WA 6150
 Australia
 e-mail: borowitz@possum.murdoch.edu.au

BRAGG Aline
 Commercial Cisandina Chile Ltda.
 Av. El Salto 4447
 Huechuraba
 Santiago
 Chile
 e-mail: cccbragg@entel.chile.net

BRAULT Dominique
 CEVA
 B.P. 3
 F-22610 Pleubian
 France
 e-mail: algue@ccva.fr

BROWN Murray T.
 Department of Biological Sciences
 University of Plymouth
 Plymouth PLL 8AA
 U.K.
 e-mail: mtbrown@plymouth.ac.uk

BUSCHMANN Alejandro
 University of Los Lagos
 Casilla 933
 Osorno
 Chile
 e-mail: abuschma@puyehue.di.ulagos.cl

CALUMPONG Hilconida
　Marine Laboratory
　Silliman University
　Dumaguete City
　Philippines

CAMBONGA Anastacio
　GENU Philippines
　6th Flr. Unit 3, Metrobank Bldg.
　Osmena Blvd.
　Cebu City 6000
　Philippines

CAMITZ Astrid
　Department of Botany
　Stockholm University
　S-106 91 Stockholm
　Sweden
　e-mail: astrid.camitz@botan.su.se

CANDIA Arturo
　Instituto de Fomento Pesquero
　Balmaceda 252
　Puerto Montt,
　Casilla 665
　Chile
　e-mail: acandia@ifop.cl

CARILLO-DOMINGUEZ Silvia
　National Institute of Nutrition
　Vasco de Quiroga # 15
　D.F. 14000
　Mexico
　e-mail: carrillo@aztlan.innsz.mx

CARLSON Lena
　LeCa Marin
　Kung Oscars vag 7
　S-222 40 Lund
　Sweden
　e-mail: lena.carlson@lund.mail.telia.com

CECERE Ester
　Instituto Talassografico
　C. N. R. V. Roma
　I-74100 Taranto
　Italy
　e-mail: cecere@istta.le.cnr.it

CHEN Chiu-Ming
　Department of Chemistry
　National Tsing Hua University
　Hsinchu
　Taiwan

CHIANG Young-Meng
　8 Alley 6, Lane 30
　Chou Shan Rd.
　Taipei
　Taiwan

CHIRAPART Anong
　Department of Fishery Biology
　Kasetsart University
　50 Phaholyothin Rd. Chatuchak
　Bangkok 10900
　Thailand
　e-mail: ffisanc@nontri.ku.ac.th

CHOPIN Thierry
　Centre for Coastal Studies and Agriculture
　University of New Brunswick
　P.O. Box 5050
　Saint John,
　N.B. E2L 4L5
　Canada
　e-mail: tchopin@unbsj.ca

CIANCIA Marina
　Dpto. de Quimica Organica
　FCEYN, UBA
　Ciudad Universitaria, Pab 2
　1428 Buenos Aires
　Argentina
　e-mail: labcer@qo.fcen.uba.ar

COCHOUD Paul
　Marinalg International
　85 Boulevard Haussmann
　F-75008 Paris
　France

COPPEJANS Eric G.
　Laboratory of Botany
　University of Ghent
　Ledeganckstraat 35
　B-9000 Ghent
　Belgium
　e-mail: eric.coppejans@rvg.ac.be

CORMACI Mario
   Dipartimento di Botanica
   University of Catania
   Via Antonio Longo 19
   I-95128 Catania
   Italy
   e-mail: cormaci@mbox.dipbot.unict.it

CORREA Juan A.
   Facultad de Ciencias Biologicas
   Departamento de Ecologia
   Pontificia Universidad Catolica de Chile
   Casilla 114-D
   Santiago
   Chile
   e-mail: jcorrea@genes.bio.puc.cl

CRITCHLEY Alan
   MRC
   University of Namibia
   Private Bag 13301
   Windhoak
   Namibia
   e-mail: acritchley@unam.na

DE CASTRO Teresa
   SEAFDEC/AQD
   5021 Tigbauan
   Iloilo
   Philippines
   e-mail: rddata@i-iloilo.com.ph

DE LARA-ISSASI Graciela
   Universidad Autonoma
   Metropolitana - Iztapalapa
   Departamento de Hydrobiologia
   Laboratorio de Ficologia Aplicada
   Apdo. Postal 55 535
   D.F. 09340
   Mexico
   email: grace@xanum.uam.mx

DE NYS Rocky
   School of Biological Science &
   Centre for Marine Biofowling & Bio-Innovation
   Sydney 2052
   Australia
   e-mail: r.denys@unsw.edu.au

DEPOLO Miguel
   Fidel Otieza
   #1956 -Piso 14
   Santiago
   Chile

DEVEAU Jean-Paul
   Acadian Sea Plants Limited
   30 Brown Avenue
   Dartmouth
   Nova Scotia
   B3B 1X8
   Canada
   e-mail: jpdeveau@acadian.ca

DEVEAU Louis
   Acadian Sea Plants Limited
   30 Brown Avenue
   Dartmouth
   Nova Scotia
   B3B 1X8
   Canada

DIAKE Sylvester
   Fisheries Division
   Ministry of Agriculture and Fisheries
   P.O. Box 913
   Honiara
   Solomon Island

DIEZ Isabel
   Lab. Botanica
   Dep. Biologia Vegetal & Ecologia
   Fac. Ciencias, U. P. V./E. H. U.
   University of the Basque Country
   Apdo. 644, Bilbao 48080
   Spain
   e-mail: gvbdisai@lg.ehu.es

DURAND Patrick
   IFREMER
   Rue de I'lle d'Yeu
   B.P. 21105
   F-44311 Nantes Cedex 3
   France
   e-mail: pduran@ifremer.fr

EDWARDS Matthew S.
Department of Biology
University of California
Santa Cruz, CA 95064
U.S.A.
e-mail: edwards@biology.ucsc.edu

EKLUND Britta
Laboratory for Aquatic Ecotoxicology
Stockholm University
Studsvik
S-611 82 Nykoping
Sweden
e-mail: brittae@system.ecology.su.se

ENGKVIST Roland
Department of Natural Sciences
University of Kalmar
P.O. Box 905
S-391 29 Kalmar
Sweden
e-mail: roland.engkvist@ng.hile.se

FALSHAW Ruth
Industrial Research Ltd.
P.O. Box 31-310
Lower Hutt
New Zealand
e-mail: r.falshaw@irl.cri.nz

FAN Xiao
Institute of Oceanology
Chinese Academy of Sciences
7 Nanhai Road
Qingdao 266071
China
e-mail: fei@ms.qdio.ac.cn

FAZAL Morty
Kingsway International
Box 9712
Dar-es-Salaam
Tanzania
e-mail: kingsway@raha.com

FEI Xiugeng
EMBLC - Institute of Oceanology
Chinese Academy of Sciences
7 Nanhai Rd.
Qingdao 266071
China

FIGUEIREDO-CREED Marcia
Instituto de Pesquisas
Jardin Botanico do Rio de Janeiro
Rua Pacheco Leao 915
Rio de Janeiro, RJ 22460 030
Brazil
e-mail: mfigueir@openlink.com.br

FOREMAN Ronald E.
Department of Botany
University of British Columbia
#3529-6270 University Blvd.
Vancouver B.C.
V6T 1Z4
Canada
e-mail: ron-foreman@bc.sympatico.ca

FORSYTH Juan Alberto
AV. Argentina
3250 Callao
Peru
e-mail: jafa@crosland.tci.net.pe

FORTES Miguel
Marine Science Institute
University of the Philippines
1101 Diliman, Quezon City
Philippines
e-mail: fortesm@msi01.cs.upd.edu.ph

FOSS David
FMC Corporation-Food Ingredients Division
1735 Market Street
Philadelphia, PA 19103
U.S.A.
e-mail: david_foss@fmc.com

FOSSAA Jan Helge
Institute of Marine Research
P.O. Box 1870
Nordnes
N-5024 Bergen
Norway
e-mail: jan.helge.fossaa@imr.no

FRALICK Richard
  Biology Department
  Plymouth State College
  USNH
  Plymouth, NH 03264
  U.S.A.
  e-mail: rfralick@psc.plymouth.edu

FREDRIKSEN Stein
  Section for Marine Botany
  Department of Biology
  University of Oslo
  P.O. Box 1069
  Blindern
  N-0316 Oslo
  Norway
  e-mail: stein.fredriksen@bio.uio.no

FREILE-PELEGRIN Yolanda
  CINVESTAV-Unidad Merida
  KM. 6 Carr. Ant. Progreso
  A.P. 73 Cordemex
  Merida,
  Yucatan C.P. 97310
  Mexico
  e-mail: freile@kin.ciemer.conacyt.mx

FRESHWATER David Wilson
  Center for Marine Science Research-UNCW
  7205 Wrightsville Ave.
  Wilmington, NC 28403
  U.S.A.
  e-mail: freshwaterw@uncwil.edu

FRIEDLANDER Michael
  Israel Oceanographic and Limnological Research
  P.O. Box 8030
  Haifa 31080
  Israel
  e-mail: michael@ocean.org.il

FUJITA Daisuke
  Toyama Prefectoral Fisheries Research Institute
  366 Takatsuka
  Namerikawa 936 8536
  Japan
  e-mail: d-fujita@nsknet.or.jp

FUJITA Yuji
  Faculty of Fisheries
  Nagasaki University
  Bunkyo Machi 1-14,
  Nagasaki 852
  Japan
  e-mail: yfujita@net.nagasaki-u.ac.jp

FUKAMI Kimio
  Laboratory of Aquatic Environmental Science
  Kochi University
  Monobe-200
  Nankoku 783-8502
  Japan
  e-mail: fukami@cc.kochi-u.ac.jp

FURNANI Giovanni
  University of Catania
  Dipartimento Di Botanica
  Dell'Universita
  via A. Longo 19
  I-95125 Catania
  Italy
  e-mail: g.furnani@mbox.dipbot.unict.it

GABRIELSEN Bjorn Olav
  Hydro Research Centre-Porsgrunn
  P.O. Box 2560
  N-3901 Porsgrunn
  Norway
  e-mail: bjorn.olav.gabrielsen@hre.hydro.com

GAN Sook Yee
  Algae Laboratory
  Institute of Postgraduate Studies and Research
  University of Malaya
  50603 Kuala Lumpur
  Malaysia
  e-mail: h0gan@umcsd.um.edu.my

GAO Kunshan
  Department of Phycology
  Institute of Hydrobiology
  The Chinese Academy of Sciences
  Wuhan 430072
  China
  e-mail: ksgao@public.wh.hb.cn

GARCIA-REINA Guillermo Blairsy
    Instituto de Algologia
    Aplicada-ULPGC Muelle de Taliarte
    s/n 35214- Telde Gran Canaria
    Spain
    e-mail: iaa@ext.step.es

GEORGE Annie
    Aquatic Laboratory
    Institute of Postgraduate Studies and Research
    University of Malaya
    50603 Kuala Lumpur
    Malaysia
    e-mail: h0annie@umcsd.um.edu.my

GERUNG Grevo S.
    Usa Marine Biological Institute
    Kochi University
    Usa Cho, Tosa Shi,
    Kochi 781-11
    Japan
    e-mail: grevo@cc.kochi-u.ac.jp

GERWICK William H.
    College of Pharmacy
    203 Pharmacy Bldg.
    Oregon State University
    Corvallis, OR 97331-3507
    U.S.A.
    e-mail: gerwickw@ccmail.orst.edu

GILJE Magne
    PRONOVA Biopolymer A. S.
    P.O. Box 2045
    N-5501 Haugesund
    Norway

GILLESPIE Russell D.
    P.O. Box 802
    Randpark Ridge 2156
    South Africa
    e-mail: isaruss@icon.co.za

GINSBURG David W.
    UOG Marine Laboratory
    University of Guam
    Mangilao, GU 96923
    U.S.A.
    e-mail: xginsbu@uog.edu

GIVERNAUD Thierry
    SETEXAM BP210
    Km 7 Route de Tanger
    14000 Kenitra
    Morocco

GLANTZ Dale A.
    KELLO Biopolymers
    2145 E. Belt St.
    San Diego, CA 92130
    U.S.A.
    e-mail: dale.glantz@monsanto.com

GOMEZ-GARRETA Amelia
    Laboratorio Botanica
    Facultad Farmacia
    University of Barcelona
    Avda. Joan XXIII
    s/n 08028 Barcelona
    Spain
    e-mail: agomez@farmacia.far.ub.es

GOROSTIAGA Jose Maria
    University of Bosque Country
    Lab. Botanica
    Dep. Biologia Vegetal & Ecologia
    Fac. Ciencias, U. P. V./E. H. U.
    Apdo. 644
    48080 Bilbao
    Spain
    e-mail: gvpgogaj@lg.ehu.es

GRAHAM Michael H.
    Scripps Institute Oceanography
    9500 Gilman Drive M/C 0208
    La Jolla, CA 92093-0208
    U.S.A.
    e-mail: mgraham@ucsd.edu

GRANBOM Malena
    Department of Botany
    Stockholm University
    S-106 91 Stockholm
    Sweden
    e-mail: malena.granbom@botan.su.se

HAFTING Jeff T.
    1359 Lorilawn Court
    Burnaby, B.C.
    V5B 4W8
    Canada
    e-mail: marbiopr@comptime.com

HAMADA Minoru
    Hamada Honten Co., Ltd.
    1864 Heianura Aiga
    Sumoto City Hygo, 656-21
    Japan

HAN Lijun
    Institute of Oceanology
    Chinese Academy of Sciences
    7 Nanhai Road
    Qingdao 266071
    China
    e-mail: fei@ms.qdio.ac.cn

HEE Torben
    FMC Copenhagen A/S
    Risingevej I
    Vallensbaek Strand
    DK-2665 Copenhagen
    Denmark
    e-mail: torben_hee@fmc.com

HELLBLOM Frida T. M.
    Department of Botany
    Stockholm University
    S-106 91 Stockholm
    Sweden
    e-mail: hellblom@botan.su.se

HEMMINGSON Jacqueline Ann
    Industrial Research Ltd.
    P.O. Box 31-310
    Lower Hutt
    New Zealand
    e-mail: jhemmingson@irl.cri.nz

HERNANDEZ Gustavo
    CICIMAR-IPN
    Gaviotas 465
    La Paz B. C.S. 23070
    Mexico
    e-mail: gcarmona@umredipn.ipn.mx

HILLIS Llewellya W.
    Coral Reef Algae Consultant
    20 Brooks Rd.
    Woods Hole, MA 02543
    U.S.A.
    e-mail: hillisl@naos.si.edu

HIRAOKA Masanori
    Usa Marine Biological Inst.
    Kochi University
    Usa cho, Tosa Shi
    Kochi 781-11
    Japan
    e-mail: uesmreef@cc.kochi-u.ac.jp

HOHLBERG Andres R.
    Av. Pedro de Valdivia Norte
    061 Providencia
    Santiago
    Chile
    email: gelymar@entelchile.net

HOMMERSAND Max H.
    Department of Biology
    Coker Hall
    University of North Carolina
    Chapel Hill, NC 27599-3280
    U.S.A.
    e-mail: mhommer.coker@mhs.unc.edu

HONDA Masaki
    Central Research Institute of Electric
    Power Industry
    1646 Abiko, Abiko City
    Chiba 270-1194
    Japan
    e-mail: m-honda@criepi.denken.or.jp

HONYA Masura
    2-3-1-129 Kaga Itabashi-ku
    Tokyo
    Japan
    e-mail: masura_honya@ihi.co.jp

HORN Svein
    Department of Biotechnology
    NTNU, 7034
    Trondheim
    Norway
    e-mail: svein@chembio.ntnu.no

HUANG Rang
   Institute of Oceanography
   National Taiwan University
   4 Roos Rd. No.1
   Taipei
   Taiwan
   e-mail: rang@ccms.ntu.edu.tw

HUANG Su-Fang
   Taiwan Museum
   48 Hsuchow Road
   Taipei 100
   Taiwan
   e-mail: sfhuang@eden.tpm.gov.tw

HUNG Wei
   Food Research Laboratories
   Marutomo Co., Ltd.
   Kominato 1696, Iyo
   Ehime 799 31
   Japan
   e-mail: huangwei@interlink.or.jp

HUNTER Hamish
   Atoll seaweed Co. Ltd.
   P.O. Box 528
   Betio, Tarawa
   Kiribati

HURD Catriona
   Department of Botany
   University of Otago
   P.O. Box 56
   Dunedin
   New Zealand
   e-mail: hurd@planta.otago.ac.nz

HURTADO PONCE Anicia
   SEAFDEC/AQD
   Tigbauan 5021
   Iloilo City
   Philippines
   e-mail: annehur@iloilo.nct

INCIONG Enrique
   MCPI Corporation
   Tugbongan, Consolacion 6001
   Cebu City, Cebu
   Philippines

INOUE Osamu
   INA Food Industry Co., Ltd.
   574 Turumaki-Cho
   Waseda, Shinnjuku 162
   Japan
   e-mail: kantennpp@valley.or.jp

ISHIBASHI Kiyohide
   NABOCUL Cosmetics Co., Ltd.
   5-29-7, Sendgaya
   Shibuya-ku
   Tokyo 151 0051
   Japan

ISTINI Sri
   BPP Teknologi Gedung II 1t.
   15 Jl. M. H. Thampin 8
   Jakarta 10340
   Indonesia
   e-mail: achil@hotmail.com

IVANAC Ivo
   Fidel Otieza
   #1956-Piso 14
   Santiago
   Chile

IWAMOTO Katsuaki
   Marine Science Co., Ltd.
   Room 1013, 1-11-7 Higashi-Kanda
   Chiyoda-Ku, Tokyo
   Japan

JENSEN Arne
   Institute of Biotechnology
   NTNU, Sem Selands V. 8
   N-7034 Trondheim
   Norway
   e-mail: arnejens@kjemi.unit.no

JONES Joanna M.
   Port Erin Marine Laboratory
   Port Erin
   Isle of Man 1M9 6JA
   U.K.
   e-mail: jokain@enterprise.net

JUPP Barry P.
   Oman Seaweed Project (Fishworld)
   P.O. Box 2583
   Ruwi PC112
   Sultanate of Oman, Muscat
   Oman
   e-mail: fworld@gto.net.om

KADOYA Isamu
   Kadoya & Co. Ltd.
   No. 123-1 Higashimachi
   Chuo-ku, Kobe
   Japan

KAISER Tim E.
   Tidal Organics Inc.
   P.O. Box 868
   Yarmouth
   B5A 4JZ
   Canada
   e-mail: scotia@fox.nstn.ca

KAKITA Hirotaka
   Shikoku National Industrial Research Inst.
   Hayashi, Takamatsu
   Kagawa 761 03
   Japan
   e-mail: kakita@sniri.go.jp

KÄLLÅKER Ulf
   c/o Froytang AS
   N-7270 Dyrvik
   Norway

KANG Siu Ming
   PT Amarta Sarilestari
   Jl. Margomulyo Indah
   Blok B-6, Surabaya
   Indonesia

KAPRAUN Donald F.
   Department of Biology
   University of North Carolina
   Wilmington, NC 28403
   U.S.A.
   e-mail: kapraun@uncwil.edu

KATZ Shlomit
   Israel Oceanoraphy and Limnological Research
   P.O. Box 8030
   Haifa 31080
   Israel

KAUTSKY Nils
   Dept. of Systems Ecology
   Stockholm University
   Frescati Backe
   Suante Arrhenivsv 21A
   S-106 91 Stockholm
   Sweden
   e-mail: nils@system.ecology.su.se

KAWAMURA Toshihiro
   Research Laboratory
   Yamata Noriten Co. Ltd.
   384-60 Asitaka-azaoue
   Umazu, Shizuoka 410
   Japan

KEATS Derek
   Botany Department
   University of Western Cape
   P. Bag X17
   Bellville 7535
   South Africa
   e-mail: derek@botany.uwc.ac.za

KENDRICK Gary A.
   Department of Botany
   The University of Western Australia
   Nedlands, WA 6020
   Australia
   e-mail: garyk@cyllene.uwa.edu.au

KIATOA Ienimoa
   Atoll Seaweed Company Ltd.
   Betio
   Tarawa
   Kiribati

KIGANE Tetsuya
   544-3 Huka-cho Mihara
   Hiroshima, 723 0001
   Japan

KIM Gwang Hoon
  Department of Biology
  Kongju National University
  Shingwandong Chungnam
  Kongjushi 314 701
  Korea
  e-mail: ghkim@knu.kongju.ac.kr

KIM Hyung Geun
  Faculty of Marine Bioscience and Technology
  Kangnung National University
  Kangnung, Kangwon-do 210 702
  Seoul
  Korea
  e-mail: kimhg@knusun.kangnung.ac.kr

KIM Jeong Ha
  Department of Biology
  Sung Kyun Kwan University
  Suwon, Kyung Ki Do 440 746
  Korea
  e-mail: jhkimbio@yurim.skku.ac.kr

KIM Kil Jae
  439-13, Sojuri, Ungsang-Up
  Yangsan City
  Kyeongnam
  Korea
  e-mail: msckorea@kotis.net

KIM Kwang Ho
  439-13, Soju-ri, Ungsang-Up
  Yangsan- City
  Kyeongnam
  Korea
  e-mail: msckorea@kotis.net

KIM Kwang Young
  Department of Oceanography
  Chonnam National University
  Kwangju 500 757
  Korea
  e-mail: kykim@chonnam.chonnam.ac.kr

KIM Nam Gil
  Department of Aquaculture
  Gyeongsang National University
  Inpyoungdong 445, Tongyoung
  Kyongnam 650-160
  Korea
  e-mail: ngkim@gshp.gsnu.ac.kr

KIM Sung-Ho
  Department of Biology
  Kongju National University
  Shingwandong, Chungnam
  Kongjushi 314 701
  Korea
  e-mail: kim9093@hanmail.net

KLOCHKOVA Nina Grigorievna
  Partizanskaya Street-6
  Petropavlovsk-Kamchatsky 683000
  Russia
  email: rector@marine.kamchatka.su

KNUTSEN Svein Halvor
  Department of Biotechnology
  Agricultural University of Norway
  P.O. Box 5040
  N-1432 Aas,
  Norway
  e-mail: svein.knutsen@ikb.nlh.no

KONG Corsica S. L.
  Department of Biology
  The Chinese University of Hong Kong
  Shatin, N. T.
  Hong Kong 852
  China
  e-mail: S970662@mailserv.cuhk.edu.hk

KRAAN Stefan
  Martin Ryan Marine Science Institute
  National University of Ireland
  Galway
  Ireland
  e-mail: stefan.kraan@seaweed.ucg.ie

KRAFT Gerald T.
  School of Botany
  University of Melbourne
  Parkville, Victoria
  Australia
  email: g.kraft@botanyunimelb.edu.au

LADAH Lydia
   Universidad Autonoma de Baja California
   Instituto de Investigaciones Oceanologicas
   Apartado Postal # 453
   Ensenada
   Baja California CP 22800
   Mexico
   e-mail: drpcl2@bahia.ens.uabc.mx

LAHAYE Marc
   National Research Institute of Agriculture
   URPOI B.P. 71627
   F-44316 Nantes
   France
   e-mail: lahaye@nantes.inra.fr

LAITE Parker S.
   P.O. Box 279
   Camden, ME 04843
   U.S.A.

LARGO Danilo B.
   University of San Carlos
   Talamban, Cebu City
   Cebu City 6000
   Philippines
   e-mail: biology@durian.usc.edu.ph

LEBBAR Rachid
   Km. 7, Route de Tanger
   B.P. 210 Kenitra
   Morocco
   e-mail: setex@atlasnet.net.ma

LEDUA Esaroma
   FAO-SPADP, Phase II
   Cl-UNDP, Private Mail Bag
   Raiwaqa, Suva
   Fiji

LEE In Kyu
   Department of Biology
   Seoul National University
   Seoul 151-742
   Korea
   e-mail: inkyulee@plaza.snu.ac.kr

LEE Ju Yeon
   Department of Biology
   Chungnam National University
   Tae jeon 305 764
   Korea

LEMOINE Yves
   Cytophysiologie ve Yves getale et Phycologie
   Bat. SN2 Universite de Lille
   F-59655 Villeneuve D'Asco
   France
   e-mail: yves.lemoine@univ-lille.fr

LEONARDI Patricia I.
   University Nacional del Sur
   Dept. Biologia y Quimica
   San Juan 670
   Bahia Blanca 8000
   Argentina
   e-mail: leonardi@criba.edu.ar

LEPAGNOL Valerie
   Goemar Laboratories
   Zac la Madeleine B.P. 55
   F-35400 Saint Malo cedex
   France
   e-mail: lepagnol@sb-roscoff.fr

LEVY Israel
   Israel Oceanographic and Limnological Research
   The National Institute of Oceanography
   P.O. Box 8030
   Haifa 31080
   Israel
   e-mail: israel@ocean.org.il

LEWIS Jane
   Seaweed Biogeography Laboratory
   Institute of Marine Biology
   National Taiwan University
   Keelung
   Taiwan
   e-mail: jlewis@lglab.imb.ntou.edu.tw

LIAO Lawrence
   Marine Biology Section
   University of San Carlos
   Cebu City 6000
   Philippines
   e-mail: uscplib@pinya.usc.edu.ph

LIAO Ming-Long
    CRC for Industrial Plant Biopolymers
    School of Botany
    University of Melbourne
    Parkville, Victoria 3052
    Australia
    e-mail: m.liao@botany.unimelb.edu.au

LIAO Woan Ru
    Institute of Oceanography
    National Taiwan University
    Roos Rd. No.1
    Taipei 100
    Taiwan
    e-mail: rang@cems.ntu.edu.tw

LIN Showe-Mei
    Department of Biology
    University of Southwestern Louisiana
    P.O. Box 42451
    Lafayette, LA 70504 2451
    U.S.A.
    e-mail: sxl6893@usl.edu

LLANA Ethel
    # 14 Villa Lourdes Subdivision
    Visayas Aveenue
    Quezon City 1101
    Philippines

LLUISMA Arturo O.
    Marine Science Institute
    University of the Philippines
    Diliman
    Quezon City 1101
    Philippines
    e-mail: lluisma@msi01.cs.upd.edu.ph

LO Jir-Mehng
    Department of Chemistry
    National Tsing Hua University
    Kuan-Fiu Rd., Hsinchu
    Taiwan
    e-mail: d797405@oz.nthu.edu.tw

LOGNONE Vincent
    Seaweed Manufacturing Technology Center
    CEVA Pres'ile Pen.Lan
    B.P. 3
    F-22610 Pleubian
    France
    e-mail: alque@ceva.fr

LUONG-VAN Jim Thinh
    Faculty of Science
    Casuarina Campus
    Northern Territory University
    B 40 Darwin, NT 0909
    Australia
    email: jluongva@darwin.ntu.edu.au

LUXTON David
    D. Luxton & Associates Limited
    70 Hamurana Road
    Omokoroa, Tauranga
    New Zealand
    e-mail: dlatganz@enternet.co.nz

MAICHER Gerd R.
    Hans Binder M'Bau GMBH
    Isarstrasse 6–8
    D-85417 Marzling
    Germany
    e-mail: grm1272@aol.com

MALM Torleif
    Kalmar University
    Box 905
    S-391 29 Kalmar
    Sweden
    e-mail: torleif.malm@ng.hih.se

MANEVELDT Gavin W.
    Botany Department
    University of the Western Cape
    P. Bag X17
    Bellville 7535
    South Africa
    e-mail: gavin@botany.uwc.ac.za

MARCOS Roberto R.
    Productos del Pacifico
    Floresta 1395 Col. Obrera
    Ensenada, B.C. 22830
    Mexico
    e-mail: vivanco@microsol.com.mx

MARTINEZ GOSS Milagros
    Institute of Biological Sciences
    College of Science
    University of the Philippines
    Los Baños, Laguna
    Philippines

MATON Michel
    DISATEC S. A. -222
    Rue Michel Carre
    F-95870 Bezons
    France

MATSUYAMA Kazuyo
    Laboratory of Phycology
    Tokyo University of Fisheries
    Konan 4, Minato-ku
    Tokyo 108 8477
    Japan
    e-mail: ad95202@tokyo-u-fish.ac.jp

McHUGH Dennis J.
    School of Chemistry
    University College
    University of New South Wales
    Canberra, ACT 2614
    Australia
    e-mail: d-mchugh@adfa.oz.au

McPEAK Ron
    Global Biological Consultants
    20101 NE 196th St.
    Battle Ground, WA 98604-3719
    U.S.A.
    e-mail: rmcpeak@worldaccessnet.com

MENESES Isabel Cabellos
    Canada 253, of. "F",
    3er.piso.Providencia
    Santiago
    Chile
    e-mail: imeneses@interactiva.cl

MILLER Ian T.
    Carina Chemical Laboratories Ltd.
    P.O. Box 30366
    Lower Hutt
    New Zealand
    e-mail: ian.miller@xtra.co.nz

MINORU Hamada
    HAMADA Honten Co., Ltd
    1864 Heianura Aiga
    Sumoto City
    Hiyogo 656-2121
    Japan

MINOURA Katsuteru
    102 Heights, Miyuki 3-5-22
    Masugata, Tamaku
    Kawasaki City
    Kanagawa
    Japan
    e-mail: maruja@ibm.net

MIYASHITA Akinori
    Laboratory of Applied Phycology
    Tokyo University of Fisheries
    Konan-4, Minato-ku
    Tokyo, 108
    Japan
    e-mail: notoya@tokyo-u-fish.ac.jp

MOLLER Torben Torsbierg
    Danisco Ingredients USA Inc.
    201 New Century Parkway
    New Century, KS 66031-913 7
    U.S.A.

MONTAÑO Nemesio
    Marine Science Institute
    University of the Philippines
    1101 Diliman, Quezon City
    Philippines
    e-mail: coke@msi01.cs.upd.edu.ph

MORENO Lautaro
    Avenida Americo Vespucio Sur
    842 Las Condes
    Santiago
    Chile
    email: lmoreno@entelchile.net

MORITA Teruo
    Faculty of Fisheries
    Nagasaki University
    Bunkyo Machi 1-14
    Nagasaki 852-8521
    Japan
    e-mail: d79702h@stcc.nagasaki-u.ac.jp

MOURADI Aziza
    Laboratoire de Biochimie et.
    Biotechnologies Marines
    B.P. 133 Faculte des Sciences
    14000 Kenitra
    Morocco

MTOLERA Matern
    Institute of Marine Sciences
    P.O. Box 668
    Zanzibar
    Tanzania
    e-mail: mtolera@zims.udsm.ac.tz

MUNDA Ivka Maria
    Centre for Scientific Research of the
        Slovene Academy of Science & Arts
    Gosposka 13
    1000 Ljubljana
    Slovenia
    e-mail: zrc@zrc-sazu.si

MYSLABODSKI David
    Israel Oceanographic and Limnological
        Research Ltd.
    Shikmona
    P.O. Box 8030
    Haifa 31080
    Israel
    e-mail: mysla@ocean.org.il

NAGAHISA Eizo
    Kitasato University
    School of Fisheries Sciences
    Sanriku-cho, Kesen-gun
    Iwate 022 01
    Japan
    e-mail: nagahisa@nnettown.or.jp

NAGAURA Kazuhiro
    Laboratory of Applied Phycology
    Tokyo University of Fisheries
    Konan-4, Minato-ku
    Tokyo, 108
    Japan
    e-mail: notoya@tokyo-u-fish.ac.jp

NANBA Nobuyoshi
    School of Fisheries Sciences
    Kitasato University
    Sanriku, Iwate 022 01
    Japan
    e-mail: nanba@nnettown.or.jp

NASSOR Makame S.
    Commission for Natural Resources-Zanzibar
    P.O. Box 774
    Masomeni
    Zanzibar
    Tanzania

NISHIDE Eiichi
    College of Bioresource Sciences
    Nihon University
    1866 Kameino, Fujisawa
    Kanagawa 252-8510
    Japan

NISHIZAWA Kazutoshi
    10 -4, Kouyama -3
    Nerima-ku
    Tokyo 176-0022
    Japan

NOSEDA Miguel Daniel
    Departamento de Bioquimica
    Setor de Cs. Biologicas Universidade
    Federal de Parana-Jd. das Americas
    P.O. Box 19046
    Brazil
    e-mail: nosedaeu@garoupa.bio.ufpr.br

NOTOYA Masahiro
    Laboratory of Applied Phycology
    Tokyo University of Fisheries
    Konan-4, Minatoku
    Tokyo, 108
    Japan
    e-mail: notoya@tokyo-u-fish.ac.jp

NUÑEZ Roberto A.
    Laboratorio de Macroalgas
    CICIMAR-IPN
    Apdo. Postal 592
    La Paz, B. C. S.
    C.P. 23000
    Mexico
    e-mail: rnunez@vmredipn.ipn.mx

OHNO Masao
  Usa Marine Biological Institute
  Kochi University
  Usa cho, Tosa Shi
  Kochi 781 1164
  Japan
  e-mail: mohno@cc.kochi-u.ac.jp

OHSUMI Yukihiro
  Research & Development Center
  Shiraho Co., Ltd.
  6-1-17, Nakakasai
  Edogawa, Tokyo 134
  Japan
  e-mail: sykk0001@alles.or.jp

OLIVA Eduardo
  Prodalmar
  Universidad Arturo Prat
  Chile

OLIVEIRA Eurico C.
  Inst. Biociencias
  University of Sao Paulo
  C. Postal 11461
  Sao Paulo SP 05422-970
  Brazil
  e-mail: euricodo@usp.br

PACHECO RUIZ Isai
  Apartado Postal 453
  Ensenada, Baja California
  Mexico
  e-mail: isai@bahia.ens.uabc.mx

PAGDILAO Cesario
  Philippine Council for Aquatic and Marine
  Research and Development
  Los Baños, Laguna
  Philippines

PANG Shaojun
  Institute of Oceanology
  7 Nanhai Road
  Qingdao 266071
  China
  e-mail: sjpang@ms.qdio.ac.cn

PATRON Jessica J.
  College of Science & Math
  Western Mindanao State University
  San Jose Road
  Zamboanga City 7000
  Philippines

PAULA Edison Jose
  Departamento de Botanica
  Instituto de Biociencias
  Universidad de Sao Paulo
  C. Postal 11.461 05422-970
  Sao Paulo, SP
  Brazil
  e-mail: ejdpaula@usp.br

PEDERSÉN Marianne
  Department of Botany
  Stockholm University
  S-10691 Stockholm
  Sweden
  e-mail: marianne.pedersen@botan.su.se

PERURENA Joaquin
  Ozmar Trading
  P.O. Box 81
  Harbord
  NSW 2096
  Australia
  e-mail: perurena@fl.net.au

PHANG Siew-Moi
  Institute of Biological Sciences
  University of Malaya
  50603 Kuala Lumpur
  Malaysia
  e-mail: hlphangs@umcsd.um.edu.my

PHILP Kevin
  Quest International Phils. Corp.
  GF/GFB#1, Mactan Export Processing Zone
  Lapu Lapu City 6015
  Cebu
  Philippines

PIANTINI Rene
  PRODALMAR Ltda.
  Sucre No. 220 OF 607
  Antofatasta
  Chile
  e-mail: rpiantin@ente-chilenet

PINO Hugo
    Av. Americo Vespucio Sur 842
    Las Condes
    Santiago
    Chile

PODKORYTOVA Antonina V.
    Pacific Scientific Research Fisheries Centre
    4 Shevchenko Alley
    Vladivostok 690600
    Russia
    e-mail: root@tinro.marine.su

POOLE Lynda
    Department of Biological Sciences
    University of Dundee
    Dundee DDI 4HN
    United Kingdom
    e-mail: l.j.poole@dundee.ac.uk

PORSE Hans
    Copenhagen Pectin A/S
    DK-4623 Lille Skensved
    Denmark

POST Steven E.
    Genu Products Phil./Hercules
    44 Guldblommevej
    DK-4000 Roskilde
    Denmark

PRUD'HOMME van REINE Willem F.
    Leiden University
    Rijks Herbarium/Hortus Botanicus
    P.O. Box 9514
    2324 BD Leiden
    The Netherlands
    e-mail: prudhomme@rulrhb.leidenuniv.nl

QIN Song
    Institute of Oceanology
    Chinese Academy of Sciences
    7 Nanhai Road
    Qingdao 266071
    China
    e-mail: sqin@ms.qdio.ac.cn

RAGAN Mark A.
    Institute for Marine Biosciences
    National Research Council of Canada &
        Canadian Inst. for Advanced Research
    Program in Evolutionary Biology
    1411 Oxford Street
    Halifax N.S.
    B3H 3Z1
    Canada
    e-mail: mark.ragan@nrc.ca

RAIKAR Sanjeev Vencu
    Faculty of Fisheries
    Nagasaki University
    Bunkyo Machi 1-14
    Nagasaki 852 8521
    Japan
    e-mail: f 1088@cc.nagasaki-u.ac.jp

RAKELS Johannes L. L.
    Department of Chemical Engineering
    University of San Carlos
    Talamban
    Cebu City 6000
    Philippines
    email: che@mangga.usc.edu.ph

RASMUSSEN John
    DANISCO Ingredients
    Edwin Rahrsvej 38
    DK-8220 Brabrand
    Denmark
    e-mail: g8lor@danisco.dk

RASMUSSEN Preben B.
    DANISCO Ingredient
    Edwin Rahrs Vej 38
    DK-8220 Brabrand
    Denmark
    e-mail:g8pbr@danisco.dk

REBELLO Jacqueline
    Saitama-ken
    Washimiya-machi
    Sakurada 3-S-1.401
    Saitama-ken, 340-02
    Japan
    e-mail: jacreb@cc.kochi-u.ac.jp

REES Alwyn
Leigh Marine Laboratory
University of Auckland
P.O. Box 349 Warkworth
New Zealand
email: ta.rees@auckland.ac.nz

REIS Renata Perpetuo
Rua Tonelero 380/1007
Rio de Janeiro
RJ.CEP 22030 000
Brazil
e-mail: rreis@graziela.jbrj.gov.br

RENOUX Aline
Laboratoire de Biologie Vegetale
Faculte des Sciences
97159 Pointe a Pitre Cedex
Guadeloupe
French West Indies

RIAD Abdelwahab
Km 7, Route de Tanger
B.P. 210 Kenitra
Morocco
e-mail: setex@atasnet.net.ma

RIBERA-SIGUAN Ma. Antonia
Laboratori de Botanica
Facultat de Farmacia
Universitat de Barcelona
Avda. Joan XXIII s/n
08028 Barcelona
Spain
e-mail: ribera@farmacia.far.ub.es

RICHMOND Amos
Jacob Blaustein Institute
Ben Gurion University
Midreshet Ben Gurion
84990
Israel
e-mail: amosr@bgumail.bgu.ac.is

RICOHERMOSO Maximo
MCPI Corporation
No. 3 Eagle St.
Sto. Niño Village
Banilad, Cebu City
Philippines

RINDI Fabio
Dipartimento Scienze
Dell'Uomo E Dell' Ambiente
University of Pisa
Via A. Volta 6
I-56126 Pisa
Italy
e-mail: fabio.rindi@discat.unipi.it

ROBERTSON Bruce L.
University of Port Elizabeth
P.O. Box 1600
Port Elizabeth 6000
South Africa
e-mail: btablr@upe.ac.za

ROBINS Rebecca
Dept. EEM- Biology
University of California
Santa Barbara, CA 93106
U.S.A.
e-mail: dchapman@descartes.ucsb.edu

ROBLEDO Daniel
CINVESTAV-Unidad Merida
Km.6 Antigua Carretera A Progreso 73
C.P. 97310 Cordemex
Merida, Yucatan
97310
Mexico
e-mail: robledo@kin.cieamer.conacyt.mx

RODRIGUEZ Deni
Dept. of Biology
University of Southwestern Louisiana
P.O. Box 42451,
Lafayette, LA 70504 2451
U.S.A.
e-mail: dxr1029@usl.edu

ROJAS Ramiro
Fidel Otieza
#1956-Piso 14
Santiago
Chile

ROTMANN Klaus W. G.
Taurus Products
P.O. Box 5534
Rivonia 2128
South Africa
e-mail: taurusc@cis.co.za

ROUSSEAU Florence
Museum National d'Histoire Naturelle
Laboratoire de Cryptogamie
12 rue Buffon
F-75005 Paris
France
e-mail: gdretudi@mnhn.fr

RUDOLPH Brian
Copenhagen Pectin A/S,
Ved Banen 16
Lille Skensved
DK-4623
Denmark
e-mail: brudolph@merc.com

RUENESS Jan
University of Oslo
Department of Biology (Marine Botany)
P.O. Box 1069
Blindern
N-0316 Oslo
Norway
e-mail: jan.rueness@bio.uio.no

RUSLI Utama
P.T. Sumberguna Makasar nusa
Jln. K. H. Achmad Dachlan No. 11
Ujung Pandang 90112
Sulawesi Selatan
Indonesia

SANCHEZ-RODRIGUEZ Ignacio
Av. Instituto Politecnico Nacional
Apdo. Postal 592
La Paz Baja California Sur
Mexico
e-mail: isanchez@vnredipn.ipn.mx

SANDER Wayne J.
Nutra Sweet Kelco Co.
8355 Aero Drive
San Diego, CA 92123
U.S.A.
e-mail: wayne.sander@monsanto.com

SANTELICES Bernabe
Pontificia Universidad Catolica de Chile
Facultad de Ciencias Biologicas
Departament de Ecologia
Casilla 114-D Santiago
Chile
e-mail: bsanteli@genes.bio.puc.cl

SANTOS Gertrudes A.
1717 Ala Wai Blvd. #1109
Honolulu
HI 96815
U.S.A.

SATO Keiichi
Riken Food Co., Ltd.
1-16-3 Tago
Sendai, Miyagi
Japan

SATO Minoru
Faculty of Agriculture
Tohoku University
Tsutsumidori, Aoba-ku
Sendai 981-8555
Japan
e-mail: msato@bios.tohoku.ac.jp

SEARLE Richard
Nutrasweet Kelco Company
Lady Burn Works, Girvan
Ayrshire KA26 9JN
U.K.
email: rsearle@kelco.com

SCHIEWER Silke
Department of Biology
Hong Kong Baptist University
Waterloo Road, Kowloon
Hong Kong
China
e-mail: silke@hkbu.edu.hk

SCHRAMM Winfrid
    Institute of Marine Science
    University of Kiel
    Dusternbrookerweg 20
    D-24105 Kiel
    Germany
    e-mail: wschramm@ifm.uni-kiel.de

SEIP William F.
    Becton Dickinson Microbiology Systems
    250 Schilling Cirle
    Cockeysville, MD 21030
    U.S.A.

SEMESI Adelaida
    Botany Department
    University of Dar es Salaam
    P.O. Box 35060
    Dar es Salaam
    Tanzania
    e-mail: amu@udsm.ac.tz

SERISAWA Yukihiko
    Laboratory of Phycology
    Tokyo University of Fisheries
    Konan-4, Minato-Ku
    Tokyo 108 8477
    Japan
    e-mail: ad95201@tokyo-u-fish.ac.jp

SHIHIRA-ISHIKAWA Ikuko
    Shimizu Laboratory
    Marine Biotechnology Institute
    1900 Sodeshi-cho
    Shimizu City
    Shizuoka 424
    Japan
    e-mail: s-ishikawa@mbio.shimizu.co.jp

SMIT Albertus
    Botany Department
    University of Cape Town
    P.O. Box Rondebosch 7700
    South Africa
    e-mail: asmit@botzoo.uct.ac.za

SNOEIJS Pauline
    Department of Ecological Botany
    Uppsala University
    Villavagen 14
    S-752 36, Uppsala
    Sweden
    e-mail: pauli.snoeijs@vaxtbio.uu.se

SOMSUEB Sutheewat
    Usa Marine Biological Inst.
    Kochi University
    Usa cho, Tosa Shi
    Kochi 781 11
    Japan
    e-mail: stwtha@cc.kochi-u.ac.jp

SONNEVELD Matt
    Hercules Inc., Hercules Plaza
    1313 North Market St.
    Wilmington, DE 1989-0001
    U.S.A.

SOTTO Filipina
    University of San Carlos
    Talamban
    Cebu City, Cebu
    Philippines
    e-mail: fili@mangga.esc.edu.ph

SOUSA-PINTO Isabel
    Department of Botany
    Faculdad de Ciencias
    University of Porto
    R. Campo Alegre 1191
    4150 Porto
    Portugal
    e-mail: ispinto@caserver.fc.up.pt

STANCIOFF Dimitri J.
    2 Spring Street
    Camden, ME 04843
    U.S.A.
    e-mail: dimitri_stancioff@me.com

STEINBERG Peter D.
    School of Biological Sciences
    UNSW
    Sydney, NSW 2052
    Australia
    e-mail: p.steinberg@unsw.edu.au

STEINNES Arild
    PRONOVA Biopolymer A. S.
    P. O. Box 2045
    N-5501 Haugesund
    Norway

STEKOLL Michael S.
    University of Alaska
    11120 Glacier Highway
    Juneau, AK 99801
    U.S.A.
    e-mail: ffmss@aurora.alaska.edu

STENGEL Dagmar B.
    Martin Ryan Marine Science Inst.
    National University of Ireland
    Galway
    Ireland
    e-mail: dagmar.stengel@seaweed.ucg.ie

SUZUKI Soichiro
    9-1 Hongo-cho
    Numazu-shi
    Shizuoka-ken 410
    Japan

SUZUKI Minoru
    Philagar Japan
    Maruto Bldg.
    2-1 Koishikawa, Bunkyo-ku
    Tokyo 112
    Japan

TADAO Iri
    PROAGAR S.A.
    Avda. Vicente Perez Rosales 800
    Llanguinhue City, X Region
    Chile

TAHER Marcel
    C.U. Sumber Rejek 1
    Jl. Mahawu 156 A
    Manado, 95239
    Indonesia

TAN Ian H.
    Royal Botanic Garden Edinburgh
    20A Inverleith Row
    Edinburgh EH3 5LR
    U.K.
    e-mail: i.tan@rbge.org.uk

TANDRA Alfred
    P.O. Box 17
    Jalan. S. Cerekang No. 16 (34)
    Ujung Pandang
    Indonesia

TAYLOR Rebecca
    Department of Biological Sciences
    University of Dundee
    Dundee DDI 4HN
    U.K.
    e-mail: r.taylor@dundee.ac.uk

TAYLOR Michael
    Leigh Marine Laboratory
    University of Auckland
    P.O. Box 349
    Warkworth,
    New Zealand
    e-mail: mw.taylo@auckland.ac.nz

TAYLOR Susan G.
    R. R. 3, 9915 Craddock Drive
    Pender Island
    British Columbia
    Canada
    e-mail: staylor@sfu.ca

TEANGANA Etera
    Atoll Seaweed Co. Ltd.
    P.O. Box 25
    Bairiki
    Tarawa
    Kiribati

TETSUYA Kigane
    TEICHU
    544-3 Fukamachi
    Miharashi
    Hiroshima 723 0001
    Japan

TINNE Michael
    Atoll Seaweed Co. Ltd.
    P.O. Box 508
    Betio
    South Tarawa
    Kiribati

TOKUDA Hiroshi
  1-46-25 Ooka
  Minami-Ku
  Yokohama 232 0061
  Japan

TROELL Max
  Department of Systems Ecology
  Stockholm University
  Suante Arrhenius V. 21A
  S-106 91 Stockholm
  Sweden
  e-mail: max@system.ecology.su.se

TRONO Gavino C. Jr.
  Marine Science Institute
  University of the Philippines
  Diliman
  Quezon City 1101
  Philippines
  e-mail: trono@msi01.cs.upd.edu.ph

TSENG Chengkui
  Institute of Oceanology
  Chinese Academy of Sciences
  7 Nanhai Road
  Qingdao 266071
  China
  e-mail: sqin@ms.qdio.ac.cn

TWIDE Povl
  Genu Products Phil./Hercules
  44 Guldblommevej
  DK-4000 Roskilde
  Denmark

UGARTE Raul
  30 Brown Avenue
  Dartmouth
  Nova Scotia
  Canada

UKU Jacqueline
  Kenya Marine & Fisheries Research Institute
  P.O. Box 81651
  Mombasa
  Kenya
  e-mail: juku@recoscix.com

UPPALAPATI RAO Srinivasa
  Faculty of Fisheries
  Nagasaki University
  Bunkyo Machi 1-14
  Nagasaki 852
  Japan
  e-mail: f1089@cc.nagasaki-u.ac.jp

USOV Anatoly
  Institute of Organic Chemistry
  Russian Academy of Sciences
  Leninsky Prospect 47
  Moscow 117913
  Russia
  e-mail: usov@ioc.ac.ru

UY Wilfredo H.
  Mindanao State University at Naawan
  Naawan
  Misamis Oriental 9023
  Philippines
  e-mail: wili@chem1.msuiit.edu.ph

VALDES Flavio
  201 New Century Parkway
  New Century, KS 66031
  U.S.A.

VARELA-ALVAREZ Elena
  National University of Ireland
  Martin Ryan Institute
  Nuig, Galway
  Ireland
  e-mail: elena.varela@seaweed.ucg.ie

VASQUEZ Julio A.
  Depto. Biologia Marina
  Universidad Catolica del Norte
  Larrondo 1281
  Coquimbo
  Chile
  e-mail: jvasquez@socompa.cecun.ucn.cl

VLACHOS Vaso
  Department of Microbiology
  University of the Witwatersrand
  Private Bag 3 Wits
  Johhanesburg 2050
  South Africa
  e-mail: vaso@gecko.biol.wits.ac.za

VOLESKY Boya
    Chemical Engineering
    McGill University and BV Sorbex, Inc.
    Montreal
    H3A 2B2
    Canada

WALKER Diana I.
    Department of Botany
    The University of Western Australia
    Nedlands, WA 6907
    Australia
    e-mail: diwalker@cyllene.uwa.edu.au

WANG Su-Juan
    Shanghai Fisheries University
    334 Jungong Road
    Shanghai 200090
    China

WANG Wei-Lung
    Department of Botany
    National Chung Hsing University
    Taichung
    Taiwan
    e-mail: wlwangtw@yahoo.com

WEINBERGER Florian
    Israel Oceanographic & Limnological Research
    Tel Shikmona
    P.O.B. 8030
    31080 Haifa
    Israel
    e-mail: florian@ocean.org.il

WERNER Astrid
    Biologische Anstalt Helgoland
    Notkestrasse 31
    D-2230 Hamburg
    Germany
    e-mail: awerner@meereforschang.de

WEST John A.
    School of Botany
    University of Melbourne
    Parkville, VIC 3052
    Australia
    e-mail: jwest@rsbens.its.unimelb.edu.au

WESTERMEIER Renato
    Instituto de Pesquerias
    Universidad Austral de Chile
    Los Pinos s/n, Casilla
    Puerto Montt 1327
    Chile

WICHMAN Jesper
    Danisco Ingredients
    Edwin Rahrs Vej 38
    DK-8220 Brabrand
    Denmark
    e-mail: g8jew@danisco.dk

WU Chaoyuan
    7 Nanhai Road
    Qingdao 266071
    China
    e-mail: cywu@ms.qdio.ac.cn

YAN Xing-Hong
    Laboratory of Phycology
    Tokyo University of Fisheries
    Konan-4 Minato-ku
    Tokyo 108 8477
    Japan
    e-mail: yxm58928@tokyo-u-fish.ac.jp

YERMARK Irina
    Pacific Institute of Bioorganic Chemistry
    FED-RAS
    Vladivostok 22
    Prospekt 100-letya 159
    Vladivostok
    Russia
    e-mail: yermak@loiboc.marine.su

YOON Hwan-Su
    Deptartment of Biology
    Chungnam National University
    Daejon 305 764
    Korea
    e-mail: s-hsyoon@hanbat.chungnam.ac.kr

YOUNG David C.W.
    Tacara Sdn. Bhd.
    P.O. Box 62502
    Tawau, Sabah
    91035
    Malaysia
    e-mail: davidyg@pop.jaring.my

YOUNG Jason
   Marine Biology Section
   University of San carlos
   Cebu City 6000
   Philippines

YVIN Jean Claude
   Laboratoires Goemar
   Zac la Madeleine
   B.P. 55
   F-35400 Saint Malo
   France
   e-mail: labo@goemar.com

ZAMORANO Jaime P.
   Gelymar S.A.
   Casilla 997
   Puerto Montt
   Chile
   e-mail: gely-pmo@entelchile.net

ZEMKE-WHITE W. Lindsey
   School of Biological Sciences
   University of Auckland
   Private Bag 92019
   Auckland
   New Zealand
   e-mail: l.zemke-white@auckland.ac.nz

ZERTUCHE Jose Antonio
   U.A.B.C. - I.I.O.
   P.O. Box 453
   Ensenada B.C.   22860
   Mexico
   e-mail: zertuche@bahia.ens.uabc.mx

ZHAO Shi Jin
   Institute of Oceanology
   The Chinese Academy of Sciences
   7 Nanhai Rd.
   Qingdao 266071
   China
   e-mail: sjzhao@ms.qdiao.ac.cn

ZHOU Zhi-Gang
   The College of Fisheries
   Shanghai Fisheries University
   334 Jungong Rd.,
   Shanghai 200090
   China

ZVYAGINTSEVA Tatiana N.
   Pacific Institute of Bioorganic Chemistry
   pr Stoletija-159
   Vladivostok 690022
   Russia
   e-mail: zjag@piboc.marine.su

Diversity of the seaweed flora of the Philippines and its utilization

Gavino C. Trono, Jr.
Marine Science Institute, College of Science, University of the Philippines, Diliman, Quezon City 1101, Philippines. E-mail: trono@msi01.cs.upd.edu.ph

Key words: seaweed flora, diversity, phycocolloids, seaweed farming, natural products, utilization

Abstract

Some 820 species of marine macrobenthic algae, including many species of Cyanophyta, are reported from the Philippines. These consist of 472 species of Rhodophyta belonging to 37 families and 11 orders, 134 species of Phaeophyta belonging to 10 families and 7 orders and 214 species of Chlorophyta belonging to 11 families and 7 orders. The Rhodophyta comprise 57.6%, the Phaeophyta 16.3% and the Chlorophyta 26.1% of the flora. Many of these species are of economic importance as food, sources of industrial products such as polysaccharides, bioactive and nutritional natural products, and growth promoting substances. Farming seaweeds is presently one of the most productive and environmentally friendly forms of livelihood for coastal populations. More than 80 000 people have been estimated to culture some 10 000 ha of the coastal area. For example, production of farmed *Kappaphycus/Eucheuma* reached 58 324 dry metric tons in 1995 and was valued at US$44 million dollars. Production areas are concentrated in the southern Philippines. Other species in limited commercial production are *Caulerpa lentillifera* through farming, *Gracilaria* species through farming and gathering of local wild stocks and *Gelidiella acerosa* from harvesting wild stocks. Today, seaweeds and their products are the third most important fishery export.

Introduction

The coralline reefs, rocky shores, coves, protected and wave-exposed coasts and estuarine areas of the Philippines represent widely diverse habitats for the marine flora and fauna. Because oligotrophic waters and high rainfall characterize the archipelago, the marine waters around the islands are enriched by nutrients leached from terrestrial sources. The advent of the typhoon season also causes the vertical mixing of nutrient rich bottom waters, which contribute further to the enhancement of the fertility of the coastal waters. Among the more conspicuous marine organisms in the nearshore and shallow areas are the macrobenthic fauna and flora. The floral components are represented by three groups of primary producers, the mangroves, seagrasses and the macrobenthic algae, the seaweeds.

Diversity of the seaweed flora

Few reports on the seaweed flora were published during the nineteenth century; the earliest was written by Blanco (1837), an Augustinian priest residing in the Philippines. Later, Montagne (1844) published on the collections of Hugh Cuming, an English naturalist, while Decaisne (1842a, b) described *Galaxaura fastigiata* and *Liagora caenomyce,* from the same collections. In 1868 von Martens published on the collections made during the Preussiche Expedition nach Ost-Asien. The collections made by H. N. Mosely during the Challenger Expedition were published by Dickie (1874, 1876a, 1876b & 1877) and Piccone (1886, 1889) published materials collected during the Vettor Pisani circumnavigation trip.

The largest and most important collections were made during the Siboga Expedition in 1899 to the Sulu Archipelago. Barton (1901) produced a monograph of the Siboga collections of *Halimeda*, Gepp & Gepp (1911) on the family Codiaceae, and Weber-van Bosse (1904) published on the family Corallinaceae. The entire collections were later reviewed and published by Weber-van Bosse (1913, 1921, 1923, 1928).

Chou (1945, 1947) published a monograph of the genus *Galaxaura* from the collections of H. H.

Bartlett in 1935–1936, while Gilbert (1942, 1943, 1947) published on the entire Chlorophyceae material, Taylor (1964, 1966) on the Chlorophyceae and Phaeophyceae, and Tseng & Gilbert (1942) published a monograph of the genus *Codium*. Taylor (1966) published on the genus *Turbinaria*.

Foreign phycologists contributed most of the records on the seaweed flora, up to this period. The surge of interest among Filipino phycologists during the 1970s has resulted in 42 publications on the seaweed flora. A compilation of 820 species of marine macrobenthic algae, which include many species of Cyanophyta, was published by Silva et al. (1987). These consist of 472 species of Rhodophyta, belonging to 37 families and 11 orders, 134 species of Phaeophyta, belonging to 10 families and 7 orders, and 214 species of Chlorophyta, belonging to 11 families and 7 orders. Of the three divisions, the Rhodophyta is the most diverse representing 57.6% of the entire flora, the Chlorophyceae 26.1% and the least diverse is the Phaeophyta representing only 16.3%. Notwithstanding their small representation in the flora, members of the Phaeophyta are the most common and abundant, in terms of standing stock. About 15–20% of the binomials appear to be incorrect identifications or synonyms; for instance, some 56 species of *Sargassum* are listed, but recent monographic studies of the genus have recognised only 28 species, eight of which were described as new species. More recently, publications by Trono & Ganzon-Fortes (1988) and Trono (1997) have contributed to the flora.

Utilization of the seaweed resources

Of the 820 species recorded, some 350 species are known to have some economic value. However, less than 5% of these are economically important in the country; most have still to be developed. Those known to be economically important can be categorized according to their main use, although many fall into two or more categories.

Food species

This group includes many species which are directly used as food (de Leon et al., 1980; Trono & Ganzon-Fortes, 1988) and are mainly prepared as fresh or blanched salad vegetables or are used as flavouring in soups and garnishing in food recipes. Seaweed salads are usually mixed with sliced tomatoes and onions, sprinkled with pepper in sweet sour dressing and salted to taste. In some areas, fermented fish paste is used to enhance the taste. Except for *Caulerpa lentillifera*, *Kappaphycus alvarezii*, *Eucheuma denticulatum* and some species of *Gracilaria*, which are produced through farming, all others are gathered from wild stocks.

The following species are used as vegetable salads:

Acanthophora spicifera	*Eucheuma denticulatum*
Caulerpa lentillifera	*Gracilaria* spp.
C. racemosa (several varieties)	*Grateloupia filicina*
C. peltata	*Halymenia durvilleai*
C. sertularioides	*Scinaia hormoides*
C. taxifolia	*Hydroclathrus clathratus*
Codium edule	*Kappaphycus alvarezii*
C. intricatum	*Laurencia* spp.
C. bartlettii	*Porphyra* sp.
Enteromorpha spp.	*Gelidiella acerosa*

Young shoots of *Sargassum* spp., *A. spicifera*, *Euchuema* spp. and *Porphyra* sp. are used in soups as flavouring and thickening agents while others are utilized in gelo-type desserts (*Eucheuma* spp., *Kappaphycus* spp., *G. acerosa* and *Gracilaria* spp.). Many of these species are sold in wet markets. In the Philippines mainly fresh stocks are used, unlike the situation in Japan where species are packaged with value added processes. The use of seaweeds as food was formerly restricted to populations living in mountainous coastal areas where supply of garden vegetables was limited.

Seaweeds as a source of phycocolloids

Amongst the most important of the seaweed resources in the country today are the phycocolloid-bearing species. These consist of the agarophytes, carrageenophytes and alginophytes, although the latter are not yet commercially utilized for alginate production.

Agarophytes

Agar, a major additive in the food industry, is extracted from several species, including the genera *Gracilaria*, *Gelidium*, *Pterocladia* and *Gelidiella acerosa*. However, because of their small size and low biomass available from natural stocks, the potential for using *Gelidium* and *Pterocladia* as raw materials for agar production is limited and therefore not economically feasible. At present, only several species of *Gracilaria* and *Gelidiella acerosa* are utilized in agar processing.

Cultured *Gracilaria* species from Mindanao sells for US$0.30 dry kg^{-1} while alkali-treated anhydrous materials sell for a much higher price of US$3.00 dry kg^{-1}.

The low price of dried *Gracilaria* materials gathered from wild stocks is mainly due to the poor quality of the agar produced by the species, and to the poor post-harvest practices applied to it. Many of the *Gracilaria* species present in the Philippines possess agar of low quality, i.e. low yield and gel strength. Stock selection studies, carried out on seven of the most abundant species and grown under the same culture conditions, showed that these produce agars of different qualities. Of the seven, *G. firma*, *G. heteroclada* and *Gracilaria* sp. from Sorsogon province were highly productive and produced good quality agar. *Gracilaria firma* is well adapted to open reef farming while *G. heteroclada* and *Gracilaria* sp. are good seedstocks for pond culture. *Gracilaria eucheumoides*, gathered from wild stocks, produces good quality sugar reactive agar. However, this species does not lend itself to farming due to its slow growth and very strict ecological farming requirements. Wild stocks of *G. tennuistipitata* and *Gelidiella acerosa* also possess good quality agar; their culture technology, however, is not yet developed.

Carrageenophytes

Several species of carrageenophytes are present in the Philippines: *Kappaphycus alvarezii*, *K. cottonii*, *K. striatum*, *K. procrusteanum*, *Euchema denticulatum*, *E. gelatinae* and *E. arnoldii*. Several species of *Hypnea* and *Acanthophora spicifera* have also been reported to contain carrageenan. The genus *Kappaphycus* produces *kappa*-carrageenan while *Euchema* produces *iota*-carrageenan. At present only *K. alvarezii* and *E. denticulatum* are produced commercially through farming.

Alginophytes

The genus *Sargassum* is the most dominant in the Philippines. It generally forms large beds in rocky wave-exposed portions of the reefs and shores. More than twenty species have been taxonomically verified in the Philippines (Trono, 1992); *S. crassifolium*, *S. cristaefolium*, *S. oligocystum*, *S. binderi*, *S. cinctum*, *S. feldmannii*, *S. hemiphyllum*, *S. polycystum*, *S. paniculatum* and *S. siliquosum* are the more common species. Other species such as *Turbinaria ornata* and *T. conoides*, *Hormophysa cuneiformis* and *Hydroclathrus clathratus* are also present. However, these do not form large biomass or produce such good quality alginates as *Sargassum*.

The manufacture of alginate is feasible because the raw material is derived from abundant wild stocks of *Sargassum*. At prese nt, the bulk of the *Sargassum* biomass is used in the production of animal feed; a portion of the seaweed meal production is exported to Japan. *Sargassum* is also used in the production of liquid fertilizer. The harvesting of local stocks is presently limited to certain areas in northern Mindanao and Visayas. Local stocks in Luzon, Palawan, southern Mindanao, Sulu and Tawi-Tawi remain untapped. The potential for development of *Sargassum* and other alginophytes is great.

Seaweed farming as a livelihood

The farming of seaweeds is presently one of the most productive and environmentally friendly forms of livelihood among the coastal populations. Major farming areas are concentrated in the south-western part of the country, specifically in the provinces of Sulu, Tawi-Tawi, southern Palawan and Zamboanga. Minor production areas are located in northern Bohol, Cebu, Cuyo Island group in the northern Sulu Sea, Calatagan in Batangas and northern Mindanao. It is estimated that about 80 000 farmers cultivate more than 10 000 ha of the shallow coastal areas. In addition, more than 300 000 people are engaged in activities related to the seaweed industry, as employees of seaweed companies, entrepreneurs and small scale business personnel catering to the needs of the seaweed farming industry.

At present seaweed farming is based on just a few species such as *K. alvarezii*, *E. denticulatum*, *Caulerpa lentillifera* and some species of *Gracilaria*. The farming of these species is a relatively recent development. Prior to the mid-1960s, the supply of dried carrageenophytes for carrageenan production came from Indonesia through the harvesting of local stocks. In the second half of the 1960s, the supply from Indonesia stopped, due to political problems, and an American seaweed company was forced to seek alternative sources of dried seaweeds in the Philippines. However, because of the unabated harvesting, local wild stocks declined drastically. The prospect of running out of raw material was the main factor that influenced the development of farming technology. The first successful farm for *Kappaphycus* and *Eucheuma*

Table 1. Seaweed production and exports of the Philippines

Year	Quantity (dry t)	Value (in US$ 000)
Seaweed Production: 1990–1995		
1990	36 558	30 988
1991	35 654	40 470
1992	43 819	58 998
1993	47 844	59 408
1994	50 614	57 479
1995	58 324	43 655
Total	272 813	US$291 001
Average	45 468	US$48,000

Source: Bureau of Agricultural Statistics

Year	Quantity (dry t)	Value (in US$ 000)
Export of Dried Seaweeds: 1990–1996		
1990	35 346	49 883
1991	26 828	21 232
1992	20 529	18 550
1993	21 662	18 125
1994	23 558	21 983
1995	28 920	39 106
1996	26 408	41 974
Total	183 251	210 853
Average	26 178	30 121

Source: National Statistics Office

Export of Dried and Processed Seaweed Products in 1996		
Dried seaweed	26 408	41 974
Semi-refined carrageenan	18 293	81 220
Refined carrageenan	2 252	23 240
Total	46 953	146 434

Source: Seaweed Industry Association of the Philippines

was established in Tapaan Island lagoon, Siasi, Sulu in 1972 (Doty & Alvarez, 1973). From there the technology was extended to other areas in Zamboanga, Sulu and Tawi-Tawi. The high demand for dried seaweeds resulted in highly competitive buying among the new seaweed companies which led to the development of alternative farming areas in central Visayas, Batangas in Luzon and southern Palawan.

Production of carrageenophytes between 1990 and 1995 (Table 1) averaged 45 468 dry t y^{-1}, with a value of US$48 million dollars. Export of dried seaweeds between 1990 and 1996 averaged 26 178 dry t, worth some US$30.12 million dollars. In 1996, 46 953 dry t, consisting of dried seaweeds (53.8%), semi-refined (41.1%) and refined carrageenans (5.1%) and valued at US$ 146.434 million, were exported to Europe, Canada, Australia and the United States. At present seaweeds and their products are the third most important fishery export of the Philippines.

Other species produced in relatively smaller quantities are: *C. lentillifera*, *Gracilaria* spp. and *Gelidiella acerosa*. Production of *C. lentillifera* is mainly through pond and open lagoon cultures, production of *Gracilaria* spp. is through pond culture and gathering of local stocks while production of *G. acerosa* and *Sargassum* spp. is mainly from harvesting local stocks.

Other uses of seaweeds and their natural products

Seaweed species such as *Acanthophora muscoides*, *Gracilaria arcuata*, *Halimeda macroloba*, *Hydroclathrus clathratus*, *Laurencia* spp. and *Sargassum* spp. are known to contain growth regulators. Extracts from *Sargassum* spp. are utilized in liquid fertilizers; application of the fertilizer on vegetables and grain crops (Montaño & Tupas, 1990) enhances their production. Many genera like *Sargassum*, *Ulva*, *Hydroclathrus* and *Kappaphycus* spp. are used as soil conditioners (Chidambaram & Unny, 1953; Michanek, 1979; Mshigeni, 1982).

Other species are known to contain bioactive products that exhibit antibacterial, antiviral and antifungal properties. Some such species are: *Asparagopsis taxiformis* (Fenical et al., 1979), *Chondria armata*, *Laurencia* spp., *Cladophora* spp., *Dictyosphaeria cavernosa*, *Gelidium* sp. and several species of *Laurencia*. *Digenea simplex* is an effective anthelmintic agent, vermifuge and laxative (Michanek, 1979; Hoppe, 1979). Many species (e.g. *Gracilaria* spp.) are used in the treatment of stomach disorders such as dysentery and diarrhoea (Aguilar-Santos & Doty, 1968). Others are used for the treatment of high blood pressure (*Caulerpa* spp.), for lowering plasma cholesterol (*Ulva pertusa*, *Enteromorpha compressa*, *Porphyra* sp.) (Lahaye & Jegou, 1993), for renal problems (*Acetabularia major*, *Gracilaria* spp.), and for the control of heavy metal pollution (*E. denticulatum*, *Sargassum* spp., *K. alvarezii*). *Turbinaria* spp. are used as insect repellent agents. Others (*Sargassum* spp., *Turbinaria* spp., *Ulva* spp., *Caulerpa* spp.) are used as sources of vitamins and minerals (Diaz-Piferrer, 1979; Baker, 1984).

Large seaweed species such as *Sargassum* spp., *Kappaphycus* spp., *Gracilaria* spp., *H. clathratus*, *Hypnea* spp. and *Ulva* spp. are commercially used as components in animal feeds, as sources of carotenoids and as binders (de Guzman, 1978; Ragan, 1981; Mshigeni, 1982). Some species contain moderate amounts of protein (Diaz-Piferrer, 1979; Hoppe, 1979 Luistro et al., 1987) and are used as substitutes in animal feed (Bersamin et al., 1967). Several species, such as *Carpopeltis* spp., *Grateloupia filicina*, *Ceramium* spp. and *Gymnogongrus* spp., are also used as raw material in the manufacture of paste.

Many species contain important chemicals such as tannins, phenols, folic and folinic acid, gelan and tocopherols. These are reported in many species of brown algae, several species of the Ulvaceae and some species of *Hypnea* and *Laurencia* (Hoppe, 1979; Ragan, 1981; Lewis et al., 1988).

A more recent use is in biofiltration. For example *Gracilaria* spp. have been shown to be efficient biofiltration agents in highly fertilized waters associated with aquaculture areas (Gao & McKinley, 1994; Buschmann et al., 1996).

The large variety of natural products, including amino acids, minerals, nitrogenous compounds, polysaccharides, pigments, steroids, phytohormones, terpenes, vitamins, carbonyl compounds, fatty acids, diterpene, mannitol, carboxylic acids, phenols and lipids, found in more than 200 species of the seaweed flora and their applications have been documented in two books (Trono & Ganzon-Fortes 1988; Trono 1997).

From this report it should be apparent that only a few of the many economically important species are currently utilized to their full potential, the majority remaining undeveloped.

References

Aguilar-Santos, G. & M. S. Doty, 1968. Chemical studies of 3 species of the marine algal genus *Caulerpa*. In Freudenthal, H.D. (ed.), Drugs from the Sea: Transactions of the Drugs from the Sea Symposium. University of Rhode Island, 27–29 August 1967. Marine Technology Society, Washington: 173–176.
Baker, J. T., 1984. Seaweeds in pharmaceutical studies and applications. Hydrobiologia 116/117: 29–40.
Barton, E. S., 1901. The genus *Halimeda*. Siboga Exped. Monogr. 60: 1–32.
Bersamin, S. V., R. B. Benania & R. Rustia, 1967. Protein from seaweeds for animal feed substitutes. Phillip. J. Sci. 96: 159–175.
Blanco, M., 1837. Flora de Filipinas. Manila, 887 pp.
Buschmann, A. H., M. Troell, N. Kautsky & L. Kautsky, 1996. Integrated tank cultivation of salmonids and *Gracilaria chilensis* (Gracilariales, Rhodophyta). Hydrobiologia 326/327: 75–82.
Chidambaram, K. & M. M. Unny, 1953. Note on the value of seaweeds as manure. Proc. Int. Seaweed Symp. 1: 67–68.
Chou, R. C.-Y., 1945. Pacific species of *Galaxaura*, I: Asexual types. Pap. Michigan Acad. Sci., Arts Lett. 30: 35–56.
Chou, R. C.-Y., 1947. Pacific species of *Galaxaura*. II: Sexual types. Pap. Michigan Acad. Sci., Arts Lett. 31: 3–24.
Decaisne, J., 1842a. Essais sur une classification des algues et des polypers calciferes de Lamoroux. Ann. Sci. nat. Bot. ser. 2, 17: 297–380,
Decaisne, J., 1842b. Essais sur une classification des algues et des polypers calciferes de Lamoroux. Memoire sur les corallines ou polypiers calciferes. Ann. Sci. nat. Bot. ser. 2, 18: 96–128.
Diaz-Piferrer, M., 1979. Contributions and potentialities of Caribbean marine algae in pharmacology. In Hoppe, H. A., T. Levring & Y. Tanaka (eds), Marine Algae in Pharmaceutical Science. Walter de Gruyter & Co., Berlin: 149–164.
Dickie, G., 1874. On the algae of Mauritius. J. linn. Soc. Bot. 14: 190–202.
Dickie, G., 1876a. Contributions to the botany of the expedition of H.M.S. 'Challenger.' – Algae, Chiefly Polynesian. J. linn. Soc. Bot. 15: 235–246.
Dickie, G., 1876b. Notes on algae collected by H. N. Moseley, M. A., of H.M.S. 'Challenger', chiefly obtained in Torres Straits, Coasts of Japan, and Juan Fernanadez. J. linn. Soc. Bot. 15: 446–455.
Dickie, G., 1877. Supplemental notes on algae collected by H. N. Mosely, M. A. of H.M.S. 'Challenger', from various localities. J. linn. Soc. Bot. 15: 486–489.
Doty, M. S. & V. B. Alvarez, 1973. Seaweed farms: A new approach for the U.S. Industry. In Proceedings of the 9th Annual Conference of Marine Technology Society, Washington D.C. 1873: 701–708.
Fenical, W. O., J. McConnel & A. Stone, 1979. Antibiotic and antiseptic compounds from the family Bonnemaisoiaceae (Florideophyceae). Proc. int. Seaweed Symp. 9: 387–400.
Gao, K. & K. H. McKinley, 1994. Use of macroalgae for marine biomass production and CO_2 remediation: a review. J. appl. Phycol. 6: 45–60.
Gepp, A. & E. S. Gepp, 1911. The Codiaceae of the Siboga Expedition including a monograph of Flabellariaceae and Udoteae. Siboga-Exped. Monogr. 62, 150 pp.
Gilbert, W. J., 1942. Notes on *Caulerpa* from Java and the Philippines. Pap. Michigan Acad. Sci., Arts Lett. 27: 7–26.
Gilbert, W. J., 1943. Studies on Philippine Chlorophyceae, I: Dasycladaceae, Pap. Michigan Acad. Sci., Arts Lett. 28: 15–35.
Gilbert, W. J., 1947. Studies on Philippine Chlorophyceae, III: The Codiaceae. Bull. Torrey Bot. Club 74: 73–79.
Guzman, L. B. de, 1978. Carrageenan as heavy metal binders. B.S. Thesis, College of Arts and Sciences, University of the Philippines, Diliman, Quezon City, 43 pp.
Hoppe, H. A., 1979. Marine algae and their products and constituents in pharmacy. In Hoppe, H. A., T. Levring & Y. Tanaka (eds), Marine Algae in Pharmaceutical Science. Walter de Gruyter & Co., Berlin: 25–120.
Lahaye, M. & D. Jegou, 1993. Chemical and physical-chemical characteristics of dietary fibers from *Ulva lactuca* (L.) Thuret and *Enteromorpha compressa* (L.) Grev. J. appl. Phycol. 5: 195–200.
Leon, S. Y. de, R. M. Guirriec & L. N. Panlasiqui, 1980. Alternative foods for the growing population. National Bookstore, Inc. Metro Manila, Philippines: 50–89.
Lewis, J. G., N. F. Stanley & G. G. Guist, 1988. Commercial production and applications of algal hydrocolloids. In Lembi, C. A. & J. R. Waaland (eds), Algae and Human Affairs. Cambridge University Press, New York: 205–236.

Luistro, A. H., N. E. Montaño, G. J. B. Cajipe & E. C. Laserna, 1987. Protein content of some Philippine seaweeds with notes on the amino acid content of *Ulva lactuca* L. and *Gracilaria verrucosa* (Huds.) Papenfuss. Philipp. J. Sci. 17: 23–28.

Martens, G. von, (1866) 1868. Die tange. In Die Preussische Expedition nach Ost-Asien; Nach amtlichen Quellen: Botanischer Theil, Berlin, 152 pp.

Michanek, G., 1979. Seaweed resources for pharmaceutical uses. In Hoppe; H.A., T. Levring; Y. Tanaka (eds), Marine Algae in Pharmaceutical Science. Walter de Gruyter & Co., Berlin: 203–235.

Montagne, C., 1844. Plantae cellulares quas in insulis philippinensibus a cl. Cuming collectae recensuit, observationibus non nullis descriptionibusque illustravit C. Montagne. D.M. London J. Bot. 3: 658–662.

Montaño, N. E. & L. M. Tupas, 1990. Plant growth hormonal activities of aqueous extracts from Philippine seaweeds. SICEN Leaflet, Marine Science Institute, University of the Philippines, Diliman, Quezon City 1101: 1–5.

Mshigeni, K. E., 1982. Seaweed resources in Tanzania: a survey of potential sources for industrial phycocolloids and for other uses. In Hoppe, H. A., & T. Levring (eds), Marine Algae in Pharmaceutical Science, vol. 2. Walter de Gruyter & Co., Berlin: 131–174.

Piccone, A., 1886. Alghe del viaggio di circumnavigazione della Vettor Pisani, Genova, 97 pp.

Piccone, A., 1889. Nuove alghe del viaggio di curcumnavigazione della 'Vettor Pisani.' Atti della Reale Accademia dei Lincei, Memorie di Classe di Scienze Fisiche, Matematiche e Naturale, series 4, 6:10–63.

Ragan, M. A., 1981. Chemical constituents of seaweeds. In Lobban, C. S. & M. J. Wynne (eds), The Biology of Seaweeds. Blackwell Scientific, Oxford: 589–626.

Silva, P., E. G. Menez & R. L. Moe, 1987. Catalog of the benthic marine algae of the Philippines. Smithsonian Contributions to the Marine Sciences No. 27, Smithsonian Institution Press, Washington D.C., 179 pp.

Taylor, W. R., 1964. The genus *Turbinaria* in Eastern Seas. J. linn. Soc. Bot. 58: 475–490.

Taylor, W. R., 1966. Records on Indo-Pacific *Turbinarias*. Hydrobiologia 28: 91–100.

Trono, G. C. Jr., 1992. The genus *Sargassum* in the Philippines. In Abbott, I. A. (ed.), Taxonomy of Economic Seaweeds with Reference to some Pacific and Western Atlantic Species (vol. 3). California Sea Grant College Program, University of California, La Jolla, California: 43–94.

Trono, G. C. Jr., 1997. Field Guide and Atlas of the Seaweed Resources of the Philippines. Bookmark, Inc., Manila, Philippines, 303 pp.

Trono, G. C. Jr. & E. T. Ganzon-Fortez, 1988. Philippine Seaweeds. National Bookstore, Inc., Manila, Philippines, 330 pp.

Tseng, C. K. & W. J. Gilbert, 1942. On new algae of the genus *Codium* from the South China Sea. J. Wash. Acad. Sci. 32: 291–296.

Weber-van Bosse, A., 1904. Corallinae verae of the Malay Archipelago. In Weber-van Bosse, A. & M. Foslie (eds), The Corallinaceae of the Siboga-Expedition. Siboga-Exped. Monogr. 61: 78–110.

Weber-van Bosse, A., 1913. Liste des algues du Siboga, I. Myxophyceae, Chlorophyceae, Phaeophyceae avec le concours de M. Th. Reinbold. Siboga-Exped. Monogr. 59a: 1–186.

Weber-van Bosse, A., 1921. Liste des algues du Siboga, II: Rhodophyceae, Premiere partie: Protoflorideae, Nemalionales, Cryptonemiales. Siboga-Exped. Monogr. 59b: 187–310.

Weber-van Bosse, A., 1923. Liste des algues du Siboga, II: Rhodophyceae, Seconde partie. Ceramiales. Siboga-Exped. Monogr. 59c: 311–392.

Weber-van Bosse, A., 1928. Liste des algues du Siboga, II: Rhodophyceae. Troisieme partie. Gigartinales et Rhodymeniales et tableau de la distribution des Chlorophycees, Phaeophycees et Rhodophycees de l'Archipel Malaisien. Siboga-Exped. Monogr. 59d: 393–533.

Red algal polysaccharide industry: economics and research status at the turn of the century*

Donald F. Kapraun
Department of Biological Sciences, University of North Carolina–Wilmington, 601 S. College Rd., Wilmington, NC 28403, U.S.A.

Key words: co-production, cyclic production, mariculture, sustainable development

Abstract

Commercial research priorities target production cost reduction and expansion of new applications for established products to increase profitability. Cost effective seaweed production for phycocolloids relies on: (1) species and strain selection, (2) vegetative reproduction, (3) improvements in cultivation, harvesting and drying and storage. Increasingly, present linear technologies based on consumption of raw materials and production of product, and waste, will be replaced by cyclic technologies which maximize utilization of waste as raw material resources. Co-production and synergistic product development are not only green, but cost effective. Transportation of resources and products is energy expensive, but regional production promises to reduce transportation costs. Marine phycologists have important roles to play in the phycocolloid industry, especially in the development of co-generation and co-production technologies, in the domestication of seaweed cultivars to remove harvest pressure on target species and sensitive ecosystems, and in the development of regional low technology production.

Introduction

It is axiomatic that basic research really is basic, and provides the essential foundation for application and technology (Magne, 1993). It is also true that when headlines pronounce 'currency crisis deepens' and 'monetary austerity promoted' that even less support can be expected for basic research. I don't make policy, so don't blame me. I merely report the obvious, and make suggestions for surviving with some grace and dignity. Since interactions between marine phycologists and the phycocolloid industry are based on the premise that co-operation can increase their net profit, we need to consider how best we can enhance their profitability, while directing their efforts toward sustainable development. The following is intended as a brief status report, followed by predictions of motivating forces and directions for development in the 21st century.

Linear production versus sustainable development

Traditionally, production has been seen as a linear process, with raw materials entering at one end and then being transformed into products which are distributed and sold at the other end (Figure 1). Increasingly, the 'green' or environmental movement has helped us recognize that (1) all processes on earth are cyclical, not linear and (2) there is no 'away' where anything can be disposed of or dumped. The concept of sustainable development (Hawken, 1993) has profound implications for both marine phycologists and for ecologically balanced production in the phycocolloid industry (Bodvin et al., 1996) as we begin the 21st century.

Let's now address the questions of how marine phycologists can interact with the phycocolloid industry to enhance sustainable development by replacing linear processes with cyclic production? And how we can turn 'black' (the accountant's profit line) into 'green' (ecologically sound practice)? And what's in it for me? My specific comments are based on my experience with red algal phycolloids, both agars and carrageenans. However, many of these remarks will

---

* This paper is dedicated to my friend and colleague Dr Kimon T. Bird, 30 October 1951–29 October 1996.

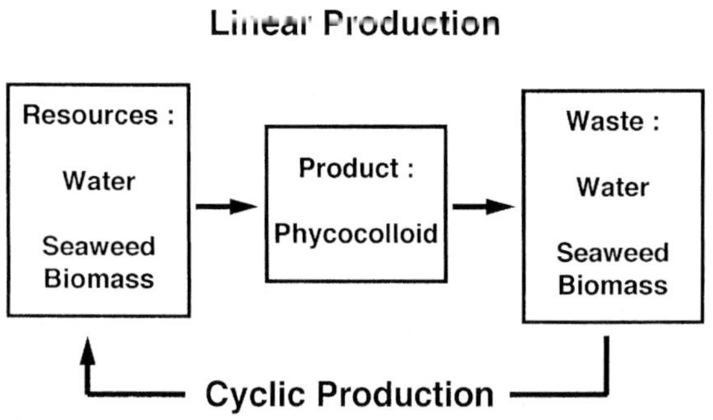

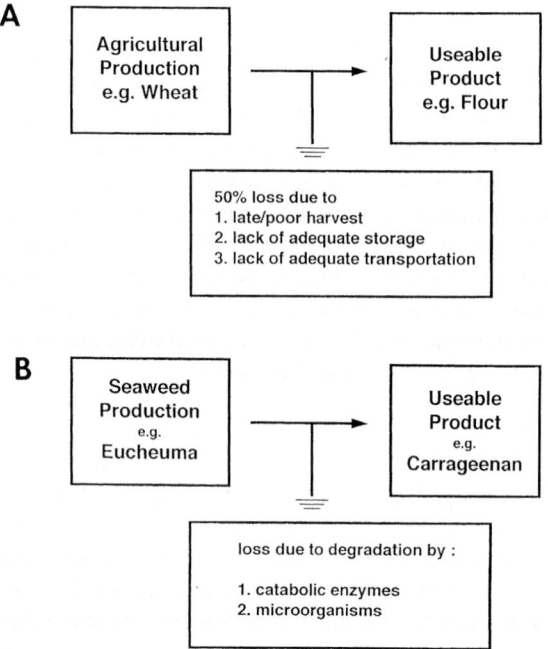

Figure 1. Linear production results in waste while cyclic production utilizes waste as resource for new products.

Figure 2. Increased production does not necessarily result in increased product when there is significant loss due to poor harvest and storage techniques, and lack of transportation. (A) Agricultural model, (B) Seaweed model.

have application to the brown algal (alginate) industry as well.

With the deconstruction of the USSR (CCCP) we learn that it was not uncommon for 50% of agricultural production in any given year to be lost due to poor harvesting, primitive storage and inadequate transportation (Figure 2A). So wheat and potatoes would rot in the field or in open air storage on the ground for lack of transport to processing facilities. Conventional wisdom makes all of us economic advisors in this simplistic scenario. Both distance and hindsight give us 20/20 vision. We realize that in this specific case increased agricultural production would not appreciably increase amounts of food. We understand that excellent solutions are worthless unless they match specific, immediate problems. In the former USSR, attempts to improve harvest efficiency, crop storage without loss, and transportation of crops for processing should have been priorities, while any attempt to increase agricultural production would have been an excercise in futility. When we superimpose this schematic of poor agricultural practices on phycocolloid production, we note an uncomfortable fit (Figure 2B). Many of us pursue research programs aimed at a marginal increase in phycocolloid production even though adequate biomass exists to produce most carrageenan and agar products. Profitability for many industries is constrained by variables other than biomass production, including (Figure 3):

(1) Poor quality control due to variability in harvest and storage of dried biomass

Unless seaweed is quickly and thoroughly dried, and kept dry, phycocolloids are irreversibly degraded both by catabolic enzymes and microorganisms. Bacteria (Jaffray et al., 1997) as well as fungi (Melo et al., 1997) have been implicated in the degradation of phycocolloid polysaccharides during storage. Under tropical field conditions, rain and humidity are inextricably connected to mariculture sites. We wonder if the effective drying and storage technologies associated with the leaf tobacco industry cannot be modified for the phycocolloid industry?

It is possible that good things come to those who wait. But for those of us who are less patient, I recommend a meeting of field agents representing the phycocolloid and tobacco industries with microbiologists to discuss microbial degradation problems associated with long-term storage and shipment of seaweed biomass.

(2) Long distance transportation of biomass for phycocolloid extraction.

When coal was discovered in Australia, entrepreneurs envisioned huge ships exporting coal to Europe. This business venture was quickly derailed by reality when an engineer calculated that a coal freighter would burn more coal for its own fuel than could be carried on such a long trip around South America. In general, for any bulky (low density) commodity, it is cost effective to pre-process prior to shipping. This gen-

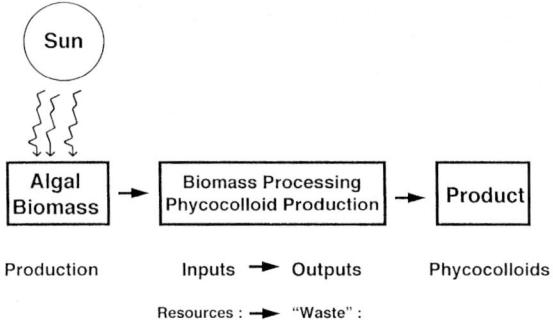

Figure 3. In the phycocolloid industry, profitability is constrained by variables in addition to biomass production.

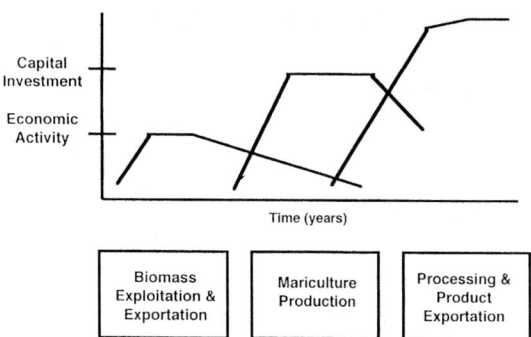

Figure 4. Relationship of capital investment and economic activity increase to the sequence of phycocolloid exploitation.

eralization will become dramatically more significant in economic equations as:
(a) the artificially low price of fuel is adjusted upward; and
(b) producing countries insist on a greater profit share which always comes from value added products rather than from commodities.

Only government beaurocracy would consider it rational and cost effective to grow biomass in southeast Asia, ship the biomass to America or Europe for processing, and then ship the value-added product back to southeast Asia for distribution and sales. In Indonesia, local processing of *Kappaphycus* began in 1988. Now alkali-treated cottonii (ATC), Philippine natural grade (PNG), semi-refined carrageenan (SRC), seaweed flour (SF) and natural washed carrageenan (NWC) are among the products developed for export and further refinement (Luxton, 1993).

(3) A conspicuous corollary to the economic liability of long distance transport of biomass is the problem of long distance transport of product.

In the 21st century, long distance separation of production and markets will be an unsustainable luxury. The phycocolloid industry must work with producers to develop semi-refined products for low cost transportation to refinement sites in Europe and America. Eventually, production and consumption must be localized, if not nationally, then at least regionally. Generalizations always lack impact and tend to be less than memorable. We identify with the specific and personal. Therefore, I will cite a specific example by way of illustration. Since commercial cultivation of *Gracilaria* began in Chile in 1982, production, freed of reliance on diminishing wild populations, has expanded dramatically (Zertuche Gonzales, 1993). Now Chile processes high-grade agar sufficient for the domestic market and for export (Oliveira & Alveal, 1990; Avila & Sequel, 1993; Alveal et al., 1997), with more than 70% of the *Gracilaria* harvest supporting local refinement (Norambuena, 1996). The present status and evolution of phycocolloid production in Chile have been summarized by Santelices (1996). It seems significant that the initial stage characterized by collection of wild populations potentially results in both the greatest environmental degradation, food web disruption and the least economic activity (Figure 4).

(4) Water, biomass and wastes management

The phycocolloid industry requires huge inputs of water and biomass, and produces huge outputs of waste water and biomass residue. There are several solutions to the associated problems: first, large amounts (as opposed to small amounts) of waste are a much bigger problem than simple mathematical increase. That is, disposing of 100 000 metric tons of biomass residue is not just 1000 times more difficult than disposing of 1000 metric tons. The relationship is probably more geometric than mathematical. Second, waste water and biomass residue are much greater problems in industrialized, urban settings than in rural, agricultural environments. I will remind you that animal manure is fertilizer on the farm, but waste and pollution in the city.

The solution suggested by both of the above problems is the same: decentralized, small scale processing in rural, agricultural areas where both waste water and biomass residue can be incorporated into agricultural production (Figure 5). My example here is Thailand. In order to reduce its dependency on imported agar, which is vital to sustaining a micropropagation industry of orchids and other horticultural plants,

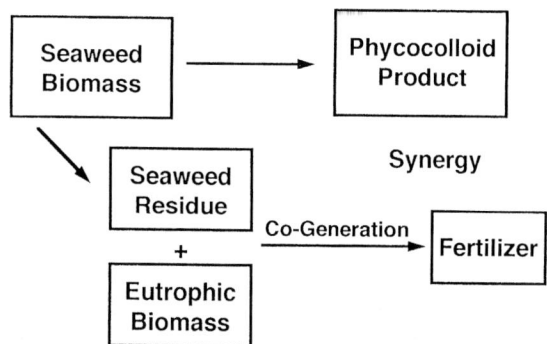

Figure 5. Co-production and synergy increase profitability and transform waste into renewable resources.

Thailand initiated a program to produce and process seaweed locally (Chardrkrachang & Chinadit, 1988; Chinadit & Chandrkrachang, 1986). This system is elegant for its simplicity and reliance on low technology as opposed to high-tech processes. It is based on small-scale co-operatives processing *Gracilaria* for agar using materials at hand such as coconut presses. Semi-refined agar can be used locally, or sold to national processors for the manufacture of high-grade agar. The seaweed residue can be composted for fertilizer, and the processing water routed to the village irrigation system.

(5) Co-production and synergy

The solution to the above problems leads logically to the concept of co-production and synergy — the process by which the interaction of two or more components results in an effect of which each is individually incapable and/or which is greater than the sum of the individual components (Figure 5).

As an example of co-production, organic wastes from salmon farming can be used as resources for new products. Specifically, production of both shellfish and seaweeds is enhanced when they are placed 'downstream' in an aquaculture system (Bodvin et al., 1996). Bioconversion of biomass by fermentation to useful products is not generally cost effective. However, transformation of algal biomass with other organic waste (wood and manure) results in a competitive fertilizer (Morand et al., 1990). In many areas of the world with polluted coastal ecosystems, eutrophication results in significant production of biomass (Rosenberg, 1985) and resultant ecological impact (Chassany de Casabianca, 1984). A bioconversion facility established to process phycocolloid biomass residue into fertilizer (co-production) could synergize by accepting eutrophication biomass, with twin positive environmental results: increased production of fertilizer and removal of nutrients which typically recycle to support rounds of biomass production (Morand et al., 1990).

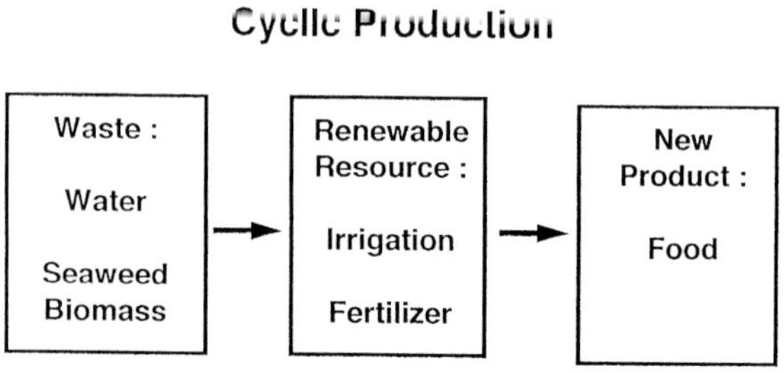

Figure 6. In cyclic processes, production design transforms waste into renewable resources.

Value of biosphere components

Once we accept that all processes are cyclic and not linear, production design will seek to transform waste into renewable, recycled resources (Figure 6). The next logical step is to understand the hidden value of biosphere components in maintaining a liveable environment. Presently, more than 40% of the Earth's primary productivity is redirected to human consumption (Hawken, 1993). We cannot save endangered species, protect habitat and redirect the bulk of solar-based production to our own selfish ends (Figure 6).

We can't have our cake and eat it too. As biologists and scientists, we should be concerned about the potential our research has to encourage and aid environmental degradation. Therefore, as we identify new target species for the phycocolloid industry, every effort should be made to shorten the transition time during which the species is collected from naturally occurring stocks and all biomass comes from mariculture. *Gelidiella* is representative of the agarophytes that many of us have helped develop as an attractive resource for the phycocolloid industry (Kapraun et al., 1994; Roleda et al. 1997a, b). More than a decade ago, it was proposed that a management scheme be developed prior to commercial harvest of *Gelidiella* natural stocks (Trono & Ganzon-Fortes, 1985), or at a minimum, that pruning/cutting harvest replace whole plant collection. Not surprisingly, over-exploitation and over-harvesting of natural beds and collection of whole plants of this slow growing species remain the norm. World-wide, most biomass continues to come from wild populations (Ganzon-Fortes, 1994) even though we know that this alga plays a significant role in coral reef ecology.

Resource utilization

At this point in any discussion of resource utilization, a distinction should be made between pragmatism and idealism. Many would agree that habitat destruction is bad, but what realistic alternative exists to foster economic development? We can just as easily ask if it is pragmatic to over-harvest and destroy a resource in a few years, or if it is idealistic to impose a management program with restraints for collecting procedure and harvest season to prolong the economic viability of a resource? It is no accident that countries with the strongest environmental laws also have the strongest economies! It has always been this way. Our English words for economy and ecology are both derived from the Greek *oikos* (household), and reflect the intimate relationship between sustainable development and husbandry of resources.

No resource exists solely and exclusively for the pleasure and benefit of man. All resources serve important functions as food, habitat, nursery and substrate stabilizers (Santelices, 1996). All of these functions have value and must be included in all cost–benefit calculations. There ain't no free lunch. Now we return to our rhetorical questions: How can we turn 'black' into 'green'? What's in it for me? We collectively have tremendous resources of information, knowledge and experience which can help the phycocolloid industry.

(1) We can help reduce production variability and degradation of stored biomass by creative use of technology for drying and storage of seaweed. Better quality seaweed translates into less biomass required for extraction, and reduced demands on water and energy resources.

(2) We can help develop co-generation and synergistic technologies. Together is better, not just more fun!

(3) We can encourage and support co-operatives based on small scale, local production–extraction of semi-refined product. This move should not be viewed as a source of competition to the big producers. They can guarantee supplies of semi-refined product by contracting with co-operatives.

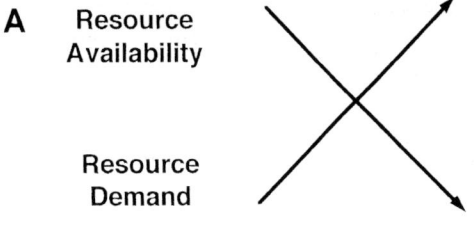

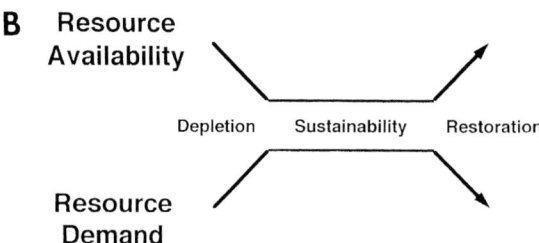

Figure 7. Non-sustainable and sustainable development compared. (A) Exploitation of natural populations results in decreased resource in response to increased demand. (B) Reliance on mariculture permits sustainability and an opportunity to restore previously damaged habitat.

(4) We can renew our efforts to turn from reliance on harvest of wild populations, which leads to destruction of ecosystems, by developing resources which are amenable to mariculture (Figure 7A). Resource sustainability, and ecosystem restoration both rely on successful mariculture (Figure 7B). The literature is replete with the consistent, predictable sequence of events that can be associated with attempts to utilize natural seaweed populations as sustainable, renewable resources for the phycocolloid industry (Norambuena, 1996). Although careful management is possible, especially with brown algae utilized for alginates (Avila & Seguel, 1993; Barilotti & Zertuche-Gonzales, 1990), typically natural wild populations are severely reduced in a region, and the harvest moves on, leaving the coastal subsistence inhabitants in predictable poverty (Fortes, 1993; Oliveira, 1990; Oliveira & Berchez, 1993). The human and ecological losses are seldom calculated in economic equations as long as new biomass sources are available to the industry. This waste of resources is unnecessary and, therefore, largely inexcusable as long-term profitability is highly correlated with sustainable development (Figure 4).

In contrast, the Philippine mariculture experience for *Eucheuma* remains a realistic model for seaweed farming in developing countries (Doty & Alvarez, 1975; Ricohermoso & Deveau, 1978; Trono, 1992;

Yarish & Wamukoya, 1990). For those of you who may have been on another planet for the last 25 years, I will summarize as follows:

(a) The Philippines carrageenan industry is based on the mariculture of *Eucheuma* and *Kappaphycus* cultivars, not wild populations (Trono, 1992).
(b) Mariculture results in less environmental impact and degradation than occurs with harvest of wild populations.
(c) Specific cultivars are grown vegetatively, making the system responsive to quality control and cultivar improvement (Dawes & Koch, 1991; Dawes et al., 1993, 1994).

My recommendation is this: if it can't be grown in a low-impact mariculture system, it should be used with great caution. As long as marine resources are viewed as common property, they will typically become part of the tragedy of the commons, overexploited with no one taking responsibility (Hardin, 1968). For coastal regions determined to develop a phycocolloid industry, we marine biologists and scientists can help in several ways. We can encourage the development and implementation of cultivation technologies for *Mazzaella (Iridaea)*, *Gigartina*, and *Gelidium*, *Pterocladia* and *Gelidiella*, all of which are desperately needed (Akatsuka, 1986; Avila & Seguel, 1993; Barilotti & Zertuche-González, 1990; Gallardo et al., 1990; Macler & Zupan, 1991; Oliveira, 1981; Oliveira & Berchez, 1993; Rueness & Fredriksen, 1990). More of us need to approach this seemingly hopeless task although, like marriage, it could be viewed by the pessimistic as a triumph of hope over reason. There is reason for cautious optimism. *Hypnea* in many respects is typical of these commercially exploited seaweeds in that its natural populations are generally insufficient to sustain economic harvest pressure (Mshigeni & Chapman, 1994; Schenkman, 1989). Fortunately, determined attempts to domesticate this seaweed and make it amenable to mariculture are starting to produce encouraging results (Berchez et al., 1993; Camaro Neto, 1987). Efforts to propagate *Gelidium* spp., which are characterized by relatively fragile natural populations and strong cyclic crop fluctuations, are showing promise (Rojas et al., 1996).

Surely someone among us can develop a system for growing other seaweeds commercially. In the mean time, we should consider phycocolloid production based on introduction of cultivars as an alternative to depletion of natural populations of target species. Problems associated with *Codium fragile* in the north Atlantic (Fralick & Mathieson, 1972; Kapraun et al., 1988) and *Caulerpa taxifolia* in the Mediterranean (Meinez et al., 1993) should not reflexively prompt us to equate introduction of seaweed cultivars with spread of the Ebola virus. "Although there are sound ecological arguments to avoid cultivar introduction, some are more passionate than scientific. The potential in many cases has been exaggerated and careful introduction can be successful provided it occurs under a set of reasonable criteria" (Oliveira, 1990). *Eucheuma* and *Kappaphycus* cultivars have been widely transported and introduced throughout the Indo-Pacific. *Gracilaria chilensis* has been safely introduced in Brazil (Plastino & Oliveira, 1988) and *Porphyra* from Japan is now grown in New England and Puget Sound (Mumford 1990, Bergdahl, 1990).

(5) We can help to develop new, high value products including agarose which has proven indispensable to biotechnology (Renn, 1990), agars with medical and dental applications (Kasloff, 1990) and alginates for high-performance bio-paper (Kobayashi, 1990). These products promote a healthy bottom line (black) with high per unit values, rather than profit by shear volume, and so qualify as 'green'. One approach to the development of new phycocolloid resources is based on genetic transformation and the creation of novel genome combinations (Cheney *in litt.*; Huang et al., 1996; Sivan et al., 1995). This basic research deserves our support and encouragement. Recent comprehensive reviews of contributions from biotechnology are available both with an industry perspective (Renn, 1990, 1997) and with a view from the field (Bird, 1995). There is little I can say in a brief time that could add to their insights, comments and recommendations.

In summary, I remind you that the marine phycocolloid industry has come a long way since Doty pioneered *Eucheuma* farming (Doty & Alvarez, 1975). Now a world-wide industry can impact on major ecosystems and affect emerging economies. I don't think that it is reasonable to blame coastal subsistence fishermen who are uneducated, or industry representatives who may be uninformed or ill-advised for environmental and economic problems which are emerging, when we marine biologists and scientists do not take the lead in doing the right thing. Sustainable development: it's good for the industry, and it's good for me!

Acknowledgements

The author gratefully acknowledges financial support from the United States Agency for International Development (HRN-5600-G-2023-0). Gratitude and appreciation for logistical support, research collaboration and friendship are expressed to Drs A. Hurtado-Ponce, L. Liao and G. Trono and to numerous other colleagues at the UP Institute of Marine Science and the University of San Carlos Marine Botany Program. Dr Don Renn is acknowledged for critical comments on the manuscript and for useful insights into the phycocolloid industry. This paper represents contribution number 182 from the Center for Marine Science Research, UNC-Wilmington.

References

Akatsuka, I., 1986. Japanese Gelidiales, especially *Gelidium*. Ocean. mar. Biol. ann. Rev. 24: 171–263.

Alveal, K., H. Romo, C. Werlinger & E. C. de Oliveira, 1997. Mass cultivation of the agar-producing alga *Gracilaria chilensis* (Rhodophyta) from spores. Aquaculture 148: 77–83.

Avila, M. & M. Sequel, 1993. An overview of seaweed resources in Chile. J. appl. Phycol. 5: 133–139.

Barilotti, D. C. & J. A. Zertuche González, 1990. Ecological effects of seaweed harvesting in the Gulf of California and Pacific Ocean off Baja California and California. Hydrobiologia 204/205: 35–40.

Berchez, F. A. S., R. T. L. Pereira & N. F. Kamiya, 1993. Culture of *Hypnea musciformis* (Rhodophyta, Gigartinales) on artificial substrates attached to linear ropes. Hydrobiologia 260/261: 415–420.

Bergdahl, J. C., 1990. Nori (*Porphyra* C. Ag.: Rhodophyta) mariculture research and technology transfer along the northeast Pacific Coast. In Akatsuka, I. (ed.), Introduction to Applied Phycology. SPB Academic Press, The Hague: 519–551.

Bird, K. T., 1995. Biotechnology of marine plants: reshaping our world now. SeaTechnology, April: 58–62.

Bodvin, T., M. Indergaard, E. Norgaard, A. Jensen & A. Skaar, 1996. Clean technology in aquaculture – a production without waste. Hydrobiologia 326/327: 83–86.

Camaro Neto, C., 1987. Seaweed culture in Rio Grande do Norte, Brazil. Hydrobiologia 151/152: 363–367.

Casabianca, M. L. de, E. Marinho-Soriano & T. Laugier, 1997. Growth of *Gracilaria bursa-pastoris* in a Mediterranean lagoon: Thau, France. Bot. mar. 40: 29–37.

Chardrkrachang, S. & U. Chinadit, 1988. Seaweed production and processing, a new approach. INFOFISH International No. 4|88, 50728 Kuala Lumpur, Malaysia: 22–25.

Chassany de Casabianca, M.-L., 1984. Analyse de problèmes écologiques liés à la récolte de biomasse algale en milieu lagunaire. Rapport de contrat no. 81.G.0983. Action concertée: écologie et aménagement rural. M.R.T., Paris, France, 80 pp.

Chinadit, U. & S. Chandrkrachang, 1986. Simplified method for agar extraction from some agarophytes in Thailand. Bull. Phamacognosy Soc. Thailand 2. 11–13.

Dawes, C. J. & E. W. Koch, 1991. Branch, micropropagule and tissue culture of the red algae *Eucheuma denticulatum* and *Kappaphycus alvarezii* farmed in the Philippines. J. appl. Phycol. 3: 247–257.

Dawes, C. J., G. C. Trono & A. O. Lluisma, 1993. Clonal propagation of *Eucheuma denticulatum* and *Kapaphycus alvarezii* for the Philippine seaweed farms. Hydrobiologia 260/261: 379–383.

Dawes, C. J., Lluisma, A. O. & G. C. Trono, 1994. Laboratory and field growth studies of commercial strains of *Eucheuma denticulatum* and *Kappaphycus alvarezii* in the Philippines. J. appl. Phycol. 6: 21–24.

Doty, M. S. & V. B. Alvarez, 1975. Status, problems, advances and economics of *Eucheuma* farms. Mar. Tech. Soc. J. 9: 30–35.

Fortes, M. D., 1993. Assessment of the natural stocks of *Gelidiella acerosa* (Forssk.) Feld. *et* Hamel in Matigue Is., Camiguin, Southern Philippines. In Calumpong, H. P. & E. G. Menez (eds), Proc. 2nd RP-USA Phycological Symposium/Workshop. Philippine Council for Aquatic and Marine Research and Development (PCAMRD) Los Baños, Laguna: 247–256.

Fralick, R. A. & A. C. Mathieson, 1972. Winter fragmentation of *Codium fragile* (Suringar) Hariot ssp. *tomentosoides* (van Goor) Silva (Chlorophyceae, Siphonales) in New England. Phycologia 11: 67–70.

Gallardo, T., M. A. Cobelas & A. Alvarez de Meneses, 1990. Current state of seaweed resources in Spain. Hydrobiologia 204/205: 287–292.

Ganzon-Fortes, E. T., 1994. *Gelidiella*. In Akatsuka, I. (ed.), Biology of Economic Seaweeds. SPB Academic Publishing, The Hague: 149–184.

Hardin, G., 1968. The tragedy of the commons. Science 162: 1243–1248.

Hawken, P., 1993. The Ecology of Commerce. Harper Business, NY: 250 pp.

Huang, X., J. C. Weber, T. K. Hinson, A. C. Mathieson & S. C. Minocha, 1996. Transient expression of the GUS reporter gene in the protoplasts and partially digested cells of *Ulva lactuca* L. (Chlorophyta). Bot. mar. 39: 467–474.

Jaffray, A. E., R. J. Anderson & V. E. Coyne, 1997. Investigation of bacterial epiphytes of the agar-producing red seaweed *Gracilaria gracilis* (Stackhouse) Steentoft, Irvine & Farnham from Saldanha Bay, South Africa and Lüderitz, Namibia. Bot. mar. 40: 569–576.

Kapraun, D. F., M. G. Gargiulo & G. Tripodi, 1988. Nuclear DNA and karyotype variation in species of *Codium* (Codiales, Chlorophyta) from the North Atlantic. Phycologia 27: 273–282.

Kapraun, D. F., E. Ganzon-Fortes, K. T. Bird, G. Trono & C. Breden, 1994. Karyology and agar analysis of the agarophyte *Gelidiella acerosa* (Forsskål) Feldmann *et* Hamel from the Philippines. J. appl. Phycol. 6: 545–550.

Kasloff, Z., 1990. The medical and dental uses of algae and algal products. In Akatsuka, I. (ed.), Introduction to Applied Phycology. SPB Academic Press, The Hague: 401–406.

Kobayashi, Y., 1990. High-performance papers from seaweeds. In Akatsuka, I. (ed.), Introduction to Applied Phycology. SPB Academic Press, The Hague: 407–411.

Luxton, D. M., 1993. Aspects of the farming and processing of *Kappaphycus* and *Eucheuma* in Indonesia. Hydrobiologia 260/261: 365–371.

Macler, B. A. & J. R. Zupan, 1991. Physiological basis for the cultivation of the Gelidiaceae. Hydrobiologia 221: 83–90.

Magne, F., 1993. Importance of basic research in applied phycology. Hydrobiologia 260/261: 25–29.

Meinesz, A., J. de Vangelas, B. Hesse & X. Mari, 1993. Spread of the introduced tropical green alga *Caulerpa taxifolia* in northern Mediteranean waters. J appl. Phycol. 5: 141–147.

Melo, V. M. M., D. A. Medeiros, F. J. B. Rios I., I. M. Castelar & A. de F. F. U. Carvalho, 1997. Antifungal properties of proteins (agglutinins) from the red alga *Hypnea musciformis* (Wulfen) Lamouroux. Bot. mar. 40: 281–284.

Morand, P., R. H. Charlier & J. Mazé, 1990. European bioconversion projects and realizations for macroalgal biomass: Saint-Cast-Le-Buildo (France) experiment. Hydrobiologia 204/205: 301–308.

Mshigeni, K. E. & D. J. Chapman, 1994. *Hypnea* (Gigartinales, Rhodophyta). In Akatsuka, I. (ed.), Biology of Economic Algae. SPB Academic Publishing, The Hague: 245–281.

Mumford, T. F., 1990. Nori cultivation in North America: growth of the industry. Hydrobiologia 204/205: 89–98.

Norambuena, R., 1996. Recent trends of seaweed production in Chile. Hydrobiologia 326/327: 371–379.

Oliveira, E. C. de, 1981. Marine phycology and exploitation of seaweeds in South America. Proc. int. Seaweed Symp. 10: 97–112.

Oliveira, E. C. de, 1990. The rationale for seaweed cultivation in South America. In Oliveira E. C. de & N. Kautsky (eds), Cultivation of Seaweeds in Latin America. IFS Workshop Proceedings, Universidade de São Paulo: 135–141.

Oliveira, E. C. de & K. Alveal, 1990. The mariculture of *Gracilaria* (Rhodophyta) for the production of agar. In Akatsuka, I. (ed.), Introduction to Applied Phycology. SPB Academic Publishing, The Hague: 553–564.

Oliveira, E. C. de & F. A. S. Berchez, 1993. Resource biology of *Pterocladia capillacea* (Gelidiales, Rhodophyta) populations in Brazil. Hydrobiologia 260/261: 255–261.

Plastino, E. M. & E. C. de Oliveira, 1988. Sterility barriers among species of *Gracilaria* (Rhodophyta, Gigartinales) from the São Paulo littoral, Brazil. Br. phycol. J. 23: 267–271.

Renn, D. W., 1990. Seaweeds and biotechnology: inseparable companions. Hydrobiologia 204/205: 7–13.

Renn, D. W., 1997. Biotechnology and the red seaweed polysaccharide industry: status, needs and prospects. Tibtech 15: 9–14.

Ricohermoso, M. A. & L. E. Deveau, 1978. Review of commercial propagation of *Eucheuma* (Florideophyceae) clones in the Philippines. Proc. int. Seaweed Symp. 9: 525–531.

Rojas H., R., N. Leon M. & R. Rojas O., 1996. Practical and descriptive techniques for *Gelidium rex* (Gelidiales, Rhodophyta) culture. Hydrobiologia 326/327: 367–370.

Roleda, M. Y., E. T. Ganzon-Fortes & N. E. Montaño, 1997a. Agar from vegetative and tetrasporic *Gelidiella acerosa* (Gelidiales, Rhodophyta). Bot. mar. 40: 501–506.

Roleda, M. Y., E. T. Ganzon-Fortes, N. E. Montaño & F. N. de los Reyes, 1997b. Temporal variation in the biomass, quantity and quality of agar from *Gelidiella acerosa* (Forsskål) Feldmann *et* Hamel (Rhodophyta: Gelidiales) from Cape Bolinao, NW Philippines. Bot. mar. 40: 487–495.

Rosenberg, R., 1985. Eutrophication. The future marine coastal nuisance? Mar. Pollut. Bull. 6: 227–231.

Rueness, J. & S. Fredriksen, 1990. Field and culture studies of species of *Gelidium* (Gelidiales, Rhodophyta) from their northern limit of distribution in Europe. Hydrobiologia 204/205: 419–424.

Santelices, B., 1996. Seaweed research and utilization in Chile: moving into a new phase. Hydrobiologia 326/327: 1–14.

Schenkman, R. P. F., 1989. *Hypnea musciformis* (Rhodophyta): ecological influence on growth. J. Phycol. 25: 192–196.

Sivan, A., J.-C. Thomas, J.-P. Dubacq, D. van Moppes & S. Arad, 1995. Protoplast fusion and genetic complementation of pigment mutations in the red microalga *Porphyridium* sp. J. Phycol. 31: 167–172.

Trono, G. C., 1992. *Eucheuma* and *Kappaphycus*: taxonomy and cultivation. Bull. mar. Sci. Fish., Kochi Univ. 12: 51–65.

Trono, G. C. & E. T. Ganzon-Fortes, 1985. The economic potentials of seaweeds. Philipp. Fish. Ann. 1985: 62–68.

Yarish, C. & G. Wamukoya, 1990. Seaweeds of potential economic importance in Kenya: field survey and future prospects. Hydrobiologia 204/205: 339–346.

Zertuche González, J. A., 1993. Situacion actual del cultivo de algas agarofitas en America Latina y el Caribe. In Zertuche González, J. A. (ed.), Situacion actual de la industria de macroalgas productoras de ficocoloides en America Latina y el Caribe. FAO Tech. Pc. 13, Roma, 57 pp.

A conceptual framework for marine agronomy*

B. Santelices
Departamento de Ecología, Facultad de Ciencias Biológicas, P. Universidad Católica de Chile, Casilla 114-D, Santiago, Chile. Tel: [+56] 2 6862648. Fax: [+56] 2 6862621. E-mail: bsanteli@genes.bio.puc.cl

Key words: agronomic ordination, farming productivity, multi-step farming, seaweeds, site fertility

Abstract

Between the late 1960s and the early 1980s, several generations of phycologists in Hawaii and the Philippines, associated with M. S. Doty, contributed to developing a new approach, and to advance concepts in marine agronomy. This study reviews the approach and the main concepts contributed. Integrating these contributions with others, a basic conceptual framework for marine agronomy is presented.

Introduction

Compared to the 8 000–10 000 year history of terrestrial agronomy, marine agronomy has a short history. Less than 250 years ago, Japanese fishermen began sticking brush on the seashore to expand the grounds where *Porphyra* would settle, and less than 50 years ago, Drew (1949) provided scientific information on its life cycle that eventually allowed Japanese fishermen and farmers to artificially seed their nets, accelerating the expansion of the *Porphyra* farms.

In the 50 years following Drew's findings, close to a hundred economic seaweed taxa have been tested for their field farming potential, and nearly a dozen are being commercially cultivated today (see Ohno & Critchley, 1993 for listing). A growing body of information on the chemical structure of commercial compounds, the physiological ecology of the various species, and production technologies for different types of algae is accumulating. Furthermore, basic concepts on marine agronomy are gradually being developed, which allow the agronomic nature of seaweed to be understood and farming to be approached with greater, scientifically based, predictive capability.

Among the several scientists that have contributed to the recent development of marine agronomy, M. S. Doty played an outstanding role. Between the late 1960s and the early 1980s, Doty was able to generate a very productive scientific environment in Hawaii that attracted over 30 students and several dozen co-workers from all over the world (see Kraft, 1997 for greater details). This group contributed to the development of a new approach to marine agronomy. Such an approach led to new concepts and generalizations of widespread application in seaweed farming. This study reviews this approach and some of the most important concepts put forward by Doty and his associates. By integrating these contributions with others, a basic conceptual framework for marine agronomy is developed.

Methodological approach to farming

When Doty started his field cultivation trials (Doty, 1970), extensive seaweed farming was already being developed in Japan, China and a few other countries in south east Asia (Tseng, 1981). In addition, various laboratories in Canada (e.g. Neish & Fox, 1971; Neish et al., 1977), the U.S.A. (e.g. North, 1971; Neushul, 1972) and Europe (e.g. Pérez, 1972) were developing experimental cultivation of various types of seaweeds. Departing from the above scientific efforts, Doty and co-workers developed a product-oriented, multi-disciplinary approach to seaweed farming. Since the ultimate objective of farming was an optimization of production, potential farmers needed to know how the production function could be optimized beforehand. This involved knowing the production possibilities of the target species (e.g. biomass, carrageenan, fine

---
* Presented in a special minisymposium in honour of Maxwell S. Doty.

chemical), and the pattern of temporal changes in the various products. Further, the maximum production of various compounds derived from a given species might not necessarily coincide in time, and the farmer may need to look at options in order to get the best product from farming efforts.

Since the chemical nature, quality and quantities of many seaweed products vary among different taxonomic units (Stoloff & Silva, 1957; McCandless, 1981; Jensen 1993), taxonomy is always required as an auxiliary discipline. In seaweeds, the value of 'good taxonomy' is often recognized, not only by other taxonomists but also by the commercial companies using the species as raw materials for specific applications.

To optimize production, the farmer needs to know the interactions of factors which increase production. Thus, experimental studies on production ecology are needed to understand the relationships between the target product and the abiotic environment. Once these are known, field sites with positive interactions can be identified and prepared for target species farming. Farming is thus conceived (Santelices & Doty, 1989) as any artificial expansion of the habitat, mainly resulting from increasing the area over which the desired seaweed grows naturally. Ultimately this process increases the production of the target product. Experimental testing of potentially useful farming areas, cultivation technologies and routines are needed before farms are expanded to full scale.

Productivity of an established farm depends on the quality of the genome being cultivated, on the frequency and intensity of the abiotic and biotic disturbances, and on management efficiency. Therefore genetics, community ecology and management also are considered auxiliary disciplines needed to successfully farm seaweed crops.

Conceptual approach to farming

Data accumulated on seaweed farming resulted in the development of a few basic principles that, so far, seem of widespread applicability in marine agronomy. In this review, these are arranged as successive steps, leading to studies and activities needed in order to farm a previously selected target species.

Agronomic ordination of seaweeds

External morphology still appears to be the main character allowing seaweeds to be categorized. It also permits a first prediction on potential farming sites and agronomic requirements of the target species to be made.

Several types of findings support this view. The first finding indicates that external morphology integrates several algal functions and, therefore, is simultaneously related to several environmental factors (Doty, 1971; Neushul, 1972). The second, well documented by taxonomists and morphologists (e.g. Dawson, 1966), indicates that although seaweeds differ in external morphology, certain morphologies are repeated among phylogenetically different algal groups. Morphological similarities are now understood (Littler & Littler, 1980; Littler et al., 1983) as convergent adaptations to critical environmental factors, while differences between morphologies would represent divergent responses to such selection factors. Thus, environment and habitat requirements of species with convergent morphologies will be more similar than the requirement of species with divergent morphologies. Since the essence of farming is habitat expansion (Santelices & Doty, 1989), it is expected that seaweed morphology could be used as a first clue to define the type of habitat to be expanded and the type of farming to be developed.

The number of commercially farmed seaweeds (see Ohno & Critchley, 1993 for a review) is not diverse enough to provide a full agronomic classification scheme of seaweeds, but the present farming practices for different morphologies (Figure 1) suggest that this is a promising concept. Future increments in the number of farmed species would undoubtedly help to better define agronomic groups. The resulting classification scheme may or may not agree with other classification schemes based on other ecological relations (e.g. grazing; Steneck & Watling, 1982). From the agronomic point of view, some of the groups already defined in those schemes (e.g. 'corticated macrophytes' *sensu* Steneck & Watling, 1982) seem agronomically heterogeneous, while agronomically important morphologies (e.g. 'broad blades' such as *Gigartina* or *Sarcothalia*) have not been considered. Similarly, characters such as plasticity to change morphology under various water movement regimes, regeneration capacities of various plant parts and clonal versus unitary organization seem agronomically more important to classify seaweeds than the characters now used in the above classification schemes (photosynthesis, nutrient uptake, grazer susceptibility).

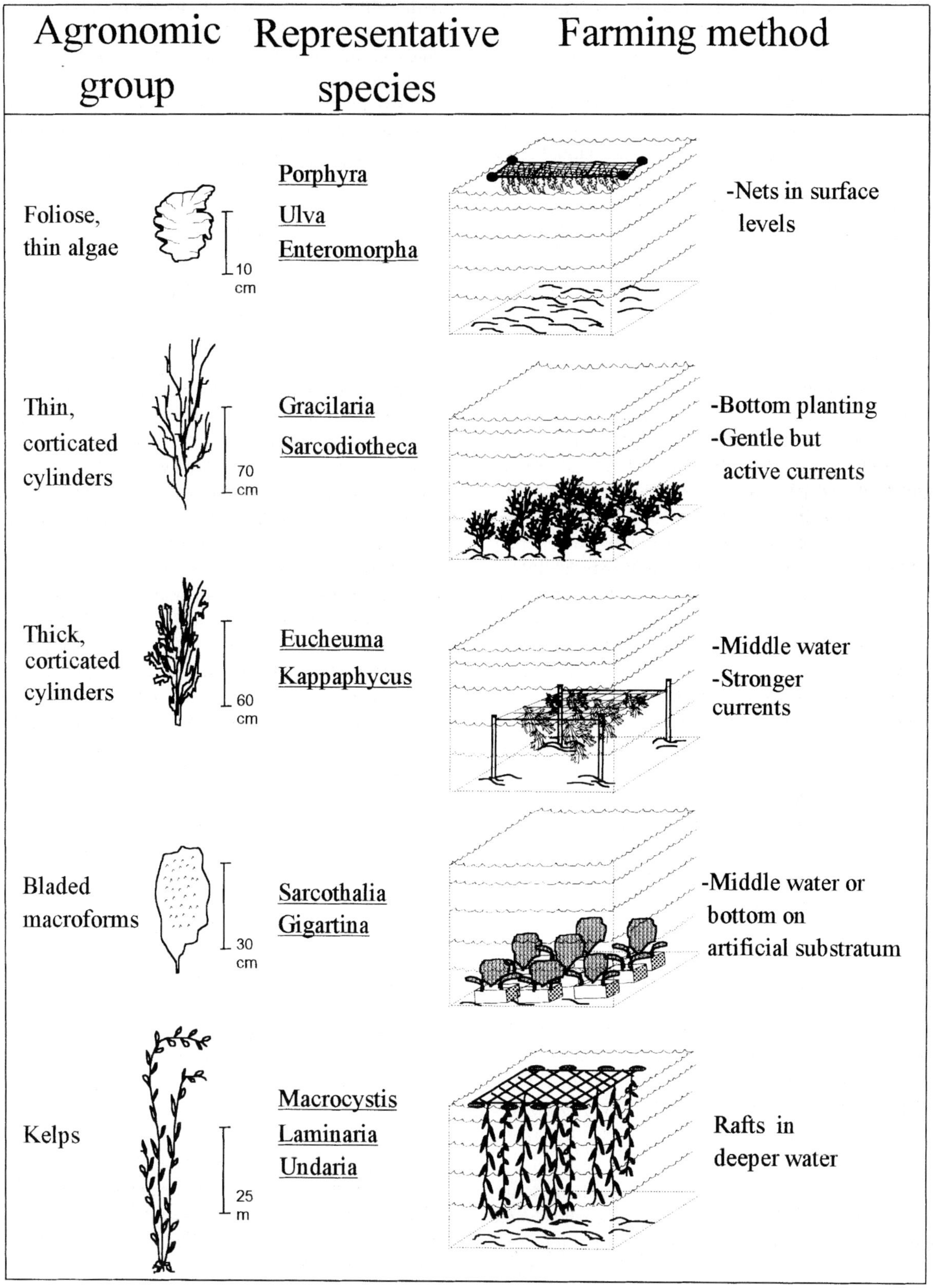

Agronomic group		Representative species		Farming method
Foliose, thin algae	~10 cm	Porphyra Ulva Enteromorpha		-Nets in surface levels
Thin, corticated cylinders	~70 cm	Gracilaria Sarcodiotheca		-Bottom planting -Gentle but active currents
Thick, corticated cylinders	~60 cm	Eucheuma Kappaphycus		-Middle water -Stronger currents
Bladed macroforms	~30 cm	Sarcothalia Gigartina		-Middle water or bottom on artificial substratum
Kelps	~25 m	Macrocystis Laminaria Undaria		Rafts in deeper water

Figure 1. Agronomic classification of seaweeds, emphasizing the relations between external morphology and farming technique.

Clonal and unitary seaweeds, one-step and multi-step farming

Seaweeds can have clonal or unitary organization. Clonal seaweeds can be grown and propagated by self-replication of genetically identical units. Such units function and survive on their own if separated from one another by natural processes or injury (Santelices, 1992). Fragments of unitary seaweeds, on the other hand, cannot survive and grow, and their propagation and farming has to be started from spores.

Clonal seaweeds are best suited for one-step farming. This consists (Figure 2) of regrowing adult thalli directly from fragments. The fragment needs to have a minimum size to successfully compete in the adult environment, and in time it will regenerate the adult form.

In contrast, farming of unitary seaweeds needs to be started from spores. However, ontogenetic development of anatomically complex seaweeds may pass through various morphologies. Since the various developmental stages use the environment differently (Neushul, 1972), to be successfully farmed, each stage may require different farming practices (Figure 2). Depending on the diversity of ontogenetic forms involved, this may lead to a two-step or to a multi-step farming practice.

Early, small-size developmental stages are often farmed in nursery facilities. Depending on the life cycle of the target species, such a nursery facility may need to support the growth of the propagules and juvenile stages only or to include also the alternate, microscopic, phase of heteromorphic species. Competition with undesired microforms and early developmental stages of other macroforms is solved by pre-emption of the available substratum through enhanced recruitment of the target species. Enhanced recruitment is attained by incubation of a large number of spores. Juvenile forms may respond better when cultivated at farming facilities of species with similar habitat requirements. However, for economic reasons they are often transplanted to the adult habitat, where they may outcompete other morphologically similar taxa because of a density effect.

Habitat partitioning and the abiotic environment

Even though external morphology is, without doubt, useful for characterizing the way a seaweed uses the environment, additional ecological information is required before farming. Many morphologically similar species may share a habitat, although they may differ in their fine grained responses to one or several environmental factors. To expand the niche of the species during farming, it is necessary to characterize the most important niche dimensions regulating or controlling production. This is normally done by experimentally testing the effects of key ecological factors on growth rates of the desired species. Results often give clues on the significant abiotic factors controlling growth.

Extensive laboratory and cultivation experiments with many species (see Lobban & Harrison, 1994 for a review) have suggested that growth is often regulated by a complex interaction of factors. This is the case, for example, in Gelidiales (Santelices, 1978; Frederiksen & Rueness, 1989) and Gracilariales (Hoyle, 1976, 1978), where growth is regulated by a complex interaction of irradiance, temperature, nutrients and water movement (Figure 3). For example, when light intensity or water movement are limiting, the effects of nutrients on growth are not evident. Nutrients can compensate for inadequate water movement when the latter factor is comparatively low. In turn, enhanced diffusion resulting from high water movement or nutrient additions, results in a more effective use of higher levels of irradiance and temperature, which leads to faster growth and higher pigment concentration. The above interactions indicate (Santelices, 1975) that factors very different in nature (e.g. chemical, such as nutrients, and physical, such as irradiance), may interact regulating growth of the target species and that a major decline of one factor (e.g. nutrients) could be compensated by other factors (e.g. water movement). Therefore, Liebig's law of the minimum does not appear to be directly applicable to marine agronomy.

Site fertility

Interactions, such as the above, are likely to occur in the field, but with some variations. In shallow water, where most seaweed crops are grown, environmental conditions may change very rapidly over space and time. This means that interactions are likely to be occurring at all times. In some places, water movement could, within a certain range, compensate for nutrient deficiencies. In others, variation in water movement could be compensated by adjustments in water quality, temperature and light.

The above variability could result in the occurrence, on a small scale, of places with different degrees of fertility for a given crop. Fertility represents the potential capacity of production in a given site due to its combination of favourable environmental factors

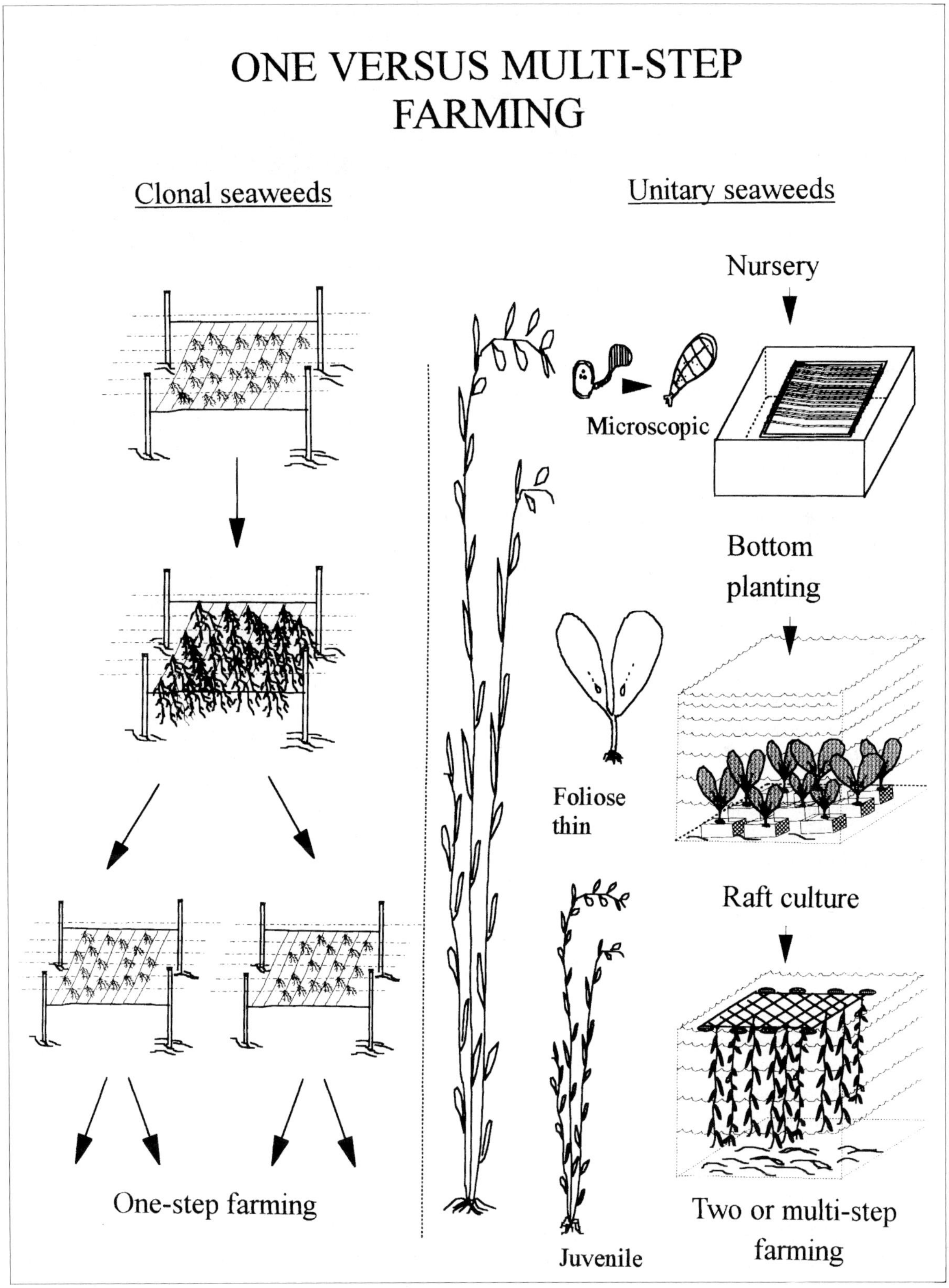

Figure 2. Relationship between seaweed organization (clones and unitary organisms) and one-step or multi-step farming.

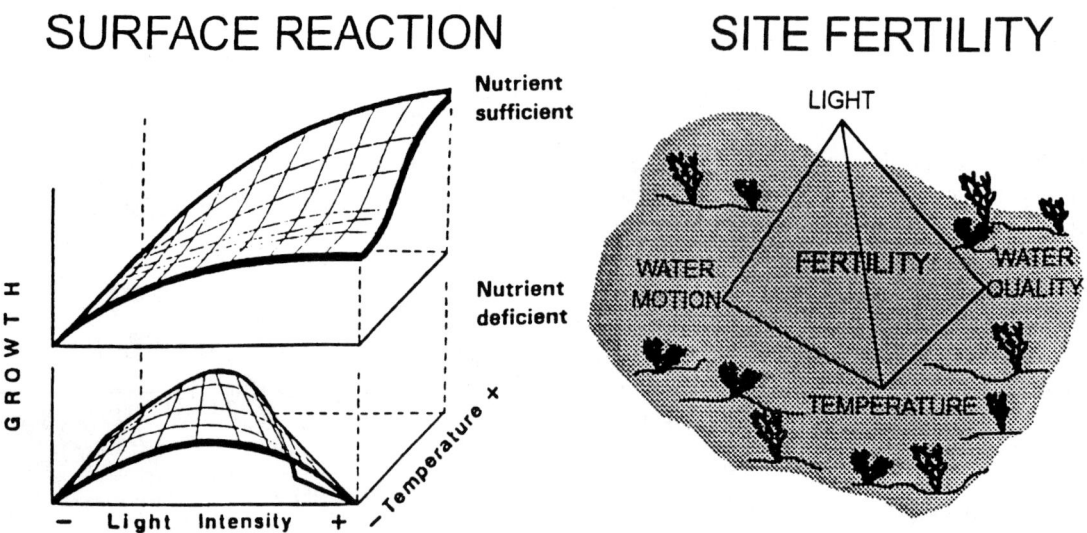

Figure 3. Comparison of diagrammatic representation of interacting factors. Left, surface reaction from laboratory experiments. Right, fertility site concept, applied to field situations.

(Doty, 1979). This can be estimated from observations of standing crop or measured by growth rates or production rates.

The site fertility concept was represented (Doty, 1979) by a tetrahedron (Figure 3). Each axis of the tetrahedron corresponds to one of the four ecologically important factors determining algal growth (irradiance, temperature, water movement, water quality), while the resulting volume represents the fertility of the given site. Any change in the physical environment modifying the above factors would change the position of the respective vertex in the tetrahedron. However, the volume representing fertility may or may not change, depending on the occurrence of compensatory changes in the other factors.

After its original formulation, Doty (1979) recognized that many more than only four environmental variables should be considered when explaining algal growth. However, for simplicity, the hypothesis included only these four components. Others could be added later if experimental and empirical results recommended their inclusion.

The site fertility hypothesis has the merit of conceptually explaining the short-term, almost random changes in standing crop and production potential often found in natural habitats. It anticipates that seaweed production will be heterogeneous in any given habitat due to the natural variations of the abiotic components. In addition, the concept allows controlled laboratory results on the effects of environmental factors on growth to be contrasted with the variability often found in the field. Many laboratory data are of limited use because they consider factor ranges which are totally unrealistic compared to field variation. Also, laboratory experiments normally maintain the different levels of a given factor constant, while in the field they may vary significantly through time. It should be noted that careful characterization of the abiotic environment determining fertility is lacking for most seaweed crops, including some of those which are currently being farmed. How to measure these factors, and what kind of statistical method to apply in order to have realistic results, are among the most complex problems facing marine farming today.

The essence of farming

The site fertility concept easily leads to an understanding of the essence of farming. If the site is fertile, the main activity will be to harvest and manage the crop, re-populating areas that might become less productive due to extensive harvest. If there is an abundance of fertile sites for the target species, but they are being occupied by a different crop, selective harvesting of the competitor, especially at times of the year when the competitor's recruitment is reduced (Santelices, 1990,

1996) would allow habitat expansion of the target species. If there is a fertile site which lacks the adequate substratum (e.g. fertile sites on sandy beaches), provision of the right substratum would allow the horizontal expansion of the habitat of the target species. Similarly, if any one of the key interacting factors determining the fertility of the site is found to be sub-optimal, that factor might be modified to increase fertility. Thus, farming always involves the introduction of artificialities that expand the area over which the desired crop will grow beyond where it would grow naturally. A successful expansion will result in a monoculture of the target species, freed from its natural community components, and at production costs that would compete with the production of the wild crop or with other farms producing the same seaweed elsewhere.

Although the site fertility hypothesis may, in theory, suggest the possibility of manipulating all abiotic factors, the situation in the sea is quite limiting. Farm enlargement to real scale is done horizontally to obtain cost-free sunlight. At a price, a farmer can attempt to control water quality. Handling of water movement and quantities in field farming is extremely difficult. Generally the farmer requires site testing to determine if water movement will allow growth of the target species. Similarly, there is little a farmer can do about sea water temperature. Due to the specific heat of water, it takes large amounts of energy to change and maintain the temperature of the water in which seaweeds grow. Thus, if the temperature is not right at a given site, moving a farm to where sea water temperatures are adequate to the crop is usually all a farmer can do to stay productive. All the above constraints emphasize the need to adequately select a site, to know how the environment varies from one place to another, how the different abiotic elements are interrelated and how their individual variabilities and compensatory effects modify crop growth.

Farming productivity

At present, close to a dozen seaweed species are commercially cultivated (see Ohno & Critchley, 1993 for review). However, only a few authors (Doty, 1981, 1986; Pizarro & Barrales, 1986; Santelices & Doty, 1989; Westermeier et al., 1993; Pizarro & Santelices, 1993; Santelices, 1996) have discussed the interaction of some of the factors determining farm productivity. Data suggest productivity would depend (Figure 4) on the fertility of the farming site, already discussed,

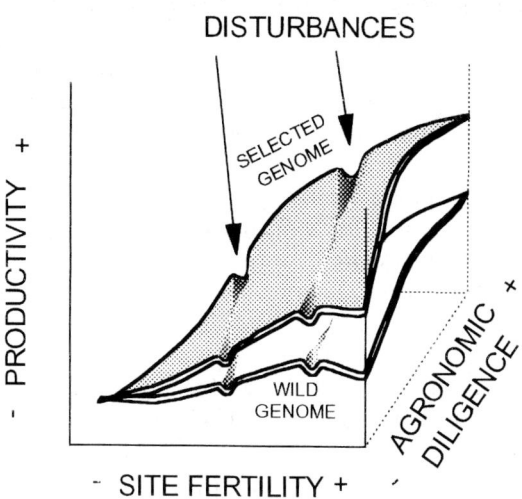

Figure 4. Interacting effects of site fertility, genome, agronomic diligence and disturbance in farm productivity.

the algal genome being cultivated, the frequency and intensity of biotic and abiotic disturbances, and the agronomic diligence.

Genetic improvement and breeding of new strains have been successful in some commercial crops such as *Laminaria* (Wu & Lin, 1987) and *Porphyra* (Miura, 1975, 1976; Ohme et al., 1986; Shin & Miura, 1990). Other crops, although characterized in terms of qualitative or quantitative genetics (see Patwary & van der Meer, 1992 for review), have not been genetically improved. Recent studies with clonal species (Santelices et al., 1995) are revealing additional sources of variability, which are likely to influence and complicate traditional strain selection practices.

Physical and biological disturbances are an unavoidable part of farming (Santelices, 1996). The habitat expansion done in farming always involves invading the natural habitat of other species, some of which may later return to the farm as competitors. On the other hand, the more distant the expanded area is from the original habitat of the target species in terms of abiotic environment, the higher the probabilities of processes that, while being normal in that habitat, would constitute disturbances for the growth of the target species. Long-term monitoring of catastrophes and major disturbances (e.g. Pizarro & Santelices, 1993) is required in all farming programmes and specially so in those involving extensive habitat modifications.

Agronomic diligence refers to the farmer's capacity to influence the other three factors determining farm productivity. Through simple selection practices the farmer can help in cultivar selection. Through farm manipulation, he can modify, to some extent, the fertility of the farming site, and through selective removal of competitors and grazers, the farmer may anticipate and reduce the negative effects of pests, consumers and other natural enemies.

The very different nature of the four types of factors determining farm productivity does not easily allow quantifications and comparisons in a single figure. Their joint inclusion (Figure 4) is intended to simultaneously outline the effects of these four factors and to call attention to the need to look at all of them if a successful farming operation is intended.

Conclusion

Farming activities with seaweeds have encouraged the development of various approaches and farming methods. As human populations and markets continue growing a larger number of seaweed species will probably be farmed in the future. It is expected that the basic concepts outlined here would help in such developments. At the same time, those new developments should serve as testing grounds for these ideas, as well as sources of new conceptual components for marine agronomy.

Acknowledgments

This study was supported by grant FONDECYT 1960646. My appreciation to D. Aedo, M. Hormazábal, A. Muñoz & P. Sánchez for their generous help during manuscript preparation, and to Dr J. Correa for reading and criticizing the manuscript.

References

Dawson, E. Y., 1966. Marine Botany. An introduction. Holt, Rinehart & Winston, Inc. New York, Chicago, San Francisco, Toronto, London, 371 pp.

Doty, M. S., 1970. The *Eucheuma* opportunity. Science Review (August 1969): 4–11.

Doty, M. S., 1971. Physical factors in the production of tropical benthic marine algae. In Costlow, J.D. (ed.), Fertility of the Sea, Vol. 1. Gordon & Breack Science Publishers, New York: 99–121.

Doty, M. S., 1979. Status of marine agronomy, with special reference to the tropics. Proc. int. Seaweed Symp. 9: 35–58.

Doty, M. S., 1981. *Eucheuma* farm productivity. Proc. int. Seaweed Symp. 8: 688–691.

Doty, M. S., 1986. The production and use of *Eucheuma*. In Doty, M. S., J. F. Caddy & B. Santelices (eds), Case Studies of Seven Commercial Seaweed Resources. FAO Fisheries Technical Paper 281. Food and Agriculture Organization of the United Nations, Rome: 123–164.

Drew, K. M., 1949. Conchocelis-phase in the life-history of *Porphyra umbilicalis* (L.) Kuetz. Nature, London 164: 748.

Frederiksen, S. & J. Rueness, 1989. Culture studies of *Gelidium latifolium* (Grev.) Born. et Thur. (Rhodophyta) from Norway. Growth and nitrogen storage in response to varying photon flux density, temperature and nitrogen availability. Bot. mar. 32: 539–546.

Hoyle, M. D., 1976. Autoecology of ogo *(Gracilaria)* and Limu Manauena (*G. coronopifolia*) in Hawaii, with special emphasis on *Gracilaria* species as indicator of sewage pollution. Ph.D. Thesis. University of Hawaii, 480 pp.

Hoyle, M. D., 1978. Reproductive phenology and growth rates in two species of *Gracilaria* from Hawaii. J. exp. mar. Biol. Ecol. 35: 273–283.

Jensen, A., 1993. Present and future needs for algae and algal products. Hydrobiologia 260/261: 15–23.

Kraft, G. 1997. In Memoriam. Maxwell Stanford Doty. Phycologia 36: 82–90.

Littler, M. M. & D. S. Littler, 1980. The evolution of thallus form and survival strategies in benthic marine macroalgae: field and laboratory tests of a functional form model. Am. Nat. 116: 25–44.

Littler, M. M., D. S. Littler & P. R. Taylor, 1983. Evolutionary strategies in a tropical barrier reef system: functional form groups of marine macroalgae. J. Phycol. 19: 229–237.

Lobban, C. S. & P. J. Harrison, 1994. Seaweed ecology and physiology. Cambridge University Press, Cambridge, 365 pp.

McCandless, E. L., 1981. Polysaccharides of the seaweeds. In Lobban, C. S. & M. J. Wynne (eds), The Biology of Seaweeds. Blackwell Scientific, Oxford: 559–589.

Miura, A., 1975. *Porphyra* cultivation in Japan. In Tokida J. & H. Hirose (eds), Advance of Phycology in Japan. Dr W. Junk Publishers, The Hague: 273–304.

Miura, A., 1976. Genetic studies of cultivated *Porphyra* (Nori) improvement. Mar. Sci. 8: 15–21.

Neish, A. C. & C. Fox, 1971. Greenhouse experiments (1971) on the vegetative propagation of *Chondrus crispus* (Irish Moss). Technical Report N° 12, Atlantic Regional Laboratory, National Research Council of Canada, Halifax, N.S., 35 pp.

Neish, A. C., P. F. Shacklock, C. H. Fox & F. J. Simpson, 1977. The cultivation of *Chondrus crispus*. Factors affecting growth under greenhouse conditions. Can. J. Bot. 55: 2263–2271.

Neushul, M., 1972. Functional interpretation of benthic marine algal morphology. In Abbott, I. A. & M. Kurogi (eds), Contributions to the Systematics of Benthic Marine Algae of the North Pacific. Japan Society for Phycology, Tokyo, Japan: 47–73.

North, W. J., 1971. Introduction and background. In North, W. J. (ed.), The Biology of Giant Kelp Beds (*Macrocystis*) in California. Nova Hedwigia Beih. 32: 1–96.

Ohme, M., Y. Kunifuji & A. Miura, 1986. Cross experiments of the color mutants in *Porphyra yezoensis* Ueda. Jap. J. Phycol. 34: 101–106.

Ohno, M. & A. T. Critchley, 1993. Seaweed cultivation and marine ranching. Japan International Cooperation Agency, Japan, 151 pp.

Patwary, M. U. & J. P. van der Meer, 1992. Genetics and breeding of cultivated seaweeds. Korean J. Phycol. 7: 281–318.

Pérez, R., 1972. Opportunité de l'implantation de l'algue *Macrocystis pyrifera* sur les côte bretonnes. Sci. Pêche 135: 1–9.

Pizarro, A. & H. Barrales, 1986. Field assessment of two methods for planting the agar-containing seaweed, *Gracilaria*, in Northern Chile. Aquaculture 59: 31–43.

Pizarro, A. & B. Santelices, 1993. Environmental variation and large scale *Gracilaria* production. Hydrobiologia 260/261: 357–363.

Santelices, B., 1975. Ecological studies on Hawaiian Gelidiales (Rhodophyta). Ph.D. Thesis, University of Hawaii, 524 pp.

Santelices, B., 1978. Multiple interaction of factors in the distribution of some Hawaiian Gelidiales (Rhodophyta). Pac. Sci. 32: 119–147.

Santelices, B., 1990. Managing the wild crop, propagating and cultivating seaweeds in Chile. In de Oliveira E. C. & N. Kautsky (eds), Proceeding Workshop on Cultivation of Seaweeds in Latin America. Universidad de Sao Paulo, Brasil: 27–34.

Santelices, B., 1992. Strain selection of clonal seaweeds. Prog. Phycol. Res. 8: 85–116.

Santelices, B., 1996. Seaweed research and utilization in Chile: moving into a new phase. Hydrobiologia 326/327: 1–14.

Santelices, B. & M. S. Doty, 1989. A review of *Gracilaria* farming. Aquaculture 78: 95–133.

Santelices, B., D. Aedo & D. Varela, 1995. Causes and implications of intraclonal variation in *Gracilaria chilensis* (Rhodophyta). J. appl. Phycol. 7: 283–290.

Shin, J. A. & A. Miura, 1990. Estimation of the degree of self-fertilization in *Porphyra yezoensis*. Hydrobiologia 204/205: 397–400.

Steneck, R. S. & L. Watling, 1982. Feeding capabilities and limitation of herbivorous molluscs: a functional form approach. Mar. Biol. 68: 299–312.

Stoloff, L. & P. Silva, 1957. An attempt to determine possible taxonomic significance of the properties of water extractable polysaccharides in red algae. Econ. Bot. 11: 327–330.

Tseng, C. K., 1981. Commercial cultivation. In Lobban, C. S. & M. J. Wynne (eds), The Biology of Seaweeds. Blackwell Scientific, Oxford: 680–725.

Westermeier, R., I. Gómez & P. Rivera, 1993. Suspended farming of *Gracilaria chilensis* (Rhodophyta, Gigartinales) at Cariquilda River, Maullín, Chile. Aquaculture 113: 215–229.

Wu, C. Y. & G. H. Lin, 1987. Progress in the genetics and breeding of economic seaweeds in China. Hydrobiologia 151/152: 57–61.

Observations on the phylogenetic systematics and biogeography of the Solieriaceae (Gigartinales, Rhodophyta) inferred from *rbc*L sequences and morphological evidence

Suzanne Fredericq[1], D. Wilson Freshwater[2] & Max H. Hommersand[3]
[1]*Department of Biology, University of Southwestern Louisiana, Lafayette, LA 70504-2451, U.S.A.*
[2]*Center for Marine Science Research, 7205 Wrightsville Avenue, Wilmington, NC 28403, U.S.A.*
[3]*Department of Biology, University of North Carolina, Chapel Hill, NC 27599-3280, U.S.A.*

Key words: biogeography, carrageenan, *Eucheuma*, Gigartinales, molecular systematics, Rhodophyta, seaweed, Solieriaceae

Abstract

A hypothesis of phylogenetic relationships inferred from analysis of plastid-encoded *rbc*L sequences is presented for members of the Solieriaceae, an economically important family containing carrageenan-type phycocolloids. Previous studies established that the Solieriaceae *sensu lato* forms a terminal clade in the Solieriaceae complex, a cluster of families sister to the Gigartinaceae complex (Gigartinaceae and Phyllophoraceae). The family Solieriaceae (including the Areschougiaceae) presently contains about 20 genera that are widely distributed in warm-temperate and tropical waters throughout the world. Molecular *rbc*L and published morphological evidence are consistent with an Australasian origin for the family and a Tethyan distribution for its tropical representatives from west to east as far as the Pacific coast of North and South America. The austral lineages *(Callophycus, Rhabdonia, Areschougia* and *Erythroclonium)* are resolved at the base of the tree. The remaining genera, comprising the tropical and subtropical cluster, form a robust terminal clade (100% bootstrap support). This assemblage can be divided into seven groups, most of which receive strong bootstrap support: (1) *Sarconema*, (2) a *Eucheuma* group, (3) an Atlantic *Solieria* group, (4) an Indo-Pacific *Solieria robusta* group, (5) a *Meristiella/Meristotheca* group, (6) *Agardhiella* and (7) a *Sarcodiotheca* group that includes '*Eucheuma*' *uncinatum* from the Gulf of California. Four clusters of species are recognized in the *Eucheuma* group that correspond to the sections: Eucheuma, Gelatiformia (= *Betaphycus*), Anaxiferae, and Cottoniformia (= *Kappaphycus*) proposed by Maxwell Doty. No decision is reached regarding the taxonomic rank appropriate to the eucheumoid taxa.

Introduction

Gene phylogenies for the red algae (Rhodophyta) inferred from plastid *rbc*L sequences contained a large terminal clade that includes nearly all the taxa known to possess both kappa- and lambda-type cell wall carrageenans (Fredericq et al., 1996; Freshwater et al., 1994). Morphologically, this group is characterized by having cystocarps (fruiting bodies) in which the gonimoblasts (diploid generation produced from the fertilized egg) develop inwardly from a special cell, the auxiliary cell, and often link to filaments of the female gametophyte by means of cell fusions or secondary pit connections (Hommersand & Fredericq, 1990; Kylin, 1956). The carrageenophytes can be subdivided into two monophyletic clades, one having cruciately divided tetrasporangia (meiosporangia of red algae) and the other zonately divided tetrasporangia. The first group, containing the families Gigartinaceae and Phyllophoraceae, has been treated elsewhere (Fredericq & Ramírez, 1996; Hommersand et al., 1994). The second group includes a large number of families: Caulacanthaceae, Cubiculosporaceae, Furcellariaceae (including *Turnerella* and *Opuntiella*), Tichocarpaceae, Rissoellaceae, Dicranemataceae, Acrotylaceae, Mychodeaceae, Mychodeophyllaceae, Cystocloniaceae, and Solieriaceae (including Areschougiaceae *sensu* Chiovitti et al., 1998). Most of these families have cystocarps in which the auxiliary cell lies in close proximity to the carpogonium (egg cell); however, in two of them, the Furcellariaceae and the Solieriaceae, auxiliary cells form in separate branch systems from those that bear the carpogonia, and the diploid nucleus is transferred to the auxiliary cell by means of long, tube-like structures called connecting filaments. The Furcellariaceae occupies a basal position in most phylogenetic trees, whereas the Solieriaceae forms a terminal clade sister to the monotypic genus *Heringia* of the Caulacanthaceae (Fredericq & Hommersand, unpubl.). Except for *Turnerella* and *Opuntiella*, which cluster with the Furcellariaceae, all the genera and species we have investigated that are presently placed in the Solieriaceae *sensu lato* appear in the terminal clade in our *rbc*L trees. In this paper we examine the phylogeny of the Solieriaceae inferred from *rbc*L sequences in comparison with published morphological observations.

Material and methods

Methods of sample preparation and DNA extraction were as previously described (Fredericq & Ramírez, 1996; Freshwater & Rueness, 1994; Hommersand et al., 1994). For automated sequencing, amplification products were cleaned of excess primer, enzyme and dNTPs by PEG precipitation (Hillis et al., 1996). Sequencing reactions were done using the Big Dye sequencing Kit and protocol (Perkin-Elmer, Foster City, CA), and run on an ABI prism 310 Genetic Analyzer (PE Applied Biosystems, Foster City, CA). The generated sequence data were compiled and aligned with Sequencher (Gene Codes Corp., Ann Arbor, MI). Parsimony analyses of *rbc*L sequence data were performed using PAUP (v. 4.0: Swofford, 1998) and MacClade (v. 3.0: Maddison & Maddison, 1992) computer programs. Parsimony trees were inferred using a two part heuristic search scheme. Initial searches, designed to increase the likelihood of swapping within the 'island' of trees leading to the most parsimonious solution (Maddison, 1991), consisted of 1000 random sequence additions, using STEEPEST DESCENT and MULPARS (but permitting only 20 trees be held at each step) options with the NNI (nearest neighbour interchange) swapping algorithm. Trees found in these initial searches were then used as starting points for further searches using STEEPEST DESCENT, MULPARS and TBR (tree bisection reconnection) swapping algorithm until swapping was completed.

Consistency (CI) and retention (RI) indices (Farris, 1989; Kluge & Farris, 1969) were calculated excluding uninformative characters. Support for nodes of parsimony trees was assessed by calculating bootstrap proportion (BP) values (Felsenstein, 1985) based on 5000 resamplings of heuristic searches with simple addition of sequences, MULPARS and TBR swapping algorithm. Neighbour joining (Saitou & Nei, 1987) and maximum likelihood (Felsenstein, 1981) analyses of the sequence data set were also done using PAUP. A neighbour joining tree was inferred based on Kimura two-parameter distances (Kimura, 1980). Bootstrap support for nodes in the neighbour joining tree were calculated based on 1000 resampling replicates. Seven maximum likelihood analyses with TBR branch swapping on initial trees built by random addition of sequences were done using empirical base frequencies and a transition/transversion ratio of 2.0. Due to time considerations, bootstrap support for nodes of the maximum likelihood tree were not determined.

Results

Nucleotide sequence data for *rbc*L was generated from 37 taxa within the red algal family Solieriaceae (Appendix). Three additional species from the Caulacanthaceae were included as outgroup taxa based on the close relationship of these families in more global analyses (Fredericq & Hommersand, unpubl.; Fredericq et al., 1996; Freshwater et al., 1994). The sequence data set contained 394 parsimony informative characters. Tree lengths of 100 000 randomly generated trees for these data had a skewed distribution ($g_1 = -0.644$, $P < 0.01$) indicating the presence of nonrandom structure (Hillis & Huelsenbeck, 1992; Hillis et al., 1993). There was no difference in the species groups resolved by parsimony, distance or maximum likelihood analyses, and bootstrap support for resolved branches was identical or nearly so in almost all cases. For the purposes of this study, all three methods gave similar results and only results of the parsimony analyses are shown here (Figures 1 and 2).

Parsimony analyses of the data set resulted in six minimal length trees of 1494 steps, CI = 0.376 and RI = 0.549. Well-supported lineages leading to *Callophycus*, *Rhabdonia*, *Areschougia* and *Erythroclonium* are resolved at the base of the trees (Figure 1). A terminal group consisting of *Sarconema*, a *Eucheuma/Betaphycus/Kappaphycus* clade, *Solieria*,

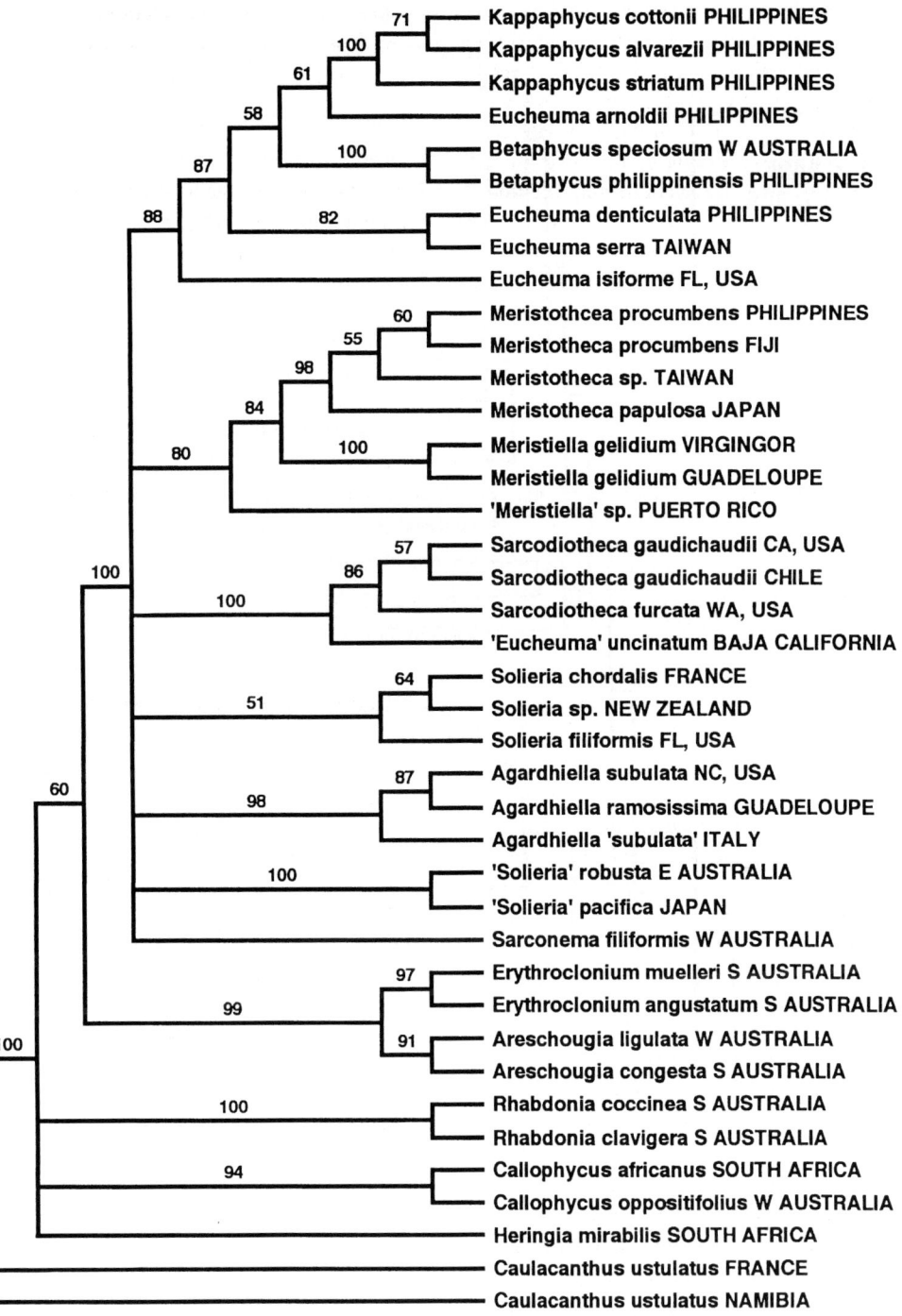

Figure 1. Fifty percent bootstrap consensus tree, based on 5000 replicates, resulting from parsimony analyses of *rbc*L sequence data for 37 Solieriaceae and two outgroup species.

a *Solieria robusta* clade, a *Meristiella/Meristotheca* clade, *Agardhiella*, and a *Sarcodiotheca* clade that included '*Eucheuma' uncinatum* is strongly supported (BP = 100, Figure 1). All of the generic groups except *Solieria* are resolved (BP = 80 to 100) as monophyletic clades on the tree. There is little to no bootstrap sup-

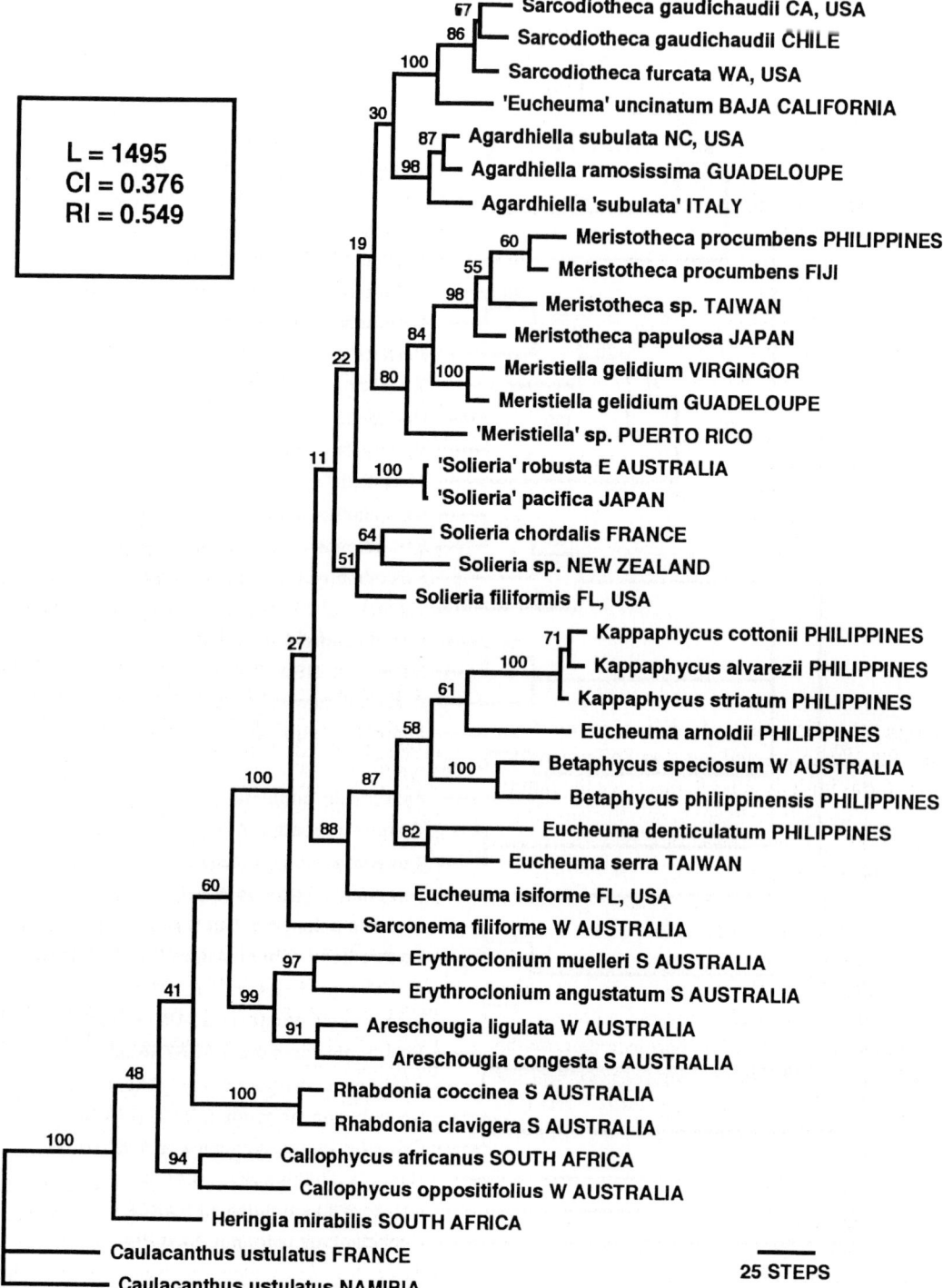

Figure 2. Non-minimal topology of length = 1495, CI = 0.376 and RI = 0.549 which is a better fit than the minimal trees to an evolutionary hypothesis based on morphological and biogeographic evidence. The lengths of branches are proportional to the number of steps. Bootstrap proportion values (5000 replicates) are shown for each branch.

port (BP = 51) for the grouping of *Solieria* species and the penalty to parsimony to bring the *Solieria* clade together with the '*Solieria*' *robusta* clade is only two steps. The sequences of '*Solieria*' *robusta* and '*S.*' *pacifica* differ at only two sites indicating that the samples sequenced refer to the same species.

Whereas the generic groups are, for the most part, well supported by bootstrap analyses, there is no support for the relationships among these groups as resolved in the minimal trees (Figure 1). A rearrangement of the groups to reflect an evolutionary hypothesis based on published morphological and biogeographical evidence is shown in Figure 2. This topology is only one step longer than minimal ($L = 1495$ vs 1494) and requires no increase in CI measured homoplasy, and has a better likelihood score than the minimal parsimony trees (LnLi = $-11\,408.319$ vs $-11\,410.067$ to $-11\,420.358$). Topologically this tree differs from the minimal tree only in the arrangement of the nodes that have no bootstrap support.

Discussion

Systematics

The status of the Solieriaceae was called into question by Silva (1980), who pointed out that Solieriaceae J. Agardh 1876 was initially superfluous since the family to which it was applied included *Caulacanthus* Kützing, the type of Caulacanthaceae Kützing 1843, and he proposed the Solieriaceae for conservation. In 1994 Womersley substituted the name Areschougiaceae J. Agardh 1876 for Solieriaceae because it had page priority and did not include the type of any previously published family. The problem was resolved by Silva (1993) who pointed out that heterotypic synonyms for family names that were specifically rejected under the Montreal *Code* of 1959 are no longer rejected under the Sydney *Code* of 1981. Thus the priority of Solieriaceae is now established for taxa that include *Solieria* J. Agardh 1842.

Among members of the Solieriaceae complex a distinction has traditionally been made between families in which auxiliary cells are borne in the same cluster of branched filaments as the carpogonial branches (procarps present) and those in which they are borne in separate clusters (procarps absent). With one exception (*Austroclonium*), genera of Solieriaceae belong to the latter group (procarps absent). Transfer of the diploid nucleus from the fertilized carpogonium to the auxiliary cell is mediated by one or more unbranched, non-septate connecting filaments that issue from a fertilized carpogonium in one system of branched cortical filaments and link to auxiliary cells in separate systems (Gabrielson & Hommersand, 1982a). Kylin (1932, 1956) placed genera that share the same type of connecting filaments in two families, a uniaxial family Rhabdoniaceae and a multiaxial family Solieriaceae. Recently, Chiovitti et al. (1998) recommended provisional reestablishment of the Rhabdoniaceae Kylin (as the family Areschougiaceae J. Agardh) to contain *Rhabdonia, Areschougia, Erythroclonium* and *Austroclonium*, in consideration of their findings of unique highly methylated carrageenans in these genera. Our *rbc*L results will be discussed in the context of a broad circumscription of the Solieriaceae, *sensu* Gabrielson & Hommersand (1982a), and in one that recognizes the family Areschougiaceae.

The Solieriaceae *sensu lato* is identified in *rbc*L trees as a terminal taxon sister to *Heringia*, a monotypic genus from southern Africa presently placed in the Caulacanthaceae (Fredericq & Hommersand, unpubl.). The family differs from *Heringia* primarily in the greater length of the unbranched connecting filament, and the fact that the gonimoblasts develop primarily inwardly rather than outwardly as in *Heringia* (Searles, 1968). Gabrielson & Hommersand (1982a,b) divided the Solieriaceae into three tribes: Solierieae, Areschougieae, and Agardhielleae based on: the number of connecting filaments, presence or absence of sterile cells on the carpogonial branches, presence or absence of an involucre or enveloping tissue around the carposporophyte, and position, whether basal or lateral, of the pit connection to the tetrasporangium. Kraft and Gabrielson (1983) concluded that the tribes were not fully defined or capable of completely consistent application, and they are not supported in trees inferred from the *rbc*L sequence data presented here.

Two major groups are identified in this study: (1) a series of basal lineages containing *Callophycus, Rhabdonia, Areschougia* and *Erythroclonium*, in addition to *Heringia*, and (2) a terminal group containing a *Sarconema* clade, a *Eucheuma/Betaphycus/Kappaphycus* clade, *Solieria* and *Solieria robusta* clades, a *Meristiella/Meristotheca* clade, an *Agardhiella* clade, and a *Sarcodiotheca* clade (Figures 1 and 2). The two groups recognized in the molecular trees can also be defined morphologically and on the basis of their biogeographic distribution. *Anatheca, Flahaultia, Placentophora*, the parasitic genera *Tikvahiella* and *Gardneriella*, and the procarpic genus *Austroclonium* were not available for inclusion in this study. Taxa belonging to the basal lineages are either uniaxial or multiaxial with an undefined number of apical cells, possibly just one, and the gonimoblasts emanate from an elongated fusion cell that includes scarcely modified proximal cells in the auxiliary cell filament. Those belonging to the terminal group are multiaxial, composed of axial filaments derived from four or more distinct apical cells, and the gonimoblast filaments either link to a modified swollen central fusion cell or to a placenta composed of a network of enveloping secondary filaments. The basal group is represented primarily by species found in southern and western Australia with a few outliers in South Africa. An exception is *Callophycus* which includes several widely distributed species in the Indo-Pacific Ocean in addition to the Australian species (Kraft, 1984). The terminal group is primarily tropical or warm-temperate and extends in a band around the globe roughly between 40° N and 30° S. Species that range further north or south apparently derive from taxa that arose within this central region.

The four putative ancestral taxa: *Callophycus*, *Rhabdonia*, *Areschougia* and *Erythroclonium* are resolved in three separate lineages in the minimal *rbc*L trees (Figure 1). It is somewhat surprising that *Areschougia* and *Erythroclonium* are so firmly associated (BP = 99). Both have well-defined central axial filaments; however, axial cells each bear two periaxial cells in *Erythroclonium* and only one in *Areschougia* (Womersley, 1994). Our observations neither support nor contradict recognition of the Areschougiaceae containing the genera: *Rhabdonia*, *Areschougia*, *Erythroclonium* and *Austroclonium*. Though paraphyletic in minimal trees, a monophyletic Areschougiaceae requires only a six-step penalty to parsimony. Chiovitti et al. (1996, 1997a, 1998) demonstrated that these four genera are distinguished from typical members of the Solieriaceae by the possession of highly methylated carrageenans composed predominantly of carrabiose 2-sulphate of α-carrageenan, carrabiose 2,4-disulphate of ι-carrageenan and the 6′-*O*-methylated counterparts of both repeating units. The galactan of *Callophycus* (Chiovitti et al., 1997b) was distinct in containing the dominant repeating disaccharide 4′,6′-*O*-(1-carboxyethylidene)caribiose 2-sulphate and minor amounts of the α-carrageenan repeating unit (carrabiose 2-sulphate). *Callophycus* was also unusual in having the highest pyruvate contents (8–10%) reported thus far for red algal galactans. An eight-step penalty to parsimony is required to create a monophyletic assemblage containing both *Callophycus* and the Areschougiaceae.

The procarpic species, *Melanema dumosum*, formerly placed in the Rhabdoniaceae (= Areschougiaceae) by Min-Thein & Womersley (1976), was removed from this assemblage by Chiovitti et al. (1995) owing to the presence of ι-, κ- and β-carrageenans without methyl or pyruvate substitutions. This type of cell wall composition is found in some eucheumoid Solieriaceae and in some Caulacanthaceae. *Melanema* lies outside the Solieriaceae in *rbc*L trees inferred from large data sets containing taxa belonging to the Solieriaceae complex (Fredericq & Hommersand, unpubl.)

Whatever the ultimate taxonomic status of the Areschougiaceae and of *Callophycus*, the remaining members of the Solieriaceae form a robust terminal clade, receiving 100% bootstrap support in *rbc*L trees. *Sarconema* occupies a basal position among genera in the terminal clade and is also the most primitive morphologically. The vegetative axis consists of a well-defined, multiaxial medulla and surrounding cortex composed of isodiametric cells. The fusion cell incorporates nutritive cells from the outer cortex as in *Solieria*; however, a special nutritive network is not formed around the developing gonimoblasts (Papenfuss & Edelstein, 1974). *Sarconema* is represented by two species, *S. filiforme* (Sonder) Kylin and *S. scinaioides* Børgesen. *Sarconema filiforme* is widespread in the Indian Ocean and may have been distributed secondarily into the eastern Mediterranean Sea (Rayss, 1963).

In the tree in Figure 2, the three clades that have cystocarps with central fusion cells (*Eucheuma/Betaphycus/Kappaphycus*, *Solieria* and '*Solieria' robusta*) are positioned basal to a terminal cluster of three clades having placental-type cystocarps (*Meristiella/Meristotheca*, *Agardhiella*, *Sarcodiotheca*). This arrangement of taxa is favoured when a larger number of outgroup taxa are used, and in analyses in which molecular and morphological data sets are combined (Fredericq & Hommersand, pers. obs.). To derive a *rbc*L tree with a terminal dichotomy of monophyletic clades containing central fusion cell-type and placental-type cystocarps requires a five-step penalty to parsimony. The Indo-Pacific Ocean is the centre of distribution for the group having cystocarps with central fusion cells, whereas the Atlantic and eastern Pacific Ocean is the centre for the group having placental cystocarps.

The large, commercially important genus *Eucheuma* contains one species, *E. isiforme* (C. Agardh) J Agardh, in the eastern tropical Atlantic Ocean and upwards of 40 described species in the Indo-Pacific Ocean, many of them poorly defined. Most of these are said to contain ι-carrageenan as their principal cell wall component; however, some species contain primarily κ carrageenan or a mixture of β, κ and ι-carrageenans (Greer & Yaphe, 1984; Santos, 1989). *Kappaphycus* Doty (1988) was established to contain species characterized by the presence of κ-carrageenan and a second genus, *Betaphycus* Doty ex Silva, Basson & Moe (1996) was informally proposed by Doty (1995) to receive species that contained β-carrageenan.

Earlier, Doty and Norris (1985) followed by Doty (1988) had recognized four sections in *Eucheuma* based on morphological characters in addition to carrageenan type. These were characterized as follows:

(1) Section Eucheuma [Lectotype = *Eucheuma denticulatum* (Burman) Collins & Hervey]. Thalli cylindrical; spines simple with narrowing bases, at first in regularly spaced pairs or whorls; branching from whorls, opposite, pectinate or irregular; axial core cylindrical, rhizoidal (axiferous); cystocarps on laterals, generally with one terminal spine; ι-carrageenan.

(2) Section Anaxiferae [Lectotype = *Eucheuma arnoldii* Weber-van Bosse]. Thalli cylindrical or dorsiventral; spines compound, scattered or in whorls; branching from whorls, often opposite; axial core absent; cystocarps on main axes without spines; ι-carrageenan.

(3) Section Gelatiformia [Lectotype = *Eucheuma gelatinum* (Esper) J. Agardh = *Betaphycus gelatinum* (Esper) Doty ex Silva, Basson & Moe (1996)]. Thalli compressed; spines simple with broadening bases, arranged in rows marginally and later dorsally and ventrally; branching from the margins, pinnate; axial core tortuous, often flattened, hyphal; cystocarps on laterals, often bearing spines; typically containing a mixture of β, κ and ι-carrageenans [*Betaphycus* Doty ex Silva, Basson & Moe; Type = *B. philippinensis* Doty].

(4) Section Cottoniformia [Type = *Eucheuma alvarezii* Doty 1985 = *Kappaphycus alvarezii* (Doty) Doty ex Silva in Silva et al., 1996]. Thalli variable, commonly cylindrical or angular; protuberances irregularly arranged, sometimes spiny; branching irregular, sometimes pinnate or falsely dichotomous; axial core usually present and vaguely cylindrical, hyphal or with thylles often present; cystocarps on main axes without protuberances or spines; κ-carrageenan.

Three of the four sections of *Eucheuma* proposed by Doty are separated by long branch lengths and received strong support. Section Eucheuma is paraphyletic with respect to the other sections, although there is questionable bootstrap support for the arrangement of sections within the larger *Eucheuma*/*Betaphycus*/*Kappaphycus* clade (BP = 58 & 61). Sampling was limited primarily to material from the Philippines, and additional taxa from other parts of the Indo-Pacific Ocean need to be investigated.

The three species included here that have a prominent rhizoidal central axis and whorled spines and branches, *Eucheuma isiforme* (C. Agardh) J. Agardh, *E. denticulatum*, and *E. serra* (J. Agardh) J. Agardh have proved difficult to tell apart (Gabrielson, 1983; Gabrielson & Kraft, 1984). Collins & Hervey (1917) first suggested that *E. isiforme* from the western tropical Atlantic Ocean is conspecific with the Indian Ocean species, *E. denticulatum*, a view expressed several times since, while *E. serra* has been maintained as a separate species by some authors and merged with *E. denticulatum* by others. Despite their morphological similarity, long branch lengths in *rbc*L trees separate the Atlantic *E. isiforme* and the Indo-Pacific *E. denticulatum*. Numerous populations covering the geographic ranges for these species will need to be investigated before their taxonomic relationships can be resolved.

Betaphycus is problematic in that the presence of β-carageenan has been confirmed in only three of the species that share the suite of morphological features that characterize the section Gelatiformia. Other species are reported to contain either ι or κ and ι-carrageenans (Santos, 1989). Weber-van Bosse (1928) placed members of this group in her section Axifera because of the presence of a prominent central axis. Doty (1988), on the other hand, interpreted the central axis of Gelatiformia as being hyphal (secondary) rather than rhizoidal (primary) in origin. If true, *Betaphycus* should be more closely related to *Kappaphycus* than to axiferous species of *Eucheuma*. The precise ontogeny of filaments of the central axis has not been illustrated or clarified and the topological position of *Betaphycus* relative to *Eucheuma* and *Kappaphycus* shown here in *rbc*L trees is unresolved (BP = 58 and 61).

Kappaphycus alvarezii is the principal commercial species cultivated in the Philippines for its κ-carrageenan. The other two species sampled here from the Philippines, *K. cottonii* and *K. striatum*, were identified based on their morphology. The three are separated by short branch lengths in *rbc*L trees; however, their true identities cannot be determined until material from the type locality of each has been sampled. For *K. alvarezii* this is Sabah, Malaysia, for *K. cottonii* it is Saya de Malha Bank, Indonesia, and for *K. striatum* it is Zanzibar, Tanzania (see Doty, 1988). Secondary hyphal filaments and thylles (cells and branched, short-celled filaments derived secondarily from medullary cells in the centre of the thallus) are best known from members of this group (Doty, 1985, 1988; Weber-van Bosse, 1928).

Two species, *Eucheuma arnoldii* and *E. amakusaensis* Okamura, which are said to contain ι-carrageenan, possess neither rhizoidal nor hyphal axial cores (Doty 1988). The position of *E. arnoldii* within the larger *Eucheuma/Betaphycus/Kappaphycus* clade is poorly supported (BP = 61).

Solieria presently contains nine species: the type species *S. chordalis* (C. Agardh) J. Agardh in Western Europe, *S. filiformis* (Kützing) Gabrielson in the western and eastern tropical Atlantic Ocean (Gabrielson, 1985; Lawson & John, 1987), and presumably as an introduced species in Wales, UK (Farnham, 1980), *S. robusta* (Greville) Kylin, which is reported to be distributed widely in the Indo-Pacific Ocean (Silva et al., 1996), *S. anastomosa* Gabrielson & Kraft (1984) from Lord Howe Island, *S. pacifica* (Yamada) Yoshida, a recent segregate from *S. robusta* from Japan, *S. dichotoma* Yoshida from Japan, *S. tenuis* Zhang & Xia from China and Japan (see Yoshida et al., 1995), *S. jaasundii* Mshigeni & Papenfuss (1981) from Tanzania and *S. dura* (Zanardini) Schmitz from the north Indian Ocean, Red Sea and eastern Mediterranean Sea (see Perrone & Cecere, 1994). At the moment, the morphological distinction between the Atlantic and Indo-Pacific groups rests solely on differences in apical growth. In members of the Indo-Pacific group the periaxial cells are disposed in orthostichous arrays (Gabrielson & Kraft, 1984), whereas in *S. chordalis* and *S. filiformis* of the Atlantic group the periaxial cells are said to rotate to the left and right along the axial filaments (Gabrielson & Hommersand, 1982a). Recently, Guimarães & Oliveira (1996) have commented that this character is difficult to observe and have questioned its utility. Bringing the Atlantic and Indo-Pacific *Solieria* clades together requires a penalty to parsimony of only two steps. Further studies are required to assess the appropriate taxonomic status of these species.

The recent discovery of a new population of a species of *Solieria* of unknown origin at Orakei Basin, Auckland, New Zealand (W. Nelson, pers. com.) is one of many examples of weedy species that have appeared in habitats modified by man around the world. Though separated by long branch lengths, the New Zealand plant has its closest affinities in *rbc*L trees to *Solieria chordalis* and *S. filiformis* from the Atlantic Ocean, rather than to any known Indo-Pacific species (Figure 1).

Three clades identified in *rbc*L sequence analyses are characterized by a placental cystocarp in which the gonimoblasts interweave amongst and link to secondary gametophytic filaments rather than forming in association with a central fusion cell. These are: *Meristiella/Meristotheca* clade with *Meristiella* in the Atlantic and *Meristotheca* in the Indo-Pacific Ocean, an *Agardhiella* clade with *Agardhiella* in the western Atlantic Ocean and the Mediterranean Sea, and a *Sarcodiotheca* clade with *Sarcodiotheca* and '*Eucheuma*' *uncinatum* Setchell & Gardner in the eastern Pacific Ocean and the Caribbean Sea. (Representatives of *Sarcodiotheca* from the Caribbean Sea are poorly known and have only recently become available for study.) Each of these clades is moderately to strongly supported by bootstrap analyses; however, their association as a group was not seen in the most parsimonious tree (Figure 1) and is represented here in a tree one step less parsimonious than minimal (Figure 2). The replacement of a central fusion cell by a placental network may be strong morphological evidence in favour of relatedness; however, not enough is known about the comparative development of the cystocarp placenta in the three clades to warrant the conclusion that they share a similar origin and common ancestor.

Bootstrap analyses provide strong support for a basal position of taxa presently placed in *Meristiella* that are found in the Atlantic Ocean and a terminal position for the Indo-Pacific species placed in *Meristotheca* (Figures 1 and 2). Little separates the two genera apart from their biogeography and the two may someday be merged. *Meristiella* was originally separated from *Meristotheca* by Gabrielson & Cheney (1987) on the grounds that an auxiliary cell complex composed of specialized cells adjacent to the auxiliary cell is present in *Meristiella* and absent in *Meristotheca*, and that tetrasporangia are nemathecial in *Meristotheca* and not in *Meristiella*. These are difficult characters to confirm, and they were recognized only tentatively in a recent study by Guimarães & Oliveira (1996) pending further investigations.

The *Agardhiella* clade presently contains only a single genus *Agardhiella*, which until recently was known only in the warm temperate and tropical Atlantic Ocean and Caribbean Sea. At the present time, cylindrical forms are placed in a single species, *A. subulata* (C. Agardh) Kraft & Wynne, ranging from New England to Brazil (Gabrielson, 1985) and compressed or flattened forms are placed either in *A. ramosissima* (Harvey) Kylin or *A. floridana* (Kylin) Gabrielson ex Guimarães & Oliviera (1996). A distinguishing feature of *Agardhiella* is the presence of short interconnecting filaments that link laterally between files of axial cells. Similar filaments are also found in both Atlantic and Indo-Pacific species of *Solieria* and this may be a feature that links the two genera phylogenetically. The discovery of *Agardhiella subulata* in the Mar Piccolo, Gulf of Taranto, Italy, has greatly extended the known range of that species (Perrone & Cecere, 1994). Despite the strong morphological similarity of the Mediterranean and western Atlantic plants, the long branch length that separates them in *rbc*L trees suggests that they have been separated for a long time (Figure 2).

Sarcodiotheca was established by Kylin (1932) upon *Anatheca furcata* Setchell & Gardner. Originally, the genus contained only flattened, more or less dichotomously branched species. Gabrielson (1982a,b) investigated the vegetative morphology and cystocarp development in *Sarcodiotheca furcata* (Setchell & Gardner) Kylin, *S. dichotoma* (Howe) Dawson, and the terete species formerly known as *Neoagardhiella gaudichaudii* (Montagne) Abbott. He redefined *Sarcodiotheca* to include both flattened and terete species in which the gonimoblasts unite with vegetative cells to produce a compact cellular central placenta bearing comparatively long, branched files of carposporangia. *Neoagardhiella gaudichaudii* was transferred to *Sarcodiotheca*. The inclusion of *Sarcodiotheca gaudichaudii* (Montagne) Gabrielson in the *Sarcodiotheca* clade is strongly supported (BP = 100). At the present time, *Sarcodiotheca* is separated from *Agardhiella* based solely on features of cystocarp development. Though the mature cystocarps are morphologically quite similar, the recognition of two genera receives support from *rbc*L sequence data. The record of *Sarcodiotheca gaudichaudii* (as *Neoagardhiella gaudichaudii*) from England (Farnham, 1980) requires confirmation.

The plant known as *Eucheuma uncinatum* Setchell & Gardner from the Gulf of California, Mexico is superficially similar to *E. isiforme* from the western Atlantic Ocean. However, the cystocarp contains a compact, parenchymatous sterile placenta, similar to that in *Sarcodiotheca*, rather than a central fusion cell as found in *Eucheuma* (Norris, 1985). '*Eucheuma*' uncinatum is an integral component of the *Sarcodiotheca* clade in *rbc*L trees (BP = 100).

Biogeography

Two genera, among the basal lineages in Figures 1 and 2, which are prominent in Australia, *Callophycus* and *Erythroclonium*, are also represented in the flora of South Africa. *Erythroclonium* is known only from Australia and South Africa, whereas *Callophycus* is more widely distributed in the Indo-Pacific Ocean (Kraft, 1984). These taxa could reflect an ancestral distribution along the northern coast of Gondwana before South Africa and Antarctica/Australia separated, or their present distribution may have resulted from long range dispersal through the tropical Indian Ocean during periods of major world-wide cooling (Hommersand, 1986).

Beginning with *Sarconema*, all the taxa in the large terminal clade (Figures 1 and 2) are present in subtropical to tropical waters either in the Indo-Pacific or the Atlantic Ocean, or by extension at an earlier time, the eastern Pacific Ocean. *Sarconema* is an Indo-Pacific genus; *Solieria* (*sensu lato*), *Meristiella/Meristotheca*, and *Eucheuma* are distributed between the Indo-Pacific and Atlantic Oceans; *Agardhiella* is primarily a genus of the western Atlantic Ocean, and *Sarcodiotheca*, a genus known primarily from the eastern Pacific Ocean, is also present in the western Atlantic and the Caribbean Sea.

When considered together, the molecular data presented here, and the many morphological observations in the published literature, support the speculation that ancestral members of the Solieriaceae are retained in the temperate waters of Australia and South Africa, and that the more widely distributed tropical and subtropical entities originated and evolved in the Tethys. They were split into two groups by the collapse of the Tethyan seaway in the vicinity of the Mediterranean Sea, beginning around 100 million years ago and ending 20 million years ago. A recent paleobiogeographical reconstruction of ocean currents during the Cretaceous suggests that a robust circumglobal tropical current flowed westward through the continental configuration of that time (Bush, 1997). Such a current could account for the early distribution of ancestral taxa that gave rise to the genera that comprise the terminal assemblage in rbcL trees. A Tethyan origin and early distribution for the Solieriaceae was first proposed by Gabrielson (pers. com.), and has been suggested to account for the presence of possible relictual Solieriaceae in the Mediterranean Sea (Perrone & Cecere, 1994). The rbc L data (Figures 1 and 2) indicate that, except for *Sarconema*, the ancestral taxa in each of the clades in the terminal group of the Solieriaceae (*Solieria*, *Eucheuma* and *Meristiella/Meristothcea* clades, as well as *Agardhiella* and *Sarcodiotheca* clades) have been conserved in the Atlantic Ocean, and that taxa found in the Indo-Pacific Ocean are more highly derived. Future morphological and molecular studies are required to test these hypotheses.

Acknowledgements

The authors extend their appreciation to Isabella Abbott and George Staples for kindly lending material of *Eucheuma* from the Maxwell S. Doty collection at the Bishop Museum, Honolulu, Hawaii, which provided a basis for our morphological concepts of species belonging to that genus. We especially thank Lawrence Liao for a timely collection of silica gel-dried material of Philippine species of *Eucheuma*. We also thank J. Cabioch, E. Cecere, E. Deslandes, J. Huisman, D. F. Kapraun, R. Lewis, A. Millar, W. Nelson, A. N'Yeurt, M. E. Ramírez, A. Renoux, C. Stoude, M. Yoshizaki, and J. Zertuche who generously provided silica gel-dried collections for this study. This project was funded in part by U.S. department of Energy grant DE-FG02-97ER12220 and Louisiana Board of Regents grants LEQSF (1997-99)-RD-A-30, and LESQSF (19797-98)-ENH-TR-86 to SF, and NSF grant DEB-9726170 to DWF.

References

Agardh, J. G., 1842. Algae maris Mediterranei et Adriatici. Fortin, Masson et Cie, Paris, x + 164 pp.

Agardh, J. G., 1876. Species genera et ordines algarum, 3(1) Epicrisis systematis Floridearum. T.O. Weigel, Leipzig, vii + 724 pp.

Benson, D. A., M. Boguski, D. J. Lipman & J. Ostell, 1994. Genbank. Nucleic Acids Res. 22: 3441–3444.

Bush, A. B. G., 1997. Numerical simulation of the Cretaceous Tethys circumglobal current. Science. 275: 807–810.

Chiovitti, A., M.-L. Liao, G. T. Kraft, S. L. A. Munro, D. J. Craik & A. Bacic, 1995. Cell wall polysaccharides from Australian red algae of the family Solieriaceae (Gigartinales, Rhodophyta): iota/kappa/beta-carrageenans from *Melanema dumosum*. Phycologia 34: 522–527.

Chiovitti, A., M.-L. Liao, G. T. Kraft, S. L. A. Munro, D. J Craik & A. Bacic, 1996. Cell wall polysaccharides from Australian red algae of the family Solieriaceae (Gigartinales, Rhodophyta): highly methylated carrageenans from the genus *Rhabdonia*. Bot. mar. 39: 47–59.

Chiovitti, A., A. Bacic, D. J. Craik, S. L. A. Munro, G. T. Kraft & M.-L. Liao, 1997a. Carrageenans with complex substitution patterns from red algae of the genus *Erythroclonium*. Carbohydr. Res. 305: 243–252.

Chiovitti, A., A. Bacic, D. J. Craik, S. L. A. Munro, G. T. Kraft & M.-L. Liao, 1997b. Cell-wall polysaccharides from Australian red algae of the family Solieriaceae (Gigartinales, Rhodophyta): novel, highly pyruvated carrageenans from the genus *Callophycus*. Carbohydr. Res. 299: 229–243.

Chiovitti, A., G. T. Kraft, A. Bacic, D. J. Craik, S. L. A. Munro & M.-L. Liao, 1998. Carrageenans from Australian representatives of the family Cystocloniaceae (Gigartinales, Rhodophyta), with description of *Calliblepharis celatospora* sp. nov., and transfer of *Austroclonium* to the family Areschougiaceae. J. Phycol. 34: 515–535.

Collins, F. S. & A. B. Hervey, 1917. The algae of Bermuda. Proc. am. Acad. Arts Sci. 53: 1–195.

Doty, M. S., 1985. *Eucheuma alvarezii* sp. nov. (Gigartinales, Rhodophyta) from Malaysia. In Abbott, I. A. & J. N. Norris (eds), Taxonomy of Economic Seaweeds with Reference to Some Pacific and Caribbean Species. La Jolla: California Sea Grant College Program [Report T-CSGCP-011]: 37–45.

Doty, M. S., 1988. *Prodromus ad systematica Eucheumatoideorum*: a tribe of commercial seaweeds related to *Eucheuma* (Solieriaceae, Gigartinales). In Abbott, I. A. (ed.), Taxonomy of Economic Seaweeds With Reference to Some Pacific and Caribbean Species. La Jolla: California Sea Grant College Program [Report T-CSGCP-018] 2: 159–207.

Doty, M. S., 1995. *Betaphycus philippinensis* gen. et sp. nov. and related species (Solieriaceae, Gigartinales). In Abbott, I. A. (ed.), Taxonomy of Economic Seaweeds With Reference to Some Pacific Species. La Jolla: California Sea Grant College Program [Report T-CSGCP-035] 5: 237–245.

Doty, M. S. & J. N. Norris, 1985. *Eucheuma* species (Solieriaceae, Rhodophyta) that are major sources of carrageenan. In Abbott, I. A. & J. N. Norris (eds), Taxonomy of Economic Seaweeds With Reference to Some Pacific and Caribbean Species. La Jolla: California Sea Grant College Program [Report T-CSGCP-011]: 47–61.

Farnham, W. F., 1980. Studies on aliens in the marine flora of southern England. In Price, J. H., D. E. G. Irvine & W. F. Farnham (eds), The Shore Environment, vol. 2. Ecosystems. Academic Press, London: 875–914.

Farris, J. S., 1989. The retention index and the rescaled consistency index. Cladistics 5: 417–419.

Felsenstein, J., 1981. Evolutionary trees from DNA sequences: A maximum likelihood approach. J. mol. Evol. 17: 368–376.

Felsenstein, J., 1985. Confidence limits on phylogenies: an approach using the bootstrap. Evolution 39: 783–791.

Fredericq, S. & M. E. Ramírez, 1996. Systematic studies of the antarctic species of the Phyllophoraceae (Gigartinales, Rhodophyta) based on rbcL sequence analysis. Hydrobiologia 326/327: 137–143.

Fredericq, S., M. H. Hommersand & D. W. Freshwater, 1996. The molecular systematics of some agar- and carrageenan-containing marine red algae based on *rbc*L sequence analysis. Hydrobiologia 326/327: 125–135.

Freshwater, D. W. & J. Rueness, 1994. Phylogenetic relationships of some European *Gelidium* (Gelidiales, Rhodophyta) species, based on *rbc*L nucleotide sequence analysis. Phycologia 33: 187–194.

Freshwater, D. W., S. Fredericq, B. S. Butler, M. H. Hommersand & M. W. Chase, 1994. A gene phylogeny of the red algae (Rhodophyta) based on plastid *rbc*L. Proc. natn. Acad. Sci. U.S.A. 91: 7281–7285.

Gabrielson, P. W., 1982a. Morphological studies of members of the tribe Agardhielleae, (Solieriaceae, Rhodophyta) I. *Sarcodiotheca furcata* Phycologia 21: 75–85.

Gabrielson, P. W., 1982b. Morphological studies of members of the tribe Agardhielleae (Solieriaceae, Rhodophyta) II. *Sarcodiotheca gaudichaudii* (Montagne) comb. nov. Phycologia 21: 86–96.

Gabrielson, P. W., 1983. Vegetative and reproductive morphology of *Eucheuma isiforme* (Solieriaceae, Gigartinales, Rhodophyta). J. Phycol. 19: 45–52

Gabrielson, P. W., 1985. *Agardhiella* versus *Neoagardhiella* (Solieriaceae, Rhodophyta) another look at the lectotypification of *Gigartina tenera*. Taxon 34: 275–280.

Gabrielson, P. W. & D. P. Cheney, 1987. Morphology and taxonomy of *Meristiella* gen. nov. (Solieriaceae, Rhodophyta). J. Phycol. 23: 481–493.

Gabrielson, P. W. & M. H. Hommersand, 1982a. The Atlantic species of *Solieria* (Gigartinales, Rhodophyta); their morphology. distribution and affinities. J. Phycol. 18: 31–45.

Gabrielson, P. W. & M. H. Hommersand, 1982b. The morphology of *Agardhiella subulata* representing the Agardhielleae, a new tribe in the Solieriaceae (Gigartinales, Rhodophyta). J. Phycol. 18: 46–58.

Gabrielson, P. W. & G. T. Kraft, 1984. The marine algae of Lord Howe Island (N.S.W.): the family Solieriaceae (Gigartinales, Rhodophyta). Brunonia 7: 217–251,

Greer, C. W. & W. Yaphe, 1984. Characterisation of hybrid (beta-kappa-gamma) carrageenan from *Eucheuma gelatinae* J. Agardh (Rhodophyta, Solieriaceae) using carrageenases, infrared and ^{13}C-nuclear magnetic resonance spectroscopy. Bot. mar. 27: 473–478.

Guimarães, S. M. P. B. & E. C. Oliveira, 1996. Taxonomy of the flattened Solieriaceae (Rhodophyta) in Brazil: *Agardhiella* and *Meristiella*. J. Phycol. 32: 656–668

Hillis, D. M. & J. P. Huelsenbeck, 1992. Signal, noise and reliability in molecular phylogenetic analyses. J. Hered. 83: 189–195.

Hillis, D. M., M. W. Allard & M. M. Miyamoto, 1993. Analysis of DNA sequence data: phylogenetic inference. Methods Enzymol. 224: 456–487.

Hillis, D. M., B. K. Mable, A. Lason, S. K. Davis & E. A. Zimmer, 1996. Nucleic Acids IV: sequencing and cloning, pp. 321–381. In Hillis, D. M., C. Moritz & B. K. Mable (eds), Molecular Systematics. Sinauer Associates, Inc., Sunderland, MA, 655 pp.

Hommersand, M. H., 1986. The biogeography of the South African marine red algae: a model. Bot. mar. 29: 257–270.

Hommersand, M. H. & S. Fredericq, 1990. Sexual reproduction and cystocarp development. In Cole, K. M. & R. G. Sheath (eds), Biology of the Red Algae, Cambridge University Press, New York: 305–345.

Hommersand, M. H., S. Fredericq & D. W. Freshwater, 1994. Phylogenetic systematics and biogeography of the Gigartinaceae (Gigartinales, Rhodophyta) based on sequence analysis of *rbc*L. Bot. mar. 37: 193–203.

Kimura, M., 1980. A simple method for estimating evolutionary rates of base substitutions through comparative studies of nucleotide sequences. J. mol. Evol. 16: 111–120.

Kluge, A. G. & J. S. Farris, 1969. Quantitative phyletics and the evolution of anurans. Syst. Zool. 18: 1–32.

Kraft, G. T., 1984. Taxonomic and morphological studies of tropical and subtropical species of *Callophycus* (Solieriaceae, Rhodophyta). Phycologia 23: 53–71.

Kraft, G. T. & P. W. Gabrielson, 1983. *Tikvahiella candida* gen. et sp. nov. (Solieriaceae, Rhodophyta), an new adelphoparasite from southern Australia. Phycologia 22: 47–57.

Kützing, F. T., 1843. Phycologia generalis. F.A. Brockhaus, Leipzig. xxxii + 458 pp, 80 pls.

Kylin, H., 1932. Die Florideen ordnung Gigartinales. Lunds Univ. Årsskr., NF., Avd 2, 28(8): 88 pp.

Kylin, H., 1956. Die Gattungen der Rhodophyceen. Lund: C.W.K. Gleerups Förlag. xv + 673 pp., 458 figs.

Lawson, G. W. & D. M. John, 1987. The marine algae and coastal environment of tropical West Africa, 2nd edn. Beihefte zur Nova Hedwigia 93: 1–415.

Maddison, D. R., 1991. The discovery and importance of multiple islands of most-parsimonious trees. Syst. Zool. 40: 315–328.

Maddison, W. P. & D. R. Maddison, 1992. MacClade: Analysis of Phylogeny and Character Evolution. Sinauer Associates, Sunderland, MA, 398 pp.

Min-Thein, U. & H. B. S. Womersley, 1976. Studies on southern Australian taxa of Solieriaceae, Rhabdoniaceae and Rhodophyllidaceae (Rhodophyta). Aust. J. Bot. 24: 1–166.

Mshigeni, K. E. & G. F. Papenfuss, 1981. *Solieria jaasundii*, a new species of red algae (Gigartinales, Solieriaceae) from Tanzania. Bot. mar. 24: 1–7.

Norris, J. M., 1985. Observations on *Eucheuma* J. Agardh (Solieriaceae, Rhodophyta) from the Gulf of California, Mexico. In Abbott, I. A. & J. N. Norris (eds), Taxonomy of Economic Seaweeds With Reference to some Pacific and Caribbean Species. La Jolla: California Sea Grant College Program [Report T-CSGCP-011]: 63–65.

Papenfuss, G. F. & T. Edelstein, 1974. The morphology and taxonomy of the red alga *Sarconema* (Gigartinales: Solieriaceae). Phycologia 13: 31–44.

Perrone, C. & E. Cecere, 1994. Two solieriacean algae new to the Mediterranean: *Agardhiella subulata* and *Solieria filiformis* (Rhodophyta, Gigartinales). J. Phycol. 30: 98–108.

Rayss, T., 1963. Sur la présence dans la Méditeranée orientale des algues tropicales de la famille des Soliériacées. Acta Bot. Horti. Bucharest 1961/1962: 91–106.

Rueness, J., 1997. A culture study of *Caulacanthus ustulatus* (Caulacanthaceae, Gigartinales, Rhodophyta) from Europe and Asia. Cryptogamie Algol. 18: 175–185.

Saitou, N. & M. Nei, 1987. The neighbor-joining method: a new method for reconstructing phylogenetic trees. Mol. biol. Evol. 4: 406–425.

Santos, G. A., 1989. Carrageenans of species of *Eucheuma* J. Agardh and *Kappaphycus* Doty (Solieriaceae, Rhodophyta). Aquat. Bot. 36: 55–67.

Searles, R. B., 1968. Morphological studies of red algae of the order Gigartinales. Univ. Calif. Publ. Bot. 43: 1–86.

Silva, P. C., 1980. Names of classes and families of living algae with special reference to their use in the Index Nominum Genericorum (Plantarum). Regnum Vegetabile 103. [iii +] 156 pp.

Silva, P. C., 1993. Report of the committee for algae: 1. Taxon 42: 699–710.

Silva, P. C., P. W. Basson & R. L. Moe, 1996. Catalogue of the benthic marine algae of the Indian Ocean. Univ. California Press, Berkeley, xiv + 1259 pp.

Swofford, D. L, 1998. PAUP∗: Phylogenetic analysis using parsimony (and other methods). Version 4.0b.1. Sinauer Associates, Sunderland, MA.

Weber-van Bosse, A., 1928. Liste des algues du Siboga IV. Rhodophyceae. Pt. 3, Gigartinales et Rhodymeniales et tableau de la distribution des Chlorophycées, Phaeophycées et Rhodophycées de L'archipel Malaisien. Siboga-Expeditie Monographie 59. Leiden: 393–533, figs 143–213, pls. XI–XVI.

Womersley, H. B. S., 1994. The marine benthic flora of southern Australia. Rhodophyta: Part IIIA. Bangiophyceae and Florideophyceae (Acrochaetiales, Nemaliales, Gelidiales, Hildenbrandiales and Gigartinales *sensu lato*). Canberra: Australian Biological Resources Study 508 pp., 167 figs, 4 pls., [4] maps.

Yoshida, T., K. Yoshinaga & Y. Nakajima, 1995. Checklist of the marine algae of Japan (revised in 1995). Jpn. J. Phycol. (Sorûi) 43: 115–171.

Appendix 1. Species, authorities, collection information, percent of *rbc*L sequence produced and GenBank accession number (see Benson et al., 1994) for taxa analyzed in this study.

Agardhiella ramosissima (Harvey) Kylin, Isle des Saintes, Basse-terre, Guadeloupe, coll. S. Fredericq, 11.vi.1995, 90% (AF099680).

Agardhiella subulata (C. Agardh) Kraft & Wynne, Federal Basin, New Hanover Co., NC, U.S.A., coll. D. W. Freshwater, iii.1991, 96% (U04176).

Agardhiella sp., Mar Piccolo basin, Gulf of Taranto, Italy, coll. E. Cecere, 12.i.1993, 92% (AF099681).

Areschougia congesta (Turner) J. Agardh, Warrnambool, Victoria, Australia, coll. M. H. Hommersand, 13.vii.1995, 93% (AF099682).

Areschougia ligulata Harvey ex J. Agardh, Cervantes, West Australia, coll. M. H. Hommersand, 20.ix.1995, 80% (AF099683).

Betaphycus philippinensis Doty, Dancalan, Bulusan, Sorsagon Prov., Luzon, Phillipines, coll. L. Liao & J. G. Sisican, 17.iii.1998, 94% (AF099684).

Betaphycus speciosum (Sonder) Doty, Tarcoola Beach. Geraldton, West Australia, coll. M. H. Hommersand, 21.ix.1995, 95% (AF099685).

Callophycus africanus (Schmitz) Hewitt, Palm Beach, Natal, South Africa, coll. M. H. Hommersand, 23.vii.1993, 91% (U21591).

Callophycus oppositifolius (C. Agardh) Silva, Cervantes, West Australia, coll. M. H. Hommersand, 20.ix.1995, 94% (AF099686).

Caulacanthus ustulatus (Turner) Kützing, Carentec, Brittany, France [introduced species, see Rueness (1997)], coll. J. Cabioch, 22.vi.1993, 69% (U04181).

Caulacanthus ustulatus (Turner) Kützing, Swakopmund, Namibia, coll M. H. Hommersand, 6.vii.1993, 93% (AF099687).

Erythroclonium angustatum Sonder, D'Estrées Bay, Kangaroo I, Australia, coll. M. H. Hommersand, 7.xi.1995, 94% (AF099688).

Erythroclonium muelleri Sonder, D'Estrées Bay, Kangaroo I, Australia, coll. M. H. Hommersand, 7.xi.1995, 93% (AF099689).

Eucheuma arnoldii Weber-van Bosse, Siquijor Town, Siquijor Prov., Philippines, coll. L. M. Liao, 20.xii.1997, 94% (AF099690).

Eucheuma denticulatum (Burman) Collins & Harvey, Danajan Reef, northern Bohol, Philippines, coll. D. F. Kapraun, i.92, 92% (U04177).

Eucheuma isiforme (C. Agardh) J. Agardh, Spanish Harbor Key, FL, U.S.A., Coll. D. W. Freshwater, 26.ii.1994, 98% (AF099691).

Eucheuma serra (J. Agardh) J. Agardh, Magang Harbor, N.E. Taiwan, Coll. S. Fredericq, 6.vii.1994, 98% (AF099692).

'Eucheuma' uncinatum Setchell & Gardner, Bahia de Los Angeles, Baja California, Mexico, Coll. J. Zertuche, from culture, 93% (AF099693).

Heringia mirabilis (C. Agardh) J. Agardh, The Kowie, Port Alfred, Cape Prov., South Africa, Coll. M. H. Hommersand, 19.vii.1993, 96% (U21601).

Kappaphycus alvarezii (Doty) Doty ex Silva, Caragasan Beach., W. coast of Zamboanga City, Mindanao, Philippines, Coll. L. M. Liao, 13.xii.1997, 97% (AF099694).

Kappaphycus cottonii (Weber-van Bosse) Doty ex Silva, N. side of Santa Cruz, Piqueño I., Zamboanga City, Mindanao, Philippines, Coll. L. M. Liao, 11.i.1998, 95% (AF099695).

Kappaphycus striatum (Schmitz) Doty ex Silva, Punta Engaño, Mactan I., Cebu Prov., Philippines, coll, L. M. Liao 28.xii.1997, 64% (AF099696).

Meristiella gelidium (J. Agardh) Cheney & Gabrielson, Baie Olive, Haute Terre, Guadeloupe, coll. A. Renoux, 10.iii.1994, 97% (AF099697).

Appendix 1. contd.

Meristiella gelidium (J. Agardh) Cheney & Gabrielson, Cockroach Isls, Virgin Gorda, British West Indies, Coll. S. Fredericq, 5.vi.1995, 92% (AF099698).

'Meristiella' sp., off Culebra I., Puerto Rico, coll. S. Fredericq, 3.vi.1995, 92% (AF099699).

Meristotheca papulosa (Montagne) J. Agardh, Okonoshima, Tateyana, Chiba Pref., Japan, coll. S. Fredericq & M. Yoshizaki, 3.ix.1993, 98% (AF099700).

Meristotheca procumbens Gabrielson & Kraft, Faniva, Itumuta, Totuma, Fiji, coll. A. N'Yeurt 29.xii.94, 64% (AF099701).

Meristotheca procumbens Gabrielson & Kraft, Punta Engaño, Mactan I., Cebu Prov., Philippines, coll. L. M. Liao, 28.xii.1997, 91% (AF099702).

Meristotheca sp., Ye Lieu, N. Taiwan, Coll. S. Fredericq & S.-M. Lin, 7.vii.1994, 67% (AF099703).

Rhabdonia clavigera J. Agardh, D'Estrées Bay, Kangaroo I, South Australia, coll. M. H. Hommersand, 9.xi.1995, 93% (AF099704).

Rhabdonia coccinea (Harvey) Harvey, Bales Beach, Seal Bay, Kangaroo I., South Australia, coll. M. H. Hommersand, 8.xi.1995, 92% (AF099705).

Sarcodiotheca furcata (Setchell & Gardner) Kylin, Falt Pt., Lopez I, Puget Sound, Washington, USA, coll. C. Stoude, 30.x.1993, 61% (AF099706).

Sarcodiotheca gaudichaudii (Montagne) Gabrielson, Pigeon Pt., San Mateo Co., CA, U.S.A., coll. M. H. Hommersand, 21.xii.1992, 65% (U04184).

Sarcodiotheca gaudichaudii (Montagne) Gabrielson, Playa Changa, Coquimbo, Chile, Coll. S. Fredericq & M. E. Ramírez, 19.i.1995, 93% (AF099707).

Sarconema filiforme (Sonder) Kylin, Drift, Cottleshoe Beach, Perth, West Australia, coll. J. Huisman, 23.v.1993, 81% (AF099708).

Solieria chordalis (C. Agardh) J. Agardh, Rade de Brest, Brittany, France, coll. E. Deslandes, 3.i.1995, 94% (AF099709).

Solieria filiformis (Kützing) Gabrielson, Harbor Branch, Indian River Co., FL, U.S.A., coll. R. Lewis, x.1992, 95% (U04185).

Solieria sp., Orakei Basin, Auckland, New Zealand, coll. W. Nelson, vii.1992, 95% (AF099712).

'Solieria' pacifica (Yamada) Yoshida, Okinoshima, Tateyana, Chiba Pref., Japan, coll. S. Fredericq & M. Yoshizaki, 3.i.1993, 87% (AF099710).

'Solieria' robusta (Greville) Kylin, At 14 m, Port Stephens, W. side of Cabbage Tree I., New South Wales, Australia, coll. A. Millar & P. Richards, 22.iii.1994, 74% (AF099711).

Development of the extant diversity in *Halimeda* is linked to vicariant events

Wiebe H. C. F. Kooistra, Magnolia Calderón & Llewellya W. Hillis[1,*]
Smithsonian Tropical Research Institute, Unit 0948, APO AA 34002-0948, U.S.A. E-mail: halimeda@mbl.edu
[1]*Present address: 20 Brooks Rd., Woods Hole, MA 02543, U.S.A.*

Key words: 18S rDNA, biogeography, Bryopsidales, evolution, *Halimeda*, phylogeny, Tethys, vicariance

Abstract

Partial 18S rDNA sequences, including a 102 base pair insertion, were used to infer a phylogeny among 48 samples across all sections in *Halimeda* Lamouroux, 1812. The phylogeny reveals a separation of the monophyletic section Rhipsalis into a western Atlantic and a western Pacific clade. Consequently, morphologically similar species within this section such as *H. monile* (Ellis & Solander) Lamouroux (western Atlantic), and *H. cylindracea* Decaisne (western Pacific), are not sister taxa. Vicariant events that separated the tropical regions of the Atlantic and Indo-Pacific Oceans can explain the observed biogeographical pattern in section Rhipsalis.

Introduction

Halimeda Lamouroux 1812, in the siphonous order Bryopsidales (see Hillis-Colinvaux, 1984) is one of the most prominent genera in benthic marine communities throughout the tropics. The segmented thalli of *Halimeda* are architecturally complex (Barton, 1901; Hillis-Colinvaux, 1980; Hillis et al., 1998). Members of this genus and its putative ancestors possess extensive fossil records (Elliott, 1960, 1965, 1982; Johnson, 1969; Dragastan et al., 1997) as a consequence of their calcified thalli (Wilbur et al., 1969; Borowitzka & Larkum, 1976a, b, 1977). The thalli form a large part of the limestone in reefs and lagoons (Milliman, 1974; Hillis, 1991, 1997; Drew, 1993) allowing fossilization. The oldest reported *Halimeda* fossils are late Cretaceous (Flügel, 1988), and were widely distributed in the pre-Miocene Tethyan realm (Elliott, 1960, 1980/81; Johnson, 1969).

Microscopic anatomical features indicate similarities among fossil and extant *Halimeda* species, and suggest that extant species may be living fossils. Their origin may well predate vicariant events that separated the Tethys Sea in what is now Syria and Iraq, ca. 12–15 million years ago (Rögl & Steininger, 1984), and the Isthmus of Panama, ca. 3.5 million years ago (Coates et al., 1992) into the tropical Atlantic and the tropical Indo-Pacific Oceans. In that case, one would expect a phylogeny of *Halimeda* with Atlantic and Indo-Pacific geminates or sister species. Alternatively, the extant diversity of the genus may be much younger, and the vicariant events could predate the origin of the extant species. In that case the phylogeny of *Halimeda* would reveal Atlantic and Indo-Pacific sister clades, each with their own sets of species.

Many authors have used hypotheses inferred from molecular characters obtained from ribosomal DNA sequences to study taxonomic relationships and to discriminate among biogeographic hypotheses (eg. Zechman et al., 1990; Olsen et al., 1994; Peters et al., 1997). Hillis et al. (1998) showed congruence between a morphological phylogeny of *Halimeda* and a molecular one inferred from partial 18S rDNA sequences including an insertion. Herein we compare our biogeographic hypotheses with a phylogeny inferred from the same DNA region as in Hillis et al. (1998) but with collections of samples which represent more thoroughly the morphological diversity and the geographic range of this genus.

---
*Author for correspondence

Materials and methods

Forty eight samples, including 15 described species of *Halimeda*, and five samples (4 taxa) not conforming to described species, were used (Table 1).

We sampled over the distribution range of species to assess their phyletic status and interspecific relationships, not to resolve intraspecific relationships. Extraction of DNA, PCR amplification and optional re-amplification were carried out as described in Hillis et al. (1998), with the modification that the cycling profile consisted of 60 sec at 94 °C (denaturation), then 30 cycles of: denaturation 35 sec at 94 °C, annealing 35 sec at 58 °C, and extension 90 sec at 72 °C, then a final extension 5 min at 72 °C. Products were run on low melting point TAE (low EDTA) agarose gels containing 1 μg/ml ethidium bromide. Bright bands were excised, heated to 65 °C for 10 min and incubated for 3 h at 45 °C with 1 μl Gelase (Epicentre Technologies Corporation, Madison, Wisconsin). Cycle sequencing PCR was performed in 10 μl volume, using 2 μl dRhodamine terminator cycle sequencing ready reaction mix (Applied Biosystems, Perkin-Elmer, Foster City, CA), 2 μl HalfTERM (Genpak Inc., Stony Brook, NY), 0.4 μm primer and 5 μl template DNA solution. Cycle sequencing PCR products were cleaned using sephadex G-50 columns and sequenced on an ABI model 377XL automated sequencer (Applied Biosystems) according to manufacturer's instructions. Forward and reverse sequences were compared with SeqEd 1.0.3 (Applied Biosystems), aligned by Se-Al (Rambaud, 1995). A neighbour joining tree was inferred from Kimura 2-parameter corrected distances (Kimura, 1980) using PAUP (Phylogenetic Analysis Using Parsimony*, version 4.0.0d64, an advance trial version distributed by Swofford. Final distribution Sinauer Assoc. Inc., Sunderland MA, U.S.A. Maximum parsimony trees were also obtained in PAUP. Decay indices were calculated according to Bremer (1988). Numbers of synapomorphies supporting clades in obtained and alternative phylogenies were evaluated using MacClade (Maddison & Maddison, 1992).

Results

The complete dataset consists of 712 base pairs, including an insertion of ca. 102 base pairs. The insertion is present in all samples included in this study, and is located in the loop at the end of stem 44 of the 18S rRNA secondary structure (van de Peer et al., 1993). The insertion comprises 14% of our sequences and 63% (53/84) of the variable sites. A 10 000 random trees run in PAUP* resulted in a g_1 value of -0.500465, indicating significant structure among the 55 informative sites (Hillis & Huelsenbeck, 1992).

We also obtained homologous partial 18S rDNA and insertion sequences from members of the stipitate calcified bryopsidalean genera *Penicillus* Lamouroux, *Rhipocephalus* Kützing, *Udotea* Lamouroux and the stipitate uncalcified genus *Flabellia* (Turra) Nizamuddin. These sequences could be aligned among one another but revealed many alternative alignment possibilities when aligned with *Halimeda* sequences. Therefore, phylogenetic analyses do not include outgroups, making the resulting trees unrooted.

A neighbour joining (NJ) phylogeny of the *Halimeda* sequences is presented in Figure 1. Maximum parsimony analysis resulted in 206 equally most parsimonious trees (MPTs, not shown) of 147 steps (consistency index, 0.673; retention index, 0.884; rescaled consistency index, 0.595). Decay indices and MP bootstrap values for clades, that are also recovered in the NJ tree, are presented in Figure 1 and agree with NJ bootstrap support.

The tree reveals a well supported lineage with members of sections Opuntia J. Agardh ex De Toni and Micronesicae Hillis-Colinvaux. The phylogenetic status of section Opuntia remains unresolved because *H. fragilis* Taylor in section Micronesicae is included within section Opuntia. The lineage shows a basal polytomy with four branches: one with a clade containing Atlantic and Pacific samples of *H. opuntia* (L.) Lamouroux and the Hawaiian sample of *H.* sect. Opuntia; a branch with a second sample of *H.* sect. Opuntia (Indo-Pacific origin); a third branch with an Atlantic clade of *H. copiosa* Goreau & Graham and *H. goreauii* Taylor, and the fourth branch with a Pacific sample of *H. fragilis*.

Section Rhipsalis J. Agardh ex De Toni is monophyletic, well supported and internally well-resolved. The first dichotomy of this lineage separates an Indo-West Pacific clade with *H. macroloba* Decaisne, three samples of *H.* section Rhipsalis and *H. cylindracea* Decaisne, from an Atlantic clade with *H. simulans* Howe, *H. monile* (Ellis & Solander) Lamouroux and *H. incrassata* (Ellis) Lamouroux.

The tree also shows a well resolved lineage with *H. gracilis* Harvey ex J. Agardh and *H. lacrimosa* Howe in section Halimeda J. Agardh ex De Toni. The

Table 1. List of samples from which the partial 18S rDNA sequence and the insertion therein were examined.

Taxon	Voucher code	Ocean	Geographical location	Lat.	Long.
Section Opuntia J. Agardh ex De Toni					
H. opuntia	98-083	W A	Smithsonian Sta., San Blas, Panama	9° 32' N	78° 57' W
H. opuntia	95-IG203	W A	I. Grande, Panama	9.6° N	79.8° W
H. opuntia	95-SAC2	W A	San Andres I., Colombia	12.6° N	81.7° W
H. opuntia	95-HON07	W A	I. Roatan, Honduras	16.3° N	86.3° W
H. opuntia	96-IO002	C I O	Chagos Archipelago	5.1° S	71.5° E
H. opuntia	95-GUAM1	W P	Agat Bay, Guam	13.5° N	144.8° E
H. section Opuntia	97-059	C P	Honaunau, Oahu, Hawaii	19° 18' N	155° 40' W
H. section Opuntia	98-031	I P	Tukang Besi I., Indonesia	5° 56' S	123° 48' E
H. goreauii	97-068	W A	I. Escudo de Veraguas, Panama	9°06 N	81°30 W
H. goreauii	95-SB1	W A	Icacos I. San Blas, Panama	9.32° N	78.57° W
H. goreauii	98-037	W A	Puerto Morelos, Mexico	21° 30' N	86° 30' W
H. copiosa	97-080	W A	Cayo Zapatilla, Panama	9° 14' N	82° 00' W
H. copiosa	95-SAC15	W A	San Andres I., Colombia	12.6° N	81.7° W
H. copiosa	97-481	W A	Discovery Bay, Jamaica	18° 40' N	77° 20' W
Section Micronesicae Hillis-Colinvaux					
H. fragilis	95-GUAM8a	W P	Double Reef, Guam	13.5° N	144.8° E
Section Rhipsalis J. Agardh ex De Toni					
H. macroloba	475x4	W P	Townsville, GBR, Australia	19° S	147° E
H. macroloba	97-486	E I O	Mangrove Bay, W. Australia	32° S	116° E
H. section Rhipsalis	95-GUAM12	W P	Cocos I., Guam	13.5° N	144.8° E
H. section Rhipsalis	98-001	W P	Guiguiwanen, Bolinao, Philippines	16.38° N	119.90° E
H. section Rhipsalis	97-494	W P	Nukubuco reef, Viti Levu, Fiji	18° 10' S	178.29' E
H. incrassata	97-485	W A	I. Grande, Panama	9.6° N	79.8° W
H. incrassata	95-HRS1	W A	Smithsonian Sta., San Blas, Panama	9° 32' N	78° 57' W
H. incrassata	98-040	W A	Laguna del Madre, Texas	26° 08' N	97° 14' W
H. simulans	95-HRS4	W A	Smithsonian Sta., San Blas, Panama	9° 32' N	78° 57' W
H. simulans	98-036	W A	Puerto Morelos, Mexico	21° 30' N	86° 30' W
H. simulans	94-G9	W A	Galeta, Panama	9.6° N	78.6° W
H. monile	97-072	W A	I. Escudo de Veraguas, Panama	9° 06' N	81° 44' W
H. monile	98-034	W A	Puerto Morelos, Mexico	21° 30' N	86° 30' W
H. cylindracea	475x2	W P	Townsville, GBR, Australia	19° S	147° E
Section Halimeda J. Agardh ex De Toni					
H. tuna	97-484	W A	I. Grande, Panama	9.6° N	79.8° W
H. tuna	97-096	W A	I. Colon, Panama	9° 20' N	82° 14' W
H. tuna	97-078	W A	Cayo Zapatilla, Panama	9° 14' N	82° 10' W
H. tuna	97-069	W A	I. Escudo de Veraguas, Panama	9° 06' N	81° 44' W
H. tuna	98-038	W A	Puerto Morelos, Mexico	21° 30' N	86° 30' W
H. tuna	98-073	M	Malta	35.8° N	14.5° E
H. discoidea	97-547	E A	Gran Canaria, Canary Islands	15° 30' W	28° 0' N
H. discoidea	98-052	E A	Sao Vicente, Cape Verde Islands	18° 30' N	25° 0' W
H. discoidea	95-HON08	W A	I. Roatan, Honduras	16.3° N	86.3° W
H. discoidea	97-483	W A	Portobelo, Panama	79° W	9° N
H. discoidea	98-051	C P	Midway I., Hawaii	28° N	187° E
H. discoidea	97-057	C P	Kaneohe Bay, Oahu, Hawaii	21° 29' N	157° 50' W
H. discoidea	95-CH001	E P	I. Uva, Panama	7.8° N	82.0° W
H. discoidea	95-ME007	E P	La Paz, Baja Calif. Mexico	24.2° N	110.3° W
H. discoidea	95-GUAM3a	W P	Cocos I., Guam	13.5° N	144.8° E
H. lacunalis f. lata	95-GUAM7	W P	Double Reef, Guam	13.5° N	144.8° E
H. gracilis	97-089	W A	I. Escudo de Veraguas, Panama	9° 06' N	81° 30' W
H. lacrimosa	95-BA010	W A	Lee Stocking Island, Bahamas	23° 45' N	76° 05' W
Section Crypticae Hillis-Colinvaux					
H. cryptica	97-482	W A	Discovery Bay, Jamaica	18° 40' N	77° 20' W

W A, western Atlantic; E A, eastern Atlantic; M, Mediterranean Sea; W P, western Pacific; C P, central Pacific; E P, eastern Pacific; C I O, central Indian Ocean; E I O, eastern Indian Ocean; I P, Indo-Pacific

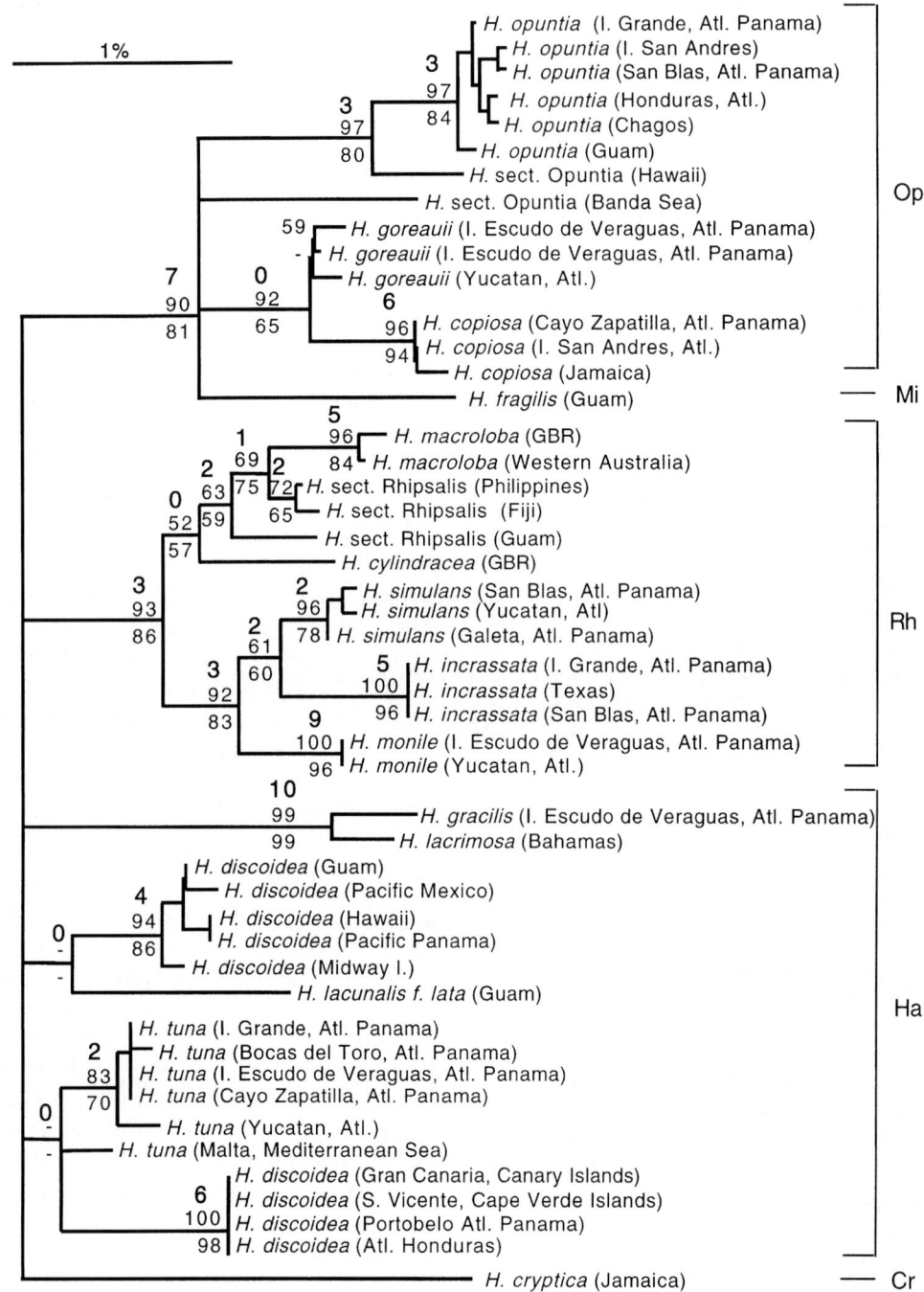

Figure 1. Neighbour joining phylogeny of partial 18S rDNA sequences and their inserts of 48 samples of *Halimeda* inferred from Kimura 2-parameter corrected distances. Branch length is indicated by the scale bar above the tree. The decay index (in bold) and neighbour joining bootstrap value (1000 replicates) of >50%, associated with a clade, are mapped above the internode leading to that clade. Maximum parsimony bootstrap values of >50% (1000 replicates, fast stepwise addition) are presented underneath internodes. "-" indicates bootstrap value of <50%. Decay index 0 indicates that the clade was recovered in all equally most parsimonious trees. No decay index is given when a clade is recovered in less that 100% of the MPTs. Dichotomies in the obtained neighbour joining phylogeny without sufficient bootstrap support and without decay index, or those that were not recovered in the consensus of the equally most parsimonious trees are presented collapsing into polytomies. Decay indices and bootstrap values associated with intraspecific clades are omitted from the figure. Abbreviations used: Op, section Opuntia; Mi, section Micronesicae; Rh, section Rhipsalis; Ha, section Halimeda, and Cr, section Crypticae.

remaining samples of this section are encountered in two weakly supported lineages: one with Pacific *H. discoidea* Decaisne and *H. lacunalis f. lata* Taylor and one with Atlantic *H. discoidea* and *H. tuna* (Ellis & Solander) Lamouroux. Tree topology does not resolve the phylogenetic status of section Halimeda.

Halimeda cryptica Colinvaux & Graham, in section Crypticae Hillis-Colinvaux, is found on a single long branch. This lineage is shown as distinct from all others.

The topology among the lineages is not well resolved. Weak resolution does not result from branch attraction (Hendy & Penny, 1989) because Kimura 2-parameter corrected distances among the insertion sequences, the most rapidly evolving region in our sequences, do not exceed 25%. A comparison of synapomorphies associated with alternative branching patterns among the basal lineages (MacClade) reveals that weak support for these ramifications is due to low numbers or absence of synapomorphies, and not to conflicting informative sites.

Most species that are represented by multiple samples are monophyletic. However, the phylogenetic status of *H. discoidea* and *H. tuna* remains unresolved. Amphi-Atlantic samples of *H. discoidea* are monophyletic and are distinct from a clade with Pacific samples of this species. The sample of *H. tuna* from the Mediterranean Sea is distinct from a clade of this species from the Caribbean.

Discussion

Quality of the data set

Since many clades, especially those of section Rhipsalis, are generally well supported, we believe that the unresolved relationships among the members of section Halimeda, and among the major lineages, are caused by a shortage of parsimony informative sites at those particular levels and not by an overall shortage. Comparisons among branch lengths in Figure 1 suggest that sequences of *H. discoidea* and *H. tuna* have relatively slow substitution rates in contrast to those of the remaining sequences; their sequences may not have acquired enough substitutions to resolve their interspecific phylogeny. The low number of synapomorphies that support the topology among the lineages could result from the relatively brief period during which all these lineages evolved.

More parsimony informative sites are needed to improve the resolution of the *Halimeda* phylogeny. We expect many such sites on the ITS regions; not primarily on the remaining 1100–1200 positions of the 18S rDNA sequence because within *Halimeda*, only 31 variable sites are encountered on the ca. 600 positions of the partial 18S rDNA sequences excluding the insertion.

All sequences included in the present data set were obtained with the Applied Biosystems dRhodamine terminator cycle sequencing kit (Perkin Elmer). All sequences in Hillis et al. (1998), were generated with the Applied Biosystems Taq dye-deoxi terminator cycle sequencing kit (Perkin Elmer) and were redone with the new kit. A few differences are apparent but they do not alter the conclusions in Hillis et al. (1998).

Monophyly of units of biogeographic comparison

Biogeographic comparisons within taxonomically perceived species, or between assumed geminates, can lead to erroneous conclusions when the species are not monophyletic or the so-called geminates result from convergence. In our comparison, Pacific and Atlantic samples of the pantropical species *H. discoidea* may not be monophyletic. In Atlantic *H. monile* and Pacific *H. cylindracea* cylindrical segments are common but these taxa are not nearest sisters.

Taxonomic status of the sections

In their preliminary phylogenies of *Halimeda*, Hillis et al. (1998) suggested monophyly for section Rhipsalis and paraphyly for section Opuntia, since the latter included a member of section Micronesicae. Our present phylogeny supports these findings. In Hillis et al. (1998) section Halimeda was only represented by *H. lacunalis f. lata* and Pacific isolates of *H. discoidea*; moreover these entities appeared to be nearest sisters. Although, the molecular phylogeny in Figure 1 includes three more species within this section as well as Atlantic samples of *H. discoidea*, the taxonomic status of section Halimeda remains unresolved. Given the morphological diversity contained in this large section (Hillis-Colinvaux, 1980), paraphyly or polyphyly would not be surprising. Finally, the lineage containing *Halimeda cryptica* appears not to be particularly related to any of the other lineages in the data set. This finding supports placement of this species in the monospecific section Crypticae (Hillis-Colinvaux, 1980).

Vicariance

Section Rhipsalis is separated into a western Pacific clade and a western Atlantic clade. This pattern could result from vicariant events that separated the tropical Atlantic from the tropical Indo-Pacific: the closure of Tethys in the Middle East (ca. 15–12 million years ago, Rögl & Steininger, 1984) and the final emergence of the Isthmus of Panama (ca. 3.5 million years ago, Coates et al., 1992). If the above inferences are correct, then cladogenesis into the extant diversity of section Rhipsalis must have occurred somewhere in the last ca. 12 million years, because the presumed origin of the diversity is not much deeper than the subdivision into Atlantic and Indo-Pacific clades. The estimated age of the appearance of modern species contrasts with the Cretaceous age of the genus (Elliott, 1960, 1980/81; Johnson, 1969; Flügel, 1988), and suggests that *Halimeda* experienced one or more earlier radiations followed by widespread extinction.

Conclusions

The observed pattern in the molecular phylogeny agrees with the hypothesis that the section Rhipsalis was separated into a western Atlantic and a western Pacific clade by vicariant events separating the tropical Atlantic from the tropical Indo-Pacific.

Acknowledgements

Research was carried out at the Smithsonian Tropical Research Institute, Republic of Panama. We thank K. Arano, J. Coleson, P. A. Colinvaux, E. A. Drew, J. A. Engman, Y. de Jong, A. Dominici, P. W. Glynn, O. Gussmann, R. Haroun, I. Hendriks, S. Holzwarth, J. Jara, N. Knowlton, J. Kowalski, G. Llewellyn, J. Maté, J. L. Olsen, F. Parrish, W. F. Prud'homme van Reine, E. Serviere-Zaragoza, H. van der Strate, J. Woodley, R. Zechman & J. Zucarello for sample collections. We gratefully acknowledge financial support to L.W.H. from the Smithsonian Institution's Scholarly Studies Program and the Andrew W. Mellon Foundation. For permission to collect in the Republic of Panama, we thank the Government of Panama (Instituto Nacional de Recursos Naturales Renovables and Recursos Marinos) and the Kuna Nation.

References

Barton, E. S., 1901. The Genus *Halimeda*. Monograph LX of: Uitkomsten op zoologisch, botanisch, oceanographisch en geologisch Gebied verzameld in Nederlandsch Oost-Indie 1899-1900 aan boord H. M. Siboga onder commando van Luitenant ter zee 1e kl. G. F. Tydeman, uitgegeven door Dr. Max Weber, Prof. in Amsterdam, Leider der Expeditie. E. J. Brill, Publishers and Printers, Leiden: 32 pp. + 4 pls.

Borowitzka, M. A. & A. W. D. Larkum, 1976a. Calcification in the green alga *Halimeda*. II. The exchange of Ca^{2+} and the occurrence of age gradients in calcification and photosynthesis. J. exp. Bot. 27: 864–878.

Borowitzka, M. A. & A. W. D. Larkum, 1976b. Calcification in the green alga *Halimeda*. III. The source of inorganic carbon for photosynthesis and calcification and a model of the mechanism of calcification. J. exp. Bot. 27: 879–893.

Borowitzka, M. A. & A. W. D. Larkum, 1977. Calcification in the green alga *Halimeda*. I. An ultrastructure study of thallus development. J. Phycol. 13: 6–16.

Bremer, K., 1988. The limits of amino acid sequence data in angiosperm phylogenetic reconstruction. Evolution 42: 795–803.

Coates, A. G., J. B. C. Jackson, L. Collins, T. M. Cronin, H. J. Dowsett, L. M. Bybell, P. Jung & J. A. Obando, 1992. Closure of the Isthmus of Panama: the near-shore marine record of Costa Rica and Western Panama. Geol. Soc. Am. Bull. 104: 814–828.

Dragastan, O., D. K. Richter, K. Barbel, P. Mihai, S. Anca & C. Ion, 1997. A new family of Paleo-Mesozoic calcareous green siphons-algae (order Bryopsidales, class Bryopsidophyceae, phylum Siphonophyta). Rev. Esp. Micropaleontol. 29: 69–135.

Drew, E. A., 1993. *Halimeda* biomass, growth rates and sediment generation on reefs in the central Great Barrier Reef province. Coral Reefs. 2: 101–110.

Elliott, G. F., 1960. Fossil calcareous algal floras of the Middle East with a note on a Cretaceous problematicum, *Hensonella cylindrica gen. et sp. nov.* Quarterly J. geol. Soc. Lond. 115: 217–232.

Elliott, G. F., 1965. The interrelationships of some Cretaceous Codiaceae (calcareous algae). Paleontology 8: 199–203.

Elliott, G. F., 1980/81. The Tethyan dispersal of some Chlorophyte algae subsequent to the Paleozoic. Paleogeogr. Paleoclim. Paleoecol. 32: 341–352.

Elliott, G. F., 1982. A new calcareous green alga from the middle Jurassic of England: its relationships and evolutionary position. Paleontology 25: 431–437.

Flügel, E., 1988. *Halimeda*: paleontological record and palaeoenvironmental significance. Coral Reefs 6: 123–130.

Hendy, M. D. & D. Penny, 1989. A framework for the quantitative study of evolutionary trees. Syst. Zool. 38: 297–309.

Hillis, D. M. & J. P. Huelsenbeck, 1992. Signal, noise, and reliability in molecular phylogenetic analyses. J. Heredity 83: 189–195.

Hillis, L., 1991. Recent calcified Halimedaceae. In Riding, R. (ed.), Calcareous Algae and Stromatolites. Springer Verlag, Berlin, Heidelberg, New York: 167–188.

Hillis, L. 1997. Coralgal reefs from a calcareous green alga perspective, and a first carbonate budget. Proc. 8th Int. Coral Reef Sym. 1: 761–766.

Hillis, L. W., J. A. Engman & W. H. C. F. Kooistra, 1998. Morphological and molecular phylogenies of *Halimeda* (Chlorophyta, Bryopsidales) identify three evolutionary lineages. J. Phycol. 34: 669–681.

Hillis-Colinvaux, L., 1980. Ecology and taxonomy of *Halimeda*: primary producer of coral reefs. Adv. mar. Biol. 17:1–327.

Hillis-Colinvaux, L., 1984. Systematics of the Siphonales. In Irvine D. E. G. & D. M. John (eds), Systematics of the Green Algae. Systematics Ass., Special Vol. 27, Academic Press, London & Orlando: 271–296.

Johnson, J. H., 1969. A Review of the Lower Cretaceous Algae. Professional Contribution Colorado School of Mines, No. 6, Colorado School of Mines, Golden, Colorado, 180 pp.

Kimura, M., 1980. A simple method for estimating evolutionary rates of base substitution through comparative studies of nucleotide sequences. J. mol. Evol. 16: 111–120.

Lamouroux, J. V. F., 1812. Extrait d'un mémoire sur la classification des polypes coralligènes non entièrement pierreux. Nouv. Bull. Sci. Soc. Philom. 3: 181–8.

Maddison, W. P. & D. R. Maddison, 1992. MacClade: Analysis of Phylogeny and Character Evolution, Version 3. Sinauer Associates Inc., Sunderland, Massachusetts, 398 pp.

Milliman, J. D., 1974. Recent sedimentary carbonates, part I: Marine biocarbonates. Springer Verlag, NY, 378 pp.

Olsen, J. L., W. T. Stam, S. Berger & D. Menzel, 1994. 18S rDNA and evolution in the Dasycladales (Chlorophyta): modern living fossils. J. Phycol. 30: 730–744.

Peters, A. F., M. J. H. van Oppen, C. Wiencke, W. T. Stam & J. L. Olsen, 1997. Phylogeny and historical ecology of the Desmarestiaceae (Phaeophyceae) support a southern hemisphere origin. J. Phycol. 33: 294–309.

Rambaud, A., 1995. Se-Al, Sequence alignment program v1-d1. Department of Zoology, University of Oxford. (andrew.rambaud@zoology.ox.ac.uk)

Rögl, F. & F. F. Steininger, 1984. Neogene Paratethys, Mediterranean and Indo-Pacific seaways. Implications for the paleobiogeography of marine and terestrial biotas. In Brenchley P. J. (ed.), Fossils and Climate, John Wiley & Sons, Chichester: 171–200.

van de Peer, Y., J. M. Neefs, P. De Rijk & R. De Wachter, 1993. Reconstructing evolution from eukaryotic small-subunit RNA sequences: calibration of the molecular clock. J. mol. Evol. 37: 221–332.

Wilbur, K. M., L. Hillis-Colinvaux & N. Watabe, 1969. Electron microscope study of calcification in the alga *Halimeda* (order Siphonales). Phycologia 8: 27–35.

Zechman, F. W., E. C. Theriot, E. A. Zimmer & R. L. Chapman, 1990. Phylogeny of the Ulvophyceae (Chlorophyta): cladistic analysis of nuclear encoded rRNA sequence data. J. Phycol. 26: 700–710.

Phylogeny of Alariaceae (Phaeophyta) with special reference to *Undaria* based on sequences of the RuBisCo spacer region

Hwan Su Yoon & Sung Min Boo*
Department of Biology, Chungnam National University, Daejon 305-764, Korea,
E-mail: smboo@hanbat.chungnam.ac.kr

Key words: Alariaceae, gene tree, Phaeophyta, phylogeny, RuBisCo spacer, *Undaria*

Abstract

In order to investigate the phylogenetic relationships of the family Alariaceae, we determined the complete sequences of the plastid RuBisCo spacer region for fourteen taxa of the Alariaceae and for two reference species. The RuBisCo spacer sequences showed that the Alariaceae forms two clades: one comprises *Alaria, Pterygophora* and *Undaria*, while the other comprises *Ecklonia, Eckloniopsis, Eisenia, Egregia* and *Laminaria*. These results favour the narrow concept of the Alariaceae, in which *Alaria, Pterygophora* and *Undaria* only may be placed. *Ecklonia, Eckloniopsis, Eisenia, Egregia* and *Laminaria* appear not to have a common ancestor. The RuBisCo spacer sequences of the three *Undaria* species also indicate that the species are very closely related, with an intermediate relationship at the DNA level; we hypothesize that there might be a reticulate evolution among the species. In this scenario, *U. undarioides* might be an ancient hybrid species from the parental species of *U. peterseniana* and *U. pinnatifida*. The RuBisCo spacer region, including its flanking areas, is useful for inferring phylogenetic relationships within the family Alariaceae.

Introduction

The family Alariaceae (Laminariales) was established by Setchell & Gardner (1925) on the basis of midrib and sporophyll of the sporophytes. Three tribes, nine genera and thirty-three species are recognized in this family (Papenfuss, 1951; Estes & Steinberg, 1988; Druehl et al., 1997). Most of the members occur in the North Pacific, where they have been suggested to have radiated late in the Cenozoic (Estes & Steinberg, 1988).

The Alariaceae is typified by the genus *Alaria*, which is closely related to *Pterygophora* and *Undaria* in having sporophylls (Setchell & Gardner, 1925). The concept of the family has, however, been controversial in the light of morphological as well as molecular data (Saunders & Druehl, 1992; Kawashima, 1993; Druehl et al., 1997). *Ecklonia, Eckloniopsis, Eisenia* and *Egregia* have traditionally been placed in the Alariaceae (Setchell & Gardner, 1925; Papenfuss, 1951; Miyabe, 1957; Bold & Wynne, 1985). On the other hand, Okamura (1936) and Kawashima (1993) treated them within the Laminariaceae, based on the presence of mucilage ducts and absence of sporophylls. Recently, Bolton & Anderson (1994) considered that *Ecklonia* should be placed in the Alariaceae, based on production of secondary blades from the lateral meristems.

The genus *Undaria* includes three species, and it is endemic to China, Japan and Korea (Okamura, 1916; Kang, 1966). Although the type species, *U. pinnatifida*, is distinguished by the *Alaria*-type sporophylls and midrib, the other members are very diverse in these characters (Okamura, 1916; Kawashima, 1993). Interspecific hybrids of *Undaria* species have been successful in artificial crossings (Saito, 1972), as is often the case in kelps (Druehl, 1989; Druehl & Saunders, 1992). Kawashima (1993) postulated that the genus is less specified in the course of evolution, and further commented that *U. undarioides* is difficult to distinguish from *U. peterseniana*.

The large and small subunits of the ribulose-1,5-biphosphate carboxylase/oxigenase (RuBisCo) gene

---
* Author for correspondence.

are separated by a short spacer within the plastid genome in brown and red algae (Destombe & Douglas, 1991). The spacer region has been used to compare specific and generic divergence (Goff et al., 1994; van Oppen et al., 1995; Stache-Crain et al., 1997). These studies showed that the region is more conserved than the ribosomal internal transcribed spacer (ITS 1 & 2), but more variable than the coding genes of RuBisCo and the 18S rDNA. Although nuclear DNA sequence data have contributed to the understanding of kelp molecular phylogeny (Saunders & Druehl, 1992, 1993a, b; Druehl et al., 1997), RuBisCo spacer sequences have rarely been used for kelp systematics.

The aims of this study were to circumscribe the family Alariaceae and to provide a hypothesis of phylogenetic relationships within the family based on molecular data. We investigated the RuBisCo spacer region of the fourteen taxa to cover all the genera in the Alariaceae. The related members of the Laminariales, *Laminaria japonica* and *Chorda filum* were included as references.

Materials and methods

The taxa selected for this study were representatives of all the genera within the Alariaceae (Setchell & Gardner, 1925; Papenfuss, 1951) from the North Pacific, including reference species that are related to this family (Table 1). All the voucher specimens have been deposited in the herbarium of Chungnam National University, Daejon, Korea.

Thalli dried in silica gel were washed by filtered seawater and distilled water before grinding. Total DNA were extracted using Chelex 100 resin (BioRad) or 2 × CTAB (2% hexadecyltrimethylammonium bromide) buffer (Stache-Crain et al., 1997). The quality and quantity of extracted DNA were estimated on 0.8% agarose gels with EtBr.

To amplify the RuBisCo spacer fragment, polymerase chain reaction (PCR) was performed in an automated thermocycler (MJ Research, Watertown). Reaction cocktails followed the recipes of Tan & Druehl (1996). A pair of RS1 + RS2 primers was used for each reaction (RS1: 5′ GCC AAA TGC ACC AAC TTC TT 3′, RS2: 5′ AGA CCC CAT AAT TCC C 3′). A negative control without target DNA template was included in each set of reactions. The initial cycle of PCR was denaturation at 95 °C for 4 min, annealing at 45 °C for 1 min, and elongation at 72 °C for 1 min. It was followed by 28 cycles of denaturation at 95 °C for 30 sec, annealing at 45 °C for 30 sec, extension at 72 °C for 1 min, and the final extension at 72 °C for 10 min. To check the yield and size of amplified DNA, PCR products were separated through electrophoresis on 0.8% agarose gels with EtBr. Amplified DNA were purified by GeneClean II kit (Bio 101, La Jolla, CA).

The amplified and purified DNA of the RuBisCo spacer region was directly used for sequencing by the dideoxy method using Sequenase 2.0 (U.S. Biochemicals, OH). A nested primer, RS3 (RS3: 5′ AAA GCG GCT TTA GAT TTA TG 3′) and the two PCR primers were used for sequencing. The sequencing cocktail included 150 ng of purified DNA template, 5 ng of primer, and mixtures according to manufacturer's recommendation. Each primer was annealed to the DNA using the snap-cooling technique for brown algae (Tan & Druehl, 1996). Ten percent dimethyl sulfoxide (DMSO) was added to the annealing and labelling mixes to enhance the sequencing signals. DNA sequences of the RuBisCo spacer were separated on 8% polyacrylamide gels following the manufacturer's protocol. The gels were fixed for 15 min with 5% glacial acetic acid and 15% methanol; autoradiographs were exposed for 96 h.

The RuBisCo spacer sequences were aligned using SeqPup, a multisequence editing program (Gilbert, 1995). Final alignment was visually done by reference to the published RuBisCo spacer sequences for *Ectocarpus* species (Stache-Crain et al., 1997). Maximum parsimony (MP) analysis was done with PAUP program (ver. 3.1.1, Swofford, 1993). All nucleotides were equally weighted and gaps were treated as missing characters. Branch and bound search option was used to find the shortest trees. Bootstrap and decay indices were used to assess the robustness of each clade on the trees. Bootstrap analyses with heuristic search was performed using 1000 replicates with TBR and MULPARS option (Felsenstein, 1985). Decay analyses were performed for 10 steps (Morgan, 1997). A tree length frequency distribution was determined by evaluating 100 000 trees sampled randomly from the set of all possible trees (Hillis & Huelsenbeck, 1992).

Sequence divergence was calculated by DNADIST program of PHYLIP (Felsenstein, 1993) with Kimura's (1980) two parameter model of nucleotide changes. The divergences were converted to produce the neighbour-joining (NJ) tree by NEIGHBOR program, and bootstrapping of 1000 replicates was given. The maximum likelihood (ML) tree was constructed using DNAML program of PHYLIP and bootstrapping of 100 replicates was also given.

Table 1. List of taxa used for the RuBisCo spacer sequencing study with source localities and herbarium vouchers with accession numbers (Chungnam National University, Daejon, Korea)

Family Species & collection localities	Vouchers
Alariacae	
Alaria crassifolia Kjellman, Hokodate, Hokkaido, Japan	CU-97081
A. esculenta (L.) Graville, Ketch Harbor, Nova Scotia, Canada	CU-97071
A. marginata Postels & Ruprecht, Bamfield, B.C., Canada	CU-97072
A. praelonga Kjellman, Akkeshi, Hokkaido, Japan	CU-97082
Eckonia cava Kjellman, Sungsan, Cheju, Korea	CU-96101
Ec. stolonifera Okamura, Geojedo, Tongyoung, Korea	CU-9602
Eckloniopsis radicosa (Kjellman) Okamura, Wakayama, Koji, Japan	CU-9802c
Egregia menziesii (Turner) Areschoug, Bamfield, B.C., Canada	CU-97073
Eisenia arborea Areschoug, Carmel Bay, California, U.S.A.	CU-97074
Ei. bicyclis (Kjellman) Setchell, Ulreungdo, East Sea, Korea	CU-96103
Pterygophora californica Ruprecht, Carmel Bay, California, U.S.A.	CU-97075
Undaria peterseniana (Kjellman) Okamura, Sungsan, Cheju, Korea	CU-9607
U. pinnatifida (Harvey) Suringar, Padori, Taean, Korea	CU-9612
U. undarioides (Yendo) Okamura, Wakayama, Koji, Japan	CU-97083
Chordaceae	
Chorda filum (L.) Stackhouse, Padori, Taean, Korea	CU-9608
Laminariaceae	
Laminaria japonica Areschoug, Shinnam, Samcheok, Korea	CU-9604

Results

Complete sequences of the RuBisCo spacer region were determined for 16 taxa, mainly the family Alariaceae, but also *Chorda filum* (Chordaceae) and *Laminaria japonica* (Laminariaceae). The total sequences determined here were from 611 to 630 bp in the taxa of the Alariaceae, including the flanking regions. The sequence length of *Chorda* was 493 bp with a big indel and that of *Laminaria* 612 bp. The flanking regions in all the taxa treated here were determined for 244 bp from the RuBisCoL 3′ end and 99 bp from the RuBisCoS 5′ end. The actual sequence length of the RuBisCo spacer ranged from 268 to 287 bp in the Alariaceae treated here. There were no differences in spacer size and gap positions among the *Undaria* species.

All 16 sequences were aligned with 646 positions (Figure 1). The variable nucleotides were observed at 185 positions (29%). The intrageneric divergence was very low with 0.004 ± 0.001 within *Undaria*, while high with 0.016 ± 0.011 within *Alaria*. The interfamilial divergence was 0.102 ± 0.013 between the alariacean and the laminariacean clades (data not shown).

Ninety-six variable positions were uninformative and 89 positions were parsimony-informative in MP analysis. The analysis produced 6 equally parsimonious trees with a tree length of 265 steps, a consistency index of 0.853, and a retention index of 0.866. The frequency distribution of tree length showed a skew to the left ($g1 = -0.764$). These descriptions of the parsimonious trees indicated supportable phylogenetic signals on the data set of the RuBisCo spacer sequences for the Alariaceae. The topology of the MP trees was identically reflected on the strict consensus tree with bootstrap and decay indices (Figure 2). The Alariaceae was separated into two clades. One, so-called 'alariacean' clade, comprised *Alaria, Pterygophora,* and *Undaria* with 51% bootstrap value, and 1 step of decay index. Within this clade, the branch of *Alaria* and *Undaria* was produced with below 50% bootstrap value and 1 step of decay index. Bootstrap and decay supports were, however, very strong within *Alaria* (97%, 5 steps) and *Undaria* (100%, 9 steps), respectively. The other, so-called 'laminari-

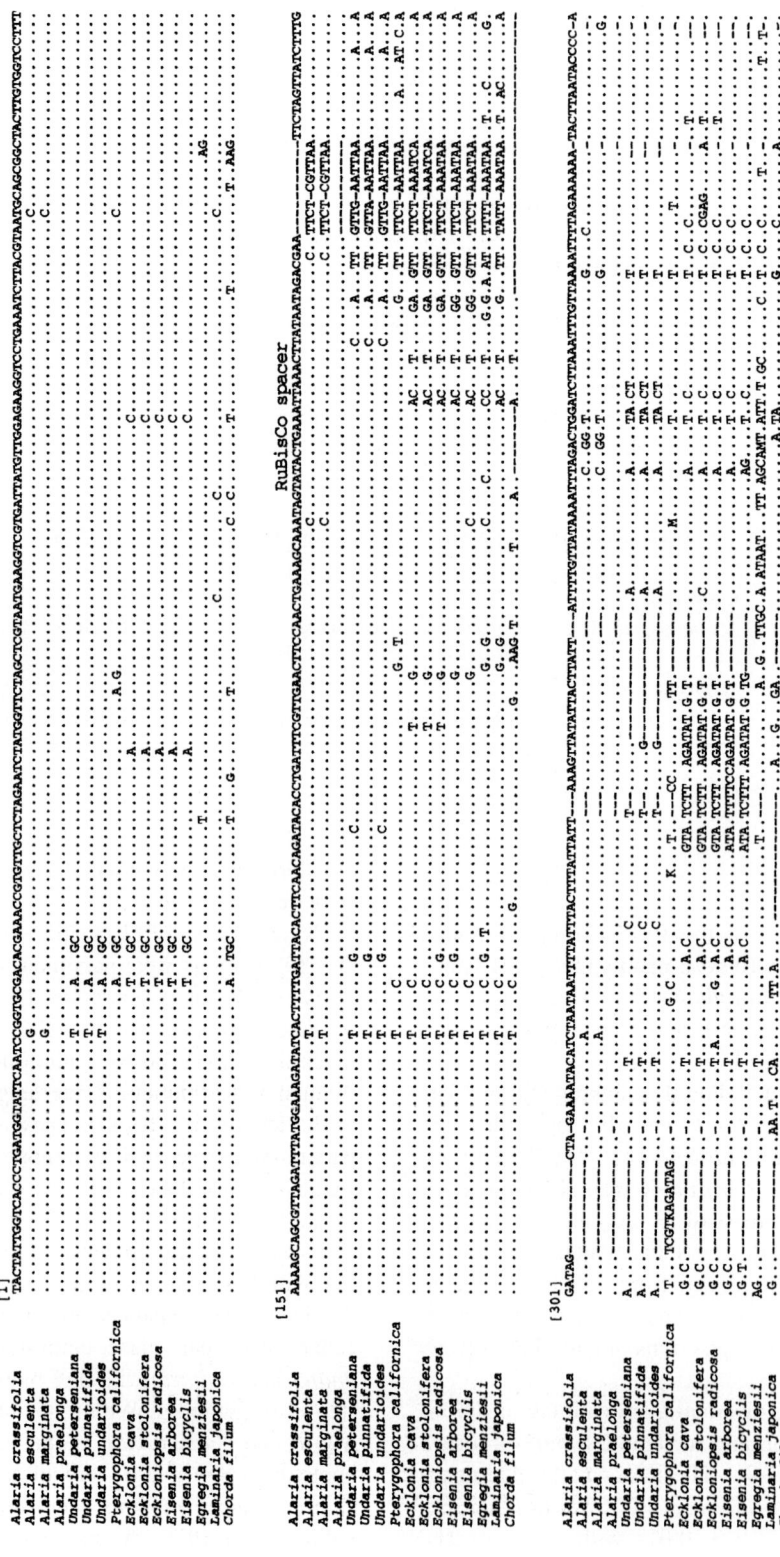

Figure 1. Sequence alignment of the RuBisCo spacer region, including the flanking areas, of the family Alariaceae *sensu lato* and the reference species. Numbers on the head of the sequences are alignment position, (-) represents alignment gap, and (.) position identical to first line.

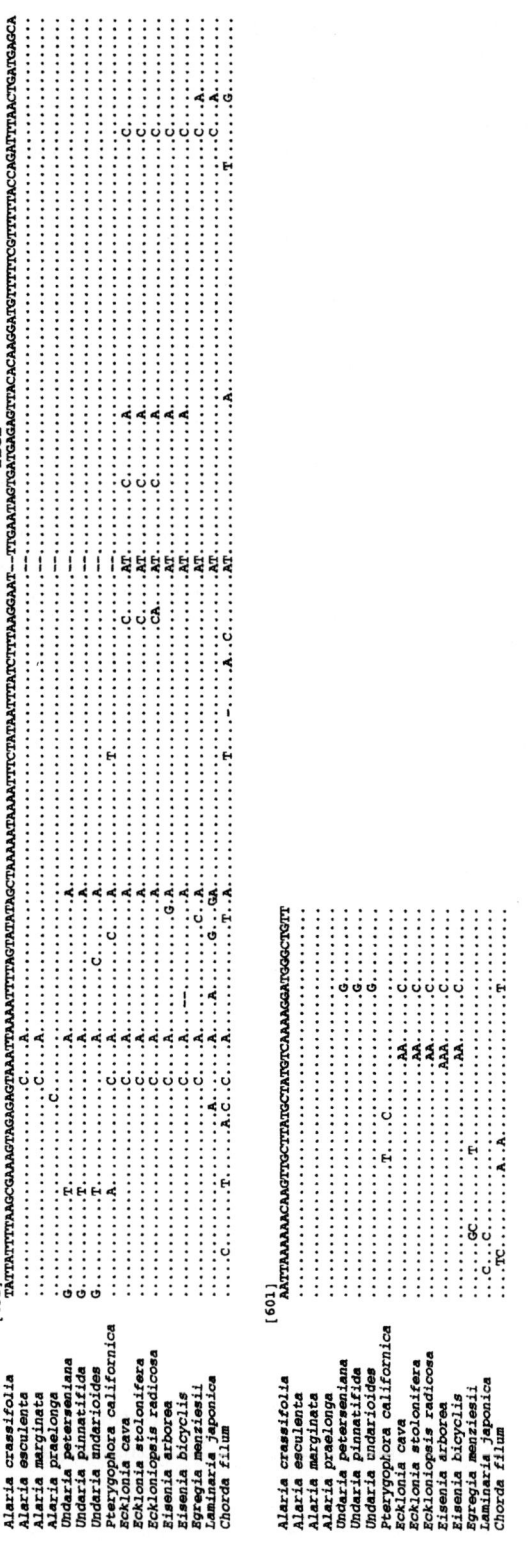

Figure 1. (continued).

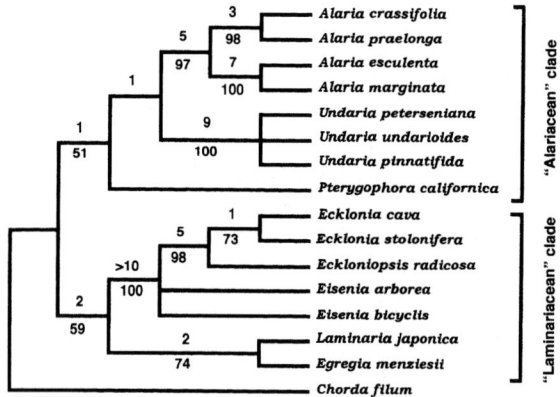

Figure 2. Strict consensus of six equally parsimonious RuBisCo trees of the Alariaceae *sensu lato*, with a length of 265 steps, CI of 0.853 and RI of 0.866. Bootstrap values >50% are given above the branches and decay indices are below the branches.

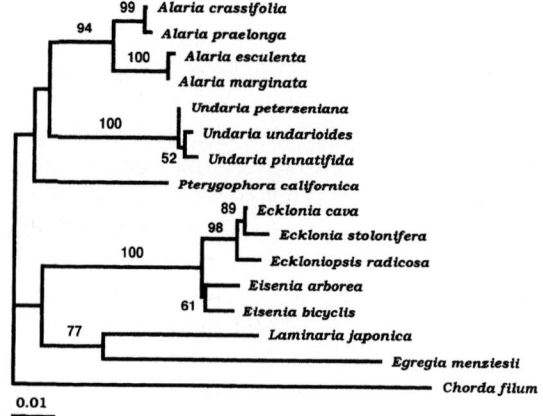

Figure 3. The RuBisCo ML tree of the Alariaceae *sensu lato*. Bootstrap values >50% are given above the branches.

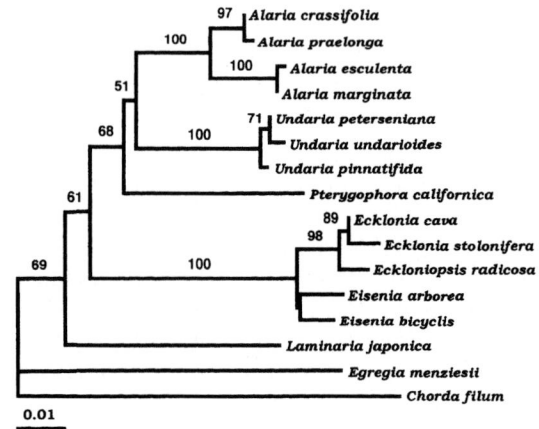

Figure 4. The RuBisCo NJ tree of the Alariaceae *sensu lato*. Bootstrap values >50% are give above the branches.

acean' clade, comprised two subclades: one subclade included *Ecklonia*, *Eckloniopsis* and *Eisenia* with 98–100% bootstrap values and 5–10 steps of decay index, the other included *Egregia* and *Laminaria* with 74% bootstrap value and 2 steps of decay index. There was, however, no association between two *Eisenia* species.

The topology of the ML tree (Figure 3) was nearly the same as that of the strict consensus MP, but with slight changes in the internal branches and bootstrap values. Two *Eisenia* species formed a clade with 61% bootstrap value, so did two *Ecklonia* species with 89% bootstrap. In this point, the topology of the ML tree was similar to that of the NJ tree. The topology of the NJ tree (Figure 4) was, however, different from those of the MP and ML trees in forming a single large clade included all taxa of the Alariaceae and *Laminaria*, except *Egregia*.

Three *Undaria* species produced a strong branch with 100% bootstrap value in the MP, ML and NJ trees. The internal branching pattern within the genus was different in all three trees. A relationship of metataxon was shown in the strict consensus MP tree, but *U. peterseniana* was a sister in the ML tree, and *U. pinnatifida* in the NJ tree.

Discussion

General characteristics of RuBisCo spacer in Alariaceae

The RuBisCo spacers of all taxa of the Alariaceae show similar sizes of 268 – 287 bp in this study. The RuBisCo spacer ranges from 130 to 190 bp in the primitive brown algae like *Ectocarpus* and *Kuckuckia* (Stache-Crain et al., 1997). The sizes are 64–120 bp in *Gracilaria* and *Gracilariopsis* (Goff et al., 1994), 77 bp in *Porphyra* (Brodie et al., 1996) and 113 bp in *Phycodrys* (van Oppen et al., 1995) of red algae. The sequences of the spacer in the Alariaceae is richer in A + T than ITS, as seen in previous studies (Goff et al., 1994; Stache-Crain et al., 1997).

The RuBisCo sequence divergence among species of *Undaria* ranges from 0.008 in the spacer to only 0.004 in the region including the spacer, RuBisCoL 3' and RuBisCoS 5' ends. The divergence values of the RuBisCo spacer are lower than 0.024 of the ITS in *Undaria* (Yoon & Boo, unpubl.), as is the case within the genus *Alaria*. This amount of sequence divergence accords with that of previous studies of red (Goff et al., 1994) and brown (Stache-Crain et al., 1997) algae.

The RuBisCo spacer sequences are highly conserved but sufficiently variable to use as a good indicator for understanding relationships at the generic and specific level in the Alariaceae.

Circumscription of Alariaceae

Our sequence data of the RuBisCo spacer region support the presence of two major clades in the traditional group of the Alariaceae, as suggested by morphological data by Okamura (1936) and Kawashima (1993), the 18S rDNA (Saunders & Druehl, 1992), and ITS sequences also (Saunders & Druehl, 1993a, b; Druehl et al., 1997). The 'alariacean' clade corresponds closely to the narrow interpretation of the Alariaceae (Kawashima, 1993; Saunders & Druehl, 1993a). The 'laminariacean' clade includes taxa belonging to the Laminariaceae *sensu* Kawashima (1993) and Saunders & Druehl (1993a) based on the absence of sporophylls and ITS data.

Data of the RuBisCo spacer region, therefore, contrast with the classical circumscription of the family Alariaceae *sensu lato* Setchell & Gardner (1925). The Alariaceae *sensu lato* involves all taxa with sporophylls, either kinetic or potential, arising as outgrowths from either stipe or blade and arising at the transition place. The wide concept of the Alariaceae was adopted by later phycologists (Papenfuss, 1951; Bold & Wynne, 1985; Estes & Steinberg, 1988). Bolton & Anderson (1994) supported this concept of the Alariaceae on the basis of a diagnostic feature of the production of secondary blades from the lateral meristems, rather than the position of sorus on the plant. The genera *Ecklonia, Eckloniopsis, Eisenia* and *Undaria* usually occur in warm waters in Australia, Japan, Korea, New Zealand and South Africa (Kang, 1966; Kawashima, 1993), where the selection pressure of warm waters might make the phylogenetically different genera look similar. This speculation leads to the convergent evolution of the genera. It is guessed that it was due to convergent evolution that the genera *Ecklonia, Eckloniopsis, Eisenia, Egregia* and *Undaria* together had been placed in the Alariaceae.

The RuBisCo data provide a further corroboration of the view of Kawashima (1993) and Saunders & Druehl (1993a, b) that *Alaria, Pterygophora, Lessoniopsis* and probably *Undaria* should be placed in the Alariaceae. Kawashima (1993) circumscribed the family as having mucilage cells and sporophylls, while the Laminariaceae have mucilage ducts and sori on blades. This taxonomic system has been adopted by Chihara (1997). Saunders & Druehl (1993a) emended the Alariaceae based on the ITS data and placed *Alaria, Pterygophora* and *Lessoniopsis* within the family. The same grouping is shown in the recent work of Druehl et al. (1997), which also mentioned that the genus *Undaria* is potentially characterized in the Alariaceae.

Although *Alaria* and *Undaria* are well placed in the family, there is still a conflict on the putative members between molecular and morphological data. *Pterygophora* has sporophylls with mucilage ducts instead of mucilage cells, as is the case in *Lessoniopsis*. Molecular data of the 18S, rDNA and ITS sequences showed that the Lessoniaceae members clustered with the Alariaceae (Saunders & Druehl, 1993a, b; Tan & Druehl, 1996). Therefore, it is probable that the narrow grouping of the Alariaceae will be changeable with more samplings of the Lessoniaceae.

Our RuBisCo sequence data indicate that *Ecklonia, Eckloniopsis, Eisenia*, and *Egregia* may be placed in the family Laminariaceae (Figure 2). They share diagnostic characters of mucilage ducts and blade-bearing sori. The view of the taxonomic position of these genera within the Laminariaceae is supported by morphological (Kawashima, 1993) and the ITS data (Saunders & Druehl, 1993a; Druehl et al., 1997). The genus *Egregia* appears to be closely related to *Laminaria*, as seen in the ML and MP trees or each member evolved along its own lineage, as seen in the NJ tree. Therefore, the laminariacean clade based on current molecular data may be polyphyletic.

Phylogenetic relationships within Undaria

The RuBisCo spacer sequences of the genus *Undaria* show that the species have the same divergence per every pair of the species, and the divergence value is lower than that of other genera. The close relationship within *Undaria* is supported by high bootstrap and decay supports, but the topologies of the intrageneric clades are differently produced according to the MP, NJ or ML trees. As seen in Figure 2, *U. undarioides* always forms an ingroup with *U. pinnatifida* in the ML tree, but with *U. peterseniana* in the NJ tree. These results indicate that the *Undaria* species are very closely related with an intermediate relationship at the DNA level. The close and intermediate relationship among species is also supported by morphological and cross experiment data. *Undaria pinnatifida* has sporophylls and blade with a midrib, while *U. peterseniana* has sori on the blade without a midrib. *Undaria undarioides* is intermediate in having both sporophyll and sori on the lower blade, entire blade and weak development of a midrib (Okamura, 1916; Kawashima, 1993).

The interspecific hybrids of *Undaria peterseniana* male and *U. pinnatifida* female were successful in the laboratory (Migita, 1967). Gametophytes from the hybrids were almost similar to the parental species in morphological and ecological aspects. The chromosome number of the hybrids was 30, identical to the parental species. Nishibayashi & Inoh (1960) already observed the same number of chromosomes in zoospores of *U. undarioides*. Artificial hybrids between the *Undaria* species were also successfully produced by Saito (1972). All of these studies indicate that a mixing of the gene pool among species of *Undaria* may be possible in the field.

Undaria pinnatifida has a wide distribution from the warm waters to the cold, while *U. peterseniana* is often distributed in southern Korea and Japan. *Undaria undarioides* has a very narrow distribution in southern Japan, where *U. peterseniana* occurs together (Kawashima, 1993). This biogeographical overlapping may provide a high opportunity of crossing among species in the field.

Natural hybridization, as well as artificial hybrids, often occurs in kelps (Druehl & Saunders, 1992). Although it is difficult to demonstrate natural hybrid species, intermediate morphology, the successful hybrids between two species, the same chromosome number of the hybrids to that of the parental species, and the biogeographical overlapping lead us to speculate that natural hybrids occur within the genus *Undaria*. Furthermore, the very low sequence divergences among the RuBisCo spacer sequences provided for their support. We hypothesize that the process of reticulate evolution might be more suitable to explain all the scientific observations that we have made so far. In this scenario, *U. undarioides* might be an ancient hybrid species from the parental species of *U. peterseniana* and *U. pinnatifida*. Further genetic analysis of these species will provide more details and concrete data to support this hypothesis of reticulate evolution hypothesis.

Acknowledgements

The authors thank Handoyo Kusumo, Marao Ohno and Jim Watanabe for supplying specimens, Charlene Mayes for techniques and papers, Ian Tan for primer design, Louis Druehl and Shoiji Kawashima for valuable comments on kelp, Ki-Joong Kim for data analysis, and Kee-Jeong Ahn for reading draft. This work was supported by KOSEF 96-04-01-03. Dr Joanna M Jones crtically read and improved the earlier draft of this paper.

References

Bold, H. C. & M. J. Wynne, 1985. Introduction to the Algae. 2nd. edn. Prentice-Hall, Inc., N.J.

Bolton, J. J. & R. J. Anderson, 1994. *Ecklonia*. In Akatsuka, I. (ed.), Biology of Economic Algae. SPB Academic Publlishing, The Hague: 385–406.

Brodie, J., P. K. Hayes, G. L. Barker & L. M. Irvine, 1996. Molecular and morphological characters distinguishing two *Porphyra* species (Rhodophyta: Bangiophycidae). Eur. J. Phycol. 31: 303–308.

Chihara, M., 1997. Biology of Algal Diversity. Uchida Rokakuho Publ. Co. Ltd, Tokyo, 386 pp.

Destombe, C. & S. E. Douglas, 1991. Rubisco spacer sequence divergence in the rhodophyte alga *Gracilaria verrucosa* and closely related species. Curr. Genet. 19: 395–398.

Druehl, L. D., 1989. Molecular evolution in the Laminariales: A review. In Garbary D.J. & G.R. South (eds), Evolutionary Biogeography of the Marine Algae of the North Atlantic. NATO ASI series Vol. G22, Springer-Verlag. Berlin: 205–217.

Druehl, L. D. & G. W. Saunders, 1992. Molecular explorations in kelp evolution. In Round F.E. & D.J. Chapman (eds), Progress in Phycological Research Vol. 8. Elsevier, Amsterdam: 47–83.

Druehl, L. D., C. Mayes, I. H. Tan & G. W. Saunders, 1997. Molecular and morphological phylogenies of kelp and associated brown algae. In Bhattacharya, D. (ed.), Origins of Algae and their Plastids. Springer, Wien: 221–235.

Estes, J. A. & P. D. Steinberg, 1988. Predation, herbivory and kelp evolution. Paleobiology 14: 19–36.

Felsenstein, J., 1985. Confidence limits on phylogenies: An approach using the bootstrap. Evolution 39: 783–791.

Felsenstein, J., 1993. PHYLIP (Phylogenetic Inference Package) Version 3.572. Department of Genetics, University of Washington, Seattle.

Gilbert, D., 1995. SeqPup, a biological sequence editor and analysis program for Macintosh computers. Published electronically on the Internet available via anonymous ftp to ftp.bio.indiana.edu.

Goff, L. J., D. A. Moon & A. W. Coleman, 1994. Molecular delineation of species and species relationships in red algal agarophytes *Gracilariopsis* and *Gracilaria* (Gracilariales). J. Phycol. 30: 521–537.

Hillis, D. M. & J. P. Huelsenbeck, 1992. Signal, noise and reliability in molecular phylogenetic analyses. J. Hered. 83: 189–195.

Kang, J. W., 1966. On the geographical distribution of marine algae in Korea. Bull. Pusan Fish. Coll. 7: 1–125.

Kawashima, S., 1993. Illustrated book of Japanese kelp. Revised version, North Japanese Ocean Publ. Sapporo, 206 pp.

Kimura, M., 1980. A simple method for estimating evolutionary rates of base substitution through comparative studies of nucleotide sequences. J. mol. Evol. 16: 111–120.

Migita, S., 1967. Studies on artificial hybrids between *Undaria peterseniana* (Kjellman) Okam. and *U. pinnatifida* (Harv.) Sur. Bull. Fac. Fish., Nagasaki Univ. 24: 9–20.

Miyabe, K., 1957. On the Laminariaceae of Hokkaido. J. Sapporo agri. Coll. 1: 1–50.

Morgan, D. R., 1997. Decay analysis of large sets of phylogenetic data. Taxon 46: 509–517.

Nishibayashi, T. & S. Inoh, 1960. Morphological studies in Laminariales. V. The formation of zoospores in *Undaria undarioides* (Yendo) Okamura. Biol. J. Okayama Univ. 6: 83–90.

Okamura, K., 1916. *Undaria* and its species. Bot. Mag. Tokyo 29: 266–278.

Okamura, K., 1936. Nippon Kaiso-shi. Uchida-rokakuho, Tokyo (in Japanese).

Papenfuss, G. F., 1951. Phaeophyta. In Smith, G. M. (ed.), Manual of Phycology. Chronica Botanica, Waltham, Mass.: 119–158.

Saito, Y., 1972. On the effects of enviromental factors on morphological characteristics of *Undaria pinnatifida* and the breeding of hybrids in the genus *Undaria*. In Abbott A. & M. Kurogi (eds), Contribution to the Systematics of the Benthic Marine Algae of the North Pacific. Jap. Soc. Phycol. Kobe: 117–132.

Saunders, G. W. & L. D. Druehl, 1992. Nucleotide sequences of the small-subunit ribosomal RNA genes from selected Laminariales (Phaeophyta): implications for kelp evolution. J. Phycol. 28: 544–549.

Saunders, G. W. & L. D. Druehl, 1993a. Revision of the kelp family Alariaceae and the taxonomic affinities of *Lessoniopsis* Reinke (Laminariales,Phaeophyta). Hydrobiologia 260/261: 689–697.

Saunders, G. W. & L. D. Druehl, 1993b. Nucleotide sequences of the internal transcribed spacers and 5.8S rRNA genes from *Alaria marginata* and *Postelsia palmaeformis* (Phaeophyta: Laminariales). Mar. Biol. 115: 347–352.

Setchell, W. A. & N. L. Gardner, 1925. The marine algae of the Pacific coast of North Amarica. Part III. Melanophyceae. Univ. Calif. Publ. Bot. 8: 383–898.

Stache-Crain, B., D. G. Muller & L. J. Goff, 1997. Molecular systematics of *Ectocarpus* and *Kuckuckia* (Ectocarpales, Phaeophyceae) inferred from phylogentic analysis of nuclear- and plastid-encoded DNA sequences. J. Phycol. 33: 152–168.

Swofford, D. L., 1993. PAUP: Phylogenetic Analysis Using Parsimony. Version 3.1.1, Illinois Natural History Survey, Champaign.

Tan, I. H. & L. D. Druehl, 1996. A ribosomal DNA phylogeny supports the close evolutionary relationships among Sporochnales, Desmarestiales and Laminariales (Phaeophyceae). J. Phycol. 32: 112–118.

van Oppen, M. J. H., S. G. A. Draisma & J. L. Olsen, 1995. Multiple trans-Arctic passages in the red alga *Phycodrys rubens*: evidence from nuclear rDNA ITS sequences. Mar. Biol. 123: 179–188.

A new agarophyte, *Curdiea balthazar* sp. nov. (Gracilariales, Rhodophyta), from the Three Kings Islands, northern New Zealand

Wendy A. Nelson[1], Glenys A. Knight[1] & Ruth Falshaw[2]
[1]*Museum of New Zealand Te Papa Tongarewa, P.O. Box 467, Wellington, New Zealand*
[2]*Industrial Research Ltd, PO Box 31-310, Lower Hutt, New Zealand*

Key words: Curdiea sp. nov., Gracilariales, Rhodophyta, agar chemistry, New Zealand, Three Kings Islands

Abstract

A new species of *Curdiea, C. balthazar* sp. nov. (Gracilariales, Rhodophyta) is described from the Three Kings Islands north of the North Island of New Zealand. This species is compared with the other endemic species of *Curdiea* found in New Zealand (*C. codioides, C. coriacea, C. flabellata*) and a key to the New Zealand species is presented.

Curdiea balthazar sp. nov. contains a highly methylated agar which is consistent with the other known members of this genus. The agar in this new species can be distinguished from agars found in other New Zealand members of the genus in that it is completely methylated at the 2-position of the 4-linked 3,6-anhydro-α-L-galactopyranosyl units, but has no methylation on the 6-position of the 3-linked β-D-galactopyranosyl units.

Curdiea balthazar sp. nov. is distinguished from other species in the genus by its morphology (irregularly branched thalli up to 0.8 m in length) and anatomy (large medullary cells). It shares with other *Curdiea* species the reproductive features characteristic of the genus.

Introduction

The genus *Curdiea* (Gracilariales, Rhodophyta) is restricted to Australasia and Antarctica. Three species of *Curdiea* are known to be endemic to New Zealand – *C. codioides* V.J. Chapm., *C. coriacea* (Hook.f. et Harv.) V.J. Chapm. and *C. flabellata* V.J. Chapm. Earlier studies on New Zealand representatives of this genus (Nelson & Knight, 1997) suggested that there is a further endemic undescribed species restricted to the Three Kings Islands, an offshore island chain located to the north of the North Island (Figure 1). Previously, this fourth species had been referred to the western Australian species, *C. obesa* (Harv.) Kylin (Adams, 1994; Adams & Nelson, 1985), based on four herbarium specimens (two of which were fragmentary). A recent expedition to the islands enabled subtidal algal collections to be made, including entire specimens of an undescribed species of *Curdiea*.

Members of this genus have attracted recent attention as a potential source of the phycocolloid, agar. Initial structural studies of the agars from New Zealand species of *Curdiea* have shown that they are all highly methylated and in certain cases this leads to very interesting properties (Furneaux et al., 1990; Miller et al., 1994). In a more detailed survey of polysaccharides of *Curdiea* spp. (from New Zealand, Australia and Antarctica), Falshaw et al. (1998) have shown that all the species in this genus contain agars [i.e. with a basic structure of alternating 3-linked β-D-galactopyranosyl (G) and 4-linked 3,6-anhydro-α-L-galactopyranosyl (LA) units – Figure 2] but substituted with high levels of methyl ether (M) groups at LA2 and/or G6 (shown as R^1 and R^2, respectively, in Figure 2). Three types of substitution pattern emerged: >60% methylation of the 6-position of G units only [*C. flabellata* (NZ), *C. codioides* (NZ), *C. crassa* (Aus.) and *C. angustata* (Aus.)], >95% methylation of the 2-position of LA units only [*C. irwinii* (Aus.) and *C. racovitzae* (Ant.)], or >90% methylation of both the 6-position of G units and the 2-position of LA units [*C. coriacea* (NZ) and *C. obesa* (Aus.)].

In this paper we describe a new species of *Curdiea*, based on morphological, anatomical, and chemical characteristics as well as distributional range.

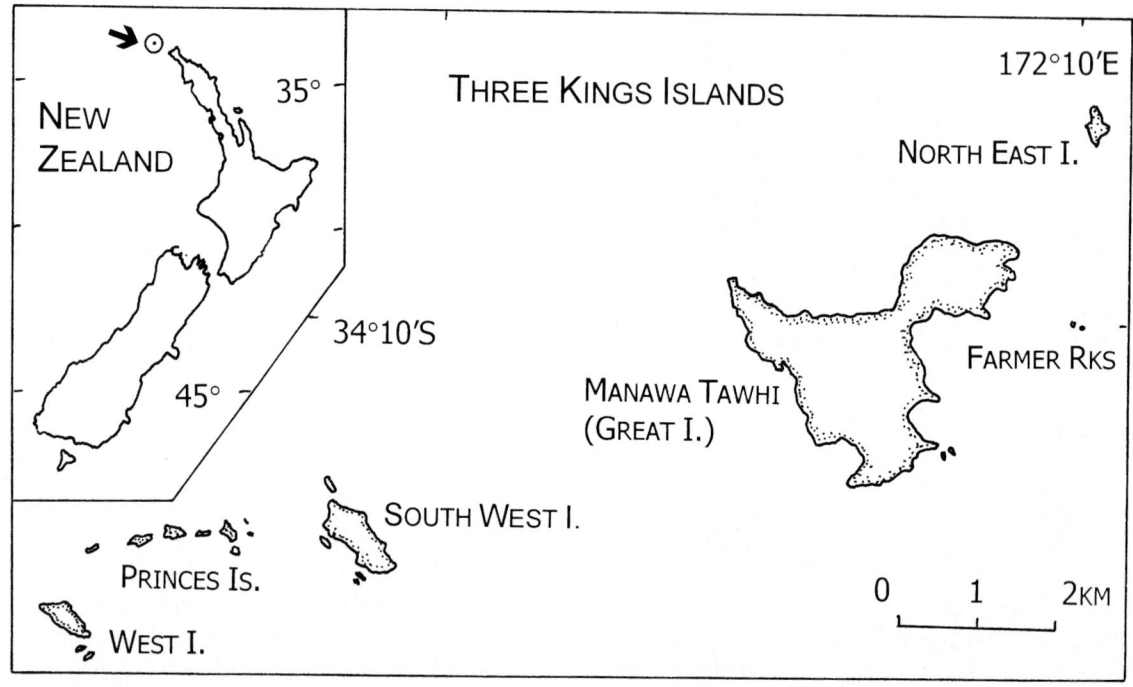

Figure 1. Map of northern New Zealand and the Three Kings Islands (57 km from the northern North I.).

Materials and methods

Locality

The Three Kings Islands lie 57 km to the north of the North Island of New Zealand. They are volcanic and have been separated from the North Island for at least 2 million years (Hayward, 1986) (Figure 1). These uninhabited islands are extremely rugged and precipitous, with difficult access, and experience strong storm and tidal conditions. The region around the Three Kings Islands is influenced by the warm East Australian Current as well as by active upwelling of cold water between the island group and mainland New Zealand (Garner, 1959; Stanton, 1973).

Collections

A field trip in March 1997 yielded collections from two sites and observations from other localities around the Three Kings Islands. Thalli were either air-dried for polysaccharide analysis and for herbarium specimens, or formalin-preserved before preparation as herbarium specimens. Colour of thalli was recorded using Horticultural Colour Chart (H.C.C.) (Wilson, 1938–1941; Stearn, 1973).

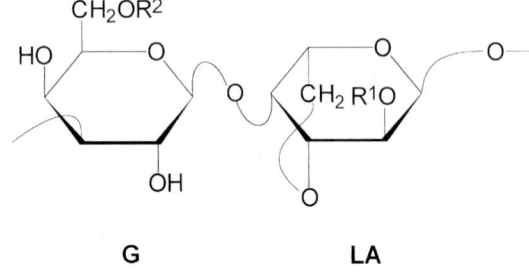

Figure 2. Structure of idealised agar with common positions of methyl ether (M) substitution (R^1 and/or R^2 = M or H).

Thalli were sectioned either by hand or using a Bright Starlight 2212 freezing microtome, either viewed directly, or stained with aniline blue, and then examined with an Olympus BH2 microscope, using both bright field and differential interference optics.

Chemistry

Samples of algal material (approx. 2 mg) were analysed directly for constituent sugars using the reductive hydrolysis method of Stevenson & Furneaux (1991). This technique combines hydrolysis (with

Figure 3. Holotype of *Curdiea balthazar* sp. nov.— WELT A22010. Scale bar = 5 cm.

CF$_3$CO$_2$H) and reduction (with 4-methylmorpholine-borane), followed by acetylation, to produce alditol acetates. Derivatized samples were analysed by GLC using a Hewlett-Packard 5890 Series II chromatograph with a Supelco SP-2330 column (15 m × 0.25 mm id, 0.22 μm film thickness) at 220 °C and flame-ionization detector (FID). Identification of components was by comparison of retention times with authentic standards.

Results

Description

Curdiea balthazar W.A. Nelson, G.A. Knight et R. Falshaw, sp. nov. (Figures 3–10)

Thalli coccinei vel carminei (lacca indica H.C.C. 8.26 vel carmineus chrysanthemi H.C.C. 8.24), usque ad 0.8 m longis, axibus erectis irregulariter ramosis loratis applanatis; rupi per stipitem centralem brevem crassum carnosum fortiter affixi. Cortex e stratis 1–2 cellularum cubicarum (c.10 μm diam.) et sub-cortex e stratis 3–4 cellularum pigmentosarum rotundatarum in amplitudine augentium medullam versus constatus. Cellulae medullosae parietibus crassis, in amplitudine variabiles, inclusionibus conspicuis saepe auricoloribus; strata centralia 1–3 cellularum maximarum (120–240μm diam.). Cystocarpia prominentia, globosa usque mamillata, ostiolata, utrinque praesentia; tetrasporangia in nematheciis utrinque disposita.

Thalli dark red to crimson, up to 0.8 m in length; upright, irregularly branched, strap-like flattened axes; strongly attached to rock by central short thick, fleshy stipe. Cortex (1–2 layers) of cuboidal cells (ca. 10 μm), sub-cortex (3–4 layers) of pigmented, rounded cells increasing in size towards medulla (Figure 4). Medullary cells thick-walled (Figure 5), variable in size with conspicuous often golden-coloured inclusions; 1–3 central layers of very large cells (120–240 μm diameter) (Figures 4–5). Cystocarps prominent, globose to mamillate, ostiolate, present on both surfaces (Figure 6); tetrasporangia in nemathecia on both surfaces (Figure 7, 8).

Holotype: WELT A22010 (Figure 3).
Isotypes: WELT A22011, A22012, A22013.
Type locality: Three Kings Is., Princes Is., Archway I., 24.5 m, 2 March 1997, *G.A. Knight.*

Etymology

The Dutch explorer Abel Tasman named the Three Kings Islands when he discovered them on Epiphany 1643 (the twelfth day of Christmas). Although none of the islands is named Balthazar after one of the Three Kings, this name has been chosen both because of the geographical location of this species and the medullary cells that bear golden cell inclusions.

Distribution

Restricted to the Three Kings Islands, northern New Zealand. Specimens examined: *C. balthazar* — West Point, Manawa Tawhi, *Mason*, Jan 1976, WELT A9400-01; North West Bay, Manawa Tawhi, *Choat*, Oct 1978, WELT A10445-6; Eastern point, Manawa Tawhi, *Knight & Duffy*, Mar 1997, WELT A22024/A21969, A22025/A21970, A22028-22038, A22042; North West Bay; Archway I., Princes Is., *Knight*, Mar 1997, WELT A22022, A22039-40. *C. codioides* – South East Bay, Manawa Tawhi, *Knight*, Mar 1997, WELT A22020. *C. coriacea* – Archway I., Princes Is., *Knight & Duffy*, Mar 1997, WELT A22021; North West Bay, Manawa Tawhi, *Francis*, Mar 1997, WELT A22023.

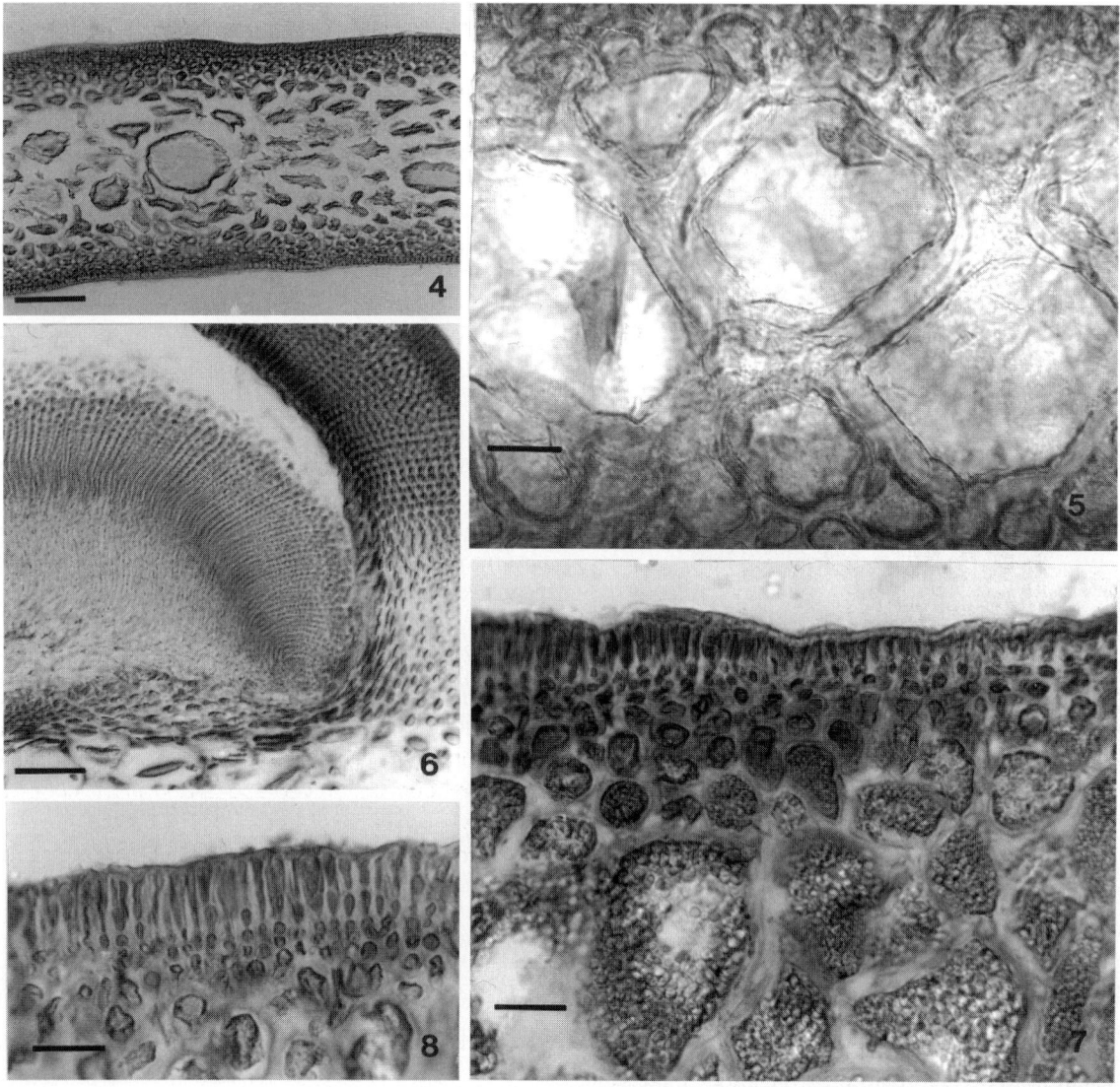

Figures 4–8. **Figure 4.** Cross-section of *C. balthazar* showing the irregularly-sized medullary cells including some very large cells. Scale bar = 150 μm. **Figure 5.** Cross section showing the thick walls of medullary cells. Scale bar = 90 μm. **Figure 6.** Cross section of a cystocarp of *C. balthazar* — parallel files of carposporangia, thick pericarp of small, isodiametric cells. Scale bar = 20 μm. **Figure 7.** Developing tetrasporangial nemathecium. Scale bar = 20 μm. **Figure 8.** Tetrasporangial nemathecium with maturing tetrasporangia. Scale bar = 23 μm.

Habit

The attachment and branching of this species are very irregular. The thalli are attached centrally to rock (Figure 9) and as the thallus expands the lower part of the frond splits, in some cases becoming markedly fenestrate (Figure 10). The lower parts of the thallus thicken and become almost terete. The splitting and thickening of the thallus in this way may produce a single very strong false stipe or a number of 'stipe-like' attachments for a single frond.

Reproduction

Cystocarps in *C. balthazar* are found on both surfaces of the thalli and are very conspicuous (up to 2 mm in diameter) and have a prominent ostiole. The outer pericarp is very thick and is made up of 20–

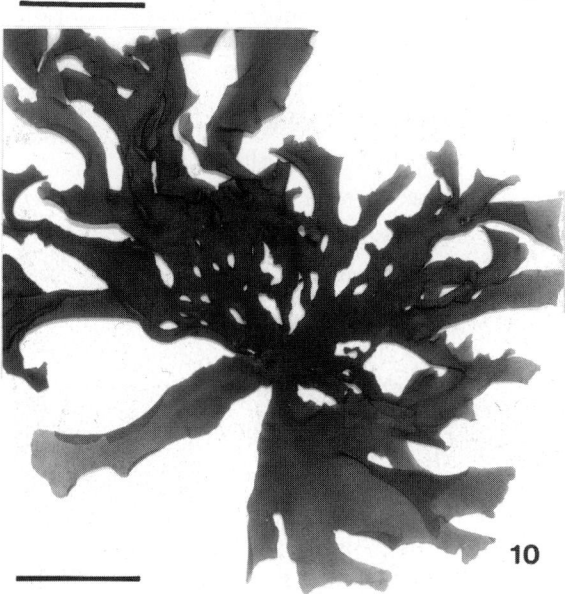

Figures 9–10. **Figure 9.** Young thallus showing central short terete stipe – WELT A22042. Scale bar = 3.5 mm. **Figure 10.** Older thallus with fenestrate basal area – WELT A22028. Scale bar = 3.5 mm.

30 rows of small, isodiametric cells. The cells of the inner pericarp lying below the gonimoblast are pigmented. The filaments of the gonimoblast produce straight, parallel chains, and mature into carposporangia. The carposporangia are small-celled (8–9 μm) and spherical on release.

Tetrasporangial nemathecia can be detected on specimens by the slightly raised surface (Figures 7 and 8) in comparison to the surrounding vegetative blade. The development of tetrasporangial nemathecia in *C. balthazar* follows the same sequence described for *C. flabellata* and *C. coriacea* (Fredericq & Hommersand, 1989; Nelson & Knight 1997). Tetrasporangia are 28–35 μm in height and cruciately divided. Spermatangia have not been observed in this species; they have been reported for only one species in the genus (Nelson & Knight, 1997).

Ecology, habitat

Thalli occur abundantly at (12-) 18–25 m, typically on boulders covered with non-geniculate coralline algae, amongst sand in open areas, in trenches and at the bases of vertical rock faces. Species growing nearby include *Sargassum johnsonii* V.J. Chapm., *Phacelocarpus labillardieri* (Turner) J. Agardh and *Curdiea coriacea*.

Chemistry

Constituent sugar analysis of six specimens collected from the Three Kings Islands (Table 1) showed sample 1 contained constituent sugars corresponding to an agar containing predominantly G6M and LA units, i.e. highly methylated only at G-6 (Figure 2). Based on the work of Falshaw et al. (1998) this would indicate *C. codioides* since *C. flabellata* is a southern New Zealand species. Samples 2 and 3 (Table 1) contained constituent sugars corresponding to an agar containing predominantly G6M and LA2M units, i.e. highly methylated at both G-6 and LA-2, corresponding to *C. coriacea*. A small amount of unmethylated galactose was also observed in these two samples. As these were algal specimens rather than purified agar extracts, other sources of galactose are likely to be present, such as floridoside [α-D-galactopyranosyl-(1$\rightarrow$2)-glycerol] (Hemmingson et al., 1996). The remaining three specimens (samples 4–6, Table 1) contained a combination of constituent sugars distinctly different from the others which corresponded to an agar containing G and LA2M units, i.e. highly methylated at LA-2 but with no methylation at G-6. This type of substitution pattern has previously been observed only in *C. irwinii* and *C. racovitzae*, neither of which occur in, or close to, New Zealand. The combination of constituent sugars in samples 4–6 indicated that these three specimens represent a new species of *Curdiea* (i.e. *C. balthazar*) and distinguished samples of *C. balthazar* from samples of other Three Kings *Curdiea* species.

Discussion

Fredericq & Hommersand (1989, 1990) have outlined the diagnostic characters of *Curdiea* and other genera

Table 1. Constituent sugar analyses of algal specimens collected from the Three Kings Islands

Constituent Sugar[a]	Deduced unit & substitution	Three Kings Islands *Curdiea* samples					
		1	2	3	4	5	6
2-Me-AnGal	LA2M	x[c]	✓	✓	✓	✓	✓
AnGal	LA	✓	x	x	x	x	x
6-Me-Gal	G6M	✓	✓	✓	x	x	x
Gal	G (+floridoside)	✓	(✓)	(✓)	✓	✓	✓
Species Identification - Voucher #[b]		*C. codioides* WELT A22020 CD539	*C. coriacea* WELT A22021 CD540	*C. coriacea* WELT A22023 CD542	*C. balthazar* WELT A22022 CD541	*C. balthazar*♀ WELT A22024 CD543	*C. balthazar*⊕ WELT A22025 CD544

[a] Determined as (naturally partially methylated) alditols acetates, i.e. AnGal as 1,2,4,5-tetra-*O*-acetyl-3,6-anhydrogalactitol, 6-Me-Gal as 1,2,3,4,5-penta-*O*-acetyl 6-*O*-methyl-galactitol, etc. [b] WELT specimens in the herbarium of the Museum of New Zealand; CD samples at Industrial Research Ltd. [c] x absent; ✓ present; (✓) traces.

within the Gracilariaceae. The new species from the Three Kings Islands clearly belongs to *Curdiea* based on the development of the gonimoblast (absence of unfused, sterile gonimoblast tissue), the carposporangia, which are aligned in long, parallel files, the gonimoblast and carposporangia, which completely fill the cystocarp cavity, the pericarp consisting of small isodiametric cells, cells of the cystocarp floor being transformed into darkly staining nutritive tissue, and cruciately divided tetrasporangia, which are produced in raised nemathecia. Although the New Zealand species of this genus are very variable in their external morphology, they all share the anatomy and reproductive development diagnostic of the genus.

The polysaccharide chemistry of *Curdiea* is very interesting and has also been shown to be taxonomically informative. Although high levels of methylation, reported here and by Falshaw et al. (1998), are not unique to agars from *Curdiea* species, the presence of a highly methylated agar in combination with other distinctive anatomical and reproductive characteristics strongly supports the placement of the new species in the genus *Curdiea*.

Key to New Zealand species of *Curdiea*:

1. Thallus dichotomously branched..............................2
1. Thallus irregularly expanded............3
2. Thallus prostrate, cushion forming, 5–8 cm diameter.... *C. codioides*
2. Thallus erect, strap-like axes from wedge-shaped base, 40–60 cm.............. *C. flabellata*
3. Thallus decumbent, astipitate, expanding from central attachment to form circular blade, becoming lobed and perforate when mature; to 30 cm diameter.............. *C. coriacea*
3. Thallus upright, with short thick, fleshy stipe, irregularly branched, strap-like axes; up to 80 cm............ *C. balthazar*

Acknowledgements

We wish to thank Clinton Duffy for his assistance with diving and collections in the Three Kings Islands; Howard Choat and Rosemary Mason for collections of *C. balthazar*; Peter Bostock (Queensland Herbarium) for translating the description; Te Papa Photography team. This research has been supported by grants from the Foundation for Research, Science and Technology MNZ602 (W.A. Nelson) and CO8607 (R. Falshaw).

References

Adams, N. M., 1994. Seaweeds of New Zealand. Canterbury University Press, Christchurch, 360 pp.

Adams, N. M. & W. A. Nelson, 1985. Marine algae of the Three Kings Islands. National Museum of New Zealand misc. ser. 13: 1–29.

Falshaw, R., R. H. Furneaux & D. E. Stevenson, 1998. Agars from nine species of red seaweed in the genus *Curdiea* (Gracilariceae, Rhodophyta). Carbohydr. Res. 308: 107–115.

Fredericq, S. & M. H. Hommersand, 1989. Reproductive morphology and development of the cystocarp in *Curdiea flabellata* Chapman (Gracilariales, Rhodophyta). N.Z. J. Bot. 27: 521–530.

Fredericq, S. & M. H. Hommersand, 1990. Diagnoses and key to genera of the Gracilariaceae (Gracilariales, Rhodophyta). Hydrobiologia 204/205: 173–178.

Furneaux, R. H., I. J. Miller & T. T. Stevenson, 1990. Agaroids from New Zealand members of the Gracilariaceae (Gracilariales, Rhodophyta) – a novel dimethylated agar. Hydrobiologia 204/205: 645–654.

Garner, D. M., 1959. The subtropical convergence in New Zealand surface waters. N.Z. J. Geol. Geophys. 2: 315–317.

Hayward, B. W., 1986. Origin of the offshore islands of northern New Zealand. In: Offshore Islands of northern New Zealand. New Zealand Department of Lands and Survey, Information Series 16: 129–138.

Hemmingson, J. A., R. H. Furneaux & V. H. Murray-Brown, 1996. Biosynthesis of agar polysaccharides in *Gracilaria chilensis* Bird, McLachlan et Olivera. Carbohydr. Res. 287: 101–115.

Miller, I. J., R. Falshaw & R. H. Furneaux, 1994. Chemical methylation of agaroid hydroxyl-groups. Carbohydr. Res. 262: 127–135.

Nelson, W. A. & G. A. Knight, 1997. Reproductive structures in *Curdiea coriacea* (Gracilariales, Rhodophyta) including the first report of spermatangia for the genus. N.Z. J. Bot. 35: 195–202.

Stanton, B. R., 1973. Hydrological investigations around northern New Zealand. N.Z. J. mar. freshw. Res. 7: 85–110.

Stearn, W. T., 1973. Botanical Latin. David & Charles, Newton Abbot, 566 pp.

Stevenson, T. T. & R. H. Furneaux, 1991. Chemical methods for the analysis of sulphated galactans from red algae. Carbohydr. Res. 210: 277–298.

Wilson, R. F., 1938–1941. Horticultural Colour Chart. 2 vols. London.

Marine benthic algae of North East Herald Cay, Coral Sea, South Pacific

Alan J. K. Millar
Royal Botanic Gardens Sydney, Mrs Macquaries Rd, Sydney, NSW 2000 Australia

Key words: Australia, Coral Sea, marine algae, taxonomy, biogeography.

Abstract

The marine benthic algae from North East Herald Cay, Coral Sea, South Pacific, are listed with taxonomic, bibliographic and biogeographic details. The checklist includes 66 species of which 23 are green, 2 are brown, and 41 are red algal species. The almost complete absence of brown algae from what is seemingly a typical tropical reef environment on which a true coral cay has developed is noteworthy. All samples were from the lagoon, which forms the concave side of a crescent-shaped reef and which ranges in depth from 0–30 m. The endemic Caribbean green alga *Chamaedoris peniculum* is recorded for the Pacific for the first time, and a possibly undescribed species of the genus *Rhipiliopsis* was also discovered. Although only preliminary, the survey shows that the marine flora is seemingly typical of coral cays for the general region of the Great Barrier Reef.

Introduction

North East Herald Cay (16° 56′ S; 149° 11′ E) is a semi-circular, crescent-shaped coral cay with its long axis oriented NE–SW. The cay itself is approximately 34 ha, 1200 m long and 500 m wide, and has a maximum elevation of 5 m, with a broad moderately sloping sandy beach with some beachrock slabs towards its centre, and a fringing coral reef. It is part of the Coringa-Herald National Nature Reserve, which lies in the central region of the Coral Sea Plateau in the Western Coral Sea, approximately 440 km due east of the city of Cairns on the north Queensland coast of Australia. The Plateau is separated from the Great Barrier Reef by the Queensland Trough, and is flanked by the Coral Sea Basin to the north-east and by the Townsville Trough to the south (Orme, 1977). The Herald Cays are bathed by the East Australian Current, which has its origins in the South Equatorial Current, and sea-surface water temperatures rarely drop below 24 °C in winter. Because it is a National Nature Reserve, it has been the subject of intensive, multi-disciplinary scientific expeditions (mostly by the Australian National Parks and Wildlife Service – NPWS), the most recent of which was in June 1997, when some 20 scientists and volunteers spent 14 days researching the cay. This was organized by the Royal Geographical Society of Queensland with various logistical support from NPWS. Part of this team included the author and Karlene L. Christian, and our principal task was to collect, document and describe the marine benthic algae of the entire cay. However, due to prevailing weather conditions, only the extensive lagoon area, which forms the entire, concave north west side of the cay, was examined in detail. This lagoon, however, is very extensive in area and has many (ca. 30) large, and often cavernous bommies (large coral and rock pinnacles), rising out of the surrounding coral-sand seabed, which is 2–30 m deep. This offers a wide range of habitats for the algae; so a reasonable indication of the marine flora was possible.

Methods

All collections were made using SCUBA and preserved in 4% formaldehyde, stored in a 30 l black plastic barrel, then transported back to Sydney for sorting, pressing, and hand-sectioning. The collections are filed in the Royal Botanic Gardens Sydney, National Herbarium of New South Wales (NSW). Herbarium abbreviations follow Holmgren et al. (1990).

List of species

Chlorophyta
   Ulvales
      Ulvaceae

Enteromorpha Link

Enteromorpha intestinalis (Linnaeus) Nees 1820:, index 2. Type locality: 'in Mari omni'. Voucher: west side of Lagoon, *A. J. K. Millar & K. L. Christian*, 24.vi.1997, NSW Slide 20–79.

Cladophorales
Anadyomenaceae
Anadyomene Lamouroux

Anadyomene stellata (Wulfen) C. Agardh, 1822: 400. Type locality: Adriatic Sea. Voucher: south-west side of lagoon, *A. J. K. Millar & K. L. Christian*, 25.vi.1997, NSW 418137.

Microdictyon Decaisne

Millar & Kraft (unpublished) have determined that the genus may be represented by only four of its 18 included species. These are differentiated by the anatomy of the anastomosing cell wall segments and certain stable differences in primary filament branching patterns. Characters defined by Setchell (1929), by which the majority of species are separated today (mesh size, filament width, frond and cell-wall colour, and stellate vs cruciate primary filament branching patterns), have been found to be untenable. The four species can be separated as follows:

1. Anastomosing segments of thickened wall rings.2
1. Anastomosing segments with crenellated walls. . 3
   2. Primary filament branches stellate or cruciate . .
   . . . . *M. umbilicatum*
   2. Primary filament branches secund . . . .
   . . . . *M. vanbosseae*
   3. Crenellated walls on unmodified cells . . . .
   . . . . *M. setchellianum*
   3. Crenellated walls on modified cells, tenaculae . . .
   . . . . *M. montagnei*

Microdictyon montagnei was transferred to the genus *Boodlea* by Egerod (1952), primarily because of its tenaculae, which are attachment cells (with crenellated walls) considerably smaller than the joined parental cells. This single feature also links it with the genera *Chamaedoris* (Okamura, 1932), *Cladophoropsis* (Egerod, 1952), *Struvea* and *Phyllodictyon* (Kraft & Wynne, 1996), which are otherwise separated on major morphological differences, as well as the occurrence of segregative cell division in the former three. *Boodlea* species are defined by their cell divisions, which occur in random planes giving the resultant plant a three-dimensional habit. *Microdictyon montagnei* would be unique in the genus *Boodlea* because of its cells that are branched strictly in a single plane, which thus form flattened blades typical of the genus *Microdictyon*.

Microdictyon setchellianum Howe, 1934: 38. Type locality: Hawaii. Voucher: north side of lagoon, *A. J. K. Millar & K. L. Christian*, 22.vi.1997, NSW 418073.

Siphonocladales
Siphonocladaceae
Chamaedoris Montagne

Chamaedoris peniculum (Ellis & Solander) Kuntze, 1898: 400. Basionym: *Corallina peniculum* Ellis & Solander, 1786: 127, pl. 7, Figures 5–8, pl. 25, Figure 1. Synonyms: *C. annulata* (Lamarck) Montagne, 1842: 261. *Penicillus annulatus* Lamarck 1813: 299. *Nesea annulata* (Lamarck) Lamouroux, 1816: 256. *Scopularia annulata* (Lamarck) Chauvin 1842: 122. Type locality: Florida. Voucher: north side of lagoon, *K. L. Christian & A. J. K. Millar*, 26.vi.1997, NSW 415041.

The felt-like capitulum is generally peltate but occasionally may be acentrically placed on the stalk. In all plants, however, the capitulum is flat and curved downwards like an umbrella, and the plants are found only in deep water (18 m plus). This differs from Caribbean material in that the capitulum is cup-shaped, ie. curved upwards, and plants are said to be restricted to shallow water (above 10 m; Littler et al., 1989: 84). The Herald Cay specimens are similar to the West Indian species in that both have simultaneously produced whorls of capitulum filaments arising from distal ends of the distal stalk-cells (Børgesen, 1912: Figure 17). This same capitulum filament production also occurs in the Indian species *C. auriculata* Børgesen, and the Japanese species *C. orientalis* Okamura & Higashi (Okamura, 1932: 68, pl. 284, Figures 8–15) and thus differs substantially from the other species of the genus [*C. delphinii* (Hariot) Feldmann & Børgesen] in which capitulum filaments arise randomly from a single, terminal stalk-cell and are not cut off as separate cells (Børgesen 1940: 16, Figure 5). Since even the stalk is seemingly of one cell (i.e. monosiphonous), the generic placement of *C. delphinii* is questionable. Lastly, tenaculae used to attach neighbouring filaments are present in all but *C. delphinii*. Critical study is required to ascertain the specific differences between species of *Chamaedoris*. If the identity of our plants is confirmed, this would constitute the first record for the Pacific.

Valoniaceae
Dictyosphaeria Decaisne ex Endlicher

Dictyosphaeria cavernosa (Forsskål) Børgesen, 1932: 2, pl. 1, Figure 1. Syntype localities: Saudi Arabia and Yemen. Voucher: north side of lagoon, *A. J. K. Millar & K. L. Christian*, 20.vi.1997, NSW 417366.

Dictyosphaeria versluysii Weber-van Bosse, 1905: 144. Syntype localities: various in Indonesia. Voucher: north side of lagoon, *A. J. K. Millar & K. L. Christian*, 20.vi.1997, NSW 417365.

Valonia C. Agardh

Valonia aegagropila C. Agardh, 1822: 429. Type locality: Venezia, Italy. Voucher: south-west side of lagoon, *A. J. K. Millar & K. Christian*, 25.vi.1997, NSW 418177.

Valoniopsis Børgesen

Valoniopsis pachynema (Martens) Børgesen, 1934: 10, Figures 1, 2. Syntype localities: Bengkulu and Pulau Tikus, Sumatra, Indonesia. Voucher: west side of lagoon, *A. J. K. Millar & K. L. Christian*, 24.vi.1997, NSW.

Ventricaria Olsen & West

Ventricaria ventricosa (J.Agardh) Olsen & J. West, 1988: 104. Type locality: Guadeloupe, French West Indies. Voucher: north side of lagoon, *A. J. K. Millar & K. L. Christian*, 22.vi.1997, NSW 417359.

Bryopsidales
Caulerpaceae
Caulerpa Lamouroux

Caulerpa cupressoides (Vahl) C. Agardh, 1817: xxiii. Type locality: St Croix, Virgin Islands. Voucher: west side of lagoon, *A. J. K. Millar, K. L. Christian*, 24.vi.1997, NSW 417358.

Caulerpa racemosa (Forsskål) J. Agardh, 1873: 35. Type locality: Suez, Egypt. Voucher: west side of Lagoon, *A. J. K.Millar & K. L. Christian*, 24.vi.1997, NSW 417356.

Caulerpa taxifolia (Vahl) C. Agardh, 1817: xxii. Type locality: St Croix, Virgin Islands. Voucher: north-west side of lagoon, *A. J. K. Millar & K. L. Christian*, 17.vi.1997, NSW 418047.

Codiaceae
Codium Stackhouse

Codium ovale Zanardini, 1878: 37. Type locality: Sorong, Irian Barat, Indonesia. Voucher: north side of lagoon, *A. J. K. Millar & K. L. Christian*, 22.vi.1997, NSW 418039.

Halimedaceae
Halimeda Lamouroux

Halimeda fragilis W. R. Taylor, 1950: 88, pl. 48, Figure 2. Type locality: Eniwetok Atoll, Marshall Islands. Voucher: north side of lagoon, *A. J. K. Millar & K. L. Christian*, 22.vi.1997, NSW 418018.

Halimeda tuna (Ellis & Solander) Lamouroux, 1816: 309, pl. XI, Figure 8. Type locality: Mediterranean Sea. Voucher: north side of lagoon, *A. J. K. Millar & K. L. Christian*, 22.vi.1997, NSW 418016.

Halimeda opuntia (Linnaeus) Lamouroux, 1816: 308. Type locality: Jamaica. Voucher: north side of lagoon, *A. J. K. Millar & K. L. Christian*, 22.vi.1997, NSW 418036.

This species is often credited with many varieties of which the Herald Cay collections could be placed in at least two (*opuntia* and *hederacea*). The morphological variation within single plants makes this task almost impossible, and the variations are here considered as one single species.

Halimeda cylindracea Decaisne, 1842: 103. Type locality: Nosy-bé, Malagasy Republic. Voucher: north side of lagoon, *A. J. K. Millar & K. L. Christian*, 17.vi.1997, NSW 418026.

Udoteaceae
Avrainvillea Decaisne

Avrainvillea lacerata Harvey ex J. Agardh, 1887: 54. Type locality: Tonga. Voucher: west side of lagoon, *A. J. K. Millar & K. L. Christian*, 24.vi.1997, NSW 417352.

Three specimens of Harvey's Friendly Islands Exsiccatae No. 86 (isotypes) are filed in NSW. The Herald Cay plants match these very closely.

Rhipilia Kützing

Rhipilia orientalis A. & E. S. Gepp, 1911: 57, 140, pl. XVI, Figures 134–136. Syntype localities: Pulu Sebangkatan, Borneo Bank and Fau Island. Voucher: north side of lagoon, *A. J. K.Millar & K. L. Christian*, 20.vi.1997, NSW 418055.

Rhipilia tenaculosa A. & E. S. Gepp, 1911: 56, 140, pl. XV, Figures 130–132, pl. 16, Fig. 133. Type locality: off Barra Grande near Penambuco, Brazil. Voucher: north side of lagoon, *A. J. K. Millar & K. L. Christian*, 22.vi.1997, NSW 418053.

Larger, flattened and distinctly flabellate in comparison to the small spongy plants of *R. orientalis*.

Rhipiliopsis A. & E. S. Gepp

Rhipiliopsis sp. nov. Voucher: south-west side of lagoon, *A. J. K. Millar & K. L. Christian*, 25.vi.1997, NSW 418170.

This species is undescribed and will be dealt with in a forthcoming publication.

Rhipiliopsis gracilis Kraft, 1986: 55, Figures 17–21. Type locality: The Canyons, Heron Island, Great Barrier Reef, Queensland. Voucher: south-west side of lagoon, *A. J. K. Millar & K. L.Christian*, 25.vi.1997, NSW 418136.

Udotea Lamouroux

Udotea orientalis A. & E. S. Gepp, 1911: 119, 142, pl I, Figures 1, 4. Syntype localities: Indian and Pacific Oceans. Voucher: north side of lagoon, *A. J. K. Millar & K. L. Christian*, 22.vi.1997, NSW 418064.

This species grows extensively throughout the Great Barrier Reef.

Phaeophyta

Dictyotales
Dictyotaceae
Dictyota Lamouroux

Dictyota dichotoma (Hudson) Lamouroux, 1809: 42. Type locality: Walney Island, Lancashire, England. Voucher: north side of lagoon, *A. J. K. Millar & K. L. Christian*, 22.vi.1997, NSW 418078.

Fucales
Sargassaceae
Turbinaria Lamouroux

Turbinaria ornata (Turner) J. Agardh, 1848: 266. Type locality: unknown. Voucher: north side of lagoon, *A. J. K. Millar & K. L. Christian*, 20.vi.1997, NSW 417371.

Rhodophyta

Nemaliales
Galaxauraceae
Galaxaura Lamouroux

Galaxaura marginata (Ellis & Solander) Lamouroux, 1816: 264. Type locality: Bahama Islands. Voucher: north side of lagoon, *A. J. K. Millar & K. L. Christian*, 22.vi.1997, NSW 418085.

Galaxaura rugosa (Ellis & Solander) Lamouroux, 1816: 263. Type locality: Jamaica. Voucher: north side of lagoon, *A. J. K. Millar & K. L. Christian*, 20.vi.1997, NSW 417383.

Liagoraceae
Liagora Lamouroux

Liagora ceranoides Lamouroux, 1816: 239 Type locality: St Thomas, Virgin Islands. Voucher: north side of lagoon, *A. J. K. Millar & K. L. Christian*, 20.vi.1997, NSW 417401.

Gelidiales
Gelidiaceae
Gelidiella Feldmann & Hamel

Gelidiella pannosa (J.Feldmann) J.Feldmann & G.Hamel, 1934: 534, Figures 1–2. Type locality: Biarritz, France. Voucher: north side of lagoon, *A. J. K. Millar & K. L. Christian*, 22.vi.1997, NSW 418176.

This species has been recorded from the GBR by Price and Scott (1992), whose description and illustrations of *G. pannosa* match those of the Herald Cay specimens. Hatta & Prud'homme van Reine (1991), and more recently Kraft & Abbott (1998) recognize both *G. pannosa* and *G. lubrica* as distinct species and tabulate the differences. Again, our specimens fit their interpretations of *G. pannosa* more closely than *G. lubrica*.

Corallinales
Corallinaceae
Amphiroa Lamouroux

Amphiroa crassa Lamouroux in Quoy & Gaimard, 1824: 627. Type locality: Shark Bay, Western Australia. Voucher: north side of lagoon, *A. J. K. Millar & K. L. Christian*, 20.vi.1997, NSW 418195.

Being strictly cylindrical or terete, this taxon could easily pass for a *Jania* species, but the presence of cortically derived conceptacles distinguishes it from *Jania* which has axillary conceptacles.

Gigartinales
Peyssonneliaceae
Peyssonnelia Decaisne

Peyssonnelia capensis Montagne, 1847: 177. Type locality: Port Natal, South Africa. Voucher: west side of lagoon, *A. J. K. Millar & K. L. Christian*, 24.vi.1997, NSW 418107.

Peyssonnelia cf. *evae* Weber-van Bosse, 1921: 279, Figure. 95. Type locality: Pearl Bank, Tawitawi Prov., Sulu Archipelago. Voucher: north side of lagoon, *A. J. K.Millar & K. L. Christian*, 20.vi.1997, NSW 417391.

Plants consist of fine, pink and undulate blades with obovoid cells in erect filaments, multicellular rhizoids and large (40 μm diam) isodiametric hypobasal cells. Calcification occurs in both the hypobasal layer and in erect filaments. Herald Cay specimens differ from Weber-van Bosse's description in having multicellular rhizoids and in having fewer cells in erect filaments. The obovate shape of the perithalial cells would seem to be distinctive.

Peyssonnelia inamoena Pilger, 1911: 311, Figures 24, 25. Type locality: Gross-Batanga, Cameroon, West Africa. Voucher: south-west side of lagoon, *A. J. K. Millar & K. L. Christian*, 24.vi.1997, NSW 418107.

Plants consist of a very thin, lightly calcified and green-yellow thallus. Rhizoids are unicellular and the erect filaments are only about 3–5 cells long.

Peyssonnellia neocaledonica Kützing, 1869: 32, pl. 90, Figures c, d. Type locality: New Caledonia. Voucher: south-west side of lagoon, *A. J. K. Millar & K. L. Christian*, 25.vi.1997, NSW 418104.

Plants consist of anastomosing, imbricate, leathery blade masses, that are supple and mottled brown-red in colouring. Rhizoids are multicellular and erect filaments arise at 45° angles from the hypothallial layer. Ultimate and penultimate cells of erect filaments are refractive (cystoliths) and the filaments are up to 30 cells long.

Peyssonnelia sp. Voucher: south-west side of lagoon, *A. J. K. Millar & K. L. Christian*, 24.vi.1997, NSW 418109.

Plants are a very bright red, have a very thick (to 3 mm), and heavily calcified thallus. Rhizoids are muticellular and erect filaments are at a 90° angle to the hypobasal layer. It looks like half a dinner plate sticking out from the rocks and there seem to be no described species that match this large and distinctive taxon.

Acrosymphytaceae
Acrosymphyton Sjoestedt

Acrosymphyton taylorii Abbott, 1962: 845. Type locality: Oahu, Hawaii. Voucher: north side of lagoon, *A. J. K. Millar & K. L. Christian*, 20.vi.1997, NSW 417386.

Dumontiaceae
Gibsmithia Doty

Gibsmithia hawaiiensis Doty, 1963: 458, Figures 1–17. Type locality: Waikiki, Oahu, Hawaii. Voucher: north side of lagoon, *A. J. K. Millar & K. Christian*, 18.vi.1997, NSW 418091.

Nemastomataceae
Titanophora (J. Agardh) J. Feldmann

Titanophora weberae Børgesen, 1943: 39, Figure 13. Type locality: Salee Strait, Irian Barat, Indonesia. Voucher: south-west side of lagoon, *A. J. K. Millar & K. L. Christian*, 25.vi.1997, NSW 418117.

Halymeniales
Halymeniaceae
Halymenia C. Agardh

Halymenia sp. Voucher: south-west side of lagoon, *A. J. K. Millar & K. L. Christian*, 25.vi.1997, NSW 418202.

Measuring only 10 mm in diameter, this fully mature flattened blade displays all the essential criteria for inclusion in the genus *Halymenia*, including filaments linking the opposing cortices, darkly staining arachnoid cells and auxiliary cell ampullae. Only two specimens have been found to date so that a true indication of the habit of this possibly undescribed species is not possible at this stage.

Rhodymeniales
Champiaceae
Champia Desvaux

Champia caespitosa Dawson, 1944: 311, pl. 46, Figures 3–4. Type locality: Isla Pond, Baja California, Mexico. Voucher: north-west side of lagoon, *A. J. K. Millar & K. L. Christian*, 27.vi.1997, NSW Slide 20–80.

See Millar (1999) for a discussion on the recognition of this species as separate from *C. parvula*.

Rhodymeniaceae
Asteromenia Huisman & Millar

Asteromenia peltata (W.R.Taylor) Huisman & A.Millar, 1996: 139 (Figure 35). Type locality: Tortuga Island, Venezuela. Voucher: north side of lagoon, *A. J. K. Millar & K. L. Christian*, 26.vi.1997, NSW 417014.

The majority of the plants collected were cystocarpic, which is only the third instance of such for the genus and species. Interestingly, Huisman & Millar (1996) reported that North Carolina plants were cystocarpic in June (boreal summer) yet the Herald Cay and Abrolhos plants were fertile in June and September (austral winter and spring).

Halichrysis Huve & Huve

Halichrysis cf. *micans* (Schmitz in Schmitz & Hauptfleisch) Huvé & Huvé, 1977: 106. Type locality: southern coast of Java (*fide* Weber-van Bosse, 1928). Voucher: west side of lagoon, *A. J. K.Millar & K. L. Christian*, 24.vi.1997, NSW 418096.

This species has a habit superficially similar to *Asteromenia peltata* in that individual, stellate blades (3–5 cm across) emanate from central stalks and anastomose together into large stands. The Herald Cay species, however, has repeatedly dichotomously branched lobes, and the texture is firm and leathery, not soft and flaccid as in *Asteromenia*. Until fertile plants are found the identification must remain tentative.

Ceramiales
Ceramiaceae
Anotrichium Nägeli

Anotrichium tenue (C. Agardh) Nägeli, 1862: 399. Type locality: Venezia, Italy. Voucher: north side of lagoon, *A. J. K. Millar & K. L. Christian*, 20.vi.1997, NSW Slide 20–81 (tetrasporic); 22.vi.1997, NSW Slide 20–82 (male and female).

The tetrasporangia in the Herald Cay plants are borne strictly terminally, but appear to become only slightly adaxially placed when mature. This would suggest that the feature that separates *A. tenue* from *A. secundum* is not entirely stable (see comments in Millar & Abbott, 1997).

Antithamnion Naegeli

Antithamnion makroklonion Athanasiadis, Garbary & Vandermeulen, 1990: 560, Figures 1–8, 9A–C. Type locality: Eilat, Red Sea, Sinai Peninsula. Voucher: north side of lagoon, *A. J. K. Millar & K. L. Christian*, 17.vi.1997, NSW Slide 20–83.

Crouania J. Agardh

Crouania minutissima Yamada, 1944: 40. Type locality: Ant atoll, Caroline Is. Voucher: north side of lagoon, *A. J. K. Millar & K. L. Christian*, 22.vi.1997, NSW Slide 20–84.

Centroceras Kützing

Centroceras clavulatum (C. Agardh) Montagne, 1846: 140. Type locality: Calloa, Peru. Voucher: north side of lagoon, *A. J. K. Millar & K. L. Christian*, 22.vi.1997, NSW Slide 20–85.

Ceramium Roth

Ceramium cf. *caudatum* Setchell & Gardner, 1924: 776, pl. 27, Figures 55–57. Type locality: Eureka, near La Paz, Pacific Mexico. Voucher: northwest side of lagoon, *A. J. K. Millar & K. L. Christian*, 17.vi.1997, NSW Slide 20–86.

The present material differs from that described in the protologue only in that the basipetal cortical cells never undergo further division or only rarely so.

Ceramium flaccidum (Harvey ex Kützing) Ardissone, 1871: 40. Type locality: Kilkee, County Clare, Ireland. Voucher: west side of lagoon, *A. J. K. Millar & K. L. Christian*, 24.vi.1997, NSW 417404.

Ceramium marshallense Dawson, 1957: 120, Figure 27a–b. Type locality: Rigili Island, Eniwetok Atoll, Marshall Islands Voucher: north side of lagoon, *A. J. K. Millar & K. L. Christian*, 22.vi.1997, NSW Slide 20–94.

Haloplegma Montagne

Haloplegma duperreyi Montagne, 1842: 258, pl. 7, Figure. 1. Type locality: Martinique. Voucher: north side of lagoon, *A. J. K. Millar & K. L. Christian*, 22.vi.1997, NSW Slide 20–87.

Wrangelia C. Agardh

Wrangelia argus (Montagne) Montagne, 1856: 444. Type locality: Roque del Gando, Islas Canarias. Voucher: north side of lagoon, *A. J. K. Millar & K. L. Christian*, 20.vi.1997, NSW 417372.

Dasyaceae
Heterosiphonia Montagne

Heterosiphonia crispella (C. Agardh) Wynne var. *laxa* (Børgesen) Wynne, 1985: 87. Type locality: St Croix, Virgin Islands. Voucher: north-west side of lagoon, *A. J. K. Millar & K. L. Christian*, 17.vi.1997, NSW Slide 20–88.

Delesseriaceae
Hypoglossum Kützing

Hypoglossum cf. *geminatum* Okamura, 1908: pl. 32. Type locality: Enoshima, Sagami Province, Japan. Voucher: north side of lagoon, *A. J. K. Millar & K. L. Christian*, 20.vi.1997, NSW Slide 20–89.

Although there are instances of paired lateral bladelets arising from the midrib, there are equally many single blades arising from the midrib. Tetrasporic material, which is lacking in the Herald Cay populations, is needed for a complete identification of this species.

Martensia Hering

Martensia australis Harvey, 1855: 537. Type locality: King George Sound, Western Australia. Voucher: south-west side lagoon, *A. J. K. Millar & K. L. Christian*, 25.vi.1997, NSW 418125.

Rhodomelaceae
Chondria C. Agardh

Chondria simpliciuscula Weber-van Bosse, 1913: 125, pl. 12, Figures 9–10. Type locality: Aldabra Island, Indian Ocean. Voucher: west side of lagoon, *A. J. K. Millar & K. L. Christian*, 24.vi.1997, NSW 417377.

Chondria succulenta (J. Agardh) Falkenberg, 1901: 205, t. 22, Figures 22, 23. Type locality: King George Sound, Western Australia. Voucher: south-west side of lagoon, *A. J. K. Millar & K. L. Christian*, 25.vi.1997, NSW 418112.

Although smaller than plants of this species growing on mainland Australia (Coffs Harbour; Millar, 1990), they match in all essential details.

Chondria (?) **sp. 1.** Voucher: north side of lagoon, *A. J. K. Millar & K. L. Christian*, 22.vi.1997, NSW 20–94.

As this material is sterile, a true indication of its generic identity is difficult. The plants are small (to 3–4 cm in length), entangled and decumbent. The pericentral cells are visible through the thin cortex, but the most striking feature of this species is the compressed nature of its branches. Apices are sunken, trichoblasts are rare, and branches are not constricted at their base.

Chondria (?) **sp. 2.** Voucher: north-west side of lagoon, *A. J. K. Millar & K. L. Christian*, 17.vi.1997, NSW Slide 8–42.

As with the previous species, this material is sterile. Features include pericentral cells which are not visible, a cortex consisting of distinctly elongate cells in tiled rows, an abundance of trichoblasts arising from depressed apices and markedly constricted branch bases.

Herposiphonia Naegeli

Herposiphonia arcuata Hollenberg, 1968: 538, Figure 5. Type locality: Kailua, Oahu, Hawaii. Vouchers: south-west side of lagoon, *A. J. K. Millar & K. L. Christian*, 25.vi.1997, NSW Slide 20–90.

Laurencia Lamouroux

Laurencia perforata (Bory) Montagne, 1840: 155. Type locality: Santa Cruz de Tenerife, Canary Islands. Voucher: north side of lagoon, *A. J. K. Millar & K. L. Christian*, 20.vi.21997, NSW 418094.

Lophocladia (J. Agardh) Schmitz

Lophocladia kipukaia Schlech, 1990: 174, Figures 7–12. Type locality: Kipukai, Kauai, Hawaii. Voucher: south-west side of lagoon, *A. J. K. Millar & K. L. Christian*, 25.vi.1997, NSW Slide 20–91.

Schlech (1990) points out the distinctive feature of this species as being its staggered pericentral cells, which lie halfway between the ends of the central cells. This feature is evident in the Herald Cay collections, which are, however, much smaller than the dimensions given for this species. *Lophocladia minor* Itono (1973) is a similarly diminutive species, but both it and Hawaiian *L. kipukaia* are at least three times taller than our collections.

Polysiphonia Greville

Polysiphonia sphaerocarpa Børgesen, 1918: 271. Type locality: St Thomas Island, Virgin Islands. Voucher: north side of lagoon, *A. J. K. Millar & K. L. Christian*, 22.vi.1997, NSW 418095.

Veleroa Dawson

This distinctive genus contains three species, the type *V. subulata* Dawson (1944: 335, pl. 72, Figure 2) from the Gulf of Mexico, *V. karulvalensis* (Varma) Krishnamurthy & Thomas from India, and *V. elongata* Saenger (1982) from Queensland, Australia. The Herald Cay species differs from all of these in being strictly ascendent, smaller, thinner and with monosiphonous filaments up to 18 cells in length and with blunt or rounded apical cells. *Veleroa karulvalensis* has monosiphonous filaments in which the suprabasal cell is markedly inflated (Saenger, 1982, Figure 1C), and in the Australian species, the cells are elongate (Saenger, 1982, Figure 1B). The type species has distinctly mucronate tips of laterals which are only 8 cells in length and closely match what Abbott (1996) described as *Micropeuce setosus* from Hawaii. Abbott states that there are four pericentral cells in her species, which would preclude it from the genus *Micropeuce* which has five pericentral cells. I suggest that the Hawaiian species most likely represents the type species *V. subulata*.

***Veleroa* sp. nov.** Voucher: north side of lagoon, *A. J. K. Millar & K. L. Christian*, 20.vi.1997, NSW Slides 20–92, 20–93.

Genus of uncertain position
Ethelia Weber-van Bosse

Ethelia biradiata (Weber-van Bosse) Weber-van Bosse, 1921: 297–298. Type locality: Seychelles. Voucher: south west lagoon, *A. J. K. Millar & K. L. Christian*, 25.vi.1997, NSW 418222.

Small (to 1 cm diameter) crustose blades growing entangled among branches of *Amphiroa crassa* were a common component of this marine flora.

Discussion

Perhaps the most striking aspect of the marine algae of this coral cay is the almost complete absence of phaeophytes. Apart from a few fragments of *Turbinaria* and some torn plants of *Dictyota dichotoma*, not even any *Sargassum* species were noted, yet this is a genus characteristic of tropical marine environments. Both green and red algal divisions were well represented, but the lack of brown algae is noteworthy. One would normally expect a fairly good coverage of genera and species of the brown algal order Dictyotales, especially the calcified member *Padina*, which is a regular inhabitant of coral cay lagoons such as those in the Capricorn section of the Great Barrier Reef (GBR) and elsewhere in the tropics and subtropics. A new record for the Pacific was discovered here in the form of the green alga *Chamaedoris peniculum*. Thought to be restricted to the tropical western Atlantic and the Caribbean, this distinctive species was found in 18 m depths and deeper. Sharing morphological features with both the Dasycladales, in its simultaneous whorls of capitulum filament swellings at apices of distal stalk cells, and the Siphonocladales, with its filaments undergoing segregative division, *C. peniculum* is an excellent candidate for further critical study, which was beyond the scope of the present project.

The discovery of a new species of *Rhipiliopsis* with a monosiphonous stalk, monostromatic blade and capitulum filaments that are sinusoidal and fused along their entire length also merits further critical research.

As is typical of tropical reef lagoons, members of the green algal family Udoteaceae predominate, including *Udotea*, *Rhipilia*, *Rhipiliopsis* and *Avrainvillea*. The closely related genus *Halimeda* of the Halimedaceae is also well represented. And for the red algal contingent, such genera as *Acrosymphyton*, *Gibsmithia*, *Liagora* and *Titanophora* are classic examples of tropical reef environments. No algae were found that would normally be expected in temperate or cool temperate localities, such as the mixture occurring at Norfolk Island further south in the Coral Sea (Millar, pers. obs.). The Herald Cays are clearly a typical coral cay environment as far as its marine algae are concerned.

Because of the very restricted sampling regime, a comparison with other similar latitudinally situated Pacific Islands would not be particularly enlightening. To the east of the Herald Cays, Garrigue & Tsuda (1988) have catalogued the marine algae from New Caledonia, N'Yeurt et al. (1996) have documented that of Fiji which lies on almost the same latitude, N'Yeurt (1996) has started documenting Rotuma Island, and that of French Polynesia has been examined by Payri & N'Yeurt (1997). To the northwest, Coppejans et al. (1995a,b) have been systematically documenting the marine flora of Papua New Guinea, and Womersley & Bailey (1970) have detailed the species from the

Solomon Islands. Tsuda & Wray (1977), although considerably further away in the northwestern Pacific, have listed the marine algae from Micronesia; and to the southwest of the Herald Cays, knowledge exists on the marine algae of the GBR (Cribb, 1983; Price & Scott, 1992). Thus, knowledge of the marine algae of the Herald Cays will fill a large gap in the Coral Sea, west of New Caledonia and east of the GBR. Suffice it to say that based on the present species composition, one would expect there to be an equally diverse marine algal flora at the Herald Cays as is found on these nearby coral cays. This list of 66 species, therefore, is but a fraction of the true diversity of the area.

Acknowledgments

Special thanks go to my diving partner, Karlene Christian (proprietor of Bounty Divers, Norfolk Island), for her experienced help, keen algal eye and for her delicious fish-head soup whilst on the Cay. Karlene was generously sponsored by the Norfolk Island Government and Agnes Hains of 'Walk in the Wild' on Norfolk Island. Ian Loch (Australian Museum) helped with transportation of the barrel of seaweeds from Cairns to Sydney, and the Royal Geographical Society of Queensland are thanked for the invitation to participate in the North East Herald Cay expedition.

References

Abbott, I. A., 1962. Morphological studies in a new species of *Acrosymphyton* (Rhodomelaceae). Am. J. Bot. 49: 845–849.
Abbott, I. A., 1996. New species and notes on marine algae from Hawai'i. Pac. Sci. 50: 142–156.
Agardh, C. A., 1817. Synopsis Algarum Scandinaviae. Berling, Lund, 135 pp.
Agardh, C. A., 1822. Species Algarum. Volume 1, part 2. Berling, Lund. 169–531 (Reprint A. Asher and Co., Amsterdam 1969).
Agardh, J. G., 1848. Species Genera et Ordines Algarum. Volumen Primum: Algas Fucoideas Complectens. Gleerup, Lund, 363 pp.
Agardh, J. G., 1873. Till algernes systematik, Nya bidrag. Acta Univ. Lund 9(8): 1–71.
Agardh, J. G., 1887. Till algernes systematik, Nya bidrag (Femte afdelningen). Acta Univ. Lund 23(2): 1–174.
Ardissone, F., 1871. Revista dei Ceramii della flora Italiana. Nuova Giorn. Bot. Ital. 3: 32–50.
Athanasiadis, A., D. J. Garbary & H. Vandermeulen, 1990. *Antithamnion makroklonion* sp. nov. (Rhodophyta, Ceramiales) from the Gulf of Aqaba (Red Sea). Nordic J. Bot. 10: 557–564.
Børgesen, F., 1912. Some Chlorophyceae from the Danish West Indies. Part 2. Bot. Tidsskr. 32: 241–273.
Børgesen, F., 1918. The marine algae of the Danish West Indies. Part 3: Rhodophyceae (4). Dansk bot. Ark. 3: 241–304.
Børgesen, F., 1932. A revision of Forsskål's algae mentioned in Flora Aegyptiaco-Arabica and found in his herbarium in the Botanical Museum of the University of Copenhagen. Dansk bot. Ark. 8(2): 1–14.
Børgesen, F., 1934. Some Indian Rhodophyceae, especially from the shores of the Presidency of Bombay. 4. Kew Bull. 1934: 1–30.
Børgesen, F., 1940. Some marine algae from Mauritius. 1. Chlorophyceae. Biol. Meddel. Kongel. Danske Vidensk. Selsk. 15(4): 1–81.
Børgesen, F., (1943). Some marine algae from Mauritius, III. Rhodophyceae, Part 2: Gelidiales, Cryptonemiales, Gigartinales. Biol. Meddel. Kongel. Danske Vidensk. Selsk. 19(1): 1–85.
Chauvin, J. F., 1842. Recherches sur l'Organisation, la Fructification et la Classification du Plusieurs Genres d'Algues. Caen. 132 pp.
Coppejans, E., O. De Clerck & C. Van den Heede, 1995a. Annotated and illustrated survey of the marine macroalgae from Motupore Island and vicinity (Port Moresby area, Papua New Guinea). I. Chlorophyta. Biol. Jaarb. 62: 70–108.
Coppejans, E., O. De Clerck & C. Van den Heede, 1995b. Annotated and illustrated survey of the marine macroalgae from Motupore Island and vicinity (Port Moresby area, Papua New Guinea). II. Phaeophyta. Belgian J. Bot. 128: 176–197.
Cribb, A. B., 1983. Marine Algae of the Southern Great Barrier Reef. Part 1. Rhodophyta. Australian Coral Reef Society, Handbook No. 2. Brisbane. 173 pp.
Dawson, E. Y., 1944. The marine algae of the Gulf of California. Allan Hancock Pacif. Exped. 3: 189–454.
Dawson, E. Y., 1957. An annotated list of marine algae from Eniwetok Atoll, Marshall Islands. Pac. Sci. 11: 92–132.
Decaisne, J., 1842. Memoire sur les corallines ou polypiers calciferes. Ann. Sci. nat., Bot. 2, 18: 96–128.
Doty, M. S., 1963. *Gibsmithia hawaiiensis* gen. n. et sp. n. Pac. Sci. 17: 458–465.
Egerod, L. E., 1952. An analysis of the siphonous Chlorophycophyta, with special reference to the Siphonocladales, Siphonales and Dasycladales of Hawaii. Univ. Calif. Publ. Bot. 25: 325–454.
Ellis, J. & D. Solander, 1786. The Natural History of Many Curious and Uncommon Zoophytes Collected from Various Parts of the Globe. London, 208 pp.
Falkenberg, P., 1901. Die Rhodomelaceen des Golfes von Neapel und der Angrenzenden Meeres-Abschnitte. Fauna and Flora des Golfes von Neapel, Monographie 26. Friedlander, Berlin, 754 pp.
Feldmann, J. & G. Hamel, 1934. Observations sur quelques Gélidiacées. Rev. Gén. Bot. 46: 528–549.
Garrigue, C. & R. T. Tsuda, 1988. Catalog of marine benthic algae from New Caledonia. Micronesica 21: 53–70.
Gepp, A. & E. S. Gepp, 1911. The Codiaceae of the Siboga Expedition including a monograph of Flabellarieae and Udoteae. Siboga-Exp. Mon. 62, 150 pp.
Harvey, W. H., 1855. Some account of the marine botany of the colony of Western Australia. Trans. r. Irish Acad. 22: 525–566.
Hatta, A. M. & W. F. Prude'homme van Reine, 1991. A taxonomic revision of Indonesian Gelidiales (Rhodophyta). Blumea 35: 347–380.
Hollenberg, G. J., 1968. An account of the species of the red alga *Herposiphonia* occurring in the central and western tropical Pacific Ocean. Pac. Sci. 22: 536–559.
Holmgren, P. K., N. H. Holmgren & L. C. Barnett, 1990. Index Herbariorum. Part. I: The Herbaria of the World. 8th edn. International Association for Plant Taxonomy, New York Botanical Garden, Bronx, New York, 693 pp.
Howe, M. A., 1934. Hawaiian algae collected by Dr Paul C. Galtsoff. J. Wash. Acad. Sci. 24: 32–42.
Huisman, J. M. & A. J. K. Millar, 1996. *Asteromenia* (Rhodymeniaceae, Rhodymeniales), a new red algal genus based on *Fauchea peltata*. J. Phycol. 32: 138–145.

Huvé, P. & H. Huvé, 1977. Notes de nomenclature algale. 1. – Le genre *Halichrysis* (J. Agardh, 1851 emend. J. Agardh, 1876) Schousboe mscr. in Bornet 1892 (Rhodyméniales, Rhodyméniacées). Bull. Soc. phycol. France 22: 99–107.

Itono, H., 1973. Notes on marine algae from Hateruma Island, Ryukyu. Bot. Mag. Tokyo 86: 155–168.

Kraft, G. T., 1986. The green algal genera *Rhipiliopsis* A. & E. S. Gepp and *Rhipiliella* gen. nov. (Udoteaceae, Bryopsidales) in Australia and the Philippines. Phycologia 25: 47–72.

Kraft, G. T. & I. A. Abbott, 1998. *Gelidiella womersleyana* (Gelidiales, Rhodophyta), a diminutive new species from the Hawaiian Islands. Bot. mar. 41: 51–61.

Kraft, G. T. & M. J. Wynne, 1996. Delineation of the genera *Struvea* Sonder and *Phyllodictyon* J. E. Gray (Cladophorales, Chlorophyta). Phycol. Res. 44: 129–142.

Kuntze, O., 1898. Revisio Generum Plantarum. Part 3(2). Leipzig, 576 pp.

Kützing, F. T., 1869. Tabulae Phycologicae, Vol.19. Nordhausen, 36 pp.

Lamarck, J. B., 1813. Sur les polypiers empâtés. Ann. Mus. natn. Hist. nat. Paris 20: 294–312, 370–386, 432–458.

Lamouroux, J. V. F., 1809. Exposition des caractères du genre *Dictyota*, et tableau des espèces qu'il renferme. J. Bot. (Desvaux) 2: 38–44.

Lamouroux, J. V. F., 1816. Histoire des Polypiers Coralligènes Flexibles, Vulgairement Nommés Zoophytes. Published by author, Caen, 560 pp.

Littler, D. S., M. M. Littler, K. E. Bucher & J.N. Norris, 1989. Marine Plants of the Caribbean, a Field Guide from Florida to Brazil. Smithsonian Institution Press Washington, D.C., 263 pp.

Millar, A. J. K., 1990. Marine red algae of the Coffs Harbour region, northern New South Wales. Aust. Syst. Bot. 3: 293–593.

Millar, A. J. K., 1999. Marine benthic algae of Norfolk Island, South Pacific. Aust. Syst. Bot. 12 (in press).

Millar, A. J. K. & I .A. Abbott, 1997. The new genus and species *Ossiella pacifica* (Griffithsieae, Rhodophyta) from Hawaii and Norfolk Island, Pacific Ocean. J. Phycol. 33: 88–96.

Montagne, J. F. C., 1840. Plantae Cellulaires, Sect 4. In Webb P. B., & S. Berthelot (eds) Histoire Naturelle des Iles Canaries. Paris. 3(2): 1–208.

Montagne, C., 1842. Troisième centurie de plantes cellulaires exotiques nouvelles, Décade V, VI, VII, et VIII. Ann. Sci. Nat., Bot., series 2, 18: 241–282.

Montagne, C., 1846. Phyceae. In Durieu de Maisonneuve, M.C. (ed.), Exploration Scientifique de l'Algérie Pendant les Annees 1840, 1841, 1842, Sciences Naturelles: Botanique. Paris: 1–197.

Montagne, C., 1847. Enumeratio fungorum quos a cl. Drege in Africa meridionali collectos et in herbario migueliano servatos descripsit C. Montagne. Ann. Mus. Hist. nat. Marseille 1843, 7, Bot. sect. 3, 7: 166–181

Montagne, C., 1856. Sylloge Generum Specierumque Cryptogamarum. Bailliere, Paris, 498 pp.

Nägeli, C., 1862. Beiträge zue Morphologie und Systematik der Ceramiaceae. Sitzungsber. Königl. Bayer. Akad. Wiss. München (1861) 2: 297–415.

Nees, C. G., 1820. Horae Physicae Berolinensis. Bonn, 123 pp.

N'Yeurt, A. D. R., 1996. A preliminary floristic survey of the benthic marine algae of Rotuma Island. Aust. syst. Bot. 9: 361–490.

N'Yeurt, A. D. R., G. R. South & D. W. Keats, 1996. A revised checklist of the benthic marine algae of the Fiji Islands, South Pacific (including the island of Rotuma). Micronesica 29: 49–98.

Okamura, K., 1908. Icones of Japanese Algae. 1(7), Published by the author, Tokyo, pls 31–35.

Okamura, K., 1932. Icones of Japanese Algae. 6(70), Published by the author, Tokyo, pls 281–285.

Olsen, J. L. & J. A. West, 1988. *Ventricaria* (Siphonocladales-Cladophorales complex, Chlorophyta), a new genus for *Valonia ventricosa*. Phycologia 27: 103–108.

Orme, G. R., 1977. The coral plateau – a major reef province. In Jones, O. A. & R. Endean (eds), Biology and Geology of Coral Reefs Vol. 4. Geology 2. Academic Press, New York.

Payri, C. E. & A. D. R. N'Yeurt, 1997. A revised checklist of Polynesian benthic marine algae. Aust. syst. Bot. 10: 867–910.

Pilger, R., 1911. Die Meeresalgen von Kamerun nach der Sammlung von C. Ledermann. Engl. Bot. Jahrb. 46: 294–323.

Price, I. R. & F. J. Scott, 1992. The Turf Algal Flora of the Great Barrier Reef. Part I. Rhodophyta. James Cook University Press, Townsville, 266 pp.

Quoy, J. R. C. & P. Gaimard, 1824. Zoologie. In L. de Freycinet, Voyage autour du monde..... exécuté sur les corvettes de S.M. l'Uranie et la Physicienne, pendant les années 1817, 1818, 1819 et 1820. Paris, 713 pp.

Saenger, P., 1982. A new species of *Veleroa* (Rhodophyta: Rhodomelaceae) from eastern Australia. Proc. r. Soc. Qd 93: 65–69.

Schlech, K. E., 1990. *Eupogodon iridescens* and *Lophocladia kipukai*, two new species of Rhodophyta from Hawaii. Brit. phycol. J. 25: 169–77.

Setchell, W. A., 1929. The genus *Microdictyon*. Univ. Calif. Publ. Bot. 14: 453–588.

Setchell, W. A. & N. L. Gardner, 1924. Expedition of the California Academy of Sciences to the Gulf of California in 1921. The marine algae. Proc. Calif. Acad. Sci., ser. 4, 12: 695–949.

Taylor, W. R., 1950. Plants of Bikini and Other Northern Marshall Islands. University of Michigan Press, Ann Arbor, 227 pp.

Tsuda, R. T. & F. O. Wray, 1977. Bibliography of marine benthic algae in Micronesia. Micronesica 13: 191–198.

Weber-van Bosse, A., 1905. Note sur le genre *Dictyosphaeria* Dec. Nouva Notarisia 16: 142–144.

Weber-van Bosse, A., 1913. Marine algae, Rhodophyceae, of the 'Sealark' Expedition, collected by J. Stanley Gardiner, M.A. Trans. Linn. Soc. London, Bot. ser. 2, 8: 105–142.

Weber-van Bosse, A., 1921. Liste des algues du Siboga. II. Rhodophyceae. Première partie. Protoflorideae, Nemalionales, Cryptonemiales. Siboga-Exped. Mon. 59b: 187–310.

Weber-van Bosse, A., 1928. Liste des algues du Siboga. IV. Rhodophyceae. 3. Gigartinales et Rhodymeniales. Siboga-Exped. Mon. 59b: 393–533.

Womersley, H. B. S. & A. Bailey, 1970. Marine algae of the Solomon Islands. Phil. Trans., Ser. B. Biol. Sci. 259: 257–352.

Wynne, M. J., 1985. Concerning the names *Scagelia corallina* and *Heterosiphonia wurdemannii* (Ceramiales, Rhodophyta). Crypt. Algol. 6: 81–90.

Wynne, M. J., 1995. Benthic marine algae from the Seychelles collected during the R/V Te Vega Indian Ocean Expedition. Contr. Uni. Michigan Herb. 20: 261–346.

Yamada, Y., 1944. A list of the marine algae from the Atoll of Ant. Sci. Pap. Inst. algol. Res. Fac. Sci. Hokkaido Imp. Univ. 3: 31–45, pls 6–7.

Zanardini, G., 1878. Phyceae papuanae novae vel minus cognitae a cl. O. Beccari in itinere ad Novam Guineam annis 1872-75 collectae. Nuovo Giorn. Bot. Ital. 10: 34–40.

Changes of the benthic algal flora of the Tremiti Islands (southern Adriatic) Italy

Mario Cormaci & Giovanni Furnari
Dipartimento di Botanica dell'Università, Via Antonino Longo 19, 95125 Catania, Italy

Key words: algal flora, Adriatic Sea, environmental changes, the Tremiti Islands

Abstract

During the last two decades, the benthic algal flora of the northern Adriatic Sea has changed as a result of increased pollution by sewage, agricultural drainage and industrial discharges. In order to verify if pollution has also influenced the benthic algal flora of the southern Adriatic, a study was undertaken of the Tremiti Islands and the results compared with previously published floristic data of the area. The results indicate changes in the benthic flora. Although the total number of species had increased slightly (301 species compared with 265 from the literature), the floristic composition was quite different, with 109 species not now found and 145 species new to the area. Since several sensitive species, like *Cystoseira* spp. and *Sargassum* spp., had disappeared and a greater number of opportunistic species was recorded, it can be concluded that the southern Adriatic is nowadays also influenced by pollution impact.

Introduction

Pollution-induced changes of the benthic algal flora and vegetation in the northern Adriatic Sea were pointed out by Munda (1993). They were related to pollutants that have been discharged for the last two decades from direct urban effluents as well as via the rivers Po, Adige, Brenta, Piave, Tagliamento and Isonzo. In order to verify if any changes in the benthic algal flora have also occurred in the southern Adriatic, a floristic study of the marine park around the Tremiti Islands, which is not subject to terrestrial pollution, was carried out. Moreover, this area was chosen since data on its benthic flora are available from studies carried out about 30 years ago (Pignatti et al., 1967; Giaccone, 1969; Rizzi Longo, 1972; Giaccone, 1978). From such literature data (appropriately updated from both taxonomic and nomenclature points of view), the benthic algal flora consisted of 265 taxa at specific and infraspecific rank (158 Rhodophyceae, 60 Fucophyceae and 47 Chlorophyceae).

Investigated area

The Tremiti Islands are a small archipelago situated in the southern Adriatic Sea at a distance of about 22 km from the Italian coast (Figure 1). There are three main islands: St. Domino (2.08 km^2), St. Nicola (0.41 km^2), Caprara (0.44 km^2) and an islet named 'Il Cretaccio' (0.03 km^2). Because of the distance from the coast, as well as the lack of waste waters due to the small number of inhabitants (about 400 on the whole), the sea around these islands shows both temperature and salinity values similar to those found in the open sea (Rizzi Longo, 1972). In the last few decades such values have shown wide annual variations and tendencies to increase (Zore-Armanda, 1991; Zore-Armanda et al., 1991).

Material and methods

Sampling was carried out on rocky substrata in the second half of May 1997 at 16 stations: six at St. Domino, four at St. Nicola, five at Caprara and one at Cretaccio. A total of 116 samples were collected by SCUBA diving along 16 transects perpendicular to the shore from the midlittoral zone to a maximum depth of 40 m. The collected material was preserved in seawater–formalin (5%) for later study in the laboratory. Voucher specimens are kept in the Herbarium of

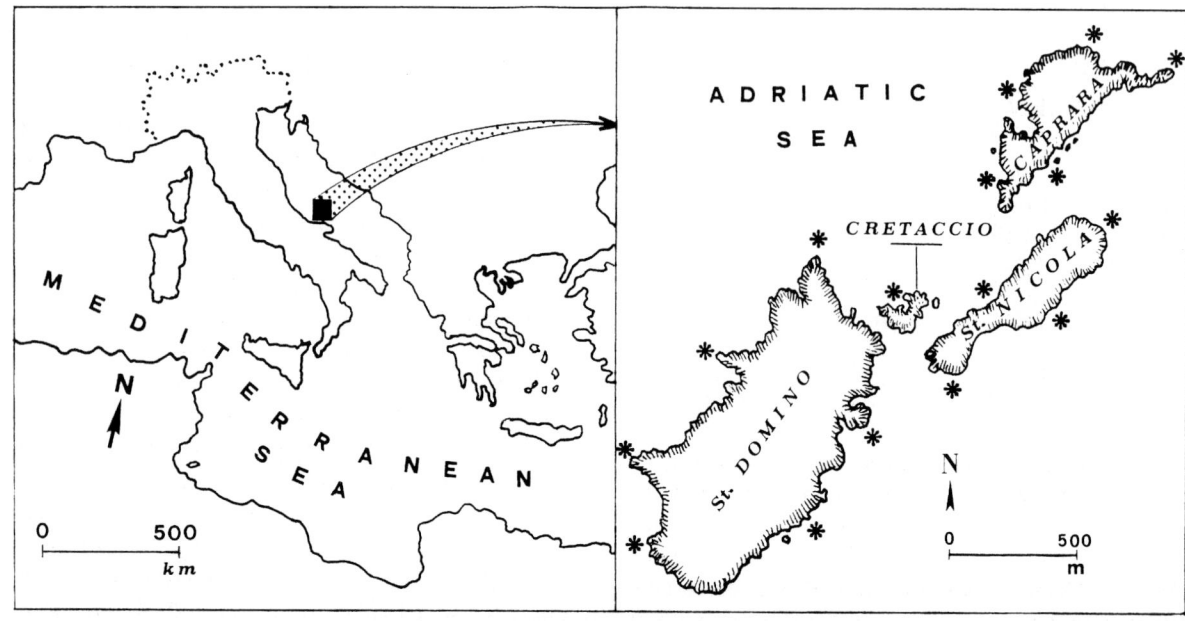

Figure 1. Map of the Tremiti Islands. Asterisks indicate sampling sites.

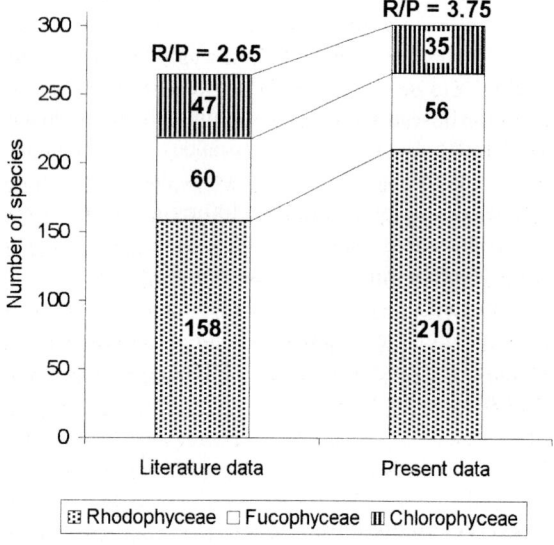

Figure 2. Comparison between the floristic richness of the previous and present floras. (R/P = Rhodophyceae/Fucophyceae).

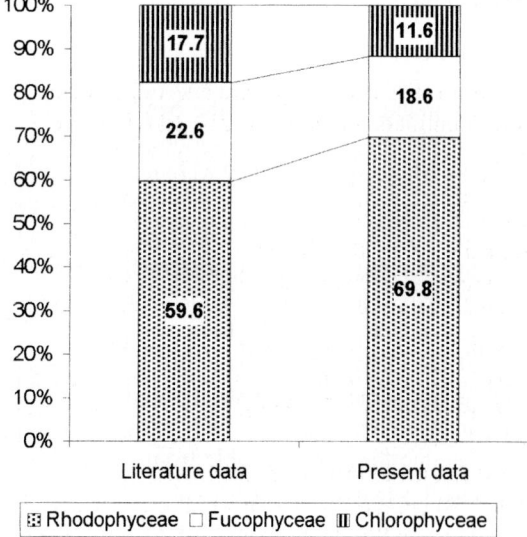

Figure 3. Comparison between the floristic richness of the previous and present floras (in per cent values).

the Department of Botany of the University of Catania (CAT).

Floristic lists (available on request) of:

1. taxa present in both floras;
2. taxa previously reported and not found in this study;
3. taxa newly reported, were made.

In the lists of each taxon the phytogeographic element, named according to Cormaci et al. (1982), is reported.

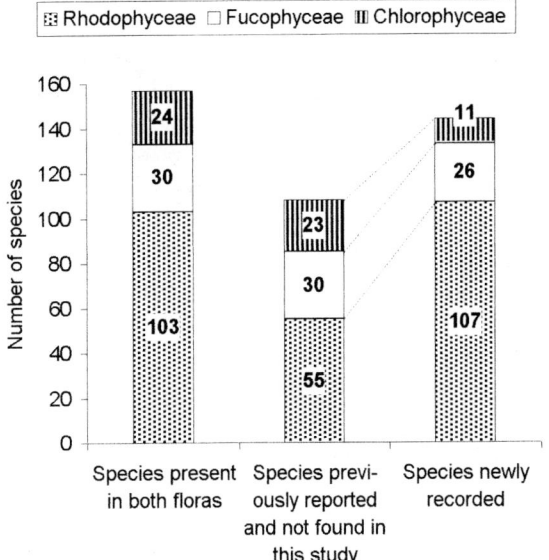

Figure 4. Comparison between the floristic richness of previous and present floras showing the number of common, disappeared and newly recorded taxa.

Results and discussion

A total of 301 taxa at specific and infraspecific rank were identified, comprising 210 Rhodophyceae, 56 Fucophyceae and 35 Chlorophyceae. Even though our data cannot be considered as exhaustive being based only on late spring samples, they can be safely compared with the literature data which also refer to samplings carried out in late spring and summer months. From the comparison, it appears that the present flora is slightly richer than that reported previously (301 taxa compared with 265), showing a minor decrease of both Fucophyceae and Chlorophyceae (4 and 12 species less, respectively) but a substantial increase in the number of Rhodophyceae (210 compared with 158) (Figure 2). These variations correspond to an increase in Rhodophyceae of 10.2% and a decrease in Fucophyceae and Chlorophyceae of 4% and 6.1%, respectively (Figure 3).

Only 156 taxa (102 Rhodophyceae, 30 Fucophyceae and 24 Chlorophyceae) of the present flora have been reported previously, while 145 (108 Rhodophyceae, 26 Fucophyceae and 11 Chlorophyceae) were new to the area. Conversely, 109 taxa (56 Rhodophyceae, 30 Fucophyceae and 23 Chlorophyceae) recorded previously were not found this time (Figure 4).

From a qualitative point of view it should be pointed out that: 1) several sensitive species of Fucales like *Cystoseira crinita* Duby, *C. humilis* Kützing, *C. schiffneri* Hamel, *C. spinosa* Sauvageau, *Sargassum acinarium* (Linnaeus) Setchell, *S. hornschuchii* C. Agardh which characterize the Mediterranean Sea photophilic communities on hard substrata in non-polluted environments (Giaccone & Bruni, 1973; Ros et al., 1985), had disappeared; 2) numerous deep water Rhodophyceae e.g. *Aglaothamnion tenuissimum* (Bonnemaison) Feldmann-Mazoyer, *Antithamnion heterocladum* Funk, *Balliella cladoderma* (Zanardini) Athanasiadis, *Nitophyllum tristromaticum* Rodriguez *ex* Mazza, *Rodriguezella pinnata* (Kützing) Falkenberg, *Sebdenia rodrigueziana* (J. Feldmann) Parkinson had appeared.

From a phytogeographic point of view, the present flora has a R/P (Rhodophyta/Phaeophyta) Index value (Feldmann 1937) of 3.75, much higher than that obtained from the published floral data (2.63). Such a high value, according to Feldmann (1937), should be evidence of nearly tropical characteristics of the present flora. That, in our opinion, is not the case in the Tremiti Islands even though the water temperature has increased in the last few decades. In fact: (1) the increase of the R/P Index value is mainly due to the increase in the number of Rhodophyta but not to a considerable decrease in the number of Phaeophyta (Figure 4); (2) from the comparison between chorological spectra of present and previous floras (Figure 5), the only substantial difference is the higher incidence in the present flora of the Mediterranean element (24% against 18%) coincident with a decrease in both Atlantic and Cosmopolitan elements; there are no changes to the Circumboreal and Circumboreoaustral elements (species with cold water affinity) (2% in both floras) nor to the Indo-Pacific (4%) and Circumtropical (4%) elements (species with warm water affinity). The increase in the Mediterranean element is mainly due to the high number of newly reported skiophilous or deep water Rhodophyta belonging to that element. This is probably related to a reduced transparency of water rather than to the increase in water temperature.

Conclusions

In our opinion, changes to the benthic algal flora of the Tremiti Islands are mainly related to the reduction in water transparency. Such a reduction was recorded by Morovic & Domijan (1991) who, on the basis of

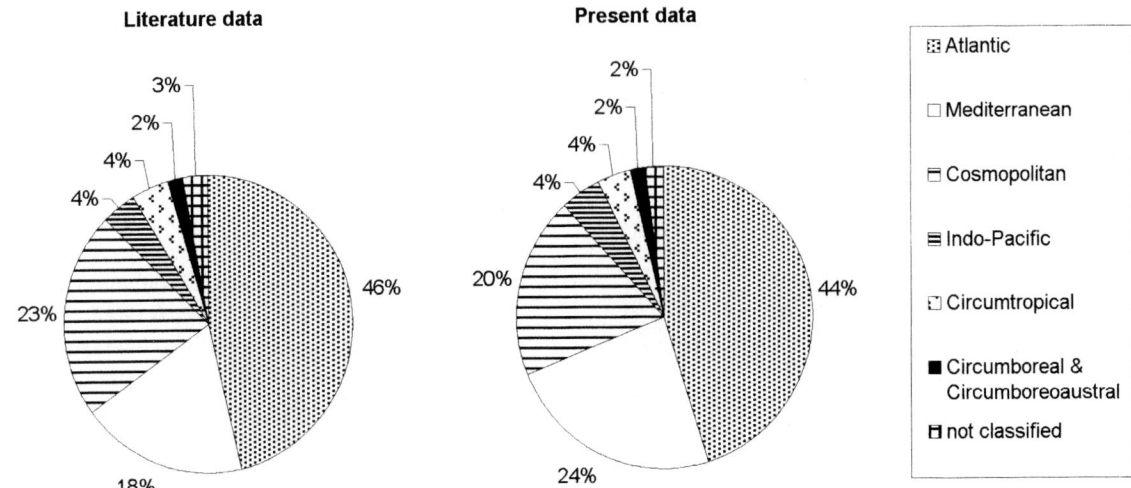

Figure 5. Comparison between chorological spectra of previous and present floras. Chorological elements from Cormaci et al. (1982).

irradiance measurements, concluded that ". . . the euphotic zone depth has considerably decreased in the southern Adriatic from sixties to eighties in both the coastal and open waters".

According to Zore-Armanda (1991), water turbidity is most probably due to the eutrophication of the Adriatic waters. However, while the occurrence of eutrophic deep waters around the Tremiti Islands cannot be ruled out, the almost total lack of communities with Ulvales in both midlittoral and upper sublittoral zones, as well as the persistence near the surface of a well structured community with *Cystoseira amentacea* (C. Agardh) Bory v. *stricta* Montagne (to which should be referred *C. spicata* Ercegovic v. *elegans* Ercegovic as reported by Rizzi Longo 1972), would seem to indicate the lack of eutrophication at least in shallow waters. In contrast to the reports by Rizzi Longo (1972) of an extraordinary transparency of water, because of the almost total lack of suspended materials, during our dives a noticeable turbidity, due to presence of suspended inert terrigenous materials, was observed. This might explain both the increase of skiophilous (mostly Rhodophyta) and the decrease of photophilic (*Cystoseira* and *Sargassum* spp.) species in the sublittoral zone.

Finally, even though the floristic richness is higher in 1997, the disappearance of both sublittoral species of *Cystoseira* and *Sargassum*, shows a considerable structural deterioration of phytobenthic communities of the Tremiti Islands, similar to that recorded by Munda (1993) in moderately polluted sites in the northern Adriatic.

Acknowledgements

This work was supported by a grant from the Italian C. N. R. within the project PRISMA 2, sub-project 3. We are grateful to M. T. Brown for correcting the text.

References

Cormaci, M., A. Duro & G. Furnari, 1982. Considerazioni sugli elementi fitogeografici della flora algale della Sicilia. Naturalista sicil. 6 (suppl. 1): 7–14.

Feldmann, J., 1937. Recherches sur la végétation marine de la Méditerranée. La côte des Albères. Rev. algol. 10: 1–340.

Giaccone, G., 1969. Raccolte di fitobenthos sulla banchina continentale italiana. Giorn. Bot. Ital. 103: 485–514.

Giaccone, G., 1978. Revisione della flora marina del mare Adriatico. Annuario del WWF. Parco Marino di Miramare, Trieste 6: 5–118.

Giaccone, G. & A. Bruni, 1973. Le Cistoseire e la vegetazione sommersa del Mediterraneo. Atti Ist. Veneto Sci., Lett. ed Arti, Venezia. 131: 59–103.

Morovic, M. & N. Domijan, 1991. Light attenuation changes in the middle and southern Adriatic. Acta Adriatica 32: 621–635.

Munda, I. M., 1993. Changes and degradation of seaweed stands in the northern Adriatic. Hydrobiologia 260/261: 239–253.

Pignatti, S., P. De Cristini & L. Rizzi, 1967. Le associazioni algali della grotta delle Viole nell'isola di S. Domino (Is. Tremiti). Giorn. Bot. Ital. 101: 117–126.

Rizzi Longo, L., 1972. La flora sottomarina delle Isole Tremiti. Atti Ist. Veneto Sci., Lett. ed Arti, Venezia 130: 329–376.

Ros, J. D., J. Romero, E. Ballesteros & J. M. Gili, 1985. Diving in blue water. The benthos. In Margalef R. (ed.), Key Environments: Western Mediterranean. Pergamon Press, Oxford: 233–295.

Zore-Armanda, M., 1991. Natural characteristics and climatic changes of the Adriatic sea. Acta Adriatica 32: 567–586.

Zore-Armanda, M., M. Bone, V. Dadic, M. Morovic., D. Ratkovic, L. Stojanoski & I. Vukadin, 1991. Hydrographic properties of the Adriatic Sea in the period from 1971 through 1983. Acta Adriatica 32: 5–547.

Cell–cell recognition during fertilization in the red alga, *Aglaothamnion oosumiense* (Ceramiaceae, Rhodophyta)

Sung-Ho Kim & Gwang Hoon Kim*
*Department of Biology, Kongju National University, Shingwandong, Kongjushi, Chungnam 314-701, Korea;
Tel: [+82]-416-50-8504. Fax: [+82]-416-50-8479*

Key words: Aglaothamnion oosumiense, cell-cell recognition, confocal microscope, fertilization, lectin, Rhodophyta

Abstract

The binding of fluorescein isothiocyanate (FITC) conjugated lectins to gametes of *Aglaothamnion oosumiense* Itono during fertilization was studied by the use of confocal microscopy. The physiological effects of lectins and carbohydrates on gamete binding were also examined. Four different lectins, concanavalin A (ConA), soybean agglutinin (SBA), *Dolichos biflorus* agglutinin (DBA) and wheat germ agglutinin (WGA) bound to the surface of spermatia, but each lectin labelled a different region of the spermatium. SBA and DBA bound only to the spermatial appendages but ConA bound to all the spermatial surface except the spermatial appendages. WGA labelled a narrow region that connects the spermatial body and appendages. During fertilization, the ConA and WGA specific substances on the spermatial surface moved towards the area contacting the trichogyne and accumulated on the surface of the fertilization canal. Spermatial binding to trichogynes was inhibited by pre-incubation of spermatia with SBA or ConA, while trichogyne receptors were blocked by the complementary carbohydrates, D-glucose or *N*-acetyl-galactosamine, respectively. WGA and DBA as well as their complementary carbohydrates had little effect on gamete binding. The inhibitory effects of ConA and SBA were increased when the two lectins were applied simultaneously. The inhibitory effects of both lectins were partially reversed (to 80–90% of controls) by addition of complementary carbohydrates at the same time. The results suggested that SBA and ConA receptors on the spermatial surface are involved in gamete recognition in *Aglaothamnion oosumiense*.

Introduction

In red algae, fertilization begins with gamete-gamete contact between the trichogyne cell wall of the female carpogonium and the spermatial coverings (Pickett-Heaps & West, 1998). The gamete binding is highly selective suggesting the presence of recognition factors along their surfaces (Kim et al., 1996).

Cell surface carbohydrates have been reported as primary markers for cell-cell recognition events in many organisms (Wassarman, 1987; Sharon & Lis, 1989; Karlsson, 1991; Kim, 1997). Such recognition systems depend on complementary binding between carbohydrate moieties of a glycoconjugate on one cell with a specific sugar-binding proteins (lectin-like protein) on another cell (Sharon & Lis, 1989; Chrispeels & Raikhel, 1991; Kim, 1997).

Gamete recognition in some algae is mediated by such a carbohydrate-receptor system (Wiese & Shoemaker, 1970; Bolwell et al., 1979, 1980; Catt et al., 1983; Schmid, 1993; Kim & Fritz, 1993a, b; Mine & Tatewaki, 1994; Kim et al., 1996). In these studies, pre-treatment of one gamete with specific carbohydrates or treatment of the other with complementary lectins was demonstrated to inhibit gamete binding and fertilization.

It is difficult to imagine, however, that the high specificity requirement for cell recognition during fertilization is based on a single lectin-like receptor and its complementary carbohydrate. Therefore, it was hypothesized that multiple carbohydrate-receptor systems may operate in gamete recognition and that those

* Author for correspondence.

molecules may play an additional role during the fertilization process.

In this study, we examined the binding of FITC conjugated lectins to the surfaces of spermatia and trichogynes of *Aglaothamnion oosumiense*. The distribution of each lectin receptors on the spermatial surface was observed at each step of the fertilization process by confocal microscopy.

Materials and methods

Tetrasporic plants of *Aglaothamnion oosumiense* were collected from 5–10 m depth from Auchungdo, Korea and were maintained in modified f/2-enriched seawater (Guillard & Ryther, 1962). The plants were kept at 15 °C under a 16:8 h light:dark (LD) photons cycle with 10 μmol photons cm^{-2} s^{-1} cool-white fluorescent light. Tetraspores were released and developed into male and female plants four weeks after germination. The sexual plants were maintained separately.

Application of fluorescein-labelled lectins

FITC-labelled lectins (Vector laboratories, Burlingame, CA) were added with plants in seawater to give a final concentration of approximately 10 μg ml^{-1} and were rotated mildly during incubation. Unbound lectin was removed by three washes in culture media for 5 min each. Sugar specificity of the lectin-ligand interactions was assayed as described previously (Kim & Fritz, 1993c).

Binding assay

Dense suspensions of spermatia were obtained from male plants with actively developing spermatial clusters as described previously (Kim & Fritz, 1993a). Briefly, male plants were incubated in 10 ml of fresh culture media with mild shaking for 12 h. Aliquot samples (10 μl) were collected several times and the number of spermatia was counted. After male plants were removed, the solution was diluted to various concentrations (10^2–10^4 spermatia ml^{-1}) by adding culture medium. Spermatial binding to trichogynes was assayed with ca. 100 trichogynes placed in varing concentrations of spermatia (10–400 spermatia per carpogonium), in a total volume of 5 ml for 2 h at 20 °C with mild rotation. The total number of trichogynes and the percentage with one or more attached spermatia were recorded. The average number of spermatia attached to trichogynes was also counted.

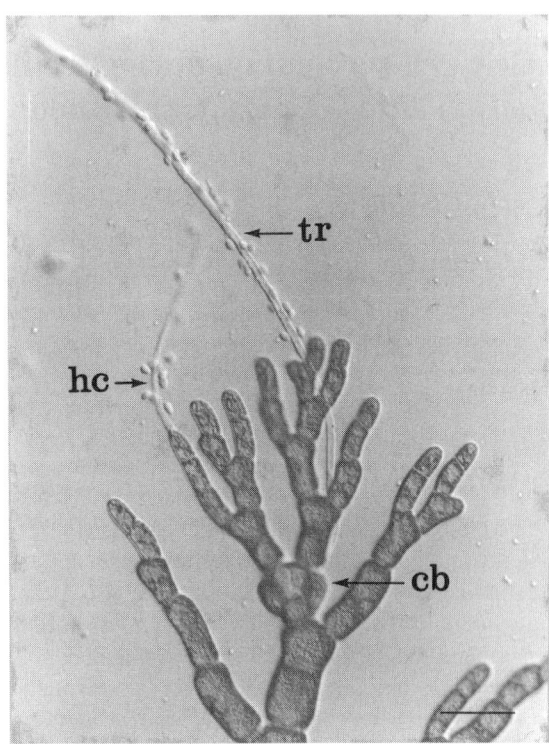

Figure 1. Female plant of *Aglaothamnion oosumiense* developing carpogonia and hair cells. Note the spermatia attached to the trichogyne and hair cell (cb, carpogonial branch; hc, hair cell; tr, trichogyne; scale bar: 20 μm).

Assays were conducted in triplicate, and a minimum of 100 trichogynes was counted from each replicate. Variation between replicates was usually within 5%.

Blocking of fertilization by lectins and carbohydrates

Apical portions of female plants and released spermatia were preincubated for 1 h at 20 °C in a solution containing various lectins (25 μg ml^{-1}) or carbohydrates (10 mm) dissolved in seawater. About 100 trichogynes were incubated with ca. 100 spermatia per trichogyne, in a total volume of 1 ml for 1 h at 20 °C with mild rotation. The assay method and quantification was the same as in the binding assay.

Microscopy

All specimens were examined with an Olympus Fluoview confocal microscope equipped with epifluorescence illumination and differential interference optics. Micrographs were taken with Kodak TMAX 100 film.

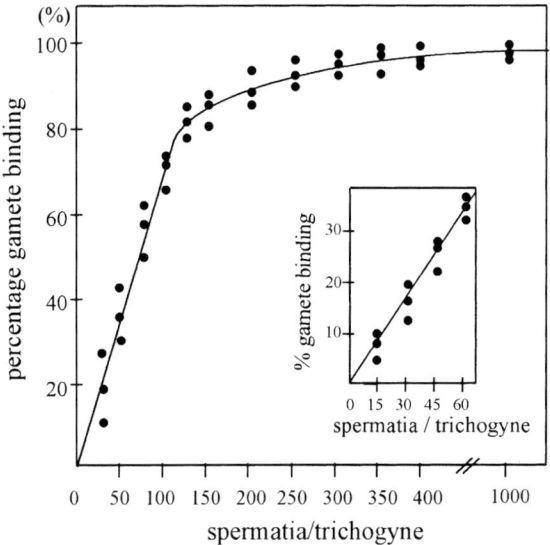

Figure 2. Spermatial attachment to trichogynes according to spermatia/trichogyne ratio in *Aglaothamnion oosumiense*.

Results

The female reproductive structure of *Aglaothamnion oosumiense* is composed of four-celled carpogonial branches with a single long trichogyne (i.e., the spermatium-receptive structure) terminating the carpogonium (Figure 1). In the presence of female plants, spermatia of *Aglaothamnion oosumiense* bound only to trichogynes and hair cells, and very few spermatia (<3%) were observed in contact with any other parts of the female thallus. Trichogynes and hair cells were often located close to each other, and spermatia bound to both cells.

Spermatial attachment could be detected within a few seconds of adding female plants, and was completed within 30 min – 1 h according to spermatial concentration. The percentage of spermatial binding to trichogynes was proportional to the relative concentration of spermatia (Figure 2). Maximum percentage binding (96.9%) was attained at approximately 200 spermatia per trichogyne (2×10^4 spermatia ml^{-1}), and half maximum binding occurred at a relative concentration of 100 spermatia per trichogyne. There was little increase of binding percentage at over 150 spermatia per trichogyne.

The spermatium was composed of two distinct parts, an ovoid spermatial body and two spermatial appendages projecting from each distal end of spermatial body (Figures 3A, B). The carbohydrate moieties on spermatial surfaces were visualized using fluorescein-labelled lectins (Table 1; Figures 3C–F). Four lectins, concanavalin A (ConA), soybean agglutinin (SBA), *Dolichos biflorus* agglutinin (DBA) and wheat germ agglutinin (WGA) bound to the surface of released spermatia (Figures 3C–F). Each lectin labelled a different region of the spermatium. ConA bound mainly to the surface of the spermatial body but did not bind to the spermatial appendages (Figure 3C). SBA and DBA bound only to the spermatial appendages but SBA labelling was more intensive than DBA (Figures 3D, E). This difference in labelling intensity may be due to slight difference in sugar moiety of the two lectins; SBA bound to *N*-acetyl-galactosamine while DBA bound more specifically to α-*N*-acetyl-galactosamine (Table 1). WGA labelled the spermatial body but was different from ConA in binding more strongly to the narrow region connecting the spermatial body and appendages (Figure 3F). Thus, each lectin receptors had a unique distribution. The carpogonium and trichogyne did not bind to the lectins, but thay stained with calcofluor white (Table 1). Non-specific binding was not observed in any of the lectins applied (Figures 1G–J).

Changes in the distribution of lectin receptors were observed according to stages of fertilization (Figure 4). Sixty min after spermatial attachment to trichogynes, the spermatial wall protruded toward the trichogyne and spermatia changed from ovoid to mushroom-shaped (Figures 4A, B). At this time, the ConA receptors on the spermatial surface accumulated at the area contacting the trichogyne (Figure 4B′), while the SBA receptors did not change their distribution (Figure 4E′). The distribution of WGA receptors was of interest because they appeared to move towards the contacting area with trichogyne (Figures 4G, H). The fertilization canal (FC) between spermatia and trichogyne developed around 120 min after spermatial binding (Figure 4C). The ConA receptors distributed along the surface of the fertilization canal as well as the spermatial body (Figure 4C′), but WGA receptors were found only on the fertilization canal (Figure 4I′). The SBA and DBA receptors did not show any change in distribution during fertilization (Figures 4D′–F′).

The percentage of spermatial attachment to trichogynes was assayed after pre-incubation of spermatia or trichogynes with various carbohydrates and lectins (Table 2). The binding of spermatia to trichogynes of *Aglaothamnion oosumiense* was inhibited by the lectins ConA and SBA, and their complementary carbohydrates, D-glucose and *N*-acetyl-galactosamine, respectively. SBA inhibited gamete binding more

Table 1. List of fluorescent probes and their binding sites. SC = spermatangial clusters, RS = released spermatia, VW = vegetative cell wall, TR = trichogyne; – = no labelling, * = trace labelling, +, ++ and +++ = increasing intensity of labelling

Name	Specificity	Binding site *Aglaothamnion oosumiense*			
		SC	RS	VW	TR
ConA	α-D-glucose, α-D-mannose	+++	+++	–	–
SBA	N-acetyl-galactosamine	–	++++	–	–
DBA	α-N-acetyl-galactosamine	–	++	–	–
PNA	β-D-galactose, D-galactose-β-galactosamine	–	*	–	–
WGA	N-acetyl-glucosamine	–	++	–	–
UEA	L-fucose	–	–	–	–
RCA	β-D-galactose	–	–	–	–
Calcofluor-white ST	Cellulose Chitin fiber	–	–	+++	+

strongly than ConA (Table 2). When spermatia were pretreated with both lectins at the same time, spermatial binding to trichogynes was further reduced. WGA and DBA were less inhibitory for spermatial binding to trichogynes than ConA and SBA (Table 2). The inhibitory effects of the latter lectins were partially reversed (to 80-90% of controls) by addition of complementary carbohydrates.

Trichogynes preincubated with D-glucose and N-acetyl-galactosamine had inhibition of spermatium-trichogyne binding, but the inhibitory effect increased when the two sugars were used simultaneously (Table 2). Although ConA binds more strongly to D-mannose than to D-glucose, spermatium-trichogyne binding was not inhibited by D-mannose.

The average number of spermatia attached to trichogynes was also affected by preincubation of spermatia or trichogynes with lectins and sugars. When spermatia were preincubated with both ConA and SBA at the same time, the average number of attached spermatia was reduced to almost one sixth of control. The results were consistent with the percentage of spermatial attachment to trichogynes.

Discussion

Our data suggest that gamete recognition in *Aglaothamnion oosumiense* is mediated by a double-docking recognition process based on the interaction of two carbohydrate moieties with their complementary receptors (lectin-like molecules). Surfaces of spermatial body and appendages have two types of glycoconjugates specific to ConA and SBA, respectively, while those of trichogynes possess the complementary lectin-like receptors. Therefore, there are at least two sets of carbohydrate-receptor systems involved in gamete recognition in this species.

The distribution of SBA receptors did not change during the fertilization process suggesting that the molecule serves only for gamete recognition and binding. The ConA receptors, however, accumulated along the fertilization canal (FC) implying their involvement in FC development. The behavior of WGA receptors during the fertilization is of interest because they appeared to move from the distal ends to the center of the spermatia, and accumulate on the fertilization canal. This movement of WGA receptors may be necessary for the development of the fertilization canal and for gamete membrane fusion.

The lectin SBA which bound to spermatial appendages showed a stronger inhibitory effect on gamete binding than ConA. This result may support our previous suggestion that the primary role of spermatial appendages is in spermatial binding to trichogynes (Kim et al., 1996). Magruder (1984) reported selective spermatial binding to trichogynes and vegetative hair cells in *Aglaothamnion neglectum* and suggested that spermatia of this species may first attach to hair cells and later transfer to a nearby trichogyne. Although

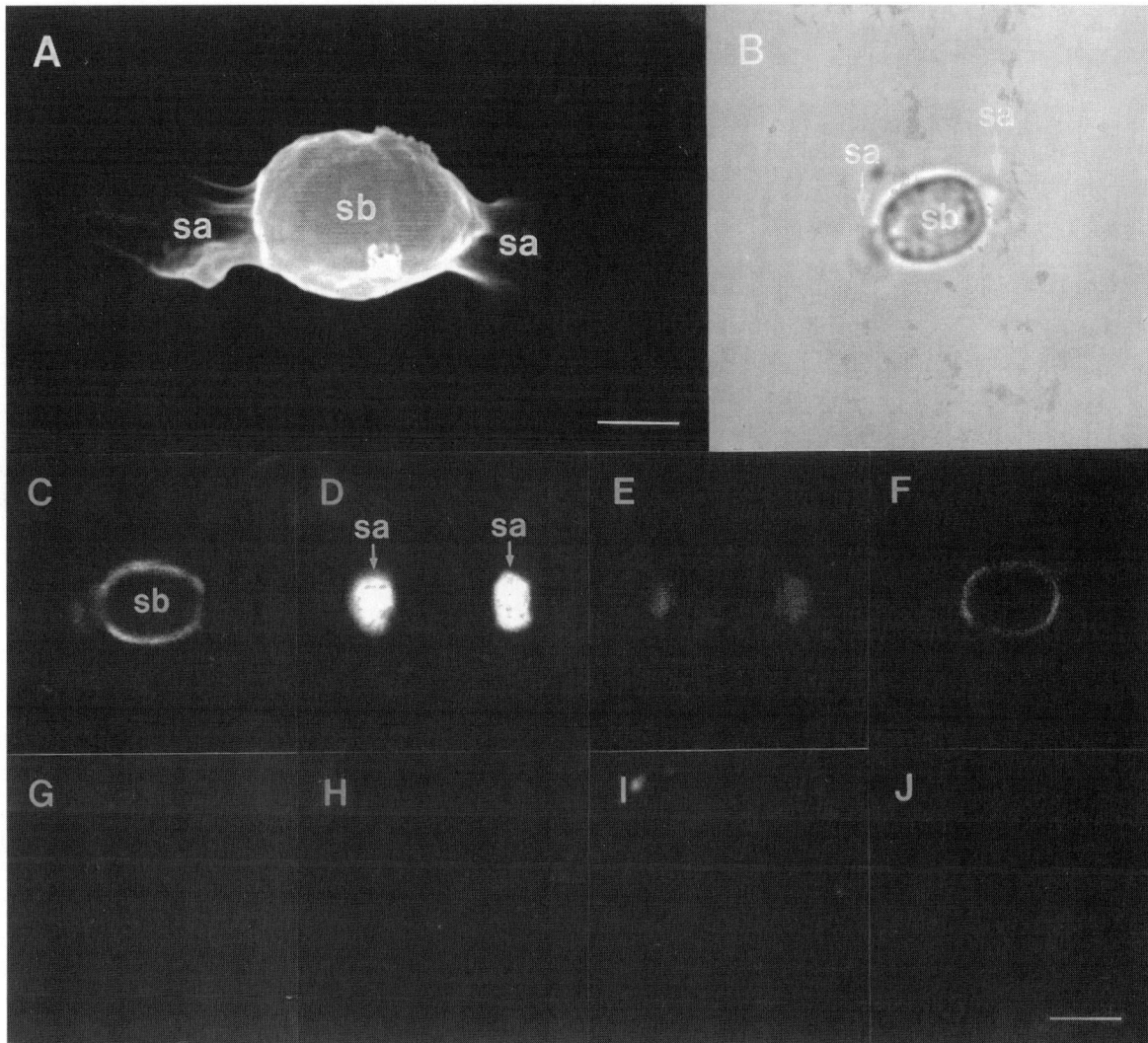

Figure 3. Released spermatia of *Aglaothamnion oosumiense* (sa, spermatial appendage; sb, spermatia body; scale bars: A–B, 1 μm, C–J, 2 μm). A. Scanning electron micrograph showing ovoid spermatia body and two spermatial appendages at each distal end; B. Differential interference (DIC) micrograph; C. Fluorescence micrograph of a FITC-ConA labelled spermatium. The labelling occurred only on the body; D. FITC-SBA bound only on spermatial appendages; E. FITC-DBA labelling on spermatial appendages. The labelling is much dimmer than with SBA; F. FITC-WGA labelling mainly occurs at the distal ends of the spermatium body; G. Heat denatured FITC-ConA labelling on spermatium. No labelling was observed; H. FITC-SBA labelling after pre-incubation of spermatia with *N*-acetyl-galactosamine; I. Heat denatured FITC-DBA labelling on spermatium; J. FITC-WGA labelling after pre-incubation of spermatia with *N*-acetyl-glucosamine.

hair cells developed rarely in *Aglaothamnion oosumiense*, no significant preference of spermatial binding toward either trichogynes or hair cells was observed.

We reported a similar gamete recognition mechanism based on carbohydrate-receptor system in a related genus, *Antithamnion* (Kim & Fritz, 1993a, b; Kim et al., 1996). In antithamnioid species, however, different carbohydrates, D-mannose and L-fucose, were involved in gamete recognition. We observed the commonality of a D-mannose carbohydrate-receptor system operating in two different *Antithamnion* species (Kim et al., 1996). Magruder (1984) reported that spermatia of *Aglaothamnion* spp. bound to the trichogynes of closely related species. The carbohydrate-receptor based gamete recognition in red algae, therefore, may work on more broad (i.e., generic) level rather than being species specific.

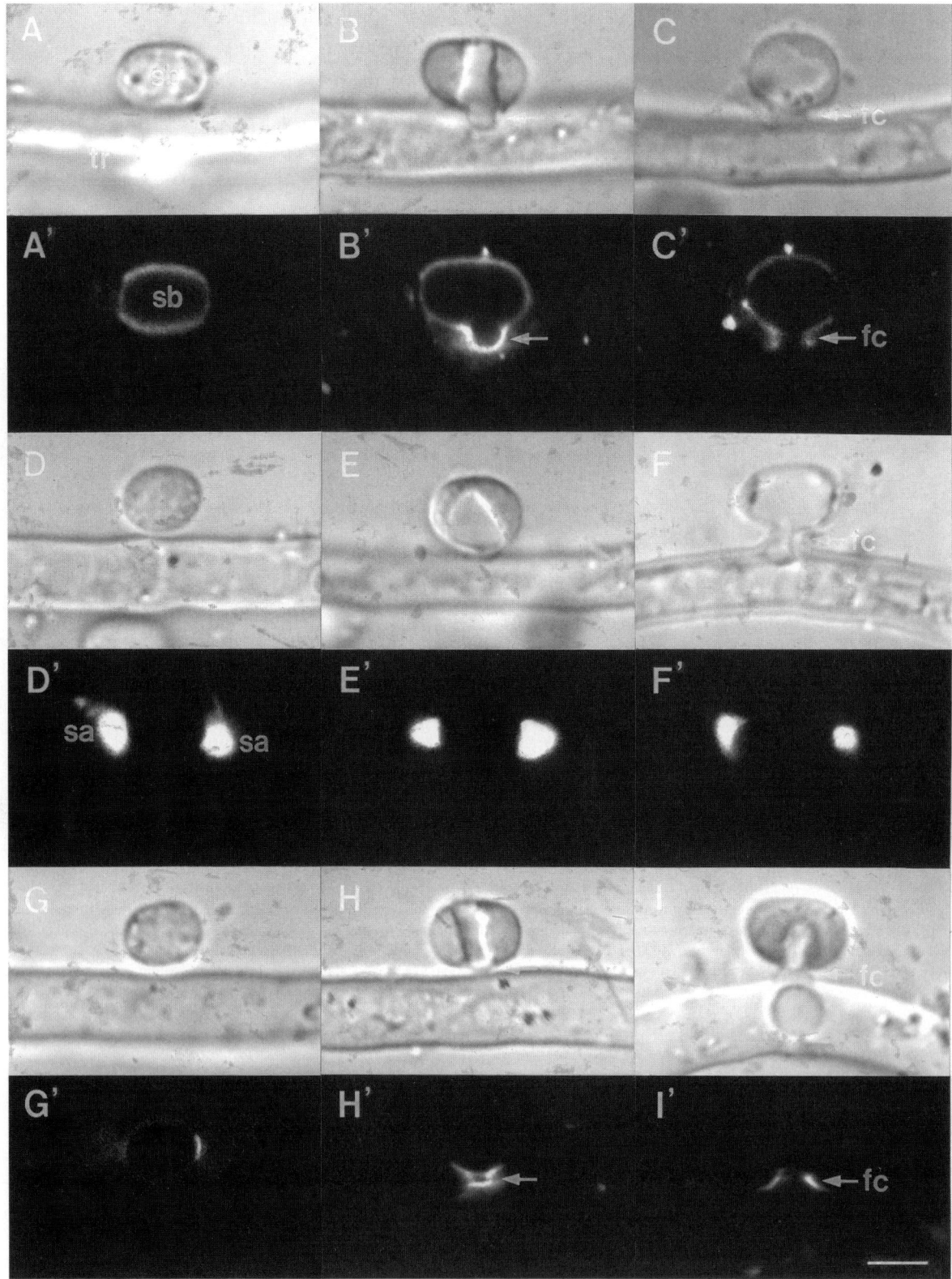

Figure 4.

Table 2. Inhibition of spermatial attachment to trichogynes by lectins ($\times$ 25 μg ml^{-1}) and carbohydrates (10 mM) in *Aglaothamnion oosumiense* (mean $\pm$ standard deviation, t-test, $n = 20$, $p<0.01$).

Treatment	Spermatial attachment	
	% of trichogynes with attached spermatia	Average number of attached spermatia per trichogyne
Sea water	96.9 $\pm$ 3.1	6.6
Preincubation of spermatia:		
Soybean Agglutinin (SBA)	54.6 $\pm$ 3.4	2.1
Heat denatured SBA	79.5 $\pm$ 4.2	5.2
SBA + D-galactosamine	80.1 $\pm$ 2.3	5.3
Concanavalin A (ConA)	77.9 $\pm$ 4.8	3.1
ConA + D-glucose	92.3 $\pm$ 3.7	6.0
ConA + SBA	30.8 $\pm$ 4.6	1.3
Heat denatured ConA + SBA	74.1 $\pm$ 3.1	4.8
ConA + SBA + D-glucose + D-galactosamine	81.9 $\pm$ 4.9	5.3
Dolichos biflorus Agglutinin (DBA)	93.2 $\pm$ 3.8	5.6
ConA+DBA	81.4 $\pm$ 3.5	5.0
Peanut Agglutinin (PNA)	92.6 $\pm$ 4.7	5.5
Wheat germ Agglutinin (WGA)	90.1 $\pm$ 2.4	4.7
Ulex europaeus Agglutinin I (UEA I)	91.5 $\pm$ 3.5	5.1
Ricinus communis Agglutinin I (RCA I)	96.2 $\pm$ 3.4	5.3
Preincubation of trichogyne:		
N-acetyl-D-galactosamine	76.4 $\pm$ 3.3	3.0
D-glucose	85.1 $\pm$ 2.7	5.2
D-glucose + *N*-acetyl-D-galactosamine	67.3 $\pm$ 2.1	2.4
D-mannose + *N*-acetyl-D-galactosamine	80.5 $\pm$ 5.3	4.6
D-mannose	98.1 $\pm$ 1.7	6.4

Cell-cell recognition mechanisms involving surface carbohydrates during fertilization have been reported for a few species of green and brown algae (Bolwell et al., 1980; Adair, 1985; Callow, 1985; Samson et al., 1987; Jones et al., 1988, 1990; Green et al., 1990; Stafford et al., 1992; van den Ende et al., 1992; Schmid, 1993). The glycoproteins and agglutinins involved in gamete binding have been isolated and purified for several species of *Chlamydomonas* (Musgrave, et al., 1981; Adair et al., 1982) and *Ectocarpus* (Schmid et al., 1994).

In red algae, many other cases of species specific cell recognition processes have been reported: parasite infection (Goff, 1982), secondary pit-connection

Figure 4. DIC and fluorescence micrograph of spermatia and trichogynes during the fertilization process of *Aglaothamnion oosumiense* (fc, fertilization canal; sa, spermatial appendages; sb, spermatial body; tr, trichogyne; scale bar: 2 μm). A–C. DIC micrograph of spermatia and trichogynes after 10 (A), 60 (B) and 120 min (C) of spermatia binding to trichogynes; A$'$–C$'$. Fluorescence micrograph of the same cells as in A-C with FITC-ConA labelling. Labelling first observed only on the spermatium body (A$'$), accumulates at the contacting area with the trichogyne (B$'$) and fertilization canal (C$'$); D–F. DIC micrograph of spermatia and trichogynes after 10 (D), 60 (E) and 120 min (F) of spermatia binding to trichogynes; D$'$–F$'$. Fluorescence micrograph of the same cells as in D-F with FITC-SBA labelling. The SBA receptors were observed only on spermatial appendages throughout the fertilization process; G–I. DIC micrograph of spermatia and trichogynes after 10 (G), 60 (H) and 120 min (I) of spermatia binding to trichogynes; G$'$–I$'$. Fluorescence micrograph of the same cells as in G-I with FITC-WGA labelling. The WGA receptors were first observed at distal ends of spermatia body (G$'$) and moved to the contacting area with trichogynes (H$'$) and later to the fertilization canal (I$'$).

formation (Goff & Coleman, 1985) and the wound-healing response (Waaland, 1990; Kim & Fritz, 1993c). Little is known, however, about the means whereby cell specificity and recognition is determined. To date, only one satisfactory explanation for cell-cell signalling in red algae has been reported, i.e. that for the wound-healing response (Waaland, 1990). A signal glycoprotein with D-mannose residues has been identified in the wound-healing process of *Griffithsia pacifica* Kylin (Watson & Waaland, 1983, 1986; Kim et al., 1995), *Antithamnion sparsum* Tokida (Kim & Fritz, 1993c) and *Antithamnion nipponicum* Yamada and Inagaki (Kim et al., 1995). It is noteworthy that glycoconjugates with the same sugar moiety (D-mannose) are involved in both wound-healing and gamete recognition in *Antithamnion* spp.

Our studies utilizing lectins and their complementary carbohydrates have produced indirect evidence that gamete recognition in filamentous red algae is based upon the interaction between glycoconjugates and their complementary receptors on the surfaces of spermatia and trichogynes. Further work is required to isolate and characterize the individual molecules of this gamete recognition mechanism. Such findings will be relevant to many interesting cell-cell interactions in red algae.

Acknowledgements

The authors extend sincere thanks to O.-K. Chah and Drs I. K. Lee and S. M. Boo for their careful review and very useful comments. This work has been partially supported by Ministry of Education, BSRI-97-4416 to G. H. Kim and Korean Science & Engeneering Foundation, KOSEF 951-0506-007-1 to G. H. Kim. Research Institute of Natural Sciences, Kongju National University.

References

Adair, W. S., B. C. Monk, R. Cohen, C. Hwang & W. Goodenough, 1982. Sexual agglutinins from the *Chlamydomonas* flagellar membrane. J. Biol. Chem. 257: 4593–4602.

Adair, J., 1985. Characterization of *Chlamydomonas* sexual agglutinins. J. Cell. Sci. (Suppl.) 2: 233–260.

Bolwell, G. P., J. A. Callow, M. E. Callow & L.V. Evans, 1979. Fertilization in brown algae. II. Evidence for lectin sensitive complementary receptors involved in gamete receptors involved in gamete recognition in *Fucus serratus*. J. Cell Sci. 36: 19–30.

Bolwell, G. P., J. A. Callow & L. V. Evans, 1980. Fertilization in brown algae. III. Preliminary characterization of putative gamete receptors from eggs and sperm of *Fucus serratus*. J. Cell Sci. 43: 209–224.

Callow, J. A., 1985. Sexual recognition and fertilization in brown algae. J. Cell. Sci. (Suppl.) 2: 219–232.

Catt, J. W., H. I. M. V. Vithanage, J. A. Callow, M. E. Callow & L. V. Evans, 1983. Fertilization in brown algae. V. Further investigation of lectins as surface probes. Exp. Cell Res. 147: 127–133.

Chrispeels, M. J. & N. V. Raikhel, 1991. Lectin, lectin genes, and their role in plant defense. Plant Cell 3:1–9.

Goff, L. J., 1982. The biology of parasitic algae. In Round, F .E. & D. J. Chapman (eds), Progress in Phycological Research I. Elsevier Biomedical Press, Amsterdam: 289–370.

Goff, L. J. & A. W. Coleman, 1985. The role of secondary pit connections in red algal parasitism. J. Phycol. 21: 483–508.

Green, J. R., J. L. Jones, C. J. Stafford & J. A. Callow, 1990. Fertilization in *Fucus*: exploring the gamete cell surface with monoclonal antibodies. In Dale, B. (ed.), Mechanism of fertilization. NATO ASI series, Vol. H. 45, Springer-Verlag, Berlin, Heidelberg: 189–202.

Guillard, R. R. L. & J. H. Ryther, 1962. Studies of marine planktonic diatoms. I. *Cyclotella nana* Hustedt and *Detonula confervacea* (Cleve). Gran. Can. J. Microbiol. 8: 229–239.

Jones, J. L., J. A. Callow & J. R. Green, 1988. Monoclonal antibodies to sperm surface antigens of the brown alga *Fucus serratus* exhibit region, gamete, species and genus preferential binding. Planta 176: 298–306.

Jones, J. L., J. A. Callow & J. R. Green, 1990. The molecular nature of *Fucus serratus* sperm surface antigens recognised by monoclonal antibodies FS1 to FS12. Planta 182: 64–71.

Karlsson, K. A., 1991. Glycobiology: a growing field for drug design. TIBS. 12: 265–272.

Kim, G. H., 1997. Gamete recognition and signal transduction during fertilization in Red Algae. Algae (Kor. J. Phycol.) 12: 263–268.

Kim, G. H. & L. Fritz, 1993a. Gamete recognition during fertilization in a red alga *Antithamnion nipponicum*. Protoplasma 174: 69–73.

Kim, G. H. & L. Fritz, 1993b. Ultrastructure and cytochemistry of early spermatangial development in *Antithamnion nipponicum* (Ceramiaceae, Rhodophyta). J. Phycol. 29: 797–805.

Kim, G. H. & L. Fritz, 1993c. A signal glycoprotein with D-mannosyl residues is involved in the wound-healing response of *Antithamnion sparsum* (Ceramiales, Rhodophyta). J. Phycol. 29: 85–90.

Kim, G. H., I. K. Lee & L. Fritz, 1995. The wound-healing responses of *Antithamnion nipponicum* and *Griffithsia pacifica* (Ceramiales, Rhodophyta) monitored by lectins. Phycol. Res. 43: 161–165.

Kim, G. H., I. K. Lee & L. Fritz, 1996. Cell-cell recognition during the fertilization in a red alga, *Antithamnion sparsum* (Ceramiaceae, Rhodophyta). Plant Cell Physiol. 37: 621–628.

Magruder, W., 1984. Specialized appendages on spermatia from the red alga *Aglaothamnion neglectum* (Ceramiales, Ceramiaceae) specially bind with trichogynes. J. Phycol. 20: 436–440.

Mine, I. & M. Tatewaki, 1994. Gamete surface and attachment during fertilization of *Palmaria* sp. (Palmariales, Rhodophyta). Jpn. J. Phycol. 42: 291–299.

Musgrave, A., E. van Eijk, R. Welscher, R. Broekman, P. Lens, W. Homan & H. van den Ende, 1981. Sexual agglutination factor from *Chlamydomonas eugametos*. Planta 153: 362–369.

Pickett-Heaps, J. D. & J West, 1998. Time-lapse video observations on sexual plasmogamy in the red alga *Bostrychia*. Eur. J. Phycol. 33:43–56.

Samson, M., F. M. Klis, W. L. Homan, P. van Egmond, A. Musgrave & H. van den Ende, 1987. Composition and properties of the sexual agglutinins of the flagellated green alga *Chlamydomonas eugametos*. Planta 170: 314–321.

Schmid, C., 1993. Cell-cell recognition during fertilization in *Ectocarpus siliculosus* (Phaeophyceae). Hydrobiologia 260/261: 437–443.

Schmid, C., N. Schroer & D. Muller, 1994. Female gamete membrane glycoproteins potentially involved in gamete recognition in *Ectocarpus siliculosus* (Phaeophyceae). Plant Sci. 102: 61–67.

Sharon, N. & H. Lis, 1989. Lectins as cell recognition molecules. Science 177: 949–959.

Stafford, C. J., J. A. Callow & J. R. Green, 1992. Isolation and characterization of plasma membranes from *Fucus serratus* eggs. Br. phycol. J. 27: 429–434.

van den Ende, H., M. L. van den Briel, R. Lingeman, P. van der Gulik & T. Munnik, 1992. Zygote formation in the homothallic green alga *Chlamydomonas monica* Strehlow. Planta 188: 551–558.

Waaland, S. D., 1990. Development. In Cole, K. M. & G. R. Sheath (eds), Biology of the Red Algae. Cambridge University Press, New York: 259–273.

Wassarman, P. M., 1987. The biology and chemistry of fertilization. Nature 235: 553–560.

Watson, B. A. & S. D. Waaland, 1983. Partial purification and characterization of a glycoprotein cell fusion hormone from *Griffithsia pacifica*, a red alga. Plant Physiol. 71: 327–332.

Watson, B. A. & S. D. Waaland, 1986. Further biochemical characterization of a cell fusion hormone from the red alga, *Griffithsia pacifica*. Plant Cell Physiol. 27: 1043–1050.

Wiese, L. & D. W. Shoemaker, 1970. On sexual agglutination and mating-type substances (gamones) in isogamous heterothallic *Chlamydomonas*. II. The effect of concanavalin A upon the mating-type reaction. Biol. Bull. 138: 88–95.

Morphology and reproduction of the adventive Mediterranean rhodophyte *Polysiphonia setacea*

Fabio Rindi[1], Michael D. Guiry[1] & Francesco Cinelli[2]

[1]*Department of Botany and Martin Ryan Marine Science Institute, National University of Ireland, Galway, Ireland*
[2]*Dipartimento di Scienze dell'Uomo e dell'Ambiente, Università di Pisa, Via A. Volta 6, 56126 Pisa, Italy*
Current address: Dipartimento di Scienze dell'Uomo e dell'Ambiente, Università di Pisa, Via A. Volta 6, 56126 Pisa, Italy
E-mail: fabio.rindi@discat.unipi.it

Key words: *Polysiphonia setacea*, Rhodophyta, culture, growth, biogeography, Mediterranean Sea

Abstract

The red alga *Polysiphonia setacea* Hollenberg (=*Womersleyella setacea* (Hollenberg) R. Norris), described originally from the Hawaiian Islands and later reported for other tropical localities, has recently become widespread in the Mediterranean. In several localities it forms a dense, almost monospecific turf but, despite its abundance, it appears to reproduce only vegetatively; neither sporangia nor sexual reproductive structures have thus far been found. In order to elucidate its life history, plants were cultured in a variety of conditions of temperature, daylength and photon irradiance, and the upper thermal limit was also determined. Isolates of *P. setacea* grew well in culture, but were more tufted and branched than wild plants. No reproduction by spores or gametes was observed and only an unusual form of vegetative regeneration was found. Some pericentral cells became darker and larger than the others and produced proliferations from which new plants arose. Plants grew best at 15 and 20 °C, and relatively poor growth was observed at 10 and 25 °C. At 20 °C growth was better in long days than in short days. The upper thermal limit was 28 °C, and plants were able to tolerate a temperature as low as 5 °C for 4 weeks without any damage. These observations show that *P. setacea* is well adapted to the environmental conditions of the western Mediterranean, and there is good agreement with the phenology of the species in the wild. Although this entity has been treated as an introduction from a tropical area, our results suggest that the Mediterranean entity would not be able to survive or grow in surface waters of tropical areas whence this species has been reported. However, the occurrence of thermal ecotypes in *P. setacea* is a possibility and further studies, based perhaps on molecular data, are necessary to assess the origin of the Mediterranean populations.

Introduction

Hollenberg (1968), in describing the red alga *Polysiphonia setacea* Hollenberg (Rhodomelaceae, Ceramiales), based the species on plants from Oahu, Hawaiian Islands, but also reported it from a number of other areas in the tropical Indian and Pacific Oceans, together with a single collection from the Atlantic. This alga was later recorded in other tropical localities by Oliveira-Filho & Cordeiro-Marino (1970), Egerod (1971), Schnetter & Bula-Meyer (1982) and Wynne (1993). Norris (1992) proposed the transfer of *P. setacea* to another rhodomelacean genus, *Womersleyella*. However, Abbott and Millar (pers. comm.) consider that the morphological features characterizing the type species of *Womersleyella*, *W. pacifica* Hollenberg, are not stable enough to justify its separation as a genus; *Womersleyella* may therefore be regarded as a heterotypic synonym of *Polysiphonia* (type species: *P. urceolata* (Lightfoot ex Dillwyn) Greville; syn. tax.: *P. stricta* (Dillwyn) Greville).

Recently Verlaque (1989) reported for the first time the occurrence of *P. setacea* in the Mediterranean Sea, and subsequently regarded it as an introduction from tropical seas (Verlaque, 1994). Later reports indicated that *P. setacea* was spreading rapidly in the Mediterranean (Rodriguez Prieto et al., 1993; Cormaci et al., 1994; Airoldi et al., 1995; Rindi & Cinelli,

Table 1. Temperature and daylength conditions in culture

Temperature (°C)	Daylength (light: dark, h)
10	8:16, 12:12, 16:8
15	8:16, 11:13, 12:12, 13:11, 16:8
20	8:16, 11:13, 12:12, 13:11, 16:8
25	12:12, 16:8

1995; Athanasiadis, 1997). In many Mediterranean localities it is exceedingly abundant, forming a thick, persistent turf that completely covers deep, sublittoral, rocky substrata (Airoldi et al., 1995; Athanasiadis, 1997). Despite its abundance, Mediterranean populations seem to show only vegetative reproduction; neither sexual structures nor sporangia have been recorded, even with weekly collections (Airoldi et al., 1995). It should be noted that tetrasporangia are the only reproductive structure thus far reported for *P. setacea* (Hollenberg, 1968); spermatangial and carpogonial branches, and carposporangia remain unknown worldwide.

In order to elucidate the life history and the responses of Mediterranean *P. setacea* to environmental factors, a culture study was initiated; the results are reported here.

Materials and methods

Specimens of *Polysiphonia setacea* were collected at Calafuria (Livorno, north-western Italy; 43°30′ N, 10°20′ E) on 14 July 1996 by SCUBA, at depths of 12–14 m. In this locality *P. setacea* is a dominant species on deep sublittoral rocky substrata, forming a dense turf between 10 and 30 m (Airoldi et al., 1995). Collections were made at several sites, at least 20 m far from each other. The material was transported to Ireland in a cooler and placed at 15 °C, 16:8 h light:dark (LD) at low photon irradiances (2–4 μmol m^{-2}s^{-1}) for a month. A stock culture, maintained at 15 °C, 16:8 h LD, 10-20 μmol m^{-2}s^{-1}, was obtained by excision of a vegetative tip from each separate collection. A 25% Von Stosch enriched seawater medium, as modified by Guiry & Cunningham (1984), was used.

Plants, obtained from stock cultures by tip excision, were grown in a variety of temperatures and photoperiodic regimes (Table 1), with photon irradiances varying from 2–50 μmol m^{-2}s^{-1}. For each combination at least 4 plants were used; in each experiment, these were from the same stock culture. Plastic dishes containing approximately 30 ml of medium were used; a plant was placed in each dish. Medium was replaced every week in order to avoid depletion of nutrients. Growth of the plants was followed for up to 7 months.

In order to quantify the growth of *P. setacea*, plants grown under different combinations of temperature/photon irradiance and temperature/daylength were dried (after 12 weeks in culture in the first case, 14 weeks in the second one); their surface area (estimated in mm^2 by a scanner and an image analysis system) was used as an approximate estimate of the size of the plants. Surface area data were analysed by 2-way ANOVAs. In a first analysis temperature (4 levels: 10 °C, 15 °C, 20 °C and 25 °C) and photon irradiance (4 levels: 3, 10, 20 and 30 μmol m^{-2}s^{-1}) were fixed orthogonal factors (plants were cultured in a regime 16:8 h LD); in a second one the fixed factors were temperature (2 levels: 15 °C and 20 °C) and daylength (5 levels: 8:16 h LD, 11:13 h LD, 12:12 h LD, 13:11 h LD and 16:8 h LD). The photon irradiances were 30 μmol m^{-2}s^{-1} for 8:16 h LD, 15 μmol m^{-2}s^{-1} for 16:8 h LD and correspondingly compensated for the other daylengths. For each combination of factors there were 3 replicated measurements. The assumption of homogeneity of variances was verified using Cochran's test (Winer, 1971); Student-Newman-Keuls (SNK) tests were used for *a posteriori* multiple comparisons of means. For all statistical analysis a probability level of 0.05 was regarded as significant; the analysis were performed using the STATISTICA package.

Upper thermal limits were established by incubating *P. setacea* in water baths ($\pm$ 0.2 °C), at 26, 28, 30 and 32 °C in a long-day (16:8 h LD) regime at 6–12 μmol m^{-2}s^{-1}. Fully-grown plants (2-cm-wide tufts) were used. Plants were incubated in glass dishes containing approximately 250 ml of medium (3 plants in each dish). At each temperature, 6 plants were incubated for 2 weeks and 6 plants for 4 weeks. Temperature was gradually raised to the experimental values in steps of no more than 1 °C per day; after the incubation period, temperature was similarly gradually brought back to 20 °C and the health of plants noted. Damage was defined as an extensive loss of pigmentation. If plants appeared severely damaged (more than 50% of the plant tissue was bleached), they were cultured for a further 6 weeks at 20 °C, to see if recovery or regeneration would occur. During incubation, the medium was changed every 2 weeks, using medium previously preheated to the experimental temperature. A similar study could not be carried out to determine the lower thermal limit, but plants were cultured experimentally for 4 weeks at 5 °C, 8:16 LD, 10–12 μmol photons m^{-2}s^{-1}.

Representative specimens of plants used for cultures are deposited in the Phycological Herbarium, Department of Botany, National University of Ireland, Galway (GALW).

Additionally, several collections (herbarium sheets, permanent slides and liquid-preserved specimens) of *P. setacea* from the tropics and from the Mediterranean were examined in order to evaluate morphological variation of field material. These are as follows:

Hawaii Islands: (**1**) ABBOTT 20775. Akiki Bay, Maui Island, 1/8/92. Intertidal, on eroded coral; det. R.E. Norris. (**2**) BPBMH587433 (GJH 63-132). (**3**) BPBMH 587401 (MSD 14696, Isotype), 14/11/1954. On rocks in algal turf in a high tide pool near Koko Head parking area, Eastern Oahu; det. G. J. Hollenberg. (**4**) BPBMH 587392 (MSD 22301.2). (**5**) BPBMH 387446 (MSD 22301.1). Tidepools, northeast bench of Hanauma Bay, Oahu; det. G. J. Hollenberg. **6**) A collection by G. F. Papenfuss, Hanauma Bay, Oahu, 30/3/1941. **Indonesia:** Rijksherbarium, Leiden, L. 10210, No. 3. Behind Gunung Monjet, Maura-Arau, Mid Sumatra. On basalt rock mixed with sand, at low water mark; det. G.J. Hollenberg. **Mediterranean Sea:** (**1**) Comino (Malta), 12/9/94, – 33 m; det. M. Cormaci & G. Furnari. (**2**) Lampedusa Island (Sicily Channel), 16/3/93, – 40 m; det. M. Cormaci & G. Furnari. (**3**) Pantelleria Island (Sicily Channel), 23/6/96, – 5 m; det. M. Cormaci & G. Furnari. (**4**) Isola delle Femmine, Sicily, 16/6/95, – 17 m; det. F. Rindi. (**5**) Galeria Bay, Corsica, 18/9/89, – 15 m; det. M. Verlaque. (**6**) Portquerolles, Var, France, April 1987; det. M. Verlaque. (**7**) Giens Bay, Var, France, October 1987, – 12 m; det. M. Verlaque. (**8**) Giens Bay, Var, France, 19/11/87, – 12 m; det. M. Verlaque. Furthermore, several collections from Calafuria (years 1994–1996, between −4 and −30 m); det. L. Airoldi & F. Rindi.

Results

Cultures

Polysiphonia setacea grew well at 15 and 20 °C, even at very low photon irradiances. Cultured material basically showed the morphology of field plants, as described by several authors (Hollenberg, 1968; Verlaque, 1989; Athanasiadis, 1997), but it appeared to differ in some features. Wild plants are usually branched very little or not at all, whereas cultured plants bore numerous cicatrigenous branches, produced by division of scar cells. In plants maintained in long days or at high photon irradiances, a branch was usually produced every 2–4 segments. Cultured material showed a tufted habit and usually the distinction between creeping axes and erect branches was not as clear as in material collected in the wild. Trichoblasts occurred in all plants, although they seemed to be less abundant at lower photon irradiances.

Under all culture conditions, plants showed rhizoidal outgrowths which have never been reported in the wild; they were more abundant at 10 and 15 °C than at 20 and 25 °C. Every outgrowth was produced as an outgrowth of the wall of a pericentral cell (Figure 1), that later became divided from the rest of the cell (Figure 2). A single pericentral cell could produce 2–3 such outgrowths; often adjacent cells of different segments formed outgrowths, so groups of 5–6 contiguous rhizoidal outgrowths were commonly observed (Figure 3).

No gametangia or sporangia were apparent. Variations in the strength of the medium, variations in salinity, cultivation in pure red light, and several types of physical stress were ineffective in stimulating the production of gametangia or sporangia.

The only form of reproduction was an unusual kind of vegetative regeneration. Some pericentral cells (or groups of pericentral cells in contiguous segments) became darker than the rest of the thallus (Figure 4) and occasionally produced multicellular rhizoids (Figure 5). In some cases these cells appeared swollen and larger than neighbouring ones (Figure 6). They usually started to divide, producing proliferations (Figures 7–9) from which new erect plants eventually arose (Figures 10, 11). This phenomenon was particularly evident in damaged plants, where old parts were almost completely bleached (in contrast with the pigmented, regenerating plants); but it could be observed under virtually all available culture conditions.

The ANOVA performed on the surface area data of plants grown in different combinations of temperature and photon irradiance revealed a strong interaction between these factors (Table 2). In long days (16:8 h LD) at photon irradiances between 10 and 30 μmol m^{-2}s^{-1} *P. setacea* grew best at 15 °C and 20 °C (Figure 12); the SNK test showed that the size

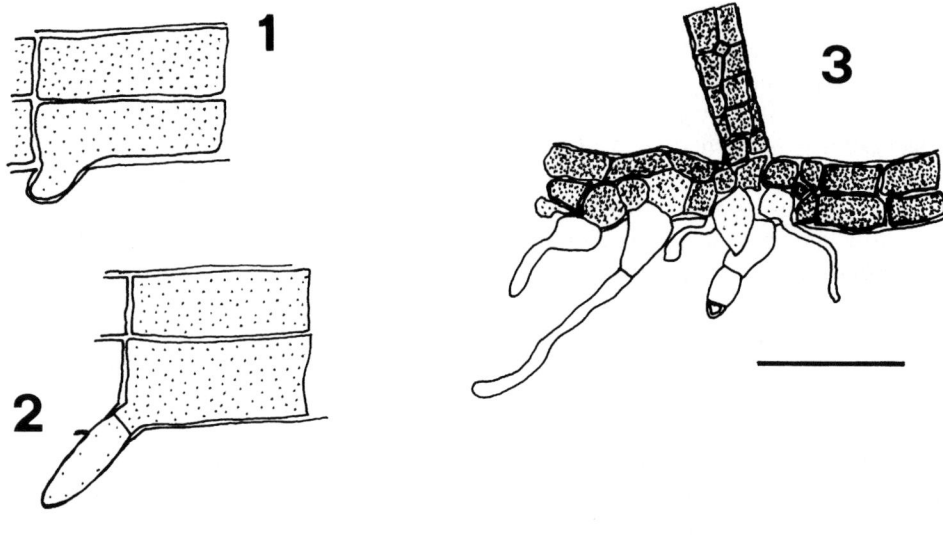

Figures 1–3. Development of rhizoidal outgrowths from pericentral cells in *Polysiphonia setacea*. Figures 1–2. Scale bars = 50 μm. Figure 3. Scale bar = 100 μm.

Table 2. Results of the ANOVA on the effect of temperature and photon irradiance on the size of plants of *P. setacea*

	df	MS	F	p
Temperature	3	91931.41	942.0811	<0.00001
Photon irradiance	3	8078.91	82.7898	<0.00001
Temper. × photon irrad.	9	4068.02	41.6877	<0.00001
Error	32	97.58334		

Table 3. Results of the ANOVA on the effect of temperature and daylength on the size of plants of *P. setacea*

	df	MS	F	p
Temperature	1	364762.1	348.4434	<0.00001
Daylength	4	1838.1	1.7559	0.177443
Temper. × daylength	4	89925.1	85.9021	<0.00001
Error	20	1046.833		

of plants cultured under these conditions was significantly larger than in other cases. Plants grown at 10 °C and 25 °C were not significantly different and they were significantly smaller than at 15 °C and 20 °C; at 10 °C and 25 °C no significant differences occurred also between different photon irradiances. Plants grown with 3 μmol m^{-2}s^{-1} at 15 °C and 20 °C showed an intermediate size, being significantly larger than plants at 10 °C and 25 °C but significantly smaller than plants at 15 °C and 20 °C in higher photon irradiances.

A significant interaction between temperature and daylength was also found (Table 3). At 15°C good growth was observed under all daylengths tested (Fig-

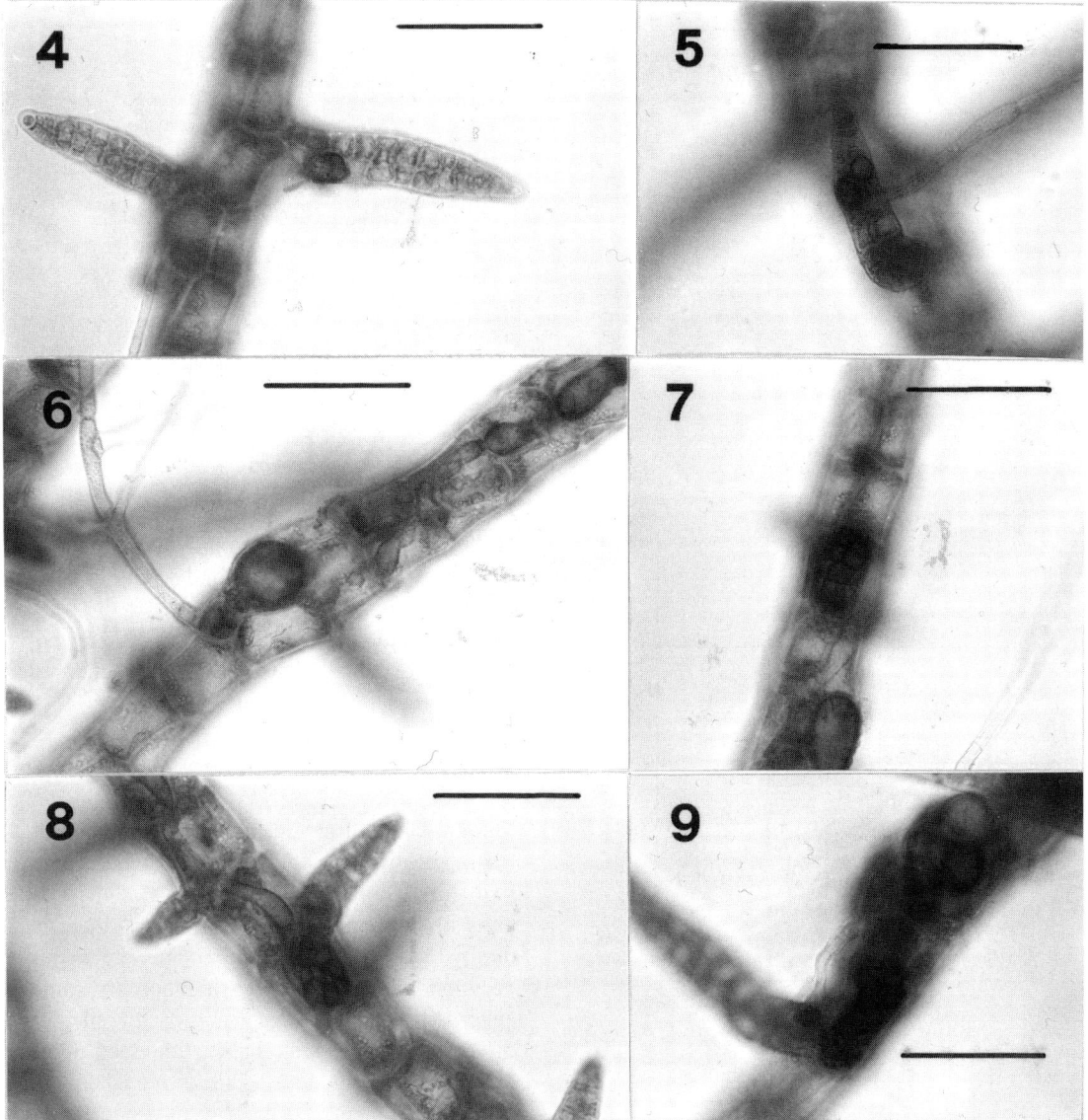

Figures 4–9. Development of new plants of *Polysiphonia setacea* by vegetative regeneration from pericentral cells. Scale bars = 100 μm. Figures 4–5. Darkening of pericentral cells and rhizoid initiation . Figure 6. A large, swollen pericentral cell, with pluricellular rhizoid. Figures 7–8. Division of enlarged pericentral cells and initiation of proliferations. Figure 9. A proliferation at a later stage.

ure 13); the SNK test showed that the size of plants was significantly larger in short days (8:16 LD and 11:13 LD) than for the other daylengths. The opposite was found at 20 °C: in this case plants grown in short days were significantly smaller. For long days (16:8 LD and 13:11 LD) and neutral days (12:12 LD) no significant differences were detected between 15 °C and 20 °C. Not all such daylengths could be tested for the other temperatures, but at both 10 and 25 °C (see Table 1 for the daylengths used) no clear differences were detectable between different photoperiods.

Results of the upper thermal tolerance experiments are given in Table 4. At 25 and 26 °C plants showed some bleached parts; but after 2 and 4 weeks damage was not extensive and new branches were continuously produced. After 2 weeks at 28 °C, plants appeared severely damaged, most of the thallus being bleached. Some cells did not lose pigmentation, however, and when returned to 20 °C were able to re-

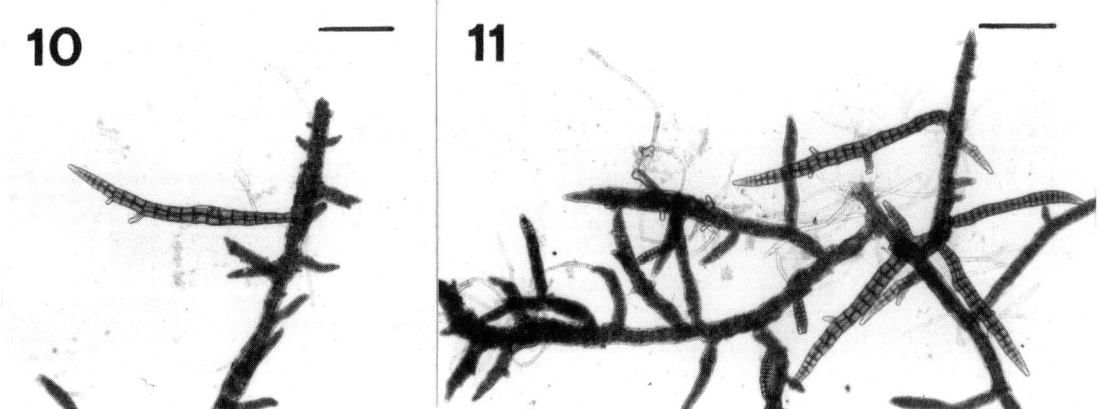

Figures 10–11. Vegetative regeneration by cultured *Polysiphonia setacea.* Scale bars = 100 μm.

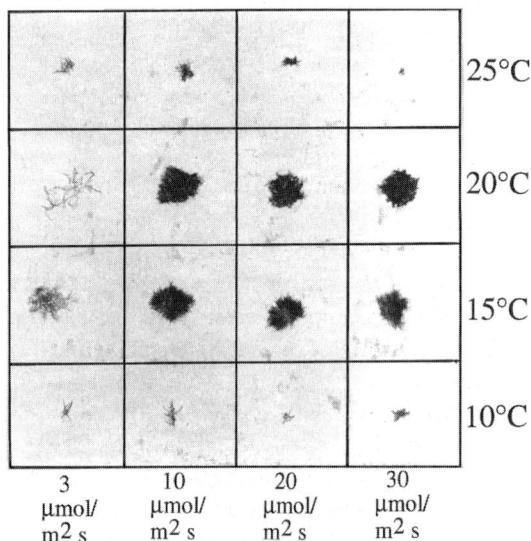

Figure 12. Growth of *Polysiphonia setacea* after 12 weeks at the temperatures and photon irradiances indicated.

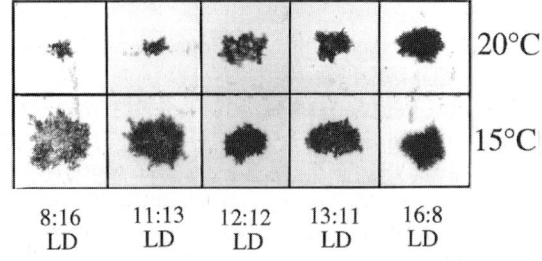

Figure 13. Growth of *Polysiphonia setacea* after 14 weeks at the temperatures and daylengths indicated.

generate new plants as described above. After 4 weeks at 28 °C plants were irreversibly damaged and no regeneration was found. At 30 and 32 °C plants were not able to survive for 2 weeks.

Plants cultured at 5 °C, 8:16 h LD did not show any signs of growth, but survived without any signs of damage for a period as long as 4 weeks.

Herbarium specimens

Variation was observed in some quantitative features (diameter of prostrate axes, length/width ratio of the segments), both in different plants and in different parts of the same plant. For features that are commonly regarded as most important in the taxonomy of *Polysiphonia* and other polysiphonous rhodomelacean genera (number of pericentral cells; complete absence of cortication; prostrate habit with erect, unbranched branches; cicatrigenous origin of branches; scar-cells occurring on every segment in prostrate axes, absent in some segments in the erect branches; abundant, rapidly deciduous trichoblasts; rhizoids with multicellular tips, cut off from the distal end of pericentral cells), the morphology of this species appeared fairly stable and no actual differences were found between Mediterranean and tropical populations of *P. setacea*. The only apparent difference was in the shape of trichoblasts. In Mediterranean specimens trichoblasts, when present, were well developed, up to 900 μm long, with 3–4 dichotomies. In tropical plants trichoblasts were rarely observed; if present, they were coarse, small, with 2–3 dichotomies, no more than 300 μm long.

Discussion

It was not possible to elucidate the sequence of morphological phases of Mediterranean *P. setacea* because of cultured plants' inability to reproduce sexually or by spores. Production of well-defined asexual propagules, reported for other species of *Polysiphonia* (Kapraun, 1977; Koch, 1986; Maggs & Hommersand, 1993), was not observed either. As in the wild, Mediterranean plants only reproduce vegetatively in culture. It is not easy to establish if the lack of reproduction is due to a specific requirement for a very narrow range of one or several environmental factors or to an actual loss of sexual reproduction. In culture, species of *Polysiphonia* and other rhodomelacean genera with a low number of pericentral cells complete their life history relatively rapidly (e.g., Edwards, 1968; Rueness, 1971; Kapraun, 1978; Kudo & Masuda, 1986; Aponte & Ballantine, 1987), producing abundant sporangia, procarps and spermatangial branches. It is therefore uncertain whether or not sexual or sporangial reproduction actually play a rôle in the spread of this alga in the Mediterranean. In any case, the occurrence of an unusual kind of vegetative regeneration (observed under virtually all culture conditions tested) suggests that mechanisms different from the simple vegetative growth may have contributed to the wide and rapid spread of *P. setacea* in recent years.

Cultural observations agree with the phenology shown by the species in the field and show a good adaptation of *P. setacea* to the environmental conditions in the western Mediterranean. Here surface water temperature ranges approximately between 12 °C in March and 25 °C in August (U.S. Navy, 1981; Lüning, 1990) and, for most of the year, it is 15–20 °C. Showing the best growth in this range, *P. setacea* is not limited by temperature in vegetative development. In this thermal interval *P. setacea* was able to grow under all the photoperiodic conditions tested, although a combination of 20 °C and short daylengths seemed to be the least favourable. These conditions occur in early autumn and, in fact, this is the period when *P. setacea* turfs appear least dense and abundant, although the supposed influence of daylength on field phenology must be treated with caution. Occupying mainly deep subtidal habitats, *P. setacea* experiences effective daylengths throughout the year that can be much shorter than those on land or in shallow water. This is true particularly for the rocky shore where the plants of this study were collected; in that locality the water is usually more turbid than in many other parts of the Mediterranean and sediment deposition is an important factor influencing composition and structure of the benthic algal communities (Airoldi et al., 1996). Probably after periods of rough seas (especially near its lower limit of distribution, about 30 m) this species must survive for some time with daylengths and light levels that are too low for growth, and plants may often be in virtual darkness for considerable periods. In this regard, it is noteworthy that *P. setacea* was able to grow at very low photon irradiances; independent of daylength, growth was observed at photon irradiances as low as 3–4 μmol m^{-2}s^{-1}.

Our results have some interesting biogeographical implications. The importance of temperature as a major factor determining the geographical boundaries of seaweeds has been stressed by many authors (for a comprehensive review, see Breeman, 1988). In recent years experimental work has revealed that temperature responses in culture can explain the geographical distribution of many species of benthic algae (van den Hoek, 1982; Breeman 1988). It is generally evident that tropical populations of marine algae grow best between 25 and 30 °C and survive undamaged for considerable periods at 30 °C. They are usually able to tolerate temperatures as high as 32–33 °C, while their lower survival limit is usually close to 10 °C (McLachlan & Bird, 1984; Cambridge et al., 1987, 1991; Pakker et al., 1994, 1995). Being unable to survive permanently at 28 °C and with a lower limit <5 °C, the Mediterranean plants of *P. setacea* exhibit the thermal characteristics of a temperate seaweed rather than a tropical one. Published records report *P. setacea* for several areas of the tropical Indopacific: Hawaii, Johnston Is., American Samoa, Society Is., Fiji Is., Marshall Is., Caroline Is., Philippines (Hollenberg, 1968), Thailand (Egerod, 1971), Indonesia and the Maldive Is. (Wynne, 1993, Silva et al., 1996, p. 556) and Colombia (Schnetter & Bula-Meyer, 1982).

Table 4. Tolerance to high temperatures in cultures of Mediterranean *Polysiphonia setacea*

Time (weeks)	Temperature (°C)				
	25	26	28	30	32
2 weeks	++	++	+	−	−
4 weeks	++	++	−	−	−

++ = no damage (healthy plants or no more than 50% of the plant bleached). + = reversible damage (more than 50% of the plant bleached, but ability to recover if brought back to favourable conditions). − = irreversible damage: plants died.

Two records exist for the tropical Atlantic: Bermuda (Oliveira-Filho & Cordeiro-Marino, 1970) and Costa Rica (Hollenberg, 1968). The other record reported by this author for the Atlantic Ocean is incorrect, as El Salvador does not border the Atlantic. An examination of surface-water isotherms (U.S. Navy, 1981) indicates that in many of these zones surface temperature for most of the year is very close to (or reaches) 28 °C, the upper limit for Mediterranean *P. setacea*. It seems very unlikely that such a population can survive or grow actively in most parts of the tropics where this species has been reported. It should also be noted that, in contrast to its subtidal Mediterranean habitat, *P. setacea* has been collected in intertidal habitats and in intertidal pools in the tropics. For tropical species tolerance to high temperatures is reported to be greater in intertidal species than subtidal ones (Pakker et al., 1995). Accordingly, tropical *P. setacea* might be expected to have much higher upper thermal limits than those of Mediterranean plants.

Morphological investigations of herbarium and preserved material did not reveal clear differences between Mediterranean and tropical specimens; the only apparent difference was in the morphology of trichoblasts. However, these structures are shed easily, and observations on cultured tropical plants were not possible. Further observations are therefore necessary to establish the taxonomic value of this feature. At this stage, it seems probable that Mediterranean and tropical populations do not represent a single taxon. Although thermal tolerances of tropical plants could not be examined, the occurrence of thermal ecotypes should not be dismissed as a possibility. Evidence for the existence of thermal ecotypes in seaweeds involves differences in optimum temperatures for growth and reproduction, differences in thermal ranges over which growth and reproduction proceed and differences in tolerance to high or low temperatures, or both (Breeman, 1988). Thermal ecotypes have already been found in species of seaweeds with a tropical-warm temperate distribution (Pakker & Breeman, 1996). Because thermal tolerance is a stable feature in seaweed populations and ecotypic differentiation is regarded as a sign of divergence on a wide temporal scale (Novaczek et al., 1989, 1990; Breeman & Pakker, 1994; van Oppen et al., 1996), our observations appear to be contrary to the widespread belief that in the Mediterranean *P. setacea* is a recent introduction from the tropics. Considering its widespread abundance, it may be that *P. setacea* occurred in the Mediterranean long before it was first noted (Verlaque, 1989).

Its occurrence in relatively deep subtidal habitats and its superficial resemblance to other species occurring in the Mediterranean, such as *Lophosiphonia scopulorum* (Harvey) Womersley, suggest that it may have been overlooked for some time. It is equally possible that, if *P. setacea* is really an introduction to the Mediterranean, it was introduced from a temperate zone rather than the tropics. Its thermal responses indicate that this alga has considerable potential for expansion to other temperate areas, including the north-eastern Atlantic.

Our results highlight the necessity of further studies in order to understand the origin of the Mediterranean strain (and indicate the importance of similar observations for other Mediterranean populations of algae regarded as recent introductions, such as species of *Antithamnion*, *Caulerpa*, and so on). Because of the impossibility of hybridization in culture (an approach that has been successful with other species of *Polysiphonia*: Rueness, 1973), comparative molecular methods are probably the only means of clarifying the relationships of this entity. Observations on the occurrence of *P. setacea* in the most easterly parts of the Mediterranean would also be informative. Our observations indicate that surface waters of the warmest areas (southern coast of Turkey and the Levant States) are the only part of the basin where this species might be limited by high summer temperatures; in July and August surface temperatures in these areas reach 28 °C, which is the upper thermal limit for Mediterranean *P. setacea*. It might be significant that this species has not been recorded in these areas to date.

Acknowledgements

We wish to thank Professor Isabella Abbott and Dr Alan J. K. Millar, who kindly provided us with slides of *P. setacea* from Hawaii and with much useful information. Professors Mario Cormaci, Giovanni Furnari and Marc Verlaque are gratefully acknowledged for loan of liquid-preserved specimens from several localities in the Mediterranean. The Curator of Rijksherbarium, Leiden, The Netherlands and the Collections Manager of the Bernice P. Bishop Museum, Honolulu, Hawaii, allowed us to examine material from their collections by sending us herbarium specimens and permanent slides.

Massimo Menconi helped us in the statistical analyses; Wendy Guiry in the maintenance of cultures.

Stefano Acunto and Michele Magri provided SCUBA assistance during the collection of the plants used for culture.

References

Airoldi, L., M. Fabiano & F. Cinelli, 1996. Sediment deposition and movement over a turf assemblage in a shallow rocky coastal area of the Ligurian Sea. Mar. Ecol. Prog. Ser. 33: 241–251.

Airoldi, L., F. Rindi & F. Cinelli, 1995. Structure, seasonal dynamics and reproductive phenology of a filamentous turf assemblage on sediment influenced, rocky subtidal shore. Bot. mar. 38: 227–237.

Aponte, N. E. & D. L. Ballantine, 1987. The life history and development of *Murrayella periclados* (C. Agardh) Schmitz (Rhodophyta, Rhodomelaceae) in culture. Cryptogamie, Algol. 8: 29–39.

Athanasiadis, A., 1997. North Aegean marine algae IV. *Womersleyella setacea* (Hollenberg) R. E. Norris (Rhodophyta, Ceramiales). Bot. mar. 40: 473–476.

Breeman, A. M., 1988. Relative importance of temperature and other factors in determining geographic boundaries of seaweeds: experimental and phenological evidence. Helgoländer Meeresunters. 41: 199–241.

Breeman, A. M. & H. Pakker, 1994. Temperature ecotypes in seaweeds: adaptative significance and biogeographic implications. Bot. mar. 37: 171–180.

Cambridge, M. L., A. M. Breeman & C. van den Hoek, 1991. Temperature responses and distribution of Australian species of *Cladophora* (Cladophorales: Chlorophyta). Aquat. Bot. 40: 73–90.

Cambridge, M. L., A. M. Breeman, S. Kraak & C. van den Hoek, 1987. Temperature responses of tropical to warm temperate *Cladophora* species in relation to their distribution in the Atlantic Ocean. Helgoländer Meeresunters. 41: 329–354.

Cormaci, M., G. Furnari, G. Alongi & D. Serio, 1994. On three interesting marine red algae (Ceramiales, Rhodophyta) from the Mediterranean Sea. Giorn. bot. ital. 128: 1001–1006.

Edwards, P., 1968. The life history of *Polysiphonia denudata* (Dillwyn) Kützing in culture. J. Phycol. 4: 35–37.

Egerod, L., 1971. Some marine algae from Thailand. Phycologia 10: 121–142.

Guiry, M. D. & E. M. Cunningham, 1984. Photoperiodic and temperature responses in the reproduction of north-eastern Atlantic *Gigartina acicularis* (Rhodophyta: Gigartinales). Phycologia 23: 357–367.

Hollenberg, G. J., 1968. An account of the species of *Polysiphonia* of the central and western tropical Pacific Ocean. Pacific Sci. 22: 56–98.

Kapraun, D. F., 1977. Asexual propagules in the life history of *Polysiphonia ferulacea* (Rhodophyta, Ceramiales). Phycologia 16: 417–426.

Kapraun, D. F., 1978. Field and cultural studies on selected North Carolina *Polysiphonia* species. Bot. mar. 21: 143–153.

Koch, C., 1986. Attempted hybridization between *Polysiphonia fibrillosa* and *P. violacea* (Bangiophyceae) from Denmark; with culture studies primarily on *P. fibrillosa*. Nord. J. Bot. 6: 123–128.

Kudo, T. & M. Masuda, 1986. A taxonomic study of *Polysiphonia japonica* Harvey and *P. akkeshiensis* Segi (Rhodophyta). Jap. J. Phycol. 34: 293–310.

Lüning, K., 1990. Seaweeds. Their environment, biogeography and ecophysiology. John Wiley & Sons, New York: 527 pp.

Maggs, C. A. & M. Hommersand, 1993. Seaweeds of the British Isles. Vol. 1 Rhodophyta. Part 3A Ceramiales. HMSO, London: 444 pp.

McLachlan, J. & C. J. Bird, 1984. Geographical and experimental assessment of the distribution of *Gracilaria* species (Rhodophyta: Gigartinales) in relation to temperature. Helgoländer Meeresunters. 38: 319–334.

Norris, R. E., 1992. Ceramiales (Rhodophyceae) genera new to South Africa, including new species of *Womersleyella* and *Herposiphonia*. S. Afr. J. Bot. 58: 65–76.

Novaczek, I., A. M. Breeman & C. van den Hoek, 1989. Thermal tolerance of *Stypocaulon scoparium* (Phaeophyta, Sphacelariales) from eastern and western shores of the North Atlantic Ocean. Helgoländer Meeresunters. 43: 183–193.

Novaczek, I., G. W. Lubbers & A. M. Breeman, 1990. Thermal ecotypes of amphi-Atlantic algae. I. Algae of Arctic to cold-temperate distribution (*Chaetomorpha melagonium*, *Devalerea ramentacea* and *Phycodrys rubens*). Helgoländer Meeresunters. 44: 459–474.

Oliveira Filho, E. C. de & M. Cordeiro-Marino, 1970. On the identity of *Lophosiphonia bermudensis* Collins and Hervey and *Dipterosiphonia rigens* (Schousboe) Falkenberg. Phycologia 9: 1–3.

Oppen, M. J. H. van, H. Klerk, J. L. Olsen & W. T. Stam, 1996. Hidden diversity in marine algae: some examples of genetic variation below the species level. J. mar. biol. Ass. U.K. 76: 239–242.

Pakker, H. & A. M. Breeman, 1996. Temperature responses of tropical to warm-temperate Atlantic seaweeds. II. Evidence for ecotypic differentiation in amphi-Atlantic tropical-Mediterranean species. Eur. J. Phycol. 31: 133–141.

Pakker, H., W. F. Prud'homme van Reine & A. M. Breeman, 1994. Temperature responses and evolution of thermal traits in *Cladophoropsis membranacea* (Siphonocladales, Chlorophyta). J. Phycol. 30: 777–783.

Pakker, H., A. M. Breeman, W. F. Prud'homme van Reine & C. van den Hoek, 1995. A comparative study of temperature responses of Caribbean seaweeds from different biogeographic groups. J. Phycol. 31: 499–507.

Rindi, F. & F. Cinelli, 1995. Contribution to the knowledge of the benthic algal flora of the Isle of Alboran, with notes on some little-known species in the Mediterranean. Cryptogamie, Algol. 16: 103–114.

Rodriguez Prieto, C., C. F. Boudouresque & M. Verlaque, 1993. Nouvelles observations sur les algues marines du Parc Naturel Regional de Corse. Trav. sci. Parc nat. rég. Rés. nat. Corse, Fr. 41: 53–61.

Rueness, J., 1971. *Polysiphonia hemisphaerica* Aresch. in Scandinavia. Norw. J. Bot. 18: 65–74.

Rueness, J., 1973. Speciation in *Polysiphonia* (Rhodophyceae, Ceramiales) in view of hybridization experiments: *P. hemisphaerica* and *P. boldii*. Phycologia 12: 107–109.

Schnetter, R. & G. Bula-Meyer, 1982. Marine algen der Pazifikküste von Kolombien. Chlorophyceae, Phaeophyceae, Rhodophyceae. Bibliotheca Phycol. 60: 1 287.

Silva, P. C., P. Basson & R. L. Moe, 1996. Catalogue of the benthic marine algae of the Indian Ocean. University of California Press, Berkeley, 1259 pp.

U.S. Navy, 1981. Marine climatic atlas of the world. Vol. 9. Worldwide means and standard deviations. U.S. Government Printing Office, Washington D.C.

van den Hoek, C., 1982. The distribution of benthic marine algae in relation to the temperature regulation of their life histories. Biol. J. linn. Soc. 18: 81–144.

Verlaque, M., 1989. Contribution à la flore des algues marines de Méditerranée: espéces rares ou nouvelles pour les côtes françaises. Bot. mar. 32: 101–113.

Verlaque, M., 1994. Inventaire des plantes introduites en Mediterranée: origines et répercussions sur l'environment et les activités humaines. Oceanol. Acta 17: 1–23.

Winer, B. J., 1971. Statistical principles in experimental design. McGraw-Hill Kogakusha, Tokyo, 907 pp.

Wynne, M. J., 1993. Benthic marine algae from the Maldives, Indian Ocean, collected during the R/V Te Vega expedition. Contr. Univ. Mich. Herb. 19: 5–30.

Long-term and diurnal carpospore discharge patterns in the Ceramiaceae, Rhodomelaceae and Delesseriaceae (Rhodophyta)

J. A. West & D. L. McBride[1]
School of Botany, University of Melbourne, Parkville, Victoria, 3052, Australia
E-mail: jwest@rubens.its.unimelb.edu.au
[1]*Present address: College of the Rockies, Cranbrook, B.C. V1C5L7 Canada;*
E-mail: mcbride@cotr.bc.ca

Key words: diurnal, long-term, carpospore-discharge, Ceramiaceae, Delesseriaceae, Rhodomelaceae

Abstract

In laboratory culture, controlled fertilization resulted in mature carposporophytes of the Ceramiales indicated below. Carpospore discharge was observed daily. In the Ceramiaceae *Spyridia filamentosa* cystocarp-pairs released spores twice in succession (range: 32–537 total spores/carposporophyte). Discharge by all cystocarp-pairs occurred during the dark period of the daily light:dark cycle (12:12 LD) and was completed in 12 d. In reverse cycle (12:12 and 16:8 DL) spore release reversed in 3d. In the Rhodomelaceae *Bostrychia moritziana*, *Pterosiphonia pennata* and *Murrayella periclados* discharged spores from isolated cystocarps over periods up 58 d (ranges: 318–4112, 1051–2271 and 451–3162 total spores/carposporophyte respectively) without any diurnal or long-term rhythmicity. In the Delesseriaceae cystocarps of *Caloglossa leprieurii* and *Caloglossa ogasawaraensis* released spores for up to 31 d (ranges: 271–3050 and 565–1286 total spores/carposporophyte respectively). Discharge peaks occurred at 5–7 d intervals with viable cystocarps and spore numbers gradually declining. Thus, in the plants studied, there are at least three patterns of carpospore release from individual mature carposporophytes: a relatively short-term dual release pattern, a long-term non-rhythmic release pattern and a long-term rhythmic release pattern. Results also indicated that excised cystocarps without associated vegetative branches showed a much reduced spore production.

Introduction

In red algae the carposporophyte generation typically arises directly from the haploid female gametophyte generation as a result of gametic fusion. These diploid carposporophytes release carpospores which, in turn, germinate to form a second diploid generation, the tetrasporophyte. The production of the carposporophyte generation can be thought of as a mechanism of zygote amplification (Hawkes, 1990) which dramatically increases reproductive potential. Thus, total carpospore production is critical to efficient reproduction in red algae.

Most investigations treating the post-fertilization development and carposporophyte formation (review by Hommersand & Fredericq, 1990) and reproductive strategies (Searles, 1980; Hawkes, 1990; Santelices, 1990) do not consider the total number of spores produced from individual carposporophytes or females during a reproductive season. West & Crump (1975) provide a brief analysis of the reproductive potential of the laboratory cultured plants in *Mastocarpus papillatus* (C. Agardh) Kützing indicating that a single plant had the extraordinary potential to produce over 10×10^7 spores. Similarly, Boney (1978) estimated a potential of 8.3×10^7 carpospores per reproductive plant of *Rhodymenia pertusa* (Postels et Ruprecht) J. Agardh, based on examination of one specimen. Seasonal studies of carpospore shedding include those done by Guzman-del Proo et al. (1971) on *Gelidium robustum* (Gardner) Hollenberg et Abbott and by Bhattacharya (1985) on *Chondrus crispus* Stackhouse, who estimated 9.61×10^{10} carpospores per m^2 of population stand.

Table 1. Species, collection sites and dates, isolate numbers, culture conditions

Specis	Collection site and date	Isolate No.	Condition
Spyridia filamentosa (Wulfen) Harvey	Calatagan, Batangas, Philippines, 11 v 1987	2846	1
Bostrychia moritiziana (Sonder ex Kützing) J. Ag.	Tooradin, Western Port Bay, Victoria, Australia, 5 x 1986	2748	1
Murrayella periclados (C. Ag.) Falkenberg	Agana, Guam, 30 viii 1989	2999	1
Pterosiphonia pennata (C. Ag.) Falkenberg	Williamstown, Port Phillip Bay, Victoria, Australia, 16 iii 1997	3780	2
Caloglossa leprieurii (Montagne) J. Ag.	Hawkesbury R., New South Wales, Australia, 12 iii 1993	3356	1
Caloglossa leprieurii	Fraser I., Queensland, Australia 29 x 1995	3560	1
C. ogasawaraensis Okamura	Darwin, Northern Territory, Australia, 4 vi 1989	2994	1

Culture conditions: 1. 24–25 °C, 12:12 LD, 75 rpm shaker, 10–25 μmol photons·m^{-2}s^{-1}, 30 ppt Provasoli's Seawater (PES/2), Pyrex No. 3250 100 × 80 mm storage dishes (200 ml). 2. 20 °C, 12:12 LD, stationary culture, 10–25 μmol-photons·m^{-2}s^{-1}, 30 ‰ Provasoli's Enriched Seawater (PES/2), Pyrex No. 3 250 100 × 80 mm storage dishes (200 ml).

Many species of the Ceramiales can be easily grown in laboratory culture to determine reproductive potential in defined conditions. *Spyridia filamentosa* (Wulfen) Harvey is a ubiquitous cold-temperate to tropical ceramiaceous species, often growing freefloating in quiet bays. Carposporophyte development is rapidly completed in culture (West & Calumpong, 1989), making it ideal for experimentation on patterns of spore development and release. Several rhodomelaceous genera are also useful for laboratory research. The mangrove inhabiting *Bostrychia*, especially *B. moritziana* (Sonder *ex* Kützing) J. Agardh, has been extensively used for studies on ecophysiology, reproductive biology and molecular systematics (e.g. Karsten et al., 1993; West et al., 1993; Zuccarello & West, 1997) and because of the rapid development of the carposporophyte (Pickett-Heaps & West, 1998) is valuable for assessing sporic potential. Another mangrove inhabitant is the pantropical *Murrayella periclados* (C. Agardh) Falkenberg. The complete life history of a Puerto Rican isolate of *M. periclados* in culture was observed by Aponte & Ballantine (1987). The temperate rhodomelacean species *Pterosiphonia pennata* (C. Agardh) Falkenberg was also selected for assessing reproductive potential. Among the Delesseriaceae there have been many genera successfully cultured through complete sexual life histories and *Caloglossa* [especially *C. leprieurii* (Montagne) J. Ag., *C. continua* (Okamura) King *et* Puttock, *C. apomeiotica* West et Zuccarello and *C. ogasawaraensis* Okamura] have proven the most reliable in long term culture investigations (Kamiya & Tanaka, 1993; West et al., 1994; Kamiya et al., 1995, 1997, 1998).

The availability of these isolates offered a good opportunity to evaluate carpospore release in the Ceramiales, the largest and, perhaps, the most complex order of the Rhodophyta, thus yielding useful information pertaining to overall red algal reproductive strategy.

Materials and methods

The algae used in this research are listed in Table 1 and methods of isolation are described by West & Calumpong (1988). Stock cultures of male and female gametophytes were maintained separately in the following conditions: 22–26 °C, 12:12 h LD (light:dark) daily photoperiod, 10–20 μmol photons m^{-2} s^{-1}, in Pyrex (No. 3250) 500 ml storage dishes containing 250 ml Modified Provasoli's Enriched Seawater (Table 2) adjusted to 30 ‰ salinity with MilliQ-filtered water. The medium and culture dishes were changed at 1–3 month intervals. For fertilization experiments subcultures of 5–10, 1-cm long shoot tips of females and males were

Table 2. Modified Provasoli's Enrichment. Directions for preparing 2-liter stock in glass distilled or milliQ water

Constituent	Amount in each stock	Amount of each stock solution to add to make 2 ℓ
Disodium DL-b-glycerophosphate pentahydrate	50 g/l	16 ml
NaNO$_3$	35 g/l	220 ml
Iron-EDTA (1:1 molar):		
combine Fe(NH$_4$)$_2$(SO$_4$)$_2$•6H$_2$0	700 mg/l	
with Na$_2$EDTA	600 mg/l	
Vitamin B$_{12}$*	25 mg/l	7 ml
Thiamine*	500 mg/l	16 ml
Biotin*	50 mg/l	16 ml
PII Trace Metals Mix	see below	400 ml

Add all solutions to 2-liter flask, bring to volume with glass-distilled or MilliQ water.

Steam sterilize (do not autoclave) and store at 4 °C.

*Biotin, thiamine and B$_{12}$ stock solutions should be frozen for storage.

P II Trace Metals Mix (1-litre stock)

Na2EDTA		1.0 g/l
Boron	Boric Acid (H3BO3)	1.12 g/l
Manganese	Manganese sulfate (MnSO4•H2O)	120 mg/l
Zinc	Zinc sulfate (ZnSO4•7H2O)	22 mg/l
Cobalt	Cobalt sulfate (CoSO4•7H2O)	5 mg/l

Combine in the above order to 500 ml glass distilled or MilliQ water. Mix.

Bring to 1 liter. Steam sterilize. Store at 4 °C.

P II Metals Mix was originally designed from a synthetic seawater medium.

transferred to 75 ml of fresh medium in Pyrex (No. 3140) 100 ml crystallizing dishes one week before use. For *Bostrychia* on Day 0 of fertilization male plants were placed in 10 ml of 20 ‰ medium to induce spermatial discharge by mild osmotic shock. The male plants were then removed and the spermatial suspension was swirled to mix uniformly. Individual female shoots bearing procarps on lateral branches were then dipped in the suspension several times before placing them in fresh medium in culture dishes on a New Brunswick Laboratory Gyratory™ G2 rotary shaker at 75 rpm.

For *Spyridia*, *Pterosiphonia*, *Murrayella* and *Caloglossa* the males were placed with females for 24 h on the rotary shaker and then removed. The fertilized females were then placed in new dishes on the shaker and within 7–20 d were prepared for carpospore discharge observations. Branches bearing mature carposporophytes were excised with microdissection scissors. In some instances cystocarps were excised with minimal vegetative tissue to determine their viability relative to those with vegetative tissue attached. Each was placed individually in polystyrene Cell Wells™ (Corning 25820). Each of the 24 wells was 16 mm diam. and contained 1.5 ml of medium. Spore counts were done by direct observation at 20–30 × with a stereoscope daily at 0600 and the carposporophytes were transferred daily to new wells with fresh medium.

For the diurnal discharge experiments with *Spyridia* and *Bostrychia*, growth cabinets were set at 12:12 LD (lights on 0600, off 1800); 12:12 DL (dark:light – reverse, lights off 0600, on 1800); 16:8 DL (reverse, lights off 0600, on 1400) and spores counted at 0600, 1200 and 1800 daily.

Results

Ceramiaceae

Spyridia filamentosa

The 2, 3 or 4 cystocarps on each fertile segment arise from fertilization of one procarp with transfer of the diploid nuclei to 2, 3 or 4 auxiliary cells on the same segment. For the following experiments the determinate laterals bearing two cystocarps were selected because this is the most common pattern.

Carposporophytes reached maturity 7 days after fertilization (Figure 1a–d) and, when grown in a normal 12:12 LD cycle, spores discharged between 2200 and 0600 (Figure 2). Between 20 and 150 spores were released on the same day by each cystocarp on the same segment; however, occasionally discharge of the adjacent cystocarp occurred on the second day. Following the initial discharge, gonimoblast regeneration occurred once, followed by a second discharge in 2–3 d. The regenerating carposporangial mass was smaller and more irregular than the first (Figure 1d). The spore number in the second discharge was about 50% lower than in the first. Carpospore release from fertile branches was completed in 10–12 d. The range and mean number of released carpospores is given in Table 3.

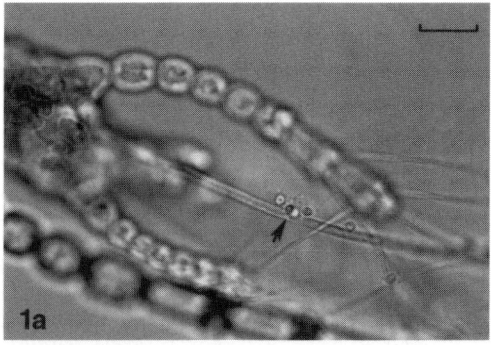

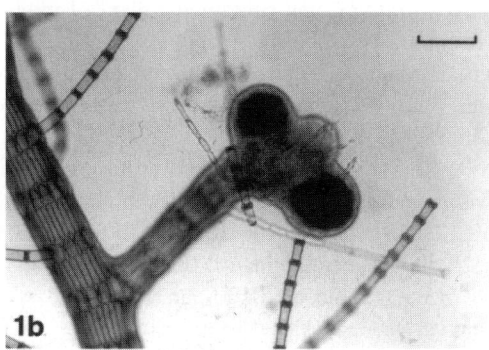

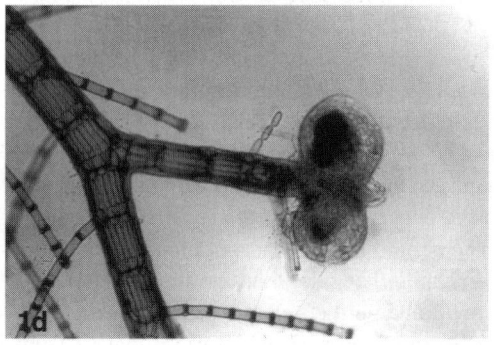

Figure 1. Spyridia filamentosa. (a) Trichogyne with attached spermatia (arrowhead). Scale bar: 25 μm. (b) Six-day-old cystocarp. Scale bar: 200 μm. (c) Branch with mature 7-day-old cystocarp-pairs on determinate laterals. Scale bar: 1.5 mm. (d) Nine-day-old cystocarp-pair with regenerating gonimoblasts. Scale: as 1b.

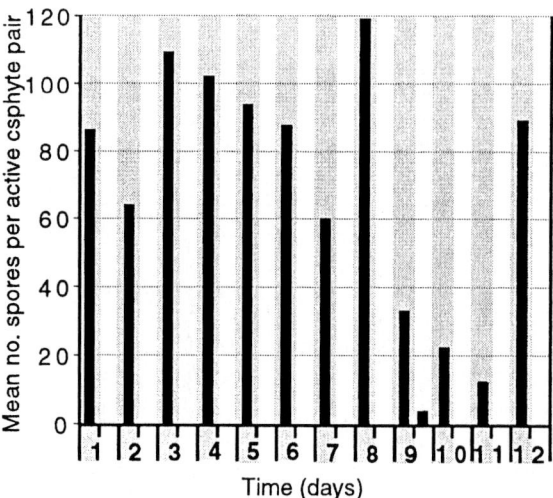

Figure 2. Spyridia filamentosa. Diurnal discharge patterns for 12 cystocarp pairs over 12 days in normal 12:12 light-dark cycle (on 0600, off 1800). Discharge occurs in daily dark period (stippled vertical bars).

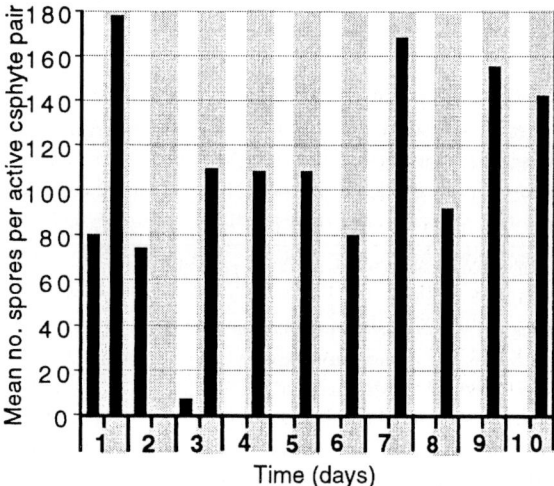

Figure 3. Spyridia filamentosa. Diurnal discharge patterns for 12 cystocarp pairs over 12 days in reversed 12:12 dark-light cycle (on 1800, off 0600). Discharge occurs in the dark cycle (stippled vertical bars).

To determine if there is a dark requirement for discharge and if it is fixed or variable, a set of cystocarpic plants were placed in a reverse LD cycle for 10–12 d. Within 3 days discharge shifted from the 2200–0600 to 1200–1800 in the dark cycle (Figure 3). As seen in Figure 4 the same pattern was evident in cystocarpic plants placed in a short-day reverse cycle (16:8 DL).

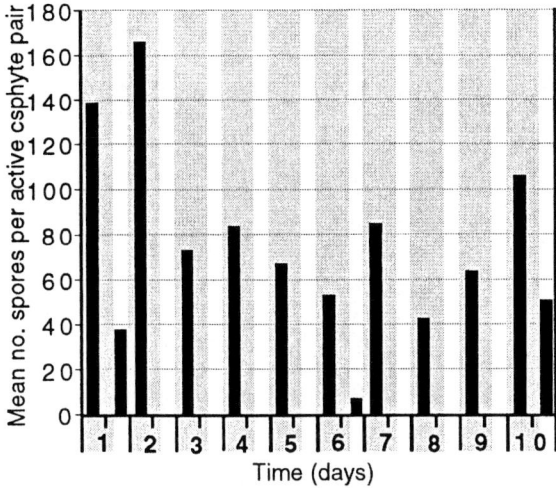

Figure 4. Spyridia filamentosa. Diurnal discharge patterns for 12 cystocarp pairs over 12 days in reversed 16:8 dark-light cycle (on 2200, off 0600). Discharge occurs in the dark cycle (stippled vertical bars) after one day of adjustment from the normal cycle in Day 1.

Rhodomelaceae

Bostrychia moritziana

In experiments with low spermatial levels the cystocarps matured 16 d after fertilization. Each lateral branch generally bore only one large cystocarp about 500 μm diam (Figure 5a). During most of the experiment individual cystocarps liberated 50–150 spores each day, but by 35 d the majority had ceased release although sporadic release occurred until Day 58 (Figure 6). These few cystocarps which continued to release, even though the pericarp began to break down, contained viable carposporangia which were still forming and releasing (Figure 5c). The range and mean number of released carpospores is given in Table 3.

In another experiment the female was overloaded with spermatia to determine the impact of this on the abundance and size of crowded cystocarps and their spore discharge (high density fertilization, Figure 6). This resulted in each lateral branch developing as many as 4–5 cystocarps (Figure 5b). These cystocarps were smaller overall (220–280 μm diam.) compared to those that developed in the first experiment (450–520 μm diam.; Figure 5a) when a more dilute suspension of spermatia resulted in usually one cystocarp on each lateral. The smaller cystocarps also produced smaller numbers of spores per day and fewer spores in total (Table 3). Diurnal spore discharge was followed in *B. moritziana* in normal 12:12 LD cycles over 12 d (Figure 7). Discharge occurred in both light and dark; consequently there is no clear pattern visible like that seen in *Spyridia* (Figures 2–4).

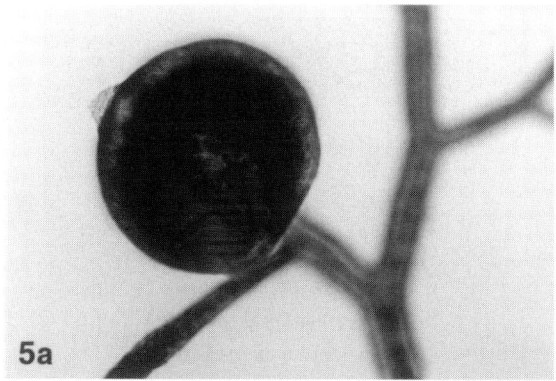

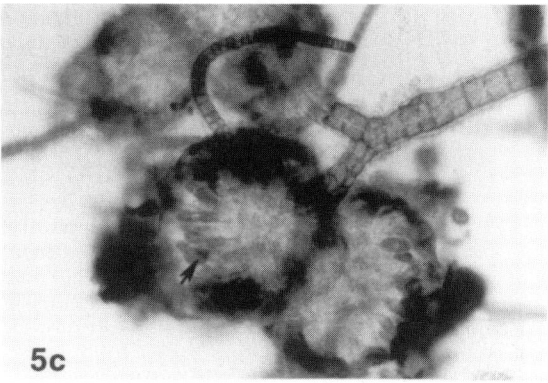

Figure 5. Bostrychia moritziana. Scales: as 1b. (a) Large single cystocarp in low spermatial density experiment. (b) Small multiple cystocarps in high spermatial density experiment. (c) Four 44-day-old cystocarps with disintegrating pericarp and exposed viable carposporangia (arrowhead).

Table 3. Number of carposporophytes monitored; ranges and means of carpospore production

Genus and species	No. csphytes	Range of total no. spores/csphyte	Mean no. spores/ csphyte ± SD	Mean no. spores/ csphyte/day ± SD
Spyridia filamentosa	72	32–537	157 ± 101	78 ± 50
Bostrychia moritziana (low density)	21	847–4117	2511 ± 998	73 ± 28
Bostrychia moritziana (high density)	15	318–1184	969 ± 424	30 ± 11
Murrayella periclados	10	451–3162	2015 ± 748	40 ± 15
Pterosiphonia pennata	8	1051–2271	1738 ± 372	45 ± 12
Caloglossa leprieurii 3356	13	271–2341	1389 ± 638	218 ± 107
Caloglossa leprieurii 3560	12	493–3050	1164 ± 806	142 ± 51
Caloglossa ogasawaraensis	7	565–1286	974 ± 294	81 ± 27

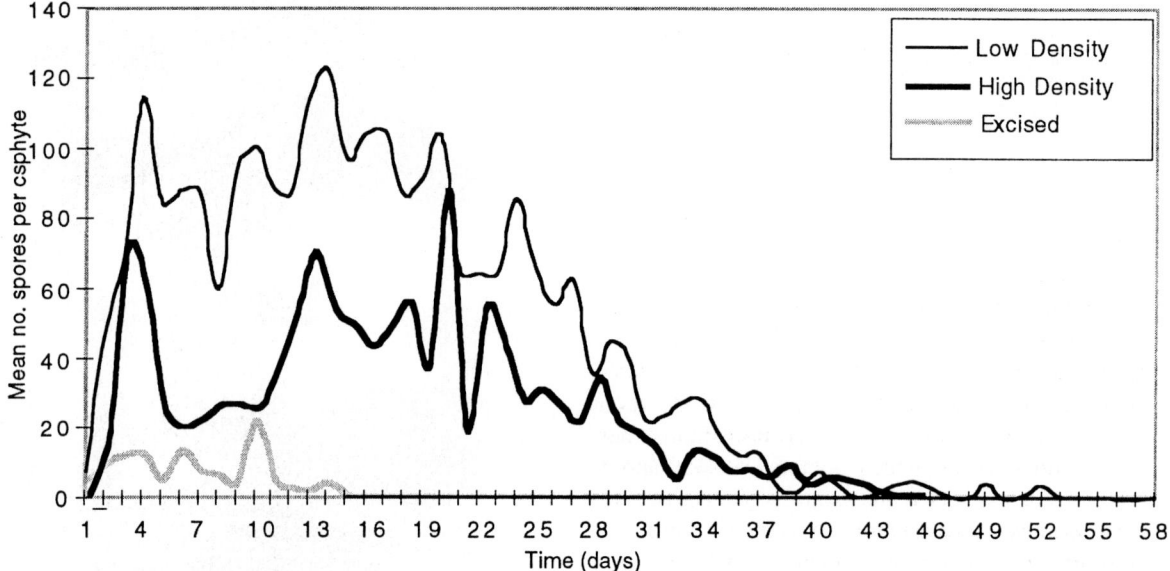

Figure 6. Bostrychia moritziana carpospore release patterns. Daily pattern of carpospore discharge in 24 carposporophytes over 57 days (7 Sept.–4 Nov. 1997) resulting from low spermatium density fertilization (low density). Also shown is the daily pattern of carpospore discharge in 19 carposporophytes over 45 days (5 Oct.–18 Nov. 1997) resulting from high spermatium density fertilization (high density). In the latter experiment the female was overloaded with spermatia to determine the impact of this on the abundance, size and spore discharge when crowded cystocarps developed. In addition, the carpospore release pattern for excised cystocarps is illustrated.

Murrayella periclados

These cystocarps matured in 14–15 d (Figure 8a–b) and carpospore discharge continued with a gradual decrease evident at 58 d when observations were terminated (Figure 9). Mature cystocarps ranged in diameter from 250–340 μm. The range and mean number of released carpospores is given in Table 3.

Pterosiphonia pennata

First carpospore discharge was observed 20 d after fertilization (Figure 8c–d). At maturity the carposporophytes were quite large (595–615 μm diam.) with uniform gonimoblast development. Carpospore discharge in unexcised cystocarps was steady for about 35 d followed by a gradual decline to zero (Figure 9). The range and mean number of released carpospores is given in Table 3.

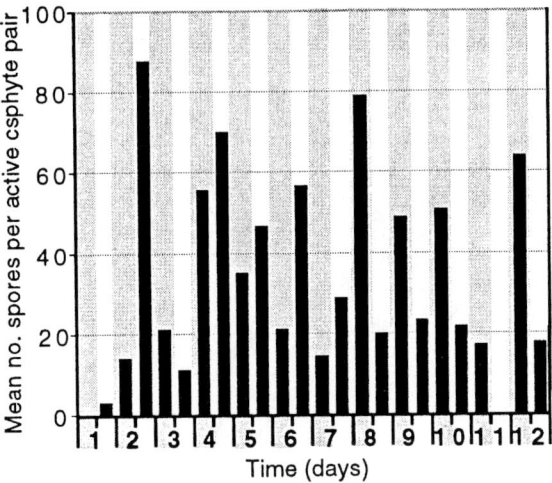

Figure 7. Bostrychia moritziana. Diurnal carpospore discharge in 12 carposporophytes over 12 days in normal 12:12 light-dark cycle (on 0600, off 1800). There is no clear diurnal pattern. Stippled area indicates dark period.

Delesseriaceae

Caloglossa leprieurii

The carposporophytes of *C. leprieurii* (isolate 3356) matured 13 d after fertilization (Figure 10a) which is several days quicker than *Bostrychia* and *Pterosiphonia*. Most of the carposporophytes discharged all their spores on the first day of discharge (day 14 after fertilization). This was clearly indicated by the complete absence of the dark carposporangial mass. The range and mean number of released carpospores is given in Table 3. On one blade two cystocarps developed side-by-side and both discharged all the spores on the same day (Figure 10b). There was a 5–7 d interval between discharges of individual cystocarps (Figure 11). Isolate 3560 showed a pattern of carpospore discharge similar to that of isolate 3356, but had a more sustained discharge period of 45 days (Figure 11). The range and mean number of released carpospores is given in Table 3.

Caloglossa ogasawaraensis

First spore discharge was evident 17 days after fertilization when carposporophytes reached 300–400 μm diam. (Figure 10c). The range and mean number of released carpospores is given in Table 3. Discharge from each cystocarp also occurred at 5–7 day intervals (Figure 11). During these intervals the dark gonimolobes were initially not visible but became larger and darker 1–2 d prior to the next discharge.

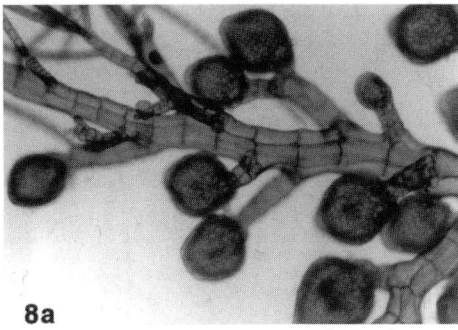

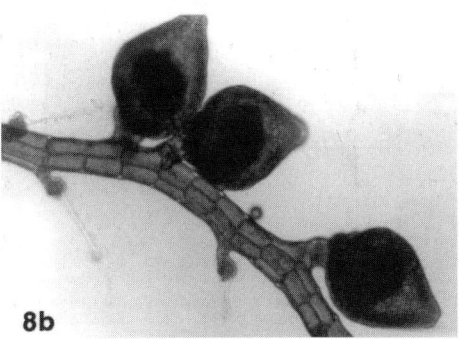

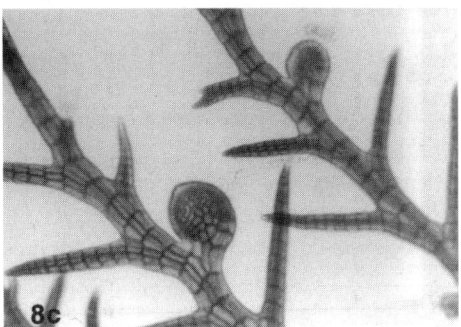

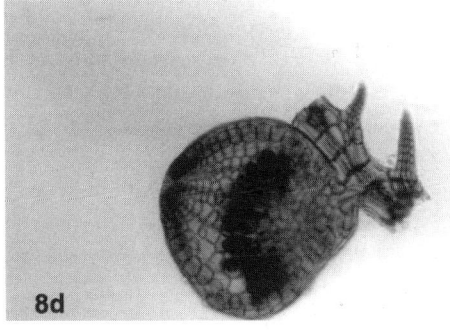

Figure 8. Murrayella and *Pterosiphonia.* Scales: as 1b. (a) *Murrayella periclados.* Developing 12-day-old cystocarps. (b) *Murrayella periclados.* 15-day-old mature cystocarps. (c) *Pterosiphonia pennata.* Developing 7-day-old cystocarps. (d) *Pterosiphonia pennata.* 28-day-old excised cystocarp.

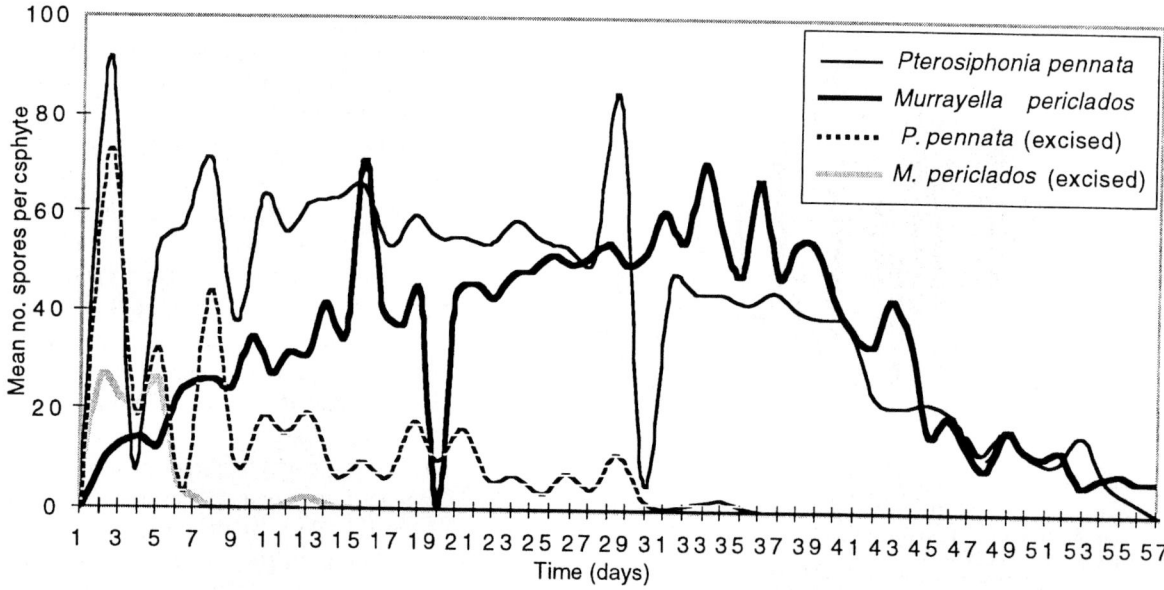

Figure 9. Daily pattern of carpospore discharge from *Pterosiphonia pennata* over 43 days (29 October–16 December 1997). Also shown is the daily pattern from *Murrayella periclados* over 50 days (15 December 1997–14 January 1998). In addition, the carpospore release pattern for excised cystocarps is illustrated.

As adjunct experiments in *B. moritziana*, *M. periclados*, *P. pennata*, *C. leprieurii* 3356 and *C. ogasawaraensis*, a variable number of cystocarps were excised in each with minimal attached vegetative tissue to determine their viability relative to those with large branch or blade tissue attached. All trials indicated a decrease in duration of release (see Figures 6, 9 for examples), carpospore production and spore size. Total carpospore production decreased to 40% of normal in *C. leprieurii* 3356, 35% in *C. ogasawaraensis*, 20% in *P. pennata*, 11% in *B. moritziana* and to a mere 3% of normal production in *M. periclados*. Also in *C. leprieurii* 3356 spore size decreased from a mean of 47 μm to 37 μm. Usually the pericarp and internal carposporophyte tissue was yellow and lysing in all the excised cystocarps long before similar unexcised cystocarps had stopped discharge.

Discussion

In order to estimate the total spore output of individual female red algae and carposporophytes it is necessary to monitor these in the field or laboratory culture. With the majority of red algae in the field, especially those in mangrove habitats, the reproductive plants are quite infrequent and only briefly present. This emphasizes the benefit of using laboratory cultures where the environment is more protective and the full potential can be expressed without interference by the many physical, chemical and biological agents. Our observations on the Ceramialaes carpospore discharge patterns may provide some new insights into the phylogeny and ecology of the order.

Ceramiaceae

Carpoblepharis, *Ceramium*, *Gulsoniopsis*, *Microcladia* and *Reinboldiella* (Hommersand, 1963), *Aglaothamnion* (as *Callithamnion*, O'Kelly & Baca, 1984) and most other Ceramiaceae appear to have successional formation of gonimolobes which implies that at least two sets of spore discharge occur during the life of each cystocarp. *Spyridia* is probably not typical of the Ceramiaceae because of its distinct overall vegetative and reproductive morphology (Feldmann & Feldmann, 1943; Hommersand, 1963; West & Calumpong, 1989). Consequently, it may not reflect the patterns seen in other Ceramiaceae, however it is the only one we have available at present

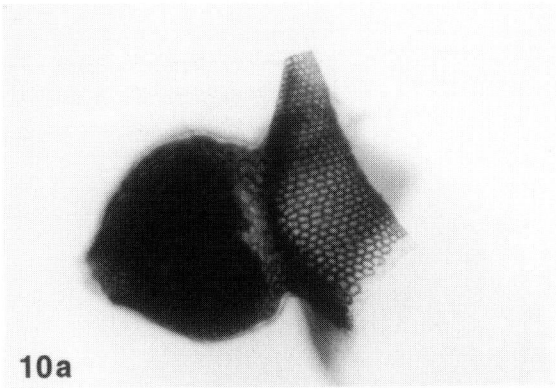
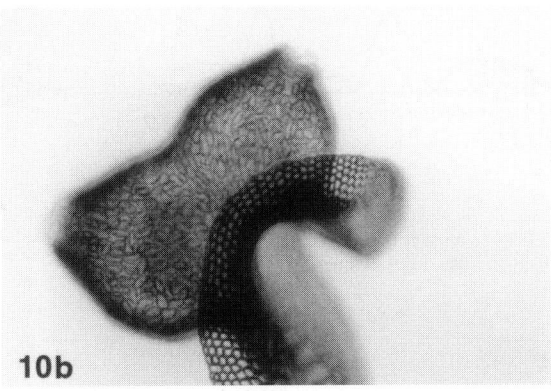
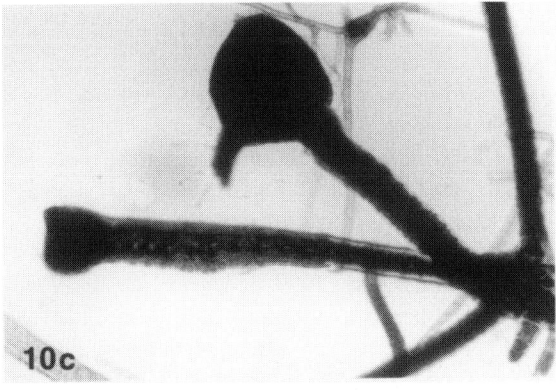

Figure 10. Caloglossa. Scales: as 1b. (a) *Caloglossa leprieurii.* 14-day-old excised cystocarp with fully developed dark gonimoblast. (b) *Caloglossa leprieurii.* Empty double cystocarps after first spore discharge series. Gonimoblast lightly visible. (c) *Caloglossa ogasawaraensis.* 14-day-old cystocarp.

which performs satisfactorily in laboratory culture for controlled experiments on fertilization and carpospore discharge. Nonetheless, it is clear that the *Spyridia* carposporophytes discharge only two successive sets of carpospores rather than several repeated sets as in the Delesseriaceae tested and continuous release in the Rhodomelaceae tested.

A marked diurnal periodicity was also associated with carpospore release in *Spyridia*. Sagromsky (1961) showed that tetraspore discharge by *Nitophyllum punctatum* (Stackhouse) Greville occurred mainly during the daily light periods and ceased after several days of continuous light or dark. Tetraspore discharge by *Gelidiella acerosa* (Forsskål) Feldman et Hamel occurred primarily between 1400 and 1600 (Umamaheswara Rao, 1974). Ngan & Price (1983) investigated the diurnal periodicity of carpospore and tetraspore discharge in a number of tropical red algae including *Caloglossa* and *Bostrychia*. The discharge peak in many species was correlated with daily floodtide although there was no consistent pattern. Kurogi et al. (1962) indicated that monospore liberation by the *Conchocelis*-phase of *Porphyra tenera* Kjellman peaked at 1200 in 10:14 LD , at 0800 and 1600 in 12:12 LD and at 1000 in a 15:9 LD.

Ngan & Price (1983) suggested that the various diurnal spore release periods of different species would reduce space competition. It is also possible that the release periods occur during minimal herbivore activity to maximize settlement, adhesion and germination although no evidence exists at present to support either of these hypotheses.

Rhodomelaceae

In most species of *Bostrychia* the mature cystocarps from field-collected plants have 30–120 carposporangia (King & Puttock, 1989) although this does not appear to reflect the long term potential spore output. These observations are quite different from those seen in the genera considered here. Also, it is of interest that we could not demonstrate evidence of diurnal periodicity of carpospore release in *Bostrychia moritziana* in contrast to Ngan & Price (1983) who reported a daily pattern of carpospore and tetraspore release in various *Bostrychia* species. In the one experiment with a very high spermatial suspension, multiple fertilizations on each branch resulted in the development of up to 5 cystocarps per branch. Compared with low spermatial suspension the number of cystocarps per branch increased, the diameter of individual cystocarps decreased and, subsequently, the number of spores released by each cystocarp declined. There was a 62% decrease in mean spore production in the smaller crowded cystocarps compared with the single large cystocarps. It is unlikely that spermatia occur in such

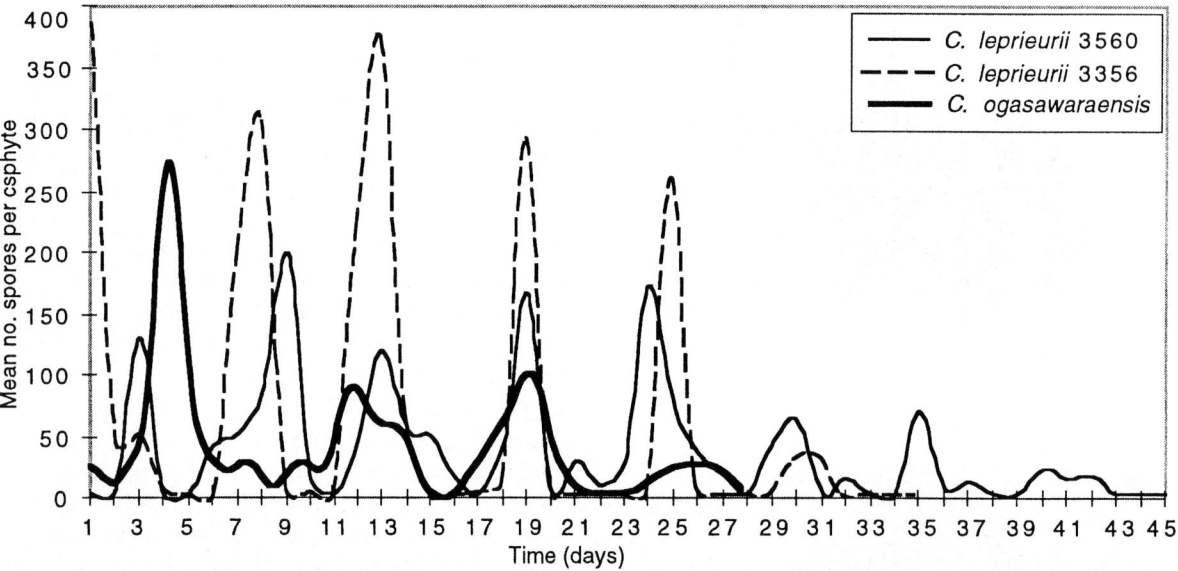

Figure 11. Daily pattern of carpospore discharge from *Caloglossa leprieurii* (3356) in 31 days (17 November–20 December 1997). Also shown are daily patterns of carpospore discharge from *Caloglossa leprieurii* (3560) in 45 days (30 December 1997–8 February 1998) and from *Caloglossa ogasawaraensis* in 26 days (8 January–2 February 1998). Note rhythmicity of production in all patterns.

high levels in nature, but again the reduced size and diminished spore output of the crowded cystocarps certainly implies that nutritional dependence affects the overall success of each fertilization.

It is difficult to construct a model of *B. moritziana* for estimation of total spore production but, with reference to the hypothetical plant illustrated by Figure 12, if one assumes that a 'typical' female is 10 mm tall and has a 'normal' bilateral branching pattern with alternate primary laterals uniformly spaced about 400 μm apart, each of these having about 4–10 (mean 6) alternate secondary laterals and each of these bearing 3–10 procarps (mean 5). In low spermatium density fertilization 1 cystocarp would be present on each secondary lateral and in high density about 4 cystocarps, the resulting calculations would be as follows:

LD – 25 primary laterals × 6 secondary laterals × 1 cystocarp (2500 spores) = 375,000 carpospores per plant.

HD – 25 primary laterals × 6 secondary laterals × 4 cystocarps (1000 spores/cystocarp) = 600,000 carpospores per plant.

On this basis it would appear that there is a slight advantage to high spermatium density fertilization but it is also clear from field observations that males and females are seldom (less than 1%) seen in the wild populations of *B. moritiana*. In 10 years of field observations we have never collected a male and female *together* and therefore the likelihood of HD fertilization occurring is extremely unlikely. Also there is a very small chance of cystocarps surviving for more than a week in field conditions so the total spore potential is greatly reduced. Finally, the low percentage germination and survival to reproductive maturity in nature further reduce the odds.

The Guam isolate (2999) of *Murrayella periclados* investigated here was the most impressive of all the Rhodomelaceae in long-term carpospore output, extending for 58 d. Further release was quite likely, but the experiment was discontinued due to space limitations. The number of spores discharged per day was about half that seen in *Bostrychia* (Table 3).

In *Pterosphonia* Hommersand (1963) describes the development of the gonimoblast as successional suggesting that spore release occurs repeatedly as successive sporangia mature. In a life history study of *P. pennata* in culture, Masuda (1973) observed that the time required from fertilization to initial carpospore discharge was 30 d, but no information on sustained spore release was provided. From the results described

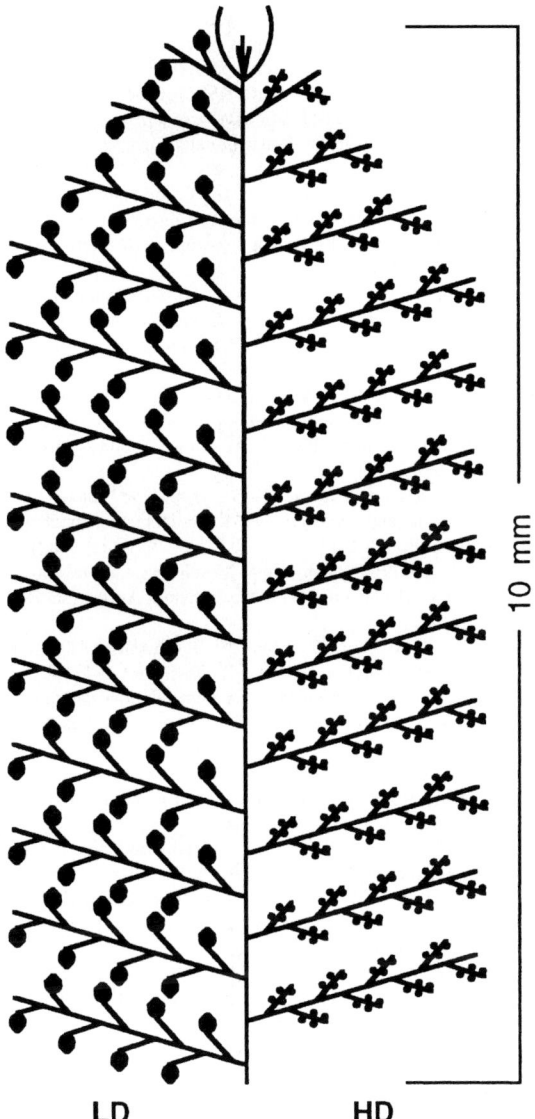

Figure 12. Hypothetical model of a *Bostrychia moritziana* female 10 mm high with low density cystocarp disposition (LD) on the left and high density disposition (HD) on the right.

here it is evident that long term spore discharge, over 40 d, is achieved by this species when the cystocarp is well supported by female vegetative tissue, but is drastically curtailed in cystocarps without this support. However, the very large cystocarp size in *P. pennata* (Figure 10d) did not offer a greater spore output over a longer time when compared with the output by *Bostrychia*.

Delesseriaceae

As noted by King & Puttock (1994) the mature carposporophytes of *Caloglossa* are subspherical, ostiolate and subsessile on the female blade. Kamiya et al. (1997) indicated that the cystocarps of *C. continua* (Okamura) King et Puttock, a closely related species derived from *C. leprieurii*, were 300–600 µm diameter. This compares well with the cystocarps (450–525 µm) in our isolates of *C. leprieurii*. It was unexpected to see, in these two isolates, the 5–7 d intervals of spore discharge and the long series lasting 35–45 d.

The life history of *C. ogasawaraensis* has been described by Kamiya & Tanaka (1993) although the carpospore discharge pattern was not considered. The cystocarps were quite large (500–800 µm diam.) compared with those (300–400 µm diam) observed in our isolate. The small cystocarp size may account for the smaller number of carpospores compared with *C. leprieurii* (Table 3). The sustained period of discharge up to 26 d, was shorter than that for *C. leprieurii* as well. The 5–7 d cycle of carpospore discharge in *Caloglossa* appears similar to the 4–5 d cycle noted in carpospore shedding of *Gracilaria edulis* (Gmelin) Silva (Rama Rao & Thomas, 1974).

Hommersand (pers. comm.) stated that 'progressive release of carpospores in monopodially developed Delesseriaceae gonimoblasts and sympodially developed Rhodomelacae gonimoblasts would represent a significant evolutionary advancement over the typical batch-wise pattern in Ceramiaceae indicating that a nutrient feeding strategy has evolved to sustain productivity of a diploidized auxiliary cell'. Present results indicate that excised cystocarps without associated vegetative branches are quickly impaired and do not achieve prolonged gonimoblast formation. This certainly adds verification to the long held assumption that the carposporophytes are nutritionally dependent throughout their development on the female gametophyte. It is evident that the reduced nutrient flow strongly impaired the total reproductive potential of these cystocarps: the amount of supporting vegetative tissue is a determining factor in (1) the number of spores released overall by each cystocarp, (2) the number of spore discharge events and (3) the spore size.

Of ecological interest is the long term potential of carpospore release in many of the Ceramiales studied here. Kain & Norton (1990) provided a discussion on high levels of released spore density which may combat high mortality and low germination rates resulting in low recruitment rates in red algae. Thus longer term release patterns may also allow circumvention of environmental periods which are unfavourable for germination. From our studies it is evident that in the Ceramiales there are at least three patterns of carpospore release: a finite dual release pattern (e.g. *Spyridia*), a long-term non-rhythmic release pattern (e.g. *Bostrychia*, *Pterosiphonia* and *Murrayella*) and a long-term rhythmic release pattern (e.g. *Caloglossa*), all of which would serve to increase reproductive efficiency.

Acknowledgements

Partial support for this research has been provided by Australian Research Council small grants SG0935526 (1994) and S19812824 (1998). Louise Phillips and Sarah Wilson graciously did spore counts on several days when we were not present. Louise also provided the specimens of *Pterosiphonia pennata* for culture. Mitsunobu Kamiya confirmed identifications of *Caloglossa* specimens.

References

Aponte, N. & D. Ballantine, 1987. The life history and development of *Murrayella periclados* (C. Agardh) Schmitz (Rhodophyta, Rhodomelaceae) in culture. Cryptogamie Algol. 8: 29–39.

Bhattacharya, D., 1985. The demography of fronds of *Chondrus crispus* Stackhouse. J. exp. mar. Biol. Ecol. 91: 217–231.

Boney, A. D., 1978. The liberation and dispersal of carpospores of the red alga *Rhodymenia pertusa* (Postels et Rupr.) J. Ag. J. exp. mar. Biol. Ecol. 32: 1–6.

Feldmann, J. & G. Feldmann, 1943. Le développement des spores et le mode de croissance de la fronde chez le *Spyridia filamentosa* (Wulf.) Harv. Bull. Soc. Histor. nat. Afr. Nord. 34: 213–221.

Guzman-del Proo, S. A., S. de la Campa-de Guzman & J. Pineda-Barrera, 1971. Shedding rhythm and germination of spores in *Gelidium robustum*. Proc. int. Seaweed Symp. 7: 221–228.

Hawkes, M., 1990. Reproductive strategies. In Cole, K. M. & R. G. Sheath (eds), Biology of Red Algae. Cambridge University Press, Cambridge: 455–476.

Hommersand, M. H., 1963. The morphology and classification of some Ceramiaceae and Rhodomelaceae. Univ. Calif. Publ. Bot. 35: 165–366.

Hommersand, M. H. & S. Fredericq, 1990. Sexual reproduction and cystocarp development. In Cole, K. M. & R. G. Sheath (eds), Biology of Red Algae. Cambridge University Press, Cambridge: 305–347.

Kain, J. M. &. T. A. Norton, 1990. Marine ecology. In Cole, K. M. & R. G. Sheath (eds), Biology of Red Algae. Cambridge University Press, Cambridge: 377–422.

Kamiya, M. & J. Tanaka, 1993. Reproductive structure of *Caloglossa ogasawaraensis* Okamura (Ceramiales, Rhodophyceae) in nature and culture. Jap. J. Phycol. 41: 113–121.

Kamiya, M., J. Tanaka & Y. Hara, 1995. A morphological study and hybridization analysis of *Caloglossa leprieurii* (Ceramiales, Rhodophyta) from Japan, Singapore and Australia. Phycol. Res. 43: 81–91.

Kamiya, M., J. Tanaka & Y. Hara, 1997. Comparative morphology, crossability, and taxonomy within the *Caloglossa continua* (Delesseriaceae, Rhodophyta) complex from the Western Pacific. J. Phycol. 33: 97–105.

Kamiya, M, J. A. West, R. J. King, G. Zuccarello, J. Tanaka & Y. Hara, 1998. Evolutionary differentiation in the red algae *Caloglossa leprieurii* and *C. apomeiotica*. J. Phycol. 34: 361–370.

Karsten, U., J. A. West & E. K. Ganesan, 1993. Comparative physiological ecology of *Bostrychia moritziana* (Ceramiales, Rhodophyta) from freshwater and marine habitats. Phycologia 32: 401–409.

King, R. J. & C. F. Puttock, 1989. Morphology and taxonomy of *Bostrychia* and *Stictosiphonia* (Rhodomelaceae/Rhodophyta). Aust. syst. Bot. 2: 1–73.

King, R. J. & C. F. Puttock, 1994. Morphology and taxonomy of *Caloglossa* (Delesseriaceae, Rhodophyta). Aust. syst. Bot. 7: 89–124.

Kurogi, M., K. Akiyama & S. Sato, 1962. Influence of light on the growth and maturation of *Conchocelis*-thallus of *Porphyra*. (I) Effect of photoperiod on the formation of monosporangia and liberation of monospores. Bull. Tohoku reg. Fish. Res. Lab. 20: 121–126.

Masuda, M., 1973. The life history of *Pterosiphonia pennata* (Roth) Falkenberg (Rhodophyceae, Ceramiales) in culture. J. Jap. Bot. 48: 122–127.

Ngan, Y. & I. Price, 1983. Periodicity of spore discharge in tropical Florideophyceae (Rhodophyta). Br. phycol. J. 18: 83–95.

O'Kelly, C. J. & B. J. Baca, 1984. The time course of carpogonial branch and carposporophyte development in *Callithamnion cordatum*. Phycologia 23: 407–417.

Pickett-Heaps, J. & J. West, 1998. Time-lapse video observations on sexual plasmogamy in the red alga *Bostrychia*. Europ. J. Phycol. 33: 43–56.

Rama Rao, K. & P. C. Thomas, 1974. Shedding of carpospores in *Gracilaria edulis* (Gmel.) Silva. Phykos 13: 54–59.

Santelices, B., 1990. Patterns of reproduction, dispersal and recruitment in seaweeds. Oceanogr. mar. Biol. ann. Rev. 28: 177–276.

Searles, R., 1980. The strategy of the red algal life history. Am. Nat. 115: 113–120.

Sagromsky, H., 1961. Durch Licht-Dunkel-Wechsel induzierter Rhythmus in der Entleerung der Tetrasporangien von *Nitophyllum punctatum*. Pubbl. Staz. Zool. Napoli 32: 29–40.

Umamaheswara Rao, M., 1974. Observations on fruiting cycle, spore output and germination of tetraspores of *Gelidiella acerosa* in the Gulf of Mannar. Bot. mar. 17: 204–207.

West, J. & H. Calumpong, 1988. Mixed-phase reproduction of *Bostrychia* (Ceramiales, Rhodophyta) in culture. I. *B. tenella* (Lamouroux) J. Agardh. Jap. J. Phycol. 36: 292–310.

West, J. A. & H. Calumpong, 1989. Reproductive biology of *Spyridia filamentosa* (Wulfen) Harvey (Rhodophyta) in culture. Bot. mar. 32: 379–387.

West, J. A. & E. Crump, 1975. Carpospore discharge periodicity in excised cystocarpic papillae of *Gigartina-Petrocelis*. (Rhodophyta). J. Phycol. 11 (supplement): 17.

West, J. A., G. Zuccarello, U. Karsten & H. P. Calumpong, 1993. Biology of *Bostrychia*, *Stictosiphonia* and *Caloglossa*

(Rhodophyta, Ceramiales). In Calumpong, H. & E. Menez (eds), Proceedings of the Second RP-USA Phycology Symposium/Workshop. 6–19 January 1992. Cebu & Dumaguete. PCAMRD, Los Banos: 145–162.

West, J. A., G. C. Zuccarello, F. Pedroche & U. Karsten, 1994. *Caloglossa apomeiotica* sp. nov. (Ceramiales, Rhodophyta) from Pacific Mexico. Bot. mar. 37: 381–390.

Zuccarello, G. C. & J. A. West, 1997. Hybridization studies in *Bostrychia*: 2. Correlation of crossing data and plastid DNA sequence data within *B. radicans* and *B. moritziana* (Rhodophyta, Ceramiales). Phycologia 36: 293–304.

Porphyra sp. (Bangiales, Rhodophyta): reproduction and life form

A. Candia[1], S. Lindstrom[2] & E. Reyes[3]

[1]*División de Fomento Acuicultura, Instituto de Fomento Pesquero, Puerto Montt, Chile*
[2]*Department of Botany, University of British Columbia, Canada*
[3]*Departamento de Biología Molecular, Universidad de Concepción, Chile*

Key words: archeospores, asexual, epiphyte, *Porphyra*, reproduction

Abstract

The reproductive pattern of *Porphyra* sp. was observed in laboratory culture. The thalli obtained from both natural populations and from cultures were monostromatic and reniform, obovate to linear-oblanceolate. Under the different culture conditions, no differentiation of sexual cells, spermatia or carpogonia, was observed; propagation occurred by archeospore formation (*sensu* Magne, 1991), with bipolar germination. Thirty successive generations have been obtained in culture, at temperatures of 12 °C and 15 °C, and each time a laminar gametophytic thallus was produced. It is concluded that *Porphyra* sp. has a strictly asexual reproductive pattern. This is interpreted as a reproductive adaptation to its epiphytic life form.

Introduction

Species of the genus *Porphyra* C. Agardh possess a biphasic life history with a macroscopic laminar gametophytic phase, and a microscopic sporophytic filamentous phase, known as the conchocelis (Drew, 1949: Miura, 1975; Cole & Conway, 1980; West & Hommersand, 1981). The gametophytic phase reproduces sexually through fertilization of the carpogonium by the spermatium and subsequent carpospore formation (Hawkes, 1978); the development of these spores gives rise to the conchocelis phase.

The gametophytic thallus may possess an alternative method of asexual reproduction, by the formation of monospores or archeospores (*sensu* Magne, 1991) and aplanospores or neutral spores (*sensu* Nelson & Knight 1996), which germinate in a bipolar way to reproduce the gametophytic phase (Bird, 1973; Miura, 1975; Conway et al., 1975; Kapraun & Lüster, 1980; West & Hommersand, 1981; Hymes & Cole, 1983; Polne-Fuller & Gibor, 1984; Li, 1984; Nelson & Knight, 1996). This asexual multiplication of the gametophytic thallus is common in some species during the winter months, under short photoperiod conditions (Hawkes, 1977; Cole & Conway, 1980; Hymes & Cole, 1983); it may also take place when the thallus cells are stimulated by enzymatic treatment (Polne-Fuller & Gibor, 1984), or when irradiance is increased in culture (Li, 1984). Usually, asexual propagation of the gametophytic thallus has been observed in *Porphyra* by the formation of aplanospores or neutral spores (spore bundles: 2, 4, 16, 32 spores), as in the case of *P. subtumens* J. Agardh ex Laing. (Conway & Wylie, 1972; Nelson & Knight, 1996), *P. perforata* (= *P. sanjuanensis*) (Krishnamurthy, 1969; Conway et al., 1975; Lindstrom & Cole, 1990) and *P. argentinensis* (Piriz, 1981).

Observations conducted on gametophytic *Porphyra* populations in the littoral of Bahía San Vicente, Chile, showed the presence of specimens whose reproductive features differed from those of the monoecious species *P. columbina*, the only species of this genus cited for the Chilean central coast (Levring, 1960) which, under laboratory conditions, possesses the typical life cycle of the genus (Etcheverry & Collantes, 1977; Avila et al., 1986). The purpose of this study was to determine the reproductive pattern and the life form of this *Porphyra* species in laboratory culture.

Materials and methods

Porphyra sp. thalli were collected seasonally (January, April, August, October) in the locality of Lenga,

Table 1. Culture conditions of thalli and spores of Porphyra sp.

Temperature (°C)	Photoperiod (hours light:dark)	Irradiance (μmol m^{-2}s^{-1})
10	16:8	10
12	16:8 (8:16)	12 (25)
15	16:8	10
20	24 light	20

Bahía San Vicente (36° 45′ S; 73° 10′ W), and taken to the laboratory in boxes maintained at ca. 4 °C in a large volume of seawater. Collections and cultures were made as follows: mature thalli were separated and washed in sterilized seawater; 15 thalli were selected and a section of 1 mm² was cut from the margin of each thallus. For each experimental temperature, 5 pieces of tissue from 5 different thalli, were placed in each of three Petri dishes (20 × 100 mm), containing filtered sea water that had been sterilized and enriched with Provasoli solution (Provasoli, 1968), in order to obtain a spore solution. These cultures were kept at 10 °C, 16 h light:8h dark and 10 μmol m^{-2} s^{-1}. After one day, a spore solution was formed in each Petri dish. A 1 ml aliquot of each spore solution was inoculated into 3 new Petri dishes containing filtered seawater and enriched with Provasoli solution. The cultures were kept under different conditions of temperature, photoperiod and irradiance (Table 1). Light was provided by cool-white 20 watt fluorescent tubes. After 6 months, the cultures maintained at 12 °C, 16 h daylength and 12 μmol m^{-2} s^{-1} were transferred to 8 h daylength and 25 μmol m^{-2}s^{-1}. Cultures were periodically monitored for spore development and thalli maturity.

From the successive generations, obtained from the initial cultures established each season, 15 thalli were selected to initiate new cultures and the steps to obtain spore solutions repeated. Three Petri dishes containing 1 ml of spore solution from each thallus were then cultured at 12 °C and 15 °C.

The vegetative and reproductive anatomies of the thalli from natural populations and from cultures, either fresh or preserved in 4% formalin, were observed in semi-permanent preparations of hand-cut sections and set in 60% Karo. Cell measurements were recorded under the microscope and photomicrographs and photographs were taken.

Results

Porphyra sp. has a monostromatic laminar thallus, ovobate to linear-oblancelolate (Figure 1), with a plain margin of a reddish to pink colour. The thallus produces rhizoids, which aggregate to form an adhesive disc (holdfast) when growing on a hard substrate (Petri dishes). The mature fronds are from 2 mm to 50 mm long. The thickness of the thallus ranges from 16 to 23 μm in its mid-part. The cell shape in cross-section is rectangular to ovate, 20 to 30 μm in length and 15 to 26 μm in width (Figure 2).

In both cultured and field collected thalli, reproductive cell formation occurred without cell division, in the apical margin of the thallus. No sporangial formation was observed, the whole protoplasm of the vegetative cell functioning as the reproductive cell (Figure 3). In keeping with this form of spore production, in *Porphyra* sp. we adopted the term 'archeospore' (*sensu* Magne, 1991). Spore generation on cultured thalli may occur gradually from apical cells to basal cells of the thallus, or the cells of the marginal and mid-part of the thallus may be massively released and germinate. In cultures, very early propagation was observed with thalli of 1–3 mm in length generating and massively liberating spores.

At both temperatures, spore release from incubated frond sections occurred after 2–36 hours. Mean spore diameter was 17 μm. Under the initial culture conditions, and after changes in photoperiod and irradiance, the spores always germinated bipolarly, giving rise to gametophytic thalli which were similar to the parents. Maturity of these cultured fronds occurred after 2–3 weeks at the different temperatures used.

Porphyra sp. grows as an epiphyte on gametophytic and tetrasporic thalli of *Gracilaria chilensis* Bird, McLachlan and Oliveira, *Grateloupia* sp. and *Mazzaella membranacea* J. Agardh. The attachment to the host thallus and subsequent development of spores are similar for all three basiphyte species. Spores are retained in a stratum formed by mucilage and microorganisms; they then initiate their bipolar germination, and one to four rhizoids perforate the cuticle and penetrate between the cortical cells. In *G. chilensis*, the hole increases in size with rhizoidal growth. *Grateloupia* sp. and *M. membranacea* produce a lump formed by the cuticle and cortical cells of the host thallus. Attachment of thalli into the basiphyte results from the growth and elongation of rhizoids; they penetrate between cortical and subcortical cells and toward the external medullar cells of the host thallus

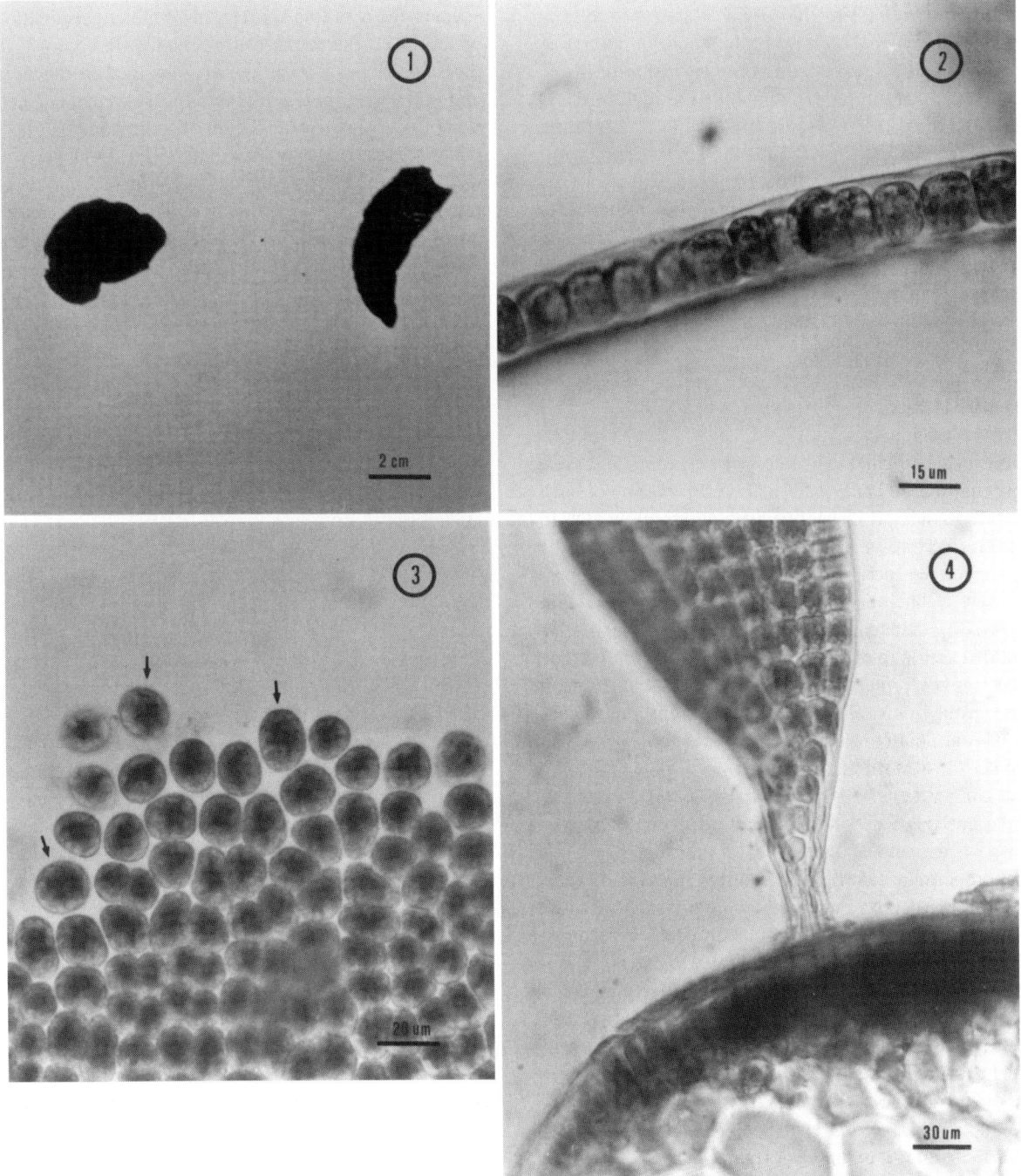

Figures 1–4. **Figure 1.** Morphology of *Porphyra* sp. thalli collected from a natural population. **Figure 2.** Cross section of the monostromatic gametophytic thallus. **Figure 3.** Apical portion of the gametophytic thallus of *Porphyra* sp. with archeospore release (arrows). **Figure 4.** *Porphyra* sp. growing epiphytically on the gametophytic thallus of *Gracilaria chilensis*.

(Figure 4). On the fronds of *Grateloupia* and *Mazzaella*, thalli of *Porphyra* sp. are mainly distributed in the mid sub-apical part of the frond (about 40–50 thalli per cm^2), whereas in *Gracilaria*, they are distributed throughout the length of the thallus, but concentrated in the mid-part of the branches. Densities ranging from 3 to 150 thalli per cm were observed on thalli collected during autumn and spring.

Porphyra thalli produced viable spores in all seasons of the year. This was verified both in laboratory cultures and in collections of basiphytes with developing (20–40 μm long) and mature (3–50 mm) thalli. In laboratory culture, spore production was initiated when the thalli reached 0.5–6 mm.

After culturing *Porphyra* sp. under different conditions of temperature, photoperiod and irradiance, or with fluctuations in these parameters, and having repeated the cycle for 30 successive generations at temperatures of 12 °C and 15 °C, a laminar gametophytic thallus was always obtained.

Discussion

The results obtained indicate that *Porphyra* sp. has an asexual reproductive pattern. The observation of such a pattern over 30 successive generations allowed us to determine that in this species reproduction could be considered as obligatory and not facultative. Its reproductive behaviour might be due to its epiphytic life form on other macroalgae and, therefore, represents an adaptation to the habitat in which it develops. The early spore formation and the massive colonization on the basiphytes would also explain this reproductive adaptation. Thalli from this species were not observed on rocky substrata in the study area.

The presence of *Porphyra* thalli throughout the year and the observation of viable spores in any season of the annual cycle would indicate a lack of photoperiodic control in the differentiation of their reproductive cells, a common occurrence in some species with a facultative asexual cycle (Kapraun & Lüster, 1980; Cole & Conway, 1980). *Porphyra* sp. possesses reproductive features that differentiates it from the other species of the genus that have been reported to have a single asexual reproductive pattern. These differences are related to the origin of the spores in the foliose thallus: *P. perforata* (= *P. sanjuanensis*) and *P. argentinensis* grow in the intertidal on rocky substrate, spores originate in the same way as carpospores, with 'bundles' of spores in numbers of 16 to 32, and of a similar size (Krishnamurthy, 1969; Piriz, 1981), *P. subtumens* grows epiphytically on *Durvillaea* thalli, spores are formed by a vegetative cell as an archeospore or by neutral sporangia development of a protothallus structure (Nelson & Knight, 1996), *P. fucicola* (= *P. maculosa*), spores are formed in bundles of 2–4 which, once released, develop in a bipolar way (Hymes & Cole, 1983). In all these species, the resulting spores show bipolar germination which gives rise to the foliose thallus. In contrast, in the specimens of *Porphyra* sp. obtained in the field and in culture, spores originate from a vegetative cell without divisions or formation of spore bundles.

The dimensions of the thallus, its life form and its reproductive form, together with the fact that these characteristics are not found in other *Porphyra* species cited for the Chilean littoral, all support the view that *Porphyra* sp. is a new, undescribed, species.

Acknowledgments

Financial support was provided by Projects FONDECYT 2960070 granted to A. Candia. We acknowledge the two anonymous reviewers who have improved this manuscript.

References

Avila, M., B. Santelices & J. McLachlan, 1986. Photoperiod and temperature regulation of the life history of *Porphyra columbina* (Rhodophyta, Bangiales) from central Chile. Can. J. Bot. 64: 1867–1872.

Bird, C. J., 1973. Aspects of the life history and ecology of *Porphyra linearis* (Bangiales, Rhodophyceae) in nature. Can. J. Bot. 51: 2371–2379.

Cole, K. M. & E. Conway, 1980. Studies in the Bangiaceae: Reproductive modes. Bot. mar. 23: 545–553.

Conway, E. & P. Wylie, 1972. Spore organization and reproductive modes in two species of *Porphyra* from New Zealand. Proc. int. Seaweed Symp. 7: 105–107.

Conway, E., T. F. Mumford Jr. & R. F. Scagel, 1975. The genus *Porphyra* in British Columbia and Washington. Syesis 8: 185–244.

Drew, K. M., 1949. Conchocelis-phase in the life history of *Porphyra umbilicalis* (L.) Kützing. Nature 164: 748–751.

Etcheverry, H. & G. Collantes, 1977. Cultivo artificial del luche, *Porphyra columbina* Montagne (Rhodophyta, Bangiaceae). Rev. Biol. Mar., Valparaíso 16: 195–202.

Hawkes, M. W., 1977. A field, culture and cytological study of *Porphyra gardneri* (Smith & Hollenberg) comb. nov. (= *Porphyrella gardneri* Smith & Hollenberg), (Bangiales, Rhodophyta). Phycologia 16: 457–469.

Hawkes, M. W., 1978. Sexual reproduction in *Porphyra gardneri* (Smith et Hollenberg) Hawkes (Bangiales, Rhodophyta). Phycologia 17: 329–353.

Hymes, B. J. & K. M. Cole, 1983. Aplanospore production in *Porphyra maculosa* (Rhodophyta). Jap. J. Phycol. (Sorui) 31: 225–228.

Kapraun, D. F. & D. G. Lüster, 1980. Field and culture studies of *Porphyra rosengurtii* Coll et Cox (Rhodophyta, Bangiales) from North Carolina. Bot. mar. 23: 449–457.

Krishnamurthy, V., 1969. On two species of *Porphyra* from San Juan Island, Washington. Proc. int. Seaweed Symp. 6: 225–234.

Levring, T. 1960. Contributions to the marine algae flora of Chile. Lunds Univ. Arsskr., adv.2, 56: 1–85.

Li, S. Y., 1984. The ecological characteristics of monospores of *Porphyra yezoensis* Ueda and their use in cultivation. Hydrobiologia 116/117: 255–258.

Lindstrom, S. C. & K. M. Cole, 1990. An evaluation of species relationships in the *Porphyra perforata* complex (Bangiales, Rhodophyta) using starch gel electrophoresis. Proc. int. Seaweed Symp. 13: 179–183.

Magne, F., 1991. Classification and phylogeny in the lower Rhodophyta: a new proposal. J. Phycol. 27 (suppl.): 46.

Miura, A., 1975. *Porphyra* cultivation in Japan. In Tokida J. & H. Hirose (eds), Advance of Phycology in Japan. VEB Gustav Fischer Verlag Jena: 273–304

Nelson, W. A. & G. A. Knight, 1996. Life history in culture of the obligate epiphyte *Porphyra subtumens* (Bangiales, Rhodophyta) endemic to New Zealand. Phycol. Res. 44: 19–26.

Piriz, I. M., 1981. A new species and a new record of *Porphyra* (Bangiales, Rhodophyta) from Argentina. Bot. mar. 24: 599–602.

Polne-Fuller, M. & A. Gibor, 1984. Developmental studies in *Porphyra*. Blade differentiation in *Porphyra perforata* as expressed by morphology, enzymatic digestion and protoplast regeneration. J. Phycol. 20: 609–616.

Provasoli, L., 1968. Media and prospect for the cultivation of marine algae. In Watanabe A. & A. Hattori (eds), Cultures and Collections of Algae. Proc. U.S. Japan Conf. Hakone, Sept. 1966. Jap. Soc. Plant Physiol: 63–75.

West, J. A. & M. H. Hommersand, 1981. Rhodophyta: life histories. In Lobban C. S. & M. J. Wynne (eds), The Biology of Seaweeds. Blackwell, Oxford: 133–193.

Life history, in culture, of the obligate epiphyte *Porphyra moriensis* (Bangiales, Rhodophyta)

Masahiro Notoya & Akinori Miyashita
Laboratory of Applied Phycology, Tokyo University of Fisheries, Konan-4, Minato-ku, Tokyo, 108 Japan

Key words: Bangiales, *Chorda filum*, life history, obligate epiphyte, *Porphyra moriensis*, Rhodophyta

Abstract

The life history of the obligate epiphyte *Porphyra moriensis* Ohmi, collected from the type locality in Japan, was demonstrated in culture. Conchocelis filaments were cultured at temperatures between 5 and 30 °C under photon irradiances of 10–80 μmol m^{-2}s^{-1} and photoperiods of 10L:14D or 14L:10D. Conchosporangial branches developed at 5–20 °C under both photoperiods, and were abundant when the conchocelis was cultured at 20 °C, 20 μmol m^{-2}s^{-1}, 14L:10D and 15 °C, 40 μmol m^{-2}s^{-1}, 10L:14D. No foliose thalli survived at 25 or 30 °C and conchocelis failed to survive at 30 °C. Released conchospores attached very poorly to synthetic strings and glass slides but settled very well on *Chorda filum* thalli and thereafter germinated to develop into foliose thalli. Thalli grew well at 15 °C and 14L:10D or 10L:14D and at 20 °C and 10L:14D. No archeospores were observed on the thallus under any culture conditions. Spermatangial and zygotosporangial sori were formed in patches along the upper margins of mature thalli. Anatomical examination revealed that the mature spermatangial packets were 128 (a/4, b/4, c/8) and zygotosporangia were 16 (a/2, b/2, c/4) according to the Hus formula. Released zygotospores gave rise to conchocelis filaments. Neither monospores nor protothalli were produced by the conchocelis in culture. The life history of this plant is a simple biphasic cycle, with the thallus stage appearing to require attachment to *Chorda filum*.

Introduction

Twenty-eight species of *Porphyra* have been reported from Japan (Yoshida et al., 1995); a few of these are obligate epiphytes, endemic to Japan. *Porphyra moriensis* Ohmi, epiphytic on *Chorda filum* (Laminariales), was first described by Ohmi (1954), and later, Fukuhara (1968) studied its morphology and distribution along the coasts of Uchiura Bay, Hokkaido and the Pacific coast of southern Hokkaido.

The present study attempted to demonstrate, in culture, the life history of *P. moriensis* and to elucidate the influence of various environmental conditions (temperature, irradiance and photoperiod) on the growth and maturation of the foliose thallus and conchocelis, as well as the relationship between this species and *C. filum*.

Materials and methods

Mature foliose thalli of *Porphyra moriensis* Ohmi epiphytic on *Chorda filum* (Figure 1) were collected on 9 April 1997 from the type locality, Mori near Hakodate, Hokkaido, Japan. Pieces of about 1 × 1 cm were excised from the upper portion of the mature thallus, and the surfaces cleaned with an artist's brush in sterilized seawater. The pieces were kept overnight in an antibiotic solution (Polne-Fuller & Gibor, 1987) with 5 ppm of GeO$_2$ at room temperature in the dark and then transferred into Petri dishes containing sterilized seawater for several hours to obtain zygotospores (following the terminology of Guiry, 1990). Liberated spores were collected by glass pipette and transferred to a new Petri dish containing sterilized seawater. The spores were washed three to four times. After the spores were firmly attached to glass slides, they were transferred to glass tubes containing 50 ml of medium and cultured at 5, 10, 15, 20, 25 and 30 °C at photon irradiances of 10, 20, 40 and 80 μmol m^{-2}s^{-1} and photoperiods of 14L:10D or 10L:14D.

The conchocelis colonies were examined weekly for conchosporangial branches. The growth of conchocelis was measured at the densest part of ten

Figure 1. Mature foliose thalli of *Porphyra moriensis* epiphytic on *Chorda filum*, collected on 9 April 1997 from Mori, Hokkaido, Japan. Scale bar, 5 cm.

clumps. Mature filaments with conchosporangial branches were cultured at 15 °C under a photon irradiance of 80 μmol m^{-2}s^{-1} and 14L:10D, together with a few young thalli of *C. filum, Laminaria japonica* and *Undaria pinnatifida* (1 cm blade length) which had been cultured for a month at 10°C, 80 μmol m^{-2}s^{-1} and 14L:10D. Released conchospores germinated and gave rise to plantlets, which were then cultured for 5 weeks at 15 °C, 80 μmol m^{-2}s^{-1} and 14L:10D in a 300 ml flask with aeration. After these thalli reached 1 cm in length, they were cultured at a range of temperatures (5, 10, 15, 20, 25 and 30 °C) under a photon irradiance of 80 μmol m^{-2}s^{-1} and photoperiod of either 14L:10D or 10L:14D to determine their growth and maturation. All cultures were examined weekly. Modified Grund medium (McLachlan, 1973) was used throughout the study and the medium was replaced weekly. Voucher specimens of the material are deposited in the Laboratory of Applied Phycology, Tokyo University of Fisheries.

Results

Zygotospores from the foliose thallus germinated to give rise to conchocelis filaments, which grew to a colony of about 3 mm diameter at 20 °C after 5 weeks in culture. The cells of the conchocelis filaments were 10 μm in diameter and had parietal chloroplasts. The conchosporangial branches were about 16–20 μm in diameter and were produced from the conchocelis filaments at 5–20 °C under both long and short days, after 5–9 weeks in culture (Figure 2A, B). Conchospores were liberated under these same culture conditions.

Released conchospores attached very poorly to synthetic string, glass slides and young thalli of *L. japonica* and *U. pinnatifida* and failed to develop into thalli. In contrast, spores attached very well to *C. filum* thalli and developed into foliose thalli (Figure 2, C–E). Young foliose thalli removed from *C. filum* thalli grew well and matured after some weeks in culture.

Conchospore germlings died at temperatures of 25 and 30 °C under all photon irradiances and photoperiods. At 15 °C, foliose thallus grew well and matured quickly. Patches of spermatangia or zygotosporangia developed along the distal margin of the thallus at 5–20 °C under both photoperiods, within 12 weeks (Figure 2F, G). The formation of the spermatangia, carpogonia and zygotosporangia was observed anatomically: 128 (a/4, b/4, c/8) for spermatangia and 16 (a/2, b/2, c/4) for zygotosporangia according the Hus (1902) formula.

Growth and maturation of conchocelis under various culture conditions are shown in Figure 3. The conchocelis filaments did not survive at a temperature of 30 °C, with most of the colonies dying after 7 days under both photoperiods. However, the conchocelis filaments survived at 5–25 °C, with good growth at 15–25 °C and photon irradiances of 20–80 μmol m^{-2}s^{-1} under both photoperiods.

Under long days and temperatures of 5 and 10 °C, more conchosporangial branches developed at lower than at higher irradiances, whereas at higher temperatures (15 and 20 °C) and short days the reverse was true. The formation rate of conchosporangial branches reached 100% within 8–9 weeks at 15 °C, 40–80 μmol m^{-2}s^{-1} and 10L:14D and at 20 °C, 20 μmol m^{-2}s^{-1} and 14L:10D (Figure 4).

Foliose thallus growth at various temperatures and both photoperiods is shown in Figure 5. Fastest growth and maturation occurred at 15 °C under long days (Figure 5); the thalli grew to 15–16 cm length within 3 weeks. Under short days, faster growth of thalli occurred at 15 and 20 °C. Thalli grown at 25 °C died within 1 week. The earliest maturation of the foliose thallus, after only 2 weeks in culture, was found at 20 °C under long days (Figure 6). At 10–20 °C the mature thalli eroded and became shorter after the release of reproductive cells, whereas thalli continued to grow during the entire culture period of ten weeks at 5 °C.

Liberation of spermatia preceded the release of zygotospores by almost 1 week, and reproductive

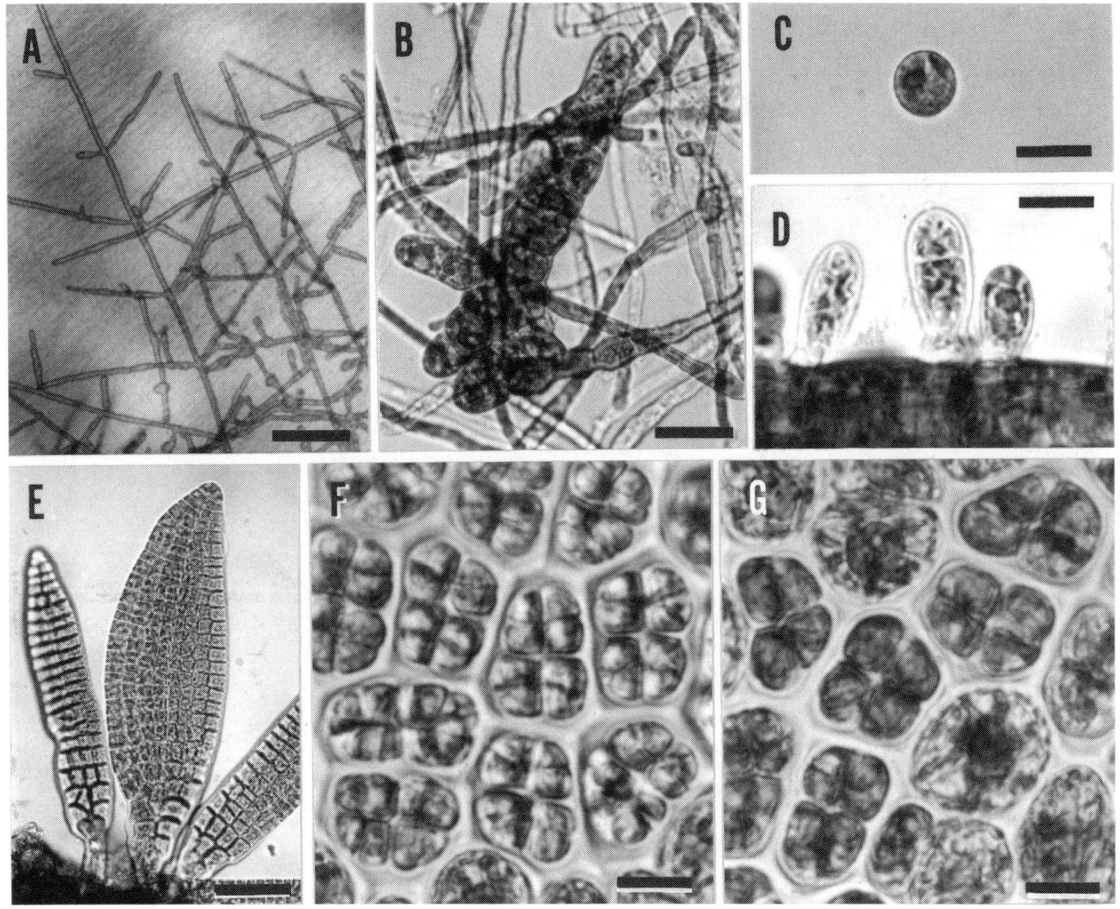

Figure 2. Life history of *Porphyra moriensis*. A: Conchocelis growing in a shell at 20 °C and a photon irradiance of 40 μmol m^{-2}s^{-1} under 14L:10D. B: Conchosporangial branches on 7-week-old conchocelis at 15 °C, 40 μmol m^{-2}s^{-1}, 10L:14D. C: Liberated conchospore. D: one-to-two celled stage of conchospore (3-day-old) at 10 °C, 80 μmol m^{-2}s^{-1}, 14L:10D attached to the host, *Chorda filum*. E: Juvenile foliose thalli at 2 weeks of age at 15 °C, 80 μmol m^{-2}s^{-1}, 10L:14D. F: Surface view of spermatangia. G: Surface view of zygotosporangia. Scale bars, 80 μm in A; 40 μm in B, E; 20 μm in C, D, F, G.

cells generally matured faster under long days and higher temperatures (Figure 6). Under both photoperiods at 5 °C few reproductive cells were evident and discharged.

Discussion

The life history of *P. moriensis* from Mori, Hokkaido, an epiphyte of *C. filum*, has a typical biphasic life history of the *P. lacerata*-type (Notoya et al., 1993b). However, archeospores were not formed on the foliose thallus under any culture conditions. Asexual reproduction by the foliose thallus (e.g. agamospore (Kornmann, 1994), neutral spore (Hollenberg, 1958), endospore (Nelson & Knight, 1996)), or by the conchocelis (e.g. monospore (Chen et al., 1970), protothallus (Cole & Conway, 1980), spherical cell (Notoya et al., 1993a)) was not observed.

The present culture studies demonstrated that liberated conchospores settled on the thallus of *C. filum* and germinated to form the foliose thallus, which matured at between 5 and 20 °C. Only a few conchospores settled and germinated on glass slides and synthetic strings. This result suggests that this species has an obligate epiphytic habit, and may have a specific relationship with *C. filum*. However, young foliose thalli grew well and reproduced when unattached. Thus, it seems that the *C. filum* thallus is necessary for settlement but not for growth or maturation of the foliose thallus. Further investigations are required to understand the requirements of conchospore settlement.

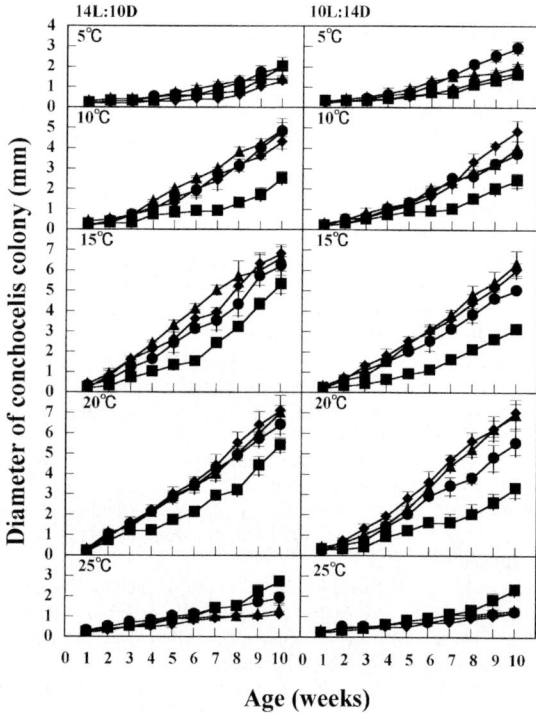

Figure 3. Growth of conchocelis of *Porphyra moriensis* at different temperatures, irradiance and photoperiods. Squares, 10 μmol m^{-2}s^{-1}; circles, 20 μmol m^{-2}s^{-1}; triangles, 40 μmol m^{-2}s^{-1}; diamonds, 80 μmol m^{-2}s^{-1}. Vertical bars are standard deviations.

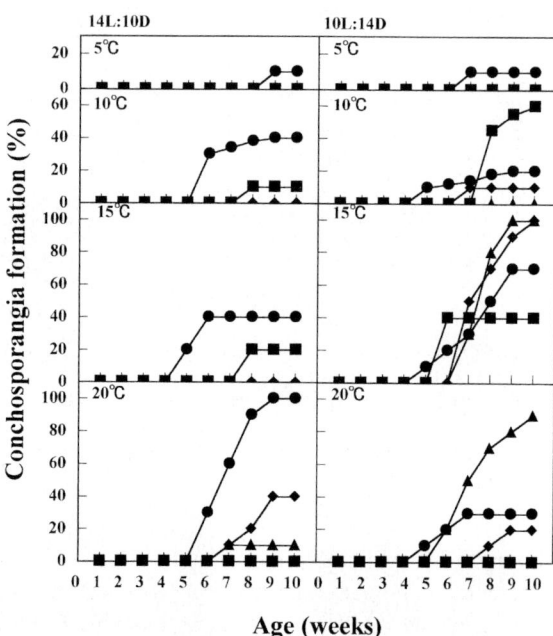

Figure 4. Formation of conchosporangial branches in *Porphyra moriensis* at different temperatures, irradiance and photoperiods. Squares, 10 μmol m^{-2}s^{-1}; circles, 20 μmol m^{-2}s^{-1}; triangles, 40 μmol m^{-2}s^{-1}; diamonds, 80 μmol m^{-2}s^{-1}. Vertical bars are standard deviations.

Hawkes (1977, 1978) and Nelson & Knight (1996) have reported on culture studies of the obligate epiphytic species *P. gardneri* and *P. subtumens*, respectively. Wound sites were found to be important for conchospore establishment of *P. gardneri* on the host (Hawkes, 1977) but not for conchospores or archeospores of *P. subtumens* to settle on *Durvillaea* (Nelson & Knight, 1996). Our studies on *P. moriensis* found that wound sites were not important for settlement of the conchospores.

The conchocelis grew between 5 and 25 °C under both long and short days, but optimum temperatures for growth and maturation were 15–20 °C. The range of temperatures for reproduction of conchosporangial branches was slightly lower than those of other temperate species of *Porphyra* from Japan (Iima & Migita, 1990; Iwasaki, 1961; Kurogi & Saito, 1962; Migita & Ito, 1987; Notoya et al., 1992, 1993a,b).

The foliose thalli grew at 5–20 °C under both photoperiods, with maturation occurring earlier at higher temperatures. However, the results suggest that the time of release of reproductive cells is related to thal-

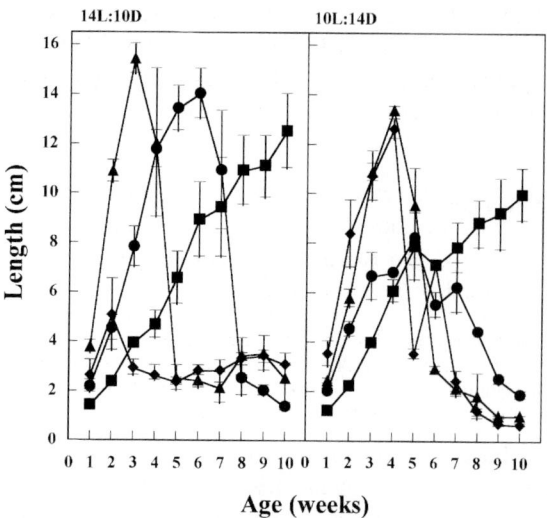

Figure 5. Growth of foliose thalli in *Porphyra moriensis* at different temperatures and photoperiods at a photon irradiance of 80 μmol m^{-2}s^{-1}. Squares, 5 °C; circles, 10 °C; triangles, 15 °C; diamonds, 20 °C. Vertical bars are standard deviations.

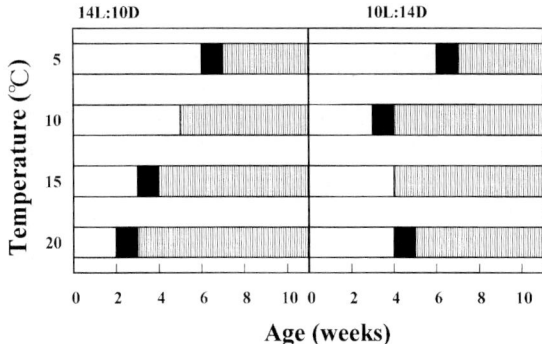

Figure 6. Liberation of spermatia and zygotospores in relation to the age of cultured thalli of *Porphyra moriensis* at different temperatures and photoperiods at a photon irradiance of 80 μmol m^{-2}s^{-1}. Open space, neither spermatia nor zygotospores liberated; solid space, spermatia liberated; vertical stripes, both spermatia and zygotospores liberated.

lus size. Thus the largest thalli will be obtained at 5°C and either photoperiod after a more prolonged culture period, which is similar to material found in the field.

Hus' formulae for spermatangia and zygotosporangia on culture thalli agree with earlier reports by Ohmi (1954) and Fukuhara (1968) from field material.

Acknowledgements

We are very grateful to Dr L. M.-C. Chen for his critical reading of the manuscript. We also thank Mr R. Terada of Faculty of Fisheries, Hokkaido University for providing specimens of *P. moriensis* and *C. filum* for culture material from Mori.

References

Chen, L. C.-M., T. Edelstein, E. Ogata, & J. McLachlan, 1970. The life history of *Porphyra miniata*. Can. J. Bot. 48: 385–389.

Cole, K. M. & E. Conway, 1980. Studies in the Bangiaceae: Reproductive modes. Bot. mar. 23: 545–553.

Fukuhara, E., 1968. Studies on the taxonomy and ecology of *Porphyra* of Hokkaido and its adjacent water. Bull. Hokkaido Reg. Fish. Res. Lab. 34: 40–99 (in Japanese).

Guiry, M. D.,1990. Sporangia and spores. In Cole, K. M. & R. G. Sheath (eds), Biology of the Red Algae. Cambridge University Press, New York: 347–376.

Hawkes, M. W., 1977. A field, culture and cytological study of *Porphyra gardneri* (Smith & Hollenberg) comb. nov. (=*Porphyra gardneri* Smith & Hollenberg), (Bangiales, Rhodophyta). Phycologia 16: 457–469.

Hawkes, M. W., 1978. Sexual reproduction in *Porphyra gardneri* (Smith et Hollenberg) Hawkes (Bangiales, Rhodophyta). Phycologia 17: 326–353.

Hollenberg, G. J., 1958. Culture studies of marine algae. III. *Porphyra perforata*. Am. J. Bot. 45: 653–656.

Hus, H. T. A., 1902. An account of the species of *Porphyra* found on the Pacific coast of North America. Proc. calif. Acad. Sci., 3rd Ser. 2: 173–240.

Iima, M. & S. Migita, 1990. Laboratory culture of *Porphyra lacerata* (Rhodophyta, Bangiales). Bull. Fac. Fish. Nagasaki Univ. 68: 13–20 (in Japanese).

Iwasaki, H., 1961. The life-cycle of *Porphyra tenera* in vivo. Biol. Bull. (Woods Hole). 121: 173–187.

Kornmann, P., 1994. Life histories of monostromatic *Porphyra* species as a basis for taxonomy and classification. Eur. J. Phycol. 29: 69–71.

Kurogi, M. & S. Saito, 1962. Influences of light on the growth and maturation of *Conchocelis*-thallus of *Porphyra*. III. Effect of photoperiod in the different species. Bull. Tohoku Reg. Fish. Res. Lab. 20: 138–156 (in Japanese).

McLachlan, J., 1973. Growth media – marine. In Stein, J. R. (ed.), Handbook of Phycological Methods. Cambridge University Press, New York: 25–51.

Migita, S. & R. Ito, 1987. The life history of *Porphyra tanegashimensis* (Rhodophyta, Bangiales) in laboratory culture. Bull. Fac. Fish. Nagasaki Univ. 61: 7–14 (in Japanese).

Nelson, W. A. & G. A. Knight, 1996. Life history in culture of the obligate epiphyte *Porphyra subtumens* (Bangiales, Rhodophyta) endemic to New Zealand. Phycol. Res. 44: 19–25.

Notoya, M., N. Kikuchi, Y. Aruga & A. Miura, 1992. *Porphyra kinositae* (Yamada et Tanaka) Fukuhara (Bangiales, Rhodophyta) in culture. Jap. J. Phycol. 40: 273–278 (in Japanese).

Notoya, M., N. Kikuchi, Y. Aruga & A. Miura, 1993a. Life history of *Porphyra tenuipedalis* Miura (Bangiales, Rhodophyta) in culture. La Mer 31: 125–130 (in Japanese).

Notoya, M., N. Kikuchi, M. Mastuo, Y. Aruga & A. Miura, 1993b. Culture studies of four species of *Porphyra* (Rhodophyta) from Japan. Nippon Suisan Gakkaishi 59: 431–436.

Ohmi, H., 1954. A new species of *Porphyra*, epiphytic on *Chorda filum* from Hokkaido. Bull. Fac. Fish. Hokkaido Univ. 5: 231–234.

Polne-Fuller, M. & A. Gibor, 1987. Calluses and callus-like growth in seaweeds: Induction and culture. Hydrobiologia 151/152: 131–138.

Yoshida, T., K. Yoshinaga & Y. Nakajima, 1995. Check list of marine algae of Japan (revised in 1995). Jap. J. Phycol. 43: 115–171 (in Japanese).

Culture studies of *Porphyra dentata* and *P. pseudolinearis* (Bangiales, Rhodophyta), two dioecious species from Korea

Kim Nam-Gil
Department of Aquaculture, Gyeongsang National University Tongyoung Kyoungnam 650-160, Korea

Key words: Bangiales, culture, dioecious species, *Porphyra dentata*, *P. pseudolinearis*, Rhodophyta

Abstract

The life cycle of two dioecious species, *Porphyra dentata* and *P. pseudolinearis*, from Korea was completed in the laboratory. Conchocelis filaments were cultured at temperatures of 5–30 °C with a photon irradiance of 10–80 μmol m^{-2} s^{-1} under a photoperiod of 14L:10D. In both species, the fastest growth of the conchocelis was observed at 20 °C and 40 μmol m^{-2} s^{-1}, and conchosporangial branches were produced at 10–25 °C and 10–80 μmol m^{-2} s^{-1}. Foliose thalli of both species were cultured at 5–30 °C, an irradiance of 40 μmol m^{-2} s^{-1} and a photoperiod of 10L:14D. The best growth of *P. dentata* and *P. pseudolinearis* was at 10–15 °C and 5 °C, respectively. Archeospores were not formed on either, under any culture condition. Zygotospores of both species were released at 15 °C. They germinated to form the filamentous conchocelis. Neither monospores nor protothalli were produced by the conchocelis of either species in culture. For both species, growth and maturation of the two phases were influenced more by temperature than by irradiance.

Introduction

Porphyra dentata Kjellman (1897) and *P. pseudolinearis* Ueda (1932), originally described from Japan, were recorded in a checklist of marine algae from Korea by Lee & Kang (1986). *Porphyra dentata* is widely distributed in warm temperate waters (Ueda, 1932; Tseng, 1938, 1948; Fukuhara, 1968), while *P. pseudolinearis* is found in cold temperate waters of the North Pacific (Ueda, 1932; Tanaka, 1952; Kang, 1966; Fukuhara, 1968; Wynne, 1972; Perestenko, 1982; Scagel et al., 1989; Lindstrom & Cole, 1992a, b, 1993). In Korea, *P. dentata* and *P. pseudolinearis* occur along the southern and eastern coasts, respectively.

The life histories of several *Porphyra* species from Japan (e.g. *P. dentata, P. kinositae, P. lacerata, P. suborbiculata, P. tanegashimensis, P. tenera, P. tenuipedalis, P. yezoensis*) have been completed in laboratory culture (Iwasaki, 1961; Kito, 1978; Migita & Ito, 1987; Iima & Migita, 1990; Notoya et al., 1992, 1993; Matsuo et al., 1994). However, until now, there have been no studies on the life histories or the effects of temperature, irradiance and photoperiod on the life histories of Korean species of *Porphyra*. Here we report on experiments to determine the influence of these factors on the growth and maturation of the foliose and conchocelis phases of two *Porphyra* species from Korea.

Material and methods

Mature thalli of *Porphyra dentata* and *P. pseudolinearis* were collected from the upper intertidal zone at Wando, Chonnam Prefecture, and Pohang in Kyongpuk Prefecture (Figure 1), on 11 and 28 February 1996, respectively (Figure 2A, Figure 3A). From these thalli, zygotospores (Guiry, 1990) were obtained to start free-living conchocelis cultures. Zygotospores were cultured at six temperatures (5, 10, 15, 20, 25 and 30 °C) and four photon irradiances (10, 20, 40 and 80 μmol m^{-2} s^{-1}) under a photoperiod of 14L:10D (light:dark). The growth of conchocelis was determined by measuring the diameter of colonies at weekly intervals.

Conchospore liberation in both species was induced by reducing the temperature from 20 to 15 °C. The resulting foliose thalli were cultured at temperatures of 5–30 °C, an irradiance of 40 μmol m^{-2} s^{-1} and a photoperiod of 10L:14D. Growth was determined from measurements of blade length and width.

Modified Grund medium (McLachlan, 1973) was used in all cultures and was renewed weekly.

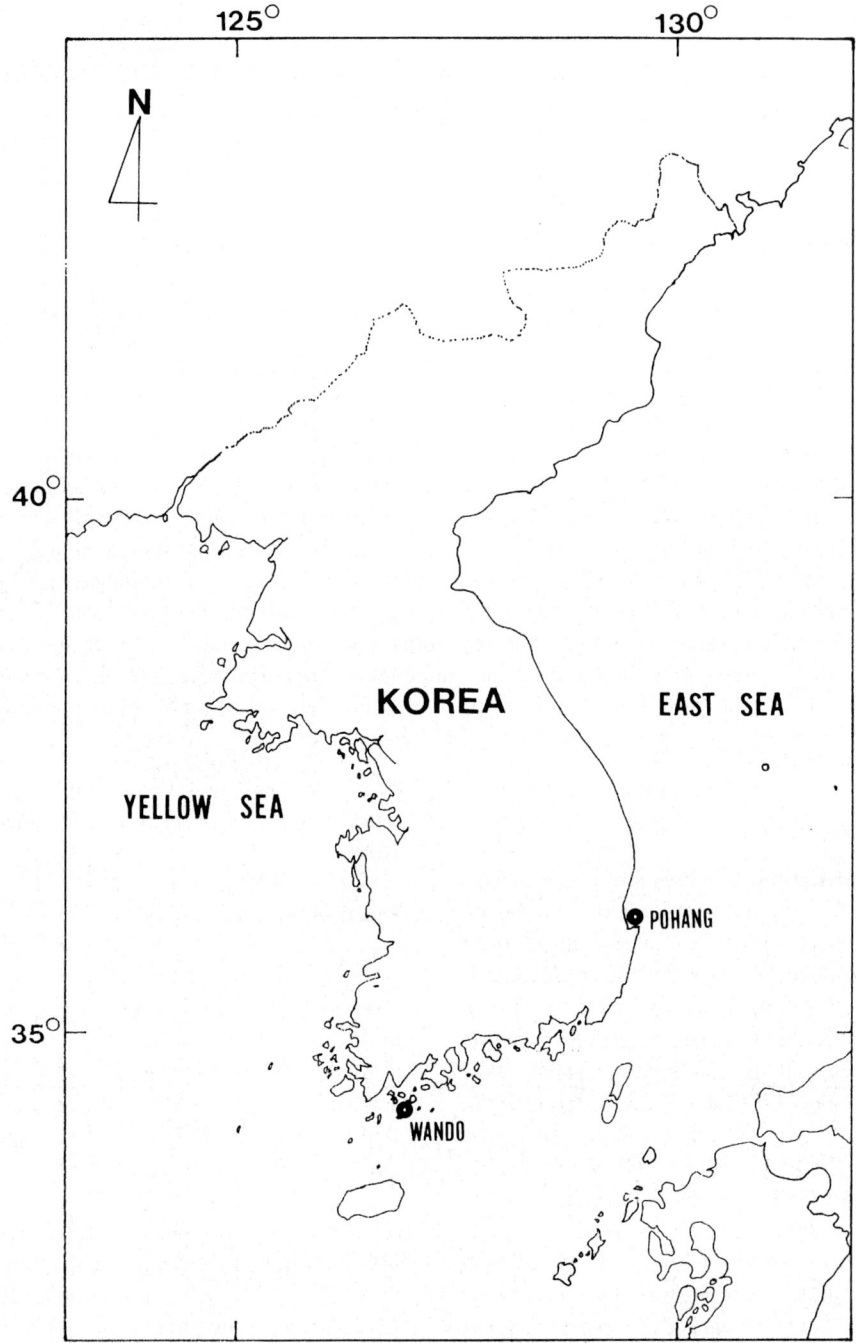

Figure 1. Map of the two sampling sites on the eastern and southern coasts of Korea.

Results

Water temperature in sampling sites

Water temperature ranged between 9.6 and 20.5 °C at Pohang, and 13 and 22.9 °C at Wando. Annual means were 14.6 °C and 16.6 °C at Pohang and Wando, respectively (Figure 4).

Growth and maturation of the conchocelis phase

Zygotospores of *P. dentata* and *P. pseudolinearis* ger-

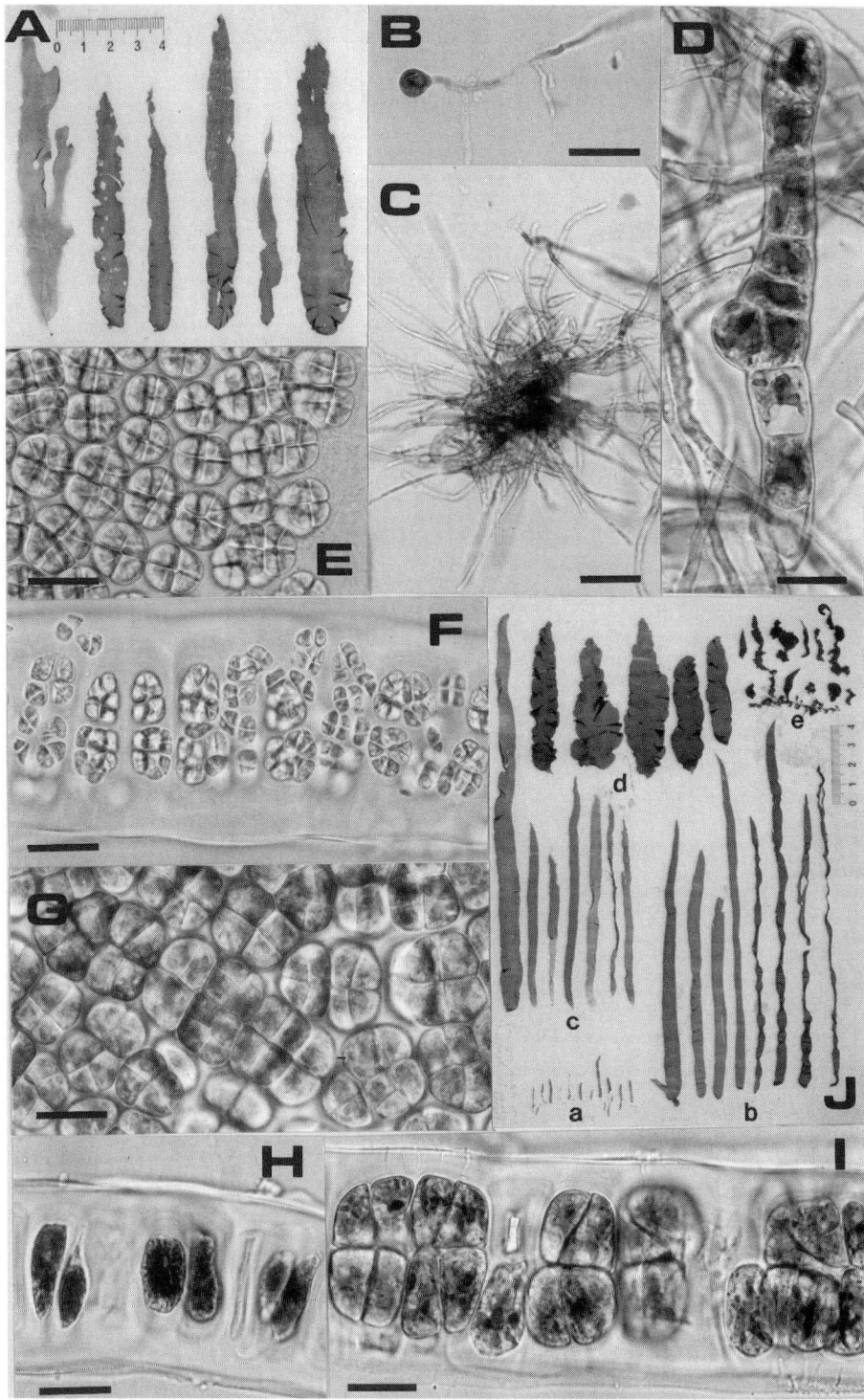

Figure 2. Life history of *Porphyra dentata* Kjellman. (A) Mature foliose thalli of *Porphyra dentata* collected on 11 February 1996 at Wando, Chonnam, Korea. (B) Germinating spore showing developing conchocelis from zygotospore. (C) Free-living conchocelis colony grown at 20 °C and 40 μmol m^{-2} s^{-1}. (D) Conchosporangial branches after 5 weeks at 20 °C and 40 μmol m^{-2} s^{-1}. (E) Surface view of spermatangia. (F) Spermatangia in cross-section. (G) Surface view of zygotosporangia (H) Cross-section showing carpogonia with prototrichogynes. (I) Zygotosporangia in cross section. (J) Foliose thalli of 15 weeks old grown in culture showing different thallus shape; (a) 5 °C, (b) 10 °C, (c) 15 °C, (d) 20 °C, (e) 25 °C. Scale bars, 80 μm in B; 200 μm in C; 20 μm in D, E, F, G, H and I.

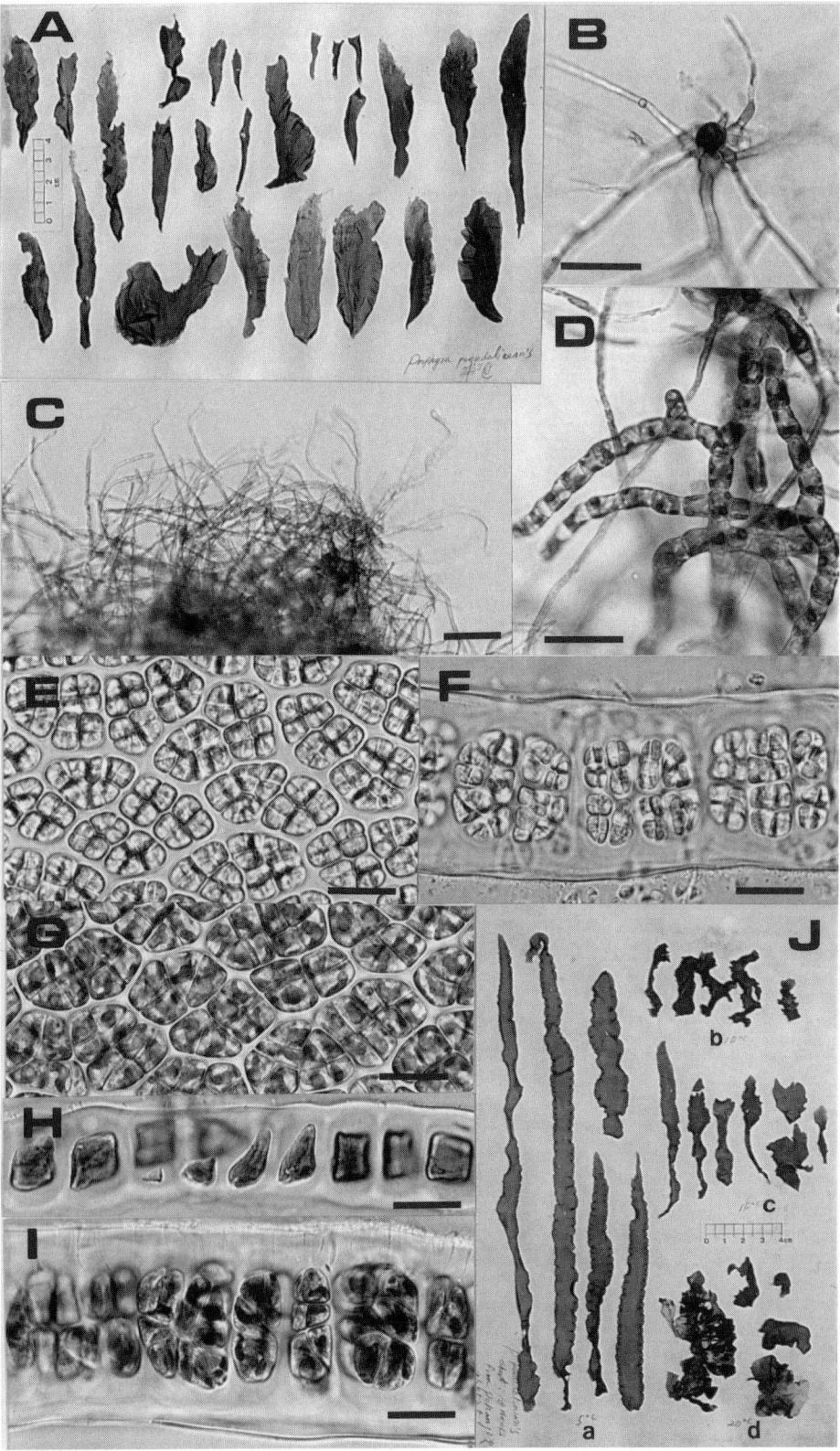

Figure 3.

minated to produce conchocelis colonies at 5–30 °C and 10–25 °C, respectively (Figures 2B,C, 3B,C). Maximum growth of conchocelis in both species was observed at 20 °C and 40 µmol m^{-2} s^{-1}. Conchocelis colonies of *P. dentata* and *P. pseudolinearis* grew to about 6.0 mm and 4.3 mm, respectively within 10 weeks (Figure 5).

Conchosporangial branch formation (Figures 2D, 3D) under various conditions of temperature and photon irradiance is presented in Figure 6. In *P. dentata*, conchocelis colonies produced conchosporangial branches under all conditions except 5 °C, within 6 weeks. Conchosporangial branches of *P. pseudolinearis* were produced at 10–25 °C within 7 weeks but not at 5 or 30 °C. In both species, more conchosporangial branches formed with increasing temperature. The highest production occurred at 25 °C and 20 µmol m^{-2}s^{-1} in *P. dentata*, and at 25 °C and 40 µmol m^{-2} s^{-1} in *P. pseudolinearis*. Within 10 weeks at temperatures of 20 and 25 °C, 100% of colonies had elongated conchosporangial branches (Figure 6). Conchospores were not liberated under any of the culture condition used continuously, in either species. Release of the conchospores was induced by reducing the temperature from 20 to 15 °C. No difference in growth was detected under different irradiances.

Growth and maturation of the thallus phase

Germlings of *P. dentata* grew well at 5–25 °C, with fastest growth occurring at 15 °C. Growth of foliose thalli at 5 and 25 °C was retarded and thalli failed to mature (Figure 7). Germlings died at 30 °C. After 4 weeks in culture, foliose thalli grew to 130 × 20 µm (length by width) at 10 °C, and produced denticulae along the basal margin. Three monoecious foliose thalli were produced at 15 °C, after 15 weeks in culture. After carpogonia production (Figures 2H, 3H), spermatia and zygotospores were liberated at 15 °C after 11 and 15 weeks, respectively.

Germlings of *P. pseudolinearis* reached their largest size at 5 °C after 15 weeks in culture, whereas initially thalli grew faster at 15 °C. At 25 °C, thalli showed almost no growth (Figure 7). Spermatia were liberated at 10 and 15 °C within 15 weeks. Zygotospores were liberated only at 15 °C. Only dioecious thalli were obtained for this species.

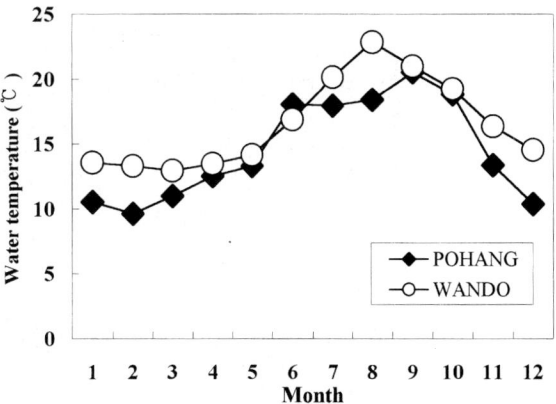

Figure 4. Monthly means of water temperature in 1996 at the two sampling sites.

Archeospores (Magne, 1991) were not produced in either species.

The linear shape of foliose thalli, similar to that in nature, was observed in both species under conditions that produced rapid growth (Figure 8). At 20 and 25 °C, thalli were elliptical or irregular in shape (Figures 2J and 3J).

In *P. dentata*, mature spermatangia had a Hus' (1902) formula of 128, (a/4, b/4, c/8) (Figure 2E, F), and zygotosporangia 16, (a/2, b/2, c/4) (Figure 2G, I). In *P. pseudolinearis*, mature spermatangia had a formula of 128, (a/4, b/4, c/8) (Figure 3E, F), and zygotosporangia 32, (a/2, b/4, c/4) (Figure 3G, I).

Discussion

The life histories of the two species studied were the same as that of *P. dentata* from Japan (Notoya et al., 1993). These authors reported that foliose thalli grew fastest at 20 °C and that conchosporangial branches were formed at between 20 and 25 °C. In the present culture study, growth and maturation of the conchocelis of *P. dentata* occurred at 10–25 °C and formation

←

Figure 3. Life history of *Porphyra pseudolinearis* Ueda. (A) Mature foliose thalli of *Porphyra pseudolinearis* Ueda collected on 28 February 1996 at Pohang, Kyongpuk, Korea. (B) Germinating spore showing developing conchocelis from zygotospore. (C) Free-living conchocelis colony grown at 15 °C and 40 µmol m^{-2} s^{-1}. (D) Conchosporangial branches after 6 weeks at 20 °C and 40 µmol m^{-2} s^{-1}. (E) Surface view of spermatangia. (F) Spermatangia in cross-section. (G) Surface view of zygotosporangia. (H) Carpogonia with prototrichogynes and vegetative cells in cross-section. (I) Zygotosporangia in cross-section. (J) Foliose thalli of 15 weeks old grown in culture showing different thallus shape; (a) 5 °C, (b) 10 °C, (c) 15 °C, (d) 20 °C. Scale bars, 50 µm in B and D; 200 µm in C; 20 µm in E, F, G, H and I.

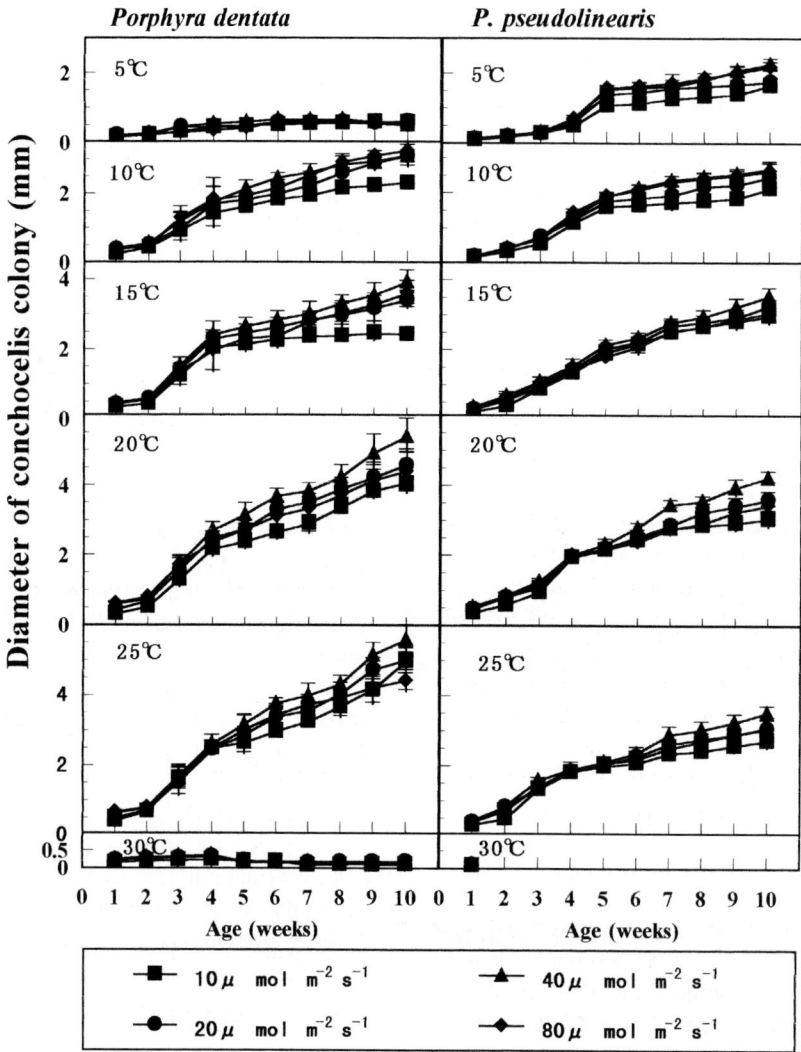

Figure 5. Growth of conchocelis colonies of *Porphyra dentata* and *P. pseudolinearis* at six temperatures (5–30 °C), four photon irradiances (10–80 μmol m^{-2} s^{-1} and 14L:10D. Vertical bars are standard deviations.

of conchosporangial branches occurred most abundantly between 20 and 25 °C. The foliose thalli grew fastest at 15 °C followed by 10 °C, although good growth was also observed at 20 and 25 °C. Our results suggest that this species should grow well at the wide range of temperatures they encounter in nature.

The growth of *P. pseudolinearis* conchocelis occurred between 5 and 25 °C under all photon irradiances used. Conchosporangial branches were formed between 10 and 25 °C under all light levels, but conchospore liberation was not observed under any of the continuous culture conditions. These results are similar to those obtained for Japanese *P. pseudolinearis* by Kurogi & Akiyama (1966). Blades of

P. pseudolinearis did not grow well at 25 °C, and thalli died after 10 weeks culture. These results and those from other field studies (Yoshida, 1966), suggest that growth and maturation of *P. pseudolinearis* are particularly affected by temperature.

Typically, *P. dentata* and *P. pseudolinearis* are dioecious species (Ueda, 1932; Tanaka, 1952; Fukuhara, 1968; Kito, 1978), but monoecious thalli of *P. dentata* have been reported in field and culture observations from Japan by Kikuchi (1992). Hwang (1994) and Funano (1961) reported monoecious thalli of *P. pseudolinearis* from Japan and Korea, respectively. *Porphyra dentata* thalli matured into monoecious blades at 15 °C under short days (10L:14D) by the end

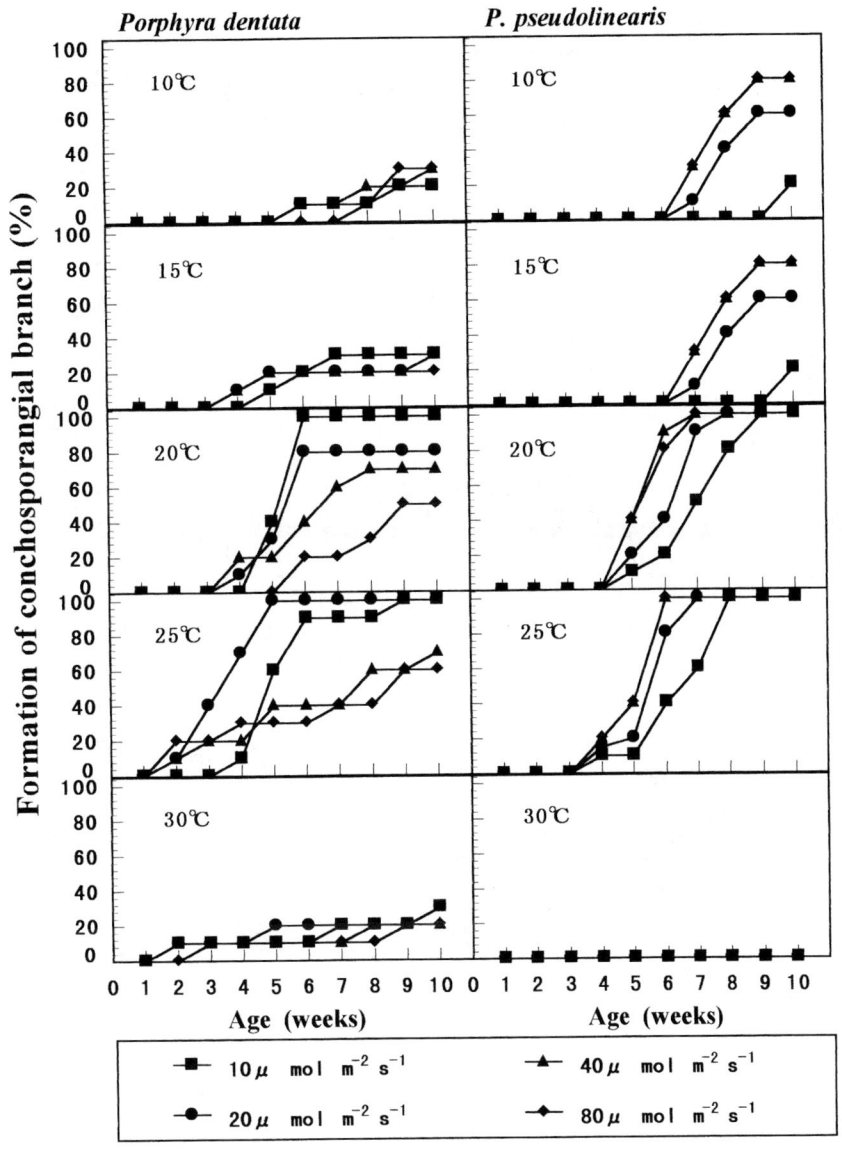

Figure 6. Formation rates of conchosporangial branches in *Porphyra dentata* and *P. pseudolinearis* at six temperatures (5–30 °C), four photon irradiances (10–80 μmol m^{-2} s^{-1}) and 14L:10D. Percentage of conchocelis colonies with conchosporangial branches.

of the experiment. Other authors have also reported monoecious thalli of dioecious *Porphyra* species e.g. *P. haitanensis* (Chang & Zheng, 1960), *P. lanceolata, P. pseudolanceolata P. linearis, P. mumfordii, P. pseudolinearis* and *P. purpurea* (Lindstrom & Cole, 1992a, b, c; 1993).

In this study, carpogonia on blades of both species were observed. They had the typical prolongation of prototrichogynes similar to that reported by Kornmann & Sahling (1991) for *Porphyra* species from Helgoland. Typically, it has been very difficult to observe carpogonia with prototrichogynes in dioecious species of *Porphyra* (Kito, 1978).

Anatomical examination of both species, revealed divisional formulae (Hus, 1902) for spermatangia and zygotosporangia that agreed with those reported by Ueda (1932) and Tanaka (1952), respectively.

Based on the results of this study and on previous culture (Kurogi & Akiyama, 1966; Kikuchi, 1992; Notoya et al., 1993) and field investigations (Kang, 1966; Fukuhara, 1968; Wynne, 1972; Scagel et al., 1989; Lindstrom & Cole, 1992a, b), the temperat-

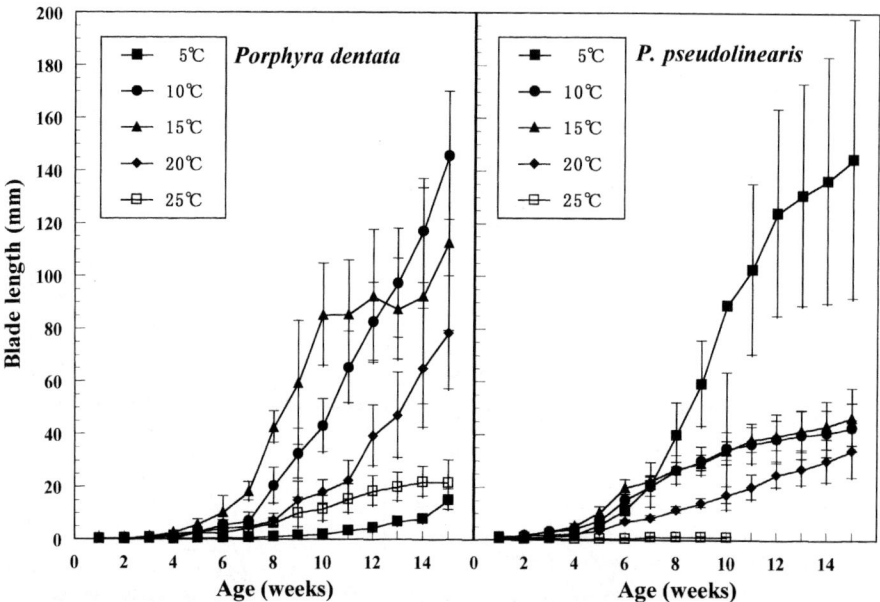

Figure 7. Growth of foliose thalli of *Porphyra dentata* and *P. pseudolinearis* at five temperatures (5–25 °C), 40 μmol m^{-2} s^{-1} and 10L:14D. Vertical bars are standard deviations.

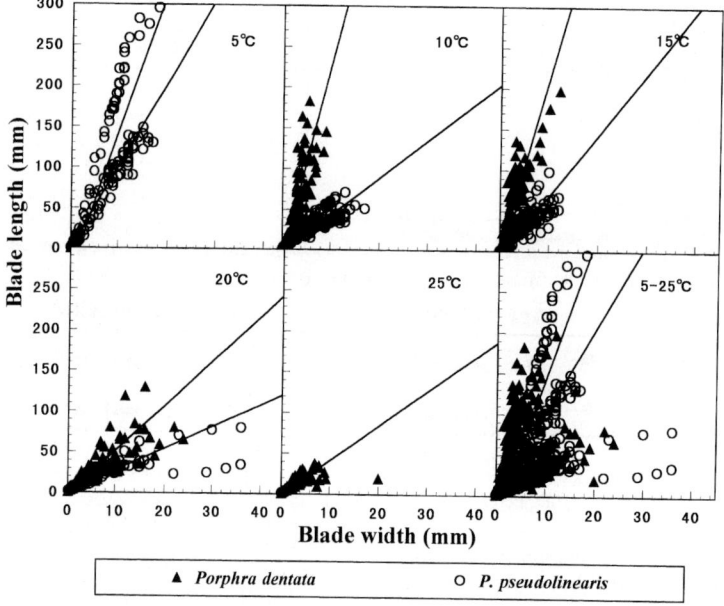

Figure 8. Relationship between blade length and width in *Porphyra dentata* and *P. pseudolinearis* grown at five temperatures (5–25 °C), 40 μmol m^{-2} s^{-1} and 10L:14D.

ure requirements for growth and maturation of the foliose and conchocelis phases are much lower for *P. pseudolinearis* than for *P. dentata*.

The life history of *P. dentata* and *P. pseudolinearis* obtained in our study corresponds to a typical biphasic cycle (Conway et al., 1975) without asexual reproduction in either phase, which commonly occurs in other species (Krishnamurthy, 1969; Conway & Wylie, 1972; Bird, 1973; Piriz, 1981; Notoya et al., 1993). Gall et al. (1993) reported that monosporangia produced on *P. dentata* conchocelis could be either stalked or sessile and that released monospores germinated into new filaments. However, in our culture, neither monospores nor protothalli were produced by the conchocelis phase of either species.

In conclusion, these culture studies indicate that the growth and maturation of both the foliose and conchocelis phases of *P. dentata* and *P. pseudolinearis* are more influenced by temperature than irradiance level. Furthermore it is apparent that *P. dentata* has monoecious and dioecious thalli.

References

Bird, C. J., 1973. Aspects of the life history and ecology of *Porphyra linearis* (Bangiales, Rhodophyceae) in nature. Can. J. Bot. 51: 2371–2379.

Chang, T. & B. Zheng, 1960. *Porphyra haitanensis*, a new species of *Porphyra* from Fukien. Acta Bot. Sinica 9: 32-36.

Conway, E., T. F. Mumford, Jr. & R. F. Scagel, 1975. The genus *Porphyra* in British Columbia and Washington. Syesis 8: 185–244.

Conway, E. & A. P. Wylie, 1972. Spore organization and reproductive modes in two species *Porphyra* from New Zealand. Proc. int. Seaweed Symp. 7: 105–107.

Fukuhara, E., 1968. Studies on the taxonomy and ecology of *Porphyra* of Hokkaido and its adjacent waters. Bull. Hokkaido reg. Fish. Res. Lab. 34: 40–99 (in Japanese with English abstract).

Funano, T., 1961. A monoecious individual of *Porphyra pseudolinearis* Ueda. J. Hokkaido Fish. Exp. St. (Hokusuishi Geppo) 18: 23–27 (in Japanese).

Gall, E. A., Y.-M. Chiang & B. Kloareg, 1993. Isolation and regeneration of protoplasts from *Porphyra dentata* and *Porphyra crispata*. Eur. J. Phycol. 28: 277–283.

Guiry, M. D., 1990. Sporangia and spores. In Cole, K. M. & R. G. Sheath (eds), Biology of the Red Algae. Cambridge University Press, New York: 347–376.

Hus, H. T. A., 1902. An account of the species of *Porphyra* found on the Pacific coast of North America. Cal. Acad. Sci., 3rd Series Botany 2: 173–241.

Hwang, M. S., 1994. Taxonomic studies of genus *Porphyra* of Korea. Ph.D. thesis, Seoul Natn. Univ. 277 pp (in Korean with English abstract).

Iima, M. & S. Migita, 1990. Laboratory culture of *Porphyra lacerata* (Rhodophyceae, Bangiales). Bull. Fac. Fish. Nagasaki Univ. 68: 13–20 (in Japanese with English abstract).

Iwasaki, H., 1961. The life-cycle of *Porphyra tenera in vitro*. Biol. Bull. 121: 173–187.

Kang, J. W., 1966. On the geographical distribution of marine algae in Korea. Bull. Pusan Fish. Coll. 7: 1–125.

Kikuchi, N., 1992. Study on the systematics and life history of Florideophycidac. M.S. thesis, Tokyo Univ. Fish. 105 pp (in Japanese).

Kito, H., 1978. Cytological studies on genus *Porphyra*. Bull. Tohoku. Reg. Fish. Res. Lab. 39: 29–84 (in Japanese with English abstract).

Kjellman, F. R., 1897. Japanska arter af slagtet *Porphyra*. Bihang Till Kongl. Sv. Vet-Akad. Handl. 23: 1–29.

Kornmann, P. & P.-H. Sahling, 1991. The *Porphyra* species of Helgoland (Bangiales, Rhodophyta). Helgoländer. Meeresunters. 45: 1–38.

Krishnamurthy, V., 1969. The Conchocelis phase of three species of *Porphyra* in culture. J. Phycol. 5: 42–47.

Kurogi, M. & K. Akiyama, 1966. Effects of water temperature on the growth and maturation of conchocelis-thallus in several species of *Porphyra*. Bull. Tohoku. Reg. Fish. Res. Lab. 26: 77-89 (in Japanese with English abstract).

Lee, I. K. & J. W. Kang, 1986. A check list of marine algae in Korea. Kor. J. Phycol. 1: 311–325 (in Korean).

Lindstrom, S. C. & K. M. Cole, 1992a. Relationships between some North Atlantic and North Pacific species of *Porphyra* (Bangiales, Rhodophyta): evidence from isozymes, morphology, and chromosomes. Can. J. Bot. 70: 1355–1363.

Lindstrom, S. C. & K. M. Cole, 1992b. A revision of the species of *Porphyra* (Rhodophyta: Bangiales) occurring in British Columbia and adjacent waters. Can. J. Bot. 70: 2066–2075.

Lindstrom, S. C. & K. M. Cole, 1992c. The *Porphyra lanceolata-P. pseudolanceolata* (Bangiales, Rhodophyta) complex unmasked: recognition of new species based on isozymes, morphology, chromosomes and distributions. Phycologia 31: 431–448.

Lindstrom, S. C. & K. M. Cole, 1993. The systematics of *Porphyra*: character evolution in closely related species. Hydrobiologia 260/261: 151–157

Magne, F., 1991. Classification and phylogeny in lower Rhodophyta: a new proposal. J. Phycol. (Suppl.) 27: 258.

Matsuo, M., M. Notoya & Y. Aruga, 1994. Life history of *Porphyra suborbiculata* Kjellman (Bangiales, Rhodophyta) in culture. La mer 32: 57–63 (in Japanese with English abstract).

McLachlan, J., 1973. Growth media - marine. In Stein, J. R. (ed.), Handbook of Phycological Methods, Cambridge University Press, New York: 25–51.

Migita, S. & R. Ito, 1987. The life history of *Porphyra tanegashimensis* (Bangiales, Rhodophyta) in laboratory culture. Bull. Fac. Fish. Nagasaki Univ. 61: 7–14 (in Japanese with English abstract).

Notoya, M., N. Kikuchi & Y. Aruga, 1992. *Porphyra kinositae* (Yamada et Tanaka) Fukuhara (Bangiales, Rhodophyta) in culture. Jap. J. Phycol. 40: 273–278 (in Japanese with English abstract).

Notoya, M., N. Kikuchi, M. Matsuo, Y. Aruga & A. Miura, 1993. Culture studies of four species of *Porphyra* (Rhodophyta) from Japan. Nippon Suisan Gakkaishi 59: 431–436.

Perestenko, L. P., 1982. Species generis *Porphyra* Ag. in Maribus orientaris extremi URSS. I. *Novitates systematicae plantarum non vascularium tomus* 19: 16–29 (in Russian).

Piriz, M. L., 1981. A new species and a new record of *Porphyra* (Bangiales, Rhodophyta) from Argentina. Bot. mar. 24: 599–602.

Scagel, R. F., P. W. Gabrielson, D. J. Garbary, L. Golden, M. W. Hawkes, S. C. Lindstrom, J. C. Oliveira & T. B. Widdowson, 1989. A synopsis of the benthic marine algae of British Columbia, Southeast Alaska, Washington and Oregon. Phycological Cont. 3, Department of Botany, Univ. British Columbia. Vancouver. 532 pp.

Tanaka, T., 1952. The systematic study of the Japanese Protoflorideae. Mem. Fac. Fish. Kagoshima Univ. 2: 1–91.

Tseng, C. K., 1938. Notes on some Chinese marine algae. Lingnan Sci. J. 17: 591–604.

Tseng, C. K., 1948. Marine algae of Hongkong VII. The order Bangiales. Lingnan Sci. J. 22: 121–131.

Ueda, S., 1932. Taxonomic studies on the Japanese *Porphyra*. J. imp. fish. Instit. 28: 1–45 (in Japanese).

Wynne, M. J., 1972. The genus *Porphyra* at Amchitka Island, Aleutians. Proc. int. Seaweed Symp. 7: 100–104.

Yoshida, T., 1966. Species of *Porphyra* and their growth in the cultivation ground of Shukunohe, Iwate Pref., Japan. Bull. Tohoku. Reg. Fish. Res. Lab. 26: 109–116 (in Japanese with English abstract).

Biological basis for the management of *Gigartina skottsbergii* (Gigartinales, Rhodophyta) in southern Chile

R. Westermeier, A. Aguilar, J. Sigel, J. Quintanilla & J. Morales
Instituto de Acuicultura, Facultad de Pesquerías y Oceanografía Universidad Austral de Chile, Casilla 1327, Puerto Montt, Chile

Key words: Gigartina skottsbergii, population studies, demography

Abstract

Environmental and biological parameters were analysed for populations of *Gigartina skottsbergii* (Setchell & Gardner) at Calbuco, Llanquihue Province (sheltered environment) and Ancud, Chiloe Province (exposed environment) in the south of Chile, with the aim of determining the biological basis useful in guiding management efforts for this species. In both populations, the biomass, density, recruitment, growth and survival showed maximum values in spring–summer. Most plants were sterile, and while cystocarps were present throughout the year, they were more abundant in autumn–winter. It was concluded that at Ancud, this species presents a synchronized and predictable population with high recruitment and low survival, while Calbuco showed low recruitment, higher survival and higher growth rates. Calbuco is, therefore, recommended as a growth (culture) area, and Ancud as an area for recruitment. As a management strategy, it is recommended that the harvest be mainly restricted to the end of spring (November–December), reducing towards the end of summer (February–March). Cutting should be selective and done by hand, removing fronds of at least 600 cm^2, corresponding to a frond length of a minimum of 25 cm.

Introduction

Gigartina skottsbergii is endemic to the extreme south of South America (Setchell & Gardner, 1936; Pujals, 1963; Kühnemann, 1972; Westermeier, 1981), and is of high commercial value for its biomass and carrageenan content. In 1989, carrageenophyte production reached 26 t, and within six years this figure had increased by over 4000% (Schnettler et al., 1996). About 80% of the national production originates in this region, and success achieved in its export has resulted in over-exploitation in some localities, producing the necessity to look for alternative areas for collection (Westermeier et al., 1995, 1996, 1997).

In the area of Ancud, this species dominates the subtidal zone, where it is found growing principally together with *Sarcothalia crispata*, *Macrocystis pyrifera* and calcareous algae (Zamorano & Westermeier, 1996). Most of the plants were found to be vegetative, with maximum production of biomass in September (spring) and March (Autumn). Piriz (1996) working in the Atlantic (Chubut, Argentina), recorded values similar to those reported for Chile by Zamorano & Westermeier (1996), although the periods of maximum biomass differed. In studies carried out by Westermeier et al. (1995, 1996, 1997) between 41° S and 53° S in areas where harvesting pressures were supposedly less, biomass values were obtained which were similar to those reported for the Ancud area. This implies that harvesting was either greater than previously supposed, or that the population was able to regulate its biomass. This factor, added to the dynamics of the phases (tetrasporophyte/cystocarpic balance) being different from that reported by Zamorano & Westermeier (1996) and Piriz (1996), makes it necessary to evaluate a series of factors in the population dynamics of this species with greater precision by analysing populations occurring in distinct environmental conditions within this geographic area.

In view of the above considerations, while previous studies were centred principally on phenology, production of biomass and chemical composition, the present study has as its objective the development of a basis for management of this species-resource.

For this, over 32 months period, we determined the dynamics of the phases of the life cycle, biomass, density, size structure, recruitment, demography and growth.

Materials and methods

Study area

The work was carried out on natural populations of *Gigartina skottsbergii* located near San Agustin in Calbuco (41° 43′ S; 73° 05′ W) and Ancud Bay (41° 51′ S; 73° 49′ W) of the X Region in Chile (Figure 1). The Calbuco bed has an average extent of 1 ha, at a depth of 15 ± 1 m, where the attachment substratum for the species consists of boulders of metamorphic origin. This bed is in a sheltered environment. In contrast, the bed at Ancud, being approximately 0.5 ha, is more exposed to waves and currents. It is located in a zone of broken coastline at a depth of 9 ± 1 m, upon large formations of metamorphic rock, known as 'slab rock formation' (Illies, 1970).

Methodology and experimental design

Research was carried out monthly between July 1995 and February 1998. Plants were collected by destructive sampling from five 1 m^2 quadrats for determination of biomass m^{-2}, density m^{-2}, length (cm) of fronds, size structure (see Table 1) and phenology. The phase of the life cycle was determined by the resorcinol method described by Garbary & De Wreede (1988). (This test was checked by using control plants whose reproductive phase was known.) After this determination, gametophytic plants were examined with a microscope and divided into fertile (cystocarpic) and vegetative types. Tetrasporic plants were not distinguished from sterile tetrasporophytes.

Recruitment was measured using artificial blocks cleared of plants and placed on the bottom.

Demography and growth were evaluated using non-destructive sampling following the methods of Bhattacharya (1985). Four size classes were determined, using the blade surface as the discriminating factor (Table 1). Forty plants were marked in each class, using plastic labels attached to boulders; these were followed over time, measuring blade area from drawings on acrylic sheets. Survival and mortality were determined and the relative percentage growth rate in area each month was calculated from: (final area – initial area)/initial area × 100.

Environmental factors measured included atmospheric temperature, solar radiation and precipitation. In the water column, we determined temperature, salinity, pH, CO_2, HCO_3, O_2, NO_3, NH_4 and PO_4. Data on current intensity was obtained from the Meteorological Division of he Chilean Navy.

Results

Phenology

In each of the 32 months at both sites most of the plants were sterile. Overall 15.1% of the plants were cystocarpic at Calbuco and 9.8% at Ancud. Reproduction was minimal in early summer, cystocarpic plants showing marked seasonality with a maximum in autumn/winter at both sites (Figure 2).

The tetrasporophytic phase was present throughout the year, with a slight maximum in spring/summer (Figure 2). The overall balance between the phases was strongly in favour of gametophytes with only 11.1% of the population being tetrasporophytes at Cabulco and 11.3% at Ancud.

Biomass, density and length

Overall, the biomass was similar at the two sites but it was more variable at Calbuco (Figure 3A). At both sites the variability obscured what might have been seasonality; if anything, the biomass was higher in spring or early summer. There was a drop in biomass after the first 12 months of study.

There was a lower density of fronds at Calbuco, with an average of 7.3 plants m^{-2}, without marked seasonality, whereas in Ancud Bay, the mean number of fronds was 26 m^{-2} (Figure 3B). Again, possible seasonality was masked by high variability. Concurrently, and similar to biomass values, after the first 12 months the number of fronds decreased considerably at the latter station.

Frond lengths of *G. skottsbergii* were consistently greater at Calbuco (Figure 3C). Here again variability masked possible seasonality.

Size structure

All size classes 1–10 were present in both populations. At Ancud there was a dominance of class 1 (30.7% overall) when compared with Calbuco, which,

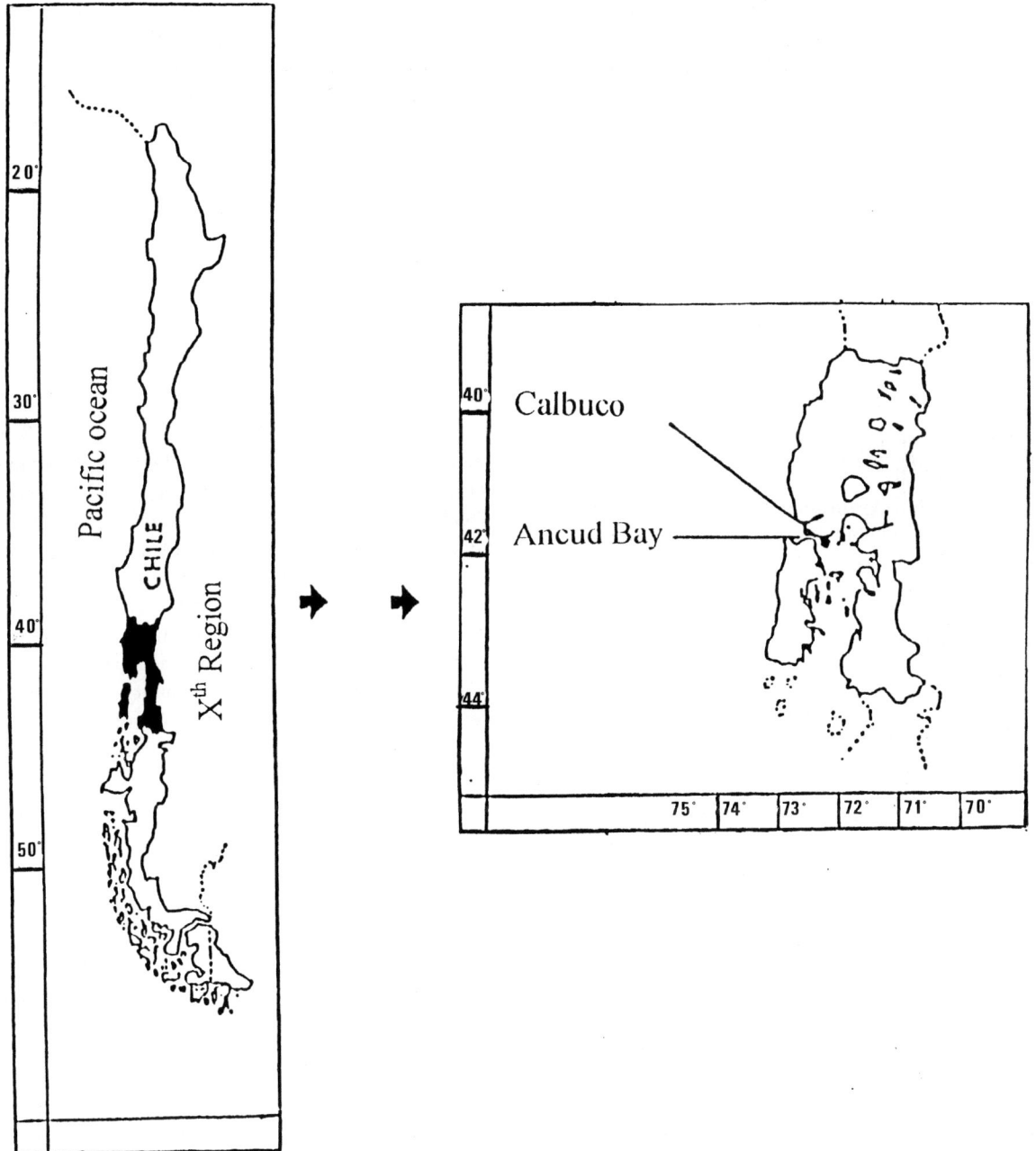

Figure 1. Gigartina skottsbergii study sites in Calbuco (Llanquihue) and Ancud Bay (Chiloe), in southern Chile.

although presenting an important proportion of this class (18.8%), showed a more homogeneous distribution of frequency among the different size classes (Figure 4A, B).

All classes I–VIII were present in the Calbuco population with class I (<1000 cm^{-2}) most abundant (62.9%) (Figure 4C, D). At Ancud, only the first six classes were present, with an average for class I of 93.3%.

Table 1. Definition of size classes used in the study on *G. skottsbergii*, based on thallus area

Size structure observations				Demography/growth observations	
Class	Area in cm^2	Class	Area in cm^2	Class	Area in cm^2
1	<100	I	<1000	A	<50
2	100–199	II	1000–1999	B	50–99
3	200–299	III	2000–2999	C	100–149
4	300–399	IV	3000–3999	D	>150
5	400–499	V	4000–4999		
6	500–599	VI	5000–5999		
7	600–699	VII	6000–6999		
8	700–799	VIII	7000–7999		
9	800–899				
10	900–999				

Recruitment, demography and growth

There were notable differences between populations in monthly recruitment; an average of 2.8 fronds m^{-2} was found at Calbuco, and 12.9 fronds m^{-2} at Ancud. Recruitment showed seasonality at both sites with clear maxima in the spring/summers of 1995/1996 and 96/97 (Figure 5).

G. skottsbergii showed seasonal behaviour, mortality being highest in the months of autumn/winter. Of 160 marked plants observed over a period of 28 months, about 20% of them survived at each site. The pattern was similar at the two sites; at the beginning of the second year of observation the survival curve changed from descending to stable (Figure 6). Overall mean mortality rates were 5% per month at Calbuco and 6% at Ancud, but there was no significant difference between the sites in the total number of plants 27 months later. By this time, however, at Calbuco there were significantly more original Class D plants than at Ancud Bay ($p < 0.025$, G-test of Independence) so large plant mortality had been greater at Ancud.

Growth showed seasonality, with an increase in rates in spring/summer at both sites (Figure 7). At Cabulco, the size groups remained discreet and their respective mean blade areas increased by 3 to 11-fold by the end of the year. In Ancud Bay, there was less growth in spring, more loss in autumn and little difference in the mean areas of the size classes after a year.

In both populations of *G. skottsbergii*, fronds with a blade area greater than 200 cm^{-2} were reproductive.

Environmental factors

Atmospheric temperature, solar radiation, water temperature and salinity presented a clearly seasonal pattern, with values increasing during spring/summer and decreasing in autumn/winter (Figure 8). An inverse behaviour was reported for the parameters of precipitation, carbon dioxide, nitrate and phosphate, with highest values reached in autumn/winter, lowering in spring/summer. The pH, oxygen, bicarbonate and ammonium values showed no seasonal pattern.

Currents were under the primary influence of the semi-diurnal tidal oscillation. Current magnitude showed a bi-seasonal pattern. At Calbuco in summer (October–March), currents measured 2–16 cm s^{-1}; in winter (April–September) this value reached 10–45 cm s^{-1}. Corresponding values for Ancud, where prevailing winds from the fourth quadrant brought torrential rains, were 9–23 cm s^{-1} (summer) and 16–75 cm s^{-1} (winter).

Discussion

During 32 months of research, most of the plants were sterile. Cystocarpic plants were abundant in autumn/winter. Tetrasporophytes were always in a minority, with an overall level of 11%, but were slightly more common in spring/summer. Zamorano & Westermeier (1996) and Piriz (1996) found a similar pattern for cystocarpic plants, but not for tetrasporophytes. Westermeier (1995, 1996, 1997), studying populations of the XI Region (Melinka, 44° S), and the XII Region (Otway Sound 53 ° S), in spring/summer, found notable differences in these abundances: in the XII Region, 14% of the plants bore cystocarps at Charles Island, but 50% at Charles III Island and the proportions of tetrasporophytes were 26% and

141

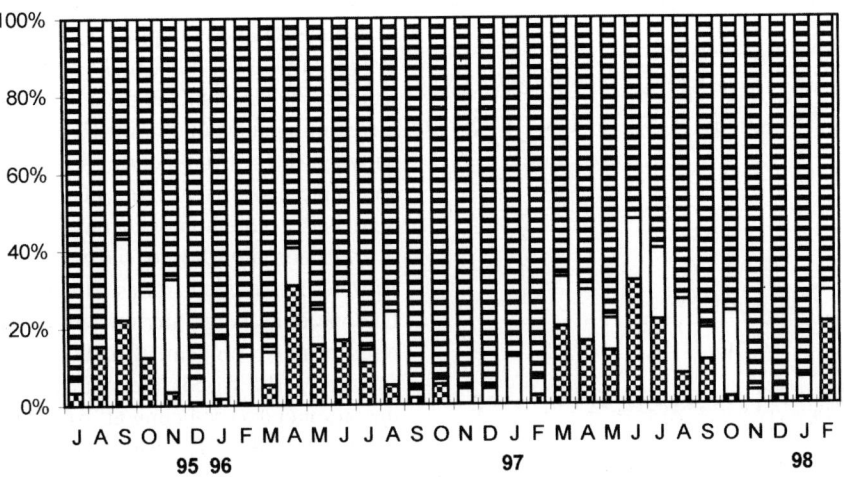

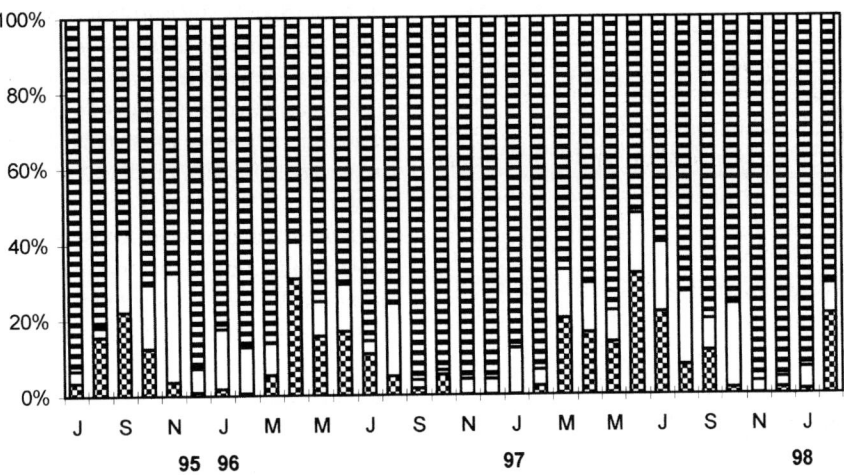

Figure 2. Percentages in relative frequency of the life history of *Gigartina skottsbergii* in the study areas. Horizontal bars, sterile gametophytes; white, tetrasporophytes; chequered, cystocarpic. Bars, SD.

33%, respectively. At some stations in the XI Region, although the differences were not marked as this, tetrasporophytes outnumbered cystocarpic plants. It is possible that differences in latitude, recruitment, harvesting pressure, and/or innate dynamics of the phases of the life cycle triggered these differences. We believe that *G. skottsbergii* is a long-lived species, having observed plants as old as 4 years, which implies that there may be cohorts maturing out of phase which could produce such changes. Variations in phase abundance have been described by Dyck et al. (1985) for *Mazzaella splendens* from British Columbia and by Westermeier et al. (1987) for *Iridaea laminaroides* in southern Chile.

In comparing the life cycle phase between Ancud (exposed) with Calbuco (sheltered), we found no dif-

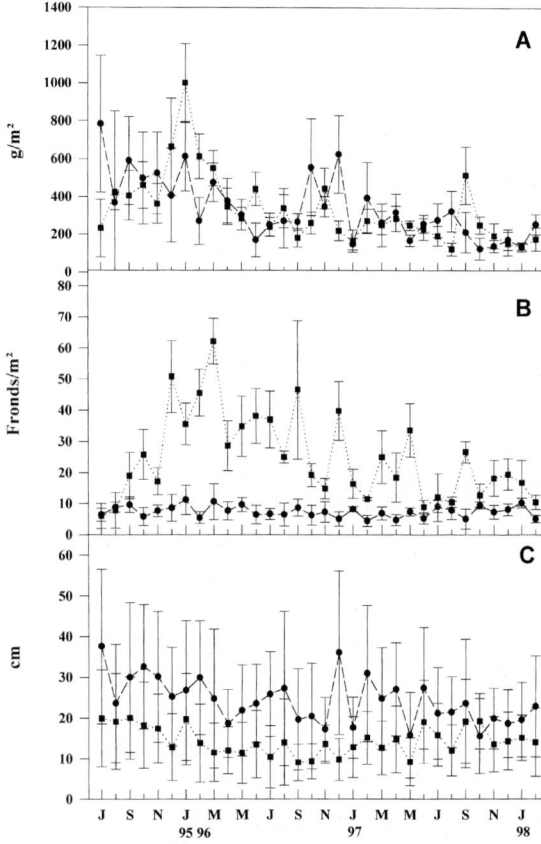

Figure 3. Monthly variation in (A) biomass, (B) density, and (C) thallus length of *Gigartina skottsbergii* in the study areas. —●—, Calbuco; ...■..., Ancud Bay.

ference in the proportion of tetrasporophytes in the populations. On the other hand, 11% signifies a considerable contribution by the diploid phase which until now had appeared to be of little importance. The dominance of gametophytes could be a response to the greater reproductive efficiency of the tetrasporophyte phase. In the laboratory, Westermeier & Sigel (1997) showed that tetraspores are superior in settlement, viability, germination and growth. Other studies point to the abundance of haploid plants as occurs with *Rhodoglossum affine* and *Gigartina* sp which may be the consequence of apomixis (Abbott, 1980). Our results suggest that this process is not involved in *G. skottsbergii*, and that this is likely a natural variation similar to that occurring in other Gigartinales in southern Chile (Westermeier et al., 1987; Westermeier et al., 1996). It is therefore necessary to concentrate on studying the gametophyte proportion (male and female plants) and with this clarify the phase balance.

In the study by Zamorano & Westermeier (1996), biomass was related to light and temperature; however, with more detailed sampling, no clear pattern emerges. This signifies that other biological and environmental factors (such as depth, exposure, substratum) also regulate timing and values. Size class distribution differed between sites. At Ancud, there was a gradient of abundance, with few large plants, while at Calbuco there was a more homogeneous distribution. Although overall mortality did not differ between sites, at Ancud the larger sized fronds showed greater mortality. Undoubtedly the greater movement of water and currents produce movement of boulders, and the large surface of the fronds incur greater abrasion, resulting in complete detachment, or partial destruction of fronds. Norall et al. (1981) reported this situation as 'storm pruning' in subtidal populations of *Phycodrys rubens* and *Callophyllis variegata* during winter. Bhattacharya (1985) observed a similar situation where the greater mortality of *Chondrus crispus* occurred in the larger size classes and those with major branching.

Nevertheless, recruitment at both stations showed some similarity in temporal variation, notably in the third year it was less, and occurred in mid-winter. This could have been due to a cohort temporally out of phase in germination and growth or the recruitment may have been produced by tetraspores liberated in winter of the same year which, due to some environmental factor were caused to recruit massively. Nevertheless, the plants are mature and viable at approximately 200 cm^2 which suggests that perhaps cohorts from the preceding year produced the recruitment, and that tetraspores need more time to germinate, recruiting later. This would suggest two periods of recruitment, one by way of carpospores in winter (with plants in summer), and another by way of tetraspores in summer (with plants in winter). This is consistent when it is considered that the carpospores are viable in the laboratory only in winter (Westermeier & Sigel, 1997).

This also suggests that the tetraspores are better adapted for growth in winter and carpospores to spring/summer, thus tetrasporophytic (2N) plants would be more abundant in summer, while gametophytes (N) would be present in winter, terminating the cycle with the production of carpospores. Guiry & Coleman (1982) found the maximum liberation of carpospores in winter in *G. stellata*, which corresponds with *G. skottsbergii* in southern Chile. Another carra-

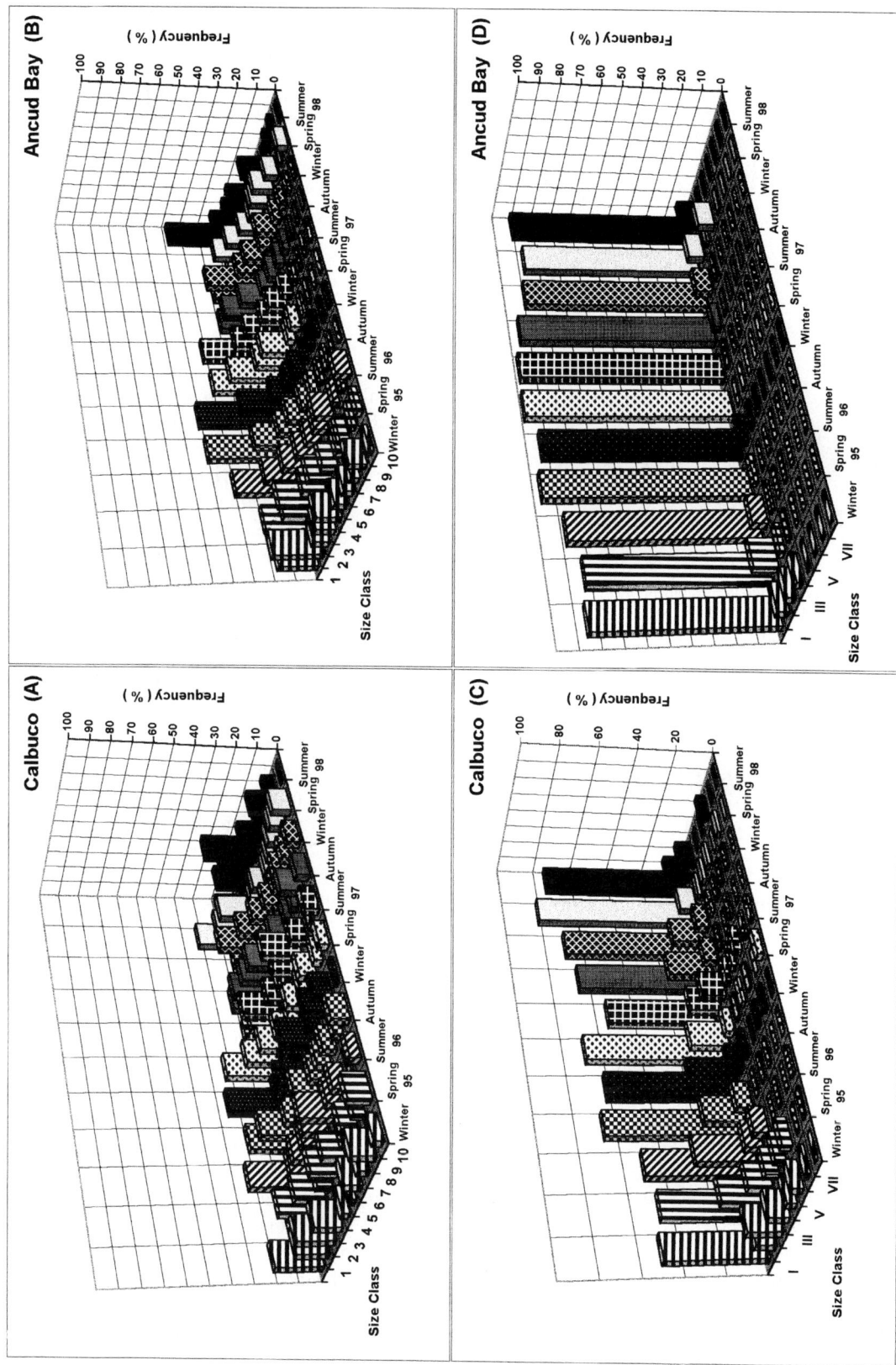

Figure 4. Seasonal variation in frond size structure of *Gigartina skottsbergii* in the study areas. (A) and (B), < 1000 cm²; (C) and (D) all sizes. See Table 1 for size class definitions.

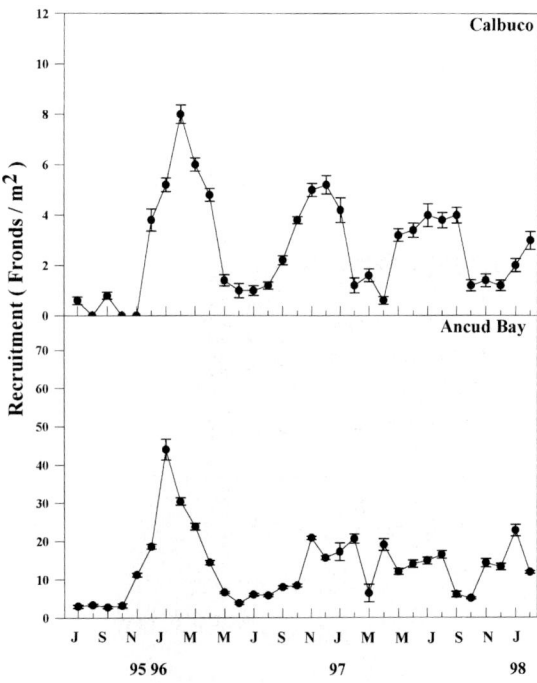

Figure 5. Monthly variation in recruitment of *Gigartina skottsbergii* in the study areas. Bars, SE.

Table 2. Mortality and growth rates in thallus area of *G. skottsbergii* over a periods of 27 and 12 months respectively

Class	Mortality (% month^{-1})		Growth rate (% month^{-1})	
	Calbuco	Ancud	Calbuco	Ancud
A	7.81	5.64	22.57	29.81
B	5.22	5.25	17.17	10.75
C	3.99	4.87	14.23	6.84
D	4.24	7.71	8.77	3.44
Mean	5.3	5.9	15.7	12.7

genophyte, *Eucheuma*, showed maximum recruitment upon increase of turbulence in winter periods, which would be equivalent to the tetrasporic phase in the areas studied.

G. skottsbergii showed a growth rate average of 15% area/mo (0.5% area day^{-1}), which included all size classes at both stations. In *Chondrus crispus*, Taylor (1972) obtained a rate of 1.2% area day^{-1}, while the rate for *Constantinea subulifera* was 3.1% area day^{-1} (Powell, 1964). This indicated that *G. skottsbergii* has a lower growth rate, related to its morphological characteristics. Its maximum rates were 2.3% area day^{-1} at Ancud (Oct) and 1.4% area day^{-1} at Calbuco (Dec). In *Iridaea laminaroides* the major rates of net growth occurred in winter months, except when a maximum was reported for January with 54.7% (Gómez & Westermeier, 1991).

Other species such as *Gracilaria*, *Ulva* (Rosenberg & Ramos, 1982), *Chordaria flageliformis* (Probyn & Chapman, 1982, 1983) and *Macrocystis pyrifera* (Westermeier & Moller, 1990) showed highest growth rates in spring/summer. Piriz (1996), working with *G. skottsbergii* in the Atlantic, reported the highest biomass at the end of spring, relating his results to the length of day and increase in radiation.

In contrast to its close relatives, the holdfast of *G. skottsbergii* is formed of haptera. It appears that a germination disc forms only in the early stages, after which haptera originate towards the substratum, where the attachment surface is increased. A possible method of management of this species/resource is postulated via maintenance of haptera and pruning of these, hopefully cultivating vegetative growth (Westermeir in prep.). At the same time we have observed *in situ* appearance of new fronds sharing the niche in the basal zone of *G. skottsbergii*. These observations differ from those of Piriz (1996), who suggests it may not be possible to obtain the renewal of fronds using this system of attachment. This undoubtedly requires further research; it may be that new fronds arise from previously settled spores.

In processing factories, the minimum frond size utilized is 20 cm^2, which implies lowering the potentially harvestable biomass in subsequent periods, and reduction of the possibility of reproductive generations of new thalli. From our results, it can be concluded that the strategy for management of this resource should include restriction to the sole removal of plants exceeding 600 cm^2 (25 cm length) between December and February. This would signify a substantial contribution of reproductive material to the different cohorts, which would ensure later recruitment. It is possible to suggest permanent monitoring, except in winter, to extract plants of this size. The type of harvesting to which this resource is exposed may produce major reductions of biomass over a short time period, which suggests the need for more conservative harvesting. It is also recommended that the harvest be done by pruning (cutting), conserving the basal fixation system, as the thallus is capable of regenerating new tissue by action of the meristems involved (Westermeier in prep.).

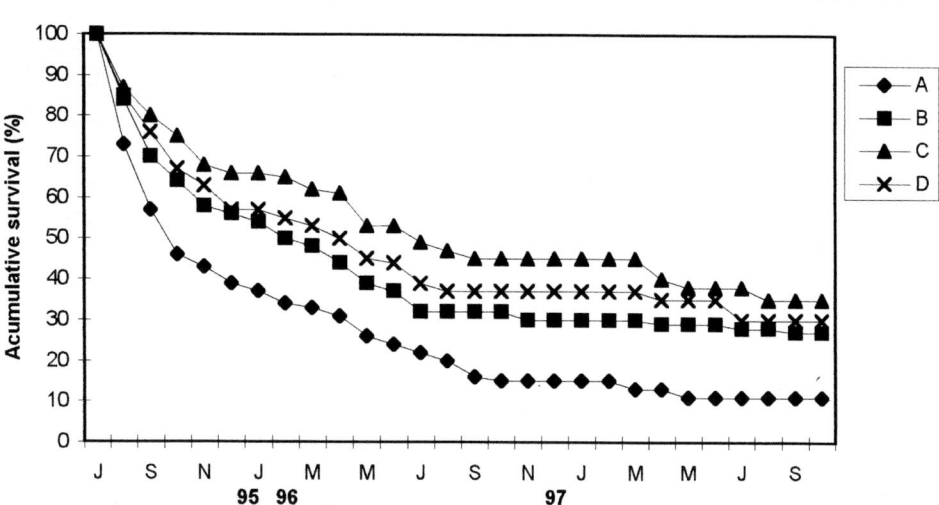

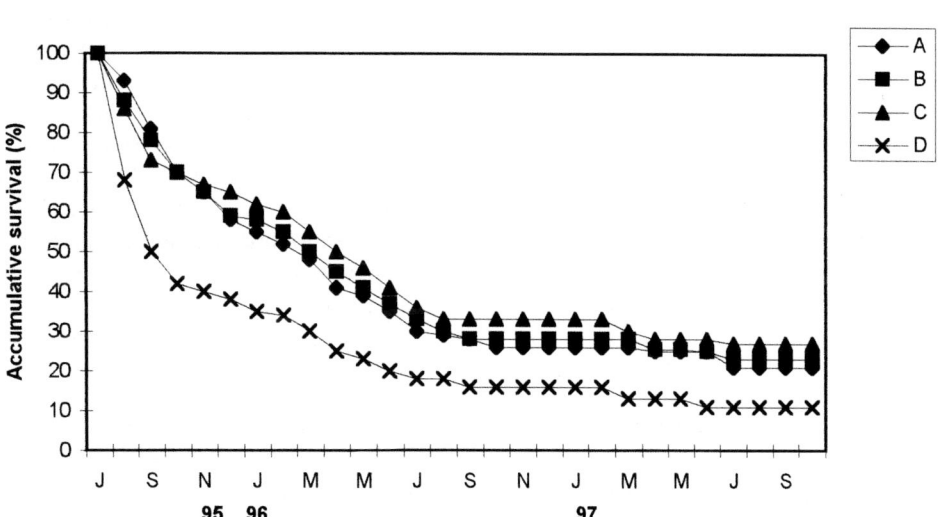

Figure 6. Cumulative survival curves of *Gigartina skottsbergii* in the study areas. See Table 1 for definitions of size classes A–D.

In future activities related to the culture of this resource, it is proposed that exposed areas such as the Gulf of Ancud be used as areas for spore capture, where the recruits, after reaching sizes of 100 cm^2 (10 cm length) are transplanted to protected areas such as Calbuco to complete their growth.

Acknowledgements

This work was financed by FONDECYT 1951203 (first author). Thanks to a contribution from this program and from the FONDAP Algae program of CONICYT, this work was presented at Cebu, Phillippines. We thank Dr Juan Correa and two anonymous reviewers for critical reading of the MS and David Patiño, Bile Vera, Carlos Atero, and Jose Seron of the Universidad Austral de Chile Aquaculture Station

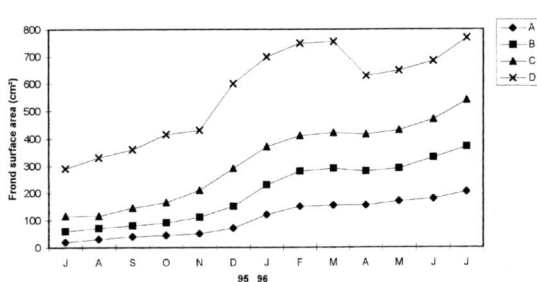

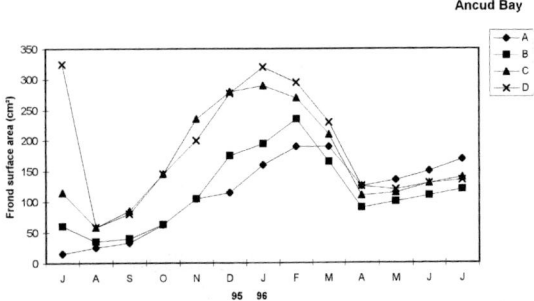

Figure 7. Growth curves of *Gigartina skottsbergii* in the study areas. See Table 1 for definitions of size classes A–D.

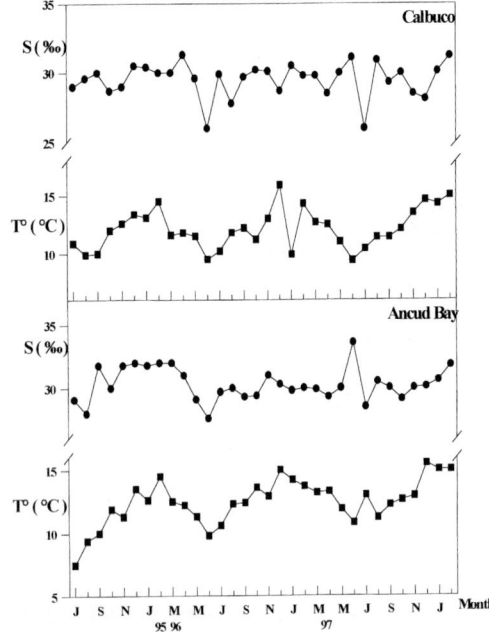

Figure 8. Monthly variation in temperature (°C) and, salinity (‰) of the water column in the study areas.

in Maullin for their work in the field and handling of samples. We are grateful to Lewis O. Veale for help with printing e-mailed figures.

References

Abbott, I. A., 1980. Some field and laboratory studies on colloid-producing red algae in Central California. Aquat. Bot. 8: 255–266.

Bhattacharya, D., 1985. The demography of fronds of *Chondrus crispus* Stackhouse. J. exp. mar. Biol. Ecol. 91: 217–231.

Dyck, L., R. E. De Wreede & D. Garbary, 1985. Life history phases in *Iridaea cordata* (Gigartinaceae): relative abundance and distribution from British Columbia to California. Jap. J. Phycol. 33: 225–232.

Garbary, D. & R. E. De Wreede, 1988. Life history phases in natural populations of Gigartinaceae (Rhodophyta): quantification using resorcinol. In Lobban C. S., D. J. Chapman & B. P. Kremer (eds), Experimental Phycology, A Laboratory Manual. Cambridge University Press, Cambridge: 174–178.

Gómez, I. M. & R. C. Westermeier, 1991. Frond regrowth from basal disc in *Iridaea laminarioides* (Rhodophyta, Gigartinales) at Mehuín, southern Chile. Mar. Ecol. Prog. Ser. 73: 83–91.

Guiry, M. D. & M. M. Coleman, 1982. Observations on the phenology and life history of a monoecious strain of *Gigartina stellata* (Stackh.) Batters (Rhodophyta). Mar. Ecol. Prog. Ser. 22: 291–303.

Illies, H., 1970. Geología, Volcanismo y Tectónica en márgenes del Pacífico en Chile Meridional. Publicación, Inst. Geología y Geografía, Universidad austral de Chile, Valdivia.

Kühnemann, O., 1972. Bosquejo fitogeográfico de la vegetación marina del litoral argentino. Phisis 31: 295–325.

Norall, T. L., A. C. Mathieson & J. A. Kilar, 1981. Reproductive ecology of four subtidal red algae. J. exp. mar. Biol. Ecol. 54: 119–136.

Piriz, M. L., 1996. Phenology of a *Gigartina skottsbergii* Setchell et Gardner population in Chubut Province (Argentina). Bot. mar. 39: 311–316.

Powell, J. H., 1964. The life-history of a red alga, *Constantinea*. Ph.D. thesis, University of Washington, Seattle, 154 pp.

Probyn, T. A. & A. R. O. Chapman, 1982. Nitrogen uptake characteristics of *Chordaria flagelliformis* (Phaeophyta) in batch mode and continuous mode experiments. Mar. Biol. 71: 129–133.

Probyn, T. A. & A. R. O. Chapman, 1983. Summer growth of *Chordaria flagelliformis* (O. F. Muell.) C. Ag.: physiological strategies in a nutrient stressed environment. J. exp. Mar. Biol. Ecol. 73: 243–271.

Pujals, C., 1963. Catálogo de Rhodophyta citadas para Argentina. Revista del Museo Argentino de Ciencias Naturales 'Bernardino Rivadavia', Botánica 3: 1–139.

Rosenberg, G. & J. Ramus, 1982. Ecological growth strategies in the seaweeds *Gracilaria foliifera* (Rhodophyceae) and *Ulva* sp. (Chlorophyceae): the rate and timing of growth. Bot. mar. 24: 583–589.

Schnettler, P., M. Avila & E. Bustos, 1996. Pesquería de la luga roja *Gigartina skottsbergii* Set. et Gard. (Rhodophyta, Gigartinales) en la Bahía de Ancud. VI Symposium de Algas Marinas Chilenas y IV Encuentro de Microalgólogos, Resúmenes: 84.

Setchell, W. A. & N. L. Gardner, 1936. *Iridophycus* gen. nov. and its representation in South America. Proc. Nat. Acad. Sci. 22: 469–473.

Taylor, A. R. A., 1972. Growth studies of *Chondrus crispus* in Prince Edward Island. Proc. Mtg. Can. Atl. Seaweeds Industry Ind. Dev. Can., Fish. Serv. Environ. Can. 1972: 29–36.

Westermeier, R., 1981. The marine seaweed of Chile's Tenth Region (Valdivia, Osorno, Llanquihue y Chiloé). Int. Seaweed Symp. 10 Göteborg: 215–220.

Westermeier, R. & P. Möller, 1990. Population dynamics of *Macrocystis pyrifera* (L.) C. Agardh in the rocky intertidal of Southern Chile. Bot. mar. 33: 363–367.

Westermeier, R. C. & J. A. Sigel, 1997. Reproductive patterns of *Gigartina skottsbergii* (Rhodophyta) in southern Chile. Phycologia 6 (Suppl): 123.

Westermeier, R. C., D. J. Patiño & J. E. Morales, 1995. Prospección de algas con especial referencia en Iridaea ciliata y *Gigartina skottsbergii* en la X y XI Región. Publicación ocasional. Universidad Austral de Chile, 100 pp.

Westermeier, R. C., D. J. Patiño & J. E. Morales, Prospección de algas con especial referencia en *Iridaea ciliata* y *Gigartina skottsbergii* en la XII Región. Publicación ocasional. Universidad Austral de Chile, 60 pp.

Westermeier, R. C., D. J. Patiño & J. E. Morales, 1997. Prospección de algas con especial referencia en *Iridaea ciliata* y *Gigartina skottsbergii* en la XI y XII Región. Publicación ocasional. Universidad Austral de Chile, 150 pp.

Westermeier, R., P. Rivera, M. Chacana & I. Gómez, 1987. Biological bases for management of *Iridaea laminarioides* Bory in southern Chile. Hydrobiologia 151/152: 313–328.

Zamorano, J. & R. Westermeier, 1996. Phenology of *Gigartina skottsbergii* (Rhodophyta, Gigartinales) in Ancud Bay, southern Chile. Hydrobiologia 326/327: 253–258.

Reproductive biology of *Gigartina skottsbergii* (Gigartinaceae, Rhodophyta) from Chile

M. Avila[1], A. Candia[1], M. Núñez[1] & H. Romo[2]
[1]*División de Fomento de la Acuicultura, Instituto de Fomento Pesquero, Casilla 665, Puerto Montt, Chile*
[2]*Departamento de Oceanografía, Facultad de Ciencias Naturales y Oceanográficas, Universidad de Concepción, Chile*
E-mail: mavila@ifop.cl

Key words: cystocarps, *Gigartina*, phenology, reproductive phases, reproductive potential, spores, tetrasporangial sori, viability

Abstract

Reproductive phenology and spore viability were studied in a natural bed of *Gigartina skottsbergii* at the locality of Ancud, Chile. Monthly sampling of biomass and density disclosed a decrease of both parameters in time, from 40 g m^{-2} in June 1996 to 1 g m^{-2} December 1997 and from almost 4 thalli m^{-2} to 1 thallus m^{-2} in the same period, respectively. Mature reproductive structures, cystocarps and tetrasporangial sori, were observed over the whole study period. Greatest cystocarp densities occurred in October through February (16 to 29 cystocarps cm^{-2}) and of tetrasporangial sori between July and October (66 to 88 sori cm^{-2}). In both types of reproductive structures, sporulation is more frequent in winter and early spring. High mortality of carpospores and tetraspores was observed in laboratory experiments performed between June and September.

Introduction

Gigartina skottsbergii Setchell & Gardner is a subtidal carrageenophyte of commercial importance in Chile (Avila & Seguel, 1993; Norambuena, 1996; Bixler, 1996). Its distribution is from Niebla (39° 53′ S) to Cape Horn (54° 56′ S) and it is endemic to the southernmost part of South America (Ramírez & Santelices, 1991). It is also found on the southern coasts of Argentina and the sub-antarctic islands (Piriz, 1988; 1996). In Chile this species has been exploited since the late 1980s reaching up to 8000 dry tons in 1996 (Avila et al., 1997), and the annual quantity of carrageenan extracted has gradually increased, up to 1700 t in 1996. The landing zone of this resource extends from Faro Corona (41° 47′ S; 73° 53′ W) to Puerto Aguirre (45° 10′ S; 73° 32′ W) (Schnettler, pers. com.).

The information available on this species refers mostly to its geographical distribution (Ramírez & Santelices, 1991; Santelices, 1989), vegetative frond morphology (Santelices, 1989; Alveal et al., 1990), reproductive structures (Kim, 1976) and chemical composition (Schnettler, pers. com). Other recent studies deal with the fluctuation of biomass and density in natural populations (Piriz, 1988; 1996; Zamorano, pers. com.) as well as with the species phenology (Zamorano & Westermeier, 1996; Avila et al., 1997; Candia, pers. com.).

The reproductive biology of this species is little documented, although sexual reproduction is known to occur seasonally (Zamorano & Westermeier, 1996; Westermeier & Sigel, 1997); an alternation in the dominance of the reproductive phases, as described for other species of the Gigartinales (e.g. *Iridaea splendens*, Ang et al., 1990) also occurs seasonally. Kim (1976) indicated that the species has a triphasic life cycle with alternation of isomorphic generations.

In contrast to other Gigartinales where the population is maintained by perennial holdfasts which produce or regenerate new blades (May, 1986; Santelices & Norambuena, 1987; Gómez & Westermeier, 1991), *G. skottsbergii* adheres to the substratum by means of secondary haptera which develop in juvenile stages (Piriz, 1996). Studies conducted at the Island of Chiloé showed that natural recruitment occurs in winter time between June and August, being strongly associated to the period of greatest abundance of mature reproductive phases (Avila et al., 1997).

The relative importance between the abundance of cystocarpic and tetrasporic phases, spore production (carpospores and tetraspores) and recruitment has not been established for *G. skottsbergii*. The aim of this work was to determine plant fecundity and spore viability in order to establish the importance of the reproduction through spores in the maintenance of *G. skottsbergii* natural beds.

Materials and methods

To determine the abundance of the tetrasporophyte, gametophyte and carposporophyte phases in *G. skottsbergii*, a subtidal population was sampled at the locality of San Antonio, Ancud (41° 52′ S; 73° 51′ W), Chiloe Island, Chile. The depth in the sampling area was between 6 and 10 m below mean tide level, the subtidal algal community was dominated by *G. skottsbergii* and covered approximately an area of 12 ha. Monthly samplings were done from June 1996 to February 1998 by hooka diving. On each occasion two non-permanent transects of 100 m each were established, perpendicular to the coastline. A 1×1 m quadrat was sampled every 10 m along the transect, making a total of 20 samples in each month. Fronds were collected with the substrate and placed in polyethylene bags and labelled, subsequently analyzed in the laboratory, reproductive phases and sizes were separated, and wet biomass (g m^{-2}) and frond density (fronds m^{-2}) were determined. Plants below 1 cm were directly counted on the substratum under a stereomicroscope in the laboratory.

Since mature fronds were scarce in the San Antonio locality, reproductive thalli of the tetrasporophyte and carposporophyte phases to study plant fecundity were sampled in a nearby natural bed. Two wet kg of each mature tetrasporophyte and cystocarpic were sampled for 21 months from June 1996 until February 1998, at Bahía de Ancud, Chiloé (41° 55′ S; 73° 51′ W). Of these, 10 fronds bearing cystocarps and 10 thalli with tetrasporangial sori were separated in the laboratory. Wet biomass was determined for each frond. Subsequently, the density of reproductive structures (cystocarps or tetrasporangial sori) was determined in the 10 thalli and from each phase, by counting under the microscope the structures present in ten 1 cm^2 random locations on each frond. In tetrasporophytic thalli, the number of mature, immature and empty tetrasporangial sori present in each sampled area was determined. Maturity in sori was determined by colour scale (Santelices & Martínez, 1997), as follows: mature sori, brown to black: immature sori, orange to light brown, and empty sori, white (Figure 1).

Gigartina skottsbergii cystocarps originate in papillae which jut out from the surface of the female gametophytic thallus. Due to this arrangement, we first counted the number of papillae cm^{-2}, then the number of cystocarps per papilla and finally the total number of cystocarps cm^{-2}. With these data, we estimated the density and abundance of reproductive structures in an area of 1 cm^2 of the tetrasporophytic and carposporophytic phases.

Carpospore and tetraspore cultures were started to determine their viability during June, July and September, when mature tetrasporophyte and carposporophyte are abundant. Sections of 2×2 cm were cut from gametophytic thalli with cystocarps and from tetrasporophytic thalli with tetrasporangial sori. Three pieces of each phase were placed into Petri dishes (20×100 mm) with filtered seawater enriched with Provasoli solution (Provasoli, 1968). After the liberation of carpospores and tetraspores, 1 ml aliquots of each spore type were drawn separately. They were put in dishes with culture medium to obtain germination and development of carpospores and tetraspores.

These cultures were kept at 10 °C and 16:8 h light: dark cycle, with a 10 μmol m^{-2} s^{-1} and observed after 5 days. Two Petri dishes with carpospores and tetraspores were sampled under an inverted microscope, with 10 ocular fields ($10 \times 10 \times$) corresponding to an approximate area of 2.443 mm^2. Live and dead spores were counted in each field. The percentage of mortality and of survival was estimated for each spore type.

Data of density of reproductive structures and biomass were tested using a correlation analysis. Mortality rates were tested using a two-way ANOVAs with *a posteriori* Tukey HSD for multiple comparisons (Sokal & Rohlf, 1981).

Results

During the study period (June 1996 to February 1998), a decrease was observed in the total and reproductive biomass (reproductive phases) in the bed studied at San Antonio. This biomass fell from about 40 g m^{-2} in winter (June, 1996) to under 1 g m^{-2} the subsequent

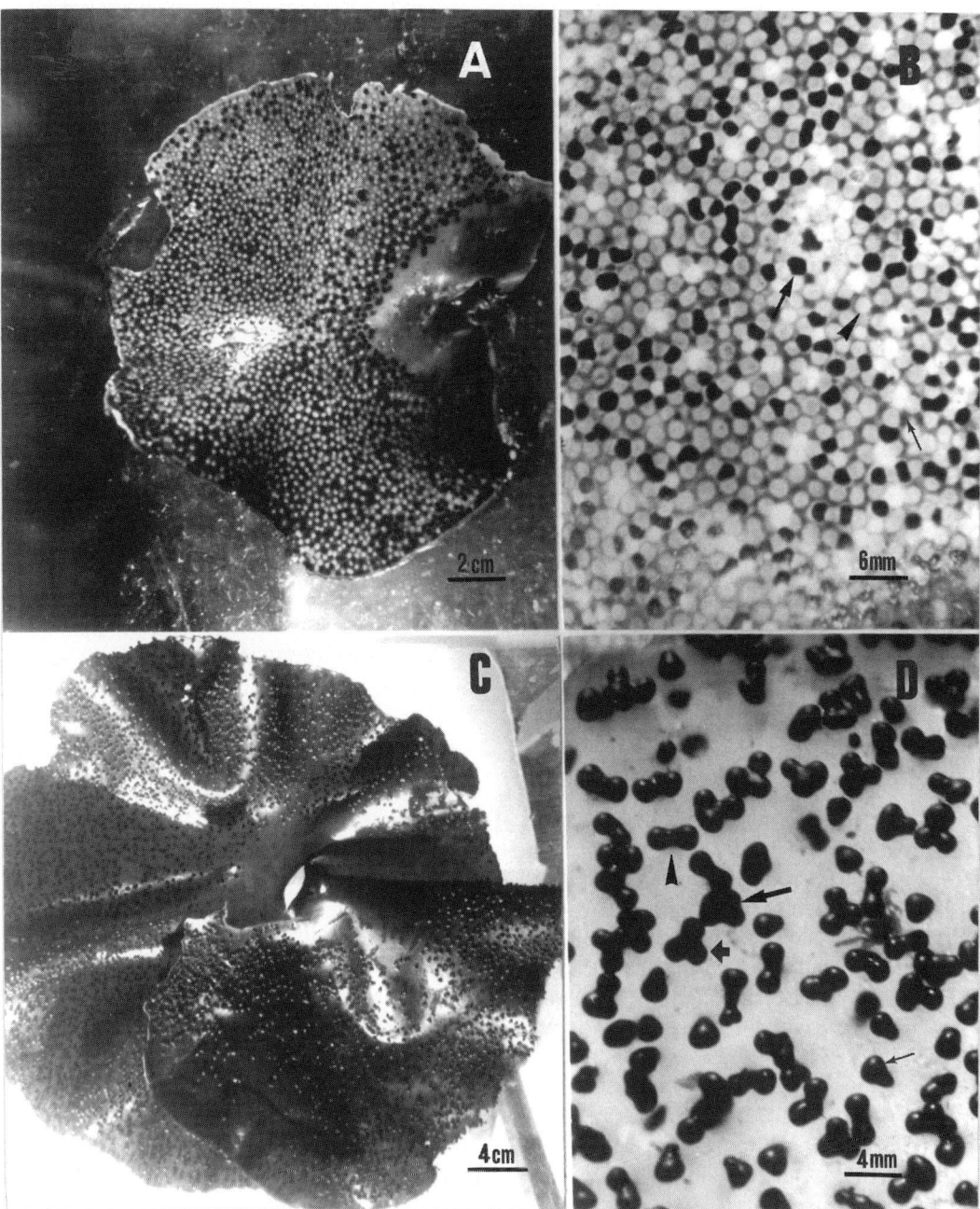

Figure 1. Morphology of the female gametophytic and tetrasporophytic thallus of *G. skottsbergii* and cystocarps and tetrasporangial sori disposition in a gametophytic and tetrasporophytic thallus. A: Tetrasporophytic thallus with tetrasporangial sori distribution. B: Mature sori (➡), immature (→) and sporulated sori (▶). C: Female gametophytic thallus with papillae and cystocarps. D: Papillae and cystocarps distributions, one papilla with one cystocarp (→), with two (▶), three (⬧) or four (➡).

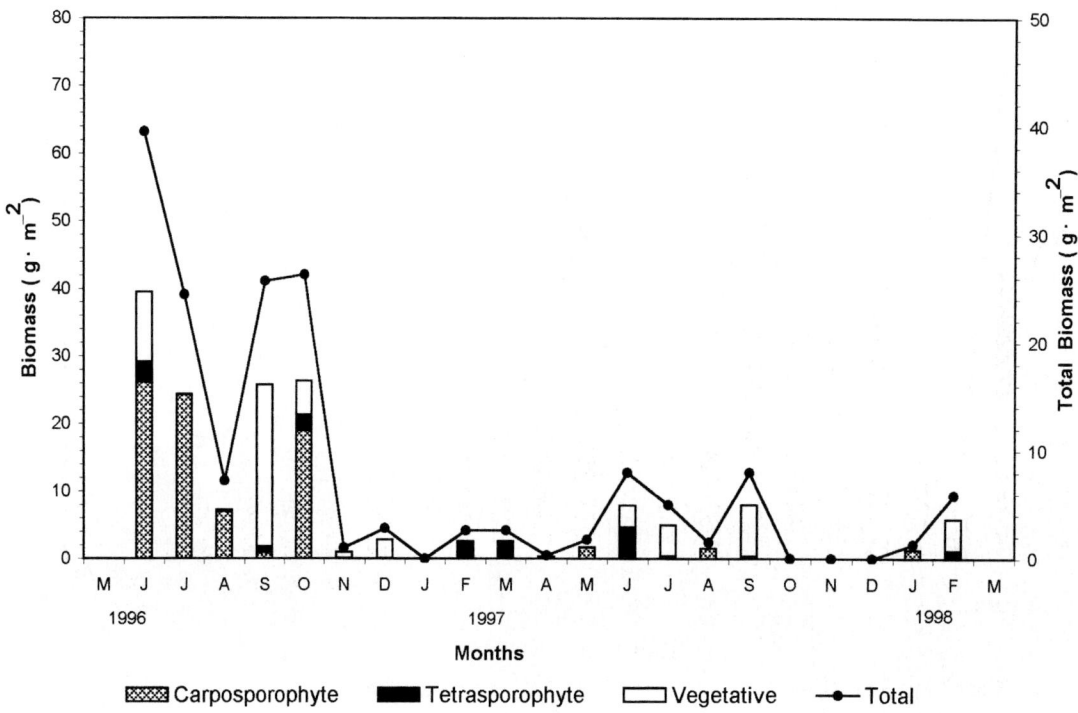

Figure 2. Monthly fluctuations in biomass (g m^{-2}) of *G. skottsbergii*.

autumn (Figure 2). A similar trend was observed in the density of reproductive thalli which strongly fluctuated from less than 4 thalli m^{-2} early winter (June, 1996) to less than 1 thallus m^{-2} in summer (February to March, 1997) and from late winter 1997 to summer 1998 (Figure 3). A clear dominance of carposporophytic biomass was recorded during winter/spring 1996 (except September), from about 26 to 8 g m^{-2}, followed by 16 months in which the reproductive biomass was less than 6 g m^{-2} until the end of the sampling period (Figure 2). Contrasting with the dominance of reproductive biomass, a very high density of vegetative fronds dominated by young / small thalli was observed during 1996–1997 (Figure 3). Notwithstanding, a simultaneous and continuous drop of plant density was recorded throughout the study period (Figures 2 and 3).

Over the whole study period, gametophytic thalli with cystocarps and tetrasporophytic thalli with tetrasporangial sori were observed. The abundance of these reproductive structures fluctuated during the study. The total density of tetrasporangial sori in the tetrasporophytic fronds remained relatively constant over the annual cycle, with a mean of 60 sori cm^{-2} (Figure 4). A higher density of sori was observed in winter and early spring. The highest density of mature sori was seen in autumn and winter, with 10 to 28 sori cm^{-2}, and a lower abundance in spring-summer with 1 to 10 sori cm^{-2}. As for immature sori, density was high in spring and summer, ranging between 15 to 58 sori cm^{-2} The density of dehisced sori showed great fluctuations, with a peak at the end of winter (August and September).

In the female gametophytic thallus, the density of papillae and cystocarps in winter months was less than 6 papillae cm^{-2} with less than 10 cystocarps cm^{-2} (Figure 5). In both cases mean values fell to 0 in early spring. Between October and November a rapid increase was observed in the density of papillae and cystocarps, with mean values of 8 to 13 papillae cm^{-2} and of 16 to 29 cystocarps cm^{-2} over the spring-summer months (Figure 5). The minimal density observed in winter-spring may have derived from the massive carpospore liberation and subsequent necrosis of the papillae, which caused their detachment and perforations, which subsequently healed.

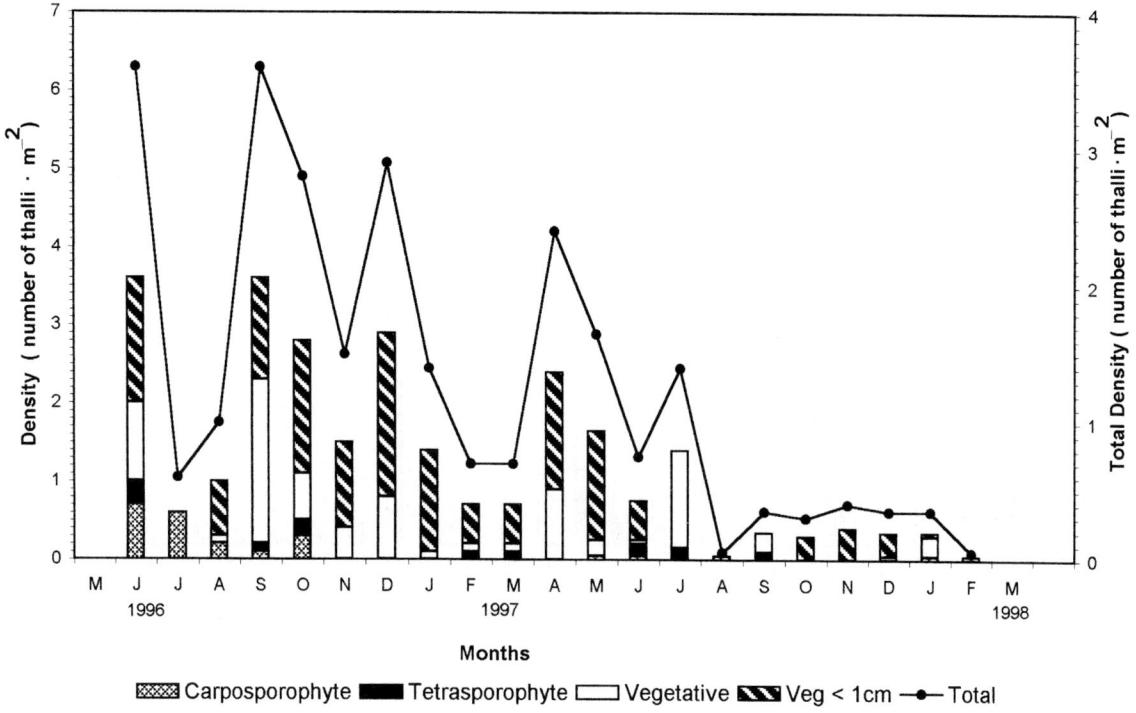

Figure 3. Monthly variations in thallus density (thalli m^{-2}) of *G. skottsbergii*.

When comparing the wet biomass of blades with tetrasporangial sori with the density of reproductive structures, we did not find any correlation between the biomass and density of reproductive structures, $r = 0.17$, while there was a better correlation between density of cystocarps and wet biomass of gametophytic fronds, $r = 0.504$. Figures 6 and 7 show monthly variations of number of tetrasporangial sori or cystocarps cm^{-2} and the variation of the biomass of the corresponding reproductive blades.

Based on our culture studies of the viability of carpospores and tetraspores performed during June, July and September, a high natural mortality for both types of spores was observed. No significant differences were found between months in mortality of tetraspores and carpospores ($p < 0.05$; two-way ANOVA). Comparisons among types of spores, indicated significant ($p < 0.05$; two-way ANOVA) differences in mortality between types of spores. For inoculated tetraspores mortality ranged from 79 to 88% and for inoculated carpospores ranged from 87% to 92%. The spores that survived germinated and developed holdfasts.

Discussion

Fluctuations of biomass and density over the study period indicated that the natural bed of *G. skottsbergii* is decreasing its standing stock. This is shown by the progressive decrease in biomass even during the growing season (spring-summer). In Chiloé Island commercial exploitation has increased since 1993 due to the great importance of Chilean raw materials for the production of carrageenans (Zamorano & Westermeier, 1996). Results show that there is no evidence of recuperation on the biomass specially during spring and summer, probably as a consequence of commercial exploitation of the resource. In relation to commercial harvesting there are 2 important aspects to consider: (1) harvesting pressure on natural beds has been increasing; and (2) harvest of attached plants by diving is permitted since there are no regulations for this species by law.

The biomass of mature tetrasporic fronds was under 5 g m^{-2} over the whole study period, so the availability of this type of spore for recruitment is quite scarce. The biomass of cystocarpic fronds fluc-

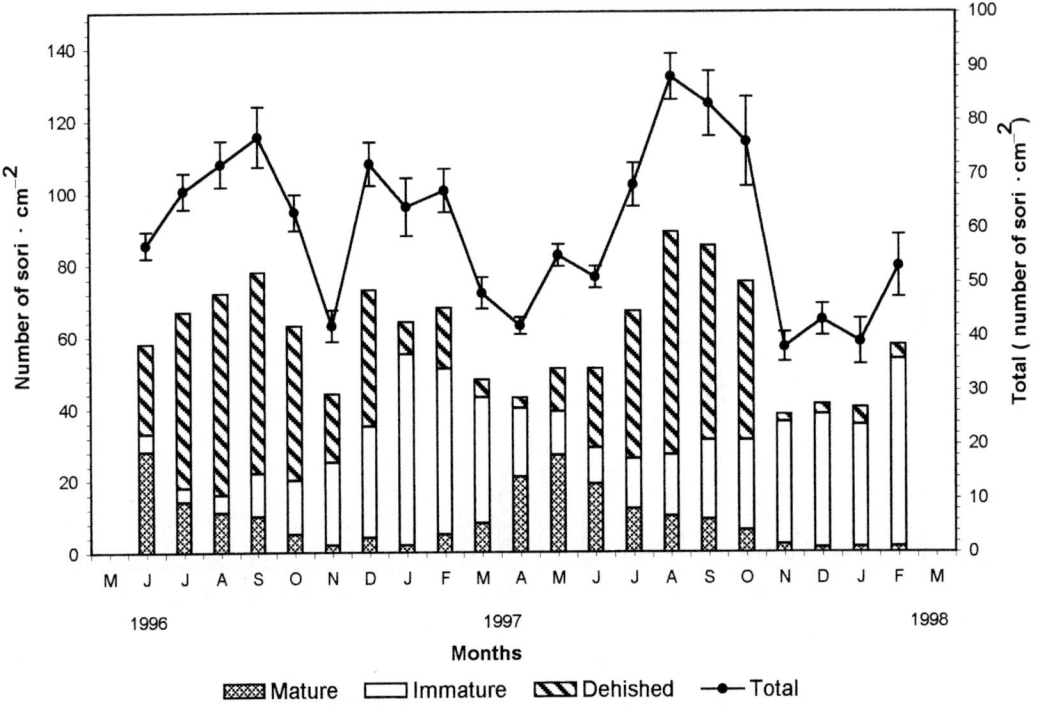

Figure 4. Monthly fluctuations in density of tetrasporangial sori in *G. skottsbergii* thalli; sori were identified as mature, immature and dehisced sori over the study period (± standard error).

tuated during the study period between 26 g m^{-2} in June 1996, to less than 1 g m^{-2} from November 1996 to February 1998.

Reproductive phenology is similar to that obtained by other authors for *Gigartina skottsbergii* (Piriz, 1996; Zamorano & Westermeier, 1996). Vegetative fronds are found throughout the year where the peak of biomass is recorded in September (spring). Nevertheless carposporophytes were found only at the beginning of this study recording the highest biomass between June and October of 1996 (winter and early spring), on Argentina coasts this phase is found to be abundant in autumn (Piriz, 1996). Zamorano & Westermeier (1996) has described a bimodal cycle for cystocarpic plants, with a maximum in late winter and late summer. This second peak was not observed in this study. All authors agree that tetrasporophytes may be found throughout the year but in small proportions (Piriz, 1996; Zamorano & Westermeier, 1996; Avila et al., 1997).

The density of vegetative, tetrasporophyte and carposporophyte phases is low, under 2 thalli m^{-2} over the study period. However there was a continuous presence of *Gigartina* small fronds, under 1 cm. This can be only explained as a result of germination of tetraspores or carpospores in spite of the low density of mature reproductive plants and the fact that spore mortality is high. The density of small thalli fluctuated along the study period and was also under 2 thalli m^{-2}. Regrowing from holdfasts was not observed during the study period, in any of the samples, probably due to harvesting methods. In other Gigartinaceae, like *Iridaea cordata*, *Mazzaella laminarioides* and *Sarcothalia crispata* it has been found that fronds are able to propagate through regeneration of the perennial holdfast (May, 1986; Santelices & Norambuena, 1987; Poblete et al., 1984).

Although all stages of tetrasporangial sori were present throughout the year, the proportions at different seasons showed that new sori developed mainly in the summer, matured mainly in autumn/winter and dehisced mainly in spring. For female gametophytes a cycle was also observed in the development of cystocarps. Papillae were more abundant from December to April (summer and early autumn), the same as cystocarps. The number of cystocarps slowly decreases

155

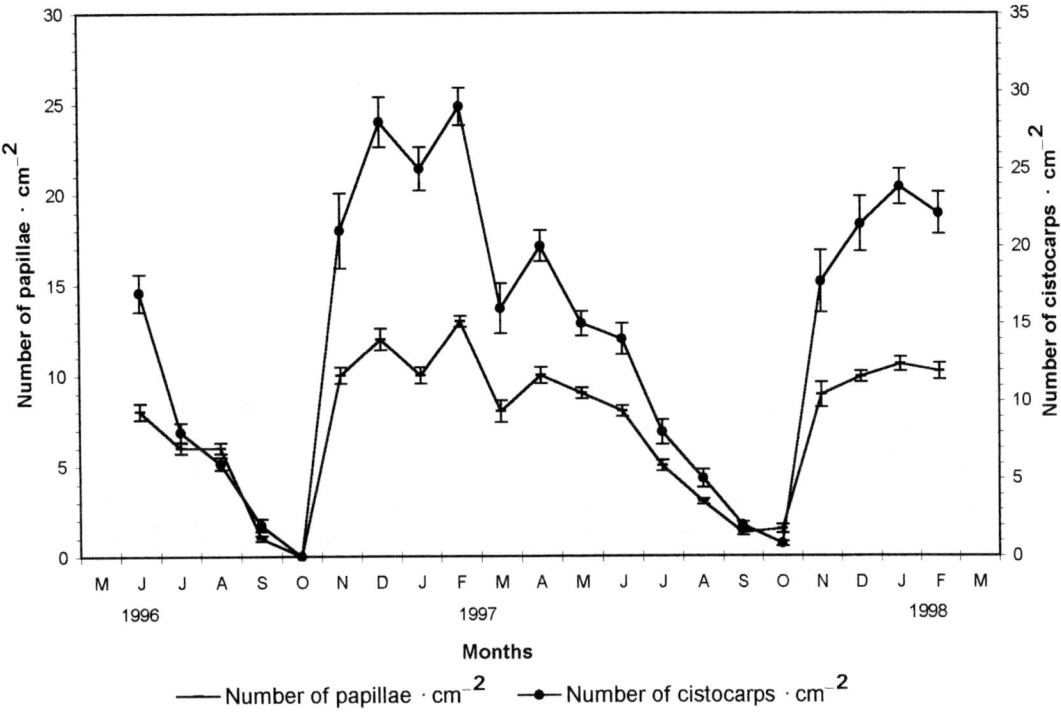

Figure 5. Monthly fluctuations in number of papillae and cystocarps in *G. skottsbergii* thalli, over the study period 1996–1998 (± standard error).

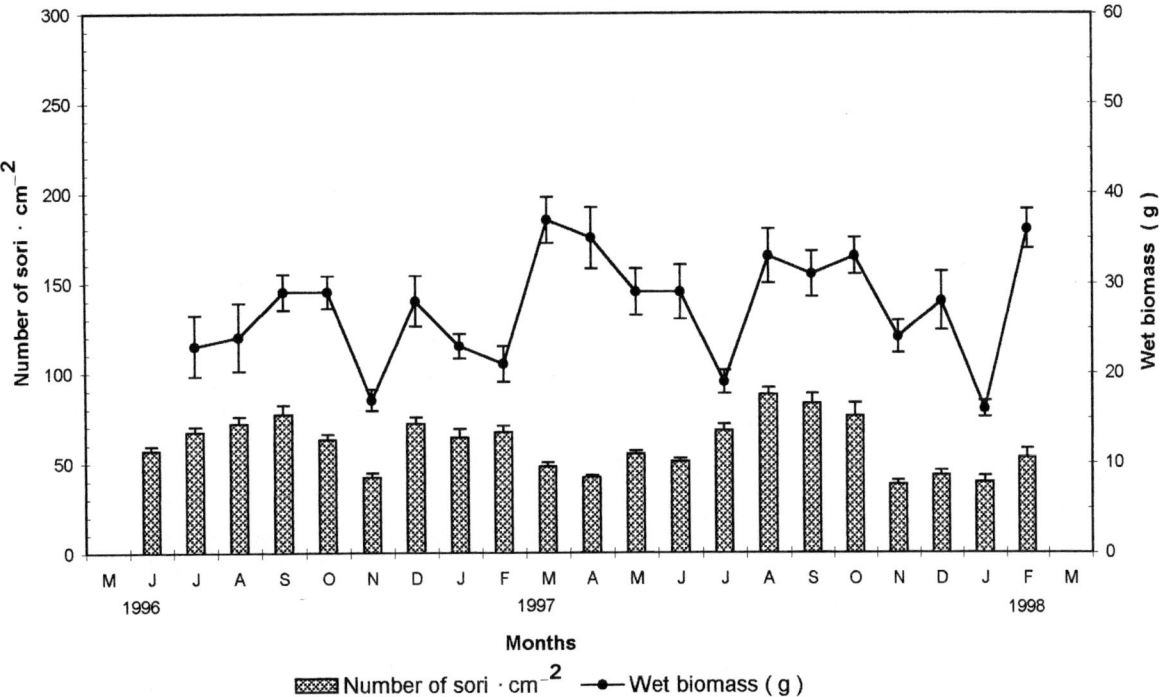

Figure 6. Monthly fluctuations in number of sori and biomass of the fronds in *G. skottsbergii* (± standard error).

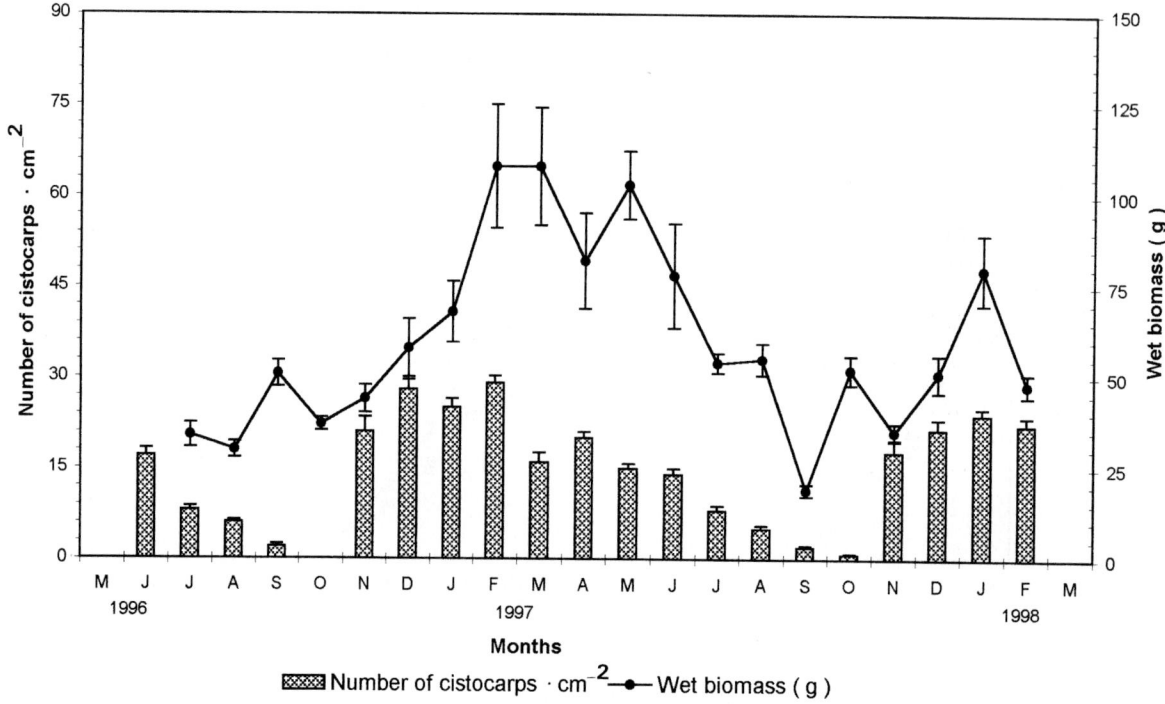

Figure 7. Monthly fluctuations in number of cystocarps and biomass in *G. skottsbergii* female gametophytic thalli (± standard error).

until June (winter), after July carpospores are released and the number of cystocarps decreases to low values during September-October (spring).

Field observations suggest that regrowth of the holdfast does not occur in this species. Fronds of < 1 cm that occurred during the study period could correspond to new fronds that have been developed from spores. We suggest that propagation and dispersal in a *Gigartina skottsbergii* natural bed is mainly due to spores. Results obtained under controlled conditions in the laboratory indicated that mortality in tetraspores and carpospores is high during the period of abundance of mature reproductive structures. Although viability of carpospores and tetraspores was low, spores that survive do germinate and develop holdfasts.

Acknowledgements

The authors acknowledge the funding for this project from the Fondo Nacional de Desarrollo Regional (FNDR). We are also grateful to FONDEF for supporting the attendance of A. Candia to the XVIth International Seaweed Symposium in Cebu, Phillipines. Many thanks to J. M. Jones, and two anonymous reviewers who have improved this manuscript.

References

Alveal, K., A. Candia, G. Collantes, M. Edding, E. Fonck, C. Melo, A. Poblete, P. Rivera, H. Romo & R. Westermeier, 1990. In Guía de Algas Marinas Chilenas de Importancia Económica. Red de Algas Marinas, Chile. 114 pp.

Ang, P. Jr., R. E. De Wreede, F. Shaughnessy & L. Dyck, 1990. A simulation model for an *Iridaea splendens* (Gigartinales, Rhodophyta) population in Vancouver, Canada. Hydrobiologia 204/205: 191–196.

Avila, M. & M. Seguel, 1993. An overview of seaweed resources in Chile. J. appl. Phycol. 5: 133–139.

Avila, M., M. Núñez, A. Candia & R. Norambuena, 1997. Reproductive phenologic patterns of a population of *Gigartina skottsbergii* (Gigartinaceae, Rhodophyta), from Ancud, Chile. 1997. Gayana Oceanol. 5: 21–32.

Bixler, J. H., 1996. Recent developments in manufacturing and marketing carrageenan. Hydrobiologia 326/327: 35–57.

Gómez, I. & R. Westermeier, 1991. Frond regrowth from basal disc in *Iridaea laminariodes* (Rhodophyta, Gigartinales) at Mehuín, southern Chile. Mar. Ecol. Prog. Ser. 73: 83–91.

Kim, D. H., 1976. A study of the development of cystocarps and tetrasporangial sori in Gigartinaceae, (Rhodophyta, Gigartinales). Nova Hedwigia 27: 1–146.

May, G. 1986. Life history variations in a predominantly gametophytic population of *Iridaea cordata* (Gigartinaceae, Rhodophyta). J. Phycol. 22: 448–445.

Norambuena, R., 1996. Recent trends of seaweed production in Chile. Hydrobiologia, 326/327: 371–379.

Piriz, M. L., 1988. Marine phycology in Argentina, a review. Gayana, Bot. 45: 83–89.

Piriz, M. L., 1996. Phenology of a *Gigartina skottsbergii* Setchell et Gardner population in Chubut Province (Argentina). Bot. mar. 39: 311–316.

Poblete, A., A. Candià, I. Inostroza & R. Ugarte, 1984. Crecimiento y fenología reproductiva de *Iridaea ciliata* Kutzing (Rhodophyta, Gigartinales), en una pradera submareal. Biol. Pesq. 14: 23–31.

Provasoli, L., 1968. Media and prospects for the cultivation of marine algae. In Watanabe, A. & A. Hattori (eds), Culture and Collection of Algae. Jap Soc. Plant Physiol. 63–75.

Ramírez, M. & B. Santelices., 1991. Catálogo de las algas marinas bentónicas de la costa templada del Pacífico de Sudamérica. Monografías Biológicas 5, Pontificia Universidad Católica, Santiago, 437 pp.

Santelices, B., 1989. Algas Marinas de Chile. Ediciones Universidad Católica de Chile, Santiago, 400 pp.

Santelices, B. & E. A. Martínez, 1997. Hierarchical analysis of reproductive potential in *Mazzaella laminarioides* (Gigartinaceae, Rhodophyta). Phycologia 36: 195–207.

Santelices, B. & R. Norambuena, 1987. A harvesting strategy for *Iridaea laminariodes* in central Chile. Hydrobiologia 151/152: 329–333.

Sokal, R. & J. Rohlf, 1981. Biometry. W.H. Freeman & Co, New York, 859 pp.

Westermeier, R. & J. Sigel, 1997. Reproductive patterns of *Gigartina skottsbergii* (Rhodophyta) in southern Chile. Phycologia. 36 (Suppl.): 123.

Zamorano, J. & R. Westermeier, 1996. Phenology of *Gigartina skottsbergii* (Gigartinaceae, Rhodophyta) in Ancud Bay, southern Chile. Hydrobiologia, 326/327: 253–258.

Population structure and reproduction of the carrageenophyte *Chondracanthus pectinatus* in the Gulf of California

Isaí Pacheco-Ruíz & José A. Zertuche-González.
Instituto de Investigaciones Oceanológicas, UABC, P.O. Box 453, Ensenada, Baja California, Mexico
E-mail: isai@bahia.ens.uabc.mx

Key words: Chondracanthus pectinatus, recruitment, reproduction, ecology, California Gulf, Mexico.

Abstract

Population structure and dynamics of the carrageenophyte *Chondracanthus pectinatus* (Dawson) L. Aguilar & R. Aguilar, an endemic species from the Gulf of California, were studied from November 1994 to December 1995. Plant maximum size and weight were reached in May (80 cm, 480 g dry wt), when new recruits were approximately 0.6 mm in length. In April, the first plants that washed ashore were observed at densities of 25 g dry wt m^{-1} of shore line. In summer (June–July), plants in the water became fragmented into small pieces. Gametophytes were always more numerous than tetrasporophytes (7:3). Male plants were not observed *in situ*. Reproduction was observed as early as December in small plants (6 cm long), with a low cystocarp and tetrasporangia density. However, the density of reproductive tissue increased exponentially in spring (384 000 cystocarps, 111 000 tetrasporangia per plant). Although reproduction by spores is significant, vegetative tissue remaining submerged is capable of re-attaching and generating new plants.

Introduction

Chondracanthus pectinatus (Dawson) L. Aguilar & R. Aguilar, is an annual endemic species of the Gulf of California, appearing in autumn (November), most abundant in spring, and disappearing completely in summer (August) when surface water temperature reaches 30–31 °C (Zertuche-González, 1988). The maximum biomass (398 ± 119 g dry wt m^{-2}) and growth (10.6% d^{-1}) occurs in spring (Zertuche-González, 1988; Pacheco-Ruíz et al., 1992). However, the strategy of its persistence *in situ* during the critical periods of summer-fall is still unknown.

The life history in Gigartinales has been studied extensively in the past decades using controlled culture conditions (Chen & McLachlan, 1972, 1980; West & Guiry, 1982). However, the life history in natural populations has been neglected (May, 1986). Some of the most studied aspects are reproductive periods, phase proportion and reproductive effort (Hansen & Doyle, 1976; Bhattacharya, 1985; May, 1986; Ballesteros et al., 1990, 1991). With respect to phase proportion, there are mixed reports with either gametophyte dominance over sporophytes or the opposite, and in some cases where the male stage is absent from natural populations (Melo & Neushul, 1993). There are no studies or references in *C. pectinatus* on these aspects although the gametophytes and tetrasporophytes are known to be isomorphic. Female gametophyte plants with cystocarps are recorded, but male plants and the spermatagia are unknown. Tetrasporophytes bear nemathecia at the base of the spines. The majority of plants have many reproductive structures (tetrasporangia and cystocarps) in spring (Dawson, 1961; Norris, 1975; Zertuche-González et al., 1995). From what little is known about reproduction in the plant, the hypothesis formed is that the generation of propagules is the main factor contributing to its permanence *in situ* and is the strategy for survival in an extreme ecological environment such as the north of Gulf of California (summer-autumn); rejecting the possibility that new generations of plants come from vegetative growth because *C. pectinatus* is a strict annual plant (Zertuche, 1989; Pacheco-Ruíz et al., 1992).

Because of this, we collected *C. pectinatus* to determine the following:

1. The seasonal variation of a population in time.
2. The exact period of reproduction.
3. The proportion of phases.
4. The reproductive effort.
5. Seasonality of tissue disintegration and recruitment.

Materials and methods

Bahía de Los Ángeles is located between 28° 53′ 00″ and 29° 10′ 00″ N, and 113° 25′ 00″ and 113° 36′ 30″ W on the northeast coast of the Gulf of California. It is an important upwelling zone from deep waters of the Canal de Ballenas, and is distinguished by strong tidal mixing (Roden & Groves, 1959; Bray & Robles, 1991), resulting in a high nutrient concentration all year long (Alvarez-Borrego et al., 1978). The collection area was located in La Sílica (Figure 1) between 1 and 5 m deep where *C. pectinatus* is located. The intertidal zone is composed of volcanic rocks. The subtidal zone is sandy, with random volcanic boulders and mollusc shells. Seawater surface temperature ranges between 14.0 and 20.0 °C in winter–spring and between 25.0 and 31.6 °C in summer–autumn. Ambient surface irradiance is 1050–1750 μmol m^{-2} s^{-1} in winter–spring and 1800–1500 μmol m^{-2} s^{-1} during summer–fall. At 4 m depth, the irradiance is between 300 and 500 μmol m^{-2} s^{-1} in winter–spring and 100–600 μmol m^{-2} s^{-1} during summer–autumn (Zertuche-González, 1988).

During the period 1986–94, 26 samplings were performed collecting fragments of *C. pectinatus* (n = 60) haphazardly by SCUBA diving. In the laboratory each fragment was separated by phase directly or indirectly with a compound microscope or stereoscope. Between November 1994 and December 1995, monthly random collections of complete plants attached to the bottom ($n > 40$) were performed. Plants of both samples were preserved in 4% formaldehyde in dark plastic bags and tagged for later analysis.

In the laboratory each plant was measured, weighed, and separated by stage, following the previous methodology. The sterile plants were evaluated for carrageenan by the resorcinol test (Garbary & DeWreede, 1986).

Reproductive effort (number of cystocarps or tetrasporangia per plant) was calculated in each cystocarpic and tetrasporic plant that was collected during the 1994–95 period. For plants collected between November and March, all reproductive structures were

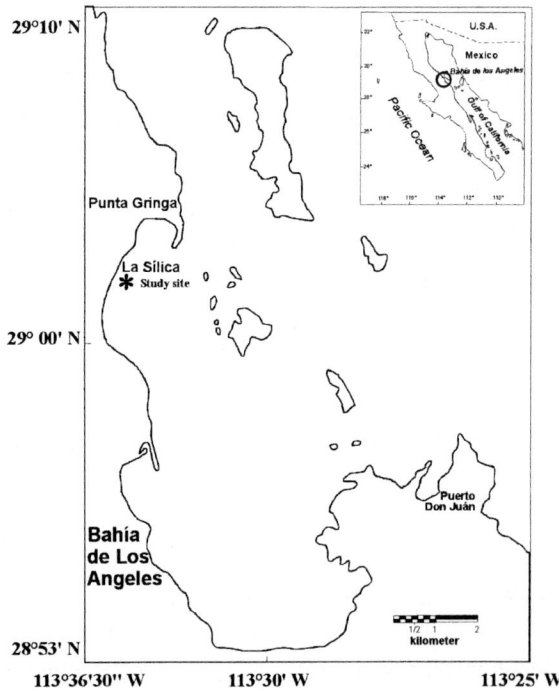

Figure 1. Study site in Bahía de Los Ángeles, Baja California.

counted using a stereoscopic microscope providing the cystocarps and tetrasporangia density per plant. Between April and June, each plant was divided into three fractions; basal, middle and apical, because there was a large concentration of structures in each plant. In each fraction, a 30 gram (wet weight) sample was separated, and for each gram the reproductive structure density was calculated. Results were extrapolated to the whole plant.

To determine the recruitment period *in situ*, on February 27, 1997 a nylon rope was installed below zero tide, the rope was perpendicular to the shoreline, facing east, and crossing the *C. pectinatus* bed. The rope was labelled every 10 m. The rope was 200 m long, and at every 20 m a 3 m length string of shells, with 15 oyster shells each 20 cm apart, was vertically positioned. The total number of shell strings installed on March 8, 1997 was 11. Each was tagged, and anchored to the bottom with a concrete dead weight with a buoy on the surface to keep it vertical in the water column. Six new strings were installed on May 28, one every 40 m over the same transect. Finally, in August 14, three strings at 40, 120 and 200 m were installed. In May 1997, random collections of the shell strings set out in March were performed (20, 100 and 180 m). In August, three strings from March (60, 140 and 200 m) and three from May (40, 120 and 200 m) were collected. Finally in October, three from the March deployment (0, 80 and 160 m), three from May (0, 80 and 160 m) and three from August (40, 120 and 200 m) were collected. For transportation the shell strings were kept in seawater inside a dark plastic bag and placed in a cooler with ice then to be analyzed with a microscope to detect recruits.

When present, the biomass of *C. pectinatus* washed to shore was estimated per m of coast. Plants were cleaned of sediments and dried and weighed *in situ*.

A thermograph was installed at 7 m, and was located south of the bed. Water temperature was recorded daily.

Results

Maximum plant length (80 ± 5.5 cm) and weight (480 ± 92 g dry weight) appeared in spring (May) and minimum (0.5 ± 0.05 cm; 0.008 ± 0.0006 g dry weight) 'quantifiable *in situ*' in autumn (October; Figure 2). The first fragments of *C. pectinatus* were found on the beach in the spring (April) (25 g m^{-1} dry weight shore line). In summer (June), most of the plants broke into small pieces, and material of *C. pectinatus* (<2 cm length) was located in suspension in the water column, decomposing above the bottom or deposited on the beach with biomass ≈ 4.6 kg m^{-1} dry weight shore line (Figure 3).

Between 1986 and 94, and 1994 and 95, the highest percentage of carposporic plants was in spring and summer (March and July) (25–62%). Tetrasporic plant percentage was always less. There were no tetrasporic *in situ* plants in 1990. During 1986–94, 40 – 50% of the plants were non reproductive and in 1994, 30 – 35% were non reproductive (Figures 4, 5a).

With the resorcinol test, applied to sterile plants in the period of 1994–95, the percentage with κ-carrageenan (male and female) was greater than recorded with ξ-carrageenan. Complete analysis showed 70% of κ-carrageenan plants, and 30% of ξ-carrageenan plants in spring (May) (Figure 5b). The gametophyte:tetrasporophyte ratio was 7:3.

The first reproduction was detected in winter (December), in plants of 6.0 ± 0.3 cm length, with low cystocarp and tetrasporangia density (1.8 ± 0.6 vs 4 ± 0.6 plant^{-1}) which then increased in spring (April), with a maximum density in May (384,170

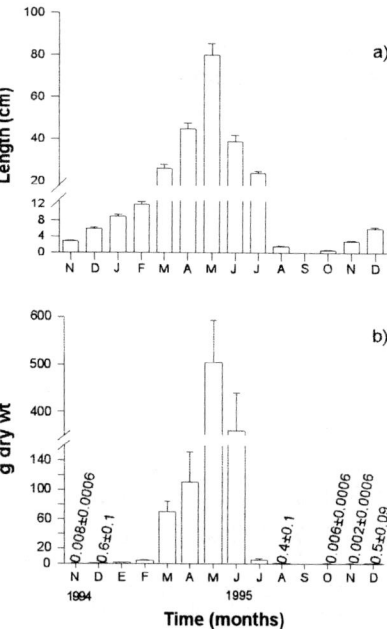

Figure 2. Plant length (a) and weight (b) of *C. pectinatus* in Bahía de Los Ángeles.

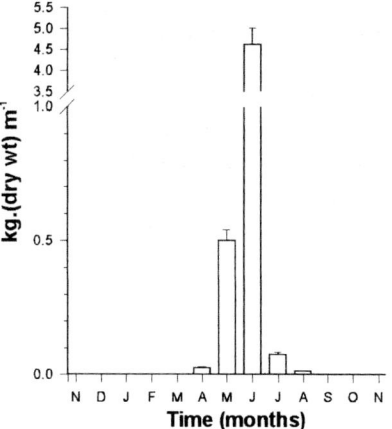

Figure 3. *C. pectinatus* biomass washed ashore in La Sílica (± = SE, n = 30).

± 34,200 cystocarps and 111,270 ± 12,033 tetrasporangia plant^{-1}) (Figure 6).

First micro-recruitment was detected in May on shell collectors placed in early spring (March). There were plantlets 0.6 ± 0.05 mm long (n = 4), crust phases 0.5 ± 0.03 mm in diameter (n = 18) and apical fragments of *C. pectinatus* <4 mm in length (n = 38) fixed on shells. Plantlets and crust phases resulted from spores while apical fragments resulted from re-attachment of plants.

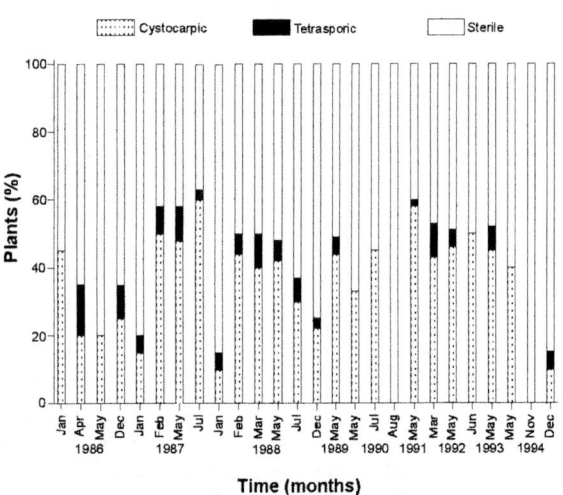

Figure 4. Seasonal reproduction of *C. pectinatus* in La Sílica bed, Bahía de Los Ángeles, during 9 years of sampling ($n = 60$).

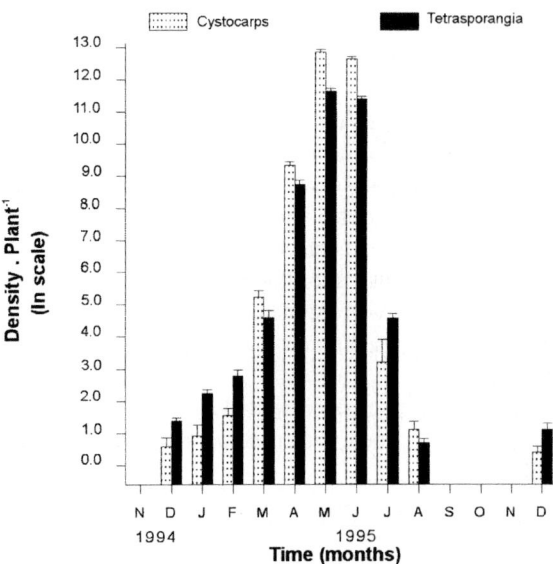

Figure 6. Reproductive effort of *C. pectinatus*, number of cystocarps and tetrasporangia per plant ($\pm = $ SE, $n = 40$).

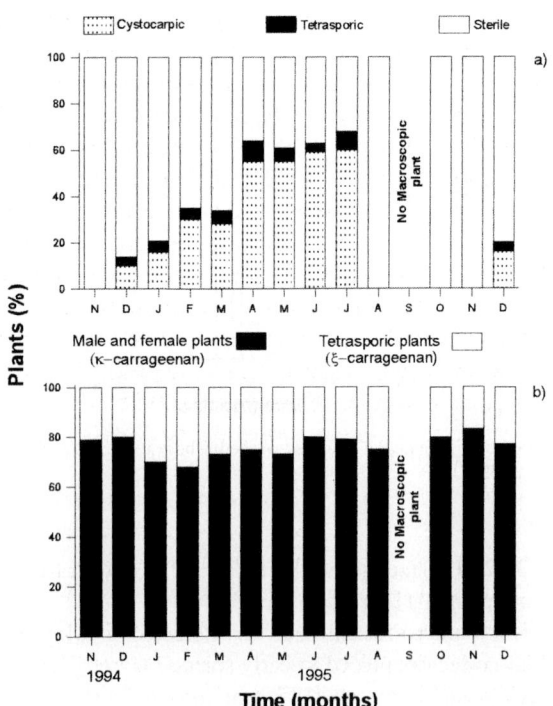

Figure 5. Seasonal variation of reproductive phase from Nov. 94 to Dec. 95 (a), and haploid/diploid presence using resorcinol (b) in *C. pectinatus* ($n = 40$).

The fragments were decomposing except for the attached portion. This mass of vegetative tissue appeared to be a perforated crust in the centre (ring) with a 1.06 ± 0.11 mm diameter ($n = 14$). After a while, on the periphery of this vegetative tissue mass, four developing plantlets were detected (1.15 ± 0.3 mm in length) identified as *C. pectinatus*.

In August, plantlets (1.527 ± 273 mm length, $n = 22$) were detected on shells placed in March, some showing a first branch (0.554 ± 0.087 mm length, $n = 5$). Crusts generated by spores with 0.91 ± 0.21 mm diameter ($n = 3$) and vegetative rings of 0.98 ± 0.17 mm diameter ($n = 4$) were also detected, some showing up to four buds of 637 ± 92 μm length ($n = 8$). Also, in August small plants (1.13 ± 0.21 mm length, $n = 19$) were detected in shells placed in May, of which only one had an erect shoot (0.55 mm length); also vegetative crusts were located with 0.83 ± 0.14 mm diameter ($n = 3$) and three buds 0.96 ± 0.11 mm in length.

In October small plants (5.84 ± 0.54 mm length, $n = 38$) were detected on shell collectors placed in March with erect thalli from vegetative material 5.89 ± 1.5 mm long ($n = 13$). Forty percent of previous material showed between 1 and 6 ramifications per plantlet, with 4.5 ± 0.6 mm length. Also small plants were detected on shells placed in May, with 4.2 ± 0.83 mm length ($n = 29$), 4 of which presented ramifications of 0.371 ± 0.46 mm length, and plantlets produced

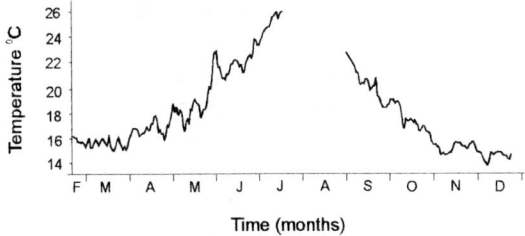

Figure 7. Water temperature at La Sílica, Bahía de Los Ángeles.

vegetative growth with buds of 3.66 ± 0.69 mm in length ($n = 5$). On the shell collectors placed in August, no spore recruitment or vegetative material of *C. pectinatus* was found. Maximum sea water temperature was in summer (26–29 °C), and the minimum in winter (14–17 °C) (Figure 7).

Discussion

For years *Chondracanthus pectinatus* was considered an annual plant, because of its apparent disappearance *in situ*, attributed to high temperatures (29–31 °C) in summer (August) and showing the appearance of the new generation at the end of Autumn (November), when the temperature is less than 22 °C (Zertuche, 1989; Pacheco-Ruíz et al., 1992). This study demonstrates that macroscopic plants of 0.5 cm are present in autumn, but the first small recruits (600 μm) are found in the spring (May). Zertuche-González (1988) and Pacheco-Ruíz et al. (1992), mention that the first recruits appear at the end of Autumn (November). However, what they detected were the first juveniles observed directly *in situ*; plants of *C. pectinatus* from 2 to 4 cm long are the ones that can be observed directly and quantified by percent cover and biomass.

Although cystocarps and tetrasporangia were found in December, there were few per plant. By April there was a big increase and recruitment from spores was evident, but by July there were few reproductive structures and in August adult plants disappeared. New recruits from this sporulation appeared from May to August but not after that. The fact that small plants were found in October, on shells deployed in March, shows that the unfavorable season, when macroscopic plants are absent, is survived by the early stages. This implies that individual plants can have a longevity of at least 15 months.

The new generation provided from propagules was detected in May and disappeared after ≈15 months (July–August), after the majority of the population had reproduced. Temperature is considered a regulating factor in the liberation of spores and fertility in many species (Katada, 1955; Christie & Evans, 1962; Rao & Subbarangaiah, 1981; Santelices, 1990; Rao & Kaliaperumal, 1983). Photoperiod has been shown to also be a regulating factor in fertility of red algae, correlating frequently with irradiance and temperature (Rao & Kaliaperumal, 1983; Ávila et. al., 1985). Even though these factors have not been measured in an experimental manner, for *C. pectinatus*, it is notable that the reproductive effort increases as the temperature increases. Oza & Krishnamurthy (1968), indicate that besides temperature, the maturity of the algae influences reproduction significantly. In accordance with this, we can suppose that there is a difference in the density of reproductive structures because the temperature and photoperiod influence the maturation process of plants, therefore contributing to a gradual increase in the number of reproductive structures per plant. This fact is considered the reproductive strategy that *C. pectinatus* uses *in situ*, between winter and spring before the plants pass into a state of senescence and death.

In *C. pectinatus*, the generation of new plants from small portions of vegetative tissue of course make up part of the population year after year *in situ*, and is pseudoperennial. The development of these adhesion structures occurred when the temperature was ≥22 °C (June–July) and plants of *C. pectinatus* fragmented (Pacheco-Ruíz et al., 1992). During this period the major part of the tissue was deposited on the beach (3.5 ± 0.5 kg dry wt m^{-2}) and small pieces and ramifications *C. pectinatus* (<2 cm) maintained adrift in the water column activating secondary mechanisms of attachment in the apex ('suckers'), something similar to the secondary mechanisms that the plant develops in spring (Dawson, 1960; Norris, 1975; Zertuche-González et al., 1995), permitting the plants to obtain a better attachment to the substrate.

The vegetative tissue that is generated in the apices of the fragments attaches to any hard surface that exists *in situ* (rocks or shells) and as the temperature continues to rise (>24 °C), the part of the thallus attached to the substrate degrades until it disappears and only the vegetative tissue remains in the form of a small microscopic ring of 0.45 ± 0.05 mm in diameter. This tissue survives the critical period of summer and autumn when the temperature *in situ* can reach 29 °C or more. From these peripheral rings of vegetative tissue develop small buds which in summer (August) can reach 0.4 ± 0.03 mm, and with time develop into adult plants which integrate into the population as pseudoperennials.

The presence of pseudoperennial macroalgae in the Gulf has been documented for *Sargassum herporhizum* Setch. et Gardn., *S. johnstonii* Setch. et Gardn. and *S. sinicola* Setch. et Gardn., in these plants the recovery of the biomass *in situ* is attributed to their organ of attachment and/or psuedoperennial stipes (McCourt, 1984a,b). *Eucheuma uncinatum* (Setch. et Gardn.) Dawson was described as an annual plant (Dawson, 1960; Norris, 1975) that disappeared in summer when temperatures were greater than 28 °C (Polne et al., 1981). However, it is shown now that this carageenophyte is pseudoperennial, and survives *in situ* by vegetative fragments, which stay embedded in the sand and regenerate to form part of the new generation (Zertuche-González, 1988). This phenomenon is also observed in populations of *Gracilariopsis lemaneiformis* (Bory) Dawson, Acleto & Foldvik in Bahía de Las Animas (Pacheco-Ruíz et al., 1999).

The permanence *in situ* of macroalgae from vegetative tissue is detected in many parts of the world (Knight & Parke, 1931). Dixon (1973) reported this permanence for *Gelidium* sp, *Pterocladia capillacea* (S. G. Grielin) Santelices & Hommersand and *Pterosiphonia complanata* (Clemente y Rubio) Foldkenberg. Perennation also can be attributed to the crust base in *Mazaella splendens* (Setchell & Gardner) Fredericq (Hansen & Doyle, 1976; Hansen, 1977, 1981). A high ratio of perennial thalli of *Bornetia secundiflora*, (J. Agardh) Thuret was a product of basal fragments (Dixon, 1965). The same occurs for *Chondrus crispus* Stackhouse (Taylor et al., 1981; Bhattacharya, 1985; Lazo et al., 1989) and *Solieria filiformis* (Kützing) Gabrielson (Perrone & Cecere, 1997). This demonstrates that the vegetative growth of *C. pectinatus* from secondary mechanisms of attachment (vegetative tissue), is part of the strategy that the plant uses to maintain *in situ* during critical periods of summer in the Gulf of California. It is not known if this strategy or a similar one is utilized by other annual red algae in the Gulf, but considering *Eucheuma uncinatum* (Setchell & Gardner) Dawson, *Gracilariopsis lemaneiformis* (Bory) Dawson, Acleto & Foldvik and *C. pectinatus*, it is possible that many other annual red and endemic algae may also be pseudoperennials. This modification in the life history of algae previously known, such as annuals in the Gulf, is in accordance with Dixon (1973); "the apparent disappearance of thallus *in situ*, for whatever adverse factor, does not necessarily signify that the plant has disappeared completely, and many red algae persist in the environment as minute hidden fragments, and would cause consideration of plants as annual when in reality their life history is pseudoperennial". The strategy performed by *C. pectinatus*, *G. lemaneirformis* and *Eucheuma uncinatum* fit the argument discussed by Chapman (1986) and Santelices (1990), where early developmental stages with suspended growth and propagules in different degrees of development provide an unseen population not usually identifiable that act as a bank of microscopic stages of seaweeds. These stages survive unfavourable conditions and provide the new generation of macroscopic plants when conditions improve.

In the natural population of *C. pectinatus*, the overall ratio of fertile gametophytes to sporophytes was 6:1 but sterile plants were abundant. When the resorcinol test was applied to the sterile plants, the gametophyte:sporophyte ratio became 7:3. This anomaly in the ratio of the phases cannot be attributed to a greater fertility in tetrasporic plants but the early stages of gametophytes may be more tolerant of the high temperatures that are common in this region of the Gulf. Alternatively an apomictic process, as reported for other red algae (Polanshek & West, 1977; West et al., 1992), may be taking place.

Acknowledgements

This research was supported by the regional program Sistema de Investigación del Mar de Cortés (SIMAC/94/CM-11), and the Universidad Autónoma de Baja California (UABC). The authors thank M. Sc. Lydia B. Ladah for English editing, M. Sc Rafael Blanco and Biol. Alberto Gálvez for valuable assistance during the field work.

References

Alvarez-Borrego, S., J. A. Rivera, G. Gaxiola-Castro, M. J. Acosta-Ruíz & R. A. Schwartzlose, 1978. Nutrientes en el Golfo de California. Cienc mar. 5: 53–71.

Avila, M., A. J. Hoffman & B. Santelices, 1985. Interacciones de temperatura, densidad de flujo fotónico y fotoperíodo sobre el desarrollo de etapas microscópicas de *Lessonia nigrescens* (Phaeophyta, Laminariales). Rev. chil. Hist. nat. 58: 71–82.

Ballesteros-Grijalva, G., G. Chauvet-Allard & E. Durazo-Beltrán, 1991. Estimación de la abundancia de *Gigartina canaliculata* Harvey, en Bahía San Quintín, Baja California, México. Cienc. mar. 17: 99–108.

Ballesteros-Grijalva, G., J. U. Labastida-Woods & E. Durazo-Beltrán, 1990. Abundancia de *Gigartina canaliculata* Harvey, en el ejido Eréndira y Popotla B.C., México. Cienc. mar. 16: 23–34.

Bhattacharya, D., 1985. The demography of fronds of *Chondrus crispus* Stackhouse. J. exp. mar. Biol. Ecol. 91: 217–31.

Bray, N. A. & J. M. Robles, 1991. Physical Oceanography of the Gulf of California. In: Dauphin J. P., & B. R. T. Simoneit (eds), The Gulf and Peninsular Province of the California. Tulsa, Okla.: Am. Assoc. Petrol. Geol.: 511–533.

Christie, A. O. & L. V. Evans, 1962. Periodicity in the liberation of gametes of zoospores of *Enteromorpha intestinalis* Link. Nature 193: 193–194.

Chapman, A. R. O., 1986. Populations and community ecology of seaweeds. Adv. mar. Biol. 23: 1–161.

Chen, L. C-M. & J. McLachlan, 1972. The life history *Chondrus crispus*. Can. J. Bot. 50: 1055–1060.

Chen, L. C-M. & J. McLachlan, 1980. *Rhodoglossum affine* (Harv.) Kylin (Gigartinaceae, Rhodophyta) in culture. Syesis 12: 113–116.

Dawson, E. Y., 1960 A review of the ecology, distribution and affinities of the benthic flora, In Symposium on the biogeography of Baja California and Adjacents Seas. Syst. Zool. 9: 93–100.

Dawson, E. Y., 1961. Marine red algae of Pacific Mexico (Gigartinales). Pac. Nat. 2: 191–375.

Dixon, P. S., 1965. Perennation, vegetative propagation and algal life histories, with special reference to *Asparagopsis* and other Rhodophyta. Bot. Gothoburg. 3: 67–74.

Dixon, P. S., 1973. Biology of the Rhodophyta. Oliver & Boyd, Edinburgh, 285 pp.

Garbary, D. J. & R. E. DeWreede, 1986. Life history phases in the natural populations of Gigartinaceae (Rhodophyta): quantification using resorcinol. In Lobban, C. S., D. C. Chapman & B. P. Kremer (eds), Experimental Phycology a Laboratory Manual. Cambdrige Univ. Press, New York: 174–178.

Hansen, J. E., 1977. Ecology and natural history of *Iridaea cordata* (Gigartinales, Rhodophyta) growth. J. Phycol. 13: 395–402.

Hansen, J. E., 1981. Studies on the population dynamics of *Iridaea cordata* (Gigartinales, Rhodophyta). Proc. int. Seaweed Symp. 8: 336–341.

Hansen, J. E. & W. T. Doyle, 1976. Ecology and natural history of *Iridaea cordata* (Rhodophyta; Gigartinnaceae): population structure. J. Phycol. 12: 273–278.

Katada, M., 1955. Fundamental studies on the propagation of Gelidiaceous algae with special reference to shedding and adhesion of the spores, germination, growth and vegetative reproduction. J. Shim. Coll. Fish. 5: 1–87.

Knight, M. & M. W. Parke, 1931. Manx Algae. Liverpool University Press, Liverpool, 260 pp.

Lazo, M. L., M. Greenwell & J. McLachlan, 1989. Population structure of *Chondrus crispus* Stackhouse (Gigartinaceae, Rhodophyta) along the coast of Prince Edward Island, Canada: distribution of gametophytic and sporophytic fronds. J. exp. mar. Biol. Ecol. 126: 45–58.

McCourt, R. M., 1984a. Niche differences between sympatric *Sargassum* species in the northern Gulf of California. Mar. Ecol. Prog. Ser. 18: 139–148.

McCourt, R. M., 1984b. Seasonal patterns of abundance, distributions, and phenology in relation to growth strategies of three *Sargassum* species. J. exp. mar. Biol. Ecol. 74: 141–156.

May, G., 1986. Life history variation in a predominantly gametophytic population of *Iridaea cordata* (Gigartinaceae, Rhodophyta). J. Phycol. 22: 448–455.

Melo, A. R. & M. Neushul, 1993. Life history and reproductive potential of the agarophyte *Gelidium robustum* in California. Hydrobiologia 260/261: 223–29.

Norris, J. N., 1975. The marine algae of the northern Gulf of California. Ph. D. diss., Univ. Calif., Santa Barbara, 575 pp.

Oza, R. M. & V. Krishnamurthy, 1968. Studies on carposporic rhythm of *Gracilaria verrucosa* (Huds.) Papenf. Bot. mar. 11: 118–121.

Pacheco-Ruíz, I., J. A. Zertuche-González, A. Cabello-Passini & B. H. Brinkhuis, 1992. Growth responses and seasonal biomass variation of *Gigartina pectinata* Dawson (Rhodophyta) in the Gulf of California. J. exp. mar. Biol. Ecol. 157: 263–274.

Pacheco-Ruíz, I., J. A. Zertuche-González, F. Correa-Díaz, F. Arellano-Carbajal & A. Chee-Barragán, 1999. *Gracilariopsis lemaneiformis* beds along the west coast of the Gulf of California, Mexico. Hydrobiologia 398/399 (Dev. Hydrobiol. 137): 509–514.

Perrone, C. & C. Cecere, 1997. Regeneration and mechanisms of secondary attachment in *Solieria filiformis* (Gigartinales, Rhodophyta). Phycologia 36: 120–127.

Polanshek, A. R. & J. A. West, 1977. Culture and hybridization studies on *Gigartina papillata* (Rhodophyta). J. Phycol. 13: 141–149.

Polne, M., M. Neushul & A. Gibor, 1981. Studies in the domestication of *Eucheuma uncinatum*. Proc. int. Seaweed Symp. 10: 619–624.

Rao, U. M. & N. Kaliaperumal, 1983. Effects of environmental factors on the liberations of spores from some red algae of Visakhapatnam coast. J. exp. mar. Biol. Ecol. 70: 45–53.

Rao, U. M. & G. Subbarangaiah, 1981. Effects of environmental factors on the diurnal periodicity of tetraspores of some Gigartinales (Rhodophyta). Proc. int. Seaweed Symp. 10: 209–214.

Roden, G. I. & G. W. Groves, 1959. Recent oceanographic investigations in the Gulf of California. J. mar. Res. 18: 10–35.

Santelices, B, 1990. Patterns of reproduction, dispersal and recruitment in seaweeds. Oceanog. Mar. Biol. Ann. Rev. 28: 177–276.

Taylor, A. R. A., L. C.-M. Chen, B. D. Smith & L. S. Staples, 1981. *Chondrus* holdfasts in natural populations and in culture. Proc. int. Seaweed Symp. 8: 140–143.

West, J. A. & M. D. Guiry, 1982. A life history study of *Gigartina johnstonii* (Rhodophyta) from the Gulf of California. Bot. mar. 25: 205–11.

West, J. A., G. Zuccarello & H. P. Calumpong, 1992. *Bostrychia bispora* sp. nov. (Rhodomelaceae, Rhodophyta), an apomictic species from Darwin, Australia: reproduction and development in culture. Phycologia 31: 37–52.

Zertuche, J., 1989. Strategies for continuous culture of nonperennial carrageennophytes from the Gulf of California. In Oliveira, E. C. de & N. Kautsky (eds), Cultivation of Seaweeds in Latin America. Univ. S. Paulo/Int. Foundation for Science: 95–100.

Zertuche-González, J. A., 1988. *In situ* life history, growth and carrageenan characteristics of *Eucheuma uncinatum* (Setchell and Gardner) Dawson from the Gulf of California. Ph. D. diss., State Univ. New York, Stony Brook, 162 pp.

Zertuche-González, J. A., I. Pacheco-Ruíz & J. González-González, 1995. Macroalgas. In Ficher, W., F. Krupp, F. Schneider, W. Sommer, K. E. Carpenter V. H. Niem (eds), Guía FAO para la Identificación de las Especies para los Fines de la Pesca. Pacifico centro-oriental. Organizacion de las Naciones Unidas para la Agricultura y la Alimentación. Plantas e invertebrados., Roma: 1: 9–82.

Seasonal variations of growth and agar composition of *Gracilaria multipartita* harvested along the Atlantic coast of Morocco

Thierry Givernaud[2], Abderrazak El Gourji[1], Aziza Mouradi-Givernaud[1], Yves Lemoine[3] & Nadia Chiadmi[1]

[1]*Laboratoire de Biochimie et de Biotechnologies Marines, Faculté des Sciences, 14000, Kénitra, Morocco*
[2]*SETEXAM, Usine El Assam, Route de Tanger, B.P. 210, 14000, Kénitra, Morocco*
[3]*Laboratoire de Cytophysiologie Végétale et Phycologie, Bât. SN2, UST Lille, F-59655 Villeneuve d'ascq Cedex, France*

Key words: Gracilaria multipartita, seasonal variation, growth, chemical composition, agar

Abstract

The biology and agar composition and properties of *Gracilaria multipartita*, a common species along the coasts of Morocco, have been studied on samples collected monthly for one year. Growth of the alga was maximum in spring and autumn, and the seaweed partially decayed after its maximum fertility was reached in June and October. The agar content and composition showed seasonal variations. The agar content was maximal in winter (30% dw), and decreased during the growth periods to minima in June and October (25% dw) which also corresponded to the maxima of fertility. The agar composition was characterized by high 6-*O*-methyl galactose (38–59 mol%) and 3,6 anhydrogalactose (24–39%) contents together with galactose (12.6–25.7 mol%) and sulphate (24–5.0% dw). The gel strength varied between 246 and 511 g cm^{-2} and increased after alkali treatment to reach a maximum of 880 g cm^{-2}. The gel strength decreased after the alga reached its maxima of fertility, indicating a possible relationship between growth, fertility and agar metabolism. The content and quality of agar from *G. multipartita* growing in Morocco are suitable for an industrial use of the seaweed for the production of food-grade agar.

Introduction

Agar is a polysaccharide extracted principally from members of Gelidiales and Gracilariales and has numerous applications in the food industry and in biotechnologies (Armisen & Galatas, 1987). It is a family of polymers based on the repeating disaccharide unit [->3)-β-D-galactopyranosyl-(1->4)-3,6-anhydro-α-L-galactopyranosyl-(1->] (Anderson et al., 1965; Araki, 1966). These galactans have various charge densities and substitution patterns by sulphate half-ester, methoxyl and pyruvate (Anderson & Rees, 1965; Duckworth & Yaphe, 1971). Members of the Gracilariales are among the most economically important agarophytes (Dawes, 1987), and support a significant industry based on both cultivated and wild plants (Santelices & Doty, 1989).

In Morocco, the agar industry is well developed and the main source of raw material is *Gelidium sesquipedale* (Clem.) Bornet et Thuret which is collected by divers all along the Atlantic coast. *Gracilaria multipartita* (Clemente) Harvey, a common species along the Atlantic coast of Morocco, is not exploited, although Humm (1951) stated that a similar species was once used together with *Gracilaria verrucosa* (Harvey.) and some other species of red algae for agar manufacture in North Carolina and Florida. Some studies have been done on this species to investigate its economic potential for agar extraction. For example, the seasonal variation of the alga composition and agar content have been reported by Pennimann (1977).

Agar composition is known to vary greatly depending on the season (Craigie & Wen, 1984; Bird, 1988; Luhan, 1992), while correlations between biomass and agar yield would appear to be species dependent in both temperate and tropical zones, or may be a function of different life phases of *Gracilaria* species (Whyte et al., 1981; Lahaye & Yaphe, 1988). Variations in the quality of agar have been demonstrated to differ from species to species, and also to be greatly influenced by environmental factors (Cote & Hanisak, 1986; Lahaye et al., 1986; Duckworth et al., 1971; Whyte & Englar, 1980, 1981). Bird & Ryther (1990) found that agar yield was lowest whereas gel strength was highest in the summer months. According to seasonality studies, summer appears as the most suitable season for harvesting *Gracilaria* (Whyte et al., 1981; Lahaye & Yaphe, 1988).

Before any commercial exploitation of *Gracilaria multipartita* is undertaken in Morocco, knowledge and understanding of its biology and the chemical and gelling properties of its agar are required. The present study was aimed at following these parameters over the growth cycle of the alga over one year.

Material and methods

Algal material

Gracilaria multipartita was collected monthly from January to December 1995, from mid and lower littoral rocks on the coast of Nations beach (20 km to the north of Rabat). The seaweeds were thoroughly cleaned with filtered seawater to remove undesirable material attached to the thalli.

One hundred thalli were used for the biological survey. The number of reproductive cystocarpic thalli was recorded for each collection. Algal size was measured by recording the weight and the total number of branches.

Agar extraction and treatment

The thalli were dried for 48 h at 60 °C to determine dry weight, ground to a powder and extracted using a method adapted from Craigie & Leigh (1978). Ten grams of powder were hydrated overnight in distilled water (500 ml). Then, the agar was extracted twice at 100 °C for 2 h, in distilled water with continuous stirring. The residue of algae was ground and re-extracted at 100 °C for 2 h and once again at 121 °C for 1 h. The hot agar solution was purified by filtration (0.45 μm), gelled at room temperature, freeze-thawed twice to remove impurities and non-gelling materials, washed with distilled water, resolubilized and dialyzed before freeze-drying.

Part of the seaweed powder was treated for 1 h at 80 °C in 6% NaOH solution. The resulting chemically treated powder was rinsed with deionized water before extraction of agar.

Physical properties of agar

The melting temperature was measured according to Whyte & Englar (1976) on a 1.5% (w/v) gel.

The agar gelling temperature was obtained using the apparatus designed by O. Sandren (Innovest AB, Ronnang, Sweden). Hot agar solution (1.5% w/v, 15 ml) was poured into plastic vials (4.5 cm diameter, 3.5 cm height) and placed directly in a beaker containing water at room temperature.

Gel strength was measured according to Kim (1970) on a 1.5% solution using a Nikansui Shiki apparatus.

The measurement of physical properties of agar have been done in accordance with ISO 9000 norm requirements. The results were given $\pm$ 20 g.cm^{-2} for gel strength and $\pm$ 1 °C for gelling and gel fusion temperatutre.

Chemical analyses

Ash content was determined according to Larsen (1978).

The constituent sugars were quantified as previously described (Mollet et al., 1995) by GLC on an instrument equipped with a flame-ionization detector (Szhimadzu CR14A). The alditol acetates derivatives of the constituent sugars prepared using the reductive hydrolysis method of Stevenson & Furneaux (1991) were separated on a capillary column (OV-1701, 25 m, 0.32 mm, Pierce, Illinois, U.S.A.) using helium as carrier gas, and identified by their retention times using *myo*-inositol hexacetate as internal standard. The initial temperature of the oven was 120 °C. The temperature was raised at 10 °C min^{-1} up to 160 °C, then, at 2 °C min^{-1} to reach a final temperature of 210 °C which was maintained for 5 min. The injector and detector temperatures were kept at 240 °C. Molar response factors were determined using commercial sources (Sigma), except for the 3,6-anhydrogalactose derivative for which the molar response factor was derived from analysis of the agarose Sea Kem (Serva) and for 4-*O*-methylgalactose obtained by courtesy of Dr J. S. Craigie (Atlantic research Laboratory, Halifax, Canada).

Sulphate content was determined from agar (1 mg) hydrolyzates (HCl 2N, 2 h at 100 °C). The acidic solution was evaporated to dryness and sulphates were dissolved in purified H$_2$O (UHQ, Elgastat). After filtration (0.22 μm, Whatman) sulphate ions were separated by HPLC (Maas et al., 1986) using a Wescan

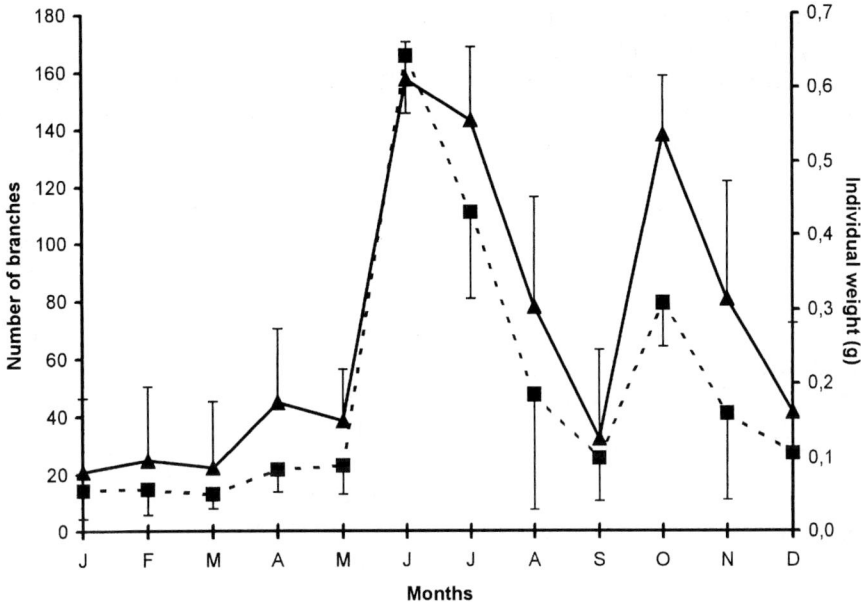

Figure 1. Seasonal variation of the thallus weight and total number of branches (measurement were done every month on 100 thalli). Number of branches (■—■), weight (▲—▲).

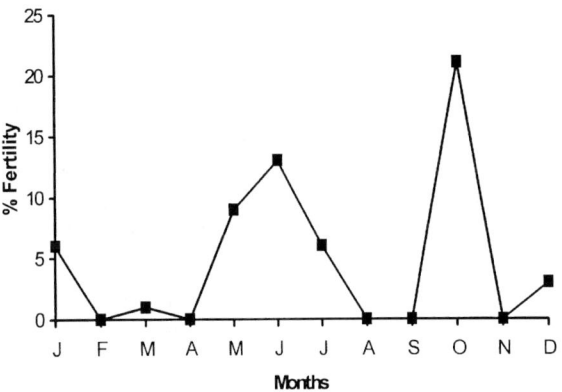

Figure 2. Percentage of reproductive structures as cystocarps (measurements were done every month on 100 thalli).

anion exchange column (4.6 mm × 250 mm, Alltech) and detected with a conductimonitor (Milton Roy).

Results

Growth and reproduction

Analysis of the variation of thallus weight and number of branches in samples collected monthly for one year showed two periods of active growth (Figure 1). The first one occurred between May and June, the second, less important, between September and October.

The percentage of thalli that bore cystocarps was the highest in June and October (Figure 2) during the two growth maxima. After these events, a relative decay of thalli was observed. Winter thalli (from late December to March) were reduced to small axes that were composed of old decayed fronds and new shoots.

Seasonal changes of seaweed composition

The percentage of dry matter was maximum in winter and declined in spring and early summer to a minimum of 11.9% in July (Figure 3). From July, the dry matter increased to reach a peak of 17.3% in October. The variation in ash content followed that of the dry matter content as it was its main component (Figure 3).

Agar content was high from January to March (around 30%, Figure 3). During these months, the seaweed was reduced to its perennating structure (fixation disk) and new young axes. From April to June, the agar decreased to a minimum of 25% (Figure 3) and then increased from July to August to about 34%. From August to October, there was a decrease in the agar content, while from October, as the seaweed decayed, the agar content increased. The two agar minima (June and October) corresponded to the maxima of the alga fertility (Figure 2) and size (Figure 1).

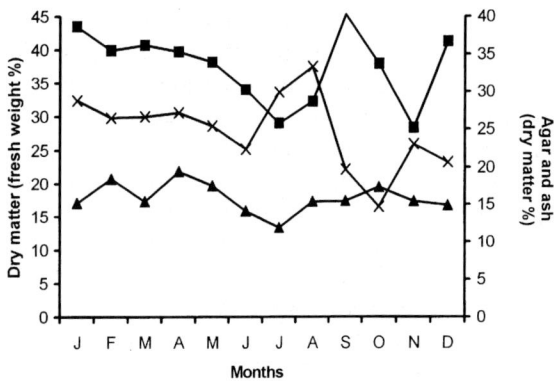

Figure 3. Seasonal variation of dry matter (▲ —▲), ash (■—■) and agar yield (X—X) in the thallus of *Gracilaria multipartita*. The results are expressed as function of dry matter (DM).

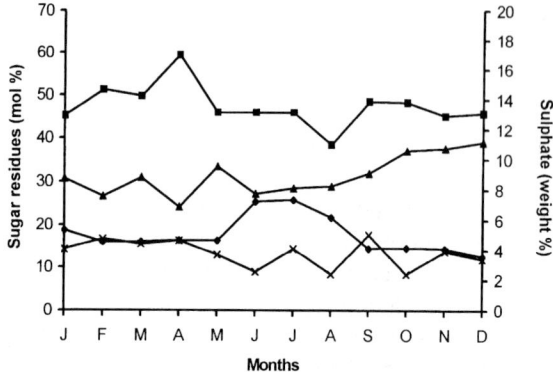

Figure 4. Seasonal variation of the main constituents of *Gracilaria multipartita* (sugar residues expressed as mol% of total sugar content, and sulphates as percentage of agar dw); 3,6-anhydrogalactose (▲—▲), D-galactose (♦—♦), 6-O-methyl galactose (■— ■) and sulphates (X—X).

Composition of agar

The agar of *G. multipartita* was characterized by high 6-O-methyl galactose (about 45 mol%) and 3,6-anhydrogalactose (about 30 mol%) contents together with galactose (about 20 mol%), glucose that never exceeded 6 mol% and traces of mannose (Figure 4). Glucose probably originated from floridean starch which was not totally eliminated during extraction and purification of agar. The proportions of the major sugars varied with season: between 24.1 and 39.0 mol% for 3,6-anhydrogalactose, 12.6% and 25.7 mol% for galactose and 38.4% and 59.5 mol% for 6-O-methylgalactose. The sulphate content ranged between 2.4 and 5.0% (Figure 4). There was an inverse relationship between the galactose and anhydrogalactose contents which indicated that part of the galactose was probably present in the L-form if an ideal repeating structure of agar is assumed. However, there was no relationship between the variations in sulphate and 3,6-anhydrogalactose contents (Figure 4) which would have been expected if sulphate was mainly under the form of L-galactose-6-sulphate. Thus, unsulphated L-galactose may be present in this agar.

It was not possible to establish any relationship between the variations in the different constituents and the different biological events because both growth and reproduction influenced polysaccharide metabolism.

Physical characteristics of agar gel

The gel strength of the native agar extracted from *G. multipartita* varied from 246 g cm^{-2} in July to 511

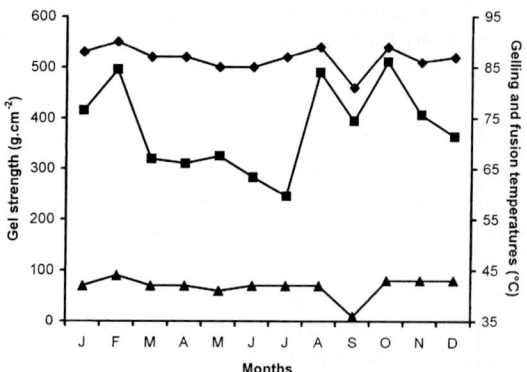

Figure 5. Seasonal variation in gel strength (■- -■), gelling temperature (▲ —▲) and melting temperature (♦—♦) in native agar from *Gracilaria multipartita*.

g cm^{-2} in October (Figure 5). After alkali pretreatment of the alga, the gel strength of agar increased to 880 g cm^{-2} for the sample harvested in June. Gel strength, at a maximum in winter (500 g cm^{-2}), decreased before and during the spring growth period to reach a minimum of 250 g cm^{-2} in July. From July, the gel strength increased to 511 g cm^{-2} after which its values remained high. Gel strength decreased after the two maxima of fertility.

The variations of melting temperature followed that of the gelling temperature (Figure 5) and were not related to variations in 3,6-anhydrogalactose and sulphate contents.

Discussion

Biology and chemical composition of thallus

The biological cycle of G. multipartita was characterized by two growth periods. During the first one, between March and June, growth mostly resulted from water absorption as the dry matter and agar contents of thalli decreased. Growth stopped from July to September during which months the alga reached its maximum of fertility and was then partially fragmented due to the disruption of the cystocarps. From July to August, in response to low nutrients and high photon irradiance, agar was actively synthesized. Then, in September and October, with the second growth period and fertility maximum, the agar content decreased. After October and in winter, the seaweed partly decayed due to the disruption of reproductive structures and to storms. The associated increase in agar content reflected the high proportion of agar-rich old basal parts in the heterogeneous thalli populations collected; similar observations were made from the perennating parts of *Gelidium latifolium* thalli (Mouradi-Givernaud et al., 1993). The mean agar yield over the year compared well with values reported for other *Gracilaria* (Orosco et al., 1992; Phang et al., 1996). The seasonal variations in the chemical composition and agar content of G. multipartita followed similar trends as described for *Gelidium latifolium* (Grév.) Thuret et Bornet (Mouradi-Givernaud et al., 1993) and *Gracilaria multipartita* from tank cultures (Lapointe & Ryther, 1978; Lapointe, 1981). These variations with a shift of two months were also observed on G. multipartita collected in 1996 (data not shown). In particular, the agar content of the seaweed was inversely proportional to growth and agar synthesis was enhanced by low nutrient levels. However, the annual biochemical cycle was also complicated by the fertility periods which probably affected the cell wall organization and led to a decrease in agar content and an increase in thalli fragmentation after reproduction.

Composition and physical properties of agar

The agar of *Gracilaria multipartita* was characterized by a high 6-O-methyl galactose content (around 45 mol%) which is found in other Gracilariales such as, *Gracilariopsis lemaneiformis* (Bory) Dawson, Acleto et Foldvik (Chirapart et al., 1995), *Gracilaria domingensis* and *Gracilaria mammillaris* (Mont.) Howe (Valiente et al., 1992). The variations in the rheological properties showed that the seasonal physiological cycle of this species and particularly its agar metabolism was influenced by the reproduction process. In July and November, just after the highest percentage of fertility, the low gel strength could result from an enzymatic hydrolysis of agar in the cell wall as already proposed by Cosson et al. (1995). In support of this interpretation is the decrease in agar yield before the two periods of reproduction of the alga. It will then be important to verify if agar can play a role as reserve polysaccharide for the seaweed. The melting and gelling temperatures were relatively stable over the year, except a small decrease in September. The mean gelling temperature at about 44 °C is in agreement with the high 6-O-methyl galactose content of this agar (Guiseley, 1970).

Conclusion

Gracilaria multipartita produces an agar of gel strength after alkali modification of 880 g cm^{-2} suitable for food applications. The best harvest period of the alga appears to be in summer when the biomass, agar content and gel strength are close to their maximum.

Acknowledgments

We are grateful to SETEXAM (Morocco) for technical and financial assistance. We are also grateful to Mr S. Rahaoui, Laboratoire de Cytophysiologie Végétale et Phycologie, UST de Lille, France, for his assistance in the determination of the chemical composition of agar by GLC.

References

Anderson, N. S. & D. A. Rees, 1965. Porphyran: a polysaccharide with a masked repeating structure. J. Chem. Soc.: 5880–5887.

Anderson, N. S., T. C. S. Dolan & D. A. Rees, 1965. Evidence for a common structural pattern in the polysaccharide sulfates of the Rhodophyceae. Nature (London), 205: 1060–1062.

Araki, C., 1966. Some recent studies on the polysaccharides of agarophytes. Proc. int. Seaweed Symp. 5: 3–17.

Armisen, R. & F. Galatas, 1987. Production, properties and uses of agar. In McHugh, D. J. (ed.), Production and Utilisation of Products from Commercial Seaweeds. FAO Technical paper 288: 1–57.

Bird, K. T., 1988. Agar production and quality from *Gracilaria* sp. Strain G-16: Effects of environmental factors. Bot. mar. 31: 33–39.

Bird, K. T. & H. Ryther, 1990. Cultivation of *Gracilaria verrucosa* (Gracilariales, Rhodophyta) strain G-16 for agar. Hydrobiologia 204/205: 347–351.

Chirapart, A., M. Ohno, H. Ukeda, M. Sawamura & H. Kusunose, 1995. Chemical composition of agars from the newly reported Japanese agarophyte, *Gracilariopsis lemaneiformis*. J. appl. Phycol. 7: 359–365.

Cosson, J., E. Deslandes, M. Zinoun & A. Mouradi-Givernaud, 1995. Carrageenans and agars, red algal polysaccharides. Prog. Phycol. Res. 11: 269–324.

Cote, G. L. & M. D. Hanisak, 1986. Production and properties of native agars from *Gracilaria tikvahiae* and other red algae. Bot. mar. 29: 359–366.

Craigie, J. S. & C. Leigh, 1978. Carrageenans and agars. In Hellebust, J. A. & J. S. Craigie (eds), Handbook of Phycological Methods. Cambridge University Press: 110–131.

Craigie, J. S. & Z. C. Wen, 1984. Effects of temperature and tissue age on gel strength and composition of agar from *Gracilaria tikvahiae* (Rhodophyceae). Can. J. Bot. 62: 1665–1670.

Dawes, C. J., 1987. The biology of commercially important tropical marine algae. In Bird, K. T. & P. H. Benson (eds), Seaweed Cultivation for Renewable Resources. Elsevier, Amsterdam: 155–190.

Duckworth, M. & W. Yaphe, 1971. The structure of agar. I. Fractionation of complex mixture of polysaccharides. Carbohyd. Res. 16: 189–197.

Duckworth, M., K. C. Hong & W. Yaphe, 1971. The agar polysaccharides of *Gracilaria* species. Carbohydr. Res. 19: 1–9.

Guiseley, K. B., 1970. The relation between methoxyl content and gelling temperature of agarose. Carbohydr. Res. 13: 247–256.

Humm, H. J., 1951. The seaweeds resources of North Carolina. In Taylor, H. F. (ed.), Survey of Marine Fisheries of North Carolina. University of North Carolina Press, Chapel Hill: 231–250.

Kim, D. H., 1970. Economically important seaweeds in Chile-I *Gracilaria*. Bot. mar. 13: 140–162.

Lahaye, M. & W. Yaphe, 1988. Effects of seasons on the chemical structure and gel strength of *Gracilaria pseudoverrucosa* agar (Gracilariaceae, Rhodophyta). Carbohydr. Polymers 8: 285–301.

Lahaye, M., C. Rochas & W. Yaphe, 1986. A new procedure for determining the heterogeneity of agar polymers extracted in the cell wall of *Gracilaria* spp. (Gracilariaceae, Rhodophyta). Can. J. Bot. 64: 579–585.

Lapointe, B. E., 1981. The effect of light and nitrogen on growth, pigment content, and biochemical composition of *Gracilaria foliifera* v. *angustissima* (Gigartinales, Rhodophyta). J. Phycol. 17: 90–95.

Lapointe, B. E. & J. H. Ryther, 1978. Some aspects of the growth and yield of *Gracilaria tikvahiae* in culture. Aquaculture 15: 185–193.

Larsen, B., 1978. Brown seaweeds: analysis of ash, fiber, iodine and mannitol. In Hellebust, A. & J. Craigie (eds), Handbook of Phycological Methods: Physiological and Biochemical Methods. Cambridge University Press: 182–188.

Luhan, R. J., 1992. Agar yield and gel strength of *Gracilaria heterocladia* collected from Iloilo, Central Philippines. Bot. mar. 35: 169–172.

Maas, F. M., I. Hoffmann, M. J. Van Harmelen & L. J. de Kok, 1986. Refractometric determination of sulfate and others anions in plants separated by H.P.L.C. Plant Soil 91: 129–132.

Mollet, J. C., M. C. Verdus, R. Kling & H. Morvan, 1995. Improved protoplast yield and cell wall regeneration in *Gracilaria verrucosa* (Huds.) Papenfuss (Gracilariales, Rhodophyta). J. exp. Bot. 46: 239–247.

Mouradi-Givernaud, A., T. Givernaud, H. Morvan & J. Cosson, 1993. Annual variations of the biochemical composition of *Gelidium latifolium* (Greville) Thuret et Bornet. Hydrobiologia 260/261: 607–612.

Orosco, C. A., A. Chirapart, M. Nukaya, M. Sawamura & H. Kusunose, 1992. Yield and physical characteristics of agar from *Gracilaria chorda* Holmes: Comparison with those from Southeast Asian species. Nippon Suisan Gakkaishi 58: 1711–1716.

Pennimann, C. A., 1977. Seasonal chemical and reproductive changes in *Gracilaria foliifera* (Forskal.) Boerg. from Great Bay, New Hampshire (U.S.A.) J. Phycol. (Suppl.) 13: 53.

Phang, S. M., S. Shaharuddin, H. Noraishah & A. Sasekumar, 1996. Studies on *Gracilaria changii* (Gracilariales, Rhodophyta) from Malaysian mangroves. Hydrobiologia 326/327: 347–352.

Santelices, B. & M. Doty, 1989. A review of *Gracilaria* farming. Aquaculture 78: 95–133.

Stevenson T. T. & R. H. Furneaux, 1991. Chemical methods for the analysis of sulfated galactans from red algae. Carbohydr. Res. 210: 277–298.

Valiente, O., L. E. Fernandez, R. M. Perez, G. Marquina & H. Velez, 1992. Agar polysaccharides from the red seaweeds *Gracilaria domingensis* Sonder *ex* Kützing and *Gracilaria mammilaris* (Montagne) Howe. Bot. mar. 35: 77–81.

Whyte, J. N. C. & J. R. Englar, 1976. Fisheries and Marine Service, Canada. Tech. rep. N° 623.

Whyte, J. C. & J. R. Englar, 1980. Chemical composition and quality of agars in the morphotypes of *Gracilaria* from British Columbia. Bot. mar. 23: 277–283.

Whyte, J. C. & J. R. Englar, 1981. Agar from an intertidal population of *Gracilaria* sp. Proc. int. Seaweed Symp. 10: 537–542.

Whyte, J. N. C., J. R. Englar, R. G. Saunders & J. C. Lindsay, 1981. Seasonal variations in biomass, quantity and quality of agar, from the reproductive and vegetative stages of *Gracilaria* (*verrucosa* type). Bot. mar. 24: 493–501.

Biomass and agar assessment of three species of *Gracilaria* from Negros Island, central Philippines

H.P. Calumpong[1,2], A. Maypa[1], M. Magbanua[2] & P. Suarez[3]
[1]*Silliman University Marine Laboratory, Dumaguete City 6200, Philippines*
E-mail: mlsucrm@mozcom.com
[2]*Department of Biology, Silliman University, Dumaguete City 6200, Philippines*
[3]*Department of Chemistry, Silliman University, Dumaguete City 6200, Philippines*

Key words: agar, biomass, *Gracilaria*, Philippines, seasonality, yield

Abstract

Biomass, cover and agar quality of three species of *Gracilaria* were monitored monthly in three sites in Negros Oriental, central Philippines, during the period July 1992–June 1993. The biomass per unit area of *G. arcuata* peaked around March while *G. salicornia* was maximal around November. Seasonality in *G. blodgettii* is not obvious from our data as this is a highly harvested species and we may have sampled after gleaners have harvested the algae.

Agar yield values were for *G. arcuata* from Bais, 2.9 ± 7% to 21.7 ± 0.7% while the same species from Siaton yielded 2.6 ± 0.6% to 18.3 ± 0.5%; *G. salicornia* yielded 2.9 ± 0.1% to 15.7 ± 1.3% and *G. blodgettii*, 0.1% to 20.7 ± 5.9%. Gel strength and viscosity were variable. Gel strength values were: *G. arcuata* (Bais), 17-260 g cm^{-2}; *G. arcuata* (Siaton), 58-270 g cm^{-2}; *G. salicornia*, 29.5–147 g cm^{-2}; *G. blodgettii*, 29.6–235 g cm^{-2}, and related to the 3,6-anhydrogalactose and sulphate content of the algae.

Introduction

Because the major seaweed product in the Philippines is carrageenan from *Eucheuma* and *Kappaphycus* species, much of the phycocolloid research and product development have been geared towards this compound. However, carrageenan has limited applications and the need to develop mariculture technologies for agarophytes is high. Prerequisite to mariculture is basic research on the biology of the source algae and their agar content.

This study aimed at gathering information on *Gracilaria* beds in three sites on eastern Negros Island, central Philippines, specifically on the standing stock of their most abundant *Gracilaria* species, and their agar yield and quality. This is part of a larger study dealing with the assessment of *Gracilaria* beds in the Philippines, with a view towards mariculture and recommending options for the management of existing stocks. *Gracilaria* species are currently harvested from wild populations in the Philippines for food, animal feeds, fertilizers and phycocolloids.

Materials and methods

Study sites

Based on preliminary surveys of potential study sites, three areas were selected in Negros Oriental: (1) Lag-it, Capiñahan, South Bais Bay, (2) Canday-ong Dumaguete City, and (3) Malo, Siaton.

Lag-it, Bais City (9° 36′ N, 123° 10′ E) is a large intertidal flat of seagrasses covered most of the year by algae. Landward is a mangrove forest and seaward is a coral reef. The dominant seagrass is *Enhalus acoroides* (L.f.) Royle. The major substratum consists of a mixture of sand and silt. Species of *Gracilaria* found were *G. arcuata* Zanardini, *G. salicornia* (C. Ag.) Dawson and *G. eucheumoides* Harvey. The last occurred in negligible quantities and was not included in the monitoring. The area is highly fished and gleaned by people collecting fishes, invertebrates and algae.

Canday-ong, Dumaguete City (9° 25′ N, 123° 20′ E) is a seagrass-algal bed inside a back reef. It is located near the mouth of Banica River. Only one species, *G. blodgettii* Harvey, was found growing on pebbles. The area is highly fished and gleaning occurs, including the harvest of *G. blodgettii*.

Malo, Siaton (9° 6′ N, 122° 54′ E) is a narrow and rocky intertidal area exposed to the southwest monsoon. Surge channels are present as a result of the strong monsoon winds. Seaward is a coral reef. Species found were *G. arcuata* and *G. salicornia* but the latter occurred in negligible amounts and was not monitored. The area is highly fished but no gleaning occurs.

Biomass, cover and zonation

To determine abundance, cover and zonation, systematic sampling was employed. Two to four transect lines, 20 m apart, were laid from the shore seaward. Transect line length varied among sites, between 50 m and 150 m, depending on the width of the *Gracilaria* bed. In Lag-it, two 150 m transects were used; in Canday-ong, four 50 m transects and in Malo, two 100 m transects were used. Permanent metal quadrats measuring 0.5 m × 0.5 m, divided into 25 subquadrats, were set up at intervals of 10 m along each transect for monthly monitoring of cover and associated macroflora and fauna.

Biomass was quantified using 0.25 m^2 quadrats randomly thrown within the *Gracilaria* bed. Ten replicates were taken monthly from August 1992 to July 1993. All algae within the quadrat were collected and brought back to the laboratory for sorting to species, cleaning and weighing. Wet weights were obtained after blotting with cheesecloth to remove the excess water. The algae were then oven-dried at 60-70 °C to constant dry weight. Adhering dirt particles and other organisms are considered as contaminating material (admixture).

Environmental parameters

Salinity was measured using a hand-held temperature-compensated AO refractometer (American Optical), pH with an Orion portable pH meter (Model SA 250, Orion Research Incorporated, U.S.A.) and temperature using an ordinary mercury quick reading thermometer. Phosphate-phosphorus (PO_4^{-3}-P) and nitrate-nitrogen (NO_3^--N) were determined using ascorbic-molybdate method for the former (Koroleff, 1983) and cadmium reduction method for the latter (Grasshoff, 1983). Rainfall data were taken from the local Philippine Atmospheric Geophysical and Astronomical Services Administration (PAGASA). All measurements were made monthly.

Agar yield

For agar yield, about 1 kg was harvested monthly for extraction. Prior to extraction, the algae were sorted and cleaned in the laboratory. Extraction of agar followed the method modified from Hurtado-Ponce & Umezaki (1988) by treating 25 g dry algae with 5% NaOH solution for one hour at 90 °C. The sample was then washed and soaked for another hour in 0.5% acetic acid. After soaking, the sample was again washed. Extraction was done with boiling distilled water for one hour. The sample was then blended with an ordinary kitchen blender before squeezing through a 3-ply cheesecloth. The extract was then frozen overnight at 0 °C, thawed the following day, oven-dried to constant weight and stored in plastic bags prior to quality testing. The following agar qualities were determined: viscosity of 1.0% agar solution at 65 °C using a Brookfield dial reading viscometer (Spindle 2, 60 rpm, Brookfield Engineering Laboratories, Inc., MA., U.S.A.), gel strength of 1.5% agar solution set overnight using an MCD gel tester (Model G741, No. 181, Marine Colloids Division, FMC Corporation, Rockland, ME, U.S.A.), gelling and melting temperatures of 1.0% agar solution (modified from Whyte & Englar, 1980), sulphate content using the extraction method of Jackson & McCandles (1978) and turbidimetric method of determination (Golterman et al., 1978), and 3,6-anhydrogalactose content using the colorimetric method of Craigie & Leigh (1978).

Results

Biomass, abundance and cover

The mean monthly and cover biomass (Figure 1) expressed in per cent of three *Gracilaria* species in three sites exhibited a seasonal trend.

In Lag-it, the mean dry biomass of *G. arcuata* ranged from 1.7 ± 1.6 g m^{-2} in April 1993 to 25.7 ± 24.6 g m^{-2} in March 1993 (Figure 1). There was a gradual increase in biomass from November 1992, peaking in March 1993 and a sudden decline. Cover followed the same trend. On the other hand, the seasonal trend in the mean biomass of *G. salicornia* was not as distinct and did not coincide with cover data (Figure 1). Highest mean biomass was in November

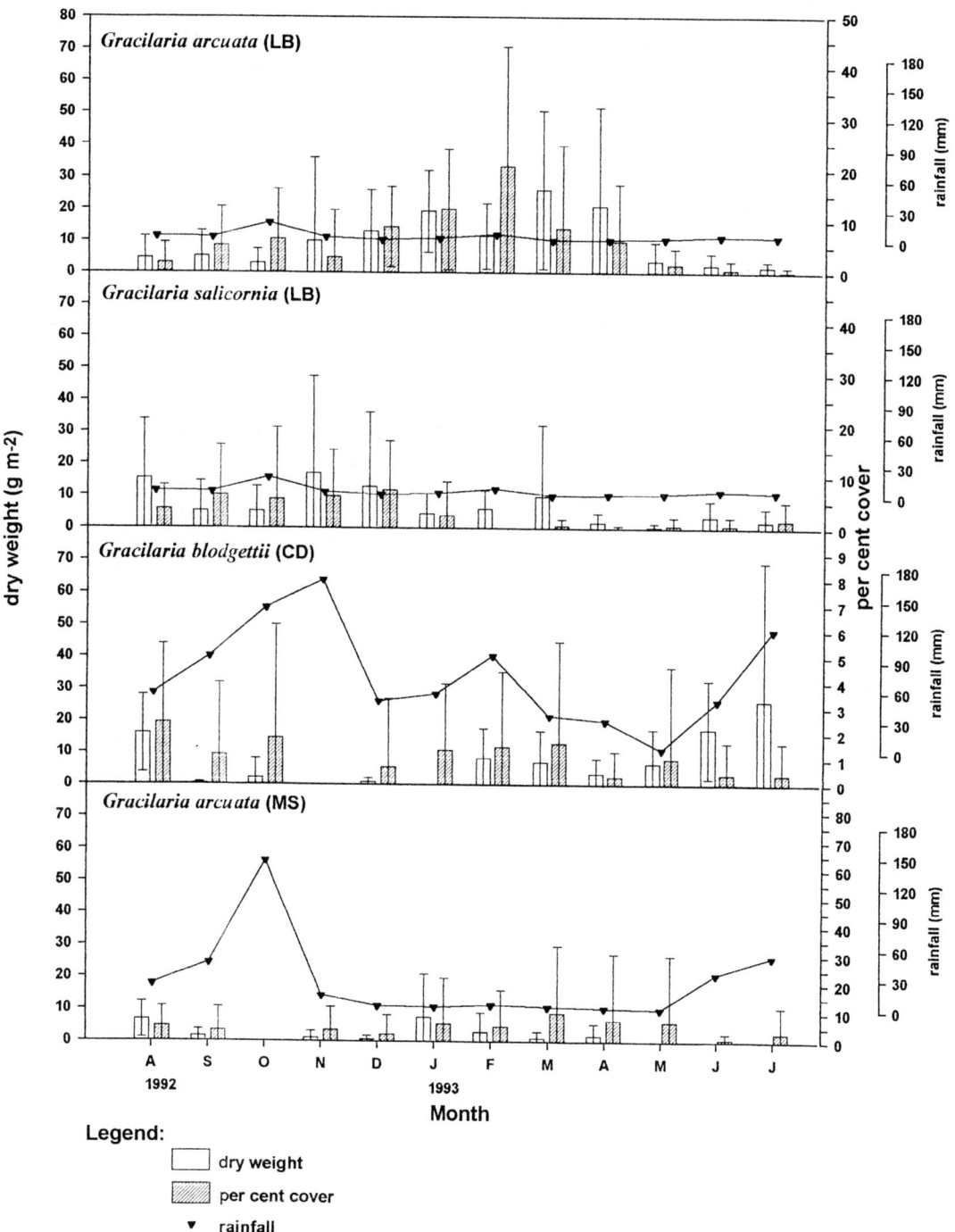

Figure 1. Monthly mean dry biomass in g m^{-2} ($n = 10$) and mean per cent cover ($n = 30$ for Lag-it; $n = 20$ for Canday-ong and Malo) of three species of *Gracilaria* from August 1992 to July 1993 in three sites plotted against rainfall (mm). (LB) Lag-it, Bais; (CD) Canday-ong, Dumaguete; (MS) Malo, Siaton. Bars represent standard deviation.

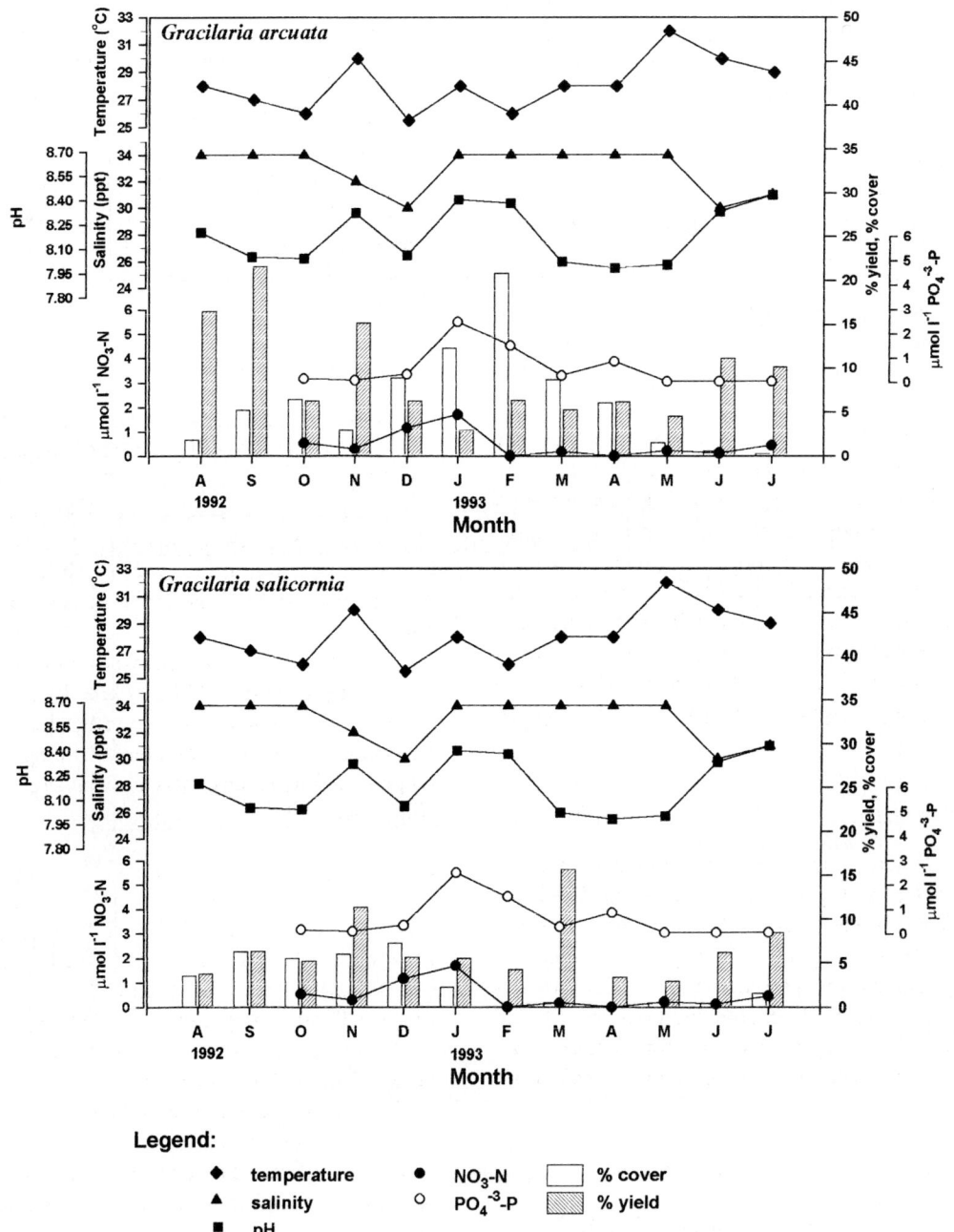

Figure 2. Monthly mean per cent cover and agar yield of *G. arcuata* and *G. salicornia* in Lag-it, Bais plotted with chemical parameters (pH, salinity, temperature, PO_4^{-3}-P and NO_3^--N).

1992 (22.3 ± 30.6 g m^{-2}) and lowest was in May 1993 (0.3 ± 1.1 g m^{-2}). Relatively high values were obtained in August and December with highest cover in December 1992. Occasional harvesting of *G. salicornia* was observed but not for *G. arcuata*. Rainfall in Bais during August 1992 to July 1993 was relatively uniform, below 10 mm per month, except for October 1992 when rainfall reached 18.3 mm. Water temperature ranged from 25 to 32 °C, salinity, 30 to 34 ‰, pH, 7.98 to 8.43, NO_3^--N from not detectable (ND) to 1.16 μmol l^{-1} and PO_4^{-3}-P, 0.05 to 2.5 μmol l^{-1} (Figure 2).

Gracilaria blodgettii biomass in Canday-ong was the lowest during the rainy months of September 1992 to January 1993 (0 to 0.2 ± 6.0 g dry wt m^{-2}) and the highest during the months of August 1992 and June–July 1993 (15.9 ± 12.2 to 26.1 ± 43.0 g dry wt m^{-2}). Rainfall was relatively high (above 25 mm) except for February–May 1993 (Figure 1). Water temperature ranged from 28 to 32 °C, salinity, 24 to 34 ‰, pH, 8.07 to 8.45, NO_3^--N, ND to 5.56 μmol l^{-1} and PO_4^{-3}-P, ND to 5.23 μmol l^{-1} (Figure 3). This species is heavily harvested for food by gleaners.

The biomass and cover data of *G. arcuata* in Malo suggest that the population starts to grow in January through May and starts to decline by June (Figure 1). No data were collected for October due to heavy rains and strong winds which resulted in zero underwater visibility. Water temperature ranged from 27 to 31 °C, salinity, 3 to 34 ‰, pH, 7.98 to 8.66, NO_3^--N, ND to 2.71 μmol l^{-1} and PO_4^{-3}-P, 0.55 to 2.25 μmol l^{-1} (Figure 3).

Reproduction

During the monthly monitoring of permanent quadrats over a year period, cystocarpic thalli of *G. blodgettii* in Canday-ong were observed only in August 1992, January 1993 and July 1993. In August 1992 and July 1993, cystocarps were present in the intertidal population only. In January 1993, both intertidal and subtidal populations bore cystocarps. No reproductive structures were observed on *G. arcuata* and *G. salicornia* in Bais.

Water content and contaminating material

Among the three species, *G. salicornia* had the highest water content, taking up 75% of its wet weight. The two species, *G. arcuata* and *G. salicornia* had about 50% moisture content each. There was no significant monthly variation among species and between sites.

Gracilaria blodgettii was the 'cleanest' among the three species. Much of its admixture was sand and adhering dirt. No significant monthly variation was discernible. No samples were obtained from Malo in October 1992.

Yield

Monthly variations in agar yield occurred in all species in all sites (Figure 4) but was not found significant (2-Way ANOVA, between months, $p = 0.97043$; between species, $p = 0.54711$, alpha = 0.05). Agar yield for *G. arcuata* from Lag-it ranged from 2.9 ± 7% to 21.7 ± 0.7% (Figure 2) while the same species from Malo yielded 2.6 ± 0.6% to 18.3 ± 0.5% (Figure 3). *G. salicornia* yielded 2.9 ± 0.1% to 15.7 ± 1.3% (Figure 2) while *G. blodgettii* yielded 0.1% to 20.7 ± 5.9% (Figure 3).

Gel strength and viscosity

The following gel strength values were obtained: *G. arcuata* (Lag-it), 17–260 g cm^{-2}; *G. arcuata* (Malo), 58–270 g cm^{-2}; *G. salicornia*, 29.5–147 g cm^{-2}; *G. blodgettii*, 29.6–235 g cm^{-2} (Figure 4). Monthly variations within species and between species were seen but were not significant (2-Way ANOVA, between months, $p = 0.045687$; between species, $p = 0.02194$, alpha = 0.05).

Agar extracted from *G. arcuata* from Lag-it ranged in viscosity from 0 to 42 cps while the same species from Malo had an agar viscosity range of 0 to 28.8 cps. *G. blodgettii* from Canday-ong had an agar viscosity range of 2.5 to 15 cps and *G. salicornia* 0 to 47 cps (Figure 4).

Gelling and melting temperatures

Gelling temperatures of the extracted agars were between 38 °C and 49.5 °C for the Bais species, *G. arcuata*, and between 44 °C and 53 °C for *G. salicornia*. For Canday-ong agar, it was between 42 °C and 49 °C for *G. blodgettii* and for Malo agar between 42 °C and 55.5 °C for *G. arcuata*.

Melting temperatures of the Bais agars were 59 °C – 81 °C for *G. arcuata* and 64 °C – 77 °C for *G. salicornia*; of the Malo agar it was 54 °C – 84.5 °C for *G. arcuata* and of the Canday-ong agar it was 52.5 °C – 78 °C.

Sulphate and 3,6-anhydrogalactose contents

Among the three species, *G. salicornia* recorded the lowest mean sulphate content (4.2% ± 1.5; $n = 21$) compared with *G. arcuata* (5.2 ± 1.1; $n = 24$) and *G. blodgettii* (5.3 ± 1.7; $n = 21$) (Figure 5).

As in sulphate content, the same pattern was found for 3,6-anhydrogalactose. *G. salicornia* recorded the

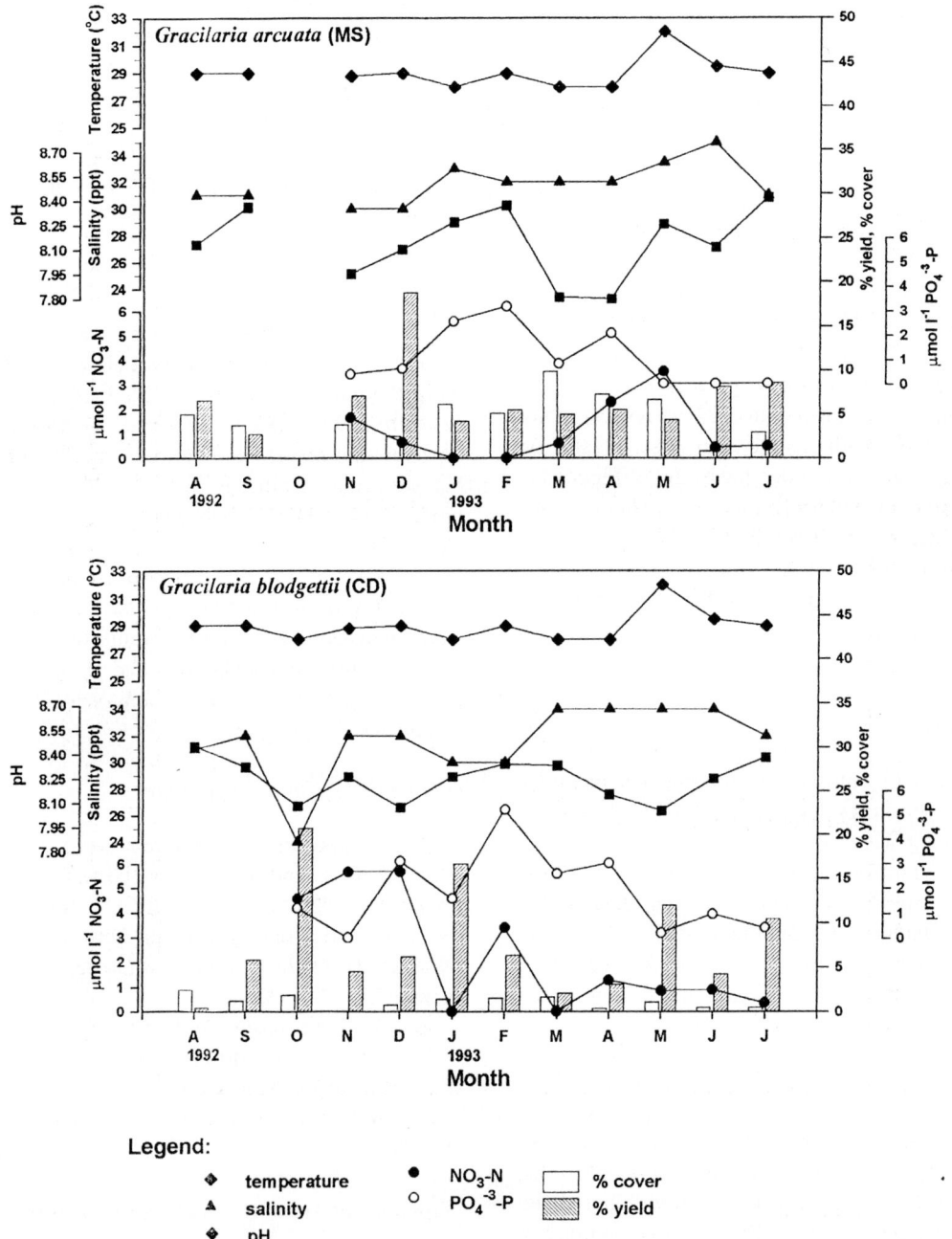

Figure 3. Monthly mean per cent cover and agar yield of *G. arcuata* in Malo, Siaton (MS) and *G. blodgettii* in Canday-ong, Dumaguete (CD) plotted against chemical parameters (pH, salinity, temperature, PO_4^{-3}-P and NO_3^--N).

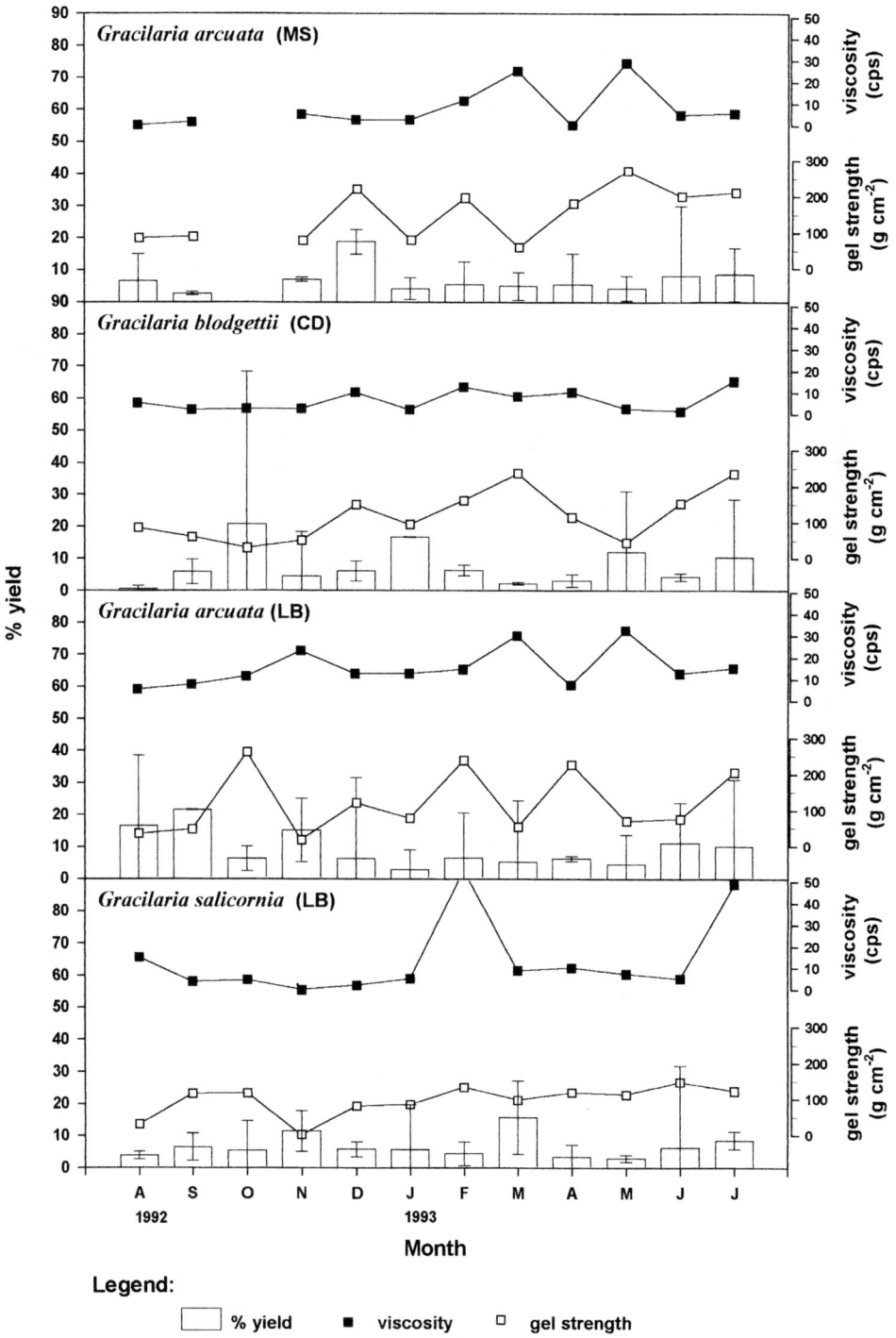

Figure 4. Monthly mean agar yield ($n = 3$), gel strength and viscosity of *G. arcuata* and *G. salicornia* in Lag-it, Bais, *G. arcuata* in Malo, Siaton (MS) and *G. blodgettii* in Canday-ong, Dumaguete (CD) from August 1992 to July 1993. Bars represent standard deviation.

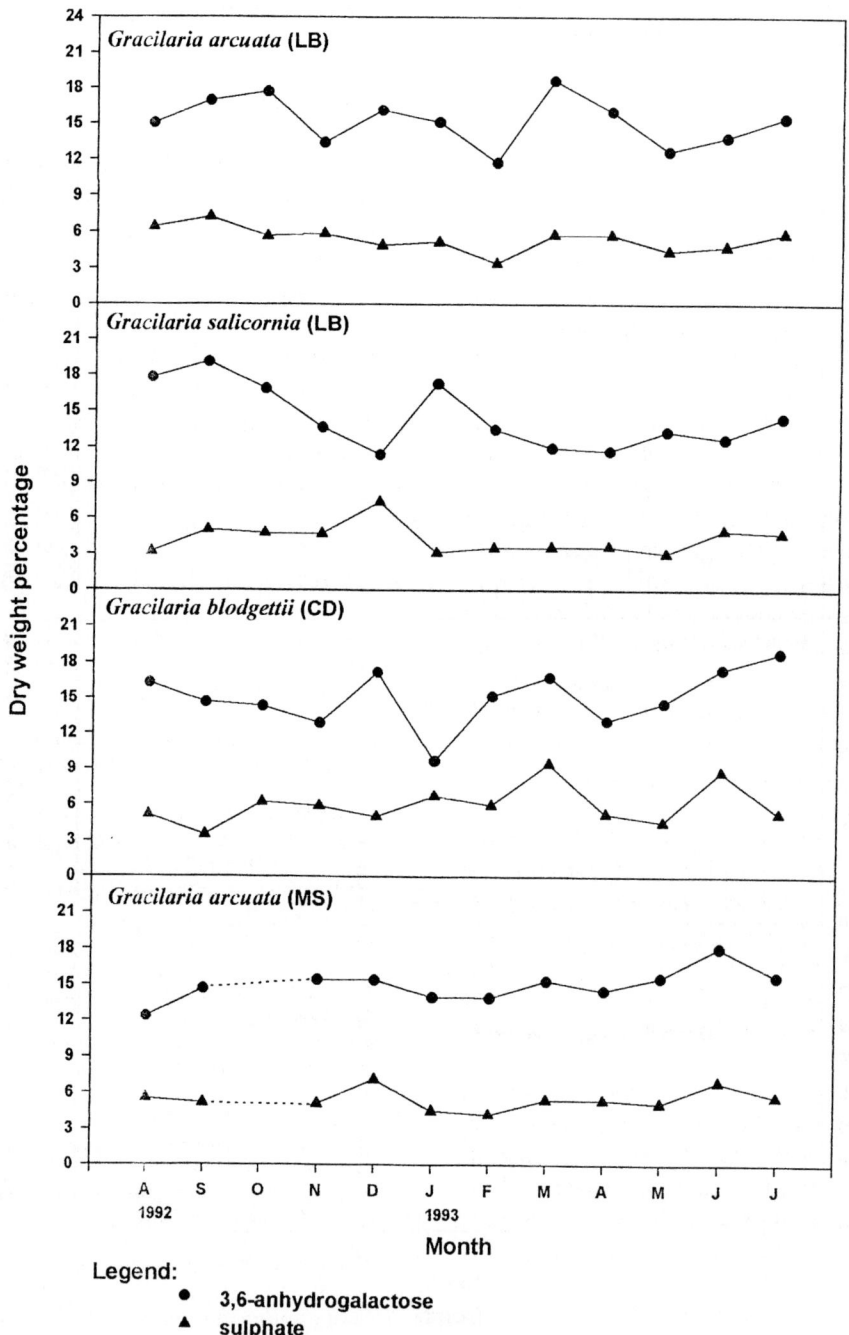

Figure 5. Dry weight percentages of agar 3,6-anhydrogalactose and sulphate contents of *G. arcuata* and *G. salicornia* in Lag-it, Bais (LB) and *G. arcuata* in Malo, Siaton (MS) and *G. blodgettii* in Canday-ong, Dumaguete (CD) from August 1992 to July 1993.

lowest mean anhydrogalactose content (13.9–2.4; $n = 22$) compared with *G. arcuata* (15.3 ± 1.8; $n = 24$) and *G. blodgettii* (15.1 ± 2.6; $n = 20$) (Figure 5).

Discussion

The two *Gracilaria* species in Lag-it, Bais seemed to reach growth peaks at different times with *G. arcuata* peaking at months when *G. salicornia* was declining (Figure 1) which may be a result of interspecific competition. Trono & Azanza-Corrales (1981) found that the biomass of *G. verrucosa* (Huds.) Pappenf. in Manila Bay declined when the biomass of *Acanthophora spicifera* (Vahl) Børg. was high.

For *Gracilaria arcuata* at Lag-it, Bais, a growth peak occurred after a period of higher PO_4^{-3}-P and NO_3^--N levels in the water (Figure 2). Sousa-Pinto et al. (1996) demonstrated that growth of *Gelidium robustum* (Gardn.) Hollen. & Abb. was directly proportional to increasing concentrations of phosphates and Macler & West (1987) found growth rate of *Gelidium coulteri* Harv. reduced to nearly zero after three weeks of nitrogen starvation. Seasonality in *G. blodgettii* is not obvious from our data as this is a highly harvested species and we may have sampled after gleaners have harvested the algae.

Other factors may have contributed to the seasonal growth. Lag-it Bais and Canday-ong, Dumaguete are both exposed to the northeast monsoon which occurs during the months of November through April while Malo, Siaton is exposed to the southwest monsoon. Trono & Azanza-Corrales (1981) found low biomass of *G. verrucosa* in Manila Bay during the southwest monsoon.

It was interesting to note that no reproductive thalli were collected during the whole sampling period, except for *G. blodgettii*. In Manila Bay, Trono & Azanza-Corrales (1981) found reproductive thalli of *G. verrucosa* the whole year round with a preponderance of reproductive over vegetative ones during months of high salinities.

Agar yield also showed variation but was not obviously correlated with other factors. This is in contrast to the findings of other workers (Macler & West, 1987; Mouradi-Givernaud et al., 1992; Chirapart & Ohno, 1993; Sousa-Pinto et al., 1996), where it has been correlated with the nutrient levels in the water.

Gels produced from the three species had monthly variations in gel strength and viscosity. Rees (1969) and Lahaye & Rochas (1991) have shown that the gel strength of agar is related to its 3,6 anhydrogalactose content and sulphate.

Acknowledgments

The authors thank FMC Cebu and Mr. Farley Bariquatro for the use of their viscometer and gel tester. We also gratefully acknowledge the assistance of Jacinta Radones-Lucañas in the laboratory and in the field, Grace Ozoa and Pablina Cadiz for their help in the field and Jay Naalam and Mannix Balawang for assisting P. Suarez with the chemical analyses. Jasper Leif Maypa's help in the field and in the generation of graphs is also gratefully acknowledged. This research was funded by the Philippine Council for Aquatic and Marine Research and Development.

References

Chirapart, A. & M. Ohno, 1993. Seasonal variation in the physical properties of agar and biomass of *Gracilaria* sp. (*chorda* type) from Tosa Bay, southern Japan. Hydrobiologia 260–262: 541–547.

Craigie, J. S. & C. Leigh, 1978. Carrageenans and agars. In Hellebust, J.A. & J.S. Craigie (eds), Handbook of Phycological Methods, Physiological and Biochemical Methods. Cambridge University Press, Cambridge: 109–131.

Golterman, H. L., R. S. Clymo & M. A. M. Ohnstad, 1978. Methods for Physical and Chemical Analysis of Fresh Waters. IBP Handbook No. 8. Blackwell Scientific Publications, Oxford, Edinburgh, London, Melbourne, 82 pp.

Grasshoff, K., 1983. Determination of nitrate. In Grasshoff, K., M. Ehrhardt & K. Kremling (eds), Methods of Seawater Analysis, Second, Revised and Extended Edition, Verlag Chemie GmbH, D-6940, Weinheim: 143–150.

Hurtado-Ponce, A. & I. Umezaki, 1988. Physical properties of agar gel from *Gracilaria* (Rhodophyta) of the Philippines. Bot. mar. 31: 171–174.

Jackson, S. G. & E. L. McCandles, 1978. Simple, rapid turbidimetric determination of inorganic sulfate/or protein. Anal. Biochem. 90: 802–808.

Koroleff, F., 1983. 9.1. Determination of phosphorus. In Grasshoff, K., M. Ehrhardt & K. Kremling (eds), Methods of Seawater Analysis, Second, Revised and Extended Edition, Verlag Chemie GmbH, D-6940, Weinheim: 125–138.

Lahaye, M. & C. Rochas, 1991. Chemical structure and physico-properties of agar. Hydrobiologia 221: 137–148.

Macler, B. A. & J. A. West, 1987. Life history and physiology of the red alga, *Gelidium coulteri*, in unialgal culture. Aquaculture 61: 281–293.

Mouradi-Givernaud, A., T. Givernaud, H. Morvan & J. Cosson, 1992. Agar from *Gelidium latifolium* (Rhodophyceae, Gelidiales): biochemical composition and seasonal variations. Bot. mar. 35: 153–159.

Rees, D. A., 1969. Structural conformation and mechanism in the formation of polysaccharide gels and networks. Adv. Carbohydr. Chem. Biochem. 24: 267–332.

Sousa-Pinto, I., R. Lewis & M. Polne-Fuller, 1996. The effect of phosphate concentration on growth and agar content of *Gelidium robustum* (Gelidiaceae, Rhodophyta) in culture. Hydrobiologia 326/327: 437–443.

Trono, Jr. G. C. & R. Azanza-Corrales, 1981. The seasonal variation in the biomass and reproductive states of *Gracilaria* in Manila Bay. Proc. int. Seaweed Symp.10: 743–748.

Whyte, J. N. C. & J. R. Englar, 1980. Chemical composition and quality of agar in the morphotypes of *Gracilaria* from British Columbia. Bot. mar. 23: 277–283.

The effects of a simulated harvest on *Porphyra* (Bangiales, Rhodophyta) in South Africa

N. J. Griffin[1], J. J. Bolton[1] & R. J. Anderson[2]
[1]*Botany Department, University of Cape Town, Private Bag, Rondebosch 7701, South Africa*
Fax: [+ 27] 21 650 4041. E-mail: griffin@botzoo.uct.ac.za
[2]*Seaweed Unit, Sea Fisheries Private Bag X2, Roggebaai 8012, South Africa*

Key words: Porphyra, Bangiales, Rhodophyta, harvest trials, harvest impact, nori, South Africa, resource management, abalone

Abstract

In South Africa, *Porphyra* has, until recently, been little exploited, having been harvested only for a small health-food market. However, the advent of land-based abalone farming has increased the pressure on wild *Porphyra* populations, as *Porphyra* is in demand for abalone fodder. This paper reports on the effects of a simulated harvest on *Porphyra* populations and those of sympatric fauna. Harvesting, starting in autumn, was found to reduce the biomass of *Porphyra*, an effect detectable up to six months later. *Porphyra* had a patchy distribution, with patches having a mode of approximately 300 thalli m^{-2}. The main effect of harvesting was the removal of patches, as mean thallus size changed little in response to harvesting. Nine months after the start of the experiment, control populations had been reduced, through loss of patches, to the level of treatment populations. Although harvesting *Porphyra* reduced populations of some sympatric fauna (amphipods, isopods and littorinid snails), natural *Porphyra* population decreases had a comparable effect. Some recommendations are discussed for the management and controlled harvesting of *Porphyra* populations in South Africa.

Introduction

Porphyra is an important resource world-wide, due to its cultivation and harvesting for the food market. With the exception of some modest exports totalling 18.5 tonnes (dry weight) between 1965 and 1978 (Anderson et al., 1989), samples of South African *Porphyra* sent for export have been rejected as inferior, in particular for being too tough (*P. capensis* Kützing can be 150 μm or more thick). Otherwise, harvesting of local *Porphyra* has been restricted to small-scale and relatively isolated harvests of wild material to be sold locally in the health-food market. These harvests continue, but are still on a small scale.

Feeding trials by abalone farmers found *P. capensis*, in association with the kelp *Ecklonia maxima* (Osbeck) Papenfuss, to be a valuable supplementary feed for the South African abalone *Haliotis midae* Linneaus (Simpson, 1994; Stepto & Cook, 1996). Of the macroalgae tested, *H. midae* preferred *Porphyra*, which also significantly improved the growth rate and shell elongation rate of *H. midae*. As a result of these findings, and the growth of the abalone farming industry, pressure on South African *Porphyra* is increasing.

The taxonomy of South African *Porphyra*, like that of *Porphyra* species in many parts of the world, is in a state of flux. Two species, *P. capensis* and *P. augustinae* (nom.illeg.), were initially described by Kützing (1843), but were later synonymized as *P. capensis* by Agardh (1890), a decision supported by Isaac (1957) and Graves (1969), though both observed and recorded different forms of *Porphyra*. Recent work has described two new species and recorded two more (Stegenga et al., 1997), and we have discovered that other apparently undescribed species are present. Only two of the five species so far recorded are likely to be harvested (the others are small, and two are kelp epiphytes and so relatively inaccessible to harvesters), and of these, *P. capensis* (*sensu* Stegenga et al., 1997) is by far the commonest. It is, however, evident that *P. capensis* is itself a species complex.

Porphyra species have been recorded from most of the coastline of the Western and Eastern Cape, South Africa (Graves, 1969; Anderson et al., 1989; Stegenga et al., 1997), but harvestable quantities are restricted to more temperate waters, especially Western Cape and Cape Peninsula shores, influenced by the cold, nutrient-rich waters of the Benguela upwelling system. No *Porphyra* has been recorded north of Port Edward on the east coast, where the algal flora has many tropical elements. Generally, South African *Porphyra* populations are seasonal (though thalli may be found throughout the year), with recruitment peaks in autumn and spring leading to dense summer and winter populations (Griffin et al., 1998). Typically, thalli growing in the upper two-thirds of the eulittoral are epilithic, while those in the lower third are often epizoic or epiphytic. Gametophytes may give rise to sporophytes by zygotospores and to gametophytes by archeospores (Graves, 1969).

This study was initiated to determine the effect of harvesting on *Porphyra* populations and associated fauna, as part of a larger project that aimed to draw up a resource management plan for South African *Porphyra*.

Materials and methods

We assessed the effect of three harvesting regimes (three-, six- and twelve-monthly harvests) on populations of *Porphyra* at Slangkoppunt (34° 08′ 40″ S, 18° 19′ 10″ E) on the Cape Peninsula. The experiment was started in autumn (24 April 1995), and ran for twelve months. Six replicate, non-abutting 2 × 1 m permanent quadrats were haphazardly placed in apparently homogenous populations of *Porphyra* in an approximately 50 m broad stretch of the mid-upper eulittoral. The slope of the shore at the study site is shallow, and quadrat orientation varied. Each 2 × 1 m permanent quadrat was subdivided into eight square 0.5 × 0.5 m quadrats. Three of the 0.5 × 0.5 m quadrats were randomly assigned to one of three harvest treatments, and the remainder were randomly assigned as controls. Every three months from the start of the experiment, one control quadrat per replicate was harvested, along with appropriate treatment quadrats. All plants greater than ca. 3 cm in length were hand-picked; if thalli tore, the holdfast and torn remnants were left if they were shorter than ca. 3 cm. It was common that thalli tore, and holdfasts were frequently left behind; however, thallus fragments longer than ca. 1 cm in length were relatively rare. While the entire quadrat was cleared, only material with holdfasts in the central 0.25 × 0.25 m was retained, to eliminate edge effects. The wet mass of retained plants was determined after soaking plants for 15 min in seawater, and then removing superficial water first by spinning plants in a salad spinner until no more water was collected, then by blotting them with paper towels. During the harvest, the presence, if any, of fauna in the harvested quadrats was noted, and fauna removed with the harvested thalli were collected, identified and counted. No fauna were collected beyond those trapped among *Porphyra* thalli and unable to escape, in an attempt to simulate the action of commercial harvesters.

The effect of treatments on average thallus mass, thallus density, and quadrat biomass was evaluated using analysis of variance. Null hypotheses of no difference between treatment and control populations, no change in control populations over time, and no difference between treatments after 12 months were examined using contrast analysis (Keppel, 1991). Density in control populations was further analyzed to determine whether *Porphyra* in the mid-upper eulittoral shows patchy growth by plotting frequency distributions on various scales. The effect of patchiness on thallus growth was assessed by testing for correlation between thallus density and mean thallus size and stand biomass using Spearman's rank correlation (Zar, 1984).

The frequency of occurrence of fauna associated with *Porphyra* in various controls and treatments was examined for change with time and treatment using cluster analysis (UPGMA, Sokal & Sneath, 1963) and non-metric multidimensional scaling (NMDS), minimizing global stress, of similarity matrices (after Field et al., 1982; Minchin, 1987) derived using the Dice similarity coefficient (Dice, 1945). Species abundances from the shore were grouped to higher taxonomic levels (mussels, limpets, snails, chitons, anemones, starfish, barnacles, amphipods, isopods and polychaete worms) before calculating similarity indices as accurate identification of all fauna in field conditions was not possible. The Dice index, which uses presence/absence rather than abundance data, was chosen as a measure of similarity as accurate and consistent enumeration was not always possible for all fauna. Data were aggregated to taxa and abundances transformed to presence/absence in order to include information on isopods and amphipods, which were abundant in dense *Porphyra* patches but were difficult to count and identify in the field. The number of taxa, Margalef's richness (d), Shannon diversity (H) and Simpson's dominance (D) were calculated, using frequency data transformed as above, for all quadrats (Shannon & Weaver, 1949; Simpson, 1949; Clifford & Stephenson, 1975). The correlation of these indices with *Porphyra* biomass was tested using Spearman's rank correlation test (Zar, 1984). A Mantel-type Monte Carlo analysis, or ANOSIM, was used to test hypotheses of treatment and time effect on community structure (Clarke, 1993).

Results and discussion

All the thalli harvested during this experiment were *P. capensis* (*sensu* Stegenga et al., 1997), and no representatives of *P. saldanhae* Stegenga, Bolton et Anderson or the linear (*augustinae*) form of *P. capensis* were collected in experimental or control quadrats.

Significant changes in the biomass of control populations of *Porphyra* ($p = 0.005$) over time were largely

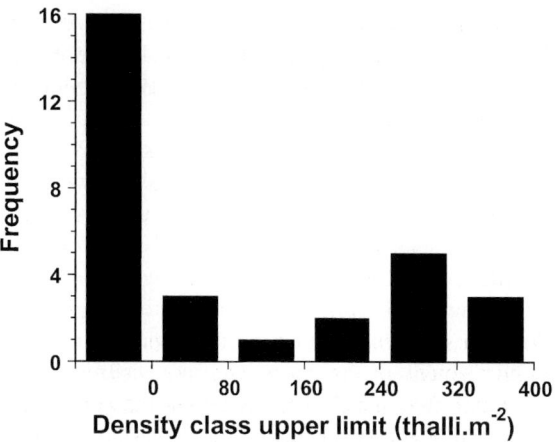

Figure 2. Frequency distribution of *Porphyra* thallus density in control quadrats pooled over time (Quadrat area 0.25×0.25 m). X axis values show inclusive upper limits of density (thalli m^{-2}) classes.

driven by changes in population density ($p < 0.001$), with no real change over time in mean thallus mass ($p = 0.783$) being detected (Figure 1). The decrease in density in control quadrats was not due to even thinning of *Porphyra* populations, but was a function of the loss of patches of *Porphyra*. Distribution of *Porphyra* was patchy; frequency analysis of *Porphyra* densities in control plots suggests a bimodal distribution with peaks at zero thalli m^{-2} and approximately 300 thalli m^{-2} (Figure 2). There was no gradual shift in the frequency distribution with time. This pattern is what would be expected from a combination of high density within-patch samples, samples where *Porphyra* is absent, and relatively few samples from thinning patches or patch margins. The zero thalli m^{-2} mode was derived mostly from samples from the later half of the experiment, when most initial populations had disappeared. The 300 thalli m^{-2} mode was mostly from samples taken from extant populations six months or less after the start of the experiment. When data from within patches and patch margins only were examined (data from quadrats with density of zero are excluded), no significant change in *Porphyra* density ($p = 0.316$) or stand biomass ($p = 0.603$) over time was detected.

Control populations may have experienced some recruitment into the ca. 3 cm harvestable size class throughout the study period, as evidenced by the growth of thalli in treatment quadrats after harvesting during most of the experiment (Figure 1). Growth in treatment quadrats could be due to recruitment from spores, growth of unharvested sporelings or regrowth from holdfasts (here thallus remnants up to ca. 3 cm).

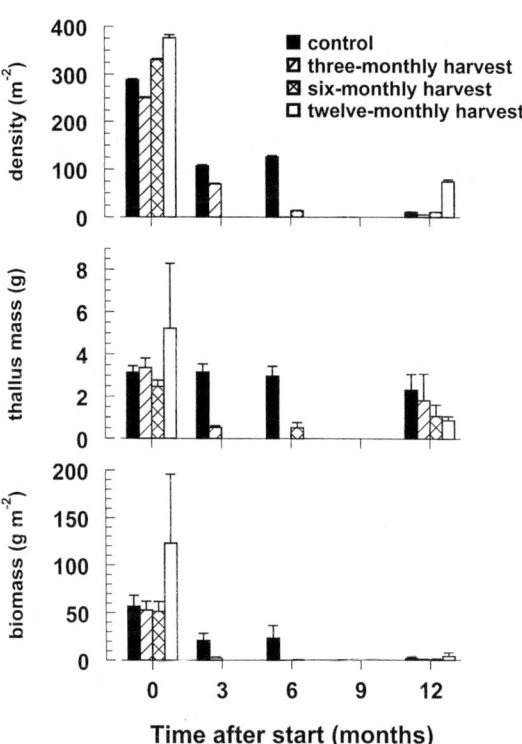

Figure 1. Density (thalli m^{-2}), mean thallus mass (g), and stand biomass (g m^{-2}) of *Porphyra* thalli in three harvest treatments and control quadrats. All data are mean + s.e. As all quadrats were examined, the absence of data from any period that should be present according to the experimental design indicates a zero and not missing data.

Nelson & Conroy (1989) found that regrowth of *P. columbina* was greatly improved if thallus stubs were left after harvesting. While they did not specifically attribute all regrowth to these thallus stubs, they did note that the stubs regenerated new tissue. When individual *P. capensis* thalli were monitored, we found no regrowth from holdfasts in ten permanent quadrats scattered haphazardly between the 2 × 1 m quadrats (Griffin et al., 1998). However, there was evidence of slow continual recruitment regardless of season, with peaks in spring and autumn (recruitment defined here as the appearance of sporelings, not visibly derived from holdfasts which were visible to the naked eye). Sporeling recruitment may at least partially account for the increase in biomass in harvest (and potentially in control) quadrats.

When compared to the controls, harvesting significantly decreased the biomass of *Porphyra* in treatment quadrats ($p = 0.005$) (Figure 1). Overall, mean thallus mass in harvest quadrats was lower than in contemporary control plots ($p < 0.001$). As harvesting removes all thalli longer than ca. 3 cm, those thalli in harvest quadrats either grew as new recruits from spores, or from a sporeling understorey that survived the harvest treatment. We consider it unlikely that regrowth from torn thallus fragments and remaining holdfasts after harvesting contributed significantly to the regrowth of *Porphyra* in treatment quadrats for the reasons discussed above. However, improved recruitment due to archeospores produced by holdfasts and torn plant remnants, and improved recruitment of sporelings in the sheltered environment provided by the holdfasts may have contributed to the regrowth.

Differences in thallus density between harvest and control plots were less significant ($p = 0.088$). Harvested populations showed a pronounced decrease in density after the first harvest; however, a parallel, delayed decrease in density was evident in control populations (Figure 1). The dramatic decrease in density of the control population, which occurred between six and nine months after the start of the experiment, can be largely attributed to the mid-eulittoral die-back of *Porphyra* that occurs in early summer of most years (Griffin et al., 1998). At this time, eulittoral algal populations, established during the winter, are exposed to hot, dry spring-summer south-easterly winds (Bolton & Joska, 1995; Stegenga et al., 1997).

During the monitored period there was insufficient recruitment to regain the densities of control or treatment quadrats encountered at the start of the experiment. This is partially a function of the experimental design; experimental quadrats were not randomly placed in the eulittoral, but were placed in areas that had large, relatively uniform populations of *Porphyra*. After twelve months, although regrowth of winter populations of *Porphyra* had begun, relatively little regrowth had occurred in the permanent quadrats. That regrowth had begun in some quadrats indicates that harvesting does not prevent recruitment and the low average recruitment after twelve months was due more to the stochastic spatial distribution of *Porphyra* patches.

Other workers (Branch et al., 1990) have remarked that *Porphyra* is restricted to spatially patchy populations. *Porphyra* species, often common or dominant in the mid- to upper eulittoral throughout their range (e.g. Lubchenco & Cubit, 1980; Roland & Coon, 1984), are tolerant of the unusual and extreme conditions there (e.g. see Lipkin et al., 1993). Although environmental stress undoubtedly acts, to some extent, in determining the upper limit of growth of many macroalgae in the mid- to upper eulittoral (e.g. see Stekoll & Deysher, 1996), we believe that the spatially patchy growth of *Porphyra* at Slangkoppunt is, at least partially, a function of disturbance patterns when thalli are sporelings (here young thalli with an area of less than 16 mm^2) and, more particularly, of grazing patterns (Griffin et al., 1998). Grazers may act to clear young sporelings from areas shortly after their recruitment; this hypothesis is supported by the observations that following dense sporeling recruitment, sporelings would commonly survive until a disturbance removed all of them. Seldom were few recruits lost; they either survived or died together. Apart from grazers, no other disturbances that might result in spatially localized removal of established sporelings are known at the study site. Also supporting this hypothesis is the observation by Branch et al. (1990) that, after freshwater floods had killed most grazers present at Steilhoogte on the West Coast, the initial patchy distribution of *Porphyra* was replaced by a *Porphyra* bloom with a cover of 100%. The high-shore grazers at Steilhoogte were dominated by *Patella granularis* Linneaus, *Oxystele variegata* Anton and *Siphonaria* spp., all of which were common at Slangkoppunt. Grazing of sporelings has been found elsewhere to be the major limitation to the vertical spread of macroalgae in the intertidal (Underwood & Jernakoff, 1984).

Survival in patches in the upper eulittoral would seem to benefit plants because there would be reduced water loss by presenting a relatively small surface area for evaporation (Levitt & Bolton, 1991), and so would potentially allow prolonged photosynthesis following emersion (see Herbert & Waaland, 1988; Hanelt et al., 1993). The extensive contact between thalli in patches should also facilitate efficient gamete transfer. That patches seem to improve the growth of *Porphyra*, despite potential competition for space and light, is supported by the positive correlation between thallus density and mean thallus size ($p < 0.001$), and stand biomass ($p < 0.001$), within the range of control densities recorded (0–384 thalli m^{-2}). The densities observed in this study were well below those predicted by the ultimate biomass-density line (Weller, 1987; Scrosati & DeWreede, 1997). However, since smaller thalli were not harvested and therefore harvesting of patches was incomplete, conclusions regarding the relationship of biomass and density in *Porphyra* patches cannot be drawn from these data. Growth of thalli longer than ca. 3 cm did seem to be favoured in patches with densities below 384 thalli m^{-2}.

The faunal taxa apparently most affected by a reduction in *Porphyra* biomass were amphipods (mostly *Hyale* spp.) and isopods (mostly *Parisocladus* spp.) (Figure 3). When the initial dense *Porphyra* stands were reduced by natural population changes or harvesting, both amphipods and isopods, previously present in very high numbers, were nearly always absent. Snails (predominantly *Nodilittorina africana* Phillipi and *Oxystele variegata*) were less frequent in treatment as opposed to control quadrats, and this may indicate a response to harvesting. Limpets appeared unaffected by harvesting. Limpet densities were generally higher in dense *Porphyra* stands, but high limpet densities may occur without seaweed cover. Other taxa occurred too infrequently and their frequencies were too variable to suggest a harvesting impact. When the abundance of fauna removed with the harvested *Porphyra* was examined, the genera most commonly found were *Hyale*, *Nodilittorina* and *Parisocladus*. The gut contents of six numerically dominant herbivores that commonly co-exist with *Porphyra* were examined and *Hyale*, *Parisocladus* and *Patella granularis* had recognizable fragments of *Porphyra* in the guts of 80% or more of the animals (Griffin et al., 1998). As the number of taxa in quadrats was positively correlated with harvested *Porphyra* biomass ($p < 0.001$), the effects of more intensive harvesting on eulittoral fauna should be monitored.

The NMDS ordination of treatment and control quadrats had relatively high global stress (0.22), which indicates that the relationships between the quadrats are essentially multivariate and cannot be completely resolved in two dimensions. A gradient of time after experiment start was visible, and for the most part the control quadrats clustered separately from the treatment quadrats. Beyond this the multivariate approach used here yielded little information on the effect of harvesting *Porphyra* on eulittoral communities. The Mantel-type Monte Carlo analysis detected no significant change with time and treatment in the Dice similarity matrix underlying the ordination plots. Somerfield & Clarke (1995), commenting on the efficacy of aggregating data to higher taxonomic levels in marine community studies, noted that the effects of aggregating data on ability of the Mantel test to discriminate between stations varied. Our ordination plots indicate that the effect of time and treatment on aggregated data is still detectable, but this approach does not seem to be the best way to assess the effect of harvesting *Porphyra* on eulittoral communities.

Eulittoral species of *Porphyra* are capable of remaining viable for some time after they have been dried (Smith et al., 1986; Lipkin et al., 1993). Any *Porphyra* collected on a commercial scale in South Africa would of necessity comprise intertidal species, as these are more easily collected and much larger and more common than subtidal forms. Thalli could therefore be collected and air- or sun-dried, after which transport and/or storage of thalli would be facilitated (air-dried thalli of *P. capensis* can remain viable for periods exceeding one year if stored at $-18\,°C$). When required, dried thalli could be rehydrated and fed to abalone. This would spread the impact of harvesting *Porphyra* over a wider area and time, and may make larger, sustainable harvests of *Porphyra* possible.

Roland & Coon (1984) and Nelson & Conroy (1989) examined the effect of harvesting on populations of mixed *Porphyra* species in British Columbia and New Zealand, respectively. Both came to the conclusion that provided harvesting was not so extensive as to significantly reduce sporophyte populations, eulittoral *Porphyra* populations were capable of providing a sustained yield of *Porphyra*. From our results, we come to a similar conclusion: harvesting eulittoral *Porphyra* should have little impact provided that sufficient gametophytes are left unharvested to maintain sporophyte populations and to provide some food and shelter for those fauna associated with eulittoral *Porphyra*. Specific management proposals for *Porphyra* in South Africa have been presented (Griffin et al., 1998). In that document, it was proposed that regularly spaced patches, and in particular dense patches, should be left throughout the eulittoral after harvesting to

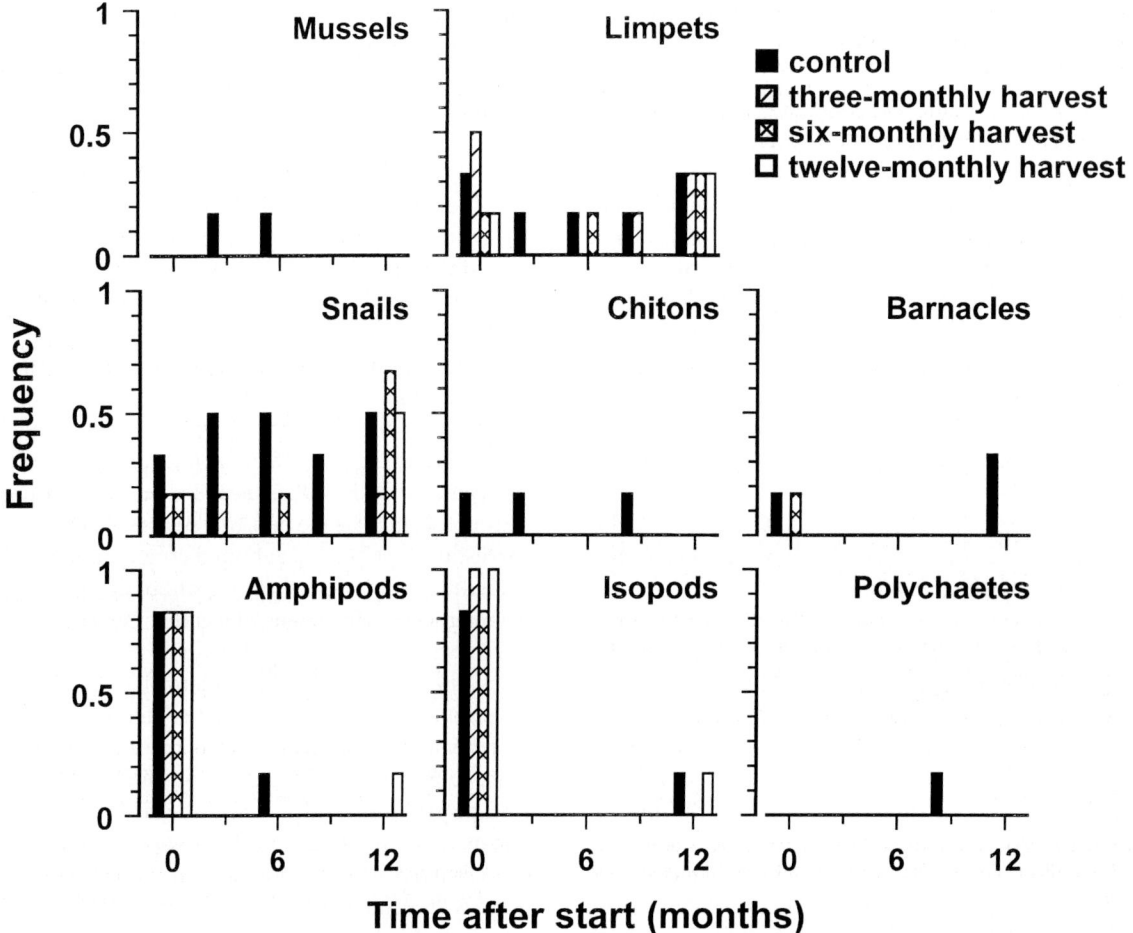

Figure 3. Frequency of eulittoral faunal taxa in experimental quadrats under three harvest and one control treatments, over time. Frequency is expressed as the proportion of quadrats containing a given taxon. As all quadrats were examined, the absence of data from any period that should be present according to the experimental design indicates a zero and not missing data.

facilitate growth and sporulation of unharvested *Porphyra* gametophytes, and survival of associated fauna, and, moreover, that the frequency of harvesting on any stretch of shore should be limited.

Until the taxonomy of *Porphyra* in South Africa is properly resolved, effective management of *Porphyra* populations will at best be difficult. Once the taxonomy is resolved it will be possible to more properly investigate the ecology and response to harvesting of the genus. Until these longer-term goals are achieved, ongoing monitoring of harvested sites will be necessary to assess the effect of widespread harvesting on eulittoral communities.

Acknowledgements

We would like to thank the Sea Fisheries Research Institute for funding this project, and the Foundation for Research Development and the University of Cape Town for additional support. We would also like to thank two anonymous referees for their comments on the manuscript.

References

Agardh, J. G., 1890. Till algernes systematik. Nya bidrag. Lunds Universitets Årsskrift, Afd. 2, 26: 1–125.

Anderson, R. J., R. H. Simons & N. G. Jarman, 1989. Commercial seaweeds in southern Africa: a review of utilization and research. S. Afr. J. mar. Sci. 8: 277–299.

Bolton, J. J. & M. A. P. Joska, 1995. Population studies on a South African carrageenophyte: *Iridaea capensis* (Gigartinaceae, Rhodophyta). Hydrobiologia 260/261: 191–195.

Branch, G. M., S. Eekhout & A. L. Bosman, 1990. Short-term effects of the 1988 Orange River floods on the intertidal rocky-shore communities of the open coast. Trans. r. Soc. S. Afr. 47: 331–354.

Clarke, K. R., 1993. Non-parametric multivariate analyses of change in community structure. Aust. J. Ecol. 18: 117–143.

Clifford, H. T. & W. Stephenson, 1975. An Introduction to Numerical Classification. Academic Press, London, 229 pp.

Dice, L. R., 1945. Measures of the amount of ecological association between species. Ecology 26: 297–302.

Field, J. G., K. R. Clarke & R. M. Warwick, 1982. A practical strategy for analysing multispecies distribution patterns. Mar. Ecol. Prog. Ser. 8: 37–52.

Graves, J. M., 1969. The genus *Porphyra* on South African coasts: I. observations on the autecology of *Porphyra capensis* sensu Isaac (1957), including a description of dwarf plants. J. s. afr. Bot. 35: 343–362.

Griffin, N. J., J. J. Bolton & R. J. Anderson, 1998. Potential for harvest of *Porphyra* species in the south western Cape. Unpublished Sea Fisheries Research Institute Report, Sea Fisheries Research Institute, Roggebaai, 48 pp.

Hanelt, D., K. Huppertz & W. Nultsch, 1993. Daily course of photosynthesis and photoinhibition in marine macroalgae investigated in the laboratory and field. Mar. Ecol. Prog. Ser. 97: 31–37.

Herbert, S. K. & J. R. Waaland, 1988. Photoinhibition of photosynthesis in a sun and shade species of the red algal genus *Porphyra*. Mar. Biol. 97: 1–7.

Isaac, W. E., 1957. The distribution, ecology and taxonomy of *Porphyra* on South African coasts. Proc. linn. Soc. Lond. 168: 61–65.

Keppel, G., 1991. Design and Analysis: a Researcher's Handbook. Third edition. Prentice-Hall, Englewood Cliffs, New Jersey, 594 pp.

Kützing, F. T., 1843. Phycologia Generalis. Brockhaus F. A, Leipzig, 458 pp., 80 pl.

Levitt, G. J. & J. J. Bolton, 1991. Seasonal patterns of photosynthesis and physiological parameters and the effects of emersion in littoral seaweeds. Bot. mar. 34: 403–410.

Lipkin, Y., S. Beer & A. Eschel, 1993. The ability of *Porphyra linearis* (Rhodophyta) to tolerate prolonged periods of desiccation. Bot. mar. 36: 517–523.

Lubchenco, J. & J. Cubit, 1980. Heteromorphic life histories of certain marine algae as adaptations to variations in herbivory. Ecology 61: 676–687.

Minchin, P. R. 1987. An evaluation of the relative robustness of techniques for ecological ordination. Vegetatio 69: 89–107.

Nelson, W. A. & A. M. Conroy, 1989. Effect of harvest method and timing on yield and regeneration of Karengo (*Porphyra* spp.) (Bangiales, Rhodophyta) in New Zealand. J. appl. Phycol. 1: 277–283.

Roland, W. G. & L. M. Coon, 1984. Postharvest recovery of beds of the edible red alga *Porphyra perforata*. Can. J. Bot. 62: 1968–1970.

Scrosati, R. & R. E. DeWreede, 1997. The dynamics of the biomass-density relationship and frond biomass inequality for *Mazzaella cornucopiae* (Gigartinaceae, Rhodophyta): implications for the understanding of frond interactions. Phycologia 36: 506–516.

Shannon, C. E. & W. Weaver, 1949. The Mathematical Theory of Communication. University of Illinois Press, Urbana, Illinois, 117 pp.

Simpson, B. J. A., 1994. An investigation of diet management strategies for the culture of the South African abalone, *Haliotis midae*. M. Sc. thesis. University of Cape Town, 80 pp.

Simpson, E. H., 1949. Measurement of diversity. Nature 163: 688.

Smith, C. M., K. Satoh & D. C. Fork, 1986. The effects of osmotic tissue dehydration and air drying on morphology and energy transfer in two species of *Porphyra*. Plant Physiol. 80: 843–847.

Sokal, R. R. & P. H. A. Sneath, 1963. Principles of Numerical Taxonomy. W. H. Freeman and Company, San Francisco, 359 pp.

Somerfield, P. J. & K. R. Clarke, 1995. Taxonomic studies, in marine community studies, revisited. Mar. Ecol. Prog. Ser. 127: 113–119.

Stegenga, H., J. J. Bolton & J. J. Anderson, 1997. Seaweeds of the South African west coast. Contributions from the Bolus Herbarium Number 18, Bolus Herbarium, University of Cape Town, Cape Town, 655 pp., 61 pl.

Stekoll, M. S. & L. Deysher, 1996. Recolonisation and restoration of the upper intertidal *Fucus gardneri* (Fucales, Phaeophyta) following the Exxon Valdez oil spill. Hydrobiologia 326/327: 311–316.

Stepto, N. K. & P. A. Cook, 1996. Feeding preferences of the juvenile south African abalone *Haliotis midae* (Linneaus, 1758). J. shellfish Res. 15: 653–657.

Underwood, A. J. & P. Jernakoff, 1984. The effects of tidal height, wave-exposure, seasonality and rock-pools on grazing and the distribution of intertidal macroalgae in New South Wales. J. exp. mar. Biol. Ecol. 75: 71–96.

Weller, D. E., 1987. A re-evaluation of the -3/2 power rule of plant self-thinning. Ecol. Monographs 57: 23–43.

Zar, J. H., 1984. Biostatistical Analysis. Prentice-Hall, Englewood Cliffs, New Jersey, 718 pp.

Effects of seasonal growth rate on morphological variation of *Undaria pinnatifida* (Alariaceae, Phaeophyceae)

M. D. Stuart[1], C. L. Hurd[1] & M. T. Brown[2]
[1]*Department of Botany, University of Otago, P.O. Box 56, Dunedin, New Zealand*
[2]*Marine Biology and Ecotoxicology Group, Department of Biological Sciences and Plymouth Environmental Research Centre, University of Plymouth, Plymouth, PL4 8AA, Devon, U.K.*
E-mail: mstuart@doc.govt.nz

Key words: seaweed, morphology, seasonal variation, growth, phenotypic modulation, kelp, *Undaria pinnatifida*

Abstract

Undaria pinnatifida (Harvey) Suringar is currently divided into two morphological forms, f. *typica* Yendo. and f. *distans* Miyabe & Okamura. The objective of this study was to determine the effects of seasonal variation in growth rate on the morphology of *U. pinnatifida*, and to define the form of *U. pinnatifida* growing in Otago Harbour, New Zealand. Morphological variables (stipe length, blade length, blade width, sporophyll length and degree of blade incision), growth rates (frond, blade and stipe) and blade erosion were measured each month from August 1993 to February 1995, and compared using correspondence analysis. Variation in the morphology of *U. pinnatifida* was largely accounted for by varying growth rates. Definition of the form of *U. pinnatifida* growing in Otago Harbour is equivocal because morphological characteristics of both f. *typica* and f. *distans* were exhibited at different times of the year.

Introduction

Although the laminarian kelp, *Undaria pinnatifida* (Harvey) Suringar is endemic to Japan, Korea and China, introductions throughout the world, via maritime shipping and aquaculture, have enlarged its range to include the French Mediterranean, the French Atlantic, New Zealand, Tasmania, Argentina, Venice, the Channel Islands, England and southern Australia (Pérez et al., 1981; Hay & Luckens, 1987; Sanderson, 1990; Curiel et al., 1994; Santiago Caamano et al., 1990; Floc'h et al., 1991; Fletcher & Manfredi, 1995; Casas & Piriz, 1996; Fletcher, pers. comm; Burridge, in litt.).

Undaria pinnatifida has an annual, heteromorphic life-cycle characterized by a macroscopic sporophyte and microscopic gametophyte. The appearance of the sporophytes is variable and may occur in autumn, from winter to mid-spring, or year round, whereas sporophyte degeneration occurs over a four month period during summer and early autumn (Akiyama & Kurogi, 1982; Zhang et al., 1984; Floc'h et al., 1991; Hay & Villouta, 1993). In regions with wide temperature ranges and summer temperature maxima of 23–29 °C, only the gametophyte stage is evident during the summer. However, juvenile and mature seaweeds may be present throughout the year in regions with small temperature ranges and summer temperature maxima of 15–19 °C (Stuart, 1997).

Undaria pinnatifida is currently divided into two recognized forms on the basis of morphology; f. *typica* Yendo. (= *Undaria pinnatifida* Sur., *Alaria pinnatifida* Harv.), and f. *distans* Miyabe & Okamura (= *Ulopteryx pinnatifida* Kjellm.; Saito, 1975). *Undaria pinnatifida* f. *typica* has a short stipe, shallow pinnate division of the blade and sporophyll development is often confluent with the base of the blade; whereas *U. pinnatifida* f. *distans* is characterized by an elongated stipe and a little-divided blade, with sporophyll development confined to the basal portion of the stipe (Kito et al., 1981; Ohno & Matsuoka, 1993). A third form, f. *narutensis*, was described by Yendo (1911) as containing a short stipe, with less folded sporophylls which became confluent with the lamina and grew sterile ligules from the sporophyll margin. This form was subsequently recognized as an extreme expression of f. *typica* (Okamura, 1915).

Although the morphological features used to distinguish the formae of *U. pinnatifida*, such as blade incision and stipe length, have a genetic basis (Saito, 1972), they may be modulated by environmental factors to the extent that the two recognized forms of *U. pinnatifida* are mutable (Saito, 1972; Kito et al., 1981; Taniguchi et al., 1981). Since portions of a seaweed may grow or develop at different rates, the overall dimensions of the seaweed may vary seasonally and between sites, in response to the environment. Likewise, the degree of blade incision will be affected by the balance between blade growth and the decay of blade tissue.

Following the discovery of *U. pinnatifida* in Wellington Harbour in 1987, and its subsequent spread about New Zealand, morphological differences among discrete populations inhabiting the east coast of New Zealand have become apparent. Sporophytes harvested from Oamaru and Timaru harbours, were of the large 'nambu' variety, whereas Wellington Harbour isolates were relatively small, had shorter stipes, and often contained blades which were basally sporogenous, a feature comparable to *Undaria undarioides* (Hay & Villouta, 1993). As the 'nambu' variety is synonymous with *U. pinnatifida* f. *distans* (Tamura, 1970), the work of Hay & Villouta (1993) suggests that f. *distans* was found in Oamaru and Timaru Harbours; however, isolates from Wellington Harbour exhibited variable morphological characteristics which eluded classification. It has been suggested that the 'naruto' variety occurs in New Zealand (Hay, in litt.); this is equivalent to *U. pinnatifida* f. *narutensis* (Yendo, 1911; Tamura, 1970) and was considered by Okamura (1915) to represent an extreme morphology of *U. pinnatifida* f. *typica*.

From measurements of growth rates and productivity of *U. pinnatifida* in Otago Harbour (Stuart, 1997), seasonal changes in morphology were noticed which incorporate morphological characteristics consistent with both recognized forms of *U. pinnatifida* and those described by Hay & Villouta (1993) for Wellington isolates. These observations raised several questions concerning the effect of growth rates on the morphology of *U. pinnatifida* in New Zealand:

1. Can the morphological variation exhibited by *U. pinnatifida* in Otago Harbour be accounted for by the environmental modulation of growth rates?

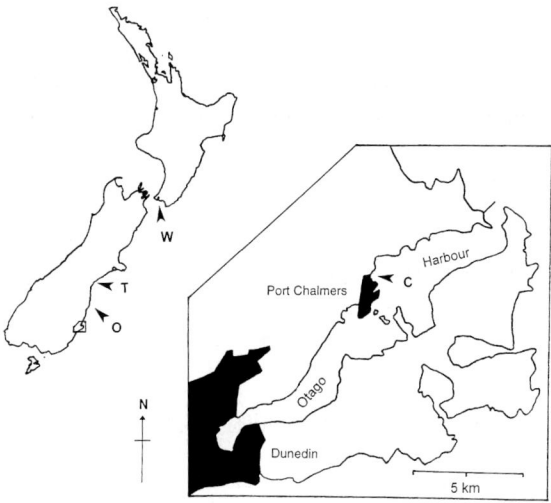

Figure 1. Location of Otago Harbour, study site and other sites mentioned in the text. W = Wellington, T = Timaru and O = Oamaru.

2. Do the morphological characteristics of Otago Harbour isolates conform to recognized forms of *Undaria pinnatifida*?
3. Or does seasonal variation in *U. pinnatifida* morphology represent a range of phenotypic expression of which the recognized forms only represent the extremes of a seasonal cline in morphological characteristics?

To answer these questions, the seasonal variation in morphology of *U. pinnatifida* in Otago Harbour, New Zealand, was compared to seasonal variation in growth rates and the results are discussed in relation to the current taxonomic classification of *U. pinnatifida*.

Description of site studied

The study site is located within Carey's Bay (170° 37′ E; 45° 49′ S), situated in Otago Harbour, south eastern New Zealand (Figure 1). The tidal range is 2.3 m and water type 9, according to the classification of Jerlov (1976). Within Carey's Bay, *Undaria pinnatifida* forms a continuous fringe about wharf pilings from MLW to a depth of 1 m, and isolated clumps of *U. pinnatifida* are found attached to the piles to a depth of 5 m, or growing upon refuse littering the muddy harbour floor, such as wire rope and wooden planks. *Undaria pinnatifida* is also present on the adjacent rocky shore, where clumps of several seaweeds grow on boulders. Dominant algae growing in association with *U. pinnatifida* are *Bryopsis plumosa* (Hudson) C. Agardh, *Cladophora feredayi*

Harvey, *Codium fragile* (Suringar) Hariot, *Desmarestia ligulata* (Lightfoot) Lamouroux, *Enteromorpha compressa* (Linnaeus) Nees, *Mediothamnion lyalli* (Harvey) Gordon, *Phycodrys quercifolia* (Bory) Skottsberg, *Rhodymenia leptophylla* J. Agardh, *R. foliifera* (Harvey) and *Ulva rigida* C. Agardh.

Materials and methods

Measurement of morphological characters and growth rates

For the purpose of this study, the following terminology is used: frond refers to the combined length of the stipe and blade, blade refers to the lamina and stipe refers to region extending from the base of the holdfast to the base of the midrib of the blade.

To measure growth rates, 30 sporophytes, each with a frond length greater than 30 cm were randomly selected on wharf pilings and labelled with PVC tape each month from August 1993 to February 1995 by divers using SCUBA. Sporophytes were distributed over depths ranging from MLW to 5 m. Frond and stipe lengths of each tagged seaweed were measured to the nearest 1 cm and two holes (0.5 cm diameter) were punched, one each side of the midrib, 10 cm from the stipe-blade junction. Tagged sporophytes were harvested 3–4 weeks later. Frond and stipe lengths of the harvested sporophytes were re-measured and frond and stipe growth rates were calculated from the difference between the initial and final measurements. The distance of the holes from the blade-stipe junction was also measured and blade growth rate estimated from the movement of the punched holes along the blade. Due to erosion of tissue from the distal portions of the blade, frond growth rate is equivalent to the sum of blade growth rate, stipe growth rate and erosion. As measurements of frond, blade and stipe growth rates were made during this study, the rate of blade erosion was calculated as the difference between frond growth rate and the combined rates of blade and stipe growth.

Relative growth rates (μ) were expressed relative to frond and stipe length using the following formula (Kain, 1987):

$$\mu = 100 \cdot \ln L_2 - \ln L_1 / T_2 - T_1,$$

where L_1 = initial length (cm), L_2 = final length (cm) and $T_2 - T_1$ = duration of experiment (days). Growth rate data are presented in Stuart (1997).

Morphometric measurements (Figure 2) were also made of all harvested sporophytes to the nearest 0.5

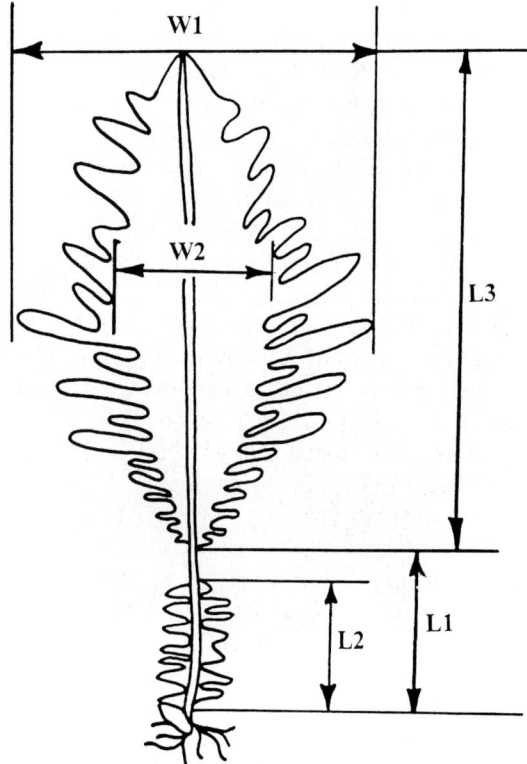

Figure 2. Morphometric measurements of *Undaria pinnatifida* from December 1993 to February 1995. L1: entire length of stipe, L2: length of sporophyllous region of stipe, L3: length of blade, W1: width of blade, W2: width of undivided blade (adapted from Hara & Akiyama, 1985).

cm and were converted to morphological characters (labelled F1–F5) using the following formulae: F1

(degree of blade incision) = W1/W2,
F2 (no. pinnae per unit blade length) = N1/L3,
F3 (stipe length:frond length) = L1/(L1 + L3),
F4 (blade length:blade width) = L3/W1,
F5 (proportion of stipe containing sporophylls) = (L2/L1) × 100,

where W1 = maximum blade width, W2 = maximum width of undivided portion of the blade, L1 = stipe length, L2 = sporophyll length, L3 = blade length, and N1 = number of pinnae on one side of the blade.

Statistical analysis

The effect of growth rate on *Undaria pinnatifida* morphology was investigated by comparing ordinations produced by detrended correspondence analysis (DCA), canonical correspondence analysis (CCA) and detrended canonical correspondence analysis

Figure 3. Specimen of *Undaria pinnatifida* collected from Carey's Bay (26/5/94) showing extension of sporophyll development into the base of the blade and along the midrib. sp = sporophyll, bl = blade, mr = midrib.

(DCCA). DCA differs from CCA and DCCA in that it is an indirect method of gradient analysis in which the environmental gradients (here equivalent to growth rates) are not studied directly, but are inferred from the species composition (here equivalent to morphological data) (Palmer, 1993). Each axis can therefore be thought of as a hypothetical environmental gradient, which is subsequently interpreted according to measured environmental factors (ter Braak, 1986). In this study, each sporophyte was placed along a morphological gradient and can be interpreted according to the measured growth rates. As the ordination axes of the CCA are constrained to be linear combinations of the growth rates (ter Braak, 1987), it is assumed that the measured growth rates account for the observed morphological variation. As the axes of DCA are not constrained by growth rates, similarity between DCA and DCCA ordinations will determine the ability of the measured growth rates to account for the morphological variation summarized by the ordination (ter Braak, 1986). Likewise, similarity between the eigenvalues and correlation coefficients of DCA and DCCA indicate that the pattern of morphological variation is explained by the measured growth rates, dissimilarity implies that other factors are responsible for the observed patterns.

Direct comparison between growth rates and morphology were made by constraining the axis of a CCA ordination to the growth rates of harvested seaweeds. Growth rates were plotted as vectors on a CCA biplot. Relationships between individual morphological characters and morphological variation were presented as a DCA biplot using the sample score centroid as the origin of vectors describing the morphological characters. As correlation between growth rates may result in instability of canonical coefficients (ter Braak, 1986), only the intraset coefficients of the CCA were compared to determine the relative importance of each of the growth rates to morphological variation in *U. pinnatifida*. All data were log transformed before correspondence analysis using CANOCO (ter Braak, 1990).

To verify that significant variation occurred between monthly measurements of morphological characteristics and growth rates, data were first tested for skewed distributions and kurtosis (Anscombe & Tukey, 1963; Snedecor & Cochran, 1980). Data were log transformed if required, and then analysed by ANOVA and Duncan's New MRT using the statistical package, Teddybear (Wilson, 1975).

Results

Seasonal morphological variation

Undaria pinnatifida sporophytes sampled in May and June were characterized by sporophyll development which extended along the entire length of a short stipe. The blade was relatively entire with few pinnae and shallow incisions. Seaweeds collected in July had a similar degree of sporophyll development but stipes were longer and the blade had more pinnae. The frond length of *U. pinnatifida* was maximal in August but declined thereafter until February. Stipe length increased from May to September, then declined from October to February.

The majority of sporophytes collected throughout the study were characterized by sporophylls which were confined to the basal portions of the stipe margin, but sporophyll development extended along the entire stipe margin and was confluent with the blade in some sporophytes. During May and June 1994, sporophyll

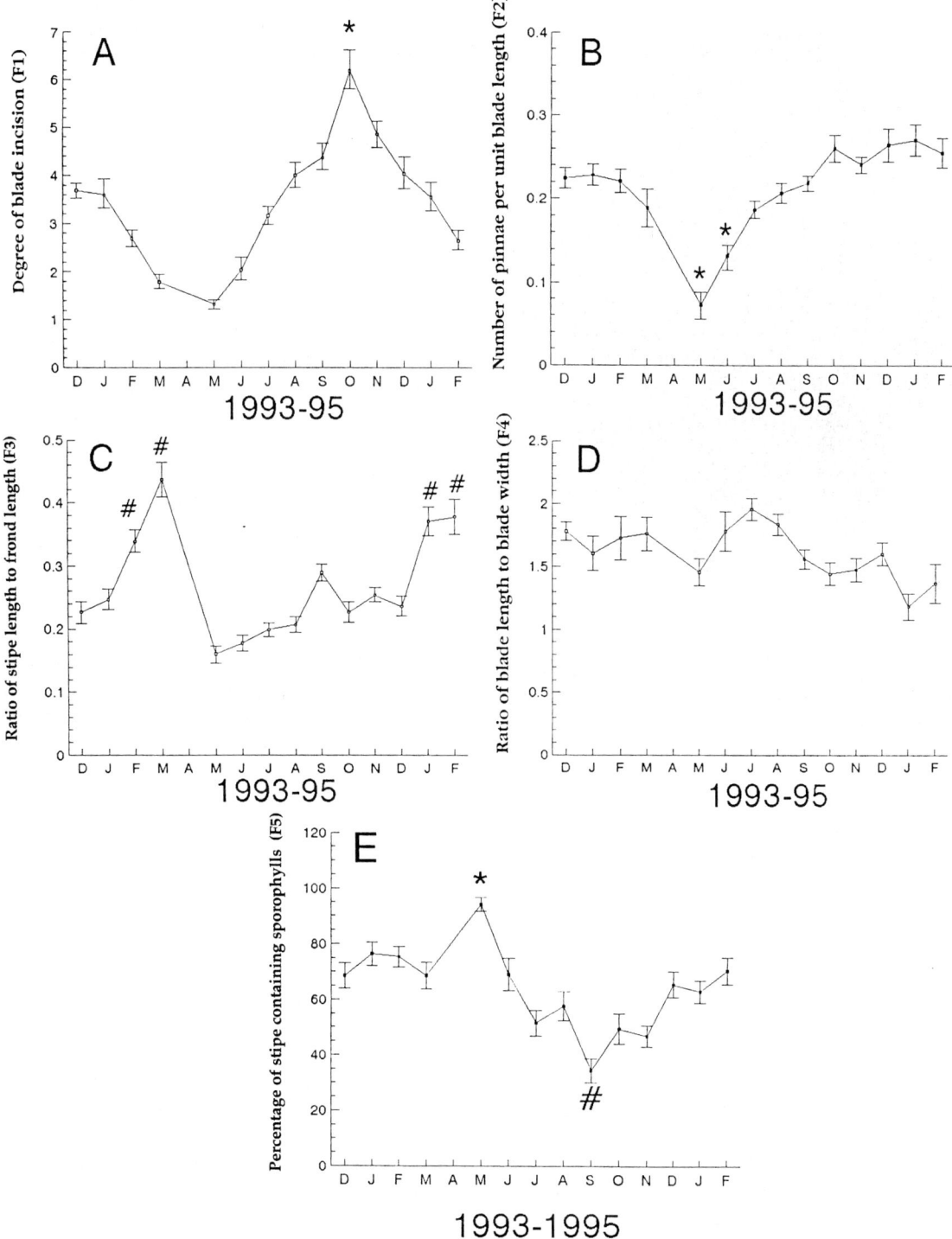

Figure 4. Seasonal variation of morphological characteristics for *Undaria pinnatifida*. Each point represents the mean ($n = 24$-30) $\pm$ 1 S.E. # indicates dissimilarity from other values at $p = 0.05$ and * represents dissimilarity from other values at $p = 0.01$.

Table 1. Intraset correlation coefficients of growth rates with axis 1 and axis 2 of the detrended correspondence analysis (DCA) and detrended canonical correspondence analysis (DCCA)

	DCA		DCCA	
Axis	1	2	1	2
Frond growth rate	−0.4467	−0.1593	−0.4540	−0.4426
Blade growth rate	−0.4196	−0.0870	−0.4258	−0.3887
Stipe growth rate	0.0396	0.0632	0.0438	0.0670
Erosion	0.1066	0.0898	0.1094	0.1273

Table 2. Intraset correlations between CCA axes defined by morphological characteristics and constrained by the growth rates of *Undaria pinnatifida* (Figure 5)

	Correlation coefficient	
Axis	1	2
Frond growth rate	−0.4494	0.1175
Blade growth rate	−0.4218	0.2812
Stipe growth rate	0.0466	0.2525
Erosion	0.1084	0.1167

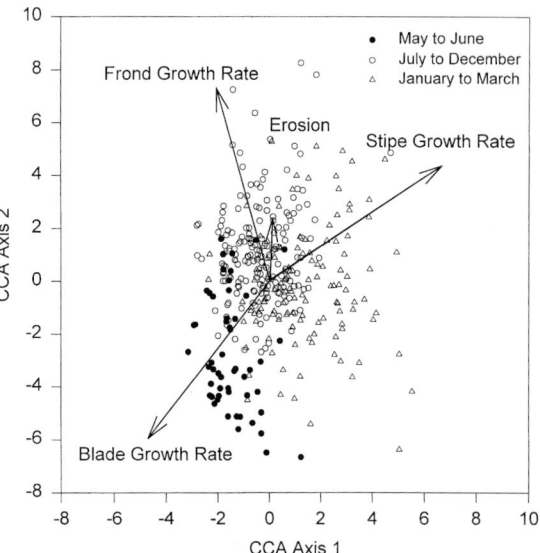

Figure 5. Canonical correspondence analysis (CCA) ordination of morphological characteristics of *Undaria pinnatifida* upon CCA axes constrained by growth rates.

development extended into the basal portions of the blade (Figure 3).

All morphological characteristics and growth rates of *U. pinnatifida* varied significantly between months ($p < 0.001$. Blade incision was lowest in May 1994 and highest in October 1994 (Figure 4A), with values during October significantly higher than for all other months ($p = 0.01$). The number of pinnae per blade length (F2) was significantly lower in May and June than for the rest of the year ($p = 0.01$, Figure 4B). The ratio of stipe length to frond length (F3) was significantly higher during February–March 1994 and January–February 1995 than for the other months ($p = 0.05$, Figure 4C). There was no seasonal trend in the ratio of blade length to blade width (Figure 4D). Sporophyll development was almost confluent with the blade margin in May 1994 ($p = 0.01$), but was confined to the basal half of the stipe during September 1994 ($p = 0.05$, Figure 4E).

Gradient analysis

The similarity between the intraset correlations of DCA and DCCA indicate that morphological variation in *U. pinnatifida* is adequately accounted for by the measured growth rates (Table 1). Comparison of the intraset correlations for growth rates and the axes of the CCA ordination indicate that the first axis is defined by frond growth rate and blade growth rate, whilst the second axis is defined by the rate of blade and stipe growth (Table 2). Growth rates accounted for 61.8% and 41.6% of the morphological variation summarized by the first and second CCA axis, respectively.

To aid in the interpretation of the CCA and DCA ordinations (Figures 5–6), the length of a vector indicates the contribution of growth rate or a morphological character to variation along each of the axes, and the direction indicates how well the growth rate or morphological character is correlated with the axes. A small angle between vectors indicates a high correlation and a large angle indicates a low correlation between growth rates or morphological characters (Palmer, 1993). A strong negative correlation between blade and stipe growth rates resulted in seaweeds with high blade growth rates and low stipe growth rates (May–June), opposing seaweeds with low blade growth rates and high stipe growth rates (January–March, Figure 5). Seaweeds harvested from July to December occur about the origin of the first axis and have similar rates of blade and stipe growth. The vector defined by frond growth rate describes a morphological cline formed by the combined effects of blade and stipe growth rate on the morphology of *U. pinnatifida*. Hence, seaweeds situated at the top

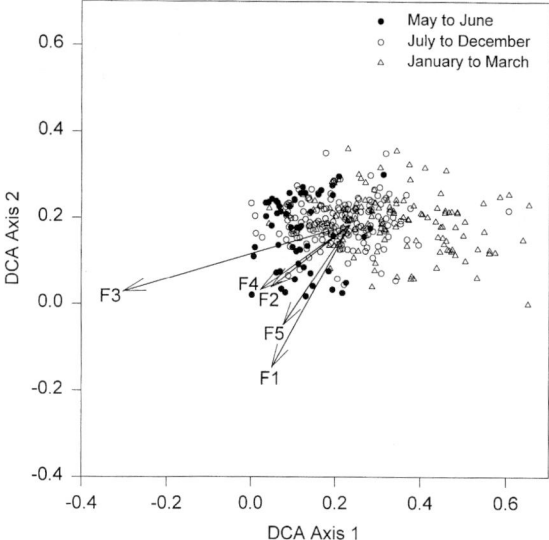

Figure 6. Detrended correspondence analysis (DCA) ordination of *Undaria pinnatifida* morphology, with morphological characters (F1–F5) plotted as vectors arising from the sample score centroid. The morphological characters are: (F1) degree of blade incision, (F2) number of pinnae per unit blade length, (F3) the ratio of stipe length to frond length, (F4) the ratio of blade length to blade width and (F5) the proportion of stipe containing sporophylls.

of the ordination have the highest frond growth rate, and those seaweeds situated at the bottom of the ordination have the lowest frond growth rate. Erosion contributes little to seasonal variation in *U. pinnatifida* morphology.

The contribution of specific morphological characters to the seasonal morphological variation in *U. pinnatifida* is presented in Figure 6. Variation in the ratio of stipe length to frond length (F3), and the degree of blade incision (F1) contribute most to the morphological variation of *U. pinnatifida* in Carey's Bay, as indicated by the greater length of the arrows. The ratio of blade length to blade width (F4), the number of pinnae per unit blade length (F2) and the proportion of stipe containing sporophylls (F5), were relatively conservative and contributed least to morphological variation in *U. pinnatifida*. The strong correlation between the degree of blade incision (F1) and the proportion of stipe containing sporophylls (F5) indicates that confluence of blade and sporophyll is more likely to occur in seaweeds with a low degree of blade incision. The general direction of the morphological vectors in Figure 6 shows that seaweeds harvested from May and June 1994 have a relatively short stipe and long blade, slight blade incision and confluence of the blade and sporophyll. *Undaria pinnatifida* harvested from January to March 1994, and January to February 1995, were characterized by relatively long stipes and short blades.

Discussion

Seasonal variation in the rates of blade and stipe growth accounts for the majority of morphological variation exhibited by *Undaria pinnatifida* in Carey's Bay, Otago Harbour. Although proportionate growth of the blade and stipe occurred from July to December and produced seaweeds with a long blade and stipe, disproportionate growth of blade and stipe either resulted in sporophytes with long blades and short stipes (May–June), or sporophytes with long stipes and short blades (January–March). Seasonal morphological variation was not attributable to variation in the rate of blade erosion, but blade growth rate was the most variable component determining a net increase or decrease in blade size. Consequent rates of frond growth, representing the summation of blade and stipe growth rate, produced a seasonal cline of frond length (y axis of CCA ordination) superimposed upon a seasonal cline defined by the relative proportion of stipe length to frond length (x axis of CCA ordination). Seasonal morphological variation therefore represents phenotypic modulation of stipe and blade morphology possibly caused by the different responses of stipe and blade growth rates to seasonal changes in the environment.

Although a significant relationship between sea surface temperature and blade growth rates has been demonstrated in Otago Harbour for *U. pinnatifida*, no single environmental factor accounts for the observed seasonal variation in stipe growth rate (Stuart, 1997). Various authors have suggested that stipe growth may be influenced by irradiance and the shading effects of aggregated kelp (Duncan, 1973; Holbrook et al., 1991), and it is possible that a similar mechanism modulates stipe elongation in *U. pinnatifida*. As with other annual kelp species, reduced blade and stipe growth rates during late summer (December–February) are most likely to be linked to the initiation of spore release and senescence (Norton & Burrows 1969; Pérez et al., 1981). However, without a direct causal relationship between specific environmental parameters and growth rates, those factors responsible for the environmental modulation of *U. pinnatifida* morphology remain unclear. Inter-correlations between seawater nitrate, sea surface temperature and

Table 3. Descriptions of *Undaria pinnatifida* and *U. undarioides* directly from Yendo (1911) and Okamura (1915)

Undaria pinnatifida
forma *distans* Stipe elongated, subequal to the length of deeply pinnated lamina with large sporophylls limited to the base of the stipe, without proliferations.
forma *typica* Stipe short, with sinuses between adjacent pinnae comparatively shallow which are more distant from the midrib, with the upper portions of large sporophylls confluent with the base of the lamina.
forma *narutensis* **(Yendo 1911) = forma** *typica* **(Okamura, 1915)** Stipe shortest, with less folded sporophylls, which become confluent with the lamina and proliferating sterile ligules from the margin of the sporophylls.

Undaria undarioides
Stipe shortest (0.5 - than 20 cm.), subcylindrical at base or compressed, soon ancipate and complanated above, more or less winged; lamina thin membranaceous, midribed, bullato-rugose, oblongo-ovate or cordate, entire or subpinnatifid. Sori on both sides of midrib at the beginning on both surfaces, afterward becoming confluent towards the base of the lamina and often at the same time continuous with those on the sporophylls.

incident PFD make it difficult to separate their individual effects from field observations without supporting experimental evaluation of the specific environmental factors influencing *U. pinnatifida* morphology.

The morphological forms exhibited by *U. pinnatifida* in Otago Harbour are not discrete and represent a seasonal cline in stipe and sporophyll morphologies which do not conform to characteristics consistent with the recognized forms of *U. pinnatifida*. Definition of the form of *U. pinnatifida* growing in Otago Harbour is therefore ambiguous because morphological characteristics of both f. *typica* and f. *distans* are exhibited. On the basis of blade morphology, the shallow blade incision of *U. pinnatifida* growing in Carey's Bay is indicative of f. *typica* (Table 3). However, the sporophyll and stipe morphology is variable and produces characteristics of both f. *distans* and f. *typica*. The presence of sporogenous sporophytes with short stipes, undivided blades and the extension of sporophyllous tissue into the basal portions of the blade is reminiscent of spring recruits of *U. pinnatifida* collected from Wellington Harbour by Hay & Villouta (1993). These morphological characteristics are shared by both *U. undarioides* and *U. pinnatifida* f. *typica* (Table 3; Okamura, 1915). The intermittent occurrence of this morphology in Otago Harbour would suggest that the genetic basis of such morphological characters is modulated by environmental variables, resulting in the observed phenotypic differences. It is equally possible that the growth of the stipe and blade may be influenced by the development of the sporophyte, such that the onset of sporophyll development affects the rate of stipe elongation. As the reproductive maturity of *U. pinnatifida* may be under endogenous and/or environmental control, it is difficult to determine the relative importance of growth and development on *U. pinnatifida* morphology.

Without direct comparisons among herbarium specimens, lectotype material and New Zealand isolates, the classification of New Zealand *Undaria pinnatifida* remains inconclusive. Likewise, an attempt should be made to compare the genetic and morphological similarity of the populations of *U. pinnatifida* about New Zealand. This could be done either by comparing parental and F1 morphologies under controlled environmental conditions, or comparing morphologies of seaweeds grown in culture with reciprocal transplants between different sites. Molecular techniques should also be employed to assess the genetic similarity of the various forms exhibited by *U. pinnatifida*.

Acknowledgements

We wish to thank Greig Funnell, Richard Gimpel, Alan Johnston, Craig Loveridge & Katrin Berkenbusch for diving assistance, and Dr Janice Lord for assistance with data analysis. We also thank Drs Joanna Jones & Cameron Hay for their helpful comments. MDS was supported by a University of Otago post-graduate scholarship.

References

Akiyama, K. & M. Kurogi, 1982. Cultivation of *Undaria pinnatifida* (Harvey) Suringar, the decrease in crops from natural plants following crop increase from cultivation. Bull Tohoku Reg. Fish. Lab. 44: 91–99.

Anscombe, F. J. & J. W. Tukey, 1963. The examination and analysis of residuals. Technometrics 5: 141.

Braak, C. J. F. ter, 1986. Canonical correspondence analysis: A new eigenvector technique for multivariate direct gradient analysis. Ecology 67: 1167–1179.

Braak, C. J. F. ter, 1987. The analysis of vegetation-environment relationships by canonical correspondence analysis. Vegetatio 69: 69–77.

Braak, C. J. F. ter, 1990. Canoco – a FORTRAN program for Canonical Community Ordination. Microcomputer Power, Ithaca, New York, U.S.A. 95 pp.

Casas, G. N. & M. L. Piriz, 1996. Surveys on *Undaria pinnatifida* (Laminariales, Phaeophyta) in Golfo Nuevo, Argentina. Hydrobiologia 326/327: 213–215.

Curiel, D., A. Rismondo, M. Marzocchi & A. Solazzi, 1994. Distribuzione di *Undaria pinnatifida* nella laguna di Venezia. Lavori-Soc. Ven. Sci. Nat. 19: 121–126.

Duncan, M. J., 1973. *In situ* studies of growth and pigmentation of the phaeophycean *Nereocystis luetkeana*. Helgoländer wiss. Meeresunters. 24: 510–525.

Fletcher, R. L. & C. Manfredi, 1995. The occurrence of *Undaria pinnatifida* (Phaeophyceae, Laminariales) on the south coast of England. Bot. mar. 38: 355–358.

Floc'h, J. Y., R. Pajot & I. Wallentinus, 1991. The Japanese brown alga *Undaria pinnatifida* on the coast of France and its possible establishment in European waters. J. Cons. perm. int. Explor. Mer. 47: 379–390.

Hara, M. & K. Akiyama, 1985. Heterosis in growth of *Undaria pinnatifida* (Harvey) Suringar. Bull. Tohoku Reg. Fish. Res. Lab. 47: 47–50.

Hay, C. H. & P. A. Luckens, 1987. The Asian kelp *Undaria pinnatifida* (Phaeophyta: Laminariales) found in a New Zealand harbour. N.Z. J. Bot. 25: 329–332.

Hay, C. M. & E. Villouta, 1993. Seasonality of the adventive Asian kelp *Undaria pinnatifida* in New Zealand. Bot. mar. 36: 461–476.

Holbrook, N. M., M. W. Denny & M. A. R. Koehl, 1991. Intertidal 'trees': consequences of aggregation on the mechanical and photosynthetic properties of sea-palms *Postelsia palmaeformis* Ruprecht. J. exp. mar. Biol. Ecol. 146: 39–67.

Jerlov, N. G., 1976. Marine Optics. Elsevier, Amsterdam. 231 pp.

Kain, J. M., 1987. Seasonal growth and photoinhibition in *Plocamium cartilagineum* (Rhodophyta) off the Isle of Man. Phycologia 26: 88–99.

Kito, H., K. Tanigushi & K. Akiyama, 1981. Morphological variation of *Undaria pinnatifida* (Harvey) Suringar - II. Comparison of the thallus morphology of cultured F_1 plants originated from parental types of two different morphologies. Bull. Tohoku Reg. Fish. Res. Lab. 42: 11–18.

Norton, T. A. & E. M. Burrows, 1969. Studies of marine algae of the British Isles 7. *Saccorhiza polyschides* (Lightf.) Batt. Br. phycol. J. 4: 19–53.

Ohno, M. & M. Matsuoki, 1993. *Undaria* cultivation 'Wakame'. In Ohno, M. & A.T. Critchley (eds), Seaweed Cultivation and Marine Ranching. JICA, Yokosuka, Japan: 41–49.

Okamura, K., 1915. *Undaria* and its species. Bot. Mag. Tokyo 29: 269–281.

Palmer, M. W., 1993. Putting things in even better order: The advantage of canonical correspondence analysis. Ecology 14: 2215–2230.

Pérez, P., J. Y. Lee & C. Juge, 1981. Observations sur la biologie de l'algue japonaise *Undaria pinnatifida* (Harvey) Suringar introduite accidentellement dans l'étang de Thau. Sci. Pêche 315: 1–12.

Saito, Y., 1972. On the effects of environmental factors on morphological characteristics of *Undaria pinnatifida* and the breeding of hybrids in the genus *Undaria*. In Abbott, I. A. & M. Kurogi (eds), Contributions to the Systematics of Benthic Marine Algae of the North Pacific. Japanese Society of Phycology, Kobe, Japan: 117–134.

Saito, Y., 1975. *Undaria*. In Tokida, J. & H. Hirose (eds), Advances of Phycology in Japan. Dr W. Junk Publishers, The Hague: 304–320.

Sanderson, J. C., 1990. A preliminary survey of the distribution of the introduced macroalga, *Undaria pinnatifida* (Harvey) Suringer on the east coast of Tasmania, Aust. Bot. mar. 33: 153–157.

Santiago Caamano, J., C. Duran Neira & R. Acuna Castroviejo, 1990. Aparicion de *Undaria pinnatifida* en las costas de Galicia (Espana). Un nuevo caso en la problematica de introduction de especies foraneas. C.I.S., Santiago de Compostela, Informes tecnicos 3: 1–43.

Snedecor, G. W. & W. G. Cochran, 1980. Statistical Methods. Edn. 7. Iowa State University Press, 593 pp.

Stuart, M. D., 1997. Seasonal ecophysiology of *Undaria pinnatifida* in Otago Harbour, New Zealand. Ph.D. thesis, University of Otago, New Zealand. 193 pp.

Tamura, T., 1970. Propagation of *Undaria pinnatifida*. In Tamura T. (ed.), Marine aquaculture. Washington D.C. National Technical Information Service, National Science Foundation (English translation of Japanese 2nd edn, 1966): 1–14.

Taniguchi, K., H. Kito & K. Akiyama, 1981. Morphological variation of *Undaria pinnatifida* (Harvey) Suringer -I. on the difference of growth and morphological characteristics of two types at Matsushima Bay, Japan. Bull. Tohoku Reg. Fish. Res. Lab. 42: 1–9.

Wilson, J. B., 1975. Teddybear – a statistical system. New Zealand Statistician 10: 36–42.

Yendo, K., 1911. The development of *Costaria*, *Undaria* and *Laminaria*. Ann. Bot. 25: 691–715.

Zhang, D. M., G. R. Miao & L. Q. Pei, 1984. Studies on *Undaria pinnatifida*. Hydrobiologia 116/117: 263–265.

Phenology of *Sargassum* spp. (Sargassaceae, Phaeophyta) from Reunion Rocks, KwaZulu-Natal, South Africa

R. D. Gillespie & A. T. Critchley[1]
Department of Botany, University of the Witwatersrand, Private Bag 3, Wits, 2050, South Africa
[1] *Joint address: Multidisciplinary Research Centre, University of Namibia, Private Bag 13301, Windhoek, Namibia and Department of Botany, University of the Witwatersrand, South Africa*
[1] *Present address: P.O. Box 802, Randpark Ridge, 2156, South Africa*
Fax: [+27] 11 795 1443. E-mail: isaruss@icon.co.za

Key words: alginate, biomass, density, *Sargassum elegans*, *Sargassum* sp., standing stock

Abstract

The density, standing stock, reproductive periodicity and alginate yields have been investigated for a mixed *Sargassum* species population at Reunion Rocks, KwaZulu-Natal, South Africa. The population consists of two dominant components, *Sargassum elegans* Suhr and a second species, here termed *Sargassum* sp. 1. *Sargassum elegans* typically makes up more than 70% of the population which is heterogeneous in both biomass and density distribution. These species share the same macro-scale distribution within the site and display the same periods of growth and senescence. Standing stock dropped in the cooler winter months and reached a maximum in the hot summer months. Biomass was correlated to one month antecedent minimum, maximum and mean air temperature. Density patterns followed a similar pattern and biomass increased with density. Alginate yield of both species peaked in the warm summer months and was lowest during the cooler winter months. Reproduction peaks were recorded in the cooler months after the peak in biomass and density.

Introduction

The genus *Sargassum* is well represented in both tropical and sub-tropical waters world-wide, often being the dominant member of both the sub- and intertidal flora. *Sargassum* spp. and their derivative products are used throughout the world (Ganzon-Fortes et al., 1993; Masuda et al., 1993; Trono & Tolentino, 1993). However, they are not utilized in South Africa, where seaweed resource utilization has focused on the large kelps, *Ecklonia* and *Laminaria* and the major agar producing red seaweeds, *Gracilaria* and *Gelidium* (Critchley et al., 1998). There may be potential for the utilization of *Sargassum* on a small-scale by local coastal communities in KwaZulu-Natal, South Africa, for example as fertilizer or animal feed supplements. While members of the genus *Sargassum* have been the subject of several stock assessments, phenological and reproductive ecology studies in many parts of the world (De Wreede, 1976; Deysher, 1984; Glenn et al. 1990; Kendrick & Walker, 1994; Largo et al., 1994; Trono & Tolentino, 1993), the genus in South Africa has been relatively ignored from an ecological point of view.

Gillespie et al. (1996) stated that there is a poor understanding of the ecology of the seaweeds along the KwaZulu-Natal (KZN) coastline of South Africa, especially with regard to population structure and function and in regard to conservation status of the seaweed species. Until 1995 there had been no detailed stock assessment or population ecology studies of the seaweeds of this coast. During 1995 two studies were initiated, one investigating the stocks and effects of harvesting on *Gelidium* spp. from Reunion Rocks, KwaZulu-Natal (Aingworth, 1997; Aingworth & Critchley, 1997), the other, this study, investigating the stocks of *Sargassum* species from the same site. The lack of ecological data has serious implications for the development of conservation strategies, which are at present non-existent for these seaweeds, and guidelines for the potential and sustainable utilization of the resource. Gillespie & Critchley (1997,

1998) discussed aspects of the taxonomy of *Sargassum* species from Reunion Rocks. This paper presents biomass data, growth strategies, reproductive phenologies, biomass–density relationships and alginate yields for *Sargassum elegans* Suhr and *Sargassum* sp. 1 from Reunion Rocks, KwaZulu-Natal, South Africa.

Materials and methods

Study site

Reunion Rocks is situated on the east coast of South Africa in the province of KwaZulu-Natal, approximately 15 km south of Durban at 29° 59′ 00″ S and 30° 58′ 40″ E. The intertidal rock of Reunion Rocks is Bluff Sandstone of the Pleistocene era (Krige, 1932). The study site has been described previously by Gillespie & Critchley (1998). The shore consists of three important areas: lagoon, lower platform and seaward belt (terminology from Eyre & Stephenson, 1938; Birkett, 1984). The lower platform slopes upwards from the lagoon side to the seaward side (Figure 1). The study site was located near the seaward edge of this platform. The lowest terrace occurs above the level of the mean low water springs (MLWS) in the lagoon, while the seaward edge of the platform is approximately 2 m above MLWS (Figure 1; Birkett, 1984).

Sampling method

A metered transect line was laid across the width of the lower platform, approximately parallel to the seaward belt. The line was split into two sections of 33.4 and 37.6 m, due to the irregularity of the seaward edge of the platform (Figure 1a). The ends of the lines were marked by brass screws drilled into the rock. A band of 1.6 m was defined on each side of this line (i.e. a sample band of 71 m × 1.6 m was used). Random placement of quadrats was determined using a random number table to indicate the distance along the line and placement distance above or below the line. Optimal quadrat area and replicate number were determined using the concept of minimal area curves (Hopkins, 1957). Ten 0.16 m^2 quadrats were used at each sample time. Data were collected over eight sample periods during periods of low spring tides from August 1995 to July 1997 and corresponding with spring, summer, autumn and winter seasons. At each sample time, all of the *Sargassum* present in the quadrats was scraped off the rocks, including as much of the holdfasts as possible and ensuring that individual thalli were kept whole. No quadrat was sampled twice. Samples were placed in plastic bags and transported on ice back to the laboratory.

Biomass and density data

The number of thalli per quadrat was enumerated. Data for the first set of seasons (August 1995 to May 1996) did not distinguish between different species present and were used to provide an estimate of total *Sargassum* biomass and density of thalli per m^2. During the second set of seasons (October 1996 to July 1997) individual thalli were divided between *S. elegans* (a clumped, prostrate species) and *Sargassum* sp 1 (an erect species) and analyzed separately, as well as combined in order to provide an overall estimate of *Sargassum* biomass and density per m^2 for the population as a whole and for the separate species. A second erect species, *Sargassum* sp 2, discussed in Gillespie & Critchley (1998), was grouped with *Sargassum* sp. 1 as the number of these individuals within the sampled quadrats was very low (less than three individuals were collected during any one sample period). *Sargassum* sp. 1 has been commonly referred to as *S. heterophyllum/S. incisifolium*, however Gillespie & Critchley (1998) suggested that this name was inappropriate for the specimens from Reunion Rock). Specimens were cleaned of epiphytes and dried in an oven at 45 °C (±1 °C) to constant weight (DW).

Maximum, minimum and mean monthly (at each hour of the day) air temperatures for Durban (approximately 15 km north of Reunion Rocks) were provided by the South African Weather Bureau. Surface sea temperatures recorded 400 m offshore in the early morning at Isipingo Beach (about 1 km south of the study site) were provided by the Natal Sharks Board. Linear regression analyses were performed using Systat 5.03 to analyze the relationship between air and seawater temperature and biomass.

Alginate yield

Specimens collected during October 1996, January, May and July 1997 were used to investigate alginate yield from *Sargassum elegans* and *Sargassum* sp. 1. The method used followed that of Trono & Tolentino (1993) with minor modification. Three replicates were used for each species for each season. The specimens from one randomly placed quadrat (as described above) represented one replicate. Five grams of dried

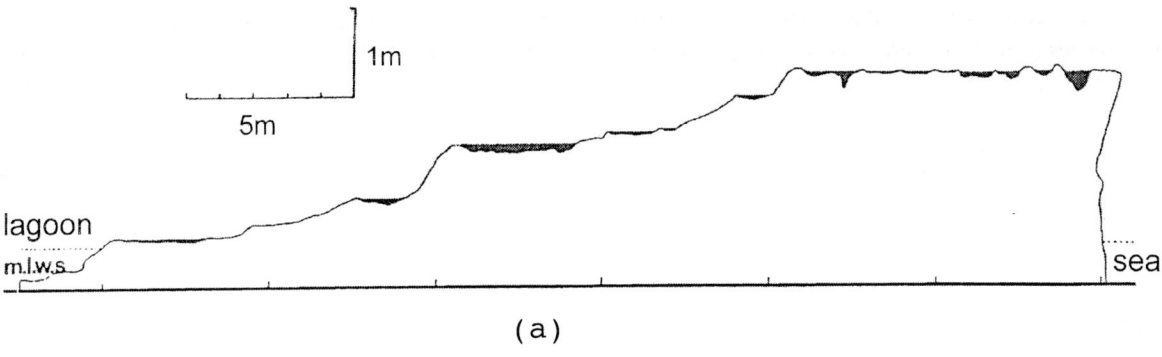

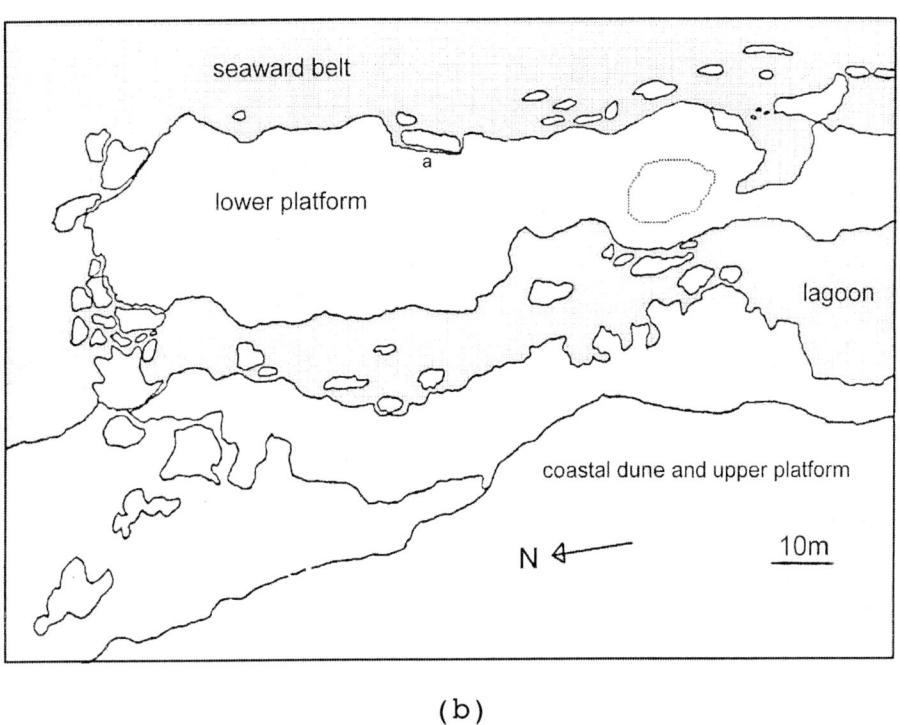

Figure 1. (a) Transect across the lower platform of the study site at Reunion Rocks. (b) Aerial View of the study site. *Sargassum* species occur on the lower platform. Shaded areas indicate permanent water (after Birkett, 1984).

thallus were soaked in 100 ml of 2% formalin-distilled water solution (distilled water pH 6.6) for 24 h. The liquid was poured off and the material was dried at 45 °C, after which it was ground to a coarse powder (individual pieces no larger than 2 mm^2) using a mortar and pestle. The sample was then placed in 100 ml of 0.2 N HCl for 24 h, after which the HCl was decanted and the sample placed in 200 ml of 2% Na_2CO_3 for 3 h. The liquid portion was filtered through Whatman filter paper. Sodium alginate was precipitated using 96% ethanol and dried in an oven at 45 °C to constant weight.

Results

The spatial pattern of *Sargassum* species biomass and density over the study site was heterogeneous. Both biomass and density declined from the seaward to the landward side of the site with highest levels in the central region. *S. elegans* and *Sargassum* sp. 1 showed peak density distribution within similar areas of the site. Density and biomass peaked in an area of shallow corrugations containing permanent water at low tide, to the south of the platform and also peaked further westwards on the northern side of the platform. This is an area with less corrugation of the substratum and fewer permanent pools than on the southern side.

Population composition

The study area was dominated by *S. elegans*, which made up approximately 75% of the *Sargassum* population at Reunion Rocks (Figure 2). The species composition of the population was relatively stable between October and May. However during the cooler winter months of July/August the contribution of *Sargassum* sp. 1 to the population declined to less than 15%. This change was consistent over the two years of study.

Standing stock and seasonality

The average monthly air temperatures recorded at 10:00 h (corresponding to the average time of daytime low spring tides for Reunion Rocks) at Durban are presented in Figure 3. The mean monthly sea surface temperatures recorded at Isipingo Beach are also presented in Figure 3. Figure 4 shows that the total *Sargassum* biomass at Reunion Rocks rose as temperature warmed (from August to January), reached a maximum in the hottest months of the year and decreased with decreasing temperature to a minimum in the coldest months of the year. Both *S. elegans* and *Sargassum* sp. 1 showed the same temporal pattern. The variation evident within a season was a result of the spatial biomass heterogeneity within the site. *Sargassum* species biomass was positively correlated to one month, antecedent mean air temperature ($r^2 = 0.58$, $p = 0.048$, $n = 7$), mean maximum air temperature ($r^2 = 0.61$, $p = 0.039$, $n = 7$) and mean minimum air temperature ($r^2 = 0.57$, $p = 0.05$, $n = 7$). Surface seawater temperature could not be related to biomass change. A regression using one month antecedent sea surface temperature yielded an insignificant r^2 value of 0.21 ($p = 0.31$, $n = 7$).

Sargassum spp. density

The density of *Sargassum* species followed the same temporal pattern as biomass, with a decrease in the cooler winter months and an increase with warming temperatures (Figure 5).

Biomass-density relationship

Figure 6 indicates a positive relationship between biomass and density (on linear scale axes) for the *Sargassum* population as a whole. Recorded biomass reached a peak of approximately 350 g DW/m^2 for the range of densities sampled. When *S. elegans* and *Sargassum* sp. 1 were considered separately (Figure 7) it can be seen that *S. elegans* reached a higher density but not a higher biomass than *Sargassum* sp. 1.

Reproductive phenology

The *Sargassum* population attained a reproductive peak in the cooler autumn/winter months and a minimum during the warmer months (Figure 8). This was also the pattern of *S. elegans* (and contrary to the pattern of biomass change; Figure 4). *Sargassum elegans* always had a greater percentage of thalli which were reproductive than *Sargassum* sp. 1 (Figure 8). The percentage of reproductive thalli of *Sargassum* sp. 1 never exceeded 25% at any time, whereas reproductive percentages of greater than 50% were recorded for *S. elegans*. The pattern of occurrence of reproductive thalli was more variable for *Sargassum* sp. 1 than for *S. elegans*.

Alginate yield

Sargassum sp. 1 consistently yielded a greater percentage of sodium alginate than *S. elegans* (Figure 9). Both species displayed the same seasonal pattern with a peak in the cool winter months and a decrease in the warm summer months (Figure 9). This pattern is the opposite to that recorded for biomass and corresponds to that recorded for reproductive thalli.

Discussion

Standing stock, density, reproduction and temperature

Seasonal variations in phenological properties such as density, biomass and reproduction have been reported for *Sargassum* by several authors and have differed between temperate and tropical areas in terms of the seasonality of the changes (e.g. Ang, 1984; De Wreede, 1976; Deysher, 1984; Glenn et al., 1990; Kendrick, 1993; Kendrick & Walker, 1994; Koh et

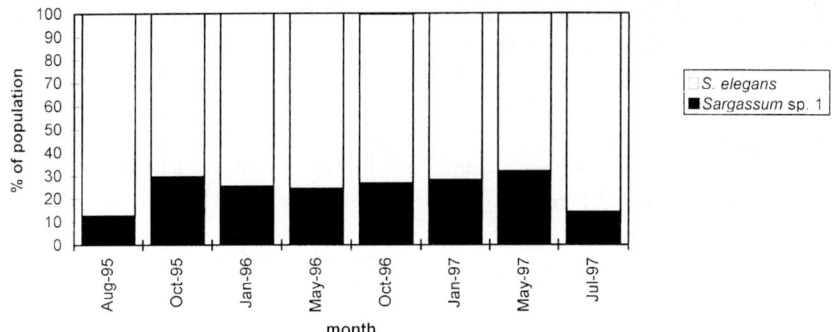

Figure 2. Relative percentage contributions of *Sargassum elegans* and *Sargassum* sp. 1 to the *Sargassum* spp. population at Reunion Rocks.

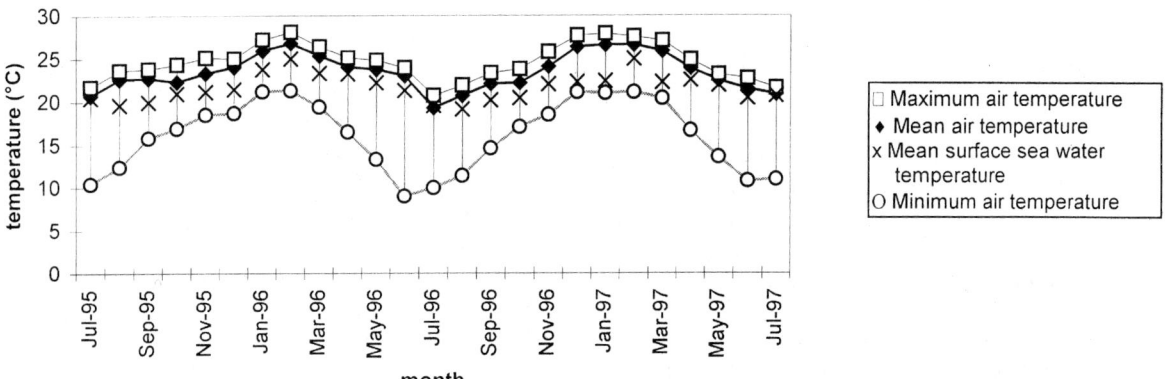

Figure 3. Maximum, minimum and mean air temperature recorded at 10:00 h at Durban, and mean surface seawater temperature recorded at Isipingo Beach.

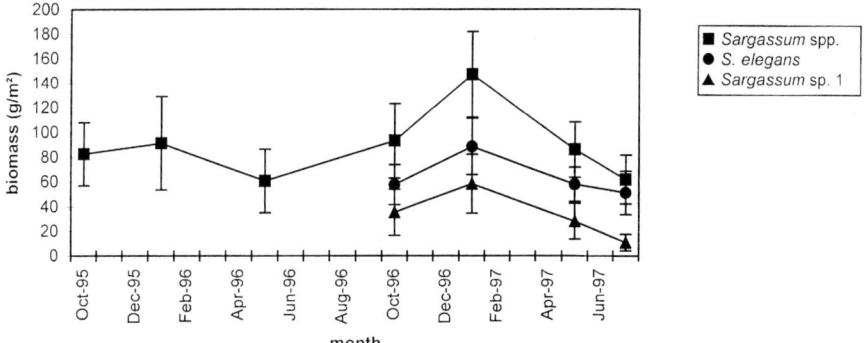

Figure 4. Seasonal variability of total *Sargassum* spp., *S. elegans* and *Sargassum* sp. 1 standing stock (± S.E.) recorded at Reunion Rocks.

al., 1993; Largo et al., 1994; McCourt, 1984a, b; Trono & Lluisma, 1990; Trono & Tolentino, 1993). In general, the biomass levels for *Sargassum* species at Reunion Rocks appear to match the lower to mid-biomass levels recorded previously for tropical species of *Sargassum*. De Wreede (1976) and McCourt (1984b) generalized that *Sargassum* species in tropical regions are usually most abundant during the coolest part of the year, while the reverse was true in temperate areas. As a general rule this is substantiated by the published literature on *Sargassum* phenology (see reference list). Based on such a generalization it would be expected that *Sargassum* from Reunion Rocks on the sub-tropical KZN coast

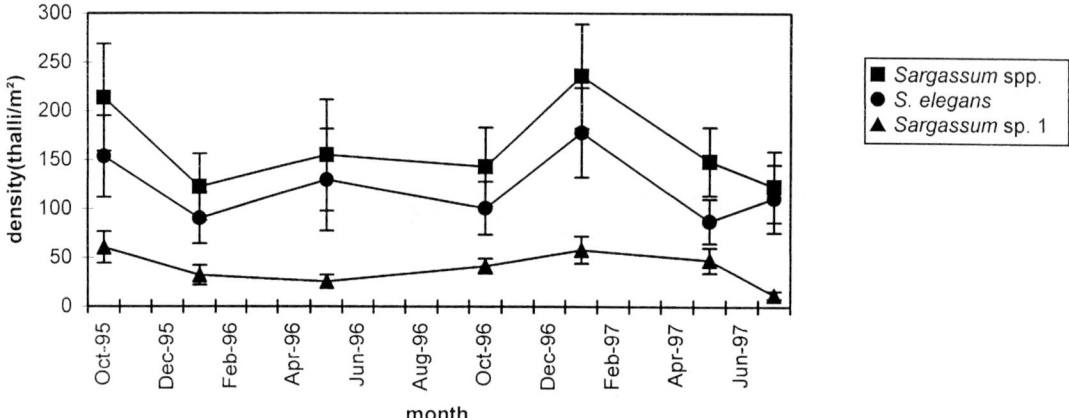

Figure 5. Seasonal variability of *Sargassum* spp., *S. elegans* and *Sargassum* sp. 1 thallus density (± S.E.) recorded at Reunion Rocks.

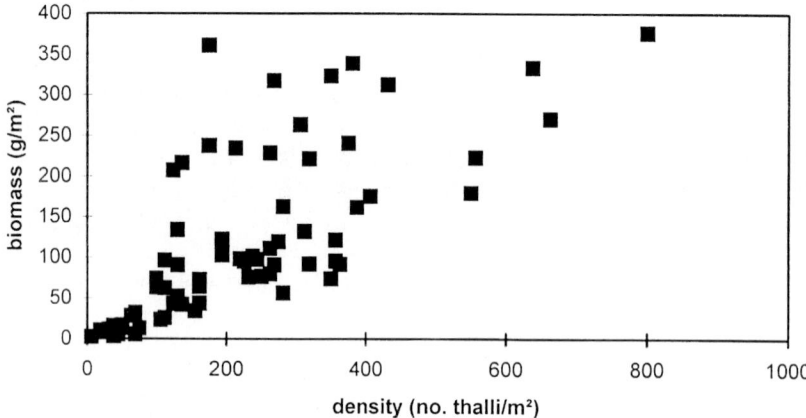

Figure 6. Bivariate plot of total *Sargassum* spp. biomass against total *Sargassum* spp. density at Reunion Rocks for the sampling period August 1995 to July 1997.

of South Africa would display biomass and density maxima between the minimum and maximum temperature months of winter and summer. The Reunion Rocks *Sargassum* population showed peak biomass during the hottest period of the year (i.e. January) and declined with lower temperatures. There have been conflicting reports of temperature effects on *Sargassum* standing stocks in tropical regions (Glenn et al., 1990). Tropical areas have been termed 'temporally uniform' and as a consequence temperature has been thought to play a less important role in regulating biomass fluctuations than in temperate areas. Storms and physico-chemical factors have been suggested as playing a more important role than temperature in regulating biomass in tropical areas (Doty, 1971; Largo et al., 1994; Trono and Lluisma, 1990; Trono and Tolentino, 1993). Glenn et al. (1990) found a strong correlation between temperature and biomass for tropical *Sargassum* species in Hawaii; less than 30% of the variation in biomass was explained by temperature variation when using the concurrent week's temperature; however, 65% was explained when using the third antecedent week's temperature. Air temperature variation was found to explain up to 61% of the biomass variation of *Sargassum* in the present study. A stronger relationship was recorded for air temperatures at 10:00 h as opposed to 12:00 h (the hottest time of the day). The former time corresponds to the average period of low spring daytime tide at Reunion Rocks (i.e. periods of exposure for the seaweeds) and is when

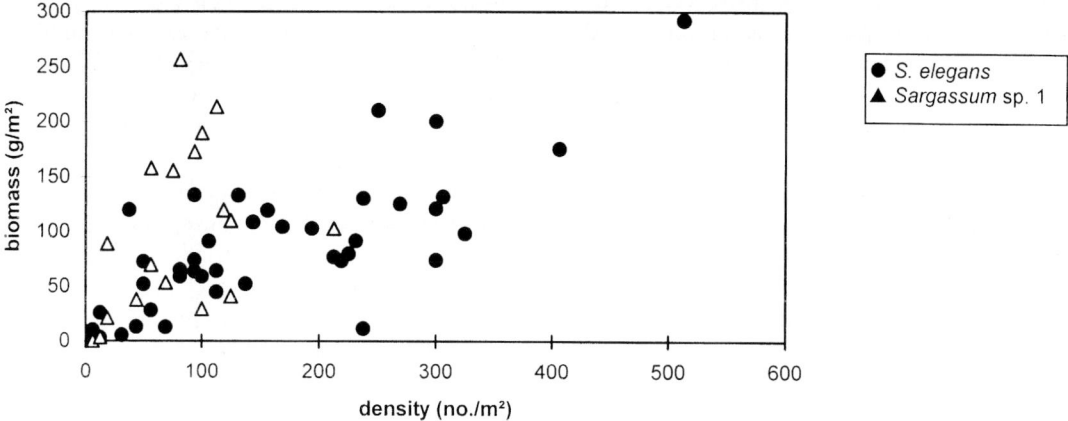

Figure 7. Bivariate plot of *Sargassum elegans* and *Sargassum* sp. 1 biomass plotted against density of each species for the sampling period October 1996 to July 1997.

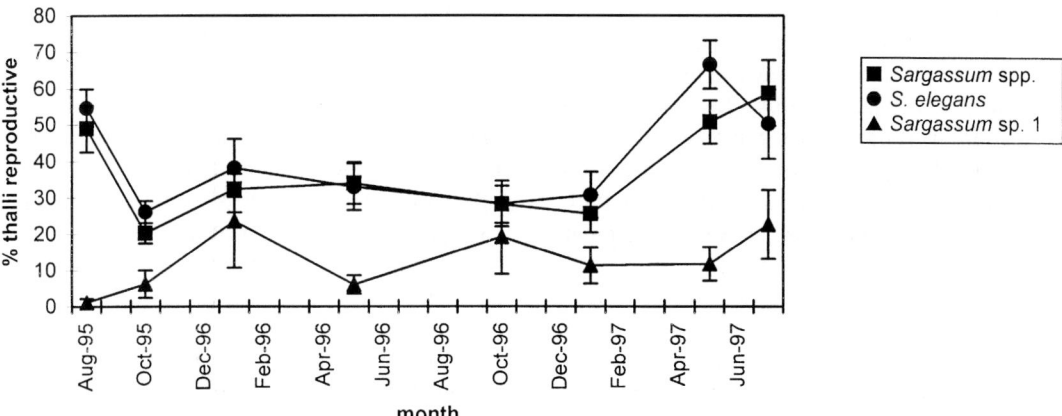

Figure 8. Seasonal variability in percentage of *Sargassum* spp., *S. elegans* and *Sargassum* sp. 1 thalli with receptacles ($\pm$ S.E.).

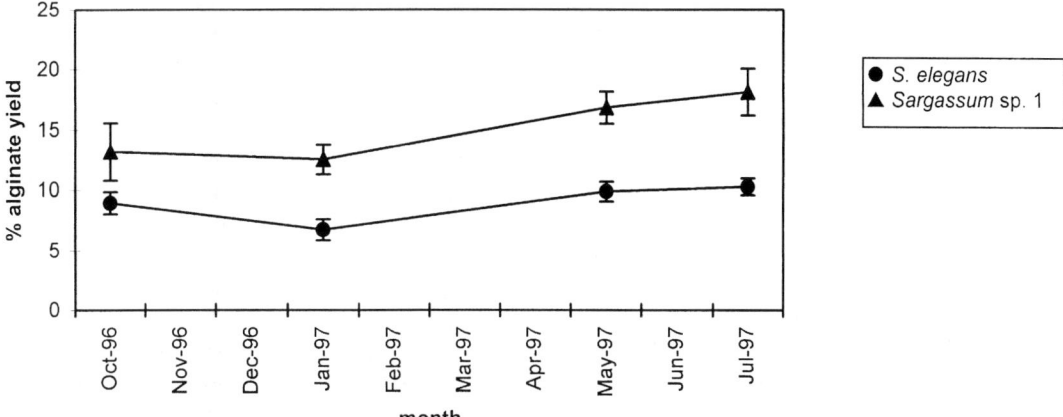

Figure 9. Seasonal variability of percentage sodium alginate yield ($\pm$ S.E.) from *S. elegans* and *Sargassum* sp. 1.

the study site is often free of extensive wave action and is without wave splash over spring tide periods. The Reunion Rocks population is only without some degree of wave splash for a period of about 2 h over low spring tides. Outside of these periods the site is exposed to constant wave action and water wash. Despite the long periods of wave-splash/submergence, sea temperature was not related to biomass change.

Temperature does not serve to increase vegetative growth alone in *Sargassum* species and the pattern of biomass change must be explained with reference to reproduction, cycles of growth and senescence of *Sargassum* populations (De Wreede, 1976; Glenn et al., 1990; McCourt, 1985). *Sargassum* populations generally have the following development phases: recruitment, growth, reproduction, senescence and regeneration. These phases can be assessed when examining the temporal variation in biomass/density along with reproductive phenology. It was indicated in De Wreede (1976) that peak abundance in *Sargassum* usually coincides with peak fertility. Kendrick (1993) showed that an intertidal species complex of *Sargassum* at Rottnest Island, Western Australia displayed peak densities during the warm summer months with a decline in the cooler months. He showed the number of reproductive thalli outside of these peak periods declining to zero or very low values. Reproduction was shown to peak just prior to peak densities and prior to the warmest months, sometime between September and December. Glenn et al. (1990) stated that when temperatures began to decline in Hawaii, *Sargassum* thalli were triggered to become fertile. McCourt (1984a) showed that three *Sargassum* species from the northern Gulf of California became fertile just after or concurrent with attainment of maximum stipe length in the spring. McCourt (1984a) suggested that peak periods of reproduction in *Sargassum* are adaptive in that the seasonal fertility pattern provides an example of adaptive timing of reproduction so that in colder temperate regions thalli are fertile in warmer months and in warmer tropical waters thalli are fertile in cooler months. Yet the present study showed peak reproductive periods (greatest proportion of thalli with receptacles) towards cooler months and at opposing times to peaks in biomass.

Sargassum elegans and *Sargassum* sp. 1 showed the same growth patterns. Individuals of *S. elegans* are on average markedly smaller than *Sargassum* sp. 1, displaying a clumped, prostrate growth form as opposed to the upright growth form of *Sargassum* sp. 1 (Gillespie & Critchley, 1998). Their higher biomass in the *Sargassum* population indicates their greater density in the population and relative contribution to the population. *Sargassum* sp. 1 has a low relative percentage of reproductive thalli when compared to *S. elegans*. Gillespie & Critchley (1997) showed that *S. elegans* had a higher reproductive investment than *Sargassum* sp. 1, yet the receptacles of *Sargassum* sp. 1 had a higher number of conceptacles per unit weight than *S. elegans*. In terms of reproduction, *Sargassum* sp. 1 is more variable than *S. elegans* which shows a clear seasonal pattern. *Sargassum* sp. 1 was found to be 100% reproductive in some quadrats and often 0% reproductive in others during the same season. This variation may have been a result of environmental factors (e.g. high wave action, higher temperatures in intertidal pools at low tides). Possibly in the study area *Sargassum* sp. 1 is close to its distribution limits (in terms of environmental conditions, e.g. wave exposure, temperature fluctuations, water availability) and consequently a lower proportion of the population becomes reproductive. De Wreede (1978) found *S. muticum* in British Columbia to be reproductive during the warm summer months. Interestingly, he found that tide pool populations displayed a higher percentage of reproductive thalli (peak at over 75%) compared to lower intertidal populations which reached a peak of about 48% in the mid zone of the lower intertidal and about 25% in the lower zone of the lower intertidal. Largo et al. (1994) showed that in the same habitat *S. polycystum* from the intertidal showed a lower percentage peak of reproductive thalli compared to *S. siliquosum* which occurred subtidally. The peak for *S. polycystum* was 43% in May. When elongate forms of *Sargassum* develop in wave exposed sites, such as the intertidal, it appears that peak reproductive percentages are reduced.

Alginate yield

Alginic acid is a structural component of the brown algae (McCandless, 1981) and as such it may be expected to reach higher levels in species living under conditions of high wave action. The fact that *Sargassum* sp. 1 yielded a greater percentage alginate, relative to dry weight, than *S. elegans*, across all seasons, may indicate that it allocated more resources to support than *S. elegans*. This is a plausible explanation in that both species grow in close association and would thus be subject to similar levels of wave action. The longer, less compact type of growth form of *Sargassum* sp. 1 is probably subjected to greater hydrodynamic forces and drag than *S. elegans* and consequently produces greater levels of alginate. Daly & Prince (1981) found that alginic acid concentrations were progressively higher in the perennial, annual and reproductive portions of *S. pteropleuron*. They concluded that the seasonal change in alginic acid concentration on a whole-plant basis coincided with the seasonal change in the composition of the *Sargassum* biomass. The fact that the seasonal trends in alginate yield are the same for both species, suggests the same controlling factors, and are also probably a reflection of the growth strategies of the species. The alginate peak for *S. elegans* matches that of peak reproduction of the *S. elegans* population. A similar result is observed for the corresponding period for *Sargassum* sp. 1. Not all studies have recorded seasonality in alginate yield. Trono & Tolentino (1993) recorded yields of 16.7–21% for *Sargassum* sp., 16–19% for *S. oligocystum* and 13–16% for *S. polycystum*. This study recorded a range of 5–11.4% for *S. elegans* and 8.5–22% for *Sargassum* sp. 1.

Biomass–density relationships

It has been suggested that positive density dependence in algal species arises from benefits associated with the closer proximity of neighbours in a manner that outweighs competitive disadvantage (Creed, 1995; Scrosati, 1996). Such positive density dependence could be due to increased protection from desiccation (Hruby & Norton, 1979), protection from physical battering (Schiel & Choat, 1980), grazing (Lubchenco, 1983) or both (Ang & De Wreede, 1992).

Neither of the *Sargassum* species in the present study was found in distinct clumpings within the site, indicating an overlap of distributions. It is generally accepted that *Sargassum spp.* with a clumped or short, compact growth form are more abundant in, or adapted for, high wave action environments, and possibly able to resist longer periods of emergence and consequent desiccation (De Paula & Oliveira, 1982; De Wreede, 1978; Largo et al., 1994; Schiel & Choat, 1980) Neither of the species studied occurred in greater abundance/density at the seaward edge of the platform (where wave action would be at its most intense) and density and biomass of both species were greatest about 80 cm away from the front edge (unpublished data). Furthermore, the overlap between distributions and peaks in abundance and biomass of both species seem to counter arguments of variation of phenotype in response to environmental factors. Both species occur in a high wave action environment, close to the front edge of the lower platform which is constantly exposed to wave action outside of periods of low spring tides. The low relative contribution of *Sargassum* sp. 1 to the population does provide some support for a suggestion that clumped forms are better suited, than more elongate upright forms, to survive in areas of wave exposure. Such observations do not preclude the existence of micro-scale factors allowing for the coexistence of the two species, such as a highly heterogeneous substratum providing for emergent and submerged habitats over distances of several centimetres. Niche differentiation is probably the ultimate cause for the coexistence of the two species in the same site. Further research to analyze niche differentiation and intra-locality variation is necessary to assess the mechanisms for the coexistence of the two species.

Acknowledgements

The authors would like to acknowledge funding received from the FRD, SANCOR and the University of the Witwatersrand Research Fund for various phases of this study.

References

Aingworth, J., 1997. An investigation of the effect of harvesting on the biomass and agar yield from *Gelidium abbottiorum* from Reunion Rocks, KwaZulu-Natal, South Africa. MSc thesis. University of the Witwatersrand, South Africa, 125 pp.

Aingworth, J. & A. T. Critchley, 1997. Simulated harvest of *Gelidium abbottiorum* R. E. Norris at Reunion Rocks, KwaZulu Natal, South Africa. S. Afr. J. Bot. 63: 404–409.

Ang, P. O. Jr., 1984. Preliminary study on the alginate contents of *Sargassum* spp. in Balibago, Calatagan, Philippines. Hydrobiologia 116/117: 547–550.

Ang, P. O. & R. E. De Wreede, 1992. Density dependence in a population of *Fucus distichus*. Mar. Ecol. Prog. Ser. 90: 169–181.

Birkett, D. A., 1984. Seasonality and growth of *Sargassum* Agardh on the Natal coast with some taxonomic notes. M.Sc. thesis, University of Natal, Durban, 266 pp.

Creed, J. C., 1995. Spatial dynamics of a *Himanthalia elongata* (Fucales, Phaeophyta) population. J. Phycol. 31: 851–859.

Critchley, A. T., R. D. Gillespie & K. W. G. Rotmann, 1998. The seaweed resources of South Africa. In Ohno, M., A. T. Critchley, D. Largo & R. D. Gillespie (eds), Seaweed Resources of the World. JICA, Japan: 413–427.

Daly, E. L. & J. S. Prince, 1981. The ecology of *Sargassum pteropleuron* Grunow (Phaeophyceae, Fucales) in the waters off South Florida. III. Seasonal variation in alginic acid concentration. Phycologia 20: 352–357.

De Paula, E. J. & E. C. Oliviera, 1982. Wave exposure and ecotypical differentiation in *Sargassum cymosum* (Phaeophyta-Fucales). Phycologia 21: 145–153.

De Wreede, R. E., 1976. The phenology of three species of *Sargassum* (Sargassaceae, Phaeophyta) in Hawaii. Phycologia 15: 175–183.

De Wreede, R. E, 1978. Phenology of *Sargassum muticum* (Phaeophyta) in the Strait of Georgia, British Columbia. Syesis 11: 1–9.

Deysher, L., 1984. Reproductive phenology of newly introduced populations of the brown alga, *Sargassum muticum* (Yendo) Fensholt. Hydrobiologia 116/117: 403–407.

Doty, M. S., 1971. Antecedent event influence on benthic marine algal standing crops in Hawaii. J. exp. mar. Biol. Ecol. 6: 161–166.

Eyre, J. & T. A. Stephenson, 1938. The South African intertidal zone and its relation to ocean currents. V. A subtropical Indian Ocean shore. Ann. Natal Mus. 9: 21–46.

Ganzon-Fortes, E. T., R. R. Campos & J. Udarbe, 1993. The use of Philippine seaweeds in agriculture. Appl. Phycol. Forum, 10: 6–7.

Gillespie, R. D. & A. T. Critchley, 1997. Morphometric studies of *Sargassum* spp. from Reunion Rocks, KwaZulu-Natal, South Africa. I. Receptacle studies. S. afr. J. Bot. 63: 356–362.

Gillespie, R. D. & A. T. Critchley, 1998. Morphometric studies of three *Sargassum* species (Sargassaceae, Phaeophyta) from KwaZulu-Natal, South Africa. II. Thallus morphology and variability. Submitted for publication.

Gillespie, R. D., J. Aingworth & A. T. Critchley, 1996. The Seaweed resources of KwaZulu-Natal, South Africa: The current state of our knowledge. In Björk, M., A. K. Semesi, M. Pedersén & B. Bergman (eds), Current Trends in Marine Botanical Research. SIDA, Sweden: 186–193.

Glenn, E. P., C. M. Smith & M. S. Doty, 1990. Influence of antecedent water temperatures on standing crop of *Sargassum* spp.-dominated reef flat in Hawaii. Mar. Biol. 105: 323–328.

Hopkins, B., 1957. Pattern in the plant community. J. Ecol. 45: 451–463.

Hruby, T. & T. A. Norton, 1979. Algal colonization on rocky shores in the Firth of Clyde. J. Ecol. 67: 65–77.

Kendrick, G. A., 1993. *Sargassum* beds at Rottnest Island: species composition and abundance. In Wells, F. E., D. I. Walker, H. Kirkman. & R. Lethbridge (eds), Proceedings of the Fifth International Marine Biological Workshop: The Marine Flora and Fauna of Rottnest Island, Western Australia. Western Australian Museum, Perth: 455–472.

Kendrick, G. A. & D. I. Walker, 1994. Role of recruitment in structuring beds of *Sargassum* spp. (Phaeophyta) at Rottnest Island, Western Australia. J. Phycol. 30: 200–208.

Koh, C.-H., Y. Kim & S.-G. Kang, 1993. Size distribution, growth and production of *Sargassum thunbergii* in an intertidal zone of Padori, west coast of Korea. Hydrobiologia 260/261: 207–214.

Krige, L. J., 1932. Peculiar little rock basins at Isipingo Beach, Natal. S. Afr. J. Sci. XXIX: 262–264.

Largo, D. B., M. Ohno & A. T. Critchley, 1994. Seasonal changes in the growth and reproduction of *Sargassum polycystum* C. Ag. and *Sargassum siliquosum* J. Ag. (Sargassaceae, Fucales) from Liloan, Cebu, in Central Philippines. Jap. J. Phycol. 42: 53–61.

Lubchenco, J., 1983. *Littorina* and *Fucus*: effects of herbivores, substratum heterogeneity, and plant escapes during succession. Ecology 64: 1116–11123.

Masuda, M., T. Ajisaka, S. Kawaguchi, H. Quang Nang & Huu Dinh Nguyen, 1993. The use of *Sargassum mcclurei* as medical tea in Vietnam. Jap. J. Phycol. 41: 39–42.

McCandless, E. L, 1981. Polysaccharides of the seaweeds. In Lobban, C. S. & M. J. Wynne (eds), The Biology of Seaweeds. Blackwell Scientific Publications, Oxford: 559–588.

McCourt, R. M., 1984a. Seasonal patterns of abundance, distributions, and phenology in relation to growth strategies of three *Sargassum* species. J. exp. mar. Biol. Ecol. 74: 141–156.

McCourt, R. M., 1984b. Niche differentiation between sympatric *Sargassum* species in the northern Gulf of California. Mar. Ecol. Prog. Ser. 18: 139–148.

McCourt, R. M., 1985. Reproductive biomass allocation in three *Sargassum* species. Oecologia 67: 113–117.

Schiel, D. R. & J. H. Choat, 1980. Effects of density on monospecific stands of marine algae. Nature 285: 324–326.

Scrosati, R., 1996. The relationship between stand biomass and frond density in the clonal alga *Mazaella cornucopiae*. Hydrobiologia 326/327: 259–265.

Trono, G. C. Jr. & A. O. Lluisma, 1990. Seasonality of standing crop of a *Sargassum* (Fucales, Phaeophyta) bed in Bolinao, Pangasinan, Philippines. Hydrobiologia 204/205: 331–338.

Trono, G. C. Jr. & G. L. Tolentino, 1993. Studies on the management of *Sargassum* (Fucales, Phaeophyta) bed in Bolinao, Pangasinan, Philippines. Korean J. Phycol. 8: 249–257.

Population and alginate yield and quality assessment of four *Sargassum* species in Negros Island, central Philippines

H. P. Calumpong[1,2], A. P. Maypa[1] & M. Magbanua[2]
[1]*Silliman University Marine Laboratory, Dumaguete City, Philippines 6200*
E-mail: mlsucrm@mozcom.com
[2]*Biology Department, Silliman University, Dumaguete City, Philippines 6200*

Key words: Philippines, *Sargassum*, growth stages, sodium alginate, yield, viscosity

Abstract

Four species of *Sargassum* in Negros Island, central Philippines, were monitored for biomass, growth stages, alginate yield and quality. All four species exhibited different zonation, fertility patterns and seasonality in growth and biomass. Mean biomass differed significantly for all species.

Each species followed different patterns in alginate yield and viscosity in relation to growth stages. Highest yields were obtained from *S. ilicifolium* during its reproductive stage, *S. feldmannii* and *S. polycystum* during secondary growth while no significant difference in yield was found between different stages of *S. cristaefolium*. Viscosity was highest in *S. ilicifolium* and *S. polycystum* during secondary growth. No significant difference in viscosity was found between the different stages of *S. feldmannii* and *S. cristaefolium*.

Introduction

Alginic acid extracted as sodium alginate is the major polysaccharide of brown seaweeds. It is used as a thickener, emulsifier, stabilizer and gel-forming agent in industries (Laserna et al., 1982; Sumera et al., 1992). In the Philippines, a potential source of alginate is *Sargassum*, an alga primarily used for fertilizer and animal feeds. Laserna et al. (1982) studied the yield and quality of alginate produced by *Sargassum polycystum* C. Ag. Monthly alginate yield and quality of different Philippine species have also been assessed in relation to seasonality: *S. crassifolium* J. Ag., *S. paniculatum* J. Ag. and two unidentified *Sargassum* species (Sumera et al., 1992). However, the growth stage of the algae used for these alginate studies was not specified.

Members of the genus *Sargassum* exhibit a high degree of age-dependent morphological variations (Ang & Trono, 1987; Chiang et al., 1992; Kilar et al., 1992). Growth stages of tropical *Sargassum* consist of embryonic, growing, reproductive and senescent or dieback stages (Ang, 1982, 1985; Trono & Tolentino, 1992).

The objectives of this study are to assess the yield and quality of alginate in relation to growth stages of *S. cristaefolium* C. Ag., *S. feldmannii* Pham-Hoang, *S. ilicifolium* (Turn.) C. Ag., and *S. polycystum*. Observations on zonation and morphological changes were made and biomass was quantified.

Study sites

Three study sites were chosen along the east coast of Negros Island. Lag-it, Bais Bay (9° 36′ N, 123° 10′ E) is a large intertidal flat in front of a mangrove forest. This area is covered with mixed seagrasses and algae. The dominant substratum is sand-silt mixture. Lala-an, San Jose (9° 25′ N, 123° 15′ E) has a narrow intertidal zone, 50 m wide extending 3 km along the shoreline. The shoreline and intertidal zone have volcanic and limestone substrates. The subtidal zone has a coral reef. Malo, Siaton (9° 6′ N, 122° 54′ E) is characterized by a narrow and rocky intertidal zone with a subtidal coral reef. Surge channels are present as a result of strong southwest monsoon winds.

Materials and methods

To determine abundance, cover and zonation, a systematic sampling method was employed. Two to four transect lines, 20 m apart, were laid from the shore seaward. Transect line length varied among sites, between 50 m to 150 m, depending on the width of the *Sargassum* bed. Permanent quadrats of 0.25 m^2 and subdivided into 25 subquadrats were laid at every 10 m interval along each transect. They were used for the monthly monitoring of the algal cover and growth stages.

Biomass was monitored monthly from August 1992 to July 1993. Ten 0.25 m^2 quadrats were randomly thrown within the *Sargassum* bed. All algae within these quadrats were collected, sorted by species, cleaned and oven dried at 60 °C to constant dry weight. Biomasses are expressed on the algal dry weight basis.

Alginate was extracted from six samples using a method modified from Laserna et al. (1982). Two hundred grams of cleaned seaweeds were soaked in 2.0% formalin for 24 h after harvest then washed and soaked again in 0.2 N HCl for another 24 h, then washed again. The samples were then soaked in 2.0% Na_2CO_3 for 3 h, blended and filtered through three layers of cheesecloth. The filtrate was poured into isopropyl alcohol to precipitate the alginate. The sodium alginate recovered was placed in an oven at 60 °C for drying to constant weight.

Sodium alginate viscosities at 1.5% (w/w) aqueous (distilled water) solutions were measured from six samples at room temperature with a Brookfield Viscometer, spindle No. 2 at 60 rpm.

One-Way Analysis of Variance was used to analyze differential biomass, alginate yield and viscosity. Bonferroni Post Hoc test (Neter et al., 1990) ranked the differences between means. Differences in biomass were analyzed monthly, while alginate yield and viscosity were analyzed according to observed differences in morphology in relation to growth stages: (1) primary growth (absence of secondary laterals); (2) secondary growth (presence of secondary laterals and bushy in appearance); (3) reproductive (presence of receptacles); and (4) senescent (absence of most 'blades' and other structures aside from stipe and holdfast and presence of epiphytes). Yield and viscosity of different species were also compared using the Mann-Whitney U test.

Ang & Trono (1987), Noro et al. (1994) and Trono (1992, 1994) were used as taxonomic references for

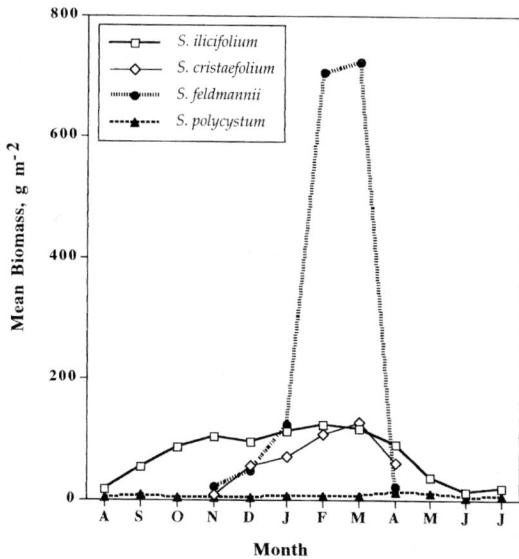

Figure 1. Monthly mean biomass of four *Sargassum* species in Negros Island at the following sites: *S. polysystem* at Lag-it, Bais Bay; *S. ilicifolium* S. and *cristaefolium* at Lala-an, San Jose; *S. feldmannii* at Malo, Siaton.

the identification of specimens.

Results

Abundance, cover, biomass and zonation

All studied *Sargassum* species exhibited different zonations, fertility patterns and seasonality in growth and biomass. Monthly mean biomass of all four species were significantly different ($p < 0.05$) indicating seasonality in biomass production.

Sargassum polycystum formed a monospecific bed extending from the upper intertidal to subtidal in Lagit, Bais Bay. Biomass and cover were highest during the months of March to May, peaking in May (11.3 ± 0.5 g m^{-2}, 10-15% cover; Figure 1). This coincided with its reproductive stage. Lowest biomass was observed in October (3.6 ± 0.2 g m^{-2}) during primary growth stage.

Lala-an, San Jose exhibited a multispecies *Sargassum* bed. *Sargassum ilicifolium* was found in the upper intertidal to the subtidal zones, *S. cristaefolium* and *S. oligocystum* co-occurred in the lower intertidal to the subtidal areas and a fourth unidentified species of *Sargassum* (not included in the alginate analysis) was limited to the upper intertidal zone, where it was totally exposed during low tide. Primary growths of *S. ilicifolium* were observed in June while *S. cristaefolium* underwent degeneration at this time. Biomass of *S. ilicifolium* reached its peak in February (Figure 1) (124.5 ± 7.3 g m^{-2}) when it was in its reproductive stage which lasted from January to April. A significant decrease in biomass was observed by late May (37.6 ± 2.6 g m^{-2}) when the population started senescence.

The growing season for *S. cristaefolium* was November to April. Peak growth was in February, as shown by the highest biomass value (109.9 ± 2.7 g m^{-2}) when the alga was reproductive (reproductive period = February to April). Primary growths of *S. oligocystum* were first observed in January and receptacles were seen in late March to late April. The fourth unknown *Sargassum* in Lala-an, San Jose on the other hand, occurred throughout the year except for the months of July and August. Receptacles were first observed in March and lasted until May.

The *Sargassum* bed of Malo, Siaton was composed of *S. polycystum* which was limited to the intertidal area, *S. cristaefolium* and *S. feldmannii* co-occurring in the lower intertidal to subtidal zone, and *S. furcatum*, strictly in the subtidal zone. *Sargassum feldmannii* first appeared in November. Its reproductive stage lasted from February to April. Biomass during these months significantly increased (Figure 1), peaking in March (722.7 ± 16.5 g m^{-2}). Lowest biomass was obtained during the months of April (22.3 ± 2.6 g m^{-2}) and November (21.5 ± 0.6 g m^{-2}) during senescent and primary growth stages, respectively. *Sargassum cristaefolium* in this area (not included in biomass and alginate extraction) followed a similar reproductive pattern. Receptacles were visible from February to April. No receptacles were seen in *S. polycystum* and *S. furcatum* for the entire sampling period.

Alginate yield and viscosity

Alginate was extracted from the dominant species, determined by their highest cover and biomass at each study site. Alginate yield (Figure 2) and viscosity (Figure 3) results of four *Sargassum* species, were grouped according to their growth stages.

Results of ANOVA indicated significant differences in alginate yield between stages of *S. polycystum*, *S. ilicifolium* and *S. feldmannii* (Table 1). The Bonferroni Post Hoc Test revealed that secondary growths of both *S. polycystum* (2.7 ± 0.2 g), and *S. feldmannii* (3.6 ± 0.3 g) yielded the highest amount

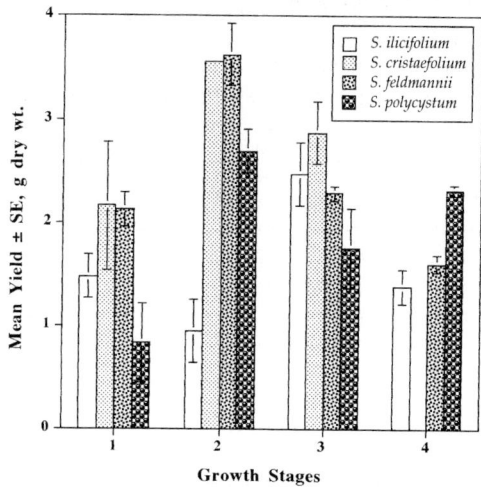

Figure 2. Mean alginate yield ($n = 6$) of four *Sargassum* species in Negros Island (sites as in Figure 1). *S. cristaefolium* was not seen during the senescent stage. Growth stages: 1 = primary growth; 2 = secondary growth; 3 = reproductive; 4 = senescent.

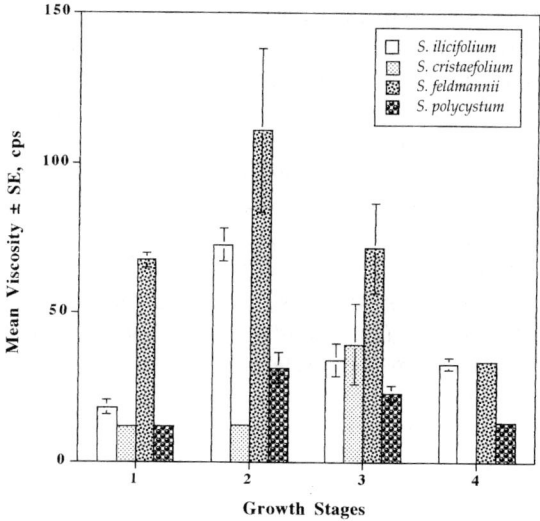

Figure 3. Mean viscosity ($n = 6$) of four *Sargassum* species in Negros Island (sites as in Figure 1). *S. cristaefolium* from Lala-an, San Jose only; was not seen during the senescent stage. Growth stages: 1 = primary growth; 2 = secondary growth; 3 = reproductive; 4 = senescent.

of sodium alginate, while primary growth of *S. polycystum* (2.7 ± 0.2 g) and senescent *S. feldmannii* (1.6 ± 0.1 g) yielded the lowest. In contrast, *S. ilicifolium* exhibited its lowest yield during secondary growth (1.0 ± 0.3 g) and highest during its reproductive stage (2.5 ± 0.3 g). No significant difference in yield was

Table 1. Results of the One Way Analysis of Variance (ANOVA) and Bonferroni Post Hoc Test (BPHT) on the yield and viscosity of sodium alginate of different *Sargassum* species in Negros I. Significant at $p < 0.05$. NS = not significant; 1 = primary growth, 2 = secondary growth, 3 = reproductive, 4 = senescence

Species	ANOVA		BPHT	
	Yield p value	Viscosity p value	Yield ranking	Viscosity ranking
S. cristaefolium	0.2846	0.5125	NS	NS
S. feldmannii	0.0003	0.1989	2>3>1>4	NS
S. ilicifolium	0.0082	0.0001	3>1>4>2	2>4>3>1
S. polycystum	0.0069	0.0093	2>4>3>1	2>3>4>1

found between all stages of *S. cristaefolium*. In addition, no significant differences were found between monthly alginate yield for each species, indicating that growth stage or morphological variation may be a more appropriate indicator of yield than month.

Mean viscosity of sodium alginate from *S. polycystum* and *S. ilicifolium* were significantly different for each stage (Table 1). Highest mean viscosity was obtained from the secondary growth of both species, with *S. ilicifolium* (72.6 ± 5.4 cps) exhibiting a higher viscosity than *S. polycystum* (31.6 ± 5.1 cps). No significant differences in viscosity were found between stages in *S. cristaefolium* and *S. feldmannii*. Mean monthly difference in viscosity was not significant within species.

Results of the comparison of the yield and viscosity of different species using the Mann-Whitney U test revealed no significant difference in yield for all four species. However, alginate from *S. feldmannii* and *S. ilicifolium* demonstrated significantly higher viscosity values than that of other species.

Discussion

The growth and development of *Sargassum* in Negros Island followed the patterns reported by Trono & Tolentino (1992), Trono & Lluisima (1990) and Ang (1982) for *Sargassum* in the Philippines. The major biomass peaks occurred when populations were fertile. The following factors were found by several workers to affect the growth stages of *Sargassum*: tide levels, temperature, (Trono & Tolentino, 1992) day length and total reactive phosphorus (Ang, 1982).

Monsoonal variation is also a major factor affecting seasonality and standing crop of algae in the tropics (Trono & Azanza-Corrales, 1981). All *Sargassum* species in our study sites were affected by the southwest monsoon which is characterized by strong winds, high wave action and very low water visibility and lasts from June to October. This was particularly evident in Malo, Siaton situated on the southwestern part of Negros Island. Low to zero biomass was recorded in this area and Lala-an, San Jose during these months. The exception was *S. polycystum* which grew year round in protected Bais Bay. All species exhibited reproductive stages during dry months.

Alginate yield and quality vary with different species (Laserna et al., 1982; Sumera et al., 1992) and different months (Sumera et al, 1992). One species, *S. cristaefolium* in Lala-an, San Jose, showed no significant difference in yield and viscosity between stages. The different stages of *S. ilicifolium* in the same site, however, exhibited differential yield and viscosity with its reproductive stage having the greatest yield and its secondary growth the most viscous alginate. *S. polycystum* in Bais Bay and *S. feldmannii* in Malo, Siaton also followed different patterns in yield and viscosity. The highest yield of both species was obtained from secondary growth. However, alginate from *S. feldmannii* did not differ in viscosity between the algal growth stages, while it was most viscous from the secondary growth of *S. polycystum*. These differences in yield and viscosity, which are clearly species-specific, may also be affected by physical and chemical factors of the area.

Among the four *Sargassum* species, a higher yield and viscosity may be obtained from *S. ilicifolium* and *S. feldmannii*. Considering biomass and phenolic content, however, *S. feldmannii* is the preferred species. Only this species had a light to almost white extract, indicating a low phenolic content. In addition, the viscosity of the alginate extracted from *S. feldmannii* was not affected by the age of the alga, and thus can be harvested any time. Harvesting, however, should be regulated in order to avoid depletion of wild stocks. The best time would be immediately after reproduction and before the thalli become senescent. Holdfasts and a few centimetres of the stipe should be left during harvest to allow the algae to regenerate.

Acknowledgements

The authors thank FMC Cebu and Mr. Farley Baricuatro for the use of their viscometer. We also gratefully acknowledge Jacinta Radones-Lucañas for her laboratory and field assistance, Grace Ozoa, Pablina Cadiz and Jasper Leif Maypa for their help in the field, and Laurie J. H. Raymundo for reviewing this paper and for her valuable advice in our data analysis. This research was funded by the Philippine Council for Aquatic and Marine Research and Development.

References

Ang, P. O., Jr., 1982. Studies on the taxonomy of the genus *Sargassum* (Phaeophyta) and the phenology and some aspects of the ecology of *Sargassum siliquosum* J. Ag. and *Sargassum paniculatum* J. Ag. in Balibago, Batangas. Master Thesis, College of Arts and Sciences, University of the Philippines, Diliman, Quezon City, Philippines.

Ang, P. O. Jr., 1985. Phenology of *Sargassum siliquosum* J. Ag. and *S. paniculatum* J. Ag. in reef flat of Balibago, Calatagan, Philippines. Proc. int. Coral Reef Symp. 5: 51–57.

Ang, P. O., Jr. & G. C. Trono Jr., 1987. The genus *Sargassum* (Phaeophyta, Sargassaceae) from Balibago, Calatagan, Philippines. Bot. mar. 30: 387–397.

Chiang, Y. M., T. Yoshida, T. Ajisaka, G. Trono Jr., C. K. Tseng & L. Baoren, 1992. Distribution and variation in *Sargassum polycystum* C.A. Agardh (Fucales, Phaeophyta). In Abbott, I. A. (ed.), Taxonomy of Economic Seaweeds with Reference to some Pacific and Western Atlantic Species. 3. California. Sea Grant College, Univ. California, La Jolla, Calif.: 211–231.

Kilar, J. A., M. D. Hanisak & T. Yoshida, 1992. On the expression of phenotypic variability: Why is *Sargassum* so taxonomically difficult? In Abbott, I. A. (ed.), Taxonomy of Economic Seaweeds with Reference to some Pacific and Western Atlantic Species. 3. California. Sea Grant College, Univ. California, La Jolla, Calif.: 95–117.

Laserna, E. C., R. L. Veroy, A. H. Luistro, N. E. Montaño & G. J. B. Cajipe, 1982. Alginic acid from brown seaweeds. *Kalikasan*, Philipp. J. Biol. 11: 51–56.

Neter J., W. Wasserman & M. H. Kutner, 1990. Applied linear statistical models. Irwin, Boston, Mass. 1171 pp.

Noro, T., T. Ajisaka & T. Yoshida, 1994. Species of *Sargassum* subgenus *Sargassum* (Fucales) with compressed primary branches. In Abbott, I. A. (ed.), Taxonomy of Economic Seaweeds with Reference to some Pacific and Western Atlantic Species. California. 4. Sea Grant College, Univ. California, La Jolla, Calif.: 23–31.

Sumera, F., M. Pepito & E. Corpuz, 1992. Sodium alginate from Philippine *Sargassum:* dependence of its viscosity and yield on seasonality and pretreatment. In Calumpong, H. P. & E. G. Meñez (eds), Proc. 2nd RP-USA Phycology Symp. Workshop. Philippine Council for Aquatic and Marine Research and Development, Los Baños, Laguna: 93–101.

Trono, G. C., Jr., 1992. The genus *Sargassum* in the Philippines. In Abbott, I. A. (ed.), Taxonomy of Economic Seaweeds with Reference to some Pacific and Western Atlantic Species. 3. California. Sea Grant College, Univ. California, La Jolla, Calif.: 43–93.

Trono, G. C., Jr., 1994. New species of *Sargassum* from the Philippines. In Abbott, I.A. (ed), Taxonomy of Economic Seaweeds with Reference to some Pacific and Western Atlantic Species. 4. California. Sea Grant College, Univ. California, La Jolla, Calif.: 3–7.

Trono, G. C., Jr. & R. Azanza-Corrales, 1981. The seasonal variation in the biomass and reproductive states of *Gracilaria* in Manila Bay. Proc. int. Seaweed Symp. 10: 743–747.

Trono, G. C., Jr. & A. O. Lluisima, 1990. Seasonality of standing crop of a *Sargassum* (Fucales, Phaeophyta) bed in Bolinao, Pangasinan, Philippines. Hydrobiologia 204/205: 331–338.

Trono, G. C., Jr. & G. L. Tolentino, 1992. The reproductive phenology of *Sargassum* spp. (Fucales, Phaeophyta) in Bolinao, Pangasinan. In Calumpong, H. P. & E. G. Meñez (eds), Proc. 2nd RP-USA Phycology Symp. Workshop. Philippine Council for Aquatic and Marine Research and Development, Los Baños, Laguna: 181–193.

The ecological effects of mining discharges on subtidal habitats dominated by macroalgae in northern Chile: population and community level studies

J. A. Vásquez[1], J. M. A. Vega[1], B. Matsuhiro[2] & C. Urzúa[2]
[1]*Dept. Biología Marina. Universidad Católica del Norte Casilla 117 Coquimbo, Chile*
E-mail: jvasquez@socompa.cecun.ucn.cl
[2]*Dept. Ciencias Químicas, Universidad de Santiago de Chile Casilla 307 Correo 2 Santiago, Chile*

Key words: coastal pollution, copper pollution, iron pollution, heavy metals, macroalgae, macroinvertebrates, *Lessonia*, *L. trabeculata*

Abstract

In 1996/97, a study was carried out to evaluate several variables related to the potential ecological effects of soluble copper and iron released as the result of direct dumping of mine tailing into the littoral zone of the Pacific Ocean off northern Chile. Variables studied included:

1. content of copper and iron in mining discharges;
2. distribution of Cu and Fe in seawater at study sites;
3. distribution of Cu and Fe in the seaweed *Lessonia trabeculata* and in its alginates (obtained from frond, stipe and holdfast);
4. alterations in *Lessonia* morphology; and
5. variability in the macroinvertebrate community associated with *Lessonia* holdfasts and the inter-plant subtidal community.

The variables were evaluated for different depths and distance from discharge sources, as well as for control areas far from any mining activity. It was observed that tailings from copper mining caused more ecological perturbation than those from iron mining; however, the lack of organisms very close to tailing discharges could be caused by stress produced by loading of fine sediments rather than by the presence of heavy metals. This work shows that the concentrations of heavy metals in seawater, plants, and alginates of *Lessonia* in contaminated and control sites were highly variable, decreasing with depth and distance from the contamination source. What were originally considered as control areas far from anthropogenic metal release, showed high concentration of heavy metal due to natural orogenetic processes occurring along the Chilean coast.

Introduction

In studies of the effects of mine tailings on littoral marine communities in the south-east Pacific, the general trend has been to refer almost exclusively to the comparative occurrence of heavy metals in different organisms (algae, invertebrates and fish) and sediments (Boré et al., 1989; Trucco et al., 1990; Vermeer & Castilla, 1991; Lecaros & Astorga, 1992; Ahumada, 1994; Vásquez & Guerra, 1996). However, most of these studies are difficult to compare due to the lack of methodological standardization and the lack of data for uncontaminated areas. This latter point is particularly relevant in northern Chile, where natural orogenetic processes, volcanic activity and climatic conditions may produce elevated levels of heavy metals in the environment, unrelated to mining activity (Vila & Sillitoe, 1991; Vásquez & Guerra, 1996).

Although the literature suggests that contamination by heavy metals may cause dramatic ecological impacts on coastal marine environments, these effects are rarely documented except in cases of overt contamination (Bryan & Langston, 1992). As suggested by Morrisey et al. (1996), the lack of evidence is partly due to the way in which these perturbations have been investigated. Traditionally, the effects of contaminants in marine environments have been approached in two ways:

1. ecotoxicological studies may identify the effects of contaminants under controlled laboratory conditions, and
2. field studies may show correlation between distribution of contaminants and the composition of benthic faunal assemblages which demonstrate environmental variability.

Unfortunately, the former type of study does not reproduce the range of potential environmental factors that regulate the magnitude of the effect. The latter type of study does not generally include the intermediate effects proper to an environmental gradient of perturbation distribution, and the eventual response of a population and/or communities to this gradient. Therefore, the actual impact of these contaminants in nature can differ widely from that reported under laboratory conditions. In this context, Bryan & Langston (1992) suggested that the effects of Cu and Zn on species distribution, although evident, were not as apparent as would be predicted under controlled laboratory conditions. For example, in the field, some deleterious effects on benthic organisms that could be directly attributed to specific effects of metallic contaminants are very scarce compared to those reported for equivalent laboratory assays.

Chile is a country of enormous mining potential, in which copper is the main export product accounting for over 60% of the total income. Between 18° and 30° S, mineral deposits include over 25% of the copper, 40% of the molybdenum, and 30% of the lithium world reserves. Moreover, in the same zone there are important deposits of Au, Fe, Ag, Mn, Co, Hg, Tu, Pb and Zn (Corvalán, 1985). Although there is evidence of serious environmental impact by the deposition of fine sediments resulting from Cu and Fe mining (e.g. Chañaral in northern Chile, 25 000 tons d^{-1} from 1939 to 1975), few data have been obtained on the effects of such discharges in marine communities (Castilla & Nealler, 1978; Castilla, 1983; Vásquez & Guerra, 1996). These authors reported changes in the species composition of intertidal areas affected by Cu tailings.

Between 18° and 42° S, subtidal rocky bottom environments to 35 m depth are dominated by *Lessonia trabeculata* (Villouta & Santelices, 1986; Vásquez, 1992). Holdfasts of *L. trabeculata*, like those of other brown macroalgae, are micro-habitats, which promote larval settlement, recruitment, and physical shelter against bottom currents and predators. A high diversity of macro-invertebrates and fish communities is associated with these habitats (Vásquez & Santelices, 1984; Ojeda & Santelices, 1984; Vásquez, 1993). Holdfasts of brown algae, a spatially and naturally delimited habitat, have been widely used as a biological model for the functioning of discrete communities (Vásquez & Santelices, 1984; Snider, 1985; Vásquez, 1993), and as study units to assess the effects of contamination (Smith & Simpson, 1993; Smith, 1996). In northern Chile, these environments have often been affected by tailings of Cu and Fe mining. Considering the logistical problems met when experimentally injecting (see Morrisey et al., 1996) heavy metals into exposed rocky environments, a valid method to experimentally study the effect of mine tailings in these communities is to determine the distribution of the contaminants over a gradient from a contamination focus, evaluating the effects on populations and/or communities related to their distance from the focus, assuming a dilution effect. This natural experiment may allow determination of *in situ* effects of mining tailings, without separating out possible co-variation effects from natural environmental variables such as temperature, salinity, pH, wave exposure and water movement. The method infers the existence of a gradient in the magnitude of the perturbations to which the biological communities are subjected to in the natural environment.

The present work characterizes some subtidal environments contaminated by Cu and Fe mining through the quantification of these cations in sea water and in *L. trabeculata* plants and alginates. It also evaluates the temporal and spatial effects of Cu and Fe tailings on *Lessonia* populations, and on rocky subtidal communities in northern Chile.

Material and methods

Rocky subtidal environments receiving solid and liquid tailings from copper (Michilla 22° 48′ S) and iron mining (Chapaco 28° 28′ S), were sampled seasonally between June 1996 and August 1997. Simultaneously, two localities 60 km north of Chapaco (Carrizal Bajo 28° 05′ S) and 100 km south of Michilla (Caleta Constitución 23° 25′ S) were evaluated as unexposed control areas (Figure 1). Samples of seawater and of the dominant subtidal populations and communities were taken in the study areas, and at permanent stations located along a distance gradient from the contamination source (0, 1, 2, 3 and 5 km) and a depth gradient (0, 10, 20 and 30 m) facing the mining discharges at Michilla and Chapaco. The subtidal localities correspond to rocky areas exposed to predominant SW winds, with coastal currents of northern direction and with bottom communities dominated by *L. trabeculata* (Vásquez, 1992).

Chemically, the study areas were characterized by the concentration of Cu^{2+} (Michilla) and Fe^{2+} (Chapaco) in bottom seawater obtained by SCUBA. The damps and outfalls that carry mining residues to the shore were simultaneously evaluated in all the study areas.

Heavy metals in seawater

The pre-concentration of the seawater samples was carried out according to Berndt et al. (1985). A 500 ml sample was filtered (0.45 μm millipore) at pH 4–5 (Na Acetate-acetic acid) and treated with 80 mg l^{-1} ammonium pyrrolidine-dithiocarbamate (APDTC) dissolved in 2 ml water. It was then vigorously stirred and filtered drop by drop through activated carbon. The metallic complexes adsorbed by the carbon were dried at 120 °C for 20 min. After the system had cooled, it was treated with 1ml HNO_3, heated and dried. Finally, the carbonaceous residue, with the metallic cations, was suspended in 1.5 ml 4.5M HNO_3 and centrifuged at 10 000 rpm for 15 min. Analytical readings were done with 200 μl aliquots of each sample by atomic absorption spectroscopy (Perkin Elmer 2380) using 1 000 μg ml^{-1} metallic Cu and Fe in 0.3 M HNO_3 (J.T. Baker-INSTRA-ANALYZED Reagent) as standards.

Heavy metals in alginates and plants of L. trabeculata

For each study area, the heavy metal content of samples of fronds, stipes and holdfasts were analysed. Additionally, the concentration of Cu and Fe were determined in alginate samples. Algae were oven-dried for 26 h at 65 °C, cut into small pieces with a plastic knife and ground in a porcelain mortar. Samples weighing approximately 2 g were ashed in a muffle furnace for 45 min at 700 °C. The ash was digested with 20 ml distilled water and filtered through Whatman N° 54 paper. The volume of the filtrate was adjusted to 50 ml with distilled water and stored in polyethylene flasks. Heavy metal concentrations were measured as above. Results are expressed as μg of metal per g of dried alga (algal content) and μg of metal per g of alginate (alginate content).

Population and community effects

In order to determine the effects of the mine tailings at the community and population level, the subtidal populations of *L. trabeculata*, the communities on hard substrata among *Lessonia* plants, and the invertebrate communities associated with *L. trabeculata* holdfasts, were sampled along the same spatial and temporal gradients where metal samples were obtained.

L. trabeculata plants were sampled *in situ*: 339 at Chapaco, 57 at Carrizal Bajo (Chapaco-control), 160 at Michilla and 50 at Caleta Constitución (Michilla control). The following measurements were taken for each plant: maximum length, holdfast basal diameter, numbers of stipes, and total weight. Frequency of reproductive plants was evaluated at each locality. Using these variables, size distribution, density, biomass, and distribution of reproductive plants representing the total of the sampled population were calculated.

The effects of Cu and Fe tailings on the structural changes of the communities associated with *L. trabeculata* holdfasts were evaluated in Michilla ($n = 53$), Caleta Constitución (100 km south of Michilla, $n = 49$), Chapaco ($n = 61$) and Carrizal Bajo (60 km north of Chapaco, $n = 57$). The holdfasts were removed by Scuba diving using crowbars, fixed in the field with 8% formaldehyde in seawater and taken to the laboratory for analysis. The fauna were extracted from the central cavity, identified when possible to the species level, counted, weighed and measured.

In each study area and throughout the distance gradient from the contamination source, the communities from hard bottoms amongst *L. trabeculata* plants were evaluated using transects of 50 m length parallel to the shore, at 10 m depth. The transects (Michilla $n = 6$, Chapaco $n = 8$) were sub-divided each 10 m, and the cover and/or density (2 replicates at random every 10 m) evaluated with 0.25 m^2 quadrats, with 100 interception points.

Spatial changes of the subtidal communities between *Lessonia* plants were evaluated by monitoring species richness, diversity, total density, total biomass and cover. Due to the lack of seasonal responses with mining discharges, population and community variables were grouped according to the distance and depth gradients. To estimate differences in the population and community parameters, one way ANOVA and *a posteriori* Tukey tests (Sokal & Rolf, 1981)

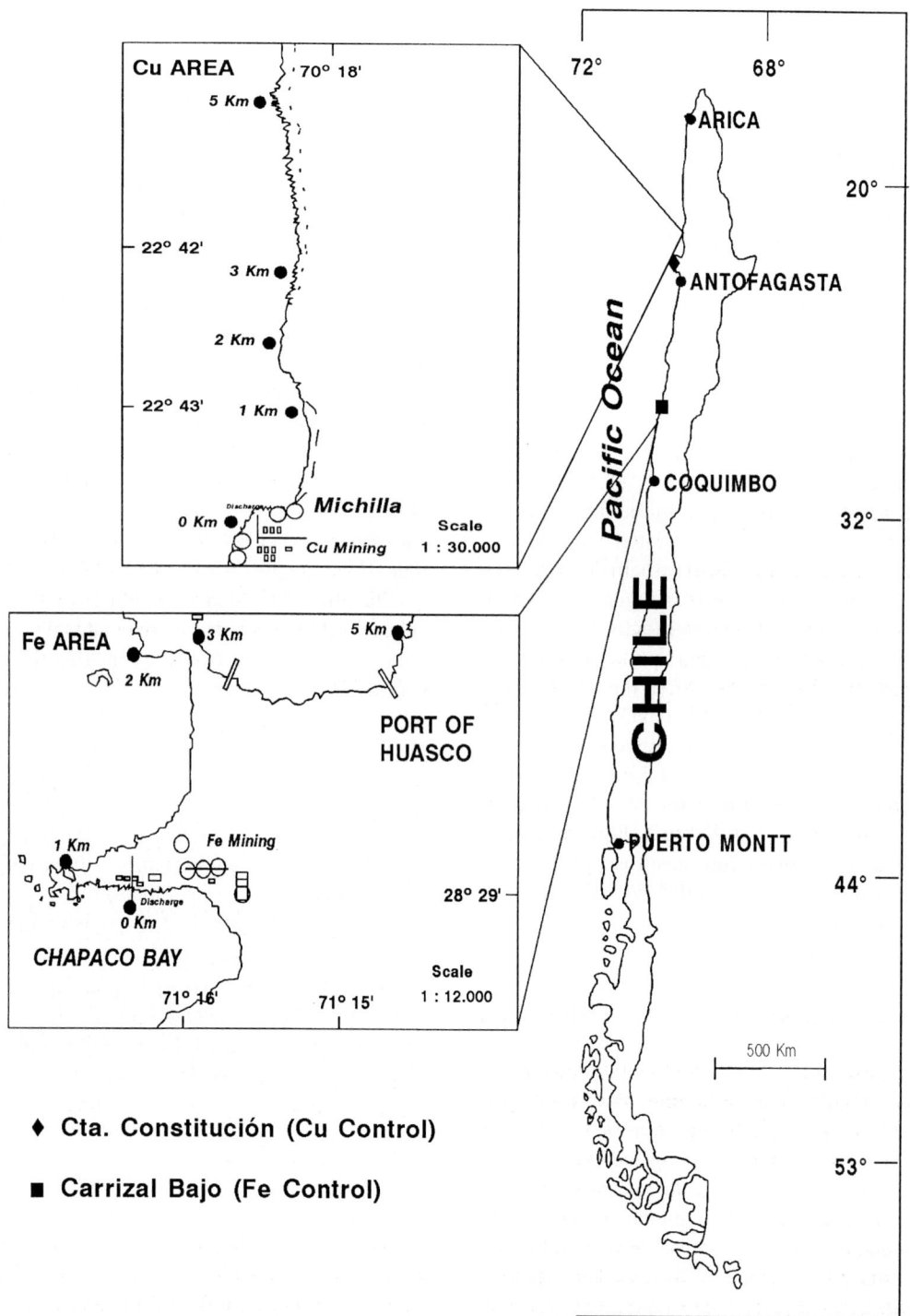

Figure 1. Location of study areas.

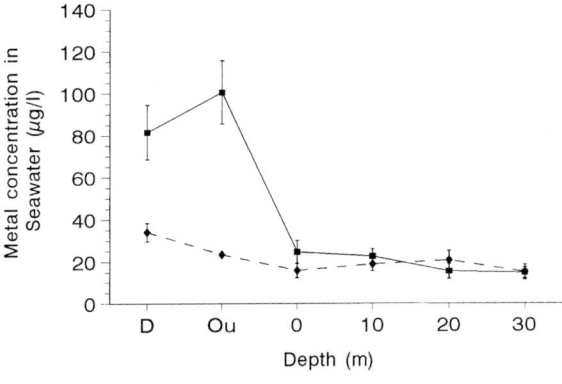

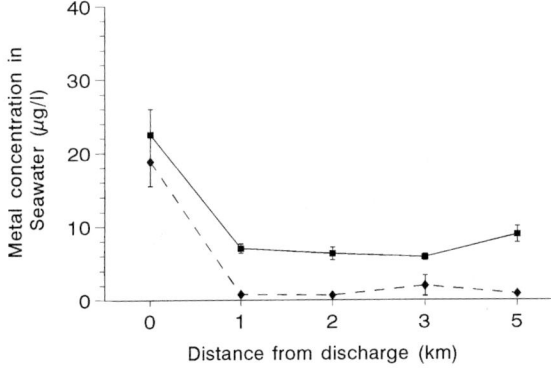

Figure 2. Variability of Cu (dashed line) and Fe (continuous line) contents in seawater along a depth (above) and distance (below) gradient from the source of mine tailings. D = dump, Ou = mining outfall. Bars = standard error.

were performed. Confidence intervals for diversity index were estimated by Jacknife (Jacksic & Medel, 1987).

Results

Heavy metals in seawater

The spatial distribution of Cu and Fe in seawater, at sites in the vicinity of mine tailings, decreased with depth (0–30 m) and with distance (0–5 km) from the contamination focus (Figure 2). At Michilla, the mean values of Cu were higher in the dam (D = 36 μg l^{-1}) and outfall (Ou = 25 μg l^{-1}) than in the intertidal areas (18 μg l^{-1}). These values did not change significantly ($p > 0.05$) until 30 m depth. Fe in seawater showed a similar pattern at Chapaco, where higher values occur in the dam and outfall (80 and 100 μg l^{-1}, respectively). These values were significantly different ($p < 0.05$) from those observed between 0 and 30 m depth, which fluctuated between 20 and 30 μg l^{-1} (Figure 2).

Between 0 and 5 km from the discharge source, Cu and Fe concentrations decreased at the two localities studied (Figure 2). Both Cu and Fe had maximum values at 0 m depth (intertidal areas), and decreased significantly ($p < 0.05$) over the first km from the contamination source. However, between 1 and 5 km from the contamination source, the values did not vary significantly ($p > 0.05$) (Figure 2).

Cu and Fe concentrations at control areas in Caleta Constitución (100 km away Michilla) and Carrizal Bajo (60 km away Chapaco) showed similar values to those of Cu at 30 m depth (23 μg l^{-1}), and were significantly higher ($p < 0.05$) than those 5 km from the contamination source. Fe concentrations (80 μg l^{-1}) in controls areas, were significantly higher ($p < 0.05$) than those observed at the limits of the depth and distance gradients at the discharge sites.

Between 0 and 30 m depth, and 0 and 5 km from the contamination source, there was no seasonal pattern of change in Cu and Fe concentrations at either Michilla or Chapaco (Figure 3). There was greater variability with depth than with distance from the mine tailings. Cu and Fe values increased in the winter months at 0 m depth and 0 km at both sites.

Heavy metals in plants and alginates

Concentrations of Cu and Fe in fronds, stipes and holdfasts of *L. trabeculata* and in alginates showed no clear pattern as a function of distance from the source of contamination (Figure 4). *Lessonia* showed variable maximum Cu concentrations in the different parts analyzed. At the control area, alginates extracted from stipes and fronds had significantly ($p < 0.05$) higher Cu values (20 μg g^{-1}) than alginates from holdfasts of the same plants (Figure 4).

At Chapaco, Fe concentrations in the different *Lessonia* structures did not vary significantly over the distance gradient studied. The high variability of metal concentrations in both the plants and extracted alginates from the study and control sites was notable (Figure 4).

Population effects

At Michilla and Chapaco, no *L. trabeculata* plants were found between 0 and 30 m depth. Within this depth range at c. 1 km from the source a fine sediment that precludes settlement and growth of algal spores and invertebrate larvae covers the rocky substratum.

The first *Lessonia* plants appeared at 3 km from Michilla, whereas a number of plants were found

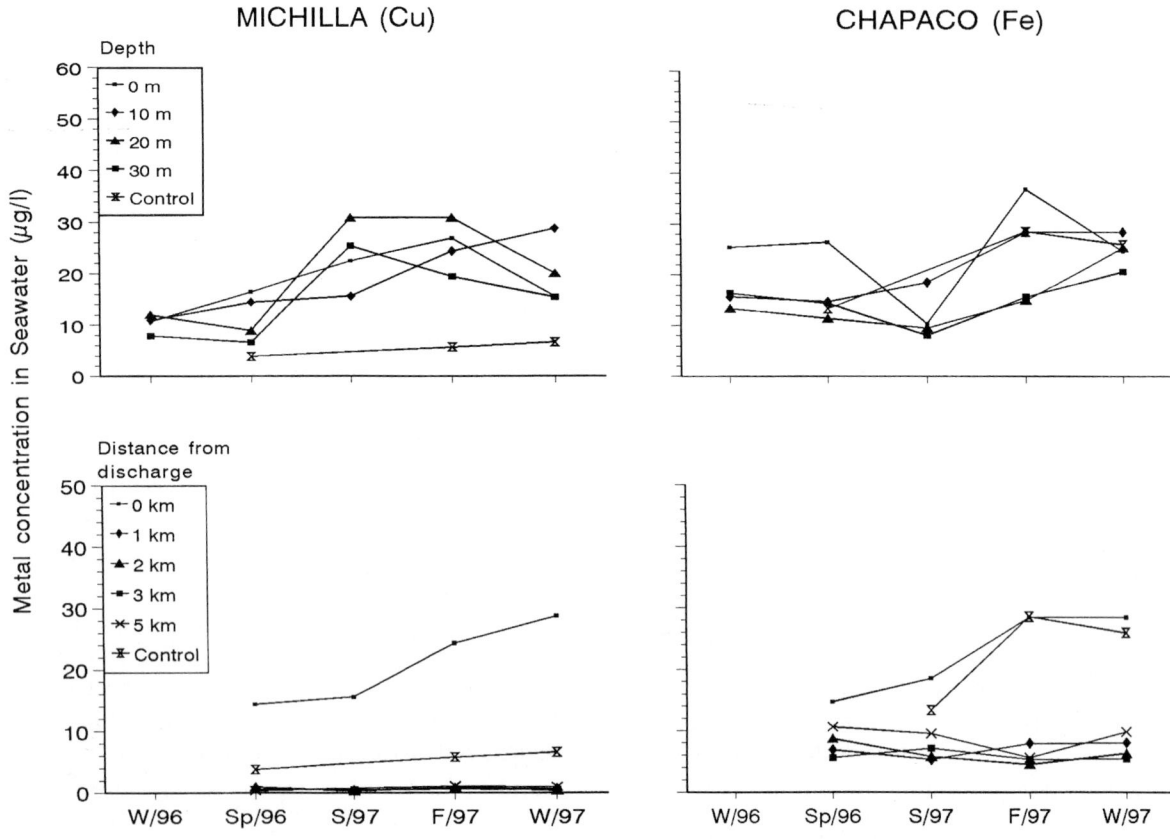

Figure 3. Seasonal variability of Cu and Fe seawater concentrations along the depth and distance gradients from the mining outfall.

only a few metres from the Fe tailing outfall at Chapaco (Figure 5). At the Michilla site, the population descriptors for *Lessonia* suggest that populations consist of individuals not exceeding 1.8 m in length and 20 cm holdfast diameter, with less than 10 stipes per plant and a mean wet weight of 5 kg. However, mean plant density is high, with over 35 plants per 10 m^2 (Figure 5).

At 5 km from Chapaco, individuals of *L. trabeculata* can exceed 2 m in length, 25 cm in mean holdfast diameter and 15 kg wet weight. The mean density increases, and the mean number of stipes per plant decreases with distance from the contamination source (Figure 5).

The higher density of juvenile plants close to the outfall of mining activities, the increase in plant size (Figure 6), and increase in the reproductive plant frequency with distance from the contamination source (Figure 7), suggest a longer survival of adult plants with decrease of the perturbation effect. Although these patterns could be produced by sediment-effect rather than heavy metal concentrations, Cu tailings produce greater perturbations than do Fe, at the population level.

Community effects

The communities of macroinvertebrates associated with *L. trabeculata* holdfasts show a significant ($p < 0.05$) reduction in number of species, density and biomass in areas closer to the Cu contamination source compared with control areas (Figure 8). Macroinvertebrate communities associated with *Lessonia* in environments contaminated by Fe tailings, show no significant differences ($p > 0.05$) with those in control areas, or with those inhabiting *Lessonia* holdfasts between 1 and 5 km from the contamination source (Figure 8). Communities on hard substrata between *Lessonia* plants showed similar responses to those observed in intra-holdfast communities at both study sites except that at 2 km from Michilla diversity on the hard bottom was high in the absence of *Lessonia* (Figure 9). Greatest effects of mining pollution were observed in areas closest to the Cu and Fe discharges.

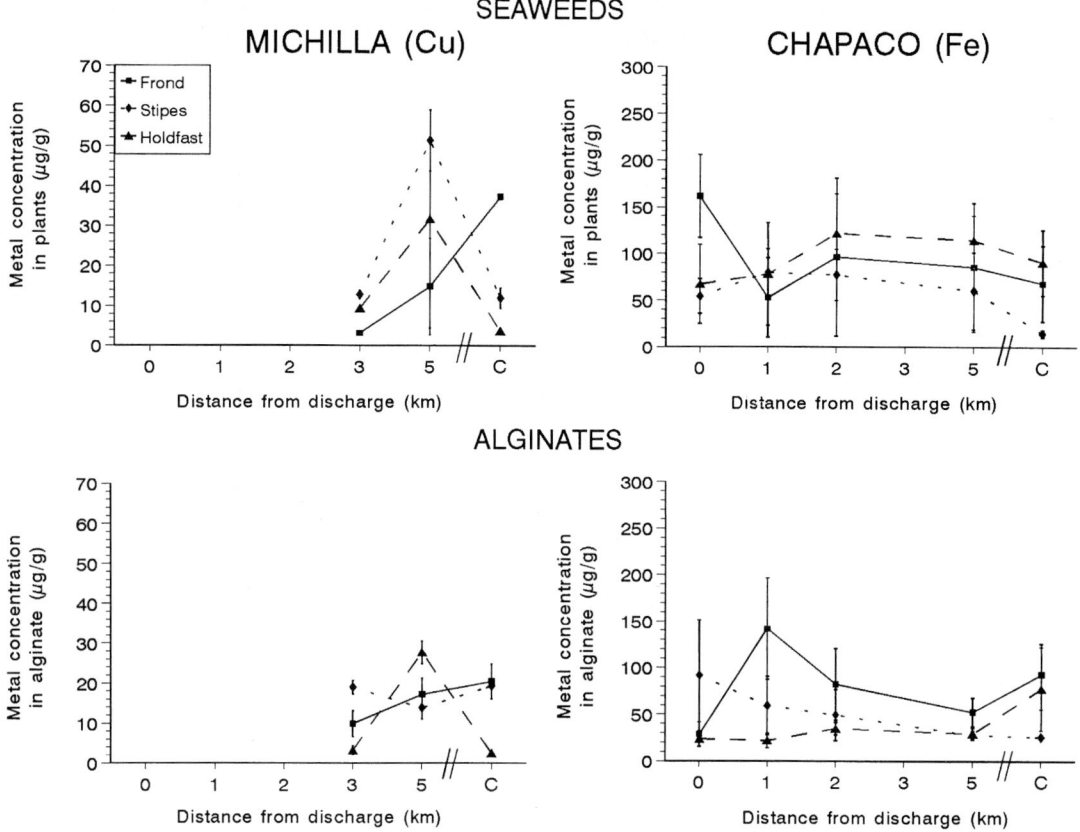

Figure 4. Content of Cu and Fe in *Lessonia trabeculata* fronds, holdfast and stipe, and alginate extracted from them, along the distance gradient. C = control area. Bars = standard error.

Discussion

This work contains the first data on the effects of mine tailings on marine subtidal populations and communities of the south-eastern Pacific. The spatial distribution of Cu and Fe in seawater, close to mine tailings, showed values that decreased with depth and with distance from the contamination source. In contrast, concentrations of Cu and Fe in fronds, stipes and holdfasts of *L. trabeculata* and in extracted alginates showed no clear pattern as a function of distance from the contamination source.

Tailings from Cu mining appear to cause more ecological perturbations than those from Fe mining. At Michilla (Cu effect) and Chapaco (Fe effect), no *L. trabeculata* plants were detected between 0 and 30 m depth. In this depth range, a fine sediment that precludes settlement and growth of benthic organisms covers the rocky substratum. The communities of macroinvertebrates associated with *Lessonia* holdfasts in the vicinity of Cu tailings show a greater reduction in species richness, density and biomass than the holdfast communities close to the Fe contamination source. Communities on hard bottoms between *Lessonia* plants showed similar responses to those observed in intra-holdfast communities at the study sites.

Even though the mining activities at Michilla between 1971 and 1994 evacuated an annual mean of 415 155 tons of solids directly to the sea, perturbations are restricted to less than 3 km distance from the contamination source. The lack of algae and other benthic organisms over the depth gradient adjacent to Cu and Fe mine tailings was probably caused by the accumulation of fine sediments rather than the heavy metal content *per se*. The 9.2×10^6 Mt of sediments dumped into the sea over 19 years at Michilla is undoubtedly the cause of the damage to the subtidal benthic communities at the study area. Castilla & Nealler (1978) have documented similar ecological damage for intertidal communities near the El Salvador copper mine in northern Chile. Resuspension of sediments during winter storms is one important

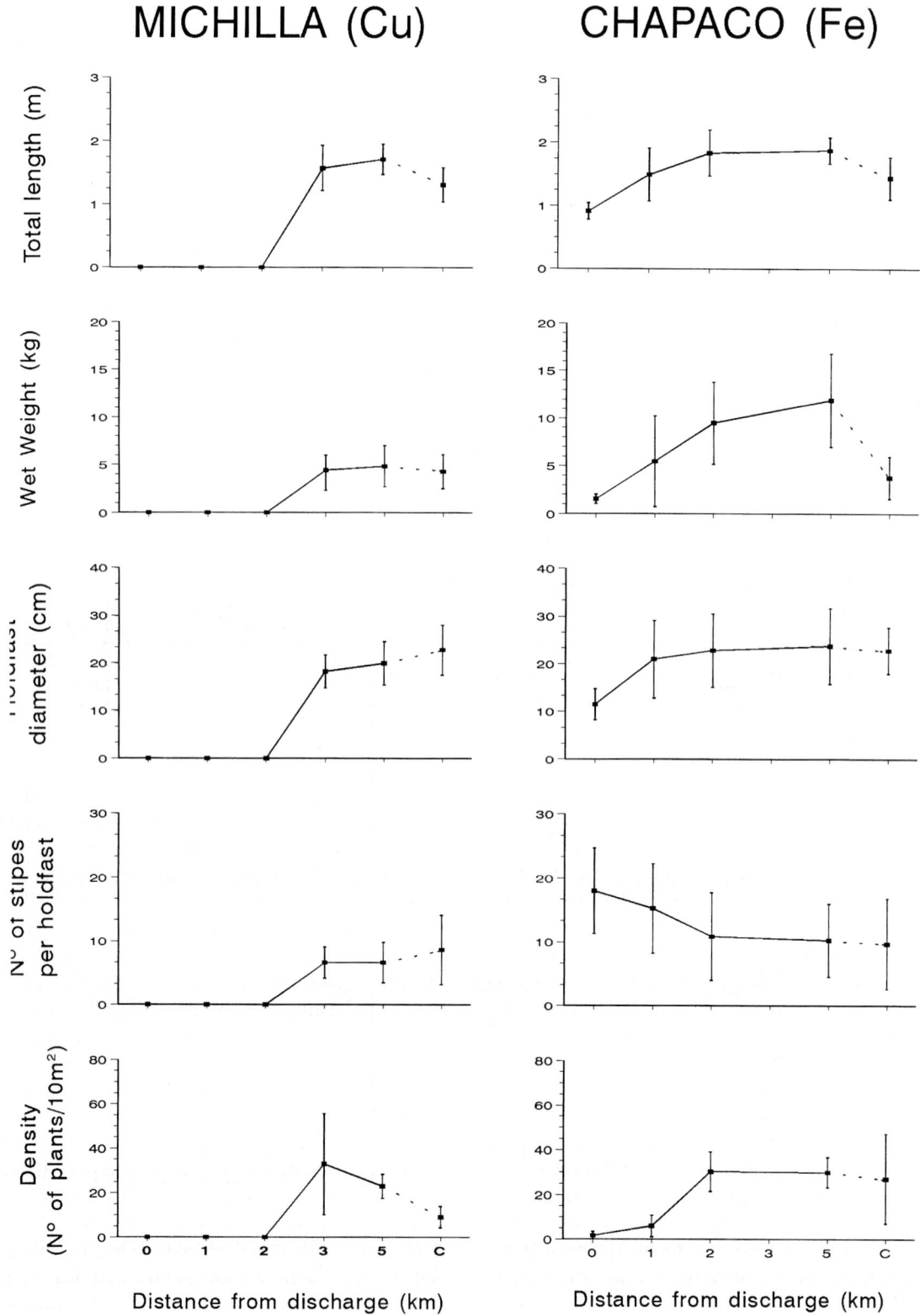

Figure 5. Spatial variability of morphological variables of *Lessonia trabeculata* populations exposed to different Cu and Fe concentrations in the field. C = control area. Bars = standard error.

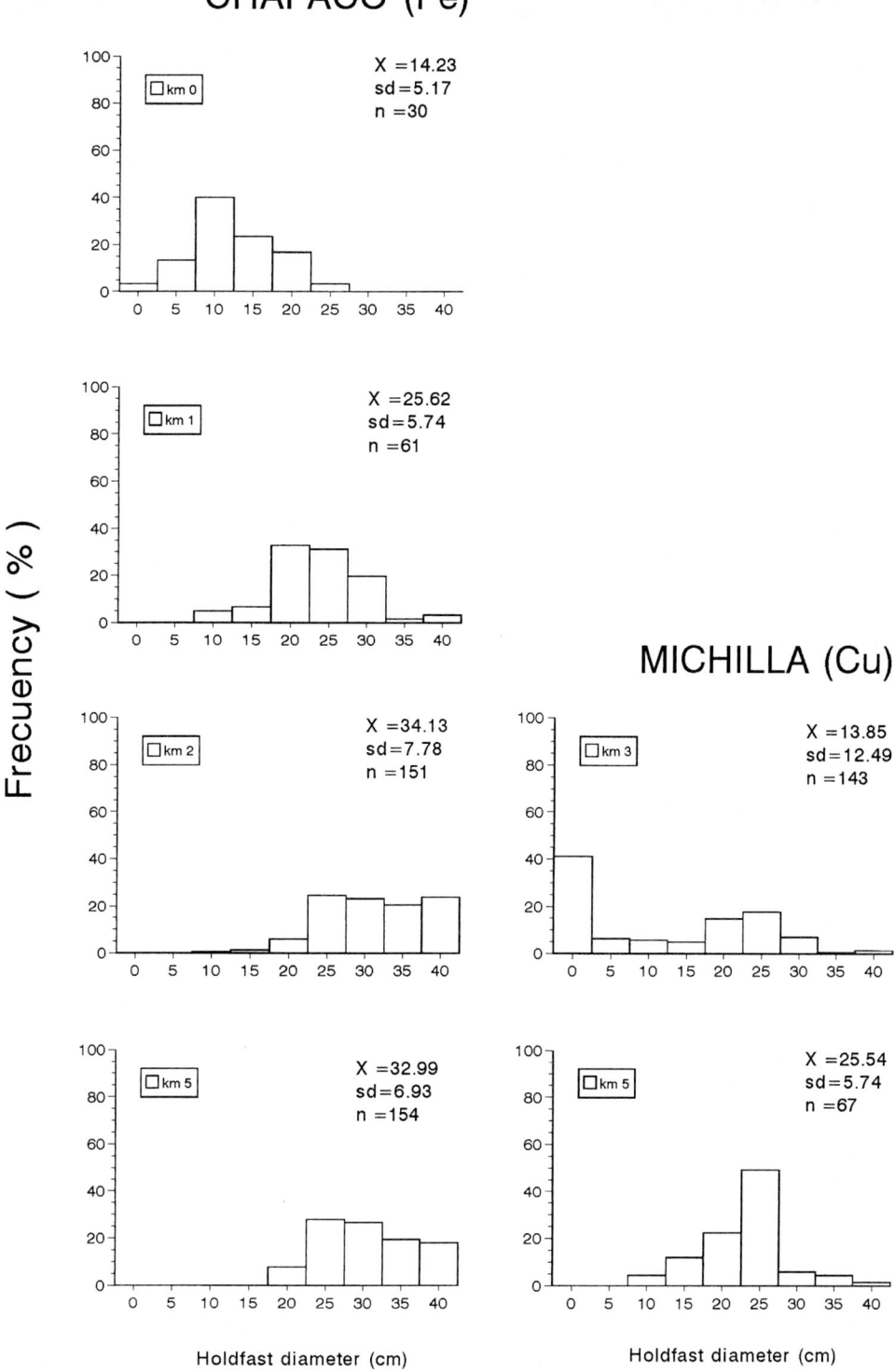

Figure 6. Spatial variability of size frequency in subtidal *Lessonia trabeculata* populations exposed to different Cu and Fe concentrations in the field.

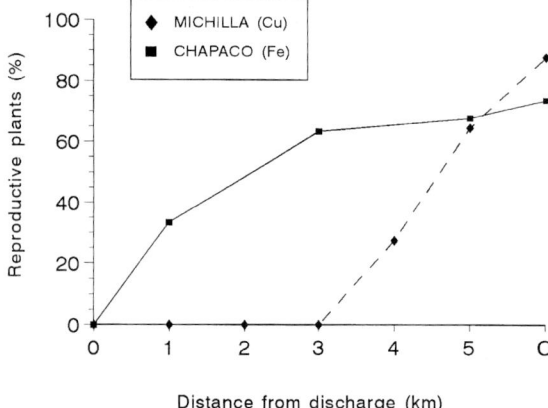

Figure 7. Frequency of reproductive plants in subtidal populations of *Lessonia trabeculata* at the study areas. C = control area. Bars = standard error.

factor that must be considered in the distribution and variability of heavy metal concentration in seawater and algae. During winter, the possibility of higher heavy metal concentration could be minimized by the greater water movement (dilution effect) caused by the predominant SW wind and surge common during this time of the year. Our data on temporal variability may reflect this phenomenon.

Tailings from Fe mining do not generate severe modifications of the subtidal populations and communities in the studied areas. Only a few meters from the tailing outfall, *Lessonia* populations have a morphology and abundance similar to those of more distant sites, and to populations in the control localities. No effects on the fauna associated with *Lessonia* holdfasts were detected in the sampling gradient. Though the concentrations of Cu and Fe exceed the 'normal' values for seawater in other latitudes (Lewis, 1994), they do not seem to generate modifications *per se*. This study shows that there are subtidal communities dominated by macroalgae (e.g. *Lessonia*) in areas with high Cu and Fe concentrations in seawater, as in the case of Carrizal Bajo (28° 05′ S), Caleta Constitución (23° 25′ S), or those reported by Vásquez & Guerra (1996) for other localities of northern Chile. The high levels of heavy metal concentration in areas without mining pollution are possibly due to orogenetic processes, high frequency of volcanic activity and climatic conditions which naturally increase its availability (Vila & Sillitoe, 1991). Other factors (not documented), which may increase the heavy metal concentrations in pristine areas of northern Chile are the runoff of summer rain from the slopes of the nearby Andes Mountains, the frequency and intensity of ENSO (El Niño Southern Oscillation), and the high degree of coastline exposure. In this context, the high levels of heavy metals all along the Chilean coast, might have resulted in adaptation in ruderal macroalgal species (*sensu* Vásquez & Guerra, 1996) allowing them to occur in coastal environments receiving high loads of anthropogenic pollution.

The sediments associated with high concentrations of heavy metals from mine tailings probably produce a greater effect than the toxic cations themselves. The kind of physical perturbations that limit the amount of light and maximize abrasion phenomena have not been assessed in the Chilean littoral, and are poorly documented for other parts of the world.

Binding of metal ions to polyphenols has been described by several authors (e.g. Ragan et al., 1979; Pedersen, 1984). Karez & Pereira (1995) found that concentrations of Zn, Cd, Pb, Cr and Cu were as much as two orders of magnitude higher in polyphenolic fractions than in whole plants of *Padina gymnospora* from south-eastern coast of Brazil coast. In general, at the most contaminated area, Cu and Pb were more concentrated in polyphenols than Zn, Cd and Cr.

Results presented here indicate that the Cu and Fe contents of extracted alginates are higher than the plants of *L. trabeculata* at the study sites. These represent the first report of heavy metal contents of alginic acid, although Paskins-Hurlburt et al. (1976) reported the metal binding properties of fucoidan.

The analysis of Cu and Fe contents in the different structures of *L. trabeculata*, particularly in alginates, shows that this macroalga is a good indicator of the heavy metal levels in seawater, although they do not disclose clear-cut variability patterns. The wide distribution range of *L. trabeculata* in subtidal communities of the south-west Pacific (Villouta & Santelices, 1986) makes this species a useful tool as study unit, not only because its individual and population characteristics may help to evaluate chemical perturbations within a distribution gradient, but also because their holdfasts contain a rich community of associated macroinvertebrates (Vásquez, 1992). These communities are discrete and biologically delimited, allowing replication over latitudinal and bathymetric gradients. Moreover, the populations occurring inside them respond differentially as a function of the perturbation to which they are exposed, and to different perturbations according to their own tolerance ranges.

This work shows that the values of heavy metals in seawater, plants and alginates of *L. trabeculata* from

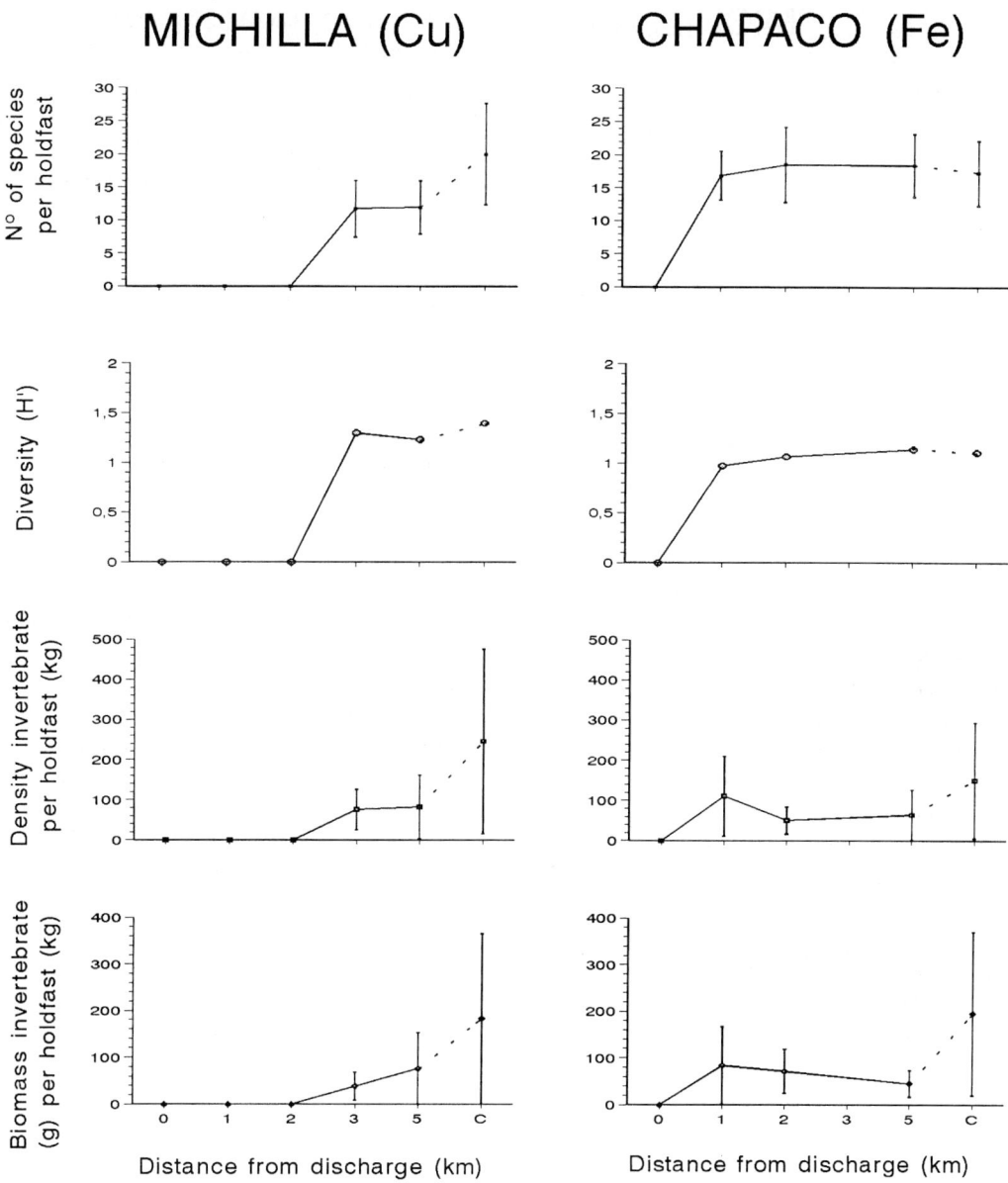

Figure 8. Spatial variability of macroinvertebrate fauna associated with *Lessonia trabeculata* holdfasts exposed to different Cu and Fe concentrations in the field. C = control area. Bars = standard error.

contaminated and control sites are highly variable. In intertidal areas adjoining the tailing ducts, the high values of Cu and Fe similar to those found in areas with the same contaminant agents in other localities of the Chilean coast (Vásquez & Guerra, 1996; Correa et al., 1996).

In laboratory experiments, exposure to the metal concentrations encountered at the study areas would result in inviability or a dramatic decrease of the reproductive and growth potentials of individuals (Bryan &

Langston, 1992; Anderson & Kautsky, 1996; Gledhill et al., 1997). It is important that the influence of environmental factors such as: temperature, wind intensity, tidal regimes, water movement, wave impact, coastal circulation, local orogenetic processes, tectonic movements, shore topography, coastal upwelling and global oceanographic phenomena like ENSO, should all be considered when assessing the intensity of the effects of contaminant agents on marine coastal communities. Future field studies should assess the effects of

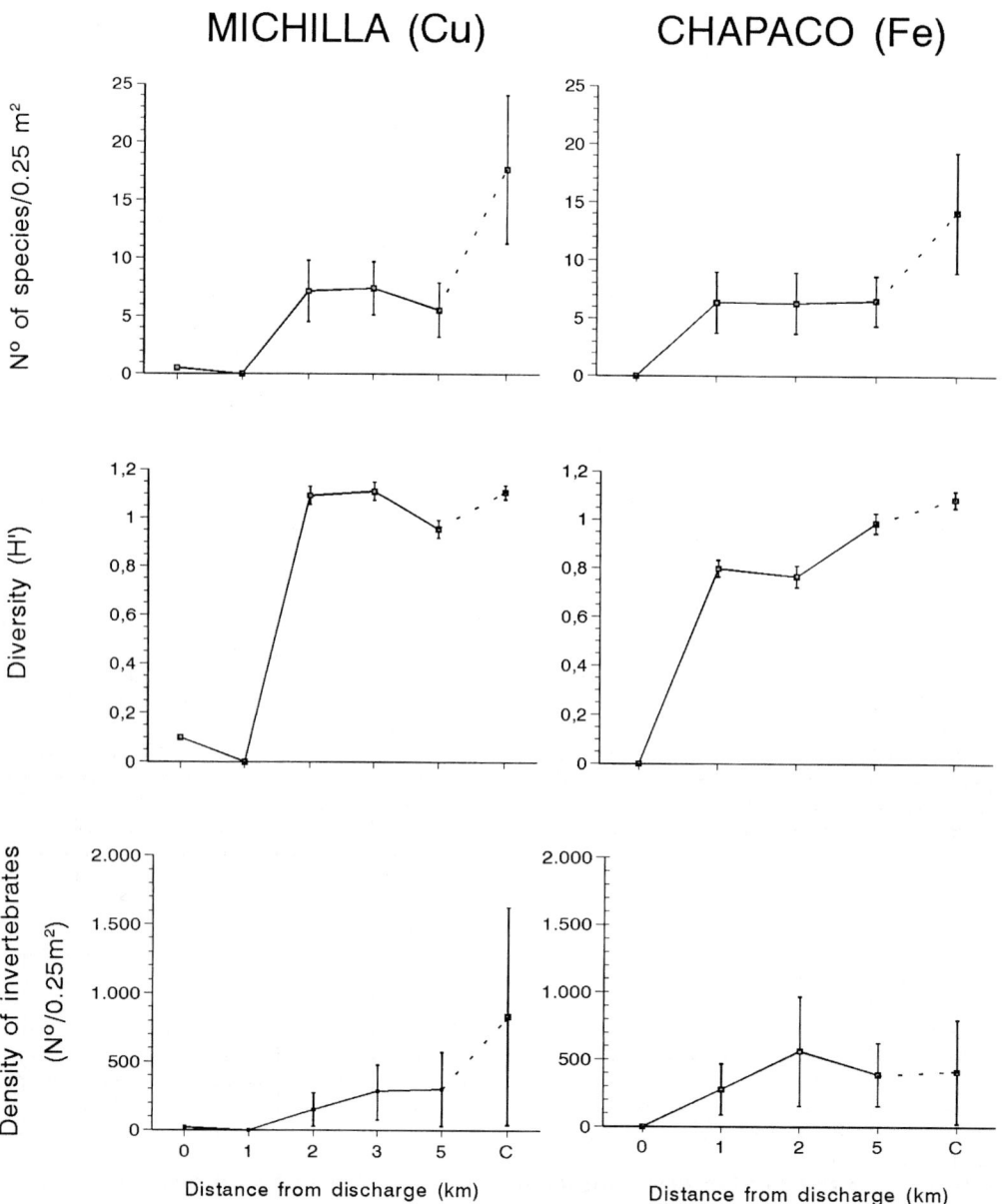

Figure 9. Spatial variability of macroinvertebrate fauna associated with hard substrata between *Lessonia trabeculata* plants exposed to different Cu and Fe concentrations in the field. C = control area. Bars = standard error.

contaminants along an intensity gradient in order to evaluate the mechanisms by which organisms, populations and communities minimize the effects of these agents of environmental perturbation.

Acknowledgements

We are grateful to N. Barroso and S. Espinoza for assistance in field and laboratory analyses. This work was funded by FONDECYT 1960202 to JAV. Sorting and evaluation (C. Cerda and F. Véliz) of macroalgae associated fauna was funded by FONDAP O & BM.

References

Ahumada, R., 1994. Niveles de concentración e índices de bioacumulación de metales pesados (Cd, Cr, Hg. Ni, Cu, Pb, y Zn) en tejidos de invertebrados bénticos de Bahía San Vicente, Chile. Rev. Biol. mar. 29: 77–87.

Anderson, S. & L. Kaustky, 1996. Copper effects on reproductive stages of Baltic Sea *Fucus vesiculosus*. Mar. Biol. 125: 171–176.

Berndt, H., U. Harms & M. Sonneborn, 1985. Multielementspurenanreicherung aus wassern an aktivkohle zur probenvorbereitung fur die atomspektroskopie (Flammen-AAS, ICP/OES) Fresenius Z. Anal Chem. 3: 329–333.

Boré, D. H. Robothan, R. Trucco, J. Inda & M. L. Fernandez, 1989. Evaluacion Preliminar de la presencia de metales pesados en recursos pesqueros de importancia comercial de la III Región de Chile. Rev. Pacífico Sur.(special number): 195–203.

Bryan, G. W. & W. J. Langston, 1992. Bioavailability, accumulation and effects of heavy metals in sediments with special reference to United Kingdom estuaries: a review. Envir. Pollut. 76: 89–131.

Castilla, J. C., 1983. Environmental inpact in sandy beaches of copper mine tailings at Chañaral, Chile. Mar. Pollut. Bull. 14: 459–464.

Castilla, J. C. & E. Nealler, 1978. Marine environmental impact due to mining activities of El Salvador Copper Mine, Chile. Mar. Pollut. Bull. 9: 67–70.

Correa, J. A., P. González, P. Sanchez, J. Muñoz & M. C. Orellana, 1996. Copper-algae interactions: inheritance or adaptation? Env. monitor. Ass. 40: 41–54.

Corvalán, J., 1985. Recursos no renovables, In Soler, F. (ed.), Medio Ambiente en Chile. Ediciones Universidad Catolica de Chile: 165–181.

Gledhill, M., M. Nimmo, S. J. Hill & M. T. Brown, 1997. The toxicity of copper (II) species to marine algae, with particular reference to macroalgae. J. Phycol. 33: 2–11.

Jaksic F. & R. Medel, 1987. El acuchillamiento de datos como método de obtención de intervalos de confianza y de prueba de hipótesis para índices ecológicos. Medio Ambiente 8: 95–103.

Karez, C. S. & R. C. Pereira, 1995. Metal contents in polyphenolic fractions extracted from the brown alga *Padina gymnospora*. Bot. mar. 38: 151–155.

Lecaros, O. & M. S. Astorga, 1992. Metales pesados en *Macrocystis pyrifera* (huiro) de la costa del Estrecho de Magallanes. Rev. Biol. mar. 27: 5–16.

Lewis, A. G., 1994. Copper In Water and Aquatic Environments. International Copper Association Ltd. New York, NY. U.S.A., 72 pp.

Morrisey, D. J., A. J. Underwood & L. Howitt, 1996. Effects of copper on the fauna of marine soft-sediments: an experimental field study. Mar. Biol. 125: 199–213.

Ojeda, F. P. & B. Santelices, 1984. Invertebrate communities in holdfasts of the kelp *Macrocystis pyrifera* from southern Chile. Mar. Ecol. Prog. Ser. 16: 65–73.

Paskins-Hurlburt, A., Y. Tanaka & S. C. Skoryna, 1976. Isolation and metal binding properties of fucoidan. Bot. mar., 19: 327–328.

Pedersen, A., 1984. Studies on phenol content and heavy metal uptake in fucoids. Hydrobiologia 116/117: 498–504

Ragan, M. A., O. Smidsrod & B. Larsen, 1979. Chelation of divalent metal ions by brown alga polyphenols. Mar. Chem. 7: 265–271.

Smith, S. D. A., 1996. The macroafaunal community of *Ecklonia radiata* holdfasts: description of the faunal assemblage and variation associated with differences in holdfast volume. Aust. J. Ecol. 21: 81–95.

Smith, S. D. A. & R. D. Simpson, 1993. Effects of pollution on holdfast macrofauna of the kelp *Eklonia radiata*: discrimination at different taxonomic levels. Mar. Ecol. Prog. Ser. 96: 199–208.

Snider, L. J., 1985. Demersal zooplankton of the giant kelp *Macrocystis pyrifera*: patterns of emergence and the population structure of three gammarid amphipod species. Ph.D. thesis, University of California, San Diego, Scripps Intitution of Oceanography, 238 pp.

Sokal, R. R. & F. J. Rohlf, 1981. Biometry. The Principles and Practice of Statistics in Biological Research 2nd Edn., W.H. Freeman & Company, New York, 859 pp.

Trucco, R. G., J. Inda & M. L. Fernandez, 1990. Heavy metal concentration in sediments from Tongoy and Herradura Bays, Coquimbo, Chile. Mar. Pollut. Bull. 21: 229–232.

Vásquez, J. A., 1992. *Lessonia trabeculata* a subtidal bottom kelp in northen Chile: a case study for a structural and geographical comparison. In Seeliger, U. (ed.), Coastal Plant Communities of Latin America. Academic Press, San Diego: 77–89.

Vásquez, J. A., 1993. Effects on the animal community of dislodgement of holdfasts of *Macrocystis pyrifera*. Pac. Sci. 47: 180–184.

Vásquez, J. A. & N. Guerra, 1996. The use of seaweeds as bioindicators of natural and anthropogenic contaminants in northern Chile. Hydrobiologia 326/327: 327–333.

Vásquez, J. A. & B. Santelices, 1984. Comunidades de macroinvertebrados en discos adhesivos de *Lessonia nigrescens* Bory (Phaeophyta) en Chile central. Rev. Chile. Historia. Natural 57: 131–154.

Vermeer, K. & J. C. Castilla, 1991. High cadmium residues observed during a pilot study in shorebirds and their prey downstream from El Salvador copper mine, Chile. Bull. envir. Contam. Toxicol. 46: 242–248.

Vila, T. & R. H. Sillitoe, 1991. Gold-rich porphyry systems in the Maricunga belt, northern Chile. Econ. Geol. 86: 1238–1260.

Villouta, E. & B. Santelices, 1986. *Lessonia trabeculata* sp. nov. (Laminariales, Phaeophyta), a new kelp from Chile. Phycologia 25: 81–86.

Obtaining absorption spectra from individual macroalgal spores using microphotometry

Michael H. Graham & B. Greg Mitchell
Scripps Institution of Oceanography, University of California San Diego, 9500 Gilman Drive, La Jolla, CA, 92093-0208, U.S.A.
E-mail: mgraham@ucsd.edu

Key words: absorption spectra, microphotometry, spores, kelp, Laminariales, Phaeophyceae, red algae, Rhodophyta

Abstract

Information on the ecophysiology of macroalgal planktonic propagules (e.g. spores) has been hard to obtain, given their small size and low concentration in the water column. Studies of the photo-physiology of macroalgal spores, for example, have been limited by the need to aggregate many spores into bulk samples for analysis. Subsequently, physiological variability among spores (e.g. pigment concentration, absorption characteristics) is lost, and taxonomic comparisons from multi-taxa samples are impossible. Here we present a technique that utilizes a spectral microphotometer to produce visible (400-800 nm) absorption spectra from individual particles; the particles in our case are macroalgal spores. The microphotometer consists of a microscope fitted with a monochromator and spectrophotometer. After mounting spores from laboratory or field suspensions onto transparent membrane filters, absorption characteristics of individual spores, or even individual plastids, can be evaluated independently from the remaining particles in the sample. Use of transparent rather than opaque membrane filters allows for determination of absorption spectra, as well as more traditional microscopic analyses (e.g. bright field, dark field, epi-fluorescence). Glutaraldehyde fixation and cold storage ($-10\ °C$) were found to be appropriate for maintaining the integrity of absorption spectra for at least 3 days. To demonstrate the utility of microphotometry for macroalgal studies, absorption spectra were obtained and analyzed from spores of various kelps and filamentous red algae.

Introduction

As typical of most benthic marine organisms, marine macroalgal life histories alternate between separate benthic and planktonic stages. Although considerable information is available concerning the ecophysiology of macroalgal benthic stages (e.g. Lobban & Harrison, 1994), laboratory studies have provided only minimal information on the ecophysiology of macroalgal planktonic propagules (Amsler & Neushul, 1991; Brzezinski et al., 1993; Beach et al., 1995), while the ecophysiology of the propagules *in situ* remains completely unknown. This dearth of information concerning macroalgal propagules is primarily due to (1) difficulties in sampling and identifying the small and often rare propagules, and (2) technological limitations that prohibit the analysis of propagules on an individual basis. As an example of the latter, recent laboratory studies of the photo-physiology of brown (Amsler & Neushul, 1991) and green macroalgal zoospores (Beach et al., 1995) required the analysis of bulk zoospore samples in order to utilize traditional phycological methods (e.g. spectrophotometry, fluorometry, HPLC, light/dark incubations). As such, variability in physiological variables (e.g. pigment concentrations, absorption spectra) among individual zoospores, of the same taxa or between taxa, is averaged away. Flow cytometry has been used to study some aspects of the physiology of individual kelp zoospores (Brzezinski et al., 1993), however, the large sample sizes required for analysis and the inability to observe directly the particles being analyzed makes this method impractical for most applications, especially those utilizing multi-taxa field samples.

Microphotometric methods have been developed for determining *in vivo* absorption spectra for individual microalgal cells (Iturriaga et al., 1988). These microphotometric systems typically consist of a microscope fitted with a spectrophotometer and scanning monochromator, and work by aligning a transparent aperture over individual particles through which spectral light transmittance (or absorbance) is measured. The result is continuous transmission (or absorption) spectra over a range of wavelengths determined by the optical characteristics of the system. Although application of such methods by phytoplankton ecologists has been described (Iturriaga et al., 1988; Iturriaga & Siegel, 1989; Carpenter et al., 1990, 1991; Bidigare et al., 1993; Robinson et al., 1995; Stephens, 1995), microphotometry has yet to be used in macroalgal studies.

Here we present a refined method for obtaining absorption spectra from macroalgal spores using microphotometry. The method was developed to provide absorption spectra for species-specific identification of individual kelp zoospores from *in situ* water samples, and is part of a larger investigation into the role of pre-settlement processes on the dynamics of subtidal kelp populations. Microphotometry, however, should prove useful to those interested in general photo-physiology of macroalgal spores and thalli. The method includes description of: (1) preparation (fixation and mounting) of macroalgal spore samples for use with the microphotometry system; (2) procedures for obtaining absorption spectra from mounted spore samples; (3) techniques for storing mounted spore samples that preserve the integrity of spore absorption characteristics. In addition, we reveal the advantage of microphotometry over traditional spectrophotometric methods by examining absorption spectra from: (1) individual zoospores of two co-occurring species of subtidal kelp (*Macrocystis pyrifera* [L.] C. Agardh and *Pterygophora californica* Ruprecht); and (2) individual tetraspores within attached tetrasporangia of two species of filamentous red algae (*Callithamnion biseriatum* Kylin and *Scagelia pylaisaei* [Montagne] Wynne).

Materials and methods

Sample preparation

Kelp zoospores
Macrocystis and *Pterygophora* sporophylls were collected from 15 m depth within the Pt. Loma kelp forest, southern California, USA in April 1997. Sporophylls with sori were trimmed, wrapped in moist paper towels, sealed in dry plastic bags, and transported to the laboratory in coolers at ambient temperature. Zoospore suspensions were created 4 to 6 hours later by re-immersing sporophylls in 500 ml of 15 °C 0.2 μm filtered seawater (FSW) under cool white fluorescent lighting. Zoospores were fixed by mixing 1 ml aliquots of zoospore suspensions with 4 ml of 2.5% glutaraldehyde in FSW (2% final concentration); aliquots were taken from the upper 1 cm of zoospore suspensions in order to sample motile zoospores only. Fixed zoospore samples were vacuum filtered onto Whatman Cyclopore transparent polycarbonate membrane filters (25 mm filter diam.; 1.0 μm pore diam.) under low pressure (< 1 cm of mercury). The last 1 mm of suspension was released from vacuum pressure and gravity filtered in order to minimize zoospore damage (Iturriaga & Siegel, 1989). Filters were air-dried until all visible moisture had evaporated, and mounted (zoospores up) between glass microscope slides and cover slips using Cargille fluorescence microscopy immersion oil (Type FF). Zoospore movement on the filters during mounting was minimized by coating slides and coverslips with immersion oil before placement of the filters. Coverslips were sealed to the slides using clear nail varnish.

Red algal tetraspores
Tetrasporangial thalli of *Callithamnion* and *Scagelia* were supplied from San Juan Islands, Washington, USA by Dr David Garbary (Garbary et al., 1999). The plants were collected subtidally from 5–10 m depth. All thalli were fixed in 2% glutaraldehyde and gravity filtered onto Whatman Cyclopore transparent polycarbonate membrane filters (25 mm filter diam.; 3.0 μm pore diam.). Filters were mounted to microscope slides as described above for kelp zoospores.

Microphotometry

Our microphotometric system consisted of a universal microscope (Olympus Provis AX70, Lake Success, New York) equipped with a Nano500 diode-array spectrophotometer and a halogen light source (Nanometrics, Sunnyvale, California). A circular aperture, through which irradiance was sampled, was placed between the objective and monochromator. Light passing through the sample and aperture was projected onto the monochromator (at the focal plane) which dispersed the spectrum onto a linear diode-array detector. This system recorded relative spec-

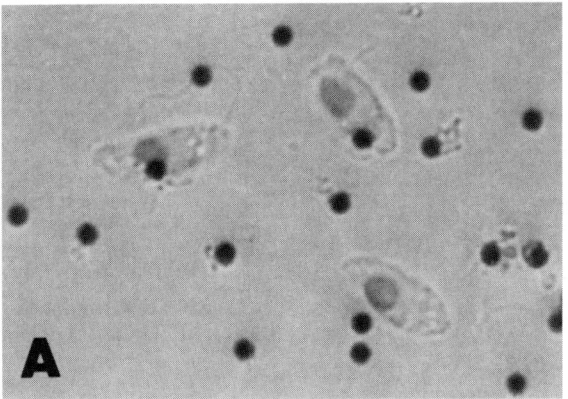

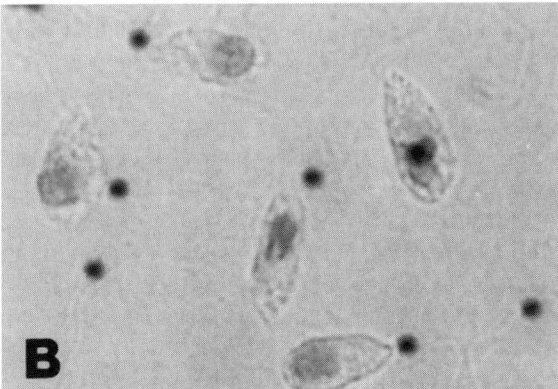

Figure 1. Light microscopy photographs of *Macrocystis* and *Pterygophora* zoospores mounted onto transparent membrane filters. A. *Macrocystis* zoospores, and B. *Pterygophora* zoospores, each showing the location of the single discoid plastids within each zoospore. Flagella are visible for some zoospores of each species; retention of these presumably fragile structures suggests that zoospores suffered little damage during mounting. *Pterygophora* zoospore plastids are larger than those of *Macrocystis*. For scale, black filled circles are the pores of the membrane filters and have diameters of 1 μm. Magnification was 2500× for both photographs.

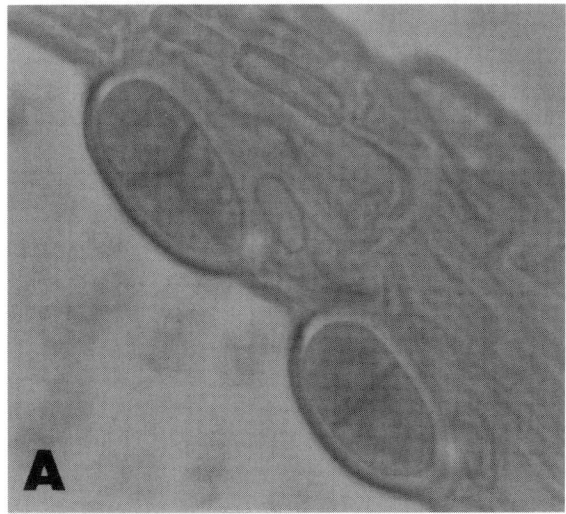

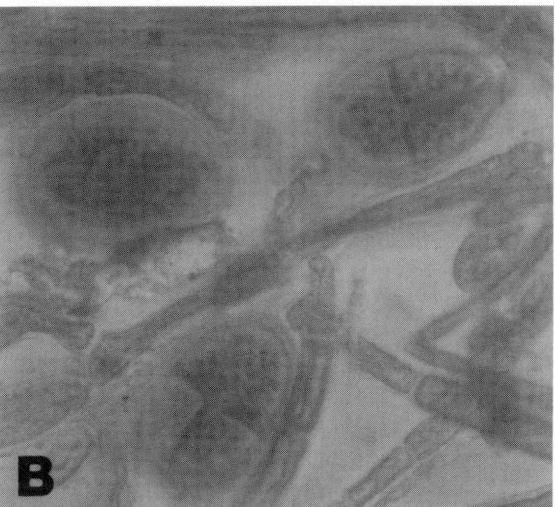

Figure 2. Light microscopy photographs of *Callithamnion* and *Scagelia* tetrasporangial thalli mounted onto transparent membrane filters. A. *Callithamnion* tetrasporangia, and B. *Scagelia* tetrasporangia, each showing the distinction between individual tetraspores. Magnification was 400× for both photographs.

tral irradiance over a range of 400–800 nm (1.2 nm intervals).

Irradiance transmitted through each spore, $I_{s(\lambda)}$, was determined by placing the aperture either directly over individual kelp zoospore plastids which are single and discoid (Figure 1; Henry & Cole, 1982) or within individual red algal tetraspores (Figure 2). To facilitate alignment, the sample image was superimposed with an image of the aperture onto a video system; a monitor allowed the aperture to be manoeuvred precisely along the sample without using the oculars. Irradiance incident on the spore (i.e. the blank), $I_{b(\lambda)}$, was determined by placing the aperture over an adjacent particle-free area without changing the focus. Background current noise was measured prior to sampling each mounted filter and was subtracted from each spectral scan of that filter (i.e. both $I_{s(\lambda)}$ and $I_{b(\lambda)}$). Absorbance (optical density) for each scan was determined using:

$$A_\lambda = -\log(I_{s(\lambda)}/I_{b(\lambda)}),$$

where A_λ is the absorbance at a specified wavelength, λ. The system required < 5 s to obtain a final absorption spectra. All measurements were made in a dark room. Measurements for kelp zoospores were made with a 100× oil-immersion objective, while measurements for red algal tetraspores were made with a 40× objective. The aperture had a 1.83 μm diam. and 2.63 μm^2 cross-sectional area when using the 100× oil-immersion objective, and a 4.53 μm diam. and 16.18 μm^2 cross-sectional area when using the 40× objective. For kelp zoospores, plastid diameter varied greatly between species (Figure 1, see Results) and A_λ was affected by the cross-sectional area of the aperture occupied by the plastids; only *Pterygophora* plastids were larger than the aperture. All red algal tetraspores were larger than the aperture. Thus, absorption spectra obtained by the above method represent true *in vivo* individual-particle (plastid) absorption spectra (*sensu* Iturriaga et al., 1988) only for *Pterygophora* and the red algae. However, since the aperture diameter was fixed and dependent only on the level of magnification, resulting absorption spectra at the same magnification for different cells can be compared within and among species.

Smoothing of each final absorption spectrum was done using the Savitsky-Golay algorithm which performs a least-squares polynomial fit to a specified number of convolution points (Savitsky & Golay, 1964; Steinier et al., 1972; Madden, 1978). We empirically determined that second-degree polynomials and 15 convolution points were best for ensuring that absorption peaks were not 'over-smoothed' (Figure 3) (see Wilson & Polo, 1981 for method). Use of 15 convolution points in the smoothing process subsequently truncated the absorption spectra from 400–800 nm to 408–792 nm. Although photon scattering was shown to be low for similar microphotometric systems (Iturriaga et al., 1988; Iturriaga & Siegel, 1989; Stephens, 1995), A_{750} was subtracted from all A_λ values to compensate for any photon loss due to scattering (Iturriaga et al., 1988). Thus, A_λ are reported from 408–750 nm only, above which absorbance was assumed to be zero (Figure 3).

Storage effects

Experiments were done to test the effect of storage duration on absorption spectra of *Macrocystis* and *Pterygophora* zoospores. These taxa were chosen for the storage duration experiments because fertile plants were readily available and preliminary observations suggested that, among brown and red macroalgae, absorption spectra of kelp zoospores were the most vulnerable to storage degradation. Zoospore suspensions from sporophylls of 1 *Macrocystis* sporophyte and 1 *Pterygophora* sporophyte were used to create 14 mounted zoospore samples for each species, as described above. Sampling from only 1 plant per species was necessary to control for potential variability among plants. Storage duration treatments consisted of mounted zoospore samples analyzed 0, 1, 2, 3, 4, 7, and 10 days after mounting. Absorption spectra for the 0 day treatments were obtained within 1 h of mounting, whereas all other treatments were stored for the allotted time duration in the dark at $-10\ °C$. For each species, absorption spectra were determined for 20 randomly chosen (undamaged) zoospores in each of 2 replicate samples per treatment.

The effect of storage duration on zoospore absorption spectra was analyzed graphically for each species by plotting average A_λ (averaged over all 40 zoospores for each of the 7 storage duration treatments) as a function of wavelength. Significant differences among treatments and controls (i.e. 0 day treatments) were determined by comparing average A_λ for each treatment with 99% confidence intervals of the average A_λ for the controls. Treatments falling outside these 99% confidence regions at any wavelength were considered significantly different from the controls and therefore unsuitable durations for sample storage. In addition, nested analysis of variance (ANOVA) was used to test the effect of storage duration (fixed factor; 7 levels) and sample replication (random factor nested within storage duration; 2 levels) on A_{438} (dependent variable). The primary use of these analyses was to determine whether A_{438} differed among replicate samples; non-significant results of the nested factor for both *Macrocystis* and *Pterygophora* (i.e. no significant difference among samples per treatment; see Results) support the use of the 40 individual zoospores per treatment as independent replicates for the initial analysis of storage duration on A_λ (Underwood, 1997). Cochran's test and analysis of residuals indicated that the assumptions of homogeneity of variances, independence, and normality of error terms were met.

Results

Mounting quality

The mounting procedure was highly effective in producing undamaged spore samples for use with microphotometry. Very few kelp zoospores (< 1%) showed signs of damage from the filtration process. The method used to apply filters and coverslips to the slides appeared to minimize movement of the kelp zoospores

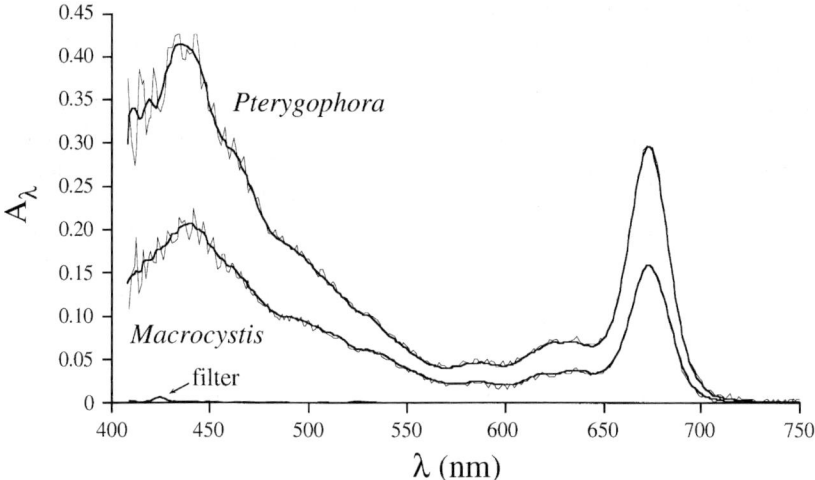

Figure 3. Typical absorption spectra (A_λ as a function of wavelength, λ) for single zoospore plastids of *Macrocystis* and *Pterygophora*. Thin lines are unsmoothed spectra and thick lines are Savitsky-Golay smoothed spectra. The thin line barely visible along the abscissa is average A_λ of an unused filter relative to air ($n = 20$). A_λ for the filter shows a small peak at 422 nm, but was negligible at all other wavelengths.

on the filters; clumping was not evident and the distribution of zoospore densities did not significantly differ from a normal distribution (mean ± SD: 55.63 ± 8.81; Kolmogorov-Smirnov chi-square: $n = 40$, maximum difference = 0.0752, Lilliefors probability = 0.86). Further, resulting sample mounts were very stable (i.e. zoospores did not move during microphotometric analyses). Mounted samples of red algal tetrasporangial thalli were also very stable, with branches and tetrasporangia lying flat and rarely overlapping. The transparent membrane filters were truly transparent (i.e. absorption between 408 and 750 nm was insignificant, Figure 3) and the addition of immersion oil between the filter and coverslip enhanced the detail of the samples. Consequently, it was very easy to visualize plastids in each zoospore for both kelp species (Figure 1) or to distinguish individual tetraspores within tetrasporangia (Figure 2), facilitating the positioning of the aperture directly over samples when obtaining absorption spectra. Occasionally, samples were located over filter pores resulting in enhanced apparent absorbance at wavelengths > 700 nm, presumably due to increased scattered light errors; such absorption spectra were discarded.

Absorption spectra

Kelp zoospores

Macrocystis and *Pterygophora* zoospore plastids sampled immediately (i.e. 0 day control treatments) exhibited absorption spectra of similar shape to those previously described for adult kelps (Smith & Alberte, 1994; Grzymski et al., 1997) (Figure 4). Presence of chlorophyll *a* was indicated by absorption peaks around 438 and 673 nm and a shoulder between 618 and 620 nm, whereas presence of chlorophyll *c* was indicated by high absorption between 460 and 470 nm and a distinct peak at 634 nm (Rowan, 1989). Fucoxanthin also could be identified by multiple shoulders between 485 and 560 nm and a small peak at 584 nm (Smith & Alberte, 1994). Although absorption spectra shape was similar between *Macrocystis* and *Pterygophora* zoospore plastids, magnitude of absorption spectra differed greatly between the two taxa (Figure 4). For example, absorbance at 438 nm was greater for *Pterygophora* (mean ± SD: 0.433 ± 0.031, $n = 20$) than for *Macrocystis* (mean ± SD: 0.191 ± 0.020, $n = 20$), and A_λ averaged between 408 and 750 nm was = 2.3 times higher for *Pterygophora* (0.144) than *Macrocystis* (0.062). Range in absorbance at 438 nm (*Macrocystis*: 0.144–0.232; *Pterygophora*: 0.377–0.506) and 673 nm (*Macrocystis*: 0.101–0.182; *Pterygophora*: 0.257–0.382), and 99% confidence regions of control treatments at wavelengths < 700 nm (Figure 4), did not overlap. In addition, plastid size varied significantly between the two species with plastid diameter (μm) being greater for *Pterygophora* (mean ± SD: 2.10 ± 0.15; $n = 20$) than for *Macrocystis* (mean ± SD: 1.30 ± 0.14; $n = 20$) (*t*-test: $t = 17.22$, df = 38, $p < 0.0001$).

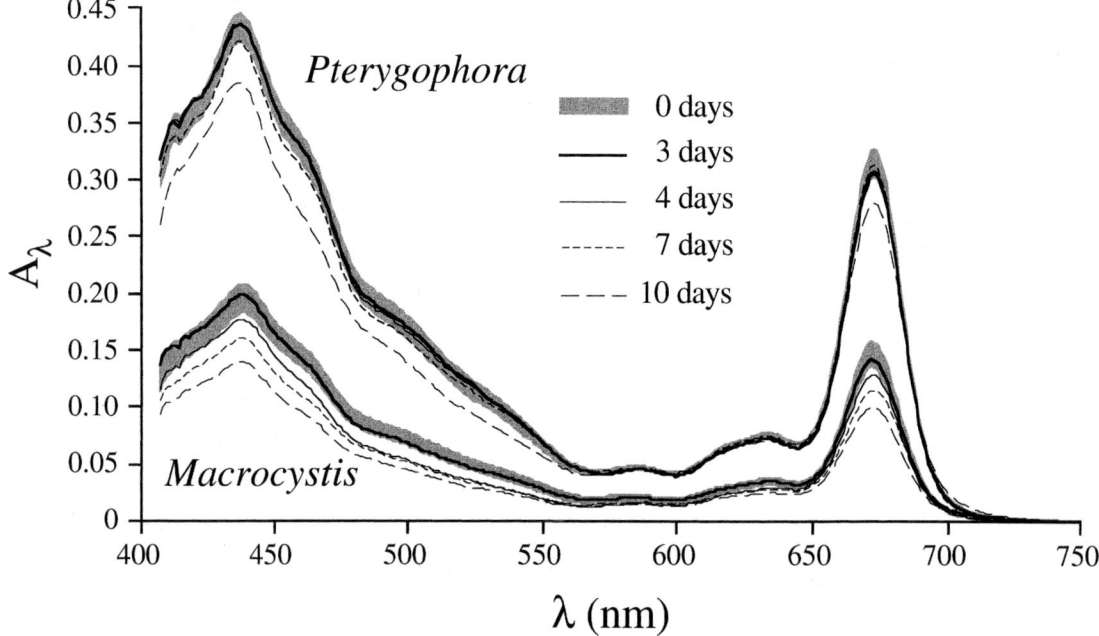

Figure 4. Effect of storage duration on absorption spectra for *Macrocystis* and *Pterygophora* zoospores. Gray shading indicates 99% confidence regions around average A_λ for those zoospores sampled immediately after fixation (controls). Lines indicate average A_λ for those zoospores sampled after various durations of storage at $-10°C$; average A_λ for the 0, 1, and 2 day treatments could not be distinguished visually and for clarity were not included in the figure. Note: average A_λ for the 3 & 4 day *Pterygophora* treatments perfectly overlap. Since average A_λ per treatment did not significantly differ among replicate samples ($n = 20$) for either species (see text), data represent average A_λ pooled for all zoospores per treatment ($n = 40$).

For both *Macrocystis* and *Pterygophora*, magnitude of absorption spectra varied significantly among zoospores sampled immediately and those stored in the dark at $-10°C$ for various durations (Figure 4). For *Macrocystis*, average absorbance of zoospores stored for 1, 2 and 3 days were well within 99% confidence intervals of those sampled immediately (i.e. 0 day control treatment) at all wavelengths, whereas average absorbance of zoospores sampled immediately were significantly higher than those sampled after 4, 7 and 10 days at wavelengths > 430 nm. For *Pterygophora*, only average absorbance from zoospores sampled after 10 days differed significantly from that of the control, although average absorbance from zoospores sampled after 7 days were at the edge of 99% confidence intervals of the control at all wavelengths (Figure 4). These results were supported by the ANOVAs used to test the effect of storage duration and sample replication on A_{438}. The effect of storage duration on A_{438} was significant for both species (*Macrocystis*: $F = 39.22$; df = 6, 7; $p < 0.001$, *Pterygophora*: $F = 10.14$; df = 6, 7; $p < 0.001$). The effect of the nested factor (sample replication) on

$A438$, however, was non-significant for both species (*Macrocystis* sample replication effect: $F = 0.49$; df = 7, 266; $p = 0.841$, *Pterygophora* sample replication effect: $F = 1.12$; df = 7, 266; $p = 0.348$).

Red algal tetraspores

The tetraspores analyzed from *Callithamnion* and *Scagelia* were within cruciate tetrasporangia still attached to the adult thalli (Figure 2); three absorption spectra were obtained within each of three tetraspores for a single tetrasporangium per species. The shape of these absorption spectra indicated absorption characteristics similar to those previously published for red algae (Smith & Alberte, 1994) (Figure 5). Presence of chlorophyll *a* was again identified by absorption peaks near 435 and 673 nm and between 612 and 620 nm. Multiple peaks were observed between 450 and 600 nm indicating the presence of rhodophycean phycoerythrin (Smith & Alberte, 1994). As found by Smith & Alberte (1994) for other rhodophyte taxa, absorbance between 450 and 600 nm was similar to absorbance at 438 nm for *Scagelia* (Figure 5), although absorbance between 450 and 600 nm for *Callithamnion* often exceeded absorbance at 438

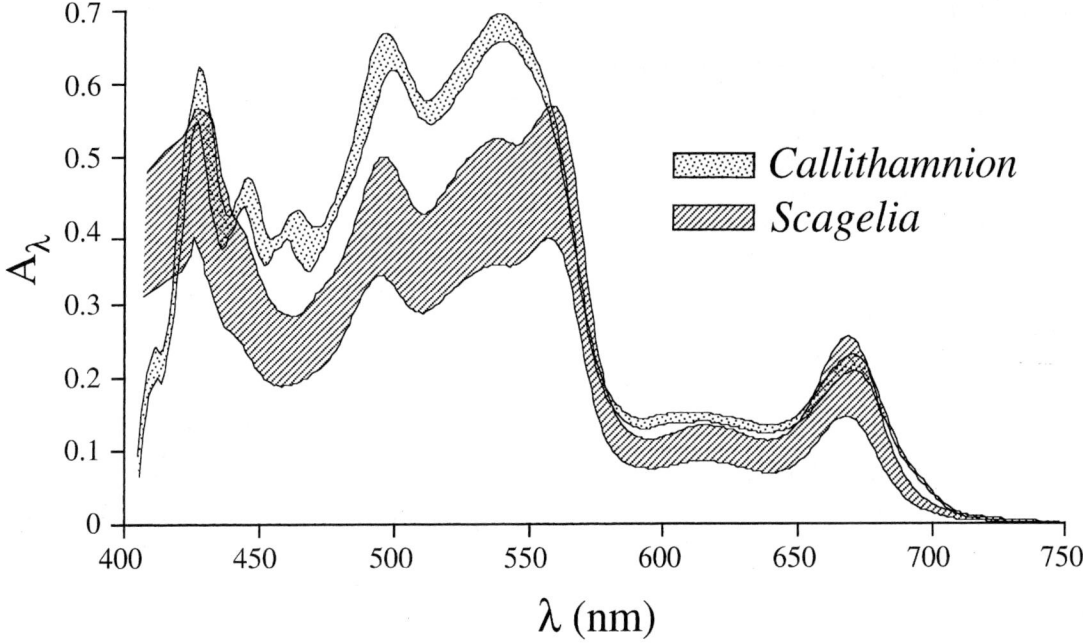

Figure 5. Absorption spectra for *Callithamnion* and *Scagelia* tetraspores. Data are 95% confidence regions around average A_λ for 3 tetraspores within a single cruciate tetrasporangium per species.

nm. Both shape and magnitude of absorption spectra varied greatly between *Callithamnion* and *Scagelia* tetraspores (Figure 5). Average absorbance ($\pm$ SD; $n = 3$) at the blue (435 nm; *Callithamnion*: 0.404 $\pm$ 0.023; *Scagelia*: 0.329 $\pm$ 0.115) and red peaks (673 nm; *Callithamnion*: 0.189 $\pm$ 0.014; *Scagelia*: 0.168 $\pm$ 0.072) was similar between the two species. Differences between *Callithamnion* and *Scagelia* in absorbance due to rhodophycean phycoerythrin (i.e. between 450 and 600 nm), however, were dramatic (Figure 5); average absorbance was higher for *Callithamnion* at all wavelengths between 450 and 550 nm. Some absorbance peaks were observed only for *Callithamnion* (e.g. 448 and 462 nm), whereas others were observed only for *Scagelia* (e.g. 552 nm). In addition, among-tetraspore variability in absorbance was greater at all wavelengths for *Scagelia* than for *Callithamnion* (Figure 5).

Discussion

The results of this study indicate that microphotometry can be used reliably to obtain absorption spectra from macroalgal spores, whether the spores have been released or are still within attached sporangia. The microphotometric system produced accurate and precise absorption spectra in a short period of time (< 5 s per spore), and the absorption spectra suggested spore pigment compositions similar to those described previously for adult macroalgae (e.g. Smith & Alberte, 1994; Grzymski et al., 1997). The transparent membrane filters were found to be extremely stable for mounting samples, such that locating and sampling individual spores was very easy. Due to the high stability of the mount, spore samples remained directly beneath the aperture for the entire duration of spectral scans. Unlike other 'transparent' membrane filters, the Whatman filters used in this study were truly transparent resulting in very high optical quality and detail, and should find use in studies other than microphotometry where visualization of filtered samples is necessary (e.g. as an alternative to the traditional method of estimating spore concentrations using hemacytometers). Finally, variability in the quality of sample mounts was shown to have little effect on resulting absorption spectra (i.e. sample replication factor in ANOVAs was non-significant), suggesting that results for any given spore suspension are reproducible among different samples.

Storage of mounted glutaraldehyde-fixed kelp zoospore samples in the dark at -10 °C was found to preserve the integrity of absorption spectra for at least 3 days for both *Macrocystis* and *Pterygophora*, similar to results previously described for storage of phytoplankton (Jeffrey et al., 1997). This is ample time to allow for timely and efficient analysis of multiple spore samples using microphotometry (i.e. samples do not have to be analyzed immediately). Although glutaraldehyde fixation was satisfactory for our purposes, recent evidence suggests that fixation with paraformaldehyde may allow for greater storage durations (Troussellier et al., 1995). Since our primary use of microphotometry will be for analysis of field samples, it was decided that storage of multiple water samples using liquid nitrogen would not be feasible given the limited space available on small boats to accommodate the relatively large storage containers. However, future efforts to utilize on-board liquid nitrogen storage may also extend the viability of kelp zoospore samples beyond that described herein (Vaulot et al., 1989; Jeffrey et al., 1997).

Among-taxa comparisons of absorption spectra indicated that microphotometry may be useful in addressing a variety of questions regarding macroalgal spore photo-physiology. Kelp zoospore absorption spectra were quantitatively different between *Macrocystis* and *Pterygophora*, two species often coexisting in subtidal habitats of the northeast Pacific. Our results are consistent with previous work showing differences in pigment concentration (chlorophyll *a*) per zoospore for different kelp species (Amsler & Neushul, 1991). Exactly why pigment concentrations and subsequent absorption spectra of kelp zoospores differ between species is not clear. Differences in plastid diameter between species probably may explain much of the variation in absorption spectra since *Macrocystis* plastids (1.34 μm^2 average area) occupied an average of only 50.9% of the total area of that sampled by the microphotometer (2.63 μm^2 area), whereas all *Pterygophora* plastids (2.72 μm^2 minimum area) occupied 100% of the aperture when perfectly centred. Still, functional significance of larger zoospore plastids in *Pterygophora* relative to *Macrocystis* is unknown. Amsler & Neushul (1991) hypothesized that among-species differences in pigment concentrations and photosynthetic capabilities of kelp zoospores may be adaptations favouring increased growth and survival of the microscopic benthic stages. Given application of our microphotometric method to kelp zoospores from field samples, additional research into the ecophysiology of kelp zoospores, and early microscopic stages, will be possible.

Among-taxa comparisons of red algal tetraspores were also interesting. Absorption spectra shape and magnitude differed greatly between *Callithamnion* and *Scagelia* tetraspores. Given that our data represent true *in vivo* individual-particle absorption spectra for these taxa, we suggest that differences observed between *Callithamnion* and *Scagelia* were due to variable pigment compositions, as well as variable pigment concentrations. It was surprising to observe such high variability in putative pigment composition among closely related taxa (both species within the family Ceramiaceae). Further, the high variability in magnitude of absorption spectra observed among *Scagelia* tetraspores suggests that pigment concentration may differ even among tetraspores within the same tetrasporangium. Additional work is necessary to critically test this hypothesis. Regardless of the conclusions that can be drawn from the data presented here, our study demonstrates the utility of microphotometry as a tool for isolating absorption characteristics of individual macroalgal cells; whether they be free-living propagules such as spores, from macroalgal thalli such as spores within attached sporangia, or even macroalgal vegetative cells.

Acknowledgements

We thank K. Riser, P. Edwards, E. Sala, and A. Hobday for help in collecting kelp sporophylls, D. Garbary for supplying the red algal thalli, and T. Moisan and C. Hewes for advice during development of sample fixation and mounting protocols. Thanks to P. Edwards, S. Rumsey, D. Reed, M. Dring, J. Jones and two anonymous reviewers for their comments on the manuscript. This research was supported in part by the Office of Naval Research via equipment grant ONR N000149410951 to C. Lange, B. Mitchell, and E. Venrick, research grant N001491J1186 to B. Mitchell, M. Graham was supported by NOAA / national sea grant NA66RG0477.

References

Amsler, C. D. & M. Neushul, 1991. Photosynthetic physiology and chemical composition of spores of the kelps *Macrocystis pyrifera*, *Nereocystis luetkeana*, *Laminaria farlowii*, and *Pterygophora californica* (Phaeophyceae). J. Phycol. 27: 26–34.

Beach, K. S., C. M. Smith, T. Michael & H. Shin, 1995. Photosynthesis in reproductive unicells of *Ulva fasciata* and *Enteromorpha flexuosa*: implications for ecological success. Mar. Ecol. Prog. Ser. 125: 229–237.

Bidigare, R. R., M. E. Ondrusek, M. C. Kennicutt II, R. Iturriaga, H. R. Harvey, R. R. Hoham & S. A. Macko, 1993. Evidence for a photoprotective function for secondary carotenoids of snow algae. J. Phycol. 29: 427–434.

Brzezinski, M. A., D. C. Reed & C. D. Amsler, 1993. Neutral lipids as major storage products in zoospores of the giant kelp *Macrocystis pyrifera* (Phaeophyceae). J. Phycol. 29: 16–23.

Carpenter, E. J., J. Chang & L. P. Shapiro, 1991. Green and blue fluorescing dinoflagellates in Bahamian waters. Mar. Biol. 108: 145–149.

Carpenter, E. J., J. Chang, M. Cotrell, J. Schubauer, H. W. Paerl, B. M. Bebout & D. G. Capone, 1990. Re-evaluation of nitrogenase oxygen-protective mechanism in the planktonic marine cyanobacterium *Trichodesmium*. Mar. Ecol. Prog. Ser. 65: 151–158.

Garbary, D. J., K. Y. Kim, T. Klinger & D. Duggins, 1999. Preliminary observations on the development of kelp gametophytes endophytic in red algae. Hydrobiologia, 398/399 (Dev. Hydrobiol. 137): 247–252.

Grzymski, J., G. Johnsen & E. Sakshaug, 1997. The significance of intracellular self-shading on the biooptical properties of brown, red, and green macroalgae. J. Phycol. 33: 408–414.

Henry, E. C. & K. M. Cole, 1982. Ultrastructure of swarmers in the Laminariales (Phaeophyceae). I. Zoospores. J. Phycol. 18: 550–569.

Iturriaga, R. & D. A. Siegel, 1989. Microphotometric characterization of phytoplankton and detrital absorption properties of the Sargasso Sea. Limnol. Oceanogr. 34: 1706–1726.

Iturriaga, R., B. G. Mitchell & D. A. Kiefer, 1988. Microphotometric analysis of individual particle absorption spectra. Limnol. Oceanogr. 33: 128–135.

Jeffrey, S. W., R. F. C. Mantoura & S. W. Wright, 1997. Phytoplankton pigments in oceanography: guidelines to modern methods. UNESCO Publishing, Paris, 661 pp.

Lobban, C. S. & P. J. Harrison, 1994. Seaweed ecology and physiology. Cambridge University Press, Cambridge, 366 pp.

Madden, H. H., 1978. Comments on the Savitsky-Golay convolution method for least-squares fit smoothing and differentiation of digital data. Anal. Chem. 50: 1383–1386.

Robinson, D. H., K. R. Arrigo, R. Iturriaga & C. W. Sullivan, 1995. Microalgal light-harvesting in extreme low-light environments in McMurdo Sound, Antarctica. J. Phycol. 31: 508–520.

Rowan, K. S., 1989. Photosynthetic Pigments of Algae. Cambridge University Press, Cambridge, 334 pp.

Savitsky, A. & M. J. E. Golay, 1964. Smoothing and differentiation of data by simplified least squares procedures. Anal. Chem. 36: 1627–1639.

Smith, C. M. & R. S. Alberte, 1994. Characterization of *in vivo* absorption features of chlorophyte, phaeophyte and rhodophyte algal species. Mar. Biol. 118: 511–521.

Steinier, J., Y. Termonia & J. Deltour, 1972. Comments on smoothing and differentiation of data by simplified least square procedures. Anal. Chem. 44: 1906–1909.

Stephens, F. C., 1995. Variability of spectral absorption efficiency within living cells of *Pyrocystis lunula* (Dinophyta). Mar. Biol. 122: 325–331.

Troussellier, M., C. Courties & S. Zettelmaier, 1995. Flow cytometric analysis of coastal lagoon bacterioplankton and picophytoplankton: fixation and storage effects. Estuar. coast. shelf Sci. 40: 621–633.

Underwood, A. J., 1997. Experiments in ecology: their logical design and interpretation using analysis of variance. Cambridge University Press, Cambridge, 504 pp.

Vaulot, D., C. Courties & F. Partensky, 1989. A simple method to preserve oceanic phytoplankton for flow cytometric analysis. Cytometry 10: 629–635.

Wilson, P. D. & S. R. Polo, 1981. Polynomial filters of any degree. J. opt. Soc. Am. 71: 599–603.

Evaluating substances that facilitate algal spore adhesion

B. Santelices & D. Aedo
Departamento de Ecología, Facultad de Ciencias Biológicas, P. Universidad Católica de Chile, Casilla 114-D, Santiago-Chile
Tel: [+56] 2 6862648. Fax: [+56] 2 6862621. E-mail: bsanteli@genes.bio.puc.cl

Key words: algal spores, ligand, nursery, polylysine, recruitment

Abstract

Non-toxic substances that enhance the adhesion of spores are of ecological and economic interest. When used as spore trappers, they may help to trace distributional changes of spore abundance in the water column. Spread over artificial substrata, they may enhance recruitment of economic seaweeds. Gastropod pedal mucus has been used as a substance that enhances adhesion, but its efficiency varies with the type and physiological state of the gastropod. In a search for adhesion promoting compounds, the attachment and germination effects of solutions of albumin (chicken), agar, gelatine (type B), glycerine and polylysine, on spores of *Mazzaella laminarioides, Lessonia nigrescens* and *Ulva rigida*, were compared. Polylysine was the only product significantly increasing the number of spores attached, as compared to uncoated controls. It did not affect germination rates of *U. rigida* or *M. laminarioides* but decreased the germination rates of *L. nigrescens*. Artificial substrata coated with polylysine retained 4–10 times more spores than uncoated controls, both in field-exposed and in nursery experiments.

Introduction

Non-toxic substances that enhance the adhesion of seaweed spores are both ecologically and economically important. When used as spore trappers, they could help document patterns of propagule abundance and dispersal within the water column. Such patterns are as yet unknown because many studies (e.g. Hruby & Norton, 1979; Amsler & Searles, 1980; Kennelly & Larkum, 1983; Zechman & Mathieson, 1985; Reed et al.,1988, 1992) have, in fact, measured recruitment, rather than spore abundance or settlement, at a given place. More recent techniques, using vital staining and venturi suction, (Kendrick & Walker, 1991) limit the study to only one or a few species whose reproductive tissues are previously treated with vital stain.

Non-toxic substances that enhance the adhesion of seaweed spores could also be used to enhance recruitment of economic seaweeds. This might be specially useful to reduce nursery time, to improve recruitment during low fertility seasons, or to increase density of propagules in species with distance-dependent fertility of microscopic stages.

Previous studies (Santelices & Bobadilla, 1996), have shown that pedal mucus of several rocky intertidal species of gastropods, *Littorina peruviana, Tegula atra, Siphonaria lessonii* and *Collisella* sp. retained and allowed germination of seaweed propagules, suggesting that mucus may contribute to seaweed propagule attachment in the field. Gastropod pedal mucus also traps microalgae and enhances their growth (Connor, 1986; Davies et al.,1990) but the quantities of mucus produced, its persistence time and its capacity to retain or stimulate microalgal growth varies with the physiological state of the gastropod, and from one to another species (Connor & Quinn, 1984; Connor, 1986; Peduzzi & Herndl, 1991). Such variability also occurs with the attachment capacity of macroalgal propagules (Santelices & Bobadilla, 1996). Given this variability, a search of adhesion promoting compounds was started. First, we compared the capacity of five natural compounds to trap spores and enhance or reduce germination rates of three algal species, under controlled laboratory conditions. Then, we tested the spore entrapping capacity in the field and in larger cultivation facilities of the most effective compound.

Materials and methods

Experimental procedures, in the laboratory and in the field, follow closely those described in Santelices & Bobadilla (1996).

Laboratory experiments

Fertile fronds of the chlorophyte *Ulva rigida* C. Agardh, the phaeophyte *Lessonia nigrescens* Bory and the rhodophyte *Mazzaella* (formerly *Iridaea*) *laminarioides*(Bory) Fredericq & Hommersand were collected at Pelancura and Las Cruces, two coastal localities near San Antonio Port in central Chile (33° 32' S; 71° 38' W). They were transported to the laboratory inside plastic bags, without water, in a refrigerated container. In the laboratory, they were desiccated for 14 h under controlled conditions of temperature (14 ± 2 °C) and darkness. Spore release was stimulated by immersing these fronds for 45 min in chilled (5–6 °C), filtered seawater. To prevent spore settlement prior to the experiment, spore solutions were maintained on top of a fast moving (100 rpm) rotary shaker until the start of the experiment (15–30 min after shedding).

In these laboratory experiments, two types of control slides without mucus were used: 'untreated control slides' never exposed to seawater, and 'incubated control slides' maintained for 48 h in (filtered) seawater.

Five adhesion promoting compounds, some in various concentrations, were tested. These included chicken albumin at 5%, agar (1.0%), gelatine, at concentrations of 2, 10 and 20%, glycerine (99%) and polylysine (0.1%). Glass slides were immersed 2 h in the respective solution, then removed and dried at room temperature for 6 h.

To evaluate their spore retaining capacity, glass slides coated with the various compounds and uncoated control were placed, each inside one 9 cm diam. Petri dish, covered with 20 ml of SWM-3 growth medium (McLachlan, 1973) and inoculated with 10 ml of spore solution. Spores were allowed to settle under low light (20 ± 5 μmol m^{-2}s^{-1}) in still water for 1 h. Thereafter, the medium with unsettled spores was discarded and replaced by another 20 ml of SWM-3 solution. Ten replicate slides per treatment were used.

Counting of the number of spores settled on coated and uncoated slides was started immediately after completion of the settlement time. In each slide, 200 microscopic fields (about 2 cm^2) of slide surface were examined under a Zeiss inverted microscope (magnification 320 x). Per cent germination was determined after 48 h of incubation under controlled conditions of temperature (14 ± 2 °C), photon irradiance (50 ± 2 μmol m^{-2} s^{-1}) and photoperiod (12 h of daily light).

Data on spore abundance and on percent germination per each algal species, on the various kinds of slides were subjected to square root and arcsine transformation, respectively. Then, differences between treatments were analyzed using ANOVA (Sokal & Rohlf, 1969) followed by an a posteriori Tukey's test.

Field experiment

This experiment was done in intertidal pools located in Las Cruces, near San Antonio Port in central Chile (33° 32' S; 71° 38' W). They were conducted to test the spore trapping capacity of polylysine in the field and to determine the minimum exposure time needed to trap spores by applying the running mean technique (Kershaw, 1964) to ten replicate slides, immersed in the pools for various times, ranging from 1 to 60 min. Ten untreated, uncoated slides were used as a control. Spore counting was done as described earlier. Data on temporal changes in spore abundance on coated and uncoated slides were subjected to square root transformation and analyzed using ANOVA and Tukey's test.

Nursing experiment

Previously washed, 1 cm diam. polypropylene strings were used as seedling substrata. Ten, 15 cm long, strings were immersed for 2 h in a 0.1% polylysine solution. Ten control strings were immersed in filtered (0.2 μm) seawater. Strings were then tagged, dried at room temperature for 24 h and wound onto a rectangular woody frame.

Fifty grams of freshly collected fertile cystocarpic thalli of *Mazzaella laminarioides* were used for seeding. Spore shedding was stimulated as described above using a tank with 3 l of filtered seawater. The spore solution was maintained with air bubbling at all times to prevent spore settlement.

The rectangular woody frame with strings was placed at the bottom of a 60 cm broad, 25 cm wide and 30 cm-long tank, with 35 l of filtered seawater which contained the 3 l of spore solution. Spore concentration, measured with hematocytometer, was 19 000 spores ml^{-1}. After 30 min of incubation with air bubbling, strings were transferred to a tank with filtered seawater.

Table 1. Significance values for the effects of adhesives on spore retaining capacity and germination rates under laboratory conditions

Spore trapping			Germination rate		
M. laminarioides	L. nigrescens	U. rigida	M. laminarioides	L. nigrescens	U. rigida
$F = 20.33$	$F = 129.96$	$F = 48.70$	$F = 2.64$	$F = 6.74$	$F = 79.85$
$p < 0.0001$	$p < 0.0001$	$p < 0.0001$	$p = 0.01$	$p < 0.0001$	$p < 0.0001$

Counting the number of spores settled on coated and uncoated strings started immediately after string transfer to filtered seawater. A total of ten microscopic fields (0.25 cm^2), randomly selected along each string, were examined under a compound microscope. Differences between treatments were analyzed using ANOVA.

This experiment was repeated three times. Finding no significant differences among the three experiments, data were pooled to increase number of replicates.

Results

Laboratory experiments

Polylysine was the only adhesion promoting substance among those tested that retained significantly more propagules than the uncoated controls (Figure 1, Table 1, Tukey's test, $p < 0.05$). Coating with glycerine, agar or albumin retained approximately similar number of spores compared to uncoated controls, while some concentrations of gelatine retained even fewer spores.

The spores of all three algal species were able to germinate after settling in the different types of coatings (Figure 1). However, some of the adhesives significantly inhibited germination of some species (e.g. *Ulva rigida* on the various solutions of gelatine). Germination rates of *Lessonia nigrescens* on polylysine coating were about half of that on the uncoated, previously incubated control, but approximately similar to the germination rate on uncoated, non-incubated control. On the other hand, germination rates of *Ulva rigida* and *Mazzaella laminarioides* propagules on polylysine did not differ significantly from that on uncoated controls (Tukey's test, $p < 0.05$).

Field experiments

Since the laboratory experiment showed that polylysine had a significantly greater capacity to retain seaweed propagules than any of the other adhesion promoting substances and the control, without significantly affecting germination of two of the three species tested, this compound was used in field experiments.

The slides coated with polylysine retained 4–10 times as many seaweed propagules (Figure 2) as compared to the uncoated slides used as control ($F = 246.4$; $p < 0.001$). Differences between coated and uncoated slides are greater under shorter field exposures (2–10 min).

In the field, coated slides could entrap spores within 2 min of immersion in the water. Application of the running mean technique indicated that 8–10 min is the sufficient exposure time to have representation of spore abundance, similar to that found under the most extended field exposure tested (60 min). No significant differences in spore abundance (Tukey test, $p < 0.05$) were found among exposure times ranging from 7 to 60 min.

Nursing experiment

Strings coated with polylysine retained 4.5 times as many propagules of *Mazzaella laminarioides* (Figure 3) as compared to the uncoated strings. Differences between both treatments were highly significant ($F = 78.76$; $p < 0.0005$).

Discussion

Results gathered in the laboratory, nursery and field experiments, all indicated that polylysine is a non-toxic substance promoting spore adhesion and that retains significantly more spores than the other natural compounds tested. The propagule retaining capacity of polylysine in the laboratory and field experiments

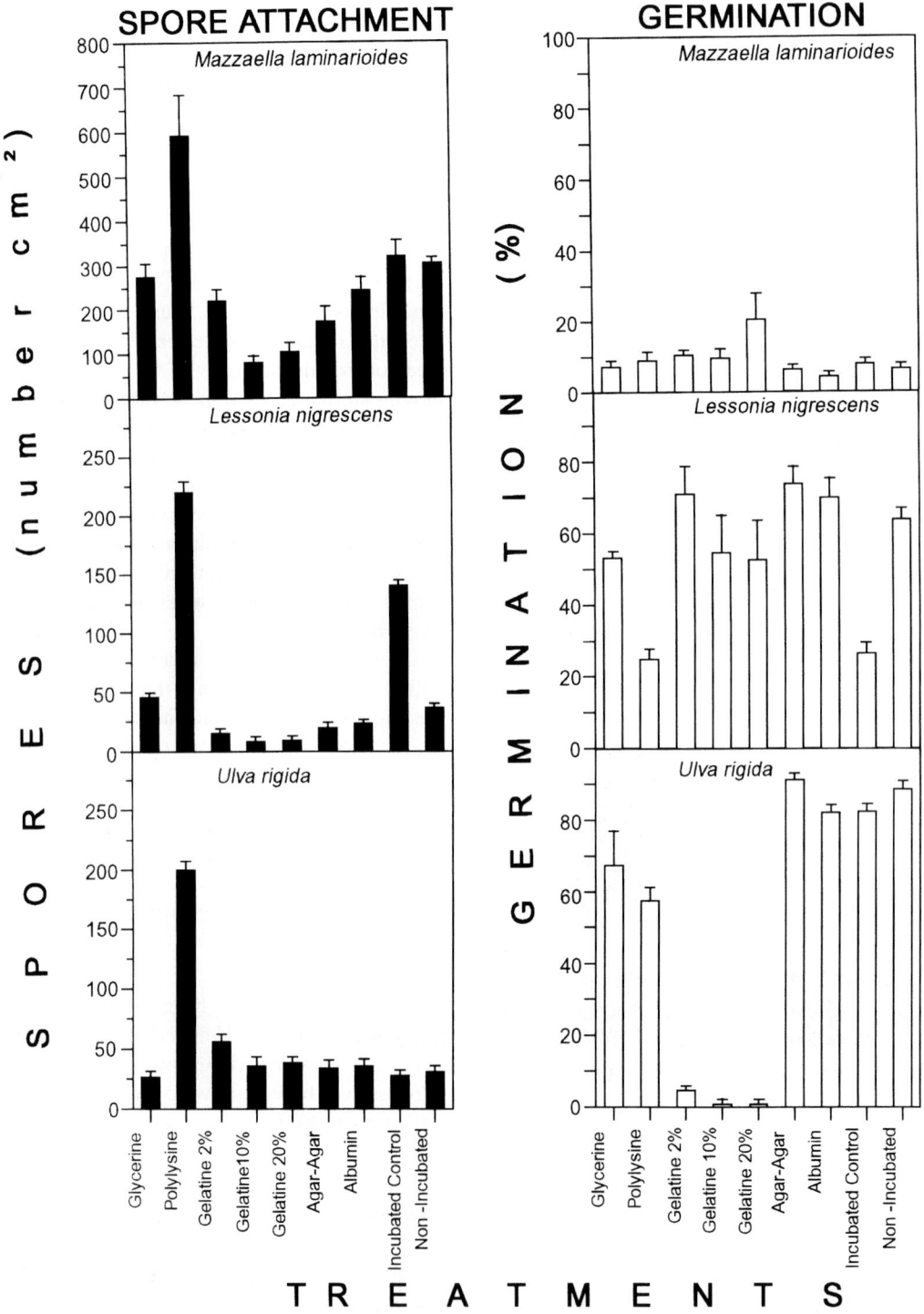

Figure 1. Spore retaining capacity and effects on germination of different adhesion-promoting compounds and evaluated in three algal species: *Mazzaella laminarioides* (Rhodophyta), *Lessonia nigrescens* (Phaeophyta) and *Ulva rigida* (Chlorophyta). Vertical lines represent 1 S.E.

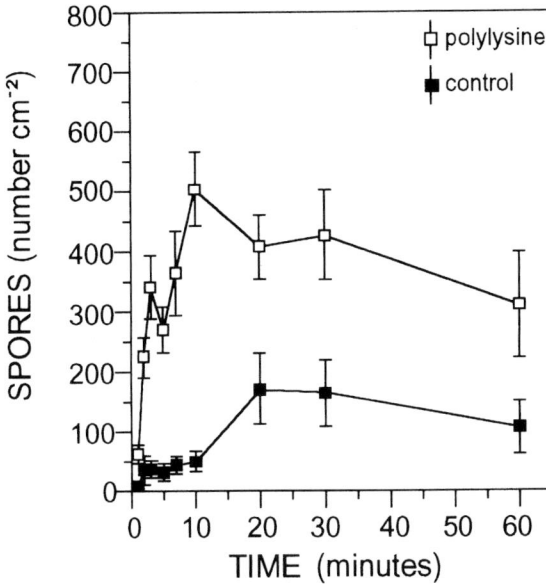

Figure 2. Spore trapping capacity of polylysine in the field for various exposure times. Vertical lines represent 1 S.E.

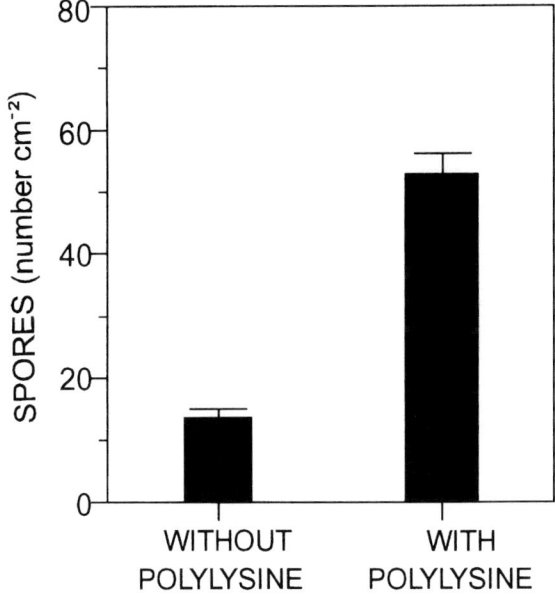

Figure 3. Spore retaining capacity of polypropylene strings coated with polylysine in the nursery experiment. Vertical lines represent 1 S.E.

was higher than that exhibited for the pedal mucus of most gastropods (Santelices & Bobadilla, 1996); its variability is smaller and it is easier to manipulate both in the field and in the laboratory. The nursery experiments suggest, in addition, that the use of polylysine could reduce the time required for spore settlement of some species due to increased recruitment. Thus, this natural compound has a significant potential for ecological studies and nursery manipulations of economic seaweeds. Future research should establish other important aspects related to its potential uses, including maximum effective time in seawater, minimum effective concentration, specific germination and early growth effects on other economic species.

The effects of polylysine enhancing recruitment of these seaweed propagules are consistent with known uses of this substrate as a cell surface ligand (Jacobson & Branton, 1977). The positive charge conferred by polylysine to the coated surfaces promote the adherence of negatively charged cells. This capacity has been shown with a wide variety of cells and organelles (Jacobson & Branton, 1977) and now it is being expanded to seaweed propagules. Fletcher & Callow (1992) had already called attention to the importance of chemical charges and surface free energy for spore settlement and attachment. While low energy, chemically inert, non-stick surfaces are important as antifouling surfaces, high energy, chemically active surfaces, such as those coated with polylysine appear now as promising to enhance recruitment of selected species.

Acknowledgments

Our appreciation to J. Correa for calling our attention to the use of polylysine as a ligand in E.M. studies, and to Marcela Hormazábal for assistance in the field experiments. This research was supported by grant FONDECYT 1960646 to B.S.

References

Amsler, C. D. & R. B. Searles, 1980. Vertical distribution of seaweed spores in a water column offshore of North Carolina. J. Phycol. 16: 617–619.

Connor, V. M., 1986. The use of mucus trails by intertidal limpets enhance food resources. Biol. Bull. 171: 548–564.

Connor, V. M. & J. F. Quinn, 1984. Stimulation of food species growth by limpet mucus. Science 225: 843–844.

Davies, M. S., H. D. Jones & S. J. Hawkins, 1990. Seasonal variation in the composition of pedal mucus from *Patella vulgata*. J. exp. mar. Biol. Ecol. 144: 101–112.

Fletcher, R. L. & M. E. Callow, 1992. The settlement, attachment and establishment of marine algae spores. Br. phycol. J. 27: 303–329.

Hruby, T. & T. A. Norton, 1979. Algal colonization on rocky shores in the Firth of Clyde. J. Ecol. 67: 65–77.

Jacobson, B. S. & D. Branton, 1977. Plasma membrane: rapid isolation and exposure of the cytoplasmic surface by use of positively charged beads. Science 195: 302–304.

Kendrick, G. A. & D. Walker, 1991. Dispersal distances for propagules of *Sargassum spinuligerum* (Sargassaceae, Phaeophyta) measured directly by vital staining and venturi suction sampling. Mar. Ecol. Progr. Ser. 79: 133–138.

Kennelly, S. J. & A. W. D. Larkum, 1983. A preliminary study of temporal variation in the colonization of subtidal algae in an *Ecklonia radiata* community. Aquat. Bot. 17: 275–282.

Kershaw, K. A., 1964. Quantitative and dynamic ecology. Edward Arnold, London, 183 pp.

McLachlan, J., 1973. Growth media-marine. In Stein, J. R. (ed.), Handbook of Phycological Methods. Culture Methods and Growth Measurements. Cambridge University Press, Cambridge: 267–274.

Peduzzi, P. & G. Herndl, 1991. Mucus trail in the rocky intertidal: a highly active microenvironment. Mar. Ecol. Prog. Ser. 75: 267–274.

Reed, D. C., C. D. Amsler & A. W. Ebeling, 1992. Dispersal in kelps: factors affecting spore swimming and competency. Ecology 73: 1577–1585.

Reed, D. C., D. R. Laur & A. W. Ebeling, 1988. Variation in algal dispersal and recruitment: the importance of episodic events. Ecol. Monogr. 58: 321–335.

Santelices, B. & M. Bobadilla, 1996. Gastropod pedal mucus retains seaweed propagules. J. exp. mar. Biol. Ecol. 197: 251–261.

Sokal, R. R. & F. J. Rohlf, 1969. Biometry. W. H. Freeman, San Francisco, CA, 776 pp.

Zechman, F. W. & A. C. Mathieson, 1985. The distribution of seaweed propagules in estuarine coastal and offshore waters of New Hampshire, U.S.A. Bot. mar. 28: 283–294.

Preliminary observations on the development of kelp gametophytes endophytic in red algae

David J. Garbary[1], Kwang Young Kim[2], Terrie Klinger[3] & David Duggins[3]

[1]Department of Biology, St. Francis Xavier University, Antigonish, Nova Scotia, B2G 2W5, Canada
E-mail: dgarbary@juliet.stfx.ca
[2]Faculty of Earth Systems and Environmental Science, Institute of Marine Sciences, Chonnam National University, Kwangju 500-757, Korea
[3]University of Washington, Friday Harbor Laboratories, Friday Harbor, WA 98250, U.S.A.

Key words: Agarum, endophytism, gametophytes, Laminariales, Phaeophyceae, Pleonosporium, reproduction, symbiosis

Abstract

The development of kelp gametophytes is described from field collections from the San Juan Islands, Washington from November, 1997 to March 1998. All gametophytes were endophytic in the cell walls of red algae, especially species with filamentous or polysiphonous construction. Gametophyte density ranged from a few to many hundreds of distinct individuals per host plant. Gametophytes formed extensive vegetative growths of irregularly branching filaments, mostly parallel to the host surface, consisting of up to 50 or more cells. Antheridia were formed at/or just above the surface of the host thallus. The stalked egg apparatus was perpendicular to the host surface. Following presumed fertilization, the zygotes developed with typical kelp embryology to form small epiphytic blades. The specific identity of the gametophytes is unknown, although the host plants were collected from three sites where the dominant kelp species were: a) *Agarum fimbriatum*, b) *Nereocystis luetkeana* and c) *Alaria marginata*, *Costaria costata* and *Laminaria groenlandica*.

Introduction

The life history pattern of Laminariales with its large, diploid sporophytes and microscopic, haploid gametophytes was established early in this century by Sauvageau (1915, 1918). Since then, numerous culture studies on most genera of Laminariales have been carried out. The basic pattern of the life history is thus well established (e.g. Hollenberg, 1939; Kain, 1964; Cole, 1968; Lüning & Neushul, 1978; Lüning, 1980). There is an apparent paradox in that although there are many billions of spores produced per year (Chapman, 1984), the gametophytic stages have only sporadically been described from nature. Examples of kelp gametophytes described from nature include Parke (1932), Sakai & Funano (1964), Funano (1969) and Kaneko (1973) (review by Kain, 1979). All of these reports are based on observations from beds of *Laminaria*, except for Parke's observations, which were made on gametophytes scraped from a floating buoy and described as *Laminaria* and *Saccorhiza*. These gametophytes were typically free-living, although Funano (1969) described the substratum for the gametophytes in his study as being coralline red algae. These observations were presumed to be sporadic because of the microscopic size of the gametophytes. In addition, the presumed rock substratum would hamper direct observation of the gametophytes.

We initially observed kelp gametophytes in situ endophytic in the cell walls of *Pleonosporium vancouverianum* (J. G. Agardh) J. G. Agardh and *Callithamnion acutum* Kylin collected in November 1997 from a bed of *Agarum fimbriatum* Harvey at 5–10 m depth off San Juan Island. This observation was extended to other hosts and sites through March 1998. Here we describe the vegetative and reproductive development of the kelp gametophytes based on the development of the gametophytes in situ.

Materials and methods

General collections of red algae were made between November, 1997 and March, 1998, at:
1. Cantilever Point (5–10 m depth),
2. Reed Rock (at 5–10 m depth), and
3. attached to the floating dock at Friday Harbor Laboratories, all in the San Juan Islands of Washington.

Red algae were epilithic, epiphytic, epizoic (on shells) or epicanthic (i.e. on car tires) (attached to dock). The primary kelp species at these sites were, respectively:
1. *Agarum fimbriatum*,
2. *Nereocystis luetkeana* (Mertens) Postels et Ruprecht, and
3. *Alaria marginata* Postels et Ruprecht, *Laminaria groenlandica* Rosenvinge and *Costaria costata* (C. A. Agardh) Saunders.

Potential host plants were returned to the laboratory and maintained in a running seawater table prior to observation. Whole thalli or portions of thalli were examined using light microscopy at 100× or 400× magnification. Kelp gametophytes were identified based on the occurrence of diagnostic oogonia and antheridia, and the presence of juvenile kelp sporophytes (e.g. Bold & Wynne 1985) attached to endophytic filaments. Numbers of individual kelp gametophytes in a host plant were determined by counting only discrete filaments as individuals.

The terms oogonium and egg are retained for female gametangial structures in Laminariales even though flagella have been reported in eggs from *Laminaria* (Motomura & Sakai, 1988).

Results

We initially observed endophytic thalli in plants of *Pleonosporium vancouverianum* that were growing as epiphytes on the holdfasts of *Agarum fimbriatum*. Subsequent inspection of other red algal species at this and other sites showed that numerous, diverse taxa served as hosts for endosymbiotic gametophytes. During a five month period, over 15 host species were observed including mostly filamentous [e.g. *Pleonosporium vancouverianum*, *Callithamnion acutum*, *Scagelia pylaisaei* (Montagne) Wynne], polysiphonous [e.g. *Herposiphonia plumula* (J. G. Agardh) Hollenberg, *Pterosiphonia* spp.] and membranous species [e.g. *Polyneura latissima* (Harvey) Kylin].

These red algal hosts supported epiphytic juvenile sporophytes (Figure 1); further inspection showed that these sporophytes were associated with microscopic filamentous gametophytes embedded in the host cell wall (Figures 2–7). The endophytic filaments were common in the thicker cell walls near the base of the plant (10–25 μm thick), and absent or poorly developed in younger parts of thalli with thin cell walls (2-4 μm thick). Inspection of numerous individual thalli of *P. vancouverianum* and *C. acutum* showed that endosymbiotic gametophytes occurred in most host plants examined from our sampled habitats (over 85% of ca. 50 plants), and in all life history phases of the host (male and female gametophytes, and tetrasporangial and polysporangial thalli). Even plants less than 1 cm tall often had kelp endosymbionts.

Each host thallus had tens to more than 200 individual endophytic kelp gametophytes. These varied from separate unbranched filaments comprising one or only a few cells, to much larger, irregularly branching filaments (Figures 2, 4, 6). Filaments did not grow through the wall into the host cell cytoplasm. In none of the hundreds of host plants observed was there apparent cell death or signs of pathology of the host in response to the presence of the endophytes. Even in host filaments with a high density of endophytic cells (e.g. Figure 4), neither the cell walls nor the protoplast of the host became modified (based on bright field microscopy).

Endophytic cells were 8–28 μm long and 4–10 μm wide. They were typically longer than wide and cylindrical to barrel-shaped or somewhat irregular. Cells associated with oogonial development were more quadrate than vegetative cells. Wider cells were conspicuously flattened in the plane of the cell wall surface. Cells of male plants were typically narrower than those in female plants (4–8 *versus* 6–10 μm). All cells except antheridia contained four to many discoid, parietal chloroplasts without pyrenoids. Hairs were absent.

Oogonial development was associated with the larger, more quadrate cells. These cells were densely pigmented and contained numerous discoid chloroplasts; they developed darkly pigmented swellings above the host surface (Figure 7). Oogonia became elongate and were oriented roughly perpendicular to the host surface (Figures 7, 8). Prior to oogonial division (and presumed fertilization) the egg moved up in the stalk leaving an empty space at the base (Figure 9). The zygote divided transversely several times to form a distinctive filament before initiating par-

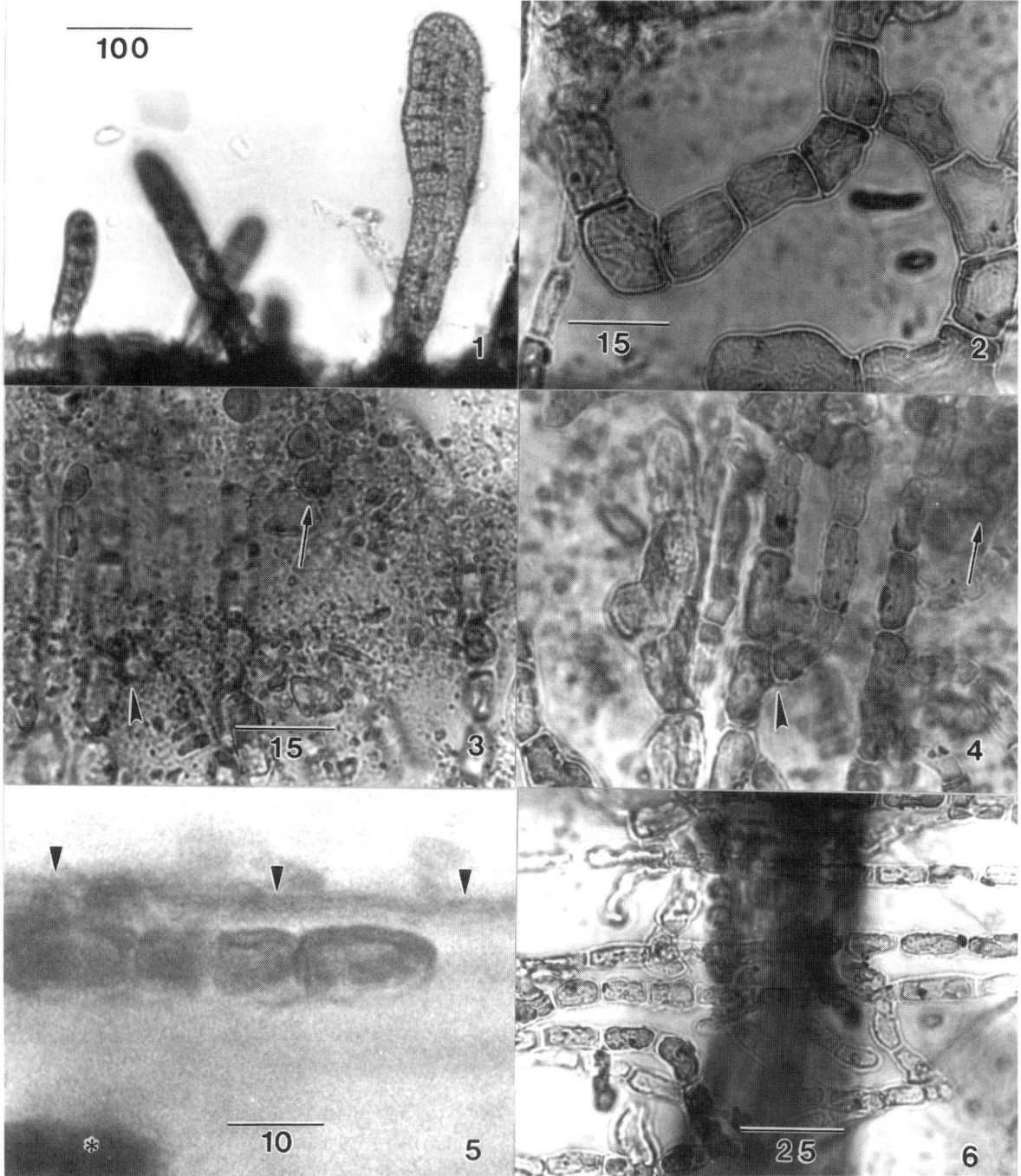

Figures 1–6. **Figure 1.** Juvenile sporophytes developing from endophytic gametophytes in cell walls of *Pleonosporium vancouverianum*. Note transverse divisions in early stage sporophyte on left. **Figure 2.** Portion of kelp gametophyte in *P. vancouverianum* showing cells with multiple discoid chloroplasts without pyrenoids. **Figures 3–4.** Optical sections showing (Figure 3) cell wall surface with epiphytic cyanobacteria (arrow) and debris and (Figure 4) gametophytic filaments within the host cell wall. Arrows and arrow heads show equivalent positions in the two focal planes. **Figure 5.** Optical section of host cell wall with gametophytic filament clearly under the surface (arrow heads) of cell wall. Asterisk indicates corner of host cell protoplast. **Figure 6.** Group of gametophyte filaments in portion of host cell. Note; dark mass in background is partially plasmolysed protoplast of host cell. Scale bars in μm.

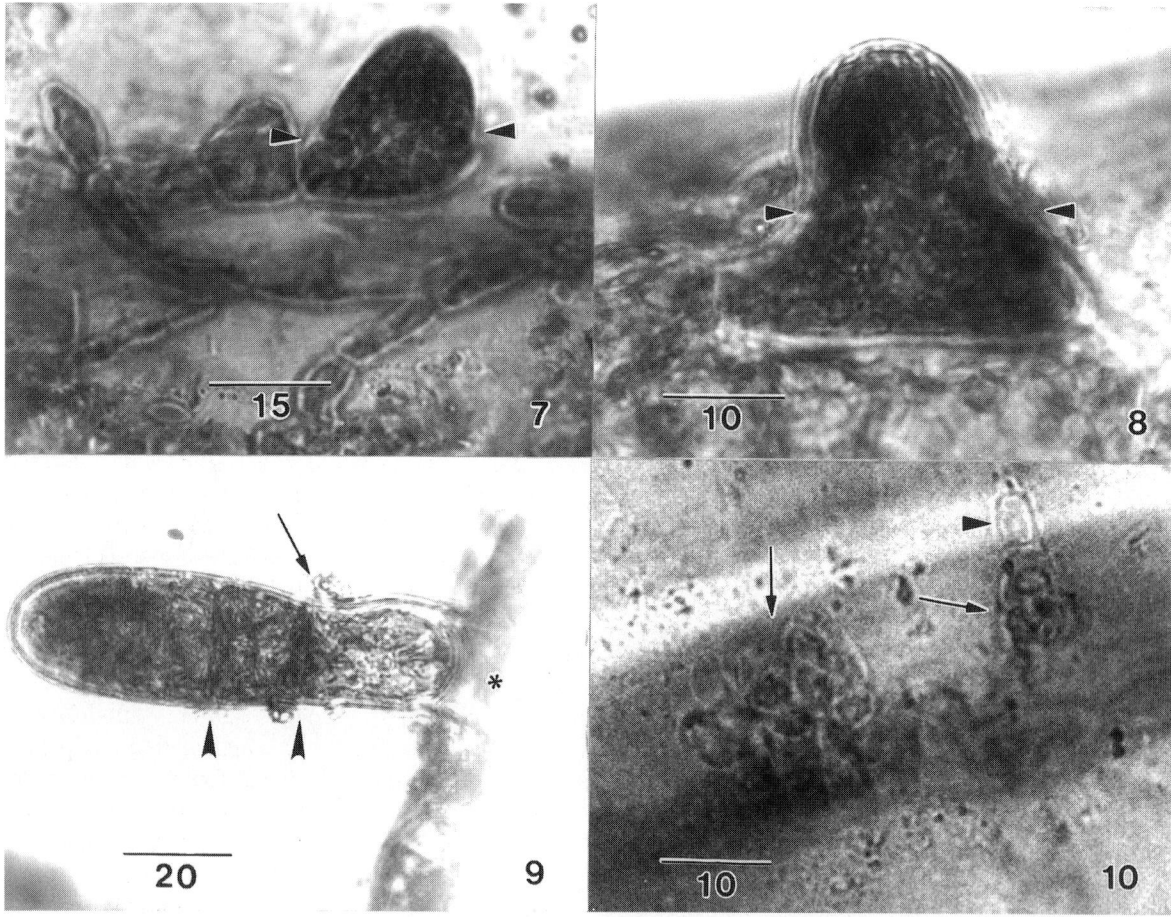

Figures 7–10. **Figure 7.** Early stage of oogonial development showing cell densely packed with chloroplasts and bulging above surface of cell wall (arrow heads). **Figure 8.** Later stage of oogonial development with dome shaped apex well above surface (arrow heads indicate upper limit of cell wall on oogonium). **Figure 9.** Cleaved zygote (arrow heads) developed from endophytic filament with distict oogonial stalk and clear space at base (asterisk). Arrow indicates apex of stalk. **Figure 10.** Optical section of host with antheridia of kelp gametophyte at host wall surface with two clusters of antheridia (arrows) and emergent antheridium (arrow head) developing from endophytic gametophyte. Clear band in upper portion of figure is transverse septum of host *Griffithsia pacifica* Abbott. Note scattered debris on wall surface.

enchymatous growth to form a flattened blade and differentiating a stipe (Figure 1). Juvenile sporophytes were commonly observed on the surface of the host (Figure 1), and where direct observation was possible, these sporophytes were always associated with endophytic filaments in the host.

Antheridia occurred singly or in clusters (Figure 10), but were often difficult to resolve because of their small size and the presence of other organisms and debris. Biflagellate motile cells 2–6 μm long were observed in the slide preparations, and some of these may have been spermatozoids. Fertilization was not observed.

Discussion

The endophytic habit is widespread in Phaeophyceae. In Ectocarpales and Chordariales, many species and even genera are endophytic [e.g. *Streblonema* (Abbott & Hollenberg, 1976), *Laminariocolax* (Burkhardt & Peters, 1998)]. Henry (1987) described *Verosphacela* (Sphacelariales) as an endophyte in *Spatoglossum*. In the Desmarestiales, an order closely related to Laminariales (Tan & Druehl, 1996), *Desmarestia antarctica* has endophytic gametophytes in the thalli of the red alga *Curdiea racovitzae* De Wildeman (Moe & Silva, 1989). In addition, Dube & Ball (1971) reported gametophytic filaments of *Desmarestia* sp. endozoic in a sea pen (*Ptilosarcus*). The prior observations of laminarialean gametophytes in no way indicates the occurrence of the endophytic habit. However, given the taxonomic distribution of endophytic species or life history stages in brown algae, it is not surprising, in retrospect, that endophytism should also occur in Laminariales.

Our observations on the vegetative and reproductive morphology of the brown algal endophytes in red algae are definitive in demonstrating that these filamentous thalli are members of the Laminariales. We know of no other group of brown algae that has the following group of features:

(a) filamentous gametophytes with multiple discoid chloroplasts devoid of pyrenoids;
(b) distinct unicellular antheridia either single and scattered or clustered in groups;
(c) an oogonium that differentiates from an intercalary (or terminal) cell in a filament to produce a stalked egg apparatus in which the egg protoplast migrates to the top of the stalk;
(d) unisexual gametophytes; and
(e) zygote embryology in which there is an initial filament that later forms a parenchymatous thallus (without the formation of hairs), that differentiates into a stipe and blade (Bold & Wynne, 1985).

The identification of our filamentous thalli as members of the Laminariales is certain; however, there is no taxonomy of laminarialean gametophytes that would allow identification at specific, generic or family level. Some species of Laminariales produce bisexual gametophytes (e.g. Chordaceae, Phyllariaceae, Pseudochordaceae: Maier, 1984; Henry, 1987; Kawai & Kurogi, 1985, respectively), however, all gametophytes that we observed were unisexual as is typical of most Laminariales (Bold & Wynne, 1985; Kawai & Kurogi, 1985). Thus it remains to be demonstrated how many species of Laminariales are involved with the endosymbiosis reported here.

The occurrence of the kelp symbiosis at multiple sites in the San Juan archipelago where different kelp species were dominant (e.g. members of Laminariaceae, Alariaceae and Lessoniaceae), allows us to infer that more than one species of Laminariales is capable of forming such a symbiosis. Using the microphotometric technique of Graham & Mitchell (1999), Graham (personal communication) has suggested to us that the variation in our samples is also consistent with our contention that several species can occur as endosymbionts. Preliminary observations on the zoospores of *Nereocystis luetkeana* and *Laminaria groenlandica* show that zoospores can attach to red algal hosts and germinate, and that some of these become endophytic in the walls of two red algal species [*Pleonosporium vancouverianum* and *Antithamnionella pacifica* (Harvey) Wollaston] (Garbary, unpublished). The complete development of these gametophytes was not followed.

Laboratory culture studies of all members of the Laminariales studied to date (e.g. Lüning & Neushul, 1978; Henry & Cole, 1982), as well as the use of spore settlement plates in nature (Reed et al., 1988) or the outplanting of laboratory settled zoospores (Hsiao & Druehl, 1973), have demonstrated that the symbiosis is not obligatory. The relative importance of the free-living *versus* endophytic gametophytes to successful re-establishment of the sporophytic phase, remains to be demonstrated. On the one hand, it is possible that the endophytic habit is an accident of spore settlement, and that it is of little or no adaptive significance. On the other hand, endophytism may represent a facultative strategy of primary importance to gametophyte growth and reproduction, at least in some species or in some environments. The irregular branching pattern and absence of heterotrichous development in the kelp gametophytes is analogous to that found in many endophytic red algae (e.g. in Acrochaetiales, Garbary et al., 1982) or endophytic green algae [e.g. *Acrochaete*, *Endophyton*, *Phaeophila* (Nielsen & McLachlan, 1986)]. These common features suggest a morphological convergence in red, brown and green algae based on the common adaptation for endophytism.

Acknowledgments

We thank the Director and Staff of Friday Harbor Laboratories for providing the research environment in which this work was performed. Dr Dianna Padilla provided encouragement, helpful discussion and access to the 'free car'. Bruno Pernet, Brian Allen & Sarah Carter kindly provided diving assistance. This work was supported by research grants from the Natural Sciences and Engineering Research Council of Canada to DJG and the Korean Science and Engineering Foundation to KYK.

References

Abbott, I. A. & G. J. Hollenberg, 1976. Marine Algae of California. Stanford University Press, Stanford, California, 827 pp.

Bold, H. C. & M. J. Wynne, 1985. Introduction to the Algae, 2nd edn. Prentice-Hall, Englewood Cliffs, N. J., 720 pp.

Burkhardt, E. & A. F. Peters, 1998. Molecular evidence from nrDNA ITS sequences that *Laminariocolax* (Phaeophyceae, Ectocarpales *sensu lato*) is a worldwide clade of closely related kelp endophytes. J. Phycol. 34: 682–691.

Chapman, A. R. O., 1984. Reproduction, recruitment and mortality in two species of *Laminaria* in southwest Nova Scotia. J. exp. mar. Biol. Ecol. 78: 99–109.

Cole, K., 1968. Gametophytic development and fertilization in *Macrocystis pyrifera*. Can. J. Bot. 46: 777–782.

Dube, M. A. & E. Ball, 1971. *Desmarestia* sp. associated with the sea pen *Ptilosarcus gurneyi* (Gray). J. Phycol. 7: 218–220.

Funano, T., 1969. Observation on the female gametophytes and the microscopic sporophytes of *Laminaria religiosa* Miyabe. Rep. Hokkaido Fish. exp. Sta. 10: 43–50.

Garbary, D., G. I. Hansen & R. F. Scagel, 1982. The marine algae of British Columbia and northern Washington: Division Rhodophyta (red algae), Class Florideophyceae, Orders Acrochaetiales and Nemaliales. Syesis 15 (supplement): 1–102.

Graham, M. H. & B. G. Mitchell, 1999. Obtaining absorption spectra from individual macroalgal spores using microphotometry. Hydrobiologia 398/399 (Dev. Hydrobiol. 137): 231–239.

Henry, E. C., 1987. Primitive reproductive characters and a photoperiodic response in *Saccorhiza dermatodea* (Laminariales, Phaeophyceae). Brit. phycol J. 22: 23–31.

Henry, E. C. & K. M. Cole, 1982. Ultrastructure of swarmers in the Laminariales (Phaeophyceae). II. Sperm. J. Phycol. 18: 570–579.

Hollenberg, G. J., 1939. Culture studies of marine algae. I. *Eisenia arborea*. Am. J. Bot. 26: 34–41.

Hsiao, S. I. C. & L. D. Druehl, 1973. Environmental control of gametogenesis in *Laminaria saccharina*. IV. *In situ* development of gametophytes and young sporophytes. J. Phycol. 9: 160–164.

Kain, J. M., 1964. Aspects of the biology of *Laminaria hyperborea* III. Survival and growth of gametophytes. J. mar. biol. Ass. U. K. 44: 415–433.

Kain, J. M., 1979. A view of the genus *Laminaria*. Oceanogr. mar. Biol. ann. Rev. 17: 101–161.

Kaneko, T., 1973. Morphology of the female gametophyte and the young sporophyte of *Laminaria japonica* var. *ochotensis* Okamura in nature. Sci. Rep. Hokkaido Fish. exp. Sta. 15: 1–8.

Kawai, H. & M. Kurogi, 1985. On the life history of *Pseudochorda nagaii* (Pseudochordaceae fam. nov.) and its transfer from the Chordariales to the Laminariales (Phaeophyta). Phycologia 24: 289–296.

Lüning, K., 1980. Critical levels of light and temperature regulating the gametogenesis of three *Laminaria* species. J. Phycol. 16: 1–15.

Lüning, K. & M. Neushul, 1978. Light and temperature demands for growth and reproduction of laminarian gametophytes in southern and central California. Mar. Biol. 45: 297–309.

Maier, I., 1984. Culture studies of *Chorda tomentosa* (Phaeophyta, Laminariales). Brit. phycol. J. 19: 95–106.

Moe, R. L. & P. C. Silva, 1989. *Desmarestia antarctica* (Desmarestiales, Phaeophyceae), a new ligulate Antarctic species with an endophytic gametophyte. Pl. Syst. Evol. 164: 273–283.

Motomura, T. & Y. Sakai, 1988. The occurrence of flagellated stages in *Laminaria angustata* (Phaeophyta, Laminariales). J. Phycol. 24: 282–285.

Nielsen, R. & J. McLachlan, 1986. Investigations of the marine algae of Nova Scotia. XVI. The occurrence of small green algae. Can. J. Bot. 64: 808–814.

Parke, M. W., 1932. Port Erin: report of the algologist. Rep. mar. biol. St. Port Erin 45: 18–21.

Reed, D. C., D. R. Laur & A. W. Ebeling, 1988. Variation in algal dispersal and recruitment: the importance of episodic events. Ecol. Monogr. 58: 321–335.

Sakai, Y. & T. Funano, 1964. Observation on the female gametophytes and the microscopic sporophytes of *Laminaria religiosa* from Oshoro Bay, Hokkaido, Japan. Sci. Rep. Hokkaido Fish. exp. Sta. 2: 1–6.

Sauvageau, C., 1915. Sur le développement et la biologie d'une Laminaire (*Saccorhiza bulbosa*). C. R. Hebd. Séanc. Acad Sci., Paris. 160: 445–448.

Sauvageau, C., 1918. Recherches sur les laminaires des côtes de France. Mém. Acad. Sci. Inst. Fr. 56: 1–240.

Tan, I. H. & L. D. Druehl, 1996. A ribosomal DNA phylogeny supports the close evolutionary relationships among Sporochnales, Desmarestiales, and Laminariales (Phaeophyceae). J. Phycol. 32: 112–118.

Using *in situ* substratum sterilization and fluorescence microscopy in studies of microscopic stages of marine macroalgae

Matthew S. Edwards

Moss Landing Marine Laboratories, P.O. Box 450, Moss Landing, CA 95039, U.S.A.
Current address: Department of Biology, University of California, Santa Cruz, CA 95064, U.S.A.
Tel: [+1] (408) 459-2357. Fax: [+1] (408) 459-2249. E-mail: edwards@biology.ucsc.edu

Key words: fluorescence microscopy, microscopic stages, recruitment, seed banks, sterilize

Abstract

The methods currently used for examining the relative contribution of microscopic stages to the persistence of natural populations of marine macroalgae can be inappropriate for use in subtidal habitats. Also, because of their microscopic size, direct examination and obtaining an estimate of recruitment, growth and mortality of these stages in the field is difficult. A method of removing microscopic algal stages from natural rock surfaces using watertight tents and water-soluble chemicals is presented. Also discussed is the use of a previously described method of fluorescent labelling these microscopic stages that, when examined under UV light, allows for their precise identification and growth to be determined. Together, these methods can be effective in examining the ecology of algal microscopic stages in the field.

Introduction

The factors influencing recruitment and persistence of marine macroalgae have received considerable attention in the ecological literature. Mostly, these studies have focused on the macroscopic stages (see reviews in Dayton, 1985; Schiel & Foster, 1986; Santelices, 1990; Vadas et al., 1992), but recently the contribution of microscopic stages to the maintenance of natural populations has been examined (Klinger, 1984; Hoffman, 1987; Hoffman & Santelices, 1991; Santelices et al., 1995; Edwards, 1996). For some species, these microscopic stages have been referred to as 'banks of microscopic forms' (Chapman, 1986) because, much like terrestrial seed banks, they persist during periods when the macroscopic stages are absent (Santelices et al., 1995). However, in contrast to the seeds of most terrestrial plants, algal microscopic stages are typically extremely small, making it difficult to directly observe their recruitment, growth and survivorship in the field (Dayton, 1985). Although it is assumed that the microscopic stages of at least some macroalgae, especially the kelps (order Laminariales), persist as filamentous gametophytes during periods when their macroscopic stages are absent (Kain, 1975; Klinger, 1984; Dayton, 1985; Reed et al., 1997), the precise identification of these stages (spore, gametophyte or embryonic sporophyte) remains largely undescribed in the field (but see Hsiao & Druehl, 1973).

A variety of experimental methods have been used to examine the importance of microscopic stages for maintaining populations of marine macroalgae. One of the most common has been to compare recruitment of algal macroscopic stages on sterilized (− microscopic stages) and non-sterilized (+ microscopic stages) substrata, and then to attribute any differences in recruitment to the presence or absence of microscopic stages (Santelices et al., 1995; Blanchette, 1996; Edwards, 1996; Reed et al., 1997). Although sterilizing natural rock substratum has commonly been done in intertidal and subtidal habitats, the physical nature of these environments has led to the use of very different methods. A standard approach to sterilizing surfaces in intertidal habitats has been to first scrape all visible organisms and then apply chemicals such as oven cleaner, formalin, alcohol or phosphoric acid, or to burn the substratum with a propane torch (see review by Littler & Littler, 1985). These methods work quite well in intertidal habitats (Wilson, 1925; Blanchette, 1996), but they may be inappropriate in subtidal habitats because the chemicals must be both water-soluble, so as not to leave a toxic residue, and able to remain on the substratum long enough to kill all microscopic organisms. Burning these substrata is problematic because, although some torches may function quite well underwater, they burn too hot and can potentially be damaging to the substratum.

Standard approaches to sterilizing substrata in subtidal and tide pool habitats have typically involved manipulating small pieces of broken substratum or boulders. These are collected in the field, brought back to the laboratory and sterilized for several minutes using an autoclave (Santelices et al., 1995) or for several days in the sun (Reed et al, 1997). Once sterilized, the boulders are returned to the field and placed alongside unsterilized ones and monitored for algal recruitment. Although relatively simple and effective, this may be inappropriate in areas where the natural substratum is solid bedrock and boulders are absent. In these types of habitats, the top few centimetres of the substratum can be chipped away with a hammer and chisel (Reed & Foster, 1984), but this is not recommended because exposing new rock may cause leaching of toxic chemicals (Littler & Littler, 1985). Given these difficulties, outplanting of artificial substrata to the field at specified times is often used. Recruitment of macroscopic algal stages on these substrata is then correlated with the availability of nearby reproductive adults (Kain, 1975). Although using artificial substrata may be informative in some situations, they can also be misleading because typically they lack the refuges characteristic of natural substrata. This can lead to artificially high susceptibility of the microscopic stages to grazing (Dayton, 1985). In addition, because of the specific nature of their surfaces, the artificial substrata themselves can bias which species occur on them (Harlin & Lindbergh, 1976; Flavier & Zingmark, 1993). Given these difficulties, a method that will allow for the sterilization of natural, subtidal rock substrata *in situ* together with the use of a previously described method of identifying specific microscopic stages in the field are presented.

Material and methods

Substratum sterilization

All field experiments were done in a giant kelp (*Macrocystis pyrifera*) forest located in Stillwater Cove, California, U.S.A. (36° 34′ N, 121° 56′ W). The study site was a 40 m diameter low-relief terrace, located at a depth of 12 m near the middle of the cove. The substratum consisted primarily of solid, conglomerate and granite rock and a few small boulders. Three 0.25 m^2 experimental plots with relatively flat surfaces were identified and marked at the corners by nailing yellow plastic tape to the substratum. The plots were then prepared for sterilization by removing all macroscopic (> 0.1 cm) organisms with a knife and then abrading the substratum with a wire brush. Two ~5 cm deep × 0.635 cm diameter holes were drilled in the middle of two opposing sides of each plot with a pneumatic drill, and an 18 cm long × 0.635 cm diameter stainless steel threaded rod was cemented into each hole. The following day, a 0.25 m^2 sterilization tent was placed over each plot and attached to the threaded rods with stainless steel nuts. These tents were made from black plastic sheeting attached to 5 cm angle-stock PVC frames (see Figure 1). A 16 cm piece of surgical tubing penetrated the middle of each tent and served as a valve which could be opened or closed by untying or tying the tubing. A ~2.5 cm thick gasket made from modelling clay wrapped in thin plastic (to prevent it from dissolving) was attached to the base of each tent such that tightening the nuts on the threaded rod forced a water-tight seal between the tents and the substratum. To check for leaks in the tent-substratum interfaces, 50 ml of fluoricine dye (1 g fluoricine per 50 ml filtered seawater) was injected into each tent with a hypodermic needle and the tent perimeter examined for leaking dye. Leaks were easily identified and plugged with petroleum-base clay, making the tents watertight.

To remove all microscopic algal stages from the experimental plots, 1 l of household bleach was injected into each tent through its valve with a large syringe. The valve was then closed and the tents left for two days. No damage was observed to the plants immediately outside the tents and subsequently *Desmarestia* recruits were observed directly on the outside edges of the sterilized plots. At the time of tent removal, all macroscopic algae were scraped from three additional 0.25 m^2 plots, and three unmanipulated control plots were established by marking them at the corners with yellow tape. To examine the efficiency of the sterilization method, a second set of three sterilized and three 'scraped-only' plots were established using the methods described above, but immediately following removal of these tents, a ~10 cm^2 piece of substratum was removed from each plot using a hammer and chisel. These substrata were individu-

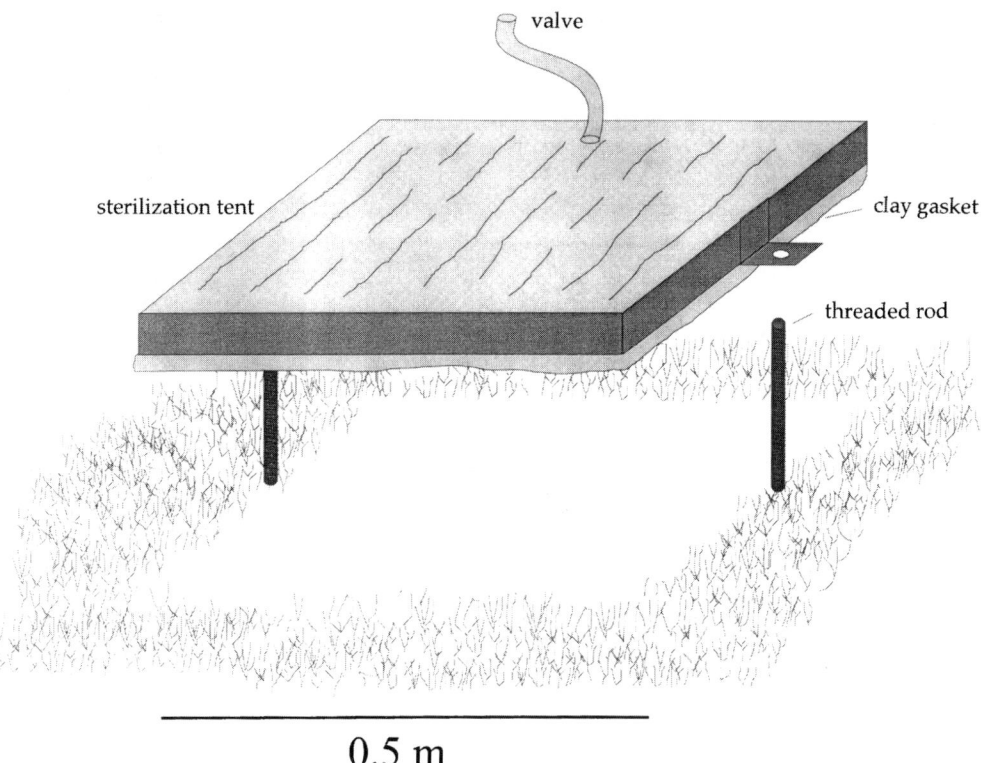

Figure 1. Diagram of sterilization tents used to remove all microscopic stages from experimental substrata. Tents measure 0.25 m² and were made out of reinforced plastic tarping attached to PVC frames ($n = 3$).

ally placed in black, plastic bags to avoid exposure to direct sunlight and transferred to the laboratory where they were placed in individual culture dishes with nutrient-enriched, 0.2 μm filtered seawater. The substrata were then incubated under an irradiance of 75 μmoles m^{-2} s^{-1}, a temperature of 9 °C and a light:dark photoperiod of 16:8 h for two months. The culture medium was changed weekly, and all substrata were monitored for the appearance of macroscopic algal stages. During this time, all experimental field plots were also monitored for recruitment of macroalgae. Monitoring of these field plots continued for two years in order to examine long-term recovery of perennial turf algae. Differences in algal densities among the field plots at both six weeks and two years after tent removal were each tested with a one-way Model I analysis of variance (ANOVA). A Bonferroni-corrected planned comparison was then used to examine differences between the sterilized and scraped-only plots six weeks after tent removal. Differences in recruitment on sterilized vs. scraped-only substrata cultured in the laboratory were examined using a t-test.

Examination of microscopic stages

To identify specific algal microscopic stages in the field, zoospores of the kelps *Macrocystis pyrifera* and *Pterygophora californica* were each settled onto glass microscope slides overnight. The following day, these slides were immersed in a 20% solution of *Fungi-Flour*™ (0.01% Cellufluor®) and filtered seawater for 24 h. *Fungi-Fluor*™ (Polysciences, Inc., Warrington, PA 18976; CAT# 17442A) is a non-lethal biostain that binds non-specifically to beta-linked polysaccharides and fluoresces (emittence 400–440 nm) when excited by UV light (240–400 nm; peak excitation 345–365 nm; Baselski & Robinson, 1989). This staining method has been used previously to stain the cell walls of laminarian (Cole, 1964; Hsiao and Druehl, 1973) and fucalean (Nakazawa et al., 1969; Serrão et al., 1996) microscopic algal stages and was found to be non-toxic and have no effects on cell growth (Nakazawa et al., 1969). After 24 h, the slides were removed from the staining solution and transferred to individual culture dishes containing nutrient enriched (70 ml Alga Grow® solution l^{-1} seawater) filtered seawater. At this time, additional unstained zoospores of *M. pyrifera* and *P. californica* were settled onto these and other slides, yielding the following treatment combinations: stained *M. pyrifera* alone, stained *P. californica* alone, stained *M. pyrifera* + unstained *P. californica*, stained *P. californica* + unstained *M. pyrifera*, unstained *M. pyrifera* alone and unstained *P. californica* alone ($n = 5$ dishes per treatment). All dishes were then cultured under 40 μmoles m^{-2} s^{-1}, 15 °C, 14:10 L:D conditions for five weeks; each week, two dishes from each treatment were randomly selected and the microscopic thalli on their slides examined under UV with fluorescence microscopy (transmittance 330–385 nm).

To determine if the above method was effective at identifying the ages and growth rates of specific microscopic algal stages in the field, zoospores of the annual alga *Desmarestia ligulata* were settled onto frosted glass microscope slides overnight and stained using the methods described above. The slides were then placed on three PVC slide racks (30 slides per rack), transferred to the field in opaque, plastic bags and bolted to the substratum at a depth of 12 m. Three slides per rack were then collected each week for six weeks and examined under UV light. After the stained thalli were identified, their size (longest axis length) was measured with an ocular micrometer using standard light microscopy at 100×. Growth rates were compared between field-grown thalli and those cultured under similar conditions (75 μmoles m^{-2} s^{-1}, 9 °C, 16:8 h) in the laboratory. Although these laboratory conditions were established on the basis of measurements made in the field at the beginning of the study, average irradiance and nutrient levels were consistently higher in the laboratory.

Results

Substratum sterilization

No macroalgae recruited to any of the substratum pieces removed from sterilized plots and then cultured in the laboratory, whereas numerous (18.33 ± 9.71 recruits 10 cm^{-2}, mean ± SE) individuals, mostly fleshy red algae and unidentified Laminariales, were observed on substratum pieces removed from scraped-only plots. These differences were not significant (t-test: $p = 0.34$), most likely due to low statistical power (> 0.1; Cohen, 1988) which resulted from insufficient replication and high variation among the scraped substratum.

Differences in initial recruitment of macroalgae among the field plots, six weeks after tent removal, were also not significant (ANOVA: $F_{2,6} = 4.38$; $p = 0.066$), although differences between sterilized and scraped-only plots were significant (Bonferroni = 0.03). No macroalgae were observed in any of the sterilized plots, while numerous individuals (15.0 ± 6.8 recruits per plot, mean ± SE), of mostly *D. ligulata*, *M. pyrifera* and *P. californica*, were observed in each of the scraped-only plots. Taken together, these results indicate that no macroalgae recruited to any sterilized substratum while significantly more macroalgae were observed on scraped-only substratum (combined probability: $p = 0.04$; Rice, 1990). This indicated that the method of sterilization effectively removed the algal microscopic stages while scraping the substratum alone did not.

A few weeks after tent removal, benthic diatoms were abundant on all scraped-only substrata, but no diatoms were observed on any of the sterilized substrata. Furthermore, three months after tent removal, macroalgae were found growing at all the field plots. Together, this suggested that the method of substratum sterilization did not leave a toxic residue that might prevent algal recruitment. Long-term monitoring of field plots further indicated that since there were no differences in turf algal abundance between the experimental and control plots, two years after tent removal (ANOVA: $F_{2,6} = 1.807$; $p = 0.243$), recovery of perennial turf algae was also not affected by the sterilization process.

Examination of microscopic stages

When examined under UV light with fluorescence microscopy, labelled microscopic stages of *M. pyrifera* and *P. californica* fluoresced blue, making them easily identifiable. Unlabelled thalli, in contrast, fluoresced red due to the excitation of their chlorophyll (Figure 2A). This allowed for easy differentiation between labelled and unlabelled stages. These microscopic stages were observed to be gametophytes. As they grew, the stain became diluted throughout their thalli and thereby decreasing in fluorescence intensity. However, they remained relatively easy to identify even after they produced sporophytes, with the female gametophytes fluorescing blue and the emerging sporophytes fluorescing red (Figure 2B).

The microscopic stages of *D. ligulata* that were labelled and outplanted to the field on slides, were easily distinguished from the microscopic stages of

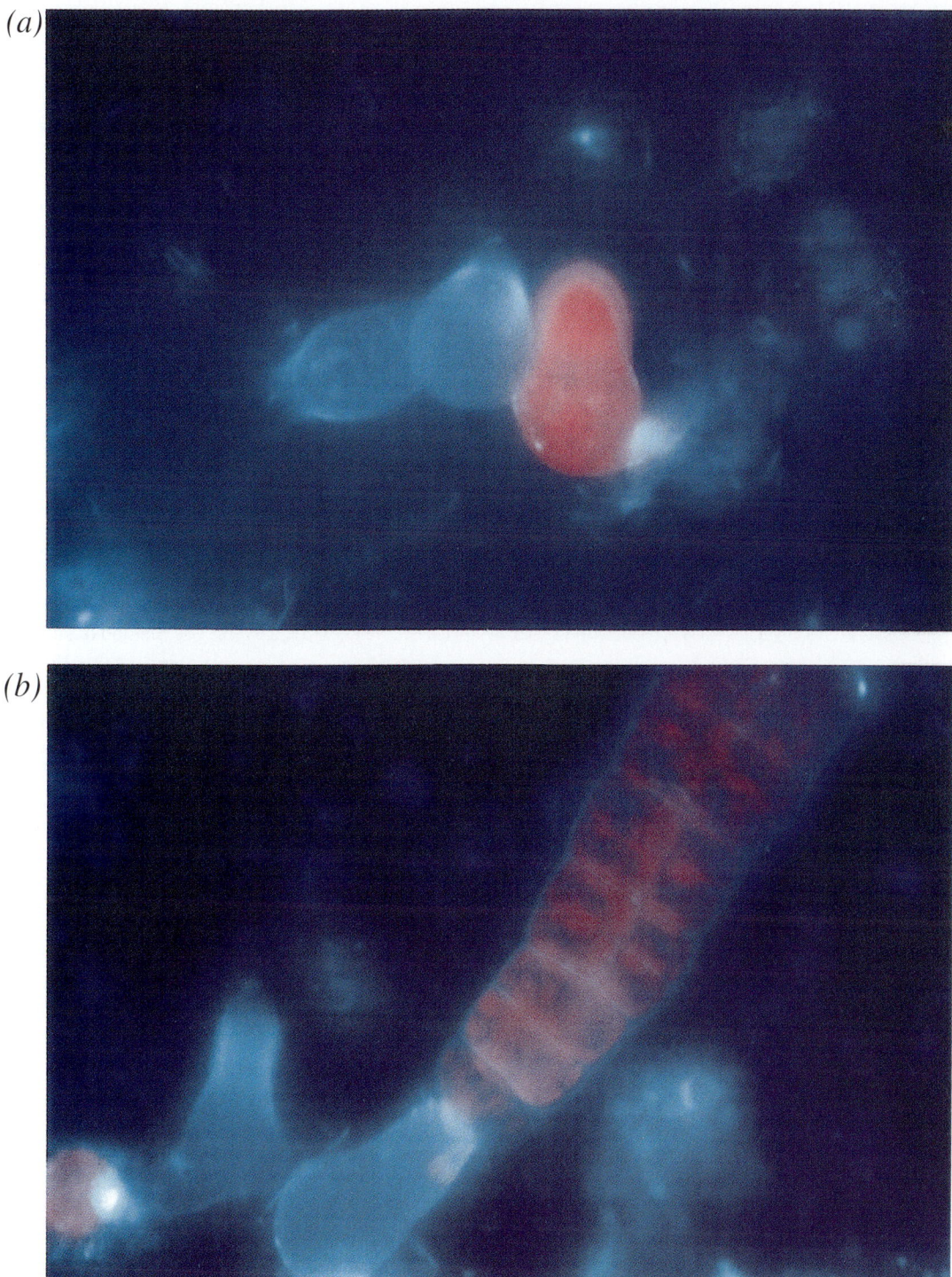

Figure 2. A. Fluorescently labelled *P. californica* gametophyte (blue) and unlabelled *P. californica* gametophyte (red) under UV light (red fluorescence is due to excitation of chlorophyll by UV light). Labelled gametophyte is 24 h older than unlabelled gametophyte. B. Single celled, labelled *P. californica* gametophyte (blue) after sporophyte (red) production (unlabelled multicellular tissue).

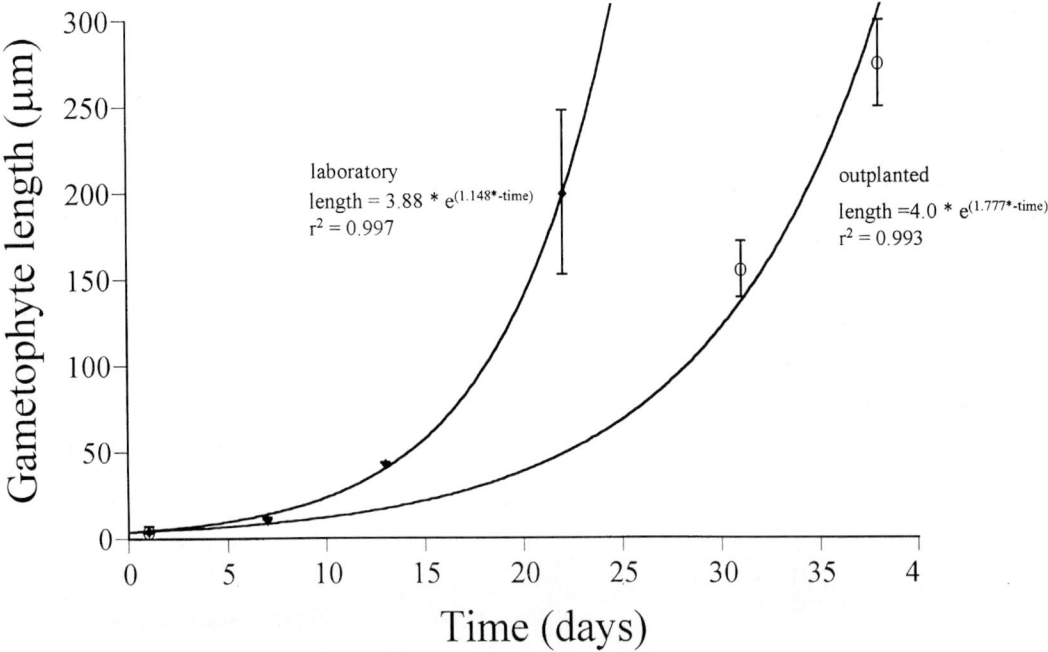

Figure 3. Growth of microscopic stages of *D. ligulata* (gametophytes) cultured in the laboratory and in the field.

other macroalgae that naturally settled on the slides when examined under UV light. The growth rates of the transplanted stages were estimated and compared with laboratory grown stages. Results indicated that *D. ligulata* microscopic stages exhibited exponential growth under both field and the laboratory conditions, with growth appearing slightly faster in the laboratory (Figure 3).

Discussion

With increasing appreciation of the contribution of banks of microscopic forms to the recruitment of natural populations of marine macroalgae, the ability to directly observe these microscopic stages in the field is becoming more important. This, however, has been problematic for subtidal species that occur on solid rock surfaces. Here, watertight tents provided an efficient way to remove the microscopic algal stages from these substrata using common household bleach. These tents allowed high concentrations of bleach to remain in continuous contact with the substratum for at least two days, thereby killing all the resident microscopic algae. By preventing the dilution of the sterilizing agent into the surrounding seawater, these tents allow for the use of a variety of water-soluble chemicals that have been available for use in intertidal habitats, e.g. formalin, phosphoric acid or alcohol. Also, since these chemicals are water-soluble, they will not leave toxic residues that may affect future settlement of algal propagules.

Fluorescent staining of algal propagules has been used to identify certain species and examine their propagule growth. The results from this study suggest that this technique can be used as a powerful tool in studying algal microscopic stages in the field. Here, it allowed for the identification of known thalli and for the precise measurement of their growth rates in the field. It can, therefore, be used to examine growth of algal microscopic stages under a variety of environmental conditions. This can be done *in situ*, whereas previously, growth of microscopic stages in various field conditions has typically been inferred from laboratory studies. This technique remained effective at identifying the microscopic stages even after production of new sporophytes, which greatly increases confidence in its ability to precisely identify labelled thalli throughout the various stages of their development. These different tools, when used together, should greatly increase our ability to examine the ecology and importance of microscopic stages to the recruitment of subtidal populations of marine macroalgae in the field.

Acknowledgements

I thank M. Foster for advice throughout this project. I also thank my field assistants: R. Clark, J. Downing, C. Roberts and especially S. Lamerdin who helped haul my tents to and from my study sites. I would like to extend my warmest appreciation to S. Tanner for her help with the fluorescence microscopy and E. Danner for his help with computer generated graphics. E. Danner, A. Boxshall, M. Foster and R. Walder reviewed this manuscript and offered helpful criticism. This project was funded in part by grants from The Dr Earl H. and Ethyl M. Meyers Oceanographic and Marine Biology Trust and the Packard Foundation.

References

Baselski, V. S. & M. K. Robinson, 1989. A staining kit for detection of opportunistic pathogens in bronchoalveolar lavage specimens. Amer. Clin. Lab. July, 1989: 36–37.

Blanchette, C. A., 1996. Seasonal patterns of disturbance influence the recruitment of the sea palm, *Postelsia palmaeformis*. J. exp. mar. Biol. Ecol. 197: 1–14.

Chapman, A. R. O., 1986. Population and community ecology of seaweeds. Adv. mar. Biol. 23: 1–161.

Cohen, J., 1988. Statistical Power Analysis for the Behavioral Sciences 2nd edn. Lawrence Erlbaum Associates, New Jersey, 567 pp.

Cole, K., 1964. Induced fluorescence in gametophytes of some Laminariales. Can. J. Bot. 42: 1173–1183.

Dayton, P. K., 1985. Ecology of kelp communities. Ann. Rev. Ecol. Syst. 16: 215–245.

Edwards, M. S., 1996. Factors regulating the recruitment of the annual alga *Desmarestia ligulata* along the central California coast. M.S. Thesis. San Francisco State University, 67 pp.

Flavier, A. B. & R. G. Zingmark, 1993. Macroalgal recruitment in a high marsh creek of North Inlet Estuary, South Carolina. J. Phycol. 29: 2–8.

Harlin, M. M. & J. M. Lindbergh, 1977. Selection of substrata by seaweeds: optimal surface relief. Mar. Biol. 40: 33–40.

Hoffman, A. J., 1987. The arrival of propagules at the shore: a review. Bot. mar. 30: 151–165.

Hoffman, A. J. & B. Santelices, 1991. Banks of microscopic forms: hypotheses on their functioning and comparisons with seed banks. Mar. Ecol. Prog. Ser. 79: 185–194.

Hsiao, S. I. C. & L. D. Druehl, 1973. Environmental control of gametogenesis in *Laminaria saccharina*. IV. *In situ* development of gametophytes and young sporophytes. J. Phycol. 9: 160–164.

Kain, J. M., 1975. Algal colonization of some cleared subtidal areas. J. Ecol. 63: 739–765.

Klinger, T., 1984. Allocation of the blade surface area to meiospore production in annual and perennial representatives of the genus *Laminaria*. M.S. Thesis, Univ. British. Columbia, Vancouver, 96 pp.

Nakazawa, S., K. Takamura & M. Abe, 1969. Rhizoid differentiation in *Fucus* eggs labelled with Calcofluor White and birefringence of cell wall. Bot. Mag. Tokyo 82: 41–44.

Littler, M. M. & D. S. Littler, 1985. Ecological field methods: macroalgae. Handbook of Phycological Methods. Cambridge University Press, New York, 617 pp.

Reed, D. C. & M. S. Foster, 1984. The effect of canopy shading on algal recruitment and growth in a giant kelp forest. Ecology, 65: 937–948.

Reed, D. C., T. W. Anderson, A. W. Ebeling & M. Anghera, 1997. The role of reproductive synchrony in the colonization potential of kelp. Ecology, 78: 2443–2457.

Rice, W. R., 1990. A consensus combined *p*-value test and the family-wide significance of component tests. Biometrics 46: 303–308.

Santelices, B., 1990. Patterns of reproduction, dispersal and recruitment in seaweeds. Oceanogr. Mar. Biol. ann. Rev. 28: 177–276.

Santelices, B., A. J. Hoffman, D. Aedo, M. Bobadilla & R. Otaíza, 1995. A bank of microscopic forms on disturbed boulders and stones in tide pools. Mar. Ecol. Prog. Ser. 129: 215–228.

Schiel, D. R. & M. S. Foster, 1986. The structure of subtidal algal stands in temperate waters. Oceanogr. mar. Biol. Ann. Rev. 24: 265–307.

Serrão, E. A., L. Kautsky & S. H. Brawley, 1996. Distributional success of the marine seaweed *Fucus vesiculosus* L. in the brackish Baltic Sea correlates with osmotic capabilities of Baltic gametes. Oecologia 107: 1–12.

Wilson, O. T., 1925. Some experimental observations of marine algal successions. Ecology 6: 303–311.

Vadas, R. L., S. Johnson & T. A. Norton, 1992. Recruitment and mortality of early post-settlement stages of benthic algae. Br. phycol. J. 27: 331–351.

The sea star *Asterina pectinifera* causes deep-layer sloughing in *Lithophyllum yessoense* (Corallinales, Rhodophyta)

D. Fujita
Toyama Prefectural Fisheries Research Institute, Namerikawa, 936-8536, Japan
E-mail: d-fujita@nsknet.or.jp

Key words: Asterina pectinifera, Corallinales, *Lithophyllum yessoense*, regeneration, sea star, sloughing

Abstract

Deep-layer sloughing is a recently described mode of surface shedding in some encrusting coralline algae. Several causative agents or ecological roles have been suggested for its occurrence, but none have been proven. During ecological studies of urchin-dominated barren grounds in southwestern Hokkaido, the dominant encrusting coralline species, *Lithophyllum yessoense*, was found to be sloughing beneath the sea star, *Asterina pectinifera*, in shallow waters. The sea stars often stayed long in one position and left body-shaped white scars on the encrusting thalli. Anatomical studies of the scars revealed that a deep layer, well below the vegetative initials and the bottom of submerged conceptacles, was being shed. The upper layer of living columnar cells in the medulla became new vegetative initials, producing new epithallial layers above them. Deep-layer sloughing also occurred on the thalli in running-water aquarium experiments, when thalli were exposed to the sea stars. Although the thalli were heavily covered with small epiphytic algae, clean surfaces were found just below the flakes of the sloughed layer. This mode of surface shedding may play an important role in recovery from damage on barren grounds where bottom feeders are abundant.

Introduction

Deep-layer sloughing is a mode of surface shedding recently described in two species of encrusting coralline algae (Corallinales, Rhodophyta) in South Africa: a species of Mastophoroideae, *Spongites yendoi* (Foslie) Chamberlain (Keats et al., 1993, 1994) and a species of Lithophylloideae, *Lithophyllum neoataleyense* Masaki (Pueschel & Keats, 1997). Keats et al. (1993) demonstrated that the deep-layer sloughing of *S. yendoi* was the surface shedding from deep within the thallus, well below the layer of initial cells, sometimes including a layer of buried conceptacles and followed by regeneration of a new layer of initial cells.

Several causative agents or ecological roles have been attributed to deep-layer sloughing on encrusting coralline algae. Keats et al. (1993, 1994) suggested that the sloughing of *S. yendoi* contributes to the alga's relative thin thallus, thereby conferring the advantages of faster growth and stronger attachment, and may also serve to discard old conceptacles or cells damaged by grazers. In *L. neoataleyense*, collected from a cove subjected to considerable sand movement, sand burial, sand abrasion or bacterial invasion was the possible source of damage inducing deep-layer sloughing, but no evidence of mechanical damage or pathogens was detected in transmission electron microscopy (Pueschel & Keats, 1997).

During ecological studies of urchin-dominated barren grounds (sensu Lawrence, 1975) along the southwestern coast of Hokkaido in the Sea of Japan, the author found that a shallow water species of the sea star *Asterina pectinifera* Müller et Troschel may have caused deep-layer sloughing in the dominant subtidal coralline alga, *Lithophyllum yessoense* Foslie. *L. yessoense* is a perennial cushion-like encrusting species with a smooth upper surface, in which epithallial shedding was the only mode of surface shedding previously reported (Masaki et al., 1981, 1984). In the present paper, results of field observations, anatomy and aquarium experiments on the deep-layer sloughing in *L. yessoense* are reported.

Materials and methods

Sampling and field observations were made in the subtidal zone near Kaitorima Fishing Port (42° 50′ N, 139° 53′ E), Taisei, Hokkaido, by SCUBA diving on June 6, September 8 and November 15, in 1994. Species composition of encrusting coralline algae and other macroalgae at the coasts of Taisei were described in Noro et al. (1983) and Fujita (1989), respectively. Densities of the sea star *A. pectinifera* and the sea urchin *Strongylocentrotus nudus*, a causative agent maintaining the urchin-dominated barren grounds (see Fujita, 1989; Agatsuma, 1997), were determined by randomly placing twenty 50×50 cm quadrats at a depth of 2–3 m on November 15 in 1994.

Thalli of *L. yessoense* were brought alive in sea water to Taisei Hatchery (TH) near the sampling sites for aquarium experiments, while others were fixed in 10% formalin and decalcified in 0.6 M HNO_3 for anatomical observations in Toyama Prefectural Fisheries Research Institute (TPFRI). Cross sections were made of paraffin-embedded specimens and stained with 1% toluidin blue or 1% DAPI.

In TH, a rock (15 cm in diameter) covered with encrusting coralline algae (largely *L. yessoense*) as well as five thalli (4 cm in diam.) of *L. yessoense* detached from rocks were placed at the bottom of an outdoor aquarium ($29 \times 48 \times 28$ cm) together with five individuals of *A. pectinifera* (ray length = 5 cm). In another aquarium (control), the same sets of corallines but no animals were included. The experiments were conducted using running fresh sea water three times in a year after collecting new materials. Close-up photographs were taken at least twice a month to check any events (e.g. occurrences of surface shedding, growth of epiphytes, necrosis) on the coralline algae in the tank. The detail of the aquarium experiments (including other animals) will be described elsewhere (Fujita et al., in prep.).

Results

Field observations

Near Kaitorima Fishing Port, many pentagonal white scars were found on encrusting coralline algae around the sea star *A. pectinifera* at depths of 1–4 m (Figures 1 and 2) in June, September and November. The mean densities of the sea star *A. pectinifera* and the sea urchin *S. nudus* in November 1994, were 2.6 m^{-2} and 9.6 m^{-2}, respectively. The sea star *A. pectinifera* was the only pentagonal species in the study area, while two other long-rayed species of sea star, *Asterias amurensis* Lütken and *Aphelasterias japonica* (Bell) were rarely found. Around the latter two species, however, no white scars were observed. Grazing marks of the sea urchin *S. nudus*, as well as the limpet *Acmaea pallida* (Gould) on *L. yessoense*, were described elsewhere (Fujita, 1992). Usually, up to ten or more pentagonal white scars were scattered around one or two individuals of *A. pectinifera* (Figure 1). Each of the white scars was of the body size and shape of the sea star (8–10 cm in diameter) (Figure 2). The white scars were only found on encrusting coralline algae covering various types of subtidal hard substrata; it was easier to find them on flat surfaces of concrete blocks or vertical walls of rocks than on the uneven surfaces of boulders.

Most of the encrusting coralline algae were *L. yessoense*, although other coralline and fleshy algae were sometimes included in the areas of white scars (Figure 2). No macroscopic animals, which were expected as prey, were found in the protruded stomach of the sea stars when fifty individuals were overturned at each sampling time. White scars on *L. yessoense* (Figure 3) resembled white patches of epithallial shedding (Masaki et al., 1984, Figure 10). The white scars, however, were exclusively pentagonal in outline (Figures 1 and 2) and composed of thicker flakes, containing conceptacles (Figure 3), although somewhat vague scars were recognized among a series of white scars around the sea stars (Figure 1). The flakes of white scars were easily removed by diver's hand and the surface below was vivid pink.

Anatomy

Cross sections revealed that many cell layers, well below the vegetative initial cells and bottom of buried conceptacles (when present), were being shed in the pentagonal white scars of *L. yessoense* (Figure 4). The sloughed layers were 100–300 μm thick, and were composed of epithallial cells, vegetative initial cells and columnar cells. These features were similar to those of deep-layer sloughing in *S. yendoi* (Keats et al., 1993). Just below the sloughed layers, a layer of unwounded columnar cells became vegetative initial cells and regenerated new epithallial cells in 2–3 layers (Figure 5). No conceptacle initiation was noticed below the sloughed layers during the present study.

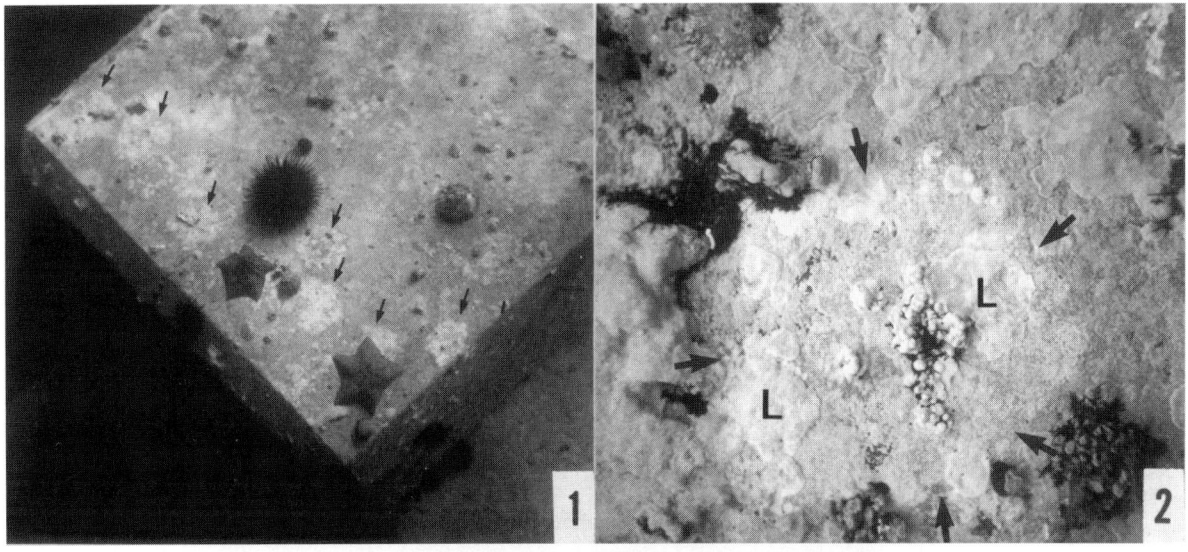

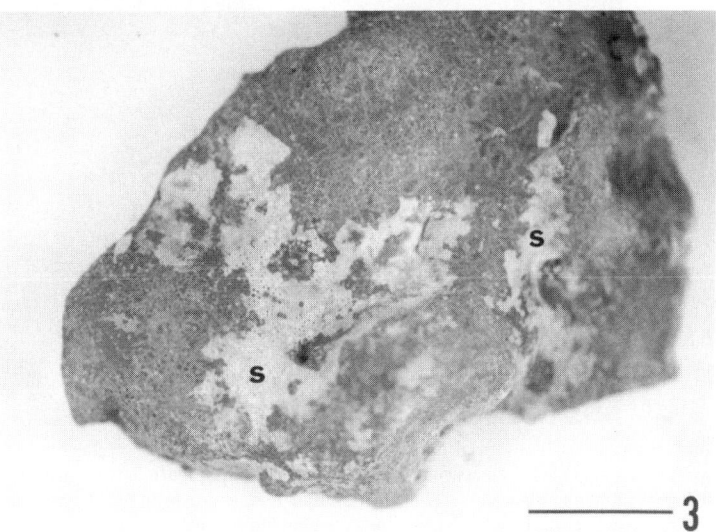

Figures 1–3. **Figure 1.** Pentagonal white scars (arrows) on encrusting coralline algae covering concrete blocks. The sea stars: *Asterina pectinifera*; the sea urchin: *Strongylocentrotus nudus*. Diameters of white scars range from 8 to 10 cm. **Figure 2.** Close-up photograph of a pentagonal white scar (surrounded by arrows) (8 cm in diameter) found on a vertical rock wall. Cushion-like crusts (L) are thalli of *L. yessoense*. **Figure 3.** White flakes (s) remained on collected samples of *L. yessoense*. Pores in white flakes are old conceptacles discarded. Bar = 1 cm.

Aquarium experiments

The sea star *A. pectinifera* caused deep-layer sloughing in *L. yessoense* in the aquarium. The deep-layer sloughing was found both on thalli covering rocks and on those detached from hard substrata (Figure 6) in every aquarium with the sea star, but not in the aquarium of control. Judging from the photographs taken twice a month, the earliest records of deep-layer sloughing were 24 d, 14 d and 4 d after commencements in June, September and November experiments, respectively. White flakes, probably removed by crawling sea stars, were sometimes scattered at the bottom of aquariums, while vivid pink-coloured surfaces were found on thalli just below the white flakes. However, thalli were more or less covered with small filamentous algae after a month. Deep-layer sloughing could not remove the epiphytes at the end of each culture (2–7 months after commencements) because

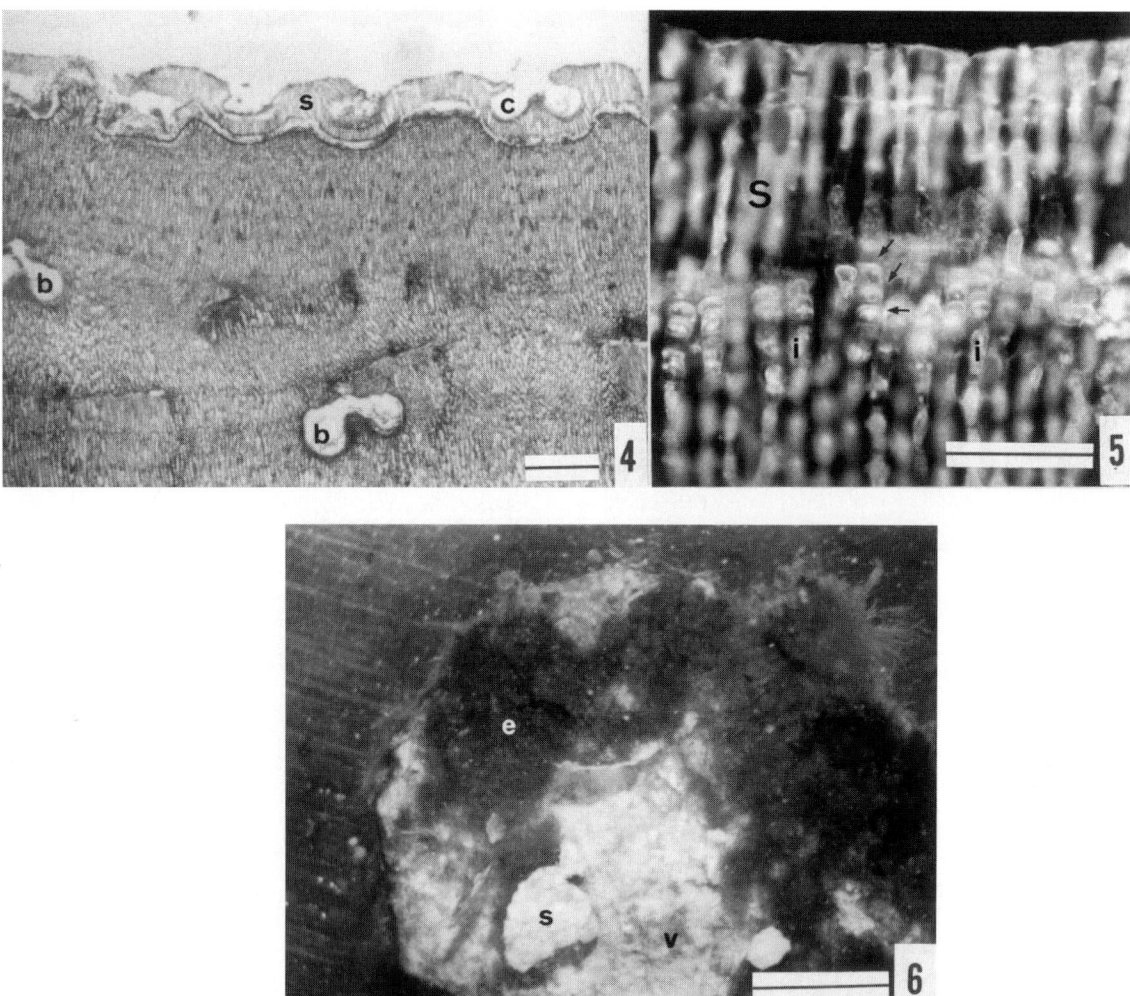

Figures 4–6. **Figure 4.** Cross section of a white-scarred thallus of *L. yessoense*, where many cell layers deep enough to contain a layer of conceptacles (c), were being shed. See also the buried conceptacles (b). Bar = 250 μm. **Figure 5.** Cross section of a white-scarred thallus of *L. yessoense*, showing new epithallial layers of two or three cells (arrows) thick regenerated from initial cells (i), just below the sloughed layers (s). Bar = 100 μm. **Figure 6.** *L. yessoense* in which deep-layer are sloughing but heavily covered with epiphytes (e) after coexisting with *A. pectinifera* in outdoor aquarium for two months. Pink-colored vivid surface (v) appeared after removing sloughed cell layers (s) by hand.

they formed compact mats and covered the whole top surfaces of *L. yessoense* (Figure 6).

Even when heavily covered with the epiphytes, pink-coloured surfaces were present under the thick accumulation of sloughed layers (Figure 6). The sea star *A. pectinifera* could not remove epiphytes and algal cover in the aquariums either.

Discussion

The sea star *A. pectinifera* is an omnivorous species inhabiting soft and hard substrata, and detritus is the most common food item (Levin et al., 1987, as *Patiria pectinifera*), while it has often been reported as a predator of released juvenile abalone shells in Japan (e.g., Hoshikawa et al., 1997). In the present study, no macroscopic prey animals were found even when the stomach of overturned sea stars were protruded. In addition, *A. pectinifera*, when given a live juvenile abalone fixed on a coralline-covered boulder fed on the abalone meat after it died, but never caused deep-layer sloughing among corallines including *L. yessoense* (D. Fujita, unpublished data). There is a possibility that *A. pectinifera* may feed on detritus or rich microscopic epibionta or inbionta (Fujita, 1994) during the stay on *L. yessoense* in nature. However, protruding the stomach for feeding is probably not the cause of deep-layer sloughing because the outline of white scars (sloughing layers) was not the amorphous shape of stomach, but the body-sized pentagon. Consequently, staying long in one position by *A. pectinifera* is the highly possible reason of deep-layer sloughing in *L. yessoense*. Cutting out light and limiting the supply of oxygen and nutrients by overlying, or the retention of digestive juices (released from the stomach) by the margin of the sea star, may cause cell damage resulting in white scars of *L. yessoense*. The time required for the sea star to cause deep-layer sloughing remained unclear but may be shorter than four days.

Keats et al. (1994) suggested that the possible reason of deep-layer sloughing in *S. yendoi* was a means of preventing the thallus from becoming thick and weakened by undercutting or from lowering the horizontal growth rate. In this relatively thin species of 1.0 mm thick (Chamberlain, 1993), sloughing deep layers had made up 26–52% of the thallus thickness (Keats et al., 1993). On the contrary, *L. yessoense*, a cushion-like species of more than 5 mm thick (Fujita, 1994) may have the advantage of surviving under the long stay of *A. pectinifera*. The fact that thin coralline crusts of *Pneophyllum* and *Titanoderma* died when covered with *A. pectinifera* in Toyama Bay (Fujita, 1996a) may support this hypothesis. The thickness of the sloughed layer is no more than 10% of the thallus of *L. yessoense*, thus, it remains thick even after sloughing.

Layers sloughed from *L. yessoense* also contained one layer of conceptacles (Figure 4) as reported in *S. yendoi* (Keats et al., 1993). In *S. yendoi*, discarding old conceptacles was one of possible reasons for deep-layer sloughing (Keats et al., 1993; 1994). However, *L. yessoense* (a perennial species) usually buries conceptacle layers in the medulla every year (Fujita, 1990, 1994), so that there seemed to be no need to discard old conceptacles, even less with the help of *A. pectinifera*.

Another hypothesized function of deep-layer sloughing was antifouling. Our early studies reported the occurrence of epithallial shedding in *L. yessoense* and suggested that this might play some role in preventing recruitment of kelp in barren grounds (Masaki et al., 1981, 1984). However, thalli of *L. yessoense* with deep-layer sloughing could not remove epiphytes in outdoor culture experiments; only grazing by sea urchins or snails could remove epiphytes on thalli of *L. yessoense* (Fujita et al., in prep.). The antifouling hypothesis was also negated in field experiments of *S. yendoi* (Keats et al., 1994). At present the author has no reason to believe in the importance of surface shedding in antifouling of *L. yessoense*.

Deep-layer sloughing of *L. yessoense* should be related to regeneration since the medulla of this species has high potential of morphogenesis (Fujita, 1992, 1996b; Fujita et al., 1992). Deep-layer sloughing and the subsequent regeneration may be advantageous for *L. yessoense* survival under frequent bottom feeding activities by the sea star *A. pectinifera* in urchin-dominated barren grounds. In such open spaces, most recruits of algae were exposed to bottom feeding by sea stars as well as grazing by sea urchins, as Dayton et al. (1984) demonstrated that kelp germlings were easily eaten by *Patiria miniata*, a closely related species of *A. pectinifera*.

Conclusion

Deep-layer sloughing was recorded in a subtidal dominant species of encrusting coralline alga *Lithophyllum yessoense* from southwestern Hokkaido, Japan. The causative agent was a bottom feeder (non-grazer) of sea star, *Asteria pectinifera*. The sea star left pentagonal white scars on the surface of *L. yessoense* after staying long in one position. Further studies are needed to clarify whether cutting resource (light, oxygen or nutrients) or chemical digestion caused the cell damages of the white scars. The deep-layer sloughing of *L. yessoense* closely resembled that of *Spongites yendoi* (Keats et al., 1993, 1994) and *Lithophyllum neoataleyense* (Pueschel & Keats, 1997) in thickness, in containing a layer of conceptacles and in regenerating new initial cells and epithallial cells. The deep-layer sloughing and the subsequent regeneration may contribute to the survival of encrusting thick thalli of *L. yessoense* under the frequent bottom feeding activities by the sea star *A. pectinifera* in open spaces such as urchin-dominated barren grounds. This is the first report on deep-layer sloughing in areas other than South Africa and on interactions between Asteroid and encrusting coralline algae.

Acknowledgements

I am grateful to Mrs K. Koyama and Y. Tsuji of Taisei Hatchery for their assistance to aquarium experiments, and to Kaitorima Branch of Hiyama Fisheries Cooperative for permitting SCUBA diving. I also thank to

Dr Derek Keats and an anonymous reviewer for improving the manuscript. A part of this research was financed by Fisheries Agency, Ministry of Agriculture, Forestry and Fishery of Japan.

References

Agatsuma, Y., 1997. Ecological studies on the population dynamics of the sea urchin *Strongylocentrotus nudus*. Sci. Rep. Hokkaido Fish. exp. Stn. 51: 1–66.

Chamberlain, Y. M., 1993. Observations on the crustose coralline red alga *Spongites yendoi* (Foslie) comb. nov. in South Africa and its relationship to *S. decipiens* (Foslie) comb. nov. and *Lithophyllum natalense* Foslie. Phycologia 32: 100–115.

Dayton, P. K., V. Currie, T. Gerrodette, B. D. Keller, R. Rosenthal & D. V. Tresca, 1984. Patch dynamics and stability of some California kelp communities. Ecol. Monogr. 54: 253–289.

Fujita, D., 1989. Marine algal vegetation in the 'Isoyake' area at Taisei, Hokkaido. Nanki Seibutsu 31: 109–114.

Fujita, D., 1990. Annual growth rate of *Lithophyllum yessoense*. Nippon Suisan Gakkaishi, 56: 1015.

Fujita, D., 1992. Grazing on the crustose coralline alga *Lithophyllum yessoense* by the sea urchin *Strongylocentrotus nudus* and the limpet *Acmaea pallida*. Benthos Res. 42: 49–54.

Fujita, D., 1994. Non-geniculate coralline algae in barren grounds. Kaiyou Monthly, 27: 60–65.

Fujita, D., 1996a. Non-geniculate coralline algae and their communities in Toyama Bay. Contr. Fish. Res. Japan Sea Block 33: 63–70.

Fujita, D., 1996b. Unusual excrescences of non-geniculate coralline alga *Lithophyllum yessoense* (Rhodophyceae, Corallinales) in culture. Bull. Toyama Pref. Fish. Res. Inst. 8: 21–24.

Fujita, D., H. Akioka & T. Masaki, 1992. Regeneration of *Lithophyllum yessoense* Foslie in culture. J. Phycol. 40: 143–149.

Hoshikawa, H., K. Tajima & C. Fujisawa, 1997. Field experiment on traps for starfish on a rocky shore to protect release abalone spats from predation. Sci. Rep. Hokkaido Fish. exp. Stn. 50: 19–26.

Keats, D. W., A. Groener & Y. M. Chamberlain, 1993. Cell sloughing in the littoral zone coralline alga, *Spongites yendoi* (Foslie) Chamberlain (Corallinales, Rhodophyta). Phycologia, 32: 143–150.

Keats, D. W., P. Wilton & G. Maneveldt, 1994. Ecological significance of deep-layer sloughing in the eulittoral zone coralline alga, *Spongites yendoi* (Foslie) Chamberlain (Corallinaceae, Rhodophyta) in South Africa. J. exp. mar. Biol. Ecol., 175: 145–154.

Lawrence, J. M., 1975. On the relationships between marine plants and sea urchins. Oceanogr. mar. Biol. Ann. Rev. 13: 213–286.

Levin, V. S., V. V. Ivin & V. I. Fadeev, 1987. Ecology of the starfish *Patiria pectinifera* (Mueller et Troschel) in Possiet Bay, Sea of Japan. Asian mar. Biol. 4: 49–60.

Masaki, T., D. Fujita & H. Akioka, 1981. Observation on the spore germination of *Laminaria japonica* on *Lithophyllum yessoense* (Rhodophyta, Corallinaceae) in culture. Bull. Fac. Fish., Hokkaido Univ.32: 349–356.

Masaki, T. ,D. Fujita & N. T. Hgen, 1984. The surface ultrastructure and epithallium shedding of crustose coralline algae in an 'Isoyake' area of southwestern Hokkaido, Japan. Hydrobiologia 116/117: 218–223.

Noro, T., T. Masaki & H. Akioka, 1983. Sublittoral distribution and reproductive periodicity of crustose coralline algae (Rhodophyta, Cryptonemiales) in southern Hokkaido, Japan. Bull. Fac. Fish. Hokkaido Univ. 34: 1–10.

Pueschel, C. M. & D. W. Keats, 1997. Fine structure of deep-layer sloughing and epithallial regeneration in *Lithophyllum neoatalayense* (Corallinales, Rhodophyta). Phycol. Res. 45: 1–8.

Spatial variability in secondary metabolite production by the tropical red alga *Portieria hornemannii*

Daniel B. Matlock, David W. Ginsburg & Valerie J. Paul
University of Guam Marine Laboratory, UOG Station, Mangilao, Guam 96923, U.S.A.
Present address: Biology Department, Seattle University, Seattle, WA 98122-4340, U.S.A.
E-mail: matlock@seattleu.edu

Key words: monoterpene, chemical variation, spatial variation, transplant, secondary metabolites, red algae

Abstract

Apakaochtodenes A and B, which are halogenated monoterpenes and the major secondary metabolites in *Portieria hornemannii*, are effective feeding deterrents toward herbivorous reef fishes on Guam. A reciprocal transplant study was conducted to determine the relative importance of environmental versus genetic factors influencing site-to-site differences in the amount of apakaochtodenes produced. The study sites were chosen for characteristically high (Anae Island) and low (Gun Beach) apakaochtodene levels. Algae collected from Anae Island and Gun Beach differed significantly in concentrations of apakaochtodene B at the start of the experiment, but by the end they had almost the same amount of the metabolite because the level had decreased in plants at Anae Island. Additionally, algae from Anae Island had relatively high levels of apakaochtodene A (60–90% of apakaochtodene B concentration), whereas this compound was rarely detected in Gun Beach algae. Transplantation to a different site had no significant effect on the levels of the apakaochtodenes, other than a decrease in concentration that might have resulted from handling the algae. Our data indicate a strong site-to-site difference in apakaochtodene levels in *P. hornemannii* on Guam, notable interplant variation in the levels of the compounds among thalli within the same site, and some evidence for temporal variation in levels of these compounds over a period of four weeks.

Introduction

Intraspecific chemical variation in the production of secondary metabolites has been demonstrated for a variety of marine algae (reviewed by Hay & Steinberg, 1992; Hay, 1996). Several studies have examined variation in chemical defenses within individual thalli (Carlton et al., 1989; Meyer & Paul, 1992, 1995; Cronin & Hay, 1996a; de Nys et al., 1996; Pennings et al., 1996). For example, de Nys et al.. (1996) found significant variation in levels of metabolites in the red alga *Delisea pulchra*, with concentrations highest at the distal tips of the thalli. Biogeographic variation in chemical defenses has also been studied for seaweeds (Steinberg, 1992; Hay & Steinberg, 1992; Targett et al., 1992; Bolser & Hay, 1996). Fewer studies have examined smaller scale site-to-site chemical variation in seaweeds or examined the causes of such interplant variation (Paul & Van Alstyne, 1988; Cronin & Hay, 1996b, c; de Nys et al., 1996; Puglisi & Paul, 1997). Temporal changes in algal metabolites have been demonstrated to be caused by induced (Van Alstyne, 1988; Yates & Peckol, 1993; Cronin & Hay, 1996b) as well as activated (Paul & Van Alstyne, 1992) defenses. Such temporal changes may be related to differences in secondary metabolites found among collections from different sites (Cronin & Hay, 1996b). The degree to which intraspecific chemical variation varies as the result of site-specific environmental influences (i.e. biotic and abiotic factors) is poorly understood.

The tropical red alga *Portieria hornemannii* (Lyngbye) Silva (Gigartinales, Rhizophyllidaceae) can be found in a variety of high wave energy habitats around Guam and was selected for this study because it has been shown to exhibit notable site-to-site variation in secondary metabolite production (Paul et al., 1987; Fuller et al., 1992, 1994; Puglisi & Paul, 1997). On Guam, the major secondary metabolite of *P. horne-*

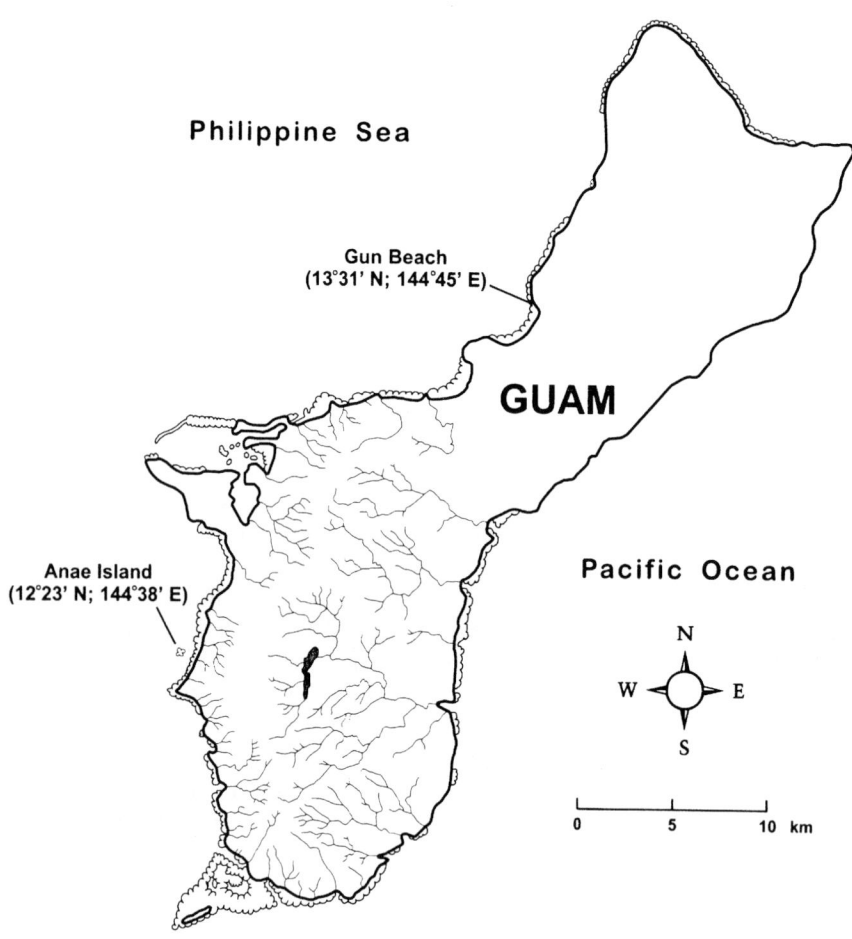

Figure 1. Map of Guam, showing experimental sites.

mannii is apakaochtodene B (Puglisi & Paul, 1997; Gunatilaka & Paul, work in progress), which is an effective deterrent against herbivores (Paul et al., 1987, 1990, 1992). Several other halogenated monoterpenes including apakaochtodene A are minor metabolites (Paul et al., 1987; Puglisi & Paul, 1997; Gunatilaka & Paul, work in progress). Puglisi and Paul (1997) showed that nutrient availability (i.e., nitrogen and phosphorus) did not influence the production of monoterpenes in *P. hornemannii;* they suggested that light as well as temporal variation might contribute to differences in the production of monoterpenes among the algae at different sites. In this study, we were interested in determining the relative importance of environmental versus genetic factors affecting the production of apakaochtodenes when individual thalli were transplanted between two sites where *P. hornemannii* varied in the concentrations of these compounds.

Materials and methods

Portieria hornemannii

Portieria hornemannii is a small red alga that grows as individual branched thalli in tufts approximately 3–8 cm tall, and 5–10 mm in diameter. They are found at 0.5–3 m intervals, attached to the reef substrate at depths of 1–10 m from the surface in areas with heavy current, surge or other water movement (Trono,

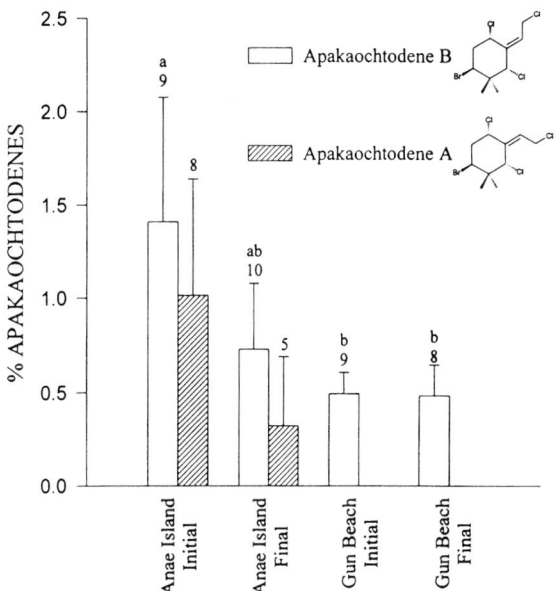

Figure 2. Comparison of apakaochtodene A and B concentrations in initial and final field samples in the transplant experiment. For both apakaochtodenes, the amount of pure compound was converted to percent yield based on total dry mass of the individual thalli. Vertical bars represent means + 1 S.D. Numbers above each histogram bar indicate samples analyzed per treatment. The number of samples was lower for apakaochtodene A because the minor metabolite was below the detectable limit in some samples. These were not considered 0 values because the compound was evident in the GC traces, but could not be quantified. Letters above each bar indicate differences among treatment means ($p > 0.05$) (analysis for apakaochtodene B only).

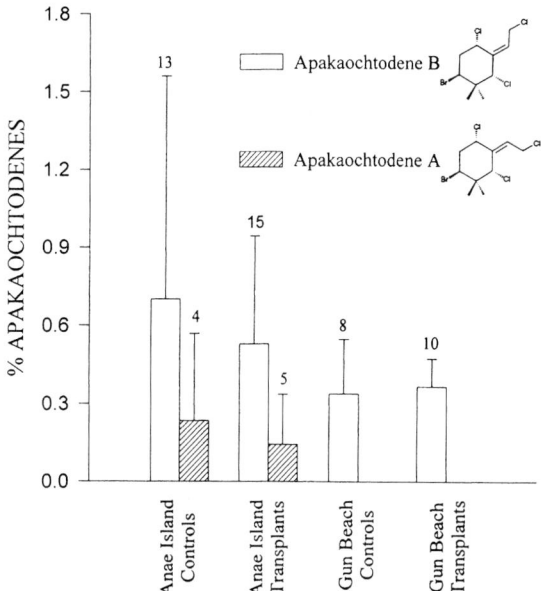

Figure 3. Comparison of apakaochtodene A and B concentrations in controls vs. transplanted specimens. Locations refer to site of origin of plants, e.g. Anae Island Transplants were collected at Anae Island and transplanted to Gun Beach. For both apakaochtodenes, the amount of pure compound was converted to percent yield based on total dry mass of the individual thalli. Vertical bars represent means + 1 S.D. Numbers above each histogram bar indicate samples analyzed per treatment. The number of samples was lower for apakaochtodene A because the minor metabolite was below the detectable limit in some samples. There was no significant difference among controls and transplants from either site (Kruskal-Wallis test, $p > 0.05$).

1969; Trono & Ganzon-Fortes, 1988; Puglisi & Paul, 1997). *P. hornemannii* is widely distributed across the Pacific from the Philippines to Hawaii, including the Marianas Islands where this study was conducted.

Chemical extraction

Fresh or frozen whole individual thalli were rinsed in freshwater, cleaned of epiphytic algae, spun 20 times in a plastic salad spinner and blotted dry. After recording fresh weights, thalli were ground using a Virtis high speed homogenizer in a small volume of solvent (dichloromethane:methanol, 1:1). The solvent and solid material were transferred to a flask and the volume brought to 50 ml with fresh solvent. After 8 h, the extract was decanted through weighed filter paper, sealed and stored in a freezer at $-20\,°C$. The solids left in the flask were resuspended in 50 ml of fresh solvent and extracted an additional 8 h. A total of three successive extractions were performed on each sample, after which all remaining solids were dried on the filters and weighed. The combined extract from each thallus was concentrated by reduced pressure rotary evaporation and weighed. These crude extracts were then dissolved in hexanes, filtered through glass wool, taken to dryness in a Speed vacuum evaporator, weighed and stored at $-20\,°C$ prior to quantitative chemical analysis of the apakaochtodene compounds.

Chemical analysis

The dried extracts were redissolved in hexanes with napthalene as an internal standard at a concentration of 100 $\mu g\ ml^{-1}$. Samples were diluted with 1 ml of hexanes (with naphthalene standard) per 1 mg extract. Gas chromatography was conducted using a Hewlett-Packard 5980 Series II Plus gas chromatograph with a cross-linked 5% methyl silicone (HP-5; 30 m × 0.25 mm). Injections were made in the splitless mode with an inlet pressure of 13.4 kPa at 70 °C. The injection port was held a 250 °C with a 70–290 °C temperature ramp at 10 °C per minute. The carrier gas was helium at a flow rate of 0.5 ml per minute.

Mass spectrometry was conducted with an HP 5972 Mass Selective Detector. Ions characteristic of the internal standard naphthalene and apakaochtodene B were monitored in the selected ion monitoring mode and were quantitatively analyzed using purified standards. Quantification was performed with a multiple point external standard method using HP Chemstation (Los Altos, CA) and Microsoft Excel software. The peak areas of apakaochtodene B and the internal standard were measured, and their ratio (compound/internal standard) calculated and converted to concentration by reference to standard curves. Calculation of concentrations of apakaochtodene A in *P. hornemannii* from Anae Island was based on the area ratios of apakaochtodene A to apakaochtodene B reported in the Chemstation analysis, because external standard curves were not run for the minor metabolite. For both apakaochtodenes, the amount of pure compound was converted to percent yield based on total dry mass of the individual thalli.

Statistical analysis

GC-MS data provided concentrations of apakaochtodene A and B as a proportion of algal dry mass. The sample variances were highly heteroscedastic (Bartlett's test for homogeneity of variances, $p < 0.0001$), and various transformations of the data did not improve the homogeneity of variances. Data were analyzed by a nonparametric Kruskal-Wallis test with a Comparisons of Mean Ranks procedure used to compare mean ranks of the different treatment groups. Statistix for Windows (ver. 1.0, Analytical Software, Tallahassee, FL) was used for all statistical analyses.

Transplant experiment

A reciprocal transplant study was conducted at two sites on the western side of Guam (Figure 1). One study site was at a depth of 5–7 m on the upper fringing reef slope at the north end of Gun Beach. The other was on the east side of Anae Island, at a depth of 5–7 m. Anae Island is a small limestone outcrop, approximately 0.45 km offshore, part of the extensive fringing reef on the western coastline of Guam. These sites were selected because the monoterpene concentration in *P. hornemannii* at Gun Beach was previously shown to be low, whereas at Anae Island concentrations were higher (Puglisi & Paul, 1997).

In total, there were four experimental groups for both sites: initial ($n = 10$) and final ($n = 10$) field samples, transplants ($n = 20$) and controls ($n = 20$). Transplant and control specimens were removed intact from the reef with a piece of the substrate to which they were attached. These were secured to labelled bricks with marine epoxy (PC-11 white epoxy paste, Protective Coating Co., Allentown, PA). This procedure was performed in the boat anchored above the site as quickly as possible to minimize handling and exposure to air. Bricks to which controls were affixed were returned to the reef after 10–15 min. Transplant specimens were immersed in saltwater in insulated containers, conveyed directly to the reciprocal location and placed on the reef at a comparable depth. Transplant and control bricks were haphazardly distributed in a small area on the reef at each site where *P. hornemannii* naturally occurred. A group of ten algae were collected at each site at the beginning and end of the experimental period as initial and final field samples. The experiment ran from April 10– May 6, 1996, and after the 28 days, transplants and controls were collected. Not all twenty transplants and controls were recovered because some were lost during the course of the experiment. All eight groups of samples were chemically analyzed as described above.

Results

The content of apakaochtodenes A and B was calculated as a percentage of total dry mass for each plant. Initial and final field samples, taken at the beginning and end of the 28-day duration of the experiment, did not differ significantly in levels of apakaochtodene B or total apakaochtodenes either at Anae Island or at Gun Beach (Figure 2), i.e. there was no significant temporal variation in concentration of apakaochtodenes over the 28-day duration of the study. The algae at Anae Island had higher levels of apakaochtodene A; this isomer was below detectable levels in specimens from Gun Beach in all of our treatments. There was initially significantly more apakaochtodene B at Anae Island than at Gun Beach.

The apakaochtodene B content of Anae Island specimens transplanted to Gun Beach did not significantly change relative to that of the controls returned to the water at Anae Island and the final field controls (Figure 3). The Anae Island transplants and controls were significantly lower than the Anae Island initial but not the final field samples. Concentrations of apakaochtodene A were highest in the initial field samples and significantly lower in transplants, and controls (Figures 2 and 3). Concentrations of apakaochtodene A in Anae Island final field samples did not differ significantly from either the initial field samples or the transplant, or control samples, so no effect of transplantation was detected.

Apakaochtodene B concentrations in Gun Beach specimens transplanted to Anae Island did not differ significantly from those left at Gun Beach (Figure 3). The Gun Beach transplants also did not differ significantly in apakaochtodene B levels from the initial or final field samples. Apakaochtodene A production was not induced in Gun Beach algae when transplanted to the Anae Island site, where it naturally occurs in *P. hornemannii*.

Our analyses detected notable interplant variation in apakaochtodene levels, even from thalli collected from the same site. This was especially true for the Anae Island algae, where some individuals had as much as 2–4% total apakaochtodenes by dry mass.

Discussion

This reciprocal transplant experiment was designed to investigate the relative importance of environmental *versus* genetic factors influencing the amount of apakaochtodenes produced. The study sites were chosen for characteristically high (Anae Island) and low (Gun Beach) apakaochtodene levels. If differences between populations of the algae are genetically predetermined, one would predict that transplanted specimens would continue to produce apakaochtodenes at concentrations equivalent to those of controls, which were treated similarly but left at their original site. There would be no significant difference between the controls and transplants from the same location. On the other hand, if the apakaochtodene content reflects environmental rather than genetic differences between the populations, one would expect the concentrations to drop when Anae Island specimens are transplanted to Gun Beach, and to increase in Gun Beach specimens transplanted to Anae Island. We found that the mean concentration of apakaochtodenes A and B did not change significantly as a result of transplantation, indicating that modifications in the environment (at least on the 4-week time scale of our study) did not alter the apakaochtodene levels in *P. hornemannii*. In particular, the production of the minor metabolite apakaochtodene A, which was below detectable levels in Gun Beach algae in our study and known to be low (only 5–10% of the apakaochtodene B levels) in many different Gun Beach collections (Puglisi & Paul, 1997; Gunatilaka & Paul, work in progress), did not increase in thalli transplanted to Anae Island. Similarly, Anae Island algae continued to contain relatively high levels of apakaochtodene A even when moved to the Gun Beach site. Consistent differences in the production of apakaochtodene A in algae from Gun Beach and surrounding reefs in the northern part of Guam and Anae Island, and surrounding reef sites in the southern part of Guam, suggest some genetic component to the biosynthesis of this minor metabolite.

We noted large variations in the values obtained within sampling groups in which all specimens were collected from the same site at the same time. For example, the standard deviation of apakaochtodene A and B content for the field samples at Anae Island was over half of the mean value (Figure 2). Moreover, the magnitude of variation was greater at Anae Island than at Gun Beach. This large interplant variation complicates the study of the effects of transplantation or any manipulative treatment on monoterpene production by this alga (see also Puglisi & Paul, 1997). The physical and biological histories of individual thalli at each site may contribute to this variation (Hay & Fenical, 1992), but these factors were not known in this study.

It is possible that there were natural factors, including seasonal cycles of algal growth and the relationship between plant size and secondary metabolite production, which affected the apakaochtodene content of the algae. For example, the highest percentages of apakaochtodenes at Anae Island were found in the smallest specimens (<100 mg dry mass). The young thalli may need the protection from grazing that these monoterpenes provide (Paul et al., 1987, 1990, 1992) more when they are small and just becoming established on the reef. If these compounds are only produced early in the life cycle of the alga, or if they are only produced in the growing tips, then one would expect to find a rapid decrease in concentration as a proportion of dry mass as the plants increase in size.

In addition to other extraneous influences discussed here, one cannot overlook the possibility that experimental handling of the transplanted specimens in the transplant experiment may have had a detrimental effect on the health of the algae, and indirectly, on their production of secondary metabolites. While the differences were not significant, both control and transplanted algae from Gun Beach and Anae Island had lower levels of compounds than any of the initial or final field collections, suggesting some effect of handling on the algae.

Conclusion

Algae from Anae Island had relatively high levels of apakaochtodene A (approximately 60–90% of the apakaochtodene B concentrations), but this compound was not found at detectable levels in the Gun Beach algae. Reciprocal transplanting of algae between sites with high and low secondary metabolite chemistry, failed to show significant changes in apakaochtodene concentrations. Levels of apakaochtodene A, which was only found in *P. hornemannii* from Anae Island, did not change in the transplanted algae. Our data indicate a strong site-to-site difference in apakaochtodene levels in *P. hornemannii* on Guam and notable interplant variation in the levels of the compounds among thalli, even within the same site. It also appears that the apakaochtodene concentration of *P. hornemannii* may vary over time within sites as well as between sites. This must be taken into account in any future experiments that attempt to determine the effect of external factors on secondary metabolite production.

Acknowledgements

The authors would like to thank Andrew Hudson, an undergraduate at Seattle University for his assistance as part of an independent study on Guam during the spring of 1996. Marine Lab marine technicians Frankie Cushing, Butch Irish and Chris Bassler provided able assistance above and below the water on the many boat trips involved in these projects. Thanks to Dr Margaret Hudson and Dr Glenn Yasuda at Seattle University for useful suggestions on the experimental set-up and handling and interpretation of results. We are grateful to the National Science Foundation for supporting this research through Grant No. HRD-9023311, to VJP and a Research Opportunity Award that made it possible for DBM to participate while on sabbatical, and to Seattle University for making the sabbatical leave possible. This is contribution #404 of the University of Guam Marine Laboratory.

References

Bolser, R. C. & M. E. Hay, 1996. Are tropical plants better defended? Palatability and defenses of temperate versus tropical seaweeds. Ecology 77: 2269–2286.

Carlton, D. J., J. Lubchenco, M. S. Sparrow & C. D. Trowbridge, 1989. Fine-scale variability of lanosol and its disulfate ester in the temperate red alga *Neorhodomela larix*. J. chem. Ecol. 15: 1321–1333.

Cronin, G. & M. E. Hay, 1996a. Within-plant variation in seaweed palatability & chemical defenses: Optimal defense theory versus the growth-differentiation balance hypothesis. Oecologia 105: 361–368.

Cronin, G. & M. E. Hay, 1996b. Induction of seaweed chemical defenses by amphipod grazing. Ecology 77: 2287–2301.

Cronin, G. & M. E. Hay, 1996c. Effects of light and nutrient availability on the growth, secondary chemistry, and resistance to herbivory of two brown seaweeds. Oikos 77: 93–106.

de Nys, R., P. D. Steinberg, C. N. Rogers, T. S. Charlton & M. W. Duncan, 1996. Quantitative variation of secondary metabolites in the sea hare *Aplysia parvula* and its host plant, *Delisea pulchra*. Mar. Ecol. Prog. Ser. 130: 135–146.

Fuller, R. W., J. H. Cardellina II, Y. Kato, L. S. Brinen, J. Clardy, K. M. Snader & M. R Boyd, 1992. A pentahalogenated monoterpene from the red alga *Portiera hornemannii* produces a novel cytotoxicity profile against a diverse panel of human tumor cell lines. J. med. Chem. 35: 3007–3011.

Fuller, R. W., J. H. Cardellina II, J. Jurek, P. J. Scheuer, B. Alvarado-Linder, M. McGuire, G. N. Gray, J. R. Steiner, J. Clardy, E. Menez, R. H. Shoemaker, D. J. Newman, K. M. Snader & M. R. Boyd, 1994. Isolation and structure/activity features of halomon-related antitumor monoterpenes from the red alga *Portieria hornemannii*. J. med. Chem. 37: 4407–4411.

Hay, M. E., 1996. Marine chemical ecology: what's known and what's next? J. exp. mar. Biol. Ecol. 200: 103–134.

Hay, M. E. & W. Fenical, 1992. Chemical mediation of seaweed-herbivore interactions. In John, D. M. S. J. Hawkins & J. H. Price (eds), Plant-Animal Interactions in the Marine Benthos. Systematics Association Special Volume No. 46. Clarendon Press, Oxford: 319–337.

Hay, M. E. & P. D. Steinberg, 1992. The chemical ecology of plant-herbivore interactions in marine versus terrestrial communities. In Rosenthal, G. A. & M. R. Berenbaum (eds) Herbivores: Their Interactions With Secondary Plant Metabolites, Vol. I. Academic Press, San Diego: 371–413.

Meyer, K. D. & V. J. Paul, 1992. Intraplant variation in secondary metabolite concentration in three species of *Caulerpa* (Chlorophyta: Caulerpales) and its effects on herbivorous fishes. Mar. Ecol. Prog. Ser. 82: 249–257.

Meyer, K. D. & V. J. Paul, 1995. Variation in secondary metabolite and aragonite concentrations in the tropical green seaweed *Neomeris annulata*: effects on herbivory by fishes. Mar. Biol. 122: 537–545.

Paul, V. J. & K. L. Van Alstyne, 1988. Chemical defense and chemical variation in some tropical Pacific species of *Halimeda* (Halimedaceae: Chlorophyta). Coral Reefs 6: 263–270.

Paul, V. J. & K. L. Van Alstyne, 1992. Activation of chemical defenses in the tropical green algae *Halimeda* spp. J. exp. mar. Biol. Ecol. 160: 191–203.

Paul, V. J., S. G. Nelson, & H. R. Sanger, 1990. Feeding preferences of adult and juvenile rabbitfish *Siganus argentus* in relation to chemical defenses in tropical seaweeds. Mar. Ecol. Prog. Ser. 60: 23–24.

Paul, V. J., K. D. Meyer, S. G. Nelson & H. R. Sanger, 1992. Deterrent effects of seaweed extracts and secondary metabolites on feeding by the rabbitfish *Siganus spinus*. Proc. 7th Internat. Coral Reef Symp. 2: 867–874.

Paul, V. J., M. E. Hay, J. E. Duffy, W. Fenical & K. Gustafson, 1987. Chemical defense in the seaweed *Ochtodes secundiramea* (Montagne) Howe (Rhodophyta): Effects of its monoterpenoid components upon diverse coral-reef herbivores. J. exp. mar. Biol. Ecol. 114: 249–260.

Pennings, S. C., M. P. Puglisi, T. J. Pitlick, A. C. Himaya & V. J. Paul, 1996. Effects of secondary metabolites and $CaCO_3$ on feeding by surgeonfishes and parrotfishes: Within-plant comparisons. Mar. Ecol. Prog. Ser. 134: 49–58.

Puglisi, M. P. & V. J. Paul, 1997. Intraspecific variation in the red alga *Portieria hornemannii*: Monoterpene concentrations are not influenced by nitrogen or phosphorus enrichment. Mar. Biol. 128: 161–170.

Steinberg, P. D., 1992. Geographical variation in the interaction between marine herbivores and brown algal secondary metabolites. In Paul V. J., (ed.), Ecological roles for marine natural products, Comstock publishing Associates, Ithaca, NY, USA: 51–92.

Targett, N. M., L. D. Coen, A. A. Boettcher & C. E. Tanner, 1992. Biogeographic comparisons of marine algal polyphenolics: evidence against a latitudinal trend. Oecologia: 89: 464–470.

Trono, G. C., Jr., 1969. The marine benthic algae of the Caroline Islands. II. Phaeophyta and Rhodophyta. Micronesica 5: 25–119.

Trono, G. C., Jr. & E. T. Ganzon-Fortes, 1988. Philippine Seaweeds. National Bookstore, Inc., Publishers, Metro Manila, Philippines: 146–147.

Van Alstyne, K. L., 1988. Herbivore grazing increases polyphenolic defenses in the intertidal brown alga *Fucus distichus*. Ecology 69: 655–663.

Yates, J. L., & P. Peckol, 1993. Effects of nutrient availability and herbivory on polyphenolics in the seaweed *Fucus vesiculosus*. Ecology 74: 1757–1766.

Influence of *Ecklonia radiata* kelp canopy on structure of macro-algal assemblages in Marmion Lagoon, Western Australia

Gary A. Kendrick[1], Paul S. Lavery[2] & Julia C. Phillips[3]
[1]*Dept of Botany, The University of Western Australia, 6907, Australia*
Fax: [+61]-8-93801001. E-mail: garyk@cyllene.uwa.edu.au
[2]*Centre for Ecosystem Management, Edith Cowan University, Joondalup, 6027, Australia*
[3]*Dept of Botany, University of Otago, Dunedin, New Zealand*

Key words: Australia, kelp, *Ecklonia radiata*, density, macroalgal assemblage

Abstract

A recent study of the influence of a wave exposure gradient on macroalgal assemblages associated with kelp stands in Marmion Lagoon, Western Australia found macroalgal communities had high spatial heterogeneity. Much of this heterogeneity was between replicate quadrats within sites and exposure level. The cause of such spatial heterogeneity is investigated further. Ninety 0.25 m^2 quadrats were sampled from 9 sites (10 quadrats from each site) during the Australian autumn (April–May) 1996. The sites were nested in groups of 3 across a gradient in wave exposure. The quadrats were also grouped by density of adult *Ecklonia radiata* thalli, which is the local canopy forming kelp. Three categories were used: <2; 2–4; and >4 kelp thalli m^{-2}. Eighty two species were observed, but only 19 species occurred in >10% of quadrats and a further 13 species in 5–10% of quadrats. This suggests that species were patchily distributed among quadrats. This patchy distribution was found to be greatest in quadrats with *Ecklonia radiata* densities <2, and least in quadrats with *E. radiata* densities >4. The major taxa contributing to this patchy distribution were the brown algae *Lobophora variegata*, *Sargassum spinuligerum* and *Sargassum* spp. juveniles, and the red algae *Amphiroa anceps*, *Chauviniella coriifolia*, *Dictymenia sonderi*, *Heterodoxia denticulata*, *Jeannerettia pedicellata*, *Pterocladia lucida* and *Rhodymenia sonderi*. Many of these species only occurred or were most abundant in areas of low kelp density. The results demonstrate that assemblage structure at local scales, between replicate quadrats, was influenced by density of the kelp canopy just as much as gradients in exposure to ocean swells between reef lines. Many species were influenced both by kelp density and exposure to swells.

Introduction

Temperate reefs in southwestern Western Australia support diverse assemblages of marine macroalgae (Huisman & Walker, 1990; Phillips et al., 1997). There are a variety of dominant canopy forming taxa, including the kelp *Ecklonia radiata*, 13 species of *Sargassum* and more than 18 species of *Cystophora*, *Scytothalia* and other large brown algae (Womersley, 1987). Eighty-two taxa of red, brown and green algae were found associated with *Ecklonia radiata* kelp forests near Perth, Western Australia (Phillips et al., 1997). Many of these taxa were rare, and only 18 occurred in >10% of samples and 13 taxa in 5–10%. Macroalgal assemblages were as variable, in the taxa present and their abundance, between replicate quadrats and sites within regions of similar exposure to oceanic swells as they were across a large gradient in exposure to ocean swells (Phillips et al., 1997). This paper assesses the influence of density of the dominant canopy former, *Ecklonia radiata*, on assemblage structure, within and across the gradient in swell exposure.

The processes that independently and in conjunction influence the structuring of seaweed assemblages on subtidal reefs in temperate Western Australia (WA) have not been studied to the same degree they have in eastern Australia and elsewhere in the world. Generally, research has focussed on individual taxa. For a few species, studies have addressed recruitment dynamics (Kendrick & Walker, 1991, 1995; Kendrick, 1994), demography (Kirkman, 1981, 1984; Kendrick & Walker, 1994) and physiological status and growth in relation to availability and quality of light (Kirkman, 1989; Wood, 1987) and temperature (Hatcher et al., 1987). Few studies have investigated the influences of competition and grazing in structuring macroalgal assemblages on subtidal reefs. Canopies of adult thalli have been found to negatively influence juvenile recruitment in stands of *Ecklonia radiata* (Kirkman, 1981) and *Sargassum* (Kendrick, 1994). Interspecific competition has not been studied. On intertidal reef platforms gastropod grazing structures the distribution and abundance of macroalgae (Wells & Keesing, 1989; Scheibling, 1994; Prince, 1995), but gastropod grazers are not as abundant in the sublittoral and their influence on subtidal macroalgae has not been studied. The influence of grazing reef fish also has received little attention, but has been shown to control macroalgal abundance on shallow subtidal reefs. At Rottnest Island, Western Australia territorial kyphosid fish have been shown to have close association to polygons of *Sargassum* and remove algae from within these polygons (Berry & Playford, 1992).

A broad scale influence of disturbance, from exposure to ocean swells, on macroalgal assemblages has also been identified from reefs at Marmion Lagoon. Onshore reefs have less biomass of kelp and more diverse assemblages of algae and sessile invertebrates than offshore reefs in Marmion Lagoon (Hatcher, 1989). We recently attempted to test Hatcher's observations, but found the variability in macroalgal assemblage structure in kelp habitat across metre scales between replicate quadrats and 100 m scales between sites of similar exposures to ocean swell swamped any trend between onshore and offshore reefs (Phillips et al., 1997). This led us to hypothesize that variable kelp density influenced the structure of macroalgal assemblages. The aim of this paper is to re-address the analysis of Phillips et al. (1997), incorporating the influence of kelp canopy on assemblage structure of offshore and onshore reefs in Marmion Lagoon. Then our intention is to generate a scale-specific conceptual model of the processes that would influence the observed macroalgal assemblage structure as a tool to generate testable hypotheses about the spatial influence of these processes.

Materials and methods

This study re-analyses data collected and described by Phillips et al. (1997). The methods of sampling and analysis are presented briefly below. Further detail can be obtained from Phillips et al. (1997).

Study area

Marmion Lagoon (31° 48′ 18″ S, 115° 42′ 11″ E) is a shallow (<15 m deep) semi-enclosed body of water situated 20 km north of Perth, Western Australia. Oceanic swells from the west and south-west dominate the local wave climate year round (Searle & Semeniuk, 1985). Swells are dampened, diffracted and refracted as they approach the coast, by a series of three parallel limestone reefs. This dissipation of energy as waves encounter each successive reef line produces a gradient of physical disturbance ranging from highly exposed sites (offshore reefs) to sites of low exposure (inshore reefs). Phillips et al. (1997) measured wave heights at each reef line and these were used to calculate the total energy per unit area of wave. Energy at the offshore, midshore and inshore reef lines was 3.89, 2.05 and 0.97 kJ m^{-2}, respectively.

Sampling design

The biomass of macroalgal communities was sampled on limestone reefs exposed to three levels of physical disturbance. Each disturbance level was replicated at three sites, giving a total of nine sampling sites. The sites were subjectively chosen on each line of reef. All sites were elevated reef > 0.5 m above surrounding substrata in a depth of 6 m and consisted of limestone with similar rugosity between sampling locations. Sites within each level of exposure were located 300–400 m apart. At each site, ten replicate macroalgal samples were collected by SCUBA divers using randomly positioned 0.25 m^2 quadrats. Sampling occurred within a three week period in the Australian autumn (April–May 1996).

Collection and processing of macroalgae samples

For each replicate, all non-crustose macroalgae were removed by hand. Small turfing and foliose algae were collected by chipping off pieces of the limestone reef that these algae were attached to with a cold chisel. Samples were stored on ice, returned to the laboratory and preserved using 4–5% buffered seawater formalin solution. Taxa were identified and then dried and weighed for dry weight before combusting for 2 h at 550 °C to determine ash-free dry weight. Due to the logistical problems of ashing large volumes of the kelp *Ecklonia radiata*, only five plants were ashed and the mean loss on ignition (23.96%) was deducted from the dry weights of remaining plants to give ash-free dry weights.

Biomass of crustose algae was calculated using a correlation between percentage cover and ash-free dry weight. Using a gridded 0.25 m^2 quadrat (5 cm × 5 cm squares giving a total of 81 intercept points), all crustose algae from a 5 cm × 10 cm area were collected. Five samples were collected which was less than we originally intended, however logistical constraints prevented the collection of a larger sample size. Samples were processed to ash-free dry weight and a regression analysis was performed on the data. The regression ($y = 0.0769x$, $r^2 = 0.8674$, $n = 5$, $p = 0.017$) was then applied to the percentage cover of crustose algae recorded for each replicate (% cover was estimated using the point-intercept method) to give an estimate of the biomass.

Data analysis

Multivariate statistical analyses were used to explore patterns in macroalgal assemblages at different kelp densities, and different levels of exposure to physical disturbance simultaneously. Each quadrat was treated as a separate data entry. This differs from the previous analysis of Phillips et al. (1997), where they summed the individual taxon biomass from each replicate quadrat for each site and analysed the sum. All data sets were log(n+1) transformed prior to analyses to reduce the weighting by abundant taxa. The multivariate statistical analysis package PATN (Belbin, 1993) was used to conduct ordination of quadrats. The dissimilarity between quadrats was first associated using the Bray-Curtis association measure. The ordination performed on the association matrix was a 2-dimensional non-metric multidimensional scaling (MDS). It was performed on all 90 quadrats. The MDS results were plotted on three separate scatter plots: low kelp densities (<2 kelp thalli 0.25 m^{-2}), mid kelp densities (2–4 kelp thalli 0.25 m^{-2}) and high kelp densities (>4 kelp thalli 0.25 m^{-2}). These divisions were entirely arbitrary, but were chosen to represent patches in the kelp canopy (low), open canopy (mid) and closed canopy (high).

To investigate which taxa were responding to the disturbance gradient, Principal Axis Correlations

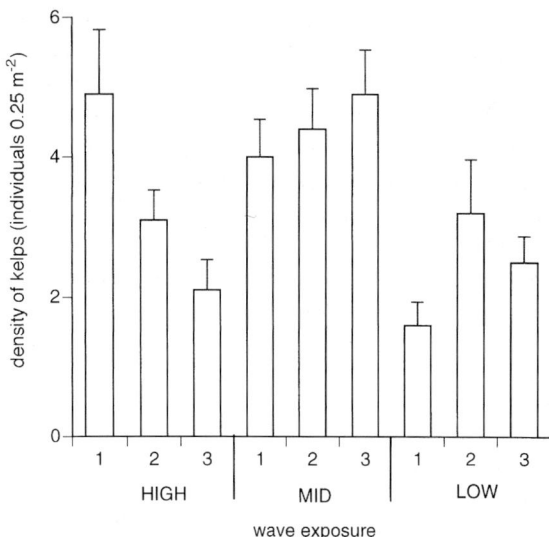

Figure 1. Variation in kelp density (mean ± s.e., $n = 10$) at 3 sites (1, 2, 3) nested within reefs lines parallel to the coast and that have high, mid and low exposure to ocean swells.

(PCC) were performed against the ordination (Belbin, 1993). PCC is a multiple-linear regression programme that determines the direction of best fit and the correlation coefficient of that fit for each taxon, or functional group in the ordination space (Belbin, 1993). The correlation coefficient was used as a rough indicator of the significance of each taxon (Belbin, 1993). Those taxa with a correlation coefficient greater than 0.5 were considered to be significantly influencing the ordination pattern.

Results

The mean density of adult kelp thalli varied more between sites within wave exposure levels than between wave exposure levels (Figure 1). The sites at the most exposed (offshore) and least exposed (inshore) locations showed greater variation than mid-exposure reefs. Densities of kelps were positively related to biomass of kelps and negatively related to biomass of understorey macroalgae across all sites (Figure 2 A, B). There was a large amount of variation, both in kelp density and biomass of kelp and understorey taxa, within sites as indicated by the x and y standard error bars ($n = 10$) on the graphs (Figure 2 A, B). This suggests there was large variation in these measures between replicate 0.25 m^2 quadrats. There were also no consistent groupings of sites into levels of wave exposure.

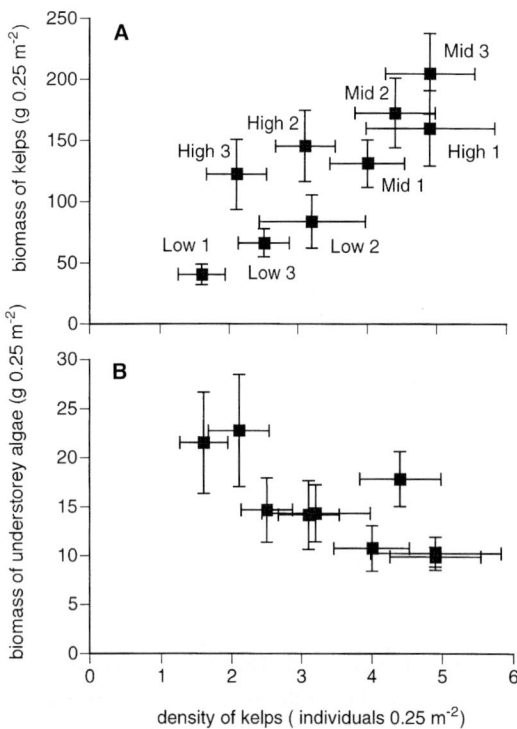

Figure 2. Influence of kelp density on A) biomass of kelps (g AFDW 0.25 m^{-2}) and B) total biomass of understorey algae (g AFDW 0.25 m^{-2}) at 3 sites nested within reefs lines parallel to the coast and that have high (High 1, 2, 3), mid (Mid 1, 2, 3) and low (Low 1, 2, 3) exposure to ocean swells. All variables are mean ± s.e., $n = 10$.

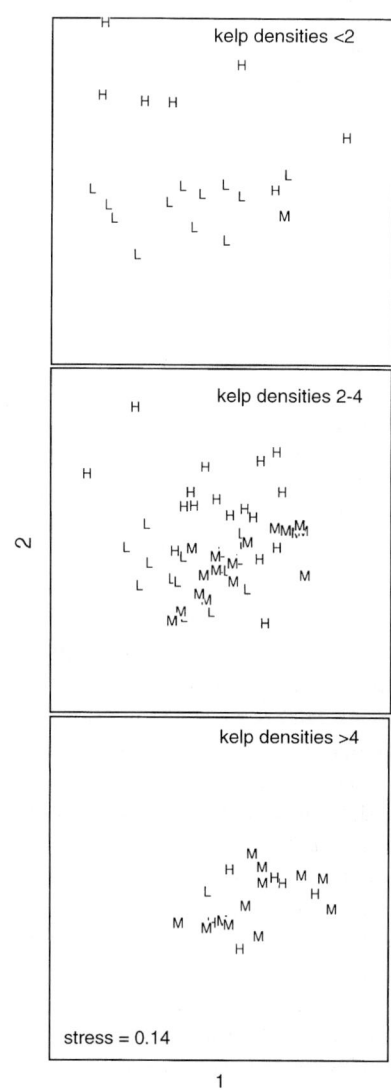

Figure 3. A two dimensional non-metric MDS of macroalgal assemblages of understorey algae associated with the kelp *Ecklonia radiata* divided into 3 plots of quadrats with densities of <2, 2–4 and > 4 kelp thalli 0.25 m^{-2}. Lettering represents reefs with different exposure to ocean swells: H = high, M = mid and L = low exposure to ocean swells.

The composition and abundance of the understorey macroalgal assemblage varied with the density of the kelp canopy. Generally, there were more taxa and greater differences in the composition of taxa between replicate quadrats and sites in areas of low kelp density (<2 kelps 0.25 m^{-2}, median = 11 taxa) than high kelp density (>4 kelps 0.25 m^{-2}, median = 6 taxa). Scatter plots of the non-metric MDS indicated decreased spread of replicate quadrats, thus decreased variation in taxa present and their abundance, with increased kelp density (Figure 3). At kelp densities less than 2 adult kelp thalli 0.25 m^{-2}, areas of high (7 quadrats) and low exposure (12 quadrats) to swells were more represented than mid exposures (1 quadrat). The trend was for high exposure quadrats to separate from low exposure quadrats along the second axis, but both exposure levels were equally spread along the first axis. A similar, but less distinct, trend was observed from quadrats with kelp densities of 2–4 adult kelp thalli.

At kelp densities greater than 4 adult kelp thalli 0.25 m^{-2} areas of mid exposure (13 quadrats) were more represented than high (5 quadrats) and low exposure (1 quadrat) and spread across both axes was reduced.

Ten taxa of understorey macroalgae substantially contributed to the pattern observed in the scatter plots: the brown algae *Lobophora variegata* and *Sargassum spinuligerum*, *Sargassum* spp. juveniles, and the red algae *Amphiroa anceps*, *Chauviniella coriifolia*,

Table 1. Understorey taxa with a correlation coefficient > 0.5 following Principal Axis Correlation of non-metric multidimensional scaling of quadrats

Taxa	Axis 1	Axis 2	Correlation coefficient
Dictymenia sonderi	− ve	+ ve	0.7730
Chauviniella coriifolia	− ve	+ ve	0.7498
Amphiroa anceps	− ve	+ ve	0.6992
Jeannerettia pedicellata	− ve	+ ve	0.6725
Sargassum spinuligerum	− ve	+ ve	0.6581
Sargassum juveniles	− ve	+ ve	0.6567
Lobophora variegata	− ve	+ ve	0.5524
Pterocladia lucida	− ve	− ve	0.8863
Rhodymenia sonderi	− ve	− ve	0.7811
Heterodoxia denticulata	− ve	− ve	0.5327

+ ve = overall trend of increased biomass with increased positive score on Axis.
− ve = overall trend of decreased biomass with increased negative score on Axis.

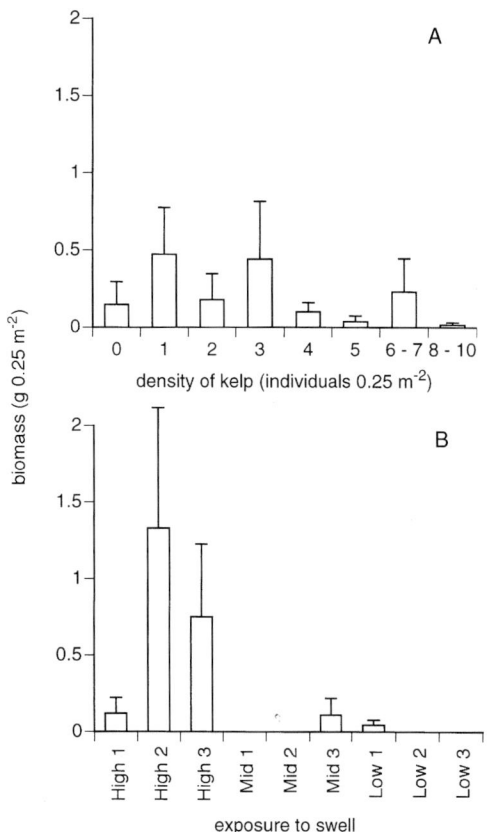

Figure 4. Biomass of *Amphiroa anceps* (g AFDW $0.25\ m^{-2}$; mean ± s.e., $n = 10$) across A) quadrats with different densities of kelps, and B) at sites with high (High 1, 2, 3), mid (Mid 1, 2, 3) and low (Low 1, 2, 3) exposure to ocean swells.

Dictymenia sonderi, Heterodoxia denticulata, Jeannerettia pedicellata, Pterocladia lucida and *Rhodymenia sonderi*. (Table 1). All of these taxa were more abundant in quadrats with low densities of kelps.

Kelp density influenced the biomass of these 10 taxa in different ways. There were taxa that were not influenced greatly by kelp density and that were common to reefs with either high or low exposure to ocean swells (eg. *Amphiroa anceps, Rhodymenia sonderi, Pterocladia lucida*). For example, *Amphiroa anceps*, which has significantly more biomass at high exposure reefs, had similar biomass across the range of kelp densities (Figure 4). *Pterocladia lucida*, in comparison, occurs in higher biomass at mid and low exposure reefs. The other taxa only occurred or were most abundant at mid to low kelp densities (e.g. *Sargassum spinuligerum, Heterodoxia denticulata, Sargassum* spp. juveniles, *Dictymenia sonderi, Lobophora variegata*). They also varied in occurrence and biomass with the level of exposure to ocean swells. For example, *Sargassum spinuligerum*, which was only found on the most exposed outer reefs (Phillips et al., 1997), only occurred in quadrats with ≤3 adult kelp thalli $0.25\ m^{-2}$ (Figure 5). In contrast, *Heterodoxia denticulata*, had greatest biomass at low kelp densities and low exposures to ocean swells (Figure 6). The examples above illustrate the range of influences of kelp density on distribution and abundance of individual taxa. They also highlight that understorey assemblage structure across the gradient in exposure was confounded by the structuring of these assemblages by kelp density.

Discussion

The distribution and abundance of understorey species in subtidal stands of the kelp *Ecklonia radiata* at Marmion Lagoon, Western Australia, is influenced by the density of kelp thalli. This influence is at the smallest scales of sampling, at metre scales between replicate quadrats. Thus variation in kelp density between quadrats confounds studies sampling between locations separated by scales of kilometres, like the study of Phillips et al. (1997) that described the influence of exposure to oceanic swells on the structure of macroalgal assemblages in kelp stands.

Stands of the kelp, *Ecklonia radiata* are identifiable to the layperson as habitats with their own struc-

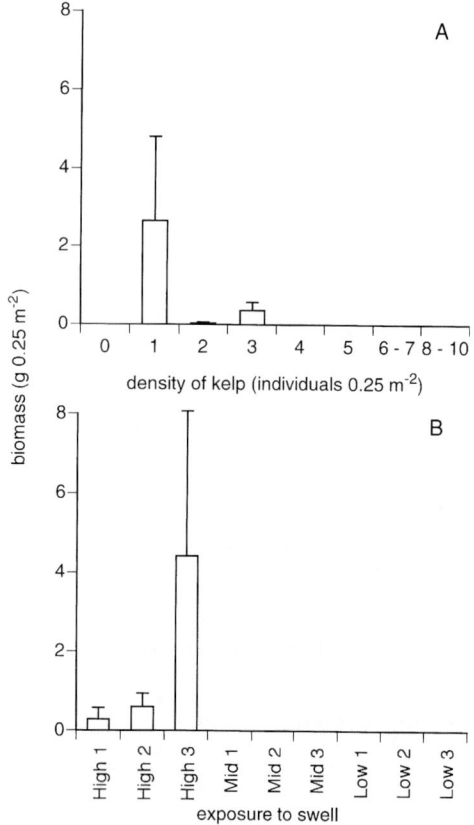

Figure 5. Biomass of *Sargassum spinuligerum* (g AFDW 0.25 m^{-2}; mean ± s.e., $n = 10$) across A) quadrats with different densities of kelps, and B) at sites with high (High 1, 2, 3), mid (Mid 1, 2, 3) and low (Low 1, 2, 3) exposure to ocean swells.

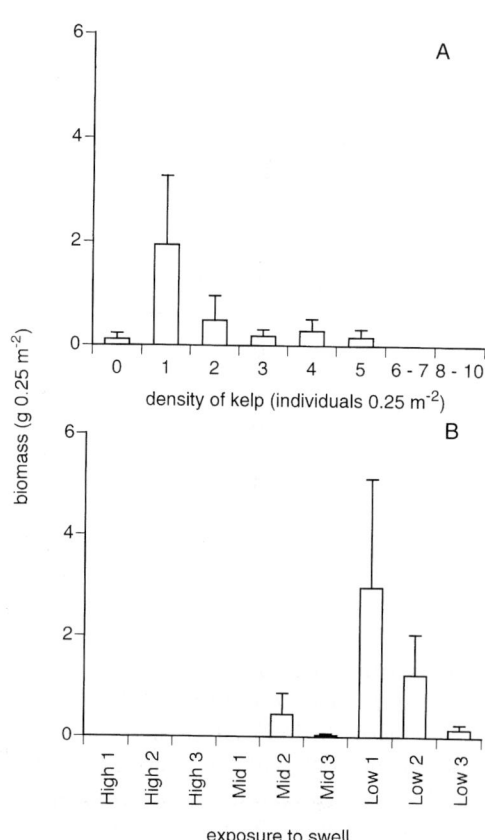

Figure 6. Biomass of *Heterodoxia denticulata* (g AFDW 0.25 m^{-2}; mean ± s.e., $n = 10$) across A) quadrats with different densities of kelps, and B) at sites with high (High 1, 2, 3), mid (Mid 1, 2, 3) and low (Low 1, 2, 3) exposure to ocean swells.

ture and assemblages of seaweeds. Kelps at Marmion Lagoon, Western Australia can account for 95% of the biomass of primary producers on nearshore reefs (Kirkman, 1984). But, as shown by Phillips et al. (1997) and Kennelly & Underwood (1992), algal assemblages associated with the kelp *Ecklonia radiata* vary within the same general location, and are even variable in the presence of taxa between stands <100 m apart within areas that have the same exposure to ocean swells. This variability is related in part to the density of the kelp canopy. Increased kelp density led to a reduction in the number of associated seaweed taxa. This pattern was most evident between quadrats within a single reef.

We propose that there is a spatial hierarchy of processes influencing the distribution and abundance of macroalgae within *Ecklonia radiata* stands. The influence of a process at small scales will be reflected in species distributions and abundances to the same degree as larger scale environmental influences (Figure 7). Regional scale influences include processes like variation in exposure to ocean swells. In this study and in Phillips et al. (1997), there are observed trends in assemblage structure and the distribution of individual taxa across the gradient in exposure to ocean swells between offshore, nearshore and onshore reefs. Examples outlined in the results are: species that are only found or most abundant on reefs of high exposure (*Amphiroa anceps*, *Sargassum spinuligerum*); and species that are only found or most abundant on reefs of low exposure (*Dictymenia sonderi*, *Pterocladia lucida*). Yet, there is equally as much structuring of these assemblages by variation in the density of the kelp canopy. Taxa that are restricted to quadrats with low kelp densities and were major contributors to the understorey assemblage structure (as demonstrated in the

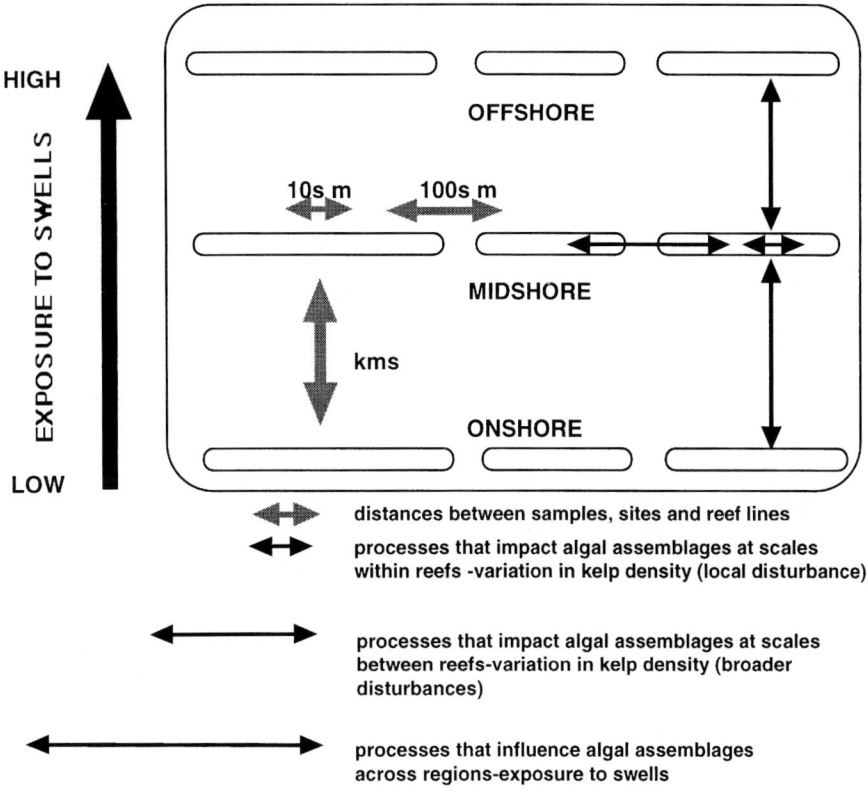

Figure 7. Conceptual model of Marmion Lagoon showing the 3 lines of reef, the distances between sampling units (quadrat, site and reef line) used in this study, and their correspondence to scales of processes that are equally important in determining the resultant structure of the understorey macroalgal assemblages common to stands of the kelp *Ecklonia radiata*.

MDS) were *Dictymenia sonderi*, *Heterodoxia denticulata*, *Jeannerettia pedicellata*, *Lobophora variegata* and *Sargassum spinuligerum*. Taxa that were equally distributed across the range in kelp densities were *Amphiroa anceps*, *Chauviniella coriifolia*, *Pterocladia lucida* and *Rhodymenia sonderi*.

These local scale patterns of species distribution and abundance probably reflect the interaction between local disturbance and colonization history, localized dispersal and recruitment in macroalgae and patchiness in grazing. Disturbance and gap creation in kelp stands occur at local (1–10 m) scales. Kennelly (1987) found in eastern Australian kelp forests turfing algae dominated substrata after kelps were removed by storms. Kennelly (1987) proposed that turfs were an alternative stable state to the *Ecklonia radiata* canopy, but warned that the spatial and temporal scales of his study heavily influenced his interpretation. The propagules of marine macroalgae, unlike pelagically dispersed larvae of other marine organisms, do not survive suspended in the water column over periods longer than a few days and even positively phototactic taxa only have a maximum dispersal shadow of 10s of kilometres (Zechman & Mathieson, 1985; Hoffman, 1987). Localized dispersal and recruitment across scales of 1–10 m has been recognized in many taxa of benthic macroalgae including kelps (Reed et al., 1991) and *Sargassum* (Kendrick & Walker, 1991, 1995). This localized pattern of dispersal results in a clumped distribution of macroalgal taxa and is an effective strategy for competition for space through localized persistence and spreading (DeWreede, 1983; Kendrick & Walker, 1995). Fish and macroinvertebrate grazers would also impact macroalgal assemblage structure at these scales due to their localized distributions. Fish and macroinvertebrate grazers can have a regional scale influence on the presence of macroalgae. In New South Wales, the influence of sea urchins can result in the complete removal of large macroalgae resulting in subtidal urchin barrens (Fletcher, 1987; Andrew, 1993). Fish grazing in benthic kelp forests can have a more localized influence on the presence and abundance of macroalgal taxa (Kennelly, 1991). No comparable work has been attempted in WA.

There is a spatial hierarchy of processes influencing understorey assemblages (Figure 7), and each is equally manifest at the lowest level of observation, the replicate quadrat. Consequently, any attempt to examine the influence of a process on understorey assemblages must take into account the very real possibility that many other processes operating at a range of spatial scales may simultaneously be affecting understorey assemblages. The sampling design must accommodate this. Our results clearly suggest that our earlier study, designed to test for the impact of disturbance on kelp stands, should have included sampling stratified by kelp density to remove its confounding effects on understorey algal assemblage structure. Even manipulative experiments studying effects of competition and grazing need to start each treatment off with similar densities of kelp. Even then, if localized dispersal and recruitment are a major cause of variation in the distribution and abundance of understorey kelp taxa, variation between replicate quadrats will remain large.

In conclusion, kelp density influences the species presence and abundance of understorey algal assemblages, with fewer species occurring under kelp canopies of greater densities. This influence is greatest within individual kelp stands and is observed as large variation in the species present and their biomass between replicate quadrats. We present a hypothesis that the structuring of understorey algal assemblages in kelp forests is influenced by a hierarchy of spatial processes from regional scale influences of wave exposure to localized disturbance, recruitment and grazing, but that this influence is manifested equally. Future planned research will experimentally test this hypothesis by manipulating kelp densities within reefs located across a wave exposure gradient.

References

Andrew, N. L., 1993. Spatial heterogeneity, sea urchin grazing and habitat structure on reefs in temperate Australia. Ecology 74: 292–302.

Belbin, L., 1993. PATN Technical Reference. CSIRO Division of Wildlife and Ecology, Canberra, 226 pp.

Berry, P. F. & P. E. Playford, 1992. Territoriality in a subtropical kyphosid fish associated with macroalgal polygons on reef platforms at Rottnest Island, Western Australia. J. r. Soc. West. Aust. 75: 67–73.

DeWreede, R. E., 1983. *Sargassum muticum* (Fucales, Phaeophyta): Regrowth and interactions with *Rhodomela larix* (Ceramiales, Rhodophyta). Phycologia 22: 153–160.

Fletcher, W. J., 1987. Interactions among subtidal Australian sea urchins, gastropods and algae: effects of experimental removals. Ecol. Monogr. 57: 197–210

Hatcher, A., 1989. Variation in the components of benthic community structure in a coastal lagoon as a function of spatial scale. Aust. J. mar. Freshwat. Res. 40: 79–96.

Hatcher, B. G., H. Kirkman & W. F. Wood, 1987. Growth of the kelp *Ecklonia radiata* near the northern limit of its range in Western Australia. Mar. Biol. 95: 63–73.

Hoffman, A. J., 1987. The arrival of seaweed propagules at the shore: a review. Bot. mar. 30: 151–165.

Huisman, J. M. & D. I. Walker, 1990. a catalogue of the marine plants of Rottnest Island, Western Australia, with notes on their distribution and biogeography. Kingia 1: 349–459.

Kendrick, G. A., 1994. Effects of settlement density and adult canopy on survival of recruits of *Sargassum* spp. (Sargassaceae, Phaeophyta) Mar. Ecol. Prog. Ser. 103: 129–140.

Kendrick, G. A. & D. I. Walker, 1991. Dispersal distances for propagules of *Sargassum spinuligerum* (Sargassaceae, Phaeophyta) directly measured by vital staining and venturi suction sampling. Mar. Ecol. Prog. Ser. 79: 133–138.

Kendrick, G. A. & D. I. Walker, 1994. Role of recruitment in structuring beds of *Sargassum* spp. (Phaeophyta) at Rottnest Island, Western Australia. J. Phycol. 30: 200–208.

Kendrick, G. A. & D. I. Walker, 1995. Dispersal of propagules of *Sargassum* spp. (Sargassaceae, Phaeophyta): observations of local patterns of dispersal and possible consequences for recruitment and population structure. J. exp. mar. Biol. Ecol. 192: 273–288.

Kennelly, S. J., 1987. Physical disturbances in an Australian kelp community. I. Temporal effects. Mar. Ecol. Prog. Ser. 40: 145–153.

Kennelly, S. J., 1991. Caging experiments to examine the effects of fishes on understorey taxa in a sublittoral kelp community. J. exp. mar. Biol. Ecol. 147: 207–230.

Kennelly, S. J. & A. J. Underwood, 1992. Fluctuations in the distributions and abundances of species in sublittoral kelp forests in New South Wales. Aust. J. Ecol. 17: 367–382.

Kirkman, H., 1981. The first year in the life history and the survival of the juvenile marine macrophyte, *Ecklonia radiata* (Turn.) J. Agardh. J. exp. mar. Biol. Ecol. 55: 243–254.

Kirkman, H., 1984. Standing stock and production of *Ecklonia radiata* (C.Ag.) J. Agardh. J. exp. mar. Biol. Ecol. 76: 119–130.

Kirkman, H., 1989. Growth, density and biomass of *Ecklonia radiata* at different depths and growth under artificial shading off Perth, Western Australia. Aust. J. mar. Freshwat. Res. 40: 169–197.

Phillips, J. C., G. A. Kendrick & P. J. Lavery, 1997. A test of a functional group approach to detecting shifts in macroalgal communities along a disturbance gradient. Mar. Ecol. Prog. Ser. 153: 125–138.

Prince, J., 1995. Limited effects of the sea urchin *Echinometra mathaei* (de Blainville) on the recruitment of benthic algae and macroinvertebrates into intertidal rock platforms at Rottnest Island, Western Australia. J. exp. mar. Biol. Ecol. 186: 237–258.

Reed, D. C., M. Neushul & A. W. Ebeling, 1991. Role of settlement density on gametophyte growth and reproduction in the kelps *Pterygophora californica* and *Macrocystis pyrifera* (Phaeophyceae). J. Phycol. 27: 361–366.

Scheibling, R. E., 1994. Molluscan grazing and macroalgal zonation on a rocky intertidal platform at Perth, Western Australia. Aust. J. Ecol. 19: 141–149.

Searle, D. J. & V. Semeniuk, 1985. The natural sectors of the Inner Rottnest Shelf coast adjoining the Swan Coastal Plain. J. r. Soc. West. Aust. 67: 116–136.

Wells, F. E. & J. K. Keesing, 1989. Reproduction and feeding in the Abalone *Haliotis roei* Gray. Aust. J. mar. Freshwat. Res. 40: 187–197.

Womersley, H. B. S., 1987. The Marine Benthic Flora of Southern Australia. Part II. The flora and fauna of South Australia Handbook Committee, 484 pp.

Wood, W. F., 1987. Effect of solar ultra-violet radiation on the kelp *Ecklonia radiata*. Mar. Biol. 96: 143–150.

Zechman, F. W., & A. C. Mathieson, 1985. The distribution of seaweed propagules in estuarine, coastal offshore waters of New Hampshire, U.S.A. Bot. mar. 28: 283–294.

Factors affecting sporulation of *Gracilaria cornea* (Gracilariales, Rhodophyta) carposporophytes from Yucatan, Mexico

Alberto Guzmán-Urióstegui & Daniel Robledo*
CINVESTAV - IPN, Unidad Merida A.P. 73 Cordemex 97310 Merida, Yucatán, México
Fax: [+52] 99 81 29 17. E-mail: robledo@kin.cieamer.conacyt.mx

Key words: carpospores, *Gracilaria cornea*, spore cultivation

Abstract

Carpospore shedding was studied in *Gracilaria cornea* in order to determine maximum spore output potential for mariculture purposes. The combined effects of temperature (23, 26 and 29 °C), daylength (8:16, 12:12 and 16:8 light:dark), photon irradiance (darkness, 20 and 40 μmol m^{-2} s^{-1}) and spore release method (spontaneous release, osmotic shock and drying) were tested. Maximum spore shedding in cystocarpic *G. cornea* occurred within the first three days depending on temperature. A reduction in spore release periodicity was more evident at 29 °C. Carpospore shedding was mainly affected by temperature and daylength. A higher number of carpospores was released per cystocarp at 26 °C than at 23 or 29 °C. Short day conditions (8:16 L:D) produced a higher number of carpospores at 26 °C, even at the lowest irradiances tested (darkness and 20 μmol m^{-2} s^{-1}). A combination of 26 °C and short days (8:16) gave the highest carpospore discharge per cystocarp. There was no significant difference between the spore release methods. These results could be applied to promote the establishment of extensive cultivation of *G. cornea* from spores as an alternative to vegetative propagation in Yucatan coast.

Introduction

Most *Gracilaria* spp. which are harvested or cultivated worldwide are utilized for the commercial production of agar, combining a fast growth rate and ease of vegetative reproduction (Kain & Destombe, 1995). The overexploitation of wild biomass of economically important agarophytes, as well as the urgent need of dependable sources of defined raw material, has led to the expansion of controlled cultivation of *Gracilaria*. Commercial cultivation is done on a very large scale in several countries, such as Chile (Avila & Seguel, 1993), China (Ren et al., 1984) and Taiwan (Chiang, 1981). Pilot scale cultivation is currently being carried out in medium sized farms mainly in Namibia, Venezuela and Malaysia (Armisen, 1995; Dawes, 1995; Alveal et al., 1997). At present, the culture methods for *Gracilaria* rely on vegetative fragments, but cultures are subject to a drop in their productivity with time (2–3 yr) and this may be due to the ageing of the thalli or excessive repeated harvesting (Buschmann et al., 1995; Glenn et al., 1996). The need for alternative methods such as spore culture is evident; on this basis, several studies have demonstrated the feasibility of growing *Gracilaria* from spores, and this coupled with algal selection, offers the possible production of genetically engineered strains of *Gracilaria* in the future (Friedlander & Dawes, 1984; Destombe et al., 1993; Dawes, 1995).

In order to compete in the biomass trade business, the cultivated *Gracilaria* must be more productive and possess special properties (e.g. better processing characteristics, agar yield, gel strength). *Gracilaria cornea* J. Agardh from the Yucatan peninsula has been recently recognized as a high quality agar producer (Freile-Pelegrin & Robledo, 1997a, b). For optimal cultivation of *Gracilaria* efficient seeding, fast growth rate and genetic improvement are of prime importance to the farmer. In this regard, seeding from spores can be achieved only if the methods of reproduction are properly understood. The use of diploid carposporophytes of *G. cornea* was considered because they are present throughout the year in natural populations, the carpospores are released in profusion from cystocarps

---
* Author for correspondence.

which could be easily recognized and collected by fishermen.

Many experiments which attempt to define the influence of abiotic factors on spore release have been restricted to testing single factor effects; for this reason, we studied the effects of temperature, light, and daylength on carpospore release on *Gracilaria cornea*, as a potential aid for intensive mariculture in the Yucatan peninsula.

Materials and methods

Gracilaria cornea is one of the 8 species of gracilarioids occurring in the Caribbean (Norris, 1985). In the Yucatan peninsula this species grows attached to rocks or dead coral, forming a narrow fringe 20 m wide and several hundred meters long parallel to the coast-line. The diploid phase that develops directly on the female thallus, the carposporophyte, is evident all year in the area, and was selected as seed stock material. Healthy mature cystocarpic thalli of *G. cornea* were collected from a natural shallow (1.5–3 m depth) bed located 3 km west of Dzilam de Bravo, Yucatan (21° 03′ N, 88° 57′ W).

Approximately 567 carposporophytic fragments (3–4 cm length) of *G. cornea* were used in the experimental design. Clean fragments bearing mature cystocarps ($n = 7$) were used for each combination of temperature, irradiance and daylength. In order to stimulate spore discharge, three sporulation induction methods were used: **Spontaneous discharge**, carposporophytic fragments were placed on glass Petri dishes (5 cm diameter) with sterile seawater (34 ‰) and left until carpospores spontaneously discharged; **osmotic shock**, carposporophytic fragments were submerged in high salinity seawater (60 ‰) for 30 min and returned to normal seawater (34 ‰) contained in glass Petri dishes; **drying**, carposporophytic fragments were dried for 1 h in the shade and re-immersed in seawater.

All Petri dishes containing the carposporophytic fragments were placed in a temperature-light cross gradient table under 81 different combinations of temperature (23, 26 and 29 °C), photon irradiance (darkness, 20 and 40 μmol m^{-2} s^{-1}), daylength [(short day 8:16, long day 16:8 and 12:12 light:dark (L:D)] and spore induction method (spontaneous release, osmotic shock and drying). Three separate experiments were made, one for each daylength, darkness being the control between experiments. Spore counts were done under a stereo microscope using a counting chamber at 24 h periods during 4 days (96 h). Each day the carposporophytic fragments were rinsed, transferred to a new Petri dish and returned to the experimental conditions described above. For each condition the mean number of carpospores released per cystocarp was recorded.

Statistical analysis were carried out using one-way MANOVA to determine differences between treatments (Johnson & Wichern, 1992). Significant levels ($p < 0.05$) between the different variables were determined in the analysis.

Results

The maximum spore shedding in cystocarpic *G. cornea* occurs within the first three days depending on temperature. Maximum carpospore release with the spontaneous discharge method occurred at 26 °C and 40 μmol m^{-2} s^{-1} under short days 8:16 (Figure 1), but release both in darkness at 29 °C and with 20 μmol m^{-2} s^{-1} under 12:12 daylength at 29 °C did not differ significantly ($p < 0.05$). Spore discharge was lowest at 23 °C in short day (8:16).

Under osmotic shock treatment maximum carpospore shedding was observed at 26 °C, 8:16 at 20 μmol m^{-2} s^{-1}, but not did differ ($p < 0.05$) from that in darkness (Figure 2). At 23 °C, spore discharge was reduced and delayed. Spore release at 29 °C, was reduced in 16:8.

For the drying method maximum spore output was found at 26 °C, short days (8:16) and 40 μmol m^{-2} s^{-1} (Figure 3), although output at the same photoperiod and temperature at 20 μmol m^{-2} s^{-1} or at 23 °C in long days (16:8) at 40 μmol m^{-2} s^{-1} did not differ significantly ($p < 0.05$).

Significant differences resulting from conditions and methods tested are summarized in Table 1. Temperature and daylength were the most effective factors though irradiance also had a significant effect. It seems that the spore induction methods were indistinguishable. The best grouping condition to induce the highest carpospore discharge per cystocarp in *G. cornea* was obtained in 26 °C under short day (8:16).

Discussion

Spore discharge in *G. cornea* was evident until the second and third day depending on the temperature.

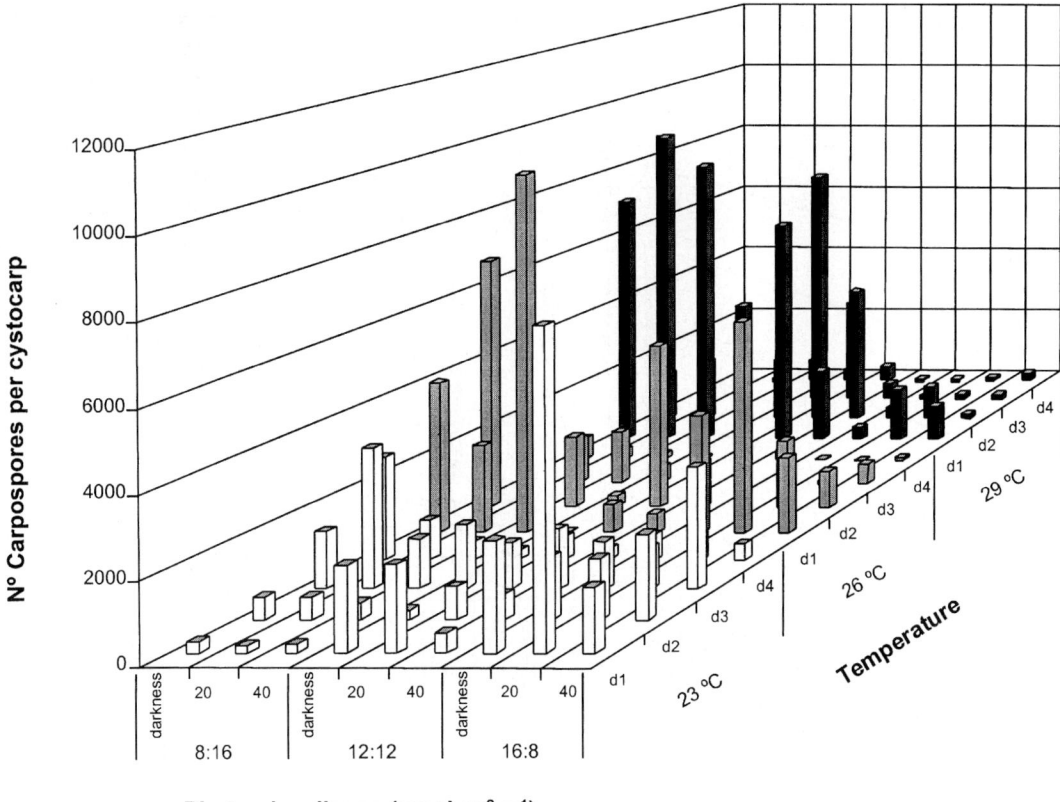

Figure 1. Number of spores released per cystocarp in *G. cornea* with spontaneous release method as a function of temperature, 23 °C (white bars), 26 °C (grey bars), 29 °C (black bars), daylength (8:16, 12:12, 16:8 light:dark) and photon irradiance (darkness, 20, 40 μmol m^{-2} s^{-1}) during 4 day period (d1, d2, d3, d4). Mean number of carpospores discharged per cystocarp for each treatment.

The decline in daily output has been also observed in *Gracilaria* spp. from India (Oza & Krishnamurthy, 1968; Rama Rao & Thomas, 1974). On the other hand, Lefebvre et al. (1987) has found that *Gracilaria gracilis* (Stackhouse) Steentoft, Irvine et Farnham shed carpospores rhythmically for about a month. This effect can be related to temperature in relation to algal origin (temperate or tropical).

Higher numbers of carpospores shed per cystocarp were obtained in *G. cornea* when compared with those in *Gracilaria corticata* (J. Agardh) J. Agardh (4911 carpospores per cystocarp per day) (Umamaheswara Rao, 1976) and *Gracilaria verrucosa* (Hudson) Papenfuss (19 700 carpospores shed per cystocarpic plant) (Oza & Krishnamurthy, 1968). Maximal spore shedding was obtained at 26 °C, which is the average water temperature in the area. A reduction in the number of spores released was observed at temperatures below or above 26 °C. The same pattern was present in other Gracilariales (Rama Rao & Thomas, 1974). According to Umamaheswara Rao & Subbarangaiah (1981) the peak of spore output is affected by temperature but not by irradiance, desiccation or salinity.

A combined effect between daylength and irradiance on carpospore release was observed in *G. cornea* with spontaneous release method. A short day period allowed a higher number of spores to be released, while in general an increase in daylength produced

Table 1. Significance levels of carpospore release for *Gracilaria cornea* (one-way ANOVA fixed factors) using pooled means (first two days) for spore induction method, temperature, daylength and photon irradiance

Variable	F	p
Spore induction method	1.015478	0.362559
Temperature	13.68927	0.000001
Daylength	12.77647	0.000003
Photon irradiance	5.002675	0.006869

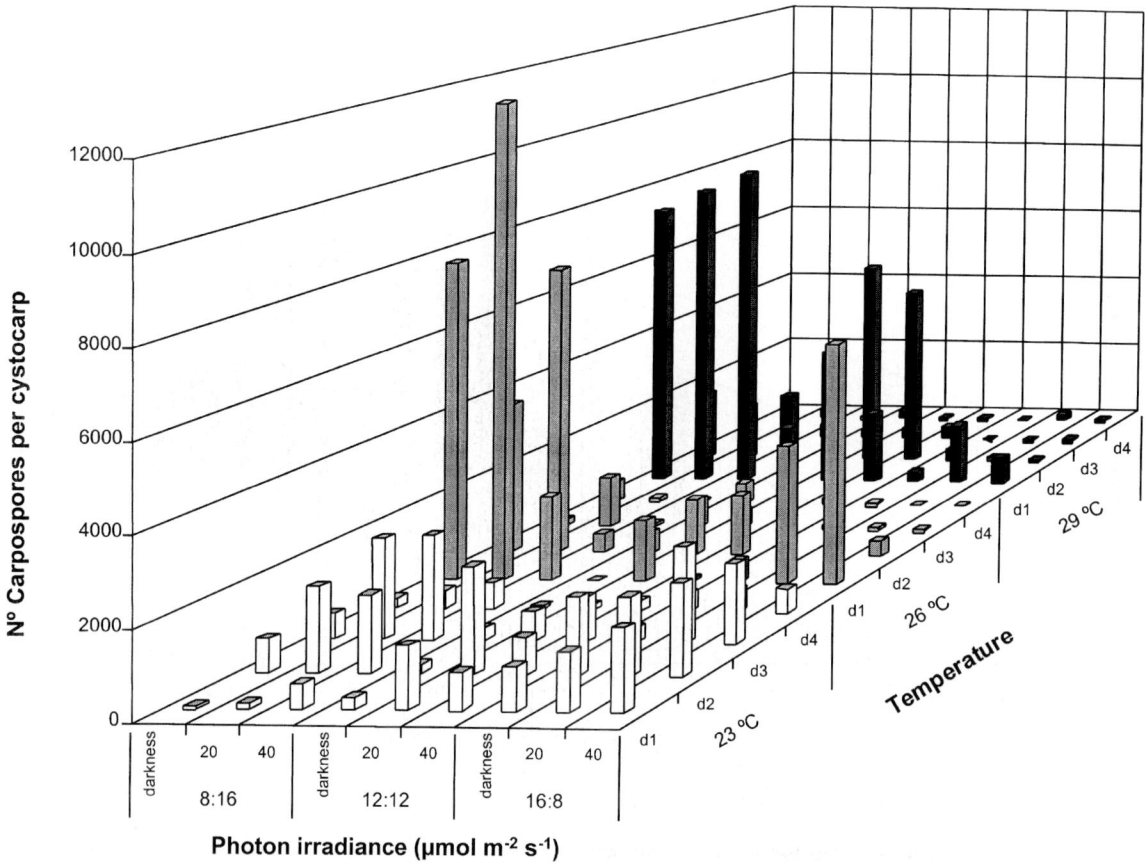

Figure 2. Number of spores released per cystocarp in *G. cornea* with osmotic shock method as a function of temperature, 23 °C (white bars), 26 °C (grey bars), 29 °C (black bars), daylength (8:16, 12:12, 16:8 light:dark) and photon irradiance (darkness, 20, 40 μmol m^{-2} s^{-1}) during 4 day period (d1, d2, d3, d4). Mean number of carpospores discharged per cystocarp for each treatment.

a decrease in the number of spores shed, except for drying at 23 °C and 40 μmol m^{-2} s^{-1}. Increasing irradiance had the same effect on spore released for *Gracilaria foliifera* (Førsskäl) Børgesen (Friedlander & Dawes, 1984) as found for *G. cornea*. These results are different from those obtained by Umamaheswara Rao & Kaliaperumal (1983) for two species of Gelidiales and other red algae, in which low irradiance and longer daylength determined the higher number of spores shed. On the other hand, spore output decreased as irradiance increased in *G. corticata* and *Gracilaria textorii* (Suringar) J. Agardh (Umamaheswara Rao & Subbarangaiah, 1981).

In relation to the spore induction method, spore shedding in red algae is stimulated by salinity changes (Reed, 1995) and desiccation followed by a reimmersion in sea water (see review article by Santelices, 1990). In this study, spore induction method had no significant effect on spore shedding (Table 1). Umamaheswara Rao (1976) reported that sometimes these induction methods do not produce any effect or even inhibit spore release. Mild desiccation followed by re-immersion in seawater was used for spore release in *Gracilaria chilensis* Bird, McLahlan et Oliveira carposporophytes (Infante & Candia, 1988). Spore output in *G. cornea* using drying treatment (1 h) at 26 °C under short day and 40 μmol m^{-2} s^{-1} was comparable to the best conditions obtained with the spontaneous release method. Nevertheless in *G. corticata*, *G. textorii* and *Gracilariopsis sjoestedtii* (Kylin) Dawson 15 min exposure to air inhibited tetraspore release (Umamaheswara Rao & Subbarangaiah, 1981). It is important to note that viability of *G. cornea* spores was not affected by the sporulation method used in this study.

The effect of environmental factors on spore output is of considerable interest to cultivators. In *G. cornea*

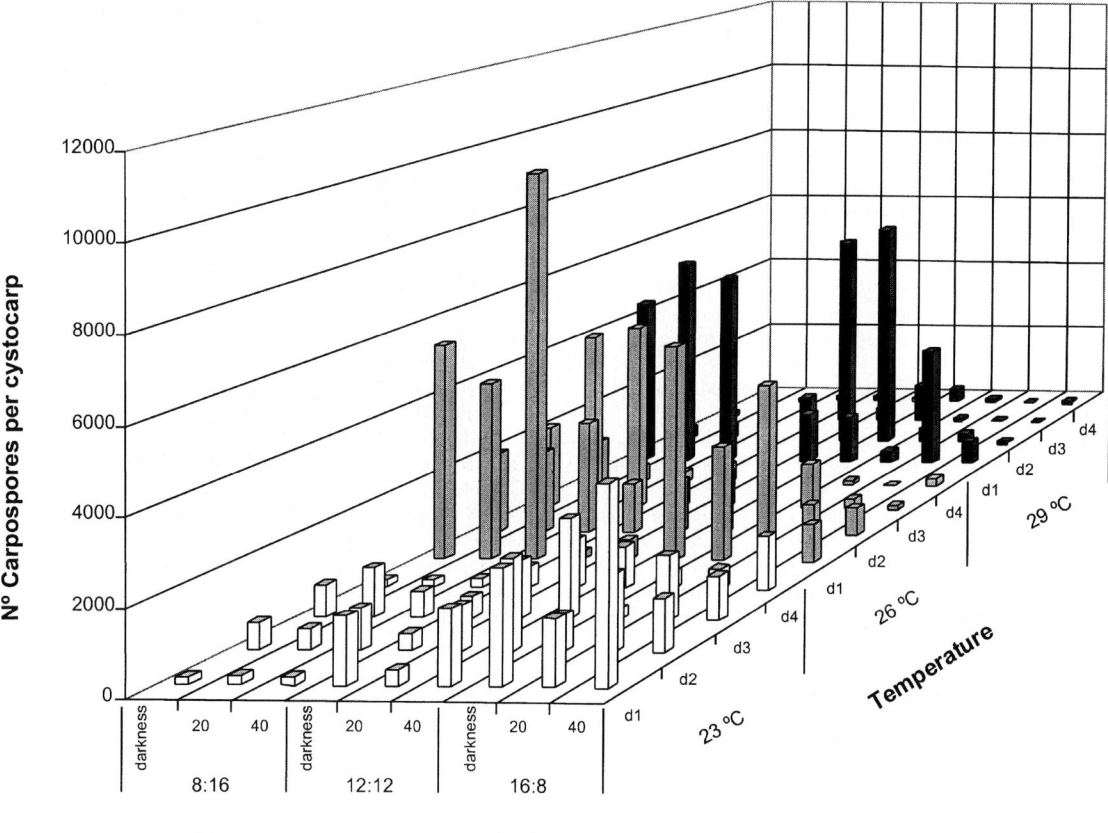

Figure 3. Number of spores released per cystocarp in *G. cornea* with drying method as a function of temperature, 23 °C (white bars), 26 °C (grey bars), 29 °C (black bars), daylength (8:16, 12:12, 16:8 light:dark) and photon irradiance (darkness, 20, 40 μmol m^{-2} s^{-1}) during 4 day period (d1, d2, d3, d4). Mean number of carpospores discharged per cystocarp for each treatment.

the best conditions to maximize carpospore shedding were 26 °C and daylength 8:16 (short day period).

Acknowledgments

This work has been supported by CONACyT (No 2198P-B) and CONABIO-McArthur Foundation (Ref. M151). We are grateful to Maria Luisa Zaldivar Romero for assistance in laboratory work. A. Guzman-Uriostegui acknowledges a Master's Degree Scholarship from CONACyT (Grant No 112873). The authors are greatly indebted to Dr Joanna M. Jones for valuable advice and critical reading of the manuscript.

References

Alveal, K., H. Romo, C. Werlinger & E. C. Oliveira, 1997. Mass cultivation of the agar-producing alga *Gracilaria chilensis* (Rhodophyta) from spores. Aquaculture 148: 77–83.

Armisen, R., 1995. World-wide use and importance of *Gracilaria*. J. appl. Phycol. 7: 231–243.

Avila, M. & M. Seguel, 1993. An overview of seaweed resources in Chile. J. appl. Phycol. 5: 133–139.

Buschmann, A. H., R. Westermeier & C. A. Retamales, 1995. Cultivation of *Gracilaria* on the sea-bottom in southern Chile: a review. J. appl. Phycol. 7: 291–301.

Chiang, Y. M., 1981. Cultivation of *Gracilaria* (Rhodophycophyta, Gigartinales) in Taiwan. Proc. int. Seaweed Symp. 10: 569–564.

Dawes, C. P., 1995. Suspended cultivation of *Gracilaria* in the sea. J. appl. Phycol. 7: 303–313.

Destombe, C., J. Godin, M. Nocher, S. Richerd & M. Valero, 1993. Differences in response between haploid and diploid isomorphic phases of *Gracilaria verrucosa* (Rhodophyta: Gigartinales) exposed to artificial environmental conditions. Hydrobiologia 260/261: 131–137.

Freile-Pelegrin, Y. & D. Robledo, 1997a. Effects of season on the agar content and chemical characteristics of *Gracilaria cornea* from Yucatan, Mexico. Bot. mar. 40: 285–290.

Freile-Pelegrin, Y. & D. Robledo, 1997b. Influence of alkali treatment on agar from *Gracilaria cornea* from Yucatan, Mexico. J. appl. Phycol. 9: 533–539.

Friedlander, M. & C. J. Dawes, 1984. Studies on spore release and sporeling growth from carpospores of *Gracilaria foliifera* (Førsskäl) Børgesen var. *angustissima* (Harvey) Taylor. I. Growth responses. Aquatic Bot. 19: 221–232.

Glenn, E. P., D. Moore, K. Fitzimmons & C. Azevedo, 1996. Spore culture of the edible red seaweed, *Gracilaria parvispora* (Rhodophyta). Aquaculture 142: 59–74.

Infante, R. & A. Candia, 1988. Cultivo de *Gracilaria verrucosa* (Hudson) Papenfuss e *Iridaea ciliata* Kutzing (Rhodophyta, Gigartinaceae) en laboratorio: esporulacion inducida y colonizacion de carposporas en diferentes sustratos. Gayana Bot. 45: 297–304.

Johnson, R. A. & D. W. Wichern, 1992. Applied multivariate statistical analysis. Prentice-Hall, New Jersey, 642 pp.

Kain, J. M. & C. Destombe, 1995. A review of the life history, reproduction and phenology of *Gracilaria*. J. appl. Phycol. 7: 269–281.

Lefebvre, C. A., C. Destombe & J. Godin, 1987. Le fonctionnement du carposporophyte de *Gracilaria verrucosa* et ses repercussions sur la strategie de reproduction. Cryptogamie Algol. 8: 113–126.

Norris, J. N., 1985. *Gracilaria* and *Polycavernosa* from the Caribbean and Florida: key and list of the species of economic potential. In Abbott I. A. & Norris J. N. (eds), Taxonomy of Economic Seaweeds. University of California, California: 101–121.

Oza, R. M. & V. Krishnamurthy, 1968. Studies on carposporic rhythm of *Gracilaria verrucosa* (Huds.) Papenf. Bot. mar. 11: 118–121.

Rama Rao, K. & P. C. Thomas, 1974. Shedding of carpospores in *Gracilaria edulis* (Gmel.) Silva. Phykos 13: 54–59.

Reed, R. H., 1995. Solute accumulation and osmotic adjustment. In Cole, K. M. & R. G. Sheath (eds), Biology of the Red Algae. Cambridge University Press, New York: 147–170.

Ren, G. Z., J. C. Wang & M. Q. Cheng, 1984. Cultivation of *Gracilaria* by means of low rafts. Hydrobiologia 116/117: 72–76.

Santelices, B., 1990. Patterns of reproduction, dispersal and recruitment in seaweeds. Oceanogr. mar. Biol. ann. Rev. 28: 177–276.

Umamaheswara Rao, M., 1976. Spore liberation in *Gracilaria corticata* J. Agardh growing at Mandapam. J. exp. mar. Biol. Ecol. 21: 91–98.

Umamaheswara Rao, M. & N. Kaliaperumal, 1983. Effects of environmental factors on the liberation of spores from some red algae of Visakhapatnam coast. J. exp. mar. Biol. Ecol. 70: 45–53.

Umamaheswara Rao, M. & G. Subbarangaiah, 1981. Effects of environmental factors on the diurnal periodicity of tetraspores of some Gigartinales (Rhodophyta). Proc. int. Seaweed Symp. 10: 209–214.

Porphyra cultivation in Alaska: conchocelis growth of three indigenous species

Michael S. Stekoll[1], Rulong Lin[1] & Sandra C. Lindstrom[2]
[1]*Juneau Center, School of Fisheries and Ocean Science, University of Alaska, 11120 Glacier Highway, Juneau, AK 99801, U.S.A.*
[2]*Department of Botany, University of British Columbia, #3529-6270 University Blvd, Vancouver, BC V6T 1Z4, Canada*

Received December 1998. Accepted December 1998

Key words: Porphyra, conchocelis, aquaculture, growth, Alaska, nori

Abstract

Experiments were performed to determine the range and optima of environmental conditions under which indigenous species of Alaskan *Porphyra* can grow. Growth of the conchocelis phase of *Porphyra abbottae, P. torta* and *P. pseudolinearis* was measured, when cultured in enriched media, under long days (16 h light) and varying conditions of irradiance (20, 40, 80 and 160 μmol photons m^{-2} s^{-1}), temperature (7, 11, 15 and 19 °C) and salinity (5, 10, 20, 30 and 40 ‰). Optimal growth (7.6% increase in volume day^{-1}) of *P. abbottae* occurred at 11 °C, 80 μmol photons m^{-2} s^{-1} and 30 ‰. *Porphyra torta* grew best (6.5% day^{-1}) at 15 °C, 80 μmol photons m^{-2} s^{-1} and 30 ‰. *Porphyra pseudolinearis* generally had higher growth rates than the other two species with optimal growth (8.8% day^{-1}) occurring at 7 °C, 160 μmol photons m^{-2} s^{-1} and 30 ‰. For all three species salinity between 20 and 40 ‰ had little effect on growth, but there was virtually no growth at salinity of 10 ‰ and below. Irradiances between 20 and 160 μmol photons m^{-2} s^{-1} generally had little effect on growth rates. However, growth of *P. abbottae* increased with irradiance at 7 °C but was inhibited at irradiances over 40 μmol photons m^{-2} s^{-1} at 15 °C and higher. *Porphyra torta* also showed growth inhibition at 15 °C and higher irradiances. *Porphyra pseudolinearis* appeared to be the most robust species with respect to tolerance to extremes of salinity and irradiance.

Introduction

Porphyra is one of the world's major aquaculture crops, and is ranked the highest valued nearshore fishery. The annual retail value for nori, the major product of *Porphyra*, is about US$2 billion on the Japanese market alone, and US imports of nori, valued at over US$25 million, have increased 10-fold in the last 10 years (Merrill, 1993; Mumford & Miura, 1988; S. Crawford, pers. commun.). There is a growing worldwide market for this and other *Porphyra* products.

Interest in *Porphyra* farming in western North America has involved both Japanese cultivars and indigenous species (Bergdahl, 1990; Mumford., 1990). Woessner and colleagues (in litt., 1977), who have conducted biological and economic studies on *P. nereocystis*, estimated that the natural crop within a 55-mile (90 km) distance along the California coastline could be worth over half a million dollars. Waaland et al. (1986) identified five native North American species with commercial potential, and they established optimal conditions for maturation and release of conchospores of these species (Waaland et al., 1987, 1990). They have also grown these species to harvestable size in experimental farms established in the waters of Puget Sound. Their studies provided the basis for our own investigations into optimal conditions for growth of species occurring in southeast Alaska. One of the species farmed by the Japanese (*P. pseudolinearis*) occurs naturally near Juneau and in other parts of coastal Alaska to the west. Four of the species studied by Waaland and colleagues grow in Alaska, and an additional ten species are known to occur in southeast Alaska (Lindstrom & Cole, 1992; Scagel et al., 1989). Although the essential technology for *Porphyra* aquaculture in Northwest America was established in Washington State, with modifications made in British Columbia and Maine, further modifications are required for Alaska because of the requirement to use native rather than an imported, but previously domesticated, species.

Mariculture of all species of *Porphyra* utilizes artificial control of the life cycle to regulate the production of spores for seeding nets. Thus, it is imperative that we understand the factors that affect growth of the conchocelis and induce conchospore production (as well as other aspects of *Porphyra* development). Such understanding will help to avoid the 'boom or bust' cycles associated with natural production. This investigation comprises the first phase of a project to domesticate Alaskan *Porphyra* species. We report here on the growth of the conchocelis phase of three species of Alaskan *Porphyra*.

Materials and methods

Unialgal cultures of each *Porphyra* species *(Porphyra abbottae* Krishnamurthy, strain PaJB03, *P. pseudolinearis* Ueda - strain PiSC06 and *P. torta* Krishnamurthy, strain PtCH03) were obtained from carpospore release. Mature blades of the gametophyte stage of each species were collected from the field. Blades were washed and scrubbed with sterile seawater to remove surface contamination. The cleaned blades were placed in sterile seawater in Petri dishes for carpospore release. After 24–36 h the blades were removed and the dishes incubated in Provasoli's enriched seawater (PES; McLachlan, 1973) under 8L:16D (light:dark) photoperiod at 15 °C. Conchocelis segments (approx. 110–250 μm) of each species were placed in cell well plates (one piece per well) and incubated at 33 ‰ and 11 °C (100–120 μmol photons m^{-2} s^{-1}) or 15 °C (140–160 μmol photons m^{-2} s^{-1}) for the culture of pure genotype conchocelis. PES enriched seawater culture medium was used.

Conchocelis growth experiments were conducted in several incubators that had been set at different temperatures and illuminated with cool-white fluorescent lamps. Irradiance gradients were obtained by wrapping the culture containers with varying layers of white paper. Autoclaved natural seawater-based PES medium, with a GeO$_2$ concentration of 1.25 mg l^{-1} added to inhibit diatom growth, was used in the growth experiments. At the beginning of the experiments, pH of the culture medium was adjusted to 7.6–7.8 (the ambient pH of the seawater in the inner waters of southeast Alaska) using 6 M HCl or 6 M NaOH. Experimental seawater with different salinities was obtained either by boiling natural seawater (for 40 ‰) or by diluting natural seawater with distilled water. Nutrients were added after adjustment of the salinity.

For growth experiments different levels of three environmental factors were employed: temperatures of 7, 11, 15, 19 °C; salinities of 5, 10, 20, 30, 40 ‰ (*P. torta* was tested at 15, 20, 30, 40, 50 ‰); irradiances of 20, 40, 80, 160 μmol photons m^{-2} s^{-1}. All conchocelis fragments died quickly at 5 ‰ and therefore no data analysis was done for growth at this salinity.

Corning cell wells (24-well with lid) were used as culture containers. About 4 ml of culture medium (PES) were placed in each cell well. A fully randomized factorial design was employed using all combinations of the three environmental factors. The growth of free conchocelis was observed and recorded under the experimental conditions. For each experimental combination, four replicate wells were used each with four to seven small, spherical conchocelis tufts per well. Culture media were changed every 15 days. Long day (16L:8D) photoperiods were used. Growth was determined from the volume increase of the filamentous tufts as calculated from the mean diameters at the beginning and end of experiments. Conchocelis specific growth rates (μ) were calculated as the mean percent volume increase per day ($\pm$ S.E.) using the formula:

$$\mu = \frac{100[\ln(V_t/V_0)]}{t},$$

in which V_t and V_0 represent respectively the mean tuft volume in every well at the end and the beginning of the experiment, and t is the number of days. The equation assumes that growth was exponential (DeBoer et al., 1978). The experiment lasted 31 days for *P. abbottae*, 26 days for *P. pseudolinearis*, and 31 days for *P. torta*.

Growth rate differences were initially analyzed by a three-factor ANOVA (growth as a function of light, temperature and salinity). Post hoc tests were performed using the Newman-Keuls multiple comparison test to identify which tested factors were important in controlling growth of the conchocelis.

Table 1. ANOVA table for growth of conchocelis of three different Porphyra species at combinations of salinity, irradiance and temperature. [a] 10, 20, 30, 40 ‰; [b] 7, 11, 15, 19 °C; [c] 20, 40, 80, 160 μmol photons m^{-2} s^{-1}; [d] 15, 20, 30, 40, 50 ‰. (*$p < 0.05$; **$p < 0.01$).

Source of variation	df	Sum of squares	Mean square	F
P. abbottae				
Salinity[a]	3	12.211	4.070	387.38**
Temperature[b]	3	2.221	0.740	70.45**
Light[c]	3	0.916	0.305	29.06**
Sal. × Temp.	9	0.469	0.052	4.96**
Sal. × Light	9	0.099	0.011	1.05
Temp. × Light	9	2.609	0.290	27.59**
Sal. × Temp × Light	27	0.483	0.018	1.70*
Residuals	192	2.017	0.011	
P. pseudolinearis				
Salinity[a]	3	71.216	23.739	238.74**
Temperature[b]	3	0.376	0.125	1.26
Light[c]	3	0.687	0.229	2.30
Sal. × Temp.	9	1.552	0.172	1.73
Sal. × Light	9	2.052	0.228	2.29
Temp. × Light	9	1.745	0.194	1.95*
Sal. × Temp × Light	27	4.996	0.185	1.86**
Residuals	192	19.091	0.099	
P. torta				
Salinity[d]	4	2.751	0.688	38.53**
Temperature[b]	3	0.532	0.177	9.93**
Light[c]	3	0.124	0.041	2.32
Sal. × Temp.	12	0.683	0.057	3.20**
Sal. × Light	12	0.293	0.024	1.37
Temp. × Light	9	0.565	0.063	3.52**
Sal. × Temp × Light	336	1.521	0.042	2.37**
Residuals	240	4.288	0.018	

Results

Porphyra abbottae

The growth of the conchocelis of *Porphyra abbottae* was influenced by all three factors (Figure 1, Table 1). At low salinity (10 ‰) growth was virtually nil, and cells became bleached after 8–10 days. Higher salinities produced growth rates of nearly 8% day^{-1}, with best growth rates at 20–30 ‰. The optimal temperature was 11 °C in the salinity range tested. Growth was significantly greater at this temperature ($P < 0.01$, Figure 4). However, growth was affected by the interaction of temperature and irradiance. Higher temperatures (15–19 °C) combined with higher irradiances (80–160 μmol photons m^{-2} s^{-1}) inhibited growth. This result is in contrast with light effects at lower temperatures, where growth was often greater at the higher irradiances (e.g. 30 and 40 ‰, Figure 1). For example, growth at 30 ‰, 19 °C, and 80–160 μmol photons m^{-2} s^{-1} was nearly zero, whereas at 20-40 μmol photons m^{-2} s^{-1} it was 5.0-5.4% day^{-1}. Maximum growth (7.6% day^{-1}) was achieved at 30 ‰, 11 °C and 80 μmol photons m^{-2} s^{-1}. *P. abbottae* was conchosporangial under all culture conditions.

Porphyra pseudolinearis

Growth of the conchocelis of *P. pseudolinearis* was also near zero at 10 ‰ (Figure 2). The ANOVA showed significant effects only with respect to salinity (Table 1). Peak growth occurred at 30 ‰. Unlike *P. abbottae*, *P. pseudolinearis* was not strongly inhibited by combinations of high temperatures and high irradiance (Figure 2). The tolerance range was greater for *P. pseudolinearis* than for *P. abbottae*. Growth was relatively high at all temperatures and irradiances tested and was independent of irradiance between 20 and 160 μmol photons m^{-2} s^{-1} (Figure 2). There was a slight trend for growth to be inversely proportional to temperature at the highest irradiance tested. Optimal growth (8.8% day^{-1}) occurred at 30 ‰, 7 °C, and 160 μmol photons m^{-2} s^{-1}. Morphologically, *P. pseudolinearis* remained in the vegetative state throughout the experiment.

Porphyra torta

Growth rates of *P. torta* were generally lower than those of the other two species (Figure 3). ANOVA results showed that growth was significantly affected by salinity and temperature but not by irradiance under the conditions tested (Table 1). *Porphyra torta* was not tested at 10 ‰ since the other species did not grow at that salinity. Growth at 15 ‰ was fairly low except at 7 °C, the lowest temperature tested (Figure 3), whereas at 30 ‰ it was significantly greater than at all other salinities tested ($P < 0.01$, Figure 4). *Porphyra torta* had significantly higher growth rates at 7 and 15 °C than at 11 or 19 °C ($P < 0.01$, Figure 4). There was some indication that *P. torta* growth was inhibited by high light at 19 °C, but the effect was not as great as that in *P. abbottae*. The optimal culture condition for the growth of *P. torta* was 30 ‰, 15 °C and 80 μmol

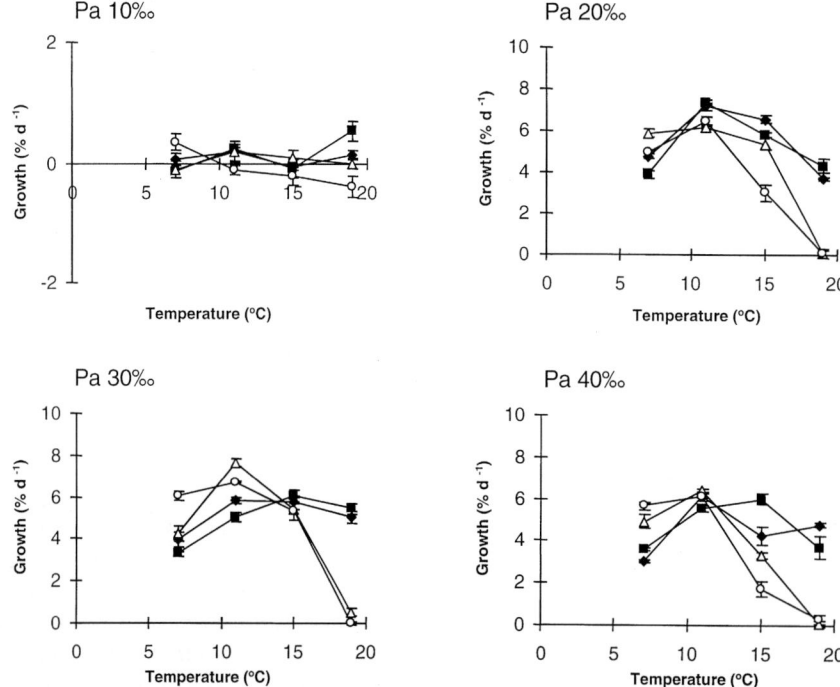

Figure 1. Porphyra abbottae (Pa). Conchocelis growth as a function of salinity (‰), irradiance (♦, 20; ■, 40; △, 80; ○, 160 μmol photons m^{-2} s^{-1}) and temperature (°C). Error bars are ± S.E. Growth is expressed as percent increase in volume per day. Note the y-axis scale for 10 ‰ is different from the others. Negative growth rates are a consequence of the sampling design.

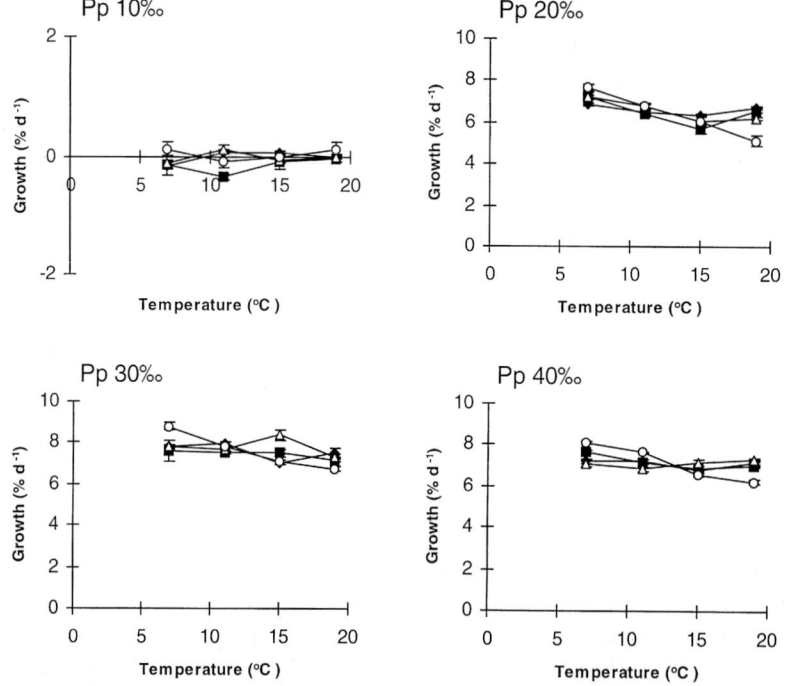

Figure 2. Porphyra pseudolinearis (Pp). Conchocelis growth as a function of salinity (‰), irradiance (♦, 20; ■, 40; △, 80; ○, 160 μmol photons m^{-2} s^{-1}) and temperature (°C). Error bars are ± S.E. Growth is expressed as percent increase in volume per day. Note the y-axis scale for 10 ‰ is different from the others. Negative growth rates are a consequence of the sampling design.

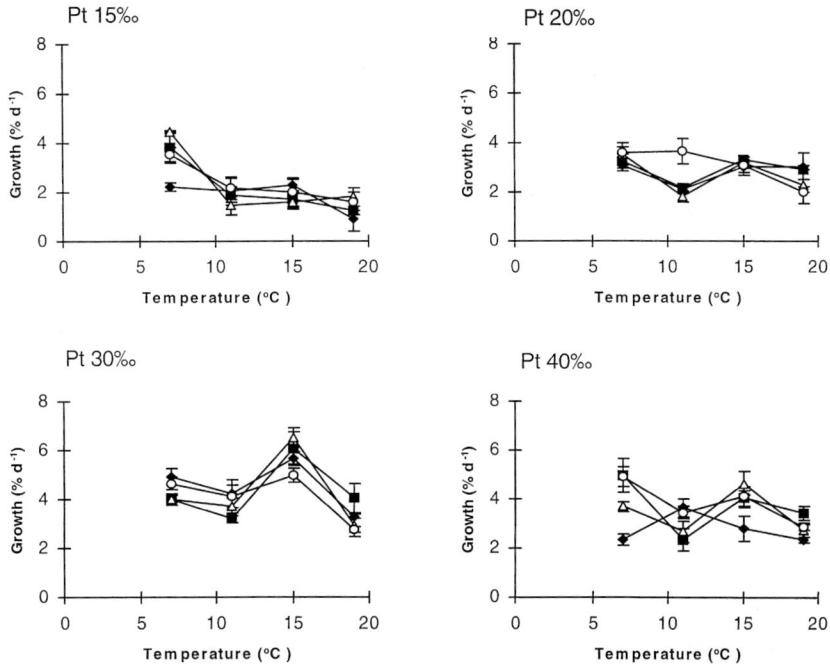

Figure 3. Porphyra torta (Pt). Conchocelis growth as a function of salinity (‰), irradiance (♦, 20; ■, 40; △, 80; ○ 160 μmol photons m^{-2} s^{-1}) and temperature (°C). Error bars are ± S.E. Growth is expressed as percent increase in volume per day.

photons m^{-2} s^{-1} (6.5% day^{-1}). *Porphyra torta* was conchosporangial throughout the experiment.

Overall, *P. pseudolinearis* had the best average growth rate, with *P. abbottae* having the second highest growth rate, except at 19 °C (Figure 4). Salinity of 30 ‰ was optimal for all three species, although they were tolerant of 20 ‰ (Figure 4). All species, in all treatments were killed under 5 ‰. Irradiance was generally not a significant factor, but the growth rate of each species differed from the others at each irradiance tested with the exception of 160 μmol photons m^{-2} s^{-1} for *P. pseudolinearis* and *P. torta* (Figure 4).

Discussion

Growth of *Porphyra* in various salinities was of interest due to the fact that the inner waters of the Alexander Archipelago in southeast Alaska have numerous freshwater inputs from streams and rivers along the coast. The high annual rainfall and snowmelt cause a marked decrease in surface water salinity in the summer and autumn (Stekoll, 1998). Most of the more accessible sites for the aquaculture of *Porphyra* in southeast Alaska are located in these inner waters. For economic and practical reasons it is necessary to understand the salinity tolerances of the various potential commercial species. Results from this study indicate that salinities of 20–35 ‰ would be optimal for the culture of *Porphyra* conchocelis, but that conchocelis growth begins to decline at salinities of 15 ‰ or less.

In these experiments, *Porphyra abbottae* conchocelis exhibited growth photoinhibition at irradiances above 40 μmol photons m^{-2} s^{-1} when grown at 15 °C and above. It is interesting to note that growth of the blade phase of this species is also reduced at higher irradiances (140 μmol photons m^{-2} s^{-1}) and higher temperatures (12 °C) (Hannach & Waaland, 1989).

The higher growth rate of *P. pseudolinearis* compared to the other two species may result from the conchocelis remaining mostly vegetative compared to the 100% conchosporangial condition of the *P. abbottae* and *P. torta* conchocelis. More energy in *P. pseudolinearis* goes into growth of filament length (which is what was measured) rather than filament width, which is significantly greater in the conchosporangial thalli.

The tolerance of the conchocelis of *P. pseudolinearis* to a wider range of salinity, irradiances and temperatures (data not shown) than the other two species corresponds with the local distribution of this species in southeast Alaska. Whereas *P. abbottae* and *P. torta* are widely and abundantly distributed on the outer coast, extending to the inner waters only up ma-

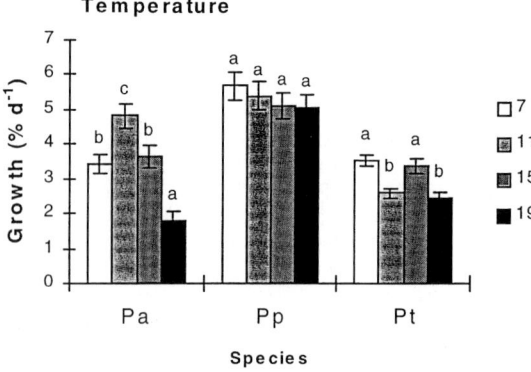

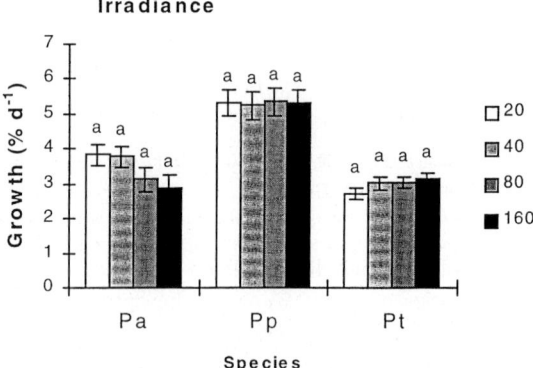

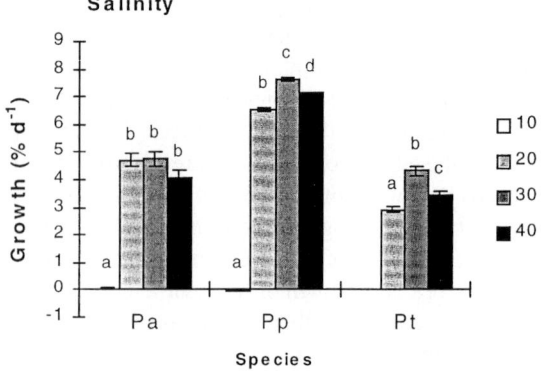

Figure 4. Comparison of pooled conchocelis growth rates of three species of *Porphyra* for each parameter tested. Error bars are ± S.E. Different letters above the bars indicate significant differences ($p < 0.01$) based on multiple comparisons using the Newman–Keuls test. Letter comparisons are relevant only within a species. Units are: Temperature = °C, Irradiance = μmol photons m^{-2} s^{-1}, Salinity = ‰

jor straits with direct connection to the outer coast, *P. pseudolinearis* has been recorded only from a limited stretch of coastline near Juneau, Alaska, several straits removed from the outer coast (Lindstrom et al., 1986). This area experiences wider temperature, salinity, and irradiance (due to reduced visibility during intense spring plankton blooms and glacial run-off in summer) fluctuations than the outer coast (Stekoll, 1998). The persistence of this species in this area obviously requires a highly tolerant cryptic phase (the conchocelis) to allow it to survive.

Measured growth rates for *P. torta* were similar to those of Waaland et al. (1987) who used strains from Puget Sound. The growth of *P. torta* from Puget Sound was significantly affected by irradiances within the range 5–300 μmol photons m^{-2} s^{-1}, whereas no significant effect of irradiance on the Alaska strain of *P. torta* was observed. However, the range of irradiances tested was narrower. Both the Alaska and Puget Sound strains had optimal growth at about 15°C at intermediate irradiances.

Porphyra mariculture offers an opportunity to develop a new industry in Alaska for seafood and seafood-based products for which markets already exist both locally and globally. The sale and barter of locally harvested species of *Porphyra* still occur among Natives in southeast Alaska, First Nations peoples in British Columbia, and Japanese–Canadians in southern British Columbia. These existing networks could provide the first market entry for a product from indigenous species (Roberts, 1993). At the national level, the primary markets for *Porphyra* are Japanese restaurants, Oriental food stores, health food outlets, and chemical companies. The research reported here is fundamental to the successful culture of *Porphyra* in Alaska. It is essential to produce spores reliably and in quantity in order to seed nets for the production phase. It is necessary to produce nets with the species and spore density that growers demand when they need them, and to be able to provide nets in the quantity needed for a commercial level of activity. Since environmental cues for reproduction are species- and possibly even population-specific, stocks selected from the wild must be manipulated in the laboratory to obtain this information.

Conclusion

Culture studies with indigenous strains of Alaskan *Porphyra* species showed the range of environmental conditions under which growth was successful. *Porphyra abbottae*, because of its existing market value in Alaska, is of special interest. However, this species is more difficult to culture, in part because of its sensitivity to certain combinations of irradiances and temperatures. Its sister species *P. torta* may be more amenable to domestication, although it quickly becomes conchosporangial in free culture, making it difficult to inoculate shells for net seeding. The tolerance of *P. pseudolinearis* to a wider range of environmental conditions and its ability to grow in or tolerate relatively low salinities are useful traits for a commercial species in southeast Alaska. Further work on domestication of this genus is necessary before the selection of a candidate species for commercialisation in Alaska.

Acknowledgements

The work reported here was supported in part by Saltonstall-Kennedy funding (NA76FD0035) administered through NOAA, U.S. Department of Commerce.

References

Bergdahl, J. C., 1990. Nori (*Porphyra* C. Ag.: Rhodophyta) mariculture research and technology transfer along the northeast Pacific coast. In Akatsuka, I. (ed.), Introduction to Applied Phycology. SPD Academic Publishing, The Hague: 519–551.

DeBoer, J. A., H. J. Guigli, T. L. Israel & C. F. D'Elia, 1978. Nutritional studies of two red algae. I. Growth rate as a function of nitrogen source and concentration. J. Phycol. 14: 261–265.

Hannach, G. & J. R. Waaland, 1989. Growth and morphology of young gametophytes of *Porphyra abbottae* (Rhodophyta): effects of environmental factors in culture. J. Phycol. 25: 247–254.

Lindstrom, S. C. & K. M. Cole, 1992. A revision of the species of *Porphyra* (Rhodophyta: Bangiales) occurring in British Columbia and adjacent waters. Can. J. Bot. 70: 2066–2075.

Lindstrom, S. C., N. I. Calvin & R. J. Ellis, 1986. Benthic marine algae of the Juneau, Alaska area. Contr. Nat. Sci. (B.C. Prov. Mus.), No. 6: 10 pp.

McLachlan, J., 1973. Growth media-marine. In Stein, J. (ed.), Handbook of Phycological Methods: Culture Methods and Growth Measurements. Cambridge University Press, London: 25–51.

Merrill, J. E., 1993. Development of nori markets in the western world. J. appl. Phycol. 5: 149-154.

Mumford, T. F. Jr., 1990. Nori cultivation in North America: growth of the industry. Hydrobiologia 204/205: 89–98.

Mumford, T. F. Jr. & A. Miura, 1988. *Porphyra* as food: cultivation and economics. In Lembi, C. A. & J. R. Waaland (eds), Algae and Human Affairs. Cambridge Univ. Press, New York: 87–117.

Roberts, W. A. Jr., 1993. An assessment of markets for *Porphyra* products. Report to Sitka Tribe of Alaska, University of Alaska Southeast, Juneau, 39 pp.

Scagel, R. F., P. W. Gabrielson, D. J. Garbary, L. Golden, M. W. Hawkes, S. C. Lindstrom, J. C. Oliveira & T. B. Widdowson, 1989 (reprinted 1993). A synopsis of the benthic marine algae of British Columbia, Southeast Alaska, Washington and Oregon. Phycological Contribution No. 3, Univ. of British Columbia, 532 pp.

Stekoll, M. S., 1998. The seaweed resources of Alaska. In Ohno, M., & A. T. Critchley (eds), Seaweed Resources of the World. Japan International Cooperation Agency, Tokyo: 258–265.

Waaland, J. R., L. G. Dickson & E. C. S. Duffield, 1990. Conchospore production and seasonal occurrence of some *Porphyra* species (Bangiales, Rhodophyta) in Washington State. Hydrobiologia 204/205: 453–459.

Waaland, J. R., L. G. Dickson, E. C. S. Duffield & G. M. Burzycki, 1986. Research on *Porphyra* aquaculture. In Barclay W. R., & R. P. McIntosh (eds), Algal Biomass Technologies: An Interdisciplinary Perspective. Nova Hedwigia 83: 124–131.

Waaland, J. R., L. G. Dickson, E. C. S. Duffield & J. E. Carrier, 1987. Conchocelis growth and photoperiodic control of conchospore release in *Porphyra torta* (Rhodophyta). J. Phycol. 23: 399–406.

Studies on the growth, in culture, of two forms of *Porphyra lacerata* from Japan

Masahiro Notoya & Kazuhiro Nagaura
Laboratory of Applied Phycology, Tokyo University of Fisheries, Konan-4, Minato-Ku, Tokyo 108, Japan

Key words: Porphyra lacerata, Bangiales, epilithic, epiphytic, life history

Abstract

Comparative studies were conducted on two forms of *Porphyra lacerata* Miura (Bangiales, Rhodophyta). Growth, morphological characteristics and maturation were compared in small epiphytic forms collected from Banda, Chiba Prefecture and large epilithic forms from Enoshima, Kanagawa Prefecture. Germlings of both types were cultured at temperatures of 10, 15 and 20 °C, an irradiance of 40 μmol m^{-2} s^{-1} and photoperiods of 14L:10D (light:dark) and 10L:14D. After 12 weeks, under identical culture conditions, the maximum size of foliose thalli of the epiphytic form was less than the epilithic form. At a temperature of 10 °C, the morphology of the Enoshima plants changed from cordate to lanceolate, whereas those from Banda retained their cordate shape. Under various culture conditions, the formation and liberation of archeospores were more abundant in the form from Banda. These observations suggest that the two forms of *P. lacerata* may represent different species, but further studies are required.

Introduction

Porphyra lacerata Miura was first described by Miura in 1967. Our previous culture studies focused on the life history of the small thallus form from Banda, Tateyama, Chiba Prefecture, Japan (Notoya & Naguara unpublished), which is epiphytic on the brown algae *Myelophycus simplex* (Harvey) Papenfuss and *Ishige okamurae* Yendo. The results indicated that temperature, irradiance and photoperiod influenced plant growth. The maximum size in culture was 22 × 25 mm (length × width), which is similar to that found in nature.

We have also reported the influence of temperature and irradiance on the life history of an epilithic form of *P. lacerata* from Enoshima, Sagami Bay, Kanagawa Prefecture, in the central part of the Pacific coasts of Japan (Notoya et al., 1993). In culture the foliose thallus of this form reached between 5.5 and 11.0 cm at 10 °C under a 14L:10D photoperiod, equivalent to the size found in nature.

The present study investigated the influence of culture conditions on the morphological characteristics, life history and growth of the foliose thallus and the conchocelis of the two forms referred to hereafter as Banda and Enoshima.

Materials and methods

Mature large epilithic foliose plants were collected on 23 February 1997 from Enoshima (Figure 1A). Mature, small, foliose thalli epiphytic on *M. simplex* were collected on 7 February 1995 from Banda (Figure 1B). To obtain zygotospores (following the terminology of Guiry, 1990) and archeospores (Magne, 1991) fragments were cut from the upper portions of thalli, the surfaces cleaned in sterilized water with an artist's brush and kept overnight in antibiotic solution containing 5 mg l^{-1} GeO$_2$ (Polne-Fuller & Gibor, 1987). Next day, these pieces were placed in Petri dishes containing sterilized seawater for several hours to induce spore release. The spores were collected by glass pipette, transferred into a new Petri dish containing sterile seawater and washed three to four times with sterilized seawater. After settling and attaching to glass slides, spores were grown at 15 °C and an irradiance of 40 μmol m^{-2} s^{-1} under a photoperiod of

Figure 1. (A) Mature foliose thalli of *Porphyra lacerata* collected on 23 February 1997 from Enoshima, Kanagawa Prefecture (scale as B). (B) Mature foliose thalli of *P. lacerata* epiphytic on *Myelophycus simplex* collected on 7 February 1995 from Banda, Taeyama, Chiba Prefecture. Scale bar is 5 cm.

14L:10D. After 1 week, the archeospores and zygotospores germinated and grew to form foliose thalli and conchocelis-phase filaments, respectively; the foliose thalli and conchocelis were separated and transferred to individual dishes. Archeospores, obtained from the young foliose thalli, were settled and attached to synthetic strings of nori nets (about 3 cm). The germlings were cultured in aerated flasks containing 300 ml of a modified Grund medium (McLachlan, 1973) at temperatures of 10, 15 and 20 °C, an irradiance of 40 μmol m^{-2} s^{-1} and photoperiods of 14L:10D and 10L:14D. The conchocelis filaments were cut into lengths of about 200 μm and attached to glass slides; they were cultured in glass tubes containing 50 ml of medium at 5, 10, 15, 20 and 25 °C, an irradiance of 40 μmol m^{-2} s^{-1} and photoperiods of 14L:10D and 10L:14D. The medium was replaced weekly during the experimental period. Three vessels were placed in each set of experimental conditions and ten thalli observed in each vessel.

Results

In culture the foliose thalli of the epiphytic (Banda) and epilithic (Enoshima) forms differed (Figure 1A, B); the maximum size of Banda plants was 22 × 25 mm (length × width) compared with 50 × 60 mm of the Enoshima plants.

Conchocelis colonies of both forms grew similarly at all temperatures. Colonies grew fastest at 15 °C under long days. No archeospores or protothalli were produced. Maturation of the conchocelis was observed

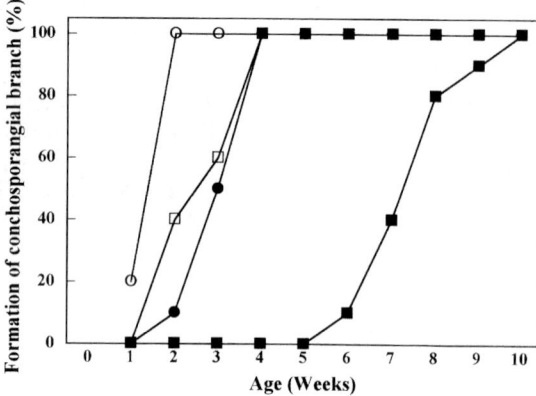

Figure 2. Formation of conchosporangial branches on Banda and Enoshima forms cultured at 25 °C, an irradiance of 40 μmol m^{-2} s^{-1} and different photoperiods. Open circles, Enoshima form under 14L:10D; open squares, Enoshima form under 10L:14D; solid circle, Banda form under 14L:10D; solid square, Banda form under 10L:14D.

only at 25 °C under both photoperiods. Conchosporangial branches formed 2 – 6 weeks later in Banda material. (Figure 2).

Growth of foliose thalli of both forms is shown in Figure 3. Similar growth was observed up to 4–7 weeks at all temperatures and both photoperiods. However, after the onset of archeospore liberation, Banda material ceased growing. The Enoshima material grew much faster at 10 °C under both photoperiods.

The maturation and liberation of archeospores, zygotospores and spermatia under the different culture conditions are shown in Figure 4. At 15 and

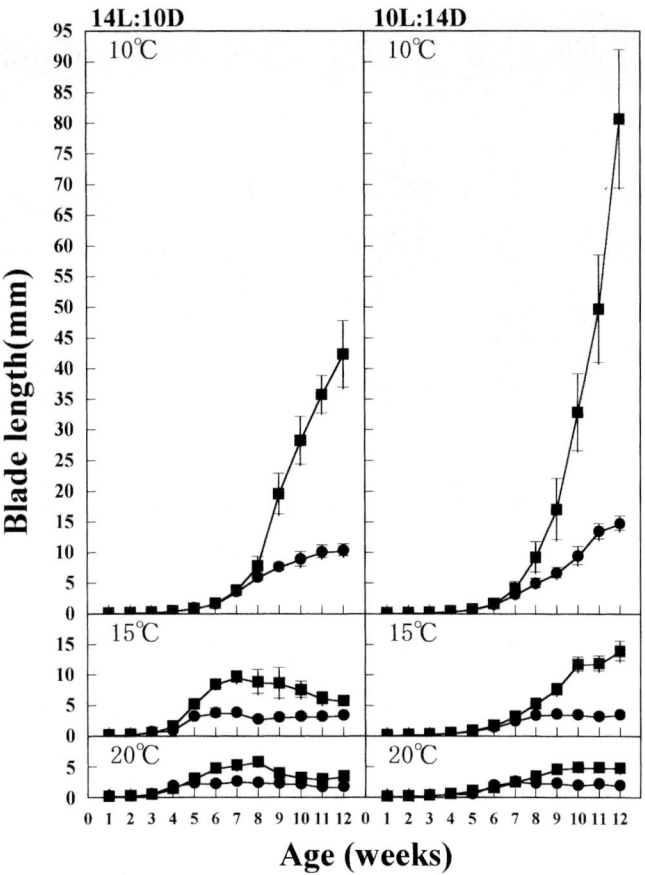

Figure 3. Growth of foliose thalli of Banda and Enoshima forms at different temperatures and photoperiods at an irradiance of 40 μmol m^{-2} s^{-1}. Solid square, Enoshima form; solid circle, Banda form. Vertical bar is ± S.D.

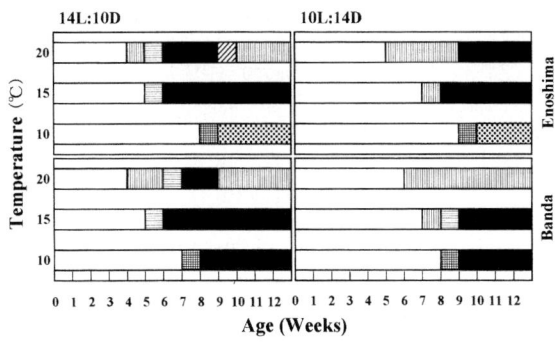

Figure 4. Liberation of archeospores, zygotospores and spermatia in relation to the age of cultured foliose thalli from Banda and Enoshima grown at different temperatures and photoperiods at an irradiance of 40 μmol m^{-2} s^{-1}. Open area, no liberation; vertical striped area, archeospores liberated; horizontal striped area, spermatia liberated; cross lined area, zygotospores liberated; solid area, archeospores, spermatia and zygotospores liberated; slanting striped area, archeospores and zygotospores liberated; dotted area, spermatia and zygotospores liberated.

20 °C, under both photoperiods, liberation of spores and spermatia were similar in both forms, except that neither spermatia or zygotospores were liberated from Banda material under short days. At 10 °C there was no liberation of archeospores from Enoshima material.

The shapes of the cultured thalli from Banda and Enoshima were different (Figure 5). At all temperatures, the thalli of Banda material were rounded and similar to field material, whereas after 12 weeks at 10 °C the Enoshima material became lanceolate.

The sizes of spores, foliose thalli and conchocelis of the Banda and Enoshima forms are compared in Table 1. For both forms in culture, the formulae for reproductive cells were: 64 (a/4, b/4, c/4) or 32 (a/4, b/2, c/4) in the spermatangium and 8 (a/2, b/2, c/2) or 4 (a/2, b/1, c/2) in the zygotosporangium.

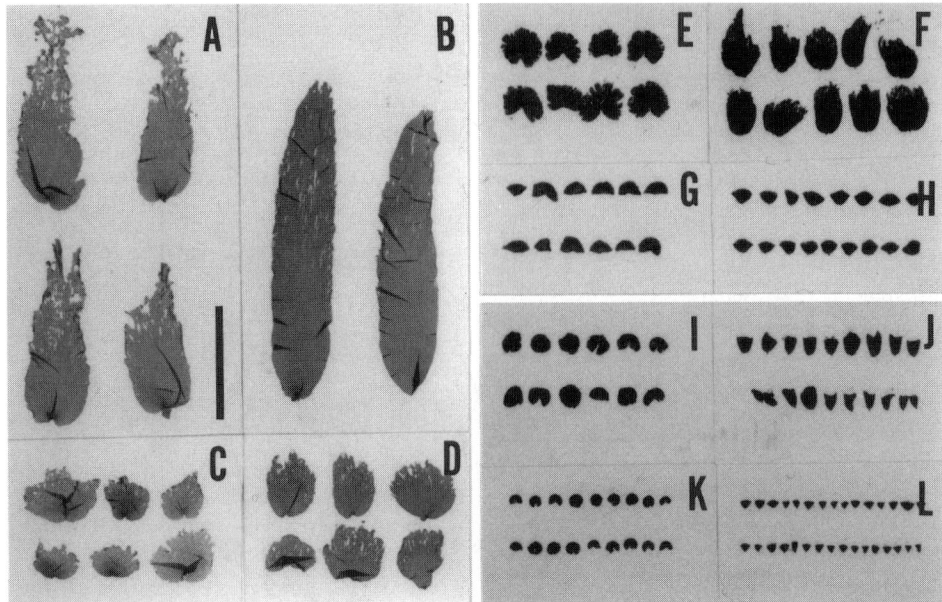

Figure 5. Foliose thalli of *Porphyra lacerata* after 12 weeks in culture at different temperatures, photoperiods and an irradiance of 40 μmol m^{-2} s^{-1}. (A) Enoshima form at 10 °C and 14L:10D; (B) Enoshima form at 10 °C and 10L:14D; (C) Banda form at 10 °C and 14L:10D; (D) Banda form at 10 °C and 10L:14D; (E) Enoshima form at 15 °C and 14L:10D; (F) Enoshima form at 15 °C and 10L:14D; (G) Banda form at 15 °C and 14L:10D; (H) Banda form at 15 °C and 10L:14D; (I) Enoshima form at 20 °C and 14L:10D; (J) Enoshima form at 20 °C and 10L:14D; (K) Banda form at 20 °C and 14L:10D; (L) Banda form at 20 °C and 10L:14D. Scale bar, 3 cm.

Table 1. A comparison of some morphological characters between natural thallus and cultured thallus of *Porphyra lacerata* from Enoshima and Banda

Character	Natural thallus		Cultural thallus	
	Enoshima	Banda	Enoshima	Banda
Foliose thallus				
Length (mm)	38.9 ± 4.5 (32–48)	12.9 ± 2.6 (9–17)	80.5 ± 11.3 (66–97)	14.6 ± 1.2 (13–17)
Width (mm)	57.0 ± 8.7 (42–69)	14.3 ± 4.6 (6–24)	18.6 ± 3.2 (16–24)	14.9 ± 2.8 (11–19)
Thickness (μm)	21.6 ± 1.8 (18–22)	20.2 ± 1.4 (18–22)	22.7 ± 2.8 (17–26)	23.2 ± 4.2 (16–30)
Shape	Cordate	Cordate	Cordate (15, 20 °C)	Cordate (10-20 °C)
			Lanceolata (10 °)	
Vegetative cell (μm)	15.9 ± 3.4 (13–22)	13.9 ± 2.2 (11–20)	21.7 ± 2.8 (17–29)	23.0 ± 2.2 (19–28)
Archeospore (μm)			15.8 ± 1.7 (13–20)	15.4 ± 1.2 (13–18)
Zygotospore (μm)			15.5 ± 1.5 (12–19)	13.1 ± 1.3 (11–15)
Spermatia (μm)			4.6 ± 0.9 (4–6)	6.0 ± 0.3 (5–7)
Hus' formula				
Spermatangium	64 (a/4, b/4, c/4)	64 (a/4, b/4, c/4)	32 (a/4, b/2 c/4)	32 (a/4, b/2, c/4)
			64 (a/4, b/4, c/4)	64 (a/4, b/4, c/4)
Zygotosporangium	8 (a/2, b/2, c/2)	8 (a/2, b/2, c/2)	4 (a/1, b/2, c/2)	4 (a/1, b/2, c/2)
			8 (a/2, b/2, c/2)	8 (a/2, b/2, c/2)

Discussion

The life history of *Porphyra lacerata* from Banda was the typical biphasic type previously reported for material from Enoshima (Notoya et al., 1993) and Nagasaki (Iima & Migita, 1990). However, the size of the Banda material was smaller, the maximum length being 10–15 mm compared with 50–90 mm for Enoshima material when cultured at 10 °C for 12 weeks. The material from Nagasaki is reported to be larger still (Iima & Migita, 1990), although the culture conditions were different. It is likely that heavy archeospore liberation from Banda material resulted in smaller thalli, indicating differences in physiology between the two forms.

The morphology of the two forms also differed. The Banda material was cordate and similar to field grown material, whereas the Enoshima material became lanceolate in form after 12 weeks in culture. Cultured thalli from Nagasaki were larger and more rounded than the material from Banda and Enoshima (Iimi & Migita, 1990).

Miura (1967) and Iimi & Migita (1990) reported only one type of Hus' formula in each type of reproductive organ: 64 (a/4, b/4, c/4) in spermatangia and 4 (a/1, b/2, c/2) in zygotosporangia. In the cultures of Banda and Enoshima material two types of reproductive cell were observed.

From these results it is clear that thalli of *P. lacerata* from Banda and Enoshima differ in their morphology, physiology and ecology. Thus, the two forms may be different species, but further studies are required.

Acknowledgements

We are grateful to Dr L. M.-C. Chen for his critical reading of the manuscript.

References

Guiry, M.D., 1990. Sporangia and spores. In Cole, K. M. & R. G. Sheath (eds), Biology of Red Algae. Cambridge University Press, New York: 347–376.

Iima, M. & S. Migita., 1990. Laboratory culture of *Porphyra lacerata* (Rhodophyta, Bangiales). Bull. Fac. Fish. Nakasaki Univ. 68: 13–20.

Magne, F., 1991. Classification and phylogeny in the lower Rhodophyta: a new proposal. J. Phycol. 27 (suppl.): 46.

McLachlan, J., 1973. Growth media – marine. In Stein, J. R. (ed.), Handbook of Phycological Methods. Cambridge University Press, New York: 25–51.

Miura, A., 1967. Two new species and new record of *Porphyra* from Enoshima, Sagami Bay. J. Tokyo Univ. Fish. 53: 65–71.

Notoya, M., N. Kikuchi, M. Mastuo, Y. Aruga & A. Miura, 1993. Culture studies of four species of *Porphyra* (Rhodophyta) from Japan. Nippon Suisan Gakkaishi 59: 431–436.

Polne-Fuller, M. & A. Gibor, 1987. Calluses and callus-like growth in seaweeds: induction and culture. Hydrobiologia 151/152: 131–138.

Effect of tissue nitrogen and phosphorus quota on growth of *Porphyra yezoensis* blades in suspension cultures

Jeff T. Hafting
Department of Botany, University of British Columbia, 6270 University Boulevard, Vancouver, B.C., Canada V6T 1Z4. E-mail: marbiopr@comptime.com

Key words: Porphyra yezoensis, suspension cultures, critical N, subsistence quota, molar N:P, nutrient status

Abstract

The effect of tissue N and P on growth of *Porphyra yezoensis* (strain U-51) blades in suspension cultures was investigated. Blades had the ability to store N in excess of requirements. The critical (0.40% fresh wt) and subsistence (0.153% fresh wt) levels of N were constant regardless of N source (NO_3^- or NH_4^+) or light level. Blades did not have the ability to store excess P over the range of P loads given. The subsistence quota for P was higher when NH_4^+ was given, suggesting a decreased ability to utilize tissue P for growth. NO_3^- was a better source of N than NH_4^+ in terms of growth. Blades became bright green in colour when N limited, suggesting a link between phycoerythrin and tissue N. The optimal molar N:P of 13–15 was constant regardless of N source (NO_3^- or NH_4^+) or light level. N:P <13–15 indicated N limitation, while N:P >13–15 indicated P limitation. P limited and light limited blades could store more N when NH_4^+ was given, than when NO_3^- was the N source, suggesting physiological mechanisms for taking advantage of this usually ephemeral N source, even when growth was limited. N and P reserves were used up relatively quickly (5 days), a characteristic of opportunistic species. Tissue analysis for N and P was a useful technique for determining nutrient status of *P. yezoensis* blades in land based tanks. As long as tissue N >0.40% fresh wt. and an N:P supply ratio of 13–15 is maintained, blade growth is not limited by N or P.

Introduction

Tank cultivation has its greatest advantage over traditional ocean-based cultivation in that one can control with greater efficiency the environment in which plants are grown (Lignell et al., 1987). Though light is the most important variable limiting growth during cultivation, it would be uneconomical to control the light environment of large-scale tanks. The next most important variable limiting growth is carbon, followed by nitrogen and phosphorus. Current mariculture technology is effective in eliminating C limitation by automatic CO_2 injection when a threshold pH is reached, and therefore carbon was not investigated in this study.

Nutrients such as N and P are not as quickly diluted and washed away in a tank (especially if in-flowing water is shut off during fertilization) as in the ocean. The optimal growing conditions occur when all nutrients added are assimilated into algal crop biomass, leaving no nutrients for epiphytes and other algae. Ideally, the crop should be grown at its maximal rate in a 'desert' of low nutrients. To accomplish this, detailed studies into algal responses to nutrients are needed. This allows the manipulation of the nutrient environment to optimize yields.

Little work has been carried out on the N and P nutrition of *Porphyra yezoensis*, even though it is farmed extensively in Asia. In traditional ocean-based farms it is difficult to maintain optimal N and P conditions, because the nutrient environment varies widely throughout the farm due to such factors as the local flow conditions. In temperate coastal waters, nutrient levels are typically high in the winter when *Porphyra* cultivation takes place (Kain, 1991). For these reasons, farmers use simple guidelines when considering algal nutrition on their farms. If the $NH_4^+ + NO_3^-$ concentration is <3 μM (50 mg m^{-3}), then the seawater is considered infertile, and fertilizer must be applied. A spraying technique is usually used to apply the fertilizer. If the $NH_4^+ + NO_3^-$ concentration is about 7 μM

(100 mg m^{-3}) it is regarded as semi-fertile. At this nitrogen concentration, growth is supported without fertilizer, but the product is of medium grade. An $NH_4^+ + NO_3^-$ concentration of about 15 μM (200 mg m^{-3}) and above is considered fertile enough to support growth and high quality without fertilizer application (Lobban and Harrison, 1994; Tseng, 1981a, b). Concentrations of phosphorus are not usually considered by farmers, and tissue analysis for this element is rarely done. Tanks allow greater control over nutrient conditions, and provide uniform growing conditions throughout the farm. In order for this advantage to be exploited, knowledge of the relationships between growth and tissue N and P is needed.

It would seem that supplying nutrients at saturating amounts for uptake rate would be desirable, but this means that nutrient levels are always high, leaving nutrients for epiphytes and weed species (Schramm, 1991). A better way of managing nutrients for optimal growth involves tissue analysis for total nitrogen and phosphorus. Tissue analysis is a direct measurement, and can provide information as to whether the algae require nitrogen or phosphorus. Most macroalgae tend to store excess nitrogen in their tissues for growth during periods of low nitrogen concentration (Thomas & Harrison, 1985).

The aim of this study was to determine if N and P are stored in the tissues of *Porphyra yezoensis*. If storage occurred, then the critical (i.e. the concentration at which growth is saturated) and subsistence (i.e. the minimum concentration at which growth occurs) levels of the nutrient were determined under both high and low light conditions. Nitrogen was given as NO_3^- or NH_4^+ because it is possible that critical nitrogen and phosphorus levels could differ depending on the nitrogen source (Wheeler & Björnsäter, 1992).

Materials and methods

Culture conditions

Porphyra yezoensis blades were grown to at least 1 cm in length before use. The blades were cultured in trough vessels, constructed of PVC pipe (20 cm diameter, 86 cm length) cut in half longitudinally, with an air tube (3 cm diameter PVC pipe) glued and sealed to the outside bottom of each trough. Small holes were drilled through the bottom of the trough into the air tube, which was connected to a large air pump. Aeration kept blades in constant motion, reduced boundary layers (Neushul et al., 1992), and also maintained temperature and nutrient distribution in each compartment (Bidwell et al., 1985). Each trough was sub-divided into four compartments by Plexiglas dividers sealed with silicon aquarium sealant. Each compartment held 3 l of culture medium. While the pump was operating, no mixing of medium among compartments occurred. Each compartment was covered with a Plexiglas lid to prevent excessive medium losses due to evaporation.

Blades were cultured under a 8L:16D (light:dark) photoperiod, 15 °C and, using fluorescent Vita-lights, 160 μmol photons m^{-2} s^{-1} in high light cultures or 50 μmol photons m^{-2} s^{-1} in low light cultures, measured just below the water level. Irradiance levels were adjusted by raising or lowering the trough vessels relative to the light source. Low light levels were achieved by using window screen to shade these compartments. All experiments were done in a walk-in growth chamber, and each treatment was replicated four times.

Seawater was filtered through 0.22 mm Millipore GS filters before use and then enriched with trace metals and vitamins following the f/2 recipe of McLachlan (1973). Nitrate ($NaNO_3$), ammonium (NH_4Cl), and phosphate (NaH_2PO_4) loads (concentration × volume given/frequency of addition) were varied according to which nutrient of interest was under investigation. When tissue N was being studied, phosphate loads were kept high to prevent phosphorus limitation. When tissue P was being considered, nitrate or ammonium loads were kept high to prevent nitrogen limitation (Table 1; note that N and P loadings, not concentrations are given in Table 1). Supply rate is more important in determining nutrient limitation, so this parameter was used instead of concentration (Lobban & Harrison, 1994). These nutrient loads compare well with ecological ranges in N and P supply. Culture medium was not changed during the experiment. Nitrogen and phosphate were added directly to the compartments from stock solutions. Culture pH was monitored and never rose above pH 8.4 in any culture throughout the experiment; carbon was assumed not to be limiting.

Experimental design

Each compartment was inoculated with 1 g fresh wt of blades. Experiments were always initiated on a Friday, and nitrogen and phosphate enrichments were done on Mondays, Wednesdays, and Fridays during the experiment. On these days, blades were removed

Table 1. Range of N and P loads given to blades during experiment.

	Nitrate experiment		Ammonium experiment
Tissue N experiments (load = μmol per day)			
NO_3^- load	2.14–257.14	NH_4^+ load	2.14–171.43
PO_4^- load	21.43	PO_4^- load	21.43
N:P supply ratio	0.1–12.0	N:P supply ratio	0.1–8.0
Tissue P experiments (load = μmol per day)			
NO_3^- load	128.57	NH_4^+ load	128.57
PO_4^- load	0.26–16.29	PO_4^- load	0.43–21.43
N:P supply ratio	7.9–494.5	N:P supply ratio	6.0–299.0

from the compartments using an aquarium fish net, excess surface water was blotted from them using paper towels, and then weighed. Before returning blades to their compartments, they were harvested back to their original 1 g fresh wt. A sample of harvested blades was saved from each compartment for tissue nutrient analysis. Blades were allowed to acclimatize to culture conditions for 7 days in high light cultures, and for 14 days in low light cultures before data were recorded. The extra 7 days acclimation time in low light cultures was necessary because growth was slower. Growth rate and tissue nutrient levels stabilized after these acclimation periods (semi-steady state conditions). The experiments ran for 21 days in total.

Growth was calculated from:

$$u = [\ln(m_2/m_1)]/(t_2 - t_1) \times 100, \quad (1)$$

where u = specific growth rate as % day^{-1} and m_2 and m_1 are culture mass at time t_2 and t_1 respectively.

Starvation experiment

To determine how quickly blades used their N and P reserves and to confirm critical and subsistence levels of N and P, a nutrient starvation experiment was run under high light conditions after 21 days in culture, using nitrogen (given NO_3^-) and phosphorus sufficient blades. Blades were given either NO_3^- or PO_4^- at the highest levels shown in Table 1. Total N and P depletion was measured using blades given high PO_4^- or NO_3^-, respectively. Blades were treated as in the previous experiment, with fresh wt measurements, harvests, nutrient additions, and tissue analysis all carried out. Growth (or death) was monitored for an additional 12 days after which time blades ceased growing and began to die.

Tissue nutrient analysis

Total tissue N and P were determined by a modified alkaline persulphate digestion technique (Björnsäter and Wheeler, 1990; D'Elia et al., 1977). This method has the advantage of allowing the measurement of total N and P from a single extraction. Algal samples were weighed (0.01–0.02 g fresh wt), and placed into 50 ml Pyrex culture tubes each containing 5 ml deionized water (DIW) and 30 ml of oxidizing reagent (3.0 g NaOH and 6.7 g low N (<0.001%) potassium persulphate dissolved in 1 l DIW). The tubes were capped tightly and autoclaved for 1 h at 110 °C and 15 psi. Tissue nutrient levels are reported as % fresh wt; results are obtained faster than if blades had been dried and this short-cut is useful in a commercial aquaculture setting (Ulrich, 1952). All surface water was removed before weighing, and samples were immediately submerged in oxidizing reagent, reducing any errors resulting from the use of fresh material.

After the tubes cooled sufficiently to be handled safely, 3 ml of 0.3 M HCl was added, to acidify the solution, followed by 4.0 ml of borate buffer (30.9 g boric acid, and 100 ml of 1 M NaOH, made up to 1 l with DIW). The volume of each tube was brought up to 51 ml with DIW. Under alkaline conditions all nitrogenous compounds were converted to NO_3^- and all phosphorus containing compounds were converted to PO_4^-. Total N and P were measured colorimetrically, phosphate by the method of Strickland & Parsons (1972) and nitrate by a spongy cadmium technique of Jones (1984), which converts all NO_3^- to NO_2^- for analysis.

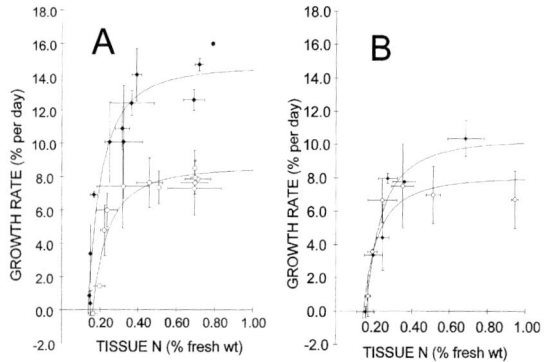

Figure 1. Relationship between tissue N and growth rate (95% confidence limits shown, $n = 4$; high light, solid diamonds; low light, hollow circles; A, NO_3^- - N source; B, NH_4^+ - N source).

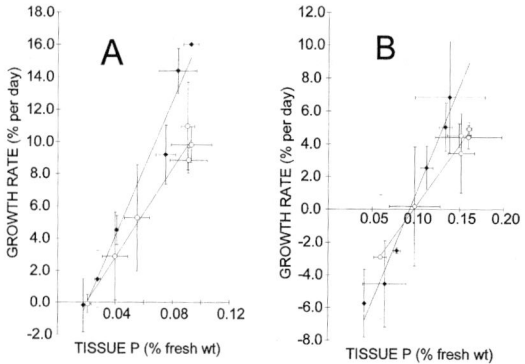

Figure 2. Relationship between tissue P and growth rate (95% confidence limits shown, $n = 4$; high light, solid diamonds; low light, hollow circles; A, NO_3^- - N source, B, NH_4^+ - N source).

Statistics

Mean growth rate and tissue N or P for each treatment were based on days 10–21 for high light conditions and on days 17–21 for low light conditions. Means were pooled for each treatment replicate ($n = 4$). All statistical analyses were done using Systat for Windows (version 5.0). Critical tissue levels of a nutrient were defined as the tissue level at which an increase of 0.01% fresh wt of tissue N or P results in less than 0.10% increase in growth rate.

Results

Tissue N

Figure 1 shows that the relationship between growth rate and internal total N is a rectangular hyperbola, regardless of N source. The data fit well ($r^2 = 0.83$–0.93, Table 2) the Droop equation (Droop, 1983) if raised to the 2nd power:

$$m = m_{max}[1 - (Q_0/Q)^2] \qquad (2)$$

where m is the growth rate (% day^{-1}), Q is the tissue concentration of the nutrient within the algal cells (% fresh wt), Q_0 is the lowest level of Q at which the alga can grow (the subsistence level, % fresh wt), and m_{max} is the maximal growth rate (% day^{-1}) at infinite Q (South & Whittick, 1987). Specific parameters for the modified Droop equation are given in Table 2.

Figure 1 and Table 2 show that regardless of the light level or N source, Q_0 and the critical N level differ very little. This allowed a mean Q_0 (0.15% fresh wt) and critical N (0.40% fresh wt) to be calculated. Growth in high light cultures was greater (mean $m_{max} = 12.6\%$ day^{-1}) than under low light conditions (mean $m_{max} = 8.4\%$ day^{-1}). NO_3^- cultures grew at a faster rate (14.7% day^{-1}) than NH_4^+ cultures (10.4% day^{-1}) under high light conditions. Blades that had a tissue N level >0.40 appeared very dark purple to black in colour but as the tissue level dropped to <0.40, the blades became more green in colour. Close to the Q_0 level, blades were bright green, with no trace of purple or reddish colouration.

Tissue P

Figure 2 shows that the relationship between internal tissue P and growth rate fits (R^2 values ranged from 0.97 to 0.99) a standard linear equation:

$$m = m(Q) + b \qquad (3)$$

where m is the growth rate (% day^{-1}), m is the slope of the line, Q is the tissue concentration of the nutrient within the algal cells (% fresh wt), and b is the y intercept (linear parameters are given in Table 3). Figure 2 and Table 3 show that the subsistence level for P was much higher when NH_4^+ was the N source (mean $Q_0 = 0.09\%$ fresh wt) than when NO_3^- was the N source (mean $Q_0 = 0.02\%$ fresh wt). High light cultures grew more quickly than low light cultures, but Q_0 did not differ between high and low light blades given the same N source. The slope of the relationship between growth and tissue P was greater when NO_3^- was the N source (Table 3). P-limited blades fell apart very easily and felt slippery. Aeration was sufficient to break P-limited blades apart, contributing to their ultimate death.

Table 2. Modified Droop equation parameters and regression statistics for the relationship between tissue N and growth rate shown in Figure 1. Modified Droop equation: $\mu = \mu_{max} [1 - (Q_0 / Q)^2]$.

Light	N source	μ_{max}	Q_0 (% per day)	Crit. N (% fresh wt)	r^2 (% fresh wt)
High	NO_3^-	14.7	0.14	0.37	0.93
High	NH_4^+	10.4	0.16	0.41	0.93
Low	NO_3^-	8.6	0.16	0.41	0.92
Low	NH_4^+	8.1	0.15	0.39	0.83
			Mean	0.15	0.40

Table 3. Parameters and regression statistics for the linear relationship between tissue P and growth rate shown in Figure 2. Linear equation: $\mu = m(Q) + b$.

Light	N source	m	b	Q_0 (% fresh wt)	r^2
High	NO_3^-	210.84	−4.30	0.02	0.99
High	NH_4^+	132.03	−12.25	0.09	0.98
Low	NO_3^-	137.70	−2.67	0.02	0.97
Low	NH_4^+	74.93	−7.30	0.10	0.98

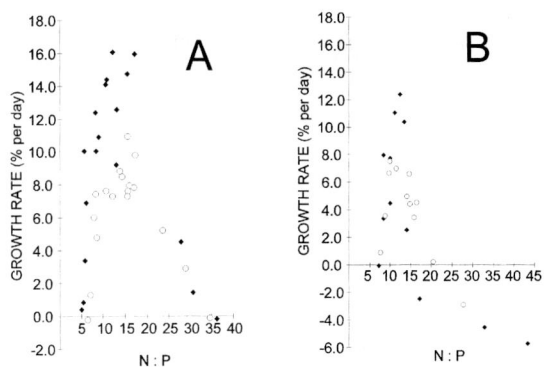

Figure 3. Relationship between molar N:P and growth rate (high light, solid diamonds; low light, hollow circles; A, NO_3^- - N source; B, NH_4^+ - N source).

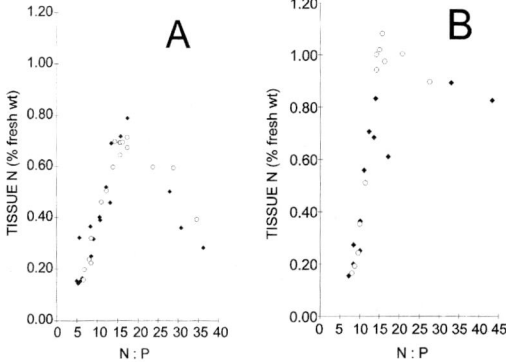

Figure 4. Relationship between molar N:P and tissue N (high light, solid diamonds; low light, hollow circles; A, NO_3^- - N source; B, NH_4^+ - N source).

Tissue N:P

Growth rate was greatest at a molar N:P of 10–17 and 10–15 when NO_3^- and NH_4^+ were supplied, respectively (Figure 3). Values below these ranges occurred in blades given low N loads (Table 1), while higher values occurred in blades given low P loads. When provided with the same N source, the N:P ranges were the same at both irradiance levels. Figure 3 also shows that reduction in growth rates was more severe in low P load blades (N:P >15) with NH_4^+ than with NO_3^-. Low P load blades given NH_4^+ began to die (i.e. growth rate < 0% per day) at N:P >17–20, while this did not occur in those given NO_3^- until N:P was 35.

Tissue N was greatest at a molar N:P of 13–17, regardless of N source or irradiance level (Figure 4). Within this range, tissue N levels were higher in blades grown under low light and given NH_4^+, than when given NO_3^-. Figure 4B shows that under both high and low light, low P load blades (N:P >17) given NH_4^+ had tissue N levels as high as blades within the N:P 13–17 range.

Tissue P was greatest at a molar N:P of 10–17, regardless of N source or irradiance level (Figure 5). Within this range, blades grown under low light given NH_4^+ had a higher tissue P content (Figure 5B) than those given NO_3^- (Figure 5A).

Starvation experiment

Figure 6A shows that the growth rates of N sufficient blades (0.67% fresh wt) when grown under high light and given only PO_4^- remained high for 5 days until tissue N fell below about 0.36% fresh wt. Growth rates then decreased quickly until the cultures began to die on day 12 and tissue N fell below about 0.15% fresh wt.

A similar pattern is seen in Figure 6B for blades with high P (0.09% fresh wt) grown under high light conditions and given only NO_3^-. Growth rates remained high for 5 days, until tissue P fell below about 0.05% fresh wt, and decreased thereafter. Cultures

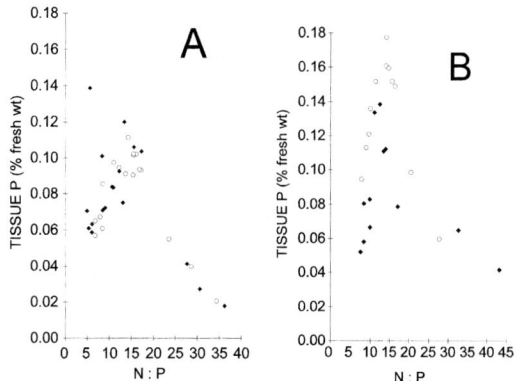

Figure 5. Relationship between molar N:P and tissue P (high light, solid diamonds; low light, hollow circles; A, NO_3^- - N source; B, NH_4^+ - N source).

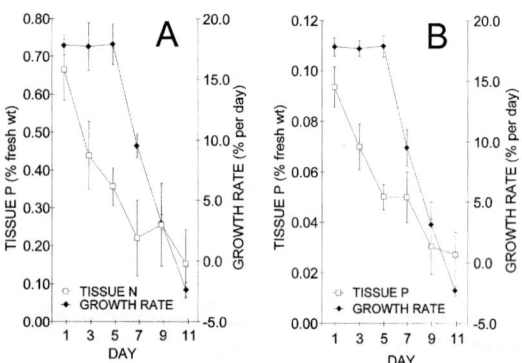

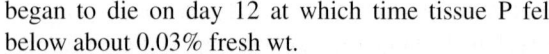

Figure 6. Results of starvation experiment. (A) Growth and tissue N of N- and P-sufficient blades, starved of N by giving them only PO_4-. (B) Growth and tissue P of N- and P-sufficient blades, starved of P by giving them only NO_3^- (95% conf. limits shown, $n = 4$).

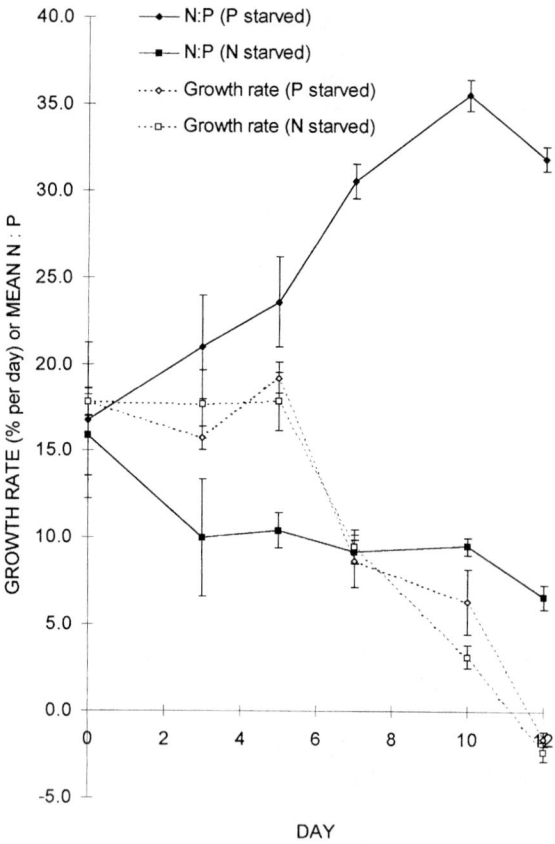

Figure 7. Results of starvation experiment, showing N:P and growth rate changes as blades were starved of N or P (95% conf. limits shown, $n = 4$).

began to die on day 12 at which time tissue P fell below about 0.03% fresh wt.

Whether blades are grown with no N, or no P, growth rates follow the same pattern; they began to drop in low N cultures once N:P <10 and in low P cultures once N:P >23 (Figure 7).

Discussion

Tissue N and P

Luxury storage of N (Droop, 1974) is indicated by the shape of the curve which describes the relationship between growth rate and internal total N (i.e., rectangular hyperbola). Under high light, NO_3^- is a better N source than NH_4^+ in terms of growth rate. This observation was somewhat unexpected since NH_4^+ is directly assimilated into amino acids, and NO_3^- must be reduced before assimilation (Lobban & Harrison, 1994). Under low light there is no advantage in using either source of N in terms of growth rate.

A critical internal tissue N level of 0.40% fresh wt and a subsistence level of 0.15% fresh wt seem to be maintained by the blades, regardless of N source or light level. Wu et al. (1984) found that the maximal growth rate of *P. yezoensis* (11.6% day^{-1}) was attained with a N content of 4.7% dry wt. Assuming a fresh to dry wt. ratio of 10, this compares well to the 0.40% fresh wt (i.e., 4.0% dry wt) critical level found in this study. The critical tissue N level is 2% dry wt for *G. tikvahiae* (Hanisak, 1990), and 1.9% dry wt for *Codium fragile* (Hanisak, 1979). This suggests that *Porphyra yezoensis* has greater N requirements than either of these species (assuming the same fresh: dry wt ratio of 10).

The fact that the light environment did not alter the critical level of N was unexpected. Light was expected to alter cell constituents such as pigment amounts, as well as the amount of N required for maximal growth rate (Lobban & Harrison, 1994). *Porphyra yezoensis* is found in the high intertidal zone, where light levels are relatively high, and perhaps is unable to increase or decrease pigment levels in response to light (Herbert & Waaland, 1988). This result differs from the interactive effect of light and N seen with *Gracilaria tikvahiae* (Lapointe & Duke, 1984), and *Macrocystis pyrifera* (Shivji, 1985). In these two algae the critical levels of N can be expected to differ depending on the light environment (Lapointe and Duke, 1984).

The two irradiances used in this experiment (50 and 160 μmol photons m^{-2} s^{-1}) were low compared to natural values *P. yezoensis* would be exposed to. If severe light limitation was occurring in high light cultures, this would prevent a meaningful discussion of how light affects critical N levels. Although the higher light level would not saturate photosynthesis (Lapointe & Duke, 1984), it is unlikely that severe light limitation was occurring. The growth rate achieved (14% day^{-1}, when given NO$_3^-$) is similar to those (15–20% day^{-1}) obtained by Imada et al. (1971) during outdoor cultivation, under irradiances levels much higher than in the present study.

Porphyra yezoensis does not have the ability for luxury storage of P (over the range of P and N:P supply given here), as indicated by a linear relationship between growth rate and tissue P. A higher subsistence level for P when NH$_4^+$, rather than NO$_3^-$, was the N source, suggests an increased requirement for P, or a decreased ability to utilize tissue P for growth. When NO$_3^-$ is the N source, *P. yezoensis* blades are more efficient at using tissue P and they will have an advantage, when light levels are high. Under low light, the need to utilize tissue P for growth is not as great and therefore no advantage is found between the different N sources. It has traditionally been assumed that N is the most important limiting nutrient in temperate oceans and during mariculture (Flores-Moya et al., 1997); however this study shows that P levels also play a very important role in determining growth rates of *P. yezoensis*.

Wu et al. (1984) concluded that NH$_4^+$, not NO$_3^-$, is a better N source for *P. yezoensis*. However, uptake rate was the only parameter considered. Even if NH$_4^+$ is taken up at a faster rate than NO$_3^-$, this does not necessarily mean that growth rate is enhanced with NH$_4^+$. The blades may be storing NH$_4^+$ as tissue N, and not using it for growth. This study goes one step further by relating internal tissue N and P to growth. Based on these parameters, it seems that NO$_3^-$ is the better N source in terms of growth. This agrees with Iwasaki (1967), who found that NO$_3^-$ was the better N source for *P. tenera*.

For *Gracilaria tikvahiae*, growth rates were identical whether NH$_4^+$ or NO$_3^-$ was given (Hanisak, 1990), whereas both *Gracilaria foliifera* and *Neoagardhiella baileyi* have higher growth rates using NH$_4^+$ rather than NO$_3^-$ (DeBoer et al., 1978). The preference is dependent on the macrophyte being considered and no general trends can be determined.

The change in colour from dark purple-black to bright green in N depleted blades suggests a loss of the pigment phycoerythrin, or the proteins associated with this pigment (Ryther et al., 1981/82). The fragmentation of *P. yezoensis* tissue given low P was also seen in *Chondrus crispus* (Neish et al., 1977). P is found in phospholipids, essential components of cell membranes (DeBoer, 1981). This suggests that P has a role in maintaining the structural integrity of *P. yezoensis* blades.

This study was performed on blades in the artificial and controlled environment of a growth chamber. It remains to be seen if blades will have a critical N level of 0.40% fresh wt under natural sunlight, and whether the critical N level varies seasonally. Further larger-scale outdoor trials are required.

Tissue N:P

The N:P for optimal growth, tissue N, and tissue P were all within the range 13–15. The fact that this relationship remains constant under different irradiances and N sources, makes N:P a particularly useful measurement for commercial purposes. A N:P <13–15 indicates N limitation and N:P >13–15 indicates P limitation. The critical N level (0.40% fresh wt) also remained constant under different N sources and light levels. If tissue N >0.40% fresh wt (the higher the N content, the higher the quality of the resulting processed nori; Johnston, 1971) and the N:P supply ratio = 13–15, blades will be neither N- nor P-limited, and should be growing at their maximum rate provided that all other factors, e.g. carbon, micronutrients are not limiting.

Marine phytoplankton have an average N:P ratio of 16:1 (the Redfield ratio) whereas marine macrophytes show a greater range of Redfield ratios, with a median of 30:1 (Lewis & Hanisak, 1996). This suggests lower

P requirements (or higher N requirements) in macrophytes. Since the optimal N:P ratio found in this study ranges from 13–15, *P. yezoensis* appears to have P requirements closer to those of marine phytoplankton. This may be due to its simple monostromatic structure (fewer structural proteins needed means lower N:P). This also illustrates the importance of P nutrition to *P. yezoensis*.

Wheeler & Björnsäter (1992) found that for *Porphyra* sp. and *Codium fragile*, N:P < 12 was indicative of N limitation and N:P > 17 suggested P limitation. Though this range is slightly wider than that found here, these results agree well with the findings of this study. Atkinson & Smith (1983) list a C:N:P ratio of 137:23:1 for *P. yezoensis*. Hernandez et al. (1993) gave a C:N:P ratio of 258:20:1 in winter, and 495:38:1 in spring for *Porphyra umbilicalis*. These ratios, obtained from blades collected from wild populations, suggest P limitation. Experiments by Flores-Moya et al. (1997) confirm that *P. leucosticta* is often P-limited in the spring.

Under P limitation, growth rate is affected much more when NH_4^+ was given. This result fits with the suggestion that blades given NH_4^+ cannot use tissue P for growth as efficiently as blades given NO_3^-. Under P limitation (N:P > 13–15), blades given NH_4^+ are able to store tissue N at levels as high as N and P sufficient blades (N:P = 13–15). Perhaps blades are physiologically adept at utilizing this usually ephemeral N source when available, even if blades cannot use the N for growth because of P limitation.

Under low light, N- and P- sufficient blades given NH_4^+ are able to store more tissue N than blades given NO_3^-. One explanation is that blades are storing N, taking advantage of a usually patchy N source (NH_4^+) while they can, even if they cannot use the N for growth because of light limitation. Also, tissue P levels were higher in N and P sufficient blades given NH_4^+ under low light than blades given NO_3^- under low light. This increased tissue P may be in response to the increased storage of NH_4^+, as an attempt to regulate N:P, keeping it within the range 13–15, optimal for growth. The potential for N accumulation was greater with NH_4^+ than with NO_3^- in *Neoagardhiella baileyi*, *Gracilaria foliifera* (D'Elia & DeBoer, 1978), and *Ulva lactuca* (DeBusk et al., 1986), for the same reasons mentioned above.

Starvation experiment

The results of the starvation experiment serve to confirm both the critical and subsistence levels of N and P, and the optimal N:P ratio. Growth rates were equally affected by N and P limitation, and this illustrates the importance of both elements in regulating growth.

N and P reserves were used up relatively quickly; growth rates remained high for only 5 days before N and P reserves fell to limiting levels. In contrast, *Gracilaria tikvahiae* was able to grow on its nutrient reserves for 2 weeks before limitation occurred (Hanisak, 1990). Low nutrient-storage capability is characteristic of opportunistic species (Lobban & Harrison, 1994), and this may pose a problem in nutrient management so that other weedy species are left with little nutrient for uptake and growth. It appears that *P. yezoensis* has a nutrient storage capacity similar to that of other opportunistic species, e.g. *Enteromorpha* and *Ulva*. There are alternative weed control methods that may be effective, e.g. exposure to air.

Conclusion

The two forms of nitrogen (NO_3^- and NH_4^+) given to *P. yezoensis* differ in their abilities to promote growth. In general, growth was higher with NO_3^- and this may be due to the increased efficiency of tissue P utilization by blades when NO_3^- is the N source. However, the potential for N accumulation is higher when NH_4^+ is the N source. Blades can store more N during light and phosphorus limitation when NH_4^+ is given, but this increased N storage does not translate into increased growth. In these experiments small blades were used and it remains to be seen whether similar results would be obtained using larger blades. However, because *Porphyra* has a simple monostromatic structure it is not expected that results would differ.

Tissue analysis is a valuable tool and should become standard practice in the cultivation of *P. yezoensis* and other macrophytes. It is the most direct way of determining the nutritional status of blades. It is used extensively with agricultural plants, and has contributed to the success of that industry (DeBoer, 1981; Hanisak, 1979). The critical nutrient concentration is the key for assessing the nutrient status of a crop (Ulrich, 1952). Maintaining a N:P supply ratio of 13–15, and a tissue N of $\geq$ 0.40% fresh wt should result in growth that is neither N- nor P-limited. Because the quality of the final product has been linked to tissue N content, it is probably desirable to maintain tissue N close to 0.70% fresh wt (Mencher et al., 1983).

Acknowledgements

This work was funded in part by a Science Council of BC, Graduate Research, Engineering and Technology (GREAT) award. Additional funding was provided by Marine BioProducts Inc. The author would also like to thank Dr R. E. Foreman and Dr P. J. Harrison at the University of British Columbia.

References

Atkinson, M. J. & S. V. Smith, 1983. C:N:P ratios of benthic marine plants. Limnol. Oceanogr. 28: 568–574.
Bidwell, R. G. S., J. McLachlan & N. D. H. Lloyd, 1985. Tank cultivation of Irish Moss, *Chondrus crispus* Stackh. Bot. mar. 28: 87–97.
Björnsäter, B. R. & P. A. Wheeler, 1990. Effect of nitrogen and phosphorus supply on growth and tissue composition of *Ulva fenestrata* and *Enteromorpha intestinalis* (Ulvales, Chlorophyta). J. Phycol. 26: 603–611.
DeBoer, J.A., 1981. Nutrients. In Lobban, C.S. & M.J. Wynne (eds), The Biology of Seaweeds, University of California Press, Los Angeles: 356–392.
DeBoer, J. A., H. J. Guigli, T. L. Israel & C. F. D'Elia, 1978. Nutritional studies of two red algae. I. Growth rate as a function of nitrogen source and concentration. J. Phycol. 14: 261–266.
DeBusk, T. A., M. Blakeslee & J. H. Ryther, 1986. Studies on the outdoor cultivation of *Ulva lactuca* L. Bot. mar. 29: 381–386.
D'Elia, C. F. & J. A. DeBoer, 1978. Nutritional studies of two red algae. II. Kinetics of ammonium and nitrate uptake. J. Phycol. 14: 266–272.
D'Elia, C. F., P. A. Steudler & N. Corwin, 1977. Determination of total nitrogen in aqueous samples using persulfate digestion. Limnol. Oceanogr. 22: 760–763.
Droop, M. R., 1974. The nutrient status of algal cells in continuous culture. J. mar. biol. Ass. U.K. 54: 825–855.
Droop, M. R., 1983. 25 years of algal growth kinetics – a personal view. Bot. mar. 26: 99–112.
Flores-Moya, A., M. Altamiran, M. Cordero, M. E. González & M. G. Pérez, 1997. Phosphorus-limited growth in the seasonal winter red alga *Porphyra leucosticta* Thuret et Le Jolis. Bot. mar. 40: 187–191.
Hanisak, M. D., 1979. Nitrogen limitation of *Codium fragile* ssp. *tomentosoides* as determined by tissue analysis. Mar. Biol. 50: 333–337.
Hanisak, M. D., 1990. The use of *Gracilaria tikvahiae* (Gracilariales, Rhodophyta) as a model system to understand the nitrogen nutrition of cultured seaweeds. Hydrobiologia 204/205: 79–87.
Herbert, S. K. & J. R. Waaland, 1988. Photoinhibition of photosynthesis in a sun and a shade species of the red algal genus *Porphyra*. Mar. Biol. 97: 1–7.

Hernandez, I., A. Corzo, F. J. Gordillo, M. D. Robles, E. Saez, J.A. Fernandez & F. X. Niell, 1993. Seasonal cycle of the gametophytic form of *Porphyra umbilicalis*: nitrogen and carbon. J. exp. mar. Biol. Ecol. 173: 181–196.
Imada, O., Y. Saito, K. I. Teramoto, 1971. Artificial culutre of laver. Proc. int. Seaweed Symp. 7: 358–363.
Iwasaki, H., 1967. Nutritional studies of the edible seaweed *Porphyra tenera*. II. Nutrition of conchocelis. J. Phycol. 3: 30–34.
Johnston, H. W., 1971. A detailed chemical analysis of some edible Japanese seaweeds. Proc. int. Seaweed Symp. 7: 427–435.
Jones, M. N., 1984. Nitrate reduction by shaking with cadmium: alternative to cadmium columns. Wat. Res. 18: 643–646.
Kain (Jones), J. M., 1991. Cultivation of attached seaweeds. In Guiry, M.D. & G. Blunden (eds), Seaweed Resources in Europe: Uses and Potential, John Wiley & Sons, Chichester: 309–376.
Lapointe, B. E. & C. S. Duke, 1984. Biochemical strategies for growth of *Gracilaria tikvahiae* (Rhodophyta) in relation to light intensity and nitrogen availability. J. Phycol. 20: 488–495.
Lewis, R. J. & M. D. Hanisak, 1996. Effects of phosphate and nitrate supply on productivity, agar content and physical properties of agar of *Gracilaria* strain G-16S. J. appl. Phycol. 8: 41–49.
Lignell, Å., P. Ekman & M. Pedersén, 1987. Cultivation technique for marine seaweeds allowing controlled and optimized conditions in the laboratory and on a pilot scale. Bot. mar. 30: 417–424.
Lobban, C. S. & P. J. Harrison, 1994. Seaweed Ecology and Physiology. Cambridge University Press, 366 pp.
McLachlan, J., 1973. Growth media-marine. In Stein, J.S. (ed.), Handbook of Phycological Methods, Cambridge University Press: 25–51.
Mencher, F. M., R. B. Spencer, J. W. Woessner, S. J. Katase & D. K. Barclay, 1983. Growth of nori (*Porphyra tenera*) in an experimental OTEC-aquaculture system in Hawaii. J. World Maricult. Soc. 14: 456–470.
Neish, I. C. & L. B. Knutson, 1977. The significance of density, suspension and water movement during commercial propagation of macrophyte clones. Proc. int. Seaweed Symp. 9: 451–461.
Neish, A. C., P. F. Shacklock, C. H. Fox & F. J. Simpson, 1977. The cultivation of *Chondrus crispus*. Factors affecting growth under greenhouse conditions. Can. J. Bot. 55: 2263–2271.
Neushul, M., J. Benson, B. W. W. Harger & A. C. Charters, 1992. Macroalgal farming in the sea: water motion and nitrate uptake. J. appl. Phycol. 4: 255–265.
Ryther, J. H., N. Corwin, T. A. DeBusk & L. D. Williams, 1981/82. Nitrogen uptake and storage by the red alga *Gracilaria tikvahiae*. Aquaculture 26: 107–115.
Schramm, W., 1991. Cultivation of unattached seaweeds. In Guiry, M. D. & G. Blunden (eds), Seaweed Resources in Europe: Uses and Potential, John Wiley & Sons, Chichester: 379–413.
Shivji, M. S., 1985. Interactive effects of light and nitrogen on growth and chemical composition of juvenile *Macrocystis pyrifera* (L.) C. Ag. (Phaeophyta) sporophytes. J. exp. mar. Biol. Ecol. 89: 81–96.
South, G. S. & A. Whittick, 1987. Introduction to Phycology. Blackwell Scientific Publications, Oxford, 278 pp.
Strickland, J. D. H. & T. R. Parsons, 1972. A Practical Handbook of Seawater Analysis, 2nd edn. Fish. Res. Bd. Canada, Ottawa, 310 pp.
Thomas, T. E. & P. J. Harrison, 1985. Effect of nitrogen supply on nitrogen uptake, accumulation and assimilation in *Porphyra perforata* (Rhodophyta). Mar. Biol. 85: 269–278.
Tseng, C. K., 1981a. Commercial cultivation. In Lobban, C. S. & M. J. Wynne (eds), The Biology of Seaweeds, University of California Press, Los Angeles: 680–725.

Tseng, C. K., 1981b. Marine phycoculture in China. Proc. int. Seaweed Symp. 10: 123–152.

Ulrich, A., 1952. Physiological bases for assessing the nutritional requirements of plants. Ann. Rev. Pl. Physiol. 3: 207–228.

Wheeler, P. A. & B. R. Björnsäter, 1992. Seasonal fluctuations in tissue nitrogen, phosphorus, and N:P for five macroalgal species common to the Pacific Northwest coast. J. Phycol. 28: 1–6.

Wu, C.-Y., Y.-X. Zhang, R.-L. Li, Z.-S. Penc, Y.-F. Zhang, Q.-C. Liu, J.-P. Zhang & X. Fang, 1984. Utilization of ammonium-nitrogen by *Porphyra yezoensis* and *Gracilaria verrucosa*. Hydrobiologia 116/117 (Dev. Hydrobiol. 22): 475–477.

Effects of nitrogen source, N:P ratio and N-pulse concentration and frequency on the growth of *Gracilaria cornea* (Gracilariales, Rhodophyta) in culture

Leonardo Navarro-Angulo & Daniel Robledo*
CINVESTAV-IPN, Unidad Mérida A.P. 73 Cordemex 97310 Mérida, Yucatán México
Fax: [+52] 99 81 29 17. E-mail: robledo@kin.cieamer.conacyt.mx

Key words: cultivation, *Gracilaria cornea*, nitrogen metabolism

Abstract

The effects of nitrogen source, nitrogen:phosphorus ratio, nitrogen pulse concentrations and pulse frequency on *Gracilaria cornea* growth were investigated under laboratory cultures. No significant differences in growth rate were detected between nitrogen sources, the mean growth rate decreased from ca. 14 to 11% d^{-1} over 8 weeks. Our results indicate that *G. cornea* can efficiently grow either with inorganic (NH_4-N, NO_3-N, NO_3NH_4) or organic (urea) nitrogen. The N:P ratio had a significant effect on *G. cornea* specific growth rate at 10:1 treatment (8.53% d^{-1}) when compared with ambient phosphate concentration (10:0), which produced the lowest growth rate (2.88% d^{-1}). Neither nitrogen pulse concentration nor pulse frequency showed a significant effect on the specific growth rate, however, pulse frequency significantly affected biomass increase at 50 μM nitrogen ($p < 0.05$). Nitrogen sources containing NH_4–N produced the highest phycoerythrin and protein contents being the most important N storage in *G. cornea*. The nitrogen storage capacity of *G. cornea* allows it to grow over a 7 day period with low nitrogen concentrations (< 50 μM). The understanding of nitrogen enrichment in *G. cornea* cultivation can be applied to manipulate pigment content or agar synthesis, and give the basis for its use in on-land biofiltering systems.

Introduction

Seaweeds belonging to the genus *Gracilaria* are being increasingly used in the production of food grade agar (Armisén, 1995). Its availability has greatly increased mainly through the development of cultivation techniques in several countries (Critchley, 1993). Successful large scale cultivation followed laboratory studies where the physiological characteristics of *Gracilaria* were studied (DeBoer, 1979; Lignell & Pedersén, 1987; Friedlander et al., 1990). *Gracilaria* has the capacity to take up and store nitrogen in excess of immediate requirements, and use it to sustain growth during subsequent periods of nutrient deficiency (Lapointe, 1985). Storage can be in the form of inorganic nitrogen (Chapman & Craigie, 1977) and/or metabolites such as proteins and pigments (Vergara &

*Author for correspondence.

Niell, 1993). This capacity has been utilized in cultivation of seaweeds to minimize the growth of epiphytes and provide the physiological basis for pulse feeding (Lapointe, 1985).

Nutrient supply is an important operating parameter in the management of seaweed cultivation systems (Lignell & Pedersén, 1987). The development of a separate management strategy is required for each species under cultivation since their physiological requirements differ. Nutrient limitation in temperate algae may occur at a different level than in tropical species which experience oligotrophic conditions (Hanisak, 1990).

Gracilaria cornea J. Agardh has been recently considered as a maricultural candidate in the Yucatan Peninsula due to its high agar yield and quality (Freile-Pelegrín & Robledo, 1997a,b). Since nitrogen limitation affects both phycocolloid content and growth rate, the nitrogen requirements of cultured seaweeds have to be determined, depending on the end use of the seaweed. The present study was designed to address the dynamics of nitrogen in *G. cornea* as a function of nitrogen source, nitrogen:phosphorus ratio and nitrogen concentration and frequency during pulse feeding.

Material and methods

Plant material

Unialgal cultures of *G. cornea* were obtained from carpospores under laboratory conditions. Tetrasporophytic thalli were grown in 1 l flasks at a density 1 g l^{-1} with filtered sterilized seawater (35 ‰) under controlled laboratory conditions of temperature (27 ± 1.3 °C), photon irradiance (100 μmol photons m^{-2} s^{-1}), and photoperiod (12:12, L:D). These conditions were previously found to optimize growth of *G. cornea* tetrasporophytes (Orduña-Rojas, 1996). Algae from the culture were preincubated 7 days before each experimental treatment under the same conditions described above. Continuous aeration was applied to the culture medium throughout the study. The fresh weight of algae was determined weekly for each treatment, adjusted to initial density and culture medium renewed. Relative daily growth rate ($R = \%$ d^{-1}) was calculated as

$$R = \frac{\ln W_t - \ln W_0}{t} \cdot 100,$$

where W_0 is the initial biomass and W_t the biomass at day t (Evans, 1972). Three replicates were used for each tested condition.

Nitrogen source

To test the effect of nitrogen enrichment on *G. cornea* growth rate, plant material was cultured with the following sources NO$_3$–N (added as NaNO$_3$), NH$_4$–N (added as NH$_4$Cl), a combination of both NO$_3$ and NH$_4$ (in the form of NH$_4$NO$_3$) and organic nitrogen source (added as urea - NH$_2$CONH$_2$). Nitrogen was added to the other Provasoli Enriched Seawater media (PES) ingredients to give a final concentration of 824 μM (Starr & Zeikus, 1993). Each treatment was carried out under conditions described above for eight weeks. Medium was changed every 7 days.

N:P ratio

The effect of N:P ratio on growth was studied in laboratory cultures under defined conditions. Enriched seawater (PES) stock solution without phosphate was used. Phosphate was added separately as Na$_2$HPO$_4$ to give four different concentrations 824, 412, 82.4 and < 0.01 μM (ambient phosphorus concentration), corresponding to the following N:P ratios 10:10, 10:5, 10:1 and 10:0, respectively. Growth rate was recorded each week during a three week period.

Nitrogen pulse concentration and frequency

This experiment determined the growth rate of *G. cornea* at four nitrogen fluxes using a factorial design with two levels of nitrogen concentrations (50 and 150 μM NH$_4$–N) and two pulse frequencies, one every two weeks (each 14 day = 1) and one every week (each 7 day = 2). This combination gave a total amount of N available to the plants equal to 3.57, 7.14, 10.71 and 21.42 μM NH$_4$–N g fresh wt d^{-1}. Each experimental treatment was maintained under unenriched seawater after a 24 h pulse. The experimental period was extended under conditions described above for ten weeks.

Analytical methods

Chemical and biochemical analyses on triplicate samples were performed as follows: chlorophyll *a* (Jeffrey & Humphrey, 1975); phycobiliproteins (Beer & Eshel, 1985) were determined in all experimental procedures. Protein content (Lowry et al., 1951) and carbohydrate content (Dubois et al., 1956) were determined for the nitrogen source trial.

Statistical analysis

Significant differences for R between treatments were determined with an ANOVA. Homogeneity of variances (Bartlett's test) was tested and transformation applied when necessary (log x). A Tukey HSD test at 95% of significance was used to assess the effect of nitrogen source on growth rate. Significant differences between each combination of nitrogen pulse concentration and frequency were determined with a Duncan test (Zar, 1984).

Results

The effect of nitrogen source on growth is shown in Figure 1. No significant difference was found for relative growth rate and biomass increase; the growth rates ranged from 11.13 to 12.77% d^{-1}. In general, growth

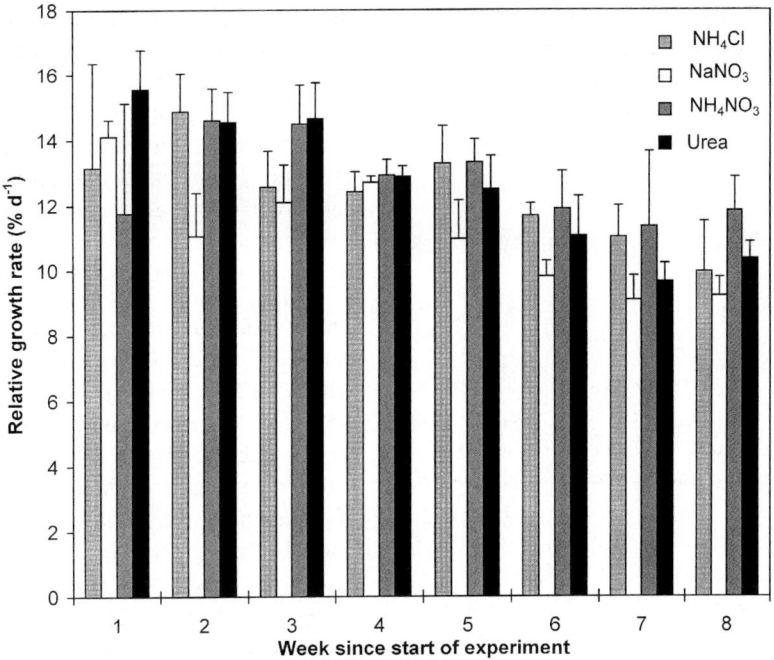

Figure 1. Relative growth rate (% d^{-1}) of *Gracilaria cornea* as a function of nitrogen source during eight weeks of cultivation. Bars show mean values ± SD.

rates decreased from 13.64 ± 1.59 to 10.34 ± 1.11% d^{-1} between treatments at the end of the experiment.

Phycobiliproteins and chlorophyll *a* contents were not affected ($p > 0.05$); both maximum phycoerythrin and phycocyanin contents were detected for NH$_4$Cl and NH$_4$NO$_3$, respectively (Table 1). In general, cultures supplemented with urea had lower pigment content. The concentration of pigments did not seem to be influenced by nitrogen source since no significant difference was found among treatments.

The only significant effect of nitrogen source on chemical composition was that protein content was highest in the presence of the NH$_4$ ion (Table 1). Maximum protein was found in NH$_4$Cl and NH$_4$NO$_3$, while it was significantly lower for NaNO$_3$ and urea. Carbohydrate did not differ significantly between any of the nitrogen source treatments ($p > 0.05$).

The nitrogen to phosphorus ratio (N:P) had a significant effect on the relative growth rate (Figure 2) which ranged from 2.88 ± 0.70 to 8.53 ± 2.95% d^{-1}. The best growth rates were obtained for N:P ratios of 10:1 and 10:5 ($p < 0.05$). The treatment with the N:P ratio of 10:0 had the lowest growth, and did not differ significantly from the 10:10 ratio. Two weeks after starting the experiment plant coloration changed for each treatment. For the 10:1 ratio, deep

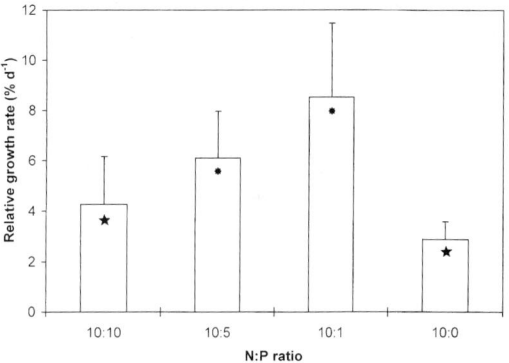

Figure 2. Relative growth rate (% d^{-1}) of *Gracilaria cornea* as a function of PO$_4^{-3}$ level in N:P experiment. Each symbol corresponds to a significant different mean in accordance with Tukey HSD test. Bars show mean values ± SD.

red fronds were observed, with 10:5 and 10:10 they showed a brownish colour, and a yellowish coloration was evident in treatment 10:0.

Neither nitrogen pulse concentration nor pulse frequency had a significant effect on the relative growth rate of *Gracilaria cornea* (Table 2); however pulse frequency significantly affected ($p < 0.05$) weight increase at 50 μM nitrogen concentration (Figure 3) with a mean value of 3.52 ± 1.01 g at 2 pulses per

Table 1. Chemical composition of G. cornea cultured with different nitrogen sources. Means with different superscript are significantly different at $p < 0.05$ (ANOVA, Tukey HSD). Standard deviation indicated, $n = 8$.

Variables	Treatments			
	NH$_4$Cl	NaNO$_3$	NH$_4$NO$_3$	Urea
Phycoerythrin (mg g dry wt^{-1})	9.70 ± 3.95^a	7.97 ± 4.29^a	8.81 ± 3.25^a	6.52 ± 2.00^a
Phycocyanin (mg g dry wt^{-1})	1.44 ± 0.71^a	1.28 ± 0.90^a	1.45 ± 0.75^a	1.12 ± 0.45^a
Chlorophyll a (μg g dry wt^{-1})	300 ± 58.4^a	304 ± 56.1^a	291 ± 61.1^a	267 ± 22.1^a
Protein (% dry wt^{-1})	16.01 ± 1.86^a	13.47 ± 1.50^b	15.58 ± 2.23^a	13.7 ± 1.51^b
Carbohydrate (% dry wt^{-1})	32.77 ± 4.84^a	35.57 ± 4.02^a	33.39 ± 4.56^a	37.02 ± 3.44^a

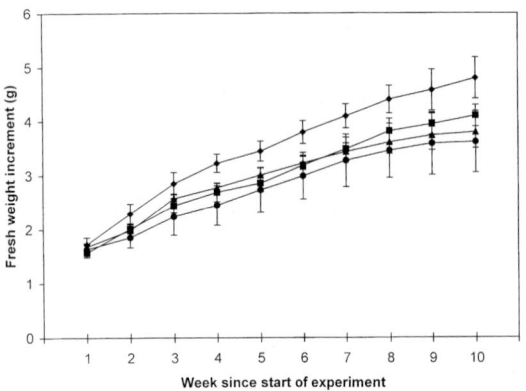

Figure 3. Fresh weight increment (g) of Gracilaria cornea as a function of nitrogen pulse concentration (NH$_4$-N) and pulse frequency. One pulse per 2 weeks at 50 μM (▲) and 150 μM (■); and 2 pulses per 2 weeks pulse at 50 μM (♦) and 150 μM (●). Bars show mean values ± SD.

2 weeks. Phycobiliprotein content was not affected, whereas the chlorophyll a content was only affected by pulse frequency (Table 2). A highly significant correlation was found between the nitrogen flux and phycoerythrin content ($r = 0.995, p < 0.01$).

Discussion

Gracilaria cornea can grow with inorganic (NH$_4$–N, NO$_3$–N) or organic (urea) nitrogen sources. The preference for NH$_4$-N has been described for other Gracilaria species and red algae (DeBoer et al., 1978).

Efficient use of different nitrogen sources at different absorption rates with preference for NH$_4$-N rather than NO$_3$-N have been shown for Gracilaria foliifera (Førsskäl) Børgesen (D'Elia & DeBoer, 1978), Gracilaria tikvahiae McLachlan (Hanisak, 1990) and Gracilaria tenuistipitata Chang et Xia (Haglund & Pedersén, 1993); however relative growth rates in G. cornea did not change significantly between different nitrogen sources in the present study. Similarly, G. tikvahiae showed similar growth rates when cultured in NH$_4$–N or NO$_3$–N enriched seawater (Hanisak, 1990). Haglund & Pedersén (1993) did not find any significant difference when using KNO$_3$ or NH$_4$NO$_3$ in the culture media, with growth rates between 4 and 9% d^{-1}. DeBoer et al. (1978) reported similar growth rates for Chondrus crispus Stackh. and G. tikvahiae, with either NH$_4$–N or NO$_3$–N, being higher for the former.

The preference for nitrogen source depends on absorption rate and is influenced by the nitrogen status of the algae (Hanisak, 1990). In Gracilaria sp. absorption rate for NH$_4$ is higher than that for NO$_3$ since NH$_4$ is directly incorporated into the amino acid pool (Haglund & Pedersén, 1993; Lobban & Harrison, 1994). Inhibition of NO$_3$ or NO$_2$ absorption have been observed, depending on ammonium concentration (Hanisak, 1990).

Phycobiliproteins, together with free amino acids, are important as nitrogen storage in red algae (Lapointe, 1981; Ryther et al., 1981), phycoerythrin being the pigment which responds faster to nitrogen availability in the medium (Vergara & Niell, 1993). In

Table 2. Growth and chemical composition of *G. cornea* cultured under different treatments of nitrogen pulse concentration (50, 150 μM) and pulse frequency (1, 2). Each combination gave the following NH$_4$-N flux (μM N g fresh wt d^{-1}): 1 × 50 = 3.57; 2 × 50 = 7.14; 1 × 150 = 10.71 and 2 × 150 = 21.34. Means with different superscript are significantly different at $p < 0.05$ (ANOVA, Tukey HSD). Standard deviation indicated, $n = 10$.

Variables	Treatments			
	1 × 50	1 × 150	2 × 50	2 × 150
R (% d^{-1})	3.33 ± 2.07^a	3.17 ± 1.79^a	4.34 ± 1.69^a	3.66 ± 1.43^a
W increase (g)	2.99 ± 0.73^a	2.79 ± 0.72^a	3.52 ± 1.0^b	3.01 ± 0.85^a
Phycoerythrin (mg g dry wt^{-1})	0.90 ± 0.29^a	0.97 ± 0.16^a	1.10 ± 0.26^a	1.52 ± 0.67^a
Phycocyanin (μg g dry wt^{-1})	38.7 ± 11.3^a	64.8 ± 51^a	41.9 ± 31^a	73.5 ± 39^a
Chlorophyll a (μg g dry wt^{-1})	78.2 ± 4.2^a	77.1 ± 9.6^a	122.4 ± 21.7^b	125.4 ± 53.2^{ab}

Gracilaria cornea phycoerythrin constituted approximately 85–87% of total phycobiliproteins. On the other hand, chlorophyll a is not considered a N pool, representing approximately 2% of total nitrogen in tissue (McGlathery et al., 1996). Increases in chlorophyll a content, with increases in cellular nitrogen, are also well known (Bird et al., 1982).

The best growing conditions for *G. cornea* were found for the N:P ratio of 10:1. Lower inorganic nutrient concentration including phosphorus reduces the algal photosynthetic capacity in tropical waters hence limiting their growth and productivity as has been shown for *G. tikvahiae* (Lapointe, 1987). In this regard, ambient phosphate concentrations supplemented with nitrogen (10:0) limited *G. cornea* growth rate. The reduction of growth observed in *G. cornea* with 10:10 ratio could be related to a detrimental effect of high phosphate concentrations. Phosphate levels of about 1 mM were found to inhibit growth of *Gracilaria conferta* (Schousboe ex Montagne) Feldmann et Feldmann (Friedlander & Ben Amotz, 1991) while an increase of N:P ratio (2.5–20) in *G. conferta* caused a significant growth rate enhancement (Friedlander & Levy, 1995).

The nitrogen storage capacity of *G. cornea* allows it to grow over 7 day period with low nitrogen concentration. Similar to our results, the internal nitrogen pool is used for growth in *G. chilensis* Bird, McLachlan et Oliveira (Pickering et al., 1993) and *G. gracilis* (Stackhouse) Steentoft, Irvine et Farnham (Smit et al., 1997). Slow growth rate in *G. cornea* at one N pulse may suggests that nitrogen reserves were used for metabolic maintenance with lower chlorophyll a content, while with two pulse treatments a higher chlorophyll a content was found suggesting that the nitrogen pool was maintained with respect to the same growth level. Smit et al. (1997) showed similar results for *G. gracilis* using two pulses per week. Phycoerythrin constituted the major nitrogen reserve in *G. cornea* representing up to 60% of total soluble protein. In *G. gracilis* nitrogen reserves in order of importance are protein, phycoerythrin, chlorophyll a and carotenoids (Smit et al., 1997). Increasing the nitrogen flux has been shown to increase protein content and the inverse relationship has been described for carbohydrates and agar content in *G. tikvahiae* (Bird et al., 1982).

Though the present results were obtained under laboratory conditions the knowledge on the regulation of the N concentration and the utilization of indicators to monitor the nutrient status of *G. cornea* should increase the efficiency and success of its mariculture and the capability of increasing phycocolloid production.

Acknowledgements

This work has been supported by CONACYT (N° 2198P-B). L. Navarro-Angulo acknowledges CONACYT (Scholarship N° 90037) and a fellowship from Fondo Yucatan. Both authors are grateful to Ma. Zaldivar Romero and Eloy Gil Trava for their assistance in laboratory analysis.

References

Armisén, R., 1995. World-wide use and importance of *Gracilaria*. J. appl. Phycol. 7: 231–243

Beer, S. & A. Eshel, 1985. Determining phycoerythrin and phycocyanin concentrations in aqueous crude extracts of red algae. Aust. J. mar. Freshwat. Res. 36: 785–792.

Bird, K. T., C. Habig & T. DeBusk, 1982. Nitrogen allocation and storage patterns in *Gracilaria tikvahiae* (Rhodophyta). J. Phycol. 18: 344–348.

Chapman, A. R. O. & J. S. Craigie, 1977. Seasonal growth in *Laminaria longicruris*: relations with dissolved inorganic nutrients and internal reserves of nitrogen. Mar. Biol. 40: 197–205.

Critchley, A. T., 1993. *Gracilaria* (Rhodophyta, Gracilariales): an economically important agarophyte. In Ohno, M. & A. T. Critchley (eds), Seaweed Cultivation and Marine Ranching. JICA. Kanagawa International Fisheries Training Center. Japan: 89–111.

D'Elia, C. F. & J. A DeBoer, 1978. Nutritional studies of two red algae. II. Kinetics of ammonium and nitrate uptake. J. Phycol. 14: 266–272.

DeBoer, J. A., 1979. Effects of nitrogen enrichment on growth rate and phycocolloid content in *Gracilaria foliifera* and *Neoagrdhiella baileyi* (Florideophyceae). Proc. int. Seaweed Symp. 9: 263–271.

DeBoer, J. A., H. J. Guigli, T. L. Israel & C. F. D'Elia, 1978. Nutritional studies of two red algae. I. Growth rate as a function of nitrogen source and concentration. J. Phycol. 14: 261–266.

Dubois, M., F. A. Gilles, J. K. Hamilton, D. A. Rebers & F. Smith, 1956. Colorimetric method for the determination of sugar and related substances. Anal. Chem. 28: 350–356.

Evans, G. C., 1972. The quantitative analysis of plant growth. Blackwell, Oxford, 734 pp.

Freile-Pelegrín, Y. & D. Robledo, 1997a. Effects of season on the agar content and chemical characteristics of *Gracilaria cornea* from Yucatán, México. Bot. mar. 40: 285–290.

Freile-Pelegrín, Y. & D. Robledo, 1997b. Influence of alkali treatment on agar from *Gracilaria cornea* from Yucatán, México. J. appl. Phycol. 9: 533–539.

Friedlander, M. & A. Ben-Amotz, 1991. The effects of outdoor culture on growth and epiphytes of *Gracilaria conferta*. Aquat. Bot. 39: 315–333.

Friedlander, M. & I. Levy, 1995. Cultivation of *Gracilaria* in outdoor tanks and ponds. J. appl. Phycol. 7: 315–324.

Friedlander, M., N. Galai & H. Farbstein, 1990. A model of seaweed growth in an outdoor culture in Israel. Hydrobiologia 204/205: 367–373.

Haglund, K. & M. Pedersén, 1993. Outdoor pond cultivation of the subtropical marine red alga *Gracilaria tenuistipitata* in brackish water in Sweden. Growth, nutrient uptake, co-cultivation with rainbow trout and epiphyte control. J. appl. Phycol. 5: 271–284.

Hanisak, M. D., 1990. The use of *Gracilaria tikvahiae* (Gracilariales, Rhodophyta) as a model system to understand the nitrogen nutrition of culture seaweeds. Hydrobiologia 204/205: 79–87.

Jeffrey, S. W. & G. F. Humphrey, 1975. New spectrophotometric equations for determining chlorophylls a, b, c_1 and c_2 in higher plants, algae and natural phytoplankton. Biochem. Physiol. Pflanzen 167: 191–194.

Lapointe, B. E., 1981. The effects of light and nitrogen on growth, pigment content, and biochemical composition of *Gracilaria foliifera* var. *angustissima* (Gigartinales, Rhodophyta). J. Phycol. 17: 90–95.

Lapointe, B. E., 1985. Strategies for pulse nutrient supply to *Gracilaria* cultures in the Florida Keys: interactions between concentration and frequency of nutrient pulses. J. exp. mar. Biol. Ecol. 93: 211–222.

Lapointe, B. E., 1987. Phosphorus- and nitrogen-limited photosynthesis and growth of *Gracilaria tikvahiae* (Rhodophyceae) in the Florida Keys: an experimental field study. Mar. Biol. 93: 561–568.

Lignell, A. & M. Pedersén, 1987. Nitrogen metabolism in *Gracilaria secundata* Harv. Hydrobiologia 151/152: 431–441.

Lobban, C. S. & P. J. Harrison, 1994. Seaweed ecology and physiology. Cambridge University Press, 366 pp.

Lowry, O. H., N. J. Rosebrough, A. L. Farr & R. J. Randall, 1951. Protein measurement with the Folin phenol reagent. J. biol. Chem. 193: 265–275.

McGlathery, K. J., M. F. Pedersen & J. Borum, 1996. Changes in intracellular nitrogen pools and feedback controls on nitrogen uptake in *Chaetomorpha linum* (Chlorophyta). J. Phycol. 32: 393–401.

Orduña-Rojas, J., 1996. Efecto de la radiación y la temperatura en la liberación y desarrollo de las carposporas del alga roja *Gracilaria cornea* J. Agardh (Gracilariales, Rodofita). MSc.Thesis CINVESTAV-Unidad Mérida, Yucatán, México, 43 pp.

Pickering, T. D., M. E. Gordon & L. J. Tong, 1993. Effect of nutrient concentration and frequency on growth of *Gracilaria chilensis* plants and level of epiphytic algae. J. appl. Phycol. 5: 525–533.

Ryther, J. H., T. Corwin, A. DeBusk & L. D. Williams, 1981. Nitrogen uptake and storage by the red alga *Gracilaria tikvahiae* (McLachlan, 1979). Aquaculture 26: 107–115.

Smit, A. J., B. L. Robertson & D. R. du Preez, 1997. Influence of ammonium-N pulse concentrations and frequency, tank condition and nitrogen starvation on growth rate and biochemical composition of *Gracilaria gracilis*. J. appl. Phycol. 8: 473–481.

Starr, R.C. & J.A. Zeikus, 1993. UTEX. The culture collection of algae at the University of Texas at Austin. J. Phycol. Suppl. 29: 1–106.

Vergara, J. J. & F. X. Niell, 1993. Effects of nitrate availability and irradiance on internal nitrogen constituents in *Corallina elongata* (Rhodophyta). J. Phycol. 29: 285–293.

Zar, H., 1984. Biostatistical Analysis. Prentice-Hall, Inc., Englewood Cliffs, N.Y., 675 pp.

Photosynthetic and respiratory responses of the agarophyte *Gelidiella acerosa* collected from tidepool, intertidal and subtidal habitats

Edna T. Ganzon-Fortes
Marine Science Institute, C.S., University of the Philippines, Diliman, Quezon City 1101, Philippines
E-mail: ednaf@msi01.cs.upd.edu.ph

Key words: photosynthesis, respiration, *Gelidiella acerosa*, culture, tidal habitat, salinity, temperature

Abstract

Several samples of the red seaweed, *Gelidiella acerosa* (Forssk.) Feldmann & Hamel occurring in tidepools, high intertidal rocks, and shallow subtidal areas on a reef flat in Ilocos Norte, northern Philippines were studied in terms of their photosynthetic and respiratory responses (μl O_2 gDW^{-1} h^{-1}) to four salinity (22, 28, 34, 40‰) and three temperature (22, 28, 34 °C) combinations. The upper intertidal plants tolerated low salinities (22–28‰) better than high salinities (34–40‰), while tidepool and subtidal plants were not affected. Temperatures of 22 through 34 °C resulted in a one-fold increase in their photosynthetic rates and insignificant differences in their respiratory rates while tidepool and subtidal plants almost doubled their photosynthetic rates and their respiration rates increased by about 5–50 times. There were no interaction effects. Therefore, intertidal plants appeared to be more tolerant to wide temperature fluctuations and low salinity levels; while tidepool and subtidal plants were least affected by salinity variations but were quite sensitive to temperature fluctuations. Vegetative and tetrasporic plants had similar photosynthetic and respiratory responses to salinity and temperature variations, although vegetative plants had significantly higher net photosynthesis under the minimum and maximum temperatures tested (22 and 34 °C). Reproductive *G. acerosa* showed greater tolerance to temperature fluctuations. These responses indicated that physiological changes may have occurred when the species became reproductive.

Tolerance of *G. acerosa* to low salinities suggests that lowering the salinities in culture tanks could be used to eradicate contaminants, i.e., dinoflagellates and filamentous green algae. Temperature of 28 °C appeared to be optimum for all plant types as reflected by their high photosynthetic and low respiratory rates.

Introduction

Gelidiella acerosa (Gelidiales, Rhodophyta) is a common component of seaweed communities in Philippine coastal waters. The species produces a high quality agar (Laserna et al., 1978; Santos & Doty, 1983; Roleda et al., 1997) that is used in the local (Trono & Ganzon-Fortes, 1988) and world (Armisen & Galatas, 1987) agar industry along with *Gracilaria*, *Gelidium*, and *Pterocladia*. However, natural populations are in danger of depletion due to commercial harvesting in many parts of the country. Because of the value of *G. acerosa* agar, cultivation technology is being developed and this study aimed to provide baseline information on aspects of its ecophysiology.

Of the primary physical factors that influence seaweed growth, water temperature is the single most important factor in the geographic distribution of tropical and subtropical seaweeds (Gessner, 1970; Lüning, 1990). In contrast, salinity is one of the most critical chemical factors affecting their local distribution (Dawes, 1981, 1987; Druehl, 1981), and both factors influence their photosynthesis and respiration (Macler, 1988; Lüning, 1990; Lobban & Harrison, 1994). Monitoring of these responses is useful in determining optimum conditions for large scale seaweed cultivation.

In Ilocos Norte, northern Philippines, *Gelidiella acerosa* occurs from the upper intertidal to the shallow subtidal areas, as well as in tidepools near the seaward margin of the fringing reef. Thalli from the different habitats vary in morphology (length and density of branches, size of cortical cells, and colour), pigment content and in their responses to increasing photon irradiance levels (P-I curve) especially during summer months (Ganzon-Fortes, 1997a). If cultivation of the species is done using stocks from different habitats, optimum cultivation conditions should be determined because environmental factors, such as salinity and temperature, vary in the different habitats and may have differential effects on the different populations.

The present study compared the photosynthetic and respiratory responses of *Gelidiella acerosa* representing the tidepool, upper intertidal and shallow subtidal habitats to 12 combinations of salinity and temperature. Except for the highest and lowest salinity values (22 and 40‰, respectively), the range of temperature and salinity values used in the experiments were based on actual field measurements monitored for two years in the study area. The goal was to determine the optimum temperature and salinity requirements of each plant type for tank and field culture, as well as to explain their tolerance to their respective habitats. In addition, vegetative and tetrasporic thalli from the same tidal zone belt were compared in order to assess their physiological fitness to the surrounding environment. Previous studies have compared growth and physiological responses of gametophytes and tetrasporophytes of macroalgae (Mathieson & Norall, 1975; Zupan & West, 1990; Britting & Chapman, 1993; Sosa et al., 1993), but only one has made physiological comparisons (P-I curves) of the vegetative and tetrasporic stages (Ganzon-Fortes, 1997a). While Hansen (1980) compared photosynthetic rates of vegetative and reproductive stages of *Gelidium coulteri* Harvey, *Rhodoglossum affine* (Harvey) Kylin and *Iridaea cordata* (Turner) Bory, she did not specify if the reproductive stage was gametophytic or tetrasporic. Her results showed that photosynthetic rates of vegetative *I. cordata* were greater than reproductive ones but this rate differential was not apparent in the two other species studied.

Materials and methods

Description of study site

The coast at Pangil, Currimao, Ilocos Norte (18° 02′ 42″ N, 120° 58′ 54″ E), in northern Philippines has a wide intertidal reef flat extending to about 150 m off-shore. The substratum is solid bedrock beset with tidepools of various shapes and sizes, and cracks that develop into deep canals near the seaward margin. Water depth ranges from 30 to 110 cm during high tide. The upper intertidal area is usually exposed for 5–8 hours during spring low water. Water is crystal clear and mixed by tidal changes and strong wave action at the seaward margin. Water temperature averages 30–33 °C between April and September (warm months) and 26–29 °C between October and March (cold months). A variation of 0.5–9.0 °C occurs across the reef, having a steeper gradient especially in summer. Salinity is stable between 32 to 35‰ throughout the year, and is uniform or may slightly vary during summer across the reef.

Collection and preparation of algal samples

Vegetative algal samples were collected by hand from tidepools, upper intertidal and shallow subtidal habitats. Fertile (tetrasporic) thalli occurred in abundance at the subtidal area and were easily recognized by numerous stichidia (swollen structures containing tetraspores) in their ultimate branchlets. Collections were placed in labelled plastic bags with ambient seawater and transported in a styrofoam box containing a minimal amount of ambient seawater. Upon arrival in the laboratory, the plants were cleaned of debris and epiphytes, washed several times with unfiltered and filtered (0.20 μm pore size) and then autoclaved seawater. Thalli without bleached parts and having dense ultimate branchlets were selected. These were cut into 20 mm segments of which only the apical and intercalary parts were chosen for experiment (representing the youngest portion of the thallus) to ensure uniformity in thallus age. These parts also showed no significant differences in their growth rates (de Venecia, 1992). The cuttings were held in salinities of 22, 28, 34 and 40‰ in deep culture dishes for 3–5 days to allow acclimation (Dawes, 1981; Kirst, 1988). Water was replaced daily. Low salinity seawater was prepared by dilution with distilled water, while high salinity seawater was produced by slow evaporation in a water bath. The plants were maintained under a 12:12 h photoperiod, irradiance of 60–80 μmol photons m^{-2} s^{-1}, and temperature of 28–32 °C.

Determination of photosynthetic and respiratory rates by manometry

A multifactorial experiment ($n = 3$) was carried out using four salinities (22, 28, 34, 40‰) and three temperatures (22, 28, 34 °C). A Gilson Differential Respirometer model TGRP-14 was used to measure the amount of oxygen evolved or consumed after a

Table 1. Statistical results: effect of salinity on the photosynthesis and respiration of Gelidiella acerosa collected from tidepool, intertidal and subtidal habitats. (Duncan grouping – means with the same letter are not significantly different; values in $\mu l\ O_2\ gDW^{-1}h^{-1}$; (n = 9–12).

Plant type	Environmental factor	Photosynthesis	Duncan grouping		p	Stat. result summary
Tidepool veg	Salinity				0.2036	Not significant
Intertidal veg	Salinity				0.0176	Significant
	22‰	4614±690	A			22 not = 34
	28‰	4303±523	A	B		22 not = 40
	34‰	3925±790	C	B		28 not = 40
	40‰	3593±673	C			22 = 28
						28 = 34
						34 = 40
Subtidal veg	Salinity				0.6961	Not significant
Subtidal tet	Salinity				0.3749	Not significant
		Respiration				
Tidepool veg	Salinity				0.0612	Not Significant
Intertidal veg	Salinity				0.0480	Significant
	40‰	−202±129	A			34 not = 40
	22	−334±163	A	B		34 = 28 = 22
	28	−373±233	A	B		40 = 28 = 22
	34	−462±214		B		
Subtidal veg	Salinity				0.4489	Not significant
Subtidal tet	Salinity				0.5935	Not significant

one-hour incubation period, as described by Ganzon-Fortes (1993) and Dawes (1981, 1985). Two 20 mm long segments were placed in each reaction flask, based on a biomass linearity experiment (Dawes, 1981). Two reaction flasks were used as 'blanks' (Dawes, 1981). Each reaction flask was provided with a reference flask in order to balance the effect of water vapor pressure and to eliminate barometric and temperature corrections. The use of blank and reference flasks controlled the problem of gas diffusion in the system brought about by the plastic and Tygon tubing materials being used in the Gilson respirometer. During photosynthetic runs, irradiance of about 235 μmol photons m^{-2}s^{-1} (which is photosynthetically saturating to Gelidiella acerosa, (Ganzon-Fortes, 1997b)) was provided by halogen lamps beneath the floor of the water bath. Irradiance was measured using an SA underwater quantum sensor (cosine collector) connected to a LI-COR 1000 data logger. At the end of the experimental run, the dry weights of the samples were determined either by actual measurements or by extrapolation based on wet weight data and previously computed ww:dw ratio. The oxygen consumed (respiration) or produced (photosynthesis) per hour was expressed in $\mu l\ O_2\ gDW^{-1}\ h^{-1}$.

Data analysis

Multivariate ANOVA (MANOVA) was used to test for significance of the differences in: (1) photosynthetic and respiratory rates of each tidal plant type per level of salinity and temperature (as single factor); (2) photosynthetic and respiratory rates of each tidal plant type per salinity-temperature combination; (3) photosynthesis and respiration of the three plant types to variations in salinity and temperature; and (4) photosynthesis and respiration of the vegetative and tetrasporic plants to variations in salinity and temperature. Initial MANOVA tests performed (i.e., Wilk's Lambda, Pillai's Trace, Hotelling-Lawley Trace, Roy's Greatest Root) showed an overall significant salinity and temperature effects if each was tested as a single factor. However, interaction of these two factors were not significant for all plant types. Therefore, no further tests were carried out to compare the responses of the different plant types to interaction of salinity and temperature. Normality was determined

Table 2. Statistical results: effect of temperature on the photosynthesis and respiration of *Gelidiella acerosa* collected from tidepool, intertidal and subtidal habitats. (Duncan grouping – means with the same letter are not significantly different; values in $\mu l\ O_2\ gDW^{-1}\ h^{-1}$, $n = 9$–12)

Plant type	Environmental factor	Photosynthesis	Duncan grouping	p	Stat. result summary
Tidepool veg	Temperature			0.0001	Significant
	34 °C	5690±876	A		22 not = 28
	28	3955±793	B		not = 34
	22	2926±596	C		
Intertidal veg	Temperature			0.0293	Significant
	34 °C	4478±631	A		22 not = 34
	28	4165±787	A B		22 = 28
	22	3684±668	B		28 = 34
Subtidal veg	Temperature			0.0001	Significant
	34 °C	5558±568	A		22 not = 28
	28	4344±616	B		not = 34
	22	3382±365	C		
Subtidal tet	Temperature			0.0001	Significant
	34 °C	4820±642	A		22 not = 28
	28	4019±634	B		not = 34
	22	2944±471	C		
		Respiration			
Tidepool veg	Temperature			0.0001	Significant
	28 °C	−32±99	A		28 not = 34
	22	−44±82	A		22 not = 34
	34	−291±212	B		28 = 22
Intertidal veg	Temperature			0.2577	Not significant
Subtidal veg	Temperature			0.0005	Significant
	28 °C	−3±304	A		28 not = 34
	22	−25±237	A		22 not = 34
	34	−402±198	B		28 = 22
Subtidal tet	Temperature			0.0000	Significant
	28 °C	−38±143	A		28 not = 34
	22	−3±6	A		22 not = 34
	34	−323±120	B		28 = 22

before MANOVA application and respiration data was transformed. Duncan's Multiple Range Test was used to determine which pair of treatment means was significant. All computations were done using Statistical Analysis System (SAS) software.

Results

Responses to salinity and temperature

Except for the intertidal plants, the four salinities did not significantly affect the photosynthetic and respiratory rates of the tidepool and subtidal plant types (Table 1). Net photosynthesis of the intertidal plants generally decreased with increasing salinity with significant differences between 22 and 40, 22 and 34, and 28 and 40‰. No pattern was observed for respiration.

Photosynthetic rates of the tidepool and subtidal plants significantly increased at each 6-degree temperature rise (Table 2), while significant responses of intertidal plants occurred only between 34 °C and 22 °C. On the other hand, respiratory rates of the tidepool and subtidal plants were similar at 22 and 28 °C and were 7–100 times lower than the rates at

Table 3. Statistical results: comparison of photosynthesis and respiration (in μl O_2 gDW^{-1} h^{-1}, $n = 9$–12) of the 3 plant types: (1) tidepool, (2) intertidal, and (3) subtidal, per level of salinity. (Duncan grouping – means with the same letter are not significantly different)

Environmental factor	Population	Photosynthesis	Duncan grouping	p	Stat. result summary
Salinity ‰					
22	1, 2, 3			0.9246	Not significant
28	1, 2, 3			0.7560	Not significant
34	1, 2, 3			0.8588	Not significant
40	1, 2, 3			0.1729	Not significant
		Respiration			
22	1, 2, 3			0.5066	Not significant
28	1, 2, 3			0.0891	Not significant
34	1, 2, 3			0.0121	Significant
	3	−109±350	A		1 not = 2
	1	−141±160	A		3 not = 2
	2	−462±214	B		1 = 3
40	1, 2, 3			0.0227	Significant
	1	−37±104	A		1 not = 2
	3	−65±142	A		3 not = 2
	2	−202±129	B		1 = 3

34 °C. Again, the intertidal plants differed with no significant differences in respiratory responses to all temperatures.

None of the salinity and temperature combinations significantly affected the photosynthesis and respiration of the different *Gelidiella acerosa* plant types.

Comparison of plants from the three habitats

Photosynthetic and respiratory responses of each plant type under varying salinities did not differ significantly except at 34 and 40‰ where respiratory rates of the tidepool and subtidal plant types were significantly lower (Table 3).

Net photosynthesis of the three plant types were similar at 28 °C (Table 4), while at 22 °C, tidepool plants were significantly lower than intertidal ones. At 34 °C, photosynthesis of intertidal plants was significantly lower and respiration of all three plant types were similar. However, at 22 and 28 °C the intertidal plants respired at higher rates than the tidepool and intertidal ones (Table 4).

Vegetative and tetrasporic plants

No significant difference was recorded in the photosynthesis and respiration of the subtidal vegetative and tetrasporic plants under the four salinity variations (Table 5). At 22 and 34 °C, vegetative plants photosynthesized at higher rates, while both life stages showed similar respiratory rates under all temperature variations.

Discussion

Responses of three plant types

The diagnostic feature of intertidal habitats is the regular alternation of emersion and submersion that accompanies the ebb and flood of the tides (Dring, 1982). This regularly exposes the intertidal zone to atmospheric conditions so that the upper intertidal plants are exposed to desiccation, while tidepool and subtidal plants are not (Lobban & Harrison, 1994). Exposure to atmospheric conditions could be adverse especially in times when solar radiation is intense, air temperature is high and daylength is longer. Even in cool temperate climates, exposed rock surfaces quickly reach temperatures of 25–30 °C, even if air temperature may be only around 20 °C and water temperature is 15 °C or less (Dring, 1982). Temperatures in shaded positions under rocks or seaweeds also climb steadily during

Table 4. Statistical results: comparison of photosynthesis and respiration (in $\mu l\, O_2\, gDW^{-1}h^{-1}$, $n = 9$–12) of the 3 plant types: (1) tidepool, (2) intertidal and (3) subtidal, per level of temperature. (Duncan grouping – means with the same letter are not significantly different)

Environmental factor	Population	Photosynthesis	Duncan grouping	p	Stat. result summary
Temperature °C					
22	1, 2, 3			0.0080	Significant
	2	3684±668	A		2 not = 1
	3	3382±365	A B		2 = 3
	1	2926±596	B		3 = 1
28	1, 2, 3			0.4415	Not significant
34	1, 2, 3			0.0003	Significant
	1	5690±876	A		1 not = 2
	3	5558±568	A		3 not = 2
	2	4478±631	B		1 = 3
		Respiration			
22	1, 2, 3			0.0015	Significant
	3	−25±237	A		1 not = 2
	1	−44±82	A		3 not = 2
	2	−263±126	B		1 = 3
28	1, 2, 3			0.0007	Significant
	3	−3±304	A		1 not = 2
	1	−32±99	A		3 not = 2
	2	−387±278	B		1 = 3
34	1, 2, 3			0.3474	Not significant

the period of emersion, although lower than in exposed positions (Dring, 1982). Salinity may increase substantially upon exposure when tide goes out, or may decrease when exposed to rain. These changes in environmental factors (e.g. temperature and salinity) associated with emersion and submersion have significant effects on the physiological adaptation or acclimation capacities of intertidal plants.

Temperature has fundamental effects on chemical-reaction rates (Lobban & Harrison, 1994); usually the rates of photosynthesis and respiration double for every 10 °C rise in temperature. The photosynthetic and respiratory responses of upper intertidal plants to the temperature treatments demonstrated their remarkable tolerance to this factor. Their photosynthetic rates increased only by one-fold per 10 °C rise in temperature so that only their responses at 22 and 34 °C differed significantly, while their respiratory rates were not affected by temperature fluctuations of 22 to 34 °C. On the other hand, tidepool plants increased their respiration rates by 5-fold compared to the 10–50-fold increased by the subtidal plants. Prevailing temperature regimes in each habitat could have influenced the differential responses (e.g. differing temperature optima) of the different plant types.

The salinity that results in minimal osmotic stress usually corresponds to the highest photosynthetic (Vosjan & Siezen, 1968), and lowest respiratory rates (Ogata & Takada, 1968). In this study, salinity range of 22 to 40‰ appeared to be well tolerated by the three ecomorphs of *Gelidiella acerosa*. However, responses of the tidepool and subtidal plants differed from the upper intertidal ones. Both photosynthesis and respiration of the tidepool and subtidal plants were not significantly affected by the salinity range, while, the intertidal plants tolerated the low salinities of 22–28‰ better than the higher ones (34-40‰). This response is not fully understood because intertidal plants were expected to exhibit greater tolerance to this factor having been exposed to precipitation and evaporation in their habitat more than the two other plant types. Yet, they also grew under foliose macroalgae or in rock crevices that could have lessened the impact of secondary effects of desiccation. It should be emphasized,

Table 5. Statistical results: comparison of photosynthetic and respiratory responses of subtidal vegetative (3) and tetrasporic (4) plants per level of salinity and temperature. (*, significant; ns, not significant; values in μl O_2 gDW^{-1} h^{-1}, $n = 9–12$)

Environmental factor	Photosynthesis	Photosynthesis p	Respiration	Respiration p
Salinity ‰				
22	(3) 4772±1130	0.1503 ns	(4) −119±203	0.2182 ns
	(4) 4036±926		(3) −286±333	
28	(3) 4433±880	0.5877 ns	(3) −113±352	0.8622 ns
	(4) 4190±979		(4) −137±191	
34	(3) 4218±1142	0.7221 ns	(4) −19±210	0.5150 ns
	(4) 4044±874		(3) −109±350	
40	(3) 4288±1077	0.1123 ns	(3) −65±142	0.5804 ns
	(4) 3441±1063		(4) −109±185	
Temperature °C				
22	(3) 3382±365	0.0186 *	(4) −3±6	0.7443 ns
	(4) 2944±471		(3) −25±237	
28	(3) 4344±616	0.2172 ns	(4) −38±143	0.6781 ns
	(4) 4019±634		(3) −3±304	
34	(3) 5558±568	0.0069 *	(4) −323±120	0.2499 ns
	(4) 4820±642		(3) −402±198	

however, that the above results were based on the context of the experiments done, and therefore, need further investigation.

The photosynthetic and respiratory responses to the above combinations of salinity and temperature by the tidepool and subtidal representatives of *G. acerosa* indicated the probable similarities in their physiological make-up. The r-phycoerythrin and r-phycocyanin concentrations and photosynthesis-irradiance curves of the summer plants are also similar, in contrast to the upper intertidal plants (Ganzon-Fortes, 1997a). This is attributed to the similarities in their habitats. Further, the upper intertidal plants showed broader tolerances to temperature, while tidepool and subtidal plants were least affected by salinity variations. The overall broad tolerances of all plant types to both salinity and temperature helped explain their presence throughout the year in the study site. Although the interaction of temperature and salinity did not affect the photosynthesis and respiration of the species, their interaction with other environmental factors, i.e., light, nutrient, or water movement, may do so.

Tetrasporic and vegetative plants

As reported by Ganzon-Fortes (1997a), tetrasporic plants of *Gelidiella acerosa* are uncommon in tidepool and upper intertidal habitats, while they are common in the subtidal area. It appeared that vegetative plants were sensitive to temperature changes above or below the normal levels when compared with the tetrasporophytes. Conversely, the tetrasporophytes appeared to have greater tolerance to temperature fluctuations. This paralleled tetrasporophyte responses to increasing light levels (P-I curve) which can be enhanced (more efficient under lowlight) when compared to its vegetative state (Ganzon-Fortes, 1997a). Likewise, significant decrease in agar quality (gel strength) of tetrasporic compared to vegetative plants has also been documented (Roleda et al., 1997). Such responses suggested that possible physiological changes may have occurred when the plant became reproductive.

Implications for mariculture

The four salinity tested (22, 28, 34 and 40‰) were tolerated by *Gelidiella acerosa* except for the intertidal plants which showed less tolerance to high salinities, e.g. 40‰. Salinities lower than 22‰ should be further tested to determine the lower limit of salinity tolerance of the species. In eradicating contaminants in culture tanks (i.e., dinoflagellates, filamentous green algae), lowering the salinity several fold could be one safe method. Temperature at 28 °C appeared to be optimum for all plant types because photosynthetic rates were high and respiration rates were low. Although highest photosynthetic rates were obtained under 34 °C for all plant types, this was also complemented by highest respiration rates.

Acknowledgement

I gratefully acknowledge the United Nation Development Program (UNDP) for providing funds for the purchase of the Gilson Differential Respirometer; also, the Philippine Council for Aquatic and Marine Research and Development (PCAMRD) for other miscellaneaous financial support. I am indebted to Dr Clinton J. Dawes for sharing the 'manometry methodology' and for greatly improving the manuscript.

This is contribution no. 285 of the Marine Science Institute, College of Science, University of the Philippines, Diliman, Quezon City, Philippines.

References

Armisen, R. & F. Galatas, 1987. Production, properties and uses of agar. In McHugh, D. J. (ed.), Production and Utilization of Products from Commercial Seaweeds. FAO Fisheries Technical Paper No. 288: 1–57.

Britting, S. A. & D. J. Chapman, 1993. Physiological comparison on the isomorphic life history phases of the high intertidal alga *Endocladia muricata* (Rhodophyta). J. Phycol. 29: 739–745.

Dawes, C. J., 1981. Marine Botany. John Wiley & Sons, Inc., New York, 628 pp.

Dawes, C. J., 1985. Respirometry and manometry. In Littler, M. M. & D. S. Littler (eds), Handbook of Phycological Methods, Ecological Field Methods: Macroalgae. Cambridge University Press, Cambridge, New York: 329–348.

Dawes, C. J., 1987. The biology of commercially important tropical marine algae. In Bird, K. T. & P. H. Benson (eds), Seaweed Cultivation for Renewable Resources. Elsevier Science Publishers, Amsterdam: 155–190.

de Venecia, M. B. B., 1992. The regenerative capacities and growth responses of *Gelidiella acerosa* (Forsskal) Feldman et Hamel branch cuttings to light and nitrates (Rhodophyta: Gelidiales). M.S. Thesis, College of Science, University of the Philippines, Diliman, Quezon City, Philippines, 90 pp.

Dring, M. J., 1982. The Biology of Marine Plants. Edward Arnold (Publishers) Limited, London, 199 pp.

Druehl, L. D., 1981. Geographical Distribution. In Lobban, C. S. & M. J. Wynne (eds), The Biology of Seaweeds. University of California Press, Berkeley and Los Angeles. Botanical Monographs 17: 306–325.

Ganzon-Fortes, E. T., 1993. Determination of photosynthetic and respiratory rates of seagrasses using a Gilson differential respirometer. In Fortes, M. D. & N. Wirjoatmodjo (eds), Seagrass Resources in Southeast Asia: Technical Papers from the Advanced Training Course/Workshop on Seagrass Resources, Research and Management (SEAGREM 2), Quezon City, Philippines, January 8–26, 1990. UNESCO-Jakarta (ROSTSEA), Indonesia: 195–202.

Ganzon-Fortes, E. T., 1997a. Influence of tidal location on morphology, photosynthesis, and pigments of the agarophyte, *Gelidiella acerosa*, from Northern Philippines. J. appl. Phycol. 7: 1–8.

Ganzon-Fortes, E. T., 1997b. Diurnal and diel patterns in the photosynthesis of the agarophyte, *Gelidiella acerosa*. Bot. mar. 40: 93–100.

Gessner, F., 1970. Temperature: plants. In Kinne, O. (ed.), Marine Ecology (vol. 1, pt. 1). Wiley, New York: 363–406.

Hansen, J., 1980. Physiological considerations in the mariculture or red algae. In Abbott, I. A., M. S. Foster & L. F. Eklund (eds), Pacific Seaweed Aquaculture. California Sea Grant College Program, Institute of Marine Resource, University of California, La Jolla, California: 80–85.

Kirst, G. O., 1988. Turgor pressure regulation in marine macroalgae. In Lobban, C. S., D. J. Chapman & B. P. Kremer (eds), Experimental Phycology – A Laboratory Manual. Cambridge University Press, New York: 203–209.

Laserna, E. C., G. J. B. Cajipe, R. L. Veroy & A. H. Luistro, 1978. Spectrofluorimetric assay of carrageenan and agar from Philippine seaweeds. Kalikasan, Philipp. J. Biol. 7: 110–116.

Lobban, C. S. & P. J. Harrison, 1994. Seaweed Ecology and Physiology. Cambridge University Press, New York, 366 pp.

Lüning, K., 1990. Seaweeds, Their Environment, Biogeography and Ecophysiology. Wiley-Interscience Publication, John Wiley & Sons, Inc., New York, 527 pp.

Macler, B. A., 1988. Salinity effects on photosynthesis, carbon allocation, and nitrogen assimilation in the red alga, *Gelidium coulteri*. Plant Physiol. 88: 690–694.

Mathieson, A. C. & T. Norall, 1975. Physiological studies of subtidal red algae. J. exp. mar. Biol. Ecol. 20: 237–247.

Ogata, E. & H. Takada, 1968. Studies on the relationship between respiration in some marine plants in Japan. J. Shimonoseki Collect. Fish. 16: 67–88.

Roleda, M. Y., E. T. Ganzon-Fortes & N. E. Montaño, 1997. Agar from vegetative and tetrasporic *Gelidiella acerosa* (Gelidiales, Rhodophyta). Bot. mar. 40: 501–506.

Santos, G. A. & M. S. Doty, 1983. Agar from some Hawaiian red algae. Aquat. Bot. 16: 385–389.

Sosa, P. A., M. Jimenez del Rio & G. Garcia-Reina, 1993. Physiological comparison between gametophytes and tetrasporophytes of *Gelidium canariensis* (Gelidiaceae: Rhodophyta). Hydrobiologia 260/261: 445–449.

Trono, G. C. Jr. & E. T. Ganzon-Fortes, 1988. Philippine Seaweeds. National Book Store, Inc., Manila, Philippines, 330 pp.

Vosjan, J. H. & R. J. Siezen, 1968. Relation between primary production and salinity of algal culture. Netherlands J. sea Res. 4: 11–20.

Zupan, J. R. & J. A. West, 1990. Photosynthetic responses to light and temperature of the heteromorphic marine alga *Mastocarpus papillatus* (Rhodophyta). J. Phycol. 26: 232–239.

The effect of light on growth and agar content of *Gelidium pulchellum* (Gelidiaceae, Rhodophyta) in culture

Isabel Sousa-Pinto[1], Erminio Murano[2], Susana Coelho[1], Ana Felga[1] & Rui Pereira[1]
[1]*CIMAR and Departamento de Botânica, Faculdade de Ciências da Universidade do Porto, R. do Campo Alegre, 1191, 4150 Porto, Portugal*
Fax: [+351] 2 6092227. E-mail: ispinto@caserver.fc.up.pt
[2]*POLYtech Research Center, Area Science Park, Padriciano 99, I-34012 Trieste, Italy*

Key words: agar, cultivation, *Gelidium pulchellum*, light, molecular weight, Rhodophyta, sulphate.

Abstract

Investigation of light conditions suitable for cultivation of *Gelidium pulchellum* (Turner) Kurtz was performed under controlled laboratory conditions at 20 °C and in the range of irradiance of 10–430 μmol photons m^{-2} s^{-1}. Growth, measured as fresh weight increment, increased with irradiance up to 130 μmol m^{-2} s^{-1} and no significant photoinhibition was observed up to 430 μmol m^{-2} s^{-1}. Maximum growth rate (10.0% day^{-1}) was obtained at 130–240 μmol m^{-2} s^{-1} under continuous light and aeration. The effect of irradiance on agar yield and quality was assessed. Agar yield varied from 31 to 38.6% of the algal dry weight, and variation was not related to irradiance. However, the yield of agar molecules soluble at 95 °C increased with increasing irradiance. A similar trend was found for sulphate content in both series of extracts, at 95 and 121 °C. On the contrary, the molecular weight and the degree of methylation of agar molecules in the 95 °C extracts decreased with increasing light intensity. As a consequence of the variations in sulphate content, molecular weight and molecular weight distribution, the gel strength was considerably lower at high light intensity. Starch content varied from 0.9 to 7.7% of the algal dry weight, and apparently was not related with irradiance.

Introduction

The most commercially important agarophyte genera include *Gelidium*, *Pterocladia* and *Gracilaria*. Although the agars from the first two genera are considered to be of higher quality and command a higher price, *Gracilaria* amounts to more than 50% of the seaweed used for agar production (Armisen, 1995), due to the farming of several species of this genus (Santelices & Doty, 1989). Few attempts (Fei & Huang, 1991) have been made to cultivate *Gelidium*, in order to provide an additional source of raw material for the extraction of high quality agar and agarose with important biomedical and industrial applications (Armisen, 1995). However, the costs involved in mariculture of macroalgae and the lack of knowledge of the optimal growth and agar production of these 'slow growing' species have prohibited profitable operations so far.

Gelidium pulchellum (Turner) Kurtz is a small intertidal species that is not used for industrial agar extraction, due to its small size and relative low biomass in the field. In order to study the physiology and biochemistry of growth and agar production of this species, it is necessary to determine its optimum growth conditions in the laboratory, as well as to evaluate the effects of environmental factors such as light, temperature, water motion, carbon, pH and nitrogen supply (Macler & Zupan, 1991).

As demonstrated for the genus *Gracilaria* (Friedlander et al., 1987; Bird, 1988; Bird et al., 1989; Chiles et al., 1989; Christeller & Laing, 1989; Ekman & Pedersén, 1990; Friedlander & Levy, 1995; Lewis & Hanisak, 1996), not only species, but also environmental and physiological factors affect growth and yield and properties of agar. However, very few studies on the effects of environmental parameters on growth and agar content have been done on *Gelidium* (Sousa-

Pinto et al., 1996). Such information obtained from algae grown under controlled conditions can be useful in selecting cultivation techniques leading to higher colloid contents of desired characteristics.

The objective of this work was to determine the effect of irradiance on growth and agar yield and quality of *Gelidium pulchellum* in laboratory culture.

Materials and methods

Culture experiments

Gelidium pulchellum was collected from Praia da Luz (41° 10′ N, 8° 39′ W), cleaned and cut in pieces of 2–3 cm and suspended in modified PES medium where no buffer was added and Na_2 glycerophosphate was replaced by Na_2HPO_4. Control of epiphytes in non-axenic cultures was achieved, either by cleaning carefully the algae before starting any experiment with a soft brush and, if necessary, 10% Betadine in seawater, or by incubating the algae for 2–4 days in a 5 ml l^{-1} solution with antibiotics (penicillin-streptomycin solution, SIGMA). The algae were briefly washed, if needed, with distilled water before being dryed and weighed.

G. pulchellum was acclimated for two weeks, before the start of the light experiments, to the following culture conditions: continuous illumination at 40 μmol photons m^{-2} s^{-1} irradiance, a temperature of 20 ± 1 °C, pH 8 and salinity 32%. Aeration was provided by bubbling sterile filtered air (3600 l h^{-1}; 0.2 μm filter) in 2 l Erlenmeyer flasks. Growth rates were measured every 3–4 days; the algae were blotted dry, weighed and transferred to fresh medium. The density of the cultures was kept between 0.5 and 1 g l^{-1} because previous experiments showed that higher densities decreased growth rates.

For the determination of the effect of varying irradiance, flasks were continuously illuminated from the side. The different irradiances were obtained by covering the light source (Phillips coolwhite confort TLD 58W/84 lamps) with layers of a neutral filter. Irradiance was measured with a Licor LI 185-B quantum-meter equipped with a LI-190 B quantum sensor.

Growth rate (G) expressed in % of increase of fresh weight per day was calculated as: g (% d^{-1}) = ln (final fresh weight/initial fresh weight) $d^{-1} \times 100$.

Three replicate flasks were used for each condition tested. Analyses of variance (ANOVA's) and a Scheffé post-hoc analyses were performed to assess the significance of differences between the growth rate obtained with each treatment.

After six weeks in culture at different irradiances, algae were air dried for agar extraction.

Agar extraction and analysis

Gelidium pulchellum (natural population), collected at Praia da Luz in December 1997, and *G. pulchellum* from the irradiance experiments at 40, 130, 240 μmol photon $m^{-2}s^{-1}$, were oven dried at 40 °C and milled (250–300 μm particle size). Dried algal powder (3 g) was treated, prior to extraction, with 100 ml of 0.1 N HCl at 4 °C for 30 min, neutralized with an equimolar amount of NaOH, and washed thoroughly with distilled water until neutral pH. Agar extraction was performed twice for 45 min at 95 °C in 200 ml of 50 mM phosphate-buffer, pH 7.2. The extracts were filtered, under pressure, at 70 °C through Whatman GF/C (1.2 mm) glass microfiber filters and combined. Algal residues were then extracted twice with 150 ml of buffer at 121 °C for 45 min and filtered as described above. Crude extracts were recovered, after isopropanol precipitation, as previously described (Murano et al., 1992).

Sulphate content was determined in triplicate with barium chloride/sodium rhodizonate according to the method of Silvestri et al. (1982), after sulphate hydrolysis by 1 N HCl, at 110 °C, for 3 h.

Gel strength was determined in triplicate on 1.5% (w/w) agar cylinders (15 mm diameter, 15 mm height) using a Stevens-LFRA texture analyser (Murano et al., 1992).

NMR experiments were performed at 80 °C on a Bruker AC 200 spectrometer equipped with a 5 mm multinuclear probe. ^{1}H-NMR (200.13 Mhz) spectra of 1% (w/v) agar solutions in D_2O were recorded with presaturation of the HOD residual signal. ^{1}H chemical shifts were measured in parts per million (ppm) from internal sodium 2,2,3,3-tetradeuterio-3-(trimethylsilyl)-proprionate (TSP). Proton composite pulse decoupled ^{13}C-NMR spectra of 3% (w/v) agar solutions in D_2O were acquired at 50.33 MHz. ^{13}C chemical shifts were referred to external tetramethylsilane (TMS) by setting the internal dimethylsulphoxide resonance to 39.6 ppm.

Starch, methoxyl and pyruvic acid contents were determined by ^{1}H-NMR spectroscopy. The starch content was estimated by the ratio between the resonance area of the H-1 of glucose at 3.35 ppm (Knutsen

& Grasdalen, 1987) and of the H-1's of 4-linked-L-galactose in the region 5.28 and 5.15 ppm (Welti, 1977; Lahaye et al., 1988). The pyruvate and the 2- and 6-O-methyl content was determined by the same procedure by using 1/3 of the area of the methyl resonances at 1.47 (Izumi, 1973) and 3.43 and 3.52 ppm (Lahaye et al., 1986), respectively, assuming a perfect alternating agar backbone.

The weight-average molecular weight ($\bar{M}_w$) and polydispersity index ($\bar{M}_w/\bar{M}_n$, where $\bar{M}_n$ is the number-average molecular weight) of agars from *G. pulchellum* were obtained, in triplicate, by means of high performance gel permeation chromatography coupled with a low angle laser light scattering detector (Martinsen et al., 1991). Agar samples were dissolved in 0.1 M tetrabutyl ammonium bromide in dimethyl sulphoxide/water 80:20 solution, and filtered through a 0.5 μm pore size membrane before injection (Murano et al., 1992).

Results

Growth rate

Growth rates depended significantly on irradiance ($p < 0.05$). The highest growth rate, 10% d^{-1}, was obtained with irradiances between 130–240 μmol photon m^{-2}s^{-1} (Figure 1). Growth decreased at lower irradiances, but the algae were healthy. A Sheffé post-hoc analysis showed no significant differences between growth rates at 240 and 430 μmol photon m^{-2} s^{-1}. However, at the high irradiance, the algae became pale after four weeks, were more epiphytized and finally two of the three cultures died after six weeks. Generally, the proliferation of cyanobacteria and fungi in the cultures was effectively inhibited by the antibiotic solution. On the contrary, green algae epiphytes were the most serious problem encountered during these experiments. One method that was found to be marginally effective in maintaining the cultures relatively free of green algae was to briefly wash the *Gelidium* with distilled water as the medium was being changed.

Agar yield and quality

The samples of *G. pulchellum* grown at 40, 130 and 240 μmol photon m^{-2} s^{-1} were investigated for their agar content, and results compared to those obtained from natural population. The total agar yield (95 and 121 °C), after correction for starch content, of all the

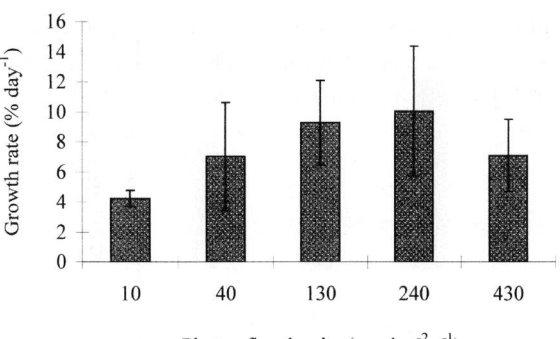

Figure 1. Effect of irradiance on growth of *Gelidium pulchellum*. Three flasks were used for each tested condition and values represent the average growth rate (% growth d^{-1}) ± SD measured during 6 weeks ($n = 3$).

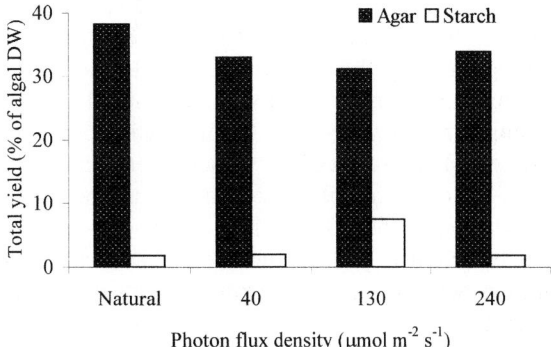

Figure 2. Total agar and starch yield, as percentage of algal dry weight, in natural population of *G. pulchellum* and *G. pulchellum* cultured at different irradiances.

cultured *G. pulchellum*, was slightly lower than that of the natural population which accounted for 38.6% of the algal dry weight (Figure 2). Although the total agar yield showed little differences in the range of irradiance 40–240 mmol photon m^{-2}s^{-1}, agar content in the 95 °C extracts increased with the increasing of irradiance from 40 to 240 μmol photon m^{-2}s^{-1} (Figure 3). On the other hand, the opposite trend was observed for the material soluble at 121 °C, which showed a marked decrease in agar content with the increasing of photon irradiance.

^{13}C NMR analysis of 95 and 121 °C extracts showed the typical signal pattern for agarose or modified agarose (see selected spectra in Figure 4). Signals at 102.3, 70.1, 82.1, 68.7, 75.3 and 61.3 ppm and at 98.1, 69.3, 80.0, 77.2, 75.6 and 69.8 ppm were attributed to carbons 1–6 in the D-galactose and 3,6-anhydro-L-galactose residues, respectively (Usov et al., 1983). Low intensity signals were also present at 98.7 (C-1) and 78.8 (C-2), ppm and at 73.5 (C-5)

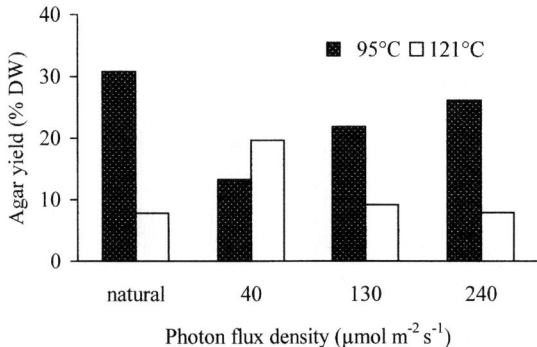

Figure 3. Agar yield, as percentage of algal dry weight, in natural population of *G. pulchellum* and *G. pulchellum* cultured at different irradiances.

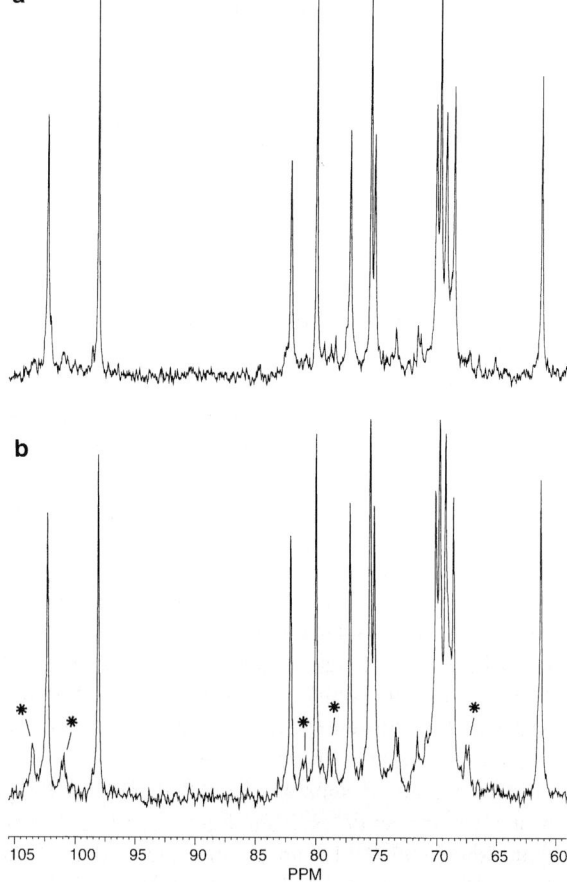

Figure 4. ^{13}C spectra of agars extracted at 95 °C from natural population of *G. pulchellum* (a) and from *G. pulchellum* grown at 240 μmol m^{-2} s^{-1} (b). ∗ refers to H-1 of 6-sulphated galactose residues.

and 71.7 (C-6) ppm, characteristic for agarobiose containing 2-*O*-methyl-3,6-anhydro-L-galactose and 6-*O*-methyl-D-galactose, respectively (Usov et al., 1980; Nicolaisen et al., 1980). A series of additional signals was detected, with increasing signal intensity from 40 to 240 μmol photons m^{-2}s^{-1}. These resonances, at 103.7, 101.1, 81.1, 78.9 and 67.7 ppm, are characteristic for agarobiose 'precursor' repeating units containing 4-*O*-L-galactose-6-sulphate (Lahaye et al., 1985). The occurrence of 6-sulphated and 2- and 6-*O*-methylated galactose residues in the agarobiose repeating units of *G. pulchellum* was confirmed by the presence, in the ^{1}H NMR spectra, of signals at 5.28 ppm (Lahaye et al., 1988) and 3.52 and 3.43 ppm (Lahaye et al., 1986), respectively (Figure 5). Distinctive signals, assigned to the methyl protons of the pyruvic ketal (1.47 ppm) and to H-1 of the L-galactose residue having a pyruvylated D-galactose residue towards the reducing end (5.24 ppm) (Izumi, 1973), were present in the ^{1}H spectra of all agar samples.

An extra signal at 5.35 ppm, attributed to H-1 of (1->4) linked α-D-glucose in floridean starch (Knutsen & Grasdalen, 1987), was detected in the ^{1}H spectra of the all extracts at 121 °C but only in the 130 μmol photons m^{-2}s^{-1} sample of the 95 °C series (Figure 5c). Characteristic signals for carbons of D-glucose in floridean starch, at 100.4, 73.5 and 72.0 ppm (Lahaye et al., 1986), were also found in the corresponding ^{13}C spectra of these samples (spectra not shown). The ratio of the resonance area of the H-1 of glucose *versus* the resonance area of H-1 of 3,6-anhydro-L-galactose can give an estimation of the starch to agar ratio in the crude extracts and, therefore, of the starch content in the algal tissue. The highest total content of glucose (7.7% of algal dry weight) was found for *G. pulchellum* cultured at 130 μmol photons m^{-2} s^{-1}, whereas the lowest values were found in the natural population (0.9%) and in *G. pulchellum* grown at 240 μmol photon m^{-2}s^{-1} (1.0%) (Figure 2). However, in the series of extracts at 95 °C, floridean starch was detected only in *G. pulchellum* grown at 130 μmol photons m^{-2} s^{-1} sample (9.1% of the crude extract, corresponding to 2.2% of the algal dry weight). The starch to agar ratio in the 121 °C extract from *G. pulchellum* cultured at 130 μmol photon m^{-2} s^{-1} was more than three times that of the others, and the starch content reached 37.5% of the extract.

The degrees of substitution with pyruvate and methoxyl groups are reported in Table 1. The amount of pyruvate ketal, expressed as mole per mole of galactose residue, was in the range 0.024–0.072 (corresponding to 0.96–2.88% of agar), with the highest values found for *G. pulchellum* grown at 130 μmol

Table 1. Pyruvate and methoxyl content of agar extracts from *Gelidium pulchellum* grown at different irradiance

		Pyruvate mol/galactose	2-O-Me-l-gal DS[a]	6-O-Me-d-gal DS[a]
natural population	95 °C	0.031	0.08	0.10
	121 °C	–	0.07	0.11
40 μmol m^{-2} s^{-1}	95 °C	0.045	0.11	0.14
	121 °C	0.060	0.05	0.08
130 μmol m^{-2} s^{-1}	95 °C	0.056	0.10	0.12
	121 °C	0.072	0.06	0.10
240 μmol m^{-2} s^{-1}	95 °C	0.024	0.09	0.09
	121 °C	0.056	0.08	0.11

[a]DS: 6-O-methyl-β-D-galactose to β-D-galactose and 2-O-methyl-α-3,6 anhydro-L-galactose to α-3,6-anhydro-L-galactose ratios. –: not determined.

photons m^{-2}s^{-1}. For all the cultured samples, the amount of 4,6-O-(1-carboxyethylidene)-D-galactose was higher in the agar fraction soluble at 121 °C. These values are similar to those reported for agar (0.2–2.5%) (Armisen & Galatas, 1987) and no direct relation was found with irradiance.

The degree of substitution (DS) with methoxyl groups in the samples from light experiments was quite low (0.05–0.14) and comparable to that of the field collected *G. pulchellum*. For both the 2-O and 6-O-methylated galactose the DS showed an inverse relation in the 95 °C and a direct one in the 121 °C extracts, with the *light intensity*.

The sulphate ester content of agar extracts from natural population was in the range 1.2–1.4%. In the light experiments, the amount of sulphate ester enhanced with increasing irradiance, and the highest value (4.8%) was found for the agar extracted at 95 °C from *G. pulchellum* grown at 240 μmol photons m^{-2}s^{-1} (Figure 6).

The weight-average molecular weight ($\bar{M}_w$) and the polydispersity index ($\bar{M}_w/\bar{M}_n$) of agar extracted from the natural population of *G. pulchellum* were 265 000 g mol^{-1} and 1.7, respectively. The corresponding $\bar{M}_w$ and $\bar{M}_w/\bar{M}_n$ of the agars extracted at 95 °C from cultured algae were in the range 159 000–204 000 g.mol^{-1} and 2.0–2.4, respectively.

Gel strength values of agar obtained from laboratory grown *G. pulchellum* were considerably lower than that of agar isolated from the natural population (Figure 7) and decreased with increasing irradiance. No appreciable difference in gel strength was found between the two series of extracts (95 and 121 °C) with the exception of agars from natural populations.

Discussion

Growth rate

The objectives of this study were to determine the effect of light intensities on growth and agar content, and characteristics of *G. pulchellum* cultivated in the laboratory under defined conditions. The free floating algal cultures at 20 °C kept in motion by filtered air bubbling were conditions that were shown in previous studies *Gelidium* species in the laboratory to give high growth rates. Green algal epiphytes were the most serious problem faced during growth experiments, but cleaning the alga before their transfer to fresh medium maintained these contaminants at a low concentration.

Growth was significantly ($p < 0.05$) affected by light intensities for the range of irradiances tested. This was true for all the *Gelidium* species studied so far (Macler & West, 1987; Fredriksen et al., 1993; Sousa-Pinto et al., 1996; Patway & Meer, 1997). Photoinhibition of growth was not significant, and optimal irradiance was found to be in the range of 130–240 μmol photons m^{-2}s^{-1}, though growth was also high at 40 μmol photons m^{-2} s^{-1}. Some *Gelidium* species showed the highest growth at higher irradiances: 300 μmol photons m^{-2} s^{-1} for *G. latifolium* (Greville) Bornet & Thuret (Fredriksen & Rueness, 1989), 250 μmol photons m^{-2}s^{-1} for *G. robustum*

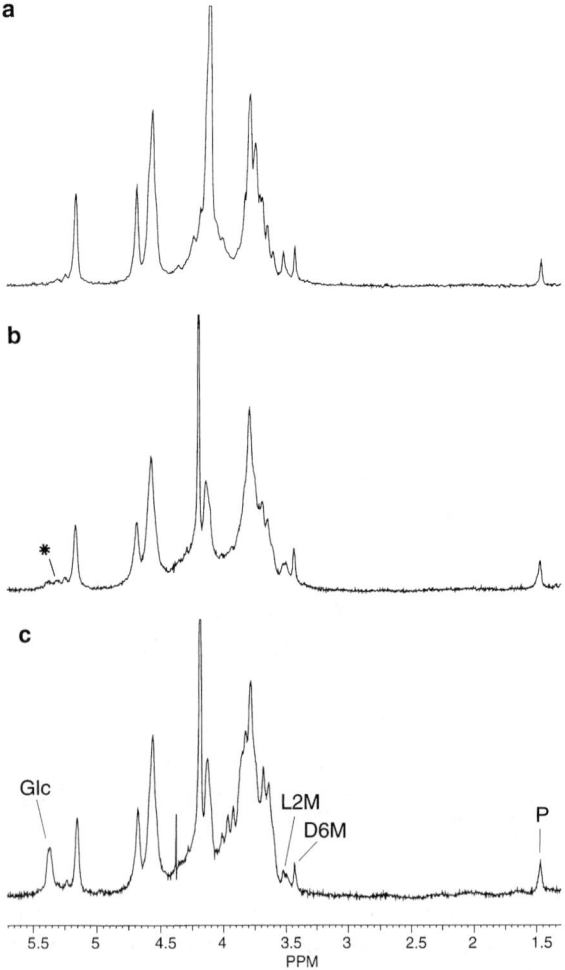

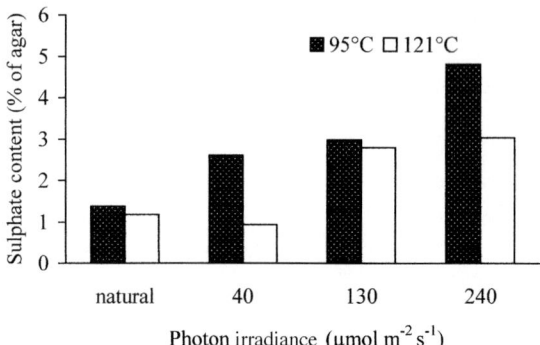

Figure 6. Sulphate content, as percentage of agar, in agar polymers extracted from natural population of *G. pulchellum* and from *G. pulchellum* cultured at different irradiances.

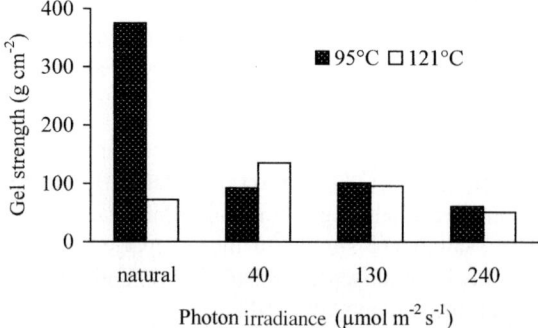

Figure 7. Gel strength of agar extracted form natural population of *G. pulchellum* and from *G. pulchellum* cultured at different irradiances.

Figure 5. ^{1}H spectra of agars from *G. pulchellum*: (a) agar extracted at 95 °C from natural population; (b) agar extracted at 95 °C from algae grown at 130 μmol m^{-2} s^{-1}; (c) agar extracted at 121 °C from algae grown at 130 μmol m^{-2} s^{-1}. ∗ refers to H-1 of 6-sulphated galactose residues; **Glc** refers to H-1 of glucose residues in floridean starch; **L2M** and **D6M** refer to H-1 of 2-*O*-methyl-3,6-anhydro-L-galactose and 6-*O*-methyl-D-galactose, respectively; **P** refers to H-1 of 3,6-anhydro-L-galactose having a pyruvylated D-galactose towards the reducing end.

(Gardn.) Hollenb. & Abb. (Sousa-Pinto, 1994). However, studies showed that some *Gelidium* and *Pterocladia* species in the field tend to grow better in shaded habitats (Seoane-Camba, 1964, 1965; Santelices, 1978, 1988, 1991). In addition others demonstrated growth and photosynthesis inhibition under irradiances higher than 400 μmol photons m^{-2}s^{-1} (Torres et al., 1991; Mouradi-Givernaud et al., 1992). Although *G. pulchellum*, an intertidal alga, did not show any adaptation for using high irradiances, since growth saturated at 130 μmol photons m^{-2} s^{-1}, it also did not show any significant growth inhibition at irradiances up to 430 μmol photons m^{-2} s^{-1}.

Agar yield and quality

Unlike other reports on *Gelidium* and *Gracilaria* (Oza, 1978; Bird et al., 1981; Macler & West, 1987; Christeller & Laing, 1989; Torres et al., 1991), the total amount (95 and 121 °C) of agar extracted from *G. pulchellum* grown under different light intensities was not clearly affected by irradiance, in the range 40–240 μmol photons m^{-2}s^{-1}. Starch content was not obviously related to irradiance in cultured *G. pulchellum*. In fact, it showed a maximum at 130 μmol photon, m^{-2}s^{-1}, which is the irradiance of growth saturation, and was remarkably lower at 240 μmol photons m^{-2}s^{-1}. Previous studies on *Gracilaria* sp. (Rotem et al., 1986) demonstrated that, when growth is limited by light, starch accumulation was directly proportional to irradiance in the range 10–168 μmol photons m^{-2}s^{-1}. It is known that floridean starch, the major carbon storage polymer in red algae, and floridoside, the principal low molecular weight photoassimilate, are accumulated during the light phase of photosynthesis and utilized, during the dark phase, for the synthesis of cell wall polysaccharides (Craigie et al., 1967; Preiss & Levi, 1980; Macler, 1986; Beck & Ziegler, 1989). Although it might be expected that, with enhanced growth and photosynthesis due to high irradiances, photoassimilates are directed preferentially towards the synthesis of starch, rather than cell-wall materials, in *G. pulchellum* the accumulation of reserve material above a certain limit of irradiance and growth rate, might be balanced or inhibited by the immediate requirement of building blocks (UDP-D-galactose and GDP-L-galactose) for the synthesis of new cell wall polysaccharides. Further studies are needed to investigate this aspect in *G. pulchellum*, in terms of relationship between floridean starch and floridoside pools, and degradation of floridean starch for the production of agarobiose precursors UDP-D-galactose and GDP-L-galactose (Yu, 1992; Hemmingson et al., 1996).

The opposite trends observed for the agar yield at 95 and 121 °C may be explained by a different solubility with temperature of agar molecules having both a different degree of substitution with sulphate and, to a minor extent, methoxyl groups, and a different molecular weight distribution. The relationship between the solubility with temperature of agar molecules and molecular weight and type, and degree of substitution is well known (Lahaye et al., 1986; Armisén & Galatas, 1987; Lahaye & Yaphe, 1988; Ji et al., 1988; Murano, 1995a). With respect to sulphate content, only the 95 °C extract of the sample grown at 240 μmol m^{-2} s^{-1} was higher than the norm for *Gelidium* agar (Armisén, 1995). In both series of extracts (95 and 121 °C) sulphate content increased with increasing irradiance and growth rates. Sulphation and methylation were found to be early processes in the biosynthesis of agar polysaccharides (Hemmingson et al., 1996). In addition, young tissue (Craigie & Jurgens, 1989) as, in general, actively growing algae (Lahaye & Yaphe, 1988) were shown to be richest in sulphate. However, data reported in the literature are not always concordant about the influence of irradiance and growth rate on the sulphate content in agarophytes (Asare, 1980; Whyte et al., 1981; Craigie, 1990). Our results with *G. pulchellum* were similar to those that found the more sulphated agars of *Gracilaria* (Cote & Hanisak, 1986; Lahaye & Yaphe, 1988). The finding that the agar polysaccharides pool is enriched in sulphated 'precursor' molecules upon increasing irradiance leads us to conclude that, also for *G. pulchellum* in culture, high irradiances determine either a higher synthesis of sulphated agar polymers or a reduced efficiency in the sulphation/desulphation mechanism.

A minor contribution to the solubility pattern may derive also from the degree of substitution (DS) with methoxyl groups. In fact, although variations were limited, the decrease of the DS in the 95 °C extracts of *G. pulchellum* with increasing irradiance may contribute to enhancing the solubility of agar molecules in the 95 °C extracts.

On the other hand, the weight-average molecular weight ($\bar{M}_w$) of agar isolated from the natural population of *G. pulchellum* (265 000 g mol^{-1}), was significantly lower than that reported for a representative number of species of *Gracilaria* (340 000–380 000 g mol^{-1}) (Murano, 1995b). It decreased with increasing light intensity, reaching the lowest value (159 000 g mol^{-1}) in the sample grown at 240 μmol photons m^{-2}s^{-1}. Since no evidence exists, to our knowledge, of depolymerisation processes of algal cell wall polysaccharides, the results on *G. pulchellum* would indicate that the polymerization rate of agar molecules is reduced at high growth rates and light intensity, thus leading to a higher amount of shorter agar chains.

The inverse relation between the amount of sulphate ester and the gel strength is well known (Rees, 1969; Christeller & Laing, 1989). In fact, the gel strength of agars obtained from cultured *G. pulchellum* was lower than that from the natural population because of the higher amount of precursor-rich molecules in the extracts. As expected, the lowest gel strength was found for the *G. pulchellum* cultivated at the highest photon flux density. These samples showed also the highest amount of sulphate, the lowest molecular weight and the highest polydispersity index.

Although for cultured *G. pulchellum*, the extent of concurrent sulphated agar blocks seems to play a major role, also the $\bar{M}_w$ and the polidispersity index ($\bar{M}_w/\bar{M}_n$) affected to some extent the solubility and the gelling ability of the final extract.

Since the gel forming ability of agar is influenced, by type and degree of sulphation, and by molecular weight and molecular weight distribution of agar molecules (Watase & Nishinari, 1983), it was reasonable to expect that the lowest gel strength was found for *G. pulchellum* grown at 240 μmol photons m^{-2}s^{-1}, which had the lowest $\bar{M}_w$ (159 000 g mol^{-1}), and the highest $\bar{M}_w/\bar{M}_n$ (2.4) and sulphate content (4.8%). Moreover, the higher $\bar{M}_w/\bar{M}_n$ reflected a higher dispersion of agar molecules with different molecular weight in the 240 μmol photons $m^{-2}s^{-1}$ sample, and the consequent interaction of chains of largely differing length negatively affected the formation of the gel network.

The effect of increasing irradiance in cultured *G. pulchellum* was that of increasing the growth rate and algal biomass. As already found for *Gracilaria* (Lahaye & Yaphe, 1988), also in *G. pulchellum* the decreasing of mechanical restrictions for the growing and dividing cells of the actively growing tissues was obtained by modifying the physico-chemical and macromolecular properties of the agarocolloid components of the cell wall. Such modifications implied a rearrangement of the solubility features as well as the cell wall interactions between agar molecules and β-D-glycans (Dea & Rees, 1987).

As for carrageenans (Knutsen et al., 1995), a further fractionation of the agar extracts from the light experiments will possibly lead to a better characterization of the different agar populations in terms of distribution of sulphated, pyruvated and methoxylated residues. This may contribute to the understanding of the mechanisms of agar deposition in the cell wall and modification in response to light intensity variations.

Conclusions

From the light experiments, the highest growth rate (10% d^{-1}) for *G. pulchellum* was obtained with continuous illumination of white light of 130–240 μmol photons m^{-2} s^{-1}, continuous aeration, pH 8.0 and a temperature of 20 °C. This species grew also very well at low irradiances (40 μmol photons m^{-2} s^{-1}) which can be beneficial if light is limiting. This culture system could be used to provide seed-stock for experiments or cultivation on a pilot scale.

The analysis of agar content showed that sulphate and molecular weight were the most affected parameters by variations of light intensity in the range 40–240 μmol photons m^{-2} s^{-1}. As a consequence of the increased sulphate content and decreased molecular weight, the gel strength was drastically reduced. A further characterization of agar extracts after alkali modification could establish the potential of agar from cultured *G. pulchellum* for a large-scale production.

Acknowledgements

This work was financed by Junta de Investigação Científica e Tecnológica (JNICT), and grants BIC/14700/97, BIC/14701/97 to A. F. and R. P. The authors wish to thank M. Bosco and F. Picotti (Trieste, Italy) for the technical assistance in the NMR analysis.

References

Armisén, R., 1995. World-wide use and importance of *Gracilaria*. J. appl. Phycol. 7: 231–243.

Armisén, R. & F. Galatas, 1987. Production, properties and uses of agar. In McHugh, D.J. (ed.), Production and Utilisation of Products from Commercial Seaweeds. FAO Fish. Tech. Pap. 288: 1–57.

Asare, S. O., 1980. Seasonal changes in sulphate and 3,6-anhydrogalactose content in phycocolloids from two red algae. Bot. mar. 23: 595–598.

Beck, E. & P. Ziegler, 1989. Biosynthesis and degradation of starch in higher plants. Ann. Rev. P. Physiol. Plant Mol. Biol. 40: 95–117.

Bird, K. T., 1988. Agar production and quality from *Gracilaria* sp. strain G-16: Effects of environmental factors. Bot. mar. 31: 33–39.

Bird, K. T., M. D. Hanisak & J. Ryther, 1981. Chemical quality and production of agar extracted from *Gracilaria tikvahiae* grown in different nitrogen enrichment conditions. Bot. mar. 24: 441–444.

Bird, K. T., K. Pendoley & F. Koehn, 1989. Variabilty in agar gel behaviour and chemistry as affected by algal growth under different environmental conditions. In Crescenzi, V., I. C. M. Dea, S. Paoletti, S. S. Stivala & I. W. Sutherland (eds), Biomedical and Biotechnological Advances in Industrial Polysaccharides. Gordon & Breach Science Publishers, New York: 365–374.

Chiles, T. C., K. T. Bird & F. E. Koehn, 1989. Influence of nitrogen availability on agar-polysaccharides from *Gracilaria verrucosa* strain G-16: structural analysis by NMR spectroscopy. J. appl. Phycol. 1: 53–58.

Christeller, J. T. & W. A. Laing, 1989. The effect of environment on the agar yield and gel characteristics of *Gracilaria sordida* Nelson (Rodophyta). Bot. mar. 32: 447–455.

Cote, G. L. & M. D. Hanisak, 1986. Production and properties of native agars from *Gracilaria tikvahiae* and other red algae. Bot. mar. 29: 359–366.

Craigie, J. S., 1990. Cell walls. In Cole, K.M. & R.G. Sheath (eds), Biology of the Red Algae, Cambridge University Press, Cambridge: 221–257.

Craigie J. S. & A. Jurgens, 1989. Structure of agars from *Gracilaria tikvahiae* Rhodophyta: location of 4-*O*-methyl-L-galactose and sulphate. Carbohydr. Polymers 11: 265–278.

Craigie, J. S., J. McLachlan & R. D. Tocher, 1967. Some neutral constituent of Rhodophyceae with special reference to the occurrence of the floridosides. Can. J. Bot. 46: 605–611.

Dea, C. M. & D. A. Rees, 1987. Affinity interaction between agarose and β-1,4-glycans: a model for polysaccharides associations in algal cell walls. Carbohydr. Polymers 7: 183–224.

Ekman, P. & M. Pedersén, 1990. The influence of photon irradiance, day length, dark treatment, temperature, and growth rate on the agar composition of *Gracilaria sordida* W. Nelson and *Gracilaria verrucosa* (Hudson) Papenfuss (Gigartinales, Rhodophyta). Bot. mar. 33: 483–495.

Fei, X. G. & L. J. Huang, 1991. General principles of on-shore cultivation of seaweeds: effects of light on reproduction. Hydrobiologia 221: 125–135.

Fredriksen, S. & J. Rueness, 1989. Culture studies of *Gelidium latifolium* (Grev.) Born. et Thur. (Rhodophyta) from Norway. Growth and nitrogen storage in response to varying photon flux density, temperature and nitrogen availability. Bot. mar. 32: 539–546.

Fredriksen, S., J. M. Rico & J. Rueness, 1993. Comparison of *Gelidium latifolium* (Grev.) Born. et Thur. (Gelidiales, Rhodophyta) isolates from Spain and Norway. J. appl. Phycol. 5: 117–12.

Friedlander, M. & I. Levy, 1995. Cultivation of *Gracilaria* in outdoor tanks and ponds. J. appl. Phycol. 7: 315–324.

Friedlander, M., R. Shalev, T. Ganor, S. Strimling, A. Ben-Amotz, H. Klar & Y. Wax, 1987. Seasonal fluctuations of growth rate and chemical composition of *Gracilaria* cf. *conferta* in outdoor culture in Israel. Hydrobiologia 151/152: 501–507.

Hemmingson J. A., R. H. Furneaux & V. H. Murray-Brown, 1996. Biosynthesis of agar polysaccharides in *Gracilaria chilensis* Bird, McLachlan et Oliveira. Carbohydr. Res. 287: 101–115.

Izumi, K., 1973. Structural analysis of agar-type polysaccharides by NMR spectroscopy. Biochim. Biophys. Acta 320: 311–317.

Ji, M., M. Lahaye & W. Yaphe, 1988. Structural studies on agar fractions extracted sequentially from Chinese red seaweeds: *Gracilaria sjeostedtii*, *G. textorii* and *G. salicornia* using ^{13}C-NMR and IR spectroscopy. Chin. J. Oceanol. Limnol. 6: 87–103.

Knutsen, S. H. & H. Grasdalen, 1987. Characterisation of water-soluble polysaccharides from norwegian *Furcellaria lumbricalis* (Huds.) Lamour. (Gigartinales, Rhodophyceae) by IR and NMR spectroscopy. Bot mar. 30: 497–505.

Knutsen, S. H., E. Murano, M. D'Amato, R. Toffanin, R. Rizzo & S. Paoletti, 1995. Modified procedure for extraction and analysis of carrageenan applied to the red alga *Hypnea musciformis*. J. appl. Phycol. 7: 565–576.

Lahaye, M. & W. Yaphe, 1988. Effects of seasons on the chemical structure and gel strength of *Gracilaria pseudoverrucosa* agar (Gracilariaceae, Rhodophyta). Carbohydr. Polymers 8: 285–301.

Lahaye, M., C. Rochas & W. Yaphe, 1985. ^{13}C NMR analysis of sulphated and desulphated agar type polysaccharides. Carbohydr. Res. 143: 240–245.

Lahaye, M., C. Rochas & W. Yaphe, 1986. A new procedure for determining the heterogeneity of agar polymers in the cell walls of *Gracilaria* spp. (Gracilariaceae, Rhodophyta). Can. J. Bot. 64: 579–585.

Lahaye, M., J. F. Revol, C. Rochas, J. McLachlan & W. Yaphe, 1988. The chemical structure of *Gracilaria crassissima* (P. et H. Crouan in Schramm et Mazé) P. et H. Crouan in Schramm et Mazé and *G. tikvahiae* McLachlan (Gigartinales, Rhodophyta) cell-wall polysaccharides. Bot. mar. 31: 491–501.

Lewis, R. & D. Hanisak, 1996. Effects of phosphate and nitrate supply on productivity, agar content and properties of *Gracilaria* strain G-16S. J. appl. Phycol. 8:41–49.

Macler, B.A., 1986. Regulation of carbon flow by nitrogen and light in the red alga *Gelidium coulteri*. Plant Physiol. 82: 136–141.

Macler, B. A. & J. A. West, 1987. Life history and physiology of the red alga *Gelidium coulteri*, in unialgal culture. Aquaculture 61: 281–293.

Macler, B. A. & J. R. Zupan, 1991. Physiological basis for the cultivation of the Gelidiaceae. Hydrobiologia 221: 83–90.

Martinsen, A., Skjåk-Bræk, G., Smisdrød O., Zanetti, F. & S. Paoletti, 1991. Comparison of different methods for determination of molecular weight and molecular weight distribution of alginates. Carbohydr. Polymers 15: 171–193.

Mouradi-Givernaud, A., T. Givernaud, H. Morvan & J. Cosson, 1992. Agar from *Gelidium latifolium* (Gelidiales, Rhodophyta), biochemical composition and seasonal variations. Bot. mar. 35: 153–159.

Murano, E., 1995a. Agar from *Gracilaria* species. Ph.D. Thesis, University of Portsmouth, Portsmouth, England.

Murano, E., 1995b. Chemical structure and quality of agars from *Gracilaria*. J. appl. Phycol. 7: 245–254.

Murano, E., R. Toffanin, F. Zanetti, S. H. Knutsen, S. Paoletti & R. Rizzo, 1992. Chemical and macromolecular characterisation of agars polymers from *Gracilaria dura* (C. Agardh) J. Agardh (Gracilariaceae, Rhodophyta). Carbohydr. Polymers 18: 171–178.

Nicolaisen, F. M., I. Meyland & K. Schaumburg, 1980. ^{13}C NMR spectra at 69.9 Mhz of agarose solutions and partly 6-*O*-methylated agarose at 95 °C. Acta Chem. Scand. Ser. B 34: 103–107.

Oza, R.M., 1978. Studies on Indian *Gracilaria*. IV. Seasonal variations in agar and gel strength of *Gracilaria corticata* J. Ag. occuring on the coast of Veraval. Bot. mar. 21: 165–167.

Patwary, U. M. & J. P. van der Meer, 1997. Construction of back-crossed *Gelidium* male-sterile and male-fertile lines and their growth comparison. J. appl. Phycol. 8: 483–486.

Preiss, J. & C. Levi, 1980. Starch biosynthesis and degradation. In Preiss, J. (ed.), The Biochemistry of Plants, Vol. 3, Academic Press, San Diego: 371–423.

Rees, D. A., 1969. Structure, conformation, mechanisms in the formation of polysaccharides and networks. Adv. Carbohydr. Chem. Biochem. 24: 267–332.

Rotem, A., N. Roth-Bejerano & S. M. Arad, 1986. Effect of controlled environmental conditions on starch and agar content of *Gracilaria* sp. (Rhodophyceae). J. Phycol. 22: 117–121.

Santelices, B., 1978. The morphological variation of *Pterocladia caerulescens* (Gelidiales, Rhodophyta) in Hawaii. Phycologia 17: 53–60.

Santelices, B., 1988. Synopsis of biological data on the seaweed genera *Gelidium* and *Pterocladia* (Rhodophyta). FAO Fisheries Synopsis 145: 1–55.

Santelices, B., 1991. Production ecology of *Gelidium*. Hydrobiologia 221: 31–44.

Santelices, B. & M. S. Doty, 1989. A review of *Gracilaria* farming. Aquaculture 78: 95–133.

Seoane-Camba, J., 1964. L'effect de l'intensité lumineuse et de la température sur la concentration de la chlorophylle dans quelques algues marines bentiques. C. r. hebd. Séanc. Acad. Sci. Paris 259: 1432–1435.

Seoane-Camba, J., 1965. Estudios sobre las algas bentonicas en la costa sur de la Peninsula Iberica. Invest. Pesq. Barc. 29: 3–216.

Silvestri, L. J., R. E. Hurst, L. Simpson & J. M. Settin, 1982. Analysis of sulphate in complex carbohydrates. Anal. Biochem. 123: 303–309.

Sousa-Pinto, I., 1994. Ecophysiology and growth of *Gelidium robustum* in culture. PhD Dissertation, University of California, Santa Barbara CA (USA).

Sousa-Pinto, I., R. Lewis & M. Polne-Fuller, 1996. The effects of phosphate concentration on growth and agar content of *Gelidium robustum* (Gelideaceae, Rhodophyta) in culture. Hydrobiologia 326/327: 437–443.

Torres, M., F. X. Niell & P. Algarra, 1991. Photosynthesis of *Gelidium sesquipedale*: effects of temperature and light on pigment concentration, C/N ratio and cell-wall polysaccharides. Hydrobiologia 221: 77–82.

Usov, A.I., E. G. Ivanova & A. S. Shashkov, 1983. Polysaccharides of algae. XXXIII. Isolation and ^{13}C-NMR spectral study of some new gel-forming polysaccharides from Japan sea red seaweeds. Bot. mar. 26: 285–294.

Usov, A. I., S. V. Yarotsky & A. S. Shashkov, 1980. ^{13}C-NMR spectroscopy of red algal galactans. Biopolymers 19: 977–990.

Watase, M. & K. Nishinari, 1983. Rheological properties of agarose gels with different molecular weights. Rheol. Acta 22: 580–587.

Welti, D., 1977. Carrageenans. Part 12. The 300 MHz proton magnetic resonance spectra of methyl -D-galactopyranoside, methyl 3,6-anhydro-D-galactopyranoside, agarose, kappa-carrageenan and segments of iota-carrageenan and agarose sulphate. J. Chem. Res. (S): 312–313.

Whyte, J. N. C., J. R. Englar, R. G. Saunders & J. C. Lindsay, 1981. Seasonal variations in the biomass, quantity and quality of agar, from the reproductive and vegetative stages of *Gracilaria* (*verrucosa* type). Bot. mar. 24: 493–501.

Yu, S., 1992. Enzyme of floridean starch and floridoside degradation in red algae. Ph.D. Thesis, Uppsala University, Uppsala, Sweden.

Effects of environmental factors and plant growth regulators on growth of the red alga *Gracilaria vermiculophylla* from Shikoku Island, Japan

Nair S. Yokoya[1], Hirotaka Kakita*, Hideki Obika & Takao Kitamura
Marine Resources Department, Shikoku National Industrial Research Institute, Hayashi, Takamatsu, Kagawa 761-03 Japan
Fax: [+81]-878-693553. E-mail: kakita@sniri.go.jp
[1]*Present address: Instituto de Botânica, Secretaria de Estado do Meio Ambiente, C. Postal 4005, 01061-970 São Paulo, Brazil*

Key words: Gracilaria, growth, irradiance, plant growth regulators, salinity, temperature

Abstract

Growth and tolerance of *Gracilaria vermiculophylla* (Ohmi) Papenfuss from Shikoku Island were investigated under a variation of temperature (5–30 °C), salinity (5–60‰), and photon irradiance (20–100 μmol photons m^{-2} s^{-1}) in unialgal culture. *G. vermiculophylla* showed wide tolerances for all factors tested, characterizing a euryhaline and eurythermal species. Two clones, one of a tetrasporophyte and the other of a female gametophyte, showed different growth rates, attributable to the difference either in phase or in genotype. The optimum temperature for the growth of the tetrasporophyte clone was 15–25 °C while that of the gametophyte clone was 20–30 °C. Maximum growth of both phases was observed at 80–100 μmol m^{-2} s^{-1}. *G. vermiculophylla* presented higher growth rates in low salinities (15–30‰). Tissue cultures were established in solid ASP 12-NTA medium supplemented with plant growth regulators (PGR), 0.5% agar, 1.0% sucrose and 0.5% inositol. Effects of two auxins (indole-3-acetic acid (IAA), and 2,4-dichlorophenoxyacetic acid (2,4-D)), and one cytokinin (6-benzylaminopurine (BA)) were tested in concentrations ranging from 0.1 to 10.0 mg l^{-1}. Growth of apical segments was significantly stimulated by the majority of treatments supplemented with PGR, while maximum growth of calluses was observed in treatments with low concentration of auxins or BA (1.0 mg l^{-1}). All treatments supplemented with PGR significantly promoted the growth of intercalary segments, except for IAA (1.0 mg l^{-1}) in combination with BA (1.0 mg l^{-1}). Growth of calluses originating from intercalary segments was observed in treatments with IAA (0.1 mg l^{-1}), 2,4-D (10.0 mg l^{-1}) or IAA (1.0 mg l^{-1}) in combination with BA (0.1 mg l^{-1}). Treatments with high concentration of IAA and BA (10.0 mg l^{-1}) were lethal for apical and intercalary segments. These results show that auxin and cytokinin play a regulatory role on the growth of *G. vermiculophylla* in tissue culture. Furthermore, results on the effects of temperature, salinity and irradiance indicate that *G. vermiculophylla* could be cultivated in brackish temperate environments with potential for economic purposes and for pollution management.

Introduction

The genus *Gracilaria* Greville is widely distributed along Japanese coast, and the occurrence of 19 species has been reported (Yamamoto, 1984). Species of *Gracilaria* have been harvested in Japan as a commercial source of agar and for food additives (Tokuda et al., 1986). Thus, information about factors controlling growth and development of Japanese species of *Gracilaria* is becoming essential to develop cultivation methods.

Environmental factors including temperature, salinity and light play an important role in the growth, reproduction and distribution of marine macroalgae (Gessner, 1970; Gessner & Schramm, 1971; Lüning, 1981; Lobban & Harrison, 1994). Temperature requirements for survival and growth in *Gracilaria* species have been extensively studied (Bird et al., 1979; Santelices & Fonck, 1979; McLachlan & Bird, 1984; Laing et al., 1989; Yokoya & Oliveira, 1992a).

---
* Author for correspondence

Tolerance and growth of *Gracilaria* species under salinity variations have been investigated by some authors (Munõz et al., 1984; Bird & McLachlan, 1986; Yokoya & Oliveira, 1992b). There are reports describing the effect of photon irradiance on the growth of *Gracilaria* spp., with some species requiring less than 100 μmol m^{-2} s^{-1} for optimal growth (Bird et al., 1979; Beer & Levy, 1983), while others require higher irradiances (Lapointe, 1981; Lapointe et al.,1984). However, few data on the effects of such environmental factors on the growth of Japanese species of *Gracilaria* in controlled conditions are available. Exceptions are two studies on the effect of temperature on growth rates of five species of *Gracilaria* from Tosa Bay, southern Japan (Orosco & Ohno, 1992), and on growth and agar quality of *Gracilaria* sp. (chorda-type) (Chirapart et al., 1994).

Tissue culture techniques have been used in many species of seaweed aquacultural crop by providing a means of propagating desired strains (Aguirre-Lipperheide et al., 1995). Growth, callus induction and regeneration on tissue culture of different species of red algae have been controlled by plant growth regulators (Bradley & Cheney, 1990; Dawes & Koch, 1991; Liu & Kloareg, 1991, Dawes et al., 1993; Kaczyna & Megnet, 1993; Yokoya & Handro, 1996, 1997; Huang & Fujita, 1997). Such information is necessary for successful DNA transformation in multicellular algae (Stevens & Purton, 1997).

The aim of this work was to characterize physiological responses of *Gracilaria vermiculophylla* to temperature, salinity and irradiance variations, assessing tolerance and optimal values for growth in unialgal cultures. Effects of plant growth regulators on callus formation and growth in tissue culture were also investigated in order to evaluate the potential of this species for further biotechnological applications.

Material and methods

Fertile plants of *Gracilaria vermiculophylla* (Ohmi) Papenfuss were collected in Takamatsu Bay, northern region of Kagawa Prefecture, Shikoku Island, Japan, in July 1996. The habitat from which plants were collected presented a wide range of temperature variation, from 4.8 to 29.5 °C, while salinity varied from 29.1 to 33.7‰, as shown in Table 1.

Specimens of *G. vermiculophylla* are deposited in the Culture Collection of Shikoku National Industrial Research Institute, Japan, and culture collection numbers are SNIGVT 3102 and SNIGVF 3203 for the tetrasporophyte and female gametophyte, respectively.

Unialgal cultures

Unialgal cultures of tetrasporophytes and female gametophytes were originated from spores liberated by fertile plants collected in nature. Only one plant of each life-cycle was selected from these unialgal cultures, and provided apical segments (5–10 mm) to start clonal cultures. Clonal cultures of the tetrasporophyte and female gametophyte were cultivated in 200 ml conical flasks, with approximately 150 ml Von Stosch's enriched seawater (VSES) at full strength (Edwards, 1970), with modifications (reduction of 50% in the concentrations of thiamine HCl, biotine and cyanocobalamine, as proposed by Yokoya (1996)). Cultures were incubated in a growth chamber (Itoh Seisakusho, Ltd., model AGP-150RL) with agitation (50 rpm), at a temperature of 20 ± 0.5 °C, 14:10 h light:dark cycle, photon irradiance of 40–50 μmol m^{-2} s^{-1} provided by cool-white fluorescent lamps, and salinity of 30–32‰. Medium renewal was carried out weekly during the first month, and at intervals of 10–14 days in subsequent months. These actively growing clonal cultures have provided inocula for subsequent experiments.

Temperature, salinity and photon irradiance

Variations of temperature (from 5 to 30 °C), salinity (from 5 to 60‰), and photon irradiances (from 20 to 100 μmol m^{-2} s^{-1}) were tested, while other experimental conditions were the same as described above for unialgal cultures.

In order to produce different salinities, seawater (30–32‰) was concentrated by freezing. Gradual melting produced seawater with different salinities, and mixtures among them provided all the desired salinities. The final salinity was checked with a salinometer (Shibuya, model S-10), and seawater was sterilized by filtering (0.20 μm, Advantec). Distilled water was not used to avoid changes on the chemical composition of the seawater.

Photon irradiances were measured with LI-250 quantum photometer equipped with LI-193SA spherical quantum sensor (LI-COR, Inc.).

Temperature and light experiments were incubated in growth chambers (Itoh Seisakusho, Ltd.), and cultures were aerated during 3 h, alternated with 3 h without aeration, while the salinity experiment was incubated in growth chamber with agitation (50 rpm).

Table 1. Temperature and salinity variations in seawater from 1.0 m of depth at collecting area, Shikoku Island, southern Japan. Monthly mean, and the minimum and the maximum values of temperature and salinity recorded from 1991 to 1997 are presented. Temperature and salinity data were collected daily and monthly, respectively

Months	Temperature (°C)			Salinity (‰)		
	Mean ± SD	Minimum	Maximum	Mean ± SD	Minimum	Maximum
January	10.1 ± 0.6	7.2	12.5	32.6 ± 0.7	31.7	33.5
February	8.3 ± 0.6	4.8	9.7	32.8 ± 0.6	32.0	33.7
March	9.2 ± 0.5	7.7	11.8	33.0 ± 0.6	32.0	33.8
April	12.0 ± 0.6	9.2	14.8	32.3 ± 0.5	31.0	32.9
May	15.8 ± 0.7	13.1	18.4	32.4 ± 0.7	31.5	33.2
June	19.2 ± 0.6	16.8	22.0	32.2 ± 0.6	31.4	33.0
July	22.6 ± 0.9	20.0	26.0	31.6 ± 0.8	30.2	32.8
August	25.8 ± 1.3	23.0	29.5	31.2 ± 0.9	29.6	32.2
September	26.1 ± 1.0	24.0	29.0	31.3 ± 1.2	29.1	32.5
October	22.8 ± 0.7	19.8	25.9	31.4 ± 1.0	29.4	32.4
November	18.5 ± 0.8	15.2	21.9	31.7 ± 1.0	30.0	32.7
December	13.5 ± 0.7	9.9	17.7	31.9 ± 1.0	30.1	32.9

SD = standard deviation.

For each treatment, three replicates of six apical segments (5 mm long) were inoculated into 200 ml-conical flasks containing 150 ml of VSES at full-strength.

Fresh weight was measured weekly or fortnightly, at intervals coinciding with the renewal of culture media. The growth rate (k) was calculated as $k = \log_2(W_f.W_0^{-1}).t^{-1}$, where W_0 is the initial fresh weight, W_f is the fresh weight after t days, and t is the number of days (Brinkhuis, 1985). This rate (k) is approximately 1.44 times the rate calculated from natural logarithms (base e). After completing the experiments, plants were dried in an oven at 60 °C for 36–48 h to determine the percentage of water content.

Data were analyzed by single factorial analysis of variance (ANOVA), and Tukey's multiple comparison test was conducted to distinguish significantly different results following the ANOVA test (Winer, 1971).

Tissue culture and plant growth regulators (PGR)

Effects of indole-3-acetic acid (IAA), 2,4-dichlorophenoxyacetic acid (2,4-D) and 6-benzylaminopurine (BA) in concentrations ranging from 0.1 to 10.0 mg l^{-1} were tested alone or in combination. PGR (Sigma) were added in filter-sterilized (0.20 μm) ASP 12-NTA synthetic medium (Iwasaki (1961), modified with the addition of 100 μg of thiamine HCl).

The protocol to obtain axenic explants was determined based on preliminary tests. Apical and intercalary segments, each 10 mm long, were cut from female gametophytes growing in clonal cultures, and incubated for one week in VSES medium supplemented with Provasoli's antibiotic solution (Sigma, with 60 000 units penicillin G, 250 μg chloramphenicol, 1500 units polymyxin B and 300 μg neomycin per liter) and 100 μg L^{-1} nystatin (Sigma). After this period, these segments were cut (5 mm long), weighed, and washed for 20 seconds in a solution of sterile seawater (filter-sterilized and autoclaved for one hour at 121 °C) with 0.5% sodium hypochlorite and 200 μl of detergent, then washed in sterile seawater to remove this solution. These explants were inoculated into 25 ml ASP 12-NTA solid medium supplemented with PGR, 1.0% sucrose, 0.5% inositol and 0.5% agar (these compounds from Wako Pure Chemical Industries, Ltd, Japan). The medium was sterilized by autoclaving for 15 min at 121 °C.

Experiments were conducted at a temperature of 20 ± 0.5 °C, 14:10 h light:dark cycle, photon irradiance of 40–50 μmol m^{-2} s^{-1}, provided by cool-white fluorescent lamps, and salinity of 30‰ treatment. Each treatment was tested with three replicates of six explants each. Observations were made weekly, and explant growth was based on fresh weight variation after six weeks. Callus growth was assessed as the length from the cut surface of the explant measured under a stereomicroscope (model SZ-PT, Olympus, Japan) with a colour camera (model CS 220, Olympus, Japan) connected to a video micrometer (model VM-30, Olympus, Japan).

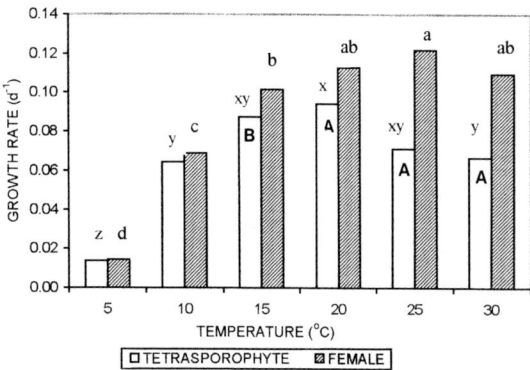

Figure 1. Growth rates for tetrasporophytes and female gametophytes of *G. vermiculophylla* cultured in different temperatures for four weeks. Each data point is the mean of three replicates. A and B indicate fertile plants and liberation of tetraspores after incubation period of 14 and 21 days, respectively. For each phase, values of bars marked by the same letter are not significantly different according to the Tukey's multiple comparison test ($p = 0.05$).

Data were analyzed by single factorial analysis of variance (ANOVA), and Tukey's multiple comparison test was conducted to distinguish significantly different results following the ANOVA test (Winer, 1971).

Anatomical observations by light microscopy were made on explants preserved in 4% formaldehyde-seawater; sections were cut in freezing microtome (model MCE-802A, Komatsu Electronics Inc.) and stained in 1.0% aqueous aniline blue acidified by the addition of 1.0 N HCl.

Results

Effects of temperature, salinity and photon irradiance

G. vermiculophylla grew in all treatments tested, showing a wide range of tolerance to temperature, salinity and photon irradiance variations. Temperature had a significant effect on growth rates of the tetrasporophyte clone ($F = 27.30$, $p < 0.001$) and the female gametophyte clone ($F = 75.48$, $p < 0.001$). Growth rates of the tetrasporophyte clone ranging from 0.014 (5 °C) to 0.094 (20 °C), and higher growth rates were observed in the female gametophyte clone, ranging from 0.014 (5 °C) to 0.122 (25 °C) (Figure 1). The tetrasporophyte clone became fertile and liberated tetraspores in temperatures from 15 to 30 °C.

Maximum growth of both phases was observed in high irradiance (80–100 μmol m^{-2} s^{-1}) and plants showed slow growth in low irradiance (20 μmol m^{-2}

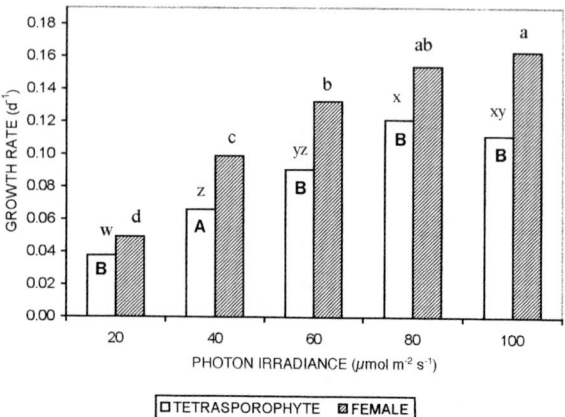

Figure 2. Growth rates for tetrasporophytes and female gametophytes of *G. vermiculophylla* cultured in different photon irradiances for four weeks. Each data point is the mean of three replicates. A and B indicate fertile plants and liberation of tetraspores after incubation period of 14 and 21 days, respectively. Values of bars marked by the same letter are not significantly different according to the Tukey's multiple comparison test ($p = 0.05$).

s^{-1}) (Figure 2). Irradiance had a significant effect on growth rates of the tetrasporophyte ($F = 38.22$, $p < 0.001$), and female gametophyte ($F = 52.88$, $p < 0.001$) clones. Growth rates varied from 0.005 to 0.010 and from 0.002 to 0.017 for the tetrasporophyte and female gametophyte clone, respectively (Figure 2). Formation of tetrasporangia and liberation of tetraspores were observed in the tetrasporophyte clone cultured in all levels of irradiance.

The female gametophyte clone of *G. vermiculophylla* tolerated a wide range of salinities, from 5 to 60‰ (Figure 3). Highest growth rates were observed in salinities ranging from 15 to 30‰ (Figure 3), and growth rates observed in different salinities were significantly different ($F = 171.43$, $p < 0.001$). Percentage of water content in female plants showed an inverse relation with salinity variation (Figure 3).

Effects of plant growth regulators on tissue culture

After three weeks in culture, development of callus from the basal pole of both apical and intercalary segments was observed in all treatments tested. Calluses presented a light-red colour, and consisted of filaments originating from divisions of cells that were produced as a wounding response on sectioned surface of segments (Figure 4A, B). Intercalary segments regenerated adventitious shoots on sectioned surface corresponding to the apical pole of the seg-

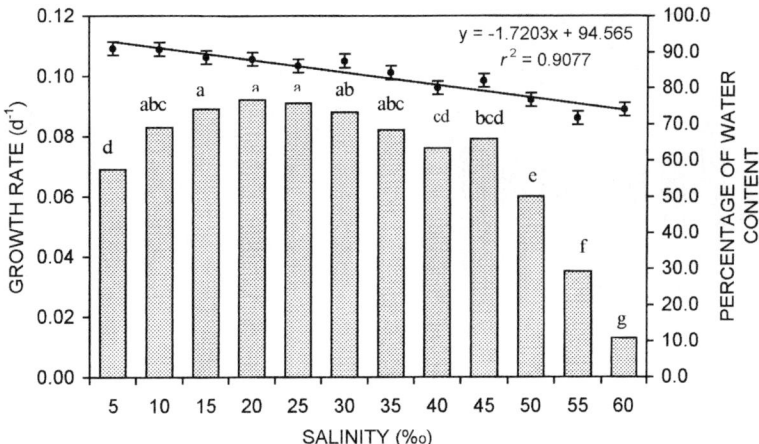

Figure 3. Growth rates and percentage of water content for female gametophytes of *G. vermiculophylla* cultured in different salinities for eight weeks. Each data point is the mean of three replicates. Equation and r^2 value are presented for linear trendline. Values of bars marked by the same letter are not significantly different according to the Tukey's multiple comparison test ($p = 0.05$).

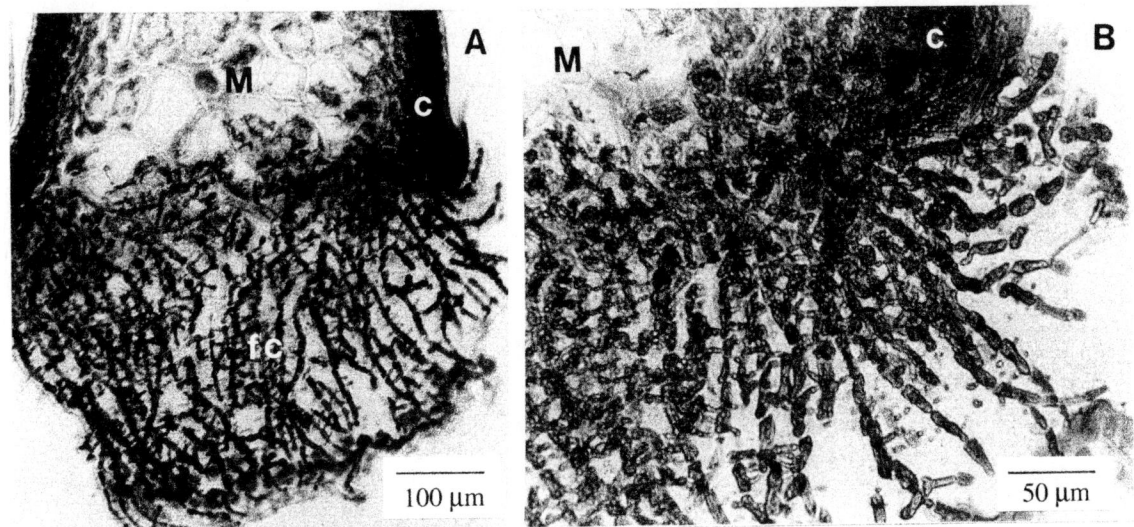

Figure 4. Longitudinal section of the basal portion of an intercalary segment isolated from a female gametophyte of *G. vermiculophylla* cultured at IAA + BA (0.1 + 1.0 mg.L^{-1}) over six weeks. Aspect (A) and detail (B) showing cortical (c) and medullary region (M) of thallus, and filamentous callus (fc).

ment. The adventitious shoots originated from cortical cells located at sub-apical region of the segment.

Treatments with high concentration of IAA and BA (10.0 mg l^{-1}) caused a progressive necrosis on the thallus, and both apical and intercalary segments died after four weeks. When compared with treatment control (without addition of PGR), growth of apical segments was significantly stimulated by the majority of treatments supplemented with PGR ($F = 3.73$, $p < 0.01$), except for the treatment with IAA (1.0 mg l^{-1}) alone or in combination with BA (1.0 mg l^{-1}) (Figure 5A). Significant differences were observed on growth of calluses originating from apical segments subjected to different treatments with PGR ($F = 3.65$, $p < 0.01$). Callus growth was inhibited in treatments with IAA (1.0 mg l^{-1}) alone, or in combination with BA (Figure 5B), while maximum growth was observed in treatments with low concentration of auxins or with BA (1.0 mg l^{-1}). All treatments supplemented with PGR significantly promoted the growth of intercalary segments ($F = 3.55$, $p < 0.01$), except for IAA (1.0 mg l^{-1}) in combination with BA (1.0 mg l^{-1}) (Figure

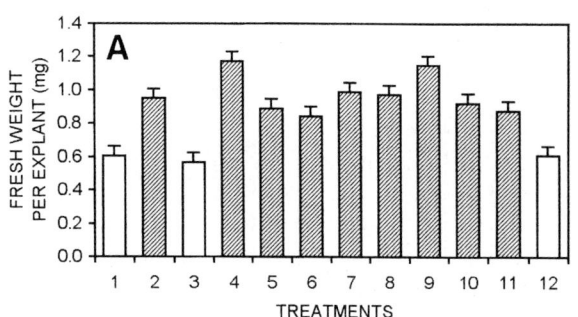
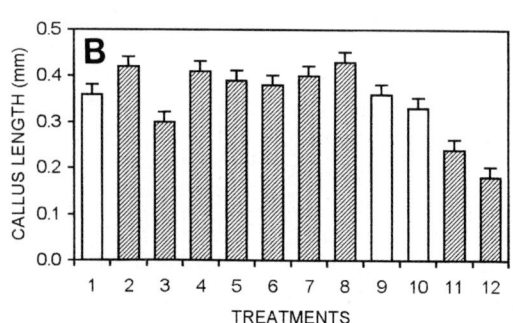
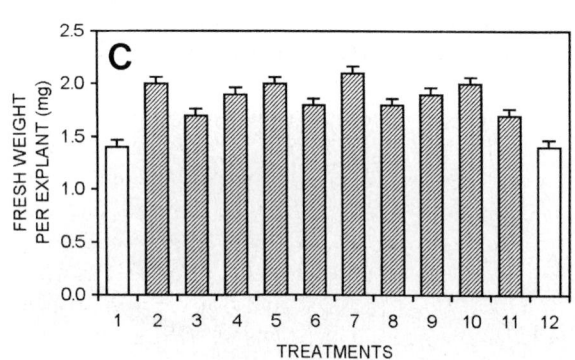
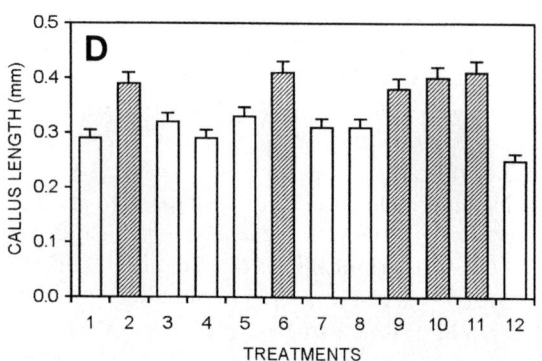

TREATMENTS												
Number	1	2	3	4	5	6	7	8	9	10	11	12
mg l^{-1}	0.0	0.1	1.0	0.1	1.0	10.0	0.1	1.0	0.1+0.1	0.1+1.0	1.0+0.1	1.0+1.0
PGR	-	IAA		2,4-D			BA		IAA+BA			

Figure 5. Effects of plant growth regulators (PGR) on growth of thallus segments (A, C) isolated from female gametophytes of *G. vermiculophylla*, and their calluses (B, D) cultured in ASP 12-NTA solid medium over six weeks. Each data point is the mean of three replicates. Error bars represent standard errors. Striped bars correspond to treatments significantly different from the control (Treatment 1), as assessed by Tukey's multiple comparison test ($p = 0.05$). IAA = indole-3-acetic acid; 2,4-D = 2,4-dichlorophenoxyacetic acid; BA = 6-benzylaminopurine.

5C). Growth of calluses originating from intercalary segments was significantly ($F = 4.77, p < 0.001$) stimulated by few treatments as IAA (0.1 mg l^{-1}), 2,4-D (10.0 mg l^{-1}), and three combinations of IAA and BA; effects of other treatments were not significantly different from control (Figure 5D).

Discussion

Effects of temperature, photon irradiance and salinity variations

G. vermiculophylla tolerated a wide range of temperature variation, from 5 to 30 °C, which correlates with the extreme values of temperature recorded at the collecting site (from 4.8 to 29.5 °C, as shown in Table 1). The eurythermal response of *G. vermiculophylla* is in accordance with observations that *Gracilaria* species from temperate waters tended to be eurythermal (Bird et al., 1979; Santelices & Fonck, 1979; McLachlan & Bird, 1984). Maximum growth of *G. vermiculophylla* was observed at 15–20 °C, and similar results have been observed in *G. lemaneiformis* (Bory) Weber-v.-Bosse (Santelices & Fonck, 1979), *G. tikvahiae* McLachlan (Bird et al., 1979), some species of *Gracilaria* from temperate coasts of the Atlantic and eastern Pacific Oceans (McLachlan & Bird, 1984), *G. chilen-*

sis Bird, McLachlan & Oliveira (Laing et al., 1989 (as *G. sordida* Nelson); Yokoya & Oliveira, 1992a), and *Gracilaria* sp. (chorda-type) from Japan (Chirapart et al., 1994). *G. vermiculophylla* has temperature responses similar to *G. tikvahiae*, growing well in temperatures as high as 30 °C, and tolerating low temperatures without necrosis of the thallus (Bird et al., 1979; McLachlan & Bird, 1984).

Maximum growth of *G. vermiculophylla* occurred at irradiances of 80–100 μmol m^{-2} s^{-1}, and these responses could be related to its intertidal distribution along Japanese coasts. Beer & Levy (1983) observed similar results in *Gracilaria* sp. (textorii-type). These irradiances are higher than those observed for *G. tikvahiae*, which was light-saturated for growth at less than 50 μmol m^{-2} s^{-1}, and became necrotic at about 65 μmol m^{-2} s^{-1} (Bird et al., 1979). However, optimum growth in higher light levels was reported in *Gracilaria foliifera* v. *angustissima* (Harvey) Taylor (Lapointe, 1981) and *G. chilensis* (Laing et al., 1989, as *G. sordida*).

No noticeable differences were found in responses between gametophytes and tetrasporophytes in *Gracilaria* species (McLachlan & Bird, 1984) and other macroalgal species with isomorphic generations (van den Hoek, 1982). However, the *G. vermiculophylla* tetrasporophyte clone appeared to have an optimum temperature range (15–25 °C) for growth which was lower than that of the female gametophyte clone (20–30 °C). This difference could be related to a seasonal variation in growth of different reproductive phases in nature or it could be due to genetic differences between the clones. Further work would be necessary to explain the difference.

A comparison of growth rates recorded in temperature and light experiments showed that the female gametophyte clone presented higher growth rates than the tetrasporophyte clone. This result could reflect the development of tetrasporangia and liberation of tetraspores in plants cultured in temperatures above 15 °C and irradiances from 20 to 100 μmol m^{-2} s^{-1} or, again, to genetic differences between the clones. McLachlan & Bird (1984) observed that reproduction in *Gracilaria* spp. occurs only when plants are growing well, and temperature, nutrients and light have no effects beyond their influence on growth. In the present study, temperature responses of *G. vermiculophylla* confirm this observation, but not light responses, since tetrasporophytes grew more slowly and became fertile at 20 μmol m^{-2} s^{-1}. These results suggest that temperature could be a limiting factor, while irradiance seems not to limit reproduction in *G. vermiculophylla*.

Wide salinity tolerance has been reported in several species of *Gracilaria* (Bird et al., 1979; Bird & McLachlan, 1986; Yokoya & Oliveira, 1992b). Similarly, *G. vermiculophylla* tolerated a wide range of salinity variation, from 5 to 60‰, although a narrow range of salinity variation was recorded at the collecting site (29.1–33.7‰, as shown in Table 1). The salinity response of *G. vermiculophylla* could reflect its intertidal distribution, since these plants are subjected to osmotic stress caused by variable periods of emersion.

The highest growth rates of *G. vermiculophylla* were observed in hypohaline conditions with an optimum at 20‰, a low value that would not be encountered in the species' habitat. These results indicate that salinity responses of *G. vermiculophylla* are genetically determined, which supports previous observation about the absence of ecotypes in *Gracilaria* species (McLachlan & Bird, 1984). Maximum growth of *G. vermiculophylla* occurred at 15–30‰, and similar responses were observed in *Gracilaria* sp. growing in estuaries in south-central Chile (Bird & McLachlan, 1986), and in *G. tenuistipitata* Zhang *et* Xia growing in ponds in southern China (Haglund & Pedersén, 1992).

Our results on the effects of temperature, photon irradiance and salinity variations on growth of *G. vermiculophylla* could be utilized as an indicator of tolerance and optimal conditions for cultivation. However, the limitation of these experiments should be considered, since they were performed in a defined set of standard culture conditions with alteration of one environmental variable at a time.

Effects of plant growth regulators on tissue culture

Apical and intercalary segments of *G. vermiculophylla* developed calluses on the sectioned surface as a wounding response and PGR regulated their growth. Similarly, wounding processes are reported to have induced calluses in other seaweed species (Polne-Fuller & Gibor, 1987; Yokoya et al., 1993), and PGR enhanced callus formation and growth in different species of red algae (Chen & Taylor, 1978; Gusev et al., 1987; Liu & Kloareg, 1991, Kaczyna & Megnet, 1993; Yokoya & Handro, 1996; Huang & Fujita, 1997).

Our results showed that PGR had a stimulatory effect on growth of thallus segments and calluses. These results could be related to regulatory roles of auxin and cytokinin on cell division and enlargement in higher plants (Evans, 1984) as well as in red algae (Bradley & Cheney, 1990; Dawes & Koch, 1991; Yokoya & Handro, 1996, 1997). However, Huang & Fujita (1997) reported no effects of IAA or BA on callus formation and growth in *Gracilaria textorii* Hariot, although they obtained different results in other red algal species.

The synthetic auxin (2,4-D) was more effective than IAA to stimulate growth of intercalary segments and calluses in *G. vermiculophylla*. This is also true for *Gracilariopsis tenuifrons* (Bird *et* Oliveira) Fredericq *et* Hommersand (Yokoya, 1996). IAA is degraded by plant cells faster than other synthetic auxins (Yeoman & Forche, 1980), and light could induce the photoinactivation of IAA. However, high concentration of IAA (10.0 mg l^{-1}) was lethal for *Gracilaria vermiculophylla* and a concentration of 1.0 mg l^{-1} inhibited the growth of apical segments and calluses. Similar results were observed in *Caulerpa* spp. (Jacobs, 1970; Dawes, 1971), and in *Gracilariopsis tenuifrons* and *Solieria filiformis* (Kützing) Gabrielson (Yokoya, 1996).

Apical and intercalary segments of *G. vermiculophylla* presented some different responses to added PGR. Intercalary segments were not inhibited by IAA (1.0 mg l^{-1}), and generally were more responsive to PGR than apical segments. These differences might be related to biochemical and/or physiological gradients along the thallus caused by the presence of apical cells. The origin of calluses (apical or intercalary segments) of *G. vermiculophylla* cultured in liquid medium with PGR has also influenced on the potential for plant regeneration (unpublished data).

In conclusion, our results on the effects of PGR in tissue culture of *G. vermiculophylla* support previous reports (Bradley & Cheney, 1990; Dawes & Koch, 1991; Kaczyna & Megnet, 1993; Yokoya & Handro, 1996, 1997; Huang & Fujita, 1997) that auxins and cytokinins can regulate growth in red algae. Formation and growth of calluses in tissue culture indicate that *G. vermiculophylla* could be a potential species for biotechnological applications. Furthermore, our results show that *G. vermiculophylla* from Shikoku Island is tolerant to wide variations on temperature and salinity, and could be cultivated in temperate brackish regions with potential for economic purposes and pollution management.

Acknowledgements

The authors thank Mr Yukitoshi Abe from Kagawa Prefectural Fisheries Experimental Station for providing data of salinity and temperature variations at the collecting area. N.S.Y. is grateful to the Science and Technology Agency of Japan for awarding a STA fellowship.

References

Aguirre-Lipperheide, M., F. J. Estrada-Rodrigues & L.V. Evans, 1995. Facts, problems, and needs in seaweed tissue culture: an appraisal. J. Phycol. 31: 677–688.

Beer, S. & I. Levy, 1983. Effects of photon fluence rate and light spectrum composition on growth, photosynthesis and pigment relations in *Gracilaria* sp. J. Phycol. 19: 516–522.

Bird, C. J. & J. McLachlan, 1986. The effect of salinity on distribution of species of *Gracilaria* Grev. (Rhodophyta, Gigartinales): an experimental assessment. Bot. mar. 29: 231–238.

Bird, N. L., L. C. M. Chen & J. McLachlan, 1979. Effects of temperature, light and salinity on growth in culture of *Chondrus crispus*, *Furcellaria lumbricalis*, *Gracilaria tikvahiae* (Gigartinales, Rhodophyta) and *Fucus serratus* (Fucales, Phaeophyta). Bot. mar. 22: 521–527.

Bradley, P. M. & D. P. Cheney, 1990. Some effects of plant growth regulators on tissue culture of the marine red alga *Agardhiella subulata* (Gigartinales, Rhodophyta). Hydrobiologia 204/205: 353–360.

Brinkhuis, B. H., 1985. Growth patterns and rates. In Littler, M. M. & D. S. Littler (eds), Handbook of Phycological Methods – Ecological Field Methods: Macroalgae. Cambridge University Press, Cambridge, 461–477.

Chen, L. C. M. & A. R. A. Taylor, 1978. Medullary tissue culture of the red alga *Chondrus crispus*. Can. J. Bot. 56: 883–886.

Chirapart, A., M. Ohno, M. Sawamura & H. Kusunose, 1994. Effect of temperature on growth rate and agar quality of a new member of Japanese *Gracilaria* in Tosa Bay, southern Japan. Jap. J. Phycol. (Sôrui) 42: 325–329.

Dawes, C. J., 1971. Indole-acetic acid in the green algal coenocyte *Caulerpa prolifera* (Chlorophyceae, Siphonales). Phycologia 10: 375–379.

Dawes, C. J. & E. W. Koch, 1991. Branch, micropropagule and tissue culture of the red algae *Eucheuma denticulatum* and *Kappaphycus alvarezii* farmed in the Philippines. J. appl. Phycol. 3: 247–257.

Dawes, C. J., G. C. Trono & A. O. Lhuisma, 1993. Clonal propagation of *Eucheuma denticulatum* and *Kappaphycus alvarezii* for Philippine seaweed farms. Hydrobiologia 260/261: 379–384.

Edwards, P., 1970. Illustrated guide to the seaweeds and seagrasses in the vicinity of Porto Aransas Texas. Contr. mar. Sc. Univ. Texas, Austin, 15: 1–228 (suppl.).

Evans, M. L., 1984. Function of hormones at the cellular level of organisation. In: Scott, T. K. (ed.), Encyclopedia of Plant Physiology, Springer-Verlag, Berlin: 23–80.

Gessner, F., 1970. Temperature. Plants. In: Kinne, O. (ed.), Marine Ecology. Wiley-Interscience, London: 363–406.

Gessner, F. & W. Schramm, 1971. Salinity. Plants. In: Kinne, O. (ed.), Marine Ecology. Wiley-Interscience, London: 705–820.

Gusev, M. V., A. H. Tambiev, N. N. Kikova, N. N. Shelyastina & R. R. Aslanyan, 1987. Callus formation in seven species of agarophyte marine algae. Mar. Biol. 95: 593–597.

Haglund, K. & M. Pedersén, 1992. Growth of the red alga *Gracilaria tenuistipitata* at high pH. Influence of some environmental factors and correlation to an increased carbonic-anhydrase activity. Bot. mar. 35: 579–587.

Huang, W. & Y. Fujita, 1997. Callus induction and thallus regeneration in some species of red algae. Phycol. Res. 45: 105–111.

Iwasaki, H., 1961. The life cycle of *Porphyra tenera* in vitro. Biol. Bull. 121: 173–187.

Jacobs, W. P., 1970. Developmental and regeneration of the algal giant coenocyte, *Caulerpa*. In: Fredrik, J. F. & R. M. Klein (eds), Phylogenesis and Morphogenesis in the Algae. Ann. N. Y. Acad. Sci. 175: 732–747.

Kaczyna, F. & R. Megnet, 1993. The effects of glycerol and plant growth regulators on *Gracilaria verrucosa* (Gigartinales, Rhodophyceae). Hydrobiologia 268: 57–64.

Laing, W. A., J. T. Christeller & B. E. Terzaghi, 1989. The effect of temperature, photon flux density and nitrogen on growth of *Gracilaria sordida* Nelson (Rhodophyta). Bot. mar. 32: 439–445.

Lapointe, B. E., 1981. The effects of light and nitrogen on growth, pigment content and biochemical composition of *Gracilaria foliifera* v. *angustissima* (Gigartinales, Rhodophyta). J. Phycol. 17: 90–95.

Lapointe, B. E., K. R. Tenure & C. J. Dawes, 1984. Interactions between light and temperature on the physiological ecology of *Gracilaria tikvahiae* (Gigartinales: Rhodophyta). Mar. Biol. 80: 161–170.

Liu, X. W. & B. Kloareg, 1991. Tissue culture of *Porphyra umbilicalis* (Bangiales, Rhodophyta). I. The effects of plant hormones on callus induction from tissue explants. C. R. Acad. Sci. Paris. Ser. 312: 517–522.

Lobban, C. S. & P. J. Harrison, 1994. Seaweed Ecology and Physiology. Cambridge University Press, Cambridge, 366 pp.

Lüning, K., 1981. Light. In: Lobban, C. S. & M. J. Wynne (eds), The Biology of Seaweeds. Blackwell Scientific Publications, Oxford: 326–355.

McLachlan, J. & C. L. Bird, 1984. Geographical and experimental assessment of the distribution of *Gracilaria* species (Rhodophyta, Gigartinales) in relation to temperature. Helgoländer. Meeresunters. 38: 319–334.

Muñoz, M. A., H. Romo & K. Alveal, 1984. Efecto de la salinidade en el crecimiento de tetrasporofitos juveniles de *Gracilaria verrucosa* (Hudson) Papenfuss (Rhodophyta, Gigartinales). Gayana 41:119–125.

Orosco, C. A. & M. Ohno, 1992. Growth rates of *Gracilaria* species (Gracilariales, Rhodophyta) from Tosa Bay, southern Japan. Jap. J. Phycol. (Sôrui) 40: 239–244.

Polne-Fuller, M. & A. Gibor, 1987. Calluses and callus-like growth in seaweeds: induction and culture. Hydrobiologia 151/152: 131–138.

Santelices, F. & E. Fonck, 1979. Ecologia y cultivo de *Gracilaria lemanaeformis*. In Santelices, B. (ed.), Actas I Symp. Algas Mar. Chilenas. Subsecretaria de Pesca, Ministerio de Economia, fomento y Reconstrucción: 165–200.

Stevens, D. R. & S. Purton, 1997. Genetic engineering of eukaryotic algae: progress and prospects. J. Phycol. 33: 713–722.

Tokuda, H., M. Ohno & H. Ogawa, 1986. The Resources and Cultivation of Seaweeds. Monographs on Aquaculture Science, 354 pp. (in Japanese).

van den Hoek, C., 1982. The distribution of benthic marine algae in relation to the temperature regulation of their life histories. Biol. J. Linn. Soc. 18: 81–144.

Winer, B. J. 1971. Statistical Principles in Experimental Design. 2nd edn. MacGraw-Hill, New York.

Yamamoto, H., 1984. An evaluation of some vegetative features and some interesting problems in Japanese population of *Gracilaria*. Hydrobiologia 116/117: 51–54.

Yeoman, M. M. & E. Forche, 1980. Cell proliferation and growth in callus cultures. In: Vasil I. K. (ed.), Perspectives on Plant Cell and Tissue Culture. Academic Press, New York: 1–24.

Yokoya, N. S., 1996. Controle do crescimento e da morfogênese por auxinas e citocininas em três espécies de rodofíceas: *Gracilariopsis tenuifrons*, *Grateloupia dichotoma* e *Solieria filiformis*. PhD thesis, São Paulo University, São Paulo, 202 pp.

Yokoya, N. S. & W. Handro, 1996. Effects of auxins and cytokinins on tissue culture of *Grateloupia dichotoma* (Gigartinales, Rhodophyta). Hydrobiologia 326/327: 393–400.

Yokoya, N. S. & W. Handro, 1997. Thallus regeneration and growth induced by plant growth regulators and light intensity in *Grateloupia dichotoma* (Rhodophyta). In Kitamura, T. (ed.), Proceedings of I.T.I.T. International Symposium on New Technologies from Marine-Sphere, Takamatsu: 83–86.

Yokoya, N. S. & E. C. Oliveira, 1992a. Temperature responses of economically important red algae and their potential for mariculture in Brazilian waters. J. appl. Phycol. 4: 339–345.

Yokoya, N. S. & E. C. Oliveira, 1992b. Effects of salinity on the growth rate, morphology and water content of some Brazilian red algae of economic importance. Cienc. mar. 18: 49–64.

Yokoya, N. S., S. M. P. B. Guimarães & W. Handro, 1993. Development of callus-like structures and plant regeneration in thallus segments of *Grateloupia filiformis* Kützing (Rhodophyta). Hydrobiologia 260/261: 407–413.

Carbon acquisition strategies of the red alga *Eucheuma denticulatum*

Malena Granbom* & Marianne Pedersén
Department of Botany, Stockholm University, S-106 91 Stockholm, Sweden
Fax: [+46]-8-165525. E-mail: malena.granbom@botan.su.se

Key words: AZ, bicarbonate uptake, carbonic anhydrase, DIDS, *Eucheuma denticulatum*, Rhodophyta

Abstract

Several species of *Eucheuma* (Rhodophyta) are commercially important and are cultivated in many places around the world for their content of carrageenan. In the present study a wild, native strain of *E. denticulatum* from Zanzibar, Tanzania, was investigated for the presence of an external carbonic anhydrase (CA) and a potential membrane-bound HCO_3^- transport protein.

The algae were brought to Sweden and cultivated under laboratory conditions. Photosynthetic activity was measured by observing changes in the pH of the media. The presence of CA and a HCO_3^- transport protein was investigated using the inhibitors acetazolamide (AZ) and 4,4′-diisothiocyanatostilbene-2,2′-disulphonic acid (DIDS), respectively. The results indicate that *Eucheuma denticulatum* has both an active external CA which was inhibited by the addition of AZ and a DIDS sensitive mechanism for anion exchange across the cell membrane. Both inhibitors could be washed away and the algae regained their full photosynthetic capacity. When AZ was washed away, the rate of pH increase was higher than in the control indicating the induction of carbon acquisition systems during the AZ treatment. The results also suggests that external CA dehydrates HCO_3^- to CO_2 below a pH of 8.5–9.0. Above this pH, direct uptake of HCO_3^- is required. The presence of both of these mechanisms for utilization of HCO_3^- are essential for the growth of *Eucheuma* which normally encounters low concentrations of CO_2 in natural sea water. This is the first report on an active DIDS sensitive HCO_3^- transport mechanism in a red alga, which needs no induction.

Abbreviations: AZ – acetazolamide; CA – carbonic anhydrase; DIDS – 4,4′-diisothiocyanatostilbene-2,2′ – disulphonic acid; C_i – inorganic carbon; DBAZ – dextran–bound acetazolamide; EZ – ethoxyzolamide

Introduction

The total concentration of inorganic carbon (C_i) in natural seawater (33‰ S), at pH 8.2 is approx. 2.2 mM and 92% of the total C_i is in the form of HCO_3^- when the seawater is in equilibrium with air. Since the diffusion rate of CO_2 in water is low and the diffusion resistance of HCO_3^- through the cell membrane is high (Gutknecht et al., 1977; Sültemeyer & Rinast, 1996) the low amount of CO_2 that diffuses into the alga may be limiting for photosynthesis. However, utilization of HCO_3^- has been suggested for a large number of macroalgae (e.g. Brechignac & Andre, 1985; Cook et al., 1986; Axelsson & Uusitalo, 1988; Smith & Bidwell, 1989; Björk et al., 1992; Haglund et al., 1992) which includes two possible mechanisms.

One possible mechanism is a periplasmic carbonic anhydrase (CA) converting HCO_3^- to CO_2 which readily diffuses across the cell membrane. This has been suggested for e.g. *Ulva rigida* (Björk et al., 1992), *Gracilaria tenuistipitata* (Haglund et al., 1992), *Ulva lactuca* (Axelsson et al., 1995) and *Enteromorpha intestinalis* (Larsson et al., 1997). CA is a ubiquitous enzyme catalyzing the interconversion of CO_2 and HCO_3^- which is also found within the cell. The external CA can be inhibited by the membrane impermeable sulphonamide, AZ (Moroney et al., 1985).

Another possibility is the presence of a bicarbonate transport system. The exact mechanism of this system is not fully known although several models have been suggested (see Lucas, 1983; Smith, 1988; Beer, 1994 for review). The potent membrane impermeable inhibitor of anion exchange (Beer, 1994), DIDS, previously applied in mammalian systems (Jennings, 1985; Falke & Chan, 1986) was used to identify a bicarbonate transport system in *Ulva* sp. (Drechsler & Beer, 1991), *Ulva lactuca* (Axelsson et al., 1995) and *Enteromorpha intestinalis* (Larsson et al., 1997). The presence of a similar anion/bicarbonate transport system in *Eucheuma denticulatum* following induction by addition of negative ions is described by Mtolera & Pedersén (1995). Bicarbonate transport has also been suggested in many other red algae (e.g. Cook et al., 1986; Cook et al., 1988). In the present study the red alga *E. denticulatum* has been investigated for the presence of both an external CA and direct uptake of HCO_3^- via an anion exchange protein. The effect of the inhibitors AZ and DIDS on the rate of pH increase was studied. The results indicate that *E. denticulatum* has both an active external CA and a direct uptake mechanism for HCO_3^-. The presence of active systems for HCO_3^- use can be important for the growth and survival of *E. denticulatum* in cultivation farms.

Materials and methods

Algal cultivation

Eucheuma denticulatum (N. L. Burman) Collins et Hervey (native strain) was collected at Fumba, Zanzibar, Tanzania and brought to Sweden. Cultures were kept in natural filtered sea water gently bubbled with air, at 22 °C, with 16 h irradiance of 5–10 μmol photons $m^{-2} s^{-1}$ and 8 h of darkness. Sea water was exchanged every second week or if the cultures looked unhealthy. Cultures were maintained without nutrient additions.

pH drift experiment

Algal samples of 2.0 g FW were used in all experiments. Measurements of pH drift were made in a closed system (recorded with a PHM 250 pH-meter, Radiometer Copenhagen), which consisted of an Erlenmeyer-flask containing 65 ml seawater sealed with a rubber stopper, with the electrode tip immersed in the water. All experiments were conducted with natural filtered sea water (34‰ S). The photosynthetic irradiance was 500 μmol photons $m^{-2} s^{-1}$ provided by fluorescent tubes. The temperature was maintained at 24 °C. Circulation was provided by a magnetic stir bar that was separated from the algae by a plastic mesh. Each set of experiments was conducted on three consecutive days on the same algal sample. Each treatment encompassed one single day, starting with the control-measurements on the first day, followed by addition of inhibitor/s on the second day. On the third day the washed algae were re-measured for changes in pH. The algal samples were carefully washed in natural sea water after each daily treatment. During the night, between measurements, the algal samples were kept in natural sea water bubbled with air under normal cultivation conditions. This experimental regime was repeated at least in triplicate for all inhibitors (AZ, DIDS and the combination of AZ and DIDS).

Inhibitors

Stock solution of AZ (Sigma, 50 mM) was prepared in 0.5 M NaOH. The final concentration in the experiment was 200 μM. The stock solution of DIDS (Sigma, 20 mM) was made with distilled water, added to a final concentration of 300 μM. In the experiments where AZ or a combination of AZ and DIDS was added, the start pH was adjusted to 6.1 with HCl. Immediately after preparing the medium, the vessel was sealed to prevent gas exchange with air, thus keeping the total amount of inorganic carbon the same. The medium was allowed to equilibrate for at least 15 minutes prior to the experiment. The experiments with DIDS were started at pH 8.0.

Statistics

In order to compare the data from the different treatments, the measuring times were recalculated to every twentieth minute with the assumption that the rate of pH increase between observations was linear. Recalculated pH values were converted into proton concentration before the calculation of means and were then converted back to pH values. The treatments, inhibitor additions and washed samples, were compared with the control using a paired t-statistic because the same algal sample was used for control and the treatments. A formal test was not used. A t-test on these data would be inappropriate according to Crowder & Hand (1990), due to repeated measurements. However, Crowder & Hand (1990) state that the t-statistics gives a good indication of the difference between the measurements. Difference was marked when the t-statistic was larger than the t-value corresponding to the significance level of 5%.

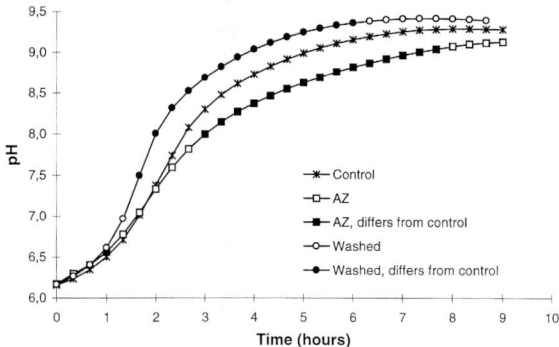

Figure 1. Drift curves of pH for *Eucheuma denticulatum* in a closed system showing mean values for control ($n = 3$), AZ (acetazolamide) addition ($n = 3$) and after washing away the inhibitor ($n = 3$). Final concentration of AZ was 200 μM. The same algal sample was used in a set up consisting of control, AZ-treatment and washed sample. Paired t-statistics were used to compare the AZ-treated and washed samples with the control every twentieth minute. Filled symbols indicate when the t-statistic was larger than the t-value corresponding to the significance level of 5%.

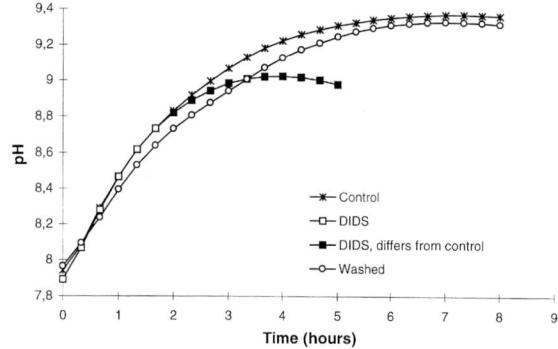

Figure 2. Drift curves of pH for *Eucheuma denticulatum* in a closed system showing mean values for control ($n = 3$), DIDS (4,4′-diisothiocyanatostilbene-2,2′-disulphonic acid) addition ($n = 3$) and after washing away the inhibitor ($n = 3$). Final concentration of DIDS was 300 μM. The same algal sample was used in a set up consisting of control, DIDS-treatment and washed sample. Paired t-statistics were used to compare the DIDS-treated and washed samples with the control every twentieth minute. Filled symbols indicate when the t-statistic was larger than the t-value corresponding to the significance level of 5%.

Table 1. Effect of inhibitors on the pH compensation points in pH drift experiments with *Eucheuma denticulatum* in a closed system. Compensation points for control, inhibitor and washed sample (after treatment with inhibitor) are calculated as the geometric mean of [H$^+$] (min value – max value). Final concentration of inhibitor/s; AZ (acetazolamide) 200 μM; DIDS (4,4′-diisothiocyanatostilbene-2,2′-disulphonic acid) 300 μM and AZ 200 μM + DIDS 300 μM. The same algal sample was used in a set up consisting of control, inhibitor/s and washed sample. The compensation points of addition of inhibitor/s and washed sample, respectively were compared with the control in a paired t-test. Significant difference from the control ($p < 0.05$) is indicated by ∗

Treatment	Control	Inhibitor	Washed
AZ ($n = 3$)	9.30 (9.27–9.34)	9.14 (9.04–9.21)	9.42 (9.38–9.47)
DIDS ($n = 3$)	9.37 (9.33–9.41)	9.02∗ (9.00–9.06)	9.33 (9.25–9.39)
AZ + DIDS ($n = 4$)	9.15 (9.07–9.25)	8.17∗ (8.02–8.36)	9.08 (8.83–9.29)

Results

In our experiments, a pH increase was observed only in the light, with acidification due to respiration occurring in darkness. The rate of pH increase was dependent on the photon irradiance. This is in agreement with results from Cook et al. (1988), Haglund et al. (1992) and Mercado et al. (1997). Spontaneous pH drift in the seawater was negligible (data not shown).

The effect of AZ, DIDS or a combination of the two on the light dependent pH increase is shown in Figures 1, 2 and 3, respectively. The inhibition of external CA by AZ reduced the rate of pH increase in a broad pH interval (Figure 4) but the alga finally reached nearly the same pH compensation point as the control, pH 9.14 and 9.30 (Table 1, Figure 1). The inhibitory effect was observed in the pH interval from 7.0 up to the pH compensation point (Figure 4). Addition of DIDS gave a significantly lower pH compensation point, pH 9.02, than the control value, pH 9.37 (Table 1, Figure 2). Below the pH compensation point a reduction in the rate of pH increase was only found above pH 8.8 (Figure 2). Additions of AZ and DIDS together resulted in a pH compensation point of 8.17 (Table 1). The rate of pH increase was also lowered in the pH interval from 6.5–7.0 up to the pH compensation point (Figure 5).

A stimulation of the rate of pH increase was seen when AZ was washed away from the algal thalli. This was most apparent in the pH interval 7.0 to 8.0 (Figure 4). The inhibition caused by AZ could be washed away completely; AZ does not usually penetrate plasma membranes (Moroney et al., 1985). The result indicates that the AZ-induced reduction in the rate of pH increase is caused by inhibition of external CA; the pH compensation point was reached after only 6 hours (Figure 1).

No stimulation was found after removal of DIDS by washing but the algal thalli regained the same rate of pH increase as the control (Figure 2). When AZ

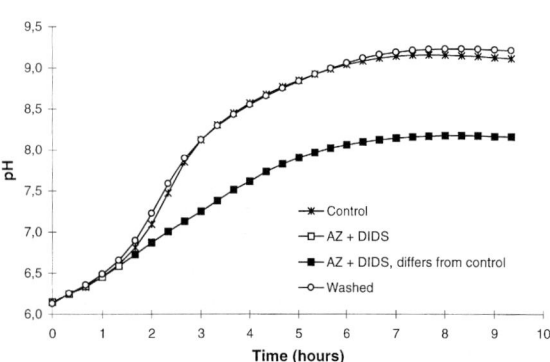

Figure 3. Drift curves of pH for *Eucheuma denticulatum* in a closed system showing mean values for control ($n = 4$), AZ (acetazolamide) and DIDS (4,4'-diisothiocyanatostilbene-2,2'-disulphonic acid) addition ($n = 4$) and after washing away the inhibitors ($n = 4$). Final concentrations of AZ and DIDS were 200 μM and 300 μM, respectively. The same algal sample was used in a set up consisting of control, AZ + DIDS-treatment and washed sample. Paired t-statistics were used to compare the AZ + DIDS-treated and washed samples with the control every twentieth minute. Filled symbols indicate when the t-statistic was larger than the t-value corresponding to the significance level of 5%.

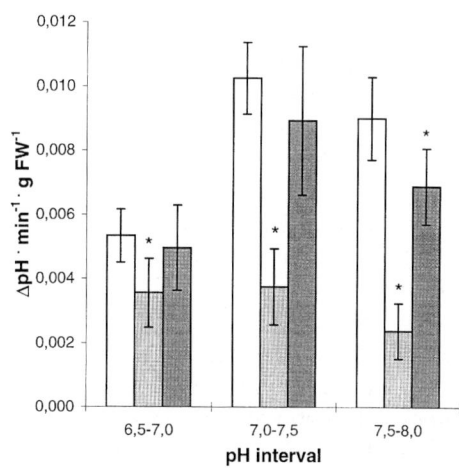

Figure 5. Drift experiments of pH with *Eucheuma denticulatum* in a closed system for control sample, after addition of the inhibitors AZ (acetazolamide, 200 μM) and DIDS (4,4'-diisothiocyanatostilbene-2,2'-disulphonic acid, 300 μM) and after washing away the inhibitors. The pH change (ΔpH) per minute and g alga (FW) is shown for pH intervals of 1.0 pH unit between 6.5 and 9.0 (S.D). The same algal sample was used in a set up consisting of control, AZ + DIDS-treated and washed sample. The AZ + DIDS-treated ($n = 4$) and washed sample ($n = 4$) were compared with the control ($n = 4$) in each pH interval using a paired t-test. Significant difference from the control ($p < 0.05$) is indicated by ∗. White symbol is control sample; grey symbol is AZ and DIDS treated sample and dark grey symbol is washed sample.

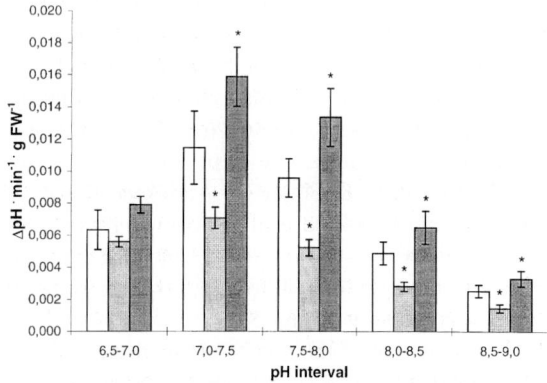

Figure 4. Drift experiments of pH with *Eucheuma denticulatum* in a closed system for control sample, after addition of the inhibitor AZ (acetazolamide, 200 μM) and after washing away the inhibitor. The pH change (ΔpH) per minute and g alga (FW) is shown for pH intervals of 1.0 unit between 6.5 and 9.0 (S.D.). The same algal sample was used in a set up consisting of control, AZ-treated and washed sample. The AZ-treated ($n = 3$) and washed samples ($n = 3$) were compared with the control ($n = 3$) in each pH interval using a paired t-test. Significant difference from the control ($p < 0.05$) is indicated by ∗. White symbol is control sample; grey symbol is AZ treated sample and dark grey symbol is washed sample.

and DIDS were added together and was washed away the algal thalli regained their capacity for bicarbonate utilization but no increase in the affinity for carbon compared to the control values was observed and the pH compensation point of 9.08 was approximately the same as for the control (pH 9.15; Table 1).

Discussion

When algae photosynthesize and take up carbon there is a change in the carbonic species in the carbonic acid buffering system ($CO_2 + H_2O \leftrightarrow H_2CO_3 \leftrightarrow HCO_3^- + H^+ \leftrightarrow CO_3^{2-} + 2H^+$). When CO_2 is removed from the medium some HCO_3^- will form CO_2 and OH^- and some CO_3^{2-} will take up a proton from H_2O forming HCO_3^- and OH^-, in order to maintain equilibrium (Uusitalo, 1996). Due to the release of OH^- ions the pH will increase in the medium. When algae take up HCO_3^- from the medium, H^+ or OH^- must be removed or added respectively, to maintain electroneutrality which also results in an increase in pH (Uusitalo, 1996). However, the carbon uptake has in certain circumstances been shown to occur without an increase in pH (Uusitalo, 1996) and pH increase has occurred without removal of carbon (Thomas & Tregunna, 1968). Both phenomena have been shown in brown algae (Thomas & Tregunna, 1968; Uusitalo, 1996).

Being able to regulate their carbon acquisition could be vital for the algae, for instance up-regulation of either external CA converting HCO_3^- into CO_2 or a protein for direct transport of HCO_3^-, during periods of low CO_2. Since it is difficult for HCO_3^- to penetrate biological membranes (Gutknecht et al., 1977; Sültemeyer & Rinast, 1996) it is possible that CO_2 diffuses across the membrane or that the CO_2 or HCO_3^- is transported directly.

Collén et al. (1992) showed that photosynthesis of a Philippine strain of *Eucheuma denticulatum* collected at Zanzibar, Tanzania was inhibited by the CA inhibitor ethoxyzolamide (EZ) which penetrates the plasma membrane and inhibits both internal and external CA. This result verified that *E. denticulatum* has an internal CA. However, addition of dextran-bound acetazolamide (DBAZ), which cannot penetrate the plasma membrane but inhibits external CA, did not reduce the photosynthetic activity of *E. denticulatum*, suggesting that external CA was not present or was inactive. Mtolera & Pedersén (1995) showed that *E. denticulatum* from Zanzibar, Tanzania (Philippine and native strain) has both an external CA and a DIDS sensitive transport protein for HCO_3^- uptake. These two systems were, however, not active in algae maintained in laboratory culture or brought in directly from the field, verifying the results of Collén et al. (1992). The presence of both systems was revealed, however, by induction either by treatment with negative ions such as NO_3^-, I^- or Br^- (700 µM, 4 mM and 4mM, respectively, in light, 10 minutes, followed by rinsing) or by cultivation at pH 9.0, followed by treatment with NO_3^- (as above).

Our present results show that the native strain of *E. denticulatum* from Zanzibar, Tanzania can have both an active DIDS sensitive ion transport protein and an active external CA without using any of the above mentioned induction procedures.

Inhibition by DIDS was not seen until the pH reached 8.8 (Figure 2) which suggests that the DIDS sensitive transport protein contributes to carbon uptake only at high pH. Addition of AZ reduced the rate of pH increase over a wide pH range providing evidence of an external CA (Figure 4). The addition of both inhibitors substantially reduced the rate of pH increase in the pH interval from 6.5–7.0 up to the compensation point, 8.17 (Figure 5) providing further evidence that *E. denticulatum* needs both systems for carbon acquisition. With inactivation of both external CA and the DIDS sensitive transport protein *E. denticulatum* cannot raise the pH above 8.17 (Table 1) which suggests that the amount of free CO_2 is limiting. The earlier conflicting results on whether or not external CA and/or a transport mechanism are present and active in *E. denticulatum* could depend on the conditions to which the thalli have been acclimated (Axelsson et al., 1995) and also may be due to investigation of different strains. Maberly (1990) also suggests that the ability to use bicarbonate is related to habitat. Both systems of bicarbonate use have also been reported for *Ulva lactuca* (Axelsson et al., 1995) and *Enteromorpha intestinalis* (Larsson et al., 1997). In both the algae the DIDS sensitive mechanism was more active at, or could be induced by, high pH, while external CA was more active at the normal pH of seawater. Mercado et al. (1997) showed that the external CA activity in *Porphyra leucosticta* was regulated by the external C_i concentration and suggested that different ways of HCO_3^- use can be activated or inactivated separately as our results also suggest. Another possible induction mechanism for HCO_3^- uptake is to allow the alga to photosynthesize and raise the pH to high values by itself, as used in these experiments. All this may explain the previous results on the presence of an active CA and an active DIDS-sensitive transport mechanisms in *E. denticulatum*.

Stimulation of the rate of pH increase after treatment with AZ may be caused by the induction of more extracellular CA or any carbon concentrating mechanisms during the AZ treatment. The results show that AZ is membrane impermeable, which is also shown by Moroney et al. (1985). The stimulation of the affinity for carbon is large in the pH interval 7.0 to 8.0 and continues up to the compensation point (Figure 4). According to our results DIDS has no influence on the rate of pH increase below pH 8.8 which means that the direct HCO_3^- uptake via the DIDS sensitive transport protein is inefficient from pH 8.0 (start pH) up to pH 8.8. From pH 9.0 the only carbon assimilation mechanism functioning is the direct uptake of HCO_3^- via the transport protein. This is shown by the complete inhibition of the pH change from pH 9.0 by DIDS. No stimulation was found after washing away the DIDS or AZ and DIDS which indicates that the two carbon assimilation systems work independently of each other. The DIDS-sensitive HCO_3^- uptake mechanism has probably evolved as a response to environmental conditions when the CO_2 concentration adjacent to the algal thalli is so low that the CA-mediated mechanism is not adequate (Larsson et al., 1997).

Our results strongly indicate that *E. denticulatum* can utilize HCO_3^- via a DIDS sensitive transport mechanism operating at pH > 8.8 and via a surface-bound CA from pH 7.0 up to 8.5-9.0. Since both inhibitors can be washed away completely it seems most unlikely that they have entered the cell and interrupted or disturbed cellular processes. Further knowledge on the carbon uptake mechanisms of *E. denticulatum* is important for the farming of this alga. When the mechanisms for inorganic carbon uptake are known, strain selection and breeding of the algae can be further improved.

Acknowledgements

This investigation was supported by Sida/SAREC (Swedish International Development Co-operation Agency/Swedish Agency for Research Co-operation with Developing Countries). Pauli Snoeijs at the Department of Ecological Botany, Uppsala University, Sweden, is thanked for the statistical help.

References

Axelsson, L. & J. Uusitalo, 1988. Carbon acquisition strategies for marine macroalgae. I. Utilization of proton exchanges visualized during photosynthesis in a closed system. Mar. Biol. 97: 295–300.

Axelsson, L., H. Rydberg & S. Beer, 1995. Two modes of bicarbonate utilization in the marine green macroalga *Ulva lactuca*. Plant Cell Environ. 18: 439–445.

Beer, S., 1994. Mechanisms of inorganic carbon acquisition in marine macroalgae (with special reference to the Cholorophyta). Prog. Phycol. Res. 10: 179–207.

Björk, M., K. Haglund, Z. Ramazanov, G. Garcia-Reina & M. Pedersén, 1992. Inorganic-carbon assimilation in the green seaweed *Ulva rigida* C.Ag. (Chlorophyta). Planta 187: 152–156.

Brechignac, F. & M. Andre, 1985. Continuous measurements of the free dissolved CO_2 concentration during photosynthesis of marine plants. Evidence for HCO_3^- use in *Chondrus crispus*. Plant Physiol. 78: 551–554.

Collén, J., M. Pedersén, Z. Ramazanov, M. Mtolera, M. Ngoile & A. Semesi, 1992. Carbon assimilation of *Eucheuma denticulatum*. In: Mshigeni, K. E., J. Bolton, A. Critchley & G. Kiangi (eds), Proceedings of the First International Workshop on Sustainable Seaweed Resource Development in Sub-Saharan Africa. K. E. Mshigeni, Winhoek, Namibia: 77–84.

Cook, C. M., T. Lanaras & B. Colman, 1986. Evidence for bicarbonate transport in species of red and brown macrophytic marine algae. J. exp. Bot. 37: 977–984.

Cook, C. M., T. Lanaras & K. A. Roubelakis-Angelakis, 1988. Bicarbonate transport and alkalization of the medium by four species of Rhodophyta. J. exp. Bot. 39: 1185–1198.

Crowder, M. J. & D. J. Hand, 1990. Analysis of Repeated Measures. Monographs on Statistics and Applied Probability 41. Chapman & Hall, London, 257 pp.

Drechsler, Z. & S. Beer, 1991. Utilization of inorganic carbon by *Ulva lactuca*. Plant Physiol. 97: 1439–1444.

Falke, J. J. & S. I. Chan, 1986. Molecular mechanisms of band 3 inhibitors. 1. Transport site inhibitors. Biochemistry 25: 7888–7894.

Gutknecht, J., M. A. Bisson & F. C. Tosteson, 1977. Diffusion of carbon dioxide through lipid bilayer membranes. Effects of carbonic anhydrase, bicarbonate and unstirred layers. J. gen. Physiol. 69: 779–794.

Haglund, K., Z. Ramazanov, M. Mtolera & M. Pedersén, 1992. Role of external carbonic anhydrase in light-dependent alkalization by *Fucus serratus* L. and *Laminaria saccharina* (L.) Lamour. (Phaeophyta). Planta 188: 1–6.

Jennings, M. L., 1985. Kinetics and mechanism of anion transport in red blood cells. Ann. Rev. Physiol. 47: 519–533.

Larsson, C., L. Axelsson, H. Ryberg & S. Beer, 1997. Photosynthetic carbon utilization by *Enteromorpha intestinalis* (Chlorophyta) from a Swedish rockpool. Eur. J. Phycol. 32: 49–54.

Lucas, W. J., 1983. Photosynthetic assimilation of exogenous HCO_3^- by aquatic plants. Ann. Rev. Plant Physiol. 34: 71–104.

Maberly, S. C., 1990. Exogenous sources of inorganic carbon for photosynthesis by marine macroalgae. J. Phycol. 26: 439–449.

Mercado, J. M., F. X. Niell & F. L. Figueroa, 1997. Regulation of the mechanism for HCO_3^- use by the inorganic carbon level in *Porphyra leucosticta* Thur. in Le Jolis (Rhodophyta). Planta 201: 319–325.

Moroney, J. V., H. D. Husic & N. E. Tolbert, 1985. Effect of carbonic anhydrase inhibitors on inorganic carbon accumulation by *Chlamydomonas reinhardtii*. Plant Physiol. 79: 177–183.

Mtolera, M. S. P. & M. Pedersén, 1995. Induction of a bicarbonate transporter and an extracellular carbonic anhydrase by addition of nitrate, iodide, or bromide ions in the red macroalga *Eucheuma denticulatum* (Rhodophyta). In Mtolera, M. S. P., Photosynthesis, Growth and Light-Induced Stress Responses in the Red Alga *Eucheuma denticulatum*, Doctoral thesis, Institute of Physiological Botany, Acta Universitatis Upsaliensis, Uppsala: I: 1–23.

Smith, R. G., 1988. Inorganic carbon transport in biological systems. Comp. Biochem. Physiol. 90B: 639–654.

Smith, R. G. & R. G. S. Bidwell, 1989. Mechanism of photosynthetic carbon dioxide uptake by the red macroalga, *Chondrus crispus*. Plant Physiol. 89: 93–99.

Sültemeyer, D. & K.-A. Rinast, 1996. The CO_2 permeability of the plasma membrane of *Chlamydomonas reinhardtii*: mass-spectrometric ^{18}O-exchange measurements from $^{13}C^{18}O_2$ in suspensions of carbonic anhydrase-loaded plasma-membrane vesicles. Planta 200: 358–368.

Thomas, E. A. & E. B. Tregunna, 1968. Bicarbonate ion assimilation in photosynthesis by *Sargassum muticum*. Can. J. Bot. 46: 411–415.

Uusitalo, J., 1996. Algal carbon uptake and the difference between alkalinity and high pH ('alkalization'), exemplified with a pH drift experiment. Sci. mar. 60 (Suppl. 1): 129–134.

Relationship of CO$_2$ concentrations to photosynthesis of intertidal macroalgae during emersion

Kunshan Gao[1,2], Yan Ji[1,2] & Yusho Aruga[3]
[1]*Marine Biology Institute, Shantou University, Shantou, Guangdong 515063, China*
[2]*Institute of Hydrobiology, The Chinese Academy of Science, Wuhan, Hubei 430072, China;*
E-mail: ksgao@public.wh.hb.cn
[3]*Tokyo University of Fisheries, Konan-4, Minato-ku, Tokyo 108-8477, Japan*

Key words: CO$_2$, emersion, macroalgae, photosynthesis, seaweeds

Abstract

In order to assess the ecological impacts of the atmospheric CO$_2$ increase on the intertidal macroalgae during emersion, the photosynthesis of *Enteromorpha linza* (a green alga), *Ishige okamurae* (a brown alga) and *Gloiopeltis furcata* (a red alga) was investigated in air as a function of CO$_2$ concentrations and water loss. Their photosynthesis was not saturated at the present atmospheric CO$_2$ level (350 μl l^{-1} or 15.6 μM), the CO$_2$ compensation point and $K_{[mCO_2]}$ increased with increasing desiccation, showing that desiccation lowers the CO$_2$ affinity of the intertidal macroalgae. It was concluded that *E. linza*, *I. okamurae* and *G. furcata*, while exposed to air, can benefit from atmospheric CO$_2$ rise, especially when the algae have lost some water.

Introduction

Marine macroalgae are distributed in the intertidal and subtidal zones of the coastal areas that are most populated, playing an important role in the coastal carbon cycle. Though macroalgae contribute less than 10% of the total marine primary production (Charpy-Roubaud & Sournia, 1990), some of them are more productive than the most productive plants on land and can be successfully cultivated on vast ocean surfaces, showing a great potential for CO$_2$ bioremediation (Gao & McKinley, 1994). They can induce large diurnal changes in the local CO$_2$ partial pressure, and turn the coral reef communities to a CO$_2$ sink, where biogenic calcification is usually a source of CO$_2$ to the atmosphere (Gattuso et al., 1997).

Atmospheric CO$_2$ concentration is increasing mainly due to industrial combustion of fossil fuels, and its subsequent ecological impacts on photosynthesis and growth of plants are of general concern (Bowes, 1993). It has been demonstrated that atmospheric CO$_2$ increase can raise oceanic primary production by phytoplankton (Riebesell et al., 1993; Hein & Sand-Jensen, 1997), though phytoplankton photosynthesis can be CO$_2$-saturated in air-equilibrium seawater (Raven, 1997). In terms of the ecological impacts of CO$_2$ on macroalgae, recent studies showed that CO$_2$ enrichment enhanced the photosynthesis and growth of *Porphyra* and *Gracilaria* plants (Lignell & Pedersen, 1989; Gao et al., 1991, 1993a) and inhibited the calcification of *Corallina pilulifera* by lowering the pH of seawater (Gao et al., 1993b). A few studies investigated the photosynthetic CO$_2$ uptake by the intertidal macroalgae while exposed during emersion (Bidwell & McLachlan, 1985; Johnston & Raven, 1986; Beer & Schragge, 1987; Madsen & Maberly, 1990; Surif & Raven 1990; Raven & Johnston 1991). For the intertidal macroalgae, when the tide is low and they are exposed to air, CO$_2$ is the only exogenous carbon source for their photosynthesis (whereas both HCO$_3^-$ and CO$_2$ are available in seawater); therefore the intertidal macroalgae may be significantly sensitive to the atmospheric CO$_2$ increase. Photosynthesis of the intertidal macroalgae during emersion has been studied previously (Brinkhuis et al., 1976; Quadir et al., 1979; Oates & Murray, 1983; Oates, 1985; 1986; Gao & Aruga, 1987). The emersed photosynthesis was enhanced at an early stage due to extracellular water loss and then reduced, probably due to intracellular dehydration, in *Porphyra* spp. (Johnson et al., 1974; Gao & Aruga, 1987), *Fucus distichus* (Quadir et al., 1979) and *Ascophyllum nodosum* (Brinkhuis et al., 1976); the maximum photosynthesis during emersion was higher than that when submerged. These studies demonstrated that the intertidal macroalgae, even dehydrated to about 15% water content, do photosynthesize in the air when they are exposed during the daytime. Nevertheless, little is known about the relationship of the photosynthesis of macroalgae during emersion to CO_2 concentrations. The present study aimed to find the relationship of the emersed photosynthesis with the CO_2 concentrations, and to assess the ecological impacts of increasing atmospheric CO_2 concentration on the intertidal macroalgae.

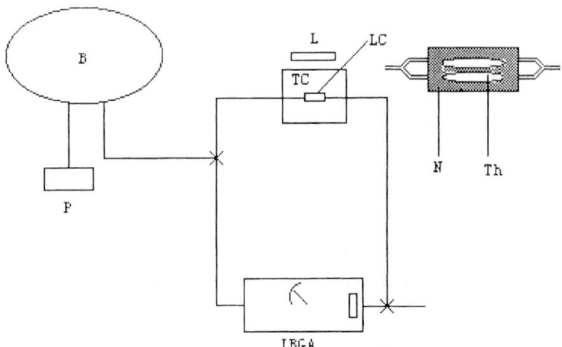

Figure 1. Outline of the system for measuring the photosynthesis. B, air bag; IRGA, infrared gas analyzer; L, fluorescent light source; LC, assimilation chamber; N, polyethylene monofilament net; P, pump; Th, thallus; TC, temperature-controlling cabinet.

Materials and methods

Three species, *Enteromorpha linza* (L.) J. Ag. (a green alga), *Ishige okamurae* Yendo (a brown alga) and *Gloiopeltis furcata* (P. et. R.) J. Ag. (a red alga) were used for the experiments. All the species are distributed in the middle to upper parts of the intertidal zone and were collected from the coast of Ibaraki Prefecture of Japan during the period from May 16 to June 10, 1997. Samples with seawater in plastic bags were transported by cooled icebox in 2–3 h to the laboratory. Wounded or unhealthy individuals were rejected, and only healthy samples were maintained in a water tank (0.1 m^3) with filtered seawater under dim light at 18–20 °C. Half of the seawater was renewed every day. Six to fifteen individuals were used for each measurement. Photosynthesis was measured at 20, 25 and 30 °C and 300 μmol photons m^{-2} s^{-1}. Photosynthesis of *E. linza, G. furcata* and *I. okamurae* in air was saturated at this light level.

The rates of photosynthetic CO_2 uptake by the algae exposed to air were determined by infrared gas analysis. The system for measuring photosynthesis consists of an infrared gas analyzer (IRGA) (Horiba ASSA-1100), a Plexiglas assimilation chamber (16.5 × 8.5 × 3 cm), light source, temperature-controlled cabinet, air bag and connecting tubes (Figure 1). The assimilation chamber with algal thalli in it was maintained in the temperature-controlled cabinet; air of known CO_2 concentration was supplied from the air bag and flushed through the assimilation chamber before the system was closed. The CO_2 concentration in the air bag was obtained by injecting pure CO_2 before pumping outdoor air into it. In the closed system, CO_2 concentration was reduced due to photosynthesis, and the photosynthetic rates changed as a function of changing CO_2 concentrations, which was recorded with time. Net photosynthetic rates were calculated with the following formula:

$$P_n = \frac{(C_j - C_i) \times V \times 60 \times 273}{(t_j - t_i) \times W \times 22.4 \times (273 + T)}, \quad (1)$$

where P_n [μmol CO_2 g(f. wt)$^{-1}$ h^{-1}] is the net photosynthetic rate at CO_2 concentration of $(C_j + C_i)/2$, C_j and C_i are the CO_2 concentrations at time t_j and t_i (min), W is the initial fresh weight (g) of samples used, V_l is the volume of the closed system, V, including the inner part of the IRGA, was determined provided that the CO_2-saturated photosynthetic rate in the closed system is consistent with that in the open system under the same conditions. It was estimated as:

$$V = \frac{(X_o - X_i) \times F \times (t_j - t_i)}{(C_j - C_i)}, \quad (2)$$

where X_i and X_o indicate the CO_2 concentrations of inlet and outlet air through the assimilation chamber at flow rate of F (l min^{-1}), determined in the open system with the same amount of samples as in the closed system; other symbols for the closed system are the same as in Equation (1).

Water loss was estimated as follows: WL (%) = (W_i − W_t)/W_i × 100, where W_i is the initial wet weight that was determined after removing excessive water drops on the algal surface by tapping or shaking the samples; W_t the instantaneous weight determined at certain time intervals during the measurement.

Results

Photosynthesis of *E. linza*, *G. furcata* and *I. okamurae* was not saturated at the present atmospheric CO_2 level (15.6 μM), and tended to be saturated at CO_2 concentrations above 30 μM CO_2 when the algae were wet (Figure 2). However, more CO_2 was required for the photosynthesis to be saturated when the algae were dehydrated: $K_{[mCO_2]}$ increased with increasing desiccation, showing that increased water loss lowers the algae CO_2 affinity. At the present level of CO_2 concentration, when the algae were wet, relative photosynthesis was 65%, 74% and 61% saturated in *E. linza*, *G. furcata* and *I. okamurae*, respectively. In *E. linza*, 70% water loss reduced the photosynthesis by 49–55%, 90% water loss resulted in CO_2 evolution at the present CO_2 level and only left 9% of the CO_2 uptake rate at doubled CO_2 concentration. In *G. furcata*, 30% water loss did not affect the photosynthesis while 60% water loss reduced it by 50–55%. Photosynthesis of *I. okamurae* also decreased with increased water loss, 40% water loss reduced it by 30% at the present CO_2 level and by 54% at doubled concentration; 60% water loss gave rise to negative and slightly positive photosynthesis at present and doubled CO_2 concentrations respectively.

Figure 3 shows the relationship of CO_2 compensation point (CO_2 CPP) to water loss. The CO_2 CPP increased with increased water loss, indicating that higher CO_2 concentrations are needed for dehydrated than for fully hydrated *E. linza*, *G. furcata* and *I. okamurae* to maintain photosynthesis positive. Fifty percent water loss did not result in negative photosynthesis at the present CO_2 level in any of the species at the temperature range of 20–30 °C except *G. furcata* at 30 °C. For *G. furcata* at 30 °C, 50% water loss brought the CO_2 CPP to 22 μM, indicating such a level of water loss at the temperature results in negative photosynthesis at the present CO_2 level. Effects of temperature on the CO_2 CPP were not obvious in *E. linza* and *I. okamurae*; however, increased temperature raised the CO_2 CPP in *G. furcata*, the CO_2 CPP at 50% water loss was 4, 8 and 22 μM, respectively, at 20, 25 and 30 °C.

Discussion

The present study showed that photosynthesis in air of *E. linza*, *G. furcata* and *I. okamurae* is CO_2-limited at the present atmospheric CO_2 concentration; higher CO_2 concentrations were needed to saturate the photosynthesis when the algae were desiccated. The CO_2 concentration for saturating photosynthesis, the CO_2 compensation point and $K_{[mCO_2]}$ increased with increasing desiccation, showing that increased water loss lowers the algae's CO_2 affinity. This pattern was true for all of the species, though they showed specific differences in the way their responses were affected by the level of water loss and temperature. Elevation of CO_2 from the present level (15.6 μM) to its doubled concentration (31.2 μM), which has been assumed to happen in the next century, brought about 25–40% increase in the photosynthesis of the three species at moderate levels of water loss while exposed. Relationship of CO_2 compensation point to water loss shows a trend that increased water loss would lead to negative photosynthetic production at the present day CO_2 level. *E. linza*, *I. okamurae* and *G. furcata* are usually exposed daily for 3 to 5 h at low tide, and their photosynthesis is subject to environmental constraints in air different from those in water, which can result in reduced or even negative production during the emersion period by the macroalgae at the present day CO_2 concentration.

It is of general concern whether the photosynthesis of the intertidal macroalgae is also CO_2-limited while submerged when the tide is high. The present day inorganic carbon composition of seawater has been regarded as not limiting the photosynthesis of marine macroalgae when submerged (Beer & Kock, 1996), albeit growth enhancement by enriched CO_2 has been reported for some red algae (Gao et al., 1991, 1993a). Most macroalgae use both HCO_3^- and CO_2 (Maberly et al., 1992; Gao & McKinley, 1994), utilization of HCO_3^- is either assisted by the dehydration via surface-bound carbonic anhydrase or HCO_3^- transport via anion exchange in some species (Larsson et al., 1997). However, for the periods of emersion at low tide when CO_2 is the only carbon source, the intertidal species may use only CO_2 when they photosynthesize in air. Maximal length of such an exposure can be as long as 5–7 h, during this period the algae lose extracellular and then intracellular water due to desiccation. Increasing atmospheric CO_2 will enhance CO_2-limited photosynthesis during such a period.

With an increase in CO_2 plants would differ in growth and competitive interactions (Bowes, 1993). We propose that CO_2 rise can benefit macroalgae, especially those suffering longer desiccation at low tide. The influence of CO_2 combined with that of sea-level rise associated with global warming can be expected

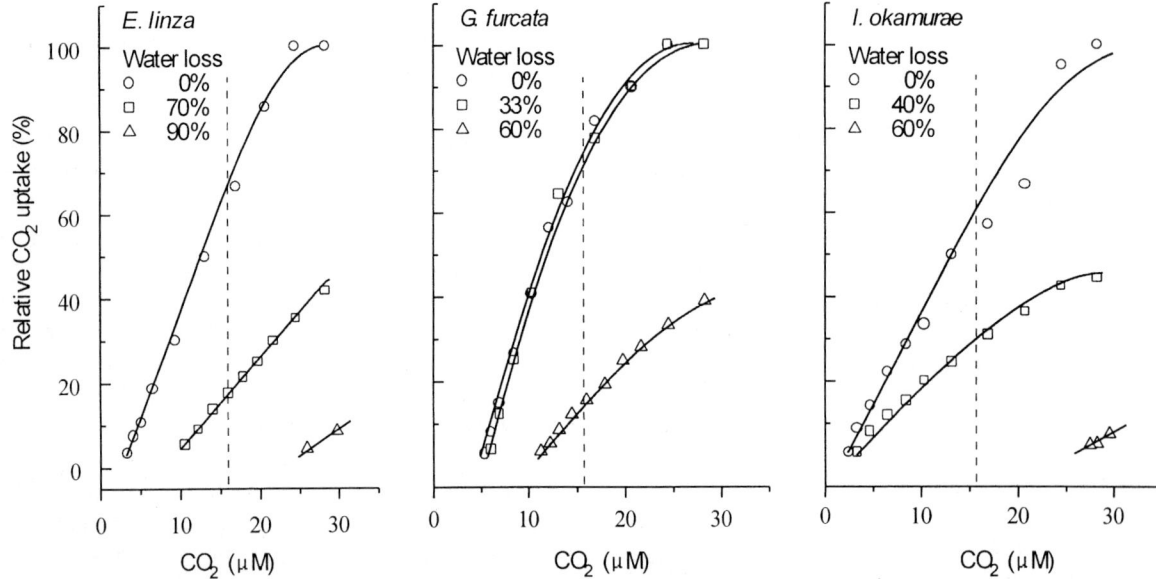

Figure 2. Photosynthetic CO_2 uptake as a function of CO_2 concentrations and desiccation (water loss) when *E. linza*, *G. furcata* and *I. okamurae* photosynthesized in air at 25°C; the maximal rates of photosynthetic CO_2 uptake were 112.6, 7.8 and 17.3 μmol CO_2 g (f.wt)$^{-1}$h^{-1} for *E. linza*, *G. furcata* and *I. okamura*, respectively. Six to fifteen individuals were used for each measurement. The dashed lines indicate the present day CO_2 level.

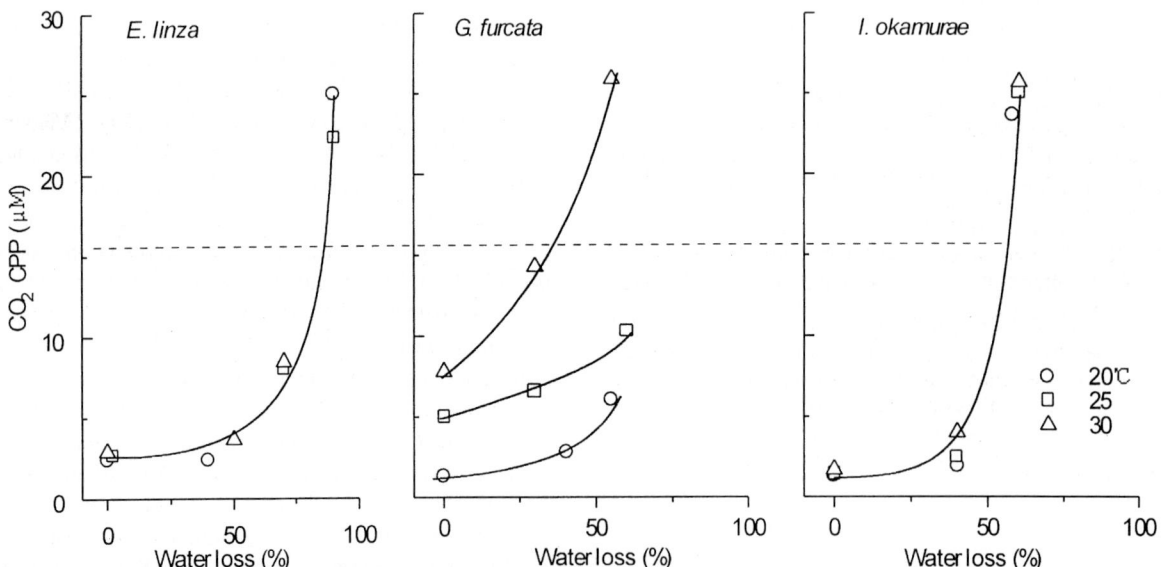

Figure 3. Relationship of CO_2 compensation point (CO_2 CPP) of *E. linza*, *G. furcata* and *I. Okamura* with desiccation (water loss), determined at 20, 25 and 30 °C. Six to fifteen individuals were used for each measurement. The dashed lines indicate the present day CO_2 level.

to bring about changes in succession and zonation of intertidal macroalgae.

Acknowledgements

This study was funded by the Natural Science Foundation of China (No. 39830060 and No. 39625002), by The Chinese Academy of Sciences ('BAIREN JIHUA', One Hundred Talents Program) and by the Research Fund for the Centennial Anniversary, Tokyo University of Fisheries of Japan. We thank Ms K. Matsuyama for her assistance in sampling.

References

Beer, S. & E. Koch, 1996. Photosynthesis of marine macroalgae and seagrasses in globally changing CO_2 environments. Mar. Ecol. Prog. Ser. 141: 199–204.

Beer, S. & B. Schragge, 1987. Photosynthetic carbon metabolism in *Enteromorpha compressa* (Chlorophyta). J. Phycol. 23: 580–584.

Bidwell, R. G. S. & J. McLachlan, 1985. Carbon nutrition of seaweeds: photosynthesis, photorespiration and respiration. J. exp. mar. Biol. Ecol. 97: 287–294.

Bowes, G., 1993. Facing the inevitable: plants and increasing atmospheric CO_2. Annu. Rev. Plant Physiol. Plant mol. Biol. 44: 309–332.

Brinkhuis, B. H., N. R. Tempel & R F. Jones, 1976. Photosynthesis and respiration of exposed salt-marsh fucoids. Mar. Biol. 34: 349–359.

Charpy-Roubaud, C. & A. Sournia, 1990. The comparative estimation of phytoplanktonic, microphytobenthic and macrophytobenthic primary production in the oceans. Mar. Microb. Food Webs 4: 31–57.

Gao, K. & Y. Aruga, 1987. Preliminary studies on the photosynthesis and respiration of *Porphyra yezoensis* under emersed conditions. J. Tokyo Univ. Fish. 47: 51–65.

Gao, K. & K. McKinley, 1994. Use of macroalgae for marine biomass production and CO_2 remediation: a review. J. appl. Phycol. 6: 45–60.

Gao, K., Y. Aruga, K. Asada & M. Kiyohara, 1993a. Influence of enhanced CO_2 on growth and photosynthesis of the red algae *Gracilaria* sp. and *G. chilensis*. J. appl. Phycol. 5: 563–571.

Gao, K., Y. Aruga, K. Asada & M. Kiyohara, 1993b. Calcification in the articulated coralline alga *Corallina pilulifera*, with special reference to the effect of elevated atmospheric CO_2. Mar. Biol. 117: 129–132.

Gao, K., Y. Aruga, K. Asada, T. Ishihara, T. Akano & M. Kiyohara, 1991. Enhanced growth of the red alga *Porphyra yezoensis* Ueda in high CO_2 concentrations. J. appl. Phycol. 3: 355–362.

Gattuso, J. P., C. E. Payri, M. Pichon, B. Delesalle & M. Frankignoulle, 1997. Primary production, calcification, and air-sea CO_2 fluxes of a macroalgal-dominated coral reef community (Moorea, French Polynesia). J. Phycol. 33: 729–738.

Hein, M. & K. Sand-Jensen, 1997. CO_2 increases oceanic primary production. Nature 388: 526–527.

Johnson, W. S., A. Gison, S. L. Gulmon and H A. Mooney, 1974. Comparative photosynthetic capacities of intertidal algae under exposed and submerged conditions. Ecology 55: 450–453.

Johnston, A. M. & J. A. Raven, 1986. The analysis of photosynthesis in air and water in *Ascophyllum nodosum* (L.) Le Jolis. Oecologia 69: 288–295.

Larsson, C., L. Axelsson, H. Ryberg & S. Beer, 1997. Photosynthetic carbon utilization by *Enteromorpha intestinalis* (Chlorophyta) from a Swedish rockpool. Eur. J. Phycol. 32: 49–54.

Lignell, A. & M. Pedersén, 1989. Effects of pH and inorganic carbon concentration on growth of *Gracilaria secundata*. Br. phycol. J. 24: 83–89.

Maberly, S. C., J. A. Raven & A. M. Johnston, 1992. Discrimination between ^{12}C and ^{13}C by marine plants. Oecologia 91: 481–492.

Madsen, T. V. & S. C. Maberly, 1990. A comparison of air and water as environments for photosynthesis by the intertidal alga *Fucus spiralis* (Phaeophyta). J. Phycol. 26: 24–30.

Oates, B. R., 1985. Photosynthesis and amelioration of desiccation in the intertidal saccate alga *Colpomenia peregrina*. Mar. Biol. 89: 109–119.

Oates, B. R., 1986. Components of photosynthesis in the intertidal saccate alga *Halosaccion americanum* (Rhodophyta, Palmariales). J. Phycol. 22: 217–223.

Oates, B. R. & S. N. Murray, 1983. Photosynthesis, dark respiration and desiccation resistance of the intertidal seaweeds *Hesperophycus harveyanus* and *Pelvetia fastigiata* f. *gracilis*. J. Phycol. 19: 371–380.

Quadir, A., P. J. Harrison & R. E. DeWreede, 1979. The effects of emergence and submergence on the photosynthesis and respiration of marine macrophytes. Phycologia 18: 83–88.

Raven, J. A., 1997. Inorganic carbon aquisition by marine autotrauphs. Adv. bot. Res. 27: 85–209.

Raven, J. A. & A. M. Johnston, 1991. Photosynthetic inorganic carbon acquisition by Prasiola *stipitata* (Prasiales, Chlorophyta) under emersed and submersed conditions: relationship to the taxonomy of *Prasiola*. Br. Phycol. J. 26: 24–25.

Riebesell, U., D. A. Wolf-Gladrow & V. Smetacek, 1993. Carbon dioxide limitation of marine phytoplankton growth rates. Nature 361: 249–251.

Surif, M. B. & J. A. Raven, 1990. Photosynthetic gas exchange under emersed conditions in eulittoral and normally submerged members of the Fucales and Laminariales: interpretation in relation to C isotope ratio and N and water use efficiency. Oecologia 82: 68–80.

A theoretical analysis and field evaluation of a light and temperature model of production by *Ecklonia cava*

Masaki Honda
Central Research Institute of Electric Power Industry, 1646 Abiko, Abiko City, Chiba 270-1194, Japan

Key words: production, mathematical model, *Ecklonia cava*, light, temperature, photosynthesis

Abstract

The dependence of photosynthesis on light and temperature is modelled through analysis of transition probabilities of photosystems. In the model, two transition probabilities are functions of light, and one transition probability is a function of temperature. The estimated light-saturated photosynthesis of *Ecklonia cava* blades at 20 °C was 0.037 mg C cm^{-2} h^{-1}. The value of the activation energy, the standard enthalpy and the standard entropy were estimated to be 56.5 kJ mol^{-1}, 204 kJ mol^{-1} and 678 J mol^{-1} K^{-1}, respectively. A production model (an integral photosynthesis model) for an *E. cava* stand was developed using the photosynthesis model. Production calculated by the model agreed well with observed data during the growing period of an *E. cava* stand at a field observation site on the west side of Miura Peninsula, Japan. Results of the analysis of the effects of irradiance and temperature on the production of the *E. cava* community by the model are:

1. Production decreased with irradiance decrease. The estimated compensation irradiance was 26.5 μmol photons m^{-2} s^{-1} when the biomass was 3 kg wet mass m^{-2} (blade:stipe ratio = 2 kg m^{-2}:1 kg m^{-2}) and the temperature was 20 °C.

2. The optimum temperature decreased when irradiance decreased and when biomass increased. The highest estimated value for the optimum temperature was 24.0 °C. The estimated optimum temperature was 18.2 °C when the biomass was 12 kg wet mass m^{-2} and the photon irradiance was 200 μmol photons m^{-2} s^{-1}.

3. The amount of biomass that resulted in the maximum production was influenced by irradiance and temperature. At 400 μmol photons m^{-2} s^{-1} and 20 °C, the estimated value of the biomass (blade:stipe = 2:1) giving the maximum production was about 5.3 kg wet mass m^{-2}. However, at 100 μmol photons m^{-2} s^{-1} and 24 °C, the estimated value was about 3.0 kg wet mass m^{-2}. The estimated values of the maximum production under the two conditions were 1.05 and 0.30 g C m^{-2} h^{-1}, respectively.

Introduction

Communities dominated by the perennial brown alga *Ecklonia cava* Kjellman, with stipes as long as 1.5 m or more, are the most typical submarine forests along the central Pacific coast of Japan. These algae are important primary producers in the coastal ecosystem. The physiology of isolated parts of *E. cava* has been previously studied, particularly with respect to the relationship between light and photosynthesis (Aruga, 1982; Maegawa et al., 1987; Sakanishi et al., 1988, 1989; Haroun et al., 1992; Aruga et al., 1997). Most ecological studies of *E. cava* are concerned with the distribution and structure of communities (Hayashida, 1977, 1986; Kida & Maegawa, 1982, 1983; Maegawa & Kida, 1987), standing crop (Ohno & Ishikawa, 1982; Yokohama et al., 1987), seasonal changes (Iwahashi, 1968; Ohno & Ishikawa, 1982; Maegawa & Kida, 1984; Yokohama et al., 1987; Haroun et al., 1989), morphological change (Kasahara & Ohno, 1983) and production (Yokohama et al., 1987; Honda, 1993, 1996).

Understanding factors regulating the production and biomass of marine macrophytes, such as *E. cava*, is especially difficult. Laboratory experiments on groups of whole plants are virtually precluded because of their large size. The ability to perform field experiments is severely restricted by variables which cannot be controlled. Models can integrate different aspects of field and laboratory work.

Calculation of the photosynthetic rate of *E. cava* is important in the estimation of production. A photosynthetic production model has been described by Monsi & Saeki (1953). Ikushima (1967, 1970) modified the model to estimate the photosynthetic rate of submerged plant communities. In their model, a rectangular hyperbola was used to simulate the photosynthesis-light response. However, the rectangular hyperbola formulation was a poor fit to the experimental photosynthesis-light response data (Jassby & Platt, 1976).

The first objective of this study is to develop a mathematical model that can be used to estimate the primary production and biomass of marine macrophytes. The dependence of photosynthesis on light and temperature is modelled through analysis of transition probabilities of photosystems. In the model, three transition probabilities are incorporated, two of these are functions of light and the other one is a function of temperature.

The second purpose of this study is to analyze the effects of irradiance and water temperature on the primary production of an *E. cava* stand. This analysis provides a variety of insights into the effects of physical factors, e.g. insolation, light-absorbing components, water depth and temperature, on *E. cava* production.

Methods

Model description

1. Biomass change

The rate of biomass change is given by

$$\frac{\Delta S}{\Delta t} = P_{\text{gross}(t)} - R_{c(t)} - D_{b(t)} + A_{(t)} - D_{i(t)}, \quad (1)$$

where $\frac{\Delta S}{\Delta t}$ is the rate of biomass change (g m^{-2} h^{-1}), $P_{\text{gross}(t)}$ is the gross photosynthetic rate (g m^{-2} h^{-1}), $R_{c(t)}$ is the respiration rate (g m^{-2} h^{-1}), $D_{b(t)}$ is the erosion rate of the blade (g m^{-2} h^{-1}), $A_{(t)}$ is the recruitment (g m^{-2} h^{-1}) and $D_{i(t)}$ is the mortality (g m^{-2} h^{-1}).

2. Gross photosynthetic rate

The oxidized reaction centre chlorophyll, P680+, is reduced by the transfer of an electron from donor Z, and the oxidized form of Z removes electrons from the water, liberating oxygen. Joliot et al. (1969) observed that oxygen production in dark-adapted chloroplasts or intact cells in a sequence of saturating flashes shows damped oscillations with a periodicity of four. As shown by Kok et al. (1970), liberation of oxygen requires the cooperation of four photochemically formed oxidizing equivalents in the individual reaction centres generated by four successive excitations of photosystem II. They proposed a four-step linear mechanism of positive charge accumulation that led to the evolution of an oxygen molecule.

This model is presented by the following scheme:

$$S_0 \xrightarrow{e^-}_{\alpha} S_1 \xrightarrow{e^-}_{\alpha} S_2 \xrightarrow{e^-}_{\alpha} S_3 \xrightarrow{e^-}_{\alpha} S_4 \xrightarrow{\beta} \quad (2)$$

where S_i are oxidized states of Z, α and β are the transition probabilities from S_i to S_{i+1} and from S_4 to S_0 respectively, and e^- is electron. As defined above, Kok's model is a Markoff process; it may be simply transcribed in matrix notation and analyzed correspondingly by

$$\begin{pmatrix} P_{0(n+1)} \\ P_{1(n+1)} \\ P_{2(n+1)} \\ P_{3(n+1)} \\ P_{4(n+1)} \end{pmatrix} = \begin{pmatrix} 1-\alpha & 0 & 0 & 0 & \beta \\ \alpha & 1-\alpha & 0 & 0 & 0 \\ 0 & \alpha & 1-\alpha & 0 & 0 \\ 0 & 0 & \alpha & 1-\alpha & 0 \\ 0 & 0 & 0 & \alpha & 1-\beta \end{pmatrix} \begin{pmatrix} P_{0(n)} \\ P_{1(n)} \\ P_{2(n)} \\ P_{3(n)} \\ P_{4(n)} \end{pmatrix} \quad (3)$$

with

$$\sum_{i=0}^{4} P_i = 1, \quad (4)$$

where $P_{i(n)}$ is the probability of S_i at the nth step, α is a constant value, when light intensity is constant, and $P_{i(n)}$ and $P_{i(n+1)}$ are the same probabilities, when the photosynthetic rate is a constant value. When the photosynthetic rate (oxygen evolving rate) is a constant value, eigenvectors of P_0, P_1, P_2, P_3 and P_4 are given by

$$t^t \left(\frac{\beta}{4\beta+\alpha} \quad \frac{\beta}{4\beta+\alpha} \quad \frac{\beta}{4\beta+\alpha} \quad \frac{\beta}{4\beta+\alpha} \quad \frac{\alpha}{4\beta+\alpha} \right). \quad (5)$$

Z in S_4, Q_4 is given by

$$Q_4 = Q_S P_{4(n)} = Q_S \frac{\alpha}{4\beta + \alpha}, \quad (6)$$

where Q_S represents the total number of reaction centres. Photosynthetic rate (oxygen evolving rate) is given by

$$v = \frac{\beta}{t_{\min}} Q_4 = \frac{\beta}{t_{\min}} Q_S \frac{\alpha}{4\beta + \alpha}, \quad (7)$$

where $t_{\min}$ is the turnover time of S_3 to S_0. α is described by the oxidizing probability of reaction centre chlorophyll, P680 and P700, in photosystem II and photosystem I respectively.

The transition states and transition probabilities of photosystem II and photosystem I are presented by the following scheme:

$$(PSII \cdot PSI)_0 \underset{\delta}{\overset{\gamma}{\rightleftarrows}} (PSII \cdot PSI)_1 \underset{\zeta}{\overset{\varepsilon}{\rightleftarrows}} (PSII \cdot PSI)_2 \xrightarrow{\eta \quad e^-} \quad (8)$$

where $(PSII \cdot PSI)_0$ is the stable state of photosystems I and II, $(PSII \cdot PSI)_1$ is photosystem I or photosystem II in the excited state and $(PSII \cdot PSI)_2$ is photosystem I and II in excited states. γ is the transition probability from $(PSII \cdot PSI)_0$ to $(PSII \cdot PSI)_1$, δ is the transition probability from $(PSII \cdot PSI)_1$ to $(PSII \cdot PSI)_0$, ε is the transition probability from $(PSII \cdot PSI)_1$ to $(PSII \cdot PSI)_2$, ζ is the transition probability from $(PSII \cdot PSI)_2$ to $(PSII \cdot PSI)_1$ and η is the transition probability from $(PSII \cdot PSI)_2$ to $(PSII \cdot PSI)_0$. Donor Z is oxidized in the process from $(PSII \cdot PSI)_2$ to $(PSII \cdot PSI)_0$. As defined above, the model is a Markoff process; it may be simply transcribed in matrix notation and analyzed correspondingly by

$$\begin{pmatrix} P_{(PSII \cdot PSI)0(n+1)} \\ P_{(PSII \cdot PSI)1(n+1)} \\ P_{(PSII \cdot PSI)2(n+1)} \end{pmatrix} = \begin{pmatrix} 1-\gamma & \delta & \eta \\ \gamma & 1-\delta-\varepsilon & \zeta \\ 0 & \varepsilon & 1-\zeta-\eta \end{pmatrix} \begin{pmatrix} P_{(PSII \cdot PSI)0(n)} \\ P_{(PSII \cdot PSI)1(n)} \\ P_{(PSII \cdot PSI)2(n)} \end{pmatrix} \quad (9)$$

where $P_{(PSII \cdot PSI)_0(n)}$ is the probability of photosystems I and II in the stable states, $P_{(PSII \cdot PSI)_1(n)}$ is the probability of photosystem I or photosystem II in the excited state and $P_{(PSII \cdot PSI)_2(n)}$ is the probability of photosystem I and II in excited states at the nth step. When the photosynthetic rate is a constant value, eigenvectors of $P_{(PSII \cdot PSI)_0}$, $P_{(PSII \cdot PSI)_1}$ and $P_{(PSII \cdot PSI)_2}$ are given by

$$\begin{pmatrix} \frac{\delta\zeta + \delta\eta + \varepsilon\eta}{\delta\zeta + \delta\eta + \varepsilon\eta + \gamma\zeta + \gamma\eta + \gamma\varepsilon} \\ \frac{\gamma\zeta + \gamma\eta}{\delta\zeta + \delta\eta + \varepsilon\eta + \gamma\zeta + \gamma\eta + \gamma\varepsilon} \\ \frac{\gamma\varepsilon}{\delta\zeta + \delta\eta + \varepsilon\eta + \gamma\zeta + \gamma\eta + \gamma\varepsilon} \end{pmatrix}, \quad (10)$$

with

$$\sum_{i=0}^{3} P_{(PSII \cdot PSI)i} = 1 \quad (11)$$

α is given by

$$\alpha = \eta \frac{\gamma\varepsilon}{\delta\zeta + \delta\eta + \varepsilon\eta + \gamma\zeta + \gamma\eta + \gamma\varepsilon}. \quad (12)$$

By substituting Equation (12) into Equation (7),

$$v = \frac{\beta}{t_{min}} Q_S, \frac{\eta\gamma\varepsilon}{4\beta(\delta\zeta + \delta\eta + \varepsilon\eta + \gamma\zeta + \gamma\eta + \gamma\varepsilon) + \eta\gamma\varepsilon}. \quad (13)$$

γ and ε are functions of excitation probabilities of reaction centre chlorophyll in photosystems, and are defined as:

$$\gamma = \varepsilon = 1 - e^{-\Gamma I} \quad (14)$$

where Γ and I are excitation probability of reaction centre of photosystems and irradiance, respectively. When most reaction centre chlorophyll molecules are in stable states, γ and ε are defined as:

$$\gamma = \varepsilon \approx \Gamma I \quad (15)$$

By substituting Equation (15) into Equation (13),

$$v \approx \frac{\beta}{t_{\min}} Q_S \frac{\eta \Gamma^2 I^2}{4\beta\delta(\zeta + \eta) + 4\beta(2\eta + \zeta)\Gamma I + (4\beta + \eta)\Gamma^2 I^2}. \quad (16)$$

η is a temperature transition probability that is expressed as the product of the probability of thermal denaturation of the electron acceptor in the photosynthetic electron transfer and the probability of an electron having higher energy than the activation energy. η is defined as

$$\eta = \frac{e^{-E_{Ap}/RT}}{1+e^{(-\Delta H_p^\circ + T\Delta S_p^\circ)/RT}} \chi, \tag{17}$$

where E_{Ap} is the activation energy (J mol^{-1}), R is the universal gas constant (J mol^{-1} K^{-1}), T is the absolute temperature (K), ΔH_p° is the standard enthalpy (J mol^{-1}), $\Delta S p^\circ$ is the standard entropy (J mol^{-1} K^{-1}) and χ is a coefficient that is not affected by temperature. By substituting Equation (17) into Equation (16) and when $\beta \gg \eta$,

$$v \approx \frac{\dfrac{Q_S}{4t_{\min}} \dfrac{e^{-E_{Ap}/RT}}{1+e^{(-\Delta H_p^\circ + T\Delta S_p^\circ)/RT}} \chi I^2}{\dfrac{\delta}{\Gamma^2}\left\{\zeta + \dfrac{e^{-E_{Ap}/RT}}{1+e^{(-\Delta H_p^\circ + T\Delta S_p^\circ)/RT}} \chi\right\} + \dfrac{\dfrac{2e^{-E_{Ap}/RT}}{1+e^{(-\Delta H_p^\circ + T\Delta S_p^\circ)/RT}} \chi + \zeta}{\Gamma} I + I^2} \tag{18}$$

In order to simplify Equation (18), we let

$$\frac{Q_S \dfrac{e^{-E_{Ap}/RT}}{1+e^{(-\Delta H_p^\circ + T\Delta S_p^\circ)/RT}} \chi}{4t_{\min}},$$

$$\frac{\delta\left\{\zeta + \dfrac{e^{-E_{Ap}/RT}}{1+e^{(-\Delta H_p^\circ + T\Delta S_p^\circ)/RT}} \chi\right\}}{\Gamma^2} \quad \text{and}$$

$$\frac{\dfrac{2e^{-E_{Ap}/RT}}{1+e^{(-\Delta H_p^\circ + T\Delta S_p^\circ)/RT}} \chi + \zeta}{\Gamma}$$

be defined as constants Φ_T, Ψ_T and Ω_T, respectively. The light-saturation curve is thus given as

$$v \approx \frac{\Phi_T I^2}{\Psi_T + \Omega_T I + I^2}. \tag{19}$$

Based on Monsi & Saeki theory (Monsi & Saeki, 1953) and the relationship among photosynthesis, light and temperature (Equation (19)), a community photosynthetic production model can be proposed as Equation (20) below:

$$P_{\text{gross}} = \int_0^F \frac{\Phi_T (K I_0 e^{-CD-KF_z})^2}{\Psi_T + \Omega_T K I_0 e^{-CD-KF_z} + (K I_0 e^{-CD-KF_z})^2} dF_Z, \tag{20}$$

where P_{gross} is the gross production (g C m^{-2} h^{-1}), F is the leaf area index (m^2 m^{-2}), I_0 is the value of photon irradiance at 0 m water depth (μmol photons m^{-2} s^{-1}), C is the vertical attenuation coefficient (m^{-1}), D is water depth (m), K is the extinction coefficient within the community (m^{-2}) and F_Z is the cumulative leaf area index between the community surface and z m depth from community surface (m^2). F is in proportion to biomass. The evaluation of Equation (20) was carried out using Mathematica® (Wolfram Research).

Respiration rate
Community respiration rate is given by

$$R_c = r_{fT} S_l + r_s S_s \tag{21}$$

and

$$r_{fT} = r_{f20} \frac{\frac{e^{-E_{Ar}/RT}}{1 + e^{(-\Delta H_r^\circ + T\Delta S_r^\circ)/RT}}}{\frac{e^{-E_{Ar}/293.15R}}{1 + e^{(-\Delta H_r^\circ + 293.15\Delta S_r^\circ)/293.15R}}} \quad (22)$$

where R_C is the community respiration rate (g C m^{-2} h^{-1}), r_{fT} is the respiration rate of blade at T (g C g^{-1} h^{-1}), S_l is the standing crop of blades (g m^{-2}), r_s is the respiration rate of stipe (g C g^{-1} h^{-1}), S_s is the standing crop of stipes (g m^{-2}), r_{f20} is the respiration rate of blades at 20 °C (g C g^{-1} h^{-1}), E_{Ar} is the activation energy (J mol^{-1}), ΔH_r° is the standard enthalpy (J mol^{-1}) and ΔS_r° is the standard entropy (J mol^{-1} K^{-1}).

Measurements of photosynthetic rate and respiration rate

Fronds of *E. cava* were taken from a coastal station during the period from November 1991 to January 1992 at 10 m depth offshore of Akiya, Yokosuka-shi, Kanagawa Prefecture on the Pacific coast of central Japan (35° 15' N, 139° 34' E). These fronds were cultured for one year in a temperature-controlled aquarium: maximum temperature and minimum temperature were 23 °C (late August) and 14 °C (late February), respectively.

An oxygen-electrode (Orbisphere Laboratories model 2714) was used to measure photosynthesis and respiration. The reaction vessel for measurement of photosynthesis and respiration had a capacity of 50 ml. A xenon-discharge lamp (SERIC XC-100) was used as the light source.

The photosynthesis-light response measurements and the respiration rate at 20 °C were made on blades in October 1993–December 1993. Discs of blades of 1.63 cm^2 were cut out of the middle portion of the longest bladelet of fronds and kept in culture bottles (2 l) for 3–4 days. Culture conditions were 110 μmol photons m^{-2} s^{-1}, 12:12 light:dark cycle, and 20 °C. The irradiance on blade discs in the reaction vessel were 2.21, 3.67, 6.09, 16.3, 26.4, 44.7, 80.1, 129, 203 and 333 μmol photons m^{-2} s^{-1}. In the experiments, results from a total of 16 measurements of the photosynthetic rate and irradiance and a total of 20 measurements of the respiration rate were obtained.

The light saturated photosynthesis (Φ_T)-temperature response measurements and the respiration-temperature response measurements were made in October 1993–December 1993 and November 1994. Discs of blades of 1.63 cm^2 were cut out of the middle portion of the longest bladelet of fronds and kept in culture bottles (2 l) for 1–4 days. Culture conditions were 110 μmol photons m^{-2} s^{-1}, 12:12 light:dark cycle, and 7, 13, 20, 24 and 28 °C.

Measurements of the respiration rate of stipe-temperature response were carried out in January, February, May, June and July 1992. For the experiments, the temperature was kept at 15 °C in January, 12 °C in February, 16 and 18 °C in May, 20 °C in June and 22 °C in July. Pieces were cut out of stipes and cut ends were covered with aluminum foil. A total of 11 measurements of the respiration rate was made.

Field production measurement

Field production measurement was based on the leaf-marking method (see Yoshida, 1970; Yokohama et al., 1987) and on observations of the relative growth of each part (stipe, primary blade and bladelet) of *E. cava* (see Honda, 1993 for details). Field data were collected from October 1988 to June 1989 at the observation site that was offshore on the west side of Miura Peninsula on the Pacific coast of central Japan (35° 15' N, 139° 34' E). The water depth at the observation site was 14 m. A permanent 1 m square quadrat was located in the *E. cava* forest. All plants enclosed within the quadrat, except for juveniles having no bladelets greater than 10 cm in length, were marked by tagging sequentially numbered plastic plates. A hole was punched in the youngest bladelet exceeding 10 cm in length and in the primary blade. Stipe length, longest bladelet length and distance between the thinnest portion of the stipe and the punched hole in the

Table 1. Seasonal changes in the number of individual of *Ecklonia cava* at the observation site

Month	Fronds	Recruitment	Mortality
Oct., 1988	9	–	–
Nov.	9	0	0
Dec.	9	0	0
Jan., 1989	10	1	0
Feb.	9	0	1
Mar.	9	0	0
Apr.	11	2	0
May	11	0	0
June	11	0	0

primary blade of all plants enclosed within the quadrat was measured. About one month later, the number of younger bladelets exceeding 10 cm in length other than the marked bladelets, stipe length, longest bladelet length and distance between the thinnest portion of the stipe and the punched hole in the primary blade of all plants was measured. The number of plants within the permanent quadrat is shown in Table 1.

The weight of each part was calculated from its length using the relative growth formula, e.g. stipe weight was calculated from stipe length, bladelet weight was calculated from bladelet length. The mass of newly produced bladelets was calculated from the weight of the longest bladelet and the number of younger bladelets exceeding 10 cm in length other than the marked bladelets. The mass of newly produced tissue of stipe was calculated from weight change during the interval. The mass of increased primary blade was calculated from the area of newly produced tissue and weight per primary blade area. The sum of the weight gained by all the individuals in a field observation interval (about 1 month), expressed in per unit square meter, was the net production for the E. cava community.

Table 2. Seasonal changes in the attenuation coefficient (C) at the observation site and in the ratio of blade production to blade and stipe production ($d_{b(t)}$), the number of bladelets eroded ($N_{e(t)}$) and the number of bladelets produced ($N_{p(t)}$) of Ecklonia cava community at the observation site

Month	C	$d_{b(t)}$	$N_{e(t)}$	$N_{p(t)}$
Oct., 1988	0.17	0.99	–	–
Nov.	0.18	1.00	3.8	3.9
Dec.	0.14	0.89	4.7	3.7
Jan., 1989	0.15	0.86	3.7	5.6
Feb.	0.15	0.89	1.9	4.9
Mar.	0.19	0.89	1.9	4.9
Apr.	0.19	0.89	1.9	4.9
May	0.18	0.98	1.9	4.9

Results

Figure 1 shows the photosynthesis-light relationship observed in E. cava blade. Using a least-squares method with linearization, I estimated the parameters for Φ_{20}, Ψ_{20} and Ω_{20}. Φ_{20}, Ψ_{20} and Ω_{20} were Φ_T, Ψ_T and Ω_T at 20 °C, respectively. The estimated mean values of Φ_{20}, Ψ_{20} and Ω_{20} with their standard deviations were 0.099 ± 0.00507 mg O$_2$ cm^{-2} h^{-1} (0.037 ± 0.00190 mg C cm^{-2} h^{-1}), 234 ± 86.4 μmol^2 m^{-4} s^{-2} and 43.6 ± 4.65 μmol m^{-2} s^{-1}, respectively.

Figure 2 shows the light saturated photosynthetic rate (Φ_T)-temperature relationship observed in a E. cava blade. The relationship of $\frac{\Phi_T}{\Phi_{20}}$ and $\frac{\eta_T}{\eta_{20}}$ is given by:

$$\frac{\Phi_T}{\Phi_{20}} = \frac{\frac{e^{-E_{Ap}/RT}}{1+e^{(-\Delta H_p^\circ + T\Delta S_p^\circ)/RT}}}{\frac{e^{-E_{Ap}/293.15R}}{1+e^{(-\Delta H_p^\circ + 293.15\Delta S_p^\circ)/293.15R}}} = \frac{\eta_T}{\eta_{20}}. \quad (23)$$

Using a least-squares method with linearization, I estimated the parameters for E_{Ap}, ΔH_p° and ΔS_p°. In this analysis, we were forced to assume that the probability of thermal denaturation of the electron acceptor at lower temperature is very small. E_{Ap} was predicted by a least-squares fit in the Arrhenius model to the data that were obtained from the experiment at a temperature lower than 16 °C. The estimated value of E_{Ap} was 56.5 kJ mol^{-1}. By a least-squares fit using Equation (23) for the whole data, ΔH_p° and ΔS_p° were predicted as 204 kJ mol^{-1} and 678 J mol^{-1} K^{-1}, respectively.

Figure 3 shows the respiration rate-temperature relationship observed in a E. cava blade. Using a least-squares method with linearization, I estimated the parameters for E_{Ar}, ΔH_r° and ΔS_r°. In this analysis, we were again forced to assume that the probability of thermal denaturation of the electron acceptor at lower temperature is very small. E_{Ar} was predicted by a least-squares fit in the Arrhenius model to the data that were obtained from the experiment at a temperature lower than 24 °C. The estimated value of E_{Ar} was 43.5 kJ mol^{-1}. By a least-squares fit using Equation (22) for the whole data, ΔH_r° and ΔS_r° were predicted as 656 kJ mol^{-1} and 2170 J mol^{-1} K^{-1}, respectively. The estimated mean value with a standard deviation of r_{f20} was 0.113 ± 0.0125 mg O$_2$ g^{-1} h^{-1} (0.0423 ± 0.00470 mg C g^{-1} h^{-1}).

There was no significant difference in the respiration rate of stipe in each temperature. The estimated mean values with a standard deviation of r_s was 0.0129 ± 0.00555 mg O$_2$ g^{-1} h^{-1} (0.00483 ± 0.00208 mg C g^{-1} h^{-1}).

Figure 4 shows the measured daily insolation. A smooth curve is of course only obtained when cloud

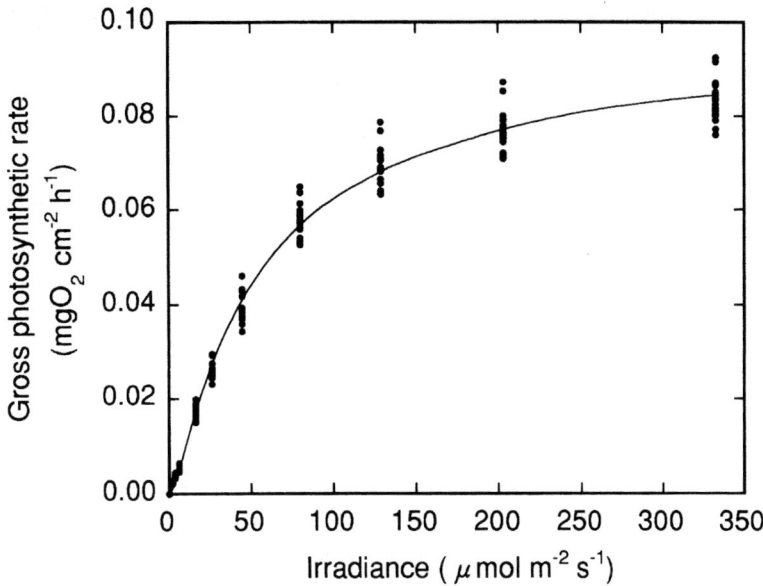

Figure 1. A light saturation curve for gross photosynthesis by the blades of *Ecklonia cava*. Curve indicates the regression line predicted by fitting Equation 19 (see text for details) to the data.

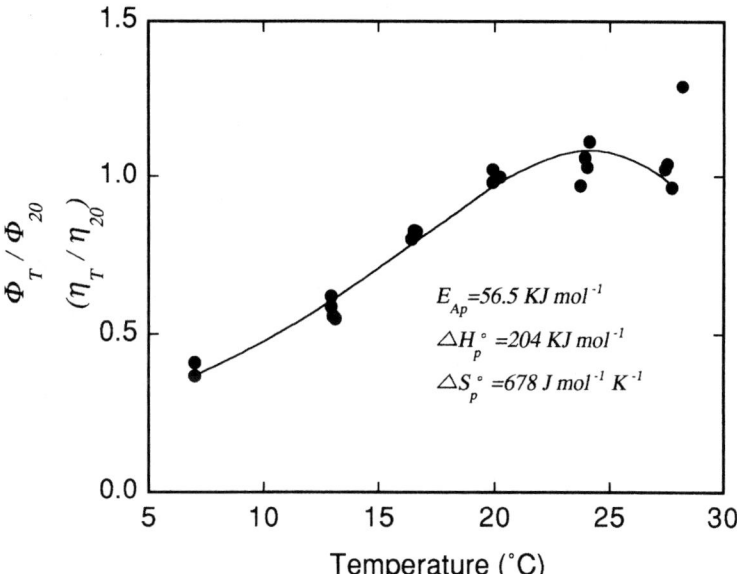

Figure 2. The effect of temperature on Φ_T and η_T. Curve indicates the regression line predicted by fitting Equation 23 (see text for details) to the data.

cover is absent, or is constant throughout the day. If $I_{0(t)}$ is the irradiance at time t h after sunrise, then by integrating $I_{0(t)}$ with respect to time we obtain the daily insolation (Q) (Monteith, 1973). The $I_{0(t)}$ is given by:

$$I_{0(t)} = \frac{\pi Q}{2N_d} \sin\left(\pi \frac{t}{N_d}\right) \tag{24}$$

where N_d is the daylength.

Table 2 shows the measured attenuation coefficient (C), and Table 3, the extinction coefficient (K) with 95% confidence limits of *E. cava* community (Honda, 1996). Figure 5 shows the measured water temperature and the non-linear regression using a sine formula.

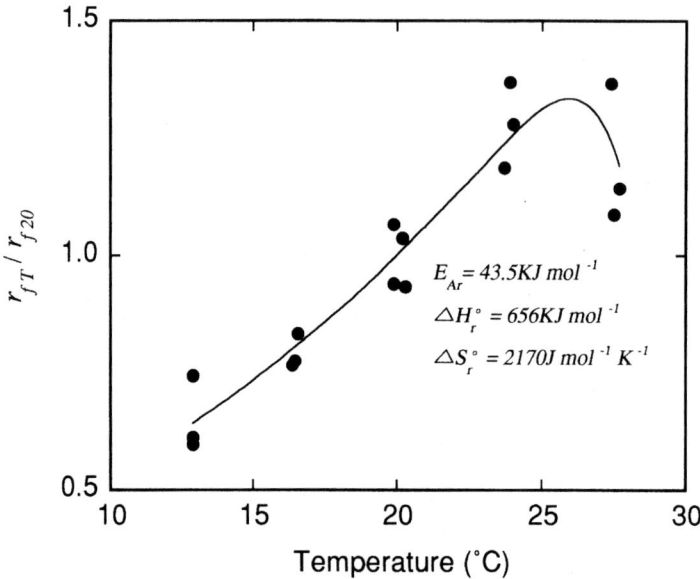

Figure 3. The effect of temperature on r_{fT}. Curve indicates the regression line predicted by fitting Equation 22 (see text for details) to the data.

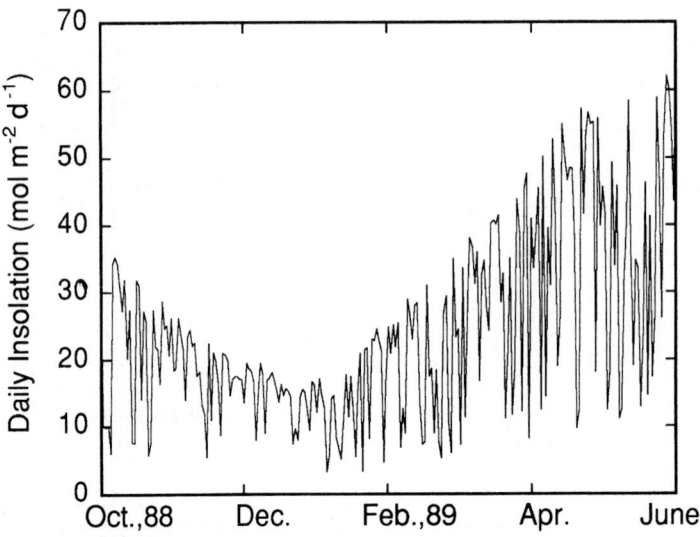

Figure 4. Seasonal changes in the daily insolation at Yokosuka Research Laboratory, CRIEPI.

The mortality, recruitment, ratio of production of blade, number of blades produced, number of blades eroded, initial biomass and production rate of *E. cava* were monitored at the field observation site from October 1988 to June 1989. The value of biomass in October 1988 was 1796 g wet mass m^{-2}. This value was the initial biomass for the production calculation. Table 1 shows changes in the number of individual of *E. cava* at the observation site. The amount of biomass changed with mortality and recruitment. In February 1989, the biomass significantly decreased with the death of one frond. Each change with recruitment in January and April 1989 resulted in less than 20 g m^{-2} in the biomass change. These biomass data were assimilated in the production calculation.

Table 3 shows the value of dry weight of blades of *E. cava* per blade area. The ratio of dry weight to wet weight of the blades was 0.15:1, and the ratio of carbon contents to dry weight was 0.3:1.

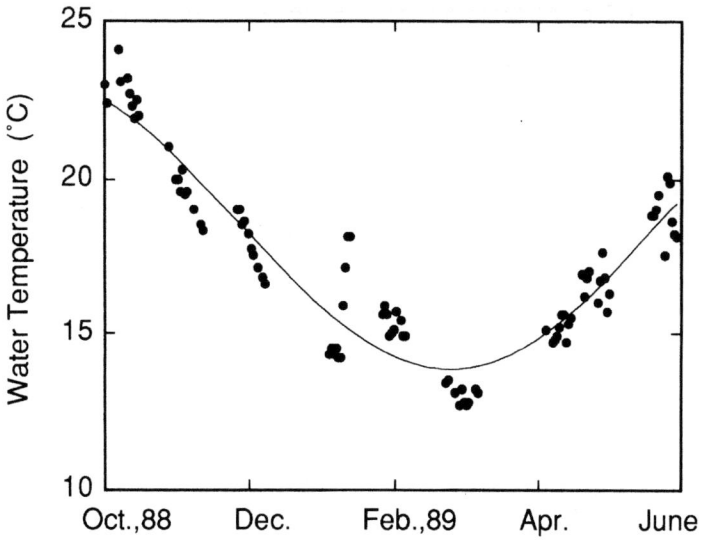

Figure 5. Seasonal changes in the water temperature at the observation site. The closed circles are the observed data. The solid line is a non-linear least-squares fit to sine formula.

Table 3. Seasonal changes in mean value of the extinction coefficient (K) with 95% confidence limits of *Ecklonia cava* community and in the value of dry weight of blades of *Ecklonia cava* per blade area

Month	K			dry weight
	Max	Mean	Min	(g m^{-2})
Oct.	0.66	0.55	0.43	55.2
Nov.	0.75	0.64	0.52	57.7
Dec.	0.64	0.53	0.42	60.4
Jan.	0.53	0.42	0.31	61.3
Feb.	0.42	0.31	0.21	62.3
Mar.	0.40	0.31	0.22	61.0
Apr.	0.37	0.30	0.24	61.0
May	0.35	0.30	0.25	59.5

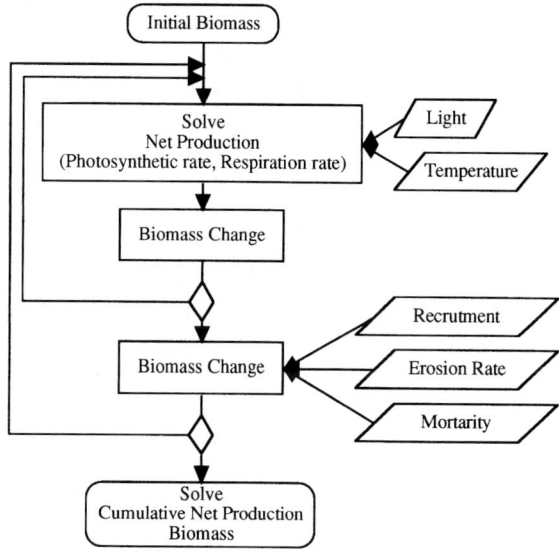

Figure 6. Flow chart for the calculation of production in *Ecklonia cava* community.

The erosion rate is given by:

$$D_{b(t)} = \{P_{\text{gross}(t)} - R_{c(t)}\}d_{b(t)}\frac{N_{e(t)}}{N_{p(t)}}, \quad (25)$$

where $d_{b(t)}$ is the ratio of blade production to blade and stipe production, $N_{e(t)}$ is the number of bladelets eroded (month^{-1}) and $N_{p(t)}$ is the number of bladelets produced (month^{-1}). The values of $d_{b(t)}$, $N_{e(t)}$ and $N_{p(t)}$ are given in Table 2.

Figure 6 shows the flow chart for the calculation of production in *E. cava* community. The time steps of calculation of photosynthetic rate and respiration rate were once every hour, and the time steps of calculation of erosion rate, recruitment and mortality were once per month. Predictions from the production model were compared with field-measured values (Honda, 1993). The relationship of predicted cumulative values to measured values of *E. cava* community production is shown in Figure 7. The result obtained agreed well with the observed data, thus indicates that the produc-

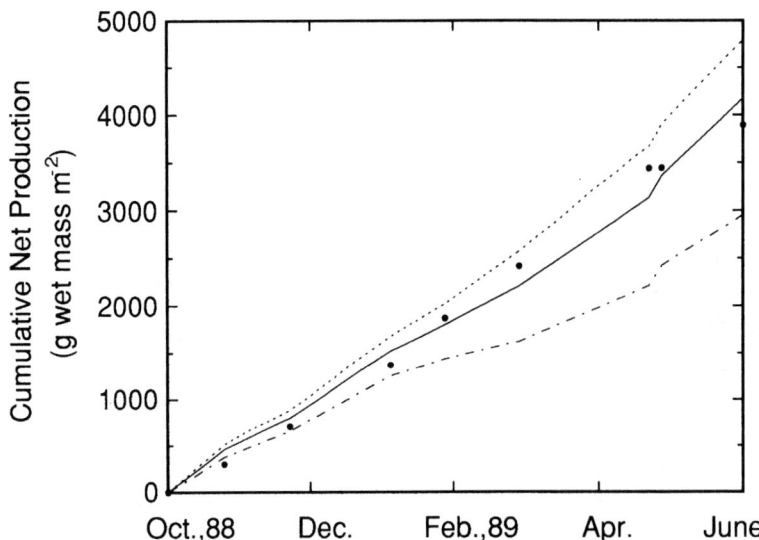

Figure 7. The relationship of predicted cumulative values to measured values of the *Ecklonia cava* community production at the observation site. The solid circles are observed data at the observation site, the solid line is cumulative value with a mean value of the extinction coefficient within the community, the dotted line and dash-dotted line are cumulative values with values of 95% confidence limits of the extinction coefficient within the community.

tion model is a successful method for analyzing the production of the *E. cava* community.

The relationships among the net production, irradiance on community, temperature and biomass are shown in Figure 8. The leaf area index (F) is in proportion to the mass of blades. In the calculation, assuming the value of dry weight of blades per blade area was 60 g m^{-2}. It is clear from the analysis that the net production changed significantly with irradiance, temperature and biomass. In Figure 8 we show that the estimated compensation irradiance increased when the biomass and temperature increased. The estimated compensation irradiance was 26.5 μmol photons m^{-2} s^{-1} when the biomass was 3 kg wet mass m^{-2} (blade:stipe = 2 kg m^{-2}:1 kg m^{-2}, $F = 5$) and the temperature was 20 °C. However, estimated value was 81.6 μmol photons m^{-2} s^{-1} when the biomass was 12 kg wet mass m^{-2} (blade:stipe = 8 kg m^{-2}:4 kg m^{-2}, $F = 20$) at 20 °C. It is clear from the analysis that the optimum temperature changed significantly with irradiance and the biomass. The estimated optimum temperatures were 23.5, 22.8, 20.3, 18.5 and 12.1 °C when the irradiance were 400, 200, 100, 80 and 50 μmol photons m^{-2} s^{-1} respectively and the biomass was 3 kg wet mass m^{-2}. The estimated optimum temperatures were 21.5, 18.2 and 12.1 °C when the irradiance were 400, 200 and 120 μmol photons m^{-2} s^{-1} respectively and the biomass was 12 kg wet mass m^{-2}. The highest estimated value for the optimum temperature was 24.0 °C when the biomass was very small. This temperature agreed with the optimum temperature of the net photosynthetic rate of blade discs when the irradiance was at a high value.

The relationship between the net production and the biomass is shown in Figure 9. The amount of biomass that resulted in the maximum production was influenced by irradiance and temperature. At 400 μmol photons m^{-2} s^{-1} and 20 °C, the estimated value of the biomass (blade:stipe = 2:1) giving the maximum production was about 5.3 kg wet mass m^{-2}. However, at 100 μmol photons m^{-2} s^{-1} and 24 °C, the estimated value was about 3.0 kg wet mass m^{-2}. The estimated values of the maximum production under the two conditions were 1.05 and 0.30 g C m^{-2} h^{-1}, respectively.

Figure 10 demonstrates the relationships among net production, the attenuation coefficient, water depth and the biomass. The maximum amount of biomass was influenced by the attenuation coefficient and the water depth.

Discussion

To analyze the dynamic nature of marine macrophyte production and biomass, a mathematical model based on the physiology of photosynthesis is a useful tool. Models can integrate different aspects of both field and

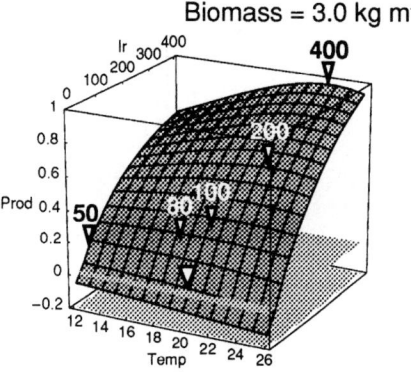

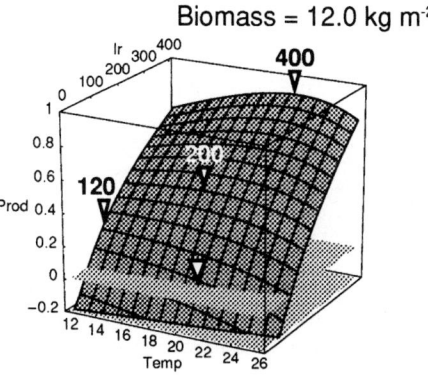

Figure 8. The relationships among the net production (Prod, g C m^{-2} h^{-1}), the irradiance on the community (Ir, μmol photons m^{-2} s^{-1}), temperature (Temp, °C) and the biomass (kg m^{-2}) of the *Ecklonia cava* community. Calculations was carried out using the production model and assuming $K = 0.4$. K is the extinction coefficient within the community. $\triangledown$, compensation points at 20 °C; $\triangledown^N$, maximum production at N μmol photons m^{-2} s^{-1}.

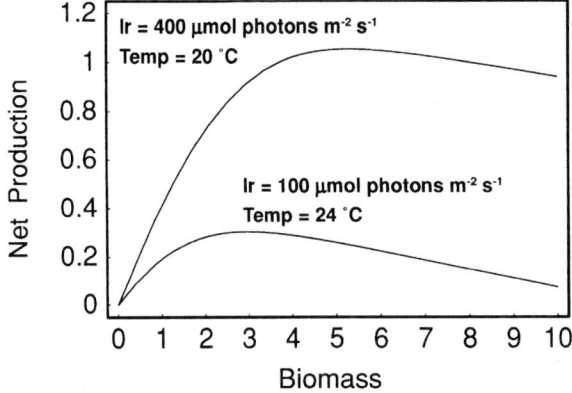

Figure 9. The relationship between the net production (g C m^{-2} h^{-1}) and the biomass (kg m^{-2}) of the *Ecklonia cava* community. Calculations were carried out using the production model and assuming $K = 0.4$. K is the extinction coefficient within the community.

laboratory works. In this study, parameters of photosynthesis and respiration were derived from laboratory work, and parameters of erosion, recruitment and mortality were obtained from field work.

Calculation of the photosynthetic rate is important in the estimation of production. The calculation results (Figures 8–10) showed the difficulty in understanding factors regulating production without computation. Jackson (1987) showed that parametric difference affected the timing of the maximum production and biomass in a year by his model. In this study, photosynthesis-light response measurements were made on the middle portion of the largest bladelet of the frond during the last three months of the year, in October–December. Sakanishi et al. (1989) showed that photosynthetic activities of grown-up bladelets of *E. cava* at 20 °C reached a maximum in January or February and a minimum in August or October. Photosynthetic rate on an area basis did not vary in different bladelets with the photosynthetic rate being highest in the apical and lowest in the basal portion of the bladelet.

In this study, the optimum temperature for light-saturated photosynthesis was 24 °C and that for respiration was about 26 °C. Sakanishi et al. (1989) showed that the optimum temperature for photosynthesis of *E. cava* bladelets was slightly higher than 25 °C in summer and slightly lower than 25 °C in winter, and respiratory rates increased with increasing temperature within the range of 5–30 °C. Kurashima et al. (1996) showed that the optimum temperature for light-saturated photosynthesis was 25–29 °C and the optimum temperatures became lower with decreasing irradiance. This small difference in the optimum temperature for photosynthesis and respiration obtained may be due to the difference in their incubation period at experimental temperature conditions and/or habitat.

Yokohama et al. (1987) showed that daily net production of an *E. cava* community reached a maximum in spring and a minimum in late summer, while standing crop was maximum in summer and minimum in winter. Data of Yokohama et al. (1987) showed that the biomass in summer was over 2 kg dry mass m^{-2} and the maximum leaf area index was 18.7. This biomass was probably over the optimum biomass for net production at 5 m water depth. Analysis by the production model estimated maximum value of optimum leaf area index for integral photosynthesis to be 15 at 24 °C. Aruga et al. (1990) showed that photosynthetic rates of *E. cava* were lower in the sorus portion than in the non-sorus portion, and respiratory rates

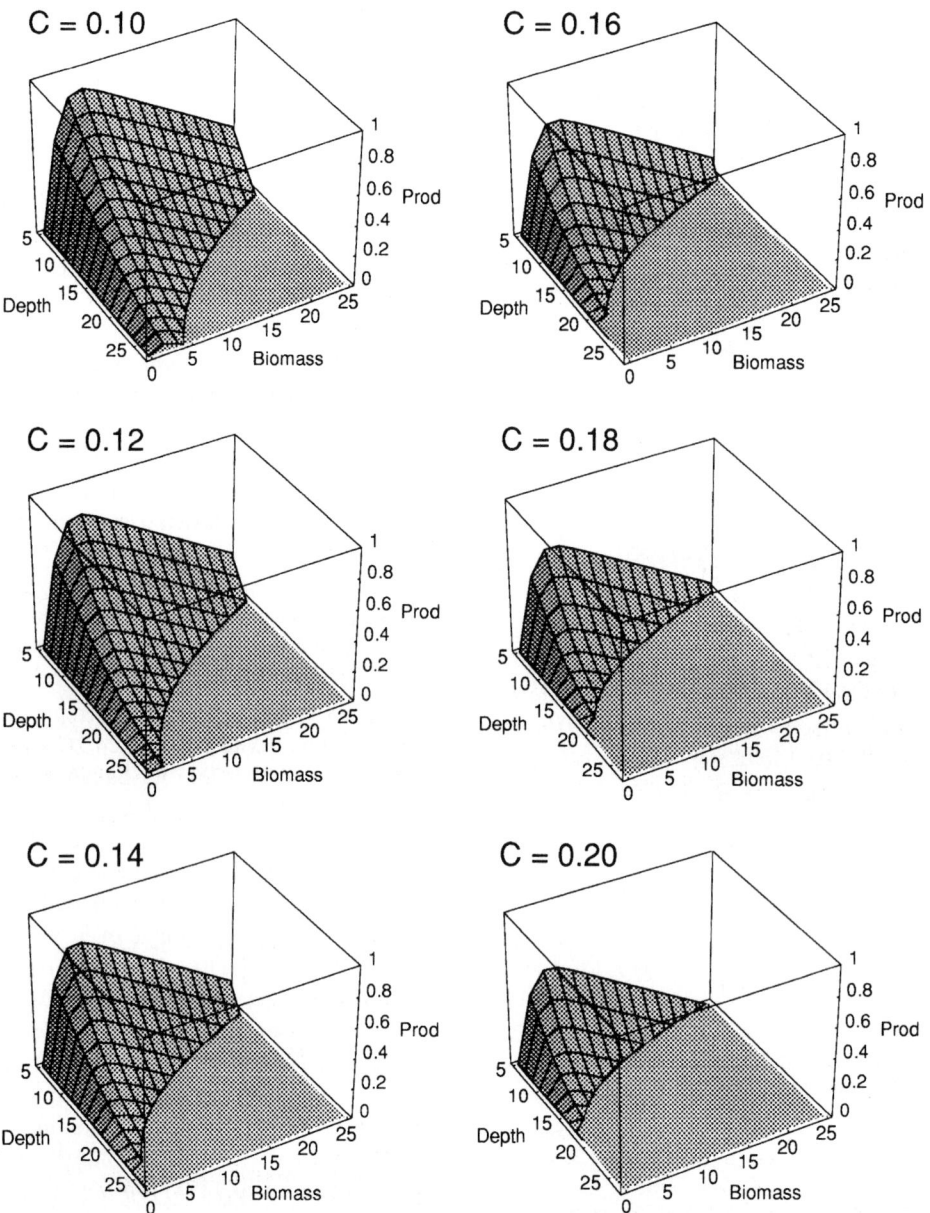

Figure 10. The relationships among the net production (Prod, g C m^{-2} h^{-1}), the attenuation coefficient (C, m^{-1}), water depth (m) and the biomass (kg m^{-2}) of the *Ecklonia cava* community. Calculations were carried out using the production model and assuming $K = 0.4$, $T = 20\,°C$ and $I_0 = 500$ μmol photons m^{-2} s^{-1}. K, T and I_0 are the extinction coefficient within the community, temperature and irradiance at 0 m water depth, respectively.

were higher in the sorus portion than in the non-sorus portion of the bladelets. Further studies (Aruga et al. 1997) showed that the ratio of sorus portion to the total area of bladelets in *E. cava* over 2 year-old was over 20%. Bladelets with zoosporangial sori occupy a greater part of the standing stock and this affects the rate of production of *E. cava* community during summer. Lower photosynthetic rate of sorus portions and large biomass were two of the causes that led to the minimum net production of *E. cava* community in summer.

This model is a simplified description of a *E. cava* community. Most of this simplification is caused by the lack of sufficient information on other important aspects of its physiology and ecology. Aspects to include in the future are:

1. Variations of photosynthetic capacity in blades.
2. Nutrient effects. From the production perspective, nutrient limitation is often considered to be the most important constraint on the growth of kelp (Jackson, 1977; Gerard, 1982; Zimmerman & Kremer, 1984).
3. A more elaborate description of tissue disappearance. Blade-tissue disappearance is also caused by fish and invertebrate grazing and by microbial infection.

Acknowledgments

The author is grateful to Mr S. Sato, Mr H. Araki, Dr N. Nakashiki, Mr M. Matsuyama, Central Research Institute of Electric Power Industry (CRIEPI), Dr T. Okuda, University of Kyushu and Prof. M. Ohno, Kochi University for their helpful comments on this paper and for their continuing guidance and encouragement, and thanks to Dr P. Ang Jr., the Chinese University of Hong Kong for critical reading of the manuscript.

References

Aruga, Y., 1982. Physiological characteristic of *Eisenia bicyclis* and *Ecklonia cava*. Marine Ranching Report 1981: 19–23 (in Japanese).
Aruga, Y., A. Kurashima & Y. Yokohama, 1997. Formation of zoosporangial sori and photosynthetic activity in *Ecklonia cava* Kjellman (Laminariales, phaeophyta). J. Tokyo Univ. Fish. 83: 103–128.
Aruga, Y., M. Toyoshima & Y. Yokohama, 1990. Comparative photosynthetic studies of *Ecklonia cava* bladelets with and without zoosporangial sori. Jap. J. Phycol. 38: 223–228.
Gerard, V. A., 1982. *In situ* rates of nitrate uptake by giant kelp, *Macrocystis pyrifera* (L.) C. Agardh: tissue differences, environmental effects, and predictions of nitrogen-limited growth. J. exp. mar. Biol. Ecol. 62: 214–224.
Haroun, R., Y. Aruga & Y. Yokohama, 1992. Seasonal variation of photosynthetic properties of *Ecklonia cava* (Laminariales, Phaeophyta) in Nabeta Bay, central Japan. La Mer 30: 339–348.
Haroun, R, Y. Yokohama & Y. Aruga, 1989. Annual growth cycle of the brown alga *Ecklonia cava* in central Japan. Scient. Mar. 53: 349–356.
Hayashida, F., 1977. On age and growth of a brown alga, *Ecklonia cava* Kjellman, forming aquatic forest. Bull. Japan Soc. sci. Fish. 43: 1043–1051 (in Japanese).
Hayashida, F., 1986. Synecological studies of a brown alga, *Ecklonia cava* Kjellman, forming aquatic forest III. Structure of *Ecklonia cava* population. J. Fac. mar. Sci. Technol., Tokai Univ. 22: 159–169.

Honda, M., 1993. New estimation method of production in kelp forest without plant collection. CRIEPI Abiko Research Laboratory Rep. U92040, 18 pp.
Honda, M., 1996. Development of a mathematical model of production in *Ecklonia cava* Kjellman community – The relationship of light and temperature to algal production. Jap. J. Phycol. 44: 149–158 (in Japanese).
Ikushima, I., 1967. Ecological studies on the productivity of aquatic plant communities III. Effect of depth on daily photosynthesis in submerged macrophytes. Bot. Mag. Tokyo 80: 57–67.
Ikushima, I., 1970. Ecological studies on the productivity of aquatic plant communities IV. Light condition and community photosynthetic production. Bot. Mag. Tokyo 83: 330–341.
Iwahashi, Y., 1968. Ecological studies on *Eisenia* and *Ecklonia* in the coast of Izu Peninsula – I. On the growth of *Ecklonia cava* Kjellman. Bull. Shizuoka Pref. Fish. exp. Sta. 1: 27–31 (in Japanese).
Jackson, G. A., 1977. Nutrients and production of giant kelp (*Macrocystis pyrifera*) determined *in situ*. Limnol. Oceanogr. 22: 979–995.
Jackson, G. A., 1987. Modelling the growth and harvest yield of the giant kelp *Macrocystis pyrifera*. Mar. Biol. 95: 611–624.
Jassby, A. D. & T. Platt, 1976, Mathematical formulation of the relationship between photosynthesis and light for phytoplankton. Limnol. Oceanogr. 21: 540–547.
Joliot, P., G. Barbieri & R. Chabaud, 1969. Un nouveau modele des centres photochimiques du system II. Photochem. Photobiol. 10: 309–329.
Kasahara, H. & M. Ohno, 1983. Physiological ecology of brown alga, *Ecklonia* on coast of Tosa Bay, southern Japan. III. Growth and morphological change. Rep. Usa mar. biol. Inst. Kochi Univ. 5: 77–84 (in Japanese).
Kida, W. & M. Maegawa, 1982. Ecological studies on *Eisenia bicyclis* and *Ecklonia cava* communities – I. Distribution and composition of the community around the coastal of Cape Goza, Shima Peninsula. Rep. Fish. Res. Lab. Mie Univ. 3: 41–54 (in Japanese).
Kida, W. & M. Maegawa, 1983. Ecological studies on *Eisenia bicyclis* and *Ecklonia cava* communities – II. Distribution and composition of the community in the coastal area of Kumanonada. Bull. Fac. Fish. Mie Univ. 10: 57–69 (in Japanese).
Kok, B., B. Forbush & M. McGloin, 1970. Cooperation of charges in photosynthetic O_2 evolution – I. A linear for step mechanism. Photochem. Photobiol. 11: 457–475.
Kurashima, A., Y. Yokohama & Y. Aruga, 1996. Physiological characteristics of *Eisenia bicyclis* Setchell and *Ecklonia cava* Kjellman (Phaeophyta). Jap. J. Phycol. 44: 87–94 (in Japanese).
Maegawa, M. & W. Kida, 1984. Ecological studies on *Eisenia bicyclis* and *Ecklonia cava* communities - IV. Seasonal changes in allometric relation of *Ecklonia* frond. Bull. Fac. Fish. Mie Univ. 11: 199–206 (in Japanese).
Maegawa, M. & W. Kida, 1987. Studies on production structures on *Eisenia bicyclis* and *Ecklonia cava* communities. Jap. J. Phycol. 35: 34–40 (in Japanese).
Maegawa, M., W. Kida, Y. Yokohama & Y. Aruga, 1987. Critical light condition for young *Ecklonia cava* and *Eisenia bicyclis* with reference to photosynthesis. Hydrobiologia 151/152: 447–455.
Monsi, M. & T. Saeki, 1953. Über den Lichtfaktor in den Pflanzengesellschaften und seine Bedeutung für die Stoffproduktion. Jap. J. Bot. 14: 22-52.
Monteith, J. L., 1973. Principles of Environmental Physics. Edward Arnold, London, 241 pp.
Ohno, M. & M. Ishikawa, 1982. Physiological ecology of brown Alga, *Ecklonia* on Coast of Tosa Bay, southern Japan. I. Seasonal

variation of *Ecklonia* bed. Rep. USA mar. biol. Inst. Kochi Univ. 4: 59–73 (in Japanese).

Sakanishi, Y., Y. Yokohama & Y. Aruga, 1988. Photosynthesis measurement of blade segment of brown alga *Ecklonia cava* Kjellman and *Eisenia Bicyclis* Setchell. Jap. J. Phycol. 36: 24–28.

Sakanishi, Y., Y. Yokohama & Y. Aruga, 1989. Seasonal changes of photosynthetic activity of a brown alga *Ecklonia cava* Kjellman. Bot. Mag. Tokyo 102: 37–51.

Yokohama, Y., J. Tanaka & M. Chihara, 1987. Productivity of the *Ecklonia cava* community in a bay of Izu Peninsula on the Pacific coast of Japan. Bot. Mag. Tokyo 100: 129–141.

Yoshida, T., 1970. On the productivity of *Eisenia bicyclis* community. Bull. Tohoku Reg. Fish. Res. Lab. 30: 107–112 (in Japanese).

Zimmerman, R. C. & J. J. Kremer, 1984. Episodic nutrient supply to a kelp forest ecosystem in Southern California. J. mar. Res. 42: 591–604.

Effects of copper pollution on the ultrastructure of *Lessonia* spp.

Patricia I. Leonardi[1] & Julio A. Vasquez[2]
[1]*Universidad Nacional del Sur. Dpto. de Biología Bioquímica y Farmacia, San Juan 670, 8000 Bahía Blanca, Argentina*
[2]*Universidad Católica del Norte, Dpto. de Biología Marina. Casilla 117, Coquimbo, Chile*
E-mail: leonardi@criba.edu.ar

Key words: coastal pollution, copper pollution, heavy metals, macroalgae, *Lessonia*, *L. nigrescens*, *L. trabeculata*, ultrastructure

Abstract

Plants of *Lessonia trabeculata* and *L. nigrescens* were studied by transmission electron microscopy in order to evaluate ultrastructural level changes in response to copper exposure. Samples of fronds, stipes, and holdfasts were collected from areas with and without copper mining discharges. Changes in cell ultrastructure observed in *Lessonia trabeculata* were related to copper concentrations in seawater, seaweeds and extracted alginates. The results strongly suggest that tolerance or adaptation of *Lessonia* to high concentrations of copper is the capacity of different plant tissues to accumulate copper as precipitates, primarily at two levels: the cell wall and periplasmalemmal space, with the vacuolar system being a third site.

Introduction

Studies of the effects of mining discharges on southeast Pacific littoral marine communities are almost exclusively evaluations of the heavy metal concentrations in different organisms and sediments (Ahumada, 1994; Boré et al., 1989; Lecaros & Astorga, 1992; Trucco et al., 1990; Vásquez & Guerra, 1996; Vermeer & Castilla, 1991). The significance of these results is difficult to infer because of: (i) the non-comparable methodologies used and (ii) the lack of similar information from non-polluted areas. This is especially important in this region where orogeny, high vulcanism and climate can increase the levels of heavy metals (Vásquez & Guerra, 1996; Vila & Sillitoe, 1991).

Despite environmental catastrophies in the region resulting from Cu and Fe mining activity, only certain aspects of changes in intertidal and subtidal populations and communities have been reported (Castilla & Nealler, 1978; Vásquez & Guerra, 1996; Vásquez et al., 1999). Changes at the ultrastructural level have never been studied.

Lessonia nigrescens and *L. trabeculata* dominate cover and biomass in intertidal and shallow subtidal areas affected by mining pollution along the Chilean coast. The main goals of this study were: (1) to observe differences in ultrastructure of individuals of both species collected from polluted and non-polluted areas and (2) to assess cell ultrastructure changes with the levels of copper in seawater, in different plant tissues and in extracted alginates. Numerous studies on macroalgal resistance to different heavy metals have been carried out (see Gledhill et al., 1997; Phillips, 1994), but only a few have focused on ultrastructural changes (Amado Filho et al., 1996a, b; Chan & Wong, 1987; Chan et al., 1981; Lignell et al., 1982; Mariani et al., 1990; Pedersén et al., 1981; Pellegrini et al., 1991, 1993; Silverberg, 1975). This is the first report of ultrastructural changes in the genus *Lessonia*.

Material and methods

Lessonia nigrescens (intertidal) and *L. trabeculata* (subtidal) fronds, stipes and holdfasts were collected from Paposo, near the Santo Domingo copper mine (ca 27° S) and from the Michilla copper mine (22° 48′ S). Control samples were collected 40 and 100 km south of Santo Domingo and Michilla, respectively.

Transmission electron microscopy (TEM)

Samples were cut into 1–2 mm long segments, fixed in 2.5% glutaraldehyde and post-fixed in 2% OsO_4, both at 5 °C. Vehicles for primary and secondary fixatives were prepared as follows: a 0.1 M sodium cacodylate buffering solution with 0.01% $CaCl_2$ was diluted 1:3 in seawater. Dehydration took place in an ascending series of acetone. Samples were embedded drop by drop with an ascending series of Spurr's low viscosity resin (Spurr, 1969), over 5 days. Sections were cut with a diamond knife and stained for 45 min with 1% aqueous uranyl acetate and 55 s with freshly prepared lead citrate. They were then examined with a J 100 CX-II electron microscope of the Centro Regional de Investigaciones Básicas y Aplicadas de Bahía Blanca, Argentina.

Cu concentrations in seawater

The pre-concentration of the seawater samples was carried out according to Berndt et al. (1985). A 500 ml sample was filtered through a 0.45 μm millipore filter at pH 4–5 (Na acetate–acetic acid) and treated with 80 mg of 1-ammonium pyrrolidine-dithiocarbamate (APDTC) dissolved in 2 ml water. It was then vigorously stirred and filtered drop by drop through active carbon. The metallic complexes absorbed by the active carbon were dried at 120 °C for 20 min. After the system had cooled, it was treated with 1 ml of HNO_3, heated and dried. Finally, the carbonaceous residue with the metallic cations was suspended in 1.5 ml of 4.5 M HNO_3 and centrifuged at 10 000 rpm for 15 min. Analyses were carried out on 200 μl aliquots of each sample by atomic absorption spectroscopy (Perkin Elmer 2380) using 1000 μg ml^{-1} metallic Cu in 0.3 M HNO_3 as a standard (J.T. Baker-INSTRA-ANALYZED Reagent).

Cu in alginates and plants of Lessonia trabeculata

For each study area, heavy metal contents of *L. trabeculata* fronds, stipes and holdfasts were analysed. Additionally, the concentration of Cu was determined in alginate samples. Algae were oven-dried for 26 h at 65 °C, cut into small pieces with a plastic knife and ground in a porcelain mortar. Samples weighing approximately 2 g were ashed in a muffle furnace for 45 min at 700 °C. Ash was digested with 20 ml of distilled water and filtered through Whatman paper No 54. The volume of the filtrate was adjusted to 50 ml with distilled water and stored in polyethylene flasks. Heavy metal concentrations were measured as before. Results are expressed as μg of metal per g of dried algae (algal content) and μg of metal per g of alginate (alginate content).

For alginate extraction, samples of dried seaweed (10.0 g) were mechanically stirred for 4 h at 50 °C with 200 ml of 3% sodium carbonate solution and filtered through muslin. The filtrate was centrifuged and the supernatant, concentrated *in vacuo*, was poured over 5 vol of ethanol. The precipitate was filtered and oven-dried for 24 h at 60 °C.

Results

Observations on L. trabeculata

Table 1 shows levels of Cu in populations of *L. trabeculata* growing in discharges with a concentration of 22.4 μg l^{-1}. Macroscopically, individuals appeared healthy but under the electron microscope, changes to organelles, cell wall, and in some cases, deterioration of cells were observed. The normal organization of the chloroplast (Figure 1) was modified, becoming fusiform with undulate, dilated lamellae, irregularly arranged (Figure 2). In some cases the chloroplast envelope was broken (Figure 2, arrowhead). The vacuolar system was filled with an electron dense granular precipitate (Figure 3) which was not apparent in control cells (Figure 4). In a few cells the plasmalemma separated from the wall (Figure 5) and in some cases was broken, allowing the movement of the dispersed lipid bodies into the periplasmalemmal space (Figures 5 and 6). Frequently, irregular electron dense precipitates also appeared in the periplasmalemmal space and in the cytoplasm (Figure 7, arrowheads). These precipitates aggregated between the layers of the inner (Figures 8 and 9) and outer (Figure 10, arrows) wall of meristoderm cells. In control cells these aggregates were lacking (Figure 11). Stipe cells showed similar modifications to those appearing in frond cells. In many cells a thin and irregular layer of an electron dense precipitate was observed between the inner and outer strata of the walls and also between contiguous cells (Figure 12, arrows, compare with control cells in Figure 13). A reduced number of holdfast cells was disrupted, mainly by the appearance of electron dense droplets in the periplasmalemmal space and cytoplasm. Electron dense precipitates were also observed in the vacuolar system and wall layers.

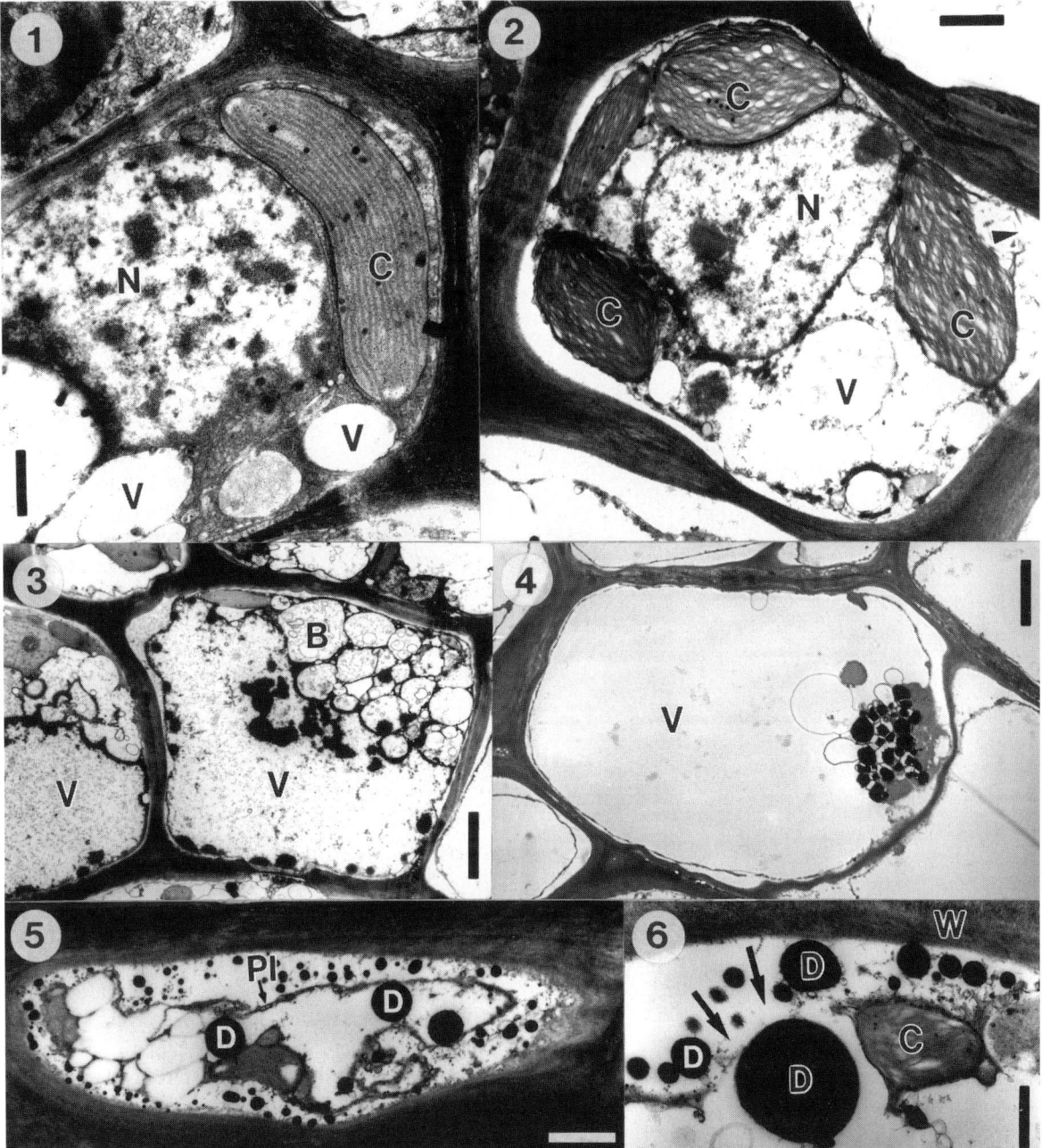

Figures 1–6. Effects of copper on frond cell structure in *Lessonia trabeculata*. **Figure 1.** Frond cell of control plant growing without copper. Portion of a meristodermic cell with characteristic organelles, scale bar = 0.7 μm. **Figure 2.** Note that the parallel thylakoid bands became undulate, dilated and irregularly arranged. The arrowhead shows a rupture of the chloroplast envelope, scale bar = 1.5 μm. **Figure 3.** The vacuole and vacuolar system are filled with an electron dense granular precipitate, scale bar = 7 μm. **Figure 4.** Frond cell of control plant growing without copper, scale bar = 5 μm. **Figure 5**. Disrupted cell, the plasmalemma is separated from the wall and broken. Note the lipid droplets in the periplasmalemmal space and cytoplasm, scale bar = 1.5 μm. **Figure 6.** Details of Figure 5 — the arrows indicate sites of plasmalemma rupture, scale bar = 1 μm. B, vacuolar system, C, chloroplast, D, electron dense droplet, N, nucleus; Pl, plasmalemma; V, vacuole; W, cell wall.

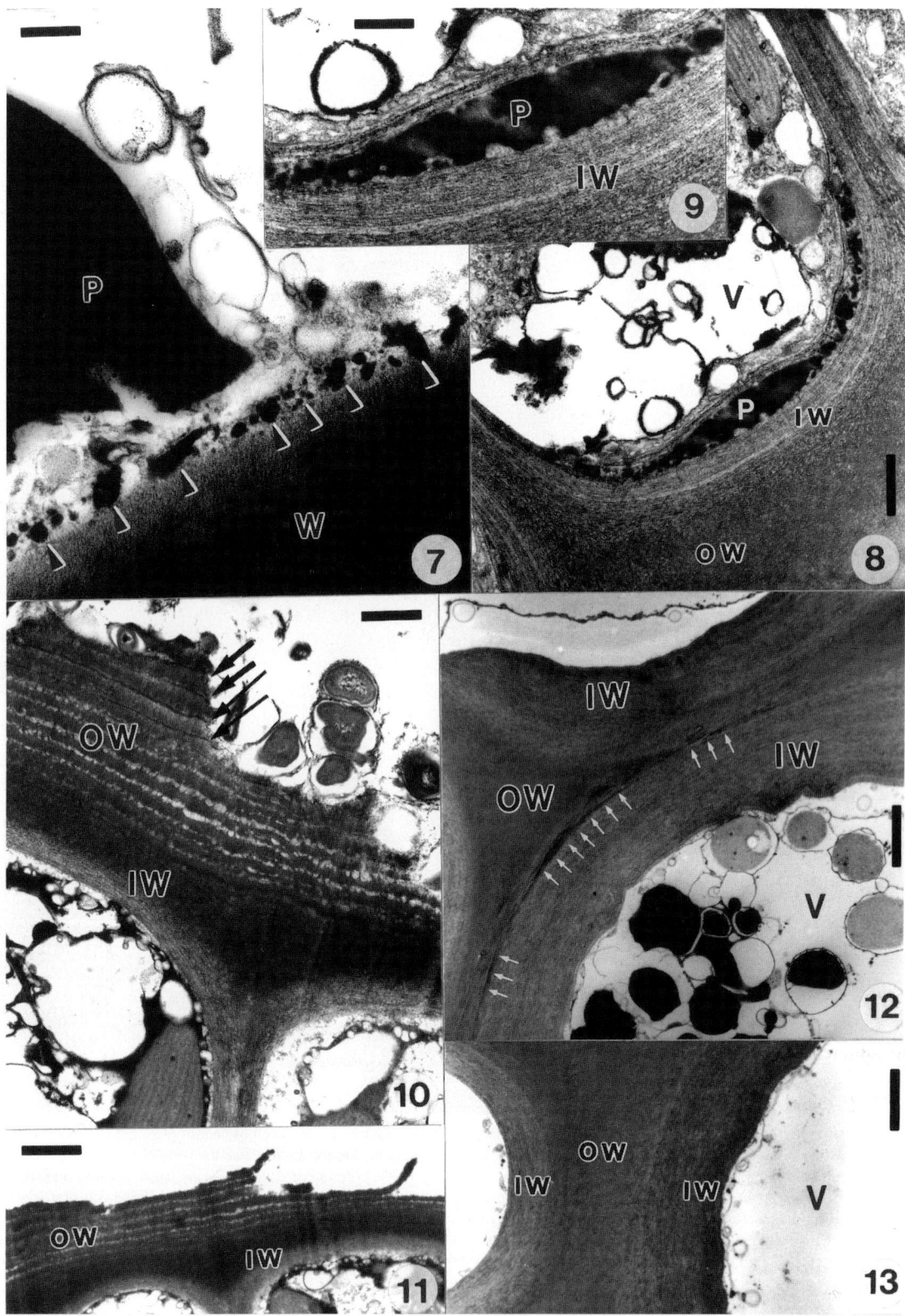

Figures 7–13.

Table 1. The range of copper concentrations in tissues and extracted alginates from different parts of plants of L. trabeculata growing in the control area 100 km (control) and 3 km from the Michilla mine discharge with a copper concentration of 22.4 μg l^{-1}.

	Control		Near discharge	
	Tissues (μg g^{-1} Cu)	Alginate (μg g^{-1} Cu)	Tissues (μg g^{-1} Cu)	Alginate (μg g^{-1} Cu)
Frond	2.92–3.55	4.63–5.38	2.39–50.96	5.56–7.22
Stipe	1.70–4.59	6.41–7.69	10.42–66.57	9.6–10.21
Holdfast	1.29–1.81	4.56–5.68	4.4–58.8	1.99–4.51

Observations on L. nigrescens

Populations of *L. nigrescens* growing in discharges with Cu concentration of 15.6 μg l^{-1} also showed ultrastructural changes. Compared to control plants (Figures 14 and 15) the cytoplasm of some meristoderm and upper cortical frond cells were greatly damaged (Figure 18). The outer layer of the external cell wall (Figures 17 and 18) showed an electron dense precipitate with irregular size and localization which was not apparent in control cells (Figure 16). Electron dense aggregates were also observed in the periplasmalemmal space (Figures 18 and 19). In stipe cells the plasmalemma withdrew from the wall and a vacuolar system appeared (Figure 20); this is not apparent in control cells (Figure 21). The holdfast meristoderm cells showed a conspicuous modification, i.e. the total or almost total replacement of the vacuolar content by a mass of electron dense material (Figures 22 and 23). In vacuoles of cortical cells this material was found as spherical or ovoid lacunae (Figure 23) or as a fine precipitate (Figure 24). Aggregates of electron dense deposits were frequently observed on the external wall and on epiphytic bacteria (Figure 25, arrowheads).

Discussion

Total Cu concentrations in seawater range between 0.002 and 0.6 μg l^{-1} (Maeda & Sakaguchi, 1990; Phillips, 1977). These values can increase considerably in coastal waters affected by copper mining discharges, as occurs in Chile. In this study the Cu values reached 4–20 μg l^{-1} (intertidal) and 20–40 μg l^{-1} (subtidal). The levels of Cu found in different tissues and extracted alginates of *L. trabeculata* growing in polluted areas (Table 1) indicated a great capacity for retaining and concentrating this heavy metal. The apparent dense precipitates appearing at different sites on the outer cell walls of these plants strongly suggest that it was one of the sites of Cu retention. Heavy metals are retained in such a way by many algae (Kloareg, 1981; Kloareg et al., 1986; Lestang & Quillet, 1974; Percival, 1979). Mariani et al. (1985) have suggested that the cell wall has a role in adaptive resistance to changes in the ionic environment during the tidal cycle by means of cation-binding to the negatively charged polysaccharides. A second form of Cu retention was found in *L. trabeculata*; the metal was accumulated as compact aggregates situated between the sheaths of the inner wall layer. Aggregates with the same appearance and position have been found in *Cystoseira barbata* exposed to Cd (Pellegrini et al., 1991).

Electron dense aggregates in the periplasmalemmal space similar to those detected in the present study have been observed in cells of the green alga *Spirogyra submargaritata* growing in media with the addition of salts of Cd, Cr, Cu, Fe, Mn, Pb and Zn (Chan et al., 1981). These authors also suggested that this space can act as a temporary reservoir for heavy metals. The dense granular precipitates located in the vacuolar system of cells from fronds, stipes and holdfasts of *L. trabeculata* and holdfasts of *L. nigrescens* have also been described as a major site of metal deposition and retention in the green alga *Stigeoclonium tenue* maintained in media with Pb (Silverberg, 1975).

←

Figures 7–13. Effects of copper on cell wall structure in fronds of *Lessonia trabeculata*. **Figure 7.** Details of a portion of meristoderm cell with irregular, electron dense precipitates in the periplasmalemmal space and cytoplasm, scale bar = 0.5 μm. **Figure 8.** General view of a meristoderm cell with an electron dense aggregate between the layers of the inner wall, scale bar = 1 μm. **Figure 9.** Details of Figure 8, scale bar = 0.5 μm. **Figure 10.** Details of the external cell wall in which electron dense, fine aggregates are observed between the layers of the outer wall (arrows), scale bar = 1 μm. **Figure 11.** External cell wall of a control frond, scale bar = 1.5 μm. **Figure 12.** Effects of copper on cell wall structure in stipe. A fine granular material can be observed between the inner and outer layers of the wall and also between contiguous cells (arrows), scale bar = 4 μm. **Figure 13.** Cell wall of a control plant stipe, scale bar = 2 μm. IW, inner wall; OW, outer wall; P, electron dense precipitates; V, vacuole; W, cell wall.

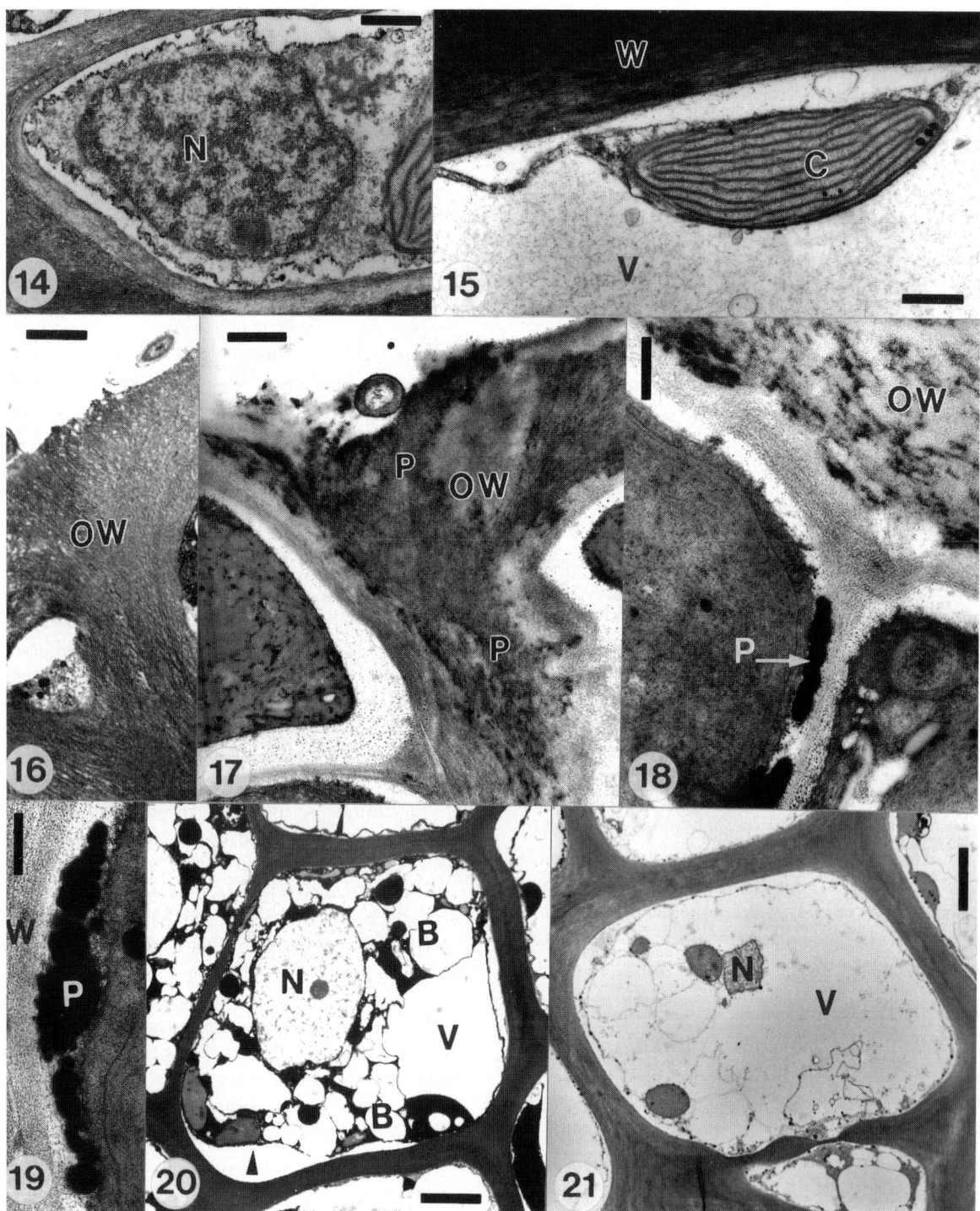

Figures 14–21. Effects of copper on frond cells and cell wall structure of *Lessonia nigrescens*. **Figures 14 and 15.** Details of two control cells in which the characteristic organelles are present, scale bar = 1 μm. **Figure 16.** Control cell with details of the external cell wall, scale bar = 1 μm. **Figure 17.** Details of copper exposed cell showing the external cell wall with an electron dense precipitate between the outer wall layers, scale bar = 1 μm. **Figure 18.** Details of extremely altered meristoderm cells exposed to copper. Organelles are not discernible. Note the presence of an electron dense precipitate between the outer layer and also the aggregates in the periplasmalemmal space, scale bars = 0.5 μm. **Figure 19.** Detail of copper exposed cell showing an aggregate of electron dense precipitate, scale bars = 0.35 μm. **Figure 20.** Effects of copper on stipe cells of *L. nigrescens*. Details of a cell in which retraction of the plasmalemma took place (arrowhead) and a vacuolar system is present, scale bar = 4 μm. **Figure 21.** Stipe cell growing without copper, scale bar = 4 μm. B, vacuolar system, C, chloroplast; N, nucleus; OW, outer wall; P, electron dense precipitates; V, vacuole; W, cell wall.

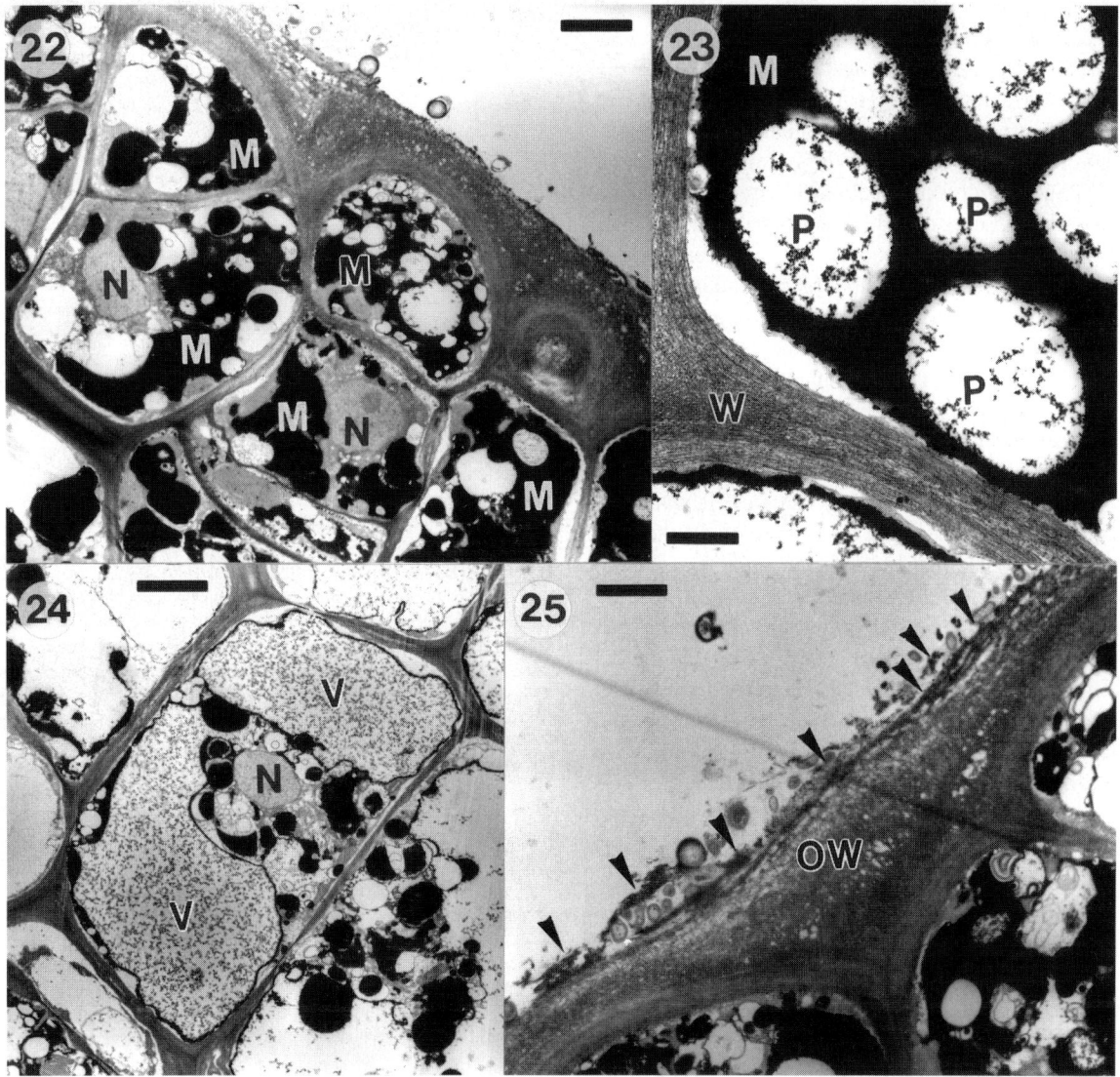

Figures 22–25. Effects of copper on holdfast cell and cell wall structure in *Lessonia nigrescens*. **Figure 22**. General view of meristoderm cells where the vacuolar content is almost all or totally replaced by an electron dense material, scale bar = 4 µm. **Figure 23.** Detail of a cortical cell showing similar vacuolar material, scale bar = 5 µm. **Figure 24.** Cortical cells with the vacuoles filled with a fine electron dense precipitate, scale bar = 1 µm. **Figure 25**. Detail of an external cell wall showing aggregates of electron dense precipitates on the wall and on epiphytic bacteria (arrowheads), scale bar = 2 µm. M, electron dense material; N, nucleus; OW, outer wall; P, electron dense precipitates; V, vacuole.

In holdfast meristoderm cells of *L. nigrescens*, vacuoles were observed to be occupied by a mass of electron dense material, not previously reported in other algae. Images of contiguous cells at different levels of the external tissues of the holdfast, showed a gradient of accumulation with the innermost cells having vacuoles filled with granular electron dense precipitates and the external cells having vacuoles filled with this mass of electron dense material.

The ultrastructural modifications found in chloroplasts of fronds and stipes of *L. trabeculata* have also been reported in *Cystoseira barbata* exposed to Cu, Cd and Zn (Pellegrini et al., 1991, 1993), and in *Padina gymnospora* growing in the presence of Zn (Amado Filho et al., 1996b). The marine filamentous green alga *Chaetomorpha brachygona,* growing on iron-ore tailings, showed a reduction in the size of the chloroplast and a large portion of the thylakoids was dilated and less compact (Chan & Wong 1987).

Large and small electron-dense lipid bodies were found in the cytoplasm and periplasmalemmal space in cells of *L. trabeculata*. The change in membrane permeability and the rupture of the plasmalemma under high Cu levels (Gledhill et al., 1997) are possible reasons for the presence of lipid droplets in the periplasmalemmal space. In *Chaetomorpha* the synthesis of large amounts of lipids is an immediate cellular response to heavy metal exposure (Chan & Wong, 1987). In the same way, the cytoplasm of the green microalga *Scenedesmus* cultured in Cu contained abundant dense bodies and lipid droplets (Silverberg et al., 1976).

In conclusion, an evaluation of a great number of photomicrographs suggests that copper is almost consistently present as precipitates in cell walls, its localization and amount in the other parts of the cell, i.e. periplasmalemmal space, vacuolar system, being, in contrast, extremely variable. The results strongly suggest that tolerance or adaptation of *Lessonia* plants growing in high concentrations of Cu is due to the capacity of the different plant tissues to sequester Cu in the form of precipitates at two main cellular levels, the wall and periplasmalemmal space, with sequestration within the vacuolar system being a third site.

Acknowledgements

This work was supported by grants of Fundación Andes (Cooperación Académica Internacional) and FONDECYT 1960202. P.I.L. is a research member of the Consejo Nacional de Investigaciones Científicas y Técnicas de la República Argentina. E. Cáceres, R. Lois and anonymous reviewers provided constructive comments on the manuscript.

References

Amado Filho, G. M., C. S. Karez, W. C. Pfeiffer, Y. Yoneshigue-Valentin & M. Farina, 1996a. Accumulation, effects on growth, and localization of zinc in *Padina gymnospora* (Dictyotales, Phaeophyceae). Hydrobiologia 326/327: 451–456.

Amado Filho, G. M., C. S. Karez, W. C. Pfeiffer, Y. Yoneshigue-Valentin & M. Farina, 1996b. Localization of zinc deposits in cells of the brown alga *Padina gymnospora* by electron probe microanalysis. Ci. Cult. J. brazilian Ass. Adv. Sci. 48: 197–200.

Ahumada, R., 1994. Niveles de concentración e índices de bioacumulación àra metales pesados (Cd, Cr, Hg, Ni, Cu, Pb, y Zn) en tejidos de invertebrados bénticos de Bahía San Vicente, Chile. Rev. Biol. mar. 29: 77–87.

Berndt, H., U. Harms & M. Sonneborn, 1985. Multielementspurenanreicherung aus wassern an aktiykhole zur probenvorbereitung fur die atmospektroskopie (Flammen-AAS, ICP/OES) Fresenius Z. anal. Chem. 3: 329–333.

Boré, D., H. Robothan, R. Trucco, J. Inda & M. L. Fernández, 1989. Evaluación preliminar de la presencia de metales pesados en recursos pesqueros de importancia comercial de la III Región de Chile. Rev. Pacífico Sur. (special number): 195–203.

Castilla, J. C. & E. Nealler, 1978. Marine environmental impact due to mining activities of El Salvador Copper Mine, Chile. Mar. Pollut. Bull. 9: 67–70.

Chan, K. & S. L. L. Wong, 1987. Ultrastructural changes of *Chaetomorpha brachygona* growing in metal environment. Cytologia 52: 97–105.

Chan, K., S. L. L. Wong & P. K. Wong, 1981. Ultrastructural changes in *Spirogyra submargaritata* growing on iron-ore tailing. Cytologia 46: 15–26.

Gledhill, M., M. Nimmo, S. J. Hill & M. T. Brown, 1997. The toxicity of copper (II) species to marine algae, with particular reference to macroalgae. J. Phycol. 33: 2–11.

Kloareg, B., 1981. Structure et rôle ecofisiologique des parois des algues littorales: contribution à la résistence aux variations de salinité. Physiol. vég. 19: 427–441.

Kloareg, B., M. Demarty & S. Mabeau, 1986. Polyanionic characteristics of purified sulphated homofucans from brown algae. Int. J. biol. Macromol. 8: 380–386.

Lecaros, O. & M. S. Astorga, 1992. Metales pesados en *Macrocystis pyrifera* (huiro) de la costa del Estrecho de Magallanes. Rev. Biol. mar. 27: 5–16.

Lestang de G. & M. Quillet, 1974. Compartement du fucoïdane sulfurylé de *Pelvetia caniculata* (Dcne & Thur.) vis à vis des cations de la mer: propriété d'echange, renouvellement des radicaux sulfuriques, coenzyme d'activation des sulfates. Intérêt fonctionnel. Physiol. vég. 12: 199–227.

Lignell, A., Roomans, G. M. & M. Pedersén, 1982. Localization of absorbed cadmium in *Fucus vesiculosus* L. by X-ray microanalysis. Z. Pflanzenphysiol. 105: 103–109.

Maeda, S. & T. Sakaguchi, 1990. Accumulation and detoxification of toxic metal elements by algae. In Akatsuka, I. (ed.), Introduction to Applied Phycology. SPB Academic Publishing The Hague : 109–136.

Mariani, P., C. Tolomino & P. Braghetta, 1985. An ultrastructural approach to the adaptive role of the cell wall in the intertidal alga *Fucus virsoides*. Protoplasma 128: 208–217.

Mariani, P., C. Tolomio, B. Balden & P. Braghetta, 1990. Cell wall ultrastructure and cation localization in some benthic marine algae. Phycologia 29: 253–262.

Pedersén, M., G. M. Roomans, M. Andrén, A. Lignell, G. Lindahl, K. Wallström & A. Forsberg, 1981. X-ray microanalysis of metal in algae. A contribution to the study of environmental pollution. Scan. Electron Microsc. 2: 499–509.

Pellegrini, L., M. Pellegrini, S. Delivopoulos & G. Berail, 1991. The effects of cadmium on the fine structure of the brown alga *Cystoseira barbata* forma *repens* Zinova et Kalugina. Br. phycol. J. 26: 1–8.

Pellegrini, M., A. Laugier, M. Sergent, R. Phan-Tan-Lu, R. Valls & L. Pellegini, 1993. Interactions between the toxicity of the heavy metals cadmium, copper, zinc in combinations and the detoxifying role of calcium in the brown alga *Cystoseira barbata*. J. appl. Phycol. 5: 351–361.

Percival, E., 1979. The polysaccharides of green, red and brown seaweeds: their basic structure, biosynthesis and function. Br. phycol. J. 14: 103–117.

Phillips, D., 1977. The use of biological indicator organisms to monitor trace metal pollution in marine and estuarine environments — a review. Envir. Pollut. 13: 281–317.

Phillips, D. 1994. Macrophytes as biomonitors of trace metals. In Kramer, K. J. M. (ed.), Biomonitoring of Coastal Waters and Estuaries. CRC Press, Boca Raton, Fl.: 85–103.

Silverberg, B. A., 1975. Ultrastructural localization of lead in *Stigeoclonium tenue* (Chlorophyceae, Ulotrichales) as demostrated by cytochemical and X-ray microanalysis. Phycologia 14: 265–275.

Silverberg, B. A., P. M. Stokes & L. B. Ferstenberg, 1976. Intranuclear complexes in a copper-tolerant green alga. J. Cell Biol. 69: 210–214.

Spurr, A. R. 1969. A low viscosity epoxy resin embedding medium for electron microscopy. J. ultrastruct. Res. 26: 31–43.

Trucco, R. G., J. Inda & M. L. Fernández, 1990. Heavy metal concentration in sediments from Tongoy and Herradura Bays, Coquinbo, Chile. Mar. Pollut. Bull. 21: 229–232.

Vásquez, J. A. & N. Guerra, 1996. The use of seaweeds as bioindicators of natural and anthropogenic contaminants in northern Chile. Hydrobiologia 326/327: 327–333.

Vásquez, J. A., J.M.A Vega, B. Matsuhiro & C. Urzúa, 1999. The ecological effects of mining discharges on subtidal habitats dominated by macroalgae in northern Chile: population and community level studies. Hydrobiologia 398/399 (Dev. Hydrobiol. 137): 217–229.

Vermeer, K. & J. C. Castilla, 1991. High cadmium residues observed during a pilot study in shorebirds and their prey downstream from El Salvador copper mine, Chile. Bull. environ. Contam. Toxicol. 46: 242–248.

Vila, T. & R. H. Sillitoe, 1991. Gold-rich porphyry systems in the Maricunga belt, northern Chile. Econ. Geol. 86: 1238–1260.

Further evaluation of the structure of the polysaccharide from *Plocamium costatum* with the use of set theory

Ian J. Miller
Carina Chemical Laboratories Ltd, P.O. Box 30366, Lower Hutt, New Zealand

Key words: algal galactan, chemical structure, NMR spectroscopy, Plocamiaceae

Abstract

The polysaccharide from *Plocamium costatum* has been examined by chemical and spectroscopic techniques. Set theory was used to confirm that the polymer chain consists of alternating 3-linked and 4-linked galactopyranosyl units, with approximately equal amounts of the 4-linked units being in the D- and L- configurations. A further analysis through set theory of the linkage analysis data, and of spectral data on methylated desulphated polysaccharide indicates that most of the 4-linked residues are 3-sulphated, most of the diads containing 4-linked L-galactosyl residues have a common substitution pattern, while there is considerable variation in substitution pattern involving the diads containing 4-linked D-galactosyl residues.

Introduction

Most red algae contain linear polymeric galactans, but not all of these have simple repeating structures. The galactan from *Pachymenia lusoria* (Miller et al., 1995) had ten significant constituent units, yet when further *Pachymenia* samples from at least three species were examined (Miller et al., 1997) all samples had common features. The differing levels of these, however, suggested that distinctions could be made between species. Thus complicated structures may contain information of taxonomic significance. Discrete mathematics, and in particular set theory, is a useful tool for handling such issues. It addresses questions such as is a feature present? Is a feature common to more than one set? It may not be necessary to understand the feature, but it is necessary that a question can always be answered yes or no. To do this it is necessary that each feature under consideration be clearly distinguishable from any other. A review (Miller, 1997) has identified taxonomic issues relating to red algal polysaccharides in terms of discrete mathematics.

The first structural report on a polysaccharide from the order Plocamiales was on *Plocamium cartilagineum* (Linnaeus) Dixon (Whyte et al., 1984). The only significant constituent sugar was D-galactose, the linkages to chain and sulphate ester substituents were {3-, 2,3,4-, 3,4-, 3,4,6-}, and since the infrared spectrum indicated the presence of 4-sulphate, it was suggested that there was a predominance of 3-linked galactosyl residues. Recently, a study on the polysaccharide from *Plocamium costatum* (C. Agardh) Hook. f. et Harv. (Falshaw et al., 1998) showed that 25% of the galactose was in the L-configuration, and the polymer backbone consisted of the expected structure of alternating 3- and 4-linked residues, however similar substitution patterns on the polymer prevented further structural information from being defined. The purpose of this paper is to report ^{13}C nuclear magnetic resonance (NMR) data on chemically modified polymers from *Plocamium costatum* and to demonstrate how set theory permits structural feature to be obtained from such spectral data.

Materials and methods

Plocamium costatum was collected at Mount Maunganui (Tauranga) in December, 1991, and air-dried. A sample of this material has been deposited in the Museum of New Zealand, Te Papa, Wellington, with the following herbarium number: WELT A 20826.

Extraction and chemical modification of the native polysaccharide

The extraction is reported in detail elsewhere (Falshaw et al., 1998). Desulphation followed the method of Miller & Blunt (1998) using arsenous oxide as the SO_3 acceptor, and methylation followed the method outlined by Miller (1998). Reactions were carried out on a 2 g scale, to give sufficient material for further chemical modification, and for nuclear magnetic resonance experiments.

Spectroscopy

The ^{13}C NMR spectra were recorded on a Varian Unity 500 spectrometer (acquisition time 1.17 s, delay 0.8 s, 90° pulse, 3000–4500 transients) on solutions 5% w/v in D_2O/H_2O at 90 °C. All signals reported below are in ppm referenced to internal DMSO at 39.4 ppm.

Symbols

The set theory analysis will use the conventions outlined previously (Miller, 1998). Polysaccharides are written in the order of the linkage position (the 3-set and the 4- set), the configuration (D- or L-), the sugar (G for galactose, AG for 3,6-anhydrogalactose) followed by the functions to the free oxygen atoms, in sets denoted by { } if there is more than one at a given position, the functions being sulphate ester (S), methyl ether (M), pyruvate acetal (P), glycosyl units such as xylose (X), and hydrogen (H) if required to complete a set. Features irrelevant at the time can be omitted in any description, and any feature can be described in terms of separable subsets. As an example, porphyran (in which the 3-linked D-galactosyl residue may be partially 6-*O*-methylated, and the 4-linked residue may be 3,6-anhydro-L-galactose or L-galactose which has predominantly, but not completely a sulphate ester at the 6-position) would be described as [3DG6{H,M}-4L(G6{S,H} ∪ AG)].

The natural galactan is termed G. A ^{13}C NMR spectral set S of the polymer G will be referred to as SG, which will contain N G elements. The amount present of any material X will be A X. The addition (or removal with a minus sign) of a substituent to a material will be represented by an operator, in vertical bars. Thus G |M,-S| would mean the extracted galactan has been methylated, then desulphated. In the absence of a number showing selectivity to the nominated positions, the operation is carried out at all available sites.

Such operators will also carry out mathematical operations to the corresponding spectral matrices (Miller, 1998).

Results and discussion

Carbohydrate linkages

While the backbone linkages have been established by conventional methods (Falshaw et al., 1998) application of set theory to the ^{13}C NMR spectrum of the desulphated polymer G | − S| permits firmer conclusions. The spectrum

$$SG|-S| = \{104.6, 103.6, 100.7, 96.2, 81.0, \\ 79.1, 78.7, 78.6, 75.5, 75.2, \\ 72.0, 70.9, 70.6, 70.4, 69.9, 69.3, \\ 68.8, 65.7, 61.4, 61.3, 61.0\} \quad (1)$$

Clearly

$$N(SG|-S|) = 21 \quad (2)$$

If a polymer consists of n hexoses, with n hexose anomeric signals

$$N = 6n - C \quad (3)$$

where C is the number of coincidences. The anomeric subset of signals is

$$SG(\text{anomeric}) = \{104.6, 103.6, 100.7, 96.2\} \quad (4)$$

which implies $n = 4$, hence from (3) $C = 3$.

From Lahaye et al. (1985),

$$S(3DG - 4LG) = \{103.7, 100.9, 81.0, 79.3, \\ 75.7, 72.2, 71.0, 69.9, \\ 69.4, 68.9, 61.4, 61.2\} \quad (5)$$

From Falshaw & Furneaux (1994),

$$S(3DG-4DG) = \{104.8, 96.3, 78.9, 78.7, \\ 75.3, 71.0, 70.6, 70.5, \\ 69.3, 65.8, 61.5, 61.3\} \quad (6)$$

∴ within a tolerance of 0.5 ppm

$$S(3DG - 4LG) \subset SG|-S| \text{ and} \\ S(3DG - 4DG) \subset SG|-S| \quad (7)$$

Now

$$S(3DG - 4LG) \cap S(3DG - 4DG) = \{70.9, \\ 69.3, 61.3\} \quad (8)$$

Table 1. Inferred linkages (mol%) for the *Plocamium costatum* polysaccharide determined from alditol acetates obtained following reductive hydrolysis

Position of linkage from galactose to chain or sulphate ester	Modified Polymer	
	$G \mid M \mid$	$G \mid -S,M \mid$
3- set		
3-	14	41
2,3-	8	–
3,6-	3	6
4- set		
4-	2	44
2,4-	4	2
3∪4- set		
3,4-	25	4
2,3,4-	24	–
3,4,6-	13	–
2,3,4,6-	4	–
Additional residues		
4-AG	1	1
Terminal X	2	1
Terminal G	–	1

The intensity of the signals of the subset defined by (8) in $\mathbf{S}\,G \mid -S \mid$ is in accord with the requirement that they are coincidences. The signal intensities of G (anomeric) are approximately equal, hence the spectrum is in accord with the structure of the desulphated *Plocamium costatum* polysaccharide containing approximately equal representation of the alternating structures required by (8).

The advantage of this mathematical approach is that all assumptions are clearly identified, it identifies checkable requirements, such as coincident signal intensities, specific signal assignments are not necessary, it works equally well as the spectrum becomes more complex, and it can be the basis of computerized spectral analysis.

Linkages to sulphate ester

The polysaccharide linkage points found previously (Falshaw et al., 1998) are also listed for convenience in Table 1. No further structural conclusion was made, as chemical analysis did not identify specific structural linkages.

Nevertheless, one further deduction can be made from the condition $A(3-) = A(4-)$ i.e. the amounts of 3-linked and 4-linked residues are equal. In Table 1 only, the substitutions {3-∼4-} (total, 25 mol%, 3G4H) and {4-∼3-} (total, 6 mol%, 4G3H) are unambiguous; the remainder, $S(u) = 3$-∩4- (chain ∪ substituent) is the subset {3G4S ∪ 4G3S}. (Note that other substitutions are not excluded because they are not there, but because there is no additional condition applicable to 2-sulphation or 6-sulphation.) Now from Table 1, $A\,\mathbf{S}(u) = 66$ mol%, $A\,3G4H = 25$ mol%, hence from $G \mid -S,M \mid$ in Table 1 $A\,3- = 47$ mol%, and since $3- = 3G4\{S \cup H\}$ then

$$A3G4S \leq 47 - 25 \leq 22 \text{mol\%} \qquad (9)$$

and

$$A4G3S \geq 66 - 22 \geq 44 \text{mol\%} \qquad (10)$$

Thus although many of the species remain unidentified and there is a level of uncertainty in the chemical analysis, it follows that most of the 4-linked residues contain 3-sulphate ester.

The ^{13}C NMR spectrum of the natural polymer is broad and provides little information, except that there are strong signals at approximately 61.4 ppm, which indicate that the primary hydroxyl groups are mainly non-substituted. Given in Table 1, a potentially useful piece of information is that there is no signal at approximately 60.5 ppm, which indicates that the combination of units (3DG2S-4DG2S3S) found in the polysaccharide from *Champia novae-zelandia* (Hook. f. et Harv.) J. Ag. (Miller et al., 1996) is not likely to be present in this polysaccharide.

^{13}C NMR analysis of methylated polysaccharides

A concept currently being developed here is to consider the equivalent of the linkage analysis through using ^{13}C NMR spectroscopy. The polymer is methylated, then desulphated. The difference between the spectrum of this and the naturally desulphated polymer is that methyl groups now occur where hydroxyl groups occurred in the natural polymer.

Although the spectra of the polysaccharides which had been methylated ($\mathbf{S}\,G \mid M \mid$) (not shown) or methylated then desulphated ($\mathbf{S}\,G \mid M,-S \mid$) (Figure 1) were too complex for full analysis, they gave three subsets of useful data: (a) A subset of the pyruvate signals $\mathbf{S}(p') = \{25.4, 65.0, 66.5\}$ Now

$$\mathbf{S}(p') \subset (\mathbf{S}G \mid M \mid) \text{ and } \{\mathbf{S}(p') \subset (\mathbf{S}G \mid M, -S \mid) \qquad (11)$$

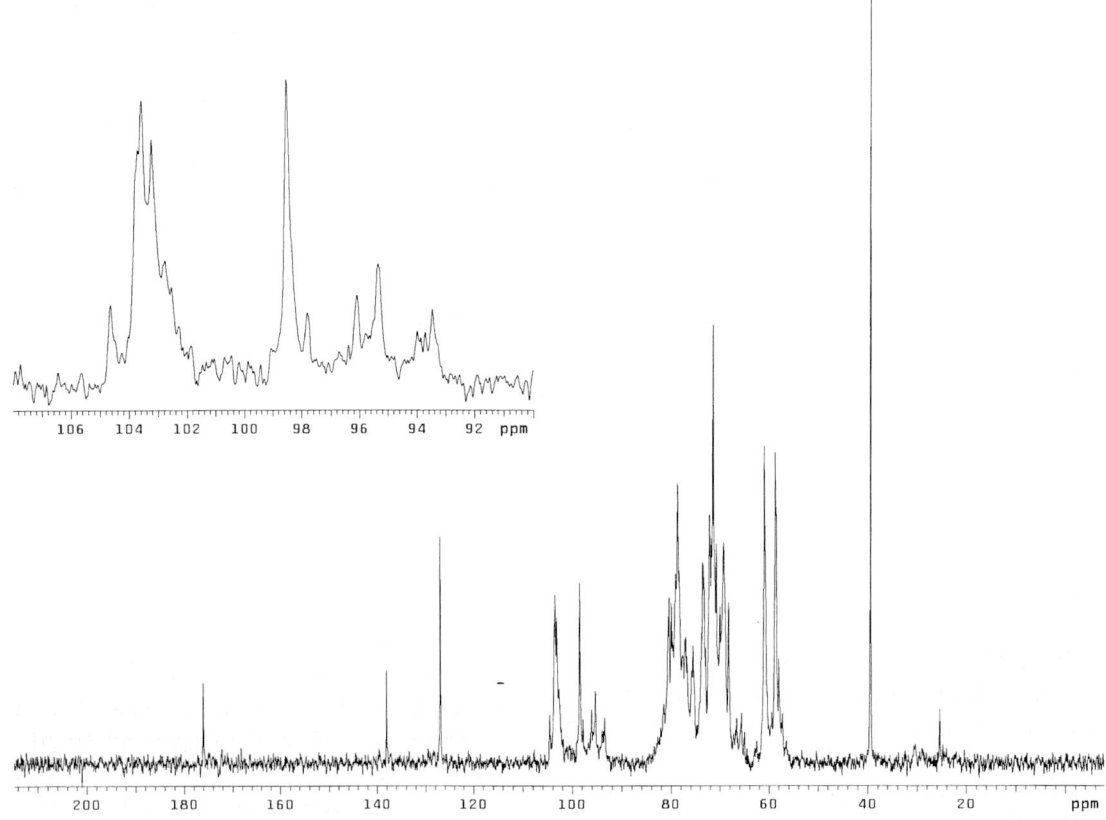

Figure 1. ^{13}C NMR spectrum of the methylated then desulphated polysaccharide from *Plocamium costatum*. The anomeric carbon signals are in the inset. Signals at 128, 138 and 177 ppm are due to pyromellitate impurity, signal at 39.4 ppm is internal DMSO.

Accordingly, pyruvate acetal appears to be present at a low level in the polymer.

(b) The subset anomeric carbon atoms of the polysaccharide G |M,-S| (anomeric) contains at least sixteen discernible signals, the main ones being

$$\mathbf{S}(\text{anomeric})(G|M, -S|) = \{104.6, 103.6,$$
$$103.3, 98.5, 98.2,$$
$$96.1, 95.3, 93.7,$$
$$93.5\} \quad (12)$$

The difference between the sets (12) and (4) is due to the presence of methyl groups on the positions devoid of sulphate ester in the natural polymer which are in a position to influence the chemical shift of the anomeric carbon atom. For 3G, these positions should be {3G2M, 4G3M}; for 4G these should be {4G2M, 3G2M, 3G4M}, and from (10), 4G3M ≈0. It is assumed that all other positions are too remote to significantly affect the shift of the anomeric carbon. Such shift changes are usually upfield, in magnitude up to 2 ppm, although they can be downfield, e.g. |2M| shifts the signal of LAG1 from 98.3 ppm to approximately 98.8 ppm (Lahaye et al., 1985).

We can make some provisional assignments.

$$\mathbf{S}(\text{anomeric})(3DG - 4DG) = \{104.6, 96.1\} \subset$$
$$\mathbf{S}(\text{anomeric})(G|M, -S|) \quad (13)$$

i.e. G contained 3DG2S4S-4DG2S and the 3-linked unit was in turn linked to a 4-linked unit with a 3-sulphate ester. By intensity, this was present at about 6–7 mol%.

If the change in shift due to methylation is upfield,

$$\mathbf{S}(\text{anomeric})4LG = \{98.5, 98.2\} \quad (14)$$

The 98.2 ppm signal was only a few% of the total, and therefore virtually all the 4LG units are likely to be in one environment, and there are probably two methyl groups about the linkage, as the change of shift (from 100.7 ppm) is 2.2 ppm.

The signals {96.1, 95.3, 93.7, 93.5} for the 4DG unit appear to encompass the full range of methylation, as the signal for λ-carrageenan |-S,M| is also at 93.5 ppm (Miller & Blunt, unpublished).

(c) The subset of methyl signals appears to be characteristic of location (Miller & Blunt, unpublished). Substitution at: G6 ≈ 59.0, 3DG2 ≈ 61.0, 3DG4 ≈ 61.5, 4D∪LG2 ≈ 58.6, 4DG3 ≈ 57.2, 4LG3 ≈ 56.5. The spectral data observed are:

$$S G|M| = \{61.2, 60.9, 60.6, 60.5 \text{ (a shoulder on } 60.6), 58.9, 58.7 \text{ (multiplet)} \\ 58.1, 57.1\} \quad (15)$$

$$S G|M - S| = \{61.0 \text{ (broad, with shoulders)} \\ 58.9, 58.8, 58.2, 57.4\} \quad (16)$$

Certain provisional conclusions can be reached. G6: identified as {58.7} in (15) and {58.9, 58.8} in (16). These are the strongest signals, hence G6 is largely unsubstituted in the natural polymer. The 58.8 signal is of approximately 80% the intensity of 58.9 in (10).

3D{G2,4}. This can only be assessed from (15) as the G6S, on desulphation, will also be in this region. There are more than 4 signals, which indicates each methyl group is in a range of environments.

4G3: the signal at ≈ 57.4 ppm is of very low intensity, indicating that most of the 4-linked residues existed as 4G3S, in agreement with (10).

4G2: the signals for this may not be properly resolved from the G6M signals.

Unfortunately, this polysaccharide has such complex substitution that the specific substitution signals have broadened and overlapped. This suggests that at least part of the polymer does not exist in significant sized blocks. Further work is required to properly evaluate its structure, nevertheless patterns do occur, and they may be of sufficient value to be of taxonomic interest.

Conclusions

The polysaccharide from *Plocamium costatum* definitely has a structure in which 3-linked and 4-linked galactopyranosyl units alternate, and the 4-linked unit comprises approximately equal amounts of D and L galactose. The environment about the D-galactosyl anomeric linkage contains most available variation in sulphation, in approximately equal amounts, the 4-linked residues are essentially totally 3-sulphated, and the environment about the L-galactosyl anomeric linkage is largely uniform, and probably contains two hydroxyl groups out of the three available positions.

Acknowledgements

I thank the Foundation for Research, Science and Technology for funding under contract CRL 501 Drs Falshaw and Furneaux for the data in Table 1, and Dr H. Wong for obtaining the NMR spectra.

References

Falshaw, R. & R. H. Furneaux, 1994. Carrageenan from the tetrasporic stage of *Gigartina decipiens* (Gigartinaceae, Rhodophyta). Carbohydr. Res. 252: 171–182.

Falshaw, R., R. H. Furneaux & I. J. Miller, 1998 The backbone structure of the sulfated galactan from P*locamium costatum* (C.Ag.) Hook. f. et Harv. (Plocamiaceae, Rhodophyta) Submitted, Bot. mar.

Lahaye, M., W. Yaphe & C. Rochas, 1985. ^{13}C NMR spectral analysis of sulphated and desulphated polysaccharides of the agar type. Carbohydr. Res. 143: 240–245.

Miller, I. J., 1997. The chemotaxonomic significance of the water soluble red algal polysaccharides. Recent Res. Devel. Phytochem. 1: 531–565 (1997).

Miller, I. J., 1998. The structure of a pyruvylated carrageenan extracted from *Stenogramme interrupta* as determined by ^{13}C NMR spectroscopy. Bot. mar. 41: 305–315.

Miller, I. J. & J. W. Blunt. 1998. Desulfation of algal galactans. Carbohydr. Res. 309: 39–43.

Miller, I. J., R. Falshaw & R. H. Furneaux, 1995. Structural analysis of the polysaccharide from *Pachymenia lusoria* (Cryptonemiaceae, Rhodophyta) Carbohydr. Res. 268: 219–232.

Miller, I. J., R. Falshaw & R. H. Furneaux, 1996. A polysaccharide fraction from the red seaweed *Champia novae-zelandiae* Rhodymeniales, Rhodophyta. Hydrobiologia 326/327: 505–510.

Miller, I. J., R. Falshaw, R. H. Furneaux & J. A. Hemmingson, 1997. Variations in the constituent sugars of the polysaccharides from New Zealand species of *Pachymenia* (Halymeniaceae) Bot. mar. 40: 119–127.

Whyte, J. N. C, R. E. Foreman & R. E. de Wreede, 1984. Phycocolloid screening of British Columbia red algae. Hydrobiologia 116/117 (Dev. Hydrobiol. 22): 537–541.

Biology and agar composition of *Gelidium sesquipedale* harvested along the Atlantic coast of Morocco

Aziza Mouradi-Givernaud[1], Lalla Amina Hassani[1], Thierry Givernaud[2], Yves Lemoine[3] & Oumaima Benharbet[1]

[1]*Laboratoire de Biochimie et Biotechnologies marines, Faculté des Sciences, B.P. 133, 14000 Kénitra, Morocco*
[2]*SETEXAM, Usine El Assam, Route de Tanger, B.P. 210, 14000 Kénitra, Morocco*
[3]*Laboratoire de Cytophysiologie Végétale et Phycologie, UST Lille, F-59655 Villeneuve d'Ascq Cedex, France*

Key words: Gelidium sesquipedale, annual growth, reproduction, rheological properties, chemical composition, agar

Abstract

Gelidium sesquipedale (Clem.) Bornet et Thuret is the main raw material used for agar production in Morocco. The biology and biochemistry of this slow growing alga collected monthly over one year has been studied. The agar content varied around 40% of algal dry weight and reached a maximum of 44.5% in November. Agar gel strength was maximum in May and July (1000 g cm^{-2}), and melting (90 °C) and gelling (35 °C) temperatures varied slightly. The agar contained a high 3,6-anhydrogalactose content (40–45 mol%) and low amounts of 6-*O*-methylgalactose (around 1 mol%) and sulphate (1.0 – 1.6% of dry weight). The reproductive status of the alga affected the agar synthesis and quality. The phycocolloid gel strength was related to the mean polysaccharide chain length but not to its chemical substitution.

Introduction

Agarocolloids are a family of gel forming galactans extracted mostly from the macroalgal orders Gracilariales and Gelidiales (Indergaard, 1983; Painter, 1983; Craigie, 1990) for industrial, cosmetic and food applications (Armisen, 1995). Agar, a fraction of this family, is composed of a repeating unit called neoagarobiose, namely [->4)-3,6-anhydro-α-L-galactopyranosyl-(1->3)-β-D-galactopyranosyl-(1->] (Araki, 1966) which can be substituted by sulphate esters, methyl ethers, pyruvate, uronic acids or 4-*O*-methyl-L-galactose (Armisen & Galatas, 1987). The chemical composition and properties of this polysaccharide vary with species (Lahaye et al., 1986; Craigie, 1990; Furneaux et al., 1990; Usov & Klochkova, 1992; Pickering et al., 1993), seasons (Cote & Hanisak, 1986; Bird & Hinson, 1992; Yenigul, 1993), age of the seaweed (Craigie & Wen, 1984; Orosco et al., 1992) and tissue (Sekkal et al., 1993). The gelling properties of agar are closely dependent on the degree and type of substitution. Sulphate residues strongly decrease the gel strength (Yaphe & Duckworth, 1972; Freile Pelegrin et al., 1995) and methyl ether groups modify the elasticity and gelling temperature (Guiseley, 1970; Whyte et al., 1981).

Agar extracted from *Gelidium sesquipedale* (Clem.) Bornet & Thuret from South European and North African Atlantic coasts is commercially and industrially well appreciated because it develops strong and rigid gels at low concentrations in water (Seoane-Camba, 1989). This species is harvested by divers from wild populations during summer months. It is the main source of raw material used for the industrial agar extraction in Morocco which is the third producer in the world.

We have studied a natural population of *Gelidium sesquipedale* over one year to ascertain whether seasonal variations occur in the yield, the rheological and physical properties and the chemical composition of its agar.

Material and methods

Gelidium sesquipedale is present along the Atlantic coasts from England to Mauritania with great abundance in Spain, Portugal and Morocco (Salinas et al., 1978). It grows on exposed rocky shores from the infralittoral zone to a depth of 20 m. The maximum biomass is found between 0 and 10 m.

Samples of *G. sesquipedale* were harvested monthly from March 1996 to February 1997 at the upper limit of the infralittoral zone at Medhia in the north of Morocco.

Biological observations

Monthly growth of the alga was monitored throughout the observation period by recording the length of the thalli and by counting, on 100 thalli, the amount of branches on the first centimetre below the apex. These two parameters gave a good estimate of the algal growth.

For each month, the total number of fertile thalli (gametophytes, tetrasporophytes and thalli bearing cystocarps) was registered.

Agar extraction

The raw material collected for agar analysis was washed in seawater and epiphytes were removed. It was then rinsed quickly with distilled water, weighed and oven-dried for 48 h at 60 °C to determine the dry weight, and ground to a powder.

Agar was extracted from the dry powder in triplicate using the method of Craigie & Leigh (1978) with the following modifications. The hot agar extract was filtered (0.45 μm), allowed to gel at room temperature, then freeze-thawed (twice) and washed with distilled water. Finally, the agar was oven-dried for 24 h at 60° C, cooled and weighed to determine the extraction yield.

Physical properties

Gel strength was measured using a Nikan-shiki apparatus according to Armisen & Galatas (1987).

Gelling temperature was obtained using the method described by Kim (1970). Gel fusion temperature was measured with the method of Whyte & Englar (1976). These two measurements were made in triplicate.

The measurement of physical properties of agar have been done in accordance with ISO 9000 norm requirements. The results were given $\pm$ 20 g for gel strength and $\pm$ 1 °C for gelling and gel fusion temperature.

Estimation of agar polymer size

The agar polymer size was estimated from the ratio of total carbohydrate to reducing sugar. Total carbohydrate was measured by the phenol sulphuric acid method of Dubois et al. (1956) using galactose as standard. Reducing sugars were quantified by the 2,5-dinotrosalicylic acid reagent Miller (1959) using galactose as standard. The analysis was made in triplicate.

Chemical composition of agar

The neutral sugars were quantified as previously described by Mollet et al. (1995) by gas liquid chromatography (GLC; Shimadzu CR14A) equipped with a flame ionisation detector and an integrator. The alditol acetates derivatives of the constituents sugars were prepared using the reductive hydrolysis method of Stevenson & Furneaux (1991), separated on a capillary column (OV-1701, 25 m, 0.32 mm, Pierce Illinois U.S.A.) using helium as carrier gas, and identified by their retention times with *myo*-inositol hexacetate as an internal standard. The initial temperature of the oven was 90 °C. The temperature was raised at 10 °C min^{-1} up to 235 °C followed by 1 °C min^{-1} up to 245 °C and 5 °C min^{-1} to reach a final temperature of 260 °C maintained for 1 min. The results were expressed as mol% of the total sugars.

For the determination of the sulphate content, samples of agar (1 mg dw) were hydrolyzed by HCl (2 N) for 2 h at 120 °C. The acidic solution was co-evaporated with ethanol and sulphate ions were dissolved in purified water (UHQ Elgastat). After filtration (0.22 μm) sulphate ions were separated by HPLC according to Mass et al. (1986) using a Wescan anion exchange column (4.6 mm × 250 mm Alltech) and detected by a conductimeter (Milton Roy).

Statistical methods

Simple linear correlation analysis was used to analyse the data.

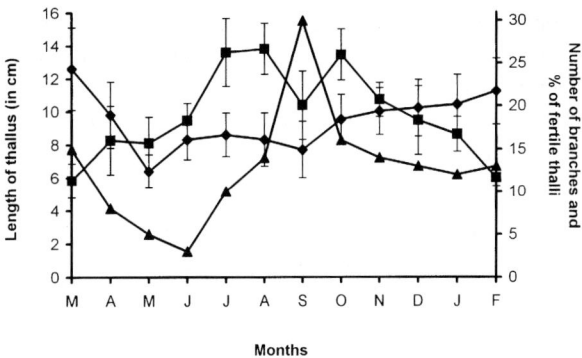

Figure 1. Seasonal variations of *Gelidium sesquipedale* fertility (▲—▲), length (♦—♦) and number of branches along the first cm of axes below the apical cell (■—■), vertical lines represent standard deviation $n = 100$.

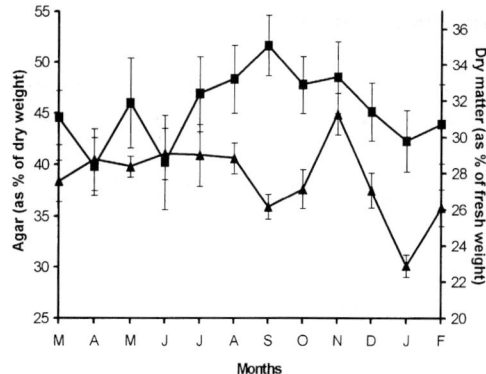

Figure 2. Annual variations of *Gelidium sesquipedale* dry weight expressed as percentage of fresh weight (■—■) and its agar content expressed as percentage of dry weight (▲—▲), vertical lines represent standard deviation, $n = 3$.

Results

Growth and reproduction

The seasonal variations in thallus length, number of branches and fertile thalli are shown on Figure 1. From March to May, the average length of thalli diminished; the old fronds decayed and were cut by storms. New fronds appeared which elongated from May to July and from September to February. The number of branches was minimum in March (11 ± 2) and increased to a maximum in July and August (26 ± 4). From September to October, the number of ramifications increased again and then decreased from October to February (12 ± 3). *Gelidium* was fertile throughout the year. The minimum percentage of fertile thalli was observed in June (2%) and the maximum in September (28%).

Chemical composition of thalli

The algal dry weight varied around 31% of fresh weight depending on the season (Figure 2). It was minimum in June ($28 \pm 2.5\%$) and maximum ($35 \pm 1.7\%$) in September. Agar yield varied slightly from March to August (around 40% of alga dw; Figure 2). It decreased to $36.0 \pm 1.6\%$ in September and then increased to a maximum of $44.5 \pm 2\%$ in November.

Physical properties

Gelling and melting temperatures varied slightly throughout the year. They ranged from 35 °C to 36 °C and from 89 °C to 92 °C, respectively. The gel strength was maximum between May and July (around 1000 g cm^{-2}), diminished to a minimum of

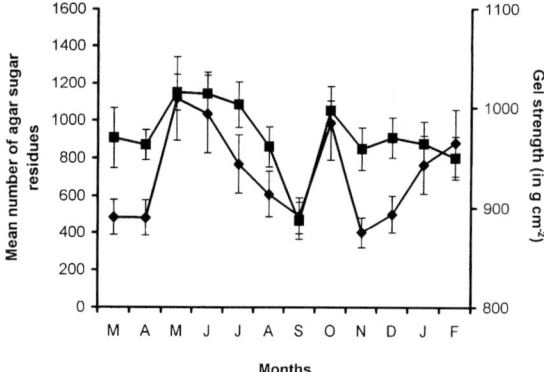

Figure 3. Annual variation of *Gelidium sesquipedale* agar polymer size as the ratio of total sugar to reducing sugar content (♦—♦) and gel strength (■—■), vertical lines represent standard deviation, $n = 3$.

890 cm^{-2} in September and then increased in autumn to reach 1000 g cm^{-2} in October. There was no correlation between melting or gelling temperature and gel strength. The seasonal change in the ratio of total sugar to reducing sugar taken as an estimate of the mean agar chain length was similar to that of the gel strength (Figure 3).

Agar composition

The agar composition was not affected by seasonal variations. Galactose (46.6–58.2 mol%) and 3,6-anhydrogalactose (40–45.1 mol%) were the main components with minor amounts of 6-*O*-methyl galactose (0.9 to 2 mol%). Glucose, rhamnose and xylose were also present but never exceeded 5 mol%. The sulphate content remained very low throughout the year, between 1.0 and 1.6%.

Discussion

The growth cycle of *G. sesquipedale* can be divided into five periods:

1. From March to May, old fronds decayed, new shoots appeared and elongated with low branching.
2. From May to July, the alga was actively growing with a regular elongation of thalli and multiplication of ramifications.
3. From July to September, growth stopped and the alga was at its maximum of fertility.
4. From September to October, there was a short period of growth with elongation of the thalli and multiplication of branches.
5. From October, there was a rapid decrease in branching and a stagnation of thallus length. This last period corresponded to ageing of the fronds and ended in March with the rapid decay of the old parts of the thalli.

This cycle was similar to that of *Gracilaria multipartita* growing in the same area (Givernaud et al., 1999). The comparison between chemical composition of the seaweed and its biological cycle showed that the development of reproductive structures between June and September resulted in a diminution in agar content ($p < 0.05$, $r = -0.89$) and gel strength ($p < 0.05$, $r = -0.87$). During this period there was also a correlation between dry matter and the percentage of fertile seaweed ($p < 0.05$, $r = 0.92$). The relationship between the agar content and reproduction was also observed on *Gracilaria multipartita* (Givernaud et al., 1999) but, in the latter species, the agar content and gel strength were also influenced by the alga's growth status. In contrast, previous work on *Gelidium latifolium* agar showed that its composition and rheological properties were mostly affected by growth and showed just a slight decrease in gel strength during the development of reproductive structures (Mouradi-Givernaud et al., 1992, 1993). Thus the reproduction and growth status influence the metabolism of the three species. During growth, there was a decrease in the dry matter and an increase in agar synthesis while during the development of reproductive structures dry matter content increased and that of agar decreased.

Hydrolysis of agar in the thallus of *Gelidium sesquipedale* seems to occur during the development of reproductive structures as the mean polysaccharide chain length was inversely correlated to fertility ($r = 0.89$, $p < 0.05$). This agar degradation could facilitate the development of reproductive structures and the release of spores and gametes but this observation needs to be supported by the isolation of the enzyme(s) responsible for this hydrolysis and the determination of the fate of the degradation products.

The general composition of *Gelidium sesquipedale* harvested in Morocco presented no major difference to that of the Spain sample (Garcia, 1988) or South of France (Mouradi-Givernaud, 1992; Vignon et al., 1994) and remained quite constant throughout the observation period. According to the compositional and rheological characteristics of *G. sesquipedale* agar determined in this study, variations of the gel strength appear to depend on the mean polysaccharide chain length and further studies should attempt to confirm this observation.

Conclusions

The results of this study confirmed the good quality of *G. sesquipedale* agar with a high gel strength and a low level of chemical substitution. The variation of the quantity and quality of the agar throughout the year was low and negligible on an industrial scale. The actual harvest period between July and September does not correspond to the maximum gel strength but to that of the maximum biomass. Further work will determine the interactions between growth, reproduction and agar metabolism of this alga.

Acknowledgements

We are greatly indebted to A. Rahaoui at the Laboratoire de Cytophysiologie Végétale et Phycologie, UST Lille for discussion and analysis of the GLC and HPLC chromatograms. Thanks are also extended to his colleagues for their valuable advice and co-operation in the experiments. We express our sincere thanks to SETEXAM, Morocco, for valuable help and financial support.

References

Araki, C., 1966. Some recent studies on the polysaccharides agarophytes. Proc. int. Seaweed Symp. 5: 3–17.

Armisen, R., 1995. World-wide use and importance of *Gracilaria*. J. appl. Phycol. 7: 231–243.

Armisen, R. & F. Galatas, 1987. Production, properties and uses of agar. In McHugh, D. J. (ed.), Production and Utilisation of Products from Commercial seaweeds. FAO Technical Paper 288: 1–57.

Bird, K. T. & T. K. Hinson, 1992. Seasonal variations in agar yields and quality from North Carolina agarophytes. Bot. mar. 35: 291–295.

Cote, G. L. & M. D. Hanisak, 1986. Production and properties of native agar from *Gracilaria tikvahiae* and other red algae. Bot. mar. 29: 359–366.

Craigie, J. S., 1990. Cell walls. In Cole, K. M. & R. G. Sheath (eds), Biology of the Red Algae. Cambridge University Press: 221–259.

Craigie, J. S. & C. Leigh, 1978. Carrageenans and agars. In Hellebust, J. A. & J. S. Craigie (eds), Handbook of Phycological Methods. Cambridge University Press: 110–131.

Craigie, J. S. & Z. C. Wen, 1984. Effects of temperature and tissue age on gel strength and composition of agar from *Gracilaria tikvahiae* (Rhodophyceae). Can. J. Bot. 62: 1665–1670.

Dubois, M., K. A. Gilles, J. K. Hamilton, P. A. Robers & F. Smith, 1956. Colorimetric method for determination of sugars and related substances. Analyt. Chem. 28: 350–356.

Freile-Pelegrin, Y., D. R. Robledo & G. Garcia-Reina, 1995. Seasonal agar yield and quality in *Gelidium canariensis* (Grunow) Seoane-Camba (Gelidiales, Rhodophyta) from Gran Canaria, Spain. J. appl. Phycol. 7: 141–144.

Furneaux, R. H., I. J. Miller & T. T. Stevenson, 1990. Agaroids from New Zealand members of the Gracilariaceae (Gracilariales, Rhodophyta). a novel dimethylated agar. Hydrobiologia 204/205: 645–654.

Garcia, I., 1988. Estudio de la variacion estcional de la calidad y el rendimiento de l'agar obtenido del alga *Gelidium sesquipedale* (Clem.) Born. et Thur. en la costa Guipuzcoana (N-de Espana). Inf. Tecn. Inv. Pesq. 146: 3–19.

Givernaud T., A. El Gourgi, A. Mouradi-Givernaud, Y. Lemoine & N. Chiadmi, 1999. Seasonal variations of growth and agar composition of *Gracilaria multipartita* harvested along the Atlantic coasts of Morocco. Hydrobiologia 398/399 (Dev. Hydrobiol. 137): 167–172.

Guiseley, K. B., 1970. The relationship between methoxyl content and gelling temperature of agarose. Carbohydr. Res. 13: 247–256.

Indergaard, M., 1983. The aquatic resource. In Cote, W. A. (ed.), Biomass Utilisation. Plenum Press, New York: 137–168.

Kim, D. H., 1970. Economically important seaweeds in Chile. – I *Gracilaria*. Bot. mar. 13: 140–162.

Lahaye, M., C. Rochas & W. Yaphe, 1986. A new procedure for determining the heterogeneity of agar polymers extracted in the cell walls of *Gracilaria* spp. (Gracilariaceae, Rhodophyta). Can. J. Bot. 64: 579–585.

Miller, G. L., 1959. Use of dinitrosalicylic acid reagent for determination of reducing sugars. Analyt. Chem. 3: 426–428.

Mollet, J. C., M. C. Verdus, R. Kling & H. Morvan, 1995. Improved protoplast yield and cell wall regeneration in *Gracilaria verrucosa* (Huds) Papenfuss (Gracilariales, Rhodophyta). J. exp. Bot. 46: 239–247.

Mouradi-Givernaud, A., 1992. Recherches biologiques et biochimiques pour la production d'agarose chez *Gelidium latifolium* (Grev.) Thuret et Bornet (Rhodophycées, Gélidiales), Thèse de Doctorat d'Etat, Univ. Caen, 350 pp.

Mouradi-Givernaud, A., T. Givernaud, H. Morvan & J. Cosson, 1992. Agar from *Gelidium latifolium* (Rodophyta, Gelidiales) Biochemical composition and seasonal variations. Bot. mar. 35: 153–160.

Mouradi-Givernaud, A., T. Givernaud, H. Morvan & J. Cosson, 1993. Annual variations of the biochemical composition of *Gelidium latifolium* (Greville) Thuret et Bornet. Hydrobiologia. 206/261: 607–612.

Orosco, C. A., M. Sawamura, M. Ohno & K. Kusunose, 1992. Effects of thallus age and alkali treatment on pyruvic acid contents of agars from the red alga *Gracilaria chorda*. Nipon Suisan Gakkaishi. 58: 1493–1498.

Painter, T. J., 1983. Algal polysaccharides. In Aspinall, G. O. (ed.), The Polysaccharides, Vol. 2. Academic Press, New York: 195–285.

Pickering, T. D., V. H. Sladden, R. H. Furneaux, J. A. Hemmingson & P. Redfearn, 1993. Comparison of growth rate in culture, dry matter content, agar content and agar quality of 2 New-Zealand red seaweeds, *Gracilaria chilensis* Bird, Mc Lachlan et Oliveira and *Gracilaria truncata* Kraft. J. appl. Phycol. 5: 85–91.

Salinas, J. M., B. Reguera & R. Gancedo, 1978. Biometria de *Gelidium sesquipedale* Rodophyta 1. Bol. Inst. Esp. Oceano. 226: 1–70.

Sekkal, M., J. P. Huvenne, P. Legrand, B. Sombert, J. C. Mollet, A. Mouradi-Givernaud & M. C. Verdus, 1993. Direct structural identification of polysaccharides from red algae by FTIR microspectrometry. 1. Localisation of agar in *Gracilaria verrucosa* sections. Mikrochim. Acta. 112: 1–10.

Seoane-Camba, J. A., 1989. On the possibility of culturing *Gelidium sesquipedale* by vegetative propagation. Proceedings of the Second Workshop of COST 48 Subgroup. 1: 59–68.

Stevenson, T. T. & R. H. Furneaux, 1991. Chemical methods for the analysis of sulphated galactans from red algae. Carbohydr. Res. 210: 277–298.

Usov, A. I. & N. G. Klochkova, 1992. Polysaccharides of algae. 45. Polysaccharide composition of red seaweeds from Kamchatka coastal waters (north western pacific) studied by reductive hydrolysis of biomass. Bot. mar. 35: 371–378.

Vignon, M. R., E. Morgan & C. Rochas, 1994. *Gelidium sesquipedale* (Gelidiales, Rhodophyta) I. Soluble Polymers. Bot. mar. 37: 325–329.

Whyte, J. N. C. & J. R. Englar, 1976. Fisheries and marine Service, Canada. Technical report N° 623.

Whyte, J. N. C., J. R. Englar, R. G. Saunders & J. C. Linsay, 1981. Seasonal variations in the biomass, quantity and quality of agar from the reproductive and vegetative stages of *Gracilaria* (*verrucosa* type). Bot. mar. 24: 493–502.

Yaphe, W. & M. Duckworth, 1972. The relationship between the structure and biological properties of agars. Proc. int. Seaweed Symp. 7: 15–22.

Yenigul, M., 1993. Seasonal changes in the chemical and gelling characteristics of agar from *Gracilaria gracilis* collected in Turkey. Hydrobiologia 261/261: 631–638.

A comparative analysis of agarans from commercial species of *Gracilaria* (Gracilariales, Rhodophyta) grown *in vitro*

Juan Macchiavello[1], Rosa Saito[2], Gracinda Garófalo[2] & Eurico C. Oliveira[3]*
[1]*Universidad Católica del Norte. Larrondo 1281, Coquimbo, Chile*
[2]*Instituto de Pesquisas Tecnológicas, São Paulo, Brazil*
[3]*Universidade de São Paulo, C. postal 11461, 05422-970 S. Paulo, Brazil*
E-mail: euricodo@usp.br

Key words: agar, agarans, 3,6-anhydrogalactopyranose, *Gracilaria*, seaweed, Brazil

Abstract

The agaran yield, 3,6-anhydrogalactopyranose (3,6-AG) and sulphate content were compared in four commercial species of *Gracilaria* grown under parallel conditions *in vitro*. *Gracilaria chilensis* Bird, McLachlan et Oliveira from Chile provided the highest agaran yield (59%), followed by *G. tenuistipitata* Zhang et Xia var. *liui* Zhang et Xia from China (53%), *Gracilaria gracilis* (Stackhouse) Steentoft, Irvine et Farnham from Namibia (34%) and from Argentina (26%), and *Gracilaria caudata* J. Agardh from Brazil (32%). The algae from Chile, China and Namibia gave higher yields after alkali treatment while those from Brazil and Argentina gave higher yields for the native agarans. Lower percentages of 3,6-AG and higher sulphate contents were found in the species from warmer waters (Brazil and China), indicating agarans of lower commercial value. The results indicate that the Chilean *Gracilaria* had a superior yield of agaran, although *G. gracilis* from Argentina presented the highest 3,6-AG content after alkali treatment compared to other species considered for commercial cultivation in Brazil.

Introduction

The genus *Gracilaria* is currently the main source of agarans in the world (Armisen, 1995), and data on agaran yields abound in the literature (c.f. Oliveira & Plastino, 1994). Nevertheless, considering that agaran composition and its rheological behaviour within the genus *Gracilaria* varies with the species, strain, seasonality, conditions of growth and many other variables (Murano, 1995), new data are welcome considering the economic importance of the genus. Besides, there is no information about the yield and the content of 3,6-AG and sulphate, obtained with the same methodology and for different species grown at the same time under the same conditions.

Despite the strong interest in the production of *Gracilaria* and/or *Gracilariopsis* biomass for agaran processing, extensive cultivation has only been achieved in Chile (Avila & Seguel, 1993), Taiwan (Chiang, 1981) and Namibia (Dawes, 1995), although promising results are being obtained in Venezuela (Rincones et al., 1992) and the Caribbean (Smith, 1992). Several pilot experiments to cultivate *Gracilaria* spp. were made in Brazil (Câmara Neto, 1987; Oliveira, 1988), but no successful commercial cultivation resulted up to now. One reason for this outcome is the lack of a suitable species that produces a good agaran and adapts well to the available cultivation techniques (Oliveira, E.C., unpublished). In a process to select a better candidate for a seaweed mariculture program in Brazil, we did a comparative study of species that have been cultivated successfully in China, Chile and Namibia, to which we added a commercial strain from Argentina and the most common *Gracilaria* species from Brazil.

Materials and methods

The studied material, places and collecting dates are given in Table 1. Small samples of each strain, consisting of vegetative tips 10 mm long were brought

---
* Author for correspondence

Table 1. Origin and collection data for the *Gracilaria* spp. studied.

Species	Collecting place	Date	Collector
G. caudata	Caraguatatuba, Brazil 23° 39′ S, 45° 26′ W	January, 1991	EC Oliveira
G. chilensis	Coquimbo, Chile 29° 57′ S, 71° 25′ W	January, 1993	J Macchiavello
G. gracilis (Argent.)	Pto. Madryn, Argentina 42° 45′ S, 65° 02′ W	November, 1991	EC Oliveira
G. gracilis (Namibia)	Luderitz, Namibia 26° 38′ S, 15° 10′ E	May, 1992	EC Oliveira
G. tenuistipitata	Haikou, China 20° 05′ N, 110° 25′ E	June, 1990	EC Oliveira

to the laboratory in São Paulo and grown *in vitro* as described by Oliveira et al. (1995). The biomass for agaran extraction was produced in sterilized seawater enriched with modified von Stosch medium (Oliveira et al., 1995). Cultures were bubbled with compressed air, temperature was kept at 22 °C under a photon irradiance of 55 μmol m^{-2} s^{-1} (Phillips daylight fluorescent lamps) and a light-dark cycle of 12 h. Agaran was extracted after the method of Craigie & Leigh (1978), 3,6-AG with the method of Yaphe & Arsenault (1965) and sulphate according to Craigie & Wen (1984). Extractions were made in triplicate.

Results

Table 2 shows that there is a large variability in agaran yield among the algae studied, ranging from 8 to 59% of dry seaweed. The maximum yield after alkali modification was obtained with *G. chilensis*, and the lowest with *G. caudata*. *G. gracilis* from Argentina presented the highest 3,6-AG content after alkali treatment, while minimum contents were obtained for *G. tenuistipitata* var. *liui* and *G. caudata*. Sulphate showed a smaller variation, from 1.5% for alkali treated *G. caudata*, to 4.2% for native *G. tenuistipitata* var. *liui*. *Gracilaria caudata* and *G. tenuistipitata* var. *liui* produced very soft gels without alkali treatment and had to be washed with isopropanol to help filtration. Lowest agaran yield after alkali treatment was obtained from *G. caudata*.

Table 2. Agaran analysis of *Gracilaria* spp. Agaran in % of dry weight of algae; 3,6-AG and sulphate in % of agaran ± standard deviation ($n = 3$). *Missing data or single sample.

Species/treatment	Agaran%	3,6-AG%	Sulphate
G. caudata Brazil			
+ alkali	8.1 ± 0.8	38.2 ± 5.3	1.5 ± 0.0
− alkali	31.5 ± 0.8	36.6 ± 0.9	3.1 *
G. chilensis Chile			
+ alkali	59.2 ± 0.9	45.1 ± 1.5	3.2 ± 0.1
− alkali	32.7 ± 1.5	36.6 ± 0.2	3.5 ± 0.5
G. gracilis Namibia			
+ alkali	34.4 ± 1.7	45.3 ± 1.3	2.3 ± 0.1
− alkali	30.9 ± 1.0	36.3 ± 0.2	1.9 ± 0.1
G. gracilis Argentina			
+ alkali	21.8 ± 1.7	50.4 ± 2.5	2.1 ± 0.1
− alkali	26.3 ± 1.0	25.3 *	2.6 ± 0.1
G. tenuistipitata China			
+ alkali	52.7 *	36.9 ± 1.9	3.3 *
− alkali	23.0 ± 1.7	22.8 ± 0.5	4.2 ± 0.5

Discussion

When we started this study in 1993, we did not know the names of the algae from Brazil or from Namibia. The latter one was identified by Bird et al. (1994) as *G. verrucosa*, later transferred to *G. gracilis* (Stackhouse)

Steentoft, Irvine et Farnham (Bird & Kain, 1995), and is therefore, taxonomically the same as the material from Argentina. The species from Brazil had gone under several names until it was finally identified as *G. caudata* J. Agardh (Plastino & Oliveira, 1997).

The comparison of agaran yields based on data from the literature is very frustrating because of the broad range of values presented for algae with the same names but which may not be similar species. This is due to the dynamic nature of agarans within the algae, varying not only inter- and intra-specifically, but also with the age and within the same individual (cf. Oliveira & Plastino, 1994). Besides all the natural variation of this complex family of colloids, and the difficulties in identifying correctly the taxa, the results are also dependent on the methods utilized by different authors, that in consequence end up extracting different fractions of the agarans (Craigie, 1990; Murano, 1995; Armisen, 1995) in addition to contaminants such as floridean starch. Another source of variation can be attributed to the difficulties in correctly identifying the species (e.g. Oliveira & Plastino, 1994). For these reasons, we decided to evaluate the agarans of some candidate species for mariculture in Brazil, grown under the same conditions, at the same time, and processed with the same methods. As an even larger variability is also found in gel strength due to all the reasons given above, including data for the same species processed by the same authors (e. g. Rebello et al., 1996), we considered that it would be better to give information on 3,6-AG and sulphate contents that present less variability within samples, yet have good correlations with the agar value. Nevertheless, we are aware that industry is more concerned with the mechanical properties of gels than with the chemical characteristics of the agarans.

In the material studied, there was a general increase in the content of 3,6-AG following alkali treatment (Table 2) as is well established in the literature (e. g. Craigie, 1990). However, *G. gracilis* from Argentina presented the highest increase (from 25 to 50%), in the content of 3,6-AG after the alkali treatment, indicating a higher amount of sulphate at carbon 6 of the L-galactose (Craigie & Leigh, 1978). On the other hand, this did not happen with *G. caudata*, where the alkali treatment did not change the content of 3,6-AG indicating that the sulphate was not at a labile position.

The agarans produced by *G. caudata*, the Brazilian species, and *G. tenuistipitata* var. *liuii* from China, both originated from tropical waters and have a lower content of 3,6-AG in comparison with agarans from the temperate water species. The data we obtained for *G. chilensis* and *G. tenuistipitata* compare well with those obtained by Matsuhiro & Urzúa (1990) and Minghou et al. (1985), respectively. The low yield of agar in alkali treated *G. caudata* and its low gel strength support the results of Lopez-Bautista & Kapraun (1995) for material from Gulf of Mexico. Recently, Rebello et al. (1997) studied the 3,6-AG and sulphate contents of *G. tenuistipitata* from China, *G. chilensis* from Chile, *G. gracilis* from Argentina and a Brazilian species identified by them as *G. gracilis*, in addition to another species from elsewhere. The results obtained with the first two species are practically the same as our results. However, the results from the Argentinean and Brazilian materials are clearly different. As Rebello et al. (1997) do not state whether their material came from Brazil or Argentina, we cannot speculate about their identification. However, in the case of Brazilian *Gracilaria* we had already shown by cross-hybridization (Plastino & Oliveira, 1989) that none of our species could be identified as *G. verrucosa* (= *G. gracilis*), a fact also confirmed by Rice & Bird (1990) utilizing DNA banding patterns. Although data obtained from material grown *in vitro* cannot always be directly extrapolated to field conditions, the results of Rebello et al. (1997) for field material *G. tenuistipitata* and *G. chilensis* are practically the same as those presented herein. There is little point in comparing our results with other data in the literature for plants referred to *G. verrucosa* (now *G. gracilis*) due to the uncertainties of species identification (Bird et al., 1982).

Considering that in the southeast coast of Brazil provides a suitable environment to grow any one of the studied species (Macchiavello, 1994), we conclude that the Argentinean strain of *G. gracilis* and *G. chilensis* from Chile are the most promising agaran producers. If a decision is taken to introduce an exotic species of *Gracilaria* into Brazil for agar production, the strain from Argentina should be preferred because it already exists in the southwestern Atlantic.

Acknowledgments

Our study was supported by a grant from FAPESP (91/3590-2). J. Macchiavello acknowledges a scholarship from Red Latinoamericana de Botánica and E. C. Oliveira a scholarship from CNPq (301436/85-1). We thank Prof. John West for valuable suggestions and for improving the English text.

References

Armisen, R., 1995. World-wide use and importance of *Gracilaria*. J. appl. Phycol. 7: 231–243.

Avila, M. & M. Seguel, 1993. An overview of seaweeds resources in Chile. J. appl. Phycol. 5: 133–139.

Bird, C. J. & J. M. Kain, 1995. Recommended names of included species of Gracilariaceae. J. appl. Phycol. 7: 335–338.

Bird, C. J., J. P. van der Meer & J. McLachlan, 1982. A comment on *Gracilaria verrucosa* (Huds.) Papenf. (Rhodophyta: Gigartinales). J. mar. biol. Ass. U.K. 62: 453–459.

Bird, C. J., M. A. Ragan, A. T. Critchley, E. L. Rice & R. R. Gutell, 1994. Molecular relationship among the Gracilariaceae (Rhodophyta): further observations on some undetermined species. Eur. J. Phycol. 29: 195–202.

Câmara Neto, C., 1987. Seaweed culture in Rio Grande do Norte, Brazil. Hydrobiology 151/152: 363–367.

Chiang, Y.-M., 1981. Cultivation of *Gracilaria* (Rhodophycophyta, Gigartinales) in Taiwan. Proc. int. Seaweed Symp. 10: 569–574.

Craigie J. S., 1990. Cell walls. In Cole K. M. & Sheath R. G. (eds), Biology of the Red Algae, Cambridge University Press, Cambridge: 221–257.

Craigie J. S. & C. Leigh, 1978. Carrageenans and agars. In Hellebust, J. A. & J. S. Craigie (eds), Handbook of Phycological Methods. Physiological and Biochemical Methods. Cambridge University Press, Cambridge: 109–131.

Craigie J. S. & Z. C. Wen, 1984. Effects of temperature and tissue age on gel strength and composition of agar from *Gracilaria tikvahiae* (Rhodophyceae). Can. J. Bot. 62: 1665–1670.

Dawes, C. J., 1995. Suspended cultivation of *Gracilaria* in the sea. J. appl. Phycol. 7: 303–313.

Lopez-Bautista, J. & D. F. Kapraun, 1995. Agar analysis, nuclear genome quantification and characterization of four agarophytes (*Gracilaria*) from the Mexican Gulf coast. J. appl. Phycol. 7: 351–357.

Macchiavello, J. E., 1994. Tolerância fisiológica e características do ágar de cinco espécies de *Gracilaria* (Rhodophyta - Gracilariales). Ph.D. thesis, University of São Paulo, São Paulo: 96 pp.

Matsuhiro. B. & C. C. Urzúa, 1990. Agars from *Gracilaria chilensis* (Gracilariales). J. appl. Phycol. 2: 273–279.

Minghou J., M. Lahaye & W. Yaphe, 1985. Structure of agar from *Gracilaria* spp. (Rhodophyta) collected in the People's Republic of China. Bot. mar. 28: 521–528.

Murano, E., 1995. Chemical structure and quality of agars from *Gracilaria*. J. appl. Phycol. 7: 245–254.

Oliveira, E. C., 1988. The cultivation of phycocolloid-producing seaweeds in the American Atlantic. Gayana 45: 55–60.

Oliveira, E. C. & E. M. Plastino, 1994. Gracilariaceae. In Akatsuka, I. (ed.), Biology of Economic Algae. SBP Academic Publishing, The Hague: 185–226.

Oliveira, E. C., E. J. Paula, E. M. Plastino & R. Petti, 1995. Metodologias para cultivos no axenicos de macroalgas marinas *in vitro*. In Alveal K., M. E. Ferrario, E. C. Oliveira & E. Sar (eds), Manual de Métodos Ficológicos. Universidad de Concepción, Concepción: 429–448.

Plastino, E. M. & E. C. Oliveira, 1989. Crossing experiments as an aid to the taxonomic recognition of the agarophyte *Gracilaria*. In Oliveira, E. C., & N. Kautsky (eds), Cultivation of Seaweeds in Latin America, Univ. São Paulo, São Paulo: 127–133.

Plastino, E. M. & E. C. Oliveira, 1997. *Gracilaria caudata* J. Agardh (Gracilariales, Rhodophyta) – restoring and old name for a common western Atlantic alga. Phycologia 36: 225–232.

Rebello, J., M. Ohno, A. T. Critchley & M. Sawamura, 1996. Growth rates and agar quality of *Gracilaria gracilis* (Stackhouse) Steentoft from Namibia, southern Africa. Bot. mar. 39: 272–279.

Rebello, J., M. Ohno, H. Ukeda, H. Kusunose & M. M. Sawamura, 1997. 3,6-anhydrogalactose, sulfate and methoxyl contents of commercial agarophytes from different geographical origins. J. appl. Phycol. 9: 367–370.

Rice, E. L. & C. J. Bird, 1990. Relationships among geographically distant populations of *Gracilaria verrucosa* (Gracilariales, Rhodophyta) and related species. Phycologia 29: 501–510.

Rincones, R. E., J. N. Rubio & E. C. Racca, 1992. *Gracilaria* pilot farming in Venezuela. In Mshigeni K. E., J. Bolton, A. Critchley & G. Kiangi (eds), Proceedings of the First International Sustainable Seaweed Resource Development in Sub-Saharan Africa. Windhoek, Namibia: 309–318.

Smith, A., 1992. Seaweed cultivation in the West Indies. In Mshigeni, K. E., J. Bolton, A. Critchley & G. Kiangi (eds), Proceedings of the First International Workshop on Sustainable Seaweed Resource Development in Sub-Saharan Africa. Windhoek, Namibia: 337–351.

Yaphe, W. & G. P. Arsenault, 1965. Improved resorcinol reagent for the determination of fructose, and 3,6-anhydrogalactose in polysaccharides. Analyt. Biochem. 13: 143–148.

Pyruvated carrageenans from *Solieria robusta* and its adelphoparasite *Tikvahiella candida*

Anthony Chiovitti[1,2], Antony Bacic[1], Gerald T. Kraft[1], David J. Craik[3] & Ming-Long Liao[1]*
[1]*CRC for Industrial Plant Biopolymers, School of Botany, University of Melbourne, Parkville, Victoria 3052, Australia*
[2]*Current address: Centre for Biomolecular Sciences, The Purdie Building, University of St. Andrews, St. Andrews, Fife KY16 9ST, United Kingdom*
[3]*Centre for Drug Design and Development, University of Queensland, St Lucia, Queensland 4072, Australia*
Fax: [+61]-3-347-1071. E-mail: m.liao@botany.unimelb.edu.au

Key words: carrageenans, chemotaxonomy, Gigartinales, *Solieria robusta*, Solieriaceae, structure determination, *Tikvahiella candida*

Abstract

Solieria, the type genus of the commercially important red algal family Solieriaceae (Gigartinales), contains seven or eight species, three of which are represented in Australia. The cell-wall galactans of the most common Australian *Solieria* species, *S. robusta* (Greville) Kylin, were analysed by a combination of compositional assays, linkage analysis, and Fourier transform infrared (FTIR) and ^{13}C nuclear magnetic resonance (NMR) spectroscopy. They are shown to be composed predominantly of carrabiose 2,4'-disulphate units (the repeating unit of ι-carrageenan) and a significant proportion of 4',6'-pyruvated carrabiose 2-sulphate units. The constituent sugars, pyruvate content, FTIR spectrum, and linkage and substitution patterns of the galactans from *Tikvahiella candida* Kraft et Gabrielson, an adelphoparasite of *Solieria robusta*, closely resemble those of its host and furnish evidence in support of a close phylogenetic relationship between the two species.

Introduction

The Solieriaceae has the most genera (ca. 20 – Kraft, 1981; Gabrielson & Hommersand, 1982a, b; Womersley, 1994) and the greatest number of commercially exploited carrageenophytes (Doty, 1988, 1995) of any family in the red algal order Gigartinales. Carrageenans are sulphated galactans composed of repeating disaccharide units of 3-linked β-D-galactopyranose (Gal*p*) and either 4-linked α-D-Gal*p* or 4-linked 3,6-anhydro-α-D-galactopyranose (AnGal*p*) residues. Most members consistently display a predominance of either the kappa (κ) or iota (ι) types through all phases of their life histories, although recent studies have revealed that several solieriaceous species produce carrageenans with exceptionally high pyruvate acetal substitutions and other unique structural modifications (Chiovitti et al., 1997a, 1998b). Polysaccharides of novel chemistry are of particular interest both for the taxonomic markers they potentially offer to the activity of framing phylogenetic hypotheses and because of the special functional properties that they can exhibit.

Solieria, the type genus of the Solieriaceae, is credited with seven or eight species. [The type species, *Solieria chordalis* (C. Agardh) J. Agardh, plus *S. anastomosa* Gabrielson et Kraft, *S. dichotoma* Yoshida, *S. dura* (Zanardini) Schmitz, *S. filiformis* (Kützing) Gabrielson [1985, = *S. tenera* (J. Agardh) *sensu* Wynne et Taylor in Wormersley, 1994], *S. jaasundii* Mshigeni et Papenfuss, *S. pacifica* (Yamada) Yoshida and *S. robusta* (Greville) Kylin]. Some, however, lack detailed morphological study and thus require taxonomic confirmation (Womersley, 1994). Gabrielson & Hommersand (1982a) have speculated that *Solieria* originated in the Indo-Pacific, probably along the coasts of southern and western Australia, and has radiated northward and westward to become widespread from

---
* Author for correspondence

tropical through to cool-temperate waters of the western Pacific, northern Indian Ocean and both sides of the Atlantic. Three of the species are recorded in Australia, the commonest and most conspicuous being *S. robusta*, which occurs along the whole of Australia south of ca. 26 ° S. Lat. in the west (Kendrick et al., 1990) and ca. 19 ° S. Lat. in the east (Gabrielson & Kraft, 1984; Womersley, 1994; Millar, 1998). The other two Australian representatives have much more restricted distributions, *S. filiformis* being a probable northern hemisphere adventive known only from the Port of Melbourne region in Victoria (Womersley, 1994), and *S. anastomosa* apparently restricted to Lord Howe and Norfolk Islands off the subtropical east coast (Gabrielson & Kraft, 1984; Millar, 1998).

Solieria robusta is the only host of *Tikvahiella candida* Kraft et Gabrielson (1983), a colourless adelphoparasite that reaches maximum lengths of 5 mm. The parasite appears to be much more restricted than its host in distribution, as it is recorded only from Kangaroo Island, South Australia, eastward to the Melbourne region of Victoria (Kraft & Gabrielson, 1983; Womersley, 1994). Adelphoparasites and their hosts by definition have a close phylogenetic relationship, being members of the same family if not the same tribe (Feldmann & Feldmann, 1958; Goff & Zuccarello, 1994). In contrast, alloparasites refer to parasites that grow on hosts to which they are not closely phylogenetically related. Kraft & Gabrielson (1983) concluded that although its closest alliances within the Solieriaceae were not clear, *Tikvahiella* appeared to be most related to the suite of genera (*Solieria, Rhabdonia, Erythroclonium*) placed by Gabrielson & Hommersand (1982b) in the tribe Solierieae. *Rhabdonia* and *Erythroclonium* have recently been removed to the family Areschougiaceae (formerly the Rhabdoniaceae), however, based on evidence from morphological and polysaccharide studies (Chiovitti et al., 1998a). Evidence from SSU nucleotide sequencing confirms both the distinctness of *Solieria* from the Areschougiaceae, and a close relationship between *Solieria* and *Tikvahiella* (G. Saunders, unpubl.). The present study examines the polysaccharides of *Solieria* and *Tikvahiella* to determine if this feature too is supportive of the morphological and molecular indications of a close relationship.

Apt (1984) has shown that the adelphoparasite *Hypneocolax stellaris* Børgesen and its host, *Hypnea musciformis* (Wulfen) Lamouroux, both members of the family Hypneaceae, have cell-wall galactans of very similar chemistry. Both are reported on the basis of IR analysis to produce mainly κ-carrageenan [composed of carrabiose 4′-sulphate units, where the term 'carrabiose' refers to the repeating disaccharide 3′-linked O-β-D-Gal$p$-(1$\rightarrow$4)-α-D-AnGal$p$]. As this is the only previous report of an analysis of host and parasite polysaccharides, we have made a similar although more detailed study of *Solieria* and *Tikvahiella*, the latter being perhaps easier to work with than *Hypneocolax*, owing to its relatively large size and reasonable abundance.

Materials and methods

Solieria robusta was collected from drift by G. & R. Kraft at Port MacDonnell, South Australia, on 20 July 1991. *Tikvahiella candida* (MELU, K8963) was collected with its host by G. Kraft and G. Saunders on 24 March 1992 in 3–5 m depths at its type locality, the jetty at Portsea, Victoria. Erumpent portions of the parasite were removed, with particular attention paid to ensure that no host tissue contaminated the sample. Although only ca. 20 mg of air-dried material was obtained for polysaccharide extraction, this was a sufficient amount on which to perform several analytical procedures.

Polysaccharides were extracted with hot water, clarified, and precipitated with 2-propanol as described previously (Chiovitti et al., 1995). The polysaccharide preparations were not subjected to further treatment prior to analysis. The sulphate content of the polysaccharide preparation from *Solieria robusta* was determined by the method of Tabatabai (1974) as modified by Craigie et al. (1984). The pyruvate content of the polysaccharides was determined by the method of Duckworth & Yaphe (1970). For the quantitative determination of constituent sugars, alditols from the polysaccharide preparations were derived by reductive hydrolysis and acetylation as described by Stevenson & Furneaux (1991). The alditol acetates were separated by gas chromatography (GC) and identified by their retention times relative to *myo*-inositol hexaacetate as described previously (Liao et al., 1993b). For quantification, molar response factors (RFs) were determined relative to *myo*-inositol for galactose (Gal), 6-O-methylgalactose (6-MeGal), glucose (Glc), and xylose (Xyl) obtained from commercial sources. The RF of 3- and/or 4-O-methylgalactose (3/4-MeGal) was assumed to be the same as that of 6-MeGal. The RF of 3,6-anhydrogalactose (AnGal) was derived by analysis of commercially available κ-carrageenan from

'*Eucheuma cottonii*' [presumably *Kappaphycus alvarezii* (Doty) Doty] known to contain Gal and AnGal in approximately equimolar proportions (Stevenson & Furneaux, 1991).

To determine their linkage and substitution patterns, the polysaccharide preparations were converted to their Me$_2$SO-soluble triethylammonium salt and methylated essentially by the protocol of Stevenson & Furneaux (1991), except that a NaOH/Me$_2$SO suspension was used to generate the alkoxide (Ciucanu & Kerek, 1984). The permethylated alditol acetates derived by reductive hydrolysis (Stevenson & Furneaux, 1991) were separated by GC on a BPX70 (SGE, Australia) capillary column, detected by electron impact ionisation-mass spectrometry (MS), and identified by their mass spectra and their retention times relative to *myo*-inositol hexaacetate (Lau & Bacic, 1993). The permethylated species were quantified directly from the reconstructed ion chromatogram with the assumption that the ion intensity of each of the permethylated alditol acetates was the same on a molar basis, except for the product derived from AnGal 2-sulphate, the RF of which was determined as described above.

Polysaccharide films for Fourier transform infrared (FTIR) spectroscopy were prepared (Liao et al., 1993a) and the FTIR spectra were recorded on a Perkin-Elmer series 2000 FTIR spectrometer in transmittance mode (8 scans, collected at a resolution of 4 cm^{-1}).

The polysaccharide from *Solieria robusta* was dissolved in D$_2$O (3% w/v) for ^{13}C nuclear magnetic resonance (NMR) spectroscopy. The proton-decoupled ^{13}C NMR spectrum was recorded on a Bruker AMX300 WB spectrometer (75.5 MHz) at 80 °C with a spectral width of 12.2 kHz, 45° pulse, an acquisition time of 0.67 s, and a relaxation delay of 0.4 s with approximately 45 000 scans. Chemical shifts are quoted relative to internal Me$_2$SO at 39.6 ppm.

Results

The yield of hot water-extractable polysaccharides from *Solieria robusta* was 40% w/w. The preparation from *S. robusta* was rich in sulphate (31.1% w/w, as SO$_3$Na) and pyruvate (2.8% w/w, as CH$_3$COCOOH). The preparation from *Tikvahiella candida* also contained pyruvate (2.0% w/w). Constituent sugar analyses (Table 1) showed that the polysaccharide preparations from both *Solieria robusta* and *Tikvahiella candida* contained mainly galactans, with Gal, AnGal,

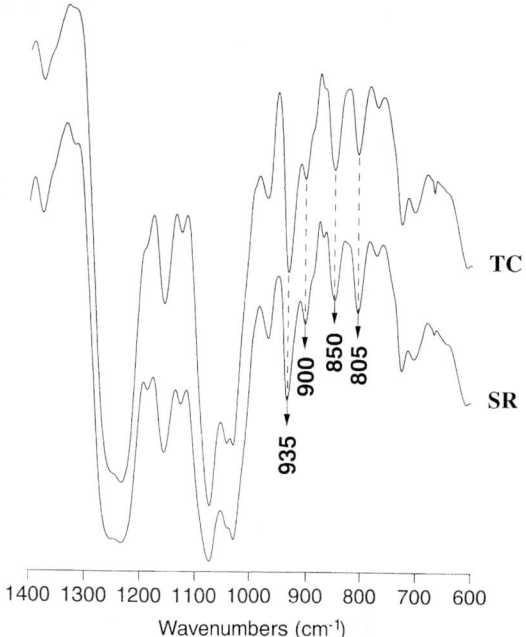

Figure 1. Fourier transform infrared spectra of the polysaccharide preparations from *Solieria robusta* (SR) and *Tikvahiella candida* (TC).

and small amounts of MeGal comprising > 90 mol% of the constituent sugars. Low levels of Xyl and Glc were also detected.

The FTIR spectra of the polysaccharide preparations from *Solieria robusta* and *Tikvahiella candida* were almost identical (Figure 1). The spectra displayed intense absorption at 1240 cm^{-1}, indicative of the presence of sulphate ester substitutions (Stancioff & Stanley, 1969). The diagnostic region of the FTIR spectra showed the characteristic absorption pattern of ι-carrageenan (Stancioff & Stanley, 1969), with intense peaks at 935 cm^{-1} (indicative of AnGal residues), 805 cm^{-1} (axial sulphate ester at O-2 of AnGal) and 850 cm^{-1} (axial sulphate ester at O-4 of 3-linked Gal residues). The FTIR spectra of both polysaccharides also exhibited small peaks at 900 cm^{-1}, suggesting that the polysaccharides contained unsulphated, 3-linked Gal residues and/or unsulphated, 3-linked residues bearing pyruvate acetal at O-4 and O-6 (Whyte et al., 1985; Zablackis & Santos, 1986; Knutsen & Grasdalen, 1987; Santos, 1989; Liao et al., 1993a; Chiovitti et al., 1997a). Notably, the intensity of the 850 cm^{-1} band was relatively greater in the spectrum of the *Tikvahiella candida* polysaccharide, whereas the intensity of the 900 cm^{-1} band was relatively greater in the spectrum of that

Table 1. Constituent sugar analysis of polysaccharide preparations from *Solieria robusta* and *Tikvahiella candida*

Sample	Monosaccharides[a] (mol%)					
	AnGal	Gal	6-MeGal	3-/4-MeGal	Xyl	Glc
Solieria robusta	31	60	1	1	4	3
Tikvahiella candida	31	64	1	-[b]	2	2

[a] AnGal = 3,6-anhydrogalactose; Gal = galactose; 6-MeGal = 6-O-methylgalactose; 3-/4-MeGal = 3-O-methylgalactose and/or 4-O-methylgalactose; Xyl = xylose; Glc = glucose. [b] - = not detected.

Table 2. Linkage analysis of constituent sugars (mol%) of the polysaccharide preparations from *Solieria robusta* and *Tikvahiella candida*

Constituent sugar[a] Deduced linkage[b]	*Solieria robusta*	*Tikvahiella candida*
AnGal$p$		
2,4-	36	32.5
Gal$p$		
3-	4	4.5
4-	1	1
3,4-	40	45
3,6-	1	1.5
3,4,6-	15.5	12
2,3,6-/2,4,6-[c]	2.5	3.5
Xyl$p$		
Terminal	tr[d]	tr

[a] AnGal$p$ = 3,6-anhydrogalactopyranose; Gal$p$ = galactopyranose; Xyl$p$ = xylopyranose. [b] 2,4-Linked AnGal$p$ deduced from 1,2,4,5-tetra-O-acetyl-3,6-anhydrogalactitol, 3-linked Gal$p$ deduced from 1,3,5-tri-O-acetyl-2,4,6-tri-O-methyl-galactitol, etc. [c] 2,3,6-/2,4,6- = 2,3,6- and/or 2,4,6-linked Gal$p$. [d] tr = trace.

extracted from *Solieria robusta*. These observations suggested that the *Tikvahiella candida* polysaccharide had a slightly higher proportion of 3-linked Gal residues bearing sulphate ester substitution at O-4 than that of *Solieria robusta*.

The results of the linkage analyses of the polysaccharide preparations from *Solieria robusta* and *Tikvahiella candida* are summarized in Table 2. All sugars were assumed to be glycosidically linked through O-1 and interpreted as pyranosyl residues. Sulphate esters and pyruvate acetals, if present, were stable during methylation but were liberated during the subsequent acid-hydrolysis, and their locations were therefore observed as 'linkages' in the results. The linkage and substitution patterns of the polysaccharides from the two species were very similar. The major components were 3,4-linked Gal$p$ and 2,4-linked AnGal$p$, which were interpreted in combination with FTIR data as mainly 3-linked Gal$p$ 4-sulphate and 4-linked AnGal$p$ 2-sulphate, respectively. The preparations also contained significant proportions of 3,4,6-linked Gal$p$. With consideration given to the high pyruvate content of the preparations of the two species, this component was likely derived mainly from 3-linked Gal$p$ bearing pyruvate acetal substitution in the form of 4,6-O-(1-carboxyethylidene)-Gal$p$ (DiNinno et al., 1979; Chiovitti et al., 1997a). This interpretation was corroborated by ^{13}C NMR spectroscopic data for the *Solieria robusta* polysaccharide (see below). The peak at 900 cm^{-1} in the FTIR spectra of the polysaccharides was therefore attributed principally to 4,6-pyruvated, 3-linked Gal$p$ residues and partly to small amounts of unsubstituted, 3-linked Gal$p$ residues which were also detected in both samples. Small amounts of 4-linked Gal$p$, 3,6-linked Gal$p$, 2,3,6- and/or 2,4,6-linked Gal$p$, and traces of terminal Xyl$p$ were also detected in the preparations from both species. The 2,3,6- and/or 2,4,6-linked Gal$p$ were most likely derived from 4-linked, Gal$p$ 2,6-disulphate, the "precursor" residue of 4-linked AnGal$p$ 2-sulphate. The former is converted to the latter biosynthetically or by chemical means (Lawson & Rees, 1970; Craigie & Leigh, 1978).

The proton-decoupled ^{13}C NMR spectrum of the polysaccharide preparation from *Solieria robusta* was recorded (Figure 2) and the resonance assignments are summarised in Table 3. The major resonances in the spectrum were assigned to carrabiose 2,4'-disulphate of ι-carrageenan by comparison with published data (Usov et al., 1980; Usov, 1984; Usov & Shashkov, 1985; Falshaw et al., 1996) and in

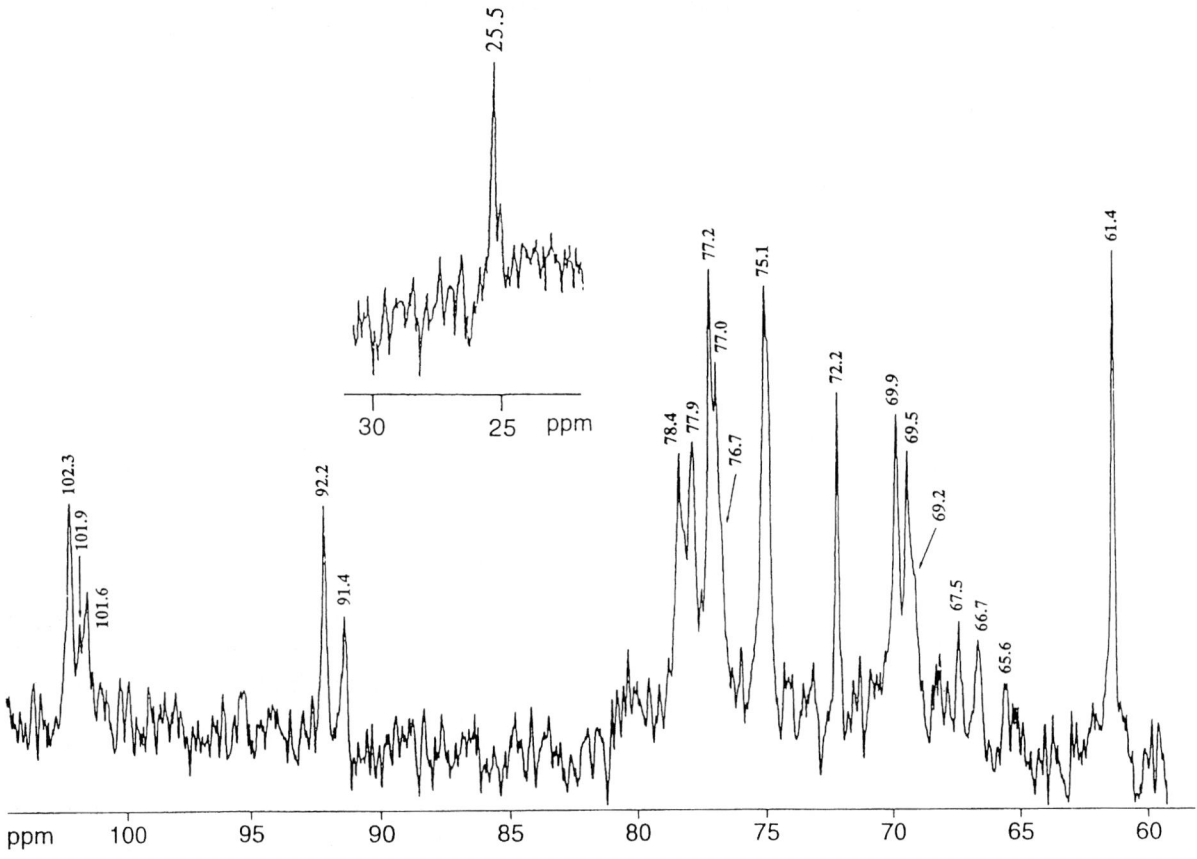

Figure 2. Proton-decoupled ^{13}C NMR spectrum of the polysaccharide preparation from *Solieria robusta*. The resonance arising from the methyl carbon (at 25.5 ppm) of the pyruvate acetal substituent is shown in the inset. The resonance for the carboxyl carbon of the pyruvate acetal substituent was not confirmed.

agreement with the data from FTIR (Figure 1) and linkage analysis (Table 2). A set of minor resonances in the spectrum corresponded to those of 4',6'-O-(1-carboxyethylidene)carrabiose 2-sulphate (Chiovitti et al., 1997a). Sulphation at O-2 of AnGal*p* residues and the occurrence of pyruvate acetal substitution probably both hindered the quantitative recovery of AnGal in the constituent sugar and linkage analyses (Tables 1 and 2; Chiovitti et al., 1997a). There was insufficient material for NMR analysis of the polysaccharide from *Tikvahiella candida*.

Discussion

The polysaccharide from Australian *Solieria robusta* is a hybrid (or mixture) of sulphated and pyruvated carrageenans comprised predominantly of carrabiose 2,4'-disulphate of ι-carrageenan and a significant proportion of 4',6'-O-(1-carboxyethylidene)carrabiose 2-sulphate (Figure 3). Small amounts of unsubstituted, 3-linked Gal*p* residues and other structural variations (Table 2) also occur. The pyruvate content, constituent sugars, FTIR spectrum, and linkage and substitution patterns of the polysaccharide from *Tikvahiella candida* are almost identical to those of the carrageenan of its host, strongly indicating the cell-wall galactans from the two species have comparable structure.

In considering the potential chemotaxonomic significance of the carrageenans of the genera *Solieria* and *Tikvahiella* within the family Solieriaceae, it is important in the light of the distinctive polysaccharides of *Solieria robusta* to first assess how consistent carrageenan chemistry is among *Solieria* species. Few *Solieria* species have been investigated in detail for their sulphated galactans. The sulphated polysaccharides from French specimens of the type species of the genus, *S. chordalis* were shown by compositional analyses, optical rotation measurements,

Table 3. Assignments of resonances[a] observed in the ^{13}C NMR spectrum of the polysaccharide preparation from *Solieria robusta*[b]

Repeating unit	Sugar	Carbon atom						Pyruvate acetal		
		C-1	C-2	C-3	C-4	C-5	C-6	Methyl	Acetal	Carboxyl
G4S-DA,2S	3-linked:	102.3	69.5	77.0	72.2	75.1[c]	61.4			
	4-linked:	92.2	75.1[c]	77.9	78.4	77.2	69.9			
GP-DA,2S	3-linked:	101.9	69.2	76.7	67.5	66.7	65.6	25.5	101.6	nd[d]
	4-linked:	91.4	75.1[c]	77.9	78.4	77.2	69.9			

Repeating units: G4S-DA,2S = carrabiose 2,4'-disulphate of ι-carrageenan; GP-DA,2S = 4',6'-O-(1-carboxyethylidene)carrabiose 2-sulphate.
[a] Chemical shifts in ppm referenced to Me$_2$SO at 39.6 ppm.
[b] Spectrum of a 3% w/v solution in D$_2$O recorded at 80°C.
[c] Coincident resonances.
[d] Carboxyl carbon resonance not determined.

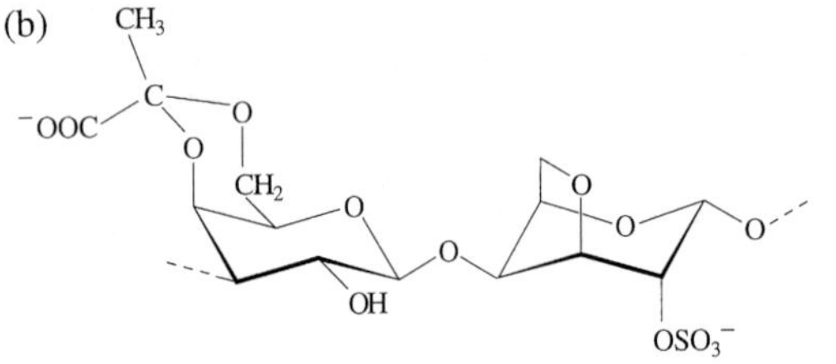

Figure 3. Proposed structures in the carrageenans from *Solieria robusta*: the dominant repeating unit (a) = carrabiose 2,4'-disulphate of ι-carrageenan; the minor repeating unit (b) = 4',6'-O-(1-carboxyethylidene)carrabiose 2-sulphate.

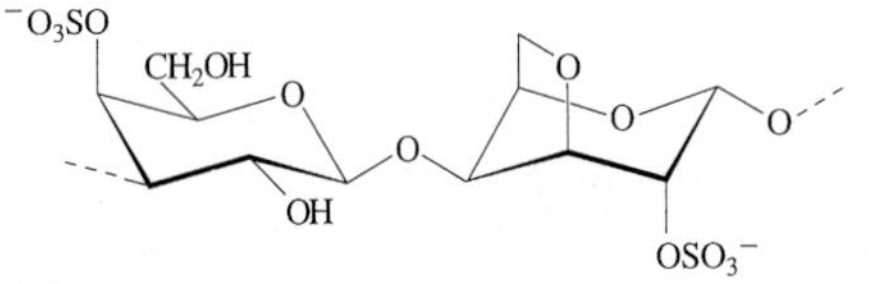

and IR and ^{13}C NMR spectroscopy to be mainly ι-carrageenan (Deslandes et al., 1985, 1990). Notably, however, the sulphate content of the *S. chordalis* carrageenan was significantly lower than that of typical ι-carrageenan from *Eucheuma denticulatum* (Burman) Collins et Hervey [as *E. spinosum* (Linnaeus) J. Agardh; Deslandes et al., 1990]. In addition, Cosson et al. (1995) remarked that the carrageenan from *S. chordalis*, as well as those from *S. filiformis* (Kützing) Gabrielson and *S. tenera* (considered synonymous taxa by Gabrielson, 1985) contained pyruvate residues. With reference also to the presence of small amounts of 6-MeGal in the *S. chordalis* carrageenan,

Cosson et al. (1995) speculated that the *Solieria* carrageenans are 'closely related to agar polymers' since pyruvate acetal and methyl ether substitutions are traditionally associated with agars. However, the data for *S. robusta* in this paper and data for polysaccharides from other species (Chiovitti et al., 1996, 1997a, b) demonstrate unequivocally that pyruvate acetal and methyl ether substitutions occur on carrabiose repeating units in the carrageenans of several Solieriaceae and related families. Indeed, the ^{13}C NMR spectrum presented for the *S. chordalis* carrageenan (Deslandes et al., 1990, Figure 2B) apparently contains some very weak resonances that could correspond to those for 4',6'-pyruvated carrabiose 2-sulphate.

On the basis of sulphate content and IR data, Saito & de Oliveira (1990) concluded that the sulphated polysaccharides from two Brazilian specimens of *Solieria filiformis* were essentially ι-carrageenan. However, the sulphate content of the polysaccharide from one of these specimens (ca. 25% w/w) was significantly lower than that of typical ι-carrageenan (> 30% w/w; Santos, 1989), suggesting the presence of minor structural variations in the polysaccharide. Murano et al. (1997) obtained FTIR, ^{13}C NMR, and ^{1}H NMR spectra of cold and hot water-extracted fractions from Mediterranean *S. filiformis*, which showed that carrabiose 2,4'-disulphate of ι-carrageenan was the dominant component of the galactans. However, numerous additional resonances in the ^{13}C spectra demonstrated the occurrence of structural variations. These additional signals were tentatively assigned mostly to 4-linked Gal*p* 2,6-disulphate (precursor) residues and their neighbouring 3-linked Gal*p* 4-sulphate residues. ^{13}C NMR spectroscopy of alkali-treated *S. filiformis* fractions would assist with resolution of the resonance assignments. Generally, however, the major components of the polysaccharides from both *S. chordalis* and *S. filiformis* are consistent with the major component of the *S. robusta* carrageenan. Verification of the presence and quantification of the amount of pyruvate acetal substitution in the carrageenans from the two former species remain topics for further investigation.

The alga *Solieria jaasundii* Mshigeni et Papenfuss, first named but not formally described as *Rhabdonia africana* by Jaasund (1977), was reported to produce ι-carrageenan largely on evidence from IR spectroscopy of the extracted polysaccharide (Mshigeni et al., 1979). However, the IR spectrum presented (Mshigeni et al., 1979, Figure 2) more closely resembled those of the hybrid α- (carrabiose 2-sulphate) and ι-carrageenans from Thai and Burmese *Catenella nipae* (a member of the Caulacanthaceae; Zablackis & Santos, 1986; Falshaw et al., 1996) and the pyruvated carrageenans from *Callophycus* spp. (Chiovitti et al., 1997a). Furthermore, the sulphate content of the *Solieria jaasundii* polysaccharide was reported (Mshigeni et al., 1979) as 18.6% w/w, considerably less than that of typical ι-carrageenan from *Eucheuma* spp. (> 30% w/w; Santos, 1989). These observations warrant more detailed study of its polysaccharide. Moreover, only the vegetative anatomy of *S. jaasundii* has been investigated (Mshigeni & Papenfuss, 1981), and studies of its reproductive anatomy are necessary to fully justify its inclusion in the genus.

Although the carrageenans of a few other *Solieria* species are superficially similar to those of *S. robusta*, further studies of the galactans of these and other *Solieria* species are required to determine whether the carrageenans of *S. robusta* are typical of the genus. It remains appropriate, however, to examine whether the galactans of *Solieria robusta* and *Tikvahiella candida* serve as chemotaxonomic markers of an alliance within the Solieriaceae. It is noteworthy that ι-carrageenan structure (carrabiose 2,4'-disulphate) seems to be a feature of carrageenans from most members of the Solieriaceae. Although the proportion of this repeat unit in different solieriaceous algae varies significantly according to source, it is the dominant structure of the carrageenans from representatives of the genera *Eucheuma*, *Anatheca*, *Sarcodiotheca* (as *Neoagardhiella*), *Agardhiella*, *Meristotheca*, and *Meristiella* (Stancioff & Stanley, 1969; Lawson et al., 1973; Mollion, 1980; Whyte et al., 1984, 1985; Doty, 1988; Santos, 1989; Chopin et al., 1990; Saito & de Oliveira, 1990; Fostier et al., 1992). The occurrence of ι-carrageenan in *Solieria robusta* and *Tikvahiella candida* is therefore of little chemotaxonomic significance within the Solieriaceae.

By contrast, polysaccharides from representatives of relatively few genera of Solieriaceae are known to contain the 4',6'-pyruvated carrabiose 2-sulphate unit. Most striking are the carrageenans of the genus *Callophycus* which, unlike those of *Solieria robusta* and *Tikvahiella candida*, are composed predominantly of the pyruvated repeating unit and are the most highly pyruvated red algal galactans yet characterized (Chiovitti et al., 1997a). The *Callophycus* carrageenans also contain carrabiose 2-sulphate, the repeating unit of α-carrageenan, as a minor component. The galactans from two Australian specimens of *Sarconema filiforme* were recently analyzed and shown to be pyruvated carrageenans (Chiovitti et al., 1998b). Although the *Sarconema filiforme* carrageenans share structural features in common with the *Solieria robusta* and *Tikvahiella candida* galactans, they differ in that the *Sarconema filiforme* carrageenans contain higher proportions of 4',6'-pyruvated carrabiose 2-sulphate as well as significant proportions of carrabiose 2,4'-disulphate and carrabiose 2-sulphate.

Carrageenans from representatives of the genera *Rhabdonia*, *Erythroclonium*, and *Areschougia* also contain small amounts of the pyruvated carrabiose repeating unit (Chiovitti et al., 1996; Chiovitti et al., 1997b; Liao et al., unpubl.). The carrageenans from these three genera differ, however, from those of *Solieria robusta* and *Tikvahiella candida* in that they contain not only 4',6'-pyruvated carrabiose 2-sulphate and carrabiose 2,4'-disulphate ('ι-carrabiose') but also significant proportions of carrabiose 2-sulphate ("α-carrabiose"), 6'-*O*-methylated ι- and α-carrabiose units, and 4-linked Gal*p* residues substituted to different extents with methyl ether at O-3. The carrageenans of *Rhabdonia*, *Erythroclonium* and *Areschougia* are unique mainly for their high methyl ether content. The three genera have recently been grouped with the genus *Austroclonium* (formerly of the Cystocloniaceae) in the Areschougiaceae on the basis of these polysaccharide features and similar vegetative and reproductive anatomies (Chiovitti et al., 1998a).

In conclusion, the cell-wall galactans of *Solieria robusta* and *Tikvahiella candida* are structurally more alike than those from either species are to the carrageenans of representatives of other solieriacean genera. This observation is congruent with 18S ribosomal subunit RNA data that support an alliance between *Tikvahiella candida* and its host within the Solieriaceae (G. Saunders, unpubl.). Although there are data to indicate the carrageenans of a few other *Solieria* species are similar to that of *S. robusta*, further investigations of the carrageenans from these and other *Solieria* species are required to validate chemotaxonomic comparisons within the genus and between *Solieria* as a whole and other solieriacean genera. In addition to nucleotide sequence and reproductive studies of *S. jaasundii*, a better understanding of the structure of its polysaccharide will assist in assessing its phylogenetic position.

Acknowledgements

We are grateful to Dr Sharon L. A. Munro (currently at the Russell Grimwade School of Biochemistry and Molecular Biology, University of Melbourne, Parkville, Victoria) for recording the ^{13}C NMR spectrum of *Solieria robusta* and to Ms Dina Chen and Ms Zofia Felton for their technical support. This research was funded by a grant from the Australian Government Co-operative Research Centre (CRC) Program to the CRC for Industrial Plant Biopolymers. D. J. Craik is a senior Australian Research Council (ARC) Fellow. A. Chiovitti acknowledges the support of an Australian Government Postgraduate Research Award.

References

Apt, K. E., 1984. The polysaccharide of the parasitic red alga *Hypneocolax stellaris* Børgesen. Bot. mar. 27: 489–490.

Chiovitti, A., G. T. Kraft, G. W. Saunders, M.-L. Liao & A. Bacic, 1995. A revision of the systematics of the Nizymeniaceae (Gigartinales, Rhodophyta) based on polysaccharides, anatomy and nucleotide sequences. J. Phycol. 31: 153–166.

Chiovitti, A., A. Bacic, D. J. Craik, S. L. A. Munro, G. T. Kraft & M.-L. Liao, 1997a. Cell wall polysaccharides from Australian red algae of the family Solieriaceae (Gigartinales, Rhodophyta): novel, highly pyruvated carrageenans from the genus *Callophycus*. Carbohydr. Res. 299: 229–243.

Chiovitti, A., A. Bacic, D. J. Craik, S. L. A. Munro, G. T. Kraft & M.-L. Liao, 1997b. Carrageenans with complex substitution patterns from the genus *Erythroclonium*. Carbohydr. Res. 305: 243–252.

Chiovitti, A., G. T. Kraft, A. Bacic, D. J. Craik, S. L. A. Munro & M.-L. Liao, 1998a. Carrageenans from Australian representatives of the family Cystocloniaceae (Gigartinales, Rhodophyta), with description of *Calliblepharis celatospora* Kraft, sp. nov., and transfer of *Austroclonium* Min-Thein et Womersley to the family Areschougiaceae. J. Phycol. 34: 515–535.

Chiovitti, A., M.-L. Liao, G. T. Kraft, S. L. A. Munro, D. J. Craik & A. Bacic, 1996. Cell wall polysaccharides from Australian red algae of the family Solieriaceae (Gigartinales, Rhodophyta): highly methylated carrageenans from the genus *Rhabdonia*. Bot. mar. 39: 47–59.

Chiovitti, A., A. Bacic, D. J. Craik, G. T. Kraft, M.-L. Liao, R. Falshaw & R. H. Furneaux, 1998b. A pyruvated carrageenan from Australian specimens of the red alga *Sarconema filiforme*. Carbohydr. Res. 310: 77–83.

Chopin, T., M. D. Hanisak, F. E. Koehn, J. Mollion & S. Moreau, 1990. Studies on carrageenans and effects of seawater phosphorus concentration on carrageenan content and growth of *Agardhiella subulata* (C. Agardh) Kraft and Wynne (Rhodophyceae, Solieriaceae). J. appl. Phycol. 2: 3–16.

Ciucanu, I. & F. Kerek, 1984. A simple and rapid method for the permethylation of carbohydrates. Carbohydr. Res. 131: 209–217.

Cosson, J., E. Deslandes, M. Zinoun & A. Mouradi-Givernaud, 1995. Carrageenans and agars, red algal polysaccharides. In Round, F. E. & D. J. Chapman (eds), Progress in Phycological Research, Vol. 11, Biopress Ltd., Bristol: 269–324.

Craigie, J. S. & C. Leigh, 1978. Carrageenans and agars. In Hellebust, J. A. & J. S. Craigie (eds), Handbook of Phycological Methods: Physiological and Biochemical Methods. Cambridge University Press, Cambridge: 109–131.

Craigie, J. S., Z. C. Wen & J. P. van der Meer, 1984. Interspecific, intraspecific and nutritionally-determined variations in the composition of agars from *Gracilaria* spp. Bot. mar. 27: 55–61.

Deslandes, E., C. Bodeau-Bellion, J.-Y. Floc'h & M. Penot, 1990. ^{13}C-NMR spectroscopy and chemical analysis of the carrageenans of four red algae (Gigartinales). Plant Physiol. Biochem. 28: 65–69.

Deslandes, E., J.-Y. Floc'h, C. Bodeau-Bellion, D. Brault & J. P. Braud, 1985. Evidence for ι-carrageenans in *Solieria chordalis* (Solieriaceae) and *Calliblepharis jubata, Calliblepharis ciliata, Cystoclonium purpureum* (Rhodophyllidaceae). Bot. mar. 28: 317–318.

DiNinno, V., E. L. McCandless & R. A. Bell, 1979. Pyruvic acid derivative of a carrageenan from a marine red alga (*Petrocelis* species). Carbohydr. Res. 71: C1–C4.

Doty, M. S., 1988. Prodromus ad Systematica Eucheumatoideorum. A tribe of commercial seaweeds related to *Eucheuma* (Solieriaceae, Gigartinales). In Abbott, I. A. (ed.), Taxonomy of Economic Seaweeds. California Sea Grant College Program Report No. CSGCP-018: 159–207.

Doty, M. S., 1995. *Betaphycus philippinensis* gen. et sp. nov. and related species (Solieriaceae, Gigartinales). In Abbott, I. A. (ed.), Taxonomy of Economic Seaweeds. California Sea Grant College Program Report No. CSGCP-035: 237–245.

Duckworth, M. & W. Yaphe, 1970. Definitive assay for pyruvic acid in agar and other algal polysaccharides. Chem. Ind. 747–748.

Falshaw, R., R. H. Furneaux, H. Wong, M.-L. Liao, A. Bacic & S. Chandrkrachang, 1996. Structural analysis of carrageenans from Burmese and Thai specimens of *Catenella nipae* Zanardini. Carbohydr. Res. 285: 81–98.

Feldmann, J. & G. Feldmann, 1958. Recherches sur quelques floridées parasites. Rev. Gén. Bot. 65: 49–129.

Fostier, A. H., J. M. Kornprobst & G. Combaut, 1992. Chemical composition and rheological properties of carrageenans from two Senegalese Solieriaceae *Anatheca montagnei* Schmitz and *Meristotheca senegalensis* Feldmann. Bot. mar. 35: 351–355.

Gabrielson, P. W., 1985. *Agardhiella* versus *Neoagardhiella* (Solieriaceae, Rhodophyta): another look at the lectotypification of *Gigartina tenera*. Taxon 34: 275–280.

Gabrielson, P. W. & M. H. Hommersand, 1982a. The Atlantic species of *Solieria* (Gigartinales, Rhodophyta): their morphology, distribution and affinities. J. Phycol. 18: 31–45.

Gabrielson, P. W. & M. H. Hommersand, 1982b. The morphology of *Agardhiella subulata* representing the Agardhielleae, a new tribe in the Solieriaceae (Gigartinales, Rhodophyta). J. Phycol. 18: 46–58.

Gabrielson, P. W. & G. T. Kraft, 1984. The marine algae of Lord Howe Island (N.S.W.): the family Solieriaceae (Gigartinales, Rhodophyta). Brunonia 7: 217–251.

Goff, L. J. & G. Zuccarello, 1994. The evolution of parasitism in red algae: cellular interactions of adelphoparasites and their hosts. J. Phycol. 30: 695–720.

Jaasund, E., 1977. Marine algae in Tanzania. V. Bot. mar. 20: 333–338.

Kendrick, G. A., J. M. Huisman & D. I. Walker, 1990. Benthic macroalgae of Shark Bay, Western Australia. Bot. mar. 33: 47–54.

Knutsen, S. H. & H. Grasdalen, 1987. Characterisation of water-extractable polysaccharides from Norwegian *Furcellaria lumbricalis* (Huds.) Lamour. (Gigartinales, Rhodophyceae) by IR and NMR spectroscopy. Bot. mar. 30: 497–505.

Kraft, G. T., 1981. Rhodophyta: morphology and classification. In Lobban, C. S. & M. J. Wynne (eds), The Biology of Seaweeds. Blackwell Scientific Publications, Oxford: 6–51.

Kraft, G. T. & P. W. Gabrielson, 1983. *Tikvahiella candida* gen. et sp. nov. (Solieriaceae, Rhodophyta), a new adelphoparasite from southern Australia. Phycologia 22: 47–57.

Lau, E. & A. Bacic, 1993. Capillary gas chromatography of partially methylated alditol acetates on a high-polarity, cross-linked, fused silica BPX70 column. J. Chromatogr. 637: 100–103.

Lawson, C. J. & D. A. Rees, 1970. An enzyme for the metabolic control of polysaccharide conformation. Nature 227: 392–393.

Lawson, C. J., D. A. Rees, D. J. Stancioff & N. F. Stanley, 1973. Carrageenans. Part VIII. Repeating structures of galactan sulphates from *Furcellaria fastigiata*, *Gigartina canaliculata*, *Gigartina chamissoi*, *Gigartina atropurpurea*, *Ahnfeltia durvillaei*, *Gymnogongrus furcellatus*, *Eucheuma cottonii*, *Eucheuma spinosum*, *Eucheuma isiforme*, *Eucheuma uncinatum*, *Aghardiella tenera*, *Pachymenia hymantophora*, and *Gloiopeltis cervicornis*. J. Chem. Soc. Perkin I: 2177–2182.

Liao, M.-L., G. T. Kraft, S. L. A. Munro, D. J. Craik & A. Bacic, 1993a. Beta/kappa-carrageenans as evidence for continued separation of the families Dicranemataceae and Sarcodiaceae (Gigartinales, Rhodophyta). J. Phycol. 29: 833–844.

Liao, M.-L., S. L. A. Munro, D. J. Craik, G. T. Kraft & A. Bacic, 1993b. The cell wall galactan of *Catenella nipae* Zanardini from southern Australia. Bot. mar. 36: 189–193.

Millar, A. J. K., 1999. Marine benthic algae of Norfolk Island, South Pacific. Australian Systematic Botany 12 (in press).

Mollion, J., 1980. Infrared and chemical studies of the carrageenan from *Anatheca montagnei* Schmitz, (Solieriaceae) from Senegal, West Africa. Bot. mar. 28: 197–199.

Mshigeni, K. E., A. K. Semesi & T. M. Ngonyani, 1979. Studies on the phycocolloid from the red seaweed *Rhabdonia africana* Jaasund (Gigartinales, Rhabdoniaceae). Bot. mar. 22: 447–450.

Mshigeni, K. E. & G. F. Papenfuss, 1981. *Solieria jaasundii*, a new species of red algae (Gigartinales, Solieriaceae) from Tanzania. Bot. mar. 24: 1–7.

Murano, E., R. Toffanin, E. Cecere, R. Rizzo & S. H. Knutsen, 1997. Investigation of the carrageenans extracted from *Solieria filiformis* and *Agardhiella subulata* from Mar Piccolo, Taranto. Mar. Chem. 58: 319–325.

Saito, R. M. & E. C. de Oliveira, 1990. Chemical screening of Brazilian marine algae producing carrageenans. Hydrobiologia 204/205: 585–588.

Santos, G. A., 1989. Carrageenans of species of *Eucheuma* J. Agardh and *Kappaphycus* Doty (Solieriaceae, Rhodophyta). Aquat. Bot. 36: 55–67.

Stancioff, D. J. & N. F. Stanley, 1969. Infrared and chemical studies on algal polysaccharides. Proc. int. Seaweed Symp. 6: 595–609.

Stevenson, T. T. & R. H. Furneaux, 1991. Chemical methods for the analysis of sulphated galactans from red algae. Carbohydr. Res. 210: 277–298.

Tabatabai, M. A., 1974. Determination of sulphate in water samples. Sulphur Inst. J. 10: 11–13.

Usov, A. I., 1984. NMR spectroscopy of red seaweed polysaccharides: agars, carrageenans and xylans. Bot. mar. 27: 189–202.

Usov, A. I. & A. S. Shashkov, 1985. Polysaccharides of algae. 34. Detection of iota-carrageenan in *Phyllophora brodiaei* (Turn.) J. Ag. (Rhodophyta) using ^{13}C-NMR spectroscopy. Bot. mar. 28: 367–373.

Usov, A. I., S. V. Yarotsky & A. S. Shashkov, 1980. ^{13}C-NMR spectroscopy of red algal galactans. Biopolymers 19: 977–990.

Whyte, J. N. C., R. E. Foreman & R. E. DeWreede, 1984. Phycocolloid screening of British Columbia red algae. Hydrobiologia 116/117: 537–541.

Whyte, J. N. C., S. P. C. Hosford & J. R. Englar, 1985. Assignment of agar or carrageenan structures to red algal polysaccharides. Carbohydr. Res. 140: 336–341.

Womersley, H. B. S., 1994. The Marine Benthic Flora of Southern Australia. Rhodophyta. Part IIIA. Australian Biological Resources Study, Canberra, 508 pp.

Zablackis, E. & G. A. Santos, 1986. The carrageenan of *Catenella nipae* Zanard., a marine red alga. Bot. mar. 29: 319–322.

Monthly changes in the content of fucans, their constituent sugars and sulphate in cultured *Laminaria japonica*

Masura Honya[1], Hiroe Mori[2], Michiko Anzai[2], Yoko Araki[2] & Kazutosi Nisizawa[3]
[1]*Research Institute, Ishikawajima-Harima Heavy Ind. Co. Ltd., 1 Sinnakahara-cho, Isogo, Yokohama 235-0031, Japan*
[2]*Tokyo Kasei Gakuin University, 2600 Aihara-cho Machida, Tokyo 194-0292, Japan*
[3]*NABOCUL Cosmetics Laboratory, NABOCUL Cosmetics Co. Ltd., 5-29-7 Sendagaya, Sibuya Tokyo 151-0051, Japan*

Key words: brown algae, *Laminaria japonica*, fucan, sulphated polysaccharide, monthly change, molar ratio

Abstract

Crude fucan was extracted monthly from makonbu, *Laminaria japonica* Areschoug, cultured from April to October at a southern site of the Hokkaido bay. The crude fucan yield tended to gradually increase from April to September, and rose markedly in October when spore formation was over. The fucans were fractionated on DEAE-Sephadex A-25 into three acidic components, A, B and C. In all the monthly extracts, fraction C was the largest. It was entirely free of laminaran and alginate, as was fraction B. Several sugars such as mannose and glucuronic acid, were identified in fraction C beside a large amount of fucose and sulphate. Only the proportion of the latter two constituents continued to increase as the alga matured to reach twice or even more than three times as much as all the other sugars in the September extract.

Abbreviations: Fuc – L-fucose; Gal – D-galactose; Xyl – D-xylose; Man – D-mannose; Glc – D-glucose; GlcA – D-glucuronic acid; Ara – L-arabinose; Rha – L-rhamnose; UA – uronic acids

Introduction

Fucans are highly heterogeneous sulphated polysaccharide families composed of various proportions of different sugar residues, such as fucose, galactose, mannose and uronic acid. The number of fucan groups and their compositional patterns differ considerably, depending on fractionation procedures as well as algal species (Mian & Percival, 1973 a, b; Abdel-Fattah et al., 1974; Medcalf, 1978; Kloareg et al., 1979; Mabeau et al., 1990).

We have formerly investigated fucans from the sporophyll of *Undaria pinnatifida* (Harvey) Suringer (Mori et al., 1982) and *Sargassum ringgoldianum* Harvey to elucidate their chemical and pharmaceutically functional properties. Three fucan fractions were obtained from these species by the same extraction procedure. All of them had similar compositions, but differed more or less in the molar ratio of their constituents. The major fucan fractions from both algae always contained the largest amounts of fucose and sulphate.

However, these studies have been performed on fucan fractions from algal fronds collected at a single growth stage. Hence, we investigated the possible variation in the monthly content and composition of fucans using blades of cultured *Laminaria japonica* Areschoug as an experimental alga.

Material and methods

Material

Germinated *Laminaria japonica* was transplanted to the Date Bay of Hokkaido in late November 1985 and cultured until October 1986 as previously reported (Honya et al.,1993).

Each mid-month, from April to October, more than 20 kelps at the same growth stage were collected, quickly transported at 4 °C to the laboratory and stored at −25 °C.

Extraction and fractionation

Frozen kelps were thawed at room temperature, and their holdfasts were removed. Fronds were quickly washed by fresh water, cut into small pieces, dried under vacuum at 15–20 °C and milled.

The extraction and fractionation of fucans were made as previously described (Mori & Nisizawa, 1982). Briefly, the milled fronds were washed with 85% ethanol at 70 °C for 3 h, then extracted thrice with 0.05 N HCl for 15–17 h. The filtered extracts were neutralized with NaHCO$_3$ and then dialyzed against running tap water. Two volumes of acetone were added to the extract to precipitate the polysaccharides. The precipitate was dissolved in distilled water, and subsequently calcium acetate and trichloroacetic acid were gently added to precipitate alginate and proteins. These were removed by centrifugation, and 4 vol. of ethanol were added to the supernatant to precipitate the sulphated polysaccharide referred to as crude fucan.

The crude fucan was fractionated by anion exchange chromatography on a DEAE-Sephadex A-25 column (3.5 × 35 cm). The sample (200 mg) was applied on the column and eluted stepwise by NaCl solutions of increasing concentration (0.5 M, 1 M, 2 M) in 0.01 N HCl. Elution of the polysaccharides was followed by the phenol-sulphuric acid method (Dubois et al., 1956). This procedure was repeated five times and the corresponding fractions were combined, dialyzed and lyophilized.

Identification and determination of neutral sugars and uronic acids

Neutral sugars and uronic acids were determined by gas chromatography (GC) and by the carbazole-sulphuric acid procedure, respectively, as described (Mori & Nisizawa, 1982). All determinations were repeated twice or thrice.

Determination of sulphate content

The sulphate content of the fucan was determined by an improved method (Ota, 1968) of the flask-combustion analysis (Schöniger, 1955). Kelp fucan powder (2–3 mg), exactly weighed on a sheet of No. 5A Toyo filter paper, was transferred into a platinum basket with the paper, and placed in a combustion flask

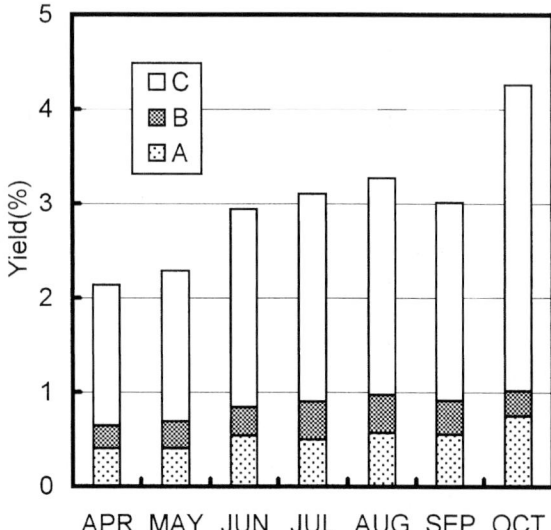

Figure 1. Monthly yield of crude fucan from kelp blades. Percent yield on a dry weight basis. A, B, and C indicate fucan fractions on the DEAE-Sephadex chromatogram as shown in Figure 2.

with 5 ml of redistilled water and two drops of 30% hydrogenperoxide. Air in the flask was replaced by oxygen gas and the fucan was completely ignited with an electric heater. The flask was then rinsed by pure ethanol; the resulting solution was adjusted to pH 6.0–6.2, and titrated with 0.005 N barium chloride solution using 0.1% dimethylsulfonazo III as indicator. The analysis was repeated twice or thrice.

Electrophoresis and paper chromatography

The homogeneity of fucan fractions was determined by electrophoresis on a cellulose acetate film, and fucan spots were detected by 0.5% toluidine blue dissolved in 3% acetic acid solution (Mori & Nisizawa, 1982).

Uronic acid residues in fucan hydrolysates prepared by the method of Haug & Larsen (1962) were identified by paper chromatography as reported by Fischer & Dörfel (1955).

Results

The yield of crude fucans extracted from kelp blades collected from April to October is shown in Figure 1. The amount of crude fucan increased almost in parallel with growth of the algae in summer. When the kelp blades were decaying in October and sporangia were

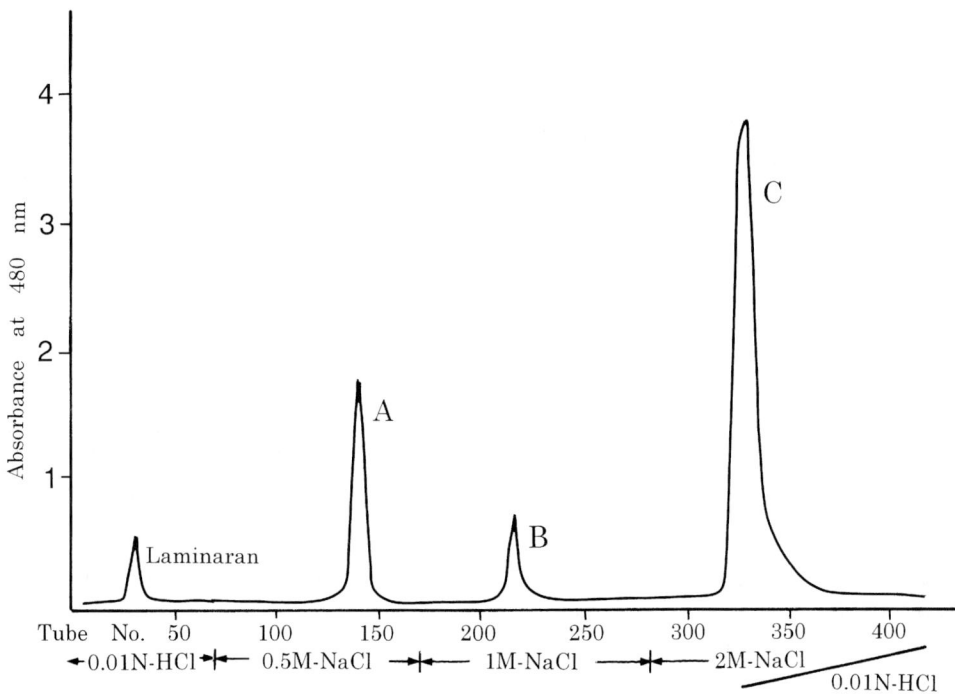

Figure 2. Fractionation on DEAE-Sephadex A-25 column chromatography of crude fucan from kelp blades collected in October. Each fraction contains 5 ml of eluate, 0.5 ml of which was used for determination of sugar content. Tube numbers in A; 130–146, B; 210–220, C; 320–350. The detection of each peak was made by colorimetry using phenol sulphuric acid method (absorbance at 480 nm). Laminaran was identified as glucose after acid hydrolysis and GC.

maturing, the fucan content was more than twice that of young algae. However, these variations in fucan content were not as high as those reported for laminaran and mannitol in *Laminaria digitata* (Haug & Jensen, 1956) and *L. japonica* (Honya et al., 1993). This may be due to different physiological functions of fucans in the kelp.

The crude fucans were fractionated by DEAE-Sephadex A-25 into a laminaran fraction eluted by 0.01 N HCl and three fucan peaks, A, B and C, eluted by NaCl solutions (Figure 2). The fucan elution pattern of the October extract represented in Figure 2 was identical to that of all of the different monthly extracts. It was also similar to that of other brown algal fucans such as those from *U. pinnatifida* and *Sargassum ringgoldianum* (Mori et al., 1982; Mori & Nisizawa, 1982; Coté, 1959; Matsuhiro & Zambrano, 1990). The elution of fucan fractions was related to their sulphate content and the major fraction contained the most sulphated polysaccharides. The yields of these frac-

Table 1. Electrophoretic analysis of the three fucan fractions from the kelp blades harvested in October. Electrophoresis was carried out on a cellulose acetate membrane under three conditions, and 0.5% toluidine blue in 3% acetic acid solution was used as detector reagent

Condition	Spot-numbers separates	Mobility from cathode (cm)		
		A	B	C
1M acetic acid-pyridine (pH3.5) 0.5 mA/cm, 30 min	1	5.1	4.8	6.0
0.1N HCl 1.0 mA/cm, 2 h	1	1.7	3.7	6.5
0.2M calcium-acetate 1.0 mA/cm, 3 h	1 or 2	0^a, 2.5^b	4.2	6.3

[a] Alginate spot identified by the carbazole-H_2SO_4 method (see 'Materials and methods').
[b] Fucan spot based on the sugar analysis.

Table 2. Monthly changes in molar ratios of sugar and sulphate residues of fucan fraction B

Month	Fuc	Gal	Xyl	Man	Glc	GlcA	SO_4^{2-}	Ara	Rha
APR	1.00	0.28	1.12	0.70	0.77	1.43	3.13	0.40	0.22
MAY	1.00	0.33	0.58	0.42	0.28	1.54	3.78	0.33	-[a]
JUN	1.00	0.27	0.78	0.52	0.12	1.13	2.03	0.08	-
JUL	1.00	0.21	0.90	0.63	0.16	1.06	2.15	0.09	-
AUG	1.00	0.19	0.49	0.46	0.13	1.01	1.98	0.12	-
SEP	1.00	0.33	1.52	1.12	0.44	2.97	2.90	0.22	-
OCT	1.00	0.07	0.24	0.17	0.08	0.48	1.44	0.07	-

[a]: Not detected

Table 3. Monthly changes in molar ratios of sugar constituent and sulphate residues of fucan fraction C

Month	Fuc	Gal	Xyl	Man	Glc	GlcA	SO_4^{2-}	Ara	Rha
APR	1.00	0.37	0.14	0.09	0.02	0.21	1.66	0.004	0.002
MAY	1.00	0.28	0.04	0.09	0.02	0.18	1.51	0.003	-
JUN	1.00	0.24	0.12	0.07	0.03	0.15	1.49	0.008	-
JUL	1.00	0.17	0.04	0.09	0.01	0.15	1.50	0.007	-
AUG	1.00	0.20	0.08	0.11	0.02	0.19	1.53	0.012	-
SEP	1.00	0.18	0.08	0.03	0.01	0.08	1.61	0.004	-
OCT	1.00	0.26	0.10	0.04	0.02	0.10	1.55	0.003	0.002

tions were 28 mg, 10 mg and 122 mg for A, B and C, respectively.

The three fucan fractions were analyzed further by electrophoresis to examine their homogeneity using three different solvent conditions (Table 1). Fraction C, being the richest in sulphate, migrated the fastest in all three conditions. Alginate was detected only in fraction A and fractions B and C migrated as single spots (Table 1). These results agree with those from other species of brown algae (Mori et al., 1982; Mori & Nisizawa, 1982).

The sugars and ester sulphate contents of fractions B and C from the kelp blades collected monthly are given in Tables 2 and 3. It is clear from Table 2 that the molar ratios (relative amounts) of ester sulphate in fucan fraction B were more than 3.0 to 1.5 times larger than that of fucose, but the absolute sulphate contents were lower than those of the fucan fraction C. This fact suggests that sulphate may occur on other sugar(s) than fucose in these fucans, particularly in young growth stages. Table 2 also shows that the molar proportion of glucuronic acid relative to fucose was about 1.5 in the young stages of the kelp while that of other sugar residues were much lower. In addition, all these molar proportions to fucose showed a tendency to decrease toward August, while they increased suddenly in the September extract, reaching a value close to 3 for glucuronic acid. These facts may be due to the possible changes in the metabolism of kelp blade such as the sporangia formation and a slight decaying of the blades. In the fucan from all the different growth stages of the alga, the proportion of sulphate in fraction C was about 1.5 times higher than that of fucose, while that of all other constituent sugars were much lower (Table 3). The variation patterns of each constituent proportions in fraction C, which are remarkably different from those in fraction B, may reflect that these fucans consist of different molecular species.

Figure 3 shows the monthly molar content of the main sugar constituents and sulphate of the major fucan fraction C. Its fucose content changed in parallel with that of sulphate as the alga matured, and reached a maximum in September. The variation of other sugars content varied less characteristically except galactose which concentration appeared inversely

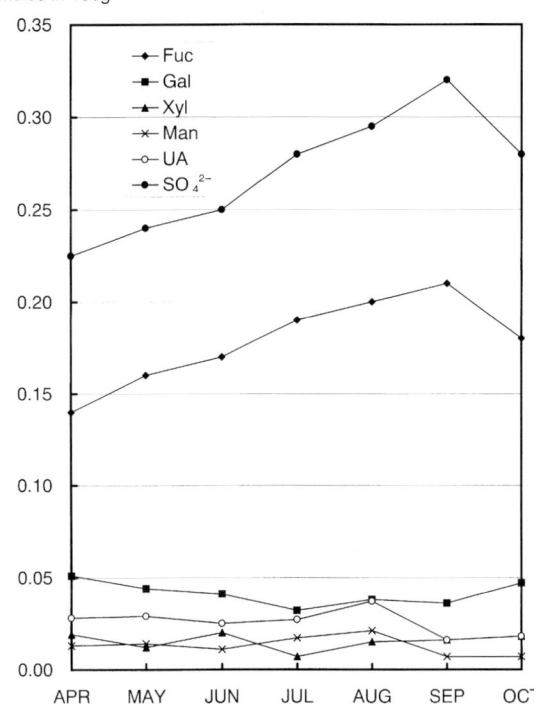

Figure 3. Monthly changes in molar contents of sugar and sulphate residues of fucan fraction C.

related to that of fucose. This fact suggests the existence of some close metabolic relationship between these two sugars.

Discussion

In the present work, the monthly changes in the amount of the individual constituents of several fucan fractions from *Laminarian japonica* fronds at different growth stages was investigated. As with the crude fucan from *Sargassum ringgoldianum* (Mori & Nisizawa, 1982), we chromatographed *L. japonica* fucan through DEAE-Sephadex into three fractions, A, B and C. These results contrasted with that obtained from the crude fucan of *Undaria pinnatifida* sporophyll which yielded four fractions of different concentrations on DEAE-Toyopearl resin chromatography (unpublished results). Thus, fucans consist of various molecular types.

Fucose and sulphate contents were the highest among all the constituents of fucan C. Their concentration increased in this fraction obtained from fronds collected from April to September, but then decreased steeply in the October collected fronds. Although the content of galactose was much smaller than that of the above two residues, its content in the different frond samples behaved almost inversely. Since *L. japonica* fronds with sori were richer in all these constituents than those without sori (unpubl. data), the reproductive status of the alga seems to be closely related with fucan metabolism.

References

Abdel-Fattah, F. M. D., M. Hussein & H. M. Salem, 1974. Some structural features of sargassan, a sulphated heteropolysaccharide from *Sargassum linifolium.* Carbohydr. Res. 33: 19–24.

Côté, R. H., 1959. Disaccharides from fucoidin. J. Chem. Soc.: 2248–2254.

Dubois, M., K. A. Gilles, J. K. Hamilton, P. A. Rebers & F. Smith, 1956. Colorimetric method for determination of sugars and related substances. Anal. Chem. 28: 350–356.

Fischer, F. G. & H. Dörfel, 1955. Die papierchromatographishe Trennung und Bestimmung der Uronsauren. Zeitschr. Physiol. Chem. 301: 224–234.

Haug, A. & A. Jensen, 1956. Seasonal variation in chemical composition of *Laminaria digitata* from different parts of the Norwegian coast. Proc. int. Seaweed Symp. 2: 10–15.

Haug, A. & B. Larsen, 1962. Quantitative determination of the uronic acid composition of alginates. Acta Chimica Scand. 16: 1908–1918.

Honya, M., T. Kinoshita, M. Ishikawa, H. Mori & K. Nisizawa, 1993. Monthly determination of alginate, M / G ratio, mannitol and minerals in cultivated *Laminaria japonica.* Bull. Jap. Soc. Sci. Fish. (Nippon Suisan Gakkaishi) 59: 295–299.

Kloareg, B., M. Demarty & M. Quillet, 1979. Extraction et purification du fucoidane de *Pelvetia canaliculata* (Dene et Thur.). Physiol. veg. 17: 731–747.

Mabeau, S., B. Kloareg & J. P. Joseleau, 1990. Fractionation and analysis of fucans from brown algae. Phytochemistry 29: 2441–2445.

Matsuhiro, B. & D. M. Zambrano, 1990. Carbohydrate constituents of *Lessonia trabeculata.* J. appl. Phycol. 2: 183–185.

Medcalf, D. G., 1978. Sulfated fucose-containing polysaccharides from brown algae: Structural features and biochemical implications. Amer. chem. Soc. Symp. Series 77: 225–244.

Mian, A. J. & E. Percival, 1973a. Carbohydrates of the brown seaweeds *Himanthalia lorea, Bifurcaria bifurcata* and *Padina pavonia.* Part I: Extraction and fractionation. Carbohydr. Res. 26: 133–146.

Mian, A. J. & E. Percival, 1973b. Carbohydrates of the brown seaweeds *Himanthalia lorea* and *Bifurcaria bifurcata.* Part II: Structural studies of the 'fucans'. Carbohydr. Res. 26: 147–161.

Mori, H., H. Kamei, E. Nishide & K. Nisizawa, 1982. Sugar constituents of some sulfated polysaccharides from sporophylls of wakame (*Undaria pinnatifida*) and their biological activities. In Hoppe, H. A. & T. Levring (eds), Marine Algae in Pharmaceutical Science, Vol.2, Walter de Gruyter, Berlin/New York: 109–121.

Mori, H. & K. Nisizawa, 1982. Sugar constituents of sulfated polysaccharides from the fronds of *Sargassum ringgoldianum.* Bull. Jap. Soc. Sci. Fish. (Nippon Suisan Gakkaishi) 48: 981–986.

Ota, S., 1968. Microdetermination of sulfur in organic compounds with modified type combustion flask. Jap. Analyst 17: 1322–1324.

Schöniger, W., 1955. Eine mikroanalytische Schnellbestimmung von Halogen in organischen Substanzen. Mikrochim. Acta 1: 123–129.

Open-water aquaculture of the red alga *Chondrus crispus* in Prince Edward Island, Canada

Thierry Chopin[1], Glyn Sharp[2], Ellen Belyea[1], Robert Semple[2] & Donald Jones[3]
[1]*University of New Brunswick, Centre for Coastal Studies and Aquaculture, Department of Biology, P.O. Box 5050, Saint John, N.B., E2L 4L5, Canada*
E-mail: tchopin@unbsj.ca
[2]*Department of Fisheries and Oceans, Bedford Institute of Oceanography, P.O. Box 1006, Dartmouth, N.S., B2Y 4A2, Canada* [3]*Miminegash Marine Research and Development Station, Miminegash, P.E.I., C0B 1Z0, Canada*

Key words: Chondrus crispus, open-water aquaculture, cold-water aquaculture, daily growth rate, carrageenans, nutrients

Abstract

The red alga *Chondrus crispus* (Irish moss) has been commercially harvested in Eastern Canada for almost 60 years. Its land-based tank aquaculture was initiated in the 1970s. In the 1990s, it became clear that production costs of these capital intensive systems were still too high for the carrageenan market but not for the production of edible seaweeds. Open-water aquaculture of cold-temperate species of carrageenophytes, and in particular of *C. crispus*, has rarely been attempted. This study re-examined the potential of the unique unattached and mostly vegetative population of *C. crispus* at Basin Head, in eastern Prince Edward Island (P.E.I.), and at 5 transplant sites in western P.E.I. Basin Head plants were successfully transplanted to other sites, providing similar or different environmental conditions, and yielding comparable, or even higher, productivity. During the peak growth periods (May to end of June and autumn), daily growth rates (DGRs) between 3 and 4% d^{-1} were recorded at Basin Head and Freeland, with some plants exceeding 6% d^{-1}. Over the whole study period (May to October), DGRs between 2 and 4% d^{-1} were lower than those reported for different species of *Eucheuma* and *Kappaphycus alvarezii*; they were, however, compensated for by extremely high carrageenan yields (between 58.1 and 71.0% DW) during the summer months when nutrients (phosphorus and nitrogen) levels in seawater and algal tissue were low. The DGRs could be increased by developing culture structures retaining fragmenting, but otherwise healthy, large distal clumps, lost with the present simple tying of plants on screens. Preliminary results demonstrated that transplantation and grow-out techniques are biologically successful, and that the Basin Head population of *C. crispus* has significant potential for open-water aquaculture in estuaries and basins of Atlantic Canada.

Introduction

Land-based aquaculture of the red alga *Chondrus crispus* (Irish moss), with air-agitated or paddle-wheel tank systems, was initiated in the 1970s in government and private facilities in both Canada and France (Neish et al., 1977; Braud & Delépine, 1981; Craigie, 1990). After 15 years of research, some believed that tank aquaculture in temperate regions could not compete with the harvest of natural populations, or tropical open-water aquaculture, mostly because of the high operation and labour costs and inadequate solar and thermal conditions (Bidwell et al., 1985). Others persevered and developed large-scale facilities (Acadian Seaplants Ltd. in Canada and Sanofi Bio-Industries in France) in the 1980s. In the 1990s, it became clear that, even if domestication of *C. crispus* had been successfully achieved by optimizing culture parameters, production costs of Irish moss for the carrageenan market were still too high and the gain in carrageenan purity and supply stability could not offset these costs (Chopin, 1998). Moreover, an evasive λ-carrageenan market and major corporate restructuring stopped these tank culture programmes in both Canada and France. Since then, Acadian Seaplants Ltd has successfully converted its facility to the culture of

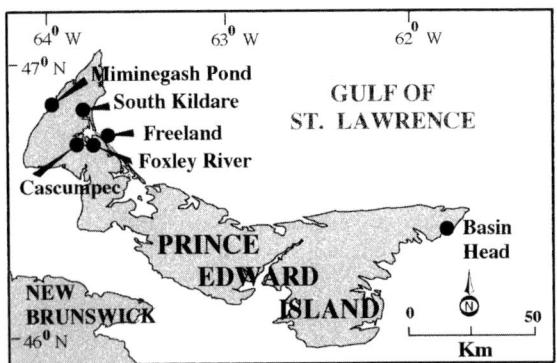

Figure 1. Map of Prince Edward Island, Canada, showing the locations of the unique unattached population of *Chondrus crispus* in Basin Head and of the 5 transplant sites (South Kildare, Freeland, Cascumpec, Miminegash Pond, and Foxley River).

C. crispus no longer for the phycolloid market, but as an edible, high added-value seaweed by manipulating the colour and the texture of selected isolates (Craigie et al., 1999).

Open-water aquaculture of carrageenophytes has been extremely successful in tropical environments, especially in the case of *Eucheuma* (Doty, 1987), but rarely attempted in temperate regions. McCurdy (1980) did some culture and transplant experiments with the unique unattached population of *C. crispus* in Basin Head, Prince Edward Island (P.E.I.), Canada (Figure 1). This isolated population grows in a shallow, sheltered arm (3000 m long, 100 to 130 m wide, and 1 m deep) extending east from the Basin proper (760 m long and 380 m wide), which has only a narrow channel (500 m long and 2.5 m deep) for exchange with open oceanic water. The plants are characterized by large, thick fronds which are rarely reproductive and mostly gametophytic [as identified by the resorcinol test (Craigie & Leigh, 1978)]. Fragmentation appears to be the main mode of reproduction in this population. Individual fronds can reach 1 kg wet weight (WW) and form large clumps by adhesion of byssal threads of *Mytilus edulis*. The total biomass of Basin Head Irish moss was estimated at between 100 and 154 t WW (McCurdy, 1980; Gallant, 1990). The potential for mariculture of this unique population was identified in the 1970s (Murchinson, 1977). McCurdy (1980) transplanted some Basin Head plants to 15 locations in basins and estuaries of eastern P.E.I. Growth was recorded at most transplant sites for the first 27–30 days in July; during August, most plants either ceased growing, fragmented or rotted. These results, in part, discouraged further experimentation and investment.

Our preliminary study re-examines the potential of *C. crispus* for open-water aquaculture at Basin Head and 5 new transplant sites in western P.E.I. Two new approaches were used: (1) an extended grow-out period from May to October, and (2) a better monitoring by measuring not only growth rates but also nutrient concentrations in seawater, and nutrient and carrageenan contents in algal tissue to help in site selection decisions.

Material and methods

Samples and transplants

On 9 May 1997, plants were collected from the east arm of Basin Head, washed clean of silt, associated animals and plants, and transported in coolers filled with ice bags to each of the 5 transplant sites (Figure 1). Sites were chosen for shelter from wave action, reduced grazing by herbivores, water circulation, and ease of access. At each site, 20 plants were weighed individually, after draining for 3 minutes in a diving bag; initial plant weight ranged from 30 to 107 g WW. They were then tied onto vinyl-coated wire mesh screens (2.5×2.5 cm mesh) and tagged with flagging tape to indicate their positions. Screens, with untagged plants (for chemical analyses) were also placed adjacent to those with tagged plants (for growth measurement) at three sites, Basin Head, South Kildare, and Freeland. Depths of the screens at mean low tide was 0.25 m to 0.5 m, and 1.25 to 1.5 m at mean high tide.

Growth measurements

At each transplant site, growth (by weight) was measured every two weeks, whereas at Basin Head, plants were measured at days 34, 70, 124 and 174 from the initial outplanting. Each plant was untied from the screen, drained for 3 min, weighed, and re-tied to the screen. Daily growth rates (DGRs) were calculated from the equation $k = 100 \ln(W_t / W_i) \cdot t^{-1}$, where k = specific growth rate expressed as % d^{-1}, W_i = initial WW, W_t = WW after t days, and t = days of growth. Cumulative percentages of growth were calculated from the equation $\sum_{t-1}^{1}(W_t - W_i) / W_i \cdot 100$, and expressed as percentages of the initial WW. The condition of the plants was noted, including the presence of epiphytes, erosion, rotting, and colour. Colour was characterized by comparison to commercial coded paint colour strips.

Seawater variables and nutrient analyses

At each site visit, salinity and temperature were recorded manually. Two seawater samples were collected, 20 cm below the surface, in 125 ml high-density polyethylene bottles and immediately frozen (−20 °C) upon returning to the laboratory. Dissolved inorganic phosphorus (DIP; as PO_4^{-3}) and nitrogen (DIN; as the sum of $NH_4^+ + NO_3^- + NO_2^{-2}$) concentrations in seawater were measured by the methods of Murphy & Riley (1962) and Grasshoff et al. (1983), respectively, using a Technicon Autoanalyzer II segmented flow analyzer.

Nutrient analysis in algal tissues

At Basin Head, South Kildare, and Freeland, some untagged plants were periodically removed from the screens, placed on ice bags in a cooler, and air freighted within a few hours to the laboratory in Saint John, N.B. for nutrient and carrageenan analyses.

Triplicate tissue samples were taken to determine tissue total phosphorus (P) and nitrogen (N) contents. Only apices were used, because they show the most changes in nutrient content (Chopin et al., 1990a). Tissue total P content was measured by the method of Murphy & Riley (1962) after acidic mineralization (H_2SO_4 and HNO_3) in Büchi 430 and 435 digester units. Tissue samples for N analyses were ground to a homogeneous powder using a Retsch Vibratory Mill Type MM-2, and dried in a forced-air oven for 72 h at 60 °C. Approximately 2 mg of powder were weighed with a Perkin Elmer Autobalance AD-6 and N contents were determined with a Perkin Elmer 2400 Series II elemental analyzer.

Extraction and content of total carrageenans

Duplicate samples for each site and date were extracted, and carrageenans were precipitated with hexadecyltrimethylammonium bromide (CTAB) (Craigie & Leigh, 1978; Chopin et al., 1990b). The coagula were dried in a forced-air oven for 72 h at 60 °C and weighed to determine the yield [=% dry weight (DW)].

Sugar analyses

Total carbohydrate was measured by the phenol-sulphuric acid method (Dubois et al., 1956), modified for microanalysis (Mollion, 1988) and using D-galactose standards. The resorcinol method, using D-fructose standards (Craigie & Leigh, 1978), was used to determine 3,6-anhydrogalactose. Sulphate was measured by the turbidimetric method after coagulum hydrolysis with 2N HCl for 2 h at 100 °C (Craigie et al., 1984).

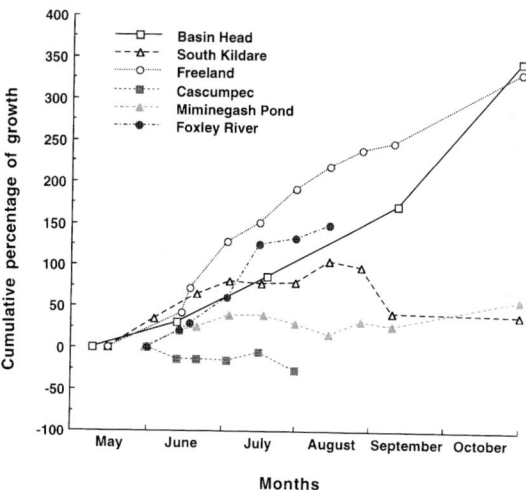

Figure 2. Cumulative percentage of growth of *Chondrus crispus* at Basin Head and the 5 transplant sites from May to October 1987.

Results

Growth was recorded at all sites (Figure 2) except Cascumpec where, after 13 days, most plants were moribund and had lost tissue through fragmentation. The screen with tagged plants at Foxley River was lost after 75 days and the experiment terminated at this site despite a promising growth rate. Fragmentation of 11 of the 20 fronds at Miminegash Pond resulted in low positive growth after 35 days. Moreover, plants were heavily covered with silt and the turbidity of seawater was high (Secchi disc readings frequently <30 cm). Seven fragmenting plants did, however, recover in late summer/early autumn and grew at a rate of 0.1 to 0.5 g WW d^{-1} in September/October. At South Kildare, like at all other sites except Foxley River, the growth rate slowed down in early July and increased in early August; fragmentation predominated at the end of August, but the growth of most plants stabilized in the autumn. At Freeland, growth rates higher than in Basin Head were sustained over the summer; however, they were similar by the end of the study. During peak growth periods (May to end of June and in the autumn), DGRs between 3 and 4% d^{-1} were recorded at both sites, with some plants exceeding 6% d^{-1}. Over

Table 1. Average salinity of seawater from May to October 1997 at Basin Head and the five transplant sites in Prince Edward Island

Location	Salinity (‰)
Basin Head	22.5 ± 3.5
South Kildare	20.3 ± 2.2
Freeland	26.6 ± 1.9
Cascumpec	17.1 ± 3.1
Miminegash Pond	23.4 ± 1.5
Foxley River	24.6 ± 1.3

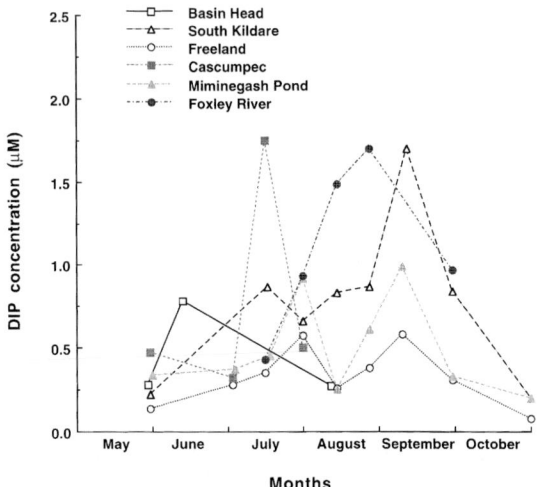

Figure 3. Variations of DIP concentration (μM P) in seawater at Basin Head and the 5 transplant sites from May to October 1997.

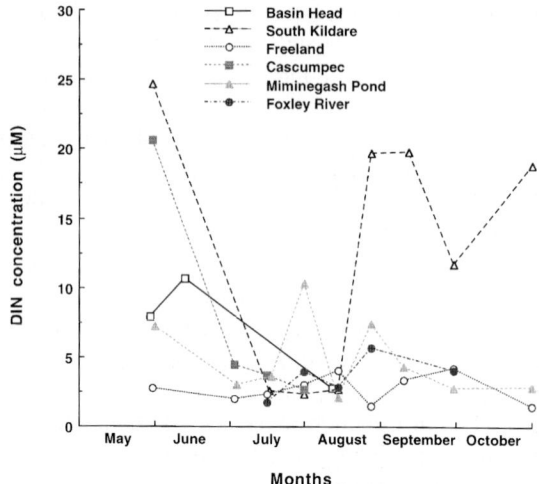

Figure 4. Variations of DIN concentration (μM N) in seawater at Basin Head and the 5 transplant sites from May to October 1997.

the whole study period, DGRs averaged 2% d^{-1} (or 0.9 g WW d^{-1} per plant) for both sites; however, the biomass of plants increased by a factor of only 6–10 because of fragmentation. The largest plant at the end of the experiment was at Basin Head, weighing 679 g WW. Growth rate was not correlated with initial plant size within the range of sizes in the study ($r^2 = 0.32$).

Plants losing weight frequently had eroding apices or sloughing tissues. Large fragments were also lost when tissues in the middle of the thallus became moribund despite healthy distal sections. Epiphytism was non-existent or negligible at all sites. Plant colour was not an indicator of plant growth or survivorship. Plants remained a dark purple colour at Basin Head and South Kildare, while their growth pattern was different and fragmentation occurred at the latter site. The fast growing Freeland plants were a light olive green in contrast to the equally productive dark purple Basin Head plants.

At all study sites, seawater exceeded 20 °C by early July. The warmest temperature was recorded at Miminegash Pond where it remained at 25.6 °C during July. Cooling began in early September and, by the end of that month, seawater temperature dropped to around 14 °C at each site. Salinity varied slightly over the study period at each site; its average was the highest at Freeland (26.6 ± 1.9‰) and the lowest at Cascumpec (17.1 ± 3.1‰) (Table 1). The Freeland site is open to wave action over 5 km from the east while all other sites have a fetch of less than 1 km. Cascumpec site has the least water movement and is the most sheltered site. Miminegash Pond is also sheltered but is subject to a strong tidal flow.

The low levels of DIP at the different sites (Figure 3) were typical of coastal cold temperate waters of the northern hemisphere at that time of the year (Chopin, 1986). Spikes of P enrichment were, however, observed at the different sites over the summer and are attributed to either summer storms putting nutrients trapped in sediments back into suspension or to agricultural run offs (Chopin et al., 1995b). Seasonal variations of DIN (Figure 4) followed the same pattern of coastal cold temperate waters: spring decline, particularly pronounced at South Kildare and Cascumpec, and low summer values with some N spikes. South Kildare regained high N contents at the end of August.

Tissue total P content was higher in Basin Head plants than in those at South Kildare and Freeland (Figure 5). At Basin Head, it remained at 3 mg P g

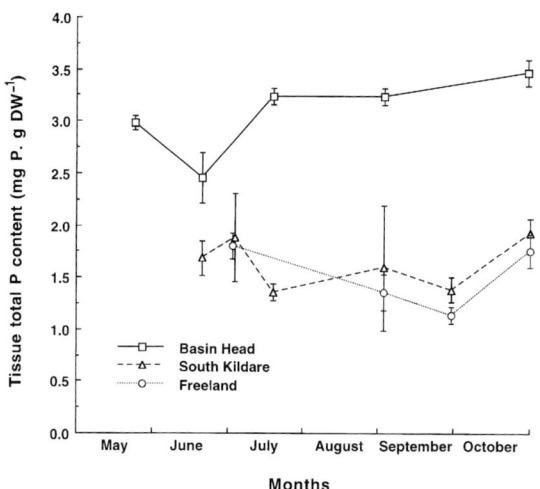

Figure 5. Variations in tissue total P content (mg P g DW^{-1}) of *Chondrus crispus* at Basin Head and the 2 transplant sites, South Kildare and Freeland, from May to October 1997. Values represent means ($n = 3$) ± SD.

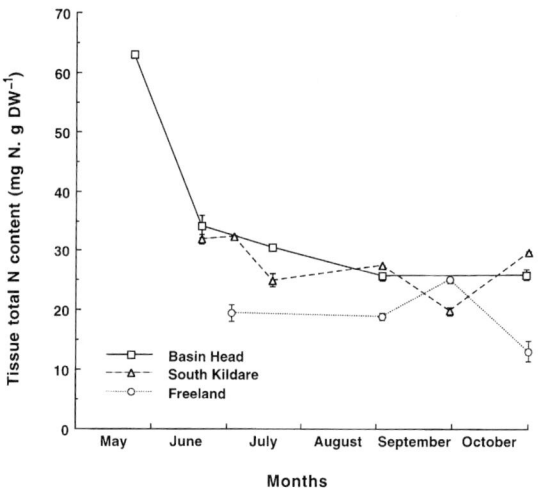

Figure 6. Variations in tissue total N content (mg N g DW^{-1}) of *Chondrus crispus* at Basin Head and the 2 transplant sites, South Kildare and Freeland, from May to October 1997. Values represent means ($n = 3$) ± SD.

DW^{-1} or above (up to 3.46 ± 0.13 mg P g DW^{-1} at the end of October), except in mid June when it reached 2.45 ± 0.24 mg P g DW^{-1}. At South Kildare and Freeland, it remained around 2.00 mg P g DW^{-1} or below (down to 1.14 ± 0.08 mg P g DW^{-1} at Freeland at the end of September). Tissue total N content decreased drastically during the spring in Basin Head plants (Figure 6). It remained low at all 3 sites dur-

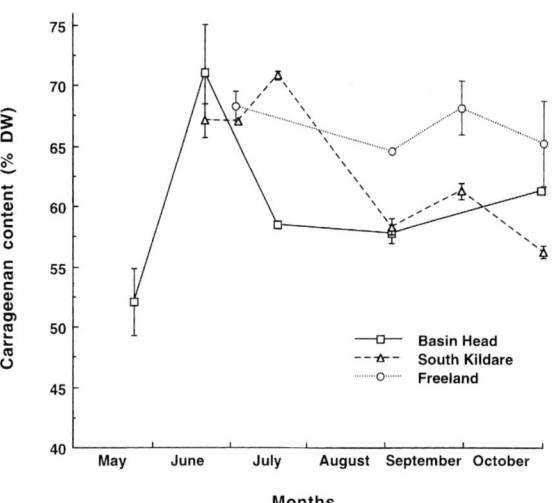

Figure 7. Variations in carrageenan content (% DW) of *Chondrus crispus* at Basin Head and the 2 transplant sites, South Kildare and Freeland, from May to October 1997. Values represent means ($n = 3$) ± SD.

ing the summer and early autumn: in Basin Head and South Kildare plants, it was between 20 and 34 mg N g DW^{-1}, while it remained at 20 mg N g DW^{-1} or below in Freeland plants, except at the end of September (25.0 ± 0.5 mg N g DW^{-1}).

Carrageenan content in Basin Head plants (Figure 7) increased drastically from the end of May ($52.1 \pm 2.8\%$ DW) to the end of June ($71.0 \pm 4.1\%$ DW). It remained at an intermediate level during the summer ($58.1 \pm 0.7\%$ DW) with a slight increase at the end of October ($61.3 \pm 0.2\%$ DW). Plants in South Kildare had high carrageenan contents in June and July (up to $70.8 \pm 0.4\%$ DW). Carrageenan contents were similar to those of the plants in Basin Head in September, but lower at the end of October ($56.2 \pm 0.5\%$ DW). Carrageenan content in Freeland plants remained high during the whole study period, between $64.5 \pm 0.1\%$ DW (early September) and $68.3 \pm 1.2\%$ DW (early July).

Sugar analyses revealed some slight variations in total carbohydrate, 3,6-anhydrogalactose, and sulphate contents at each site, even so no particular trend was evident between sites or seasons. Thus, the data were pooled and the following average contents were obtained: $76.72 \pm 5.32\%$ DW of polysaccharide for total carbohydrate content, $27.50 \pm 2.05\%$ DW of polysaccharide for 3,6-anhydrogalactose content, and $28.68 \pm 1.64\%$ DW of polysaccharide for sulphate content. This resulted in an average molar ratio galactose: 3,6-anhydrogalactose: sulphate of 1: 0.62:

1.09, the most extreme ratios obtained being 1: 0.54: 0.92 and 1: 0.76: 1.36. These ratios are indicative of carrageenans predominantly of the κ-type with some variable minor amounts of μ- and ι-carrageenans.

Discussion

The Basin Head population of unattached, large, thick, and mostly gametophytic fronds of *C. crispus*, reproducing almost entirely vegetatively through fragmentation, is unique in the Maritime Provinces of Canada. This morphotype was, however, not discriminated from six other conspecific morphotypes, representing widely contrasting forms from both sides of the North Atlantic Ocean, when compared by restriction endonuclease digestion (RFLP) of their plastid DNA, and the sequences of their internal transcribed spacers (ITS 1 and ITS 2) and the intervening 5.8S ribosomal nuclear DNA region (Chopin et al., 1996). Amplified fragment length polymorphism (AFLP), a new PCR-based fingerprinting technique, is being developed and assessed as a method for resolving population level genetic differences between phenotypically divergent strands of *C. crispus* in the Maritime Provinces of Canada (Donaldson et al., 1998). Hence, this unique colonization pattern at Basin Head remains presently unexplained. Is it due to special environmental conditions or genetic isolation, especially considering that between the closing of the east channel by sand dune accumulation in the 1930s (McCurdy, 1979) and the opening of the present southwest channel there must have been a period of physical separation of this inlet? This study demonstrated that Basin Head plants can be successfully transplanted to other sites (Freeland, Foxley River, and South Kildare), which provide environmental conditions yielding comparable or even higher productivity (Freeland, during the summer months). Moreover, the best growth rates were recorded from the two sites most diverse in physical and chemical characteristics. The relatively open water, higher salinity, cooler Freeland site contrasts with the sheltered, lower salinity, warmer Basin Head site. Plants had different nutrient statuses at these two sites. Freeland plants had lower tissue total P and N contents than Basin Head plants, reflecting lower ambient seawater P and N concentrations (Chopin et al., 1990a), associated with no immediate agricultural run off. The low nutrient levels of the Freeland plants were corroborated by their light olive green colour compared to the dark purple of the Basin Head plants (Chopin et al., 1995a).

Irrespective of their colour, plant growth at both sites was sustained and fragmentation minimal during the whole study period; conversely significant fragmentation took place in July and August at all the other sites. Fragmentation at Cascumpec occurred rapidly (13 days) and appeared to be associated with low salinity stress. At Miminegash Pond, fragmentation was associated with high siltation, turbidity, and temperature (25.6 °C during the whole month of July). Reduction of growth rate and biomass increase in July and August at the different sites followed the general pattern observed in natural beds around P.E.I. (Sharp, 1987; Chopin et al., 1992). The transplantation experiments of McCurdy (1980) were conducted during that period and explained why poor projections for aquaculture potential were reached, even if DGRs of 2 to 4% d^{-1}, similar to those of this study (3 to 4% d^{-1}), were recorded in early July. Different grow-out techniques and strategies of inoculation and harvesting are presently being tested with pilot-scale farms to develop the most efficient and commercially viable techniques.

The DGRs reported in the present study for *C. crispus*, between 2 and 4% d^{-1}, are certainly lower than those for different species of *Eucheuma* and *Kappaphycus alvarezii* recorded in tropical and subtropical regions (Table 2). These current main sources of carrageenans in the world (Chopin, 1998) generally have DGRs between 3 and 5% d^{-1}, with farms able to sustain a DGR of 7% d^{-1} being identified as highly productive. The relatively lower DGRs of this study are, however, compensated for by extremely high carrageenan yields, between 58.1 and 71.0% DW, whereas a carrageenan yield of 20 to 30% DW is generally recorded for *Eucheuma* and *Kappaphycus*. Lower and higher values reported in Table 2 could be attributed to differences in carrageenan extraction techniques (Chopin et al., 1991), state of the crop at the time of harvesting, and post-harvest treatment of the raw material (Trono & Lluisma, 1992).

Carrageenan contents measured in this study are high compared to those generally reported for *C. crispus* harvested from natural beds (between 40 and 50% DW; Chopin, 1986). Generally, studies indicate carrageenan contents for 'average' populations, i.e. samples of plants of different size and age. As Chopin et al. (1995b) suspected that conflicting reports in the literature on observations, or not, of seasonal variations in carrageenan contents in the Gigartinales, and

Table 2. Daily growth rates (% d^{-1}) and carrageenan contents (% DW) of several species of Eucheuma and Kappaphycus alvarezii in different tropical and subtropical regions

Species	Location	Daily growth rate (% d^{-1})	Carrageenan content (% DW)	Authors
Eucheuma denticulatum	Djibouti	3.3–5.4	29.7–43.3	Braud & Perez (1979)
Kappaphycus alvarezii	Indonesia	3.0–4.0	22.8	Adnan & Porse (1987)
Eucheuma denticulatum		3.0	24.2–26.8	
Kappaphycus alvarezii	Fiji	2.3–5.3		Luxton et al. (1987)
Eucheuma sp.	Philippines	4.7–9.0		Trono & Ohno (1989)
Kappaphycus alvarezii	Philippines		20.9–54.9	Trono & Lluisma (1992)
Eucheuma denticulatum				
Kappaphycus alvarezii	Philippines	5.0–5.5		Dawes et al. (1993)
Eucheuma denticulatum				
Eucheuma denticulatum	Zanzibar	5.4–7.0		Lirasan & Twide (1993)
Eucheuma denticulatum	Madagascar	up to 2.8–3.3	42.2 ± 2.4	Mollion & Braud (1993)
Eucheuma striatum			41.7	
Eucheuma uncinatum	Mexico	up to 4.9	31.5–48.0	Zertuche-Gonzalez et al. (1993)
Kappaphycus alvarezii	Japan	6.0		Ohno et al. (1994)
Kappaphycus alvarezii	Philippines	up to 3.8	4.7–11.6	Hurtado-Ponce (1995)
Kappaphycus alvarezii	Vietnam	4.0–9.0	18.8–24.6	Ohno et al. (1996)
Kappaphycus alvarezii	Venezuela	4.4–7.7	11.0–35.0	Rincones & Rubio (1999)
Eucheuma denticulatum		2.3–5.3		

in particular in C. crispus, could be a reflection of the degree of heterogeneity of the samples, they sorted their samples into five classes, based on the number of dichotomies and frond length (Chopin et al., 1988). Carrageenan contents of gametophytic plants collected in the Pleasant View bed, off Miminegash, then showed variations between 31.4 and 69.8% DW, with the larger plants of Classes 4 and 5 reaching their maximal contents of 69.8 and 66.6% DW in May and June, respectively.

The later values are within the range observed in the present study. Even if it is most likely that complex interactions of several factors are responsible for seasonal variations of carrageenan content, it appears that one of these factors, the nutrients, plays a key controlling role. The so-called 'Neish effect' (Neish et al., 1977) and 'Chopin effect' (Chopin et al., 1990b, 1995a) concerning, respectively, the impact of N and P nutrition on carrageenan production are illustrated by the present data in which high carrageenan contents were recorded during summer when levels of N and P were low.

One has to realize that DGRs reported in this study are also relatively low because fragmentation of otherwise healthy large distal sections was assimilated to loss in the present system of plants simply tied on wire mesh screens. Different suspension systems (mussel 'socks', net tubings, long-lines, enclosures, and nets) are presently being tried to improve retention of these large clumps, which are viable and, if kept in a culture structure, would certainly increase the measured DGRs.

This preliminary study not only demonstrated the possibility of successful transplantations but also that the Basin Head population of C. crispus has significant potential for open-water aquaculture in estuaries and basins of Atlantic Canada. With DGRs that could be increased and carrageenan contents that are 2 to 3 times those of Eucheuma and Kappaphycus, this approach could become economically competitive, especially at a time when the phycocolloid industry wants to diversify its sources of raw material and is showing a renewed interest in cold-water species of carrageenophytes. It could represent an alternative to a declining Irish moss fishery (Chopin, 1998) and a source of high quality carrageenans of the κ-family (instead of the κ-/λ- mixture from harvested natural beds). It could also be conceived as a source of material for the edible seaweed market or as a complementary activity integrated with mussel, oyster or finfish aquaculture for bioremediation of coastal waters and economic diversification of the present, too often, mono-aquaculture industry (Chopin and Yarish, 1998).

Acknowledgements

This study was supported by NSERC OGP46376 and EQP92706 grants to T. Chopin. We thank L. Gallant and the staff of the Miminegash Marine Research and Development Station for their assistance, and W. Morris and C. Keith for their help in the preparation of the manuscript.

References

Adnan, H. & H. Porse, 1987. Culture of *Eucheuma cottonii* and *Eucheuma spinosum* in Indonesia. Hydrobiologia 151/152: 355–358.

Bidwell, R. G. S., J. McLachlan & N. D. H. Lloyd, 1985. Tank cultivation of Irish moss, *Chondrus crispus* Stackh. Bot. mar. 28: 87–97.

Braud, J. P. & R. Delépine, 1981. Growth response of *Chondrus cripsus* (Rhodophyta, Gigartinales) to light and temperature in laboratory and outdoor tank culture. Proc. int. Seaweed Symp. 10: 553–558.

Braud, J. P. & R. Perez, 1979. Farming on pilot scale of *Eucheuma spinosum* (Florideophyceae) in Djibouti waters. Proc. int. Seaweed Symp. 9: 533–539.

Chopin, T., 1986. The red alga *Chondrus crispus* Stackhouse (Irish moss) and carrageenans — A review. Can. Tech. Rep. Fish. aquat. Sci. 1514: v + 69 p.

Chopin, T., 1998. The seaweed resources of Eastern Canada. In Critchley, A. T. & M. Ohno (eds), Seaweed Resources of the World. Japan Int. Cooperation Agency, Yokosuka: 273–302.

Chopin, T. & C. Yarish, 1998. Nutrients or not nutrients? That is the question in seaweed aquaculture . . . and the answer depends on the type and purpose of the aquaculture system. World Aquaculture 29: 31–33 and 60–61.

Chopin, T., T. Gallant & I. Davison, 1995a. Phosphorus and nitrogen nutrition in *Chondrus crispus* (Rhodophyta): effects on total phosphorus and nitrogen content, carrageenan production, and photosynthetic pigments and metabolism. J. Phycol. 31: 283–293.

Chopin, T., M. D. Hanisak & F. E. Koehn, 1991. Effects of seawater phosphorus concentration on floridean starch content in *Agardhiella subulata* (C. Agardh) Kraft *et* Wynne (Rhodophyceae). Bot. mar. 34: 369–373.

Chopin, T., J. D. Pringle & R. E. Semple, 1988. Reproductive capacity of dragraked and non-dragraked Irish moss (*Chondrus crispus* Stackhouse) beds in the southern Gulf of St. Lawrence. Can. J. Fish. aquat. Sci. 45: 758–766.

Chopin, T., J. D. Pringle & R. E. Semple, 1992. Impact of harvesting on frond density and biomass of Irish moss (*Chondrus crispus* Stackhouse) beds in the southern Gulf of St. Lawrence. Can. J. Fish. aquat. Sci. 49: 349–357.

Chopin, T., A. Hourmant, J.-Y. Floc'h & M. Penot, 1990a. Seasonal variations of growth in the red alga *Chondrus crispus* on the Atlantic French coasts. II. Relations with phosphorus concentration in seawater and internal phosphorylated fractions. Can. J. Bot. 68: 512–517.

Chopin, T., E. Whalen, G. Sharp & R. Semple, 1995b. Nutrients, growth, and carrageenans in *Chondrus crispus* and *Furcellaria lumbricalis* in a commercially harvested bed off Prince Edward Island, Canada. J. Phycol. 31 (suppl.): 15.

Chopin, T., M. D. Hanisak, F. E. Koehn, J. Mollion & S. Moreau, 1990b. Studies on carrageenans and effects of seawater phosphorus concentration on carrageenan content and growth of *Agardhiella subulata* (C. Agardh) Kraft and Wynne (Rhodophyceae, Solieriaceae). J. appl. Phycol. 2: 3–16.

Chopin, T., C. J. Bird, C. A. Murphy, J. A. Osborne, M. U. Patwary & J.-Y. Floc'h, 1996. A molecular investigation of polymorphism in the North Atlantic red alga *Chondrus crispus* (Gigartinales). Phycol. Res. 44: 69–80.

Craigie, J. S., 1990. Irish moss cultivation: some reflections. In Yarish, C., C.A. Penniman & P. Van Patten (eds), Economically Important Marine Plants of the Atlantic. Their Biology and Cultivation. Connecticut Sea Grant College Program, Groton: 37–52.

Craigie, J. S. & C. Leigh, 1978. Carrageenans and agar. In Hellebust, J. A. & J. S. Craigie (eds), Handbook of Phycological Methods, Physiological and Biochemical Methods. Cambridge University Press, Cambridge: 109–131.

Craigie, J. S., L. S. Staples & A. F. Archibald, 1999. Rapid bioassay of a red food alga: accelerated growth rates of *Chondrus crispus*. World Aquaculture 30: 26–28.

Craigie, J. S., Z. C. Wen & J. P. van der Meer, 1984. Interspecific, intraspecific and nutritionally-determined variations in the composition of agars from *Gracilaria* spp. Bot. mar. 27: 55–61.

Dawes, C. J., G. C. Trono Jr. & A. O. Lluisma, 1993. Clonal propagation of *Eucheuma denticulatum* and *Kappaphycus alvarezii* for Philippine seaweed farms. Hydrobiologia 260/261: 379–383.

Donaldson, S. L., T. Chopin & G. W. Saunders, 1998. Amplified fragment length polymorphism (AFLP) as a source of genetic markers for red algae. J. appl. Phycol. 10: 365–370.

Doty, M. S., 1987. The production and use of *Eucheuma*. In Doty, M. S., J. F. Caddy & B. Santelices (eds), Case studies of seven commercial seaweed resources. FAO Fish. Tech. Paper 281: 123–164.

Dubois, M., K. A. Gilles, J. K. Hamilton, P. A. Rebers & F. Smith, 1956. Colorimetric method for determination of sugars and related substances. Anal. Chem. 28: 350–356.

Gallant, R., 1990. Evaluation of Basin Head 'giant' Irish moss as a potential culture species. P.E.I. Dept. of Fisheries Rep., 10 pp.

Grasshoff, K., M. Ehrhardt & K. Kremling, 1983. Methods of Seawater Analysis. Verlag Chemie, Weinheim, 419 pp.

Hurtado-Ponce, A. Q., 1995. Carrageenan properties and proximate composition of three morphotypes of *Kappaphycus alvarezii* Doty (Gigartinales, Rhodophyta) grown at two depths. Bot. mar. 38: 215–219.

Lirasan, T. & P. Twide, 1993. Farming *Eucheuma* in Zanzibar, Tanzania. Hydrobiologia 260/261: 353–355.

Luxton, D. M., M. Robertson & M. J. Kindley, 1987. Farming of *Eucheuma* in the south Pacific islands of Fiji. Hydrobiologia 151/152: 359–362.

McCurdy, P. 1979. A preliminary study of the ecology of Basin Head Harbour and South Lake, P.E.I. Final report of Job Corps Project 16-01-0098. Nat. Res. Council Canada Rep., 57 pp.

McCurdy, P. 1980. Investigation of a unique population of *Chondrus crispus* in Basin Head Harbour, Prince Edward Island. Nat. Res. Council Canada Rep., 53 pp.

Mollion, J., 1988. Etude des carraghénanes de *Rissoella verruculosa*. Sur les filiations entre les différents systèmes de carraghénanes. Thèse Doctorat d'Etat, Lille, 199 pp.

Mollion, J. & J. P. Braud, 1993. A *Eucheuma* (Solieriaceae, Rhodophyta) cultivation test on the south-west coast of Madagascar. Hydrobiologia 260/261: 373–378.

Murchison, J., 1977. *Chondrus* lagoon study. P.E.I. Dept. of Fisheries Rep., 7 pp.

Murphy, J. & J. P. Riley, 1962. A modified single solution approach for the determination of phosphate in natural waters. Anal. Chim. Acta 27: 31–36.

Neish, A. C., P. F. Shacklock, C. H. Fox & F. J. Simpson, 1977. The cultivation of *Chondrus crispus*. Factors affecting growth under greenhouse conditions. Can. J. Bot. 55: 2263–2271.

Ohno, M., D. B. Largo & T. Ikumoto, 1994. Growth rate, carrageenan yield and gel properties of cultured *kappa*-carrageenan producing red alga *Kappaphycus alvarezii* (Doty) Doty in the subtropical waters of Shikoku, Japan. J. appl. Phycol. 6: 1–5.

Ohno, M., H. Q. Nang & S. Hirase, 1996. Cultivation and carrageenan yield and quality of *Kappaphycus alvarezii* in the waters of Vietnam. J. appl. Phycol. 8: 431–437.

Rincones, R. E. & J. Rubio, 1999. Introduction and commercial cultivation of the red alga *Eucheuma* in Venezuela for the production of phycocolloids. World Aquaculture 30 (in press).

Sharp, G. J. 1987. Growth and production in wild and cultured stocks of *Chondrus crispus*. Hydrobiologia 151/152: 349–354.

Trono Jr., G. C. & A. O. Lluisma, 1992. Differences in biomass production and carrageenan yields among four strains of farmed carrageenophytes in Northern Bohol, Philippines. Hydrobiologia 247: 223–227.

Trono, G. C. & M. Ohno, 1989. Seasonality of the biomass production of the *Eucheuma* strains in Northern Bohol, Philippines. In Umezaki, I. (ed), Scientific Survey of Marine Algae and their Resources in the Philippines Islands. A Technical Report of the Ministry of Education, Science and Culture, Japan: 71–80.

Zertuche-Gonzalez, J. A., I. Pacheco-Ruiz & I. E. Soria-Mercado, 1993. Carrageenan yield and properties of *Eucheuma uncinatum* (Seth. & Gard.) Daw. cultured under natural conditions. Hydrobiologia 260/261: 601–605.

Recent advances in the understanding of the biological basis for *Gigartina skottsbergii* (Rhodophyta) cultivation in Chile

Alejandro H. Buschmann[1], Juan A. Correa[2] & Renato Westermeier[3]
[1]*Departamento de Acuicultura, Universidad de Los Lagos, Casilla 933, Osorno, Chile*
[2]*Departamento Ecología, Facultad de Ciencias Biológicas Pontificia Universidad Católica de Chile, Casilla 114-D, Santiago, Chile*
[3]*Instituto de Acuicultura, Universidad Austral de Chile, Casilla 1327, Puerto Montt, Chile*

Key words: carrageenophytes, cultivation, *Gigartina skottsbergii*, population dynamics

Abstract

The demand in Chile for carrageenophytic algae has increased strongly during the last 3 years, with emphasis on *Gigartina skottsbergii*, a species representing landings of 32 438 t (wet) during 1996. Various sources of information indicate that this species is being over-exploited and therefore the development of cultivation technologies is needed to support the local carrageenan industry. In this study we summarize currently available information on laboratory, outdoor tank and open sea culture of *G. skottsbergii*. The results indicate that viable spores of *G. skottsbergii* can be obtained, mainly during winter, with germination rates of both tetraspores and carpospores, up to 40%. Germlings of *G. skottsbergii* were succesfully transplanted from the laboratory to outdoor tanks, where they displayed survival values higher than 80% during spring. Experimental trials in the field indicate that *G. skottsbergii* can be cultivated on rope systems, with tissue fragments used as inoculum. This last result suggests that regeneration from fragments is an alternative method for propagation and massive cultivation of *G. skottsbergii* in Chile.

Introduction

In Chile, the demand for carrageenophytic algae has increased steadily in the past 3 years. The most valuable species is *Gigartina skottsbergii* Setchell et Gardner, with landings of 32 438 wet t during 1996 (Buschmann et al., 1998). *G. skottsbergii* shows a seasonal variation in biomass, with a maximum recorded during summer, on both the Atlantic and Pacific coasts of southern South America (Piriz, 1996; Zamorano & Westermeier, 1996). Available information indicates that this species is being overexploited and therefore the development of cultivation technologies is urgently needed to support the local carrageenan industry. In this study we summarize the current status of a long-term research program, initiated in 1997, whose main objective is to provide the basic biological knowledge needed to allow the profitable farming of this species. In this context we concentrate on aspects related to spore availability and viability, germling suvivorship in the laboratory and in outdoor tanks, as well as alternative propagation methods in the field.

Materials and methods

All algal material used in this study was obtained from a subtidal population of *G. skottsbergii* located in Ancud (41° 51' S, 73° 49' W) in southern Chile (Figure 1). Samples for transmission electron microscopy (TEM) were fixed in 1.5% glutaraldehyde in 0.45 μm filtered seawater for 2 h at room temperature. Post-fixation was done in 1% osmium tetroxide. Dehydration in ethanol was followed by embedding in Spurr's resin. Thin sections were stained with uranyl acetate and lead citrate (Reynolds, 1963). Photographs were taken in a Siemens ELMISKOP IA TEM operated at 60 kV.

Germination experiments were done using carposporophytic and tetrasporophytic fronds collected

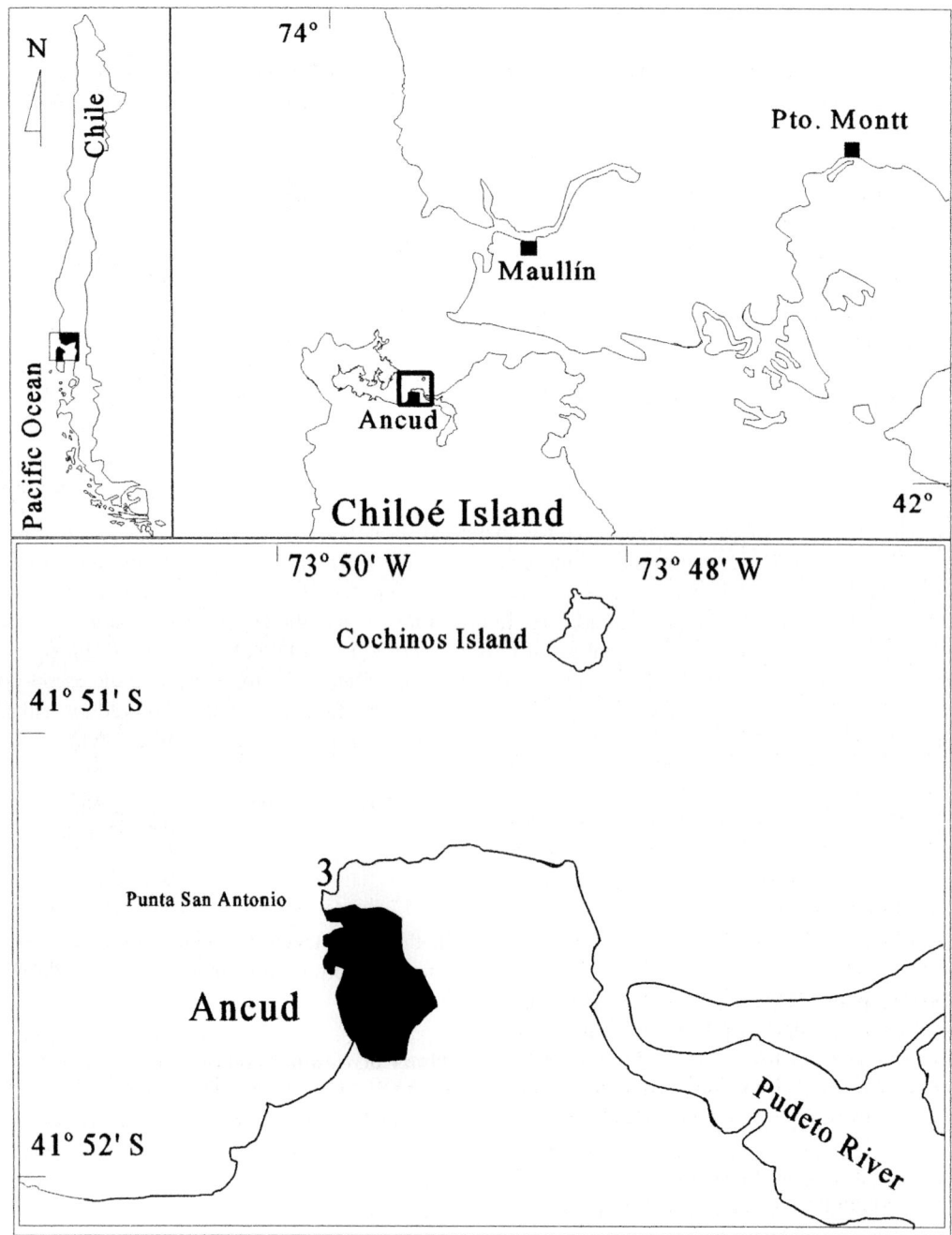

Figure 1. Map of the study area and location of the collection site (3) of *Gigartina skottsbergii* in southern Chile.

in Ancud (Figure 1). These fronds were brought to the laboratory, rinsed with sterile seawater and air dried for 10 min at room temperature. Subsequently, fragments of fertile tissues were transferred to Petri dishes with Provasoli culture medium and incubated in growth chambers at 5, 10 and 15 °C, a neutral photoperiod of 12:12 h light:dark and a photon irradiance of 48 μmol m^{-2} s^{-1}. A total of five Petri dishes were used for each temperature and karyological phase. After six days, the percentage of germination was determined by counting the number of germinated spores and the total number of spores, in three microscopic fields per dish. Positive germination was scored when a cell presented clear evidence of a complete cell division. This experimental protocol was repeated at monthly intervals during a full year to determine the existence of seasonality.

To assess the feasibiity of using laboratory-raised germlings as inoculum for an outdoor culture system, we transplanted plastic Petri dishes, 6 cm in diameter, from the laboratory to 40 l plastic outdoor containers with a water regime of a full volume replacement every 3 days. These containers were placed in 500 l tanks (see Buschmann et al., 1994, for further details) to maintain a constant temperature and were covered with black shadowing mesh to reduce to 1% and 10% the levels of irradinace reaching the germlings. At mid-day and with high solar irradiance, these treatments produce very different light environments in the culture tanks, which mimic the conditions in the field. Preliminary observations demonstrated that full irradiance levels in outdoor tanks rapidly bleached the germlings. The experiment began in November with a mean temperature of 12–14 °C, salinity of 29–31‰ and pH of 8.2–8.4. Survival of the germlings was monitored twice a month by photographic recording of fixed 1 cm^2 quadrats ($n = 6$). The final size of the plantlets was estimated by haphazardly removing a minimum of ten individuals per Petri dish ($n > 30$) and measuring individual heights. Survival and final size of the plantlets in the two light conditions were compared by one-way ANOVA after ensuring normality and homocedasticity of variances.

For assessing the wound healing responses and regeneration, vegetative plants were brought to the laboratory and rinsed with sterile seawater. Fragments of tissue, 2×2 cm each, were excised from the borders of the fronds. A total of ten fragments were transferred to 500-ml culture flasks containing 250 ml of sterile seawater. The culture conditions were 5, 10 and 15 °C at 5 μmol m^{-2} s^{-1} and neutral photoperiod. Three replicate flasks were used in each experimental condition and the culture medium was changed every 3 days. This experiment lasted 32 days, after which the number of healed fragments was recorded. Data were analyzed by a two-way ANOVA with time as a repeated measurement, as data are time dependent (Wilkinson et al., 1992). Normality and homocedasticity of variances were tested before running the standard analyses.

To test the ability of *G. skottsbergii* to regenerate in the field, 16 fragments excised from immature fronds were fastened to an 8-m nylon cord by entangling the pieces of tissue between the nylon fibers. The nylon cord was maintained at 20 cm above the bottom by fastening it to concrete blocks. The initial size of the fragments was 5×30 cm and the meristems at the edges of each portion were left untouched. Changes in size were recorded on a monthly basis by SCUBA divers.

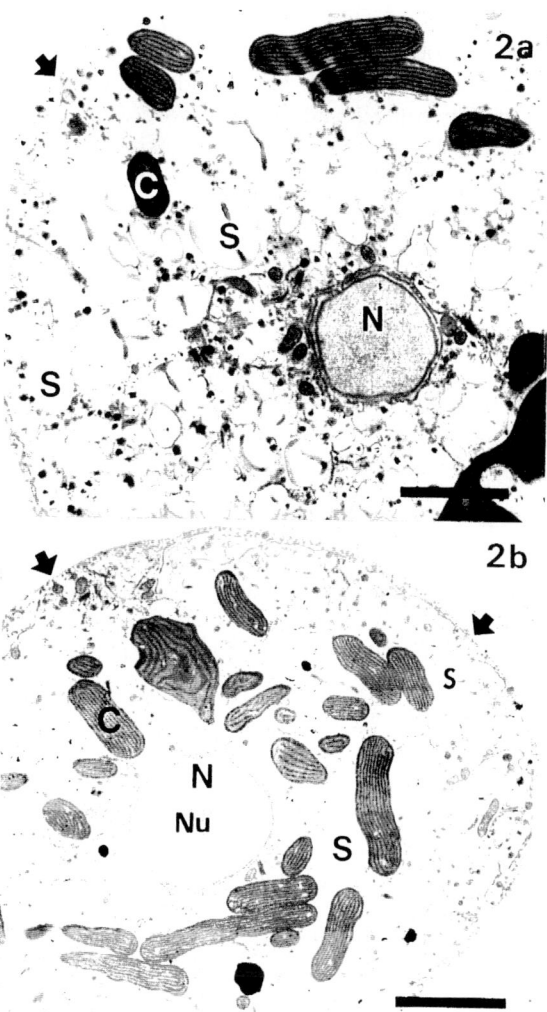

Figure 2. Ultrastructure of carpospore (a) and tetraspore (b) of *G. skottsbergii*. N, nucleus; C, chloroplast; S, starch; Nu, nucleolus. Arrows show plasmalemma without cell wall. The bar represents 5 μm.

Results

From October to April the number of spores available was very low and the germination success was zero. Carpospores and tetraspores were available in autumn to early spring, with a maximum in June. Ultrastructural information indicates that spores lack a cell wall at the moment of release (Figure 2a, b). Newly released spores are characterized by a central nucleus with a clearly distinguishable nucleolus. Abundant starch granules, plastids scattered around the nucleus, and large numbers of vesicles containing electron dense fibrillar material, likely to be the precursor of the early stages of cell wall, were present. No differences could be detected at the TEM level between carpospores and tetraspores (Figure 2a, b).

Tetraspore germination varied according to the month of frond collection in the field, with a maximum value of ca 40% observed in August (Figure 3A). Temperature modified the intensity of germination, with significantly higher values ($F = 10.086$; $p<0.001$) recorded at 5 °C as compared to 10 and 15 °C (Figure 3A). Carpospores showed similar responses to those displayed by tetraspores (Figure 3B). Both types of spores were successfully cultivated under laboratory conditions up to the stage of juvenile fronds of ca 10 mm characterized by an umbrella-like frond shape (Figure 4a–d). This level of development was achieved in 7 months under an irradiance of 25 μmol m^{-2} s^{-1}, a temperature of 10–15 °C and neutral photoperiod.

Germlings of *G. skottsbergii,* with an initial size of 0.017 mm, were transplanted from the laboratory to outdoor tanks. After 1 month of culture and regardless of the irradiance conditions, survivorship was high, ranging from 68 to 85% (Figure 5A). During this period the plantlets reached 0.012 mm in height, with no significant differences induced by the two levels of irradiance (Figure 5B). In 3 months, plantlets of both karyological phases reached 1–2 mm, with a morphology similar to that displayed in the laboratory (Figure 4d).

Laboratory wound healing responses were significantly enhanced ($F = 33.74$; $p<0.001$) by temperature, reaching mean values close to 100% at 10 and 15 °C (Figure 6). Wound healing responses also varied significantly through the year ($F = 6.125$; $p < 0.015$), although did not significantly interact ($F = 2.375$; $P > 0.11$) with temperature. At 10 and 15 °C, the maximum wound healing was reached at day 20. The histological events occuring during wound healing are by a new cortex that seals the cut and provides mer-

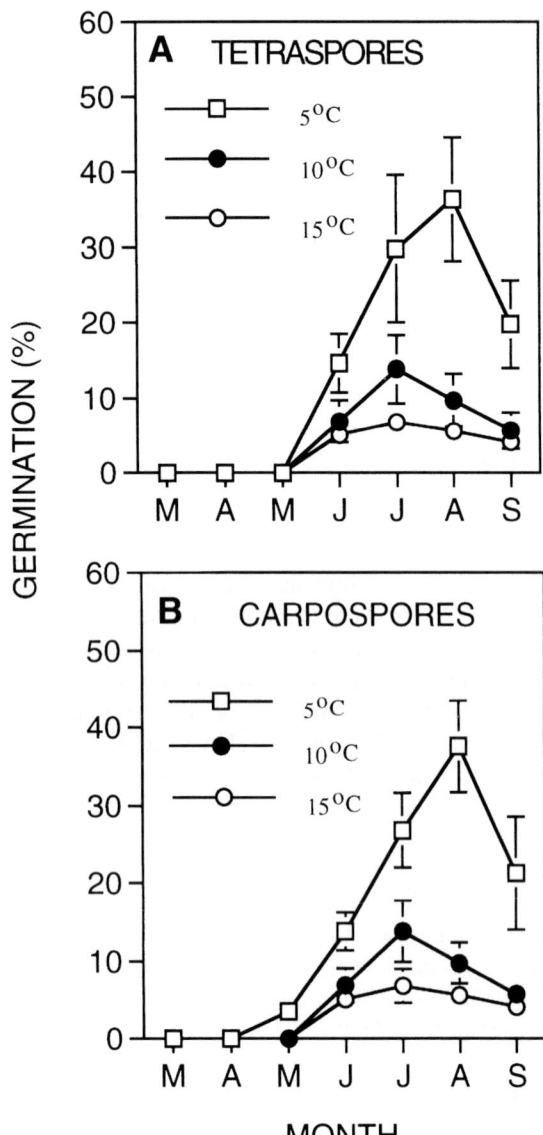

Figure 3. Monthly variation (mean ±1 S.E.) of the germination responses (%) of *G. skottsbergii* at different temperatures. A, Tetraspores and B, carpospores.

istematic cells, which eventually issue new uprights (Figure 7a–c). Following the success of the above experiments, a first trial was done to test the possible use of tissue fragments for regeneration in the field (Figure 8). The fragments showed a significant increase ($F = 4.899$; $p < 0.03$) in length, with an average increase of 44.6% for a 6-month period. The width of the fragments also increased significantly ($F = 44.394$; $p < 0.0001$) with a mean value of 48.6%. At the end of the study, the fragments of *G. skottsbergii* did not show

431

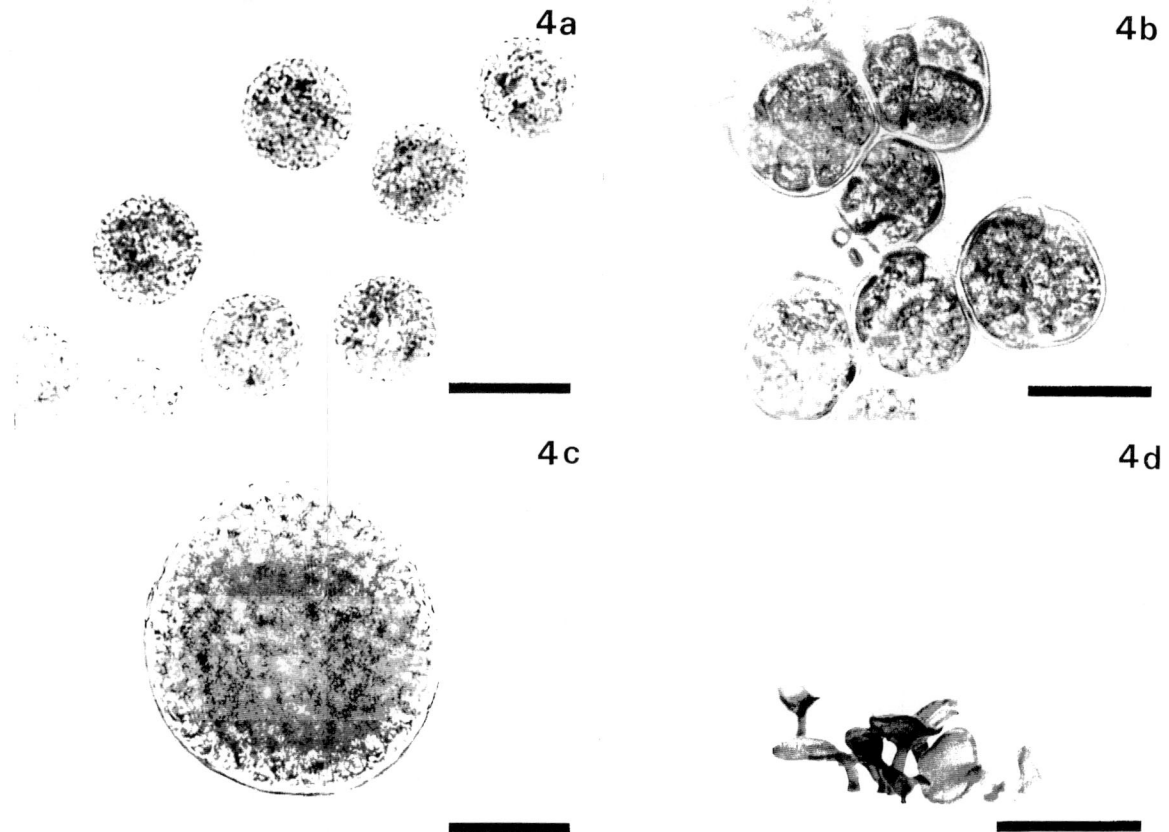

Figure 4. Development stages of gametophytes of *G. skottsbergii* in laboratory culture conditions. a, Tetraspores at time zero; bar = 30 μm; b, tetraspore germination after 4–6 days of cultivation, bar = 30 μm; c, gametophyte discs after 30 days of cultivation, bar = 40 μm; and d, gametophytic plantlets after 7 months of cultivation in the laboratory, bar = 5 mm.

any macroscopic indication of epiphytic or endophytic infections.

Discussion

Our results indicate that spores of *G. skottsbergii* do not have cell walls at the time of release. This seems to be a common feature in members of the Gigartinales and Gracilariales, as indicated by the reports on *Gracilaria* (Santelices et al., 1996), and on *Mazzaella*, *Sarcothalia*, *Chondrus*, *Ahnfeltia* and *Ahnfeltiopsis* (Santelices et al., 1997). At early stages of development (e.g. 2–4 h after release), fine structure appears normal with no indication of disruption or abnormal spatial distribution of the organelles within the cell, allowing us to individualize which spores were mature (e.g. adequate for germination). However, preliminary experiments in our laboratory testing spore viability using vital stains suggest that most spores (>90%) released under laboratory conditions are physiologically viable. This is an important finding because one of the major bottle necks preventing massive production of juveniles from spores in *G. skottsbergii* is the high mortality rates of spores before germination. It seems apparent now that spore mortality is the result of a factor, operating after the release of the spores, which triggers premature spore death. This factor remains to be found, and efforts are being invested to assess some potential candidates. One of them is temperature. It is known that *G. skottsbergii* is a cold water species, with a thermal limit reported to be 5 °C (Bischoff-Bäsmann & Wiencke, 1996). It is likely, however, that this limit is different for populations occurring in the northern extreme of the species distribution along the Pacific coast. This is supported by our results on germination, where the best responses were obtained at temperatures of 5 °C or higher. Without

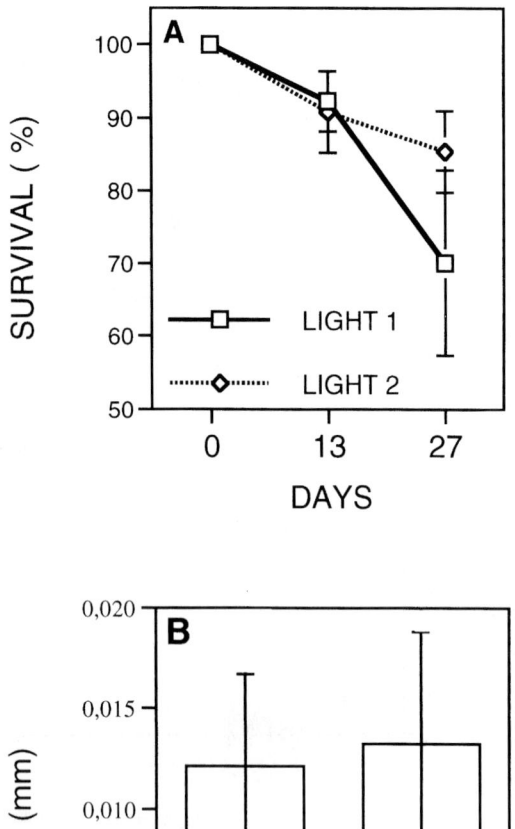

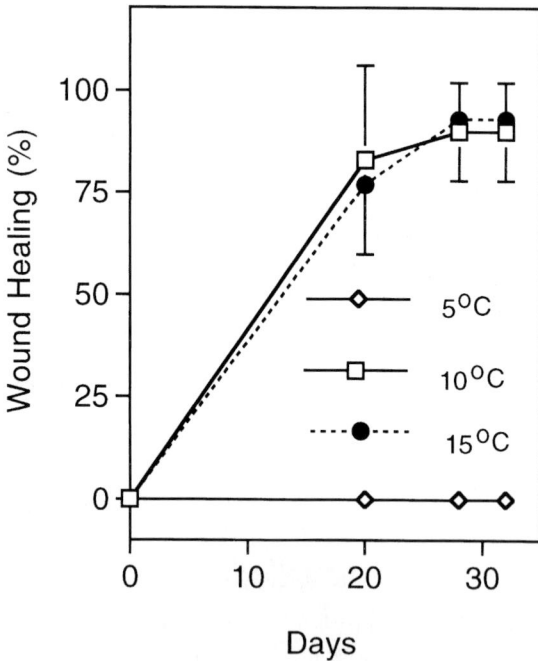

Figure 5. Mean (±1 S.D.) survivorship (A) and final size (B) of plantlets of *G. skottsbergii* transplanted to outdoor culture conditions under two light irradiances: Light 1 = 10% and Light 2 = 1% of the natural irradiance.

Figure 6. Wound healing responses (mean ±1 S.D.) of *G. skottsbergii* as a function of temperature under a low irradiance regime.

ruling out rapid adaptive responses in the laboratory as an explanation for the better responses at higher temperatures, the effect of this factor needs to be fully addressed. The importance of this issue is based on the fact that the major exploitation of *G. skottsbergii* is currently taking place in the northern extreme of its distribution, and because of logistic constraints, it is expected that farming operations will be established in the same geographic area.

Gigartina skottsbergii produces carrageenans of the kappa family in the gametophytic phase and carrageenans of the lambda family in the sporophytic phase (Piriz & Cerezo, 1991). Regardless of the methodology used to obtain spores in the laboratory, tetraspores were always more rapidly available and handled than carpospores. Post-germination development also showed better performances (e.g. growth and survivorship) in gametophytic than in sporophytic germlings (Correa et al., unpubl.). These differences may explain, at least partially, the higher abundance of gametophytic individuals found in both the Atlantic and Pacific coasts of South America (Piriz, 1996; Zamorano & Westermeier, 1996). Cultivating only plants derived from tetraspores (i.e. gametophytes) would result in a production of raw material only suitable for the extraction of carrageenans of the kappa family. As this may be an undesired feature in a farming context, alternatives must be found to allow cultivation of the two phases of *G. skottsbergii*.

Tissue culture and protoplast production have been attempted in other algae, with various degrees of success (Dawes & Koch, 1991; Kloareg et al., 1991; LeGall et al., 1990; Polne-Fuller & Gibor, 1986; Saga et al., 1986). These approaches are expensive and in most cases have been considered as unrealistic alternatives for large scale seaweed aquaculture.

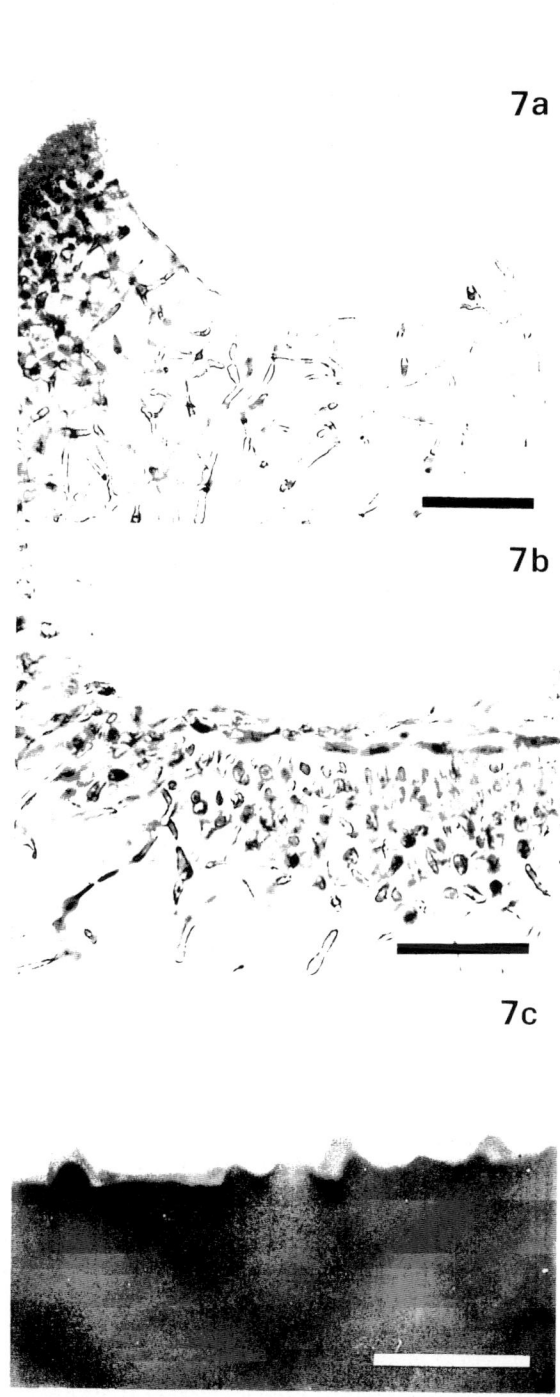

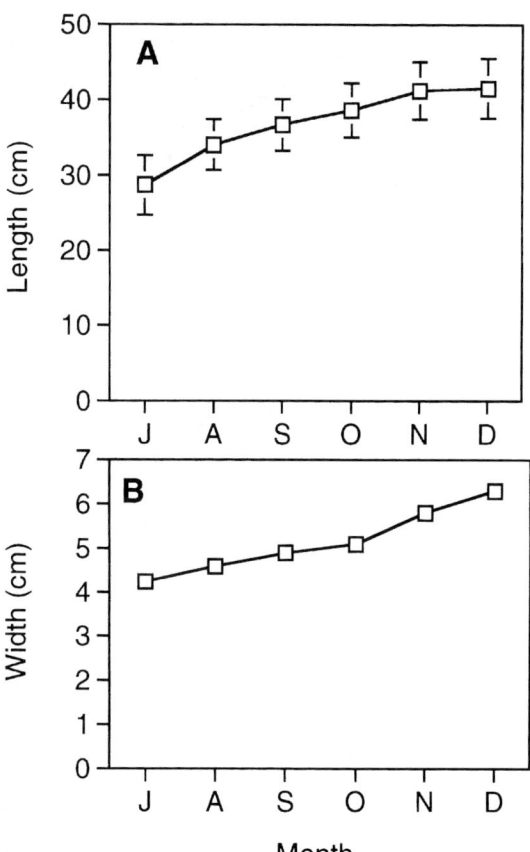

Figure 8. Growth responses of *G. skottsbergii* of tissue fragments. Data show average (±1 S.E.) of the length (A) and width (B) of the fragments.

Figure 7. Photographs showing the wound healing and regeneration responses of *G. skottsbergii* in the laboratory. a, Tissue at time zero, bar = 100 µm; b, tissue showing a wound healing response after 30 days of cultivation, bar = 50 µm; and c, regeneration tissues after 3 month in culture conditions, bar = 1.5 mm.

Still, they should not be fully ruled out for *G. skottsbergii*. A simpler approach, which has been one of the main reasons for the farming success of *Gracilaria* (Buschmann et al., 1995; Santelices & Doty, 1986), *Eucheuma/Kappaphycus* (Azanza-Corrales et al., 1996), *Grateloupia* (Iima et al., 1995) and *Chondrus* (Craigie, 1990), is to rely upon the regeneration capacity of the algal thallus. This seems to be a promising aspect in *G. skottsbergii*, which showed wound healing responses in both laboratory and in the field.

The demand for *G. skottsbergii* is likely to remain high in the near future and natural stocks appear to be quickly declining (Buschmann et al., 1998; Westermeier et al., 1997). For this reason, gaining basic biological knowledge on this resource is required, and its application is considered the only way to allow long-term sustained supply of raw material for the carrageenan industry.

Acknowledgements

This study was supported by FONDAP-Chile. The authors acknowledge the help of Juan Morales, David Patiño, Patricio Chavez, Jessica Beltrán, Verónica Flores, María Alejandra Paredes, Mariam Hernández-González, Marcos Cifuentes, Paul Güttler.

References

Azanza-Corrales, R., T. T. Alianza & N. E. Montaño, 1996. Recruitment of *Eucheuma* and *Kappaphycus* on a farm in Tawi-Tawi, Philippines. Hydrobiologia 326/327: 235–244.

Bischoff-Bäsmann, B. & C. Wiencke, 1996. Temperature requirements for growth and survival of Antarctic Rhodophyta. J. Phycol. 32: 525–535.

Buschmann, A. H., O. A. Mora, P. Gómez, M. Böttger, S. Buitano, C. Retamales, P. A. Vergara & A. Gutierrez, 1994. *Gracilaria chilensis* outdoor tank cultivation in Chile: use of land-based salmon culture effluents. Aquacult. Eng. 13: 283–300.

Buschmann, A. H., R. Westermeier & C. Retamales 1995, Cultivation of *Gracilaria* on the sea-bottom in southern Chile: a review. J. appl. Phycol. 7: 291–301.

Buschmann, A. H., J. A. Correa, R. Westermeier, M. C. Hernández-González & R. Norambuena, 1999. Mariculture of red algae in Chile. World Aquaculture 29 (in press).

Craigie, J. S., 1990. Irish moss cultivation – Some reflections. In Yarish, C., C. A. Penniman & P. van Patten (eds), Economically Important Marine Plants of the Atlantic. Their Biology and Cultivation. Connecticut Sea Grant College Program, Groton, CT: 37–52.

Dawes, C. J. & E. W. Koch, 1991. Branch, micropropagule and tissue culture of the red algae *Eucheuma denticulatum* and *Kappaphycus alvarezii* farmed in the Phillippines. J. appl. Phycol. 3: 247–257.

Iima, M., T. Kinoshita, S. Kawaguchi & S. Migita, 1995. Cultivation of *Grateloupia acuminata* (Halymeniaceae, Rhodophyta) by regeneration from cut fragments of basal crusts and upright thalli. J. appl. Phycol. 7: 583–588.

Kloareg, B, H. Benet & Y. LeGall, 1991. Applications of protoplast isolation in basic and applied phycology. In García-Reina, G. & M. Pedersén (eds) Seaweed Cellular Biotechnology, Physiology and Intensive Cultivation. Universidad de Las Palmas de Gran Canaria, Canary Islands: 17–24.

LeGall, Y., J. P. Braud & B. Kloareg, 1990. Protoplast production in *Chondrus crispus* gametophytes (Gigartinales, Rhodophyta). Plant Cell Rep. 8: 582–585.

Piriz, M. L., 1996. Phenology of *Gigartina skottsbergii* Setchell et Gardner population in Chubut Province (Argentina). Bot. mar. 39: 311–316.

Piriz, M. L. & A. S. Cerezo, 1991. Seasonal variation of carrageenan in tetrasporic, cystocarpic and 'sterile' stages of *Gigartina skottsbergii* S. et G. (Rhodophyta, Gigartinales). Hydrobiologia. 225: 65–69.

Polne-Fuller, M. & A. Gibor, 1986. Calluses, cells and protoplasts in studies towards genetic improvement of seaweeds. Aquaculture 57: 117–123.

Reynolds, E. S., 1963. The use of lead citrate at high pH as an electron-opaque stain in electron microscopy. J. Cell. Biol. 17: 208–212.

Saga, N., M. Polne-Fuller & A. Gibor, 1986. Protoplasts from seaweeds: production and fusion. Nova Hedwigia 83: 37–43.

Santelices, B. & M. S. Doty, 1986. A review of *Gracilaria* farming. Aquaculture 78: 283–290.

Santelices, B., J. A. Correa, I. Meneses, D. Aedo & D. Varela, 1996. Sporeling coalescence and intraclonal variation in *Gracilaria chilensis* (Gracilariales, Rhodophyta). J. Phycol. 32: 313–322.

Santelices, B, J. A. Correa, I. Meneses, D. Aedo, D. Varela & P. Sánchez, 1997. Coalescing Rhodophyta, life history style and adaptive traits. Phycologia 36: (suppl.) 97.

Westermeier, R., D. Patiño & J. Morales, 1997. Prospección de algas con especial referencia en *Iridaea ciliata* y *Gigartina skottsbergii* en la XI y XII Región. Publicación ocasional, Universidad Austral de Chile, Valdivia, 150 pp.

Wilkinson, L., M. A. Hill & E. Vang, 1992. SYSTAT: statistic, Version 5.2 edn. SYSTAT Inc, Evanston, IL.

Zamorano, J. & R. Westermeier, 1996. Phenology of *Gigartina skottsbergii* (Gigartinaceae, Rhodophyta) in Ancud Bay, southern Chile. Hydrobiologia 326/327: 253–258.

Economic feasibility of *Sarcothalia* (Gigartinales, Rhodophyta) cultivation

Marcela Avila[1], Erick Ask[2], Brian Rudolph[3], Mario Nuñez[1] & Ricardo Norambuena[4]
[1]*División de Fomento a la Acuicultura, Instituto de Fomento Pesquero, Casilla 665, Puerto Montt, Chile*
[2]*Marine Colloids Phil. Inc. Ouano Compound, Looc, Mandaue City, Cebu Philippines*
[3]*Copenhagen Pectin A/S Division of Hercules Incorporated. DK- 4623 Lille Skensved, Denmark*
[4]*Subsecretaría de Pesca Chile, Casilla 100V, Valparaíso, Chile*
E-mail: mavila@ifop.cl

Key words: Sarcothalia, cultivation, Gigartinales

Abstract

Studies to develop a technology for *Sarcothalia* cultivation were carried out based on results of a pilot scale farm. Mature cystocarpic and tetrasporic fronds of *Sarcothalia crispata* were collected in Chiloé island (41° S), southern Chile and sporulated in semi controlled conditions. Frames with nylon and polyfilament of different diameters were seeded and incubated in tanks of 400 l. Small gametophytes attached to ropes were maintained in the hatchery for 3 months before out-planting to the sea. Small fronds were grown in a floating system in the open sea. Plants attained 1.5 cm after 3 months cultivation in tanks in semi-controlled conditions. After out-planting to the sea, frond growth took place mainly during spring and summer. The annual timing of the cultivation and production process is described.

Introduction

In Chile, seaweeds belonging to the genus *Sarcothalia*, *Gigartina* and *Mazzaella* have been harvested from wild beds for more than three decades and used commercially for extraction of carrageenan (Avila & Seguel, 1993; Norambuena, 1996). These resources have become especially important during the last five years due to the construction of carrageenan processing plants in Chile, which provide about 20% of the world carrageenan production capacity (Bixler, 1996). These seaweeds are mainly being harvested in central Chile, from Valparaiso (36° S) to southern Chile in Chiloé Island (41° S). In 1997, around 8000 t of dry seaweed (US$ 9 200, 000) was exported as raw material and a quantity not specified was utilized together with imported seaweed by the domestic carrageenan industry (4 companies), which produced 2079 t of carrageenan, representing an income of US$ 16 400, 000. Substantial price increases during the past 5 years suggest an increase in demand of the raw material in the future (Bixler, 1996).

Sarcothalia crispata (Bory) Leister (formally *Iridaea ciliata*) is found from the intertidal to subtidal, down to 10 m and its distribution is antiboreal (Hommersand et al., 1993) from Valparaíso (33° S) to the Magellan Strait (53° S).

The thallus is foliose, with one or two oval fronds, with numerous proliferations on the blade of the gametophytes but not on the tetrasporophytes (Hommersand et al., 1993). Fronds are attached to the substratum through a disc-like holdfast, with cilia-like proliferations on the margins of the frond. This is characteristic of the species in the juvenile and adult stages. The thallus varies in length reaching up to 2 m. Its life cycle is isomorphic.

Field and laboratory studies have been conducted for cultivation of small thalli of *Iridaea cordata* and *Gigartina exasperata* (Waaland, 1973, 1976; Mumford 1977, 1979; Mumford & Waaland, 1980). Field experiments determined that the maximum growth rate was obtained at 3–5 m depth. The laboratory experiments showed that the optimum temperature for growth was between 10 and 14 °C. In semi-closed systems yield from tank cultures was determined also using different sizes of thalli. Hansen (1983), likewise, attempted to farm *Iridaea*. One methodology used to farm seaweeds commercially is the use of spores. This method has been used successfully in the commercial farms of brown seaweeds such as *Laminaria* (Kawashima, 1984), *Undaria* (Ohno & Matsuoka, 1993), red seaweeds such us *Porphyra* (Okazaki, 1971; Tseng, 1981; Oohusa, 1993) and recently in *Gracilaria* (Alveal et al., 1997) and *Eucheuma* (Azanza et al., 1996). Navarro (1991) induced sporulation of *S. crispata* and recorded the settling of viable spores on various substrata. Avila et al. (1996) inoculated artificial substrata with spores in controlled and in semi-controlled conditions and obtained fronds, but during the out-plant of the ropes to the sea, mortality rates were high (94%) and the density of discs low (5 discs/cm^2). The present study has been encouraged by the carrageenan industry to develop a technology for *Sarcothalia* cultivation, and to produce an economic model to estimate the costs of production based on results of a pilot scale farm.

Materials and methods

Mature tetrasporic fronds of *Sarcothalia crispata* were collected from Ancud (41° 52′ S; 73° 51′ W) and transported to the hatchery in insulated styrofoam boxes. Plants with mature tetrasporangia were sorted immediately upon arrival at the hatchery and kept overnight at 10 °C to increase spore production. Collections were made during autumn and winter (between March and August, 1996).

Selected fronds were rinsed in a series of four baths in filtered seawater before sporulation, blotted dry and left to dry further for 10–15 min. Mature fronds were kept in sporulation containers (3 kg seaweed/20 l capacity), and filled with filtered seawater. Sporulation containers were covered to reduce contaminants and kept still for 2 h. After this, fronds were removed and the remaining spore solution was stirred, filtered and poured into tanks (400 l capacity) where the string had been already wrapped on frames which were lying horizontally one on top of the other (Figure 1). Spore settlement and density was checked on coverslips placed on the frames.

Both sides of the frames were inoculated on two different days and on the third day filtered seawater was changed and enriched Provasoli solution (300 ml) at half strength was added to the tanks. Bubbling was started one week after seeding. Nylon and polyfilament of different diameters were used as substrata.

During winter, light was increased with fluorescent light placed 70 cm above the tanks with a long photoperiod of 18 h light and 6 h darkness to increase growth of the discs. The temperature of the tanks fluctuated as indicated in Figure 2. After the first three weeks of incubation in the tanks, effluents from a nearby scallop hatchery were added twice a week to increase nutrient availability and reduce costs. Once a week, frames were cleaned with seawater to reduce diatoms and other contaminants.

After three months in the hatchery plants reached 5–15 mm in length and 1.170 m of substrata with little fronds were transplanted to the sea. Nylon was outplanted to the sea in April, 1996 and distributed over 92 lines of 10 m in a horizontal floating system located about 2 m below the sea surface. Material of 3 mm was also out-planted and distributed in 5 lines and 50 m length and growth was evaluated. The length of about 40 plants was measured from nylon and polyfilament weekly.

For the second out-plant, lines were controlled and cleaned of epiphytes every 2 weeks in the sea but not measured. Experimental harvesting was done after 3 months in the sea, 60 individual fronds were measured in length and width. Two harvests were carried out on the pilot farm during the growth season (September–April), one in January 30, 1997 and the other in March 13, 1997. The area of the fronds was estimated using the following formula:

$$\text{Area} = a \times b \times \pi/4.$$

The design of the farm in the sea (Figure 3) was a rectangular module of 20×100 m with a 50×50 cm styrofoam float in each corner and held in place by four concrete anchors. Small anchors were placed on the sides to provide tension on the lines. A 12 mm rope was used for the anchor lines.

Results

Inoculation of substrata and incubation in tanks

The best germination and survivorship were achieved with nylon of 1 mm and polyfilament of 3 mm. Tetraspores germinated on both types of substrata but the gametophytes developed and grew at a slower rate on the polyfilament of 3 mm.

Germination of spores occurred in the 5 days following inoculation. Spores settled and became attached firmly to the substratum, starting development. Densities of discs attached on the ropes fluctuated between 56 and 80 spores mm^{-2} after inoculation and

Figure 1. Frames used to settle spores in tanks of 400 l.

after three months of incubation in tanks density decreased to an average of 10.3 ± 4.4 plants mm^{-2}. Fouling organisms such as diatoms, *Enteromorpha*, and *Ulva* were evident after two to three weeks after inoculation. After three months, plants growing on nylon attained an average length of 15 ± 6.6 mm in length while on polyfilament of 3 mm, length was 5.3 ± 5.2 mm.

Growth in the field

The first out-plant showed that fronds of less than 1 cm length were not able to grow in the sea because they became covered with epiphytes. Frond enlargement corresponding to the first out-plant during winter is shown in Figure 4. The algae *Ectocarpus*, *Giffordia* and *Polysiphonia* became abundant on the lines during spring (September–December). During spring and summer, *Ulva* was the major weed present. On the second out-plant, placed in October 1996, *Ectocarpus* and *Giffordia* were not observed but *Ulva* was present on the lines.

Production level

After three months in sea, the fronds of the second out-plant reached harvestable size >10 cm (Figure 5). In Table 1 data from the harvests are shown. In the laboratory the harvested biomass was dried in the oven to constant weight to obtain the dry weight. The first harvest was done by cutting or breaking fronds of harvestable size about 2 cm above the holdfast. After the harvest the discs remained on the line with new uprights which grew into new fronds. An average of 33 harvestable fronds were found in a linear metre of line. The mean size of the fronds of the first harvest expressed as area was 170 ± 193 cm^2. In the second harvest it decreased to 118 ± 91 cm^2.

Discussion

Preliminary surveys have shown that *Sarcothalia crispata* occurs in populations along the Chilean coast (Santelices, 1989; Westermeier et al., 1996). Natural populations in central Chile have seasonal variations in biomass (Hannach & Santelices, 1985; Poblete et

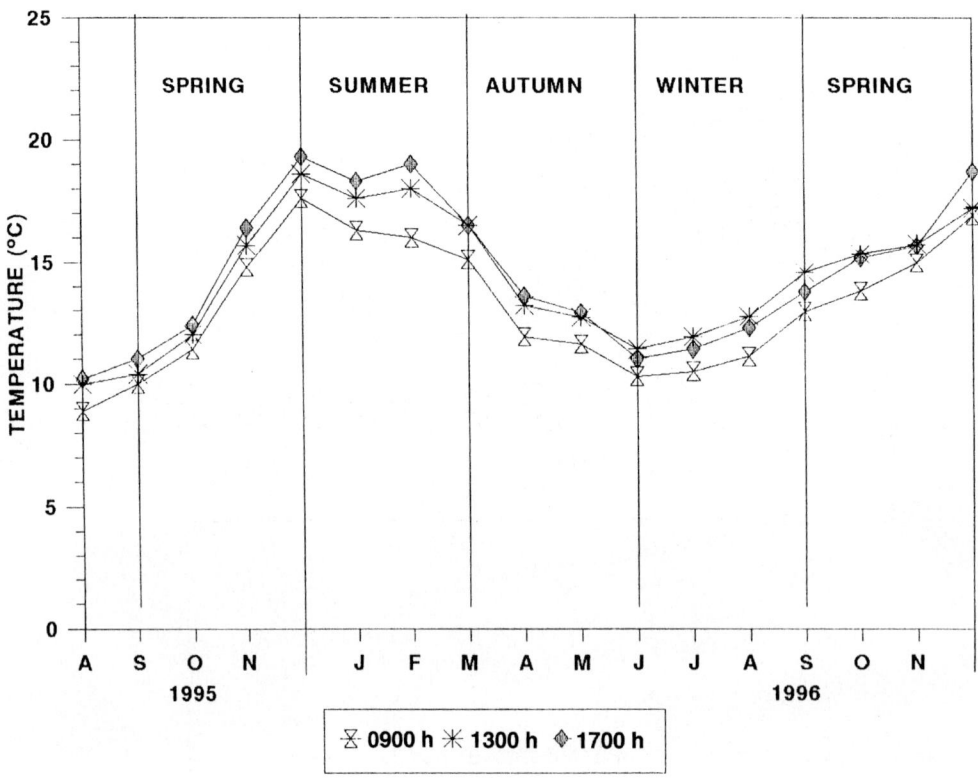

Figure 2. Fluctuations of seawater temperature in the tanks between spring of 1995 and spring of 1996, at different times of day.

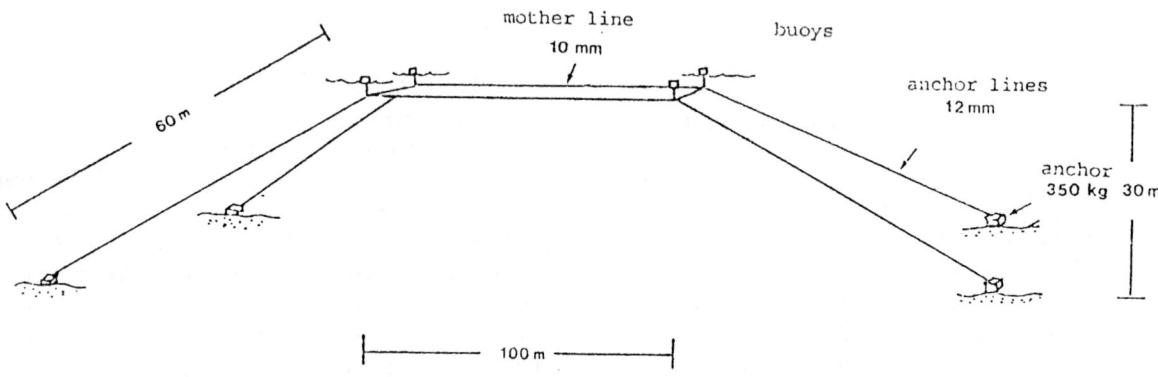

Figure 3. Design of the pilot module in the sea.

Table 1. Results of the first and second harvest in the pilot system that was installed in the sea (size is expressed as area ± S.D.)

	Wet weight, g m^{-1}	Dry weight, g m^{-1}	Number of plants m^{-1}	Size of the harvested fronds (cm^2)
First harvest	184 ± 77	37	34 ± 16	170 ± 193
Second harvest	281 ± 132	56	41 ± 23	118 ± 91
Total	465	93		

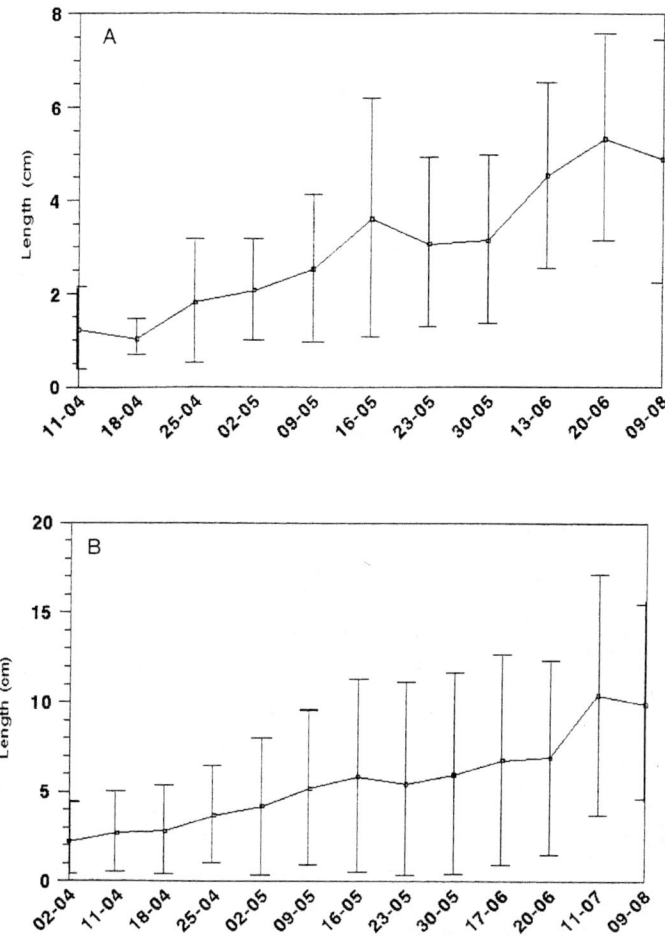

Figure 4. Frond enlargement in winter. (A) Material of 3 mm (polyfilament), (B) Monofilament. Vertical bars indicate standard error.

al., 1985) and predominance of the gametophyte over the tetrasporophyte has also been described (Otaíza, 1996). Biomass in natural beds is mostly regulated by two mechanisms: holdfast regeneration and settlement of spores (Santelices, 1989). Avila et al. (1996) showed that natural beds in southern Chile present marked seasonality both in reproduction and growth. Maximum biomass of non reproductive plants occurs in late summer. Phenological studies indicate that in late summer and early autumn (March) tetrasporic blades begin to show reproductive structures and, in the following months, appear to be the dominant reproductive phase, while cystocarpic blades reach their maximum density in winter (August) when tetrasporic fronds have already declined. Considering these phenologic data, sporulation of tetrasporic fronds was induced in winter time.

The results of this work show the technical feasibility of culturing *Sarcothalia* in the field. However, the cost of investments has to be optimized, and the productive process must be further developed in some biological aspects such as the yield per metre. *Sarcothalia* culture by spores requires, as in other red and green algae, two important phases, one in the greenhouse and the second in the sea. In the greenhouse phase, it takes about three months to obtain fronds of 1.5–2 cm. Smaller fronds are covered with epiphytes in the sea, reaching almost 100% mortality. According to our results, this size is adequate for transport to the field with low mortality; the final mean density achieved was 33 fronds m^{-1} in the sea. There is evidence that the density of the fronds can affect growth and production (Flores-Moya & Fernández, 1996). Thus it is advisable to test the effect of density on productivity.

Figure 5. Line with fronds ready to be harvested.

Culture in the greenhouse comprises sporulation, germination, adhesion to a substratum and growth up to 2 cm while culture in the sea consists in the growth of fronds until reaching a harvestable size. To induce sporulation in mature fronds we used desiccation, as described by Infante & Candia (1988). However, prior to frond stimulation, the reproductive material was kept at low temperature for at least 12 h. Using a spore suspension for seeding results in a fairly homogeneous distribution of the spores on the substratum. In the greenhouse phase, different artificial substrata were tested for spore settlement. A high settlement and good adhesion was achieved with 3 mm ropes and monofilament. The presence of epiphytic algae such as diatoms and some Chlorophyceae, which are present

Table 2. Timing of *Sarcothalia* cultivation

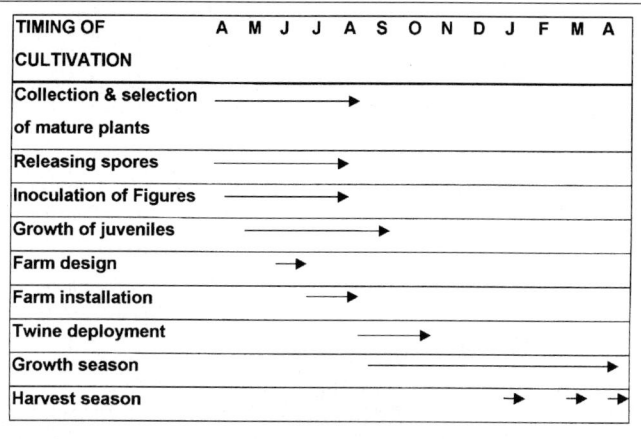

over the whole greenhouse phase, did not preclude the development of *Sarcothalia* plants.

According to the results obtained, we elaborated a calendar for *Sarcothalia* culture (Table 2), where the selection of the reproductive phase to be cultured is to be done between April and August. Sporulation and inoculation of substrata must be carried out immediately. However, the density of the plants should be optimized, since a high density could affect production. Holdfasts firmly adhered will initiate the development of small thalli after 3–4 weeks in the greenhouse, under conditions of filtered, enriched sea water and illumination. The small fronds must be kept in greenhouse culture until reaching the adequate size for their transfer to the field (1.5 cm). It is advisable to transport the ropes with plants in early spring to allow for the growth of *Sarcothalia* fronds before the occurrence of an explosive growth of unwanted species, as it has occurred in the commercial culture of other algae like *Gracilaria* (Kushel & Buschmann, 1991; Westermeier et al., 1991; Buschmann et al., 1995). Fronds longer than 10 cm may be harvested at the end of January, in March, and early May.

The cost of spore production under semi-controlled conditions may be optimized, considering that in other species, such as *Gracilaria chilensis*, this technique has been developed at low cost (Alveal et al., 1997) and implemented at commercial level.

If results obtained in the first harvests at pilot scale are extrapolated to 3 yearly harvests over the growth period (November–May), the total output would be about 700 g of fresh weight or 140 g dry weight per (linear) metre.

References

Alveal, K., H. Romo, C. Werlinger & E. C. Oliveira, 1997. Mass cultivation of the agar-producing alga *Gracilaria chilensis* (Rhodophyta) from spores. Aquaculture 148: 77–83

Avila, M. & M. Seguel, 1993. An overview of seaweed resources in Chile. J. appl. Phycol. 5: 133–139.

Avila, M., R. Otaíza, R. Norambuena & M. Núñez, 1996. Biological basis for the management of 'luga negra' (*Sarcothalia crispata* Gigartinales, Rhodophyta) in southern Chile. Hydrobiologia 326/327: 245–252.

Azanza, R., T. Aliaza & N. Montaño, 1996. Recruitment of *Eucheuma* and *Kappaphycus* on a farms in Tawi-Tawi, Philippines. Hydrobiologia 326/327: 235–244.

Bixler, H. J., 1996. Recent development in manufacturing and marketing carrageenan. Hydrobiologia 326/327: 35–57.

Buschmann, A., R. Westermeier & C. Retamales, 1995. Cultivation of *Gracilaria* on the sea bottom in southern Chile: a review. J. appl. Phycol. 7: 29–301.

Flores-Moya, A. & J. Fernández, 1996. Growth pattern, reproduction, and self-thinning in seaweeds. J. Phycol. 32: 767–769.

Hannach, G. & B. Santelices, 1985. Ecological differences between the isomorphic reproductive phases of two species of *Iridaea* (Rhodophyta: Gigartinales). Mar. Ecol. Prog. Ser. 22: 291–303.

Hansen, J. E., 1983. A physiological approach to mariculture of red algae. J. World Maricul. Soc. 14: 380–391.

Hommersand, M., M. Guiry, S. Fredericq & G. Leister, 1993. New perspectives in the taxonomy of the Gigartinaceae (Gigartinales, Rhodophyta). Hydrobiologia 260/261: 105–120.

Infante, R. & A. Candia, 1988. Cultivation of *Gracilaria verrucosa* (Hudson) Papenfuss and *Iridaea ciliata* Kutzing (Rhodophyta, Gigartinaceae), 'in vitro'. Induced shedding and carpospore colonization on different subtrata. Gayana, Bot. 45: 297–304.

Kawashima, S., 1984. Kombu cultivation in Japan for human foodstuff. Jap. J. Phycol. 32: 379/94.

Kuschel, F. & A. Buschmann, 1991. Abundance, effects and management of epiphytism in intertidal cultures of *Gracilaria* (Rhodophyta) in southern Chile. Aquaculture 92: 7–19.

Mumford, T. F., 1977. Growth of Pacific northwest marine algae on artificial substrates: potential and practice. In Krauss, R.

(ed.), The Marine Plant Biomass of the Pacific Northwest Coast. Oregon State University, Corvallis, 139–161.

Mumford, T. F., 1979. Field and laboratory experiments with *Iridaea cordata* (Floridophyceae) grown on nylon netting. Proc. int. Seaweed Symp. 9: 515–523.

Mumford, T. F. & J. R. Waaland, 1980. Progress and prospects for field cultivation of *Iridaea cordata* and *Gigartina exasperata*. In Abbott, I. A., M. S. Foster & L. F. Eklund (eds), Pacific Seaweed Aquaculture. California Sea Grant College Program, Pacific Grove, California, 92–105 & 192–197.

Navarro, R., 1991. Cultivo de *Iridaea ciliata* Kützing (Rhodophyta, Gigartinales) en el laboratorio y su factibilidad de crecimiento en terreno. Tesis de grado. Universidad Austral de Chile. Facultad de Ciencias Escuela de Biología Marina, 79 pp.

Norambuena, R., 1996. Recent trends in seaweed production in Chile. Hydrobiologia 326/327: 371–379.

Oohusa, T., 1993. The cultivation of *Porphyra* 'Nori'. In Ohno, M. & A. T. Critchley (eds), Seaweed Cultivation and Marine Ranching. Kanagawa International Fisheries Training Center, Yokosuka, 57–73.

Ohno, M. & M. Matsuoko, 1993. *Undaria* cultivation 'wakame'. In Ohno, M. & A. T. Critchley (eds), Seaweed Cultivation and Marine Ranching. Kanagawa International Fisheries Training Center, Yokosuka: 41–49.

Okazaki, A, 1971. Seaweeds and their Uses in Japan. Tokai University Press, Tokyo, 165 pp.

Otaíza, R., 1996. Respuesta a la herbivoría de discos gametofíticos y esporofíticos de *Iridaea ciliata*. VI Symposium de Algas Marinas y IV Encuentro de Microalgólogos. Puerto Montt, Chile: 70.

Poblete, A., A. Candia, I. Inostroza & R. Ugarte, 1985. Crecimiento y fenología reproductiva de *Iridaea ciliata* Kützing (Rhodophyta, Gigartinales) en una pradera submareal. Biol. Pesq. 14: 23–31.

Santelices, B., 1989. Algas Marinas de Chile. Ediciones Universidad Católica de Chile, Santiago. 400 pp.

Tseng, C. K., 1981. Commercial cultivation. In Lobban, C. S. & M. J. Wynne (eds), The Biology of Seaweeds. Blackwell Scientific, Oxford: 680–725.

Waaland, J. R., 1973. Experimental studies on the marine algae *Iridaea* and *Gigartina*. J. exp. mar. Biol. Ecol. 11: 71–80.

Waaland, J. R., 1976. Growth of the red alga *Iridaea cordata* (Turner) Bory in semi-closed culture. J. exp. mar. Biol. Ecol. 23: 45–53.

Westermeier, R., M. Morales, J. & D. Patiño, 1996. Caracterización de poblaciones de algas carragenófitas en el mar interior de la X y XI Región, con especial referencia a *Iridaea ciliata* Kutzing y *Gigartina skottsbergii* Setchell et Gardner (Rhodophyta, Gigartinales). VI Symposium de Algas Marinas y IV Encuentro de Microalgólogos. Puerto Montt, Chile: 91.

Westermeier, R., P. Rivera & I. Gomez, 1991. Cultivo de *Gracilaria chilensis* Bird, McLachlan & Oliveira, en la zona inter y sub mareal del estuario Cariquilda, Maullín Chile. Rev. chil. Hist. nat. 64: 307–321.

Comparison of the performance of the agarophyte, *Gracilariopsis bailinae*, and the milkfish, *Chanos chanos*, in mono- and biculture

Lota B. Alcantara[1,2], Hilconida P. Calumpong[2,*], Milagrosa R. Martinez-Goss[3], Ernani G. Meñez[4] & Alvaro Israel[5]

[1]*State Polytechnic College of Palawan, P.O. Box 93, Puerto Princesa City 5300, Philippines*
[2]*Silliman University Marine Laboratory, Dumaguete City 6200, Philippines*
[3]*IBS-CAS, University of the Philippines at Los Baños, College, Laguna 4301, Philippines*
[4]*Department of Botany, National Museum of Natural History, Smithsonian Institution, Washington, D.C. 20560, U.S.A.*
[5]*Israel Oceanographic & Limnological Research, National Institute of Oceanography, Tel-Shikmona, Haifa, Israel*
E-mail: mlsucrm@mozcom.com

Key words: biculture, *Chanos*, epiphytes, *Gracilariopsis*, Philippines, water quality

Abstract

The performances of the agarophyte, *Gracilariopsis bailinae*, and the milkfish, *Chanos chanos*, under monoculture and biculture conditions in aquaria and ponds were studied from May 1997 to March 1998. Water quality of both systems was monitored. The two species have reciprocal characteristics in their biological requirements and by-products. Both species attained higher growth rate in biculture: in aquaria *Gracilariopsis* obtained a mean daily growth rate of 4.72 ± 1.64% for biculture and 3.44 ± 2.74% in *Gracilariopsis* monoculture while *Chanos* had a mean daily growth rate of 4.81 ± 2.13% in biculture and 4.13 ± 2.13% in *Chanos* monoculture. In ponds, *Gracilariopsis* obtained a mean daily growth rate of 3.68 ± 0.39% in biculture and 2.46 ± 0.38% in *Gracilariopsis* monoculture while *Chanos* had a mean daily growth rate of 4.81 ± 0.33% in biculture and 2.9 ± 0.1% in *Chanos* monoculture. The growth rates for both *Gracilariopsis* and *Chanos* decreased weekly through one month of culture. Higher dissolved oxygen levels were observed in *Gracilariopsis* monoculture and in biculture and significantly lower in *Chanos* monoculture. *Chanos* did not control epiphytism in *Gracilariopsis*. There was no difference in epiphytism in either culture systems in aquaria, but a significant growth of green algae occurred in ponds with monoculture of *Gracilariopsis*.

Introduction

Monospecies aquaculture may be economically viable but its long term sustainability may be compromised. There is inefficient use of resource and production of by-products that alter the sustainability of the environment (Buschmann, 1996). For example, after a decade of intensive culture of *Penaeus monodon* Fabricius in the Philippines, many farmers experienced problems of slow growth and diseases. Introduction of feeds, fertilizers and pesticides resulting in by-products in the system led to imbalances such as deterioration of soil and water quality and disease outbreaks (Primavera, 1993).

Unlike monoculture, biculture can be designed to maximize the productivity and balance of the culture medium. Production in aquatic farming could be increased when the cultured organisms occupy different strata of the water column and have different biological requirements (Bardach et al., 1972).

Both *Gracilariopsis bailinae* Zhang and Xia and *Chanos chanos* Forsskål are economically important and grow well in brackish ponds. In the Philippines, approximately ninety per cent of the total brackish

---
*Author for correspondence

pond area is utilized for the cultivation of *Chanos* (Sumagaysay et al., 1990). Furthermore, a number of milkfish farmers now utilize high-density technology to increase production, thus necessitating supplemental feeding. The biculture of seaweed and fish has not yet been adopted commercially in the Philippines. Additional income and reduced environmental impact are some of the advantages of this culture technique (Hurtado-Ponce, 1993; Buschmann, 1996). With *Gracilariopsis*, pollution of the water in the system and its environs may be minimized.

The growth performance of the two experimental organisms in mono- and biculture is presented in this paper. The resulting effect of the two culture systems on some water quality characteristics is also discussed.

Previous studies revealed that competitor algae and epiphytes limit the harvest of *Gracilaria* and *Gracilariopsis* species (e.g., Friedlander and Ben-Amotz, 1991; Gonzales et al., 1993; Buschmann et al., 1994). This paper also reports on the epiphytes on *Gracilariopsis* and competitor algae present in both culture systems.

Materials and methods

Aquarium experiment

Three treatments were compared: treatment 1 = *Gracilariopsis* + *Chanos* (biculture), treatment 2 = *Gracilariopsis* only (*Gracilariopsis* monoculture), and treatment 3 = *Chanos* only (*Chanos* monoculture). Each treatment had four replicates distributed randomly. Twelve aquaria were used. Each aquarium measured 30 cm × 60 cm × 40 cm with water capacity of 72 l. The study lasted for one month. Stocking was one *Chanos* in each aquarium for treatments 1 and 3, and 20 g of *Gracilariopsis* in each of treatments 1 and 2. Both the alga and fish were acclimated in their new environments prior to the experiments.

Water temperature, salinity, pH, ambient temperature and photosynthetically active radiation (PAR) were monitored at 0900–1100 and 1400–1600 h daily using a mercury thermometer, temperature compensated Reichert refractometer (Japan), Orion Research Model SA 250 portable pH meter (Orion Research Incorporated, USA), and LI-COR LI 188BS quantum meter (LI-COR, Inc., Lincoln, NE. USA). Water samples for measuring dissolved oxygen (DO) were collected at 0600–0700 and 1500–1600 h weekly. The Winkler method of DO analysis was followed. Analyses of initial and final water NH_4-N, NO_3-N, and PO_4-P were carried out according to the salicylate-hypochlorite method of Bower and Holm-Hansen (1980), cadmium reduction method of Grasshoff et al. (1983), and molybdo-ascorbic acid reduction method of Grasshoff et al. (1983), respectively. Rainfall, cloud cover, and other atmospheric data were gathered from the local PAGASA (Philippine Atmospheric Geophysical and Astronomical Services Administration) office.

The non-aerated aquaria were cleaned and water changed three times a week after monitoring of water quality characteristics. *Chanos* in aquaria received supplemental commercial feed (Tateh Feeds, Manila, Philippines) three times daily at 5% of the body weight adjusted weekly. Initial mean *Chanos* body weights were 2.88 ± 0.3 g for biculture and 3.12 ± 0.1 g for monoculture. The aquaria were placed on outdoor platforms shaded with black netting.

Weights of *Gracilariopsis* and *Chanos* were measured every week with a triple beam balance (Ohaus, NJ, USA) with 0.1 g accuracy. Ten grams of *Gracilariopsis* served as sample for quantification of epiphytes and other competing algae. Epiphytes were picked by hand or by forceps from the thalli. Samples were then dried to constant weight at 60 °C.

Pond experiment

The treatments used in the ponds were the same as those in the aquarium experiment. Each of the three treatments had six replicates. Each replicate consists of one earthen pond with its own water intake or gate, a total of 18 pond compartments. Each earthen pond measured 6m × 2m × 1m and was subdivided by netting material into three equal subcompartments for serial sampling. Each subcompartment was sampled at equal intervals of two weeks, following one after another, and all organisms were removed without replacement each time.

Each subcompartment was stocked with 100 g m^{-2} of *Gracilariopsis* for biculture and algal monoculture, which translates to 1000 kg ha^{-1}, and two individuals of *Chanos* or approximately 5000 individuals ha^{-1} for biculture and fish monoculture. The milkfish received commercial supplemental feed (SEAFDEC/AQD) three times daily at 5% of initial body weight. Initial mean *Chanos* body weights were 11.0 ± 1.2 g for biculture and 14.1 ± 1.0 g for monoculture. Water monitoring was the same as in the aquarium experiment.

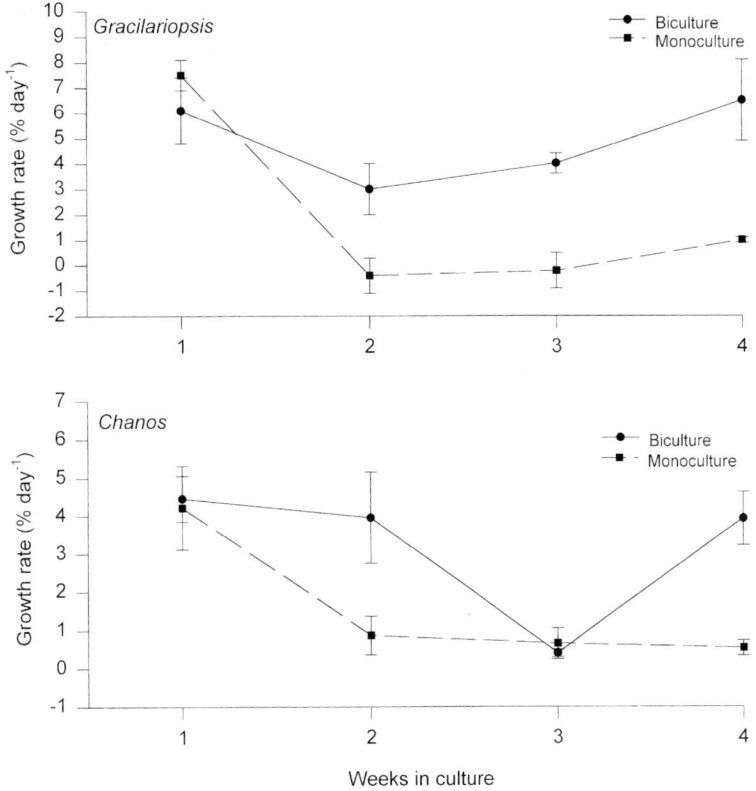

Figure 1. Growth rates of cultured organisms in aquaria computed weekly for one month. Bars: SD.

Fluctuations of water depth in pond compartments were also measured. Water replenishment took place once or twice fortnightly depending on the height of the highest high tide as water could only enter the pond when the tide was more than 1.5 m.

Data analysis

Biomass was cut back to initial values at each measurement. Determination of growth rate followed the formula:

Growth rate (% day^{-1}) = $\frac{\ln (W_f/W_i)}{t} \times 100$,

where W_f = final fresh weight at day t; W_i = initial fresh weight; t = number of culture days.

Differences in growth rate and DO levels over time were analyzed using the two-factor analysis of variance (2-way ANOVA), and the least significant difference test (LSD) or Duncan's multiple range test (DMRT). Appropriate data transformations were applied (Gomez & Gomez, 1984; Sokal & Rohlf, 1987).

Results

Aquarium experiment

The mean growth rate of *Gracilariopsis bailinae* for the duration of the culture period was 4.72 ± 1.64 (SD)% day^{-1} for the biculture and 3.44 ± 2.74% day^{-1} for the monoculture. Two-factor ANOVA revealed that *Gracilariopsis* bicultured with *Chanos* obtained a significantly higher ($p = 0.009$) growth rate than monocultured. During the first week of culture, the two treatments were similar. Using the least significant difference (LSD, $\alpha = 0.05$) test, monoculture had reduced growth during the third and fourth week of rearing (Figure 1).

The mean growth rates of *Chanos* in aquaria for four weeks were 4.46 ± 0.6% day^{-1} for biculture and 3.96 ± 2.13% day^{-1} for monoculture. ANOVA showed that the growth rate of *Chanos* bicultured with *Gracilariopsis* was significantly better ($p = 0.008$) than the one reared alone (Figure 1).

The incidence of epiphytes and other algae in the aquarium was high especially during week 4. At har-

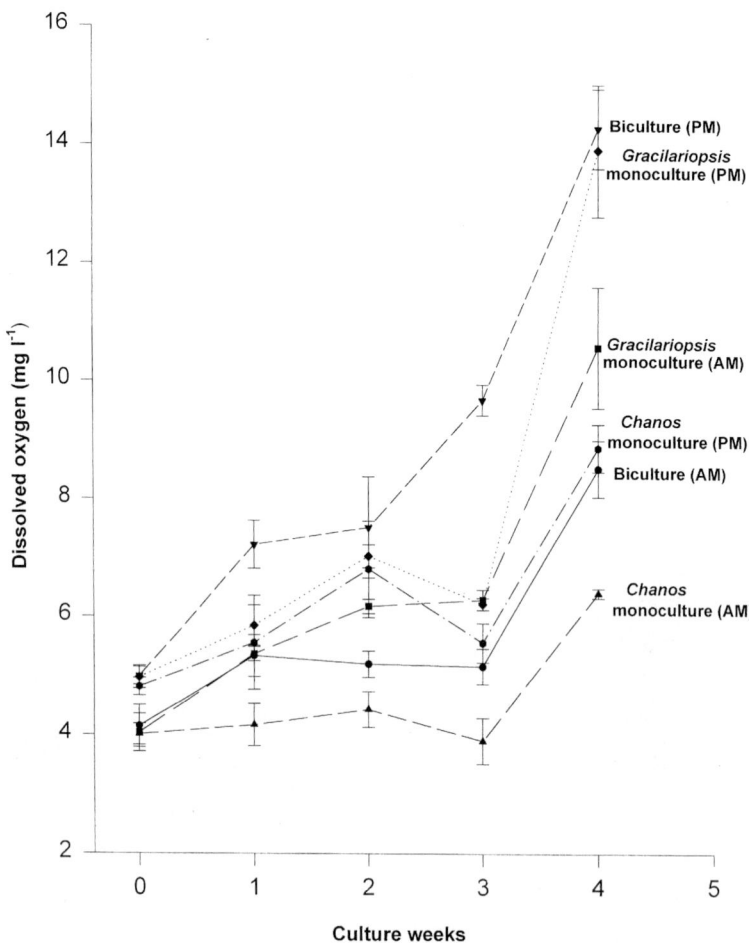

Figure 2. Dissolved oxygen in aquaria taken early in the morning and late in the afternoon once every week. Bars: SD.

vest, biculture yielded 14.08% (dry wt.) epiphytes and other entangling algae while *Gracilariopsis* monoculture yielded 18.33%. No significant difference was found in the amount of epiphytes and other algae between the two treatments using t-test ($p = 0.29$). The epiphytes and entangling algae found in the aquarium included the green algae *Enteromorpha clathrata* (Roth) Grev. and *Chaetomorpha linum* (O.F. Müller) Kützing, the cyanobacterium *Lyngbya majuscula* (Dillwyn) Harvey and diatoms belonging to the genera *Synedra*, *Navicula*, *Pinnularia*, *Mastogloia*, and *Coscinodiscus*.

Dissolved oxygen in aquaria increased through time, except for a drop at week 3 (Figure 2). Precipitation (1.4–43.4 mm) and cloudiness (3–6 okta) persisted during week 3, which may have contributed to the reduction in DO. There was less precipitation (0.8 to 8.8 mm) during the fourth week. In both morning and afternoon samplings *Gracilariopsis* monoculture had the highest DO level while *Chanos* monoculture had the lowest. Two-factor ANOVA demonstrated that a significant difference exists among treatments ($p < 0.00001$) and culture time ($p < 0.00001$). Using DMRT it was determined that *Gracilariopsis* monoculture had significantly higher DO than biculture, which had significantly greater DO than *Chanos* monoculture. For all treatments, week 4 had significantly higher DO than the earlier weeks.

Substantial amounts of NH_4-N, NO_3-N, and PO_4-P were removed from the water at the end of the four-week culture period (Table 1). However, no significant differences were found in the amounts of these nutrients removed among treatments.

Table 1. Means of available NH$_4$-N, NO$_3$-N and PO$_4$-P removed from or added (in brackets) into the water or soil in aquaria or ponds after culture period. Units are in μmol l^{-1} for water and mg/kg for soil. Values which are significantly different are in bold. Bi = biculture; *Grac* = *Gracilariopsis* monoculture; *Chanos* = *Chanos* monoculture.

	Aquarium Water			Pond Water			Pond Soil		
	Bi	*Grac*	*Chanos*	Bi	*Grac*	*Chanos*	Bi	*Grac*	*Chanos*
NH$_4$-N	24.35	41.38	46.80	1.73	1.22	**0.73**	3.68	3.27	(1.05)
	±7.5	±13.5	±4.4	±0.59	±0.73	**±0.54**	±4.08	±7.13	±9.29
NO$_3$-N	9.92	7.82	10.35	0.56	0.5	(0.02)	(7.67)	(12.86)	(18)
	±2.1	±1.94	±0.34	±0.03	±0.27	±0.36	±4.66	±5.36	±5.25
PO$_4$-P	3.44	2.98	4.62	0.72	(1.06)	(1.34)	10.66	2.87	**45.57**
	±0.68	±0.94	±0.32	±0.48	±1.36	±1.87	±5.39	±14.02	**±14.1**

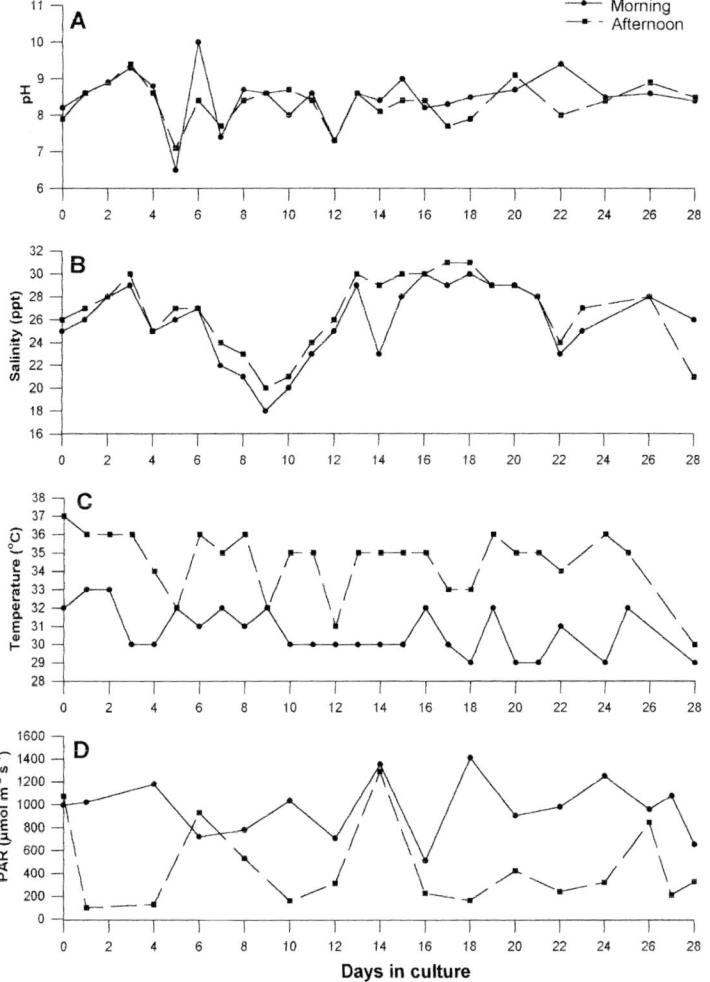

Figure 3. Characteristics of water in aquaria and ambient photosynthetically active radiation (PAR) taken twice daily during the culture period. A = pH; B = salinity, C = temperature; D = PAR.

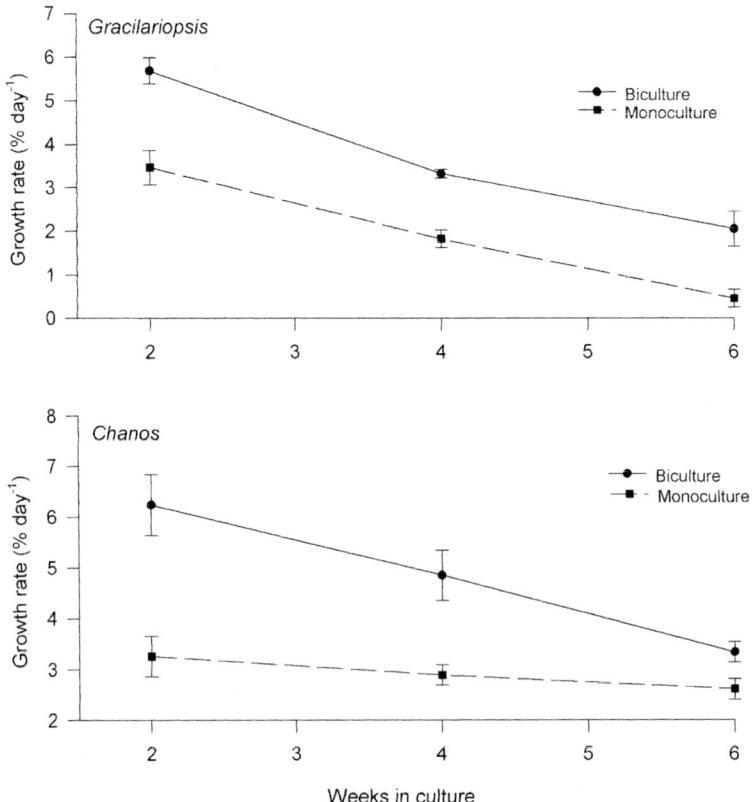

Figure 4. Growth rates of *Gracilariopsis* (A) and *Chanos* (B) in ponds computed every two weeks for six weeks. Bars: SD.

Morning and afternoon water pH ranged from 6.4 to 9.4 (Figure 3A). Salinity fluctuated between 17‰ to 34‰. Afternoon salinity was usually higher than morning salinity (Figure 3B). Precipitation and water change in the aquaria caused the decrease in salinity while evaporation increased it. Afternoon water temperatures varied between 30 °C and 37 °C, which were generally higher than morning readings of 29 °C to 33 °C (Figure 3C). Irradiance ranged from 101 μmol m^{-2} s^{-1} to 1352 μmol m^{-2} s^{-1} with generally higher morning than afternoon values (Figure 3D).

Pond experiment

Chanos had a mean daily growth rate of 4.81 ± 0.33% for biculture and 2.9 ± 0.1% for *Chanos* monoculture The growth rate of *Gracilariopsis* for six weeks in ponds is shown in Figure 4A. Its growth rate decreased with the culture period. Comparison of means using *t*-test revealed that the alga in biculture had significantly higher growth rate than the one in monoculture (p = 0.011, 0.001 and 0.001) in all the serials.

Figure 4B shows the growth rate of *Chanos* in small ponds declining through time. *Chanos* in biculture with *Gracilariopsis* performed better than in monoculture. It was indicated by t-test that the increase in per cent weight of *Chanos* in biculture in all sampling periods was significantly higher (p = 0.004, 0.013 and 0.014) than in monoculture.

There were more unattached filamentous green algae than epiphytes in experimental ponds. Four species were identified as associates first becoming epiphytes later: *Chaetomorpha linum, C. crassa, Enteromorpha clathrata* and *E. flexuosa* (Wulfen) J. Agardh. No significant difference (p = 0.64) in the amount of epiphytes and other algae was seen between biculture (2.89 ± 2.54%) and *Gracilariopsis* monoculture (3.20 ± 2.73%) but the amounts were significantly higher (2-way ANOVA) at the end of the experiment (Figure 5).

Dissolved oxygen values fluctuated with morning DO generally lower (0.11-6.5 mg l^{-1}) than afternoon DO (4.45-19.3 mg l^{-1}) (Figure 6). Two-way ANOVA

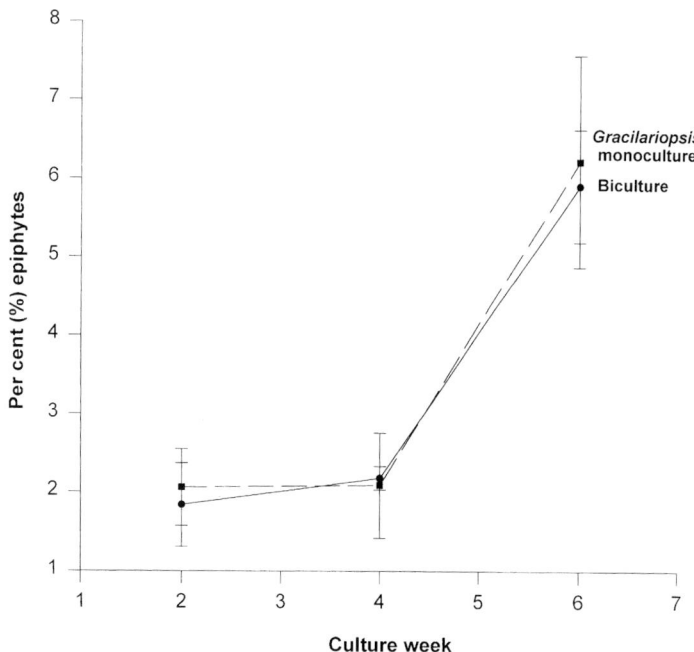

Figure 5. Relative amount of epiphytes and other algae associated with *Gracilariopsis* in ponds taken every two weeks for six weeks. Bars: SD.

revealed that morning DO in biculture and *Chanos* monoculture was significantly lower ($p = 0.00001$) than *Gracilariopsis* monoculture. The afternoon DO was significantly lower ($p = < 0.00001$) in *Chanos* monoculture than in the other treatments. Morning DO during the sixth week was significantly lower compared to that obtained during the second and fourth weeks.

A summary of water quality characteristics and PAR in ponds is given in Figure 7. The water pH occasionally fluctuated within the range of 6.78 to 8.96 (Figure 7A). Rainfall affected the variations in water salinity, depth and temperature and ambient PAR. There was lower salinity and temperature when water was deeper, which also coincided with lower PAR. Salinity during the culture time ranged from 14‰ to 40‰ (Figure 7B). Water depth fluctuated between 13 cm and 45 cm (Figure 7C). PAR was distinctly higher during mornings, which ranged from 83 to 2922 μmol s^{-1} m^{-2} (Figure 7D). Water temperature was higher in the afternoons, within the range of 22 °C to 39 °C (Figure 7E).

There was a significant difference in the amount of NH_4-N removed from pond water and PO_4-P removed from pond soil at the end of the culture period (Table 1). In pond soil, NO_3-N was added instead of removed in all treatments.

Discussion

Many environmental factors influenced growth rate of *Gracilariopsis* in both aquarium and pond experiments. In ponds and other large-scale farming, low yields were attributed to the presence of epiphytes and other competitor algae (Lapointe & Ryther, 1978; Lindsay & Saunders, 1980; Lignell et al., 1987; and Haglund & Pedersén, 1993) apparently due to shading and competition for nutrients (Lignell et al. 1987). Kuschel & Buschmann (1991) reported that dense epiphytism of *Giffordia* on *Gracilaria,* up to 30% of the latter's biomass, reduced irradiance by 94.1 to 98.5%.

In this study, the high load of epiphytes in the aquarium in the later part of the experiment probably contributed to the decline in growth of *Gracilariopsis*. McLachlan & Bird (1986) claimed that maximum available light is inversely related to production of *Gracilaria*. In addition, Rueness et al. (1987) pointed out that the abundance of epiphytic diatoms and ectocarpoid algae and reduced water motion contributed to poor growth of *Gracilaria verrucosa* (Huds.) Papenf.

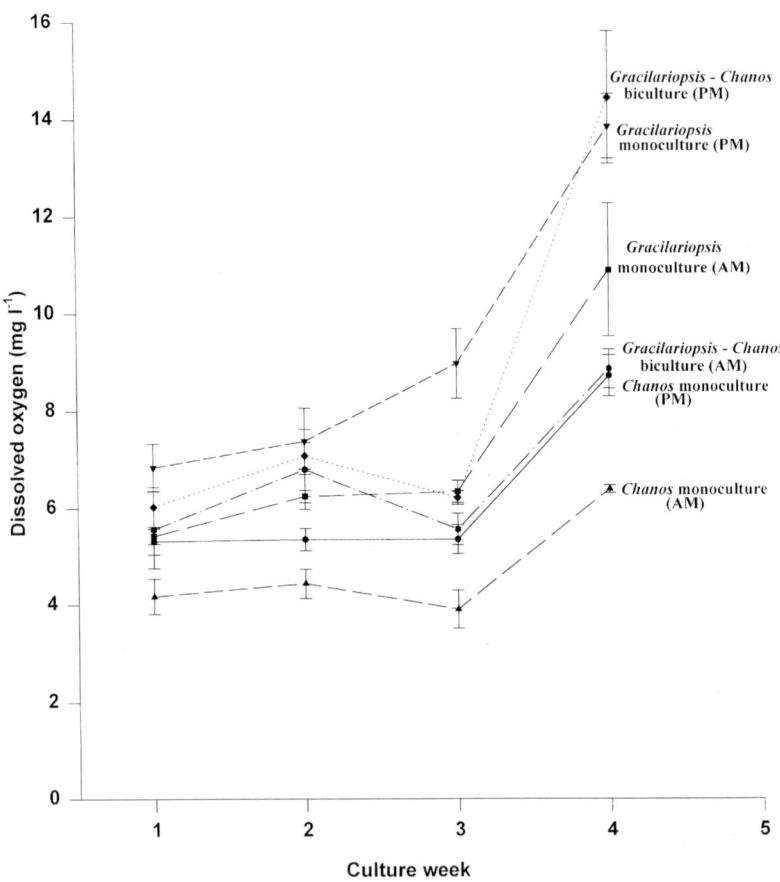

Figure 6. Dissolved oxygen in ponds taken early in the morning and late in the afternoon once every week. Bars: SD.

Haglund & Pedersén (1993) obtained the highest growth rate of *Gracilaria tenuistipitata* var. *liui* Zhang and Xia in pond culture at the beginning of the growing season but it decreased as the season advanced. Growth reduction in this study may be attributed to the increasing amount of epiphytes, especially in the aquaria. The relative epiphyte biomass (2.89%–18.33%) in this study is near the range (6.17–10.21%) found by Nelson et al. (1980), but higher than that found (<1%) by Friedlander & Ben-Amotz (1991).

Although *Chanos* probably contributed to the higher growth rate of *Gracilariopsis* in their biculture, it did not control epiphytism. This observation disagrees with the report by Santelices & Doty (1989) that epiphytes of *Gracilaria* could be controlled with 'Tilapia' and *Chanos*.

The maximum growth rates of *Gracilariopsis* in this study were lower (5.68% day^{-1} for pond and 4.72% day^{-1} for aquarium) than those attained (13% day^{-1}) in integration with *Lates calcarifer* Blóch in floating sea cages (Hurtado-Ponce, 1993). The difference may be due to location, culture method, and species of bicultured fish. In another species cultured in a pond, Haglund & Pedersén (1993) reported a growth rate of 7% day^{-1} for *Gracilaria tenuistipitata*.

Higher dissolved oxygen in both biculture and *Gracilariopsis* monoculture was apparently caused by the presence of *Gracilariopsis* and its epiphytes. The presence of *Chanos* in biculture did not affect the oxygen produced by *Gracilariopsis* and other algae. Lower DO in ponds than in aquaria early in the morning may have been caused by more respiring organisms in the former present in the sediment and water.

Nelson et al. (1980) found that *Gracilaria edulis* and *Gracilaria arcuata* have the capability of removing ammonia from seawater. Likewise, *Gracilaria tenuistipitata* was found to remove efficiently NO_3^-,

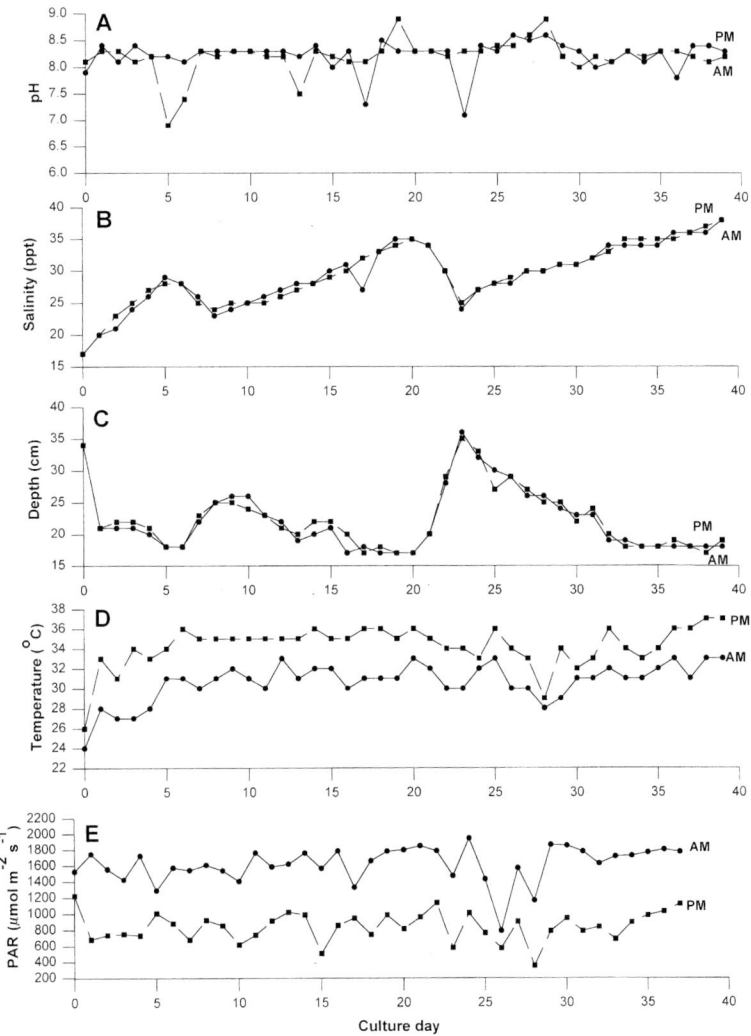

Figure 7. Characteristics of water in ponds and ambient PAR taken twice daily during the culture period. A = pH; B = salinity; C = depth; D = temperature; E = PAR.

NH_4^+, and PO_4^{3-} (Haglund & Pedersén, 1993). Results of this study show that the loss of NH_4-N, NO_3-N, and PO_4-P in the biculture of *Gracilariopsis* and *Chanos* as well as in the monocultures of *Gracilariopsis* and *Chanos* is not consistent with the mentioned reports. The regular cleaning and water change in the aquaria might have removed more of these nutrients than *Gracilariopsis* could filter. Moreover, Friedlander and Ben-Amotz (1991) reported that ammonium and nitrate did not increase growth of *Gracilaria conferta* (Schousboe) J. and G. Feldmann.

Sumagaysay et al. (1990) suggested that decreasing water temperature caused the slowing of growth rate in the last month of culture of *Chanos*. Similarly, Haglund & Pedersén (1993) reported that decreased growth of *Gracilaria tenuistipitata* was due to decreasing temperature and PAR. The observations may be true in aquarium experiments but the opposite was noted in the pond experiments of this study. With the El Niño phenomenon, pond temperatures were probably at their peak. Water pH (9 to 10) in ponds recorded by Haglund & Pedersén (1993) was higher than reported (7.5 to 8.5) in this paper. Salinity in ponds is inversely related to depth.

Conclusion

1. *Gracilariopsis bailinae* obtained a significantly higher growth rate in biculture with *Chanos chanos*. Its mean daily growth rates in aquaria for four weeks were $4.72 \pm 1.64\%$ for biculture and $3.44 \pm 2.74\%$ for *Gracilariopsis* monoculture; in ponds *Gracilariopsis* attained a mean daily growth rate of $3.68 \pm 0.39\%$ for biculture and $2.46 \pm 0.38\%$ for monoculture for six weeks.
2. *Chanos* grew better in biculture with *Gracilariopsis*. Its mean daily growth in aquaria for four weeks were $4.81 \pm 2.19\%$ for biculture and $4.13 \pm 2.13\%$ for *Chanos* monoculture while in ponds *Chanos* had a mean daily growth rate of $4.81 \pm 0.33\%$ for biculture and $2.9 \pm 0.1\%$ for *Chanos* monoculture for six weeks.
3. The growth rates for both *Gracilariopsis* and *Chanos* decreased weekly through one month of culture.
4. *Chanos* did not control epiphytism in *Gracilariopsis*. Epiphytes and other competitor algae of *Gracilariopsis* were similar in both mono- and biculture systems.
5. Dissolved oxygen in mono- and biculture of *Gracilariopsis* were similar but significantly lower in *Chanos* monoculture.

Acknowledgement

This study was funded by the University of the Philippines at Los Baños Foundation, Laguna, (US-AID CDR Project No. 96-009), the USAID-supported Center of Excellence in Coastal Resources Management at Silliman University, Dumaguete City; the Philippine Council for Aquatic and Marine Research and Development (PCAMRD) of the Department of Science and Technology; and the Silliman University Marine Laboratory Algal Fund (from Elvira Tan Award), Dumaguete City.

References

Bardach, J. E., J. H. Ryther & W. O. McLarney, 1972. Aquaculture: The Farming and Husbandry of Freshwater and Marine Organisms. John Wiley & Sons, Inc., 868 pp.

Bower, C. E. & T. Holm-Hansen, 1980. A salicylate-hypochlorite method for determining ammonia in seawater. Can. J. Fish. aquat. Sci. 27: 794–798.

Buschmann, A. H., 1996. An introduction to integrated farming and the use of seaweed as biofilters. Hydrobiologia 326/327: 59–60.

Buschmann, A. H., J. Schulz & P. Vergara, 1994. Epiphytism and herbivory in an intertidal *Gracilaria* (Rhodophyta, Gigartinales) farm in Southern Chile. In Koop, K. (ed.) Ecology of Marine Aquaculture. International Foundation for Science, Stockholm: 48–58.

Friedlander, M. & A. Ben-Amotz, 1991. The effect of outdoor culture conditions on growth and epiphytes of *Gracilaria conferta*. Aquat. Bot. 39: 315–333.

Gomez, K. A. & A. A. Gomez, 1984. Statistical Procedures for Agricultural Research, 2nd edn. International Rice Research Institute, Los Baños, Philippines 680 pp.

Gonzales, M. A., H. L. Barrales, A. Candia & L. Cid, 1993. Spatial and temporal distribution of dominant epiphytes on *Gracilaria* from a natural subtidal bed in Central-Southern Chile. Aquaculture 116: 135–148.

Grasshoff, K., M. Ehrhardt & K. Kremling (eds), 1983. Methods of Seawater Analysis, 2nd rev. edn. Verlag Chemie, D. Weinhein, 419 pp.

Haglund, K. & M. Pedersén, 1993. Outdoor pond cultivation of the subtropical marine red alga *Gracilaria tenuistipitata* in the brackishwater in Sweden. Growth, nutrient uptake, co-cultivation with rainbow trout and epiphyte control. J. appl. Phycol. 5: 271–284.

Hurtado-Ponce, A. Q., 1993. Growth rate of *Gracilariopsis heteroclada* (Zhang et Xia) Zhang et Xia (Rhodophyta) in floating cages as influenced by *Lates calcarifer* Blöch. In: Calumpong, H. P. & E. G. Meñez (eds) Proceedings of the 2nd RP-USA Phycology Symposium/Workshop, Suppl.: 13–22.

Kuschel, F. A. & A. H. Buschmann, 1991. Abundance, effects and management of epiphytism in intertidal cultures of *Gracilaria* (Rhodophyta) in Southern Chile. Aquaculture 92: 7–19.

Lapointe, B. E. & J. H. Ryther, 1978. Some aspects of the growth and yield of *Gracilaria tikvahiae* in culture. Aquaculture 15: 185–193.

Lignell, A., P. Ekman & M. Pedersén, 1987. Cultivation technique for marine seaweeds allowing controlled and optimized conditions in the laboratory and on a pilotscale. Bot. mar. 30: 417–424.

Lindsay, J. G. & R. G. Saunders, 1980. Enclosed floating culture of marine plants. In: Abbott, I., M. Foster & L. Eklund (eds), Pacific Seaweed Aquaculture. California Sea Grant College Program, La Jolla, California: 106–114.

McLachlan, J. & C. J. Bird. 1986. *Gracilaria* (Gigartinales, Rhodophyta) and productivity. Aquat. Bot. 26: 27–49.

Nelson, S., R. Tsutsui & B. Best, 1980. A preliminary evaluation of the mariculture potential of *Gracilaria* (Rhodophyta) in Micronesia: growth and ammonium uptake. In: Abbott, I., M. Foster & L. Eklund (eds), Pacific Seaweed Aquaculture. California Sea Grant College Program, La Jolla, California: 72–79.

Primavera, J. H., 1993. A critical review of shrimp pond culture in the Philippines. Rev. Fish. Sci.: 151–201.

Rueness, J., H. A. Mathisen & T. Tananger, 1987. Culture and field observations on *Gracilaria verrucosa* (Huds.) Papenf. (Rhodophyta) from Norway. Bot. mar. 30: 267–276.

Santelices, B. & M. S. Doty, 1989. A review of *Gracilaria* farming. Aquaculture 78: 95–133.

Sokal, R. & F. Rohlf, 1987. Introduction to Biostatistics, Second Edition. W.H. Freeman & Company, New York, 363 pp.

Sumagaysay, N. S., Y. N. Chiu-Chern, V. J. Estilo & M.A.S. Sastrillo, 1990. Increasing milkfish (*Chanos chanos*) yields in brackishwater ponds through increased stocking rates and supplementary feeding. Asian Fish. 3: 251–256.

Upwelling and fish-factory waste as nitrogen sources for suspended cultivation of *Gracilaria gracilis* in Saldanha Bay, South Africa

R. J. Anderson[1], A. J. Smit[2] & G. J. Levitt[1]
[1]*Sea Fisheries Research Institute, Private Bag X2, Roggebaai 8012, South Africa*
E-mail: anderson@botzoo.uct.ac.za
[2]*Department of Botany, University of Cape Town Rondebosch 7700, South Africa*

Key words: nitrogen, *Gracilaria gracilis*, cultivation, stable isotopes, $\delta^{15}N$

Abstract

In Small Bay, Saldanha, the water becomes highly stratified in summer. The cold bottom layer (of upwelling origin) is rich in nitrogen, some of which enters the surface layer by advection. However, the surface water often becomes warm and oligotrophic leading to poor growth or death of *Gracilaria gracilis* grown in experimental suspended systems. At the same time, large quantities of nitrogen-rich fish waste are released at a particular site in the bay. We tested the hypothesis that *Gracilaria* grown close to the site of waste release (1.5 km away and in the waste plume) would grow faster than at the control site 3.5 km away. In October and November (early summer) 1996, all the *Gracilaria* at the control site died, while growth at the fish waste site was good (between 8 and 10% day^{-1}). In November–December control plants grew slightly faster than those from the waste site, in February the reverse occurred, and subsequently (March–June) growth was similar at both sites as winter winds caused mixing of the water column. These results, and analyses of the C/N ratios of the *Gracilaria* tissues provide some support for our hypothesis. Also, analyses of the stable N isotope ratios in the *Gracilaria* tissues indicate that there is considerable uptake of the fish-waste N even at the control site. We conclude that while proximity to the waste site may sometimes benefit the *Gracilaria*, the fish waste would in fact provide a significant source of N for seaweed cultivated throughout the northern area of Small Bay, particularly when the water is highly stratified in summer.

Introduction

Eutrophication of marine waters, particularly bays, is a well-known phenomenon which leads to various environmental problems (Vollenweider, 1992). The Saldanha–Langebaan bay system (Figure 1) is unique in being the only deep, large embayment in the otherwise very exposed west coast of South Africa. It is also the site of numerous and sometimes conflicting human activities. While the shallow southern portion (Langebaan Lagoon) is a nature reserve, the deeper Saldanha Bay is divided into Small and Big Bays by a quay which is used for ore and oil-loading. Small Bay is used as a fishing harbour, for mariculture (it is the centre of mussel cultivation in SA) and various recreational activities. Small Bay, where most of the these activities occur, also receives about 650 tons of nitrogen annually over and above natural fluxes, in the form of liquid fish-processing waste that is discharged from two factories on its west shore (Figure 1), and which has measurable effects on benthic macrofauna (Christie & Moldan, 1977). Fish waste from the larger (pelagic processing) operation was also implicated in causing a problem bloom of the opportunistic green alga *Ulva lactuca* which reduced the benthic *Gracilaria* population and fouled commercial beach-casts of this economic agarophyte in 1993/94 (Anderson et al., 1996b). The waste is discharged at about 8 m depth, and most of it rises to the surface, where it is distributed in the surface water by the wind, often leaving a visible oily slick. In summer prevailing southerly winds drive the plume northwards to the beach, and then eastwards with the circulating current (Anderson et al., 1996b; Weeks et al., 1991). Subsequently, the use of stable isotope ratios of nitrogen confirmed that this pelagic fish-waste was the source of the nitrogen that had fed the bloom (Monteiro et al., 1997). Stable isotope ratios of carbon ($^{13}C/^{12}C$) and nitrogen

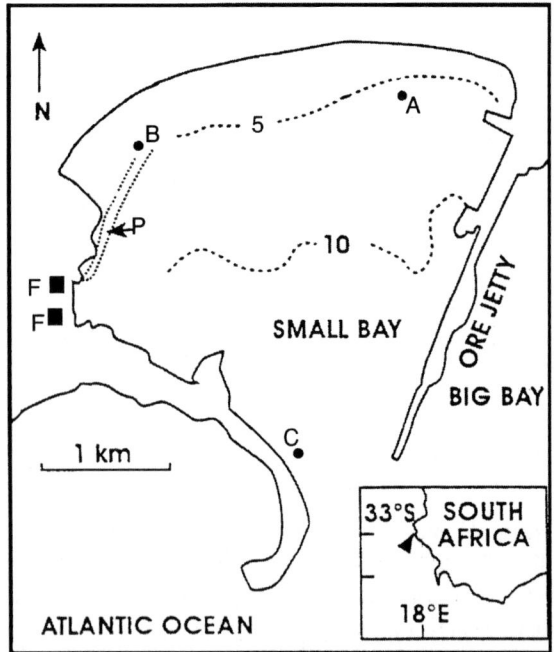

Figure 1. Map of Small Bay, Saldanha, showing positions of Sites A, B and C and fish-factories (F). The usual direction of the pelagic fish-waste plume in summer is shown (P). Depth contours (5 and 10 m) shown.

(^{15}N/^{14}N) are an accepted method of following trophic pathways in marine systems (Monteiro et al., 1991; Owens, 1987).

Features of the biogeochemistry of Small Bay that are relevant to the growth of seaweeds are summarized by Anderson at al. (1996b). In winter the water column is well-mixed, uniformly cold (12–14 °C) and nutrient-rich, However, in summer the system becomes strongly stratified, with warm, oligotrophic surface water overlying a cold, nutrient-rich bottom layer which originates from upwelling on the adjacent coast. Under natural conditions, nitrogen in the surface layer becomes depleted as it is taken up by phytoplankton, and it can only be replaced by flux across the thermocline. This is estimated at a low rate of mmol m^{-2} h^{-1} of N (P.M.S. Monteiro, pers com.). Nitrogen levels in the surface water may become too low to measure: under these conditions *Gracilaria* cultivated near the surface becomes bleached, and if the conditions persist for more than a week or two, dies (Anderson et al., 1996a). The thermocline usually varies in depth between about 5 and 10 m, on an approximately 6–7-day cycle, effectively pulsing in a manner that appears to supply sufficient nutrients to most of the benthic *Gracilaria* population (which is concentrated at depths from 3–8 m in the gently sloping north of the bay). However, sometimes in summer the southerly winds are sustained for periods of several weeks, warm oligotrophic water persists at 5–6 m depths, particularly in the NE corner of Small Bay, and even benthic *Gracilaria* dies.

The cultivation of the local species, *Gracilaria gracilis*, has been shown to be technically feasible in Small Bay, using suspended 'rafts' of rope and netting lines, and on an experimental scale yielded growth rates of 4–7% day^{-1}, suggesting a commercial yield of about 36 t dry wt ha^{-1} (Anderson et al., 1996a), similar to that obtained commercially at Luderitz in Namibia (Dawes, 1995). However, Anderson et al. (1996a) showed that over the 5 years of experimental cultivation in the NE corner of Small Bay, growth rates tended to fall very low at some stage each summer, and in 1974, all cultivated material died; these effects were a result of prolonged southerly winds and the presence of oligotrophic water.

There is thus the paradox that, while in one part of Small Bay the surface nutrient levels can in summer sometimes remain low enough to severely reduce *Gracilaria* growth or even kill the seaweed, several kilometres away there is enough nitrogen waste being released to have caused a massive bloom of *Ulva* over an area of 20 ha (Anderson et al., 1996b). Also, the maritime authority at Saldanha has received several requests for water space for the suspended cultivation of *Gracilaria*, in competition with other potential water uses such as harbour expansion, recreational sailing and fishing, and information is needed to assist in prioritizing water uses and deciding where (if at all) *Gracilaria* farming should be allowed in Small Bay. This study therefore aimed to test whether it would be preferable to cultivate *Gracilaria* in the path of the fish-waste plume, both from the point of view of improved growth and the removal of excess nitrogen from the water. Specifically, we tested the hypothesis that in summer, growth in the waste plume would be better than at the NE (control) site 3.5 km from the outfall.

Materials and methods

Growth experiments were conducted during 1996/1997 at Sites A and B in Small Bay, and information was used from part of a previous study at Site C (Figure 1). Site A was used previously for experiments between 1991 and 1995 (Site 1 of Anderson et al., 1996a) and lay in the NE corner of Small Bay in about 5.5 m deep water, about 3 km from the fish-waste outfall. Site B was adjacent to several granite reefs about 1 km N of the pelagic waste outfall, at 5.5 m depth and directly in the pelagic fish waste plume during southerly, summer winds. Data and *Gracilaria* samples from previous experiments (1995/1996) at Site C were also used: this site (Site 2 of Anderson et al., 1996a) lay adjacent to the breakwater, just inside the mouth of Small Bay, about 2 km SE of the outfall.

At each site, *Gracilaria gracilis* was grown on a rope 'raft' comprising a rectangular frame of 15 mm polypropylene rope 20 × 5 m in size, suspended horizontally about 0.4 m below the water surface, between a series of floats and anchors (see Dawes, 1995, or Anderson et al., 1996a, for details of raft construction). The seaweed lines were tied across the 5 m width of the frame, so that they were suspended horizontally below the water surface, 0.75 m apart to avoid abrasion between thalli on adjacent lines. For the experiments at Sites A and B, small tufts of *Gracilaria* were weighed out (to within 2 g of 20 g, on a spring balance) and attached to a 5.5-m length of 8-mm polypropylene rope at 0.15-m intervals using plastic cable ties ($n = 30$). After the growth period (about 1 month) the ropes were removed, the cable ties cut and each numbered tuft re-weighed. In most cases re-stocking was done with material that had just been removed from the raft, but when that was missing or in poor condition new thalli from benthic populations were used. Results are expressed as relative growth rates (RGR) here calculated from the compound interest formula

$$\text{RGR} = (n\sqrt{W_2/W_1} - 1) \times 100,$$

where n = no. of days, W_1 = initial wet wt, W_2 = final wet wt. Although epiphytes were sometimes present on thalli, they were estimated never to make up more than a few percent of the total weight, and were thus ignored.

Data from 1994 and 1995 at both sites A and C were obtained using commercial-style netting lines rather than ropes. In this method the *Gracilaria* was threaded sideways through the mesh of a plastic tube of netting ('netlon'). Between four and five replicate lines were used each month, each stocked at about 400 g f wt m^{-1} and buoyed in the middle with a plastic bottle (see Anderson et al., 1996a, for details).

Stable isotope analysis was used to assess the relative amounts of fish-derived N in *Gracilaria* grown at Site B, near the fish-waste outfall, and at Site A, as well as in thalli grown the previous year at Site C.

Gracilaria samples (three replicates) were collected off the rafts each month when plants were harvested and weighed. For reference purposes, *Gracilaria* was also collected from a relatively unpolluted site in Langebaan lagoon.

The seaweed samples were immediately placed in plastic bags and stored on ice in the dark until transfer to a freezer. For analysis the seaweed was thawed and rinsed to remove microscopic and macroscopic epiphytes and salt, then oven dried at 60 °C and ground to a fine powder using liquid N and a pestle and mortar. The powder was stored in a desiccator until analysis.

Simultaneous ^{13}C/^{12}C and ^{15}N/^{14}N ratios were determined using 0.6 mg samples (three replicates) on a Finnigan MAT 252 isotope ratio mass spectrometer according to the method of Fry et al. (1992). This also provided C/N ratios for the samples. Ratios of ^{15}N/^{14}N are expressed as ‰, and were calculated as follows:

$$\delta^{15}\text{N} = \left[\frac{^{15}\text{N}/^{14}\text{N}_{\text{smp}}}{^{15}\text{N}/^{14}\text{N}_{\text{std}}} - 1\right] \cdot 10^3,$$

where ^{15}N/^{14}N$_{\text{smp}}$ is the isotope ratio of the sample and ^{15}N/^{14}N$_{\text{std}}$ is the isotope ratio of the standard.

δ^{15}N is reported relative to atmospheric nitrogen (Mariotti, 1984). The reference gas was high purity nitrogen (99.995%) calibrated against atmospheric nitrogen. Analytical precision was to within 0.3‰ (1SD). δ^{13}C was calculated using an equation of similar form to that above, but substituting ^{13}C/^{12}C for ^{15}N/^{14}N. The reference gas was high purity carbon dioxide calibrated against Pee Dee Belemnite. Analytical precision was 0.2‰.

Water temperatures were measured continuously on submersible electronic recorders accurate to within 0.1 °C but calibrated to within 0.5 °C. They were installed at each site on the raft at 0.5 m depth. Underwater irradiance was measured several times at each site on the same days, using a Li-Cor Li-193SA Spherical Quantum Sensor and LI-100 datalogger. Because light is generally not limiting and values were similar at all sites, the results are not presented, but only referred to where necessary.

Results

The *Gracilaria* at Site A grew well in September–October, but died in November–December, when growth at Site B was good (Figure 2). It is notable that at Site A most of the benthic *Gracilaria* (at 5.5 m depth) also died during October–November. Between

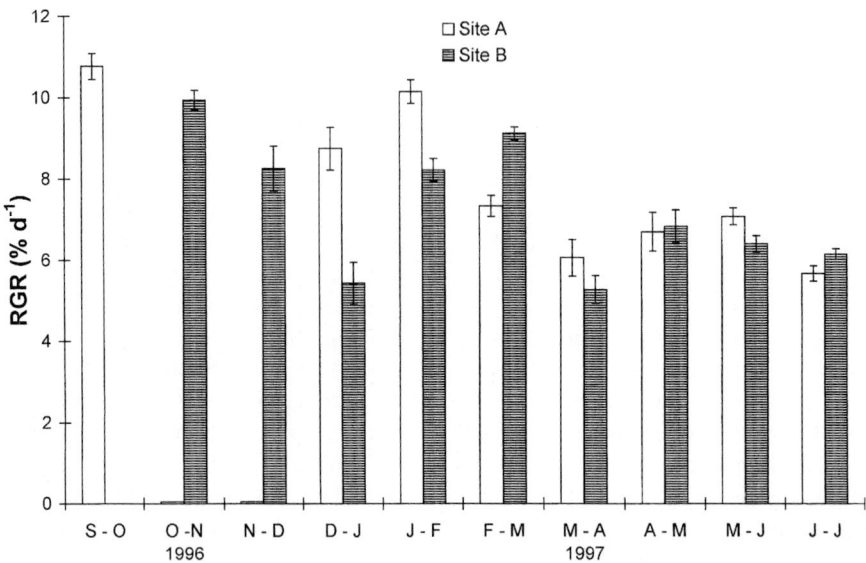

Figure 2. Mean relative growth rates (± 95% confidence limits) of *Gracilaria* tufts grown attached to ropes, at Sites A and B in Small Bay, Saldanha, from mid-September 1996 to mid-July 1997 ($n = 30$).

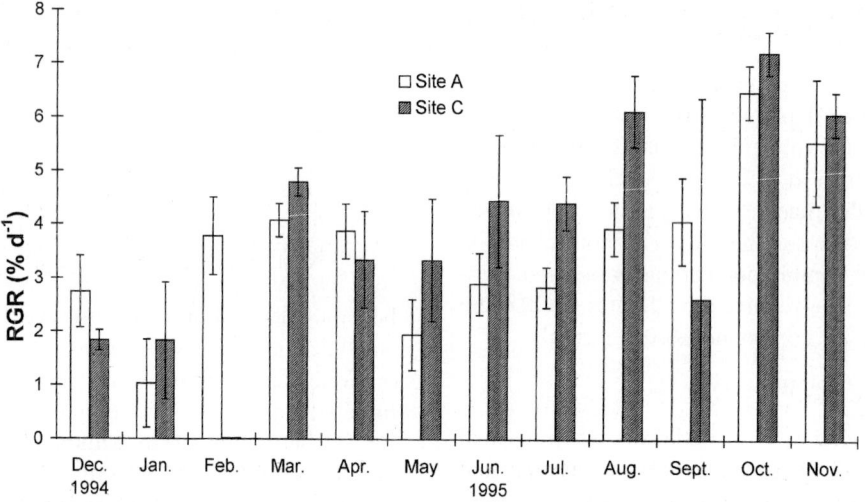

Figure 3. Mean relative growth rates (± 95% confidence limits) of *Gracilaria* grown on commercial-type netting line at Sites A and C in Small Bay, Saldanha, from December 1994 to November 1995 ($n = 4$–5).

mid-December and February the *Gracilaria* grew better (RGR => 8% day^{-1}) at Site A than Site B and in February–March it grew better at Site B. Subsequent growth rates were similar at the two sites, in autumn and winter.

In December 1994 and January 1995 growth was poor at both sites A and C (Figure 3). In February it improved at A, while the *Gracilaria* at C died. In March, growth at C was better. Subsequently, in winter (May–August), growth was similar at the two sites, except in July and August, when it was higher at Site C, near the breakwater. In September, October and November, RGR values were similar at the two sites. Relative growth rates of *Gracilaria* on the commercial-style netlon lines during 1994/1995 (Figure 3) cannot be compared directly with RGR on ropes (Figure 2), because in the former the commercial stocking weight of 400 g m^{-1} results in high net yields but relatively low RGR, while on the ropes the stocking weight is very low (less than 100 g m^{-1}),

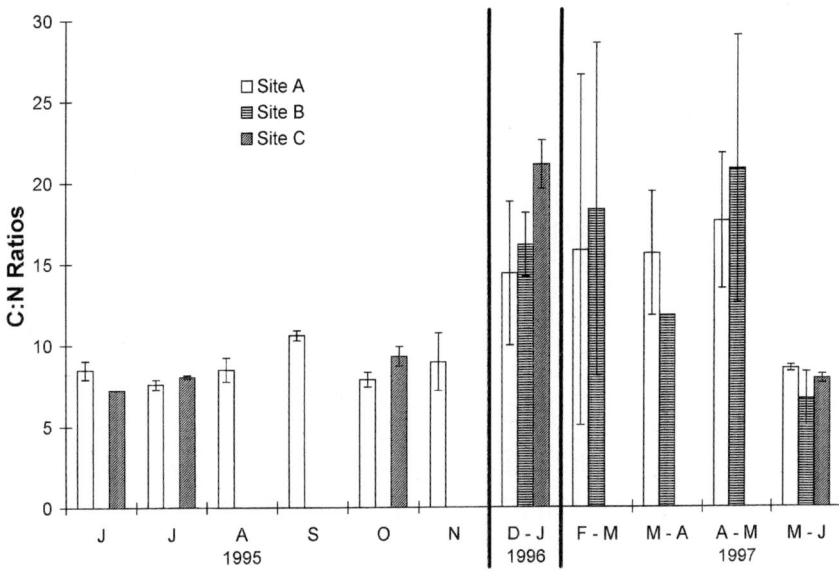

Figure 4. C/N ratios (± 95% confidence limits) of *Gracilaria* tissue cultivated at three sites in Small Bay, Saldanha, at various times (n = 3).

Table 1. δ^{15}N values in *Gracilaria* cultivated at Sites A, B and C at various times (95% confidence limits of means shown, n = 3 in all cases)

Site	Delta ^{15}N values (‰)			
	Oct '95	Dec '96–Jan '97	Feb–Mar '97	Apr–May '97
A	10.4 ± 0.7	11.4 ± 0.5	13.0 ± 0.2	11.2 ± 0.1
B		8.6 ± 0.7	11.4 ± 0.2	10.0 ± 0.3
C	8.5 ± 0.3	8.7 ± 0.4		

resulting in high RGR but low net yields (Anderson et al., 1996a).

Ratios of C/N at sites A and C (Figure 4) remained low from the start of measurements in June 1995 until November (Site B was not yet in use). In October and November 1996 all the *Gracilaria* at Site A had died, and only bleached fragments remained. The results for midsummer (December) 1996 showed raised C/N ratios at all sites, but particularly Site C. In 1997 the C/N ratios at sites A and B were variable but high between February and April–May (no data for Site C). There was no correlation between RGR values and C/N ratios for the whole data-set.

In all the samples measured (Table 1), the δ^{15}N values of *Gracilaria* tissue remained consistently high at Site A, reaching a maximum of 13.0‰ in February 1997. In tissue from Site B, this value was low (8.6) in December 1996, but higher thereafter. The two values for Site C (8.5‰ in October 1995 and 8.7‰ in a sample from a pilot commercial raft in December 1996–January 1997) were among the lowest of all the samples.

Surface (0.5 m) water temperature at Sites A and B showed a typical seasonal pattern for Saldanha Bay, with rising averages in summer, and frequent sudden drops, but more stable low temperatures in winter (May–July). Temperatures at the two sites were essentially similar (Figure 5).

Discussion

In September–October 1996, the RGR of *Gracilaria* at Site A was high because daylength was increasing, water temperature increased from 14 to 16 °C, and nutrient levels were likely to still be high after the winter mixing and before the thermocline had been present for long (as discussed by Anderson et al., 1996a). The subsequent death of all plants at Site A (and most of the benthic population) in October–November and November–December is difficult to explain. The material planted at Sites A and B came from the same benthic population. Temperatures at both sites averaged only about 16 °C, and only briefly exceeded 19 °C, and natural nutrients should have been adequate to sustain growth, especially this early in summer. However, at Site A, the completely bleached appearance of the remaining fragments of plants, and of the

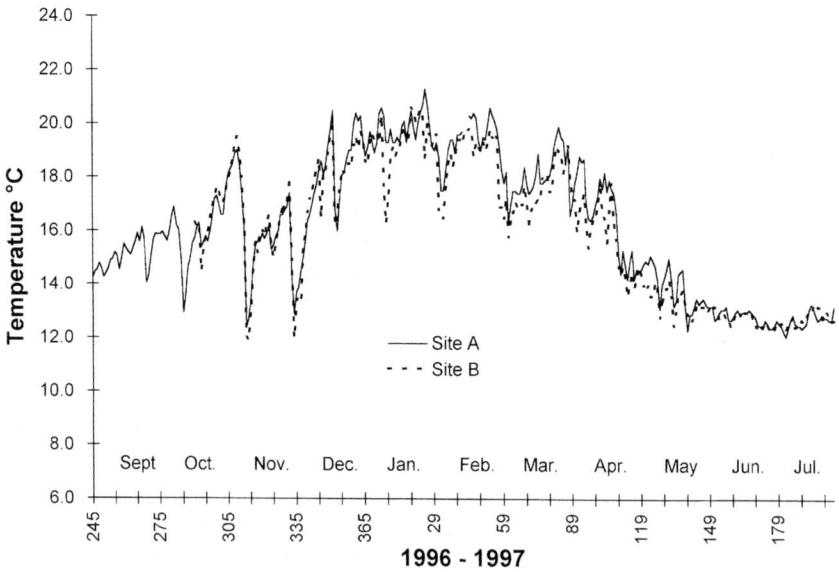

Figure 5. Daily mean water temperatures at Sites A, B and C, Small Bay, from September 1996 to July 1997, at 0.5 m depth.

benthic population, strongly indicated nutrient starvation, although the possibility of a localized disease cannot be ruled out. The important result is that despite almost identical water temperatures (and by inference natural nutrient levels) at Sites A and B, the plants at B in the fish waste plume grew well, providing strong evidence that catastrophic summer die-offs may be avoided by siting close to the fish-waste source. Unfortunately samples were not retained for C/N or $\delta^{15}N$ analysis.

In December 1996–January 1997 RGR of *Gracilaria* at Site A was fairly high ($>8\%$ day^{-1}), perhaps as a consequence of an improvement in nutrient levels, while at Site B the RGR fell to less than 6% day^{-1}, although temperatures at the two sites remained similar. The C/N values from tissue at both sites were moderately high, indicating low ambient N. However, the explanation for improved growth rates at Site A can be found in the $\delta^{15}N$ value of tissue from the two sites. Monteiro et al. (1997) showed that in *Ulva* grown in this waste plume in Small Bay, $\delta^{15}N$ values ranged from 8.5 to 13.9, with an average around 10.7‰, while control plants from relatively pollution-free Langebaan Lagoon had values ranging from about 8.0–9.6, with an average of 8.9‰. Values of anchovy muscle protein (the bulk of the fish waste) range from about 12 to 13.4 (a mean of 12.9‰ – Sholto-Douglas et al., 1991), and those of natural oceanic nitrate-N from 6 to 8‰ (Monteiro et al., 1997; Sealy et al., 1987). In December–January the $\delta^{15}N$ value of tissue of the *Gracilaria* from Site A was 11.4‰, indicating substantial fish-N input, while that from Site B was only 8.6‰, indicating minimal uptake of fish nitrogen. While it was impossible to follow daily patterns of wind and water flow, this provides evidence that even close proximity to the fish-waste is no guarantee that N from this source will reach the plants. It is possible that either waste was carried to Site A, or that considerable re-mineralization of sediment-N was occurring there. In a separate study, Smit (unpubl.) discusses the potential for re-mineralization of fish-waste N, which he shows to be distributed throughout bottom sediments in the northern area of Small Bay.

The RGRs at the two sites increased and the difference between them narrowed during January–February 1997, despite water temperatures being the highest of that summer, and natural nutrient levels by inference the lowest. In February–March the RGRs remained fairly high although Site B then performed better than Site A, despite similar water temperatures. Over the whole January–March period we would have expected low RGRs on account of low ambient nutrients (as indicated by high water temperatures), in keeping with the known biogeochemistry of Small Bay. By the end of summer nutrient levels should have been very low, since there is little mixing across the thermocline (Anderson et al., 1996a). However, tissue $\delta^{15}N$ values were high from both sites, indicating significant uptake of fish-waste N, possibly explaining the relatively good growth rates. The difference between

December–January and February–March was particularly marked at Site B (from 8.6 to 11.4‰, indicating a marked change in N-source over that period.

In March–April growth rates were relatively low at both sites, and C/N ratios moderately high, indicating some nutrient stress. Low growth was typical of this time of year (autumn) in the experiments carried out by Anderson et al. (1996a) and was ascribed to variable and weak winds, little water movement, and low ambient nitrogen levels.

In April–May, RGRs increased somewhat at both sites, to levels that were however still low relative to summer (about 6% day^{-1} compared to >8% day^{-1}). Water temperatures began dropping, indicating mixing of the water column and a possible increase in nutrient levels. However, C/N ratios were high, suggesting N-limitation, and δ^{15}N values dropped at both sites, indicating uptake of less fish-waste derived N. From May–July RGRs were steady at just over 6% day^{-1} at both sites, and water temperatures were about 13 °C, typical of winter mixed conditions, when high nutrient levels are present, for example typically 5–10 μM of NO$_3$–N (Anderson et al., 1996a). The low C/N ratios in May–June 1997 are similar to those obtained in previous samples from winter 1995 (Figure 4), and indicate high tissue N levels and a plentiful supply of N relative to growth rates.

Because we used a different cultivation method in 1994–1995, RGR values in Figure 4 cannot be compared directly to those in Figure 3. The former (commercial) method gives lower RGR values because a much higher seeding weight of *Gracilaria* was used. The relatively poor growth of *Gracilaria* at Sites A and C in December 1994 and January 1995 (compared to later in the year) was the result of the prolonged presence of warm oligotrophic surface water (see Anderson et al., 1996a). Weak southerly winds prevailed at the time, and there would have been little movement of fish-waste N to Site A and none to Site C. In February 1995 the *Gracilaria* at Site C died of apparent nutrient starvation, leaving only bleached thalli. This site is upwind of the waste source at this time of year, and would therefore have received no additional nitrogen. Site A was shown to be capable of receiving fish-waste nitrogen by the δ^{15}N ratios of the *Gracilaria* (see above). Anderson et al. (1996a) attributed better growth of *Gracilaria* at Site C than Site A in winter to the greater wind fetch (with winter northerlies), causing more wind chop and thallus movement and increasing nutrient uptake rates at the former site. However, that the northerly winds may also drive fish-waste to site C in winter is a possibility, although we have no δ^{15}N evidence to support this.

The sustained low C/N ratios in *Gracilaria* cultivated at Sites A and C from June–November 1995 indicate high N levels in the tissue (and hence in the water) relative to the growth rates of the plants. This was expected, because the water column is well-mixed and nutrient-rich in winter (see above). Although thermal stratification sets in between September and November, nutrient concentrations in the surface layer generally decline slowly over this period as they are taken up by phytoplankton (Anderson et al., 1996a; Monteiro & Brundrit, 1990). The high C/N ratios for December 1996–January 1997 indicate low tissue N levels, suggesting low N in the water at all sites, but especially Site C (the sample comes from a commercial experiment which was subsequently abandoned due to low growth rates and mussel spat settlement). The low δ^{15}N value (8.7 ‰) in December–January at Site C supports the idea of fish waste being unavailable there (south of the waste outfall) in summer.

Although we expected that the growth rates of *Gracilaria* would be higher at Site B throughout the summer and that δ^{15}N values from Site B would indicate uptake of far more fish-waste N than at Site A, this was not always the case. Except for the 2 months when cultivated *Gracilaria* died at Site A but grew well at Site B, there was no consistent difference in RGRs or in tissue C/N ratios between the two sites. There are several possible reasons. Water circulation patterns in Small Bay are not simple, and it is appears from our results that fish-waste N may be transported anywhere in the Bay at any particular time. More important, this waste clearly is spread widely enough to affect growth rates of cultivated *Gracilaria* throughout the north of the bay in summer, and even at its mouth in winter. Furthermore, Smit (unpubl.) found that fish-derived N is abundant in sediments throughout the north of Small Bay, and points out the likelihood of re-mineralization of N from these sediments, although it is not known how fast this N could enter the water column. However, re-mineralization and advection of sedimentary fish-N could explain why the δ^{15}N values of *Gracilaria* from Site A, for the four periods we measured them, were consistently higher than from Site B. Also, assumptions of nutrient concentrations from water temperatures should be treated with caution, as this study has shown that fish-waste nitrogen is widely distributed, and can strongly affect the growth of suspended seaweeds.

It is possible to estimate the overall amount of nitrogen that extensive *Gracilaria* farming in the bay would remove. Assuming a nitrogen content of 3.1% of dry weight (value from Smit, unpubl.), and a commercial net yield of 36 t ha^{-1} (from Anderson et al., 1996a), this gives a total nitrogen removal rate of 1.15 t ha^{-1} year^{-1}. In a commercial farm of 40 ha, 44.6 t of nitrogen would be removed annually, which is only about 7% of the 650 t introduced by the fish-factories. Substantial removal of nitrogen would therefore only be achieved by extensive *Gracilaria* farming.

Acknowledgements

We thank the Director, Sea Fisheries Research Institute, for support. A. J. Smit was supported by a Ph.D. bursary from the Foundation for Research Development's SANCOR Sea and Coast Programme. Thanks are due to Derek Kemp and Chris Boothroyd for their cheerful and competent technical support in the lab and in the water, and to Andre Share for assistance with some of the field work. John Bolton provided constructive criticisms of the manuscript. We are grateful to PORTNET for permission to deploy the cultivation rafts, and to Taurus Products for providing the materials for a raft.

References

Anderson, R. J., G. J. Levitt & A. Share, 1996a. Experimental investigations for the mariculture of *Gracilaria* in Saldanha Bay, South Africa. J. appl. Phycol. 8: 421–430.

Anderson, R. J., P. M. S. Monteiro & G. J. Levitt, 1996b. The effect of localised eutrophication on competition between *Ulva lactuca* and a commercial resource of *Gracilaria verrucosa* (Gracilariaceae, Rhodophyta). Hydrobiologia 326/327: 291–296.

Christie, N. D. & A. G. S. Moldan, 1977. Effects of fish factory effluent on benthic macrofauna of Saldanha Bay. Mar. Pollut. Bull. 8: 41–45.

Dawes, C. P., 1995. Suspended cultivation of *Gracilaria* in the sea. J. appl. Phycol. 7: 303–313.

Fry, B., W. Brand, F. J. Mersch, K. Tholke, & R. Garritt, 1992. Automated analysis system for coupled δ^{13}C and δ^{15}N measurements. Anal. Chem. 64: 288–291.

Mariotti, A., 1984. Natural ^{15}N abundance measurements and atmospheric nitrogen standard calibration. Nature 311: 25–252.

Monteiro, P. M. S. & G. B. Brundrit, 1990. Interannual chlorophyll variability in South Africa's Saldanha Bay system, 1974–1979. S. Afr. J. mar. Sci. 9: 281–287.

Monteiro, P. M. S., A. G. James, A. D. Sholto-Douglas & J. G. Field, 1991. The δ^{13}C trophic position isotope spectrum as a tool to define and quantify carbon pathways in marine food webs. Mar. Ecol. Prog. Ser. 78: 33–40.

Monteiro, P. M. S., R. J. Anderson & S. Woodbourne, 1997. δ^{15}N as a tool to demonstrate the contribution of fish waste-derived nitrogen to an *Ulva* bloom in Saldanha Bay, South Africa. S afr. J. mar. Sci. 18: 1–9.

Owens, N. J. P., 1987. Natural variations in ^{15}N in the marine environment. Adv. mar. Biol., 24: 389–451.

Sealy, J. C., N. J. van der Merwe, J. A. Lee Thorp & J. L. Lanham, 1987. Nitrogen isotopic ecology in southern Africa: implications for environmental and dietary tracing. Geochim. Cosmochim. Acta 51: 2707–2717.

Sholto-Douglas, A. D., J. G. Field, A. G. James & N. G. van der Merwe, 1991. 13C/12C and 15N/14N isotope ratios in the southern Benguela ecosystem: indicators of food web relationships among different size classes of plankton and pelagic fish; differences between fish muscle and bone collagen tissues. Mar. Ecol. Prog. Ser. 78: 23–31.

Vollenweider, R. A., 1992. Coastal marine eutrophication: principles and control. In Vollenweider, R. A., R. Marchetti & R. Viviani (eds), Marine Coastal Eutrophication. Elsevier, Amsterdam: 1–20.

Weeks, S. J., A. J. Boyd, P. M. S. Monteiro & G. B. Brundrit, 1991. The currents and circulation in Saldanha bay after 1975 deduced from historical measurements of drogues. S. afr. J. mar. Sci., 11: 525–535.

Outplanting of laboratory-generated carposporelings of *Gracilariopsis bailinae* off northern Philippines

Susan F. Rabanal[1] & Rhodora V. Azanza
Marine Science Institute, University of the Philippines, 1101 Diliman, Quezon City, Philippines
[1]*Present address: College of Arts and Sciences, Cagayan State University 3515 Maura, Aparri, Cagayan, Northern Luzon, Philippines*

Key words: Gracilariopsis, agarophyte, mariculture, outplanting, carpospores

Abstract

For the first time with *Gracilariopsis bailinae* Zhang *et* Xia, outplanting of laboratory-generated sporelings was undertaken. Young sporelings of the species were planted 1.0 and 2.0 m below the lowest tide level, using the monoline method (fixed to the bottom), off Amunitan, Gonzaga, Cagayan, northern Philippines from February to March 1996. After six weeks of culture at 1.0 m depth, sporelings from which the apices had been removed grew significantly faster than did intact sporelings or sporelings at 2.0 m depth.

The maximum growth rate (9.7% d^{-1}) was obtained for sporelings with cut apices grown at 1.0 m below the lowest tide level during the third week of culture period and the lowest (2.6% d^{-1}) was attained for uncut sporelings cultured 2.0 m below the lowest tide level during the first week.

Introduction

Different culture techniques have been continually developed in an attempt to meet the ever-increasing demand of the red seaweed *Gracilaria* as industrial raw materials. Bottom planting, line or net farming and floating rafts using vegetative cuttings have been tested for several species in the marine environment. *Gracilaria* could also be farmed in ponds, raceways and tanks.

In bottom farming, vegetative cuttings or fragments of *Gracilaria* may be tied to rocks, shells or polyethylene tubes, and then farmed in the open waters (Pizarro & Barrales, 1986). Portions of *Gracilaria* fragments may also be buried in the intertidal mudflats (Luxton, 1981). In line and net farming, the plants with desired length or wet weights are tied to a rope and may be enclosed with a net (Hurtado-Ponce, 1990). With floating rafts, cuttings or sporelings may be suspended horizontally or vertically by a bamboo raft (Li et al., 1984; Ren et al., 1984; Sijian et al., 1986). Cultivation of *Gracilaria* in ponds (Shang, 1976; Chiang, 1981) also involves vegetative cuttings uniformly broadcast to the bottom rather than being tied to a substratum. In raceways or in tanks, fragments of the plants are also utilized and cultured under different controlled conditions (Hanisak & Ryther, 1986; Salazar, 1996). The spore method as a source of seedstock requires a nursery unit where fertile thalli release their spores with an appropriate seeding material, such as shells or gravel for bottom farming and twine for line farming, where the spores settle and grow for outplanting purposes (Trono, 1994). Good production of *Gracilaria* from vegetative cuttings and spores has been reported in bottom farming (Pizarro & Barrales, 1986) and line farming techniques (Hurtado-Ponce, 1990), but the practice in tank cultivation at the moment appears too expensive for commercial production (Hanisak & Ryther, 1986). *Gracilariopsis bailinae* has been reported to have high growth rates (Hurtado-Ponce, 1990), and a good quality agar has been extracted (Hurtado-Ponce & Umezaki, 1988; Luhan, 1992; Pondevida & Hurtado-Ponce, 1996). There is however, limited production of the species due to a lack of farming technology. Outplanting laboratory-generated sporelings is one culture technology that could be used for the species. This paper presents results of a study along this line.

Materials and methods

Cystocarpic thalli (i.e. with well-developed reproductive structures) of *Gracilariopsis bailinae* were collected from Kulong-Kulong, Zarraga, Iloilo (123° 40′ E; 11° 39′ N). The plants were transported to the laboratory in an ice box. Stones, shells and epiphytes attached to the plants were removed using forceps and fine artist's brush. Spores were released within 24 h in plastic Petri dishes each with 30 ml enriched seawater. The spores were cultured in the Petri dishes for two months under the following conditions: 26 ± 2 °C, 65 μmol photons m^{-2} s^{-1}, 12:12 h light:dark (L:D) photoperiod, seawater enriched with 40 μM NH$_4$Cl, 4.0 μM K$_2$HPO$_4$ and at 34‰ salinity. The carposporelings (2–3 mm in length) were then detached from the Petri dishes and allowed to grow in wide-mouth large jars (2000 ml), with 1500 ml enriched medium under the following conditions: 26 ± 2 °C, 100 μmol photons m^{-2} s^{-1}, 11:13 h L:D photoperiod, seawater enriched with 25 μM NH$_4$Cl, 2.5 μM K$_2$HPO$_4$ and 25‰ salinity. The plants were aerated until they reached 7 cm in length (0.21–0.28 g fresh weight) four months after spore release.

The sporelings were then transported in an ice box to the field for outplanting experiments during the cooler months of the year i.e. February–March, 1996. Experiments lasted for six weeks off Amunitan, Gonzaga, Cagayan (Figure 1, 122° 02′ E; 18° 19′ N) at two depths (1.0 m and 2.0 m). Each sporeling (0.21–0.28 g wet weight) was tied with a plastic straw at 20 cm intervals to a 2.0 m polyethylene rope (4 mm diameter), making 9 sporelings per line. Two sets of plants were prepared, one of intact sporelings and another of sporelings with cut apices (5–6 mm removed). There were 8 lines containing both types of sporelings alternately arranged at 1 m interval in each depth.

The lines were laid perpendicular to the shore and were tied to pegs positioned 1.0 and 2.0 m below the water surface during the lowest tide. Water temperature was measured with an oxygen meter (oxi 91) and salinity with a portable refractometer. Growth rates were determined at weekly intervals using the Droop equation (Lobban & Harrison, 1994):

$$\mu = 100[\ln(N_t/N_o)]/t,$$

where: μ = specific growth rate; N_t= biomass on day t; N_o= initial biomass; t = time in days.

The data were analyzed using analysis of variance (ANOVA) and Duncan's multiple range test. Significant levels were set at 0.05.

Results

Laboratory-generated sporelings were able to survive in both areas, though the growth apparently declined after 2–4 weeks of culture period (Figure 2). One-way analysis of variance revealed that growth differed significantly with depth for sporelings with uncut apices ($p < 0.05$, with an F value of 10.13 at df = 1) and for sporelings with cut apices ($p < 0.01$, with an F value of 29.79 at df = 1) and had no significant difference with time. Duncan's multiple range test (DMRT) showed that growth of sporelings with cut apices and at 1.0 m depth were significantly different to those sporelings with uncut apices cultured in both areas, as well as those with cut apices cultured at 2.0 m depth.

Higher growth rates (5.7–9.7% d^{-1}) was observed in sporelings with cut apices cultured at 1.0 m depth. This was followed by sporelings with uncut apices (2.7–5.1% d^{-1}) cultured at the same depth. At 2.0 m depth, those with cut apices had a growth rate of 2.9–3.9% d^{-1}, while sporelings with uncut apices grew at 2.7–3.5% d^{-1}.

The highest growth rate (9.7% d^{-1}) was obtained for sporelings with cut apices on the third week, cultured at 1.0 m depth, while the lowest growth (2.6% d^{-1}) was obtained in sporelings with uncut apices cultured for one week at 2.0 m depth. The weekly wet weights and the prevailing condition in the area are shown in Table 1 and Table 2, respectively. The Secchi disc depth varied between 0.2 and 0.6 m.

The cultures were epiphytized (*Enteromorpha* and *Ulva*) during the third week culture period and onwards with more prevalence in sporelings cultured at 2.0 m depth.

Discussion

Though the growth rates (2.6–9.7% d^{-1}) of laboratory generated sporelings of *Gracilariopsis bailinae* under field conditions are lower than those obtained in the laboratory (4.5–10.3% d^{-1}) (Rabanal et al., 1997), and in the field using vegetative cuttings (10.5% d^{-1}) (Hurtado-Ponce, 1990) the results in this study are better than those obtained for other *Gracilaria* species grown in the field (Luxton, 1981; Penniman et al., 1986; Rueness et al., 1987).

The sporelings grew robustly and are morphologically similar to the mother plant except for the greater number of branches arising from the base. Superior growth of laboratory-generated sporelings was also

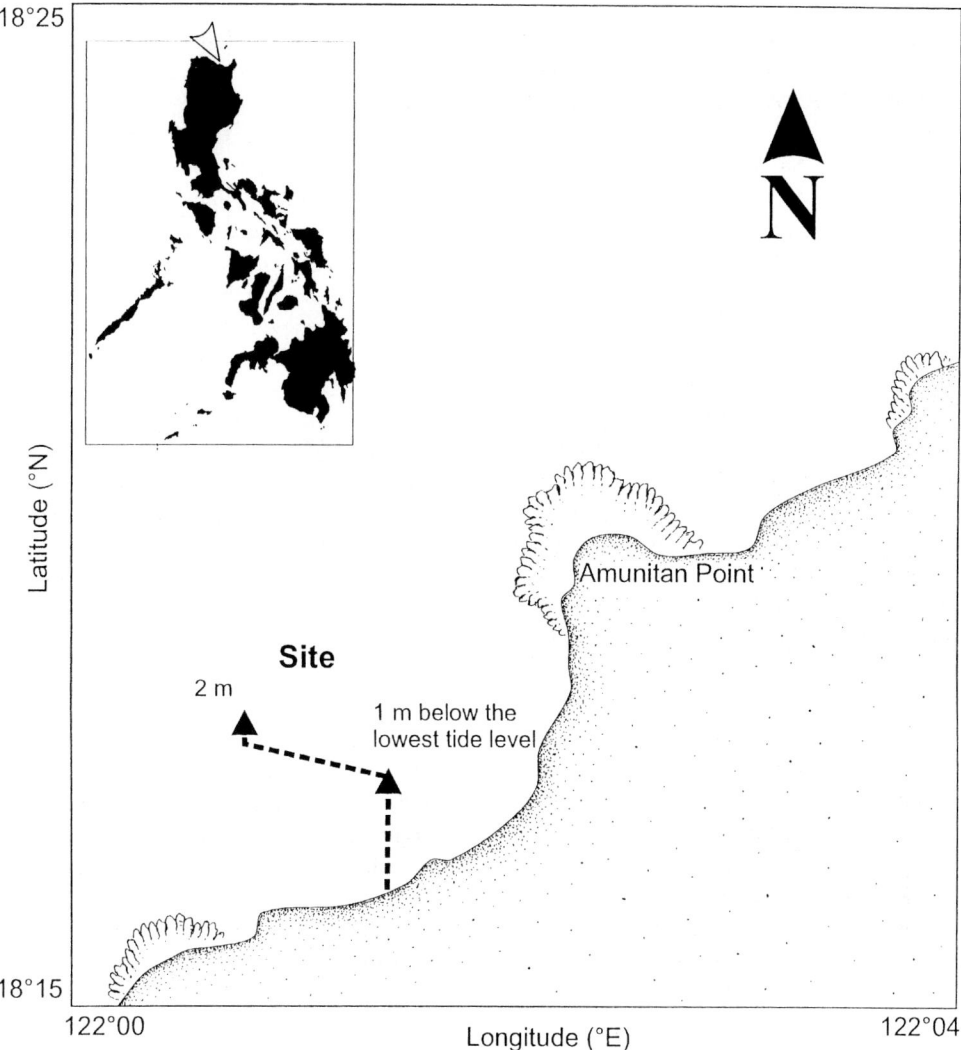

Figure 1. Map of Amunitan, Gonzaga, Cagayan, showing the outgrowing area of the sporelings of *Gracilariopsis bailinae* Zhang *et* Xia

Table 1. Average wet weights of sporelings with cut and uncut apices grown for six weeks at 2 depths (US – sporelings with uncut apices; CS – sporelings with cut apices).

Depth (m)		Average Initial WW	Weeks						Final WW
			1	2	3	4	5	6	
1.0	US	0.245	0.329	0.462	0.721	0.633	0.711	0.746	0.501
	CS	0.218	0.425	0.510	1.700	2.280	2.090	2.670	2.452
2.0	US	0.262	0.314	0.408	0.544	0.605	0.666	0.795	0.533
	CS	0.235	0.294	0.409	0.558	0.553	0.681	0.812	0.577

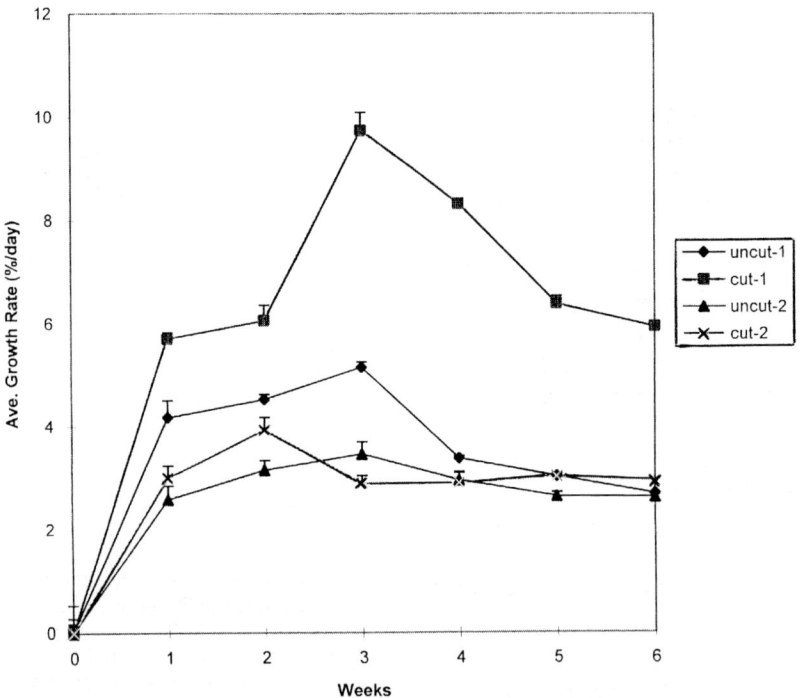

Figure 2. Average daily growth rate of *Gracilariopsis bailinae* Zhang et Xia grown for six weeks at 1.0 and 2.0 m depths off Amunitan, Gonzaga, Cagayan. Bars: SD

Table 2. Temperature and salinity at 1–2 m depth in the study area from February 1996 to January 1997

	Month	Temperature	Salinity
1996	February	24	30
	March	26–27	32–33
	April	30–31	33–34
	May	33–34	35
	June	31	33
	July	30	34
	August	33–34	35
	September	32	34
	October	30–31	32–33
	November	27–28	30–31
	December	27–28	30
1997	January	26–27	30

reported in *Gracilaria* sp. grown under greenhouse conditions (Edelstein, 1977).

The higher growth rates of plants with cut apices confirms the high regenerative capabilities of the genus (Goldstein, 1973) as evidenced by the occurrence of regenerated apices with four to five branches. It has long been observed that it is worth pruning the plants for faster growth, a proven practice of manipulating apical dominance also done for land plants. This difference in growth may be due to the greater surface area for meristematic apical growth compared to the single apical cell in uncut sporelings.

At 1.0 m depth, water movement as a result of tidal changes may have made light and nutrients more available to the sporelings. The periodic exposure of the sporelings to higher irradiance could have killed epiphytic plants that may compete for light and nutrients. Increased water motion ensures nutrient availability and removal of epiphytes (Azanza-Corrales et al., 1996). Availability of light variably affects the growth of different *Gracilaria* species. The high turbidity during the dry season in a Caribbean subtidal sand plain caused a remarkable decrease in *Gracilaria* biomass (Hay & Norris, 1984). Ohmi (1968) on the other hand, observed luxuriant growth of *G. 'verrucosa'* in shallow turbid waters in Tokyo Bay. Cultivation trials, showed relatively higher growth rate at 0.5–1.0 m depth, little growth at 2.0 m and growth limitation below 3 m off Norway (Rueness et al., 1987). Lignell et al. (1987) also described the great influence of light on the growth of *Gracilaria secundata* Harv. with a better growth rate (47% d^{-1}) in higher light and poorer growth with light limitation.

The poor growth of sporelings at 2.0 m depth may be related to epiphytism, grazing and mechanical damage. Growth of the sporelings appeared to have been affected negatively by epiphytism and grazing by siganids. Hurtado-Ponce (1990) in her culture of *Gracilariopsis heteroclada* (now *Gracilariopsis bailinae*) on vertical ropes also stressed the problem of epiphytism. Trono (1994) mentioned the damage done by epiphytes to seaweed growth. Grazing has also been considered to be an important factor in biomass production (Trono, 1988) and in resource management (Brawley & Fei, 1988).

The poor light condition at 2.0 m depth could have serious effects. Hay & Norris (1984) noted remarkable decrease in *Gracilaria* biomass when the water was turbid.

The survival of the sporelings in the area when seawater temperature was 24 °C (Feb) and 26–27 °C (March) is in agreement with the report of Wang et al. (1984) for *Gracilaria verrucosa* in Nanao Island, Guangdong Province, China. Maximum growth was attained from March to April and the plants began to disappear during summer months when seawater temperature was 30 °C or higher.

The salinity in the area (30–32‰) during the culture period favours the growth of the sporelings and is consistent with the findings in other *Gracilaria* species. *G. tenuispitata* var. *liui* (Zhang et Xia) attains optimum growth at 10–20‰ (Chen, 1991) while *G. tikvahiae* achieves best growth at 20–40‰ (Bird et al., 1979). A broad range of tolerance (15–60‰) has been observed in *Gracilaria* species, but extreme salinities could cause necrosis in the plants (Bird & McLachlan, 1986). On the other hand, Azanza-Corrales (1979) demonstrated that *Gracilaria* species grow most rapidly when salinity was higher and water was calmer.

The declining growth of the sporelings as the culture period progressed suggest that the sporelings should be harvested or pruned at least within a month in the field. Longer than this seemed detrimental to the plants as they become susceptible to fouling by epiphytes, grazing, and mechanical damage.

Results of the present investigation indicate that *Gracilariopsis bailinae* sporelings could be used potentially for maricultural purposes. Similar studies will be undertaken for longer period of time covering the warmer months of April–October to have a clearer understanding of the mariculture requirements for the species.

Acknowledgements

The authors wish to extend their gratitude to the Cagayan State University, Aparri, Cagayan and the Philippine Council for Aquatic and Marine Research and Development for the financial support, and to the Marine Science Institute, University of the Philippines for the laboratory equipment. The authors are also grateful to the editor and referees for the critical review.

References

Azanza-Corrales, R. V., 1979. The reproductive biology of *Gracilaria* in Manila Bay. Masters Thesis, College of Science, University of the Philippines, Diliman, Quezon City, 138 pp.

Azanza-Corrales, R. V., T. T. Aliaza, & N. E. Montaño, 1996. Recruitment of *Eucheuma* and *Kappaphycus* on a farm in Tawi-Tawi, Philippines. Hydrobiologia 326/327: 235–244.

Bird, C. J. & J. Mclachlan, 1986. The effect of salinity on distribution of species of *Gracilaria* Grev. (Rhodophyta, Gigartinales): An experimental assessment. Bot. mar. 29: 231–238.

Bird, N. L, L. C. M. Chen & J. Mclachlan, 1979. Effect of temperature, light and salinity on growth in culture of *Chondrus crispus*, *Furcellaria lumbricalis*, *Gracilaria tikvahiae* (Gigartinales, Rhodophyta), and *Fucus serratus* (Fucales, Phaeophyta). Bot. mar. 22: 521–527.

Brawley, S. H. & F. Xiugeng, 1988. Ecological studies of *Gracilaria asiatica* and *Gracilaria lemaneiformis* in Zhanshan Bay, Qingdao. Chin. J. Oceanol. Limnol. 6: 22–34.

Chen, C. S., 1991. The distribution of *Gracilaria tenuistipitata* var. *liu* at the estuary of Tansui River and its casual relation to salinity tolerance and character of substratum. J. Fish. Soc. Taiwan. 18: 89–96.

Chiang, Y. M., 1981. Cultivation of *Gracilaria* (Rhodophycophyta, Gigartinales) in Taiwan. In Levring, T. (ed.), XIth International Seaweed Symposium Proceedings, Walter de Gruyter, Berlin: 569–574.

Edelstein, T., 1977. Studies on *Gracilaria* sp. Experiments on inocula incubated under greenhouse conditions. J. exp. mar. Biol. Ecol. 30: 249–259.

Goldstein, M. E., 1973. Regeneration and vegetative propagation of the agarophyte *Gracilaria*. Bot. mar. 16: 226–228.

Hanisak, M. D. & J. H. Ryther, 1986. The experimental cultivation of the red seaweed *Gracilaria tikvahiae* as an energy 'crop': an overview. In Barclay, W. R. & R. P. McIntosh (eds), Algal Biomass Technologies. Nova Hedwigia Beih. 83. J. Cramer, Berlin: 212–217.

Hay, M. & J. N. Norris, 1984. Cultivation biology of *Gracilaria*: Seasonal reproduction and abundance of six sympatric species of *Gracilaria* Grev. (Gracilariaceae, Rhodophyta) on a Caribbean subtidal sand plain. Hydrobiologia 116/117 (Dev. Hydrobiol. 22): 63–94.

Hurtado-Ponce, A. Q., 1990. Vertical rope cultivation of *Gracilaria* (Rhodophyta) using vegetative fragments. Bot. mar. 33: 477–481.

Hurtado-Ponce, A. Q. & I. Umezaki, 1988. Physical properties of agar gel from *Gracilaria* (Rhodophyta) of the Philippines. Bot. mar. 31: 171–174.

Li, R. Z., R. V. Chong & Z. C. Meng, 1984. A preliminary study of raft cultivation of *Gracilaria verrucosa* and *Gracilaria sjoestedtii*. Hydrobiologia 116/117: 252–254.

Lignell, A., P. Ekman & M. Pedersén, 1987. Cultivation technique for marine seaweeds allowing controlled and optimized condition in the laboratory and on a pilotscale. Bot. mar. 30: 417–424.

Lobban, C. S. & P. J. Harrison, 1994. Seaweed Ecology and Physiology. Cambridge University Press, Cambridge, 366 pp.

Luhan, Ma. R. J., 1992. Agar yield and gel strength of *Gracilaria heteroclada* collected from Iloilo, Central Philippines. Bot. mar. 35: 169–172.

Luxton, D. M., 1981. Experimental harvesting of *Gracilaria* in New Zealand. In Levring, T. (ed.), X1th International Seaweed Symposium Proceedings, Walter de Gruyter, Berlin: 693–698.

Ohmi, H., 1968. A descriptive review of *Gracilaria* from Ghana, West Africa. Bull. Fac. Fish., Hokkaido Univ. 19: 83–86.

Penniman, C. A., A. C. Mathieson & E. C. Penniman, 1986. Reproductive phenology and growth of *Gracilaria tikvahiae* McLachlan (Gigartinales, Rhodophyta) in the Great Bay Estuary, New Hampshire. Bot. mar. 29: 147–154.

Pizarro, A. & H. Barrales, 1986. Field asssessment of two methods for planting the agar-containing seaweed, *Gracilaria*, in northern Chile. Aquaculture 59: 31–43.

Pondevida, H. B. & A. Q. Hurtado-Ponce, 1996. Assessment of some agarophytes from the coastal areas of Iloilo, Philippines. II Seasonal variations in the agar of *Gracilaria changii, G. manilaensis* and *Gracilariopsis bailinae* (Gracilariales, Rhodophyta). Bot. mar. 39: 117–122.

Rabanal, S. F., R. V. Azanza & A. Q. Hurtado-Ponce, 1997. Laboratory manipulation of *Gracilariopsis bailinae* Zhang *et* xia (Gracilariales, Rhodophyta) Bot. mar. 40: 547–556.

Ren, G. Z.., J. C. Wang & M. Q. Chen, 1984. Cultivation of *Gracilaria* by means of low rafts. Hydrobiologia 116/117: 72–76.

Rueness, J., H. A. Mathiesen & T. Tananger, 1987. Culture and field observations on *Gracilaria verrucosa* (Huds.) Papenf. (Rhodophyta) from Norway. Bot. mar. 30: 267–276.

Salazar, M., 1996. Experimental tank cultivation of *Gracilaria* sp. (Gracilariales, Rhodophyta) in Ecuador. Hydrobiologia 326/327: 353–354.

Sijian, L., L. B. Bensong, Z. Shufang, L. Bingsin & J. Zhenying, 1986. The experiment of *Gracilaria tenuistipitata* fixed to sporeling rope with float raft in tidal zone. J. Zhanjiang Fish. Coll. 1: 35–38.

Shang, Y. C., 1976. Economic aspects of *Gracilaria* culture in Taiwan. Aquaculture 8: 1–9.

Trono, G. C, 1988. Production of economically important seaweeds through culture and harvesting of natural stocks. In: Report on the Training Course of Seaweed Farming. ASEAN/SF/88/GEN/6: 59–70.

Trono, G. C., 1994. The mariculture of seaweeds in the tropical Asia-Pacific Region. In Moi, P. S., L.Y. Kun, M. A. Borowitzka & B. A. Whitton (eds), Algal Biotechnology in the Asia-Pacific Region. University of Malaya: 198–210.

Wang, Y. C., G. Y. Pan & L. C. M. Chen, 1984. Studies on agarophytes. 2. Field observations and growth of *Gracilaria* cf.*verrucosa* (Rhodophyta) in Shantou district, Guangdong, PRC. Bot. mar. 27: 265–268.

Review of genetic engineering of *Laminaria japonica* (Laminariales, Phaeophyta) in China

Song Qin, Guo-Qiong Sun, Peng Jiang, Li-Hong Zou, Yun Wu & Cheng-Kui Tseng
Institute of Oceanology, the Chinese Academy of Sciences, Qingdao 266071, China
E-mail: sqin@ms.qdio.ac.cn

Key words: genetic transformation, *Laminaria japonica*

Abstract

Progress has been made in establishing a genetic transformation model for *Laminaria japonica* (Phaeophyta, Laminariales). The model includes introduction of foreign genes by biolistic bombardment, use of promoter SV40 to drive gene expression, algal regeneration by parthenogenesis and selection by chloramphenicol or hygromycin.

Introduction

Genetic transformation of seaweeds was first proposed by Saga (1991), who highlighted potential difficulties in introducing foreign DNA, the vector to express foreign genes, and the selectable marker to screen transgenic plants. Since then, efforts have been made towards establishing transformation models for seaweeds. For example, Cheney & Kurtzman (in litt.) bombarded *uidA*, the gene which encodes β-glucuronidase (GUS), into *Eucheuma* (Rhodophyta) cells and succeeded in obtaining its transient expression by using CaMV35S and NOS promoters. Transient expression of *uidA* in protoplasts of *Porphyra miniata* (Rhodophyta) was carried out by Kuebler et al. (1994) using electroporation and CaMV35S as the promoter. Qin et al. (1994) transferred CaMV35S-*uidA* into explants of *Laminaria japonica* and *Undaria pinnatifida* (Phaeophyta) using biolistic bombardment, and transient expression was observed in rhizoids of *Laminaria japonica* and blades of *U. pinnatifida*. Here we provide a review of the work carried out since then on the genetic transformation model for *L. japonica*.

Significance of transformation studies

From the late 1950s to the early 1960s, Chinese scientists had worked on genetic breeding of kelp (*L. japonica*), and created a few highly productive strains. These new strains were partially responsible for the increase in the annual production of kelp from about 10 t dry weight in 1952 to the current 350 000 t dry weight (Tseng & Qin, 1991). Kelp production, which can be as much as 40 t of dry material per hectare is presently consumed as a subsidiary food and used as a source of low value products, such as iodine, mannitol and alginate.

Transgenic land plants are now being explored as a new source of proteins and drugs (Goddijn & Pen, 1995; Haq, 1995; Moffat, 1995). Genetic engineering is expected to be an effective means to develop kelp as a marine bioreactor to produce oral drugs such as vaccines. The basis for genetic engineering is genetic transformation studies.

Methods to introduce foreign genes

Since there is very little understanding of the genomes of bacteria or viruses associated with seaweeds, only direct modes of transformation including electroporation and microparticle bombardment should be used. Protoplasts, from sporophytes and gametophytes, of *Laminaria* are unable to regenerate (Saga & Sakai, 1984; Qin et al., 1995), and so these cannot be used as hosts. Therefore, since only intact cells can be used as gene recipients, methods such as microparticle bombardment and ultrasonication have to be used to introduce foreign DNA through cell walls.

It has been found that biolistic bombardment, using a Bio-Rad Biolistic PDS-1000/He Particle Delivery System (Bio-Rad Company, Hercules, CA,

U.S.A.), can effectively introduce foreign DNA through cell walls. No GUS or *lacZ*(β-galactosidase) background was detected by histochemical staining. A low background level (about 549.8 ng/mg total protein) of CAT (chloramphenicol acetyltransferase) was detected by ELISA (Qin et al., 1998b). Using this method sporophytes of *L. japonica*, which regenerated by parthenogenesis, exhibited CAT and *lacZ* activity, suggesting that random integration of foreign genes could occur in this process (Qin et al., 1998a, b).

Ultrasonic treatment has been developed as a new method for introducing foreign genes into higher plant cells through their cell walls (Zhang et al., 1991). Results of our study revealed that ultrasonic treatment reduced the size of filamentous female gametophytes and could partially break down cell walls. Observations under a fluorescent microscope showed that cell walls of filamentous female gametophytes of *L. japonica* were partially removed. The positive aspects of this method are lower cost and saving of time, the negative aspects are the lethal effects on gametophyte cells, characterized by the appearance of empty cells, and a decrease in the number of regenerated parthenogenetic sporophytes (Wang et al., 1998a). A method using brown algal viruses for introducing foreign DNA into algal genomes proposed by Henry & Meints (1994) is now being assessed.

Promoter study

A lack of knowledge on promoters which work effectively in macroalgae has affected transformation studies. Until powerful seaweed promoters can be isolated, promoters from land plants have to be used for establishing a model. Both CaMV35S and SV40 promoters have been tested and the former was found to work transiently in rhizoids, but the latter worked in an non-tissue-specific manner (Qin et al., 1998a, b). Results of our experiments showed that SV40 promoter induced stable expression of reporter genes after regeneration of sporophytes.

Regeneration of plants

Protoplasts, single cells and tissues are the main gene recipients in plant genetic engineering. But neither protoplasts nor single cells from sporophytes of *L. japonica* can regenerate into new plants. Tissue culture in Laminariales has been studied extensively (e.g.

Notoya et al., 1992, 1994), but as yet no efficient regeneration system has been obtained. More than two-thirds of the tissues excised from blades, stipes and rhizoids could produce calli after 4 months of culturing in enriched MS medium, but callus only differentiates in ASP12-NTA medium; when PESI medium was used calli appeared within 24 days but no further differentiation was achieved (Wang et al., 1998b). Three types of differentiation from callus have been reported in the Laminariales: formation of sporophytes directly from callus, formation of male and female gametophytes, from which sporophytes are formed by fertilization, and differentiation directly from male and female gametophyte – like filaments without fertilization. From our observations, the regeneration of sporophytes was similar to the latter type of regeneration (Wang et al., 1998b).

In genetic transformation of seaweeds, it is critical that attached bacteria do not produce any false-positive results in the gene expression test. The use of bacteria-inhibited algal materials and the establishment of controls is necessary. In transformation and culture of kelp tissues, pretreatment with 1.5% (w/v) KI produced good sterilization results (Qin et al., 1998b; Wang et al., 1998b). Both blank controls (without bombardment) and negative controls (bombarded with gold particles without DNA) are required since gold particles may introduce bacteria if they are not properly sterilized.

Fang et al. (1978) reported that female gametophytes can develop by parthenogenesis. They thought that after parthenogenesis, diploid sporophytes formed but Lewis et al. (1993) have shown that all parthenogenetic sporophytes were haploid. Female gametophytes could grow vegetatively to form a filamentous clone which can be maintained for prolonged periods in the laboratory and used to store valuable strains. Stimulated by certain conditions, these vegetative clones can develop into parthenogenetic sporophytes. This route can be used for regeneration. In our experiments, filamentous female gametophytes were first ground with glass, then captured onto a 0.2-μm membrane, and finally bombarded using the Bio-Rad Biolistic Particle Delivery System. After a few weeks new plants, with an average length of a few mm, were regenerated. Selection in an antibiotic solution could be done at this stage. Usually haploid sporophytes showed an abnormal form of thallus but, if put into specially designed containers and cultured in the sea, they matured and only female gametophytes resulted. Diploid sporophytes generate as if female gametophytes are hybridized with male gametophytes.

Screening of transgenic plants

Antibiotics and herbicides are used in genetic engineering of land crops as selectable markers. We tested the sensitivity of *L. japonica* to eight antibiotics, including lincomycin, ampicillin, streptomycin, kanamycin, neomycin, chloramphenicol (Qin et al., 1998b), hygromycin (Li, 1998), zeocin and G-418. Our result showed that *L. japonica* is sensitive only to chloramphenicol and hygromycin (Qin et al., 1998b; Li, 1998). The LD_{50} of hygromycin to parthenogenetic sporophytes was much lower than that of chloramphenicol, and was not correlated with algal length while that of chloramphenicol was (Li, 1998).

Using the biolistic particle delivery system, the SV40 promoter as transcription initiator, and female gametophytes as the gene recipients, both *cat* and *lacZ* were expressed stably in regenerated parthenogenetic sporophytes (Qin et al., 1998a, b). These results suggest that random integration of foreign genes might occur, since 2 months after transformation with pCAT-control, followed by selection in chloramphenicol-containing medium, about 0.7% of the regenerated thalli survived, while all controls died (Qin et al., 1998b). Seven months after transformation, about 6% of regenerated plants showed *lacZ* expression (% area, 0.25–10%) (Qin et al., 1998a).

In *lacZ* transformation, about 1×10^6 female gametophytes were bombarded with pSV-*lacZ* plasmid, and after about 80 days among the 435 regenerated plants (algal length including rhizoid, stipe and frond, 2–3 mm) tested for *lacZ* expression, seven plants exhibited blue spots. There were on average three to four blue spots in the meristematic zone or stipe of each plant. No blue spots were found in the 520 regenerated controls (Wu, 1996). In another experiment, the same (1×10^6) number of female gametophytes was bombarded with equivalent amounts of (pSV-*lacZ*) DNA. After 165 days of regeneration via parthenogenesis, 256 regenerated plants (algal length including rhizoid, stipe and frond, 1–3 mm) were tested and eight plants showed *lacZ* expression (% area, 0.25 – 1%). In 286 control plants no *lacZ* activity was observed. The increased number of transformants obtained in this latter experiment may be because egg-releasing gametophytes were used or because histochemical staining was carried out later, i.e. more expressing cells were produced by cell division so that transformants could be more easily found and counted. By using egg-releasing materials rather than stimulating egg release after bombardment, more gametophytes bombarded with the foreign gene regenerated into sporophytes and thus transformation efficiency was increased. The remaining regenerated plants and controls were cultured for a further 45 days, including 1 month in the sea, 30 of the 496 plants exhibited *lacZ* expression (% area, 0.25–10%) while no *lacZ* activity was detected in the 500 controls (Qin et al., 1998a). The increase in both transformants and expressed area after culturing in the sea may be due to a more suitable environment for foreign gene expression and/or more expressing cells appearing after cell division.

In a recent experiment, the same number of egg-releasing gametophytes was bombarded with the same amount of DNA. Three months later, 75.5% of regenerated parthenogenetic sporophytes exhibited *lacZ* expression and no *lacZ* activity was found in any control plant. The expression level is still low and may due to gene integration at a very late stage; the later the integration, the lower the ratio of transformed cells.

In conclusion, by using biolistic bombardment as the method of introduction, the SV40 promoter as a transcription initiator, female gametophytes as gene hosts, parthenogenesis as regeneration route and chloramphenicol (or hygromycin) as a selectable reagent, a transformation model has been established.

Environmental release of transformed algae

Environmental release of transgenic *L. japonica* should be given serious consideration. The vectors and genes used in transformation studies should be restricted in order to avoid any negative phenotype they may encode. In our opinion transgenic *L. japonica* should not cause any negative impact on natural populations or on seed stocks. *Laminaria japonica* was originally introduced from Japan and is now cultured extensively in China. In most parts of the China Sea, it cannot withstand the summer seawater temperature and plants die naturally; there are no 'natural' population of *L. japonica*. To prevent transgenic kelp being incorporated into the marine food chain or carried away by marine currents containers are needed to ensure that the kelp cannot escape or be eaten by marine animals when cultured in the sea. We have designed two types of containers to hold transformed algae in the sea. One type has a volume of about 3.6 l and is made of 0.2-

mm nylon membrane (Qin et al., 1998b); this type of container can be easily fouled but the membrane is solid enough not to be perforated. The second type is made by removing the base of a 1.25-l cola bottle, fixing a 0.2-mm nylon membrane to the bottom, and drilling many holes on the surface of the bottle. The membrane can be changed when it is fouled. To avoid the release of spores through the membrane or holes, it is necessary to harvest the kelp prior to the formation of sporangia and to collect the gametophytes indoors.

Acknowledgements

This project was sponsored by the National Climbing Plan (B6-4-1) from the China Science and Technology Committee, and by National Natural Science Grants (No.39400076 and 39670367) from the National Natural Science Foundation of China.

References

Fang, Z., J. Dai, J. Cui & Y. Ou, 1978. The first record of female sporophytes of *Laminaria japonica*. Chin. Sci. Bull. 23: 43–44.

Goddijn, O. J. M. & J. Pen, 1995. Plants as bioreactors. Trends Biotechnol. 13: 379–387.

Haq, T. A., 1995. Oral immunization with a recombinant bacterial antigen produced in transgenic plants. Science 268: 714–716.

Henry, E. C. & R. H. Meints, 1994. Recombinant viruses as transformation vectors of marine macroalgae. J. appl. Phycol. 6: 247–253.

Kuebler, J. E., S.C. Minocha & A. C. Mathieson, 1994. Transient expression of the GUS reporter gene in protoplasts of *Porphyra miniata* (Rhodophyta). J. mar. Biotechnol. 1: 165–169.

Lewis, R. J., B. Y. Jiang, M. Neushul & X. G. Fei, 1993. Haploid parthenogenetic sporophytes of *Laminaria japonica* (Phaeophyceae). J. Phycol. 29: 363–369.

Li, X., 1998. Studies on genetic transformation of *Laminaria japonica* (Phaeophyta) by using female gametophytes as gene recipients. Master Thesis, Institute of Oceanology, Chinese Academy of Sciences, 49 pp.

Moffat, A. S., 1995. Exploring transgenic plants as a new vaccine source. Science 268: 658–660.

Notoya, M., M. Nagashima & Y. Aruga, 1992. Influence of light intensity and temperature on callus development in young sporophytes of four species of Laminariales (Phaeophyta). Kor. J. Phycol. 7: 101–107.

Notoya, M., M. Nagashima & Y. Aruga, 1994. Influence of light intensity and temperature on callus development in young sporophytes of three species of Laminariales (Phaeophyta). J. mar. Biotechnol. 2: 15–18.

Qin, S., P. Jiang, X. Li & C. K. Tseng, 1998a. The expression of *lacZ* in regenerated sporophytes of parthenogenetic *Laminaria japonica*. Proceedings of the 2nd Asia-Pacific Marine Biotechnology Conference and the 3rd Asia-Pacific Conference on Algal Biotechnology: 205–208.

Qin, S., J. Wu, X. Wang, X. Li, P. Jiang & C. K. Tseng, 1998b. The expression of foreign genes in *Laminaria japonica* (Phaeophyta). In Morton, B. (ed.), The Marine Biology of South China Sea. Hong Kong University Press, Hong Kong: 209–217.

Qin, S., J. Zhang, W. Li, X. Wang, S. Tong, Y. Sun & C. K. Tseng, 1994. Transient expression of GUS gene in phaeophytes using biolistic particle delivery system. Oceanol. Limnol. Sin. 25: 353–356 (in Chinese with English abstract).

Qin, S., X. Wang, S. Tong and C. K. Tseng, 1995. Isolation and culture of female haploid protoplasts from *Laminaria japonica*. Oceanol. Limnol. Sin. 26 (suppl.): 126–129 (in Chinese with English abstract).

Saga, N., 1991. Protoplasts and somatic hybridization-controversal discussion 'No' side. Proceedings of a COST-48 Workshop, Spain: 25–30.

Saga, N. & Y. Sakai, 1984. Isolation of protoplasts from *Laminaria* and *Porphyra*. Bull. jap. Soc. Sci. Fish. 50:1085.

Tseng, C. K. & S. Qin, 1991. Marine algal biotechnology in China: present conditions and prospects. Proceeding of International Symposium on Biotechnology of Salt Ponds, Tianjin: 61–68.

Wang, X., S. Qin, X. Li, P. Jiang, C. K.Tseng & M. Qin, 1998a. Effects of ultrasonic treatment on female gametophytes of *Laminaria japonica* (Phaeophyta). Chin. J. Oceanol. Limnol. 16 (Suppl.): 62–66.

Wang, X., S. Qin, X. Li, P. Jiang, C. K.Tseng & M. Qin, 1998b. High efficient induction of callus and regeneration of sporophytes in *Laminaria japonica* (Phaeophyta). Chin. J. Oceanol. Limnol. 16 (Suppl.): 67–74.

Wu, J., 1996. Study of selectable markers and expression of foreign genes in *Laminaria japonica*, Master Thesis, Institute of Oceanology, Chinese Academy of Sciences, 39 pp. (in Chinese with English abstract).

Zhang, L., L. Cheng, J. Yuan , C. Li, S. Jia, N. Xu & N. Zhao, 1991. Ultrasonic direct gene transfer – the establishment of high efficiency genetic transformation system for tobacco. Sci. Agric. Sin. 24: 83–89 (in Chinese with English abstract).

A new method of *Laminaria japonica* strain selection and sporeling raising by the use of gametophyte clones

Dapeng Li, Zhi-gang Zhou, Haihang Liu & Chaoyuan Wu*
Institute of Oceanology, Chinese Academy of Sciences, 7 Nanhai Rd., Qingdao 266071, P. R. China
E-mail: cywu@ms.qdio.ac.cn

Key words: Laminaria japonica, gametophyte clones, crossing, strain selection, sporeling raising

Abstract

One-celled female and male gametophytes of three *Laminaria japonica* strains were isolated, cultured and gametophyte clones were formed. A technique combining strain selection with sporeling raising by the use of these female and male gametophyte clones was studied. Experiments on 9 different crossing combinations was conducted in November of 1997 in Qingdao, P. R. China. The main economic characteristics, frond length and fresh weight, of sporophytes of different crossing combinations were measured. F_1 sporophytes of No. 2 showed a higher fresh weight and longer length, therefore, No. 2 (Wh860 ♀ × Lid ♂) was selected as a good combination. Its parental female and male gametophyte clones are being mass cultured for sporeling production. By this method, the time needed for strain selection was shortened from 5–6 to 2 years. As compared with the routine method of sporeling raising by the collection of zoospores, the time of sporeling raising of this method decrease by 50%, and the production cost is also reduced by 50%. It is believed that this method will be labour and time saving and a more economic way for strain selection and sporeling raising in *L. japonica* cultivation industry.

Introduction

At present, the total annual production of cultured *Laminaria japonica* has reached 0.5 million tons dry weight in China. In the *L. japonica* cultivation industry, good strains and healthy sporelings are keys to successful cultivation. Almost all of the harvest in China comes from genetically improved strains (Wu & Lin, 1987; Patwary & van der Meer, 1992). Genetic improvement of *L. japonica* was first initiated in China in the late 1950s (Wu & Lin, 1987). At first, Chinese phycologists exploited the selection and inbreeding method to develop two improved strains, Haiqing No. 1 and Haiqing No. 2 (Fang et al., 1962, 1963, 1966). Somewhat later, two additional new varieties Nos 860 and 1170 were bred through the inbreeding and selection method (IOOMF, 1976). These strains contributed substantially to the Chinese *L. japonica* cultivation industry, but they have been losing their economic characteristics in the course of extension to farms. This is due to mixed zoospore collection from thousands of parent plants generation after generation. Moreover, the method by which they are bred takes too much time to meet the demand of production. The routine sporeling raising method of *L. japonica* is to collect zoospores in summer. The gametophytes and the resultant young sporophytes are cultured at 6–10 °C with an irradiance of 60–100 μmol m^{-2}s^{-1} in a greenhouse. After some 3 months culture in the greenhouse, the sporelings are transplanted to open sea (Tseng et al., 1955). Long term cultivation in a greenhouse at low temperature is very expensive and laborious, and the sporelings are inevitably attacked by diseases (Wu et al., 1979). In order to overcome these shortcomings, a new method of strain selection and sporeling raising of *L. japonica* by the use of gametophyte clones has been studied. The key point of this method is to combine strain selection with sporeling raising.

The method involves five steps:

1. Isolation of single female and male gametophytes and induction of gametophyte clones.
2. Crossing of female and male gametophytes of different lines.

---
*Author for correspondence

3. Selection of the resultant sporophytes.
4. Mass culture of the female and male gametophyte clones which give rise to the selected sporophyte;
5. Sporeling raising by the use of the same female and male gametophyte clones.

The main objective of this study is to find ways of *L. japonica* strains selection through crossing and sporeling raising by the use of gametophyte clones.

Materials and methods

Isolation of one-celled gametophytes and culture of gametophyte clones

Blades of three different strains with mature sporangial sori were subject to partial drying at 15 °C for several hours to induce mass discharge of zoospores. The dense zoospore suspension was diluted until several zoospores could be seen per field under microscope (100×), then slides were put into the diluted suspension to allow the zoospores to attach. The attached zoospores were cultured under 15 °C in Provasoli enriched seawater (PES) (Provasoli, 1968) with an irradiance of 15 μmol m^{-2} s^{-1}. Three days later, gametophytes were formed. Female and male gametophytes were picked up by sterilized glass pipette one by one. The isolated gametophytes grew into small, visible clones after a month's culture. These gametophyte clones were then propagated vegetatively.

Crossing of gametophytes and indoor culture of sporelings

Female and male gametophytes of three strains were crossed reciprocally, giving nine combinations (see Table 1). To avoid parthenogenesis, each pair of vegetative male and female gametophyte clone was mixed in the ratio of 5 to 1 (Zhou & Wu, 1998) by weight. The mixed gametophyte clones were cut into several-celled fragments by blender and washed thoroughly with sterilized seawater. Gametophyte fragments were suspended in seawater enriched with NO_3–N and PO_4–P to concentrations of 4×10^{-1} mol m^{-3} and 4×10^{-2} mol m^{-3} respectively, and then poured into 500 ml glass beakers and kept motionless to allow gametophytes to settle down on the palm rope substratum. The cultures were kept in a constant temperature 14 °C water bath at an irradiance of 80 μmol m^{-2} s^{-1} with a light period of L:D 10:14. Culture medium was renewed weekly. When the sporelings grew up to 0.8–1 cm in length, the palm ropes were taken out to the open sea. Selection of sporophytes was carried out afterwards.

Table 1. Crossing numbers and gametophytes used

M F	Wh860	Lid	Rong
Wh860	1	2	3
Lid	4	5	6
Rong	7	8	9

Transplantation of sporelings for open sea cultivation

After 50 days indoor cultivation from November 4 to December 23, 1997, sporelings were transplanted to Taipingjiao for 15 days precultivation to allow the sporelings to grow to 15–20 cm in length. They were detached from the palm ropes. The holdfasts of 15 sporelings of each combination were then twisted into another bigger rope of 1 m in length. The ropes were horizontally hung at 1 m in depth between two rafts.

Measurement of main economic characteristics of sporophytes

After 105 days cultivation in the open sea, sporophytes were harvested. From the economic point of view, frond length and weight of sporophytes are characteristics of importance. These are, therefore, used as strain selection standards. Frond length of each sporophyte was measured from the tip to the transition zone. Fresh weight of the intact plant was also measured immediately after harvest.

Results

Culture of gametophyte clones

Immediately after discharge, zoospores were very active, After 5 min, almost all of them attached onto the slides. Two hours later, the attached zoospores gave rise to germination tubes and the entire cell contents flowed into the apices of the germination tubes. On the third day, a cell wall appeared at the swollen apex of the germination tube. This is the sign of the formation of a gametophyte. Under the condition of L:D 16:8,

17 °C and 60 μmol m^{-2} s^{-1} irradiance, female and male gametophyte clones can propagate vegetatively without gametogenesis.

Culture of sporelings

All the cross-breeding experiments were successful and hybrid sporophytes were produced. The time needed for sporophytes to appear was somewhat different in different combinations. The sporophytes of No. 1 (see table 1) appeared only after 9 days culture, whereas that of No. 5 took 15 days. Sporelings averaged 0.8–1.0 cm after 50 days culture in the laboratory.

Main economic characteristics of sporophytes

From Figure 1, it is clearly seen that there are differences in average frond length and fresh weight among the F$_1$ sporophytes of different combinations. For instance, the average frond length of No. 6 (250 cm) strain is significantly smaller than that of No. 5 (525 cm); the highest average fresh weight was 1300 g (No. 2), twice that of the lowest one of 640 g (No. 6). Since the culture conditions for sporophytes of different combinations were almost the same both in laboratory culture and open sea cultivation, the fact that there were significant differences among combinations and similarities in the same combinations in frond length and fresh weight should be mainly attributable to genetic factors. Similar results were also obtained by the work of one of the authors, Dr Zhou Zhi-Gang, one year ago. This is the base of selection. In the course of cultivation, F$_1$ sporophyte of No. 2 showed more vigorous growth, therefore, No. 2 (Wh860 + × Lid) was selected as a good combination. Its parental female and male gametophytes are being mass cultured for sporeling production.

Discussion

Since the early 1960s, we have bred several strains of *L. japonica* through successive inbreeding and selection. Generally, it takes at least 5 years for the inbred line to approach homozygosity (IOOMF, 1976) that results in relative genetic stability. This time-consuming and laborious method can hardly meet the demand of commercial production. The use of gametophyte clones, which can be propagated vegetatively under controlled conditions as genetically identical 'seeds' (Neushul, 1987) can solve this problem. As

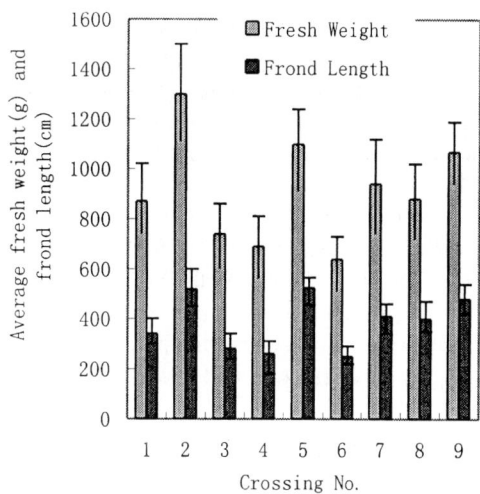

Figure 1. Average frond length and fresh weight of sporophytes of different crossing combinations. Standard errors of the mean (n = 15) measurement are represented by vertical bars.

the results in this paper show, sporophytes originating from the same crossing combination of gametophytes have very similar morphological characteristics. Through screening, the desired combination of female and male gametophytes can be purposely chosen in 2 years; the parental gametophytes can be kept as stock in the laboratory, and mass culture can be made in time of need for sporeling production on large scale. In subsequent years, the same combination of gametophytes could be used again and, therefore, the original characteristics of the sporophytes could be well conserved (Pang et al., 1997).

Sporelings used for commercial cultivation are produced by random fusion of gametes released from unrelated gametophytes derived from thousands of individual parent sporophytes. It was assumed that the cultivated strains, which had not been subjected to any further systematic selection generation after generation, would have a high degree of heterozygosity (Wu & Lin, 1987). This is taken to mean that some of the economic characteristics would be lost after several generations of extension to *Laminaria* farms. The extension work of strains Nos 860 and 1170 we bred in the 1970s proved that this is true. In the case of using gametophyte clones for strain selection and sporeling raising, the resulting F$_1$ sporophytes would have little segregation from the parents and attained more homogeneity.

The routine sporeling raising method of *L. japonica* is to collect zoospores with palm frames by the end of July in north China. It takes some 3 months for the sporelings to grow to 1 cm in length. Since the sporelings are cultured in a greenhouse and the temperature generally maintained at 6–10 °C, the electricity cost is very high. For example, the cost of a greenhouse with a capacity of producing 400 million sporelings per cultivation season will be US$50 000. In the case of culturing sporelings from gametophyte clones, the time needed is cut in half, as is the production cost. This is one of the advantages of this method. Besides, short term cultivation of sporelings in a greenhouse will decrease the possibility of being attacked by malformation and sporeling detachment diseases.

Conclusion

Experimental results showed that *L. japonica* gametophyte clones of different lines could be used for strain selection and sporeling raising. The method has the following advantages:

1. It shortens the time of strain selection from 5–6 to 2 years.
2. It decreases the time of sporeling raising by 50%.
3. It keeps selected strains from continual segregation.
4. The sporelings can be raised any time all the year round.

It is believed that this method will be a labour and time saving and more economic way for strain selection and sporeling raising in *L. japonica* cultivation industry.

Acknowledgements

This paper is contribution No. 3393 from the Institute of Oceanology, Chinese Academy of Sciences, P. R China. The study is supported by the Key Project, National Ninth-Five-Year Plan of China 96-008-01-02-06, and the Project of Bio-Engineering Center of China, SSTC 96-C01-05-01. The authors would like to thank Mr Peng Guang for help in the open sea cultivation.

References

Fang, T. C., B. Y. Jiang & J. J. Li, 1962. On the inheritance of stipe length in Haidai (*Laminaria japonica* Aresch). Acta Bot. Sinica 10: 327–335.

Fang, T. C., B. Y. Jiang & J. J. Li, 1966. The breeding of a long-frond variety of *Laminaria japonica* Aresch. Oceanol. Limnol. Sinica 8: 43–50.

Fang, T. C., C. Y. Wu, B. Y. Jiang, J. J. Li & K. Z. Ren, 1963. The breeding of a new variety of Haidai (*Laminaria japonica* Aresch). Scient. Sinica 12: 1011–1018.

IOOMF (Section of Seaweed Genetics and Breeding, Institute of Oceanology, Academia Sinica, Qingdao & Section of Seaweed Cultivation, Institute of Marine Fisheries, Qingdao. chief author, Wu C.Y.), 1976. The breeding of new varieties of haidai (*Laminaria japonica* Aresch) with high production and high iodine content. Scient. Sinica 19: 243–252.

Neushul, M. & B. W. W. Harger, 1987. Nearshore kelp cultivation yield and genetics. In: Bird, K. T. & P. H. Benson (eds), Seaweed Cultivation for Renewable Resources. Development in Aquaculture and Fisheries Science 16: 69–93.

Pang, S. J., X. Y. Hu, C. Y. Wu, A. Hirosawa & M. Ohno, 1997. Intraspecific crossing of *Undaria pinnatifida* (Harv.) – A possible time-saving way of strain selection. Chin. J. Oceanol. Limnol. 15: 227–235.

Patwary, M. U. & J. P. van der Meer, 1992. Genetics and breeding of cultivated seaweeds. Korean J. Phycol. 7: 281–318.

Provasoli, L., 1968. Media and prospects for the cultivation of marine algae. In Watanabe, A. & A. Hattori, (eds), Cultures and Collection of Algae. Proceeding of the U.S.- Japan Conference. Hakonc. Japanese Society of Plant Physiology: 63–75.

Tseng, C. K., K. Y. Sun & C. Y. Wu, 1955. On the cultivation of Haidai (*Laminaria japonica* Aresch.) by summering young sporophyte at low temperature. Acta Bot. Sinica 4: 255–264.

Wu, C. Y., N. S. Gao, D. C. Chen, B. C. Zhou, P. X. Cai, S. X. Dong, Z. C. Win & R. Y. Cong, 1979. Study on the malformation disease of *Laminaria japonica* sporeling. Oceanol. Limnol. Sinica 10: 238–253.

Wu, C. Y. & G. H. Lin, 1987. Progress in the genetics and breeding of economic seaweeds in China. Hydrobiologia 151/152: 57–61.

Zhou, Z. G. & C. Y. Wu, 1998. Clone culture of *Laminaria japonica* and induction of its sporophytes. Chin. J. Biotech. 14 (1): 99–111.

Development of commercial *Kappaphycus* production in the Line Islands, Central Pacific

David M. Luxton & Patrick M. Luxton
D Luxton & Associates Ltd., 70 Hamurana Road, Omokoroa, Tauranga, New Zealand
E-mail: dlatganz@enternet.co.nz

Key words: Kappaphycus, seaweed production, socio-economics, Kiribati, Pacific Islands

Abstract

Kappaphycus alvarezii (basionym *Eucheuma alvarezii*) was introduced to the Line Island atolls of Kiritimati (Christmas Is.) and Tabuaeran (Fanning Is.) in 1994 as an outer-island development programme in the Republic of Kiribati. Farming sites were selected, and commercial production commenced in September 1994. Production increased to 850 t y^{-1} dry weight in two years, and by 1997 over 420 people were receiving income from seaweed. On Tabuaeran seaweed has now replaced copra as the main source of income for over 70% of all households. The new seaweed-based economy has also ensured the success of the resettlement policy of the Kiribati Government. Continuous monitoring of all suppliers has revealed net incomes for a family unit as high as AUS$ 4687 per annum from a farmed area of 900–1000 m^2. On Kiritimati, a small lagoon sand-flat of 6 hectares has been developed providing income for over 100 households producing 350 t y^{-1}. Women are not only actively involved in, but are frequently the main beneficiaries of, production. The Line Islands production has been significant in raising the total Kiribati harvest to over 1200 t y^{-1} providing an important source of export earnings. The creation of a monopolistic industry and the implementation of a single-desk marketing strategy have made the development economically sustainable and competitive with S.E. Asia. The development represents a model for other isolated atoll communities in the Pacific Ocean where the economy is currently based on copra.

Introduction

Twenty years ago, Doty (1977) expressed the hope that *Eucheuma* farming would be encouraged in several Pacific Island countries, to meet increasing world demand and avoid a future reliance solely on Philippine production. Many trials have subsequently been undertaken by Pacific Island nations, but only in the Republic of Kiribati (Figure 1) and in Fiji, have trials led to commercial production (Luxton et al., 1987).

The Republic of Kiribati illustrates the development problems faced by small coral island countries in the Pacific. It has extremely limited physical resources, small land area (811 km^2), infertile soils and a scarcity of skilled I-Kiribati (indigenous people of Kiribati), as well as having the economic disadvantages of a small home market and considerable distance from other markets. Over 80% of the Kiribati workforce is engaged in subsistence agriculture and fishing, which provide most of the basic needs of the people in the outer-islands. The Republic is dependent on foreign aid to sustain its present standard of living. The country's exclusive economic zone, 3.5 million km^2 of ocean, is the most important natural resource, and the Government's objective (7th National Development Plan 1992–1995) is to further develop marine resources to improve the growth performance of Kiribati.

Kiritimati (Figure 2) became part of Kiribati in 1979 when the new republic gained independence. The population in December 1995 is recorded at 3225 (Statistics Office, 1997a). In 1983, the Kiribati Government purchased the Northern Line Island of Tabuaeran (Figure 3) from its private owner with the aim of using the island for the resettlement of people from the Gilbert Islands 3280 km to the west. At the time Tabuaeran had an I-Kiribati population of 440, but was expected to absorb an additional 3800 voluntary settlers over a ten-year period 1988–98. Due to logistical constraints and the slow development of

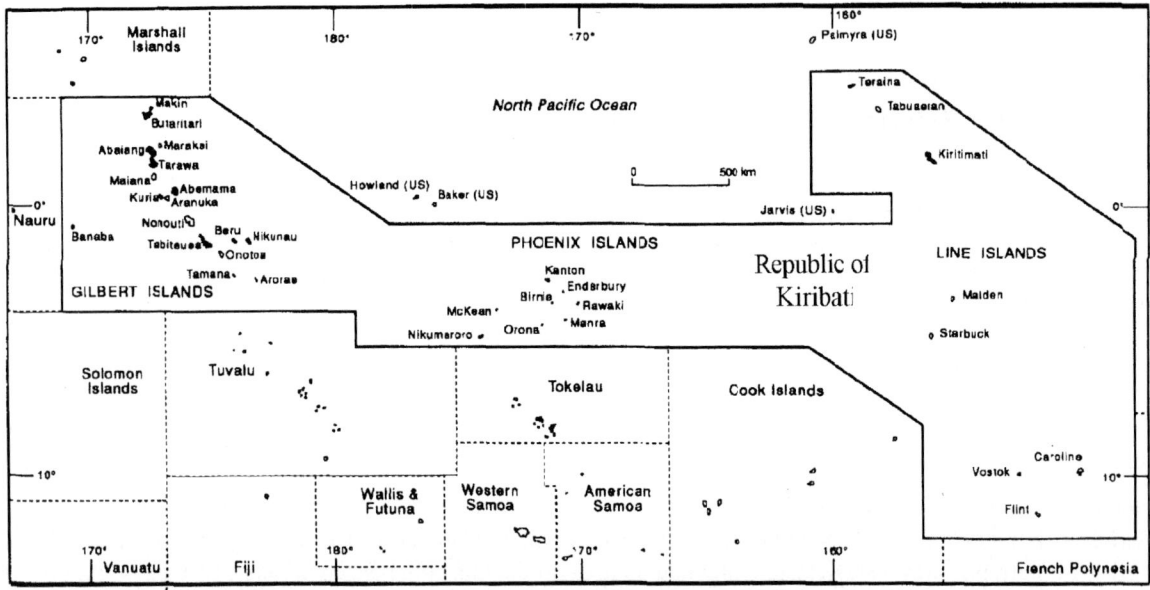

Figure 1. Location of the Republic of Kiribati, Kiritimati and Tabuaeran.

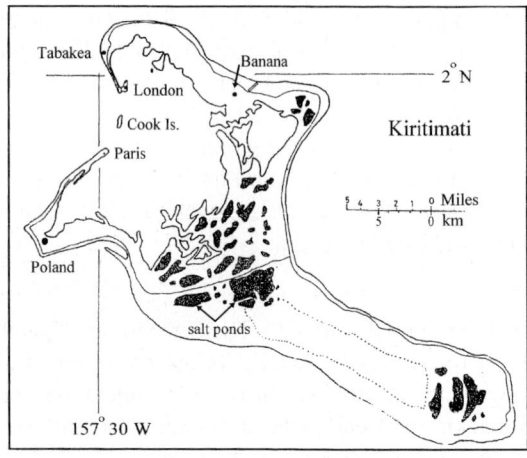

Figure 2. Kiritimati atoll.

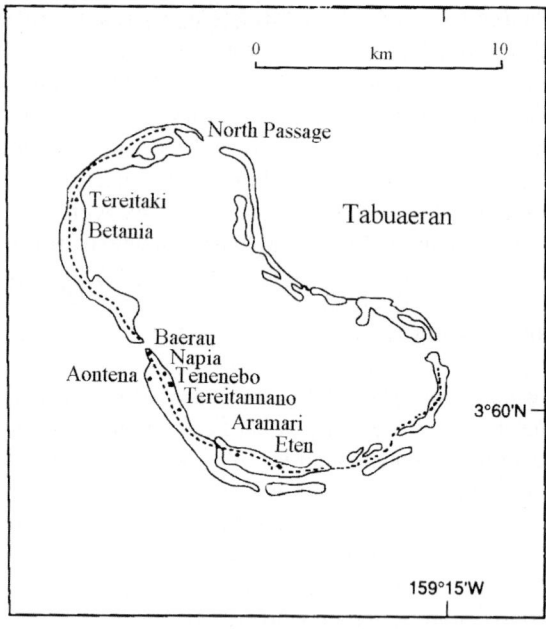

Figure 3. Tabuaeran atoll.

social infrastructure, only 800 people had been settled by 1990, and the scheme was curtailed in the following year. Furthermore, with copra cutting providing the only source of cash income, most settlers were having difficulty paying AUS$1000 for their one acre land allocation (Langston, 1993). The resettlement scheme recommenced in 1995, and the population of Tabuaeran had risen to just over 1700 by 1997.

Doty (pers. com.) first introduced Philippine *Kappaphycus alvarezii* (Doty) Doty from Hawaii to the Line Island of Kiritimati in 1977. At the same time, thalli from the Hawaiian stock were introduced to Tabuaeran by Russell (1982). This was followed by pilot farming trials on Kiritimati in 1980 by Why (pers. com.). However, it was concluded that Kiritimati was not a suitable site for commercial production due to the prevailing turbulent sea conditions at lagoon sites where thalli would grow. Russell (1982) suggested, from growth trial results, that there were about 30 ha of suitable farming area inside the North Passage of Tabuaeran lagoon. In 1981, the seaweed stock was moved from Kiritimati to the Tarawa lagoon in the Gilbert Islands, and it became the foundation stock for the future Kiribati industry.

In the Gilbert Islands, a small commercial production of *K. alvarezii* started in 1985 (Why, 1985). Dry seaweed exports commenced the following year, and after an initial increase in production (Uan, 1990), the industry declined due to a lack of business infrastructure, poor crop quality and few export markets. Annual exports declined to 339 t in 1993. The industry was restructured in 1992 with the formation of a state owned corporation, the Atoll Seaweed Company (ASC), and the following year the new company secured the future market for Kiribati seaweed by arranging a five-year forward supply agreement with a foreign processor, Copenhagen Pectin A/S.

The ASC reintroduced *K. alvarezii* to the Line Islands in March of 1994. The Line Islands development programme described in this paper commenced in May 94 with site surveys, distribution of 'seed' stock, and village workshops on farming technique and crop handling.

Methods

Lagoon farming areas were first selected based on previous Pacific island experience, observations on water movement, and reference to earlier trial work by Why (pers. com.) on Kiritimati. The six-hectare lagoon flat adjacent to the London and Port Camp shore-line was selected because of its easterly exposure to wind-driven wave action within the lagoon, and its close proximity to 2100 people in the settlements of London and Tabakea (Figure 2). On Tabuaeran the lagoon flats adjacent to the shore south-east of the main channel entrance were selected, again because of an easterly exposure and the close proximity of six villages where 73% of the island's population live (Figure 3).

Farming

Meetings were held in village maneabas (community meeting houses) to gain local approval for a new development programme, and ensure communal participation in seaweed farming. The maneaba has special significance in I-Kiribati society, representing protection of individual and collective rights, and also providing a structure for community politics where decision-making involves everyone in social and economic activities. Through workshops and demonstration field plots, prospective farmers were introduced to a high-density off-bottom farming technique, modified from that practised on Nusa Lembongan in Indonesia. A 10 m × 5 m module suitable for turbulent sea conditions contained 30 × 5 m-long polypropylene culture ropes of 3 mm diameter spaced 330 mm apart. Culture ropes were tied to a 6 mm diameter rope supported on posts 0.2 m to 0.5 m off the sea-bed. Twenty-five cuttings were tied with raffia to each culture rope to give a planting density of 15 thalli m^{-2}. No particular farm size was recommended, leaving individuals to decide on their own level of commitment in relation to time spent on subsistence living and community activities. The 30-line farm module has a recommended labour input for planting and harvesting of 2 h w^{-1}.

To ensure an adequate supply of farming materials, the development programme obtained them overseas and acted as a wholesaler to village stores. This was achieved with a revolving materials fund managed separately to the operating capital for crop purchasing and handling. New farmers provided their own posts and purchased sufficient materials for one 30-line module at a cost of AUS$7.00, with future expansions being financed from the sale of seaweed. The first 'seed' cuttings were provided free of charge from demonstration modules. Farmer credit was at the discretion of village stores. Crop drying was first accomplished on coconut-leaf mats. Once cash flow was established, farmers were encouraged to improve crop quality and minimize losses by purchasing plastic woven cloth for a drying surface, and plastic sheet covers for protection against rain-washing. More efficient harvesting was also encouraged by the provision and sale of net bags. Dry-crop storage bags were provided on an exchange basis.

Commercialization

Operating capital for the purchase and transportation of Line Islands production was provided by ASC. Two resident purchasing agents on Tabuaeran and Kiritimati were trained in quality assessment and documentation of supplies. The agents were paid a commission of AUS$15.00 t^{-1} purchased. Every supplier was issued a registration number, and a farmer data base was established, recording farmer location, gender, name, quantity and frequency of seaweed sold. The price paid to farmers was initially AUS$0.35 kg^{-1} for dry supplies free of foreign matter and with a maximum allowable moisture content of 34%. This price was subsequently increased to AUS$0.40 kg^{-1} in March 1995 by Government intervention, and remained unchanged through to March 1998. The development programme initially funded many local recurrent costs such as rents, transport, communications, and staff salaries, as well as providing management and technical training on site and overseas. These inputs were phased out over a three year period as the development became self-funding.

Kiritimati, and particularly Tabuaeran, lacked the infrastructure to handle the increasing production in 1996. Consequently the development programme funded capital items to ensure continued growth of the industry and to establish a commercial administrative centre for Line Islands seaweed on Kiritimati. A warehouse/office and laboratory facility, truck, boat, hydraulic press, and bulk handling equipment were provided on Kiritimati. Warehousing, tractor/trailer unit, HF radios, motorbike, and boat were provided on Tabuaeran.

Results

The first Line Islands farmers commenced harvesting in August 1994, and in the October the first payments were made on Kiritimati and Tabuaeran. By the end of 1995, over 200 suppliers had sold a total of 447 t in fifteen months. On Kiritimati, some farmers from Banana (Figure 2) made the daily return trip of 53 km to tend their plantings on the London lagoon-flats. A group of fourteen people from the village of Poland, on their own initiative, established farms at Paris (Figure 2) inside the southern entrance to the lagoon. Poland farmers travelled 17 km by bicycle and foot to reach Paris on a regular basis, and frequently camped overnight at their farm sites. The temporary Paris foreshore occupation brought farmers into direct conflict with the Wild Life Department and tourist bone-fishing interests. Temporary work shelters made from coconut leaves and posts also appeared on the London/Port Camp shoreline, providing shaded work areas for farmers from Tabakea and Banana. Some Government officials objected to the visual squatter appearance seaweed activities had created, but no formal evictions have so far been imposed. Most farmers have no residential land leases on Kiritimati, and London residents have not inhibited development by extending any land rights into the lagoon. The problems associated with limited farming areas envisaged

Table 1. Kappaphycus production by village on Tabuaeran 1997

Village	Number households	Number suppliers	Seaweed tonnes
Bae/Nap/Aon	52	33	78.1
Tenenebo	67	84	130.4
Tereitannano	24	32	66.8
Aramari	23	46	142.3
Eten	20	27	73.8
Betania	30	35	33.8
Tereitaki	51	60	48.9
Totals	267	317	574.1

by Tikai (1993) in the Gilbert Islands, have not been realized in the Line Islands. At the London lagoon area, there is a strong spirit of co-operation amongst farmers. Canoe passages have been willingly left open and other demands for clearway shore access have been respected. There are sufficient lagoon flats close to London to accommodate a farming interest from every resident of Kiritimati, without interfering with gillnet fishing, shell-fish collection, or tourist diving and fly-fishing interests. In 1996, the 6 ha London farming area produced 298.3 t indicating a mean production of 50 t ha^{-1} y^{-1}. Top suppliers produced at even higher rates, the best being 11.4 t from 1150 m^2.

On Tabuaeran farming started at Tenenebo, and quickly spread south to the new resettlement villages of Tereitannano, Aramari and Eten. The initial demand for 'seed' cuttings outstripped the supply and many farm units were started with just one or two culture ropes using thalli supplied by neighbours. Farmers themselves transferred plants to the northern villages of Tereitaki and Betania (Figure 3), and production started from northern lagoon flats previously considered less suitable than areas to the south of the main lagoon entrance. A second purchasing agent was established to service the two northern villages in 1995, and the following year Tabuaeran annual production increased to 494 t from 251 suppliers. Table 1 divides Tabuaeran into the seven developed farming areas in the lagoon, and shows that in 1997, production per household was highest from the most recently created resettlement villages of Aramari and Eten. Household numbers were recorded from the December 95 census, when 73% of households reported that seaweed was their main source of cash income.

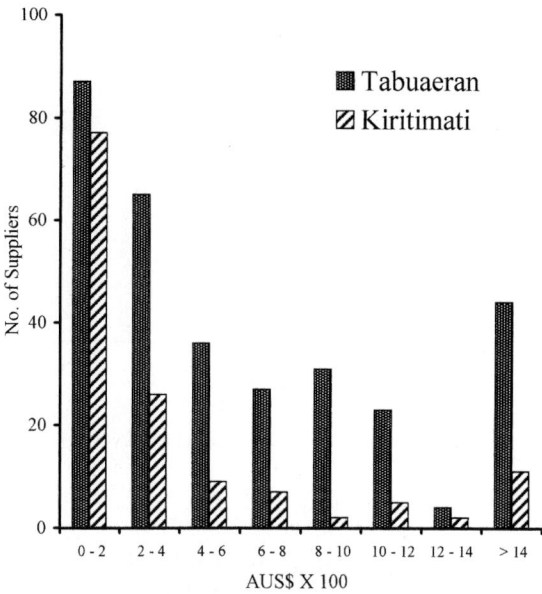

Figure 4. Annual income levels of suppliers 1997.

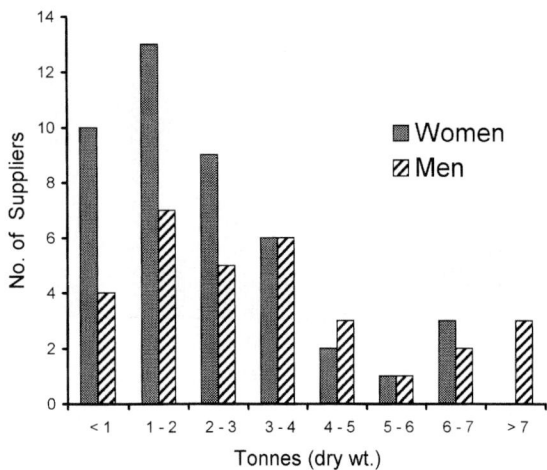

Figure 5. Gender-disaggregated suppliers 1996 (Kiritimati).

Table 2. Top farmer income statements from Tabuaeran and Kiritimati

Island	Tabuaeran	Kiritimati
Home village	Eten	Banana
Production tonnes	12.7	11.4
	AUS$	AUS$
Ropes	154	134
Other materials	148	129
Total capital costs	302	263
Depreciation (useful life)		
Ropes 3 years	51	45
Other Materials 5 years	30	26
Operating costs	312	438
Total yearly expenses	393	509
Gross yearly income	5080	4560
Less expenses	393	509
Net income	4687	4051

Annual income levels from 456 Line Islands suppliers show a wide range (Figure 4), illustrating that, for many households, there is only a small commitment to farming. On Tabuaeran, a small group of eleven farmers, mainly from recently settled villages, accounted for 16.5% of the island's 1997 production and received gross incomes of over AUS$2700 y^{-1}.

Table 2 shows that net incomes in excess of AUS$4000 y^{-1} were achieved from some family farms. The Eten supplier (Table 2) settled on Tabuaeran in 1995, and represents a husband and wife unit farming approximately 690 culture ropes throughout 1997, with some help from their school-age children. Similarly production from the top Kiritimati supplier (Table 2) comes from the efforts of a husband and wife working together, maintaining approximately 600 culture ropes throughout 1996.

All farms on Kiritimati are owned and operated by families or individuals. On Tabuaeran a small number of co-operative units are run by church and women's groups, but these are additional to the members' family farms. Women are actively involved in all production activities, and gender-disaggregated data from the London lagoon-flat farms (Figure 5) show that in 1996, 59% of suppliers were women. Women supplied 49% of the production and, hence, were the direct recipients of nearly half the total payments for seaweed, although recipients of the three highest annual incomes over AUS$3200 were men. There were more women than men supplying in the lower incomes of AUS$100–AUS1200 y^{-1}, but they accounted for 22% of all production, and hence represented an important supply group (Figure 5). Many women on both islands maintained a small part-time interest in seaweed, fitting farming activities in with domestic work and subsistence living.

The economic impact on the people and on the two atoll economies has been most significant on Tabuaeran, where the only alternative cash income is copra production. The copra price has been identical

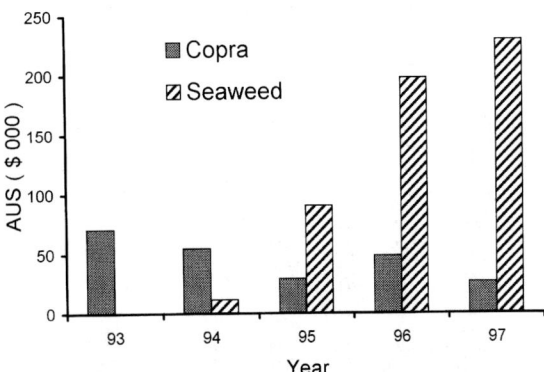

Figure 6. Annual sales from primary producers on Tabuaeran.

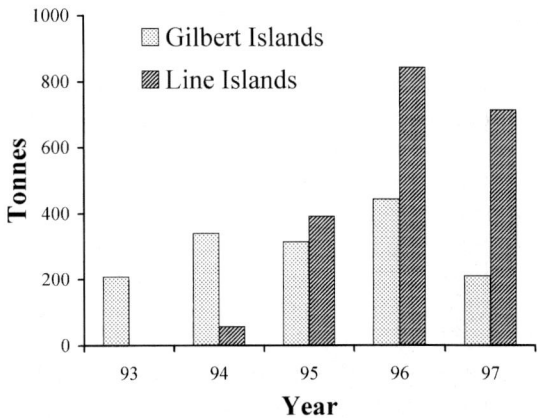

Figure 7. Annual *Kappaphycus* production (dry weight).

to the seaweed price since the introduction of farming, through to January 1998. Suppliers reported that seaweed farming was easier physically, and provided a better financial return for effort than copra. Figure 6 shows the decline in Tabuaeran copra production with the advent of seaweed production. The change in the cash economy has also increased the annual payments to commodity producers by 261% from 1993 to 1997 (Figure 6). On Kiritimati, copra still dominates the cash economy providing the main source of income for 35% of households, while seaweed was recorded as the main income for 29% of households (Dec. 1995 census). In 1996, Kiritimati copra production was 574 t, and seaweed production was 347 t. Seaweed thus added a further AUS$138 800 to payments to commodity producers. Copra production has declined since the start of seaweed farming; the 1994 Kiritimati production was 1790 t, the highest for nine years, and 865 t in 1995 (Statistics Office, 1997b).

Production from the London lagoon flat declined sharply in the second half of 1997 due to the formation of a strong El-Nino weather pattern. The reduction in water movement, due to the decrease in strong easterly winds, and the record annual rainfall (over 3600 mm) on seaweed farms, resulted in widespread die-back of thalli. Annual production from Kiritimati was reduced by 60% to 140 t in 1997, and by the end of the year most farmers had ceased operating to await the weakening of the El-Nino and a return to favourable environmental conditions for *Kappaphycus* growth. In contrast, Tabuaeran 1997 production (Table 1) increased by 16% from 494 t in 1996 although farmers in the north at Betania and Tereitaki reported a decline in plant growth in the latter half of the year. Overall the El-Nino caused a small decline in Line Islands production, with Kiritimati and Tabuaeran to-

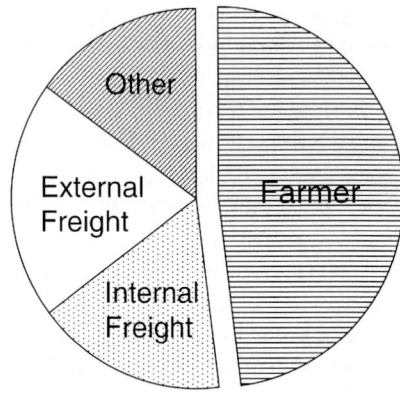

Figure 8. Line Islands variable costs breakdown.

gether accounting for 77% of the total *Kappaphycus* production from Kiribati in 1997 (Figure 7).

Figure 7 shows that the Kiribati record annual production of 1283 t in 1996 was largely due to the development of farming in the Line Islands. Although this development is part of the larger Kiribati industry, the cost of sustaining the Line Islands operation as a viable commercial business has been analysed as a separate entity (Figure 8). Production was shipped to Tarawa before exporting, but the high cost of local sea freight within Kiribati represents a constraint to profitability which was offset by cross-subsidisation from Gilbert Islands production.

Discussion

Farming

During the first survey of Tabuaeran lagoon in 1994, living *Kappaphycus* was found between Betania and

Tereitaki. The stock was assumed to have survived from the original introduction (Russell, 1982), and been left behind when the University of Hawaii closed its Pacific Equatorial Research Project in 1981. Russell (1982, Figure 1) found that plants did not survive inside the lagoon adjacent to Tereitaki, which suggests that either plants became acclimatized over time or environmental conditions have changed since 1977.

The very high productivity at Kiritimati and south Tabuaeran farming sites is attributed to the degree of water movement. Both the London and south Tabuaeran lagoon flats are subject to consistent wind-driven wave action from the prevailing easterly trade winds when normal and La Nina weather patterns prevail in the Pacific. Water motion is recognized as a key factor affecting the growth of *Kappaphycus* and the suitability of lagoon reef-flats exposed to consistent trade winds had been previously predicted (Glen & Doty, 1992). Why (pers. com.) recorded relative growth rates of 6.3% for a five week period at London, but considered thalli breakages from wave action caused unacceptable losses over ten weeks. A short harvest interval of five–six weeks, and high density planting have overcome this earlier constraint.

Both of the top Line Islands farmers (Table 2) show high returns for low capital investment. When the dollar value of their labour input (AUS$1560 y^{-1}) is added to the operating costs, the internal rates of return (IRR) for the two farms are over 900%. The IRR, also known as the discounted cash flow rate of return, represents the rate of return the farm generates, calculated in this case across ten years of cash flows. The replacement of capital items as they wear out in their respective lifetimes has been included in the calculations. High returns of over 100% for seaweed farming are common Firdausy & Tisdell (1991) show returns for seaweed farming in Bali had an IRR of 153%. Alih (1991) also indicates similar returns for seaweed farming in the Philippines, where returns on investment of 150% were recorded. One reason for the very high returns in the Line Islands is the much lower capital required to start farming. Another is the scale and nature of the type of farming carried out. The productivity yields per effort are also far higher than those achieved in the Philippines and Indonesia. High density planting in the highly fertile lagoon sites produced a maximum labour productivity of 5.7 t y^{-1} (Table 2) for the top Kiritimati farmer. The largest family farms are just over 1000 m^2, and extrapolated maximum yields per hectare on both Fanning and Kiritimati can be as high as 100–110 t ha^{-1} y^{-1}. An efficient farmer

Table 3. Aid input and financial returns to the Line Islands to 1997

Item	Year 1 (AUS$)	Year 2 (AUS$)	Year 3 (AUS$)
Consultants (2)	132 700	132 700	132 700
Capital items	18 810	45 220	317 870
Recurrent costs	11 340	24 030	11 900
Overseas business study	–	–	15 730
Revolving materials fund	22 000	–	–
Total aid input	184 850	201 950	478 200
Returns to Line Islands			
Gross farmer returns	156 241	336 552	285 558
Other returns	9602	20 687	17 709
Total returns	165 843	357 239	303 267

in Indonesia is reported as producing 48 t ha^{-1} y^{-1} (Firdausy & Tisdell, 1991), while average production at Caluya, Philippines, is reported at 27.9 t ha^{-1} y^{-1} (Hurtado-Ponce et al., 1996). Doty (1986) gives a projected yield of 31.8 t ha^{-1} y^{-1} at Pohnpei, Micronesia, and cites verbal reports of over 100 t ha^{-1} y^{-1} in the Philippines. The average I-Kiribati is probably less diligent than his or her counterpart in South East Asia, but more fortunate in having shoreline sites which require no capital items such as boats or outboards to farm. Existing copra hand-carts are utilized by some farmers to transport wet harvest from the shore to drying areas.

Commercialization

Many seaweed mariculture trials in the Pacific islands have provided valuable information, but have not led to the frequently projected commercialization. The Line Islands development described here illustrates the essential components necessary for commercialization. These are Government support and co-operation, an aid donor, foreign personnel with experience in both marine agronomy and business, and the support of a foreign processor willing to guarantee the product market. Table 3 outlines the aid input to the Line Islands development, and the actual direct returns to Kiritimati and Tabuaeran. There are also many non-quantifiable benefits from the development. Given continued support, in the form of a forward supply agreement with a foreign buyer, the financial benefits should continue to grow without any further aid input to the northern Line Islands. Table 3 does not show the operating capital input, but commercial sustainability of the operation is indicated by the ability to exceed the break even point of 629 t in the second and third year of the development. Production should continue at this level and above. The benefits of seaweed to the whole Kiribati economy are already significant. In 1996, record exports of 1204 t were achieved, making the value of seaweed exports second only to the value of copra exports.

Several other commercial factors contribute to economic viability, and distinguish the industry in the Line islands from those in the Philippines and Indonesia. The business infrastructure necessary for exporting and the provision of sufficient operating capital were already established in Tarawa before the Line Islands development commenced. Due to the small size of the country, business competition at both the seaweed buying and exporting levels was excluded by Government licence. This prevented under-capitalized and inexperienced operators entering the industry. In Kiribati, there are few entrepreneurs, and none with the necessary capital or experience to make the required investment in what is initially a high risk venture.

The creation of a monopolistic industry in the very small productive sector of Kiribati has many advantages which outweigh the disadvantages. Price control at the supplier level provides farmers with the confidence to undertake a new and previously unknown activity, with the security of knowing the future financial return for effort. Forward price-fixing with the foreign buyer provided protection for farmers against the fluctuating price cycles associated with changes in world supply and demand which have always characterized the *Kappaphycus* industry in South East Asia (Luxton, 1993). The world market *Kappaphycus* price showed large variations from 1994 to 1998. In the Philippines the farmer price can change by more than 50% within a six-month period (Hurtado-Ponce et al., 1996), and price stability is seen as a critical problem in the main production areas on Tawi-Tawi islands (Alih, 1991). In Kiribati the exporting company also has direct control of the village purchasing agents, without the need for middle-men traders, a common feature in the Philippine and Indonesian industries, where village stackers/collectors frequently have no allegiances to processors and exporters. This improves the cost structure of the industry, and allows the exporter to dictate and control the quality of farmer supplies so that no re-drying is required to meet the foreign buyer's product specifications for moisture content. Most importantly a monopoly allows for a 'single-desk' marketing strategy for all Kiribati production which, even in total, is small compared to volumes traded from South East Asia. With the world *Kappaphycus* market being dominated by the Philippines and Indonesia, the forward supply contract with a foreign processor has been critical to the development and economic sustainability of the Kiribati industry. It not only guarantees the sale of product at no marketing cost, but provides the stability and continuity to sustain both the farmers and the export company.

The variable cost structure of the Line Islands industry (Figure 8) shows that three items – farmer payments, internal and external freight – make up 85% of the variable costs. Since the variable costs make up 90% of the total running costs, these three factors have a major impact on the overall cost structure. An increase in either of the freight costs could affect the economic viability of the development. Conversely, reduced transportation costs could lead to increased growth and further returns to island communities. The construction of a warehouse/office facility, and the installation of a seaweed press on Kiritimati to service and consolidate Line Islands supplies, open up the possibility of exporting seaweed directly overseas, rather than first shipping to Tarawa. The volume of seaweed freight is currently attracting interest from shipping operators outside Kiribati.

Socio-economic impact

The replacement of copra by seaweed as the main source of income on Tabuaeran, has significantly improved the well-being of the population and raised the level of economic growth on this atoll. Copra has been heavily subsidized for a number of years and the producer price does not reflect true costs in relation to low world market prices. In fact, were it not for the subsidy from the European Commission STABEX fund, copra production would probably cease in Kiribati. Seaweed receives no such subsidy. Moreover, in February 1998 the Government again raised the producer price of copra by 12.5% to AUS$0.45 kg^{-1}. This merely encourages people to continue participating in uneconomic activities at the expense of new commercially viable enterprises.

On Tabuaeran, seaweed farming has been particularly attractive to the people resettled from the Gilbert

Islands under the Government's resettlement scheme, which provides them with cash to purchase a quarter acre house plot and a three-quarter acre bush plot for AUS$1000. Residents have recently been given the opportunity to purchase a second acre from the Government for AUS$1100. Prior to the introduction of seaweed, copra provided the only source of income, and this was limited by a predominance of low-yielding senile trees, and young nut destruction by an infestation of coconut rats. In 1993, the average household income was approximately AUS$300 y^{-1}. Some settlers have been repatriated back to their home islands, generally because of difficulties in making payments to the Land Purchase Scheme. Families that volunteer for resettlement on Tabuaeran have few possessions and tend to have no disposable income or savings. In 1995, just eight households out of 267 possessed an outboard motor, and 120 households owned a canoe (Statistics Office, 1997a). The farming of seaweed has made the cash-economy of Tabuaeran considerably larger than most of the settlers' home islands in the Gilbert group. Settlers are not only living in less crowded conditions, but many now have disposable income. Farmers give 5% of their seaweed income to the Island Council for the administration and general benefit of the whole island.

Kiritimati has a more diverse economy than Tabuaeran, and hence the socio-economic impact of seaweed farming has been less significant. As the Government centre for the Line Islands, public administration and service provide the main income for many households. Other cash-earning opportunities are copra, fishing, beche de mer (sea cucumber), pet fish and guiding for tourist fishing. Apart from copra production, all are exclusively male activities and, unlike seaweed, only relatively small groups directly benefit from the often large revenue of these activities. There are only six exporters of tropical pet fish for example. The top seaweed farmer's gross income is approximately the same as a contract diver for pet fish, and more than that of a beche de mer collector, but without the serious health risks of these occupations. Moreover, wild-cropping a limited resource, without knowledge of sustainable yields, is unlikely to lead to a viable economic activity in the future. Tourist fly-fishing and pet-fish exports are also dependent on the continuation of a regular air service to Hawaii. Seaweed farming, like copra production, involves both women and men, and consequently helps to mitigate aspects of traditional society which work against the progress of women. Traditionally, males were the food providers, planting babai and fishing from canoes. Women cared for families, prepared food, and also collected seafood from lagoon flats. It is perhaps not surprising that many men still perceive the tying of thalli to culture ropes and harvesting as women's work, and the construction of farm support structures as men's work.

Development projects in the productive sector of the Kiribati economy have largely been unsuccessful (Iuta, 1993). *Kappaphycus* mariculture in the northern Line Islands from 1994 to 1998 represents a commercially viable development of significance to the economic growth of Kiribati. Long-term sustainability will be dependent on the maintenance of good management practices, but it is considered that future sustainability will be greatly enhanced if the Government follows through on original plans to fully privatize the Atoll Seaweed Company by selling its majority share holding.

Within Kiribati, the association between the successful seaweed development and the resettlement scheme on Tabuaeran could well act as a model for further resettlements on the uninhabited Caroline and Vostok atolls in the southern Line Islands, and on Canton and Hull atolls in the Phoenix Islands. Other Pacific atoll countries, where the economy is based on copra, such as some of the northern Cook Islands and Tuvalu, could also consider developing a seaweed industry to improve economic growth. As a development, seaweed farming generates cash income in a manner harmonious with subsistence atoll life and Pacific Island culture. Professor M. S. Doty's assertion 21 years ago remains true today: "No other way has been found to achieve this sociologically very desirable end" (Doty, 1978).

Acknowledgments

The Line Islands development programme was funded by the New Zealand Official Development Assistance Programme of the Ministry of Foreign Affairs and Trade in collaboration with the Ministry of Environment and Natural Resources Development, Republic of Kiribati. The authors are indebted to Mr Mahuri Robertson and Mr Kaeti Boanereke for field operations. Co-operation and support from the European Commission's development programme to Kiribati, through Mr Michael Tinne, are also gratefully acknowledged. Staff of Fisheries Division on Tabuaeran and Kiritimati are acknowledged for their care and maintenance of the first trial plots in 1994. We also thank Kiribati Fisheries Division in Tarawa for access to the early Kiritimati trial reports of M. Doty and S. Why.

References

Alih, E. M., 1991. Economics of (*Eucheuma*) farming in Tawi-Tawi islands in the Philippines. In Hirano, R. & I. Hanyu (eds), Proceedings of the 2nd Asian Fisheries Forum. Asian Fisheries Society, Manila, Philippines: 249–252.

Doty, M. S., 1977. *Eucheuma* – current marine agronomy. In Krauss, R. W. (ed.), The Marine Plant Biomass of the Pacific Northwest Coast. Oregon State University Press, Corvallis: 203–204.

Doty, M. S., 1978. Status of marine agronomy, with special reference to the tropics. Proc. int. Seaweed Symp. 9: 35–58

Doty, M. S., 1986. Estimating farmer returns from producing *Gracilaria* and *Eucheuma* on line farms. In Santelices, B. (ed.), Usos y Funciones Ecologicals de las Algas Marinas Bentonicas. Monografias Biologicas 4: 45–62.

Glen, E. P. & M. S. Doty, 1992. Water motion affects the growth rates of *Kappaphycus alvarezii* and related red seaweeds. Aquaculture, 108: 233–246.

Firdausy, C. & C. Tisdell, 1991. Economic returns from seaweed (*Eucheuma cottonii*) farming in Bali, Indonesia. Asian Fish. Sci. 4: 61–73.

Hurtado-Ponce, A. Q., R. F. Agbayani & E. A. J. Chavoso, 1996. Economics of cultivating *Kappaphycus alvarezii* using the fixed-bottom line and hanging-long line methods in Panagatan Cays, Caluya, Antique, Philippines. J. appl. Phycol. 105: 105–109.

Iuta, T., 1993. Developments and challenges. In Van Trease, H. (ed.), Atoll Politics The Republic of Kiribati. Macmillan Brown Centre for Pacific Studies, University of Canterbury, Christchurch, New Zealand: 321–333.

Langston, P., 1993. Northern Line Islands development. In Van Trease, H. (ed.), Atoll Politics The Republic of Kiribati. Macmillan Brown Centre for Pacific Studies, University of Canterbury, Christchurch, New Zealand: 200–211.

Luxton, D. M., M. Robertson & M. J. Kindly, 1987. Farming *Eucheuma* in the south Pacific islands of Fiji. Hydrobiologia 151/152: 359–362.

Luxton, D. M., 1993. Aspects of the farming and processing of *Kappaphycus* and *Eucheuma* in Indonesia. Hydrobiologia 260/261: 365–371.

Russell, D. J., 1982. Introduction of *Eucheuma* to Fanning Atoll, Kiribati, for the purpose of mariculture. Micronesica 18: 34–44.

Statistics Office, 1997a. Report on the 1995 census of population, vol. 1. Ministry of Finance, Tarawa, Republic of Kiribati, 157 pp.

Statistics Office, 1997b. Copra Statistics up to 1996. Ministry of Finance, Tarawa, Republic of Kiribati, 15 pp.

Tikai, T., 1993. Fisheries development. In Van Trease, H. (ed.), Atoll Politics The Republic of Kiribati. Macmillan Brown Centre for Pacific Studies, University of Canterbury, Christchurch, New Zealand: 168–182.

Uan, J., 1990. Kiribati. In Adams, T. & R. Foscarini (eds), Proceedings of the Regional Workshop on Seaweed Culture and Marketing. FAO, South Pacific Agriculture Development Project. Suva, Fiji: 10–15.

Why, S., 1985. *Eucheuma* seaweed farming in Kiribati. South Pacific Commission, Noumea, Fisheries 17/WP 19, 7 pp.

The status of commercial algal utilization in New Zealand

W. Lindsey Zemke-White[1], Graeme Bremner[2] & Catriona L. Hurd[3]
[1] School of Biological Sciences, Private Bag 92019, University of Auckland, Auckland, New Zealand
[2] Ministry of Fisheries, Dunedin, New Zealand
[3] Department of Botany, University of Otago, P.O. Box 56, Dunedin, New Zealand
E-mail: l.zemke-white@auckland.ac.nz

Key words: aquaculture, harvesting, New Zealand, phycocolloid, seaweed

Abstract

In 1988, the New Zealand government instituted a moratorium on the issue of licenses to harvest wild stocks of marine macroalgae. In the intervening years, exports of algal products from New Zealand have declined while imports have increased. Exports of agar have decreased by 85%. For algal food products, exports have decreased while imports have increased by 500%. Collection of unattached rhodophytes requires no permit, and some special exemptions to the permit moratorium were made for abalone farmers, so seaweed continues to be harvested from wild stocks. In 1997, the two main rhodophyte genera harvested were *Pterocladia* and *Gracilaria*, with approximately 60 and 100 t dry weight harvested respectively. The two main phaeophyte genera harvested were *Macrocystis* and *Durvillaea*, with 51.8 and 34.5 t (wet weight) harvested respectively. Algal farming in New Zealand is still in its infancy; while there are 72 farms licensed to grow seaweed (owned by 29 different entities), only 12 of these are actively producing algae. Approximately 6 t (wet weight) was cultured in 1995, and the majority was used as feedstock for animals cultured at the same sites.

Introduction

In New Zealand's coastal waters, there are many seaweed species with potential commercial value. Since the 1940s, there has been a small industry centred around harvesting agarophytes primarily from the genus *Pterocladia* (Moore, 1944). There has also been research into harvesting wild stocks of other algal genera: *Gracilaria* (Luxton, 1981), *Porphyra*, *Durvillaea*, *Macrocystis* and *Ecklonia* (Schiel & Nelson, 1990, Schiel et al., 1997a, b). There are a number of marine farms currently licensed to grow seaweeds, and a growing number of abalone farms dependent in whole or in part on seaweeds for feed.

This paper examines the current structure of commercial algal utilization, including government policy, quantities of algae harvested and cultured, and algal imports and exports. We discuss the future prospects for utilization of New Zealand's marine macroalgae.

Material and methods

Data on the wild harvest of seaweed is derived from fishing returns submitted by permit holders to the New Zealand Ministry of Fisheries (MFish). It is mandatory to complete such a return for each day's harvesting activities. However, since no permit is required to harvest unattached rhodophytes (see below), the figures for rhodophytes are incomplete. Estimates of the illegal harvest of seaweeds have been made by MFish enforcement officers involved in policing the regulations, and are considered by them to be conservative.

Data on the production of farmed seaweeds is derived from figures collected by MFish in 1996 for reporting to the United Nations Food and Agriculture Organisation, and are considered accurate. Data on the end uses of farmed seaweeds and future intentions of seaweed farmers come from a survey of all licensed seaweed farmers undertaken by MFish in 1996. Participation was voluntary, but a 97% response was achieved.

Export and import volumes of seaweed are sourced from the databases of the New Zealand Customs Department. Unfortunately, it is not possible to identify all imported and exported seaweed products in this database, since the international customs schedule (on which the New Zealand schedule is based) lumps some products into general categories on a haphazard basis. Agar, for example, has its own category whilst carrageenan is classified as 'gums, other' along with a number of terrestrial products.

A survey of 20 companies thought to use seaweeds in their products was conducted to determine the country from which seaweeds were imported, and whether they would use a New Zealand source if it were available.

Current legislation and policy on seaweed management

Wild harvest

Seaweed harvest is currently managed under the Fisheries Act 1983. This Act requires that all commercial seaweed harvesters have a fishing permit. An exception is made for gathering unattached rhodophytes, which is completely unregulated (although there are a few small stretches of coastline where this activity is prohibited). The Fisheries Act 1996, which is enacted but not yet in force, narrows the exemption for unattached rhodophytes to any seaweeds that have been cast ashore. In future, therefore, anyone harvesting free floating red seaweeds by use of stationary nets or trawling will require a fishing permit.

The exemption of beach-cast red seaweeds from regulation has a historical rather than biological basis. During the Second World War, New Zealand was isolated from its traditional Asian suppliers of agar. Since agar was an essential material, the government encouraged the development of a domestic industry, and part-time collectors in small rural coastal communities undertook seaweed collection, primarily *Pterocladia lucida*. These part-timers would generally be available to collect seaweed only occasionally between agricultural activities. The industry and these collection arrangements have survived to the present day, and because of the number of individuals involved in collecting the seaweed and the small amount taken by each, requiring them to apply for fishing permits was considered impractical.

Permits for harvesting seaweeds other than unattached rhodophytes are rare. Those issued in the past usually carry conditions specifying the species that may be taken, the areas from which they may be taken, and a maximum annual tonnage. A moratorium on the issue of new fishing permits for seaweeds was imposed in 1988 due to environmental concerns, and under present legislation no new entrants can gain access to the fishery. Permits that were held prior to 1988 are still renewed on an annual basis, but have been gradually declining in number, as permit holders retire or leave the industry. At the time the moratorium was imposed there were 23 extant permits, but only 4 of these are still active at the time of writing.

The current supply of seaweed from the legitimate fishermen is inadequate to meet domestic consumption and the shortfall is made up through imported products and a flourishing black market.

The Minister of Fisheries decided in late 1997 to propose an amendment to the Fisheries Act, which would allow the issue of new permits for the collection of beach-cast brown and green seaweeds, but any change to the legislation is unlikely to come into force until 1999.

Another recent change to government policy that has yet to take effect resulted from the signing of the 'Ngai Tahu Deed of Settlement' at the end of 1996. This is essentially a treaty between the government of New Zealand and the largest Maori tribe in the South Island aimed at ending various grievances dating back to last century. One of the provisions of this agreement requires the government to prohibit any commercial use of the genera *Durvillaea* and *Porphyra* within the Ngai Tahu tribal area. These seaweeds are valued for traditional purposes, and under the terms of the treaty are to be reserved solely for this use. The Ngai Tahu tribal area encompasses all of the coastline on which potentially commercial quantities of *Durvillaea* is found, with the exception of several remote islands (Brown, 1998). This means that the development of any industry based on wild harvest of this genus would be extremely difficult.

Seaweed for paua farms

Until 1996 it was official government policy to foster the development of the aquaculture industry. In 1991, concerned by the effect that the permit moratorium might have on the fledgling abalone farming industry in New Zealand, the then Minister of Fisheries approved the issue of special permits for the harvest of seaweeds to aquaculturists who required seaweeds for feed. The approval covered five genera of seaweeds only (*Pterocladia, Gracilaria, Ecklonia, Macrocystis* and *Durvillaea*), and the special permits issued to abalone farmers carry conditions restricting them to taking unattached seaweed from defined areas. The permits also carry restrictions on the amounts that may be taken and the areas from which seaweed may be collected. Seaweed taken under these special permits must be used on farms and cannot be sold.

Seaweed aquaculture

To grow seaweeds in a land-based pond or facility in New Zealand requires a 'Fresh water fish farming' licence issued by the Ministry of Fisheries and a discharge consent (issued by the Regional Council) for any water being discharged from the farm. Fresh water fish farming licences are readily available, and provided that the farmer is not intending to use antibiotics or large quantities of fertilizer, discharge consents are also easy to obtain.

To grow seaweeds on a marine farm requires a marine farming permit issued by the Ministry of Fisheries, and a resource consent (to occupy coastal space) issued by the local Regional Council. Some regional councils have effectively adopted 'zoning' as a means of controlling mariculture, and intending seaweed farmers often find themselves confronted by rules and restrictions imposed primarily to avoid the water pollution problems associated with sea-caged salmon. Resource consent applications are also frequently opposed by the local populace, who are accustomed to free access to all parts of the sea. Obtaining a resource consent is therefore a lengthy and often expensive undertaking, and would normally add at least $20 000 to the start-up costs for any mariculture venture.

There are no government extension programmes to assist farmers with growing seaweeds on land or in the sea and there is no government funded research on land-based culture. There is some limited government-funded research being undertaken on seaweed mariculture.

Current harvest, production, imports and exports

Wild harvest

Apart from abalone farms, there are few businesses in New Zealand using harvested wild stocks, and one company (Coast Biologicals Ltd.) deals with the majority of collected algae. Part-time rural collectors sell dried beach-cast seaweed (mainly *Pterocladia* spp.) to

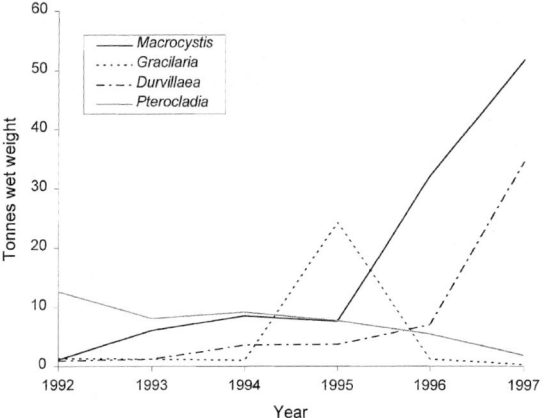

Figure 1. Wild harvest of four seaweed genera (tonnes wet weight) in New Zealand from 1992 to 1997. Figures as reported to the Ministry of Fisheries.

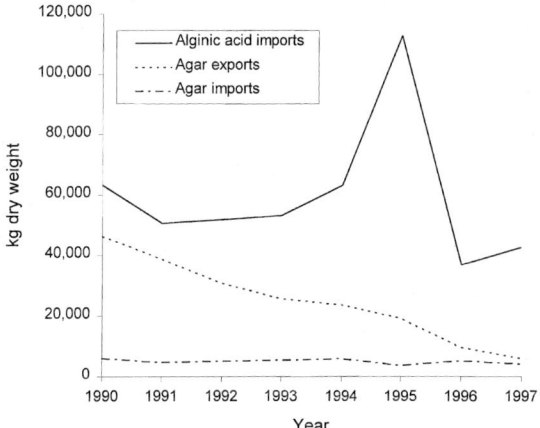

Figure 2. New Zealand imports and exports (kg dry weight) of agar and alginic acid from 1990 to 1997.

agents around New Zealand, who in turn supply this company (Schiel & Nelson, 1990). This company produces various grades of agar primarily for export, but also fertilizers, animal drenches and animal food supplements for the local market. There are also a handful of small companies that produce a variety of health products from *Ecklonia radiata* and *Porphyra* spp.

In terms of tonnage, the four main genera harvested in New Zealand are *Macrocystis, Durvillaea, Gracilaria* and *Pterocladia*. Harvest statistics for the last five years are presented in Figure 1. For *Gracilaria* and *Pterocladia* these statistics are incomplete, as there is no requirement for harvesters to report collections of unattached seaweed.

The steady increase in harvests of *Durvillaea* and *Macrocystis* and the 1995 peak in harvest of *Gracil-*

Table 1. Production of cultured seaweeds in NZ (tonnes wet weight). Figures as returned to the Ministry of Fisheries

Species	1994	1995
Porphyra columbina	1.0	0.0
Macrocystis pyrifera	3.0	6.0
Gracilaria chilensis	< 0.1	< 0.1
Asparagopsis armata	< 0.1	< 0.1

aria, are due to harvesting seaweed for abalone farms. Harvest of unattached *Gracilaria* is considered to be < 100 t yr^{-1} (wet weight). Although precise figures for the harvest of unattached *Pterocladia* are not available, we can infer from the export data for agar (Figure 2) that *Pterocladia* harvests have been decreasing steadily since 1990, and using a 1:10 return of agar from algal dry weight (Luxton & Courtney, 1987) we calculated that approximately 60 t dry weight was harvested in 1997. The illegal harvest of seaweeds is estimated to be in the region of 300 t yr^{-1}, with the bulk of this material being used in the manufacture of agricultural products.

Seaweed aquaculture

Seventy two farms are currently licensed to grow seaweed in NZ coastal waters (see Figure 3). These are owned by 29 companies, with some holding as many as nine licences. Sites are licensed for *Ecklonia, Lessonia, Ulva, Macrocystis, Gracilaria, Pterocladia* and *Porphyra*.

Despite the 72 current licences, seaweed farming in New Zealand is at an early stage. Yields from the 12 farms producing seaweeds in 1994 and 1995 (the most recent years for which statistics are available) are shown in Table 1.

Imports and exports of seaweed products

Agar and alginic acid (Figure 2), and algal food products (Figure 4) were the only three categories of algal products which were available from the New Zealand Customs database. Agar imports have been constant since 1990 at approximately 5 t yr^{-1}, while exports have decreased considerably from 46 346 kg in 1990, to 5929 kg in 1997. Alginic acid imports fluctuate annually but mean imports have changed little since 1990. Alginic acid was exported only in 1991 (120 kg) and 1997 (1150 kg). The data for algal food products included both dried and fresh seaweed so the total weight is not informative. Consequently, we have examined the dollar value of these items (Figure 4). From 1990 to 1997 imports of algal food products increased in every year except 1996. Overall there was a five-fold increase since 1990. Exports fluctuated between 1991 and 1993, but have remained low since then.

The New Zealand market for alginic acid remained fairly static, requiring between 50 and 60 t yr^{-1}. The fluctuations from 1995 to 1997 could reflect overbuying in one year and lower buying in subsequent years. From 1990 to 1997 there was no legal harvesting of phaeophytes except for use in aquaculture, and the phaeophytes cultured were used almost exclusively as feed on abalone farms. It seems, therefore, that the figures of alginic acid exports in 1991 and 1997 may reflect re-exportation of this product.

The large increase in imports of algal food products shows that the local market has grown markedly since 1990. As with alginic acid, the fluctuations in the export of algal food products may be in part due to re-export.

The processing sector

Ten companies replied to a questionnaire to assess the current domestic market and interest in using local seaweeds. There are healthy local markets for seaweed-based fertilizers (domestic and agricultural), stock food and human consumption (dietary supplements, food garnish, salt substitute and sea vegetables). All companies surveyed would use a local product if available at a similar cost and quality to imported seaweeds. Frustration was expressed at the moratorium on seaweed harvesting which is considered to be the main obstacle to developing seaweed-based industries in New Zealand.

Seaweeds for these industries were mostly imported: *Ascophyllum nodosum* from the United Kingdom, United States of America, Norway and Iceland; *Laminaria digitata* from Iceland; *Ecklonia maxima* (in liquid form) from South Africa; *Durvillaea potatorum* from Australia. Precise import data are confidential, but annual imports included several tonnes of dried *D. potatorum* and *Ascophyllum nodosum*, and several hundred litres of *E. maxima* based seaweed liquid extract.

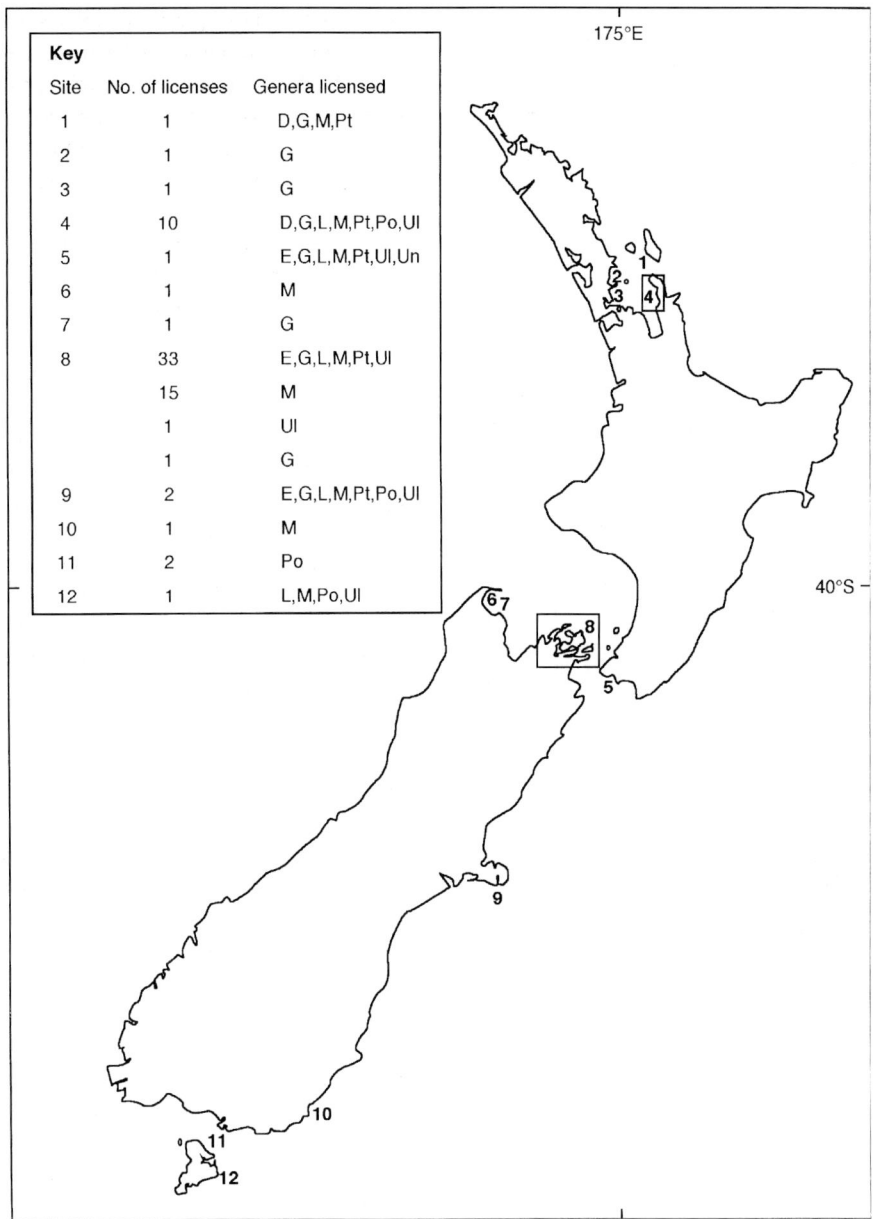

Figure 3. Sites in New Zealand where licences are held to farm seaweed and the genera which are licensed for each site. D = *Durvillaea*, E = *Ecklonia*, g = *Gracilaria*, L = *Lessonia*, M = *Macrocystis*, Pt = *Pterocladia*, Po = *Porphyra*, Ul = *Ulva*, Un = *Undaria*.

Human foods

Companies collecting drift seaweed have no problem selling their product. One company exports sea vegetables to the United States as health food. New Zealand seaweeds have a good reputation overseas (U.S.A. and Europe) for supplying high quality seaweed from (relatively) unpolluted waters. However, one of the larger New Zealand importers of seaweed had experienced New Zealand seaweed samples that were poorly harvested i.e. contained sand, and cost twice that of imported sources. This company expressed concern at the quality of drift seaweeds, and suggested that freshly cut seaweed would be preferable.

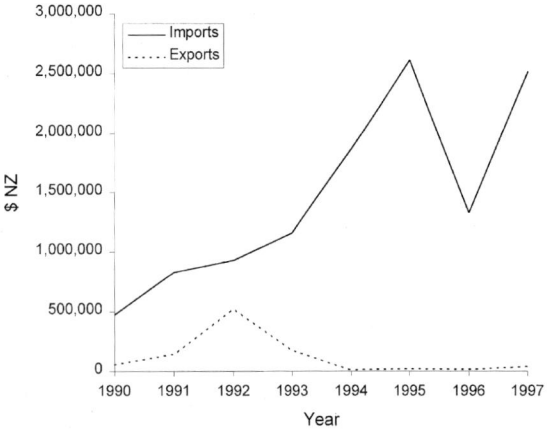

Figure 4. New Zealand imports and exports (in New Zealand dollars) of algal food products from 1990 to 1997.

Experimental ventures

Two main research teams are growing seaweeds experimentally in New Zealand: Industrial Research Limited (IRL) and the Cawthron Institute.

The IRL team, headed by Drs Richard Furneaux and Ruth Falshaw, has extracted a number of useful phycocolloids from native New Zealand seaweeds, often in collaboration with Dr I.J. Miller, Carina Chemical Laboratories Ltd (e.g. Falshaw & Furneaux, 1995; Miller & Furneaux, 1996; Miller et al., 1996; Miller et al., 1997). The growth of seven species (*Pterocladia lucida, Gelidium caulacantheum, Asparogopsis armarta, Gracilaria chilensis, Gigartina chapmanii, G. decipiens, G. circumcincta*) that contain polysaccharides with commercial potential were tested in the field on experimental lines at a mussel farm in the Marlborough Sounds in 1996. Of these, *G. circumcincta* grew well and has been selected for further laboratory and field manipulations aimed at enhancing the quality of polysaccharide produced. Concurrent ecological studies to assess the possibility of sustainable wild-harvesting are being conducted. Laboratory and field experiments to enhance the quality of agar from *Gracilaria chilensis* are also being conducted (Hemmingson et al., 1996).

Intensive research has focused on the aquaculture potential of the recently (Hay & Luckens, 1987) introduced Asian kelp, *Undaria pinnatifida*, which demands a high price in Japan. Government funding (FoRST) has been provided to the Cawthron Institute to investigate the feasibility of growing *U. pinnatifida* in New Zealand. Sporophytes have been cultured successfully on lines in the Marlborough Sounds, and gametophyte strains have been isolated from around New Zealand and maintained in chemostats (C. H. Hay, pers. com.). Other experimental ventures include growing *U. pinnatifida* on lines of different thickness in Otago Harbour (Paul Dean, University of Otago) and line cultures in Wellington Harbour (Cawthron). High quality *U. pinnatifida* can undoubtedly be grown in New Zealand. However, there is strong concern about increasing the spread of an introduced seaweed via aquaculture when we know little of its ecological impact on a diverse, but little studied, native seaweed flora.

Future prospects

Wild harvest

There are two opposing views on the utilization of attached seaweeds. One viewpoint comes primarily from the abalone farming industry and some of the industrial users of seaweed. This view holds that there are sufficient wild stocks of seaweed in New Zealand to permit a sizeable harvest on a sustainable basis, and that techniques to harvest and manage these stocks as a resource should be developed before efforts are made to farm the more common species (Beatty & Ackroyd, 1994). The other viewpoint, which has been reflected in government policy for the last decade, is that the New Zealand seaweeds are performing a valuable service *in situ* in terms of reducing coastal erosion and supporting other valuable fisheries, and that they should be left untouched. It seems unlikely that this impasse will be resolved in the near future.

A management regime involving strict controls on area, harvesting method, and timing could doubtless be developed which would allow limited harvests of attached seaweeds while still retaining the values that government policy is striving to protect (Schiel et al., 1997a, b). However, the costs of fisheries management and enforcement in New Zealand are currently passed on to the users through special government levies. The costs of developing and imposing a restrictive management regime, and the costs of undertaking the requisite preliminary impact assessment, would be high, and the resulting levy would probably financially cripple any potential harvester.

The recent government decision to increase access to beach-cast seaweeds seems likely to satisfy the needs of abalone farmers and some industrial users once it is implemented. In particular, producers of seaweed-based composts and fertilizers need access to large quantities of seaweed at low cost, and do not require raw material of a very high quality. There is a strong local market for such products, and good opportunities for both import substitution and the development of an export market. Recent research also suggests that beach-cast seaweeds are actually a better foodstuff for abalone culture than freshly harvested material of the same species (Hamid, 1997; Gay, 1998).

We predict that there will be an initial explosion of interest in the collection of drift seaweed once access to this resource is freed up, and that the market for seaweed based agricultural products and fertilizers will be taken over by locally produced products within the next decade. We also predict that there will be a continuing increase in the utilization of domestic seaweeds by the abalone farming industry.

Aquaculture

Despite considerable enthusiasm amongst the individuals involved, seaweed aquaculture has failed to thrive. This is probably due in part to marketing problems. Whilst the local market is not large, there are good prospects for import substitution of seaweed foodstuffs, especially for ground kelp products. There also appears to be an export market niche for health products. However, most seaweed farmers lack the resources to undertake a major marketing campaign, and there is no industry association established to undertake marketing on a collaborative basis.

The major consumer of raw seaweed in terms of volume is clearly the aquaculture industry itself. As abalone farms are increasing in number in New Zealand, there may be opportunities to provide seaweed as a feedstock (Bremner, 1993a, b, 1994). However, most demand appears to be being satisfied through wild harvest of seaweed via special permit, the growing of seaweed on the abalone farm itself, or by the use of artificial feeds. In future, abalone farmers will also have greater access to beach-cast seaweeds. The development of a major market for farmed seaweeds in this area is therefore unlikely. However, polyculture of seaweeds and abalone on the same farm will probably continue to expand.

The relatively high labour and export costs from New Zealand mean that international competition is a major problem facing any proposed algal culture venture. Whilst New Zealand has suitable species of algae to produce agars, carrageenans and alginates, it is difficult to see how seaweed grown as an industrial feedstock on New Zealand marine farms could ever compete in price with seaweed grown in developing nations, or harvested from the wild in the Northern Hemisphere. When the Ministry of Fisheries extends the availability of drift and beach-cast seaweeds local competition from this source will come into play. Industrial use of farmed New Zealand seaweed is likely to be viable if, and only if, a local species is discovered to contain a unique and valuable phycocolloid. There are several potential candidates (Christeller et al., 1983; Falshaw et al., 1997), and international firms have expressed interest in the production of carrageenan from several endemic species of the genera *Gigartina, Pachymenia* and *Iridaea*. However, taking these species into cultivation would require considerably greater knowledge of the plants' taxonomy, ecology and growth requirements.

Finally, there is no government assistance for potential algal culture. Many other 'new' crops and species (mainly terrestrial but including the development of mussel and oyster farming) received considerable attention from government researchers and extension advisers whilst becoming established in the past, but providing such support is now contrary to government policy. Thus the last series of extension reports was produced more than ten years ago (Bradstock & Luxton, 1984; Nelson, 1984a, b; Terzaghi et al., 1987). These reports focused on identification, biology, farming and harvesting methods. Most seaweed farmers lack the resources to undertake any sustained research programme of their own, and only limited government-funded seaweed research is being undertaken. However, groups such as those at IRL and Cawthron Institute are focusing their research efforts on field-based algal farming experiments, and working closely with seaweed farmers, mussel farmers and seaweed processors.

Conclusion

Due in part to existing government policies on wild harvests, and despite the abundance of seaweed around the New Zealand coastline, we predict that the existing phycocolloid industry is likely to remain static or decline. In terms of aquaculture, whilst there will no doubt be isolated successes, especially where seaweeds are being grown for use on-farm, we predict that seaweed farming in New Zealand will continue to struggle for the foreseeable future.

Acknowledgements

We thank all companies who replied to questionnaires and the following for providing opinion and information on the potential of the New Zealand seaweed industry: Dr Ruth Falshaw (IRL), Dr Cameron Hay (Cawthron Institute), Dr Wendy Nelson (Museum of New Zealand, Te Papa Tongarewa) & Paul Dean (University of Otago).

References

Beatty, R. & P. Ackroyd, 1994. Seaweed for farming or farming for seaweed? Seafood New Zealand February: 73–74.

Bradstock, M. & D. Luxton, 1984. Agar Seaweed: Biology, harvesting and resource management. Fishdex 27. Ministry of Agriculture and Fisheries, Wellington.

Bremner, G., 1993a. Seaweed for paua farmers. Seafood New Zealand September: 22–23.

Bremner, G., 1993b. Seaweed for paua farmers part II. Seafood New Zealand October: 27–29.

Bremner, G., 1994. Seaweed for paua farmers part III: Alternative sources of weed. Seafood New Zealand February: 29–31.

Brown, M. T., 1998. The seaweed resources of New Zealand. In Critchley, A. T. & M. Ohno (eds), Seaweed Resources of the World. Japan International Co-operation Agency, Yokosuka: 127–137.

Christeller, J. T., R. Furneaux, M. E. Gordon, W. A. Laing, I. Miller, W. A. Nelson, B. D. Shaw & B. E. Terzaghi, 1983. The potential for mariculture of seaweeds in New Zealand. Technical report, Plant Physiology Division DSIR, Palmerston North, 21 pp.

Falshaw, R. & R. H. Furneaux, 1995. Carrageenans from the tetrasporic stages of *Gigartina clavifera* and *Gigartina alveata* (Gigartinaceae, Rhodophyta) Carbohydr. Res. 276: 155–165.

Falshaw, R., G. C. Slim & W. A. Nelson, 1997. Marine hydrocolloids: commercial slimes, gums and jellies. Seafood New Zealand March: 77–80.

Gay, B., 1998. Palatability to paua of drift seaweeds in various stages of decomposition. M.Sc. thesis, Otago University, New Zealand, 250 pp.

Hamid, Z., 1997. Biochemical analysis of drift algae to be used as food in paua culture. M.Sc. thesis, Otago University, New Zealand, 70 pp.

Hay, C. H. & P. A. Luckens, 1987. The Asian kelp *Undaria pinnatifida* (Phaeophyta Laminariales) found in a New Zealand harbour. New Zealand J. Bot. 25: 329–332.

Hemmingson, J. A., R. H. Furneaux & V. H. Murray-Brown, 1996. Biosynthesis of agar polysaccharides in *Gracilaria chilensis* Bird, McLachlan et Oliveira. Carbohydr.Res. 287: 101–115.

Luxton, D. M., 1981. Experimental harvesting of *Gracilaria* in New Zealand. In Levring, T. (ed.), Xth International Seaweed Symposium. Walter de Gruyter and Co., New York: 693–698.

Luxton, D. M., W. J. Courtney, 1987. New developments in the seaweed industry of New Zealand. Hydrobiologia 151/152: 291–293

Miller, I. J. & R. H. Furneaux, 1996. A structural analysis of the polysaccharide from *Kallymenia berggrenii*. J Ag. Bot. mar. 39: 141–147.

Miller, I. J., R. Falshaw & R. H. Furneaux, 1996. A polysaccharide fraction from the red seaweed *Champia novae-zealandiae* Rhodymeniales, Rhodophyta. Hydrobiologia 326/327: 505–509.

Miller, I. J., R. Falshaw, R. H. Furneaux & J. A. Hemmingson, 1997. Variations in the constituent sugars of the polysaccharides from New Zealand species of *Pachymenia* (Halymeniaceae). Bot. mar. 40: 119–127.

Moore, L. B., 1944. New Zealand seaweed for agar-manufacture. New Zealand J. Sci.Technol. 25: 183–209.

Nelson, W. A., 1984a. *Porphyra* (Karengo): Biology, uses and harvesting. Fishdex 29. Ministry of Agriculture and Fisheries, Wellington, 1 pp.

Nelson, W. A., 1984b. *Durvillaea* (Rimurapa) bull kelp. Fishdex 30. Ministry of Agriculture and Fisheries, Wellington, 2 pp.

Schiel, D. R. & W. A. Nelson, 1990. The harvesting of macroalgae in New Zealand. Hydrobiologia 204/205: 25–33.

Schiel, D. R., J. Pirker & H. Lees, 1997a. Can giant kelp forests be commercially harvested? Part I: Natural production cycle. Seafood New Zealand October: 27–28.

Schiel, D. R., J. Pirker & H. Lees, 1997b. Can giant kelp forests be commercially harvested? Part II: Regeneration after harvest. Seafood New Zealand November: 27–28.

Terzaghi, B., W. A. Nelson & T. Hollings, 1987. *Gracilaria*: cultivation, harvest and uses. Fishdex 36. Ministry of Agriculture and Fisheries, Wellington, 4 pp.

The optimal utilization of kelp resources in the southern Cape area of South Africa

Klaus W. G. Rotmann
Taurus Products (Pty) Limited, Box 5534, Rivonia 2128, South Africa
Fax: [+27]-11-8038581. E-mail: kr21@pixie.co.za

Key words: Ecklonia maxima, kelp, production, markets

Abstract

The southern Cape area extends from the southernmost point of Africa, Cape Agulhas, to False Bay, just east of Cape Town. Large kelp beds (mainly *Ecklonia maxima*) occur on the rocky coastline. Algal utilization, which started during World War II, is currently based on the collection of natural casts, as well as a limited amount of harvesting, and offers employment in a depressed rural labour market. The resource has been researched and quantified and is utilized by low-tech methods for collection, harvesting and processing. Problems that are facing the industry are mainly of a marketing nature, but a good potential for the industry exists provided investments are made in research and development.

Introduction

The coastline of South Africa extends from about 28° S and 16° E at the Namibian border (Orange River), to about 26° S and 33° E at the Mozambique border. Two regimes determine the marine ecology: the cold Benguela current on the west coast and the warm Agulhas current on the east coast. These two systems meet at Cape Agulhas, the southernmost tip of Africa. To the west of Cape Agulhas substantial communities of kelp occur, predominantly two species: *Ecklonia maxima* and *Laminaria pallida*. These big plants occur in large beds anchored on rocky outcrops in the sea and communities extend from the Cape south coast into Namibia. *E. maxima* is the predominant species in the south whereas *L. pallida* is more predominant in the north.

This paper describes the size and usage of the kelp resource in the southern Cape area, and details some of the methods currently employed. The property rights to these seaweeds, as well as the present scope of the industry, will be identified and the future potential will be investigated (Anderson et al., 1989).

Resource

The kelp resources have been investigated by the Seaweed Research Unit of the Department of Sea Fisheries and the areas covered by kelp along the coastline between Cape Agulhas and Strand near Cape Town (Department of Sea Fisheries, in litt., 1995), estimated (Table 1). Based on Taurus' production experience since 1976 (in litt., 1997), a hectare of kelp in the sea yields about 0.6 dry tonnes of cast material per annum. The area from Cape Agulhas to Strand, just outside the city of Cape Town should therefore produce 1,172 dry tonnes of kelp per annum but the actual production figures for this area over the past 7 years look different (Figure 1). The reason for this discrepancy is the fact that not all beach cast was collected, due to fluctuations in demand.

Whilst kelps have been used in South Africa since time immemorial, for instance as wound treatment by the indigenous population, it was only during and after World War II that the economic exploitation of these large algae commenced. The first concession for the collection of kelp was issued in the 1950s at Elands Bay on the west coast. In the following years, South African kelp was mainly exported to alginate producers in Japan, Europe and the U.S.A. by traders from Cape Town. To a lesser extent kelp was also included in health products, animal feeds and a fertilizer. The industry was healthy until, in 1978, Chile sold brown seaweed on to the world market at below South Africa's cost prices, causing a collapse (Rotmann,

Table 1. Areas covered by kelp between Cape Agulhas and Strand

Depth	Hectares
0–5	1506
5–10	418
10–15	29
Total	1953

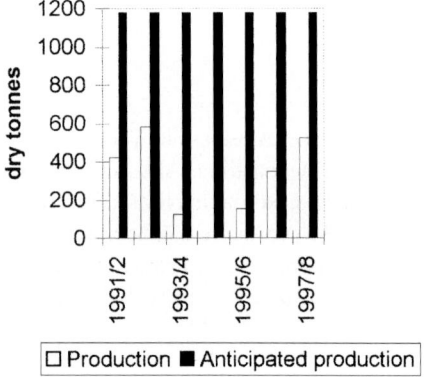

Figure 1. History of kelp stipe collections from Cape Agulhas to Strand, south coast, South Africa.

1985). Since then, production has again increased, but has yet to reach its full potential.

Production method

The brown algae are dislodged by the sea predominantly during the strong winter storms and deposited on the coast. Seaweed concession-holders and their employees visit specific areas from time to time and pull the cast material above the high water mark for drying. The wet to dry ratio is 8:1, therefore drying is organized as close as possible to the wash-up site. In certain areas, for instance near holiday resorts, however, the kelp must be removed to satisfy health regulations or aesthetic demands. Once the material is semi-dry, and after holdfasts and fronds which are unwanted in the algin production process have been removed, it is taken to the operator's final drying area. The kelp is then dried to a moisture content of not more than 17%. Once it has reached this dryness, tested by experience when the kelp particle breaks but does not not bend under pressure, the dry kelp is milled in a hammer mill, through a 2 cm screen into so-called kelp chips.

These are then put through a classified and sorted into 4 sizes:

   over 1.5 cm chip size: 75% of production
   between 1.5–1.0 cm: 15% of production
   between 1.0–0.3 cm: 5% of production
   below 0.5 cm: 5% of production

These products are then packed into woven polypropylene bags (Taurus, in litt., 1997) and stored, to be shipped in container lots to the customer.

Wet kelp is harvested by divers who cut the kelp in predetermined areas in the sea. *Ecklonia maxima* is cut just above the holdfast and allowed to drift to the surface, from where it is then cast on to the shore by the sea. This fresh product is then further processed, either into a growth stimulant or abalone feed (Levitt et al., 1992).

Property rights

The rights to the seaweeds along the South African coast are awarded by the Chief Director of Sea Fisheries. The South African coastline is divided into 17 concession areas and exclusive non-transferable permits, with a five year validity period, are issued to concession-holders.

Permit-holders must pay an annual fee together with a levy per dry tonne of seaweed collected. Collectors, who can be sub-contractors, together with their employees, must be identifiable and are the only ones allowed to collect seaweed (Levin & Share, in litt., 1997). Production figures must be submitted monthly to the Department of Sea Fisheries and severe restrictions and conditions are attached to these permits which are controlled and enforced by Sea Fisheries inspectors. The permits may be extended depending on performance.

The system, which was originally based on The Sea Fisheries Act of 1935 and which has changed frequently since that date (Rotmann, 1985), is once again under review, as quoted: 'Marine resources must be managed and controlled for the benefit of all South Africans, especially those communities whose livelihood depend on the resources of the sea. The fishing stock must be managed in such a way that promotes sustainable yield and the development of new species. The democratic Government must assist people to have access to these resources. Legislative measures must be introduced to establish democratic structures for management of sea resources' (Share et al., 1996).

A question mark has thus been put against the ownership of the resources which, in turn, militates against continuity and thus investments.

Present scope

Currently the South African kelp industry serves five distinct product markets, from algin raw material to soil improvers. It provides jobs to approximately 150 employees (Levin, pers. com.). The industry is thus important in creating work opportunities in rural areas that have otherwise no immediate potential for job creation.

Abalone Feed

South Africa is famous for its abalone, *Haliotes midae*, locally known as 'perlemoen'. The area between Cape Agulhas and Cape Point is especially known for this delicacy which is reported to have aphrodisiacal qualities and thus finds a large market with excellent prices in the Far East. This popularity has led to constant over-exploitation of the resource with a resultant significant decrease in the population. Various companies have therefore started cultivating these molluscs.

Once the juveniles are big enough (30 mm or more) they begin to feed on kelp fronds. The mariculture units, therefore, use freshly cut *Ecklonia maxima* fronds which are introduced into the on-growing tanks. On average an abalone will consume fresh kelp at a rate of between 7 and 10% of its body mass from juvenile to market size on a daily basis. It is anticipated that, in time, the abalone cultivation will demand a production of 4320 tonnes fresh *E. maxima* fronds per annum. Currently this market for fresh kelp is relatively small but, as the current development work transforms into economical operations, the demand will increase towards the aforementioned estimate by the year 2000. As the abalone feed requires fronds only, representing approximately 45% of the whole plant, this would account for about 10 000 wet tonnes total crop usage (Anderson, pers. com. 1998).

Alginate raw material

Traditionally, the alginate raw material market is the most important for the kelp industry in South Africa. Since the early 1950s, dried kelp was supplied initially to alginate manufacturers in the U.S.A., U.K., Norway, France, Japan, and lately to China and Korea. The prices of alginate raw materials have fluctuated

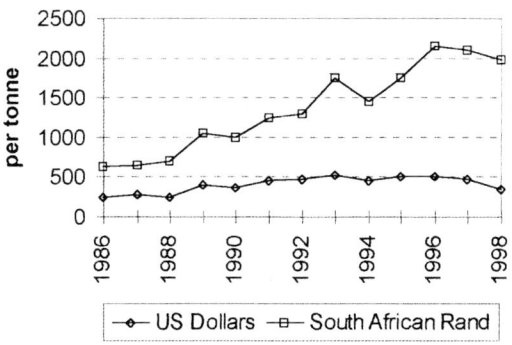

Figure 2. Price history for South African kelp chips.

widely and competition from local production in these areas as well as Chilean, Argentinean and Tasmanian raw material has been severe. Price fluctuations over time are indicated in Figure 2. Currently, however, possibly due to the *el Nino* phenomenon, prices are rising and new markets are opening.

Fish feed

Kelp powder is included in small quantities in fish feed formulations. A typical formulation is as follows:

fish meal 50%
fish oil 20%
moisture 10%
kelp powder 1–2%
rest: starch, soya bean meal, vitamins, colour, etc.

This formulation is then used in either extruded pellets or for a powder which is fed to yellowtail and other fish. The function of the kelp powder is to give a good smell and taste to the pellet and to bind it. The market size for kelp powder in aqua feed in Japan alone is estimated to be 3000–5000 tonnes. South Africa supplies the Japanese and Taiwanese markets with kelp powder for this application (Mitsui, in litt. 1998).

Artificial feed has been developed for abalone farming in South Africa (Britz et al., 1994) and a market is opening here as farmed abalone populations increase and feed demand rises.

AFRIKELP

The absence of a meaningful market for kelp products has led to research by scientists of the University of the Witwatersrand into potential other uses of kelp to support the industry (Weiersbye, pers.com.). The water-retaining properties of the algin, together with the mineral and trace element contents of the kelp, as well as the effect of growth hormones on terrestrial plants, were investigated and a product, AFRIKELP, resulted that has a clear effect on the germination, growth, production and appearance of the green plant (Weiersbye et al., in press). AFRIKELP has been shown, under drought conditions, to improve the foliar development, flowering, plant height and dry mass, as well as seed germination and establishment (Weiersbye, pers. com.).

It is claimed that AFRIKELP acts as a water and nutrient reservoir which slow-releases, thereby reducing watering times and leaching, that it increases the soil fertility by acting as a cation exchanger and increases aeration as well as aggregate soil stability and that it buffers soil temperatures and reduces soil salinity thereby acting as an environmentally friendly labour-, time- and cost-saving bio-product. The product has just been developed and thoroughly tested and will be marketed worldwide in the near future (Critchley, van Staden & Stirk, pers, com.).

Growth hormones

Natural populations of E. maxima are harvested on the west coast of the Cape peninsula for raw material to produce a liquid plant growth stimulant. The active ingredients, auxins and cytokinins, improve the general condition of green plants and in particular their root systems, resistance to drought and disease, as well as increasing their crop yield. The liquidized seaweed paste is marketed successfully in South Africa and globally (Critchley & Rotmann, 1992).

Potential

Problems facing the industry

The major factor militating against the further development of the industry is the unpredictable and generally low price obtainable on the world market for algin raw material. Severe and strong competition, together with relatively high costs in South Africa *vis-à-vis* other producing countries often price South African kelp out of the raw material market. The high costs in South Africa are caused by rapidly rising labour charges, as well as the distribution of the resource along a relatively long and rugged coastline with an insufficient infrastructure.

Recent political changes have caused some uncertainty over future property rights, leading to a reluctance to invest in research and development.

In spite of South Africa's long coastline, consumers are not generally marine-orientated and seaweeds do not command a home market, whether as a consumer product or as a raw material for hydrocolloids. This forces the South African producer to enter the world market resulting in additional transport and marketing expenses.

Future potential

South Africa has the advantage of a depreciating currency. Over the past six years the currency has depreciated from South African Rand 2.7 per US$ to South African Rand 5 per US$. This, to some extent, softens the effect of distance from market and high internal costs.

The future of the industry, however, lies in producing value-added products from the kelps. The tendency will be away from supplying commodities like algin raw material towards value-added products like growth hormones, soil improvers, and further to specialized fish feed formulations and even semi-refined or specialized alginates. There exists a potential in certain food products as well as cosmetics (Levin & Share, in litt., 1997).

Conclusion

The southern cape area of South Africa has sizeable brown seaweed resources, which are currently not fully utilized, due to market and other constraints. Development directed towards value-added products has started and is seen as a means towards expanding the industry to its optimum, thereby creating very necessary employment opportunities. Substantially more research and development is necessary to achieve this goal, but uncertainties currently discourage the necessary funding.

References

Anderson, R. J., R. H. Simons & N. G. Jarman, 1989. Commercial seaweeds in southern Africa: review of utilization and research. S. Afr. J. mar. Sci., 8: 277–299.

Britz, P. J., T. Hecht & M. G. Dixon, 1994. The development of an artificial feed for abalone farming. S. Afr. J. mar. Sci. 90: 7–8.

Critchley, A. T. & K. W. G. Rotmann, 1992. Industrial processing of seaweeds in Africa: The South African experience. In Mshigeni, M. E., J. Bolton, A. Critchley & G. Kiangi (eds), Proceedings of the First International Workshop on Sustainable Seaweed Resource Development in Sub-Saharan Africa. K E Mshigeni, Windhoek, Namibia: 85–97.

Levin, J., 1996. A socio-economic study of the seaweed industry along the west coast of South Africa: access rights and the potential for community development. Honours Thesis, University of Cape Town, South Africa.

Levitt, G., R. J. Anderson., R. H. Simons & N. G. Jarman. 1992. Past, present and future utilisation of South African Laminariales. In Mshigeni, M. E., J. Bolton, A. Critchley & G. Kiangi (eds), Proceedings of the First International Workshop on Sustainable Seaweed Resource Development in Sub-Saharan Africa. K E Mshigeni, Windhoek, Namibia: 171–187.

Rotmann, K. W. G., 1985. A strategic plan for the establishment of an integrated seaweed industry in southern Africa. M. Com. Dissertation, University of the Witwatersrand, South Africa.

Share, A., R. J. Anderson, J. J. Bolton, C. McQueen & G. Freese, 1996. South African seaweed resources: towards the development of an appropriate management policy. In Björk, M., A. K. Semesi, M. Pedersén & B. Bergman (eds), Current Trends in Marine Botanical Research in the East Africa Region: 175–185.

Weiersbye, I. M., L. B. Otter, N. S. Eccles & A. T. Critchley, in press. Constraints to the rehabilitation of degraded soils and enhanced revegetation using kelp (*Ecklonia maxima*) as soil amendment. Acta Terramis. Proceedings of the First South American Workshop on Biotechnology, Caracas, Venezuela.

Gelidium robustum agar: quality characteristics from exploited beds and seasonality from an unexploited bed at Southern Baja California, México

Y. Freile-Pelegrín[1], D. Robledo[1] & E. Serviere-Zaragoza[2]
[1]*CINVESTAV-Unidad Mérida, A.P. 73 Cordemex 97310, Mérida, Yucatán, México*
E-mail: freile@kin.cieamer.conacyt.mx
[2]*CIBNOR, A.P. 128 La Paz, Baja California Sur, México*

Key words: Gelidium robustum, agar, seasonality

Abstract

The yield and gel properties of agar from *Gelidium robustum*, harvested in Baja California for industrial production is affected by season of collection and epiphyte loading. The alga is epiphytized to various extents by the bryozoan *Membraniphora tuberculata* ('conchilla') and the resulting calcareous crust on the alga diminishes the price of the seaweed biomass. Classification of the algal biomass quality by the agar industry is based on the apparent 'conchilla' content from visual examination. The different quality classes can be categorized quantitatively into premium class (30–40% w/w of 'conchilla' load), 2nd class ($\sim$ 50% w/w) and 3rd class (> 60% w/w). For samples collected at two exploited beds, the biomass obtained from Bahía de Tortugas had a lower epiphytic coverage than that from Bahía Asunción. The agar yield from different quality classes of *G. robustum* was strongly affected by the bryozoan epiphytic coverage, while its gel characteristics were not. Algae collected at Punta Prieta, an unexploited bed not affected by 'conchilla', showed seasonality in agar yield. It ranged between 17.5 and 44.2% with two maximum values observed, one in summer and the other in winter. Gel strength ranged between 515 and 665 g cm^{-2}, reaching a maximum during autumn.

Introduction

Gelidium robustum (Gardner) Hollenberg & Abbott is the principal agar source in Mexico. This species is distributed in the cold up-welling waters of Baja California, from Punta Descanso (32° 16′ N, 117° 02′ W) to Isla Margarita (24° 26′ N, 111° 51′ W), growing from the intertidal to 15–20 m depth in areas of high water motion.

The Mexican agar industry started in 1956 when a private company got a concession to harvest *G. robustum*, gathering an average of 650 dry t annually, to produce between 40 to 75 t of agar. On the other hand, fishermen cooperatives harvest between 100–150 dry t annually, which are sold directly to the private company or exported to the U.S.A. (Robledo, 1998). *G. robustum* is harvested throughout the year, but the best period is between May and September due to optimum weather conditions (Zertuche-González, 1993). The biochemistry of agar are known to change in response to variations in environmental conditions (Mouradi-Givernaud et al., 1992; Freile-Pelegrín et al., 1995). This is especially important on the southern Baja California coast where environmental factors can be considerably affected by different phenomena (i.e. El Niño-Southern Oscillation). One problem associated with *G. robustum* beds is related with the epiphytic bryozoan, *Membraniphora tuberculata* Bosc. ('conchilla') which forms a calcareous covering on the alga and lowers its the commercial value.

Gelidium robustum is considered the second most important seaweed resource in México after *Macrocystis pyrifera* (L.) C. Agardh. Although the former is important for the Mexican agar industry, its agar content and physico-chemical properties have little been studied and no data are available on agar characteristics from more or less epiphytized algae. Espinoza & Rodríguez (1992) reported seasonal variations in yield and gel strength on *G. robustum* agar from the Mexican coast, however, the methods used in their study were not classical. On the other hand, two agar seasonality studies exist for *G. robustum* harvested in Northern California, U.S.A. (Cooper & Johnstone, 1944; Barilotti & Silverthorne, 1972) showing completely different seasonal patterns in the agar content. These contradictions may indicate that major differences can exist among geographically separated populations as well as between those in the subtidal and intertidal zones. The aim of this study was to compare the agar characteristics from three different quality classes of *G. robustum* in relation to their epiphytic covering by 'conchilla' and collected from two industrially exploited beds, Bahía Tortugas and Bahía Asunción. On the other hand, the agar characteristics from algae collected from an unexploited bed, Punta Prieta, are discussed in terms of seasonality.

Material and methods

Plant collection

The study area is located on the northwest coast of Southern Baja California. It is divided into six zones corresponding to different cooperatives (Figure 1). The local industry has classified three categories of *G. robustum* biomass based on minor to major epiphyte loading, referred to as 'conchilla': premium, second and third quality classes. Samples of the three quality classes were collected from the biomass harvested in August 1997 by fishermen cooperatives at two exploited beds, Bahía Tortugas (zone 3) and Bahía Asunción (zone 5). The harvest at both locations was done by diving between 5 to 10 m depth using an air compressor in the boat (Hooka equipment). Epiphyte loading for each quality class was expressed as a percentage of epiphyte weight present on the original sample.

The material for the agar seasonal study was collected from October 1995 to August 1996 by scuba diving at 3 m depth at the unexploited bed of Punta Prieta (zone 6). Harvested plants were transported to the laboratory, washed thoroughly with tap-water and dried in an oven at 60 °C for 48 h. Subsamples of fresh material ($n = 4$) were weighed, cleaned thoroughly of epiphytes and re-weighed in order to express agar yield as dry weight percent of epiphyte free seaweed.

Agar extraction

Dry samples (25 g) were soaked in a 0.5% w/v solution of Na_2CO_3 at 85–90 °C for 30 min prior to extraction. In order to eliminate excess alkali, the seaweeds were washed with running tap-water for 10 min. Agar extractions ($n = 3$) were carried out following the method described by Freile-Pelegrín et al. (1995).

Gel properties

Agar gel strength, melting and gelling temperatures, were measured in triplicate on 1.5% w/v solutions, made with ground dry agar (Tecator Cyclotec mill, 0.5 mm particle size) in deionized water. Gel strength (g cm^{-2}) was measured after standing overnight at room temperature, by measuring the load that causes a cylindrical plunger (1 cm^2 cross section) to break the gel within 20 s (Armisén & Galatas, 1987).

Gelling temperature was measured as follows: a hot agar solution (10 ml) was poured into a test tube containing a glass bead (5 mm diameter). The tube was tilted up and down in a water bath at room temperature until the glass bead ceased moving. The temperature in the gel was then immediately measured with a precision thermometer (0.1 °C divisions).

Melting temperature of the gel in a test tube was measured by placing a lead bead (9 mm diameter) on the gel surface. The test tube was clamped in a water bath and the temperature raised from 50 to 100 °C at 1 °C min^{-1}; the melting point was recorded with a precision thermometer when the bead sank into the solution.

Chemical characteristics

The weight percentage of sulphate was determined in triplicate by hydrolyzing 1 g of agar powder (previously dried at 105 °C) with 10 ml HNO_3 in 100 ml Kjeldahl flasks. The completely hydrolyzed ester sulphates were quantitatively precipitated with barium chloride, collected on ash-free gravimetric paper filters, dried, ignited at 600 °C for 3 h and precisely weighed. The weight of the obtained barium sulphate, multiplied by 0.4116, gave the weight equivalent sulphate.

Statistical analysis

Data were tested for normality (Kolmogorov-Smirnov), and subjected to the Bartlett's test for homogeneity of group variances using a statistical software package (Stasoft). Pearson's product moment correlation test was used to determine the linear correlation between epiphyte load, agar yield and physico-chemical properties. ANOVA comparison of means was done using a Tukey's HSD test to determine the seasonal-

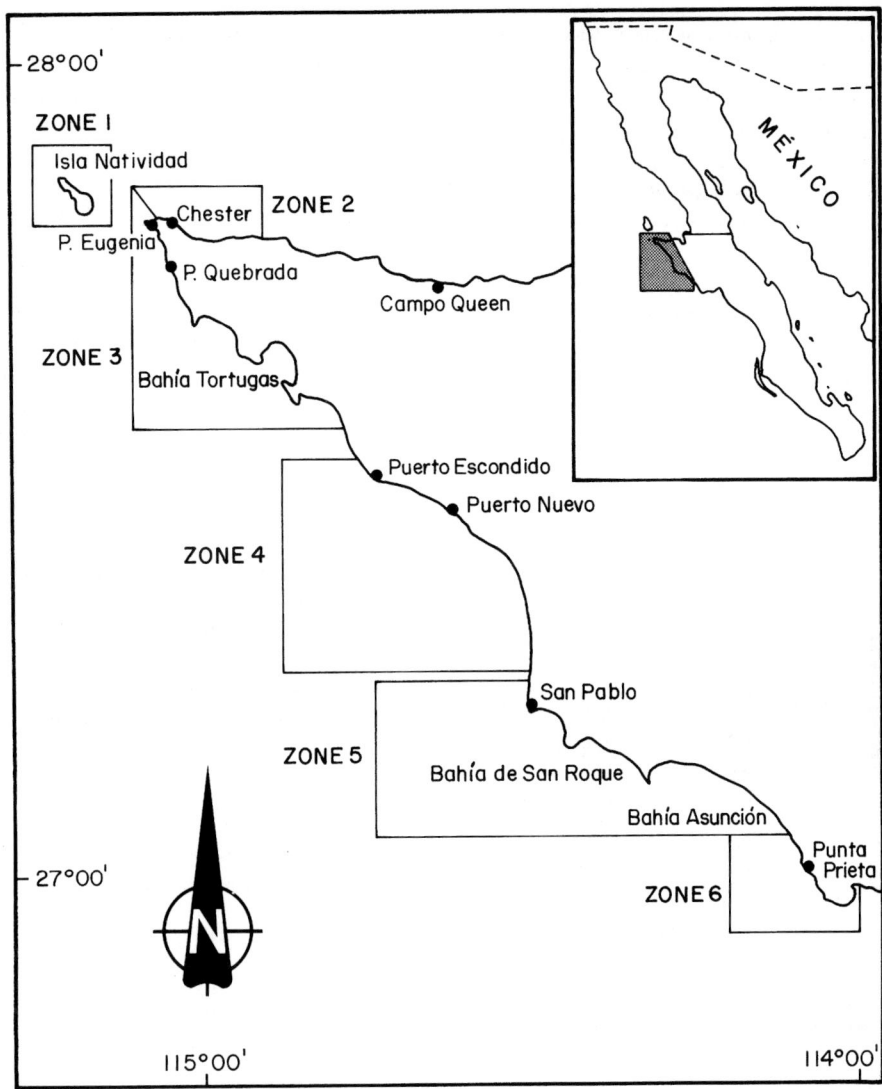

Figure 1. Map of Southern Baja California showing the distinct zones where *Gelidium robustum* is exploited.

ity and difference among the three quality classes. Groups with heterogeneous characters were tested using non-parametric Kruskal-Wallis one way analyses of variance.

Results

Quality classes at exploited beds

The epiphyte load of *G. robustum* from Bahía Tortugas and Bahía Asunción was mainly composed of the bryozoan epiphyte *Membraniphora tuberculata*. Samples collected at Bahía Tortugas showed an epiphyte content ('conchilla') varying between 31.2% w/w for premium class biomass, to 64.5% w/w for the 3rd quality class (Table 1). The epiphyte load of samples collected at Bahía Asunción was slightly higher for all quality classes, ranging from 44.5% w/w for the premium class, to 69.4% w/w for the 3rd class (Table 1).

The agar yield ranged between 21.1% for premium quality class to 13.9% for third class, from samples collected at Bahía Tortugas. The agar content in Bahía Asunción samples was slightly lower, and ranged between 19.2% for premium class to 10.5% for third class (Figure 2). ANOVA showed a significant dif-

Table 1. Epiphyte load ('conchilla') and physical properties of agar from three G. robustum quality classes at two exploited beds. Data are mean ± standard deviation.

	Epiphyte load (%)	Gel strength g cm^{-2}	Gelling temp. (°C)	Melting temp. (°C)
Bahía Tortugas				
1st class	31.2 ± 2.5	620 ± 14	35.0 ± 0.0	82.5 ± 0.7
2nd class	53.7 ± 2.5	623 ± 40	35.3 ± 0.2	89.2 ± 0.2
3rd class	64.5 ± 1.4	567 ± 22	34.4 ± 0.4	86.2 ± 1.0
Bahía Asunción				
1st class	44.5 ± 8.4	582 ± 27	34.7 ± 0.2	85.5 ± 0.0
2nd class	52.0 ± 3.1	650 ± 14	33.7 ± 0.5	86.2 ± 0.8
3rd class	69.4 ± 3.3	643 ± 41	35.0 ± 0.8	84.0 ± 0.7

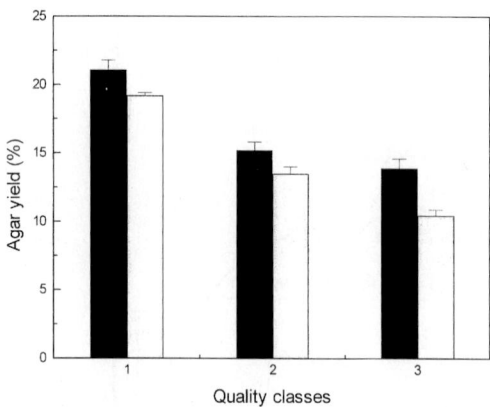

Figure 2. Agar yield of G. robustum from the three quality classes (premium, 2nd and 3rd) collected at Bahía Tortugas (black bar) and at Bahía Asunción (white bar). Bars represent standard deviation.

ference between the epiphyte load and the agar yield for the samples from the three quality classes at both localities ($p < 0.01$; Table 2). There was an inverse correlation between the epiphyte load and the agar yield for G. robustum from Bahía Tortugas ($r = -0.93$, $p < 0.01$) and from Bahía Asunción ($r = -0.97$, $p < 0.01$). However, no correlation was found between the epiphyte load and the agar gel strength, or between the former and the gelling and melting temperatures. The agar gel strength varied for all classes between 567 and 623 g cm^{-2} at Bahía Tortugas, and between 582 and 650 g cm^{-2} at Bahía Asunción, but there was no defined pattern between the two localities (Table 1). Gelling and melting temperatures of agar from the three quality classes of both localities are shown in Table 1. ANOVA showed no significant difference between the physical properties of agar from the three quality classes of both localities, except for the melting temperature at Bahía Tortugas (Table 2).

Seasonality of agar yield and characteristics from G. robustum at unexploited bed

The epiphyte load of G. robustum from Punta Prieta was mainly composed, throughout the year, of the red seaweeds *Laurencia* sp. and *Cryptopleura* spp. and by a very small amount of the bryozoan *Membraniphora tuberculata*.

The agar yield ranged between 17.5 and 44.2% reaching two maximum values, the highest in summer and another peak in winter months (Figure 3a). The agar gel strength ranged between 515 and 665 g cm^{-2}, with a maximum during autumn. In general, the gel strength decreased toward spring with a slight increase during the summer months (Figure 3b). Although the gel strength did not show any significant correlation with the agar content, the highest value of the former coincided with the minimum agar yield. The sulphate content ranged from 2.74 to 2.12% and was highest in autumn (Figure 3b). The agar gelling temperatures were higher in spring with a mean value of 35.7 °C, decreasing to a minimum in autumn (34.0 °C). The agar melting temperatures showed a similar seasonal pattern, ranging from 87.2 °C to 82.5 °C (Figure 3c). There was a positive correlation between the gelling and melting temperatures ($r = 0.75$, $p < 0.05$), and both were inversely correlated with the gel strength ($r = -0.78$, $p < 0.01$ for gelling temperature; $r = -0.85$, $p < 0.05$ for melting temperature). ANOVA showed a seasonal variation in agar content ($p < 0.01$)

Table 2. ANOVA test for significance differences among the three quality classes of G. robustum in epiphyte load and agar characteristics

Variable	df	F	p
Bahia Tortugas			
Epiphyte load (%)	2	182.73	0.0000^a
Agar yield (%)	2	70.74	0.0000^a
Gel strength (g cm^{-2})	2	1.70	NS
Gelling temp. (°C)^b	2	—	NS
Melting temp. (°C)	2	45.59	0.0006^a
Bahía Asunción			
Epiphyte load (%)	2	16.01	0.0011^a
Agar yield (%)	2	180.85	0.0000^a
Gel strength (g cm^{-2})	2	3.24	NS
Gelling temp. (°C)	2	2.62	NS
Melting temp. (°C)^b	2	—	NS

^{a}Highly significant ($p < 0.01$); NS = no significant ($p > 0.05$).
^{b}Data with heterogeneous character.

and physico-chemical properties ($p < 0.01$) for the samples collected at Punta Prieta.

Discussion

Classification of algal biomass, made by fishermen cooperatives and the agar industry, is based on the apparent 'conchilla' content from visual examination. Based on our results, the different quality classes can be categorized quantitatively into premium class with a 'conchilla' content ranging from 30 to 40%, 2nd class with approximately 50%, and 3rd class above 60% w/w bryozoan epiphyte content. Bahía Tortugas is the zone with the highest *G. robustum* harvest values, corresponding to 60% of the total agar production from southern Baja California (Casas-Valdez & Fajardo, 1990). This may be related to the lower amount of bryozoan found for all *G. robustum* biomass classes collected at this site, compared with samples from Bahía Asunción.

In general, the price of seaweeds is related to quality class which depend on 'conchilla' load. This study has shown that agar content, but not its quality is dependent on epiphytic load. This is in agreement with recent studies on the agar from *Gelidium canariensis* (Grunow) Seoane-Camba (Freile-Pelegrín et al., 1995) and *Pterocladia capillacea* (Gmelin) Bornet et Thuret (Freile-Pelegrín et al., 1996), where different

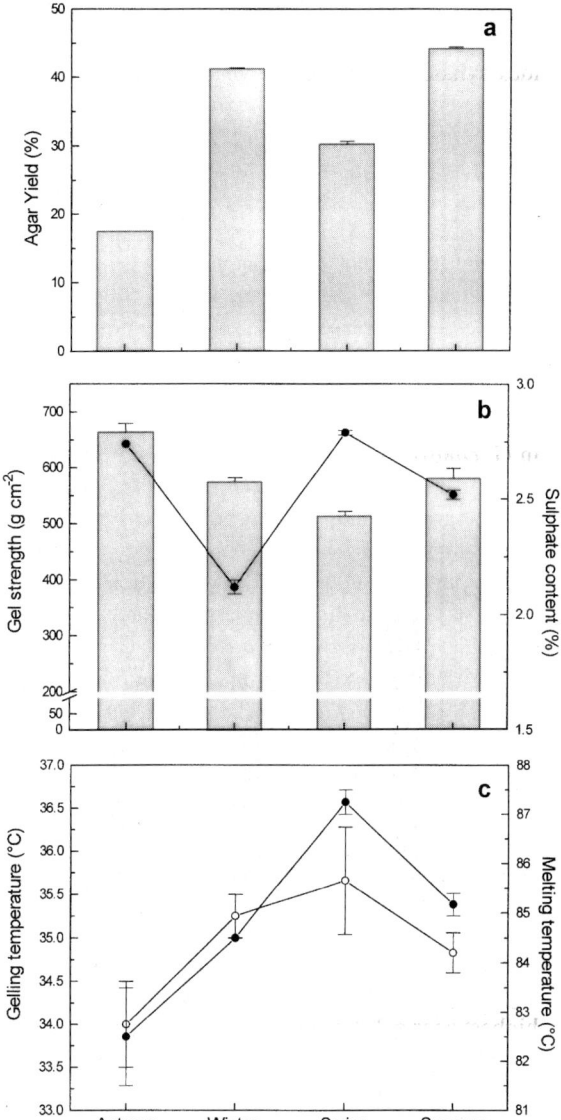

Figure 3. Seasonality of agar from *G. robustum* collected at Punta Prieta. (a) Agar yield expressed as percent of dry weight from cleaned seaweed. (b) Gel strength of 1.5% agar solution (bars) and sulphate content in agar samples (—●—). (c) Gelling (o) and melting (●) temperatures of 1.5% agar solution. Bars represent standard deviation.

epiphytic coverings of two populations did not affect quality characteristics.

On the other hand, reduction of the amount of light available for the algae can lower the agar yields (Santelices, 1988). Cancino et al. (1987) have shown that thalli of *Gelidium rex* Santelices et Abbot encrusted with *M. tuberculata* receive a lower irradiance and have a lower photosynthetic rate than non-encrusted thalli. According to Torres et al. (1991), maximal photosynthetic rates increase cell-wall polysaccharides synthesis in *Gelidium sesquipedale* (Clemente) Thuret. This may explain the differences found in *G. robustum* agar yield for the different quality classes.

In relation with the seasonal study, algae were collected at four periods over the year representing the most contrasting weather conditions in the region. A seasonal pattern for the agar yield and quality characteristics of *G. robustum* was shown. In general, the agar content increased towards the summer months, reaching values considered higher than commercial yields (17–25%). Similar results have been found for the same species (Barilotti & Silverthorne, 1972) and in *G. canariensis* collected at similar latitude (Freile-Pelegrín et al., 1995). The second peak may be related to slow growth rates influenced by the low seawater temperature during winter (Guzmán del Próo & De la Campa Guzmán, 1979). Under this condition, agar synthesis occurs at the expense of biomass production, as it has already been shown for other *Gelidium* species (Mouradi-Givernaud et al., 1992; Freile-Pelegrín et al., 1995).

During the spring months, when a decrease in temperature and high nutrient levels prevail due to up-welling conditions in the area, the lowest gel strength is found. Although nutrient enrichment has been proven to increase agar quality (Bird et al., 1981; Santelices, 1988), in *G. robustum*, the effect of high nutrient uptake is not reflected on the agar gel strength until summer when the plant reaches its highest growth (Barilotti & Silverthorne, 1972). The highest agar gel strength recorded for the autumn collected algae, could be related to an increase in water movement due to storms in this season, therefore, enhancing nutrient uptake and increasing the nitrogen status in *G. robustum*. The sulphate content of *G. robustum* agar is within the range required by industry (Armisén & Galatas, 1987), and there was no inverse correlation between this content and gel strength as other authors reported for other *Gelidium* species (Onraët & Robertson, 1987). The gelling and melting temperatures from *G. robustum* agar are similar to those of the commercially available agars obtained from *G. sesquipedale* (García, 1988). Based on our results, it is difficult to assess the optimum time for harvesting. In fact, industry collects this alga all year round, decreasing its harvest during autumn due to bad weather conditions. However, results showed that the best quality in terms of gel strength is found during these months when values are close to those of bacteriological agar (about 600 g cm^{-2}, McHugh, 1991).

G. robustum from Southern Baja California was shown to synthesize agar in quantity and quality suitable for industrial purposes. Agar yield is strongly affected by epiphytic covering, although values range within industrial requirements (17–25%). Agar gel strength is mainly affected by environmental conditions and this has to be taken into account when producing bacteriological grade agar from *G. robustum* biomass.

Acknowledgments

The authors thank Mr Ramón Franco, S.C.P.P. California San Ignacio, and Mr. Fernando López, S.C.P.P. Leyes de Reforma for providing the biomass, and Dora Uc Estrella for her technical assistance.

References

Armisén, R. & F. Galatas, 1987. Production, properties and uses of agar. In McHugh, D. J. (ed.), Production and Utilization of Products from Commercial Seaweeds. FAO Fish. Tech. Pap. 288: 1–57.

Barilotti, C. D. & W. Silverthorne, 1972. A resource management study of *Gelidium robustum*. Proc. int. Seaweed Symp. 7: 255–261.

Bird, K. T., M. D. Hanisak & J. Ryther, 1981. Chemical quality and production of agars extracted from *Gracilaria tikvahiae* grown in different nitrogen enrichment conditions. Bot. mar. 24: 441–444.

Cancino, J. M., J. Muñoz, M. Muñoz & M. C. Orellana, 1987. Effects of the bryozoan *Membraniphora tuberculata* (Bosc.) on the photosynthesis and growth of *Gelidium rex* Santelices et Abbott. J. exp. mar. Biol. Ecol. 113: 105–112.

Casas-Valdez, M. M. & C. Fajardo, 1990. Análisis preliminar de la explotación de *Gelidium robustum* (Gardner) Hollenberg et Abbott en Baja California Sur, México. Inv. Mar. CICIMAR 5(1): 83–86.

Cooper, G. L. & G. R. Johnstone, 1944. The seasonal production of agar in *Gelidium cartilagineum* a perennial red alga. Am. J. Bot. 31: 638–640.

Espinoza, J. & H. Rodríguez, 1992. Rendimiento y fuerza de gel de *Gelidium robustum* (Gelidiales, Rhodophyta) de la parte central de la península de Baja California. Rev. Inv. Cient. 3: 1–10.

Freile-Pelegrín, Y., D. Robledo & G. García-Reina, 1995. Seasonal agar yield in *Gelidium canariensis* (Grunow) Seoane-Camba (Gelidiales, Rhodophyta) from Gran Canaria, Spain. J. appl. Phycol. 7: 141–144.

Freile-Pelegrín, Y., D. Robledo, R. Armisén & G. García-Reina, 1996. Seasonal changes in agar characteristics of two populations of *Pterocladia capillacea* in Gran Canaria, Spain. J. appl. Phycol. 8: 236–246.

García, I., 1988. Estudio de la variación estacional de la calidad y el rendimiento del agar obtenido del alga roja *Gelidium sesquipedale* (Clem.) Born. et Thur. en la costa guipuzcoana (N. de España). Inf. Tecn. Inv. Pesq. 146: 3–19.

Guzmán del Próo, S. & S. De la Campa Guzmán, 1979. *Gelidium robustum* (Florideophyceae), an agarophyte of Baja California, México. Proc. int. Seaweed Symp. 9: 303–308.

McHugh, D. J., 1991. Worldwide distribution of commercial resources of seaweeds including *Gelidium*. Hydrobiologia 221: 19–29.

Mouradi-Givernaud, A., T. Givernaud, H. Morvan & J. Cosson, 1992. Agar from *Gelidium latifolium* (Rhodophyceae, Gelidiales): biochemical composition and seasonal variations. Bot. mar. 35: 153–159.

Onraët, A. C. & B. L. Robertson, 1987, Seasonal variation in yield and properties of agar from sporophytic and gametophytic phases of *Onikusa pristoides* (Turner) Akatsuka (Gelidiaceae, Rhodophyta). Bot. mar. 30: 491–495.

Robledo, D., 1998. Seaweed resources from Mexico. In Critchley, A. & M. Ohno (eds), Seaweed Resources of the World, JICA, Nagai, Yokosuka, Japan: 331–342.

Santelices, B., 1988. Synopsis of biological data of the seaweed genera *Gelidium* and *Pterocladia* (Rhodophyta). FAO Fish. Synops. 145, 55 pp.

Torres, M., F. X. Niell & P. Algarra, 1991. Photosynthesis of *Gelidium sesquipedale*: effects of temperature and light on pigment concentration, C/N ratio and cell wall polysaccharides. Hydrobiologia 221 : 77–82.

Zertuche-González, J., 1993. Situación actual de la industria de las algas marinas productoras de ficocoloides en México. In Zertuche-González, J. A. (ed.), Situación Actual de la Industria de Macroalgas Productoras de Ficocoloides en América Latina y el Caribe. D. C. No. 13. FAO-AQUILA, Mexico DF, Mexico: 33–37.

Gracilariopsis lemaneiformis beds along the west coast of the Gulf of California, Mexico

Isaí Pacheco-Ruíz[1], José A. Zertuche-González[1], Felipe Correa-Díaz[2], Fausto Arellano-Carbajal[2] & Alfredo Chee-Barragán[1]

[1]*Instituto de Investigaciones Oceanológicas, Universidad Autónoma de Baja California, P.O. Box 453, Ensenada, Baja California, México*
E-mail: isai@bahia.ens.uabc.mx
[2]*Facultad de Ciencias Marinas, Universidad Autónoma de Baja California, P.O. Box 453, Ensenada, Baja California, México*

Key words: Gracilariopsis lemaneiformis, beds, biomass, Gulf of California, Mexico

Abstract

The seasonal variation of the biomass of *Gracilariopsis lemaneiformis* (Bory) Dawson, Acleto et Foldvik was measured for 18 months in Bahía de las Ánimas. Maximum biomass per unit area (11.1 kg wet wt m^{-2}) occurred in the spring of 1995 and most of the biomass was lost by summer. Agar gel strength and yield were 891 g cm^{-2} and 14%, respectively for spring samples. Biomass per unit area was also evaluated during spring from all the beds of *G. lemaneiformis* in 850 km of the west coast of the Gulf of California. The total biomass estimated in 1995 was 5751 ± 404 dry t. The total biomass for spring of 1996 was about 30% less (4060 ± 246 dry t). Commercial exploitation of *G. lemaneiformis* started in the west coast of the Gulf of California in 1995.

Introduction

There are at least 55 species of economically important algae in the Gulf of California (Zertuche-González et al., 1995; Pacheco-Ruíz & Zertuche-González, 1996) including various endemic species (Espinoza-Avalos, 1993). *Gracilariopsis lemaneiformis* (Bory) Dawson, Acleto et Foldvik is the most abundant red alga occurring in large beds in the subtidal zone (Pacheco-Ruíz & Zertuche-González, 1996). It may be a source of agar (Zertuche-González et al., 1995). Research on certain species of *Gracilariopsis* in the Gulf of California has produced information on its taxonomy and distribution (Setchell & Gardner, 1924; Dawson, 1949; Norris, 1975); but there are no previous studies evaluating its biomass and agar characteristics or considering its commercial exploitation.

In this study, the seasonal variation of the biomass of *G. lemaneiformis* per unit area was measured during 1995–1996 (18 months) at Bahia de Las Ánimas. Additionally, the distribution and biomass per unit area of *G. lemaneiformis* beds situated along 850 km of coastline on the west coast of Baja California were determined during the spring of 1995 and 1996, when the largest biomass was present. This project is part of a programme established to determine the seaweed resources of the Gulf of California.

Study area

The study area covered the west coast of the Gulf of California from Puertecitos to Mulege, approximately 850 km of coastline (Figure 1). This region of the coast has many bays with sandy beaches where *G. lemaneiformis* is principally found. The NW region of the Gulf of California is influenced by the waters of the Ballenas Channel that are characterized by strong tidal mixing (Bray & Robles, 1991), that results in high nutrient concentrations (Alvarez-Borrego et al., 1978). Minimum temperatures occur during winter/spring (14–18 °C) and maximum temperatures during summer/autumn (29–26 °C) (Norris, 1975; McCourt, 1984a; Bray & Robles, 1991; Pacheco-

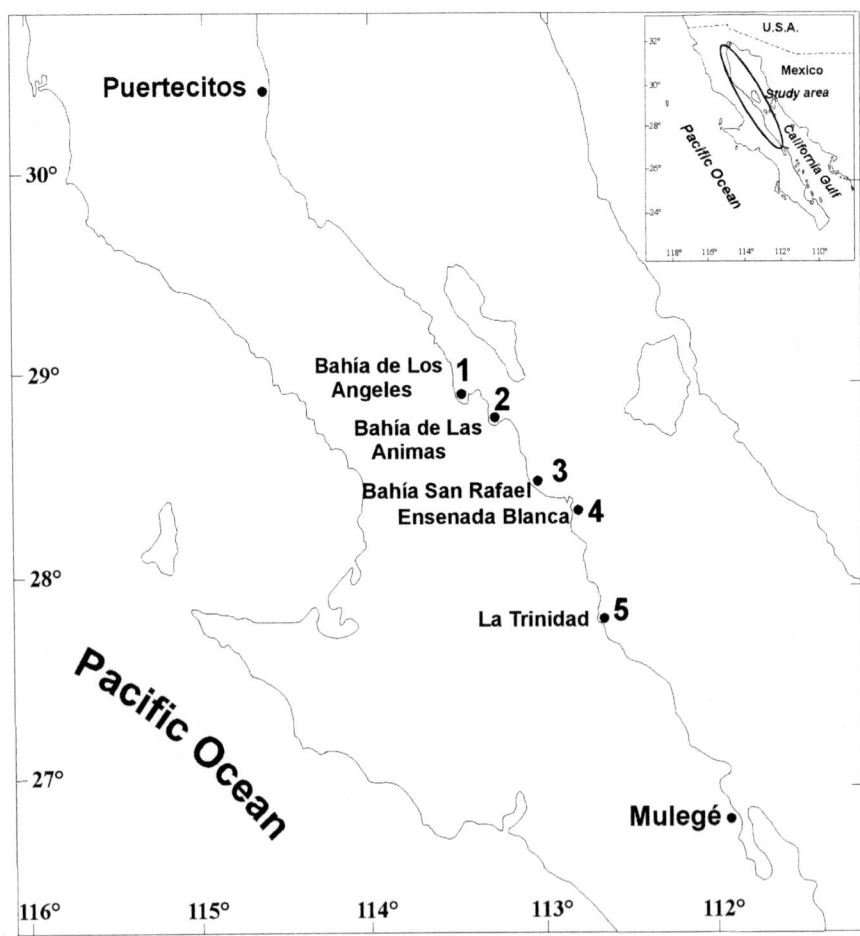

Figure 1. Gracilariopsis lemaneiformis bed localization.

Ruíz et al., 1992). In summer, nutrients are lower and the surface temperatures may reach 31 °C in the SW region (Alvarez-Borrego et al., 1978; Alvarez-Borrego, 1983). Some researchers divide this region of study into the north zone and the central zone based on oceanographic and climatic differences (Roden & Groves, 1959; Round, 1967; Espinoza-Avalos, 1993).

Materials and methods

Beds of *G. lemaneiformis* were located in late spring (May 1995) when the plants were at their greatest size and biomass (Norris, 1975) using an inflatable boat and satellite positioner (GPS, Magallan Nav 5000 DX 1500). A group of plants separated by less than 1 km and without conspicous geographic boundaries, was considered a bed. The geographic coordinates obtained were transferred to a topographic map prepared by the Instituto Nacional de Estadística Geografía e Informática (INEGI) at a scale of 1:50 000, digitized over the area of interest, and were used to determine the length of each bed. The width of each bed was variable and was measured directly using a graduated scale with transects perpendicular to the coastline at every 200 m. The area was then calculated for each bed with the computer program IDRISI ver. 4.0 [Clark University, the United Nations Environment Programme Global Resource Information Database (UNEP/GRID) and the United Nations Institute for Training and Research (UNITAR) European Office].

A quadrat of $0.25\ m^2$ was used to estimate biomass. The number of samples required to achieve 15% error for each bed was calculated using the Downing & Anderson (1985) method. The quadrats were located

Table 1. General data of beds of *Gracilariopsis lemaneiformis* in this study

Name of region	Length km	Average width (km)	Area (ha)	Beds geographic distribution N and S limits	Number of samples	Biomass dry t. May–June 1995	Biomass dry t. May–June 1996
Bahía de Los Angeles	1.5	0.1	15	29° 02′ 44″ N 113° 32′ 09″ W 28° 58′ 19″ N 113° 32′ 40″ W	40	300 ± 19	168 ± 8
Bahía de Las Animas	10.6	0.1	106	28° 54′ 08″ N 113° 22′ 14″ W 28° 48′ 28″ N 113° 21′ 39″ W	45	2064 ± 141	1376 ± 79
Bahía San Rafael	15.4	0.1	154	28° 36′ 37″ N 113° 07′ 16″ W 28° 28′ 28″ N 112° 54′ 49″ W	53	2310 ± 184	1717 ± 112
Ensenada Blanca	2.3	0.1	23	28° 25′ 08″ N 112° 50′ 55″ W 28° 24′ 08″ N 112° 51′ 34″ W	40	437 ± 26	329 ± 19
La Trinidad	8.0	0.04	32	27° 53′ 07″ N 112° 45′ 51″ W 27° 49′ 18″ N 112° 43′ 24″ W	42	640 ± 34	470 ± 28
Total			330			5751 ± 404	4060 ± 246

± = Confidence interval 95%.

haphazardly within the bed:

$$n = 5.75 x^{-0.433} A^{-0.157} p^{-2},$$

where n = number of samples, x = average biomass, A = quadrant size, p = precision or error.

Biomass (fresh weight) was obtained *in situ*. Samples were taken to the laboratory and dried at 60 °C to constant weight to determine a relationship between fresh and dry weight. Agar yield and agar strength were evaluated from a spring sample from Bahía de Las Ánimas using the methodology of Avila et al. (1989). Sulphate and 3,6-anhydrogalactose were determined using the methods proposed by Craigie & Wen (1984).

Results

Beds of *G. lemaneiformis* were located in bays and sandy beaches at depths of 2–7 m (Figure 1, Table 1). The first bed with *G. lemaneiformis* was at Bahía de Los Ángeles (29° 02′ 44″ N, 113° 32′ 09″ W), and the last bed at La Trinidad in Baja California Sur (27° 49′ 18″ N, 112° 43′ 24″ W).

The total length of *G. lemaneiformis* beds in the study region was ≈38 km, covering an area of 330 hectares. Total biomass was 5751 ± 404 dry wt t in 1995 and 4060 ± 246 dry wt t in 1996. The ratio dry:wet weight was 1:5.7.

G. lemaneiformis showed the greatest biomass per unit area in Bahía de Las Ánimas during the spring of 1995 (11.1 kg wet wt m^{-2}). Plant lengths were greater than 1.2 m. Most of the biomass was lost by summer. Gel strength and yield for spring samples were 891 g cm^{-2} and 14%, respectively (Figure 2, Table 2).

Agar yield, sulphate and 3,6-AG were; 14 ± 0.6, 0.93 ± 0.03 and 47 ± 0.8%, respectively. Gel strength was 892 ± 34 g cm^{-2}.

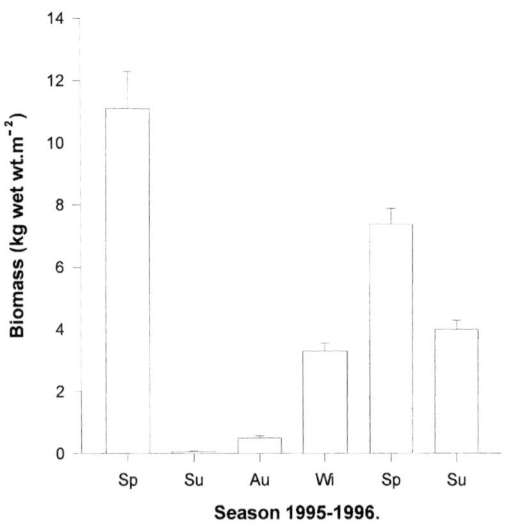

Figure 2. Biomass variation of *Gracilariopsis lemaneiformis*, in Bahía de Las Animas bed (± =SE, n= 40).

Table 2. Agar yield and quality of *G. lemaneiformis* of the bed Bahía de Las Ánimas

	Spring (May 1996)
Yield (%)	14 ± 0.6
Gel Strength (g cm^{-2})	892 ± 34
3,6-AG (%)	47 ± 0.8
Sulphate (%)	0.93 ± 0.03

(± = SE, n = 3)

Discussion

The Gulf of California is considered a subtropical region (Brusca & Wallerstein, 1979). The relatively cold water in winter and spring in north region, permits exuberant growth of macroalgae that are only present in the temperate regions of California and Baja California (Norris, 1975; Pacheco-Ruíz et al., 1992). In this study, maximum biomass of *G. lemaneiformis* occurred in the spring. This is also the case for other dominant species in the regions, such as *Sargassum johnstonii* Setch. et Gardn., *S. herporhizum* Setch. et Gardn., *S. sinicola* Setch. et Gardn., *Colpomenia sinuosa* (Roth) Derb. et Sol. and *Chondracanthus pectinatus* (Daw.) L. Aguilar et R. Aguilar in this zone (Wynne & Norris, 1976; McCourt, 1984a, b; Pacheco-Ruíz et al., 1992; Pacheco-Ruíz & Zertuche González, 1996; Pacheco-Ruíz et al., 1998). However, all of these species practically disappeared during the summer when the water temperature reached 28 °C.

The NW region of the Gulf of California is influenced by upwelling of the Ballenas Channel. As a result, the water is colder and contains higher concentrations of nutrients (Alvarez-Borrego et al., 1978; Bray & Robles, 1991) than the SW. This effect causes the presence of algal species characteristic of temperate zones (Norris, 1975), a high ratio of endemic species (Espinoza-Avalos, 1993) and seaweed biomass similar to that reported for other temperate regions of the world [e.g. *Laminaria* (4–16 kg m^{-2} wet weight) (Mann, 1972), *Ecklonia radiata* (C. Agardh) J. Agardh (6–18 kg m^{-2} wet weight) (Kirkman, 1984), *Macrocystis pyrifera* (L.) C. Agardh (4–22 kg m^{-2} wet weight) (Coon, 1982), and *Ascophyllum nodosum* (L.) Le Jolis and *Fucus vesiculosus* L. (8 kg m^{-2} wet weight) (Topinka et al., 1981)].

Gracilariopsis lemaneiformis, along with *Sargassum* sp. and *Ulva* sp., dominates the biomass along the northwest peninsula coast of Baja California (Pacheco-Ruíz & Zertuche González, 1996; Pacheco-Ruíz et al., 1998). Its agar characteristics are similar to other commercial species of *Gracilaria* (Armisen & Galatas, 1987).

The agar yield obtained in this study for *G. lemaneiformis* was relatively low compared with other species of *Gracilaria* and *Gracilariopsis* (Bird & Hinson, 1992; Chirapart et al., 1995). Agar yield differences, however, are difficult to compare due to differences in the extraction methods (McLachlan & Bird, 1986). Nevertheless, the agar quality based on agar strength, satisfied the quality required for food grade agar (Armisen & Galatas, 1987) and is similar to or better than typical commercial species of *Gracilaria* and *Gracilariopsis* (Hoyle, 1978; Whyte et al., 1981; Bird, 1988; Bird & Hinson, 1992; Luhan, 1992; Pondevida & Hurtado-Ponce, 1992).

The biomass and agar characteristics of *G. lemaneiformis* found in this study may suggest the possibility for commercial exploitation. However, it should be considered that population of *G. lemaneiformis* (as most of the benthic flora of the Gulf of the California) is subject to high variability in biomass from year to year in this subtropical region. Also, due to the annual characteristics of the plant, commercial harvesting would be limited to spring (Zertuche-González, 1994).

Based on this study, commercial trials for *G. lemaneiformis* harvesting were started in 1996 by the company PHYCOS S.A. de C.V. All the production is exported to Japan. This is the only seaweed commercially harvested from the Gulf of California and becomes the fifth seaweed species harvested commercially in Mexico along with *Gelidium robustum* (Gardn.) H. & A., *Macrocystis pyrifera* (L.) C.Ag., *Chondracanthus canaliculatus* (Harv.) Guiry and *Gracilaria pacifica* Abbott from the Pacific coast of the Baja California peninsula (Zertuche-González, 1994).

Acknowledgments

This research was sponsored by FOSIMAC (SIMAC/94/CM11) and Universidad Autónoma de Baja California (UABC). We thank Oc. Antonio Reséndiz, Oc. Alfredo Chee, Biol. Alberto Gálvez and Oc. Francisco Becerril, for logistic support in field work.

References

Alvarez-Borrego, S., 1983. Gulf of California. In Ketchum, B. H. (ed.), Estuaries and Enclosed Seas. Amsterdam, Elsevier: 427–449.

Alvarez-Borrego, S., J. A. Rivera, G. Gaxiola-Castro, M. J. Acosta-Ruíz & R. A. Schwartzlose, 1978. Nutrientes en el Golfo de California. Cienc. mar. 5: 53–71.

Armisen, R. & F. Galatas, 1987. Production, properties and uses of agar. In McHugh, D. J. (ed.), Production and Utilization of Products from Commercial Seaweeds. Roma, FAO Fish. Tech. Pap. 288: 1–57.

Ávila, M., M. J. Badilla, C. Cortes, C. Jelcez & E. Aranda, 1989. Resultados generales. Investigación, desarrollo, cultivos y uso industrial de algas *Gracilaria*. Instituto de Fomento Pesquero. Santiago de Chile, 91 pp.

Bird, K. T., 1988. Agar production and quality from *Gracilaria* sp. Strain G-16: Effects of enviromental factors. Bot. mar. 31: 33–39.

Bird, K. T. & T. K. Hinson, 1992. Seasonal variations in agar yields and quality from North Carolina agarophytes. Bot. mar. 35: 291–295.

Bray, N. A. & J. M. Robles, 1991. Physical oceanography of the Gulf of California. In Dauphin, J. P. & B. R. T. Simoneit (eds), The Gulf and Peninsular Province of the California. Tulsa, Okla.: Am. Assoc. Petrol. Geol.: 511–533.

Brusca, R. C. & B. R. Wallerstein, 1979. Zoogeographic patterns of idoteid isopods in the northeast Pacific, with a review of shallow-water zoogeography for the region. Bull. biol. Soc. Wash. 3: 67–105.

Chirapart, A., M. Ohno, H. Ukeda, M. Sawamura & H. Kusunose, 1995. Chemical composition of agars from a newly reported Japanese agarophyte, *Gracilariopsis lemaneiformis*. J. appl. Phycol. 7: 359–365.

Coon, D., 1982. Primary productivity on macroalgae in North Pacific America. In Mitsui, A. & A. C. Black (eds), CRC Handbook of Biosolar Resources. Vol. 1, P.2. Basic principles. CRC Press, Boca Raton; Fla.: 447–454.

Craigie, J. S. & Z. C. Wen, 1984. Effects of temperature and tissue age on gel strength and composition of agar from *Gracilaria tikvahiae* (Rhodophyceae). Can. J. Bot. 62: 1665–1670.

Dawson, E. Y., 1949. Estudies in the northeast Pacific Gracilareaceae. Allan Hancock Found. occ. Pap. 7: 1–104.

Downing, J. A. & M. R. Anderson, 1985. Estimating the standing biomass of aquatic macrophytes. Can. J. Fish. aquat. Sci. 42: 1860–1869.

Espinoza-Avalos, J., 1993. Macroalgas Marinas del Golfo de California. In Salazar-Vallejo, S. I. & N. E. González (eds), Biodiversidad Marina y Costera de México. Com. Nal. Biodiversidad y CIQRO, Mexico: 328–357.

Hoyle, M. D., 1978. Agar studies in two *Gracilaria* species (*G. bursapastoris* (Gmelin) Silva and *G. coronopifolia* J. Ag.) from Hawaii. II. Seasonal aspects. Bot. mar. 21: 347–352.

Kirkman, H., 1984. Standing stock and production of *Ecklonia radiata* (C. Ag.) J. Agardh. J. exp. mar. Biol. Ecol. 76: 119–130.

Luhan, Ma. R. J., 1992. Agar yield and gel strength of *Gracilaria heteroclada* collected from Iloilo, Central Philippines. Bot. mar. 35: 169–172.

Mann, K. H., 1972. Ecological energetics of the seaweeds zone in a marine bay on the Atlantic coast of Canada. II. Productivity of the seaweeds. Mar. Biol. 14: 199–209.

McCourt, R. M., 1984a. Niche differences between sympatric *Sargassum* species in the northern Gulf of California. Mar. Ecol. Progr. Ser. 18: 139–148.

McCourt, R. M., 1984b. Seasonal patterns of abundance, distribution and phenology in relation to growth strategies of three *Sargassum* species. J. exp. mar. Biol. Ecol. 74: 141–156.

McLachlan, J. & C. J. Bird, 1986. *Gracilaria* (Gigartinales, Rhodophyta) and productivity. Aquatic. Bot. 26: 27–49.

Norris, J. N., 1975. Marine algae of northern Gulf of California. Ph. D. Diss., Univ. Calif., Santa Barbara, 575 pp.

Pacheco-Ruíz, I. & J. A. Zertuche-González, 1996. The commercially valuable seaweeds of the Gulf of California. Bot. mar. 39: 201–206.

Pacheco-Ruíz, I., J. A. Zertuche-González, A. Cabello-Pasini & B. H. Brinkhuis, 1992. Growth responses and biomass variation of *Gigartina pectinata* Dawson (Rhodophyta) in the Gulf of California. J. exp. mar. Biol. Ecol. 157: 263–274.

Pacheco-Ruíz, I. & J. A. Zertuche-González, A. Chee-Barragán & R. Blanco-Betancourt, 1998. Distribution and quantification of *Sargassum* beds along the West Coast of the Gulf of California, México. Bot. mar. 41: 203–208.

Pondevida, H. B. & A. Q. Hurtado-Ponce, 1996. Assessment of some agaophytes from the coastal areas of Iloilo, Philippines; II.-Seasonal variations in the agar quality of *Gracilaria changii*, *Gracilaria manilaensis* and *Gracilariopsis bailinae* (Gracilariales, Rhodophyta). Bot. mar. 39: 123–127.

Roden, G. I. & G. W. Groves, 1959. Recent oceanographic investigations in the Gulf of California. J. mar. Res. 18: 10–45.

Round, F. E., 1967. The phytoplankton of the Gulf of California. Part 1. Its composition, distribution and contribution to the sediments. J. exp. mar. Biol. Ecol. 1: 76–97.

Setchell, W. A. & N. L. Gardner, 1924. The marine algae. Expedition of the California Academy of Sciences to the Gulf of California in 1921. Proc. Calif. Acad. Sci. 12: 695–949.

Topinka, J., L. Tucker & W. Korjeff, 1981. The distribution of fucoid macroalgal biomas along central coast of Maine. Bot. mar. 24: 11–35.

Whyte, J. N. C., J. R. Englar, R. G. Saunders & J. C. Lindsay, 1981. Seasonal variations in the biomass, quantity and quality of agar, from the reproductive and vegetative stages of *Gracilaria* (*verrucosa*) type. Bot. mar. 24: 493–501.

Wynne, M. J. & J. N. Norris, 1976. The genus *Colpomenia* Derbes et Solier (Phaeophyta) in the Gulf of California. Smithson. Contr. Bot. 35: 1–18.

Zertuche-González, J. A., 1994. Situación actual de la industria de las algas marinas productoras de ficocoloides en México. In Zertuche-González, J. A. (ed.), Situación Actual de la Industria de las Algas Marinas Productoras de Ficocoloides en América Latina y el Caribe. FAO. México: 33–37.

Zertuche-González, J. A., I. Pacheco-Ruíz & J. González-González, 1995. Macroalgas. In Fischer, W., F. Krupp, W. Schneider, C. Sommer, K. E. Carpenter & V. H. Niem (eds), Guía FAO para la Identificación de Especies Para los Fines de la Pesca Pacífico Centro-Oriental. Organización de las Naciones Unidas para la Agricultura y la Alimentación, Roma: 9–82.

Chemical composition and ^{13}C NMR spectroscopic characterisation of ulvans from *Ulva* (Ulvales, Chlorophyta)

Marc Lahaye[1]*, Enrique Alvarez-Cabal Cimadevilla[1], Ralph Kuhlenkamp[2], Bernard Quemener[1], Vincent Lognoné[2] & Patrick Dion[2]
[1]*INRA, URPOI, BP 76157, 44316 Nantes, France*
[2]*CEVA, BP 22610 Pleubian, France*

(*Author for correspondence; phone + 33-(0)2 40 67 50 63; fax + 33-(0)2 40 65 50 66; e-mail lahaye@nantes.inra.fr)

Received 1 April 1998; revised 20 August 1998; accepted 6 November 1998

Key words: ulvan, *Ulva*, HPAEC-PAD, NMR spectroscopy, iduronic acid

Abstract

The chemical composition and structures of several ulvan extracts isolated from various *Ulva* species were studied. They were all composed mainly of rhamnose, glucuronic acid, xylose, glucose and sulphate with smaller amounts of iduronic acid and traces of galactose. Proteins were also present, most likely as contaminants. Precise quantification of the uronic acid content by chemical-enzymatic hydrolysis coupled to HPAEC-PAD analysis and by colorimetry was not achieved, most likely due to the incomplete hydrolysis of glucuronan segments, inadequate HPAEC-pulsed-amperometric response factor for iduronic acid and to a possible differential colorimetric response of the two uronic acids. ^{13}C NMR spectroscopic investigation of different ulvans demonstrated that they were all based on ulvanobiuronic acid 3-sulphate A and B repeating units [β-D-Glc$p$A-(1->4)-α-L-Rha$p$3S and α-L-Ido$p$A-(1->4)-α-L-Rha$p$3S, respectively] as well as contiguous β 1->4 linked D-glucuronic acids possibly occurring either in ulvan or as a separate glucuronan. Marked variations in the content of the repeating structures were seen among the different samples. However, due to the limited number of samples studied, no conclusion was reached concerning the effects of species and ecophysiological conditions on the chemistry of ulvan.

Abbreviations: HPAEC-PAD - high performance anion exchange chromatography-pulsed amperometric detection; ^{13}C NMR spectroscopy - ^{13}C nuclear magnetic resonance spectroscopy

Introduction

The water-soluble acidic polysaccharide complex from the cell-walls of the green seaweeds *Ulva* spp., which are consumed as sea-lettuce or proliferate in eutrophicated areas, is composed of rhamnose, glucuronic acid, iduronic acid, xylose, sulphate and trace amounts of glucose, galactose and mannose (Percival & McDowell, 1967; Quemener et al., 1997). This polyelectrolyte, referred to as ulvan, is particularly resistant to biodegradation as a dietary fibre (Bobin-Dubigeon et al., 1997a, 1997b) or during biogas formation (Morand et al., 1991) and is able to gel in the presence of boron and divalent cations (Haug, 1976; Lahaye et al., 1996). In order to determine the chemical basis of its various properties, a detailed analysis of its structure was undertaken. Oligosaccharides generated by mild hydrolysis (Lahaye & Ray, 1996; Lahaye et al., 1996, 1998) and by ulvan lyase degradation (Lahaye et al., 1997) were used as model compounds in the identification of the typical ^{13}C NMR resonances of basic repeating structures of several ulvans. Two major disaccharides were identified: the ulvanobiuronic acid 3-sulphate type A and type B (Figure 1). Other minor repeating units containing

ulvanobiuronic acid 3-sulphate A

[->4)-β-D-Glc$p$A-(1->4)-α-L-Rha$p$ 3S-(1->)]$_n$

ulvanobiuronic acid 3-sulphate B

[->4)-α-L-Ido$p$A-(1->4)-α-L-Rha$p$ 3S-(1->)]$_m$

Figure 1. Chemical structure of the two major repeating units in ulvan.

glucuronic acid as a branch on O-2 of rhamnose 3-sulphate or partially sulphated xylose or contiguous glucuronic acids replacing the uronic acid in the disaccharide were also isolated. We now report on the chemical composition and structures of ulvans extracted from several *Ulva* samples in order to evaluate the variability of polysaccharide chemistry between species and in response to ecophysiological factors.

Materials and methods

Algae

Ulva rigida C. Agardh, *U. rotundata* Bliding, 'green tide' *Ulva* sp. referred to as *U. armoricana* (Dion et al., 1998), *U. olivascens* Dangeard and *U. scandinavica* Bliding were collected at several sites from the north coast of Brittany (France), from the Etang de Thau on the French Mediterranean coast and from Le Palmones estuary (Algeciras Bay, Spain). Three undefined *Ulva* samples, U2, U3 and U4 – tentatively identified as *U. gigantea* (Kütz.) Bliding – were obtained from Galicia (Spain), and commercially produced sea lettuce *Ulva* was bought from Nature-Algue (Landerneau, France, Table 1). All samples were cleaned, air dried, ground to flakes (0.5–5 mm) and stored in plastic bags at room temperature in a dry, dark place before use.

Ulvan extraction

Ulvan was extracted as described by Lahaye et al. (1998). Briefly, the algal flakes (10 g) were suspended in acidified deionized water (500 mL of ∼0.03 mol L^{-1} H$_2$SO$_4$), stirred for 30 min at room temperature and passed through a Nylon net (∼0.2 mm mesh size). The solution was discarded and the residue was resuspended in freshly prepared NaHCO$_3$ (500 mL, 0.1 mol L^{-1}), stirred 30 min in a boiling water bath and passed through a Nylon net. The solution was kept and the residue was resuspended in deionized water (500 mL) for 1 h in a boiling water bath and filtered as above. Both filtrates were combined, centrifuged (15 min, 6000 g) and filtered through a 1.2 μm membrane (Millipore). The solution was then concentrated by ultrafiltration (100 Dalton cut-off membrane) to about 100 mL, brought to pH 6.0 and stirred for 15 min at 20 °C with amyloglucosidase (Sigma 6100 units mL^{-1}) before neutralisation, addition of NaCl (0.1 mol L^{-1} final concentration) and precipitation in 4 vol. of absolute ethanol. The precipitate was recovered, washed with ethanol and dried. Before chemical analysis, ulvan samples were dissolved in deionized water, extensively dialysed, neutralised and freeze-dried. For ulvan extraction from seaweeds collected in Spain or in the Etang de Thau, the NaHCO$_3$ treatment was performed at room temperature and ultrafiltration was through a 30 KD hollow fibre cut-off cartridge (Amicon). They were repurified three times by dissolving in deionized-water at room temperature, centrifugation at 15,000 g and precipitation in 4 vol. of ethanol.

Chemical analysis

Sugar determinations were performed using the chemical-enzymatic degradation method coupled to HPAEC-PAD analysis described by Quemener et al. (1997). The different sugar response factors were calculated from standards and that of glucuronic acid was used for the iduronic acid quantification. Uronic acids were also measured by colorimetry (Thibault, 1979) using glucuronic acid as standard. Sulphate was analysed after TFA hydrolysis by HPAEC-conductimetry as described by Quemener et al. (1997). Proteins in ulvan solutions were quantified by the Bradford colorimetric assay (BioRad) with bovine serum albumin as standard. Ash was quantified gravimetrically after 12 h at 550 °C and further 4 h at 900 °C. Moisture content was determined gravimetrically after 2 h at 120 °C. All the above analyses were conducted on at least duplicate samples. PHmetric titration of ulvan was conducted on the acidic form (produced by dialysis against 0.2 mol L^{-1} HCl then against deionized water before freeze-drying). Ulvanic acid was then determined by dissolving in NaNO$_3$ (0.1 mol L^{-1}) and titration by 0.1 mol L^{-1} NaOH. ^{13}C NMR spectra were recorded at 70 °C on a Bruker ARX 400 spectrometer operating at 100.61 MHz. Chemical shifts were calculated from dimethyl sulphoxide (as internal standard) attributed to 39.6 ppm.

Table 1. Chemical composition of ulvans extracted from several species of Ulva (% dry weight).

Samples[a]	Total sugars	Sulphate	Protein	Total
Sea lettuce	74.6[b]	17.6	8.1	100.3
Ulva armoricana				
1993, St Brieuc, Brittany, France	78.0[b]	15.6	6.4[d]	100.0
9/94, St Brieuc	55.8[c]	14.0	4.1	73.9
10/94, Binic, Brittany, France	58.5[c]	14.4	3.9	76.8
4/95, St Brieuc	57.2[c]	11.7	6.5	75.4
Ulva rigida				
9/94, Pleubian, Brittany, France	75.6[c]	16.5	1.9	94.0
10/94, Roscoff, Brittany, France	70.8[c]	15.9	2.5	89.2
5/95, Etang de Thau, Languedoc, France	75.1[b]	19.8	7.2[d]	102.1
5/95, Pleubian	59.5[c]	14.3	3.5	77.3
Ulva rotundata				
9/94, Pleubian	51.2[c]	15.0	5.1	71.3
10/94, Pleubian	70.8[c]	16.5	1.5	88.8
4/95, Pleubian	49.8[c]	11.3	7.1	68.2
5/95, Pleubian	56.9[c]	13.5	3.4	73.8
10/96, Le Palmones, Algeciras Bay, Spain	65.7[b]	17.3	9.9[d]	92.9
Ulva scandinavica, 6/95, Roscoff	56.3[c]	13.1	nd[e]	
Ulva olivascens, 5/95, Pleubian	63.4[c]	13.8	4.9	82.1
U2, 10/96, Placera Peninsula, Ria de Pontevedra, Vigo, Galicia, Spain	69.3[b]	18.5	8.5[d]	96.3
U3, 10/96, Banco Placera, Ria de Pontevedra, Vigo, Galicia, Spain	91.9[b]	11.3	7.3[d]	110.5
U4 '*Ulva gigantea*', 10/96, Ria de Vigo, Vigo, Galicia, Spain	79.6[b]	11.9	5.8[d]	97.3

[a] month/year and place of collection, [b] by HPAEC for neutral sugar and colorimetry for uronic acids and [c] total sugar by HPAEC, [d] sample repurified by three cycles of resolubilisation, centrifugation and precipitation in ethanol, [e] not determined.

Results

The yields of ulvan extracts obtained from several species of *Ulva* collected at different periods of the year or at different localities were between 8 and 15% of the algal dry weight after the first ethanol precipitation, dialysis and freeze-drying. They were reduced to 4–6% for the Spanish and the Mediterranean samples purified by repeated solubilisation, centrifugation and ethanol precipitation. The chemical composition of the extracts are given in Tables 1 and 2 and indicates that the attempts to decrease the protein content by the purification method used failed. The latter was founded on the denaturation and irreversible precipitation of proteins by organic solvents such as alcohols (Scopes, 1982) and their removal by centrifugation. The low recovery yields obtained after such a purification scheme was likely to be due to losses in ulvan. The incomplete or high recoveries of the various components in this chemical analysis can be explained by the unaccounted ash fraction associated with counterions of the different charged groups and by the poor estimation of the uronic acids content. These sugars are particularly difficult to determine quantitatively by HPAEC-PAD or by colorimetry as both methods introduce errors (Table 3). Indeed, based on the pHmetric titration of ulvan from a 'green tide' sample (*U. armoricana* from 1993), 3.05 meq g^{-1} of acid groups were measured whereas 3.72 meq g^{-1} were calculated from the sum of sulphate and colorimetrically estimated uronic acids contents. In spite of these analytical limitations, all the ulvans were essentially composed of rhamnose and glucuronic acid with variable contents of xylose, iduronic acid and glucose and minor amounts of galactose (Table 2). Variations in the different sugar proportions cannot be clearly related to the species nor to ecophysiological conditions. Only the ulvan from *U. armoricana* tended to be richer in iduronic acid and that from *U. rotundata* in xylose. The slightly different ulvan extraction methods used for the Spanish and French Mediterranean *Ulva*

Table 2. Sugar molar percentages of ulvans extracted from several species of Ulva as determined by the chemical-enzymatic method coupled to HPAEC-PAD.

Samples	Rha	Gal	Glc	Xyl	GlcA	IdoA
Sea lettuce	54.8	2.1	5.6	12.4	21.5	3.7
U. armoricana						
1993, St Brieuc	48.3	3.1	17.5	10.1	15.2	5.9
9/94, St. Brieuc	51.6	1.2	8.4	9.1	22.8	7.0
10/94, Binic	53.6	1.0	6.1	7.0	25.5	6.9
4/95, St Brieuc	47.8	1.3	13.0	8.2	25.9	3.8
U. rigida						
9/94, Pleubian	50.9	1.3	6.6	7.9	28.9	4.4
10/94, Roscoff	52.2	0.9	5.3	6.5	30.4	4.7
5/95, Etang du Thau	58.3	1.7	5.0	12.0	19.0	4.0
5/95, Pleubian	52.9	1.7	7.2	6.3	29.4	2.5
U. rotundata						
9/94, Pleubian	49.6	1.5	5.4	23.8	17.8	2.0
10/94, Pleubian	53.2	1.2	5.9	8.5	25.7	5.6
4/95, Pleubian	46.8	1.5	9.5	18.7	22.5	1.0
5/95, Pleubian	54.0	1.2	7.7	5.4	28.9	2.8
10/96, Le Palmones	46.7	3.0	14.4	15.4	20.0	0.6
U. scandinavica	42.2	2.0	30.7	9.6	11.6	4.0
U. olivascens	53.8	3.0	7.6	15.1	16.7	3.8
U2	51.8	1.8	11.6	14.9	13.9	5.9
U3	27.9	1.0	38.1	16.9	12.9	3.3
U4	42.1	2.3	7.9	28.8	15.4	3.6

Table 3. Uronic acid (UA) contents (% weights) of several ulvans determined by HPAEC-PAD (GlcA, glucuronic acid; IdoA, iduronic acid) and by colorimetry (total UA) and differences (Δ) between the methods.

Samples	GlcA	IdoA	Total UA	Δ
Sea lettuce	11.6	2.0	28.1	−14.5
U. armoricana 1993	8.4	3.3	29.1	−17.4
U. rigida 5/95, Etang de Thau	10.8	3.6	25.5	−11.1
U2	8.2	3.5	18.9	−7.2
U3	7.8	2.0	34.8	−25.0
U4	6.5	1.5	37.8	−29.8
U. rotundata 10/96 Le Palmones	9.9	0.3	21.8	−11.6

samples (temperature of extraction and MW cut-off of the ultrafiltration membrane) did not markedly affect yields or composition of the extracts. The ^{13}C NMR spectra of several of these ulvans are shown in Figure 2. The signal assignments (Table 4) were done by comparison with published data for ulvan oligosaccharides and fragments (Lahaye & Ray, 1996;

Lahaye et al., 1997, 1998). In all the spectra the signals for the ulvanobiuronic acid 3-sulphate type A were observed, as were, to variable extents, those of the type B structure (Figure 2). The width or splitting observed for several signals were possible consequences of different distributions of the repeating units in the polysaccharides. The signals attributed to the iduronic acid containing sulphated disaccharide were the least intense in the *U. rotundata* (Spain) ulvan sample and were higher in those of *U. armoricana*, *U. rigida* and 'U3'. Major signals for contiguous 1,4-linked β-D-glucuronic acids, either in ulvan or as glucuronan, were observed on the spectra of 'U4', 'U3' and with lower intensities on the other spectra (Figure 2). Several signals indicated by arrows or tentatively assigned in Figure 2 may be, in part, due to the presence of xylose and glucose (X5 for Xyl C-5, X5s for Xyl2S C-5, R1x(s) for Rha3S C-1 linked to Xyl or Xyl2S, C6 for Glc C-6) in these preparations. An estimation of the iduronic acid content in ulvan from ^{13}C NMR signals clearly emphasised that the HPAEC-PAD determination of this acid sugar was underestimated. For example, iduronic acid content represented 36 or 40% of all the uronic acid in *U. armoricana* ulvan (St Brieuc, 1993) based on signal intensity or integral value, whereas HPAEC-PAD analysis yielded only 28%. Furthermore, the high intensity of the signals for contiguous 1,4-linked β-D-glucuronic acid on the spectra of U3 and U4 ulvans confirm the disagreement between the chemical-enzymatic HPAEC-PAD and colorimetric determination of glucuronic acid (Table 3).

Discussion

An in-depth knowledge of ulvan composition and chemical structure is a prerequisite to the understanding of their biological and physico-chemical properties. Furthermore, it was interesting to determine whether this information, together with their ecophysiological variability, could be used to select potential chemo-taxonomic markers for *Ulva* species. These polysaccharides, the extraction and purification of which remain to be optimised, represented about 10% of the algal dry weight and were most likely contaminated by proteins and by 1,4-linked β-D-glucuronan known to be part of these cell walls (Lahaye et al., 1996). The overall regularity of ulvan given by the great proportion of aldobiuronic acid repeating structures (Percival & McDowell, 1967)

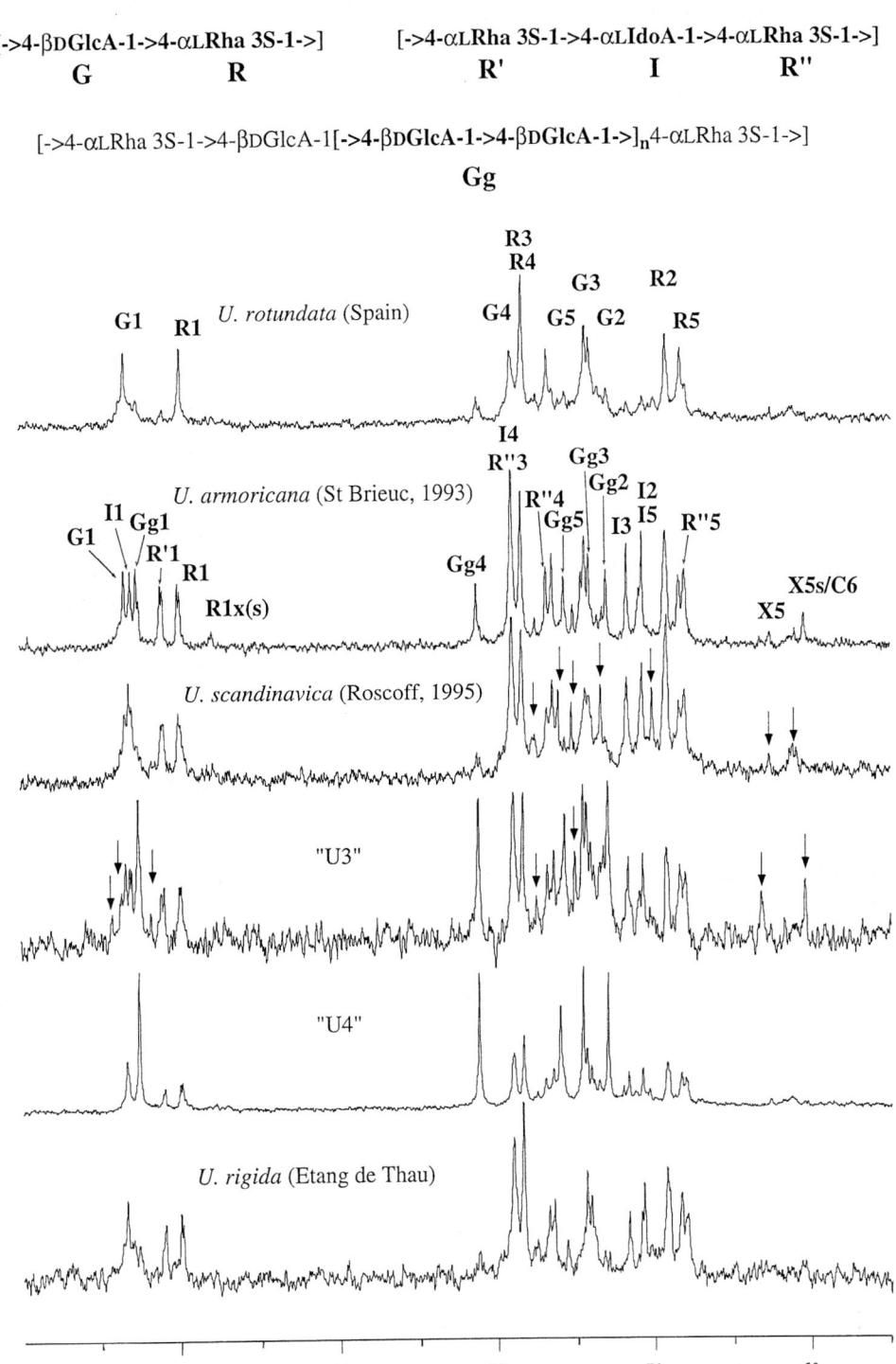

Figure 2. ^{13}C NMR spectrum of ulvan extracts from *U. rotundata* (Spain), *U. armoricana* (St Brieuc, 1993), *U. scandinavica* (Roscoff 1995), 'U3', 'U4' from Spain and *U. rigida* (Etang de Thau). Letters and numbers correspond to carbon in chemical structures depicted on top of the spectra; Gg signals correspond to carbons in glucuronan segments in ulvan as well as in contaminating 1,4-linked β-D-glucuronan.

Table 4. ^{13}C NMR chemical shifts (ppm) of main repeating units in ulvans.

Carbon	Sugar residues[a]					
	G	R	I	R'	R''	Gg
1	103.7	100.3/100.2	103.3	101.4/101.3		102.9
2	74.4	69.6	71.1		69.6	73.4
3	74.7	78.7	72.1		79.4	74.9
4	79.4	78.7	79.4		77.2	81.6
5	76.8	68.8	71.3		68.4	76.0
6	175.3	17.5	174.6		17.4	175.0

[a] letters designate sugar residues depicted in Figure 2.

suggested that a simple optimisation of the release and degradation of the acidic resistant disaccharide would be sufficient to measure precisely the ulvan sugar content (Quemener et al., 1997). However, the presence of iduronic acid complicated this analysis as there is no standard sugar available to determine its HPAEC-PAD or the colorimetric weight response factors. The presence of contiguous 1,4-linked β-D-glucuronic acid units in ulvan (Lahaye et al., 1997; Figure 2) and/or a separate 1,4-linked β-D-glucuronan constitute another limit of the HPAEC-PAD chemical-enzymatic method as the acid and enzymatic degradation conditions were most likely unable to cleave these linkages. In spite of the errors associated with the uronic acid quantification, the chemical-enzymatic method is still, in our opinion, the most suitable method for ulvan neutral sugar compositional analysis. Improvements required for a precise uronic acid content determination by HPAEC-PAD depend on the availability of standard iduronic acid and of an enzyme preparation containing both β-D-glucuronidase and glucuronanase activities.

As expected, all the ulvan samples investigated were composed mainly of rhamnose, glucuronic acid and sulphate (Percival & McDowell, 1967; Quemener et al., 1997). ^{13}C NMR spectroscopy was particularly helpful for the recognition of the major repeating units composing the different ulvan samples. These were essentially ulvanobiuronic acid 3-sulphate A and B together with 1,4-linked β-D-glucuronan as blocks within ulvan or as a separate glucuronan. Unattributed signals in these spectra also demonstrated that there are still sugar distributions and/or linkages that remain to be identified. The marked variations observed in the sugar composition and repeating structure proportions in these different ulvans cannot be related at present to seasonal or ecophysiological variations nor to species, as too few samples were studied. Medcalf et al. (1975) could not demonstrate a clear seasonality in the sulphate content and neutral sugar proportions in *U. lactuca* ulvan although they showed a lower uronic acid content in summer samples.

The presence of α-L-iduronic acid which is the epimer in C-5 of β-D-glucuronic acid is reminiscent of the mannuronic-guluronic acids and the glucuronic-iduronic acids biosynthetic filiations in alginate (Moe et al., 1995) and glycosaminoglycans (Fransson, 1985), respectively. Whether such biosynthetic pathways exist in ulvan remains to be established. Further work is also required to identify the functional role of this unusual uronic acid in the algal cell wall cohesion and in the ion exchange and gelling properties of ulvan. Interestingly, iduronic acid has been demonstrated in heparin as a more potent cation chelator than glucuronic acid (Whitfield & Sarkar, 1991, 1992; Whitfield et al., 1992).

Acknowledgements

The authors thank Prof. X. Niell (University of Malaga, Spain), Dr M. Pellegrini (Université de Marseille, France) for *Ulva* samples and Mrs C. Brunet for technical assistance. This work was financed in part by the European Union (Leonardo to EACC, RTD-programme contract ERB 4001 GT932750 to RK) and by the VANAM programme from the Région Pays de la Loire.

References

Bobin-Dubigeon C, Lahaye M, Barry J-L (1997a) Human colonic bacterial degradation of dietary fibres from sea-lettuce (*Ulva* sp.). J. Sci. Food Agric. 73: 149–159.

Bobin-Dubigeon C, Lahaye M, Guillon F, Barry J-L, Gallant DJ (1997b) Factors limiting the biodegradation of *Ulva* sp. cell wall polysaccharides. J. Sci. Food Agric. 75: 341–351.

Dion P, de Revier B, Coat G (1998) *Ulva armoricana* sp. nov. (Ulvales, Chlorophyta) from the coasts of Brittany (France). I. Morphological identification. Eur. J. Phycol. 33: 73–80.

Fransson LA (1985) Mammalian glycosaminoglycans. In Aspinall GO (ed.), The Polysaccharides Vol. 3, Academic Press, New York: 337–415.

Haug A (1976) The influence of borate and calcium on the gel formation of a sulfated polysaccharide from *Ulva lactuca*. Acta chem. Scand. B 30: 562–566.

Lahaye M, Brunel M, Bonnin E (1997) Fine chemical structure analysis of oligosaccharides produced by an ulvan-lyase degradation of the water-soluble cell-wall polysaccharides from *Ulva* sp. (Ulvales, Chlorophyta). Carbohydr. Res. 304: 325–333.

Lahaye M, Inizan F, Vigouroux J (1998) NMR analysis of the chemical structure of ulvan and of ulvan-boron complex formation. Carbohydr. Polymers 36: 239–249.

Lahaye M, Ray B (1996) Cell wall polysaccharides from the marine green algal *Ulva 'rigida'* (Ulvales, Chlorophyta). NMR analysis of ulvan oligosaccharides. Carbohydr. Res. 283: 161–173.

Lahaye M, Ray B, Baumberger S, Quemener B, Axelos MAV (1996) Chemical characterisation and gelling properties of cell wall polysaccharides from species of *Ulva* (Ulvales, Chlorophyta). Hydrobiologia 326/327: 473–480.

Medcalf DG, Lionel T, Brannon JH, Scott JR (1975) Seasonal variation in the mucilaginous polysaccharides from *Ulva lactuca*. Bot. mar. 18: 67–70.

Moe ST, Draget KI, Skjåk-Bræk G, Smidsrød O (1995) Alginates. In Stephen AM (ed.), Food Polysaccharides and their Applications. Marcel Dekker Inc., New York: 245–286.

Morand P, Carpentier B, Charlier RH, Mazé L, Orlandini M, Plunkett BA, DeWaart J (1991) Bioconversion of seaweeds. In Guiry MD, Blunden G (eds), Seaweed Resources in Europe. Uses and Potential, John Wiley, Chichester: 95–148.

Percival E, McDowell RH (1967). Chemistry and Enzymology of Marine Algal Polysaccharides. Academic Press, London. See pp. 178–186.

Quemener B, Lahaye M, Bobin-Dubigeon C (1997) Sugar determination in ulvans by a chemical-enzymatic method coupled to high performance anion exchange chromatography. J. appl. Phycol. 9: 179–188.

Scopes R (1982) Protein Purification. Principles and Practice. Springer-Verlag, Berlin. See pp. 52–59.

Thibault J-F (1979) Automatisation du dosage des substances pectiques par la méthode au méta-hydroxydiphényl. Lebensm.Wiss. Technol. 12: 247–251.

Whitfied DM, Choay J, Sarkar B (1992) Heavy metal binding to heparin disaccharides. I. Iduronic acid is the main binding site. Biopolymers 32: 585–596.

Whitfield DM, Sarkar B (1991) Metal binding to heparin monosaccharides: D-glucosamine-6-sulphate, D-glucuronic acid, and L-iduronic acid. J. inorg. Biochem. 41: 157–170.

Whitfield DM, Sarkar B (1992) Heavy metal binding to heparin disaccharides. II. First evidence for zinc chelation. Biopolymers 32: 597–619.

Direct determination of alginate content in brown algae by near infra-red (NIR) spectroscopy

Svein Jarle Horn*, Einar Moen & Kjetill Østgaard
Department of Biotechnology, The Norwegian University of Science and Technology NTNU, N-7491 Trondheim, Norway

(*Author for correspondence; phone + 47 73591687; fax + 47 73593337; e-mail svein@chembio.ntnu.no)

Received 1 April 1998; revised 23 December 1998; accepted 31 December 1998

Key words: NIR, brown algae, alginate, *Laminaria hyperborea*

Abstract

The methods used to quantify total alginate in brown algal tissue are time-consuming and may also be misleading, so faster and simpler methods for measuring alginate content would be beneficial in a variety of applications. This study reports on the use of near infra-red (NIR) analysis to monitor the alginate content of *Laminaria hyperborea* stipe during biodegradation. NIR reflectance spectra were recorded for 78 different freeze-dried samples of its stipe. The samples were collected during several biological degradation experiments and the total alginate content varied from 2.2 to 40.8% Na-alginate (w/w), determined by established methods based on ion exchange. Data analysis was performed using multivariate calibration methods in order to relate the spectral data to the alginate content. PLS2 analysis revealed some dependence on material type, probably reflecting differences in polyphenol content. In the end, a PLS1 model with 9 components was selected. The calculated model was validated both with internal data and with an external test set. Internal full cross validation explained 96.6% of the variance in alginate content. The external validation showed that the PLS1 model was able to predict the alginate concentration with a root mean square prediction accuracy of 2.1%.

Introduction

Alginate is the major structural organic component in brown algae, and may contribute up to 40% of the *Laminaria hyperborea* dry weight. This alginate may be extracted, precipitated and finally quantified by weighing (Haug, 1964). This approach may lead to underestimates due to low extraction yields and losses. Especially in the case of highly degraded alginates, as observed during biological degradation, losses will be high due to poor precipitation. As an alternative, the total alginate content may be quantified by the so called calcium-acetate method (Haug, 1964; Jensen et al., 1985). This method is based on calcium ion-exchange followed by atomic absorption analysis. However, other charged polymers may interfere with the ion-exchange analysis. This method is therefore unsuitable for analysis of brown seaweeds containing high levels of sulphated polysaccharides (Myklestad, 1968a, 1968b). Both these methods are also time consuming, and a faster and simpler method would be beneficial.

The aim of this study was to evaluate the use of near infra-red (NIR) spectroscopy of dry seaweed material as a method for direct alginate quantification. The possible advantages are obvious: NIR spectroscopy is very rapid, easily reproducible, robust and has a good penetration depth into the sample (Vårum et al., 1995). The samples originated from different biological degradation experiments (Moen et al., 1997a, 1997b) and consisted of a wide range of alginate concentrations. NIR reflectance spectra were recorded for each sample and used as a base for the data analysis.

Materials and methods

Samples and total alginate

All 78 samples to be analysed were collected during several biological degradation studies of *Laminaria hyperborea* stipe (Moen et al., 1997a, 1997b). In 37 of these samples, the thin peripheral tissue was removed by peeling prior to degradation. By doing so, most of the polyphenols, located in this outer layer, were removed, without affecting the relative alginate content of the material (Moen et al., 1997a, 1997b). Thus, our samples may be divided in two subgroups: 41 normal whole and 37 peeled stipe samples, denoted WHOLE and PEELED in this work. The samples were immediately freeze dried and then milled in a laboratory mill to pass a sieve with 1 mm holes. The total alginate content in this powder material was determined according to the calcium-acetate method, based on calcium ion exchange followed by atomic absorption analysis of the final calcium content (Haug, 1964; Jensen et al., 1985). Our samples spanned a wide range of alginate concentrations, varying from 2.2 to 40.8 g Na-alginate per 100 g dry weight of stipe.

Near infra-red spectroscopy

Visible and NIR reflectance measurements (400–2500 nm) were carried out by a scanning spectrophotometer Model 6500 from NIRSystems Inc. (Silver Springs, USA). Every second nm wavelength was recorded, giving a total of 1050 spectral variables. The whole microsample cup was filled up with seaweed powder samples to make sure that the incoming beam would be reflected. Spectra were obtained at room temperature by averaging 50 succeeding scans. Each sample was measured in three re-packed replicates, and the average value was used in the data analysis. The white ceramic plate of the sample chamber was scanned as reference between every sample measurement. Reflectance is here defined as the ratio between the intensity of the light diffusely reflected from the sample and the light reflected from the non-absorbing, diffusely reflecting reference. Thus, for reflectance measurements the Beer's law becomes: Absorbance = log(1/Reflectance).

Data analysis

Spectroscopic measurements of powders often display light scattering effects. This scattering phenomena can arise from variations in water content, surface roughness and particle size (Antti & Sjøstrøm, 1996). In this work, multiplicative scatter correction (MSC) was used as a transformation method to compensate for these effects. MSC is known to give improved performance when applied to NIR reflectance data prior to calibration (Geladi et al., 1985; Isaksson & Næs, 1988). The method separates multiplicative or scattering variations from additive or chemical information, and gives a better linear fit between spectral data and chemical composition. The MSC method used here corrects for both offset α_i and scale β_i. This is called full MSC and is given by the equation $x_{corr} = (x_{input} - \alpha_i)/\beta_i$.

The spectral data (X-data) were related to the alginate concentration and material type (Y-data) by using the multivariate calibration method partial least squares (PLS) regression. This is a linear method, which reduces the original spectral data to a few components, representing the directions of largest variance. PLS regression includes the information of the Y-data in the data analysis. In general this results in simpler models than the more common method principal component regression (PCR).

A PLS2 model was first calculated based on all the 78 samples, including both alginate content and material type (WHOLE or PEELED stipe) in the Y-data. Then, a PLS1 model with only the alginate concentration as Y-data was run. In this case, the samples were sorted according to alginate concentration, and every third sample was removed to be used as a validation set. In this way, the test samples spanned the whole experimental region. The remaining 52 samples were used for calibration of the model.

The prediction accuracy of the calibration model is described by the root mean square error of prediction (RMSEP), defined as

$$\text{RMSEP} = \sqrt{[\sum(i=1,n)(y_{i,\text{pred}} - y_{i,\text{meas}})^2 / n]}$$

where $y_{i,\text{pred}}$ is the predicted alginate concentration and $y_{i,\text{meas}}$ is the measured concentration for sample i. The symbol n is the number of cross validations or the number of samples in the test set (Martens & Næs, 1989).

All the data analyses were performed by the Unscrambler Software Version 5.5 (Camo A/S, Trondheim, Norway).

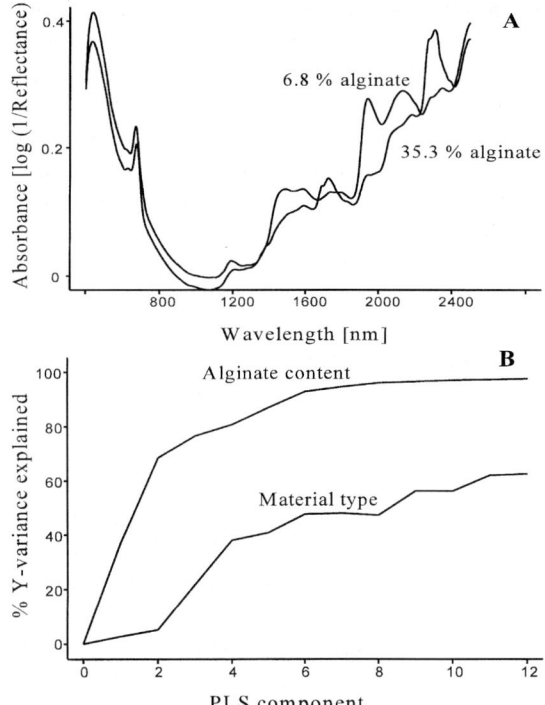

Figure 1. **A**) NIR reflectance spectra for two WHOLE stipe samples containing 6.8 and 35.3% alginate. **B**) Explained variance in the Y-data (alginate content and material type), shown as a function of the number of components in the PLS2 model.

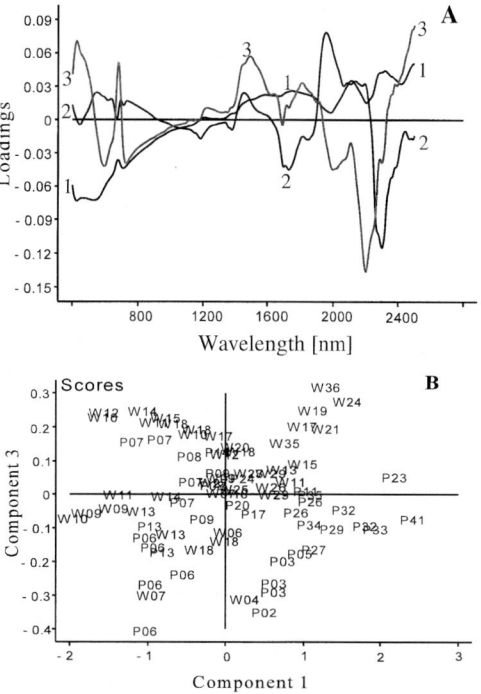

Figure 2. PLS2 model: **A**) Loading plots for the first 3 components. **B**) Score plot for component 1 (abscissa) and component 3 (ordinate). W and P refer to WHOLE and PEELED, respectively. The numbers give the alginate concentration of the samples.

Results

The spectra

Figure 1A shows the NIR reflectance spectra of two WHOLE samples after MSC transformation. Reflectance spectra of pure water have two characteristic peaks at 1442 and 1936 nm. These peaks are also found in Figure 1A, clearly showing that the water content in the samples influence the spectra.

Main predictive information

PLS2 calibration was performed on the spectra from 400 to 2500 nm with the two properties, alginate content and type of material, as Y-data. Figure 1B shows how the alginate content described the main variation in the spectra. The first two PLS components explained about 70% of the variation in alginate content, whereas a model with 8 PLS components described 95% of the variation in the Y variable.

A model extended to four components was needed to explain at least 40% of the variation in the material type. Thus, there is some correlation between the NIR spectra and the material type, but not as clear as for the alginate content. This material dependence became stronger when samples enriched in peripheral tissue, containing high concentrations of polyphenols (Moen et al., 1997a, 1997b), were included in the model (results not included).

The loading plots in Figure 2A show that the lower wavelength region 400–800 nm is important for many of the PLS components, especially component 1. Thus, many NIR instruments with spectral region only from 1100 to 2500 nm would miss out important information.

The two most important components describing the variance in alginate content and material type, were PLS components 1 and 3, respectively (Figure 1B). Figure 2B shows the two dimensional score plot for these two components, note differences in scale. It is readily seen that component 1 (abscissa) is connected to the alginate content, with low concentrations to the left and high concentrations to the right. The component 3 plot along the ordinate also illustrate a certain dependence on material type. For high alginate concentrations (to the right in Figure 2B), there is a the Materials and methods section, 26 samples were removed to be used for external validation. This 52 sample model achieved a RMSEP of 1.56%, using a model with 9 PLS components. In Figure 3A the predicted alginate content is plotted against the actual content. The correlation coefficient for the straight line is 0.986. Full internal cross validation showed that 96.6% of the variance in alginate content was explained by the model.

The test set of 26 samples was used for external validation, and the result is shown in Figure 3B. The prediction accuracy was now slightly reduced, with a RMSEP of 2.07%

Discussion

The freeze dried samples had a water content of about 3–6%. From Figure 1A it is readily seen that the water content is higher in the 6.8% sample. Large differences in water content would affect the information in the spectra, but such disturbances may again be reduced by the MSC transformation. Calculations run with those water peak spectral regions deleted gave only inferior models (results not included).

It was difficult to observe any systematic differences between spectra from WHOLE or PEELED stipe samples. The raw spectra from the WHOLE samples showed generally a little higher absorbance than the PEELED, but this difference was not visible after MSC transformation (results not included).

As shown in Figures 1B and 2B, there are probably some specific spectral characteristics that could be correlated to material type, most likely reflecting differences in polyphenol content. Thus, if reliable quantitative polyphenol data could be obtained, it might be possible to improve the PLS2 model to predict polyphenol concentrations as well. Such a study should also include samples with higher content of polyphenols, thereby facing several problems in obtaining reliable calibration data. Until then, data obtained from seaweed material richer in polyphenols than *L. hyperborea* should be treated with caution.

One alternative would be to exclude wavelengths important for the prediction of material type from a model for alginate estimates. Loading plots shown in Figure 2A revealed, however, that the same wavelength regions were important for predicting both the Y-variables, and attempted exclusions led only to inferior models (results not included).

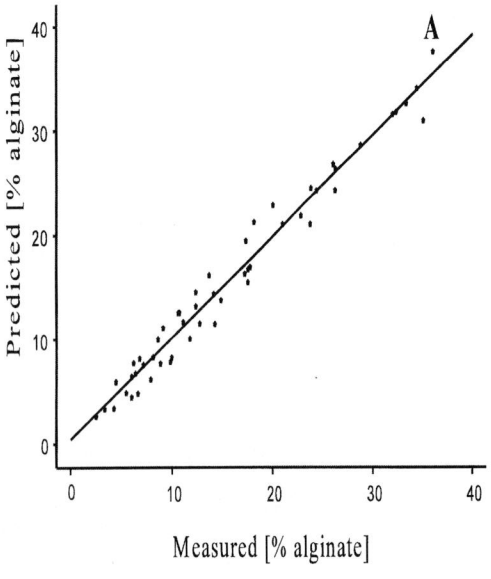

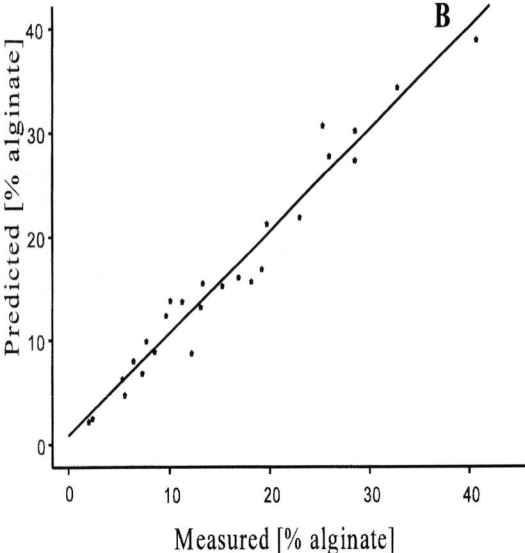

Figure 3. PLS1 model with 9 components: **A)** Calibration based on 52 samples, showing predicted *versus* measured alginate concentrations. **B)** Validation based on test set of 26 samples showing predicted *versus* measured alginate concentrations.

clear separation between WHOLE (W) and PEELED (P) samples, visualised as negative scores for PEELED and positive scores for WHOLE. The separation was less evident at low alginate concentrations.

Determination of alginate content

A separate PLS1 model was made for the prediction of alginate content in order to optimise the prediction accuracy. Using the total data set, as explained in

Although some dependence of material type was revealed, it was finally chosen to base our models on the total data set, including both material types. This gave the models a better statistical base, and also made it possible to use external test set validation. The RMSEP of 2.07% achieved in the external validation is a good prediction accuracy, especially since the total standard error in the alginate calibration data may be as high as ± 10%. Calibrations based on a more restricted type of material would probably result in even better models. However, extrapolation outside the range of the calibration samples cannot be expected to be generally valid.

The results show that NIR spectroscopy can be used for direct quantification of alginate in *L. hyperborea* stipe material. Based on NIR spectra of 56 samples, a 9-component PLS1 model was determined, and validation by a 26 sample test set showed that alginate concentration could be determined with a root mean square error of prediction of 2.1%. Seaweed material with higher and variable polyphenol content would probably be less suitable for NIR spectroscopy, unless a PLS2 model can be calibrated for both alginate and polyphenol content.

Acknowledgements

This work was supported by the Norwegian Research Council (NFR) and Norsk Hydro ASA, Porsgrunn, Norway.

References

Antti H, Sjøstrøm M (1996) Multivariate calibration models using NIR spectroscopy on pulp and paper industrial applications. J. Chemometrics 10: 591–603.

Geladi P, MacDougall D, Martens H (1985) Linearization and scatter-correction for near-infrared reflectance spectra of meat. Appl. Spectrosc. 39: 491–500.

Haug A (1964) Composition and properties of alginates. Norw. Inst. Seaweed Res. Rep. No. 30. Department of Biotechnology, Trondheim: 95–98.

Isaksson T, Næs T (1988) The effect of multiplicative scatter correction (MSC) and linearity improvement in NIR spectroscopy. Appl. Spectrosc. 42: 1273–1284.

Jensen A, Indergaard M, Holt TJ (1985) Seasonal variation in the chemical composition of *Saccorhiza polyschides* (Laminariales, Phaeophyceae). Bot. mar. 28: 375–381.

Martens H, Næs T (1989) Multivariate Calibration. John Wiley & sons, Chichester, UK: 250–253.

Moen E, Larsen B, Østgaard K (1997a) Aerobic microbial degradation of alginate in *Laminaria hyperborea* stipes containing different levels of polyphenols. J. appl. Phycol. 9: 45–54.

Moen E, Horn S, Østgaard K (1997b) Alginate degradation during anaerobic digestion of *Laminaria hyperborea* stipes. J. appl. Phycol. 9: 157–166.

Myklestad S (1968a) Ion-exchange properties of brown algae. I. Determination of rate mechanism for calcium-hydrogen ion exchange for particles from *Laminaria hyperborea* and *Laminaria hyperborea*. J. appl. Chem. 18: 30–36.

Myklestad S (1968b) Ion-exchange properties of brown algae. II. Rate mechanism for calcium-hydrogen ion exchange for particles from *Ascophyllum nodosum*. J. appl. Chem. 18: 222–227.

Vårum KM, Egelandsdal B, Ellekjær MR (1995) Characterization of partially N-acetylated chitosans by near infra-red spectroscopy. Carbohydr. Polym. 28: 187–193.

Oleic acid is the main fatty acid related with carotenogenesis in *Dunaliella salina*

H. Mendoza[1]*, A. Martel[1], M. Jiménez del Río[2] & G. García Reina[1]
[1]*Instituto de Algología Aplicada, Universidad de Las Palmas de G. C., Muelle de Taliarte s/n, 35214 – Telde, Canary Islands, Spain*
[2]*Instituto Tecnológico de Canarias, Cebrian 3, 35003 Las Palmas de Gran Canaria, Canary Islands, Spain*

(*Author for correspondence; fax +34 928 132830)

Received 1 April 1998; revised 30 November 1998; accepted 30 November 1998

Key words: carotene, *Dunaliella salina*, fatty acid, irradiance, nitrogen starvation, oleic acid (18:1)

Abstract

The variation of the fatty acid profile and the carotene content of *Dunaliella salina* in response to irradiance (80, 128, 640, 1000, 1500 μmol photon m^{-2} s^{-1}) and nitrogen starvation were analysed. The highest fatty acid content per cell and the least polyunsaturated fatty acid percentage were exhibited under 1500 μmol photon m^{-2} s^{-1}. Furthermore, the oleic acid (18:1) content maintained a positive and significant correlation with the carotene content per cell and with the irradiance. The composition of the carotene globules in *Dunaliella salina* may be the main determinant of this correlation.

Introduction

The halotolerant green microalga *Dunaliella salina* has received much attention as an ideal organism for algal biotechnology and specifically for its β-carotene production (Borowitzka et al., 1984). This microalga has the unique ability to accumulate large amounts of carotene when exposed to stress conditions such as high irradiance, high salt concentration and nutrient deficiency (Semenenko & Abdullaev, 1980; Ben-Amotz et al., 1982). β-carotene is accumulated within oily globules in the inter-thylakoid spaces of the chloroplast and it is composed mainly of the two stereoisomers 9-*cis* and all-*trans* β-carotene (Ben-Amotz et al., 1982). Irradiance and growth stress (Ben-Amotz & Avron, 1983) affect the β-carotene content and the ratio of the 9-cis to all-trans isomers. Accumulation of β-carotene represents a possible strategy of defence against damage from high irradiance under growth limiting conditions (Ben-Amotz et al., 1989; Gómez-Pinchetti et al., 1992).

Different microalgae exposed to stress growth conditions (i.e. high irradiance and nitrogen starvation) accumulate intra- or extra-plastidic lipid bodies composed of triglycerides and carotenoids (Kivic & Vesk, 1972; Roessler, 1990; Vechtel et al., 1992). The aim of the present work was to characterise the variation in the fatty acid profile and the β-carotene content in a local strain of *Dunaliella salina* Teodoresco (IAA401) grown under nitrogen starvation and different irradiances.

Materials and methods

Algae and growth conditions

Cells of *Dunaliella salina* IAA401 (Microalgae Culture Collection, Institute of Applied Algology, ULPGC, Spain) originally isolated from salt works in the north of Gran Canaria island (Gómez-Pinchetti et al., 1992), were grown in the medium described by Jiménez-Río et al. (1993) containing 1M NaCl. Cultures were bubbled with 3% CO_2 in air, at 30 °C and continuous light (200 μmol photon m^{-2} s^{-1}) provided by 20W low consumption lamps (Mazda, Eureka).

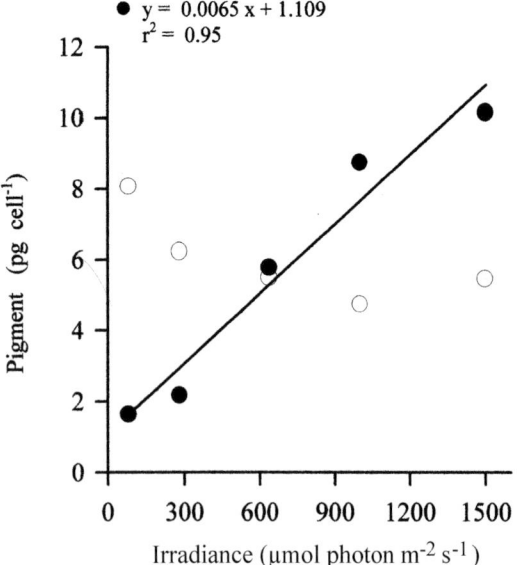

Figure 1. Relationship between carotene (●) and chlorophyll (○) contents of *D. salina* IAA 401 with irradiance.

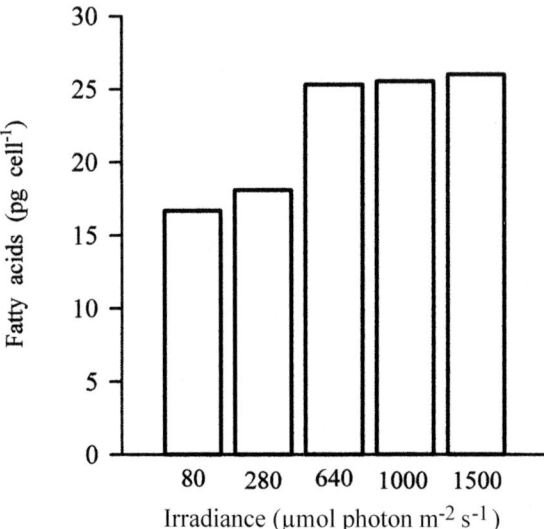

Figure 2. Effect of irradiance on total fatty acid content per cell in *D. salina* (IAA 401).

Cells were harvested at the exponential growth phase by centrifugation (5 min, 5000 × g, 5 °C), washed three times with 50mM Tris buffer containing 1M NaCl, and resuspended in fresh medium but without nitrogen, to a final density of 1.6×10^5 cells mL^{-1}. Experimental cells were then cultured under five different irradiances: 80, 280, 640, 1000, 1500 μmol photon m^{-2} s^{-1}. Samples were collected at the stationary growth phase (5 min, 5000 × g, 5 °C) for the measurement of total carotene content and fatty acid composition. Three replicates were used in all experimental conditions.

Cell growth and pigment analysis

Cell numbers were estimated with a haemocytometer. Chlorophyll and total carotene were extracted from algal pellets with cold methanol and assayed according to Wellburn (1994).

Fatty acid analysis

Total lipid extract was transesterified with methanol saturated with HCl for 90 min, at 80 °C. The fatty acids methyl esters (FAME) were separated with diethylether and purified from lipophilic substances by thin-layer chromatography 10 × 10 cm plates, coated with silica gel type g (Sigma Co.). The running solution was hexane-diethylether (10:1 v/v) and the standard used was transmethylated sunflower oil (Petkov et al., 1990). After addition of pentadecanoic acid (15:0) as internal standard, FAME were dissolved with hexane.

FAME samples were analysed using a gas chromatograph (Shimadzu, GC-15A) with a FID detector, and a polar Supelcowax 10 (Supelco) fused-silica column (30 m × 0.32 mm id.). The FID was operated at 200 °C and the column oven temperature was 195 °C. Component peaks were integrated by a Cr5A integrator (Shimatzu) and identified by comparison of relative retention times of known standards (Sigma Co.).

Statistical analysis

Statistically significant differences were established by ANOVA. Non-significant statistical differences are reported as apparent differences. Correlations were tested by regression analysis.

Results

A positive correlation ($r^2 = 0.95$) was found between irradiance and carotene content per cell (Figure 1). Cells grown at 1500 μmol photon m^{-2} s^{-1} had the highest carotene content, approximately 6 times higher than cells at 80 μmol photon m^{-2} s^{-1}. The chlorophyll content per cell was not significantly correlated ($r^2 = 0.55$) with irradiance although the content was lower in cultures grown under high radiation (Figure 1).

Table 1. Effect of irradiance on fatty acid composition of *Dunaliella salina*. (Data as percentage of total fatty acids ± sd; n = 3)

Fatty acids	Irradiance (μmol photon m^{-2} s^{-1})				
	80	280	640	1000	1500
14:0	trace	trace	trace	trace	trace
16:0	5.45 ± 1.20	7.79 ± 0.63	11.35 ± 2.57	9.34 ± 0.85	24.78 ± 6.63
16:1	3.13 ± 0.09	2.00 ± 0.46	2.17 ± 0.02	2.89 ± 0.19	2.93 ± 0.06
16:2	3.64 ± 0.27	3.13 ± 0.81	2.11 ± 0.04	3.23 ± 0.10	2.45 ± 0.05
16:3	2.41 ± 0.07	2.09 ± 0.78	1.92 ± 0.05	3.56 ± 0.25	2.91 ± 0.08
16:4	35.05 ± 6.57	25.43 ± 0.97	16.31 ± 4.41	28.05 ± 5.12	21.22 ± 4.50
18:1	0.55 ± 0.03	1.76 ± 0.47	4.25 ± 0.21	5.12 ± 0.76	6.22 ± 1.49
18:2ω6	10.99 ± 2.10	10.14 ± 2.37	16.17 ± 2.65	15.36 ± 1.68	14.39 ± 0.25
18:3ω6	1.50 ± 0.12	1.36 ± 0.06	1.10 ± 0.08	0.87 ± 0.08	0.55 ± 0
18:3ω3	36.87 ± 2.83	45.80 ± 6.26	44.33 ± 6.42	30.84 ± 6.24	24.17 ± 2.02
18:3ω4	trace	trace	trace	trace	trace
Sat.+monounsat.	9.18	11.55	17.77	17.35	33.93
Polyunsat.	90.46	87.95	81.94	81.91	65.69
Total (pg cell^{-1})	16.68 ± 0.05	18.1 ± 4.50	25.32 ± 4.73	25.55 ± 2.00	26.06 ± 0.86

The fatty acid content increased with increasing irradiance (Figure 2). Table 1 summarises the variation in the fatty acid profile under different culture conditions. High irradiance tended to increase the content of 16:0 and 18:1, and to decrease the content of 16:2, 16:4, 18:3ω6 and 18:3ω3. The percentage of unsaturated fatty acids decreased with an increase in irradiance (Table 1), while the ratio 16:0/16:4 (Figure 3) increased.

A significant correlation ($p < 0.05$), both in absolute and relative values, was found between oleic acid (18:1) and carotene content per cell (Figure 4).

Discussion

The strong correlation between irradiance and carotene content per cell and the lack of variation in chlorophyll content (Figure 1) could be attributed to the important photoprotective role of carotene in *D. salina* (IAA401). This result contrasts with other *Dunaliella* species that are unable to accumulate large amounts of carotene under high irradiance (Berner et al., 1989). An increase in fatty acid content under stress (high irradiance or nitrogen starvation) conditions has been described in *Nannochloropsis* (Sukenik et al., 1989) and *D. salina* (Cho & Thompson, 1986). In our cultures the increase in fatty acids (saturated and monounsaturated) under carotenogenic conditions (Table 1, Figure 2) can be partially attributed to the increase in carotene globules (as they are composed of 14% triacylglycerols and diacylglycerols, Fried et al., 1982). This increase in fatty acids might play an important role in the structural stability of carotene globules and thus also in the effectiveness of the photoprotective role of carotene. Katz et al. (1995) has described the importance of the structural stability of the carotene globules on its photoprotective role.

The increase in the 16:0/16:4 ratio observed in *D. salina* during adaptation to high irradiances could be related to the change in the balance between storage and photosynthetic related fatty acid during the adaptation to environmental conditions. These results are in accordance with those reported by Goes et al. (1994), who demonstrated the association between 16:0/16:4 ratio with the photosynthetic status in the green alga *Tetraselmis* sp. The increase in this ratio and the observed decrease of the polyunsaturated fatty acid content (Table 1) in response to high irradiance in *D. salina* are less marked than in other species of *Dunaliella* (Thompson et al., 1990), which might be a consequence of the photoprotective role of β-carotene.

Oleic acid (18:1) has been described as an important reserve fatty acid, mainly associated with phosphatidylethanolamine, triacylglycerols and diacylglycerols in *Dunaliella* (Fried et al., 1982; Al-Hasan, 1987); the concentration in the cell increases under

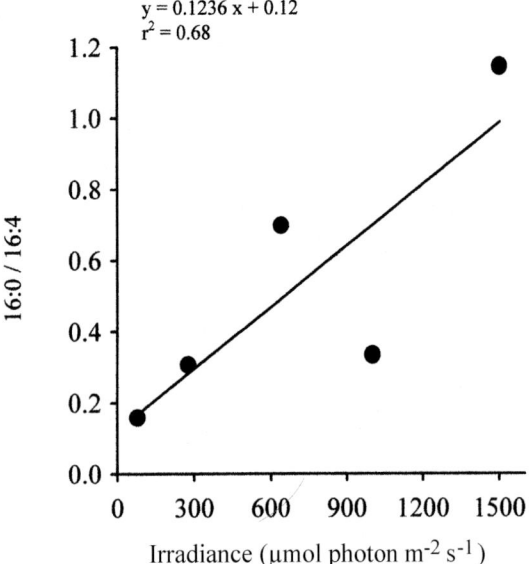

Figure 3. Relationship between 16:0/16:4 ratio of *D. salina* (IAA 401) and irradiance.

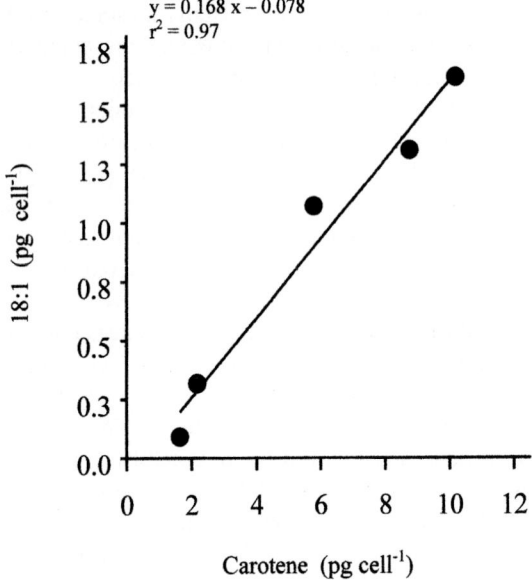

Figure 4. Relationship between oleic acid content and the carotene content per cell.

stress conditions (Mendoza et al., 1996). However, this is the first report describing the direct relation between stress, oleic acid and caroteneogenesis in *D. salina* (Figure 4). 18:1 may constitute the main fatty acid present in the β-carotene globules, which could explain the strong correlation between the oleic acid and the β-carotene contents during the adaptation to stress conditions.

The results suggest that the variations in fatty acid content and the carotenogenesis in *D. salina* could be closely related processes. This may be a consequence of the photoprotective role of carotene that determines the fatty acid profile in response to stress conditions, and to a direct relationship between oleic acid and carotenogenesis. Although there seems to be a clear correlation between carotene accumulation and oleic acid concentration, further investigations are needed to prove that the two processes are independent.

Acknowledgements

We thank Dr G. Petkov for helping to set up the fatty acid determination method in our Institute and Mrs Paula Moreno Díez for technical assistance and help. This research was partially supported by CICYT/SPAIN (Project AMB-96-0962).

References

Al-Hasan RH, Ghannoum MA, Sallal A-K, Abu-Eteen KH, Radwan SS (1987) Correlative changes of growth, pigmentation and lipid composition of *Dunaliella salina* in response to halostress. J. gen. Microbiol. 133: 2607–2617.

Ben-Amotz A, Avron M (1983) On the factors which determine massive β-carotene accumulation in the halotolerant alga *Dunaliella bardawil*. Plant Physiol. 72: 593–597.

Ben-Amotz A, Katz A, Avron M (1982) Accumulation of carotene in halotolerant algae: purification and characterisation of β-carotene globules from *Dunaliella salina* (Chlorophyceae). J. Phycol. 18: 529–537.

Ben-Amotz A, Shaish A, Mordhay A (1989) Mode of action of the massively accumulated β-carotene of *Dunaliella bardawil* in protecting the alga against damage by excess irradiation. Plant Physiol. 91: 1040–1043.

Berner T, Dubinsky A, Wyman K, Falkowski PG (1989) Photoadaptation and the 'package' effect in *Dunaliella tertiolecta* (Chlorophyceae). J. Phycol. 25: 70–78.

Borowitzka LJ, Borowitzka MA, Moulton TP (1984) The mass culture of *Dunaliella salina* for fine chemicals: From laboratory to pilot plant. Hydrobiologia 116/117: 115–125.

Cho SH, Thompson GA (1986) Properties of a fatty acid hydrolase preferentially attacking monogalactosyldiacylglycerols in *Dunaliella salina* chloroplasts. Biochim. Biophys. Acta 878: 353–359.

Fried A, Tietz A, Ben-Amotz A, Eichenberger W (1982) Lipid composition of the halotolerant alga, *Dunaliella bardawil*. Biochim. Biophys. Acta 713: 419–426.

Goes JI, Handa N, Taguchi S, Hama T (1994) Effect of UV-B radiation on the fatty acid composition of the marine phytoplankter *Tetraselmis* sp.: relationship to cellular pigments. Mar. Ecol. Progr. Ser. 114: 259–274.

Gómez-Pinchetti JL, Ramazanov ZM, Fontes AG, García Reina G (1992) Photosynthetic characteristics of *Dunaliella salina* (Chlorophyceae, Dunaliella) in relation to β-carotene content. J. Appl. Phycol. 4: 11–15.

Jiménez del Río M, Ramazanov ZM, García-Reina G (1994) Dark induction of nitrate reductase in the halophilic alga *Dunaliella salina*. Planta 192: 40–45.

Katz A, Jiménez C, Pick U (1995) Isolation and characterisation of a protein associated with carotene globules in the alga *Dunaliella bardawil*. Plant Physiol. 108: 1657–1664.

Kivic PA, Vesk M (1972) Structure and function in the euglenoid eyespot apparatus: the fine structure, and response to environmental changes. Planta 105: 1–14.

Mendoza H, Jiménez del Río M, García Reina G, Ramazanov ZM (1996) Low temperature induced β-carotene and fatty acid synthesis, and ultrastructural reorganization of the chloroplast in *Dunaliella salina*. Eur. J. Phycol. 31: 329–331.

Petkov GD, Klyachko-Gruvich GD, Furnadjieva ST, Pronina NA, Ramazanov ZM (1990) Genotypic differences and phenotypic changes of lipid and fatty acid composition in strains of *Dunaliella salina*. Sov. Plant Physiol. 3: 268–272.

Roessler PG (1990) Environmental control of glycerolipid metabolism in microalgae: commercial implications and future research directions. J. Phycol. 26: 393–399.

Semenenko VE, Abdullaev AA (1980) Parametric control of β-carotene biosynthesis in *Dunaliella salina* cells under conditions of intensive cultivation. Sov. Plant. Physiol. 27: 22–30.

Sukenik A, Carmeli Y, Berner T (1989) Regulation of fatty acid composition by irradiance level in the eustigmatophyte *Nannochloropsis* sp. J. Phycol. 24: 445–452.

Thompson PA, Harrison PJ, Whyte JN (1990) Influence of irradiance on the fatty acid composition of phytoplankton. J. Phycol. 26: 278–288.

Vechel B, Eichenberger W, Ruppel HG (1992) Lipid bodies in *Eremosphaera viridis* De Bary (Chlorophyceae). Plant Cell Physiol. 33: 41–48.

Wellburn AR (1994) The spectral determination of chlorophylls a and b, as well as total carotenoids, using various solvents with spectrophotometers of different resolution. J. Plant Physiol. 144: 307–313.

Alginate stability during high salt preservation of *Ascophyllum nodosum*

Einar Moen, Bjørn Larsen, Kjetill Østgaard* & Arne Jensen
Norwegian Biopolymer Laboratory and Department of Biotechnology, The Norwegian University of Science and Technology NTNU, N-7034 Trondheim, Norway

(*Author for correspondence; phone +47 7359 4068; fax +47 7359 1283; e-mail Kjetill.Oestgaard@chembio.ntnu.no)

Received 1 April 1998; revised 16 September 1998; accepted 28 November 1998

Key words: Phaeophyta, brown algae, alginate, preservation, salt, formaldehyde

Abstract

Formaldehyde is usually added to brown algae to avoid microbial growth during storage and to fix polyphenols in the algae before alginate extraction. Since formaldehyde is toxic, allergenic and possibly carcinogenic, dry salting of *Ascophyllum nodosum* was tested as an alternative. The seaweeds, harvested at locations with a salinity of about 30‰ from late autumn to early spring, were stored at 22 ± 2 °C under compost-like conditions. Untreated samples of seaweed lost their quality as a raw material for alginate production within 14 days. Salted (20–22%) as well as formaldehyde treated seaweed was preserved for at least 46 days. Due to the reduced water activity and oxygen saturation in the dry salted seaweed, microbial growth and brown colouring reactions were suppressed. Economic factors must also be taken into account before large-scale applications are considered.

Introduction

The brown seaweed *Ascophyllum nodosum* (Phaeophyceae) is utilised for production of alginate and seaweed extracts. In this connection, harvested material is often stored for as long as 60 days before processing. Formaldehyde in quite high concentrations (2% w/w) is widely used to prevent microbial decomposition during storage. The formaldehyde will also react with the algal polyphenols and prevent discoloration of the alginates (Haug & Larsen, 1958). Thus, the industrial application of formaldehyde has a double function (Chapman & Chapman, 1980; Jensen, 1995).

Formaldehyde is, however, toxic, allergenic and possibly carcinogenic, and should be avoided for environmental reasons. Increased concern from governments has already forced some alginate producers to reduce the use of formaldehyde, but the number of ecologically and economically acceptable alternatives is restricted.

New methods for storage must first of all be based on controlled 'first in–first out' logistics for raw material handling to identify the necessary storage period. After these criteria have been quantified, a variety of preservation techniques may be tested and evaluated. The storage environment may be manipulated to reduce microbial degradative activity in different ways, for instance by reducing the temperature, the pH or the water activity. Cooling must take into consideration the scale dependent heating that may occur due to composting if conditions are aerobic. Addition of acids may lead to loss of alginate viscosity due to chemical hydrolysis, although moderate microbial acidification may give more stable storage conditions (Moen, 1997).

Preservation by salting may have certain environmental advantages: a) leakages to a seawater recipient will be of minor harm, b) one can avoid the problems that volatile alternatives such as formaldehyde, alcohols and aldehydes may create for the workers as well as people living near the factory, c) chemical alginate degradation due to hydrolysis at extreme pH is easily avoided, d) salt is a relatively cheap chemical.

The preservation of brown seaweeds in the alginate production must not reduce the yield and viscosity of the alginate, and should also inactivate the polyphenols to secure a pure and white product. Our objective was to evaluate the stability of alginates in harvested *A. nodosum* by addition of salts instead of formaldehyde, by comparing yield, viscosity, pH and bacterial numbers.

Materials and methods

Raw material, storage and sampling

Ascophyllum nodosum was harvested at the Trøndelag coast of Norway from late autumn to early spring at locations with about 26–30‰ salinity, and arrived at the laboratory within the same day. Batches of 2–3 kg were immediately milled to pass a sieve with 4-mm holes, mixed, separated into 100 g fractions, and preservatives added (see below). Unless otherwise stated, the samples were stored at 22 ± 2 °C in polyethylene boxes ($10 \times 10 \times 4$ cm), with a cover allowing some exchange of gases with the atmosphere.

Eight separate batches of seaweed, harvested at different times and locations, were stored without additives in eight independent experiments. These experiments are summarised as a group denoted CONTROL. One particular batch of *A. nodosum* (Smøla, Norway, early spring) was separated into 25 fractions of 100 g each, to which was added: nothing, 10 g NaCl, 10 g NaCl + 1 g $CaCl_2 \cdot 2H_2O$, 20 g NaCl, 20 g NaCl + 2 g $CaCl_2 \cdot 2H_2O$, respectively. These experiments are called PRESERVATION.

To another batch was added 2 g of formaldehyde per 100 g of fresh alga, denoted as the Formaldehyde (2%) sample. Storage of seaweed was also tested in sealed containers flushed with nitrogen gas before closing, denoted as the Nitrogen sample.

Samples for analysis at zero days of storage were taken after milling, otherwise samples (2–15 g) were taken at intervals from the storage boxes. In the CONTROL and 2% Formaldehyde experiments, series of samples were taken from the same box. For each sample in the PRESERVATION experiment, the box was discarded after removing the sample. The material was analysed for bacterial numbers, pH, total solids, and yield and viscosity of extracted alginate as described below.

Analyses

In the bacteriological test, 1 g of each sample was suspended in 100 mL sterile seawater, and diluted if necessary. Bacterial numbers were estimated by counting visible colonies formed on agar Petri dishes after 72 h of incubation at 30 °C, using Tryptone Soya Agar and 1 mL of suspended sample. The numbers are given as Colony Forming Units (CFU) per g of sample, with a standard error of $\pm 60\%$. pH was measured in a 1 : 1 mixture of sample and distilled water after stirring for 30 minutes. Total solids were determined after drying the material to constant weight at 105 °C. The samples were allowed to cool in a desiccator with P_2O_5 before weighing (Larsen, 1978). Total solids have a standard error less than $\pm 5\%$. Na-alginate was determined by extraction and dialysis at room temperature (20–23 °C) according to Haug (1964). The yield is given as g Na-alginate per g dry matter. The intrinsic viscosity of the extracted alginate, $[\eta]$, was estimated according to Haug (1964) in an Ubbelohde Viscosimeter at 20 °C after filtration of the solutions through 0.8 μm Millipore AA-filters.

Results

Figure 1 summarises the total variability observed for the eight separate batches of the CONTROL samples, representing seaweed material harvested at different locations and times during the winter. The large natural variations in composition, microbial contamination and stability should always be kept in mind when single experiments are evaluated.

Some general trends were still evident, as illustrated in Figure 1 by the smooth curves obtained by second-order polynomial fitting. There was a high probability of losing more than 50% of the yield (Figure 1A) and viscosity (Figure 1B) of isolated Na-alginate within 15–20 days of storage. As the seaweed was gradually decomposed and oxidised, the colour of the material changed from greenish to brown. The isolated alginate was also reduced in quality due to an increasing discoloration, most likely because polyphenols were associated with the alginate. During storage, the pH in the algal material increased (Figure 1C) and numerous microorganisms (Figure 1D) degraded the algae. The untreated seaweed was unsuitable for alginate production within 14 days.

In the PRESERVATION experiments shown in Figure 2, the reference culture with no addition fell

Table 1. Characteristics of A. nodosum and the isolated alginate after storage at $22 \pm 2°C$. All experiments were done at compost-like conditions, except the Nitrogen sample. Values for CFU, TS, Yield and Visc. are given in percent of initial values obtained for fresh material. CFU – colony forming units; TS – total solids of algae; Yield – yield of isolated Na-alginate from the algae; Visc. – viscosity of isolated Na-alginate

Experiment	Days	CFU [%]	TS [%]	Yield [%]	Visc. [%]	Alginate colour	Seaweed colour	Seaweed aroma
No addition	46	154	72	26	77	Brown	Uniform dark brown	Wet hay, mould
10% NaCl	46	27	91	81	90	Yellow	Brown, some green	Fresh
10% NaCl + 1% CaCl$_2$	46	4	97	96	89	Yellow	Green, some brown	Fresh
20% NaCl	46	0	99	100	73	White	Green, patches brown	Fresh
20% NaCl + 2% CaCl$_2$	46	0	102	96	77	White	Green	Fresh
Formaldehyde (2%)	46	0	100	97	70	White	Green	(Toxic)
Nitrogen	27	0	102	103	63	White	Green	Sweet

within the normal range illustrated in Figure 1, and should thus be considered as a reasonably representative raw material. Samples with salts had a much higher yield of alginate (Figure 2A) than the untreated sample after 46 days. At this time, the untreated seaweed had a relatively high viscosity (Figure 2B), but the product was discoloured and the yield was only 26% of the initial value (Table 1). The colour of the isolated alginate gradually improved from brown to white as the salt concentration increased. Dry salting stabilised the pH (Figure 2C) and the bacterial numbers (Figure 2D) showed a falling trend.

According to Table 1, an acceptable quality of A. nodosum for alginate production can be chosen among greenish seaweeds with a fresh aroma. The 20% NaCl; 20% NaCl + 2% CaCl$_2$; Formaldehyde and Nitrogen samples were suitable for isolation of alginate. Dry salting will reduce both the oxygen saturation and the water activity, and thereby also possibly limit both microbial activity and chemical discoloration reactions. Low bacterial numbers in the Nitrogen sample suggested that the contamination of facultative aerobes able to grow on A. nodosum was low. The lower dose of 10% NaCl was less favourable due to higher bacterial activity, reduced dry matter and yield, and some discoloration of the alginate (Table 1).

Discussion

In addition to the sodium salt, calcium was also applied in some batches to test for additional stabilisation of the alginate. No such effect was found. The ionic status of the alginate was probably close to that of the alginate in the native algae. Evidence for this may be found from the research on making viable protoplasts from brown seaweeds with NaCl added for osmoregulation, showing that calcium chelating treatment is generally necessary to obtain dissociation (Butler et al., 1989; Rødde, 1997). Solubilisation of alginate salted at 10–20% should therefore be considered negligible. It is evident that the situation will be very different for a freely expanding alginate gel, where swelling may facilitate dissociation of calcium cross-links when exposed to ionic competition by high levels of Na$^+$.

It is not practically possible to test long-term re-use of the same storage facilities over several years in the laboratory, but selective enrichment of unwanted contaminants could possibly occur. Undesirable microbial growth may develop under a wide range of conditions. High concentrations of NaCl do not inhibit growth of all bacteria, some actually require a high NaCl concentration (Holt & Krieg, 1994). *Halomonas* species can grow in NaCl concentrations from 0.2 to 25%, while red extreme halophiles such as *Halobacterium* need at least 15% NaCl. In this study, *Halobacterium* species would not be detected by the assay used for bacterial enumeration. Most *Staphylococcus* species grow in media containing 10% NaCl. It should also be noted that human pathogens such as *S. aureus* can grow in 10% NaCl by fermentation of mannitol, which is an important constituent in A. nodosum.

When applied as the only preservation technique, dry salting at concentrations as high as 20% NaCl were necessary. This appears to be too inconvenient and expensive for the alginate industry. It is possible that an optimal solution could involve a combination of different principles including: controlled temper-

24

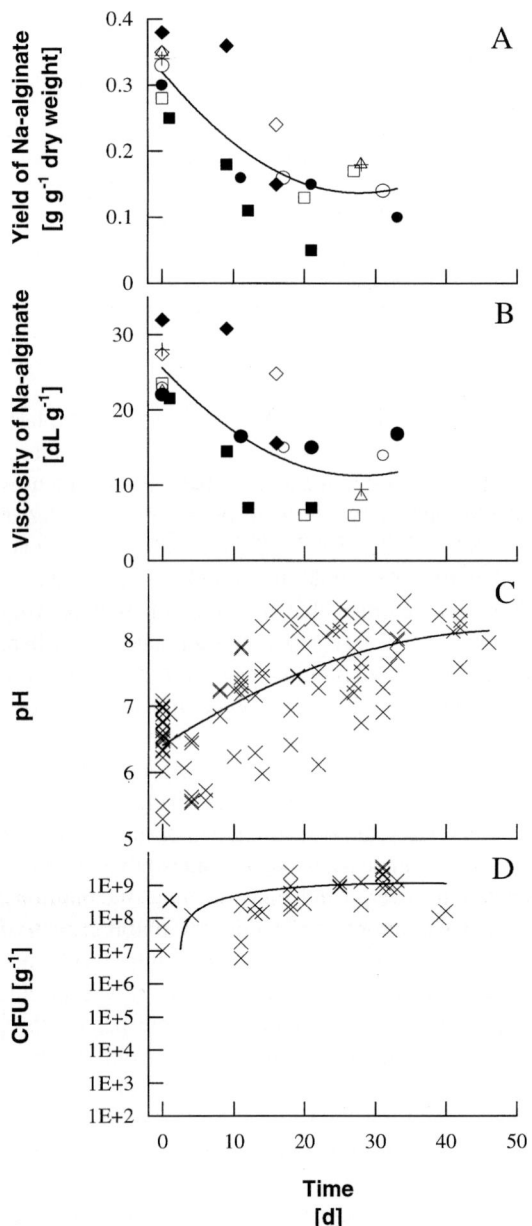

Figure 1. Natural variability of eight separate batches of untreated *A. nodosum*, shown as a function of time: (A) Yield of alginate; (B) Intrinsic viscosity of alginate (C) pH in the seaweed; (D) Bacterial counts (CFU) per g. The different batches are indicated by different symbols in A and B. Smooth curves were obtained by second-order polynomial fitting.

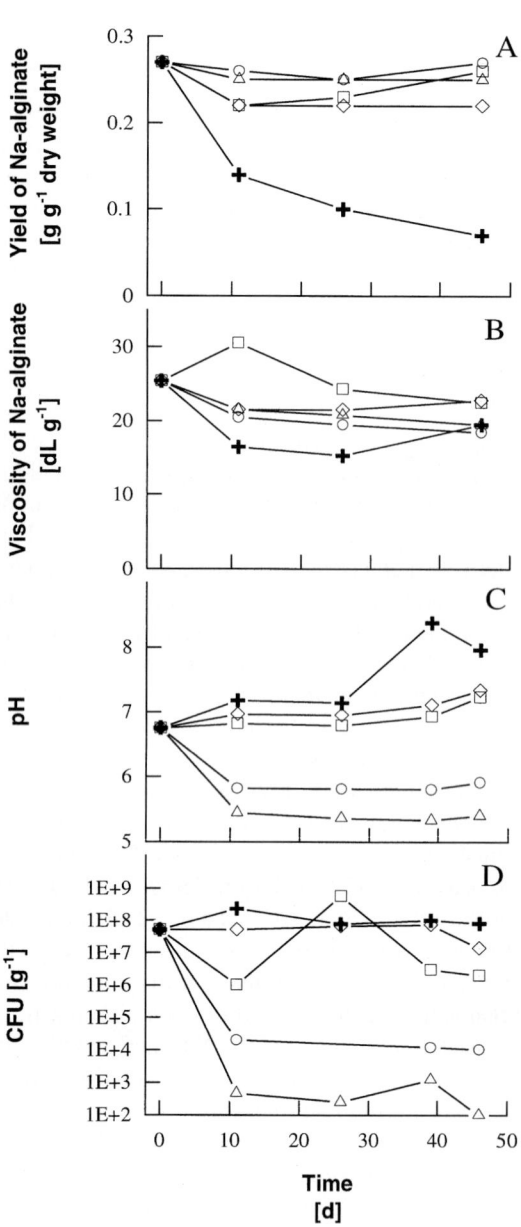

Figure 2. Characteristics of the preservation of *A. nodosum*, shown as a function of time: (A) Yield of alginate; (B) Intrinsic viscosity of alginate; (C) pH in the seaweed; (D) Bacterial counts (CFU) per g. Symbols: no addition (+); 10% NaCl (◇); 10% NaCl + 1% $CaCl_2$ (□); 20% NaCl (○); 20% NaCl + 2% $CaCl_2$ (△).

538

ature, pH, oxygen as well as salt. The suitability of a storage method can only be finally evaluated when the necessary storage period has been identified. Economic factors must also be taken into account before large-scale applications are considered.

Acknowledgements

This work was supported by the Norwegian Research Council (NFR) and Pronova Biopolymer A/S, Drammen, Norway.

References

Butler DM, Østgaard K, Boyen C, Evans LV, Jensen A, Kloareg A (1989) Isolation conditions for high yields of protoplasts from *Laminaria saccharina* and *L. digitata* (Phaeophyceae). J. exp. Bot. 40: 1237–1246.

Chapman VJ, Chapman DJ (1980) Seaweeds and their Uses. Chapman and Hall, London: 194–225.

Haug A (1964) Composition and properties of alginates. Rep. Norw. Inst. Seaweed Res. No. 30. Laboratory of Biotechnology, Trondheim: 10–25.

Haug A, Larsen B (1958) The formation of brown coloured substances in seaweed and seaweed extracts. Rep. Norw. Inst. Seaweed Res. No. 22. Laboratory of Biotechnology, Trondheim: 1–18.

Holt JG, Krieg NR (1994) Enrichment and isolation. In Gerhardt P, Murray RGE, Wood WA, Krieg NR (eds), Methods for General and Molecular Bacteriology. American Society for Microbiology, Washington DC: 179–215.

Jensen A (1995) Production of alginate. In Wiessner W, Schnepf E, Starr RC (eds), Algae, Environment and Human Affairs. Biopress Ltd., Bristol: 79–92.

Larsen B (1978) Brown Seaweeds: Analysis of ash, fiber, iodine and mannitol. In Hellebust JA, Craigie JS (eds), Handbook of Phycological Methods – Physiological and Biochemical Methods. Cambridge University Press, Cambridge: 181–188.

Moen E (1997) Biological degradation of brown seaweeds. PhD thesis, Department of Biotechnology, NTNU, Trondheim: 43–51.

Rødde RSH (1997) Chemical composition and alginate biosynthesis in protoplasts from *Laminaria digitata* and *Laminaria saccharina*. PhD thesis. Department of Biotechnology, NTNU, Trondheim: 19–31.

Chemical characteristics and gelling properties of agar from two Philippine *Gracilaria* spp. (Gracilariales, Rhodophyta)

Nemesio E. Montaño[1]*, Ronald D. Villanueva[1] & Jumelita B. Romero[1,2]
[1] *Marine Science Institute, College of Science, University of the Philippines Diliman, 1101 Quezon City, Philippines*
[2] *Mindanao State University – Tawi-Tawi, Bongao, Tawi-Tawi 1702, Philippines*

(* Author for correspondence)

Received 12 April 1998; revised 17 October 1998; accepted 20 December 1998

Key words: agar, FT-IR spectroscopy, *Gracilaria arcuata*, *Gracilaria tenuistipitata*, NMR spectroscopy, Rhodophyta

Abstract

The chemical structure of agars extracted from Philippine *Gracilaria arcuata* and *G. tenuistipitata* were determined by NMR and infrared spectroscopy. Agar with alternating 3-linked 6-O-methyl-β-D-galactopyranosyl and 4-linked 3,6-anhydro-2-O-methyl-α-L-galactopyranosyl units was isolated from *G. arcuata*, while the agar from *G. tenuistipitata* possesses the regular agarobiose repeating unit with partial methylation at the 6-position of the D-galactosyl residues. Both agars exhibit sulphate substitution at varying positions in the polymer. Chemical analyses reveal higher 3,6-anhydrogalactose and lower sulphate contents in alkali-modified than in native agar from both samples. Also, alkali modification enhanced agar gel strength and syneresis. Native *G. arcuata* agar produces a viscous solution (2000 cP at 75 °C) with a high gelling point (> 60 °C) that forms a soft gel even after alkali modification (gel strength: < 300 g cm^{-2}). On the other hand, the agar from *G. tenuistipitata* exhibits gel qualities typical of most *Gracilaria* agars.

Introduction

Agar synthesised by a number of genera in the Gigartinales constitute a heterogeneous mixture of molecules built on a disaccharide repeating unit of 3-linked β-D-galactopyranosyl (G) and 4-linked 3,6-anhydro-α-L-galactopyranosyl (LA) residues (Araki, 1966). Substitution with sulphate hemi-esters, methyl ethers and/or pyruvate ketals can occur at various sites in the polysaccharide chain, with a higher degree of substitution in *Gracilaria* agars, compared with agars from *Gelidium* and *Pterocladia* (Murano, 1995). This creates complexity in the chemical composition and consequent gel qualities of *Gracilaria* agars. Though species of *Gracilaria* generally produce agars with low gel strengths, they are considered the most important source of commercially valuable agar, especially for the food industry. Furthermore, new dimensions in the application of *Gracilaria* agars are opened by the discovery of their sugar-reactive property (Murano, 1995), hence an increase in commercial interest.

Different methods have been employed in the analysis of agar structure, e.g. chemical techniques, fractionation, enzymatic hydrolyses and physical methods, particularly nuclear magnetic resonance spectroscopy (Lahaye & Rochas, 1991). The latter is a powerful and extensively used tool in the structure elucidation of agar and other algal polysaccharides. Recently, vibrational spectroscopy, Fourier transform infrared spectroscopy in particular, has been applied to differentiate agar- and carrageenan-type seaweed galactans (Matsuhiro & Rivas, 1993), aside from its utility in sulphate detection and attachment determinations.

In this work, ^{13}C and ^{1}H NMR and FT-IR spectroscopy were used to elucidate the chemical structures of native and alkali-modified agars from *Gracilaria arcuata* Zanardini and *Gracilaria tenuistipitata* Zhang et Xia. Attention was paid to the changes in chemistry of agars upon alkali modification. In addition, data on the gelling properties of extracted agars are presented.

Materials and methods

Gracilaria arcuata and *G. tenuistipitata* were harvested in San Luis, Batangas (120° 57′ N, 13° 45′ E) and Layog, Sorsogon (124° 08′ N, 13° 50′ E), Philippines, respectively, cleaned of epiphytes, washed with tap water and dried at 60 °C. Samples were kept in a dry state until further processing was done.

Extraction of native agar was conducted by boiling 20 g dried seaweed with 800 mL water for 1 h. Diatomaceous earth was added and the mixture was blended for 5 min. Filtration was done under pressure and the filtrate frozen, thawed, dehydrated in isopropyl alcohol, and oven-dried (60 °C).

Alkali modification was performed by heating 20 g dried seaweed in 750 mL of 10% (w/w) NaOH solution at 90 °C for 2 h. After this time, the material was washed thoroughly in running water and then soaked in 600 mL of 0.5% HOAc at room temperature (28–30 °C) for 1 h. The alkali-modified seaweed was again washed and processed further as described in native agar extraction.

^{13}C and ^{1}H NMR spectroscopy were carried out using a Jeol 400 Spectrometer operating at 100.40 MHz and 399.65 MHz, respectively. Spectra of about 5% polysaccharide solutions in D_2O were obtained at 80 °C for *G. arcuata* and 90 °C for *G. tenuistipitata* samples. Chemical shifts were measured in parts per million relative to internal 3-(trimethyl)-silyl propionic acid-d4-sodium salt, TSP ($\delta = 2.4$ ppm for ^{13}C, $\delta = 0$ ppm for ^{1}H).

Degree of methylation on β-D-galactose (6-O-methyl; G) and 3,6-anhdyro-α-L-galactose(2-O-methyl; LA) residues was estimated from the ^{1}H NMR spectral data by the ratio between 1/3 of the area for methyl proton resonances at ∼ 3.43 and ∼ 3.52 ppm, respectively, and the area of the H-1 of 4-linked L-galactose in the region 5.15–5.18 ppm, assuming a perfectly alternating agar backbone (Murano et al., 1996). Degree of substitution (DS) refers to the ratio of methylated disaccharides to the total disaccharide units in the sample.

The Fourier transform infrared (FT-IR) spectra of agar samples were recorded on films using a Shimadzu 8201 PC FT-IR Spectrometer. Films were prepared by drying 5 mL of 0.5% agar solution on a Teflon-coated pan at 60 °C. Ratio of absorbance at 1250 and 930 to 2920 cm^{-1} were determined to estimate the relative amount of sulphate (total) and 3,6-anhydrogalactose, respectively to the total carbohydrate content.

The 3,6-anhydrogalactose content of the agar samples was determined colorimetrically using the resorcinol method of Yaphe & Arsenault (1965). The sulphate content was quantified by the barium chloride precipitation technique of Jackson and McCandless (1978), after sulphate hydrolysis by 1N HCl at 110 °C for 4 h. Pyruvic acid detection was done by the lactate dehydrogenase enzymatic procedure of Duckworth and Yaphe (1970).

Gel strengths of 1.5% (w/w) agar solutions were measured using a Marine Colloids Gel Tester (Model GT-1). Viscosity determination was conducted at 75 °C in a Brookfield Synchro-lectric Viscometer (Model LVF). Gelling temperature was determined using a hot solution of 1.5% agar in a test tube. A thermometer was inserted, with its bulb situated just below the solution surface, through a stopper. The solution was allowed to cool and glass beads (diam: 2.85 mm; wt: 30 mg) were dropped at intervals of 0.5 °C. The temperature at which a bead failed to drop through the agar solution was recorded as the gelling temperature. To determine the melting temperature, the gel set-up used in gelling temperature determination was refrigerated and a lead shot (d.: 4.30 mm; wt: 500 mg) was set on the gel surface. The set-up was then placed in a water bath and heated slowly. The temperature at which the lead shot dropped to the bottom of the test tube was recorded as the melting temperature. Syneresis index was measured following the method of Fiszman and Duran (1992), but utilising 39 × 39 cm cylindrical agar probes. The diameter of the exudate from the agar probe in a Whatman Filter Paper No. 1 was measured after 2 h.

Results

The ^{13}C NMR spectra of the native and alkali-treated extracts from *Gracilaria arcuata* are shown in Figure 1. The extracts exhibit similar signals which were attributed to an agar disaccharide unit with methylation occurring at O-6 (G) and O-2 (LA) (Table 1). Complete methylation was apparent at both positions as indicated by the absence of non-methylated C-6 (G) and C-2 (LA) signals at ∼ 62.5 and ∼ 71.0 ppm,

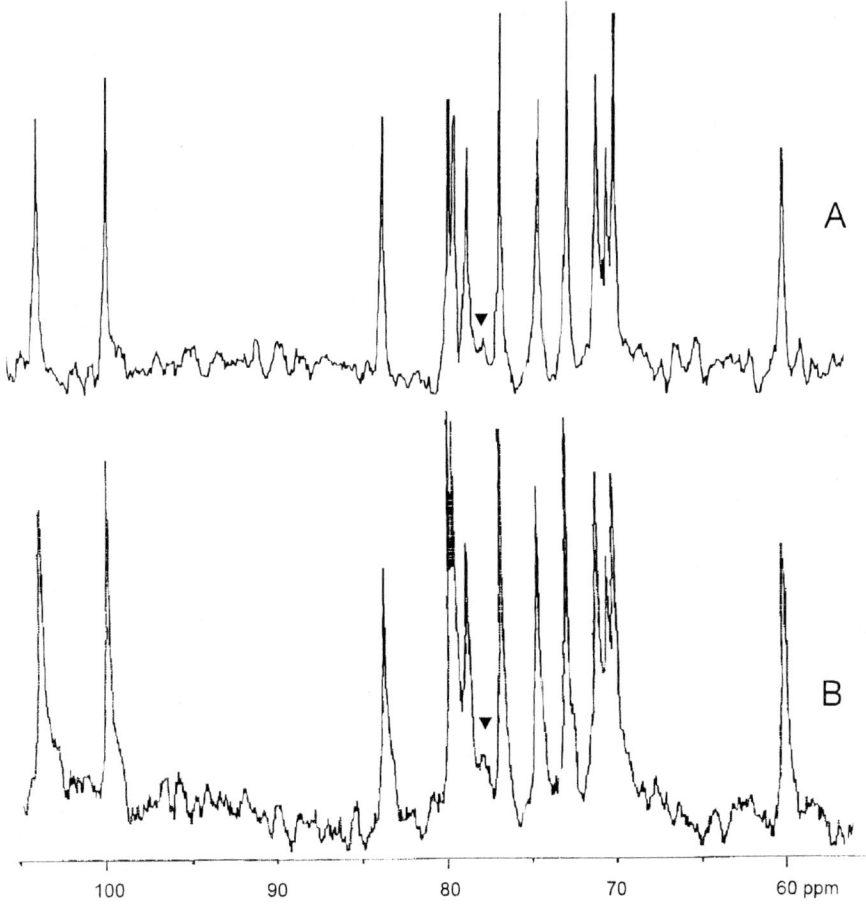

Figure 1. ^{13}C NMR spectra of (A) native and (B) alkali-treated agar extracted from *Gracilaria arcuata*. Signal marked with (▼) is attributed to C-2 (G) with sulphate ester.

respectively. Methylation pattern in both extracts was confirmed through ^{1}H NMR spectroscopy with resolution of signals at 3.43 and 3.52 ppm ascribed for protons of methyl groups attached to O-6 (G) and O-2 (LA), respectively. Complete methylation at O-6 (G) was indicated by the degree of substitution value (DS) of 1.0 measured in both extracts. At O-2 (LA), complete methylation was indicated only in the alkali-treated extract (1.0 DS) and an almost complete substitution (0.9 DS) in the native extract.

A minor signal at 77.9 ppm in the ^{13}C NMR spectra of both native and alkali-treated agar from *G. arcuata* was attributed to C-2 (G) linked to a sulphate ester (Miller et al., 1993). Furthermore, the FT-IR signal at 825 cm^{-1} (Figure 2) confirmed this sulphation (Miller et al., 1993). An FT-IR signal at 825 cm^{-1} is rather peculiar as sulphations at O-2 and O-6 of galactosyl residue were usually resolved at 830 and 820 cm^{-1}, respectively (Rochas et al., 1986). Further attribution of the 825 cm^{-1} signal to sulphate ester at O-2, instead of at O-6, is a deduction from the fact that the signal persisted after alkali modification, suggesting an alkali-stable sulphate ester.

Alkali treatment of the *G. arcuata* agar led to a fairly modest decrease in total sulphate and increase in 3,6-anhydrogalactose levels as detected by both chemical and FT-IR analyses (Table 2).

The native and alkali-treated agar from *Gracilaria tenuistipitata* possessed major ^{13}C NMR signals (Figure 3) ascribed for a regular agarobiose (G-LA) structure (see Table 1). Weak signal for methoxy group at 60.2 ppm in the native extract (at 60.1 ppm in the alkali-treated extract) indicated partial methylation at O-6 of the D-galactosyl residue and was corroborated by resolution of weak signals at 74.7, 72.9, and 70.0 ppm due to the field shift of C-5, C-6, and C-

Table 1. Chemical shifts (ppm) of major carbon resonances in the ^{13}C NMR spectra of native and alkali-treated agars from *Gracilaria arcuata* and *G. tenuistipitata*.

	Gracilaria arcuata		*Gracilaria tenuistipitata*	
	native	alkali-treated	native	alkali-treated
G-1	103.7	103.7	103.5	103.5
G-2	71.2	71.2	71.3	71.4
G-3	83.7	83.6	83.2	83.3
G-4	70.1	70.2	69.9	69.8
G-5	74.6	74.6	76.5	76.5
G-6	72.9	72.9	62.5	62.5
CH$_3$ (O-6)	60.1	60.1		
LA-1	99.8	99.8	99.3	99.3
LA-2	79.8	79.8	71.0	71.0
LA-3	79.5	79.5	81.2	81.2
LA-4	78.7	78.8	78.5	78.4
LA-5	76.8	76.8	76.8	76.7
LA-6	70.5	70.7	70.5	70.6
CH$_3$ (O-2)	60.1	60.1		

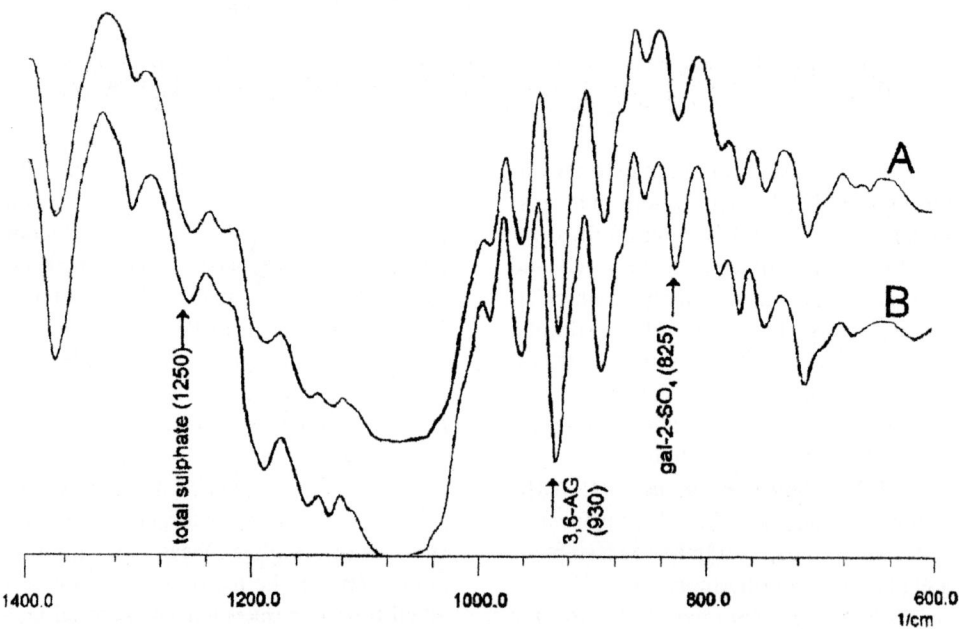

Figure 2. FT-IR spectra of (A) native and (B) alkali-treated agar extracted from *Gracilaria arcuata*.

Table 2. Sulphate and 3,6-anhydrogalactose levels of native and alkali-treated agars from *Gracilaria arcuata* and *G. tenuistipitata*. (Mean ± SE, $n = 3$)

	Gracilaria arcuata		*Gracilaria tenuistipitata*	
	native	alkali-treated	native	alkali-treated
A. Chemical Analysis				
SO_4^{-2} (% w/w)	4.17 ± 0.26	3.79 ± 0.23	3.90 ± 0.21	3.34 ± 0.08
3,6-Anhydrogalactose (% w/w)	28.5 ± 0.24	30.6 ± 0.65	34.0 ± 0.14	42.1 ± 0.37
B. FT-IR Analysis				
SO_4^{-2} (1250/2920)	1.17	1.00	1.18	0.84
3,6-Anhydrogalactose (930/2920)	1.73	1.82	1.77	1.86

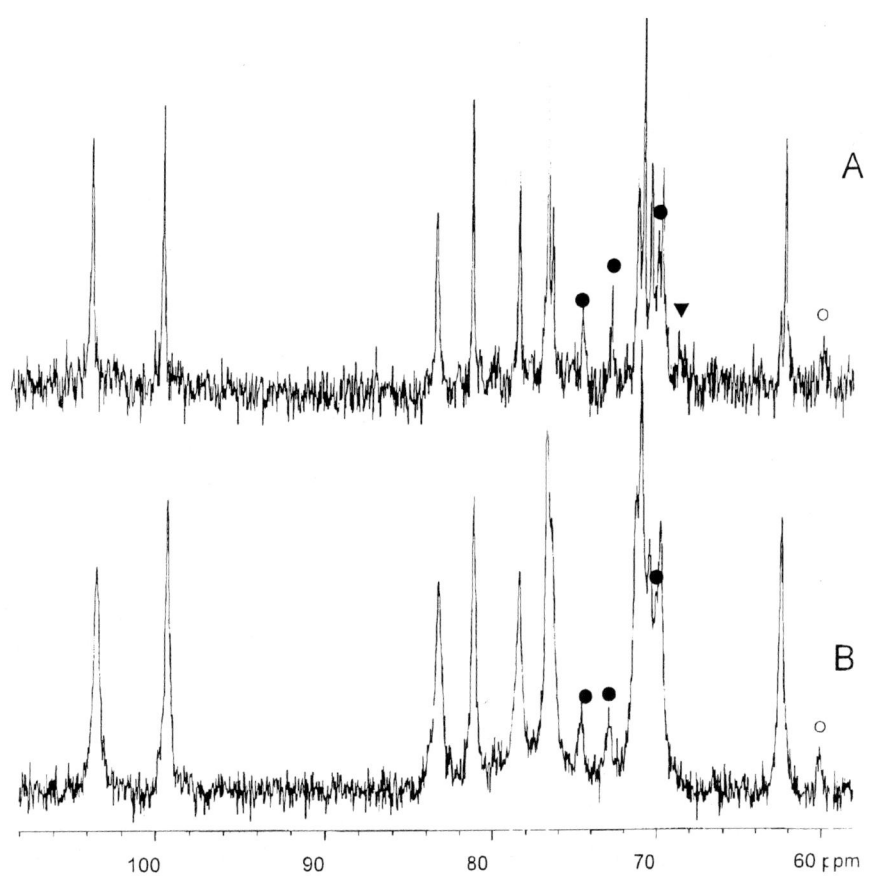

Figure 3. ^{13}C NMR spectra of (A) native and (B) alkali-treated agar extracted from *Gracilaria tenuistipitata*. Signals marked with (●), (○), and (▼) are attributed to carbons of D-galactosyl residue with O-6 methylation, methyl group at O-6 (G), and L-galactosyl residue with O-6 sulphation, respectively.

4, respectively (Murano, 1995). Methyl proton signal was resolved in the ^{1}H NMR spectra at 3.35 ppm in the native agar and at 3.45 ppm in the alkali-modified agar, with 0.4 DS in both extracts.

The presence of 3,6-anhydrogalactose precursor, galactose-6-sulphate, in the native agar of *G. tenuistipitata* was revealed by the weak signal at 68.7 ppm in the ^{13}C NMR spectrum, representing the down field shift of C-6 of the L-galactosyl residue upon sulphation. The signal disappeared after alkali modification (Figure 3), which was corroborated by a modest decrease in sulphate content and subsequent appreciable increase in the 3,6-anhydrogalactose content (Table 2). The residual sulphate content after the alkali modification could be attributed to alkali-stable sulphate at O-4 (G) as indicated by the 843 cm^{-1} band on the FT-IR spectra (Rochas et al., 1986) (Figure 4). This sulphation, however, was not detected in NMR analysis possibly due to its low level (3.3% w/w, ca. 0.1 DS).

Yield and physical properties of the extracted agars are presented in Table 3. The agar yield of *G. tenuistipitata* decreased tremendously upon alkali modification. Pyruvylation was not detected through NMR and enzymatic analysis in *G. arcuata* nor in *G. tenuistipitata* agar. Floridean starch is the major contaminant of agar extracts and could be removed by amylase treatment. Although amylase treatment was not conducted in this study, the extracts from both agarophytes were assumed to be floridean starch-free due to the absence of NMR signals attributable to starch.

Discussion

The occurrence of methyl ether at C-6 (G) and C-2 (LA) in agars extracted from different *Gracilaria* species has been widely reported, with the degree of substitution varying from species to species (Craigie, 1990; Furneaux et al., 1990). The first report where methylated agarobiose having 3,6-anhydro-2-*O*-methyl-α-L-galactose occurs as a major repeat unit in agar was that by Ji et al. (1985) for *Gracilaria eucheumoides* Harvey agar. This was supported by the findings of Lahaye et al. (1986), revealing complete methyl substitution at O-2 (LA) of an agar fraction from the alga. The agar from *G. arcuata* investigated herewith reveals complete methylation at two positions, assigning 3-linked 6-*O*-methyl-β-D-galactopyranose (G6M) and 4-linked 3,6-anhydro-2-*O*-methyl-α-L-galactopyranose (LA2M) as major repeat units. Similar methylation pattern was observed in agar synthesised by *Curdiea coriacea* (Hooker f. et Harvey) Chapman (Furneaux et al., 1990). On the other hand, the agar extracted from the *G. tenuistipitata* sample exhibits partial methylation at O-6 (G), similar to that from a sample collected in China (Ji et al., 1985). However, an additional methylation, at O-2 (LA), was elucidated by Lahaye et al. (1986).

The occurrence of a sulphate group at O-2 (G) in the *G. arcuata* sample would promote instability of the double helix conformation of agar upon gelling, as internal hydrogen bonding within the double helix is provided by a G-2 hydroxyl group (Rees, 1969). Further, axial sulphate ester at G-2 acts as a wedge, thereby preventing helix formation (Moirano, 1977). Agars from several members of the family Rhodomelaceae have been reported to exhibit such sulphation (Bowker & Turvey, 1968; Miller et al., 1993).

The decrease of sulphate in *G. arcuata* upon alkali modification indicated the presence of alkali-labile sulphate ester, at O-6 of the 4-linked L-galactosyl residue. This sulphation pattern, however, was not detected in the NMR analysis, suggesting very minimal concentrations. Galactosyl residue with sulphate substitution at O-6 is the precursor of 3,6-anhydrogalactose which is responsible for the increased gel strengths of algal extracts (Duckworth et al., 1971). An increase in the agar gel strength was observed upon alkali modification (Table 3), affirming the above contention. Despite this increase, the gel was still considered soft, reflecting the gel network effect of the sulphation at O-2 (G) and the still low 3,6-anhydrogalactose level (31%). Gelation was still possible because sulphate substitution at this position was low (3.7% w/w, ca. 0.1 DS). Weak gels, if at all, were also observed in agars possessing this sulphation pattern (Miller et al., 1993).

Similarly, alkali-stable sulphate was detected in *G. tenuistipitata* sample, but found to be linked at O-4 (G). Such sulphation is typical of κ-carrageenan, though it was reported to exist in agar from *Gracilaria bursapastoris* (Gmelin) Silva and *Gracilaria mammillaris* (Montagne) Howe (Murano et al., 1995, 1996).

The unusually high gelling temperature (> 60 °C) in *G. arcuata* agar was partly due to the very viscous solution, which resisted the sinking of glass beads used in the gelling temperature determination. Moreover, gelling temperature of agarose has been positively correlated to natural methoxyl content (Guiseley, 1970). It is due to this fact that *Gracilaria*

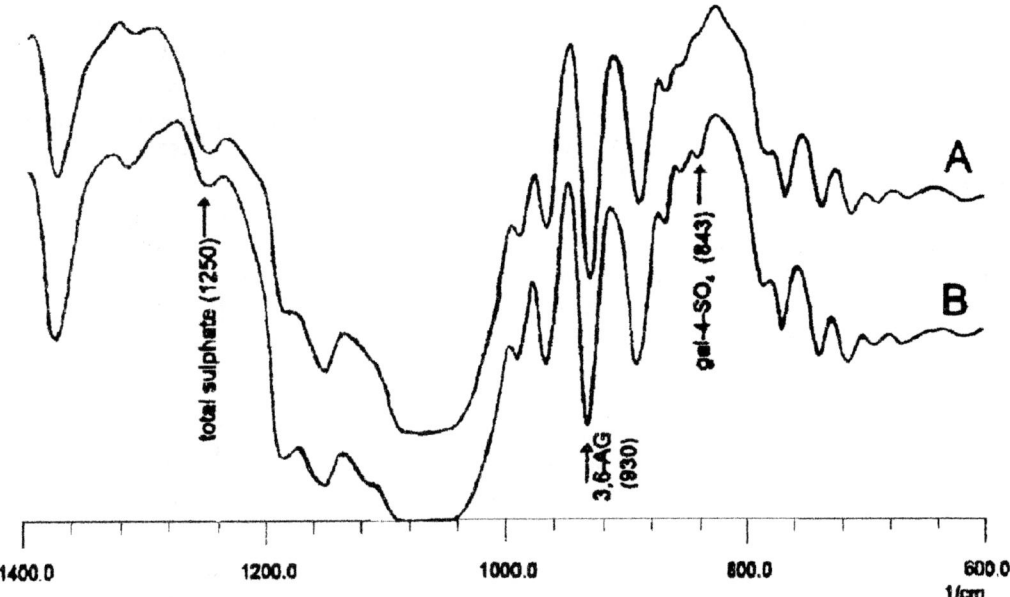

Figure 4. FT-IR spectra of (A) native and (B) alkali-treated agar extracted from *Gracilaria tenuistipitata*.

Table 3. Yield and physical properties of native and alkali-modified agars from *Gracilaria arcuata* and *G. tenuistipitata*. (Mean ± SE, $n = 3$)

	Gracilaria arcuata		*Gracilaria tenuistipitata*	
	native	alkali-treated	native	alkali-treated
Yield (%)	17.2 ± 0.9	18.8 ± 3.1	32.9 ± 2.4	15.7 ± 2.8
Gel Strength (g cm^{-2})	161 ± 4	278 ± 2	304 ± 23	606 ± 42
Viscosity (cP)	2090	538	14	15
Gelling Temp. (°C)	63.5 ± 1.3	60.7 ± 2.8	42.3 ± 0.6	42.3 ± 0.8
Melting Temp. (°C)	96.2 ± 1.8	98.7 ± 0.8	86.5 ± 1.3	86.7 ± 0.8
Syneresis Index (mm)	75 ± 7	80 ± 3	79 ± 1	127 ± 5

agars were unfit for microbiological media, as gelling temperature lower than 40 °C is generally required (Armisen & Galatas, 1987). High viscosity and melting temperature values can indicate a high molecular weight polymer (Selby & Wynne, 1973; Whyte & Englar, 1981); however, no molecular weight determination was done to prove this assumption.

The decrease in agar yield in *G. tenuistipitata* upon alkali treatment could possibly be due to leaching of agar polymers during the process, reflecting the soft thalli of the alga. On the other hand, no important yield variation was seen between untreated and alkali-treated *G. arcuata* samples (observed to possess hard thalli). Alkali modification resulted in a two-fold increase in the gel strength of *G. tenuistipitata* agar, correlating well to the 3,6-anhydrogalactose increase.

The gel network stability of *G. tenuistipitata* agar decreased upon alkali modification, as reflected by an increase in syneresis index (Fiszman & Duran, 1992). This was also observed to a lesser extent in *G. arcuata* agar. These results corroborate findings of Matsuhashi (1990), which indicated lower syneresis with higher sulphation. Viscosity, gelling, and melting temperature values of *G. tenuistipitata* agar are typical of *Gracilaria* agars (UNDP/FAO, 1990; Murano, 1995) and are lower than those of the peculiar *G. arcuata* agar.

The structural analysis demonstrated the varying methyl and sulphate substitution patterns between the agarocolloids of the two *Gracilaria* species studied. This diversity in structure and consequent gelling behaviour exemplifies the structural complexity of agar polymers synthesised by *Gracilaria*. These features, nevertheless, have made *Gracilaria* agar a vital commodity for the phycocolloid industry.

Acknowledgements

We are grateful to K. Araño and Dr E. Fortes for the seaweed identification, Dr E. Enriquez and A. Aliganga for the FT-IR analyses, Dr F. Dayrit and J. Aguilan for the NMR experiments, and the MSI-Seaweed Building staff for the laboratory assistance. This work was funded by the Philippine Department of Science and Technology through the Philippine Council for Aquatic and Marine Research and Development. This work represents UP-MSI Contribution No. 287.

References

Araki C (1966) Some recent studies on the polysaccharides of agarophytes. Proc. int. Seaweed Symp. 5: 3–19.

Armisen R, Galatas F (1987) Production, properties and uses of agar. In McHugh DJ (ed.), Production and Utilization of Products from Commercial Seaweeds, FAO Fish. Tech. Pap. 288: 1–57.

Bowker DM, Turvey JR (1968) Water-soluble polysaccharides of the red alga *Laurencia pinnatifida*. Part I. Constituent units. J. chem. Soc. (C): 983–988.

Craigie JS (1990) Cell walls. In Cole KM, Sheath RG (eds), Biology of the Red Algae, Cambridge U.P., Cambridge: 221–257.

Duckworth M, Yaphe W (1970) Definitive assay for pyruvic acid in agar and other algal polysaccharides. Chem. Ind.: 747–748.

Duckworth M, Hong KC, Yaphe W (1971) The agar polysaccharides of *Gracilaria* species. Carbohydr. Res. 18: 1–9.

Fiszman SM, Duran L (1992) Effects of fruit pulp and sucrose on the compression response of different polysaccharides gel systems. Carbohydr. Polym. 17: 11–17.

Furneaux RH, Miller IJ, Stevenson TT (1990) Agaroids form New Zealand members of the Gracilariaceae (Gracilariales, Rhodophyta) – a novel dimethylated agar. Hydrobiologia 204/205: 645–654.

Guiseley KB (1970) The relationship between methoxyl content and gelling temperature of agarose. Carbohydr. Res. 13: 247–256.

Jackson SG, McCandless EL (1978) Simple, rapid, turbidometric determination of inorganic sulfate and/or protein. Analyt. Biochem. 90: 802–808.

Ji M, Lahaye M, Yaphe W (1985) Structure of agar from *Gracilaria* spp. (Rhodophyta) collected in the People's Republic of China. Bot. mar. 28: 521–528.

Lahaye M, Rochas C (1991) Chemical structure and physicochemical properties of agar. Hydrobiologia 221: 137–148.

Lahaye M, Rochas C, Yaphe W (1986) A new procedure for determining the heterogeneity of agar polymers in the cell walls of *Gracilaria* spp. (Gracilariaceae, Rhodophyta). Can. J. Bot. 64: 579–585.

Matsuhashi T (1990) Agar. In Harris P (ed.), Food Gels, Elsevier Applied Science, England: 1–51.

Matsuhiro B, Rivas P (1993) Second-derivative Fourier transform infrared spectra of seaweed galactans. J. appl. Phycol. 5: 45–51.

Miller IJ, Falshaw R, Furneaux RH (1993) The chemical structures of polysaccharides from New Zealand members of the Rhodomelaceae. Bot. mar. 36: 203–208.

Moirano AL (1977) Sulfated polysaccharides. In Graham D (ed.), Food Colloids, The Avi Publishing Co. Inc., Connecticut: 347–381.

Murano E (1995) Chemical structure and quality of agars from *Gracilaria*. J. appl. Phycol. 7: 245–254.

Murano E, Gilli R, Navarini L, Toffanin R, D'Agnolo E, Paoletti S, Rizzo R (1995) Ion-driven gelation of highly sulfated agar. J. mar. Biotechnol. 3: 143–145.

Murano E, Toffanin R, Pedersini C, Carabot-Cuervo A, Blunden G, Rizzo R (1996) Structure and properties of agar from unexploited agarophytes from Venezuela. Hydrobiologia 326/327: 497–500.

Rees DA (1969) Structure, conformation, and mechanism in the formation of polysaccharide gels and networks. Adv. Carbohydr. Chem. Biochem. 24: 267–332.

Rochas C, Lahaye M, Yaphe W (1986) Sulfate content of carrageenan and agar determined by infrared spectroscopy. Bot. mar. 29: 335–340.

Selby HH, Wynne WH (1973) Agar. In Whistler RL (ed.), Industrial Gums, Polysaccharides and their Derivatives, Academic Press, New York: 29–48.

UNDP/FAO (1990) Training Manual on Gracilaria Culture and Seaweed Processing in China. Training Manual 6. Regional Seafarming Development and Demonstration Project (RAS/90/002) – UNDP/FAO, People's Republic of China: 1–85.

Whyte JNC, Englar JR (1981) Agar from intertidal population of *Gracilaria* sp. Proc. int. Seaweed Symp. 10: 537–542.

Yaphe W, Arsenault GP (1965) Improved resorcinol reagent for the determination of fructose, and of 3,6-anhydrogalactose in polysaccharides. Analyt. Biochem. 13: 143–148.

Polysaccharides from the red seaweed *Bostrychia montagnei*: chemical characterization

Miguel D. Noseda, Siumara Tulio & Maria E. R. Duarte*
Departamento de Bioquímica-UFPR, CP 19046, Centro Politécnico, CEP: 81531-990, Curitiba, Paraná, Brasil

(* Author for correspondence)

Received 1 April 1998; revised 1 October 1998; accepted 2 November 1998

Key words: seaweed, *Bostrychia montagnei*, polysaccharides, sulphated galactans

Abstract

Bostrychia montagnei was submitted to aqueous extraction at 25 and 85 °C. The purified polysaccharide extracts represent ~ 17% of the dried alga. Galactose is the principal monosaccharide component of these extracts (60.8–70.4 mol%). 3,6-Anhydrogalactose and its 2-*O*-methyl derivative are also present in smaller amounts (16.2–22.0 mol%), as well as other methylated sugars, such as 6-*O*- (6.5–7.8 mol%) and 2-*O*-methylgalactose (0.2–2.1 mol%). Xylose (4.1–8.1 mol%) and glucose (0.7–2.6 mol%) were also detected. The aqueous extracted polysaccharides (25 °C) were separated by anion-exchange chromatography into six sulphated galactan fractions with negative specific rotations and another two with high xylose contents and positive specific rotations. The sulphated galactans all have an agar type backbone modified by partial *O*-methyl substitution on O-6 or O-2 of the galactosyl units. The latter substitution is also present in varying degrees of 3,6-anhydrogalactose.

Introduction

Red seaweeds biosynthesize a great variety of sulphated galactans that are the major components of the extracellular matrix. Essentially, they consist of linear chains of repeating disaccharide groups of 3-linked β-galactose (A unit) and 4-linked α-galactose (B unit) (Painter, 1983). The former always belongs to the D-series, whereas the B unit (frequently as 3,6-anhydrogalactose) is in the D- and L-configuration in carrageenans and agars, respectively. This backbone is modified to varying degrees by substituents such as *O*-sulphate, *O*-methyl, 4,6-pyruvate ketal and *O*-glycosyl groups (Usov, 1992).

Bostrychia montagnei Harvey is a red seaweed of the family Rhodomelaceae (order Ceramiales) growing in the tropical Atlantic Ocean in rocky intertidal habitats and mangroves. Species of *Bostrychia* (subfamily Bostrychioideae) have been extensively studied for their polyol and low-molecular-weight contents (Kremer, 1976; Karsten et al., 1992). The Bostrychioideae produce, as their main photosynthetic metabolites, the isomeric hexitols, dulcitol and sorbitol, and for this they are unique in the red algae (Kremer, 1976). Less attention has been paid to the components of the cell wall. We now report the isolation and analysis of polysaccharides extracted with water at room temperature and then hot water, as well as the fractions obtained by anion-exchange chromatography of the first extract. As far as we know, this is the first report of a study of polysaccharides produced by a species of the genus *Bostrychia*.

Materials and methods

Specimens of *Bostrychia montagnei* were collected in the Ilha do Mel (Paraná State, Brazil). After collection, the samples were cleaned to remove undesirable material, washed with running water, sun-dried, and ground in a mill to a fine powder. The milled alga (160 g) was treated with methanol (85%, 60 °C, 800 mL) until there was an absence of colour in the organic extract (5×), and then dried in an oven at 40 °C.

Extraction procedure

The pigment-free powder from above was extracted with distilled water at room temperature with mechanical stirring for 15 h (1 L, 4×). The residue was removed by centrifugation and the supernatant poured into ethanol (3 volumes), which precipitated the polysaccharides (B-CW). The algal residue was re-extracted four times with hot water (1 L, 85 °C, 4 h) with mechanical stirring, centrifuged, and the supernatant also poured into ethanol (3 volumes), yielding the crude extract B-HW. The ethanolic supernatants of B-CW and B-HW extractions were concentrated, dialysed, and freeze-dried, yielding B-CWE and B-HWE, respectively. The four crude extracts were purified by redissolving in water, dialysis, centrifugation and lyophilization of the supernatant.

Chemical analyses

Total carbohydrate was determined by the method of Dubois et al. (1956); galactose was used to construct the standard curve. Sulphate content was determined by the turbidimetric method of Dodgson and Price (1962), hexuronic acid by the m-hydroxydiphenyl assay (Blumenkrantz & Asboe-Hansen, 1973) and protein by the Lowry method (Lowry et al., 1951). Molecular weights were determined by the method of Park & Johnson (1949). The determination of monosaccharide composition was by gas chromatography (GC), following the reductive hydrolysis procedure of Stevenson and Furneaux (1991) using N-methyl-morpholine-borane as the reductant. The relative molar GC response factors determined by Stevenson and Furneaux (1991) were used to calculate the corresponding monosaccharide content in polysaccharides.

Gelling capacity

The ion-independent gelling capacity of the polysaccharides was tested (2.5% w/v, aqueous solutions) at 4 and 25 °C. The gelling capacity in the presence of potassium chloride was determined using the fractional precipitation technique described by Matulewicz et al. (1989).

Specific optical rotation

Optical rotations of aqueous solutions of the polysaccharide samples (0.2%), were measured at 20 °C, using a 10-cm cell and the sodium D line (589.3 nm) with a Rudolph Autopol III automatic polarimeter.

Anion-exchange chromatography

The extract B-CW was fractionated by anion-exchange chromatography on a DEAE-Sephadex A-50 column. The sample (3.51g per 300 mL of water) was adsorbed onto the top of the column, first eluted with distilled water (B1), and then with solutions of increasing salt (NaCl) concentrations: 0.75 M (B2), 1.0 M (B3), 1.25 M (B4), 1.5 M (B5) and 4.0 M (B6), then 6 M urea (B7) and 0.5 M NaOH (B8). Column eluants were analysed for carbohydrates by the phenol-sulphuric acid method (Dubois et al., 1956). The fractions obtained were concentrated, dialysed against distilled water, concentrated, and freeze-dried.

Gas chromatography (GC) and GC-mass spectrometry (MS)

GC analyses were carried with a HP-5890 gas chromatograph equipped with a flame ionization detector (FID), using a fused silica capillary column (30 m × 0.25 mm) coated with DB-225. Chromatography was run isothermically at 210 °C. The temperature of both the injector and FID was 250 °C. Nitrogen was used as carrier gas at a flow rate of 1 mL min^{-1} and a split ratio of 100:1. GC-MS analyses were performed using a Varian 3300 chromatograph and a Finnigan Mat ITD spectrometer. Helium was used as carrier gas at 1 mL min^{-1}.

Gel filtration chromatography

Solutions of fractions B1, B4-B6 and B8 (2 mg per 0.5 mL of water) were applied to a Sepharose 4B column (50 × 0.8 cm) equilibrated and eluted with water at 25 °C at a flow rate of 1 mL min^{-1}. Fractions (1 mL) were collected and assayed by the reaction of Dubois et al. (1956).

Aqueous alkaline treatment

The samples (50 mg) were dissolved in water (25 mL) and sodium borohydride (5 mg) added. After 24 h at 25 °C, 3 M NaOH was added (12.5 mL), together with a further quantity of sodium borohydride

(2.5 mg), to give a final concentration of 1 M NaOH. The resulting solution was heated at 80 °C and the content of 3,6-anhydrogalactose (Yaphe, 1960) determined on samples removed at intervals; heating was continued until the 3,6-anhydrogalactose content remained constant or at least for 4 h. The reaction was terminated by cooling in an ice bath, the solution neutralized with 1 M HCl, dialysed, concentrated, and freeze-dried. The sugar composition of native and alkali-treated polysaccharides was determined by gas chromatography using the procedure of Stevenson and Furneaux (1991).

Results

The analytical data obtained for the four purified polysaccharide extracts are given in Table 1. These fractions represent 17.2% of the raw material. They are similar in terms of total carbohydrates (44.8–57.7%), sulphate (17.0–23.0%) and protein contents (1.5–3.9%). They were free of uronic acid within the analytical detection limits. Galactose is the principal monosaccharide component of all the fractions (60.8–70.4 mol%). 3,6-Anhydrogalactose and its 2-O-methyl derivative are also present in smaller amounts (16.2–22.0 mol%), as well as other methylated sugars such as 6-O- (6.5–7.8 mol%) and 2-O-methylgalactose (0.2–2.1 mol%). Xylose (4.1–8.1 mol%) and glucose (0.7–2.6 mol%) were also detected. Mannose was present in trace amount (less than 0.2 mol%), only in B-CW. Specific rotations of the purified products are negative (−35.0° up to −49.7°). Aqueous solutions of the polysaccharide fractions did not gel either at room temperature or 4 °C. All the fractions remained in solution during the fractional precipitation with KCl (up to 2.0 M).

B-CW was fractionated by anion-exchange column chromatography on DEAE-Sephadex A-50, using stepwise elution with increasing concentrations of NaCl, 6 M urea, and 0.5 M NaOH, to give eight separate fractions, with a total recovery of ∼ 77%. The yields and analyses of these fractions are reported in Table 2. All the fractions, as mentioned for the native polysaccharide, were found to be free of uronic acid. Table 2 shows that most of the product was eluted with water, 1.25 and 1.50 M NaCl, and 0.5 M NaOH (∼87% of the total recovered material). The homogeneity of the principal fractions (B1, B4, B5, B6 and B8, data not shown) was tested by gel-permeation chromatography on a Sepharose 4B column. All the polysaccharide fractions were included in the gel, each one of the first four being eluted as a symmetric peak, whereas B8 showed a broad and asymmetric peak. The protein concentrations of the B6-B8 fractions were up to three times higher than that of the native polysaccharide and accounted for all the recovered protein.

All the fractions obtained by anion-exchange chromatography had negative specific rotations (−26.0° to −51.0°), except B7 and B8 (+75.5° and +30.7° respectively). Sulphate groups are present in all the polysaccharide fractions but no correlation was found between NaCl, urea or NaOH concentrations and the degree of sulphation. Galactose was the major component sugar (Table 3) in fractions B1-B6, together with smaller amounts of 3,6-anhydrogalactose plus 2-O-methyl-3,6-anhydrolgalactose (10–36 for each 100 galactosyl residues). In addition, 6-O-methylgalactose (7–12/100 Gal), very low percentages of 2-O-methylgalactose (0.3–1/100 Gal) and xylose in variable amounts (4–43/100 Gal) were present. B7 contained xylose and galactose as the principal sugar components (molar ratio, 1:0.89), together with minor amounts of the same units which were present in the other fractions. In B8, the galactose content was much lower and almost no methylated sugars were detected, with xylose, glucose and mannose being the main sugars (molar ratio, Xyl:Glc:Man, 1:0.80:0.29).

Fractions B4 and B5 were submitted to aqueous alkaline treatment and the 3,6-anhydrogalactose content and sugar composition of the native and treated polysaccharide determined. B5 showed a decrease in its galactose (from 72.6 to 66.4 mol%) and 2-O-methylgalactose contents (from 1.1 to 0.7 mol%), accompanied by an increase in the contents of 3,6-anhydrogalactose (from 13.6 to 17.9 mol%) and 2-O-methyl-3,6-anhydrogalactose (from 4.0 to 5.3 mol%). The sugar composition of the B4 fraction was not altered after alkaline treatment. These results confirm the 3,6-anhydrogalactose determinations carried out during the alkaline treatment.

Discussion

The water-soluble polysaccharides obtained from the red seaweed *B. montagnei* are predominantly sulphated galactans. They do not show gelling properties at 25 or 4 °C, characteristic of agarose (Duckworth & Yaphe, 1971). Aqueous solutions of the polysaccharides (0.25% w/v) do not gel in the presence of

Table 1. Analyses of the polysaccharide fractions extracted from *Bostrychia montagnei*

	B-CW	B-HW	B-CWE	B-HWE
Yield (%)	6.0	6.5	2.2	2.5
Carbohydrates (%)	57.7	44.8	52.2	45.9
Sulfate (SO$_3$Na%)	23.0	17.0	21.0	18.8
Protein (%)	1.7	3.9	3.8	1.5
M_r	27,500	45,700	29,500	16,500
$[\alpha]_D$ (°)	−40.8	−49.7	−35.0	−42.5
Component monosaccharides[a] (mol%)				
Galactose	70.4	60.8	61.5	65.3
3,6-AnGal	12.1	15.4	16.8	15.1
2-*O*-Me-3,6-AnGal	4.1	6.5	5.2	4.9
6-*O*-Me-Galactose	6.5	7.8	7.2	7.0
2-*O*-Me-Galactose	0.2	2.1	0.5	1.0
Xylose	4.1	6.4	8.1	5.8
Glucose	2.6	1.0	0.7	0.9
Mannose	tr.[b]	–	–	–

[a] 3,6-AnGal = 3,6-Anhydrogalactose, 6-*O*-Me-Galactose = 6-*O*-methylgalactose, etc.
[b] tr. = Traces (< 0.2%).

Table 2. Yields and analyses of the polysaccharide fractions obtained by anion-exchange chromatography of the purified extract B-CW, from *Bostrychia montagnei*

Fraction	Eluant	Yield (%)[a]	Carbohydrate (%)	Sulphate (% SO$_3$Na)	Protein (%)	M_r	$[\alpha]_D$ (°)
B1	Water	18.8 (24.6)	60.0	13.0	0	31,300	−51.0
B2	0.75 M NaCl	0.7 (0.9)	37.0	11.2	0	5,600	nd[b]
B3	1.0 M NaCl	2.3 (2.9)	38.1	16.2	0	11,900	−26.0
B4	1.25 M NaCl	20.3 (26.5)	56.8	22.0	0	43,700	−43.9
B5	1.5 M NaCl	13.7 (17.9)	62.6	24.0	0	42,500	−42.4
B6	4.0 M NaCl	5.7 (7.4)	46.7	22.0	2.1	34,000	−44.8
B7	6.0 M Urea	1.4 (1.9)	53.5	10.1	5.2	7,000	+75.5
B8	0.5 M NaOH	13.7 (17.9)	43.4	18.1	5.3	8,300	+30.7

[a] In parentheses, percentage of the recovered material.
[b] nd = not determined.

low KCl concentrations (0.1–0.4 M), as do kappa- and iota- or high KCl concentrations (0.6–2.0 M), necessary for lambda-carrageenan gelation (Matulewicz et al., 1989, Stortz & Cerezo, 1993). All the purified extracts of *B. montagnei* contained galactose as the main sugar component, together with lesser amounts of 3,6-anhydrogalactose and its 2-*O*-methyl derivative. Other methylated sugars, such as 6-*O*- and 2-*O*-methylgalactose are also present, the latter in very low proportions. These residues were also found as components of the polysaccharides from other species of the family Rhodomelaceae, for example *Laurencia pinnatifida* (Bowker & Turvey, 1968) and *Polysiphonia lanosa* (Batey & Turvey, 1975). However, some species of this family, namely *Laurencia nipponica* (Usov & Elashvili, 1991) and *Chondria macrocarpa* (Furneaux & Stevenson, 1990), lack one or both of these methylated galactosyl units.

The purified aqueous polysaccharide extract, obtained at 25 °C from *B. montagnei* (B-CW), is composed of more than one polysaccharide structure, as demonstrated by the fractionation carried out using

Table 3. Component monosaccharides of the fractions obtained by anion exchange chromatography of the purified extract B-CW, from *Bostrychia montagnei*

	Molar ratio[a]								
Fraction	Gal[b]	3,6-AG	2-Me-3,6-AG	6-MeGal	2-MeGal	Xyl	Man	Glc	SO_3Na
B1	100	22	9	9	0.4	10	–	3	100
B2	100	7	3	7	0.3	43	6	17	85
B3	100	9	23	12	1	5	–	3	102
B4	100	21	15	10	1	5	–	–	91
B5	100	10	5	8	1	4	–	–	82
B6	100	17	5	7	–	8	–	–	97
B7	89	16	4	6	3	100	–	–	58
B8	14	2	1	0.4	–	100	29	80	136
B-CW	100	17	6	9	0.2	6	tr.[c]	4	88

[a] The value for Gal was taken as 100, except for the B7 and B8 fractions (Xyl was taken as 100).
[b] Gal, 3,6-AG, 2-Me-3,6-AG, 6-MeGal, 2-MeGal, Xyl, Man and Glc refer to galactose, 3,6-anhydrogalactose, 2-*O*-methyl-3,6-anhydrogalactose, 6-*O*-methylgalactose, 2-*O*-methylgalactose, xylose, mannose and glucose, respectively.
[c] tr. = Traces (< 0.2).

anion-exchange chromatography. From the eight fractions obtained, six were sulphated galactans with low xylose contents (eluted with water and solutions of increasing NaCl concentrations, B1-B6) and the other two having high xylose contents (eluted with 6 M urea and 0.5 M NaOH, B7 and B8 respectively). Considering the sulphate contents, the principal fractions of sulphated galactans can be separated into two groups, one with low sulphate contents (B1) and the other with higher ones (B4-B6). They have in common, M_r values, the same monosaccharide components in similar proportions, and negative specific rotations. The latter indicate that the galactosyl units are in the D- and L- forms (Cases et al., 1992). This suggests that the sulphated galactans produced by *B. montagnei* have an agar type backbone, with alternating β-D-(1→4) and α-L-(1→3) linked galactosyl units, as observed in other species of the family Rhodomelaceae (Miller et al., 1993; Miller & Furneaux, 1997). This structure is modified by partial *O*-methyl substitution on O-6 or O-2 of the galactosyl units. The latter substitution is also present in varying degrees of 3,6-anhydrogalactose. An additional complication in the substitution pattern of the backbone is introduced by *O*-sulphation.

Treatment of sulphated polysaccharides that contain 4-linked galactosyl 6-sulphate residues with hot aqueous alkali results in intra-molecular removal of the sulphate groups and formation of 3,6-anhydrogalactose (Rees, 1961). To confirm the presence of these units in *B. montagnei* polysaccharides, fractions B4 and B5 were submitted to alkaline treatment. The sulphated galactan B4 was not modified, as shown by chromatographic analysis (alditol acetates) of the native and treated polysaccharides. The small increase of 3,6-anhydrogalactose in the B5 fraction, may be due, not only to low proportions of the L-galactose 6-sulphate units, but also to the presence of substituents on O-3, such as sulphate ester or glycosyl units. This last possibility was found in the xylogalactan sulphate from *Laurencia nipponica* (Usov & Elashvili, 1991) and *Chondria macrocarpa* (Furneaux & Stevenson, 1990).

Together with the sulphated galactans, *B. montagnei* biosynthesizes polysaccharides rich in xylose, as those present in fractions B7 and B8 (see Table 3). B7 has a molar ratio of xylose:galactose close to unity, probably resulting from the presence of xylogalactans. In the order Ceramiales, methylated sulphated xylogalactans with relative high xylose content ($\sim 15\%$), have been described (Furneaux & Stevenson, 1990; Miller et al., 1993). B8 is principally composed of xylose, glucose and mannose (molar ratio, 1:0.80:0.29). Considering that this sample was heterogeneous by gel-exclusion chromatography and that glucose can arise from contaminating floridean starch (Painter, 1983), the presence of xylans, mannans and/or xylomannans in this fraction can be considered. These types of polysaccharides have been found in some species of red algae that biosynthesize neutral xylans (Usov, 1992; Matulewicz et al., 1992), as well as sulphated mannans and xylomannans (Kolender et al., 1995).

In conclusion, *Bostrychia montagnei* biosynthesizes methylated sulphated galactans with an agar type backbone modified by partial O-methyl substitution on O-6 or O-2 of the galactosyl units. The latter substitution is also present in varying degrees of 3,6-anhydrogalactose. An additional complication in the substitution pattern of these polysaccharides is introduced by O-sulphation and the presence of xylosyl units.

Acknowledgements

The authors are indebted to Miss Madalena Shirata MSc for collecting and sorting the algal material. This work was supported by a grant from PRONEX-CARBOIDRATOS (FINEP).

References

Batey JF, Turvey JR (1975) The galactan sulphate of the red alga *Polysiphonia lanosa*. Carbohydr. Res. 43: 133–143.

Blumenkrantz N, Asboe-Hansen G (1973) New method for quantitative determination of uronic acids. Anal. Biochem. 54: 484–489.

Bowker DM, Turvey JR (1968) Water-soluble polysaccharides from the red alga *Laurencia pinnatifida*. Part I. Constituent units. J. chem. Soc., C. 983–988.

Cases MR, Stortz CA, Cerezo AS (1992) Methylated, sulphated xylogalactans from the red seaweed *Corallina officinalis*. Phytochemistry 31: 3897–3900.

Dodgson KS, Price RG (1962) A note on the determination of the ester sulphate content of sulphated polysaccharides. Biochem. J. 84: 106–110.

Dubois MK, Gilles A, Hamilton JK, Rebers PA, Smith F (1956) Colorimetric method for determination of sugars and related substances. Anal. Chem. 28: 350–356.

Duckworth M, Yaphe W (1971) The structure of agar. Part I. Fractionation of a complex mixture of polysaccharides. Carbohydr. Res. 16: 189–197.

Furneaux RH, Stevenson TT (1990) The xylogalactan sulfate from *Chondria macrocarpa* (Ceramiales, Rhodophyta). Hydrobiologia 204/205: 615–620.

Karsten U, West JA, Zuccarello G (1992) Polyol content of *Bostrychia* and *Stictosiphonia* (Rhodomelaceae, Rhodophyta) from field and culture. Bot. mar. 35: 11–19.

Kolender AA, Matulewicz MC, Cerezo AS (1995) Structural analysis of antiviral sulfated α-D-$(1\rightarrow 3)$-linked mannans. Carbohydr. Res. 273: 179–185.

Kremer BP (1976) Distribution of alditols in the genus *Bostrychia*. Biochem. System. Ecol. 4: 139–141.

Lowry OH, Rosebrough NJ, Farr AL, Randall RL (1951) Protein measurement with Folin phenol reagent. J. biol. Chem. 193: 265–275.

Matulewicz MC, Cerezo AS, Jarret RM, Syn N (1992) High resolution ^{13}C-n.m.r. spectroscopy of 'mixed linkage' xylans. Int. J. biol. Macromol. 14: 29–32.

Matulewicz MC, Ciancia M, Noseda MD, Cerezo AS (1989) Carrageenan systems from tetrasporic and cystocarpic stages of *Gigartina skottsbergii*. Phytochemistry 28: 2932–2941.

Miller IJ, Furneaux RH (1997) The structural determination of the agaroid polysaccharides from four New Zealand algae in the order Ceramiales by means of ^{13}C NMR spectroscopy. Bot. mar. 40: 333–339.

Miller IJ, Falshaw R, Furneaux RH (1993) The chemical structure of polysaccharides from New Zealand members of the Rhodomelaceae. Bot. mar. 36: 203–208.

Painter TJ (1983) In Aspinall GO (ed.), The Polysaccharides Vol. 2, Academic Press, New York: 195–285.

Park JT, Johnson MJ (1949) A submicro determination of glucose. J. biol. Chem. 181: 149–151.

Rees DA (1961) Estimation of the relative amounts of isomeric sulphate esters in some sulphated polysaccharides. J. chem. Soc. 5168–5171.

Stevenson TT, Furneaux RH (1991) Chemical methods for the analysis of sulphated galactans from red algae. Carbohydr. Res. 210: 277–298.

Stortz CA, Cerezo AS (1993) The systems of carrageenans from cystocarpic and tetrasporic stages from *Iridaea undulosa*: fractionation with potassium chloride and methylation analysis of the fractions. Carbohydr. Res. 242: 217–227.

Usov AI (1992) Sulfated polysaccharides of the red seaweeds. Food Hydrocolloids 6: 9–23.

Usov AI, Elashvili MYa (1991) Polysaccharides of algae. 44. Investigation of sulfated galactan from *Laurencia nipponica* Yamada (Rhodophyta, Rhodomelaceae) using partial reductive hydrolysis. Bot. mar. 34: 553–560.

Yaphe W (1960) Colorimetric determination of 3,6-anhydrogalactose and galactose in marine algal polysaccharides. Anal. Chem. 32: 1327–1330.

Chemical structure and gel properties of carrageenans from algae belonging to the Gigartinaceae and Tichocarpaceae, collected from the Russian Pacific coast

Irina M. Yermak[1]*, Yong Hwan Kim[2], Edyart A. Titlynov[3], Vladimir V. Isakov[1] & Tamara F. Solov'eva[1]
[1]*Pacific Institute of Bioorganic Chemistry, Far East Division, The Russian Academy of Science, Vladivostok, Russia*
[2]*Dept. Food and Biotechnology, Kyonggi University, Suwon, Korea*
[3]*Institute of Marine Biology, Far East Division, The Russian Academy of Science, Vladivostok, Russia*

(*Author for correspondence; phone: +7 4232 314 855; fax: +7 4232 314 050; e-mail: yermak@piboc.marine.su)

Received 1 April 1998; revised 1 October 1998; accepted 5 January 1999

Key words: red alga, Gigartinaceae, Tichacarpaceae, carrageenan, fractionation, structure, gel properties, viscosity

Abstract

The chemical and gel characteristics of carrageenans isolated from the most abundant algal species growing on the Russian Pacific coast – *Chondrus pinnulatus*, *C. armatus* and *Iridaea cornucopiae* belonging to the Gigartinaceae and *Tichocarpus crinitus* from the Tichocarpaceae were investigated. The polysaccharides were identified by FTIR and NMR spectroscopy as predominantly κ-carrageenans with traces of ι-type (Gigartinaceae) and κ / β-type repeating structures (Tichocarpaceae) together with a small quantity of λ-carrageenan (10%). The chemical structure and the hydrodynamic properties play a determinant role on the rheology of these carrageenans. κ-Carrageenans from the Gigartinaceae displayed good gelling properties. The highest gel strength was obtained from *C. pinnulatus* (1232.7 Pa) at a 2.5% polymer concentration, while carrageenans from the Tichocarpaceae formed very weak gels (77.4 Pa) at the same concentration. Optimum gel characteristics were found with 1.0–2.0% KCl concentrations for kappa- carrageenans from Gigartinaceae and 0.75% for *T. crinitus*. The flow curves of λ-carrageenans solutions from the Gigartinaceae were similar, all between 20 and 65 °C, and characteristic of conformational disordered 'random coil' polysaccharides. Carrageenans from *T. crinitus* displayed the properties of 'random coil' only at high temperatures.

Introduction

Carrageenans are a complex family of water-soluble galactans extracted from marine red algae. These polysaccharides are composed of alternating α-(1->3) and β-(1->4) linked D-galactosyl residues and several types of carrageenan are identified on the basis of the modification of the disaccharide repeating unit by ester sulphates and by the presence of 3,6-anhydrogalactose. Native carrageenans are often hybrids of more than one type of repeating units (Craigie, 1990; Knutsen et al., 1994).

The industrial interest and economic importance of carrageenans are due to their ability to increase the viscosity of solutions or to form thermoreversible gels (Glicksman, 1983). Cations play an essential role in controlling the gelation of these biopolymers (Heyraud et al., 1990; Zhang et al., 1994) and their physico-chemical properties depend on the chemical structure and molecular weight (Rochas et al., 1990; Yermak & Khotimchenko, 1997). The relationships between chemical structure and physical properties of carrageenans are extremely complex and little understood.

Variations in carrageenan structures occur not only between polysaccharides from different species of algae, but also between different life stages of the same species and of individuals growing under different environmental conditions (McCandless et al., 1973; Craigie, 1990; Stortz & Cerezo, 1993; Falshaw & Furneaux, 1994, 1995).

While carrageenophytes are abundant in the Russian far east seas and potential sources of hydrocolloids have been identified: *Chondrus armatus*, *C. yendoi*, *C. pinnulatus* and *Tichocarpus crinitus* (Kizevetter et al., 1981), carrageenans of seaweeds from the Russian Pacific coast have not been studied in detail. Carrageenans have been isolated from several algae from the Sea of Japan (*Gigartina ochotensis*, *Rhodoglossum hemisphaericum* and *Tichocarpus crinitus*) and their chemical structures studied (Usov 1974; Usov et al., 1983), but not their physico-chemical properties.

The aim of the present study was to investigate the structure and gel properties of carrageenans isolated from the most abundant algal species on the Russian Pacific coast.

Material and methods

Algae

The following representative species of red algae were collected at Peter the Great Bay (Sea of Japan), which is near the border between the boreal and tropical zones: Order Gigartinales, Family Gigartinaceae – *Chondrus armatus* (Harv.) Okam., *C. pinnulatus* (Harv.) Okam., *Iridaea cornucopiae* P. et R. subsp. *japonica* (Yam. et Mik.) Perest., and Order Cryptonemiales, Family Tichocarpaceae – *Tichocarpus crinitus* (Gmel.) Rupr. All algae were harvested at the end of August, washed with running water and air-dried.

Extraction

Dried and milled algae (50 g) were suspended in hot water (1.5 L) and the polysaccharides extracted at 90 °C for 2 h in a boiling water bath. The suspensions were centrifuged (2500 g, 20 min, 20 °C) and the algal residues re-extracted twice with water for 2 h in a boiling water bath. The suspensions were centrifuged as above and the final algal residues recovered. The supernatants were combined, concentrated *in vacuo* to about 500 mL and solid ground KCl was added in small portions to reach a final concentration of 4% (w/v). The suspensions were left at 4 °C for 12 h and then centrifuged ($\times$ 25,000 g, 30 min, 4 °C). The pellets were suspended in hot water, dialysed against water and then against 0.15 mol L^{-1} NaCl for 4 days at 4 °C. The polysaccharides were precipitated by adding 3 volumes of 95% ethanol, washed with ethanol and dried. The samples were designated as the KCl-insoluble fractions (a). The supernatant obtained after KCl precipitation was concentrated and polysaccharides were precipitated by adding 3 volumes of 95% ethanol, washed with 95% ethanol and dried. This product was designated as the KCl-soluble fraction (b).

Carrageenan from *Eucheuma cottonii* type III and λ-carrageenan were purchased from Sigma Chemicals (St. Louis, MO, USA).

Analytical methods

The total amount of carbohydrates was estimated using the phenol-sulphuric acid method with D-glucose as standard (Dubois et al., 1956). Monosaccharides as alditol acetate derivatives (Englyst & Cummings, 1984) were identified by GLC using a Pye-Unicam-104 gas chromatograph, equipped with a flame-ionisation detector and a glass column (0.4 $\times$ 150 cm) packed with 3% QF-1(Serva) on Gas Chrom Q (100–120 mesh). The column was eluted by argon with a temperature gradient from 175 to 220 °C, at 5 °C min^{-1}.

Protein content was determined according to the method of Lowry et al. (1951) using crystalline bovine serum albumin as standard.

Ash was determined gravimetrically after incineration of samples at 550 °C for 16 h followed by 2 h at 900 °C. The cationic composition was determined by atomic absorption spectroscopy. Sodium and potassium contents were determined after mineralisation of the carrageenan samples using a microwave digestor (MLS-1200 MEGA, Italy) and analysed by ICP-AAS system (LEEMAN LABS INC, PS 1000, USA), with the following conditions: power: 10 kW, coolant gas: 13 L min^{-1}, nebulizer gas pressure: 40 Pa, pump rate: 1.0 mL min^{-1}. Sulphate was determined according to the method of Lahaye and Axelos (1993) by HPLC equipped with a conductivity detector (Waters 431) and a IC-Pack A Anion column (50 $\times$ 4.6 mm, 10 μm, Waters), eluted by 0.002 mol L^{-1} borate / gluconate eluent (1: 1; flow rate: 1.0 mL min^{-1}).

Table 1. Characterization of carrageenan fractions and content of metals in samples. (Mn was non-detectable.) a) KCl - insoluble; b) KCl – soluble.

Species	KCl solubility	Yield (%) algal dry	Content % dry weight				K	Na	Ca	Al	Cd	Cu	Zn	Pb	Cr	As	B
			sugars	sulphate	ash	protein											
Chondrus	a	30	47	22	18	1.9	0.3	0.6	2.9	0.02	ND	ND	0.11	ND	0.07	ND	ND
armatus	b	10	29	28	26	2.1											
Chondrus	a	32	43	20	17	3.5	0.6	0.8	2.6	ND	ND	ND	0.11	ND	0.09	ND	ND
pinnulatus	b	9	61	23	26	3.4											
Iridaea	a	37	53	20	24	1.7	0.2	0.5	1.5	0.04	ND	ND	0.10	ND	0.10	ND	ND
cornucopiae	b	4	34	24	27	2.1											
Mastocarpus	a	34	40	22	27	3.0	0.4	0.7	2.0	0.04	ND	ND	0.08	ND	0.09	ND	ND
pacificus	b	3.4	39	25	27	2.2											
Tichocarpus	a	32	50	22	19	2.9	0.6	0.5	2.2	ND	ND	ND	0.09	ND	0.09	ND	ND
crinitus	b	9	56	27	24	3.4											

Infrared and Nuclear Magnetic Resonance spectroscopy

Infrared (FTIR) spectra were recorded with a Perkin Elmer 580 spectrometer using carrageenan pellets in potassium bromide. ^{13}C-NMR spectra of polysaccharide solutions in D$_2$O were recorded at 60 °C with a Bruker-Physic WM-250 spectrometer operating at 62.9 MHz. Chemical shifts were determined from CD$_3$OD assigned at 50.15 ppm and used as an internal standard.

Molecular weight determinations

The molecular weight of carrageenans in solution (0.1% w/v in 0.1 mol L^{-1}NaCl) was determined by sedimentation analysis using an analytical ultracentrifuge 3130 MOM (Hungary) at 481,000 g. The apparent molecular weights were calculated by the method of Archibald (Elias, 1961).

Rheological measurements

Gel strength measurements were performed using a Sun-Rheometer Compact-100 (Japan) at 20 °C. κ-Carrageenan solutions (1% w/v) were obtained by heating the polysaccharides in water at 70 °C for 30 min. The stock solution was immediately divided in parts and KCl was added to prepare aliquots with concentrations from 0.2 to 3.0% (w/v). Solutions were placed in cylindrical glass bottles (diameter 25 mm, sample height 19 mm) and kept at 5 °C for 24 h. The gels were warmed to room temperature for approximately 2 h, before the gel strength was measured using the following conditions: plunger diameter: 12 mm, gel thickness: 0.8 mm, moving distance: 4 mm and table speed: 60 mm min^{-1}. κ-Carrageenan gels (0.5 to 2.5% w/v) in 1% (w/v) KCl were prepared as above in cylindrical bottles (diameter 15 mm).

Measurement of shear viscosity

Solutions of λ-carrageenans (1%) were prepared as above. Shear viscosity was measured for 2 min on 9 mL samples using a HAAKE RV 20 Rotovisco viscometer (UK), equipped with a NV sensor system (cylindrical type) at shear-rates ranging between 1 and 1000 s^{-1}. The measurements were performed at 20, 35, 50 and 65 °C.

Results

Polysaccharides were extracted from seaweeds by hot water and separated into 4% KCl insoluble (fraction a) and soluble (fraction b) fractions. All the fractions obtained were free of floridean starch as galactose was the only monosaccharide found in their acid hydrolysate. The yields and the chemical composition of these fractions are listed in Table 1. Fractions a and b from the four algae had similar compositions. The crude polysaccharides samples contained proteins (10% w/w) but in contrast to *Hypnea* carrageenan extracts (Knutsen et al., 1995), they could not be fractionated by KCl. Thus, the protein content was similar in both the KCl-soluble and KCl-insoluble fractions. Metal contents are given in Table 1; no man-

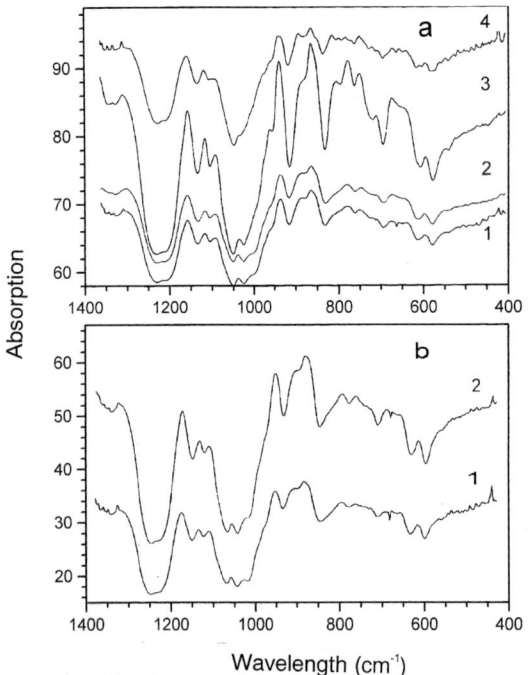

Figure 1. Infrared spectra of: **a)** KCl-insoluble fractions polysaccharide from 1) *Chondrus pinnulatus*, 2) *C. armatus*, 3) *Iridaea cornucopiae*, 4) *Tichocarpus crinitus* **b)** KCl-soluble fractions: 1) *Chondrus pinnulatus*; 2) *Iridaea cornucopiae*.

ganese, cadmium, copper, lead, arsenic and boron was detected.

Based on ^{13}C-NMR and FTIR spectra, all fractions were identified as carrageenans. The ^{13}C-NMR spectrum of the KCl-insoluble polysaccharides showed anomeric signals typical of κ-carrageenan (102.5 and 95.3 ppm). However, there was a minor signal at 92.2 ppm on the spectra of the samples from *C. armatus*, *C. pinnulatus* and *I. cornucopiae* due to the presence of ι-carrageenan repeating units (Usov et al., 1980). Additional signals were observed at 94.5 and 66.3 ppm on the spectrum of the carrageenan from *T. crinitus* and may be assigned to non-sulphated units of κ-carrageenan, defined as β-carrageenan (Usov et al., 1980; Greer &Yaphe, 1984). Thus, the carrageenan from *T. crinitus* has a κ/β hybrid structure, composed predominantly of κ-carrageenan.

Well resolved ^{13}C-NMR spectra could not be recorded, even at high temperatures, from the highly viscous KCl-soluble fractions. Hence, only FTIR spectroscopy was used for the identification of the carrageenan types in these samples.

The infrared spectra of all KCl-soluble and insoluble samples showed a broad and strong absorbance at 1240 cm^{-1}, characteristic of sulphate esters (Stancioff & Stanley, 1969). Analysis of the FTIR-spectra of the KCl-insoluble fractions confirmed that they were mainly κ-carrageenan. These spectra (Figure 1) showed absorbances at 930 cm^{-1} for 3,6-anhydro-D-galactose and at 845 cm^{-1} characteristic of the secondary axial sulphate on C-4 of galactose. The IR spectra of the carrageenans samples from the Gigartinaceae were identical to those obtained for κ-carrageenan type III (Sigma). The small absorbances at about 805 cm^{-1} (2-sulphate on 3,6-anhydro-D-galactose) and 900 cm^{-1} (unsubstituted galactose) indicated the presence of ι-carrageenan in the samples of *I. cornucopiae* (Craigie & Leigh, 1978) and β-carrageenan in *T. crinitus*, respectively (Usov et al., 1983; Reen et al., 1993).

The IR spectra of KCl-soluble polysaccharide fractions also showed a broad asymmetric band in the 800-845 cm^{-1} region due to sulphate ester groups on galactose residues (Greer & Yaphe, 1984). Carrageenan from *I. cornucopiae* had a strong absorption band at 832 cm^{-1} (equatorial secondary sulphate on C-2 of galactose) and a shoulder at 824 cm^{-1} (equatorial secondary sulphate on C-6 of galactose), which are characteristic of λ-carrageenan (Craigie & Leigh, 1978). The broad asymmetric absorbance at 836 cm^{-1} on the spectrum of the *C. pinnulatus* sample appears to result from the association of the peaks generally observed at 845, 830, 820 cm^{-1} and characteristic of λ- and κ-carrageenans. This absorbance was also observed on the spectrum of commercial λ-type of carrageenan (Sigma). The IR absorptions on the spectrum of *T. crinitus* KCl-soluble carrageenan were weaker and broader than those of the other algal polysaccharides. This spectrum exhibited the broad peak at 810 cm^{-1} which, together with the peak at 820 cm^{-1}, is also a characteristic feature of λ-carrageenan (Falshaw & Furneaux, 1994). The spectra of all the KCl-soluble samples showed the additional absorption in the 930 cm^{-1} region indicating the presence of 3,6-anhydrogalactose. Moreover, the *T. crinitus* spectrum showed a peak at about 845 cm^{-1}, indicating the presence of galactose 4-sulphate.

According to these results, the KCl-insoluble polysaccharides of *C. armatus*, *C. pinnulatus*, *I. cornucopia* and *T. crinitus* are predominantly κ-carrageenans with some ι-type and κ/β-hybrid structures, respectively. Similar carrageenans have been found in other species belonging to the Gigartinaceae (Craigie, 1990). The KCl-soluble polysaccharide frac-

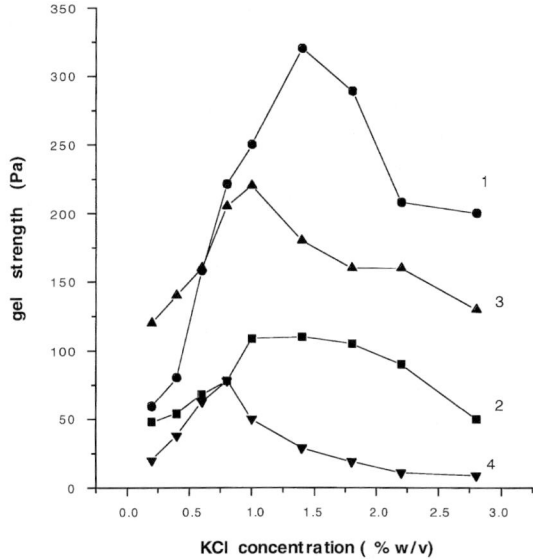

Figure 2. Effect of KCl concentrations on the gel strength of 1% solution of KCl-insoluble carrageenans from: 1) *Chondrus pinnulatus*, 2) *C. armatus*, 3) *Iridaea cornucopiae*, 4) *Tichocarpus crinitus*.

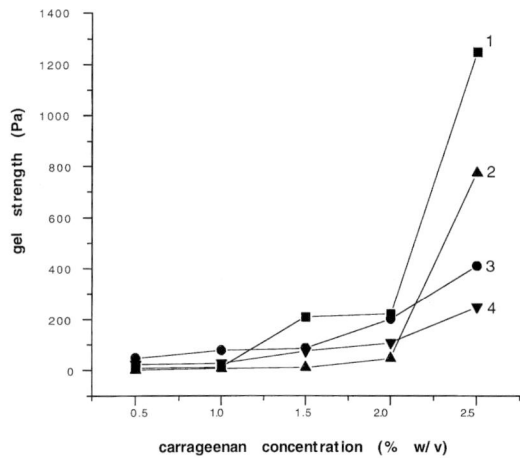

Figure 3. Effect of polysaccharide concentrations in 1% KCl on the gel strength of KCl-insoluble carrageenan from: 1) *Chondrus pinnulatus*, 2) *C. armatus*, 3) *Iridaea cornucopiae*, 4) *Tichocarpus crinitus*.

tions were mainly λ-carrageenans mixed with other carrageenan types.

The apparent molecular weights of the kappa-carrageenans, determined by sedimentation analysis, ranged from 150 to 300 kDa. The highest and lowest values were obtained with the κ-carrageenans from *C. pinulatus* and *T. crinitus*, respectively.

The rheological properties of κ-carrageenan gels from the four algae were investigated. Because gelling properties of carrageenans are strongly dependent on the type and concentration of cations present in the solution (Morris & Belton, 1982), the influence of potassium concentrations in 1% carrageenan solutions was determined. The gel strength (S) used to monitor the mechanical properties of the gels increased at first and then decreased with increasing cation concentrations (Figure 2). However, the optimal potassium concentration was distinct for the different κ-carrageenans. The gel strength of the carrageenans from *I. cornucopia* and *C. pinnulatus* rose rapidly, reaching a maximum with 1.0–1.5% KCl, but that of the carrageenans from *T. crinitus* was maximum at lower KCl concentrations (0.75%). The value of S was very low for the carrageenan from *T. crinitus* (55 Pa) and the addition of more than 1% KCl to the solution prevented gel formation.

The rheological properties of different concentrations of κ-carrageenan gels in 1% KCl are presented in Figure 3. Under identical conditions, the κ-carrageenans from *C. pinulatus*, *C. armatus* and *I. cornicopiae* produced stronger gels than that from *T. crinitus*. The highest gel strength was obtained with the carrageenan from *C. pinnulatus* (1232.7 Pa) at a 2.5% polymer concentration, whereas the carrageenan from *T. crinitus* formed the weakest gel (77.4 Pa) at the same concentration.

The viscoelastic behaviour of the KCl-soluble polysaccharides was investigated. The shear-rate dependence of the viscosity of 1% KCl-soluble polysaccharides solutions from three species of algae was observed in the range of 1 to 1000 s^{-1} at 20 °C (Figure 4). The classical rheo-thinning behaviour of carrageenans (Morris et al., 1981) was observed for all solutions of the Gigartinaceae extracts. The flow curve obtained for *T. crinitus* λ-carrageenan at 20 °C has a distinct shape (Figure 4): there is no Newtonian plateau at low shear-rates. The flow curves obtained for all the gigartinacean carrageenan solutions remain identical at 20, 35 and 65 °C (not shown) but the viscoelastic behaviour of the carrageenan from *T. crinitus* was more sensitive to temperature changes. As illustrated in Figure 5, the solution of *T. crinitus* carrageenan at 65 °C has the properties of entangled 'random coil' polysaccharide solutions. Figure 6 shows curves of the temperature dependence of viscosity obtained for carrageenans between 25–65 °C. At high shear-rate (800 s^{-1}), the viscosity of *I. cornucopiae* carrageenan was considerably lower than that of carrageenans from other seaweeds. Some systems did

46

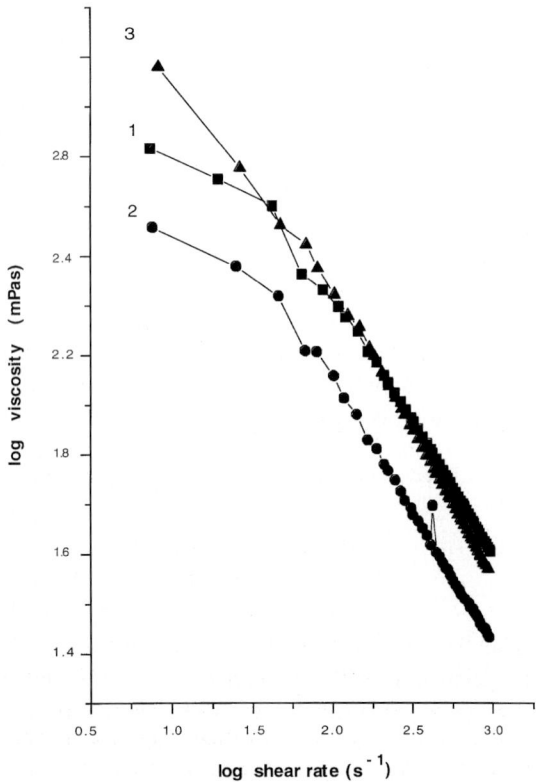

Figure 4. Double logarithmic plot of viscosity versus shear-rate at 20 °C of KCl-soluble carrageenans from: 1) *Chondrus pinnulatus*, 2) *Iridaea cornucopiae*, 3) *Tichocarpus crinitus*.

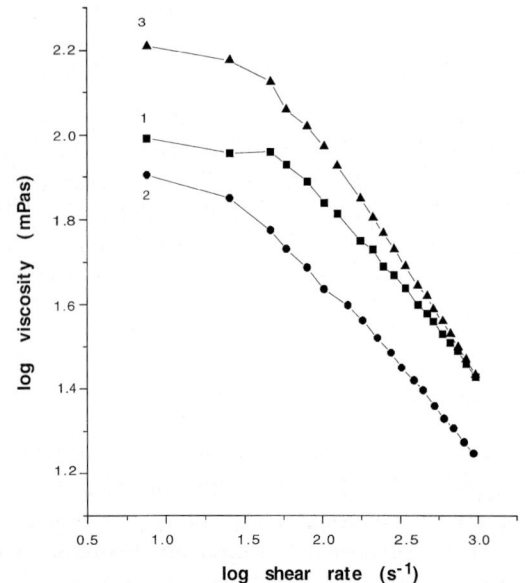

Figure 5. Double logarithmic plot of viscosity versus shear-rate at 65 °C for KCl-soluble carrageenans from: 1) *Chondrus pinnulatus*, 2) *Irideae cornucopiae*, 3) *Tichocarpus crinitus*.

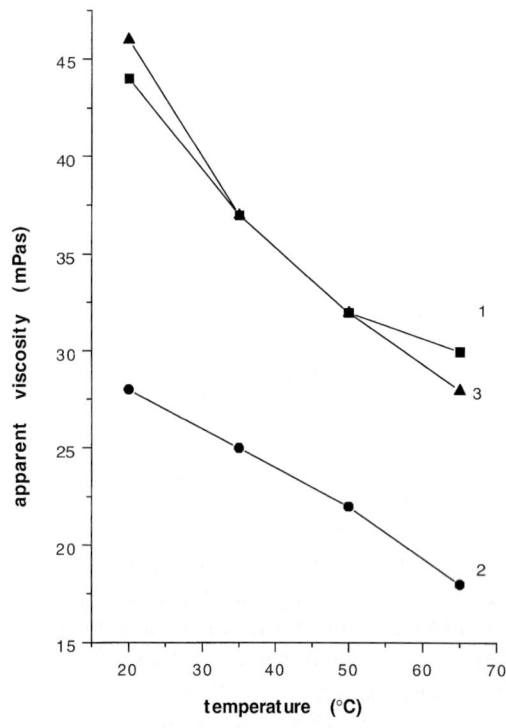

Figure 6. Dependence of apparent viscosity (at 800 s^{-1}) with temperature of KCl-soluble carrageenans from: 1) *Chondrus pinnulatus*, 2) *Irideae cornucopiae*, 3) *Tichocarpus crinitus*.

not show a conformational transition over the 20–65 °C temperature range and for these samples thermal denaturation may occur at higher temperatures.

Discussion

Carrageenophytes are abundant in the seas of the far east and on the Russian Pacific coast and potential algal sources of carrageenan are *Chondrus pinnulatus, C. armatus, C. yendoi (Iridaea cornucopiae), Mastocarpus pacificus* and *Tichocarpus crinitus* (Kizevetter et al., 1981). The purpose of this study was to characterise the chemical structure and rheological properties of polysaccharides from the most abundant species of algae of the Russian Pacific coast.

The average yield of hot-water soluble galactans from all the algal species studied was high (40–50% of the dry algae) and they had typical carrageenan composition and physical properties. They were identified as predominantly κ-carrageenan with some ι-type repeating units (Gigartinaceae) and κ/β-hybrid structures (Tichocarpaceae) together with a small quantity of λ-carrageenan (10% w/w). κ-Carrageenan fractions (more 30% of the dry algae) with high molecular weights were obtained by KCl-fractionation of the crude polysaccharides from three Gigartinaceae: *Chondrus pinnulatus, C. armatus* and *Iridea cornucopiae*. They showed good gelling properties, whereas carrageenan from the alga belonging to the Tichocarpaceae formed a very weak gel. Rheological properties of carrageenans, which strongly depend on salt concentration, showed optimum gelling characteristics with 1.0–2.0% KCl concentrations for κ-carrageenan fractions from the Gigartinaceae and 0.75% for that from the Tichocarpaceae. These different gelling behaviours may be connected with the chemical structure and molecular weights of the carrageenans and in particular those from *Tichocarpus crinitus* carrageenan which may be due to its κ/β-carrageenan hybrid structure. Unlike the ion-dependent gelling κ- and ι-carrageenans, β- carrageenan apparently does not require a sulphate ester on C-4 of the 3-linked galactose residue for gel formation, and gelling β-carrageenan has been isolated from *Eucheuma gelatinae, Eucheuma speciosa* and *Endocladia muricatum* (Reen et al., 1993). The carrageenan from *Furcellaria lumbricalis*, furcellaran, which has a κ/β-hybrid structure also shows good gelling properties (Knutsen & Grasdalen, 1987). Although the distribution of the κ- and β-segments along the κ/β-carrageenan from *Tichocarpus crinitus* and *Furcellaria lumbricalis* probably differs, furcellaran forms gels at a distinctly lower potassium-ion concentration than κ-carrageenan (Glicksmann, 1983) as does that from *T. crinitus*.

The lower gel strength of the KCl-insoluble fraction from *T. crinitus*, compared to that of the other Gigartinaceae, most likely resulted from its low average molecular weight. A relationship between the mechanical properties and the molecular weight of κ-carrageenan has been demonstrated (Rochas et al., 1990; Sing & Jacobson, 1994).

In all cases, the mechanical spectra of KCl-soluble carrageenans had a broad range of shear-rates indicating that these systems had rheo-thinning behaviour. Double logarithmic plots of shear viscosity against shear-rate of carrageenans from the Gigartinaceae in the range of temperature between 20 and 65 °C were essentially identical to those of the conformationally disordered 'random coil' polysaccharides in concentrated solutions (Morris, 1988). These plots showed horizontal Newtonian plateaux at low shear-rates and then a drastic reduction in viscosity from this maximum value at high shear-rates. This behaviour is consistent with the formation of a dynamic entangled structure in concentrated solutions of 'random coil' polymer. At low shear-rates, the disruption of polysaccharide chain entanglement, by the imposed deformation and the formation of new interactions between different chains, equilibrated, and thus no reduction of viscosity was observed (Newtonian plateau). The beginning of shear-thinning occurred at high shear-rates when the rate of disruption of existing entanglements became greater than the rate of formation of new ones, and thus the crossline density of the network was depleted and the viscosity was reduced (Morris, 1988).

The KCl-soluble carrageenan from *T. crinitus* displayed the properties of 'random coil' polymers only at high temperature. Below 65 °C, this carrageenan demonstrated a behaviour that, we suggest, is of polysaccharides forming 'weak gels' in solution (Morris, 1988), although the shear-thinning of this carrageenan in solution may begin at lower shear-rates than those we used. The different behaviours of the KCl-soluble fractions from the gigartinaceaen algae and those from the Tichocarpaceae may be due to the presence of κ-type carrageenan in the latter sample.

This study has demonstrated the relationships between chemical structures and particular rheological properties of carrageenans. The results will be useful in the selection of potential industrial sources of carrageenan from the most abundant algae growing on the Russian Pacific coast.

Acknowledgements

The authors gratefully acknowledge the Instrumental Analysis Centre of Kyonggi University, Suwon, Korea for technical help and Ms TB Tytlyanova (Institite of Marine Biology) for identification of the algae. We also acknowledge Prof. B.G. Vas'kovsky for inspiring discussions. This work was supported by a grant from the Russian Science Foundation.

References

Craigie JS (1990) Cell walls. In Cole KM, Sheath RG (eds), Biology of the Red Algae, Cambridge University Press, Cambridge: 221–257.

Craigie JS, Leigh C (1978) Carrageenans and agars. In Hellebust JA, Craigie JS (eds), Handbook of Phycological Methods, Physiological and Biochemical Methods, Cambridge University Press, Cambridge: 109–131.

Dubois M, Gilles KA, Hamilton JK, Rebers PA, Smith F (1956) Colororimetric method for determination of sugars and related substances. Anal. Chem. 28: 350–356.

Elias HG (1961) Ultracentrifugen-Methoden. Beckman Instruments, GMBH, Munchen, 71–73.

Englyst HN, Cummings JH (1984) Simplified method for the measurement of total non-starch polysaccharides by liquid chromatography of constituent sugars as alditol acetates. Analyst. 109: 937–942.

Falshaw R, Furneaux RH (1994) Carrageenan from tetrasporic stage of *Gigartina decipiens* (Gigartinacea, Rhodophyta). Carbohydr. Res. 252: 171–182.

Falshaw R, Furneaux RH (1995) Carrageenan from tetrasporic stages of *Gigartina clavifera* and *Gigartina alveata* (Gigartinacea, Rhodophyta). Carbohydr. Res. 276: 115–165.

Glicksman M (1983) Red seaweed extracts. In Glicksman M (ed.), Food Hydrocolloids 2, CRC Press Baton Rouge: 73–113.

Greer CW, Yaphe W (1984) Characterization of hybrid (beta-kappa-gamma) carrageenan from *Eucheuma gelatinae* J. Agardh (Rhodophyta, Solieriaceae) using carrageenanase, infrared and ^{13}C-nuclear magnetic resonance spectroscopy. Bot. mar. 27: 473–478.

Heyraud A, Rinaudo M, Rochas C (1990) Physical and chemical properties of phycocolloids. In Akatsuka I (ed.), Introduction to Applied Phycology. SPB Academic Publishing bv, The Hague, Netherlands: 151–176.

Kizevetter IV, Sukhoeeva MV, Shmel'kova WP (1981) Industrial marine algae and grasses from Far-Eastern seas, Legkaya and Pyshchevaya Promyshlennost: 133–140.

Knutsen SH, Grasdalen H (1987) Characterization of water-extractable polysaccharides from Norwegian *Furcellaria lumbricalis* (Huds) Lamour. (Gigartinales, Rhodophyceae) by IR and NMR spectroscopy. Bot. mar. 30: 497–505.

Knutsen SH, Murano E, D'Amato M, Toffanin R, Rizzo R, Paoletti S (1995) Modified procedures for extraction and analysis of carrageenan applied to the red alga *Hypnea musciformis*. J. appl. Phycol. 7: 565–576.

Knutsen SH, Myslabodsky DE, Larsen B, Usov AI (1994) A modified system of nomenclature for red algal galactans. Bot. mar. 37: 163–169.

Lahaye M, Axelos MAV (1993) Gelling properties of water-soluble polysaccharides from proliferating marine green seaweeds (*Ulva* spp.) Carbohydr. Polym. 22: 261–265.

Lowry OH, Rosebrough NJ, Farr AL, Randall RJ (1951) Protein measurement with the Folin phenol reagent. J. biol. Chem. 193: 265–275.

McCandless E, Craigie J, Walter J (1973) Carrageenans in the gametophytic and sorophytic states of *Chondrus crispus*. Planta 112: 201–212.

Morris ER, Cutler AN, Ross-Murphy SB, Rees DA, Price J (1981) Concentration and shear rate dependence of viscosity in random coil polysaccharide solutions. Carbohydr. Polym. 1: 5–21.

Morris VJ, Belton PS (1982) The influence of the cations sodium, potassium and calcium on the gelation of iota-carrageenan. Prog. Food Nutr. Sci. 6: 55–66.

Morris ER (1988) Polysaccharide solution properties: origin, rheological characterisation and implications for food systems. In Millane RP (ed.), Frontiers in Carbohydrate Research-1. Food Applications. Elsevier Science Publishers, London and New York: 132–163.

Reen DW, Santos GA, Dumont LE, Parent CA, Stanley NF, Stancioff DJ, Guiseley KB (1993) β-Carrageenan: Isolation and characterization. Carbohydr. Polym. 22: 247–252.

Rochas C, Rinaudo M, Landry S (1990) Role of the molecular weight on the mechanical properties of kappa carrageenan gels. Carbohydr. Polym. 12: 255–266.

Sing ST, Jacobsson SP (1994) Kinetics of acid hydrolysis of κ-carrageenan as determined by molecular weight (SEC-MALLS-RI), gel breaking strength, and viscosity measurements. Carbohydr. Polym. 23: 89–103.

Stancioff DJ, Stanley NF (1969) Infrared and chemical studies on algal polysaccharides. Proc. int. Seaweed Symp. 6: 595–609.

Stortz C, Cerezo A (1993) The systems of carrageenans from cystocarpic and tetrasporic stages from *Iridaea undulosa*. Fractionation with potassium-chloride and methylation analysis of the fractions. Carbohydr. Res. 252: 171–182.

Usov AI (1974) Polysaccharides of algae. 13. Monosaccharide compositions of polysaccharides of some red algae from the Sea of Japan. Zh. Obshch. Khim. 44: 191–196 (in Russian).

Usov AI, Yarotsky SV, Shashkov AS (1980) ^{13}C-NMR spectroscopy of red algal galactans. Biopolymers 19: 977–990.

Usov AI, Ivanova EG, Shashkov AS (1983) Polysaccharides of algae. 33. Isolation and ^{13}C-NMR spectral study of some new gel forming polysaccharides from Japan Sea seaweeds. Bot. mar. 26: 285–294.

Yermak IM, Khotimchenko YS (1997) Physical and chemical properties, applications, and biological activitues of carrageenan, a polysaccharide of red algae. Mar. Biol. 23: 129–142.

Zhang W, Piculell L, Nilsson S, Knutsen SH (1994) Cation specificity and cation binding to low sulfated carrageenans. Carbohydr. Polym. 23: 105–110.

Isolation and characterisation of a fourth hemagglutinin from the red alga, *Gracilaria verrucosa*, from Japan

Hirotaka Kakita*, Satoshi Fukuoka, Hideki Obika & Hiroshi Kamishima
Marine Resources Department, Shikoku National Industrial Research Institute, Hayashi, Takamatsu, Kagawa 761-03, Japan

(* Author for correspondence; fax 81-878-693553; e-mail kakita@sniri.go.jp)

Received 1 August 1998; revised 8 October 1998; accepted 6 November 1998

Key words: hemagglutinins, *Gracilaria*, sulphated polysaccharide, carbohydrate specificity, seaweed, electrophoretic behaviour

Abstract

Isolation and characterisation of marine algal hemagglutinins or lectins are essential for their potential industrial application as specific carbohydrate affinity ligands. The phosphate buffer extract of the red alga, *Gracilaria verrucosa* (Huds.) Papenfuss (Gigartinales, Rhodophyta) from Japan is known to contain three different hemagglutinins. The extract of the alga collected in March 1993 from Kagawa Prefecture, Japan, was purified by ammonium sulphate fractionation, ion exchange and gel filtration chromatography. Using gel filtration, two peaks were obtained (hereafter Peak 1 and Peak 2) which differed in molecular size and hemagglutinating activity against horse erythrocytes. Peak 1 corresponded to the known high molecular weight hemagglutinin, H-GVH. Peak 2 contained large amounts of hexose and sulphate along with a small amount of protein. It had a low molecular weight (gel filtration) similar to that of two of the previously reported *G. verrucosa* hemagglutinins but differed in its electrophoretic behaviour. Peak 2 is therefore a fourth hemagglutinin. Its activity was not inhibited by any of the monosaccharides tested but by the complex glycoproteins such as asialofetuin and fetuin. It had no divalent cation requirement for hemagglutination. The properties of this novel hemagglutinin could prove useful in industrial applications.

List of abbreviations: H-GVH, high molecular weight *Gracilaria verrucosa* hemagglutinin; L-GVH, low molecular weight *G. verrucosa* hemagglutinin; BSA, bovine serum albumin; SDS, sodium dodecyl sulphate; SDS-PAGE, sodium dodecyl sulphate polyacrylamide gel electrophoresis.

Introduction

Hemagglutinins are natural bioactive products, which are the focus of research aimed at developing new industrial tools using their specific carbohydrate binding abilities. In the course of such studies, we have already purified a chick 14 kDa lectin and applied it as an affinity ligand (Kakita et al., 1991). Unfortunately, this lectin had the triple disadvantage of being thermolabile, requiring a reductant such as 2-mercaptoethanol and having low carbohydrate specificity. Therefore, the search for other hemagglutinins with higher stability and carbohydrate specificities, requiring no additives, and available in a steady quality and quantity was undertaken. As marine organisms contain many different bioactive substances to those of terrestrial ones, the marine algae were surveyed for hemagglutinins as affinity ligands.

The phosphate buffer extract of the red seaweed, *Gracilaria verrucosa* has previously been reported to agglutinate rabbit erythrocytes (Hori et al., 1981, 1988a; Chiles & Bird, 1989) and three hemagglutinins have been purified. The first (named GVA-1) is a protein or glycoprotein with low carbohydrate content

(Shiomi et al., 1981), the second is a 49 kDa proteoglycan (Kanoh et al., 1992) and the third (named H-GVH) is a high molecular weight sulphated polysaccharide (Kakita et al., 1997). The first and second lectins have low molecular weights, oligomeric structures, and no disulphide bond. GVA-1 (Mr 41 000) is a tetramer constituted of two subunits (Mr 10 500 and Mr 12 000). The second hemagglutinin is a dimeric proteoglycan (Mr 49 000) consisting of two protomers (Mr 27 000 and Mr 23 000). The hemagglutinating activities of these two lectins disappeared or were reduced after treatment at 100 °C for 30 min. By contrast, the molecular weight of H-GVH is approximately Mr 480 000 and its hemagglutinating activity is heat-stable but periodate-sensitive.

Recently, we found a new low molecular weight hemagglutinin in the phosphate buffer extract of G. verrucosa and named it L-GVH. In this paper, we describe its isolation and properties.

Materials and methods

Samples of the red alga, *Gracilaria verrucosa*, were collected from Aji Peninsula, Kagawa Prefecture, Japan, in March 1993. After collection, the algae were washed with water, freeze-dried, and stored at −20 °C until use. Although the Japanese '*G. verrucosa*' was previously identified as *G. verrucosa* (Huds.) Papenfuss, several researchers objected to the classification and pointed out the necessity for re-examination (Yoshida et al., 1995). For the purpose of this paper, we refer to it as *G. verrucosa*.

Samples of animal blood were obtained from Cosmo-Bio (Tokyo, Japan). Other reagents were of analytical grade.

Purification of the hemagglutinin

The purification experiments were carried out at 4 °C. The dried alga (200 g) was homogenised with 2 L of buffer A [0.02 mol L^{-1} sodium phosphate buffer (pH 7.0), 0.15 mol L^{-1} NaCl] containing 0.005 mol L^{-1} ascorbic acid and 1% polyvinylpyrrolidone. After centrifugation to remove insoluble materials, the fraction precipitated between 35% and 70% saturation with ammonium sulphate was collected. The precipitate was dissolved in buffer A and dialysed against the same buffer. A sample of 62.0 mL was applied to a DEAE-Toyopearl 650M column (Tosoh, Japan, 29 cm × 2.6 cm I.D.) equilibrated with buffer A and eluted by stepwise increase of NaCl concentration up to 0.5 mol L^{-1} in 0.02 mol L^{-1} sodium phosphate buffer (pH 7.0). Flow rate was 2 mL min^{-1}. The eluate was collected in 10 mL fractions, which were examined for agglutination of rabbit erythrocytes. The active fractions were collected, dialysed against buffer A and concentrated to 12.5 mL by ultrafiltration (Grace amicon membrane PM10). A 500-μL volume of the concentrate was applied to a G3000PWxl column (Tosoh, Japan, 300 mm × 7.8 mm I.D.) and eluted with buffer A at a flow rate of 0.5 mL min^{-1} in 0.1 mL fractions. The gel filtration was repeated 24 times to pool sufficient fractions for testing. The active fractions (Fr. 78 and 79) were collected, dialysed against buffer A, and concentrated to 2.2 mL by ultrafiltration. A 100 μL volume of the concentrate was applied to the same column. The eluate was collected in 0.1 mL fractions and, again, this was repeated 20 times. The active fractions were collected, dialysed against buffer A, and concentrated to 0.8 mL by ultrafiltration. Thyroglobulin (Mr 669 000), ferritin (Mr 440 000), BSA (Mr 66 000), and ovalbumin (Mr 43 000) obtained from Pharmacia (Uppsala, Sweden), were used as molecular weight markers for gel filtration.

Determination of the hemagglutinin activity

Hemagglutinating activity was measured according to Hori et al. (1981). One unit of hemagglutinating activity was defined as the reciprocal of the highest dilution of the fraction (50 μL) which gave detectable hemagglutination of a rabbit erythrocyte suspension. Total activity was defined as activity (units) multiplied by the fraction volume (mL). Specific activity was defined as activity (units) divided by the quantity of hexose (mg). Chicken, cow, goose, guinea pig, horse, rabbit, sheep, and rat erythrocytes were used for hemagglutination assays. In erythrocyte specificity experiments, the minimum concentration of hemagglutinin was defined as the lowest hexose concentration (mg mL^{-1}) required to provide hemagglutination.

Sheep erythrocyte suspension treated with pronase (SETP) was prepared and the effects of dialysis, heat, pH, and divalent cations on the activity were measured according to Hori et al. (1986a, 1988a). Monosaccharides, acids, and glycoproteins were obtained from Sigma (St. Louis, MO, U.S.A.). Compounds used for the inhibition assay were: D-glucose, L-fucose, D-mannose, L-ascorbic acid, L-rhamnose, D-arabinose, D-galactose, D-ribose, D-galactosamine,

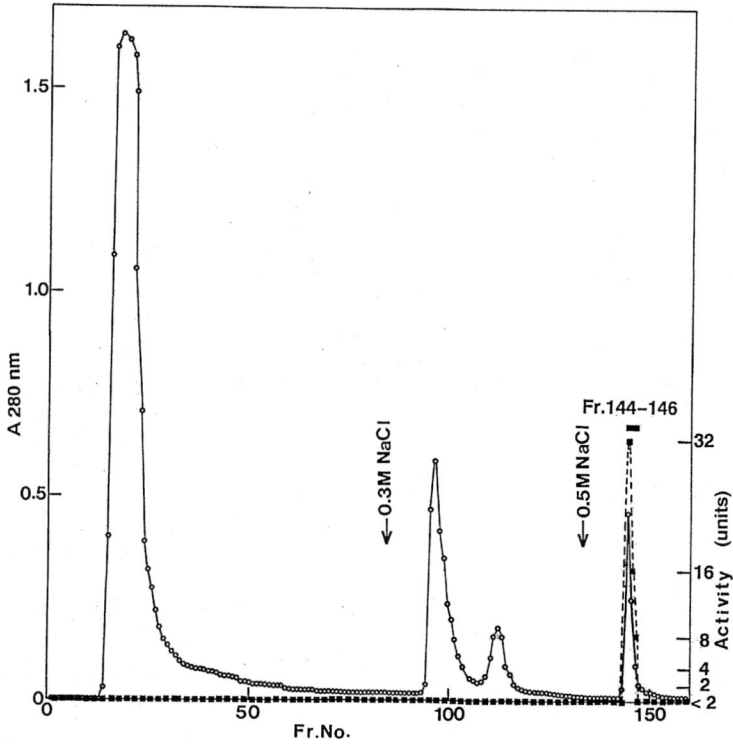

Figure 1. Ion exchange chromatography of the ammonium sulphate fractionated extract on TSKgel DEAE-Toyopearl 650M column. The arrows indicate the positions of stepwise increase of NaCl concentration. Protein elution was monitored by absorbance at 280 nm (○–○). Hemagglutinating activity against rabbit erythrocytes is represented as units (■–■). The fractions (Fr. 144–146) indicated by the solid bar were pooled for further separation.

D-glucosamine, N-acetyl-D-galactosamine, N-acetyl neuraminic acid, N-acetyl-D-glucosamine, N-acetyl-D-mannosamine, fetal calf fetuin, fetal calf asialofetuin, bovine asialomucin, bovine mucin, bovine lactoferrin, bovine ribonuclease B, *Saccharomyces cerevisiae* mannan, chicken ovalbumin, chicken egg white trypsin inhibitor, and porcine thyroglobulin. Enzymatic digestions and periodate treatment of L-GVH were carried out according to Lai et al. (1989).

Electrophoresis

SDS-PAGE was carried out by the method of Laemmli (1970). The sample was treated with 1% SDS for 5 min at 100 °C. Staining for protein was performed with a silver staining kit (Daiichi Pure Chemical, Tokyo, Japan) according to Ohsawa & Ebata (1983). The acidic polysaccharide band was stained with Alcian blue (Misevic & Burger, 1986). Acidic polysaccharide density on the stained gel was measured by absorbance at 633 nm using a Shimadzu densitometer. Rabbit myosin (Mr 200 000), *E.coli* β-galactosidase (Mr 116 000), BSA (Mr 66 000), rabbit aldolase (Mr 42 000), bovine carbonic anhydrase (Mr 30 000), and horse myoglobin (Mr 17 000) were obtained from Daiichi Pure Chemical (Tokyo, Japan) and were used as molecular weight markers.

Cellulose acetate membrane electrophoresis (Jookoo, Tokyo, Japan) was carried out according to the method of Wessler (1971). Staining of acidic polysaccharides was performed with 0.5% Alcian blue. Human umbilical cord hyaluronic acid (Seikagaku, Tokyo, Japan) was used as standard for nonsulphated polysaccharide. Shark cartilage chondroitin sulphate C and porcine intestinal mucosa heparin were obtained from Sigma (St.Louis, MO, U.S.A.) and were used as standards for sulphated polysaccharides.

Compositional analysis

Protein was determined according to Lowry et al. (1951) using BSA as standard. Hexose was determined by the phenol-sulphuric acid method using galactose as standard (Dubois et al., 1956). Sulphate content was determined by the rhodizonate method using sodium sulphate as standard (Terho & Hartiala, 1971). Sample hydrolysis was performed according to Arakawa et al. (1976). The neutral sugar composition of hydrolysates was determined by high performance liquid chromatography (Mikami & Ishida, 1983). Amino sugars were determined using a Shimadzu amino acid analyser equipped with a Shin-pack Amino-Na ion exchange column and an Amino Acid Analysis reagent kit (Shimadzu, Kyoto, Japan).

Results

Hemagglutinins extracted by phosphate buffer from *G.verrucosa* and precipitated by ammonium sulphate were purified by ion exchange chromatography (Figure 1). Gel filtration of the hemagglutinating active fractions (Figure 2A) resulted in two peaks (hereafter Peak 1 and Peak 2) with different molecular weights. Peak 1 (Fr. 68–69) corresponded to the high molecular weight hemagglutinin reported previously as H-GVH by Kakita et al. (1997). The low molecular weight hemagglutinin (L-GVH) was purified from Peak 2 (Fr. 78–79) by re-chromatography on the same column (Figure 2B). The activity recovery of L-GVH using this purification scheme is summarised in Table 1. The specific activity increased to 41.3 units mg^{-1} hexose. The total activity of L-GVH fraction was 0.2% of that of the crude extract and 6.20 mg hexose were recovered. The molecular weight of L-GVH, estimated by gel filtration to be about Mr 51 000 (Figure 2B), is similar to that of GVA-1 (Shiomi et al., 1981) and that of the 49 kDa proteoglycan reported by Kanoh et al. (1992). However, on SDS-PAGE, L-GVH (20 μL, 0.111 mg protein mL^{-1}, 3.87 mg hexose mL^{-1}) migrated as a diffuse band which stained with Alcian blue (Figure 3A) at an equivalent molecular weight of about Mr 71 000 [indicated by 71 kDa on Figure 3B]. No other band was detected by Coomassie brilliant blue or by silver staining (data not shown). This migration behaviour and the Mr value differed from the other two *G. verrucosa* hemagglutinins. On cellulose acetate membrane electrophoresis, L-GVH migrated as one spot different from that of H-GVH (Figure 4).

Chemical analysis showed that 9.0 mg of L-GVH contained 6.20 mg of hexose, 0.177 mg of protein, and 1.56 mg of sulphate. The major sugar component of L-GVH was galactose and the minor components were fucose, glucose, xylose, and glucosamine. The

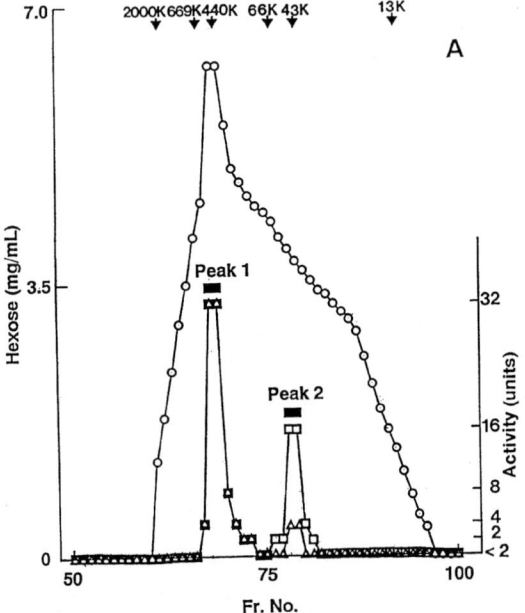

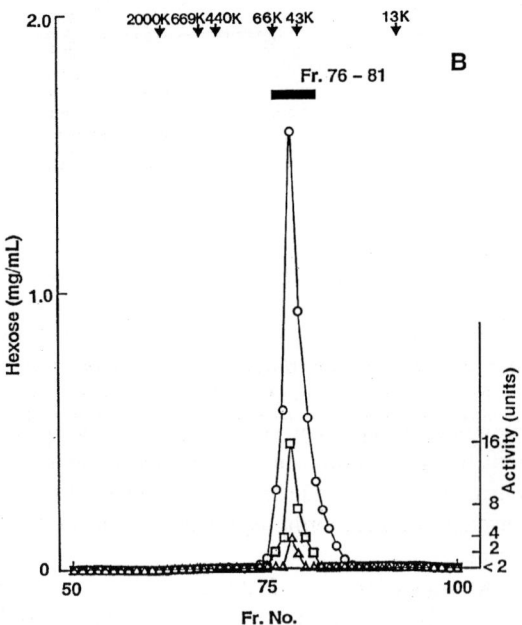

Figure 2. Gel filtration chromatographs of the active fractions on TSK gel G3000PWxl column. (A). Gel filtration chromatography of the active fraction from a TSKgel DEAE-Toyopearl 650M column. Hexose elution was determined by colorimetry (○–○) and the hemagglutinating activities against rabbit erythrocytes (△–△) and SETP (□–□) are represented as units. Peak 2 (Fr. 78–79) indicated by the solid bar were pooled for re-chromatography. (B). Re-chromatography of the Peak 2 fraction from TSKgel G3000PWxl column. The fractions (Fr. 76–81) indicated by the solid bar were pooled and concentrated. Molecular weight standards were: blue dextran (Mr > 2 000 000); thyroglobulin (Mr 669 000); ferritin (Mr 440 000); BSA (Mr 66 000); ovalbumin (Mr 43 000); ribonuclease A (Mr 13 700).

Table 1. Activity recovery of L-GVH during purification (ASppt: ammonium sulphate precipitation; RC: re-chromatography of peak 2 on G3000PWxl)

Fraction	Protein (mg)	Hexose (mg)	Total activity (units)	Specific activity (units mg^{-1})	Activity yield (%)
Crude extract	1802	9182	109900	12.0	100.0
ASppt	303	1707	39680	23.5	36.1
DEAE-Toyopearl	13.8	245	8000	37.5	7.3
G3000PWx1					
Peak 1	0.874	30.1	3072	102.2	2.8
Peak 2	0.623	14.6	352	24.1	0.3
G3000PWx1 (RC)	0.177	6.20	256	41.3	0.2

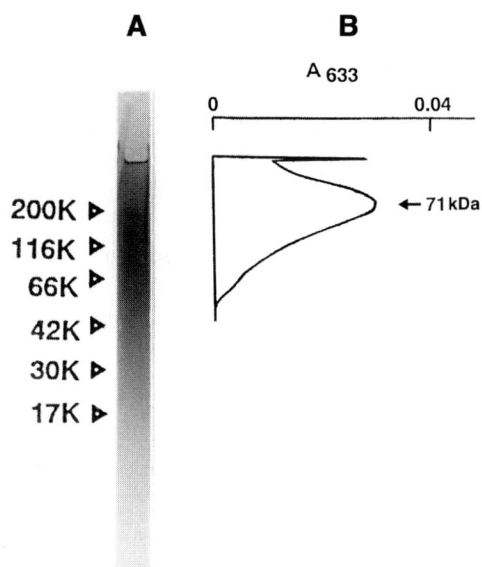

Figure 3. Sodium dodecyl sulphate electrophoresis of L-GVH on a 5–20% linear polyacrylamide gel. (A). Sodium dodecyl sulphate electrophoresis of L-GVH. Alcian blue staining was used. The open arrows indicate the position of the following molecular markers: rabbit muscle myosin (Mr 200 000), *E.coli* β-galactosidase (Mr 116 000), BSA (Mr 66 000), rabbit muscle aldolase (Mr 42 000), bovine erythrocyte carbonic anhydrase (Mr 30 000), and horse muscle myoglobin (Mr 17 000). (B). Densitometry of sodium dodecyl sulphate electrophoresis gel.

galactose content represented 97.8% of the total sugar content.

As shown in Table 2 the erythrocyte specificity of L-GVH was lower than that of H-GVH. Horse and rabbit erythrocytes and SETP were more strongly agglutinated by L-GVH than the other erythrocytes tested. Chicken erythrocytes were not agglutinated by a concentration of 7.74 mg of L-GVH hexose per mL. The erythrocyte specificity of L-GVH was different from those of the other three *G. verrucosa* hemagglutinins reported to date. The divalent cations such as 0.01 mol L^{-1} Ca^{2+}, Mg^{2+}, or Mn^{2+} did not affect agglutination (data not shown). The activity was stable at 100 °C for 30 min and between pH 5–10. It was not affected by chondroitinase ABC or pronase, but disappeared after periodate treatment.

The results of the inhibition of L-GVH activity by monosaccharides and glycoproteins showed that its activity was not inhibited by any of the monosaccharides tested at a concentration of 0.25 mol L^{-1}. Similar phenomena have been described for marine algal hemagglutinins but not from any terrestrial organism (Kamiya et al., 1980; Shiomi et al., 1981; Ferreiros & Criado, 1983; Rogers & Topliss, 1983; Hori et al., 1986a,b,c, 1987, 1988b; Okamoto et al., 1990; Kanoh et al., 1992; Kakita et al., 1997). The results of the inhibition assay also showed that its activity was inhibited by three kinds of complex type

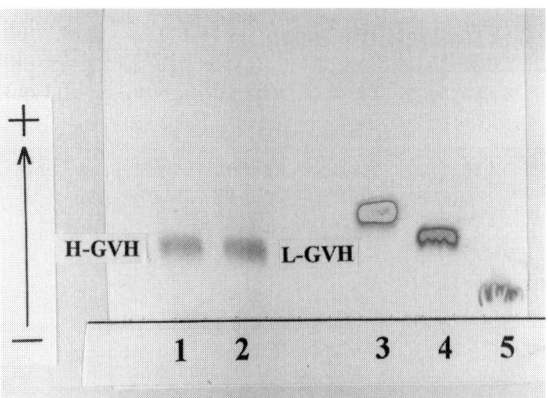

Figure 4. Electrophoresis of L-GVH on cellulose acetate membrane. Applied samples: (1) H-GVH (1 μL, 0.051 mg protein mL^{-1}, 3.89 mg hexose mL^{-1}), (2) L-GVH (1 μL, 0.111 mg protein mL^{-1}, 3.87 mg hexose mL^{-1}), (3) heparin (2 μL, 2 mg mL^{-1}), (4) chondroitin sulphate C (2 μL, 2 mg mL^{-1}), (5) hyaluronic acid (3 μL, 3 mg mL^{-1}). Alcian blue staining was used.

Table 2. Erythrocyte specificities of H-GVH and L-GVH (MC: minimum concentration; SETP: sheep erythrocyte suspension treated with pronase)

Erythrocyte	MC (mg hexose mL^{-1})	
	H-GVH	L-GVH
Chicken	> 7.78	> 7.74
Cow	> 7.78	1.93
Goose	> 7.78	1.93
Guinea pig	3.891	1.93
Horse	> 7.78	0.96
Rabbit	0.243	0.96
Sheep	> 7.78	1.93
Rat	> 7.78	3.87
SETP	0.243	0.242

glycoproteins. The minimum concentrations of these glycoproteins required to inhibit its activity were: asialofetuin, 2.48 μg mL^{-1}; fetuin, 37.5 μg mL^{-1}; thyroglobulin, 20.1 μg mL^{-1}. Its activity was not inhibited by any other glycoproteins tested at a concentration of 2 mg glycoproteins per mL. Among the glycoproteins tested, asialofetuin was the most effective inhibitor of L-GVH activity.

Discussion

The two low molecular weight *G. verrucosa* hemagglutinins reported previously have been shown to migrate as compact bands in SDS-PAGE analysis (Shiomi et al., 1981; Kanoh et al., 1992). Although the chromatographic molecular weight of L-GVH differed from that of H-GVH, it was similar to those of the two other *G. verrucosa* hemagglutinins, yet it produced a diffuse band in SDS-PAGE analysis (Figure 3A). In addition, L-GVH had a molecular weight of 71 000 by SDS-PAGE, its activity was heat-stable, periodate-sensitive. It was active against sheep erythrocytes but inactive against chicken erythrocytes. From these results, L-GVH represents a fourth hemagglutinin. Takahashi and Katagiri (1987) showed that the hemagglutinating activity of the phosphate buffer extract of *G. verrucosa* changes with seasons. Shiomi et al. (1981) purified GVA-1 from algae harvested in November 1980 from Chiba Prefecture (central east Japan). Kanoh et al. (1992) found the 49 kDa proteoglycan in algae harvested in July–August 1989 from Kagawa Prefecture (Inland Sea in southwest Japan). Kakita et al. (1997) purified H-GVH from algae harvested in June 1992 and L-GVH from algae collected in March 1993 both from Kagawa Prefecture. These findings suggest that the four *G. verrucosa* hemagglutinins differed according to the season of harvest. Thus, the difference between L-GVH and the other three *G. verrucosa* hemagglutinins may reflect seasonal variation. Hori et al. (1993) showed that novel hemagglutinins can be extracted from algae treated with pronase. That also suggests that some marine algae may possess multiple hemagglutinins.

By SDS-PAGE analysis, L-GVH produced a diffuse band, which stained with Alcian blue whereas no other band was detected using other stains. Noro et al. (1983) reported on a sulphated proteolycan that produces a diffuse band on SDS-PAGE and suggested that its broadness reflected an heterogeneous glycosylation. The migration of L-GVH as one spot on cellulose acetate membrane electrophoresis suggests that the broadness of the diffuse band observed by SDS-PAGE probably arises, not from impurities, but from electrophoretic micro-heterogeneity (Figure 4). We infer that L-GVH is not composed of a single molecular species but a group of micro-heterogeneous molecules having hemagglutinating activity. These molecules would differ in the composition or length of the sulphated polysaccharide chains (e.g. sulphated galactan) linked to a proteoglycan.

Coombe et al. (1987) suggested that marine sponge cell aggregation is caused by ionic and hydrogen interactions between sulphated polysaccharides and cell surface receptors (baseplate), and the specificity of aggregation is determined by the orientation of the sulphate groups on the polysaccharide. Our results suggest that L-GVH is not a lectin but a sulphated polysaccharide hemagglutinin, the sulphated polysaccharide moiety of which seems to be essential for hemagglutination. The chemical analysis showed that L-GVH differs from H-GVH in the molar ratio of sulphate to hexose (L-GVH, 0.46; H-GVH, 0.76). Wessler (1971) showed that the electrophoretic mobility of polysaccharide on a cellulose acetate membrane in 0.1 mol L^{-1} hydrochloric acid is proportional to the sulphate content of the polysaccharide. Because L-GVH differed from H-GVH in the migration distance on the cellulose acetate membrane, it is likely that L-GVH and H-GVH differ in their sulphate content (Figure 4). Although the hemagglutinating activities of L-GVH and H-GVH were similarly heat-stable and periodate-sensitive, the two activities differed in erythrocyte specificities. From these results, we infer that their sulphate moiety is related to their hemagglutinating property. The results from inhibition assays suggest that L-GVH has an affinity to the desialylated oligosaccharide sugar chain of complex type glycoproteins such as asialofetuin. The heat-stability and high carbohydrate specificity will prove useful in applying L-GVH as an analytical and diagnostic tool.

In a preliminary evaluation of L-GVH as an affinity ligand, we found that it met three of the five requirements (high stability, high carbohydrate specificity, and no need for any additive) but ranked only poor to fair on the other two (steady supply and steady quality). From 200 g of dried alga, we purified only 6.20 mg of L-GVH (Table 1). Thus, as for other hemagglutinins obtained from algae, *G. verrucosa* has a low L-GVH content. Furthermore, the seasonal variation of hemagglutinin types from *G. verrucosa*, discussed above, is another limitation for having a constant quality and quantity of L-GVH.

Thus, and as with other marine algae, *G. verrucosa* is a suitable source of various industrially useful hemagglutinins with some limitations that remain to be solved.

Acknowledgements

We are grateful to Dr Kanji Hori of Hiroshima University, Dr Tadaharu Watanabe of Tokushima Prefectural Industrial Technology Center, Dr Akira Ogamo of Kitazato University, Dr Hirotoshi Yamamoto of Hokkaido University, Dr Ken-ichi Kasai and Dr Jun Hirabayashi of Teikyo University, for much helpful advice.

References

Arakawa Y, Imanari T, Tamura Z (1976) Determination of neutral and amino sugars in glycoproteins by gas chromatography. Chem. Pharm. Bull. 24: 2032–2037.

Chiles TC, Bird KT (1989) A comparative study of animal erythrocyte agglutinins from marine algae. Comp. Biochem. Physiol. 94B: 107–111.

Coombe DR, Jakobsen KB, Parish CR (1987) A role for sulfated polysaccharide recognition in sponge cell aggregation. Exp. Cell Res. 170: 381–401.

Dubois M, Gilles KA, Hamilton JK, Rebers PA, Smith F (1956) Colorimetric method for determination of sugars and related substances. Anal. Chem. 28: 350–356.

Ferreiros CM, Criado MT (1983) Purification and partial characterization of a *Fucus vesiculosus* agglutinin. Rev. esp. Fisiol. 39: 51–59.

Hori K, Ikegami S, Miyazawa K, Ito K (1988b) Mitogenic and antineoplastic isoagglutinins from the red alga *Solieria robusta*. Phytochemistry 27: 2063–2067.

Hori K, Matsuda H, Miyazawa K, Ito K (1987) A mitogenic agglutinin from the red alga *Carpopeltis flabellata*. Phytochemistry 26: 1335–1338.

Hori K, Miyazawa K, Ito K (1981) Hemagglutinins in marine algae. Bull. J. Soc. Sci. Fish. 47: 793–798.

Hori K, Miyazawa K, Ito K (1986a) Preliminary characterization of agglutinins from seven marine algal species. Bull. J. Soc. Sci. Fish. 52: 323–331.

Hori K, Miyazawa K, Ito K (1986b) Isolation and characterization of glycoconjugate specific isoagglutinins from a marine green alga *Boodlea coacta* (Dickie) Murray et De Toni. Bot. mar. 29: 323–328.

Hori K, Miyazawa K, Fusetani N, Hashimoto K, Ito K (1986c) Hypnins, low-molecular weight peptidic agglutinins isolated from a marine red alga, *Hypnea japonica*. Biochim. Biophys. Acta 873: 228–236.

Hori K, Oiwa C, Miyazawa K, Ito K (1988a) Evidence for wide distribution of agglutinins in marine algae. Bot. mar. 31: 133–138.

Hori K, Shimada Y, Oiwa C, Miyazawa K, Ito K (1993) Occurrence of a novel group of hemagglutinins extractable by pronase treatment in marine algae. J. appl. Phycol. 5: 219–223.

Kakita H, Fukuoka S, Obika H, Li Z-F, Kamishima H (1997) Purification and properties of a high molecular weight hemagglutinin from the red alga, *Gracilaria verrucosa*. Bot. mar. 40: 241–247.

Kakita H, Nakamura K, Kato Y, Oda Y, Shimura K, Kasai KI (1991) High-performance affinity chromatography of a chick lectin on an adsorbent based on hydrophilic polymer gel. J. Chromatogr. 543: 315–326.

Kamiya H, Shiomi K, Shimizu Y (1980) Marine biopolymers with cell specificity-III-agglutinins in the red alga *Cystoclonium purpureum*: Isolation and characterization. J. nat. Prod. 43: 136–139.

Kanoh H, Kitamura T, Kobayashi Y (1992) A sulfated proteoglycan from the red alga *Gracilaria verrucosa* is a hemagglutinin. Comp. Biochem. Physiol. 102B: 445–449.

Lai PS, So LP, Russell CS (1989) A lipid-associated sulfated proteoglycan from *Nereis* coelomic fluid is a hemagglutinin. Comp. Biochem. Physiol. 93B: 859–865.

Laemmli UK (1970) Cleavage of structural proteins during the assembly of the head of bacteriophage T4. Nature 227: 680–685.

Lowry OH, Rosebrough NJ, Farr AL, Randall RJ (1951) Protein measurement with the folin phenol reagent. J. biol. Chem. 193: 265–275.

Mikami H, Ishida Y (1983) Post-column fluorometric detection of reducing sugars in high performance liquid chromatography using arginine. Bunseki Kagaku 32: E207–E210.

Misevic GN, Burger MM (1986) Reconstitution of high cell binding affinity of a marine sponge aggregation factor by cross-linking of small low affinity fragments into a large polyvalent polymer. J. biol. Chem. 261: 2853–2859.

Noro A, Kimata K, Oike Y, Shinomura T, Maeda N, Yano S, Takahashi N, Suzuki S (1983) Isolation and characterization of a third proteoglycan (PG-Lt) from chick embryo cartilage which contains disulfide-bonded collagenous polypeptide. J. biol. Chem. 258: 9323–9331.

Ohsawa K, Ebata N (1983) Silver stain for detecting 10-femtogram quantities of protein after polyacrylamide gel electrophoresis. Anal. Biochem. 135: 409–415.

Okamoto R, Hori K, Miyazawa K, Ito K (1990) Isolation and characterization of a new hemagglutinin from the red alga *Gracilaria bursa-pastoris*. Experientia 46: 975–977.

Rogers DJ, Topliss JA (1983) Purification and characterisation of an anti-sialic acid agglutinin from the red alga *Solieria chordalis* (C. Ag.) J. Ag. Bot. mar. 26: 301–306.

Shiomi K, Yamanaka H, Kikuchi T (1981) Purification and physicochemical properties of a hemagglutinin (GVA-1) in the red alga *Gracilaria verrucosa*. Bull. J. Soc. Sci. Fish. 47: 1079–1084.

Takahashi Y, Katagiri S (1987) Seasonal variation of the hemagglutinating activities in the red alga *Gracilaria verrucosa*. Nippon Suisan Gakkaishi 53: 2133–2137.

Terho TT, Hartiala K (1971) Method for determination of the sulfate content of glycosaminoglycans. Anal. Biochem. 41: 471–476.

Wessler E (1971) Electrophoresis of acidic glycosaminoglycans in hydrochloric acid: A micro method for sulfate determination. Anal. Biochem. 41: 67–69.

Yoshida T, Yoshinaga K, Nakajima Y (1995) Check list of marine algae of Japan (revised in 1995). Jap. J. Phycol. (Sorui) 43: 115–171.

Copper, copper mine tailings and their effect on marine algae in Northern Chile

Juan A. Correa[1]*, Juan C. Castilla[1], Marco Ramírez[1], Manuel Varas[1], Nelson Lagos[1], Sofia Vergara[1], Alejandra Moenne[2], Domingo Román[3] & Murray T. Brown[4]

[1]*Departamento de Ecología, Facultad de Ciencias Biológicas, Pontificia Universidad Católica de Chile, Casilla 114-D, Santiago, Chile*
[2]*Laboratorio de Biología Molecular, Departamento de Ciencias Biológicas, Facultad de Química y Biología, Universidad de Santiago, Chile*
[3]*Departamento de Química, Facultad de Ciencias Básicas, Universidad de Antofagasta, Casilla 170, Antofagasta, Chile*
[4]*Department of Biological Sciences, University of Plymouth, Plymouth PL4 8AA, United Kingdom*

(*Author for correspondence; e-mail: jcorrea@genes.bio.puc.cl)

Received 12 April 1998; revised 20 January 1999; accepted 21 January 1999

Key words: copper, Chile, mine tailings, intertidal diversity, algae, invertebrates

Abstract

Results are presented of a long-term research programme on the effect of copper contamination on biota in Chilean coastal waters. In spite of the magnitude of the copper mining tailings that affected Caleta Palito and surroundings in northern Chile, the effects on the intertidal assemblages remain restricted to a small geographic area. Even within the affected area, the effects are not homogeneous and there is evidence of active recovery in biological diversity in recent few years. Experimental evidence suggests that the current low algal diversity and abundance is strongly influenced by herbivory, although chronic effects of the discharges cannot be ruled out. Cellular changes in *Enteromorpha compressa* from the impacted area were characterised by abnormal granules in the cytoplasm, though these granules did not contain detectable levels of copper or other heavy metals.

Introduction

Copper is an essential micro-nutrient for aquatic primary producers and an active component in electron transport during photosynthesis, participating as co-factor in various crucial enzymatic reactions, but at elevated concentrations it can be toxic (Gledhill et al., 1997). Reported values of copper concentrations in seawater vary widely (Phillips, 1977; Haraldsson & Westerlund, 1988; Bryan & Langston, 1992; Correa et al., 1996a), but for coastal seawater with no history of copper contamination, concentrations between 0.5 and 3 μg L^{-1} are most commonly reported (Lewis, 1995). Copper concentrations also vary with latitude and depth (Correa et al., 1996b).

From both regulatory and biological perspectives, it is becoming apparent that the most meaningful concentration of copper in seawater is the bioavailable fraction of the metal; copper speciation and bioavailability in seawater have been discussed by Gledhill et al.(1997), who made it clear that information on the concentrations deleterious to marine organisms, particularly macroalgae, is scarce. This is important, because macroalgae constitute the first level in the food chain of every benthic coastal assemblage of organisms. The type and extent of the responses of marine macroalgae to copper vary according to the species under consideration. Excess copper results in toxic responses, including subtle changes in enzymatic activity to gross changes in cell structure and function.

Eventually, impacts are discernible at higher levels of biological organisation, such as depressed reproduction and growth and, ultimately, death (Brown & Depledge, 1998).

Published data on copper and its effects on biota have focussed on North America and Europe, with almost no information from South America. In Chile, in spite of some 4000 km of coastline, little research has been done on the effects of copper on the marine environment. Copper mining is of major economic importance to Chile; it produces almost 30% of the world's needs, and exports are worth ca. US$ 7000 million. Copper mining activities involve 10 major open or underground operations located between 22° and 33° S, over 2000 m above sea level in the Andes. Operations, including the dumping of tailings, usually take place around the mine pits (Castilla & Correa, 1997), but the untreated tailings of the El Salvador mine have for many years been dumped directly into the sea (Castilla & Nealler, 1978; Castilla, 1983, 1996; Correa et al., 1996a). The mine is located at ca. 120 km from the coastal city of Chañaral, the original discharge site for the untreated tailings which lasted from 1938 to 1975. During this period, the more than 150×10^6 t of untreated tailings dumped onto the coast of Chañaral Bay caused severe beach degradation (Castilla, 1983; Paskoff & Petiot, 1990). From 1976 to 1989 the dumping site was moved ca.10 km northward, to the rocky beach of Caleta Palito, which has received ca. 130×10^6 t of tailings in 13 years. Since 1990, after the building of a sedimentation dam ca. 40 km from the mine and 80 km from the coast, only sediment-free water has been channelled from the dam at a flow rate of 200–250 L s^{-1} and disposed at Caleta Palito. The maximum legal concentration of total copper can be no higher than 2000 μg L^{-1}.

The dumping of untreated tailings has resulted in beach degradation, an increase in the copper concentration of coastal water, and a decrease in biological diversity with the total loss of invertebrates and most algal species (Castilla, 1983, 1996). Following the dismantling of the benthic intertidal communities the opportunistic green alga *Enteromorpha compressa* (L.) Grev. colonized the entire intertidal fringe and even although the particulate fraction of the tailings is no longer dumped in the sea, the structure of the intertidal community continues to be characterised by the dominance of *E. compressa*, alternating with large patches of bare rock.

In 1995, a study of the coastal ecosystem around Caleta Palito was initiated to gain an understanding of the interactions between copper and the local flora and fauna. This report summarises the results of some aspects of this work.

Material and methods

Study sites, algal and invertebrate monitoring and copper seawater measurements

The study sites (Figure 1) are located at various distances from the discharge point at Caleta Palito (26° 15S′, 69° 34W′), and cover ca. 150 km of coastline, from Caleta Zenteno in the south to Caleta Huanillo in the north. A further 8 sites were established along the border of a rocky outcrop, the northern limit being the discharge point and the southern limit the artificial tailing beach of Chañaral Bay (Figure 1).

Monitoring macroalgae and sessile and mobile invertebrates in the intertidal zone was carried out at low tide (0.23–0.30 m). At each of the sites indicated in Figure 1, two rocky platforms, 30–40 m long and with slopes of 10° and 40° respectively, were selected. Each platform was divided in four intertidal fringes: low, mid-low, mid-upper, and upper (Castilla, 1996) and two independent observers walked slowly along each fringe, counting and recording the species encountered at 1-min intervals, with a maximum of 5 min per fringe. Species richness within sites and a comparison of species richness between adjacent sites were estimated using pooled data from each site following the methods outlined by Magurran (1988).

Dissolved copper concentrations of the coastal water were determined in-500 mL samples that had been filtered (0.45 μm pore Sartorius membrane) and fixed with 0.5 mL nitric acid (Merck, supra pur) by potentiometric stripping analysis in stationary solution, using a computerised Radiometer ISS 820 analyser. The certified standard CASS-2 was run simultaneously with the water samples (standard provided by the National Research Council of Canada, Division of Chemistry, Marine Analytical Chemistry Standards Program).

Copper and polluted sea water toxicity to algae

The effect of copper on algal growth was tested using 4–5 mm long apical fragments of *Centroceras clavulatum* (C. Agardh) Montagne, *Gelidium lingulatum* Kütz. (Rhodophyta) and *Halopteris hordacea* (Harvey) Sauvageau, and juvenile individuals of *Lessonia nigrescens* Bory (Phaeophyta). Ten apices or

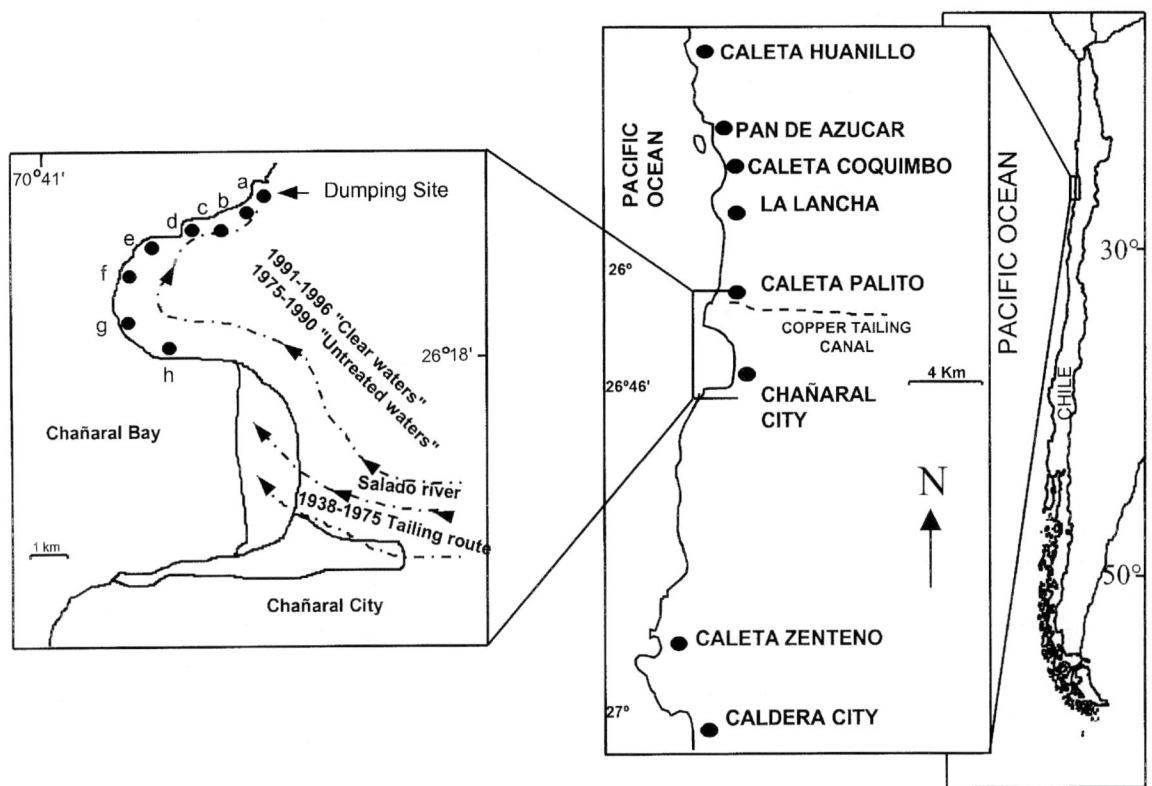

Figure 1. Map showing the various localities included in this study. The small box is a close-up of the outcrop separating Caleta Palito from Chañaral city, where 8 sites were surveyed (a–h).

10 individuals of *L. nigrescens* were cultured in 250 mL Erlenmeyer flasks containing standard SFC culture medium (Correa & McLachlan, 1991) to which was added 0, 15 or 75 μg L^{-1} copper (as CuCl$_2$). Three replicates were used for each copper concentration. The culture medium (150 mL per flask) was changed every third day and the experiments lasted 30 d. Culture conditions were 14 °C, 12:12 (L:D) photoperiod and 40 μmol m^{-2} s^{-1} irradiance. Data did not fulfil the required homocedasticity of variances, and therefore, a Kruskal-Wallis non-parametric test and *a posteriori* multiple comparison analysis were performed (Sokal & Rohlf, 1981).

Toxicity of the coastal seawater contaminated by the mine discharges was assessed using 4–5 mm long apical fragments of *Chaetomorpha linum* (Müller) Kütz. (Chlorophyta), *C. clavulatum*, *G. lingulatum*, *H. hordacea* and juvenile individuals of *L. nigrescens*. The algae were grown in individual, 15-mL glass tubes filled with 10 mL of SFC culture medium (Correa & McLachlan, 1991), using seawater collected at bare rock sites 50 m south of the discharge point of Caleta Palito. Controls were incubated in standard SFC culture medium with seawater collected from Las Cruces, a site with no history of copper enrichment, located in the central part of the country. There were 15 replicates per treatment. Culture conditions were as above and incubation lasted 20 d, with changes of the culture medium every third day. Statistical analysis of growth rate data was as in the copper toxicity experiments.

Algal propagule diversity

To determine propagule availability and their possible sources, a macroalgal census was conducted at Caleta Palito and Caleta Zenteno during winter and spring and, simultaneously, propagule diversity was indirectly assessed from water samples collected from run-off and waves (Hoffmann & Ugarte, 1985). At Caleta Palito, water was obtained from three sites, one directly at the discharge point, the others at 100 m either side of the canal. At Caleta Zenteno, two sample sites were selected. All water samples were stored in 500 mL acid-clean plastic flasks containing 12 sterile coverslips attached to the bottom, which served as sub-

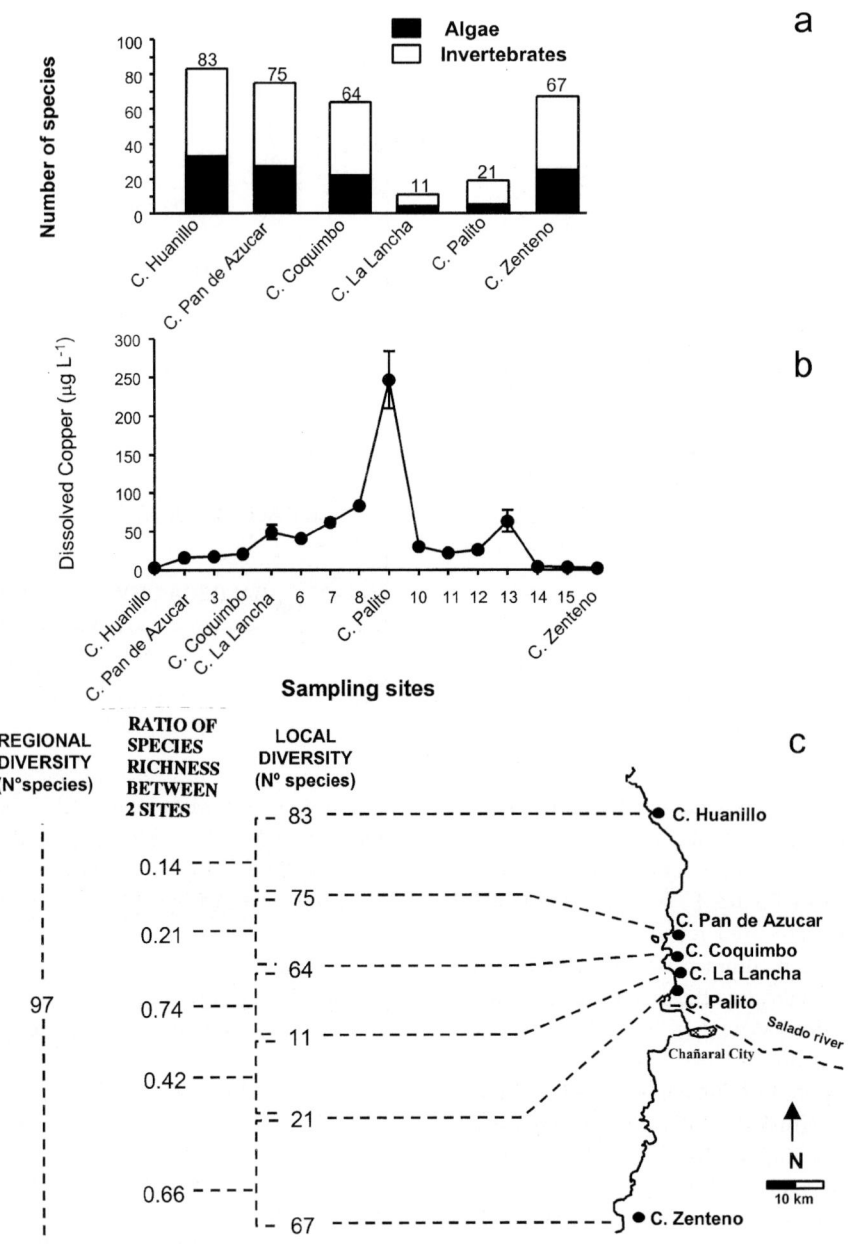

Figure 2. Patterns of biological diversity. a) Diversity profile from the northern Caleta Huanillo to the southern Caleta Zenteno. b) Profile of dissolved copper along the same coastline. The copper value in Caleta Palito was obtained at the discharge, in the mixture zone. Values for copper at sites 10, 11 and 12 correspond to water samples taken at 200, 500 and 1000 m south of the discharge, respectively. c) Detailed local diversity, values of species richness calculated for pairs of neighbouring sites.

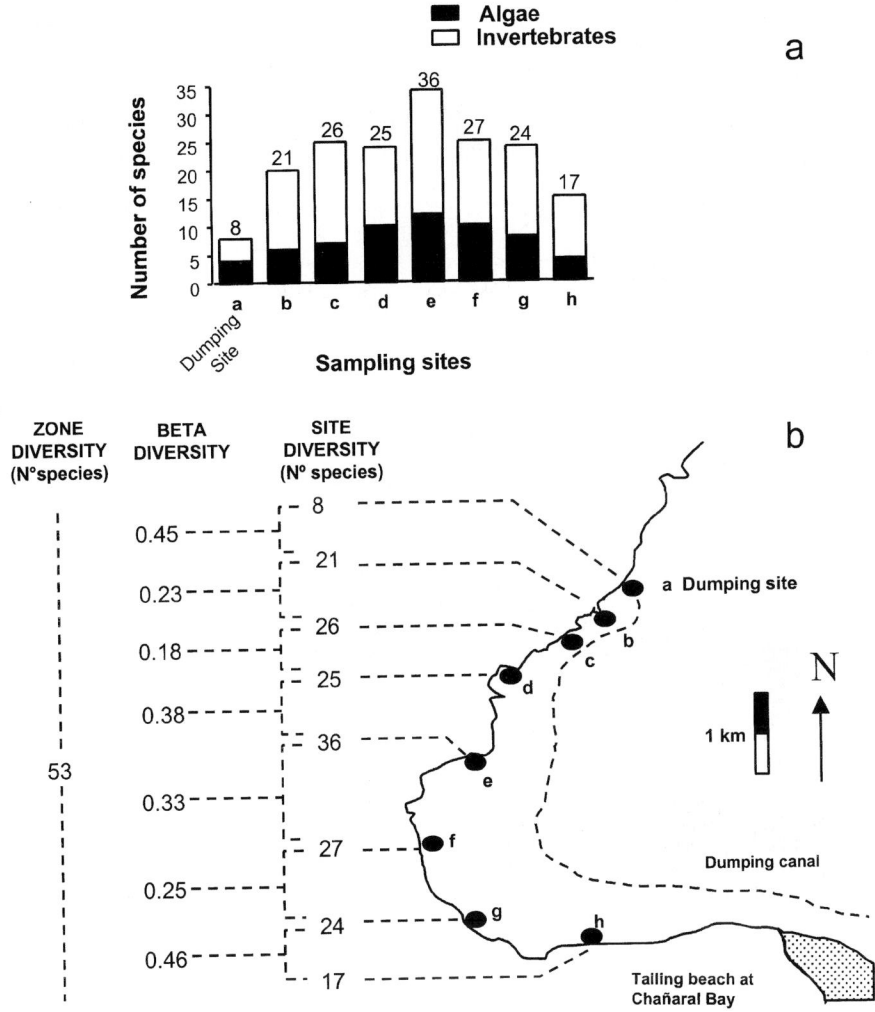

Figure 3. Small-scale changes in diversity (site diversity). a) Diversity profile along the coastline from the discharge site to Chañaral Bay. b) Detailed local diversity, values for species richness calculated for pairs of neighbouring sampling points.

strata for propagule settlement. 2.5 L were obtained at each locality and transported at 4 °C to the laboratory in Santiago. A period of 24 h was allowed to ensure propagule settlement (Santelices, 1990). Enriched seawater medium SFC (Correa & McLachlan, 1991) was used as standard culture medium. Culture conditions followed those outlined by Hoffmann and Ugarte (1985). After 5 weeks in culture, germlings were identified to the genus level, whenever possible.

Effects of grazers on algal diversity

To assess the role of grazers on the structure of the algal assemblages at Caleta Palito, artificial settlement plates were prepared using epoxy resin (Poxy Putty, Permalite Plastics, California), fringed with plastic spikes to eliminate grazers and then fastened to the rocks 3 cm above the surface. Experimental plates were left on the rocks for 3 months, after which they were removed and taken to the laboratory to determine algal diversity.

Cytological aspects of copper tolerance

For standard transmission electron microscopy, samples of *Enteromorpha compressa* from both Caleta Palito and Caleta Zenteno were fixed in 3% glutaraldehyde in 0.45 μm filtered seawater for 3 h at room temperature.

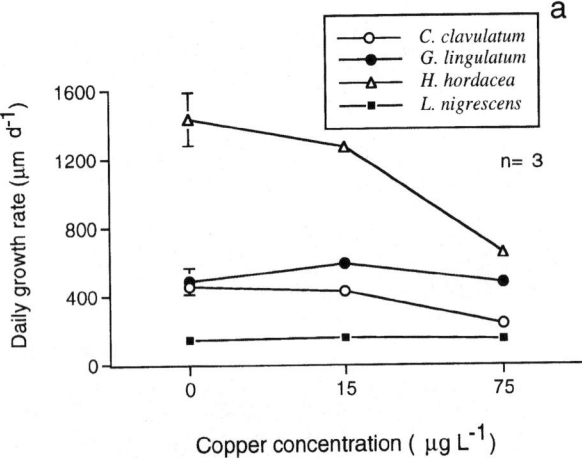

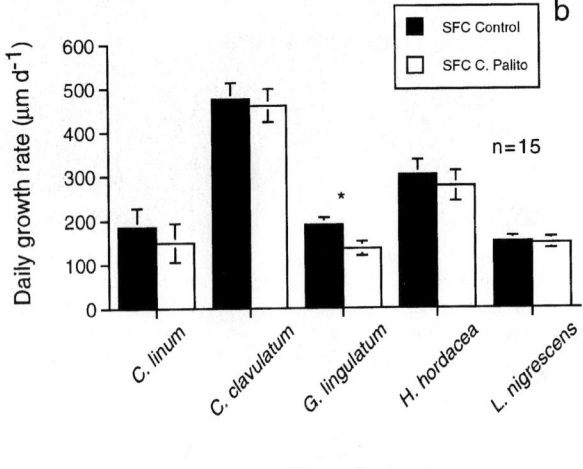

Figure 4. Growth responses of Chilean algae. a) Responses to copper enrichments. b) Responses to the seawater from the mixing zone, near to the discharge at Caleta Palito.

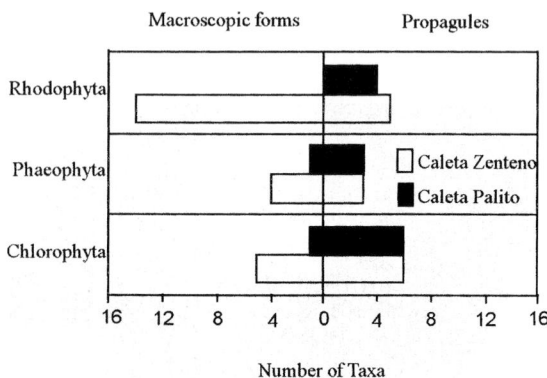

Figure 5. Propagule diversity as compared to the diversity of macroscopic algae recorded in the intertidal zones of Caleta Palito and Caleta Zenteno.

Results

Regional and local diversity associated with a copper gradient

The total number of species declined towards the impacted sites of Caleta Palito and Caleta La Lancha (Figure 2a). The greatest number of species (83) was recorded at Caleta Huanillo (the northern control site) where algae represented 36% of the total. The profile of species number was inversely related to that of the dissolved copper (Figure 2b). Highest values of beta diversity were obtained when impacted and non-impacted sites were compared (Figure 2c, e.g. Caleta La Lancha vs. Caleta Coquimbo and Caleta Palito vs. Caleta Zenteno).

On a local scale, the maximum number of species (36, 33% of which were algae) was found on the rocky outcrop at site 'e', located between the tailing beach of Chañaral and the discharge point at Caleta Palito (Figure 3a). Between-site differences were greater when one of the sites compared was close to the discharge or the tailing beach (Figure 3b).

Copper and polluted sea water toxicity to algae

Individuals of all species survived even at the highest copper concentration used. Growth of *Lessonia nigrescens* and *Gelidium lingulatum* was not significantly affected ($p > 0.05$), but growth of *Centroceras clavulatum* and *Halopteris hordacea* was reduced ($p < 0.05$) at 75 $\mu g\ L^{-1}$ (Figure 4a). Seawater mixed with the discharges from the El Salvador copper mine was not lethal to any of the algae tested. Growth rates were not significantly different ($p > 0.05$) to the controls (Figure 4b).

Postfixation was for 2 h in 1% osmium tetroxide in cacodylate buffer at pH 7.9. Samples for energy dispersive X-ray microanalysis (EDX) were not postfixed and sections were analysed using an EM 912 Omega (Carl Zeiss, Oberkocken / Germany) equipped with an EDX system (Link exL II; Oxford Instruments, High Wycombe, Buckinghamshire, UK). EDX analysis of thin sections was carried out with a spot of 100 nm diameter, 20 μA emission current and 80 kV acceleration voltage.

Figure 6. Exclusion of *Scurria* in Caleta Palito. a) Discs of artificial substratum, covered with algae at the end of the experiment. Scale: 1.5 cm. b) Mature *Polysiphonia* and c) *Antithamnion*, both from the discs. Scale: 100 μm.

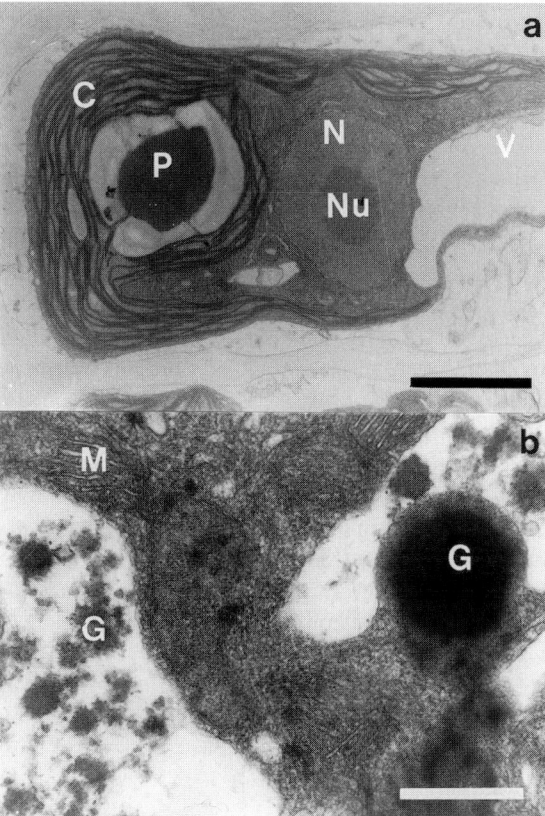

Figure 7. Ultrastructure of *Enteromorpha compressa*. a) Normal cell from Caleta Zenteno. Scale: 2 μm. b) High magnification of electron dense granules appearing only in the cytoplasm of individuals from Caleta Palito. Scale: 500 nm. C: chloroplast, P: pyrenoid, N: nucleus, Nu: nucleolus, V: vacuole, M: mitochondria, G: granules.

Algal propagule diversity

The number of taxa detected in the water column off Caleta Palito was similar to that of Caleta Zenteno (Figure 5). However, propagule diversity was not a good predictor of the diversity of macroscopic forms; at Caleta Palito the surveys of macroscopic forms indicated an absence of rhodophytes, whereas 4 red taxa appeared in the propagule component.

The effects of grazers on algal abundance and diversity

Algae rapidly colonised the artificial plates protected from herbivores, and in all cases (n = 7) almost 100% of the surface was covered by a carpet of algae (Figure 6a), including *Polysiphonia* (Figure 6b) and *Antithamnion* (Figure 6c). 8 taxa were found on the plates, including some not detected in the diversity surveys at Caleta Palito. In contrast, the surface under and around the plates remained free of algae.

Cytological aspects of copper tolerance

The observed changes in the ultrastructure and cellular components of *Enteromorpha compressa* appear to be associated with copper enrichment. Normal cells of *E. compressa* were characterised by a parietal plastid with a central nucleus and a well-developed vacuole (Figure 7a). In individuals from Caleta Palito, thylakoids were disorganised and numerous electron dense granules of diverse size, scattered within the cytoplasm, were present (Figure 7b). Copper was not detectable in the granules. EDX profiles of *E. compressa* from both Caleta Palito and Caleta Zenteno (Figure 8) were similar in terms of their copper signal and the pattern was consistent in cell wall, cytoplasm and plastid. However, a consistent copper signal was associated with the epiphytic bacteria growing on *E. compressa* from Caleta Palito, but not from Caleta Zenteno (Figure 8).

Discussion

The biological diversity data confirms earlier views that the effects of the waste from the El Salvador copper mine remain confined to a small geographic area, despite the extensive accumulation of sediments in Chañaral Bay. The lowest level of diversity was recorded at Caleta La Lancha, and not at the discharge site (Caleta Palito), even though the copper concentration at the latter site was almost five times higher. This pattern might reflect the influence of coastal currents on contaminant dispersal. At the time when discharges at Caleta Palito contained large quantities of solid residues, the northerly flowing surface waters deposited tailings at Caleta La Lancha and formed a beach similar to the one at Chañaral. Observations at Caleta La Lancha indicate that sediments are continuously resuspended by wave action, resulting in abrasion which prevents recruitment of most benthic intertidal organisms, except for fugitive, fast-growing species. Indeed, most of the organisms recorded from Caleta La Lancha were in relatively cryptic microhabitats, protected from the direct influence of the sediments.

Coastline topography may also be important. At Caleta Coquimbo, the recorded number of algal and invertebrate species was 4- and 5-fold higher than at Caleta La Lancha. The two sites are only 3 km apart but are separated by a promontory extending several hundred metres into the sea. This physical barrier seems to prevent the transport of sediments northward and to reduce the levels of dissolved copper in the water (from mean values of 50 μg L^{-1} to 20 μg L^{-1}).

At a smaller geographic scale, recovery seems to be occurring only a few hundred metres from the discharge. A number of species occur in the intertidal fringe, including *Concholepas concholepas* a carnivorous snail considered to be a top predator in the trophic webs of the Chilean intertidal assemblages. This contrasts with the total absence of algae and invertebrates during the period of untreated waste disposal (Castilla & Nealler, 1978; Castilla, 1983). This nucleus of benthic invertebrates and algae is the likely source of propagules for the areas closer to the discharge point which presently have lower biological diversity.

The green alga *Enteromorpha compressa* is the dominant organism in the intertidal zone of Caleta Palito, where it alternates with large areas of bare rock, particularly on the platforms south of the discharge point. This atypical dominance is associated with high values (range from 10 to 40.7 μg L^{-1}) of dissolved copper (Correa et al., 1996b; Castilla & Correa, 1997), and with large densities of patelloid herbivores in the areas of bare rock. It is not known what factor(s) maintains this very simple community structure.

One possibility is that the available copper or the other metals associated with the discharge (Correa, unpubl.), are highly toxic to algae other than *E. compressa*. However, the experimental evidence suggests that neither copper alone, nor the polluted coastal water off Caleta Palito, can fully explain the absence of the algae tested in the laboratory, or the poor algal diversity and abundance recorded in the intertidal zone influenced by the discharges. A second alternative is that the seawater adjacent to Caleta Palito, has a low diversity of algal propagules. The importance of these propagules to algal diversity and abundance in the intertidal zone has been discussed elsewhere (Hoffmann & Ugarte, 1985; Hoffmann & Santelices, 1991). Our results indicate that the diversity of cultivable algal propagules from the impacted area was similar to that of the control site (Caleta Zenteno) and that propagule diversity was not a good predictor of diversity of intertidal algal at any of the studied sites. Rhodophytes were present in the propagule assemblage but were absent from the intertidal close to the discharge point. A third possibility is that a factor, not directly related to the present level or quality of mining wastes, may

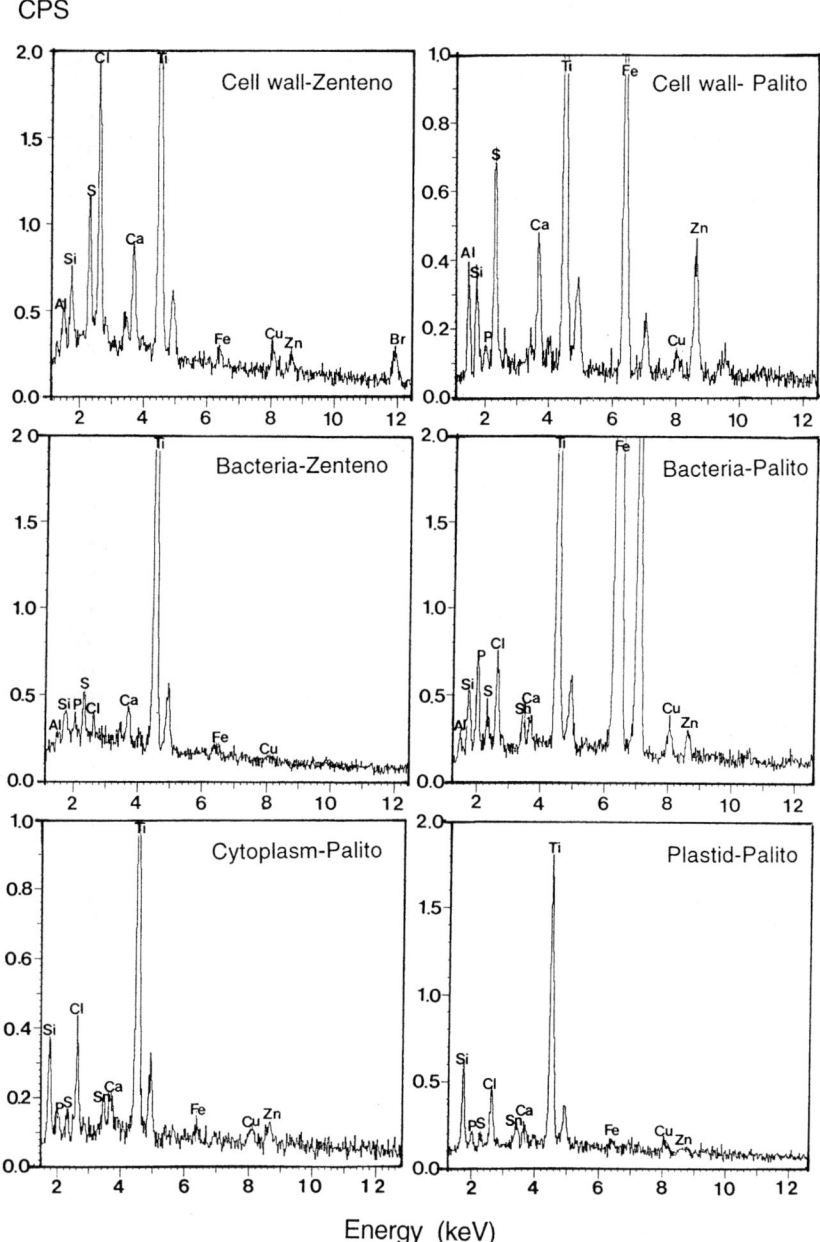

Figure 8. EDX analyses of *Enteromorpha compressa* from Caleta Palito and Caleta Zenteno. Cell wall, bacterial epiphytic film, cytoplasm and plastids are included. Controls (not included) for cytoplasm and plastids were almost identical to those from Caleta Palito.

be responsible for structuring the intertidal community at Caleta Palito. The rapid colonisation of the artificial settlement plates by a highly diverse algal turf, together with the persistence of bare rock below and around the plates, suggest that the very simple structure of the intertidal community at Caleta Palito is, at least partially, the result of grazing pressure by the herbivore *Scurria* spp. Patelloid herbivores have been recorded at exceedingly high densities (ca. 800 individuals m^{-2}) in the impacted area compared with the usual density of less than 5 individuals m^{-2} at the control sites (Correa, unpubl.). These grazers do not appear to be regulated at present by a consumer from a higher trophic level.

Enteromorpha compressa displayed clear evidence of cellular changes induced by the environment of Caleta Palito, including the accumulation of electron dense granules in the cytoplasm. In brown algae, copper is accumulated in physodes (Ragan et al., 1979; Smith et al., 1986) but for green algae the information is scarce. Intracellular deposits have been reported in the unicellular *Chlorella* (Wong et al., 1994) and *Scenedesmus* (Silverberg et al., 1976). In the latter case, these intracellular inclusions were considered to be a detoxification mechanism, as they were present only in metal tolerant strains. In our study, copper was not associated with the electron dense granules and the EDX profiles for copper were similar in material collected from the polluted and control sites. This may be an indication that copper metabolism is quite dynamic, preventing the metal from accumulating in high quantities within the cytoplasm. Alternatively, copper may not be entering the cell, but is being bound to mucilage produced by epiphytic bacteria on the thallus surface of *E. compressa* (see Holmes et al., 1991; Riquelme et al., 1997). It is known that bacterial films can bind heavy metals (Gadd & White, 1993) and that copper tolerant bacteria accumulate the metal within their cells (Brown et al., 1992; Silver & Ji, 1994). More recently, Riquelme et al. (1997) have shown that epiphytic bacteria from Caleta Palito were highly tolerant to copper and suggested that this tolerance could influence the tolerance of the algal host, *E. compressa*. This idea is consistent with the EDX profile of the mucilage deposited as part of the bacterial film on *E. compressa* from Caleta Palito, which had a much higher copper and zinc content than the controls from Caleta Zenteno.

The evidence indicates that the devastating ecological effect of the mining tailings on the coastal benthic communities of Caleta Palito during the early 1950s has not fully disappeared. However, species absent for many years from the area influenced by the discharge are now slowly appearing. The apparent simplicity of the community structure at the impacted area, which is dominated by *E. compressa*, seems under the influence of biological factors (i.e. herbivory), rather than under the absolute control of the copper concentrations or the complex mixture of today's discharges. Despite the magnitude of the early impact, ecological devastation has not spread along the coast, as originally predicted (J.C. Castilla, pers. comm.).

Acknowledgement

This study has been supported by a research grant from the ICA through CIMM to JAC.

References

Brown MT, Depledge MH (1998) Determinants of trace metal concentrations in marine organisms. In Langston WJ, Bebianno MJ (eds), Metal Metabolism in Aquatic Environments. Chapman & Hall, London: 185–217.

Brown NL, Rouch DA, Lee BTO (1992) Copper resistance systems in bacteria. Plasmid 27: 29–40.

Bryan GW, Langston WJ (1992) Bioavailability, accumulation and effects of heavy metals in sediments with special reference to United Kingdom estuaries: A review. Environ. Pollut. 76: 89–131.

Castilla JC (1983) Environmental impact in sandy beaches of copper mine tailings at Chañaral, Chile. Mar. Pollut. Bull. 14: 459–464.

Castilla JC (1996) Copper mine tailing disposal in northern Chile rocky shores: *Enteromorpha compressa* as a sentinel species. Environ. mon. Assess. 40: 41–54.

Castilla JC, Correa JA (1997). Copper tailing impacts in coastal ecosystems of northern Chile: from species to community responses. In Moore M, Imray P, Dameron C, Callan P, Langley A, Mangas S (eds), Copper. National Environmental Health Forum Monographs, Metal Series No. 3: 81–92.

Castilla JC, Nealler E (1978). Marine environmental impact due to mining activities of El Salvador copper mine, Chile. Mar. Pollut. Bull. 9: 67–70.

Correa JA, González P, Sánchez P, Muñoz J, Orellana M (1996a) Copper-algae interactions: inheritance or adaptation? Environ. mon. Assess. 40: 41–54.

Correa JA, McLachlan JL (1991) Endophytic algae of *Chondrus crispus* (Rhodophyta). III. Host-specificity. J. Phycol. 27: 448–459.

Correa JA, Ramírez M, Fatigante F, Castilla JC (1996b) Copper-algae interactions in northern Chile: The Chañaral case. In Björk M, Semesi A, Pedersén M, Bergman B (eds), Current Trends in Marine Botanical Research in the East African Region. Ord & Vetande, Uppsala, Sweden: 99–129.

Gadd GM, White C (1993) Microbial treatment of metal pollution-a working biotechnology. Tib. Tech. 11: 353–359.

Gledhill M, Nimmo M, Hill SJ, Brown MT (1997) The toxicity of copper (II) species to marine algae, with particular reference to macroalgae. J. Phycol. 33: 2–11.

Haraldsson C, Westerlund S (1988) Trace metals in the water columns of the Black Sea and Framvaren Fjord. Mar. Chem. 23: 417–424.

Hoffmann A, Santelices B (1991) Banks of algal microscopic forms: hypotheses on their functioning and comparisons with seed banks. Mar. Ecol. Progr. Ser. 79: 185–194.

Hoffmann A, Ugarte R (1985.) The arrival of propagules of marine macroalgae in the intertidal zone. J. exp. mar. Biol. Ecol. 92: 83–95.

Holmes MA, Brown MT, Loutit MW, Ryan K (1991) The involvement of epiphytic bacteria in zinc concentration by the red alga *Gracilaria sordida*. Mar. environ. Res. 31: 55–67.

Lewis AG (1995) Copper in Water and Aquatic Environments. International Copper Association, LTD., NewYork, 65 pp.

Magurran AE (1988) Ecological Diversity and its Measurement. Princeton University Press, 179 pp.

Paskoff R, Petiot R (1990) Coastal progradation as a by-product of human activity: An example from Chañaral Bay, Atacama Desert, Chile. J. Coastal Res. 6: 91–102.

Phillips DJ (1977) The use of biological indicator organisms to monitor trace metal pollution in marine and estuarine environments – a review. Environ. Pollut. 13: 281–317.

Ragan MA, Smidsrød O, Larsen B (1979). Chelation of divalent metal ions by brown algal polyphenols. Mar. Chem. 7: 265–271.

Riquelme C, Rojas A, Flores V, Correa JA (1997) Epiphytic bacteria in a copper-enriched environment in northern Chile. Mar. Poll. Bull. 34: 816–820.

Santelices B (1990) Patterns of reproduction, dispersal and recruitment in seaweeds. Oceanogr. Mar. Biol. Annu. Rev. 28: 177–276.

Silver S, Ji G (1994) Newer systems for bacterial resistances to toxic heavy metals. Environ. Health Perspectives 102: 107–113.

Silverberg BA, Stokes PM, Ferstenberg LB (1976) Intranuclear complexes in a copper tolerant green alga. J. Cell Biol. 69: 210–214.

Sokal RR, Rohlf FJ (1981) Biometry. W.H. Freeman and Co., New York, 859 pp.

Smith KL, Hann AC, Hardwood JL (1986) The subcellular localization of absorbed copper in *Fucus*. Physiol. Plant. 66: 692–698.

Wong SL, Nakamoto L, Wainwright JF (1994) Identification of toxic metals in affected algal cells in assays of waste waters. J. appl. Phycol. 6: 405–414.

Factors influencing seaweed responses to eutrophication: some results from EU-project EUMAC

Winfrid Schramm
Institute of Marine Science, University of Kiel, Düsternbrooker Weg 20, 24105 Kiel, Germany

Received 1 April 1998; revised 7 December 1998; accepted 31 December 1998

Key words: seaweed, algal blooms, eutrophication, general mechanisms, management

Abstract

Seaweed responses to eutrophication and their role in coastal eutrophication processes were compared at 8 different sites along the European coasts from the Baltic to the Mediterranean as part of the EU-ENVIRONMENT Project Marine *Eu*trophication and benthic *Mac*rophytes (EUMAC). Structural and functional changes of marine benthic vegetation typical of eutrophic waters, in particular mass development (blooms) of certain seaweeds, are not merely the result of increased nutrient loading, but must be attributed to complex interactions of primary and secondary effects during the eutrophication process. Due to species-specific physiological properties of the algae (nutrient kinetics, growth potential, light, temperature requirements), the combined effects of abiotic and biotic factors on juvenile or adult developmental stages control the development of algal blooms in different ways. In particular the role of light, temperature, water motion and oxygen depletion, as well as of grazers, on early and adult developmental stages of the algae are considered. The result are discussed in the context of coastal eutrophication control and management.

Introduction

Eutrophication is generally accepted as the principal cause of the globally observed deterioration in marine coastal environments. In coastal areas, which are the centres of population and industrialisation and the primary recipients of nutrients from land, the typical symptoms of eutrophication can be observed. Among these, changes in marine plant communities and in particular the mass development or blooms of micro- or macroalgae are probably the most conspicuous effects of increased nutrient loads. In the coastal zones of Europe, which border the most impacted sea areas in the world, such changes have been described for all coastal countries and they have become of major public concern (Schramm & Nienhuis, 1996: Figure 1).

Since primary producers are the first to directly respond to increased nutrient loads by increased production, their key position in the eutrophication process is obvious. In the search for effective and long-lasting ways of control and management it is therefore reasonable to consider the role that primary producers play in the structure and function of a system under eutrophic conditions. Within the framework of the EU-programme ENVIRONMENT, a project entitled '*Eu*trophication and marine *mac*rophytes' (EUMAC) was launched in 1993 to study the response and role of seaweeds in the eutrophication process, in particular the mechanisms of macroalgal bloom development under varying environmental and climatic conditions, at eight different sites along the European coasts from the Baltic to the Mediterranean (cf. Figure 1).

Previous investigations had already shown that primary producers generally respond to increasing nutrient levels in shallow coastal waters in a typical sequence of successive phases of increasing eutrophication (Nienhuis, 1989, 1992; Schramm & Nienhuis, 1996), which can be briefly summarised as follows (Figure 2):

a) 'Un-eutrophicated' marine or brackish shallow coastal waters can be characterised by a balanced nutrient regime i.e. over longer periods of time the nutrient imports equal exports. The dominant primary

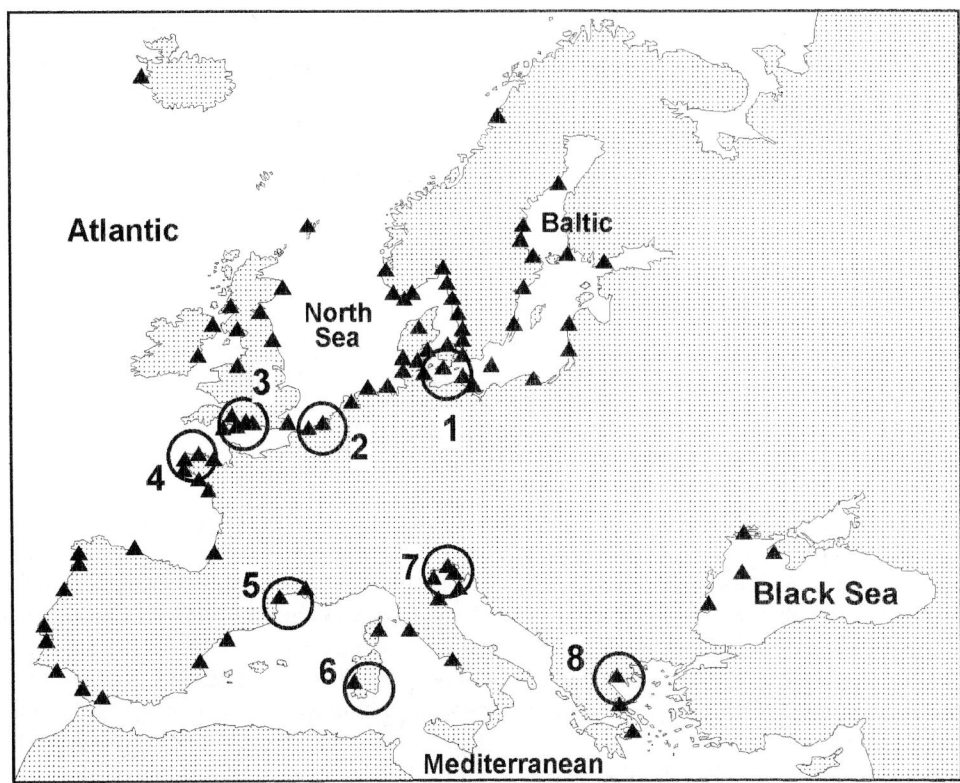

Figure 1. Locations in Europe where changes in marine benthic vegetation have been observed or studied (triangles). Study sites in the framework of the EU-ENVIRONMENT project EUMAC 1993–1996 (circles): 1) western Baltic, Schlei Fjord, G; 2) North Sea, Veerse Meer Lagoon, NL; 3) North Sea, Langstone Harbour, GB; 4) Atlantic, Bay of Lannion, F; 5) Mediterranean, Thau Lagoon, F; 6) Mediterranean, Spenia Arrubia Lagoon, I; 7) Mediterranean, Venice Lagoon, I; 8) Mediterranean, Thermaikos Gulf, GR (Schramm & Nienuis, 1996).

producers are usually perennial benthic macrophytes, such as seagrasses and other phanerogams on soft bottoms, or long-lived seaweeds on hard substrata, which successfully compete with the faster growing ephemeral benthic or planktonic algae because of their specific nutrient economy (Duarte, 1995).

b) Increased nutrient loading favours short-lived seasonal forms. From slight to medium eutrophication, increasingly blooms of 'nutrient opportunists' occur, in particular fast-growing epiphytic macroalgae and bloom-forming phytoplankton taxa. This can be explained by the specific nutrient economy and productivity potential of these forms, which take up nutrients and grow much faster than slow-growing, long-lived macrophytes (Wallentinus, 1984; Foldager-Pedersen & Borum, 1997). Phanerogams and perennial macroalgal communities gradually decline. Usually associated with this decline are changes in their structure (species composition, coverage or depth distribution limits), and of their function (production, reproduction).

c) With further increasing nutrient loading towards hypertrophic conditions, free-floating macroalgae, in particular 'green tide' forming taxa such as *Ulva* and *Enteromorpha* spp. alternating with heavy uncontrolled phytoplankton blooms, dominate and replace the perennial and slow-growing benthic macrophytes until their extinction.

d) Finally, under hypertrophic conditions with continuously high nutrient concentrations, phytoplankton constitute the dominant primary producers, and benthic macrophytes completely disappear.

From numerous field observations, as well as experimental work, it is known that this sequence of qualitative and quantitative changes in marine plant communities does not occur gradually with increasing nutrient levels, but is stepwise in sudden shifts. In addition, it has been observed that even under comparable nutrient regimes the response of primary producers, in particular the development of both phytoplankton and macroalgal blooms, may vary considerably in different locations as well as from year to

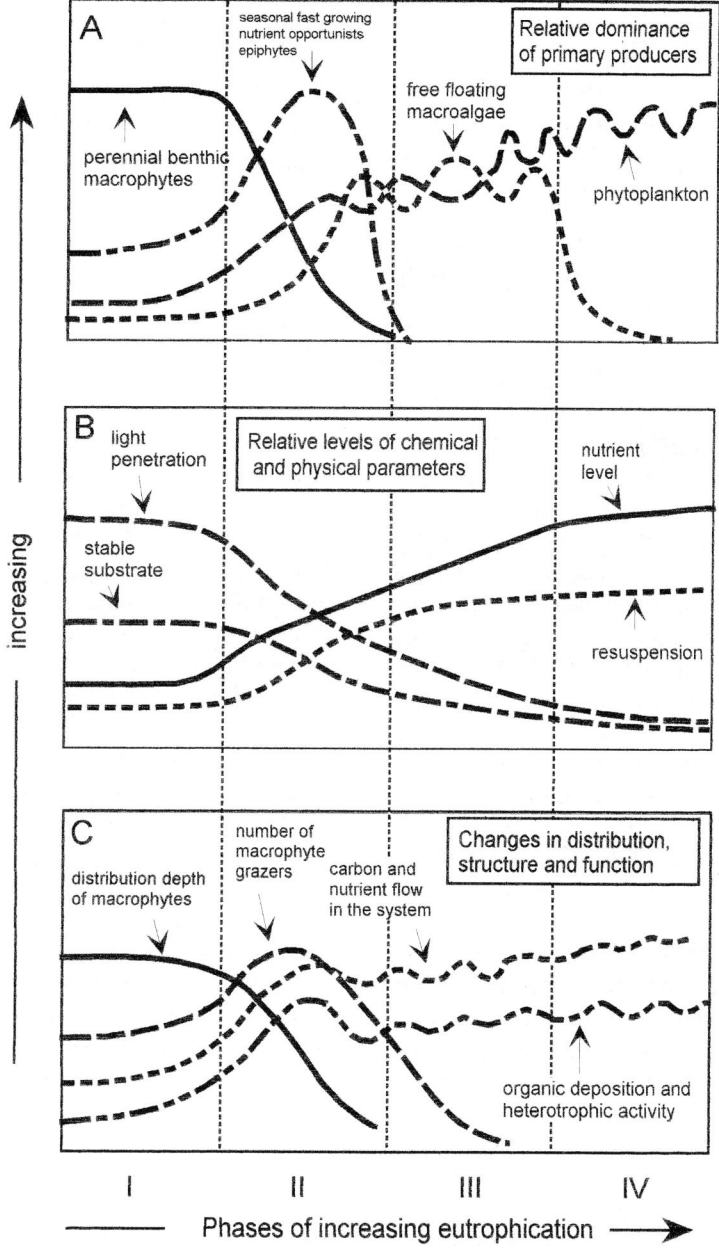

Figure 2. Schematic presentation of typical changes in dominance of primary producers and some related structural and functional parameters during phases of increasing eutrophication (Schramm, 1996).

year. These findings suggest that the observed variations cannot be attributed solely to increasing nutrient levels. Nutrients are obviously a prerequisite to trigger the proliferation of fast growing nutrient opportunists and thus the eutrophication process; however, other abiotic and biotic factors, secondary processes and feed-back mechanisms must all become effective in the initial stage and progression of the eutrophication process (e.g. Twilley et al., 1985; Vogt & Schramm, 1991; Sand-Jensen & Borum, 1991; Neundorfer & Kemp, 1993; Duarte, 1995; De Vries et al., 1996).

The primary aim of the EUMAC project was therefore not only to characterise and compare changes in marine benthic vegetation, in particular the occurrence of macroalgal blooms over the geographical and climatic range of the European coasts, but also to analyse the mechanisms causing the observed spatial and temporal variations in seaweed responses. In this paper, some results from the EUMAC project are presented to exemplify the role of some other factors besides nutrients which influence or determine the varying responses of benthic macroalgae to increased nutrient levels. In addition, implications on the ecosystem and aspects of eutrophication control and management are discussed.

Changes in light climate

Light together with temperature and nutrient supply are the principal abiotic factors, which control algal growth. Under eutrophic conditions, algal blooms of either plankton, unattached macroalgae floating in the water column, or epiphytes can considerably reduce the incident light, often to levels below the light requirements of benthic plants confined to the bottom. Long-lived benthic macrophytes, although competitively superior to short-lived nutrient opportunists under low nutrient conditions (Duarte, 1995), can therefore be outcompeted by phytoplankton and ephemeral macroalgae that are floating above or growing on top of the slow-growing plants and thus occupy positions more favourable for light harvesting.

Based on comparative studies of various eutrophicated lagoon systems, attempts were made to quantify the aspects of the light climate influencing algal bloom development (De Vries et al., 1996). It has been suggested that attached benthic primary producers can successfully compete with phytoplankton or floating macroalgae when irradiance is equal to or higher than the average irradiance in the water column. Shallow water or good transparency of the water are indicative of a high competitive ability of benthic primary producers. A typical result of the deterioration of the light climate due to eutrophication is the change in the depth distribution limits of benthic macrophytes in response to their minimum light requirements (Kautsky et al., 1986; Breuer & Schramm, 1988). In the Kiel Bight (western Baltic), for example, the lower distribution limit of *Fucus vesiculosus* communities shifted from 12 m to 2 m depth and these communities were replaced by red algal communities from the early 1950s to 1987/88. Moreover, the *Fucus* standing stock declined by 95% within this period (Vogt & Schramm, 1991). These changes were attributed to increased nutrient levels, phytoplankton production and turbidity. PAR recordings in the early 1990s showed that the annual photon irradiance below 2 m depth was insufficient for the existence of *Fucus* communities (Schaffelke, pers. comm.).

Hydrological and hydrodynamic conditions

On open coasts, as well as in enclosed coastal areas such as lagoons, fjords or estuaries, currents and tidal exchange or wind-forced flushing usually prevent the accumulation of nutrient concentration levels leading to excessive primary production. However, site specific hydrological conditions can influence the eutrophication process in other ways, and may be beneficial and even a prerequisite for the development of macroalgal blooms, particularly in their initial stages (De Vries et al., 1996). For example, the Schlei Fjord in the Baltic Sea is a severely eutrophicated inshore water. Despite heavy plankton blooms, and thus deterioration of the light climate, benthic macroalgal blooms do develop. These are favoured by wind-forced alternate flushing with nutrient loaded and turbid water from the highly eutrophic inner parts of the fjord and nutrient poor, clearer water from the Baltic Sea (Figure 3). Seasonal nutrient opportunists such as *Pilayella littoralis, Enteromorpha* spp., *Ceramium* spp., *Polysiphonia* spp., all capable of rapid uptake and accumulation of nutrients during the outflow periods, can utilise the stored nutrients under more favourable light conditions during subsequent inflow periods of clearer Baltic water.

Another example showing the importance of flushing for the development of macroalgal blooms has been reported for the Venice lagoon (Sfriso & Marcomini, 1996a,b,c), where huge masses of loose lying *Ulva* spp. covered great parts of the central lagoon in the 1980s. The standing crop was estimated to be as much as 0.6 million t wet weight (net production 1.6 million t wet weight, gross production up to 10 million t wet weight per year). From estimates of the nitrogen contained in these algae it was concluded that nitrogen import via repeated flushing is a prerequisite for the build up of the algal biomass during the main growth period of 1.5 months (De Vries et al., 1996).

The reduction or extinction of benthic macrophyte communities involves changes in the hydrodynamic conditions of the bottom layer and thus in sediment stability. Altered bottom currents or effects of wave action may result in enhanced resuspension of sediments and transfer of sediment nutrients into the water column, which further increase turbidity and nutrient

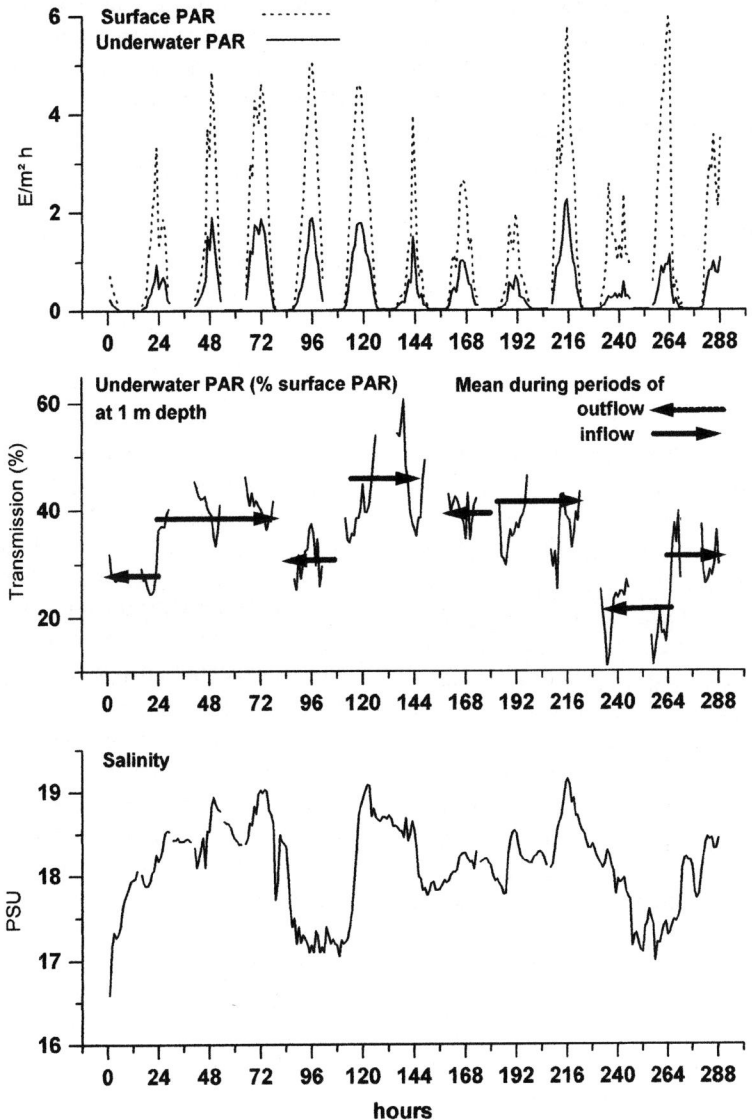

Figure 3. Light conditions in the phytobenthos as affected by wind-forced alternate flushing with water from the Baltic and from the inner parts of the eutrophic Schlei Fjord (Kiel Bight), June 1995. Daily irradiation (PAR) at the surface and at 1 m depth (top), underwater PAR (transmission in% of incident PAR, arrows = means) during periods of outflowing turbid, nutrient enriched Schlei water and of inflowing clearer, nutrient poor water from the Baltic Sea (middle), and salinity (PSU) during in- and outflow periods (bottom).

levels, respectively (Breuer & Schramm, 1988). In the Venice lagoon the dense, loose lying algal mats of 5–20 kg (wet weight) m^{-2} in the 1980s significantly reduced the resuspension of sediments. From 1990 onwards, a rapid decline of *Ulva* biomass was observed, primarily attributed to unfavourable climatic conditions during the peak growth season in consecutive years, By the mid 1990s, *Ulva* was found only in scattered populations, and the maximum standing stock had decreased to less than 0.1 million t wet weight (net production ranged between 0.3–0.4 million t per year). As a result, resuspension of the uncovered sediments increased from 20–30 g dry weight m^{-2} d^{-1} before 1990 to 500–1500 g m^{-2} d^{-1} in the mid 1990s. Increased sedimentation on the algae, turbidity and deterioration of the light climate caused further declines in *Ulva* biomass (Sfriso & Marcomini, 1996a,b,c).

The conspicuous mass development of floating green algae along the open Atlantic coast of Brittany,

France, known as 'green tides' (marées vertes), has been explained by a specific hydrological regime in combination with high nutrient loads deriving from agriculture (Menesguen, 1992; Piriou & Menesguen, 1992). Dynamic trapping of water masses in open bays with very low tidal residual drift, together with intensive turbulent mixing, allow mass development of unattached *Ulva* spp. which floats freely in the water column with no degradation; degradation occurs only after the algae have become beached. In the Bay of Lannion, which receives from the land 1500–2000 t inorganic N annually, 10 000 to 15 000 m^3 of stranded algae has had to be removed per year over the past two decades. Along the whole coastline of Brittany, nearly 100 000 t wet weight are collected annually (Dion & Le Bozec, 1996).

Oxygen depletion

In locations with reduced water exchange or mixing, oxygen depletion resulting from eutrophication may have significant secondary effects on the benthic vegetation. In eutrophic inshore waters of the Baltic, e.g. Schlei Fjord, during stagnant periods algal material from phytoplankton and epiphyte blooms accumulate in the bottom layer, enhancing oxygen-consuming heterotrophic activities. Resulting anoxia and production of toxic hydrogen sulphide in the bottom layer lead to further impairment and mortality, often resulting in the complete extinction of both the benthic flora and fauna. In addition, reducing conditions under anoxia accelerate nutrient release from the sediments, which in turn raises the eutrophication level particularly during times of increasing temperature in summer (Schramm et al., 1996).

Estimates of the nutrient budget for the Thau Lagoon (French Mediterranean coast) have shown that the greater part (approx. 85–95%) of the nutrients in the 75 km^2 lagoon is stored in the sediments (approx. 8400 t N, 1350 t P). With rising temperatures and increasing oxygen depletion, during summer, a large fraction of these nutrients and particularly P (55 t) is released from the sediments; this is the total amount of phosphorus available in the water column, and is more than double the amount (24 t P) bound by macrophytes (De Casabianca et al., 1996, 1997; De Casabianca & Posada, 1998).

The role of grazers

Interaction between primary producers and animals, in particular herbivores, suspension and filter feeders, associated with the plant communities can significantly influence the eutrophication process. Grazing may effectively reduce both the phytoplankton and phytobenthic biomass. In the Venice lagoon, for example, grazing pressure increased rapidly because anoxic conditions, which earlier frequently occurred in the dense algal mats and killed the herbivores, did not develop after the decline of *Ulva*. On an annual basis, approximately 70% of the *Ulva* production was consumed by grazers, and even exceeded the production during the summer months from June to August (Sfriso & Marcomini, 1996 a,b).

Grazing by zooplankton, but also by benthic filter and suspension feeders, can effectively reduce the phytoplankton biomass and can thus act as a natural control on effects induced by eutrophication (Officer et al., 1982; De Vries & Hopstaken, 1984; De Vries et al., 1996). The fact that at low nutrient levels bloom-forming algae usually do not flourish and outcompete the slower growing benthic perennials may be the result of grazing on epiphytes, which under nutrient limitation cannot grow fast enough to compensate for grazing losses. On the other hand, enhanced growth of epiphytic micro- and macroalgae or phytoplankton due to high nutrient levels supports the development of associated grazer and filter feeder populations, which in turn may accelerate the eutrophication process through regeneration of nutrients (Figure 4).

In red algal communities of the Swedish Baltic east coasts, for example, nutrient regeneration through associated *Mytilus edulis* exceeded considerably the nutrient demand of the benthic algae (Kautsky & Wallentinus, 1980). It was estimated that the remaining nutrients could supply the pelagic system with about 6% of its nitrogen and 17% of its phosphorus demands. Estimates for the Baltic Sea proper predicted a recycling by mussels of about 250 000 t inorganic N and 77 000 t inorganic P per year, which is higher than the input from terrestrial sources.

Early developmental stages

Comparatively little is known about the role of the early developmental stages of macroalgae in the outbreak of algal blooms. In the framework of the EU-MAC project experiments were included on juvenile stages of *Enteromorpha* spp. and *Pilayella littoralis*, two common co-occurring, bloom forming algae from the Baltic. Field observations and laboratory experiments show the importance of the mode of resting stages and overwintering, as well as of abiotic envir-

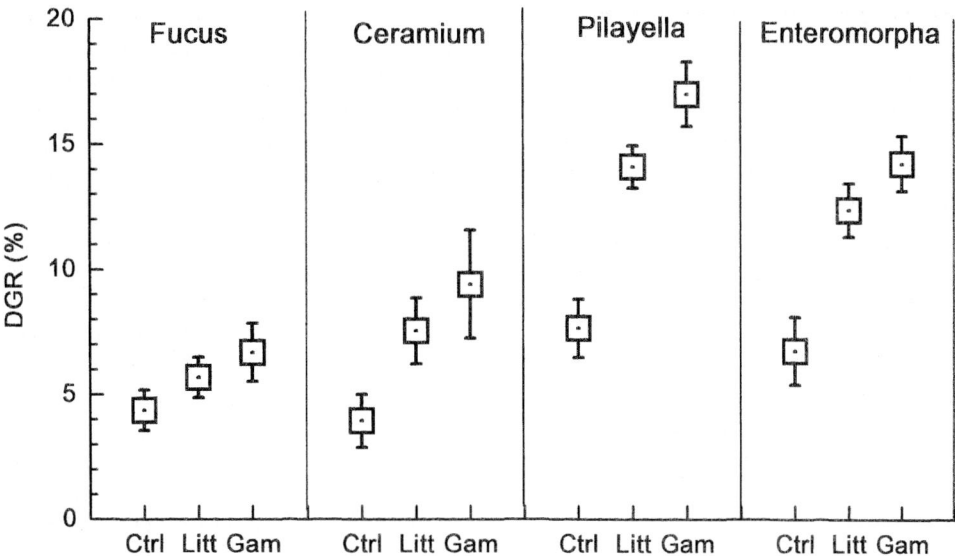

Figure 4. Effects of nutrients released from the grazers *Littorina littorea* ($n = 3$) and *Gammarus* sp. ($n = 50$) on relative daily growth rates (mean + SD, $n = 3$) of *Fucus vesiculosus* (initial wet weight 5 g each), *Ceramium nodulosum*, and *Pilayella littoralis* (initial wet weight 2 g each). During the growth period of 5 days, algae and grazers were kept separate in two aquaria connected for circulation of the seawater. Salinity 15 PSU, temperature 15 °C. Irradiance: algae 150 μmol m^{-2} s^{-1}, grazers ca 5 μmol m^{-2} s^{-1}. Grazers were fed continuously with small portions of the algae.

onmental factors, in particular light, temperature, and nutrient supply, in stimulating germination (Schories, 1995; Schramm et al., 1996; Lotze et al., 1998; Figures 5 and 6). Further, more detailed studies (Lotze 1998) revealed pronounced differences between *Enteromorpha* and *Pilayella*. The reproductive period of *Enteromorpha* extended from April to September with peak densities of settling spores of up to 60 million m^{-2} d^{-1}, and 20–40% cover on hard substrates. However, only a very small proportion reached adulthood (max. biomass 1.4 g dry weight m^{-2} in July). In contrast, *Pilayella* produced 10–30 times higher biomass (max. 15 g dry weight m^{-2}) during blooms, despite a shorter reproductive period (2–3 months) and lower densities of settled propagules (max. 15 million m^{-2} d^{-1}). The difference in timing of population development can be explained by the species-specific ecophysiological requirements for germination.

In contrast to the early life stages, no significant differences were observed in adults of the two species. Growth rates in relation to temperature, light and nutrient supply, as well as nutrient kinetics were similar. These results suggest that the beginning of macroalgal mass development in spring is determined by the ecophysiological background for timing and rate of germination, and not by productivity of the adults. The ecophysiological properties alone, however, cannot explain the further population development of *Enteromorpha* and *Pilayella*. For this reason, laboratory and field grazing experiments were carried out, which showed that the most abundant common mesograzers (*Idotea* spp., *Gammarus locusta*, *Littorina littorea*, *L. saxatilis*) at the experimental site graze heavily on both algae as germlings (64–86% reduction of germling density cm^{-2} in the field within 14 days) and adults (8–12% thallus loss d^{-1} in the field). At all life stages, *Enteromorpha* was preferred over *Pilayella*. Lotze (1998) pointed out that the relative effects exerted by the different grazers varies with the life stage of the algae, which may indicate the importance of grazer diversity for an effective control of bloom-forming algae. In further multi-factorial field experiments, a complex interaction pattern was found, which determined the resource competition between the macroalgal species:

a. In early spring, initiation of the algal populations was controlled by abiotic factors that determined the time and rate of germination out of the propagule banks. Despite lower germination temperature for *Pilayella*, and thus earlier germination, *Enteromorpha* dominated all substrates because of massive recruitment from dormant propagules.

b. This situation changed rapidly in April when grazing activity began. There was selective grazing on

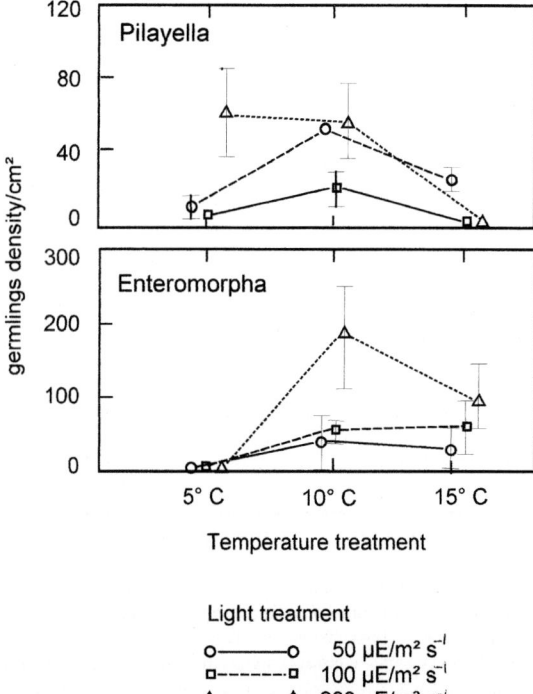

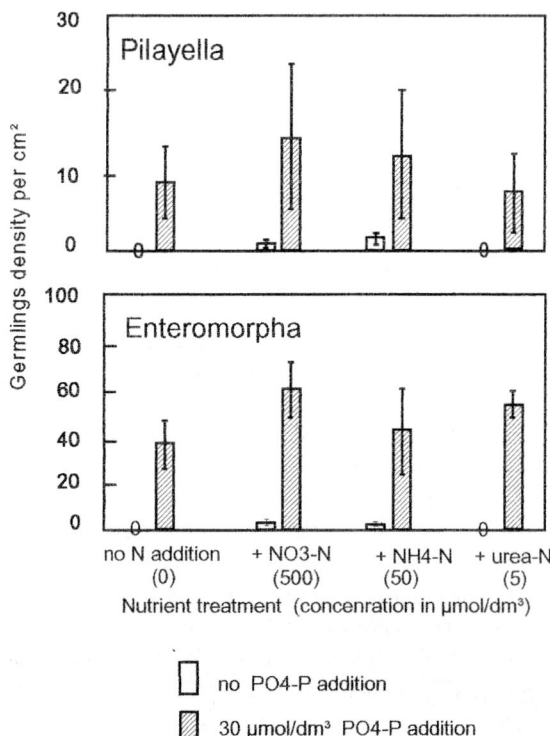

Figure 5. Germination of spores of *Pilayella littoralis* and *Enteromorpha* spp. settled on *Mytilus* shells under different temperatures fully crossed with 3 different irradiances. Number of germlings (mean + SE, $n = 5$) after 7 days cultivation in the laboratory. Daylength 14:10, light:dark (from Schramm et al., 1996).

Figure 6. Effects of single pulse P and N enrichment on germination of spores of *Pilayella littoralis* and *Enteromorpha* spp. Shells of *Mytilus* with attached overwintering spores were used as the source for germination. Density of germlings (number cm^{-2}) was counted after 10 days of cultivation (mean + SE, $n = 5$) Nutrient enrichment: 30 μmol dm^{-3} PO$_4$-P, 500 μmol dm^{-3} NO$_3$-N, 5 μmol dm^{-3} Urea-N (modified after Schramm et al., 1996).

Enteromorpha germlings, thereby indirectly favouring *Pilayella* and other macroalgal species. Exclusion of herbivores resulted in a complete dominance of *Enteromorpha*, preventing all other algae from colonising. When dormant propagules were excluded (by sterilisation of substrates), recruitment depended on new reproduction, which started in May. At this time, herbivores were already active, preventing *Enteromorpha* from settling and growing, and thereby supporting the dominance of *Pilayella*. Co-existence of both algae was observed when dormant propagules and herbivores were excluded.

Implications on the ecosystem level and aspects of eutrophication management

In summary, the changes in benthic vegetation due to eutrophication are the result of a reaction chain of direct and indirect effects, feedback mechanisms and self-accelerating processes, which are difficult to control once initiated. Duarte (1995) uses the term 'domino effect', meaning that the eutrophication process once started is maintained and amplified until the final phase of eutrophication is reached, which includes the extinction of the macrophytobenthic communities.

The progressive repression of slow-growing perennials and final replacement by fast-growing ephemeral forms alters the dynamics and pattern, but not necessarily the amount of total primary production of a system. Seagrasses and long-lived macroalgae may compensate for their slower growth rates by having much higher biomass per unit area compared to plankton and short-lived macroalgae. Probably of more importance is the fact that long-lived large macrophytes are essential components of the spatial structure of shallow water ecosystems; they provide a stable substratum for epiphytic communities, habitats for animals (physical support, shelter, spawning grounds), and strongly influence the hydrodynamic conditions by reducing turbulence and currents, resuspension and transport of sediments. Another important consequence of the change from long-lived to short-lived vegetation is the increased carbon and nutrient turnover rates, and the associated heterotrophic activities with resulting changes in energy transfer within the ecosystem.

The present knowledge of the responses of marine vegetation to eutrophication and the associated changes in the structure and function of the ecosystem, is sufficient to make general qualitative predictions and a few attempts have been made to develop simple quantitative models, e.g. De Vries et al. (1996) and Menesguen and Salomon (1988). However, we are still far from developing predictive quantitative models, in particular non-linear functional models which account for the high complexity of the eutrophication process due to feedback effects and self-acceleration. Compared to our knowledge of the effects of increased nutrient loading on marine vegetation, there is much less information available about the consequences of reduced nutrient loading, or whether changes in benthic vegetation due to eutrophication are reversible. A simple simulation model to show the time scales involved in the recovery of populations of different seagrasses in denuded areas has been used by Duarte (1995). However, comprehensive, more complex models, that would allow quantitative prediction of both eutrophication and recovery processes do not exist, as far as is known. The development of models that could provide measures for eutrophication control and management is a social demand and thus a challenge for future research.

Acknowledgements

This work was carried out under contract no. EV5V-CT93-0290.

References

Breuer G, Schramm W (1988) Changes in macroalgal vegetation in Kiel Bight (western Baltic) during the past 20 years. Kieler Meeresforsch. Sonderheft 6: 241–255.

De Casabianca ML, Laugier T, Mariho-Soriano E (1997) Seasonal changes of nutrients in water and sediment in a Mediterranean lagoon with shellfish farming activity (Thau lagoon-France). ICES J. mar. Sci. 54: 905–916.

De Casabianca ML, Laugier T, Soriano E, Posada F, Rigollet V, Pryor M (1996) Analysis of the principal macrophyte systems, *Zostera, Ulva and Gracilaria*, in a eutrophicated lagoon (Thau, France). In Rijstenbil JW, Kamermans P, Nienhuis PH (eds), EUMAC Synthesis Report and Proceedings. 2nd EUMAC Workshop, Sete, France: 139–175.

De Casabianca M-L, Posada F (1998) Analysis of environmental parameters responsible for growth of *Ulva rigida* (Thau lagoon, France). Bot. mar. 41: 157–165.

De Vries I, Hopstaken CF (1984) Nutrient cycling and ecosystem behaviour in a saltwater lake. Neth. J. Sea Res. 18: 221–245.

De Vries I, Philippart CJM, DeGroodt EG, Van der Tol MWM (1996) Coastal eutrophication and marine vegetation: a model analysis. In Schramm W, Nienhuis PH (eds), Marine Benthic Vegetation: Recent Changes and the Effects of Eutrophication. Springer, Heidelberg, Berlin: 79–113.

Dion P, Le Bozec S (1996) Eutrophication on the European coasts: the French Atlantic coasts. In Schramm W, Nienhuis PH (eds), Marine Benthic Vegetation: Recent Changes and the Effects of Eutrophication. Springer, Berlin, Heidelberg: 251–264.

Duarte CM (1995) Submerged aquatic vegetation in relation to different nutrient regimes. Ophelia 41: 87–112.

Foldager-Pedersen M, Borum J (1997) Nutrient control of estuarine algae: growth strategy and the balance between nitrogen requirements and uptake. Mar. Ecol. Prog. Ser. 161: 135–163.

Kautsky N, Kautsky H, Kautsky U, Waern M (1986) Decreased depth penetration of *Fucus vesiculosus* (L.) since the 1940s indicates eutrophication of the Baltic Sea. Mar. Ecol. Progr. Ser. 28: 1–8.

Kautsky N, Wallentinus I (1980) Nutrient release from a Baltic *Mytilus*-red algal community and its role in benthic and pelagic productivity. Ophelia (Suppl.) 1: 17–30.

Lotze HK (1998) Population dynamics and species interactions in macroalgal blooms: abiotic versus biotic control at different lifecycle stages. Dissertation, University Kiel, Germany. Ber. Inst. Meereskunde Nr. 303. 134 pp.

Lotze HK, Schramm W, Worm B (1998) Control of macroalgal blooms at early developmental stages: *Pilayella littoralis* versus *Enteromorpha* spp. Oekologia (in press).

Menesguen A (1992) Modeling coastal eutrophication: the case of the French *Ulva blooms*. In Vollenweider RA, Marchetti R, Viviani R (eds), Marine Coastal Eutrophication. Proc. Int. Conf. Bologna, Italy, 1990. Elsevier, Amsterdam: 979–992.

Menesguen A, Salomon JC (1988) Eutrophication modelling as a tool for fighting against *Ulva* mass blooms. In Schrefler BA, Zienkiewicz OC (eds), Computer Modelling in Ocean Engineering. Balkema, Rotterdam: 443–264.

Neundorfer JV, Kemp WM (1993) Nitrogen versus phosphorus enrichment of brackish waters: response of submerged plant *Potamogeton perfoliatus* and its associated algal community. Mar. Ecol. Progr. Ser. 94: 71–82.

Nienhuis PH (1989) Eutrophication of estuaries and brackish lagoons in the south-west Netherlands. In Hooghart JC, Posthumus CWS (eds), Hydroecological Relations in the Delta Waters of the South-West Netherlands. TNO Com. Hydrol. Res. Proc. Infrom. 41: 49–70.

Nienhuis PH (1992) Ecology of coastal lagoons in the Netherlands (Veerse Meer and Grevelingen). Vie Milieu 42: 59–72.

Officer CB, Smayda JH, Mann R (1982) Benthic filter feeding: a natural eutrophication control. Mar. Ecol. Progr. Ser. 9: 203–210.

Piriou JY, Menesguen A (1992) Environmental factors controlling the *Ulva* sp. blooms in Britanny (France). In Colombo G, Ferrari I, Ceccherelli VU, Rossi R (eds), Marine Eutrophication and Population Dynamics. 25th Eur. Mar. Biol. Symp. Oben & Olsen, Milwaukee: 117–122.

Sand-Jensen K, Borum J (1991) Interactions among phytoplankton, periphyton, and macrophytes in temperate freshwater and estuaries. Aquat. Bot. 41: 137–175.

Schories D (1995) Population ecology and mass development of *Enteromorpha* spp. (Chlorophyta) in the Wadden Sea intertidal at the island of Sylt (North Sea). Ber. Inst. Meereskunde, Univ. Kiel. No. 271. 145 pp.

Schramm W, Nienhuis HP (eds) (1996) Marine Benthic Vegetation: Recent Changes and the Effects of Eutrophication. Springer, Berlin, Heidelberg, 470 pp.

Schramm W, Lotze H, Schories D (1996) Eutrophication and macroalgal blooms in inshore waters of the German Baltic coasts: The Schlei Fjord, a case study. In Rijstenbil JW, Kamermans P, Nienhuis PH (eds), EUMAC Synthesis Report and Proceedings. 2nd EUMAC Workshop, Sete, France: 18–73.

Sfriso A, Marcomini A (1996a) Italy – The lagoon of Venice. In Schramm W, Nienhuis PH (eds), Marine Benthic Vegetation: Recent Changes and the Effects of Eutrophication. Springer, Heidelberg, Berlin: 339–368.

Sfriso A, Marcomini A (1996b) Macrophytes and nutrient cycles in the lagoon of Venice. In Rijstenbil JW, Kamermans P, Nienhuis PH (eds), EUMAC Synthesis Report and Proceedings. 2nd EUMAC Workshop, Sete, France: 221–248.

Sfriso A, Marcomini A (1996c). Decline of *Ulva* growth in the lagoon of Venice. Bioresource Technology 58: 299–307.

Twilley RR, Kemo WM, Staver KW, Stevensen JC, Boynton WR (1985) Nutrient enrichment of estuarine submerged vascular plant communities: I. Algal growth and effects on production of plants and associated communities. Mar. Ecol. Progr. Ser. 23: 179–191.

Vogt H, Schramm W (1991) Conspicuous decline of *Fucus* in Kiel Bay (western Baltic): what are the causes? Mar. Ecol. Progr. Ser. 69: 189–194.

Wallentinus I (1984) Comparison of nutrient uptake rates for Baltic macroalgae with different thallus morphologies. Mar. Biol. 80: 215–225.

Modelling complexation and electrostatic attraction in heavy metal biosorption by *Sargassum* biomass

Silke Schiewer
Department of Biology, Hong Kong Baptist University, Hong Kong (phone 852-23395893; fax 852-23395995)

Received 1 April 1998; revised 1 December 1998; accepted 8 December 1998

Key words: biosorption, seaweeds, ion exchange, pH, ionic strength, heavy metals

Abstract

Biosorption, the passive accumulation of metal ions by biomass, can be used for purifying metal bearing wastewater. Seaweeds represent a readily available source of biosorbent material that possesses a high metal binding capacity. For example, *Sargassum* can accumulate 2 mequiv of Cd per gram of biomass i.e. 10% of its dry weight. Binding of Cd and Cu by *Sargassum* is an ion exchange process involving both covalent and ionic bonds. The amount of cations bound covalently or by complexation can be predicted using multi-component sorption isotherms involving 2 types of binding sites, carboxyl and sulphate. A Donnan model was used to account for the effect of ionic strength and electrostatic attraction. The use of a multi-component isotherm that included one term for Na binding was less appropriate than the Donnan model for modelling ionic strength effects. It was possible to predict metal and proton binding as a function of the pH value, metal concentration and ionic strength of the solution.

List of Abbreviations

B_t	total number of binding sites B	mequiv g^{-1}
C_t	total number of carboxyl sites C	mequiv g^{-1}
I	ionic strength	mM
K_j	lumped equilibrium constant for binding of ion j	L mmol^{-1}
K_{Cj}	equilibrium constant for binding of ion j to C-sites	L mmol^{-1}
K_{Sj}	equilibrium constant for binding of ion j to S-sites	L mmol^{-1}
m	mass of biosorbent	g
pK_a	–log of acid dissociation constant	–
q_j	total amount of cation j bound (covalent + electrostatic) to all binding sites	mequiv g^{-1}
S_t	total amount of sulphate sites S	mequiv g^{-1}
V	volume of solution	L
V_m	specific cation binding volume per dry weight of biosorbent	mL g^{-1}
Y_v	fitting parameter for V_m	mL mequiv^{-1}
λ	concentration factor (intraparticle / bulk)	–
[ ]	concentration of molecular species in brackets	mM

Subscript indices: add, added for pH adjustment; app, apparent; i, initial; p, average intraparticle
Molecular species:
C^-, CH, CM$_{0.5}$ carboxyl groups ionised or bound to proton (H) or bound to metal (M)
S^-, SH, SM$_{0.5}$ sulphate groups ionised or bound to proton (H) or bound to metal (M)
H, proton; M, divalent metal ion e.g. Cu, Cd; Na, sodium ion.

Introduction

In recent years, biosorption research, which focuses on using readily available biomass that can passively accumulate heavy metals, has received growing attention. This process can be applied as a cost effective way of polishing industrial wastewater whereby drinking water quality can be reached.

Marine algae have been the focus of numerous biosorption studies and their excellent metal binding capacity has been well documented (Ramelow et al., 1992; Holan et al., 1993). The algal cell wall plays an important role in metal accumulation (Crist et al., 1988) as confirmed by TEM observations (Kuyucak & Volesky, 1989). The main components of the brown algal cell wall are cellulose, as the fibrous skeleton, alginate and fucoidan which constitute the amorphous matrix and extracellular mucilage (Lee, 1980; South & Whittick, 1987). Of these components, alginate contains carboxyl groups while fucoidan has sulphate groups, both of which are known to be able to form complexes with metals (Buffle, 1988). Kuyucak & Volesky (1989) speculated that alginate may be one of the main compounds involved in brown algal metal accumulation and more recently, Fourest and Volesky (1996) have confirmed the importance of the alginate carboxyl groups; after blocking the weakly acidic (carboxyl) groups with propylene oxide, metal binding was reduced by 80–95%. These authors also found the alginate content of *Sargassum* to be 40% of its dry weight, which corresponds to 2.25 mequiv g^{-1}, and the amount of sulphate esters was quantified as 0.27 mequiv g^{-1}.

While numerous publications have reported on screening the biosorptive properties of different types of biomass (Volesky & Holan, 1995), few have focused on developing methods to predict metal binding under different conditions (e.g. metal concentrations, pH, ionic strength). The majority of biosorption data is interpreted using simple Langmuir or Freundlich isotherms that can only reflect the influence of metal concentration on the uptake of that one particular metal, without any reference to such important parameters as pH, ionic strength, or the presence of other metals in solution.

It has long been recognised that the binding of most divalent metals (except for 'soft' or type 'b' metals or those that occur as complexes in solution) increases with increasing pH (Ferguson & Bubela, 1974; Tsezos & Volesky, 1981; Ramelow et al., 1992; Holan et al., 1993). This has been explained as an effect of decreasing competition with protons for the same binding sites (Greene et al., 1987; Crist et al., 1994). Only recently has a model for predicting this effect been developed; for example, the data for Cd binding were well predicted by the model (Schiewer & Volesky, 1995).

Fewer studies describe the influence of ionic strength on metal binding; it is generally reduced with increasing ionic strength (Ferguson & Bubela, 1974; Greene et al., 1987; Ramelow et al., 1992). A number of models have been developed for predicting the effect of ionic strength on metal binding by humic and fulvic acids (Bartschat et al., 1992; Tipping, 1993; Westall et al., 1995; Kinniburgh et al., 1996). The Donnan model has been used to interpret (but not predict) ionic strength effects in polyelectrolytes (Marinsky, 1987). For biosorption, ionic strength effects have only recently been taken into account using the Donnan model (Schiewer & Volesky, 1997a, b).

The objectives of this work were to investigate the effect of pH and ionic strength on the binding of heavy metals and protons and to provide a model able to predict these effects. The advantages of this model, compared to more conventional approaches such as the Langmuir isotherm (which is not pH sensitive) and a multi-component isotherm, which includes a term for Na binding (as a means to model ionic strength effects), shall be discussed.

Materials and methods

Beach-dried *Sargassum fluitans*, collected from Naples, Florida, was ground and sieved. The size fraction 0.84–1.0 mm was protonated in 0.1 N HCl (10 g L^{-1}), washed 10 times in the same volume of distilled deionized water, and dried in an oven at 60–80 °C. Even if auto-dissociation of some acidic groups may have occurred, it can be assumed that initially the proton binding (covalent + electrostatic) equalled the total number of binding sites. Since no other cation was present, the negative charge of any dissociated acidic groups must have been balanced by electrostatically bound protons.

The sulphate salts of $3CdSO_4 \cdot 8\,H_2O$ (AESAR) and $CuSO_4 \cdot 5\,H_2O$ (ACP Chemicals) were dissolved in distilled deionized water. Biomass (0.1–1 g) was contacted with 50 mL of metal containing solution in 125-mL Erlenmeyer flasks on a gyrotory shaker (New Brunswick Scientific, Model G2) at 2 Hz for 12 h. For experiments involving ionic strength effects, $NaNO_3$ was added, increasing the ionic strength up to 5 M in the experiments without metals. In the metal binding experiments either zero or 100 mM $NaNO_3$ were added, which is referred to as low and high ionic strength, respectively. The pH was adjusted to the desired value by frequent addition of NaOH or HCl during the biosorption process until the pH was stable. Blanks, duplicates and controls were run as appropriate. During biosorption, the biomass appeared stable, with no discoloration of the solution observed, indicative of leaching of biomass components, except at higher pH in the absence of metals. Any easily soluble polysaccharides were probably already eliminated during pre-treatment of the biomass.

Initial and final metal concentrations in solution were determined using an atomic absorption spectro-

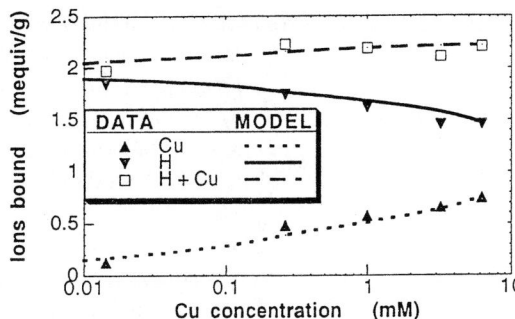

Figure 1. Biosorption isotherm for Cu and proton binding by protonated *Sargassum* biomass at pH 3: experimental data and the 2-site model.

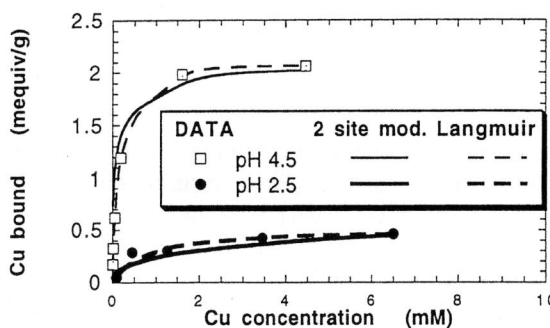

Figure 2. Biosorption isotherm for Cu binding by protonated *Sargassum* biomass at pH 4.5 and pH 2.5: experimental data, the Langmuir model and the 2-site model.

photometer (Thermo Jarrel Ash, Model Smith Hieftje II) calibrated with standard solutions prepared from 1000 mg L^{-1} certified reference solutions (Fischer). The metal bound during biosorption was calculated as:

$$q_M = 2\,([M]_i - [M])\,V\,m^{-1} \quad (\text{mequiv g}^{-1}) \qquad (1)$$

where V is the liquid volume, m the mass of the sorbent, $[M]_i$ and $[M]$ the initial and final metal concentration in solution, respectively. The factor 2 converts mmol into mequiv. The change in proton binding was calculated from the difference between the amounts of protons added $[H]_{add}\,V_{add}$ (negative values were used when base was added) and the amount of protons that accumulated in solution $([H]-[H]_i)V$

$$\Delta q_H = ([H]_{add}\,V_{add} - ([H]-[H]_i)V)\,m^{-1}$$

$$(\text{mmol g}^{-1}) \qquad (2)$$

Results and discussion

Ion exchange and pH effects: observations

The binding of Cu and protons by *Sargassum* at pH 3 is depicted in Figure 1. The amount of Cu bound to the biomass increases with increasing concentration of Cu in the solution. However, the amount of protons bound decreases with increasing Cu concentration. The sum of both Cu and protons bound remains almost constant over the whole range of Cu concentrations, as also indicated in Figure 1. This behaviour can be interpreted as an ion exchange process. While H initially saturates almost all binding sites it becomes progressively replaced by Cu with increasing Cu concentration, whereby 2 H$^+$ ions are released into the solution for each Cu^{2+} ion bound. The charge of the biomass therefore remains balanced.

The slight increase in total cation binding (protons and Cu) with increasing metal concentration can be explained and modelled as a decreasing small number of 'free' binding sites that are bound neither to protons or Cu, but whose negative charge is balanced by an electrostatic attraction of Na (which had been added for pH adjustment). At higher pH, e.g. pH 4.5, a larger number of free sites are present. When cation binding is modelled in terms of chemical binding constants, Na binding can be neglected unless present at high concentrations since it only binds weakly through electrostatic attraction and does not compete significantly with the binding of Cu and H. Electrostatic binding of Na will be considered below, in the modelling of ionic strength effects and electrostatic attraction.

Figure 2 depicts Cu binding at pH 4.5 and 2.5. At pH 4.5, Cu binding sharply increases with Cu concentration up to a maximum value of 2 mequiv g^{-1} at about 2 mM Cu. Further increase in the Cu concentration does not lead to an increasing amount of bound Cu, since all binding sites are already saturated with Cu. The binding of Cu at pH 2.5 is considerably lower than that at pH 4.5. This is a result of the above mentioned ion exchange between Cu and H$^+$. Since both Cu and H$^+$ compete for the same binding sites, higher concentrations of each ion can displace the other ion from the binding sites. Hence, at low pH, i.e. high proton concentrations, the binding of protons will increase and lead to reduced Cu binding for any given concentration of Cu.

Langmuir isotherm model

The shape of individual biosorption isotherms can be described by the Langmuir model (Langmuir, 1918):

$$q_{Cu} = K_{Cu} B_t [Cu] / (1 + K_{Cu} [Cu])$$

$$(\text{mequiv g}^{-1}) \qquad (3)$$

B_t is the total number of binding sites and equals 2.1 mequiv g^{-1} and K_{Cu} is the equilibrium constant for Cu binding and equals 8.2 L mmol^{-1}. The constants for the Langmuir isotherm determined for the Cu binding data at pH 2.5 are $B_t = 0.51$ mequiv g^{-1} and $K_{Cu} = 1.4$ L mmol^{-1}. Both values are lower than the ones for pH 4.5, reflecting the lower Cu binding at pH 2.5. The absolute mean square deviations of the model from the experimental data expressed in percent of the total binding capacity were 2.3 and 1.9% for pH 4.5 and 2.5, respectively.

However, the interpretation of the parameters for the Langmuir model is of questionable value since the number of binding sites in the biomass does not actually change with pH. Rather, the sites are less available for Cu binding when they are already occupied with H$^+$, as is the case at pH 2.5. Yet, as it is the Cu concentration that determines whether protons already occupying the sites are displaced by Cu or not, it would be more appropriate to relate the Cu binding to the true number of binding sites, which remains constant and does not change with the pH value. In that case, it is possible to account for protons occupying the binding sites by referring to the proton binding constant. Moreover, the Langmuir model assumes metal binding to free binding sites rather than ion exchange. For biosorption which involves ion exchange, it is therefore less appropriate (Crist et al., 1994).

Two site model for predicting ion exchange and pH effects

An isotherm for ion exchange, taking proton binding into account, was proposed by Schiewer and Volesky (1995). It assumes two types of binding sites, carboxyl (C) and sulphate (S) groups to be present in the biomass. The reactions for the binding of protons and metal ions to these sites are:

$$H + C^- \rightleftharpoons CH$$

$$K_{CH} = [CH] / ([H] [C^-]) \qquad (4)$$

$$H + S^- \rightleftharpoons SH$$

$$K_{SH} = [SH] / ([H] [S^-]) \qquad (5)$$

$$M^{2+} + 2 C^- \rightleftharpoons 2 CM_{0.5}$$

$$K_{CM} = [CM_{0.5}]^2 / ([M] [C^-]^2) \qquad (6)$$

$$M^{2+} + 2 S^- \rightleftharpoons 2 SM_{0.5}$$

$$K_{SM} = [SM_{0.5}]^2 / ([M] [S^-]^2) \qquad (7)$$

Divalent metal ions are assumed to bind to two monovalent C$^-$ or S$^-$ sites respectively. The binding sites are either free (ionised) or occupied by protons or metal ions:

$$C_t = C^- + CH + CM_{0.5} \quad (\text{mequiv g}^{-1}) \qquad (8)$$

$$S_t = S^- + SH + SM_{0.5} \quad (\text{mequiv g}^{-1}) \qquad (9)$$

Since at this point only *chemical* binding (covalent or complexation) is considered, Na which is only *electrostatically* attracted to the free sites is disregarded. It will be considered later in the modelling of electrostatic effects. From the above 6 equations (4–9), Schiewer & Volesky (1995) derived an isotherm equation that can predict the binding of protons and metal ions as a function of metal concentration and pH.

$$q_H = CH + SH$$

$$= C_t (K_{CH} [H]) / (1 + K_{CH} [H] + (K_{CM} [M])^{0.5})$$

$$+ S_t (K_{SH} [H]) / (1 + K_{SH} [H] + (K_{SM} [M])^{0.5})$$

$$(\text{mequiv g}^{-1}) \qquad (10)$$

$$q_M = [CM_{0.5}] + [SM_{0.5}]$$

$$= C_t (K_{CM} [M])^{0.5} / (1 + K_{CH} [H] +$$

$$+ (K_{CM} [M])^{0.5})$$

$$+ S_t (K_{SM} [M])^{0.5} / (1 + K_{SH} [H] +$$

$$+ (K_{SM} [M])^{0.5})$$

$$(\text{mequiv g}^{-1}) \qquad (11)$$

This two site model has successfully been used by Schiewer & Volesky (1995) to predict the binding of Cd at different pH values whereby the constants characterising the acidic groups in the same biomass are $C_t = 2$ mequiv g^{-1}, $S_t = 0.25$ mequiv

g^{-1}, $K_{CH} = 10^{4.8}$ (i.e. $pK_a = 4.8$) and $K_{SH} = 10^{2.5}$ (i.e. $pK_a = 2.5$) (Schiewer & Volesky, 1995). These four constants can be used for any metal binding to these sites. The only parameters that had to be determined specifically for Cu are the Cu binding constants with respect to the two sites $K_{CCu} = 20 \times 10^4$ L mol^{-1}, $K_{SCu} = 0.4 \times 10^4$ L mol^{-1}.

The number and pK_a of carboxyl sites had been determined by pH titration of the same biomass at low ionic strength (no Na salt added) which indicated that most sites will be free and easily available for metal binding at pH > 4.8. The number and pK_a of sulphate groups had been determined from metal binding data at very low pH (1–2.5) (Schiewer & Volesky, 1995). The amount of carboxyl groups was confirmed by extraction of alginate (45% of dry weight, ~ 2.25 mequiv g^{-1}) and the postulated quantity of sulphate groups matched the total amount of sulphate in the same biomass (0.27 mequiv g^{-1}) as determined by Fourest & Volesky (1996). It has to be noted that the pK_a values determined here are apparent ones which are much higher than the intrinsic pK_a values because of the contribution of electrostatic effects to proton binding. The relevance of electrostatic effects will be discussed below.

The maximum Cu binding in mequiv g^{-1} corresponds well to the total number of sulphate and carboxyl groups. This supports the assumption that one divalent ion binds to two binding sites. As shown in Figures 1 and 2, the two site model also describes the influence of pH on the binding of Cu. With the above 6 constants, the Cu binding can be predicted for any pH value or Cu concentration with the absolute mean square deviations of the model being 17 and 2.4% of the total binding capacity for pH 4.5 and 2.5, respectively. Though the errors are higher than for the Langmuir model, the advantage of the 2-site model is that it does not require the determination of a separate set of constants for each pH value. It should be noted that even though the Langmuir model can be used to describe the metal binding at a given pH value, the parameters for the model, i.e. the metal binding constant and the total number of binding sites have to be determined anew for each pH value, i.e. the pH effect can not be *predicted* with the simple Langmuir model. A model such as the Langmuir model, naturally fits better if the constants are determined individually for each isotherm but that only means that the model has less generality, not that it is the better model.

Electrostatic attraction and ionic strength effects: general considerations

Despite the success of the 2 site model in predicting pH effects in metal binding, it is only based on chemical equilibrium constants and does not take into account that the negatively charged sites (such as the carboxyl group C$^-$), that are not bound to either metals or protons, will lead to an electrostatic attraction of cations. The negative charge of these sites has to be balanced by H$^+$, heavy metal ions (Cd^2, Cu^2) or Na$^+$ (added as NaOH for pH adjustment), i.e. the fixed negative charges are largely screened by positively charged counterions. Electrostatic attraction will be especially prominent at low heavy metal concentrations and high pH values when a large number of the sites are ionised.

The contribution of electrostatic attraction to the binding of protons and divalent metal ions has been investigated by varying the ionic strength through the addition of NaNO$_3$ salt. Since Na is only bound electrostatically, it can only compete or interfere with the electrostatic (not covalent) binding of protons and divalent metal ions. In this way, it is possible to discern to what extent electrostatic and covalent binding respectively contribute to the binding of these ions. In the above equations it was assumed that the amount of metal or protons bound (q_M, q_H) is equal to the amount that is bound covalently (CH, CM$_{0.5}$), i.e. electrostatic binding was neglected. Taking both covalent and electrostatic binding into account, the proton and metal binding to carboxyl groups (C) are:

$$q_H = CH + ([H]_p - [H]) V_m \quad (\text{mequiv g}^{-1}) \quad (12)$$

$$q_M = CM_{0.5} + 2 ([M]_p - [M]) V_m$$

$$(\text{mequiv g}^{-1}) \quad (13)$$

where [H]$_p$ and [M]$_p$ are the concentrations of mobile protons or divalent metal ions in the biomass particle (not including the amount covalently bound) and V_m is the cation binding volume per mass of biosorbent. The amount of electrostatic binding equals the concentration difference between intraparticle (subscript p) and bulk concentration (no subscript) multiplied by the specific cation binding volume V_m in the biomass particle. For divalent metals the factor 2 is needed for conversion from mol to mequiv.

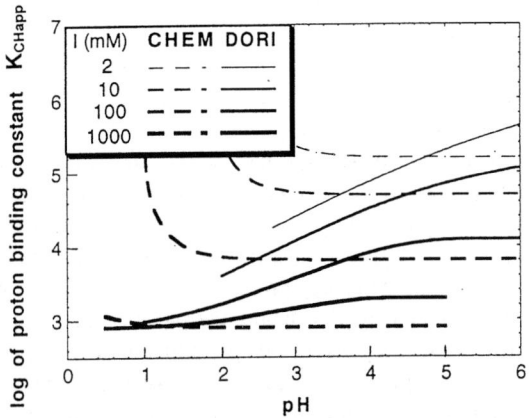

Figure 3. Variation of the apparent proton binding constant with pH for protonated *Sargassum* biomass at different ionic strengths: model predictions of the isotherm that assumes covalent binding of Na (CHEM model) and the Donnan model for constant particle volume (DORI model).

Electrostatic effects in protons binding

With increasing ionic strength ($NaNO_3$) more of the negative charge of the free binding sites is balanced by Na. Therefore, less H^+ ions are in the immediate vicinity of the binding sites. This also reduces the amount of protons covalently bound to the binding sites since the amount of H^+ bound is in equilibrium with the H^+ concentration directly at the binding sites. Due to electrostatic attraction of H^+, this concentration near the binding sites is higher than the concentration of H^+ in the bulk solution. In this context it is necessary to distinguish between apparent and intrinsic binding constants. The constants used for the two site model relating to Figures 1 and 2 are apparent binding constants since both electrostatic and covalent binding are lumped into one constant. For simplicity, the modelling of electrostatic effects will consider only one binding site C since carboxyl groups are responsible for 90% of the metal binding capacity of *Sargassum* biomass. Taking sulphate groups into account is only necessary if they significantly contribute to total metal binding which is only the case at very low pH when carboxyl groups are all protonated. At pH 3 these are not all protonated and constitute the main metal binding site. Moreover, the parameter optimisation would not yield a clear optimum if more constants are to be determined simultaneously.

The apparent proton binding constant is related to the proton concentration [H] in the bulk of the solution and is defined as

$$K_{CHapp} = CH / (C^- [H]) \quad (14)$$

The intrinsic binding constants relate to the higher intraparticle concentration of the cations near the negatively charged binding site, i.e. in this case to $[H]_p$

$$K_{CH} = CH / (C^- [H]_p) \quad (15)$$

The relation between the apparent and intrinsic proton binding constants is therefore

$$K_{CHapp} = K_{CH} [H]_p / [H] \quad (16)$$

The concentration factor $[H]_p/[H]$ can be modelled according to the Donnan equilibrium whereby the concentration of any ion is assumed to be homogeneous throughout the biomass particle and the negative charge of the biomass is balanced by counterions whose concentration factor is

$$\lambda = [H]_p/[H] = [Na]_p/[Na] = ([M^{2+}]_p/[M^{2+}])^{0.5} \ (-) \ (17)$$

One main factor determining λ is the ionic strength. λ decreases with increasing ionic strength, approaching the value 1.0. Note that even if at high ionic strengths the concentrations of H and M in the particle approach the respective concentrations in the bulk (i.e. if there is negligible electrostatic binding) there will still be covalent binding of both H and M. The concentration factor λ is also strongly dependent on the number of charges per unit volume in the biomass since it is the negative charge of the biomass that leads to the electrostatic accumulation of cations in the biomass that is reflected in $\lambda > 1$. The number of charges per unit volume (mol L^{-1}) equals the number of free ionised binding sites C^- divided by the particle volume V_m. From equation (17) and the condition for charge neutrality one can derive

$$\lambda = C^-/(2 I V_m) + ((C^-/2 I V_m)^2 + 1)^{0.5} \ (-) \ (18)$$

For *Sargassum* biomass the charge density and intrinsic proton binding constant have been determined by Schiewer & Volesky (1997a). The cation binding volume V_m equals 1.4 mL g^{-1} (if swelling is neglected, i.e. for the DORI model), the number of carboxyl binding sites is $C_t = 2.1$ mmol g^{-1} and their intrinsic pK_a value is 2.9. This is similar to the pK_a values 3.38 and 3.65 determined by Haug (1961) for mannuronic and guluronic acid monomers, respectively. From these values, the variation of the apparent proton binding constant (16) was calculated. The results are shown as the solid lines in Figure 3. The apparent pK_a value decreases with increasing ionic strength because at high ionic strength (I), Na balances

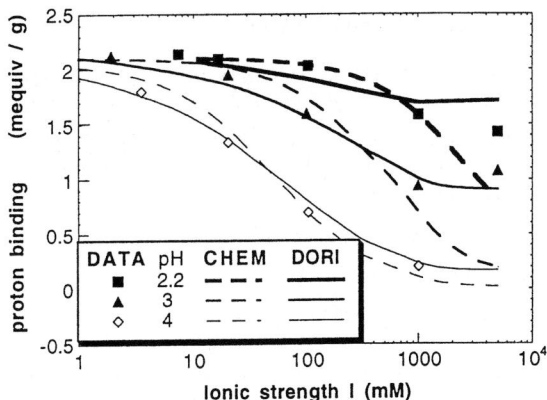

Figure 4. Proton binding by protonated *Sargassum* biomass as a function of ionic strength: experimental data and predictions of the isotherm that assumes covalent binding of Na (CHEM model) and the Donnan model for constant particle volume (DORI model).

most of the negative charge of the biomass. Consequently, the concentration factor $[H]_p/[H]$ is low. The other observable trend is that with increasing pH the apparent pK_a value increases because more sites become negatively charged due to releasing their protons. Therefore electrostatic attraction becomes more pronounced. This parallels the findings of the classical work of Marinsky and co-workers who observed an increase of the apparent pK_a with increasing pH and decreasing ionic strength for a variety of polyelectrolytes including synthetic resins, humic and fulvic acids, as well as alginate (Marinsky, 1987; Lin & Marinsky, 1993).

Isotherm model for ionic strength effects

Instead of using the Donnan model to account for ionic strength effects, one might consider treating Na as one of the cations competing for the same binding sites as protons and heavy metal ions. This approach was suggested by Westall et al. (1995) and would provide an isotherm that includes Na instead of, or in additional to, M^{2+} (equation 10)

$$q_H = C_t K_{CH} [H]/(1 + K_{CH} [H] + K_{CNa} [Na])$$

$$(\text{mequiv g}^{-1}) \qquad (19)$$

This equation (here referred to as the CHEM model) will also predict that fewer protons or divalent metals are bound to the biomass at high ionic strength, as the dashed lines in Figure 3 show. The pK_a value according to this model decreases with increasing ionic strength. However, the isotherm model (CHEM) which includes Na is not appropriate for two reasons. First, it does not reflect the commonly acknowledged electrostatic nature of Na (a 'hard' ion) binding to carboxyl groups and second, the isotherm model including Na would predict that Na can completely displace all protons from the biomass if it is present at high enough concentrations, as shown in the dashed lines of Figure 4. The experimental data in this figure rather tend to level out for the proton binding at high ionic strength. This levelling out is in accordance with approaches like the Donnan model combined with intrinsic proton binding constants that are based on electrostatic binding of Na. Electrostatically binding Na can only compete with the *electrostatic* binding of protons or heavy metals, not with their covalent binding. The maximum effect that high Na concentrations can have is to reduce the intraparticle concentrations of these ions to the level of their bulk concentrations. However, at these low concentrations, some protons (or metal ions) are still bound covalently. The divergence of the dashed lines of the isotherm (CHEM) model from the solid lines of the Donnan (DORI) model and the experimental data in Figure 4 indicate that this model is inadequate if used over a wide range of ionic strength.

A further difference between the CHEM model and the DORI model is that the former predicts an increase of the apparent pK_a with decreasing pH (i.e. with decreasing degree of ionisation) as shown by the dashed lines in Figure 3. This is not only contrary to the predictions of the DORI model (solid lines in Figure 3) but also contrary to the observed behaviour in many polyelectrolytes (Marinsky, 1987) and also in *Sargassum* (Schiewer & Volesky, 1997a).

Electrostatic effects in the presence of metals

When the Donnan model is applied for metal binding, the concentration factor λ can be calculated from the equation for charge neutrality in the particle, expressing all intraparticle concentrations in terms of their bulk concentrations and the concentration factor λ which yields (Schiewer & Volesky, 1997b)

$$\lambda = -(I-3[M])/(4[M]) + ((I-3[M])^2/(16[M]^2)$$

$$+ (I-[M] + C^-/V_m)/(2[M]))^{0.5} \quad (-) \qquad (20)$$

The calculation of the metal binding starts with evaluating Equation (20) with an estimated number of free carboxyl sites C^-. The value obtained for λ is then used to calculate the intraparticle concentrations ac-

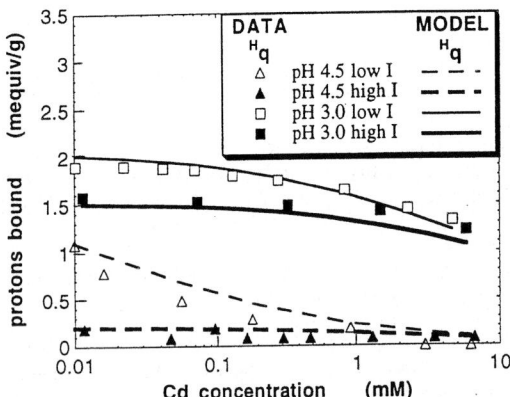

Figure 5. Proton binding by protonated *Sargassum* biomass at different ionic strength I and pH: experimental data and predictions of the Donnan model for swelling particles (DOSW).

cording to Equation (17). Subsequently the isotherm equations (e.g. (10) and (11), with $S_t = 0$ if only carboxyl group are taken into account) are evaluated at the intraparticle concentrations $[M]_p$ and $[H]_p$ instead of $[M]$ and $[H]$. This renders the calculation of C^- according to Equation (8) possible and the above routine is repeated for an iterative calculation of C^- till stable values are obtained, unless C/V_m is a constant, in which case no iterations are necessary. Finally, Equations (12) and (13) are used to add the electrostatic binding to the covalent binding calculated by (10) and (11).

The cation binding volume V_m was observed to increase with an increasing number of the free binding sites, i.e. swelling of the biomass took place at low metal concentrations and/or high pH values as well as high ionic strength (Schiewer & Volesky, 1997b). Since mechanically oriented swelling models (e.g. the one developed by Katchalsky & Michaeli, 1955) are overly complex without even including all relevant effects, swelling was neglected or a simple empirical correlation for the swelling was assumed in the modelling for ionic strength effects in biosorption by *Sargassum*. A linear increase of V_m with the number of free carboxyl sites C^- was assumed, i.e.

$$V_m = Y_V \, C^- \quad (mL\ g^{-1}) \qquad (21)$$

whereby Y_V is a fitting parameter and C^- is the number of free carboxyl groups (that do not covalently bind to any cation). With Equation (21), i.e. for the Donnan model with particle swelling (DOSW), the following equations can be derived by combining the Donnan model (17) with the isotherm equations (10, 11) for one site (Schiewer & Volesky, 1997b)

$$q_H = C_t \, (K_{CH}\,[H] + [H]\,Y_V\,(1-1/\lambda))/((1/\lambda)$$
$$+ K_{CH}\,[H] + (K_{CM}\,[M])^{0.5}) \quad (mequiv\ g^{-1}) \qquad (22)$$

$$q_M = C_t \, ((K_{CM}\,[M])^{0.5} + 2\,[M]\,Y_V\,(\lambda-1/\lambda))/((1/\lambda)$$
$$+ K_{CH}\,[H] + (K_{CM}\,[M])^{0.5}) \quad (mequiv\ g^{-1}) \qquad (23)$$

Figure 5 shows the application of the DOSW model Equation (22) to proton binding at different ionic strength and pH values as a function of the Cd concentration, using the data of Schiewer & Volesky (1997b) which have been previously interpreted assuming a rigid particle (DORI), i.e. constant V_m. Accounting for particle swelling (DOSW model) as done here, yields a better fit than previously achieved by Schiewer and Volesky (1997b). Figure 5 shows that the binding of protons is reduced both with increasing Cd concentration (similar to the above mentioned competition between Cu and H) and with increasing ionic strength (the terms low and high ionic strength, respectively, stand for the addition of zero or 100 mM $NaNO_3$). The effect of the ionic strength is especially pronounced at low Cd concentrations where most sites are ionised. Therefore it is not only competition for the same binding sites which is occurring but additionally a competition between H and Na for electrostatic (ionic) binding, whereby the electrostatic binding of protons is reduced at high ionic strength. The lower intraparticle concentrations lead to a reduction in the covalent binding of protons. This means high ionic strength *indirectly* also reduces the amount of covalent binding of protons or heavy metals.

Conclusions

Metal binding by *Sargassum* is reduced at low pH values due to increasing competition of protons for the same binding sites that metals can use. An ion exchange process takes place whereby 2 H^+ ions are released from the initially protonated biomass for each divalent metal ion bound. This behaviour can be described by a pH sensitive isotherm equation that takes ion exchange into account and is based on the assumption that both protons and metals bind to carboxyl and sulphate groups in the *Sargassum* biomass. This 2-site model successfully predicts the binding of both protons and metal at different pH values and metal concentrations. It has a larger predictive power than the Langmuir model which can predict neither pH effects nor proton binding and which is less suitable for ion exchange.

The binding of protons and metal ions is reduced with increasing ionic strength. This is due to the fact that Na, H^+ and M^{2+} compete for electrostatic binding, all acting as counterions for the negatively charged binding sites in the biomass. The effect of electrostatic attraction is quite significant; the apparent proton binding constant at low ionic strength (no $NaNO_3$ added) can be 2 orders of magnitude higher than the intrinsic one. The Donnan model was successfully used in order to account for the effect of electrostatic attraction on binding of H^+ and M^{2+}. It proved superior to a multicomponent isotherm model where Na^+ was treated as a covalently bound ion which was not appropriate for high ionic strength values.

References

Bartschat BM, Cabaniss SE, Morel FMM (1992) Oligoelectrolyte model for cation binding by humic substances. Envir. Sci. Technol. 26: 284–294.

Buffle J (1988) Complexation Reactions in Aquatic Systems: An Analytical Approach. Ellis Horwood Ltd., Chichester, UK: 156–329.

Crist RH, Martin JR, Carr D, Watson JR, Clarke HJ, Crist DR (1994) Interactions of metals and protons with algae. 4. Ion exchange vs. adsorption models and a reassessment of Scatchard plots; ion-exchange rates and equilibria compared with calcium alginate. Envir. Sci. Technol. 28: 1859–1866

Crist RH, Oberholser K, Schwartz D, Marzoff J, Ryder D, Crist DR (1988) Interactions of metals and protons with algae. Envir. Sci. Technol. 22: 755–760.

Ferguson J, Bubela B (1974) The concentration of Cu (II), Pb (II), and Zn (II) from aqueous solutions by particulate algal matter. Chem. Geol. 13: 163–186.

Fourest E, Volesky B (1996) Contribution of sulphonate groups and alginate to heavy metal biosorption by the dry biomass of *Sargassum fluitans*. Envir. Sci. Technol. 30: 277–282.

Greene B, McPherson R, Darnall D (1987) Algal sorbents for selective metal ion recovery. In: Patterson JW, Pasino R (eds), Metals Speciation, Separation and Recovery, Lewis, Chelsea, MI: 315–338.

Haug A (1961) Dissociation of alginic acid. Acta Chem. Scand. 15: 950–952.

Holan ZR, Volesky B, Prasetyo I (1993) Biosorption of cadmium by biomass of marine algae. Biotechnol. Bioeng. 41: 819–825.

Katchalsky A, Michaeli I (1955) Polyelectrolyte gels in salt solutions. J. Polym. Sci. 15: 69–86.

Kinniburgh DG, Milne CJ, Benedetti MF, Pinheiro JP, Filius J, Koopal LK, Van Riemsdijk WH (1996) Metal ion binding by humic acid: Application of the NICA-Donnan model. Envir. Sci. Technol. 30: 1687–1698.

Kuyucak N, Volesky B (1989) The mechanism of cobalt biosorption. Biotechnol. Bioengng 33: 823–831.

Langmuir I (1918) The adsorption of gases on plane surfaces of glass, mica and platinum. J. Am. Chem. Soc. 40: 1361–1403.

Lee RE (1980) Phycology. Cambridge University Press, Cambridge, UK: 224–225.

Lin FG, Marinsky JA (1993) A Gibbs-Donnan-based interpretation of the effect of medium counterion concentration levels on the acid dissociation properties of alginic acid and chondroitin sulfate. React. Polym. 19: 27–45.

Marinsky JA (1987) A two-phase model for the interpretation of proton and metal ion interaction with charged polyelectrolyte gels and their linear analogs. In Stumm W (ed.), Aquatic Surface Chemistry, Wiley Interscience, John Wiley and Sons, NY: 49–81.

Ramelow GJ, Fralick D, Zhao Y (1992) Factors affecting the uptake of aqueous metal ions by dried seaweed biomass. Microbios 72: 81–93.

Schiewer S, Volesky B (1995) Modelling of the proton-metal ion exchange in biosorption. Envir. Sci. Technol. 29: 3049–3058.

Schiewer S, Volesky B (1997a) Ionic strength and electrostatic effects in biosorption of protons. Envir. Sci. Technol. 30: 1863–1871.

Schiewer S, Volesky B (1997b) Ionic strength and electrostatic effects in biosorption of divalent metal ions and protons. Envir. Sci. Technol. 30: 2478–2485.

South GR, Whittick A (1987) Introduction to Phycology. Blackwell Scientific Publications, Oxford, UK: 61–62.

Tipping E (1993) Modelling the competition between alkaline earth cations and trace metal species for binding by humic substances. Envir. Sci. Technol. 27: 520–529.

Tsezos M, Volesky B (1981) Biosorption of uranium and thorium. Biotechnol. Bioengng 23: 583–604.

Volesky B, Holan ZR (1995) Biosorption of heavy metals. Biotechnol. Progr. 11: 235–250.

Westall JC, Jones JD, Turner GD, Zachara JM (1995) Models for association of metal ions with heterogeneous environmental sorbents: I. Complexation of Co(II) by leonardite humic acid as a function of pH and $NaClO_4$ concentration. Envir. Sci. Technol. 29: 951–959.

Ecological engineering in aquaculture: use of seaweeds for removing nutrients from intensive mariculture

M. Troell[1,2]*, P. Rönnbäck[1], C. Halling[1], N. Kautsky[1,2] & A. Buschmann[3]
[1]*Department of Systems Ecology, Stockholm University, 106 91 Stockholm, Sweden*
[2]*Beijer International Institute of Ecological Economics, Royal Swedish Academy of Sciences, Box 50005, S–10405 Stockholm, Sweden*
[3]*Universidad de Los Lagos, Departemento Acuicultura, Casilla 933, Osorno, Chile*

(*Author for correspondence; fax +46(0)8–158417; e-mail max@beijer.kva.se)

Received 1 April 1998; revised 10 November 1998; accepted 29 November 1998

Key words: ecological engineering, biofilter, aquaculture, seaweeds, mariculture, eutrophication, *Gracilaria*, shrimp farming, mangroves, ecological footprint

Abstract

Rapid scale growth of intensive mariculture systems can often lead to adverse impacts on the environment. Intensive fish and shrimp farming, being defined as throughput-based systems, have a continuous or pulse release of nutrients that adds to coastal eutrophication. As an alternative treatment solution, seaweeds can be used to clean the dissolved part of this effluent. Two examples of successfully using seaweeds as biofilters in intensive mariculture systems are discussed in this paper. The first example shows that *Gracilaria* co-cultivated with salmon in a tank system reached production rates as high as 48.9 kg m^{-2} a^{-1}, and could remove 50% of the dissolved ammonium released by the fish in winter, increasing to 90–95% in spring. In the second example, *Gracilaria* cultivated on ropes near a 22-t fish cage farm, had up to 40% higher growth rate (specific growth rate of 7% d^{-1}) compared to controls. Extrapolation of the results showed that a 1 ha *Gracilaria* culture gave an annual harvest of 34 t (d. wt), and assimilated 6.5% of the released dissolved nitrogen. This production and assimilation was more than twice that of a *Gracilaria* monoculture. By integrating seaweeds with fish farming the nutrient assimilating capacity of an area increases. With increased carrying capacity it will be possible to increase salmon cage densities before risking negative environmental effects like eutrophication and toxic algal blooms sometimes associated with the release of dissolved nutrients. The potential for using mangroves and/or seaweeds as filters for wastes from intensive shrimp pond farming is also discussed. It is concluded that such techniques, based on ecological engineering, seems promising for mitigating environmental impacts from intensive mariculture; however, continued research on this type of solution is required.

Introduction

Aquaculture, according to FAO statistics, is the fastest growing food production sector, with a yearly growth of about 4–11% (FAO, 1997; values for all aquatic products 1988–1995). Without doubt aquaculture will continue to play an important role in the global supply of fish and shellfish in the future. The present concern for continuing deterioration of coastal ecosystems and its subsequent impact on aquaculture and other uses, implies that a 'precautionary principle' be applied to any development activity that might be unsustainable. Aquaculture may contribute to the degradation of the environment, but paradoxically it is still dependent on the supply of clean waters, seed larvae supply and other ecosystem services (Rönnbäck, in press).

Traditional farming systems (e.g. extensive pond farming) dominate aquaculture production in many regions, but these are now slowly being replaced by intensive western oriented techniques (New & Wijkstrom, 1990; Ekins et al., 1994; Weber, 1996). This trend is disquieting, as it will often have negative implications for the environment. Intensive aquaculture systems have been described as 'throughput-based systems' (Daly & Cobb, 1989; Folke & Kautsky, 1992); they depend on large inputs of resources of which only a minor part is taken up by the cultured species, with the rest being released as wastes to the environment (Folke & Kautsky, 1989; Troell, 1996). Intensive farming of carnivorous fish species like yellowtail, seabream, seabass, trout and salmon in cages or tanks, and shrimp in ponds, are examples of such farming systems.

Existing solutions for aquaculture waste treatment have mainly focused on reducing particle load (discussed in Cripps, 1994), leaving the dissolved nutrient fraction untreated. However, the fact that treatment of effluents usually involves a higher degree of technology and thereby also higher costs, suggests that release of untreated water is the rule rather than the exception in intensive mariculture.

The rapid development of shrimp farming has had an adverse impact on the environment (Primavera, 1993), mainly through the conversion of large mangrove areas into shrimp ponds. Intensive shrimp farms have, compared to more extensive systems, much higher production per unit area. Thus, compared to extensive systems less mangrove forest is needed to reach a certain production level. However, outputs of wastes from intensive systems are much higher and the production depends to a larger extent upon external inputs of energy, feeds and chemicals to sustain yields.

Integration of seaweeds with fish and shrimps has been suggested as a means to counteract the release of dissolved nutrients, and at the same time convert it into a useful product (review in Brzeski & Newkirk, 1997; Kautsky et al., 1997a; Troell et al., 1999). The aim of this paper is to outline the possibilities and constraints in integrated systems where seaweeds function as biofilters in fish and shrimp mariculture. Since nitrogen is usually regarded as the limiting nutrient in coastal waters, the focus of this paper will be on nitrogen.

Characterisation of mariculture waste

The quality and quantity of wastes from aquaculture depends mainly on culture system characteristics and the choice of species, but also on feed quality and management (Iwama, 1991). The impacts on the environment will then depend on hydrodynamic conditions and the sensitivity of the receiving ecosystem. From intensive mariculture systems the principal wastes are uneaten feed and faeces, dissolved nutrients, dissolved organic compounds, chemicals and therapeutics (Beveridge et al., 1994). The release of bacteria, pathogens and farmed species escapees should also be included as waste components.

Generally, less than 1/3 of the nutrients added through feed are removed by harvesting in intensive fish farming (Gowen et al., 1991; Holby & Hall, 1991; Hall et al., 1992). For intensive shrimp pond farming it is even less, ranging between 6 and 21% (Primavera, 1994; Briggs & Funge-Smith, 1994; Robertson & Phillips, 1995). Due to differences between systems and cultured species, wastes will be partitioned differently. The rapid water exchange often associated with coastal cage farming may quickly transport released dissolved nutrients away from the farming area, or if situated in turbulent water dilute them rapidly in a large water volume. Despite this, fish cage cultures have been shown to increase dissolved nutrient concentrations in surrounding marine waters (Weston, 1986; Black & Carswell, 1987; Persson, 1991 and references cited therein). Water from land-based fish farms (e.g. salmon cultured in tanks) is usually continuously replaced due to the fish being sensitive to low oxygen concentrations and build-up of metabolic wastes. In intensive fish cage farming about 50–60% of the supplied N is released to adjacent waters in dissolved form (mainly NH_4-N). In shrimp pond farming water column ammonia is usually rapidly taken up and incorporated in phytoplankton biomass, followed by subsequent accumulation in pond sediment. As little as 10% may leave the pond in dissolved form with effluent water (Briggs & Funge-Smith, 1994; Robertson & Phillips, 1995), depending on shrimp density and water management (Hopkins et al., 1995; Lorenzen et al., 1997). Shrimps are more tolerant and pond water replacement is therefore needed less frequently. In intensive shrimp cultures, with high rearing densities, water exchange can be as high as 30–50% per day depending on phase in the production cycle (Primavera, 1993; Flores-Nava, 1995; Thongrak et al., 1997). However, sometimes it may be kept lower due to infection risks from neighbouring cultures (Lin, 1995; Thongrak et al., 1997). Hopkins et al. (1993) and Lorenzen et al. (1997) found that particulate matter and dissolved nutrients in the outflow water increase considerably with higher water exchange rates. They concluded that assimilation by phytoplankton and nitrifying bacteria attached to detrital particles are the principal process of nitrogen removal from the water column. Water exchange rates higher than 0.3 d^{-1} resulted in mainly dissolved nitrogen being discharged (Lorenzen et al., 1997).

Depending on the fate of the solids accumulated in pond bottoms i.e. either being flushed out during pond cleaning, or being dumped on land or in the sea some distance from the farm, additional dissolved nutrients will, at a later stage, enter adjacent waterways after remineralisation.

Cause and effects from nutrient discharge

For nitrogen, which is the nutrient of major concern in marine environments, there is a consensus that at least 80% of the total losses (dissolved and organically bound) from fish farms are plant available and are potentially eutrophicating substances (Håkansson et al. 1988; Persson, 1988, 1991). Due to a more rapid remineralisation of aquaculture wastes in a tropical environment (Troell & Berg, 1997), this figure may be even higher for shrimp farming. Compared to the build-up of particulate organic matter on bottoms in the vicinity of intensive mariculture operations, environmental effects from the release of dissolved nutrients may not be so obvious. Although there is general concern that coastal environments are being subjected to hypernutrification and eutrophication from mariculture activities, very few cases of increased primary production in aquaculture have been documented. It has therefore been argued that nutrient release from fish farming is of minor importance (Gowen, 1994; Ackefors & Enell, 1994; Black et al., 1997). However, the lack of direct evidence of eutrophication may be due to the fact that water exchange rates are usually high. Phytoplankton, within this enriched water, having a cell doubling time counted in hours or days, may increase in number some distance from the farm area i.e. away from where impact studies usually are being performed. Also, natural variability is often so large that eutrophication effects, while existing, cannot be proved unless extensive monitoring programmes are started with the specific aim to detect such effects (Folke et al., 1997).

Apart from increased phytoplankton production, eutrophication can cause many other effects which may be more sensitive and relevant indicators e.g. changes in: energy and nutrient fluxes, pelagic and benthic biomass and community structure, fish stocks, sedimentation, nutrient cycling, oxygen depletion, and shifts between perennial and filamentous benthic algae (Folke et al., 1997). Due to time lags and the buffering capacity of ecosystems, the eutrophication process in an area may be slow, acting over time scales of several years (Wulff & Stigebrandt, 1989). Thus, when evaluating environmental effects from increased supply of matter, not only should local factors be considered but, because we are dealing with effects on complex natural systems, short-term and long-term ecological threshold effects, on both spatial and temporal scales, must also be considered (e.g. Costanza et al., 1993).

Few publications have directly linked the release of dissolved fish farm waste with hypernutrification or eutrophication in marine waters. Ruokolahti (1988) and Rönnberg et al. (1992) found increased growth of attached filamentous algae near fish cage farms in Baltic archipelagos. Gowen and Ezzi (1992) demonstrated that intensive cage culture can cause nutrient enrichment in tidally energetic, fjordic estuaries. The effluents from fish farming have very high N/P ratios (Folke et al., 1994), which are considered as a likely cause for the development of toxic and non-toxic algal blooms (Granéli et al., 1989; Carlsson et al., 1990; Edvardsen et al., 1990; Kaartvedt et al., 1991). Kaartvedt et al. (1991) linked the blooms of the toxic plankton alga *Prymnesium parvum*, which killed 750 t salmon and rainbow trout in fish farms in a Norwegian fjord, to nutrient loading from fish farms. The cause of recent outbreaks of fish kills in the U.S.A., e.g. in the estuaries of the Mid- and South-Atlantic states, as well as along the Texas coast, is suspected to be due to blooms of *Pfiesteria piscicida* and related dinoflagellates (Burkholder & Glasgow, 1997). Potential links to fish farming need to be investigated because it has been suggested that toxin production is stimulated by fish excreta-secreta, and also by inorganic and organic phosphate (Burkholder & Glasgow, 1997).

Shrimp farming and mangroves

The quality of discharged shrimp pond water in mangrove areas can be improved by using settling and treatment ponds (Lin, 1993; Hopkins et al., 1995). However, few farmers are willing to convert growout ponds for such purposes, due to the extra costs involved and the fact that farmers have to set aside ponds that otherwise could be stocked with shrimps (Thongrak et al., 1997). Laws forcing the farmers to treat their wastewater before releasing it to the environment are applied in some countries e.g. Thailand, but the law enforcement is still rather poor, resulting in low adherence (Flaherty & Choomjet, 1995; Dierberg & Kiattisimkul, 1996; Claridge, 1996).

Few studies have investigated eutrophication effects in mangrove environments caused by shrimp farming and the literature on long-term changes in habitat functions, species shifts or biodiversity loss due to shrimp farming activity are virtually nonexistent. The impact of sewage on mangrove forests has been studied and since there are many similarities between urban and mariculture wastes with regard to their potential to cause eutrophication (Folke et al., 1997), these studies are also relevant from a shrimp farming perspective. Findings from these studies are somewhat contradictory. Thus, depending on waste load and characteristics of the mangrove system, the mangrove forest either functions as a nutrient sink, with no visible accumulation pattern (Tam & Wong, 1993), or a slow build-up of nutrients in sediment and biota (Boto & Wellington, 1983; Wong et al., 1995). It has been found that effluent discharged into a mangrove forest bypasses, to a large extent, the woody halophytes by eroding a direct path to the river systems or ocean (Temple-Banner, University of Victoria pers. com.). In addition, the water retention time can be increased by constructing effluent ponds, although this may prove fatal to mangrove species sensitive to long-term submersion.

Trott et al. (1996) studied how effluents from 5.3–11.4 ha shrimp pond farms influenced a surrounding mangrove ecosystem. Water quality, forest growth and litter fall, mangrove soil nutrient status and crab feeding activity were monitored over a two year period. Dissolved inorganic and organic nutrients, both in creek waters and in sediment porewater, together with litterfall from *Rhizophora* and *Ceriops* forests, were similar to undisturbed controls. However, during periods of shrimp farm discharge, chlorophyll a and BOD levels were significantly higher than during non-discharge periods and controls. The moderate effects on surrounding mangrove forest was explained by the fact that concentrations of suspended sediments, and dissolved organic and inorganic nutrients in the discharge waters were not significantly elevated compared to adjacent tidally flushed mangrove creeks.

In a recent study from South Australia the long-term discharge of sewage effluent into an *Avicennia* forest killed the offshore seagrasses, and as a consequence unconsolidated sediment from the seagrass beds washed into the mangroves and prevented new propagules from establishing (Peri Coleman, Delta Environmental Consulting, SA, Australia,

pers. comm.). Furthermore, the deposition of sediments in the seaward fringe of trees created land accretion, which prevented drainage of landward trees facing stagnant pools at low tide. During summer these pools occasionally become fetid and anaerobic, with sulphur-reducing bacteria moving into the water column, resulting in mass mortality of trees overnight.

As discussed previously, it is necessary to increase our understanding of long term consequences on surrounding ecosystems from the nutrient discharge by mariculture systems. Thus, effects and carrying capacity must be defined on the basis of a holistic ecosystem perspective. This can be highlighted by the capacity of mangroves to assimilate nutrients from freshwater runoff, which is of importance to the stability of coastal water quality. This function of dampening nutrient fluctuations is critical, especially during wet seasons when significant amounts of runoff reaches the coast. Due to nutrient loading from shrimp farm wastes, it is possible that the capacity of mangroves in stabilising coastal water quality is lowered. This, in turn, might lead to threshold effects detrimental to other coastal ecosystem like e.g. coral reefs which require oligotrophic waters for vigorous growth.

Previous and present use of seaweeds as biofilters

Methods for treating effluents from enclosed mariculture systems with macroalgae were initiated in the mid 1970s (Haines, 1975; Ryther et al., 1975; Roels et al., 1976; Langton et al., 1977; Harlin et al., 1979). This approach has recently gained new interest (Vandermeulen & Gordin, 1990; Cohen & Neori, 1991; Neori et al., 1991; Haglund & Pedersén 1993; Buschmann et al., 1994, 1996; in press; Jiménez et al., 1994; Krom et al., 1995; Neori, 1996; Noeri et al., 1996), verifying that wastewater from intensive and semi-intensive mariculture is suitable as a nutrient source for seaweed production, and that integration with seaweeds significantly reduces the loading of dissolved nutrients to the environment. However, in open culture systems, like fish cage farming, the continuous exchange of water makes waste disposal difficult to control, and so far, few studies have investigated the possibilities of integrating seaweeds with such cultures (Hirata & Kohirata, 1993; Petrell et al., 1993; Hirata et al., 1994 a, b; Troell et al., 1997). There is also a serious dearth of literature focusing on the feasibility or application of integrated cultures of seaweeds and shrimps (He et al., 1990; Chandrkrachang et al., 1991; Lin et al., 1992, 1993; Primavera, 1993; Flores-Nava, 1995; Enander

& Hasselström, 1994; Phang et al., 1996), although this approach seems promising for shrimp farming (Primavera, 1993; Flores-Nava, 1995; Lin 1995).

Case studies: Integrated fish farming and seaweeds

1. Land based fish tank cultivation

The agarophyte *Gracilaria chilensis* was used for removing dissolved nutrients from an outdoor intensive fish tank culture (see Buschmann et al., 1996 for details). The cultivation system consisted of eight circular 8 m^3 tanks for salmon culture (*Oncorhynchus kisutch* and *O. mykiss*), from which effluent water was channelled into decantation tanks for removal of suspended matter. The water was then lead by gravity to seaweed culture units. The water in the algal culture cells was replaced 10 times d^{-1}, had an algal inoculum maintained at 1.5 kg m^{-2} by harvesting at 15-d intervals and the algae were rotated by bubbling air. Fish production reached 30 kg m^{-3} during a 13 month production cycle, and food conversion could be maintained stable at 1.4 g food g fish^{-1} production during the entire cultivation period. Ammonium was the nutrient that increased most in the fish effluents, reaching concentrations as high as 500 μg L^{-1} in spring and summer. *Gracilaria* production was 48.9 kg m^{-2} yr^{-1} and was able to remove 50% of the dissolved ammonium in winter, increasing to 90–95% in spring.

The production of *Gracilaria* increased total income by 18% (not including production costs). If costs for nutrient emission were to be paid for by the producer, i.e. practising the 'polluter pay principle', stipulated in the Rio declaration 1992, a further saving of 4% would be possible through the integration. The final conclusion from this study and a complementary economical study (Buschmann et al., in press) is that *Gracilaria* could be used for removing dissolved nutrients from fish tank effluents, generating economic benefits to the farmers as well as the society as a whole, and permitting a diversification of the production.

2. Open cage cultivation

Rope cultures of *Gracilaria chilensis* were co-cultivated with a coastal salmon cage farm (producing 230 t a^{-1}) in Chile (Troell et al., 1997). *Gracilaria* cultivated at 10 m from the salmon cages had up to 40% higher growth rates (specific growth rate of 7% d^{-1}) than at 150 m and 1 km distance. The nutrient content of algae was also higher close to the cages.

Yield of agar per biomass ranged between 17–23% of dry weight, being somewhat lower closer to the farm but, due to higher growth rates, the accumulated agar production still peaked close to the fish cages. The degree of epiphytes and bryozoan coverage was low overall. An extrapolation of the results shows that 1 ha of *Gracilaria*, cultivated close to the fish cages, has the potential to remove at least 6.5% of dissolved nitrogen and 27% of dissolved phosphorus released from the fish farm. For nitrogen this may seem a minor reduction but, because the fish farm released nearly 16 tons of nitrogen annually, the volume assimilated by the algae is substantial. Although the size of *Gracilaria* culture used for the above calculation is small, it would give an annual harvest of 34 t (d. wt) of *Gracilaria*, valued at US$ 34,000. This Figure is twice that of a *Gracilaria* monoculture, not integrated with fish cage farming.

The conclusions from this study are that both economic and environmental advantages could be achieved by integrating algal cultivation with fish farming in open sea systems. A larger cultivation unit would increase nutrient removal efficiency and profits, but further studies focusing on full scale cultivation during different seasons are needed. The high water exchange rates that often characterise coastal fish farming will be of importance when integrating seaweeds with fish. Unlike particulate nutrients, the dissolved fraction will be transported over much greater distances. The proportion of integrated seaweeds benefiting from the surplus of dissolved nutrients will increase with the time that nutrient levels remain high in the water package passing the cages (Løland, 1993). There are two main factors that will determine the potential for seaweeds to remove nutrients: one is the capacity of the seaweeds to respond to an increased nutrient concentration, and the other is how precise the current pattern can be predicted, or how exposure to the nutrient rich water can be maximised (Troell & Norberg, 1998).

3. Footprints to visualise area required for nutrient absorption in integrated fish-seaweed farming

The spatial ecosystem area needed to take care of the released waste can be illustrated by calculating the farm's ecological footprint (Robertson & Philipps, 1995; Berg et al., 1996; Kautsky et al., 1997b; Folke et al., 1998). The ecological footprint needed for assimilating nutrients released from aquaculture indicates how densely farms can be placed in an area without risking self-pollution, formation of algal blooms, etc.

Ecological footprint analysis of intensive tilapia cage farming showed that a pelagic system with an area 115 times larger than the area of the cages is needed to take care of the phosphate released. In a semi-intensive tilapia pond the nutrients could be assimilated within the pond area itself (Berg et al., 1996). Robertson and Phillips (1995) calculated that the nitrogen and phosphorus released from semi-intensive shrimp pond farming could be entirely assimilated by a mangrove forest, which was 2.4–2.8 times larger than the pond area itself. For intensive shrimp farming this area increased to 7.4–21.6 times the pond. If seaweeds are integrated in intensive aquaculture, the footprint for waste assimilation as well as the strain on the environment will be reduced. We can calculate this reduction for the above example of open cage salmon-*Gracilaria* culture in Chile. The salmon cage will release 5.52 kg N and 0.924 kg P m^2 a^{-1} (recalculated from Troell et al., 1997). Assuming an average pelagic primary production of 200 mg C m^2 d^{-1} for the area and an atomic ratio of 80 C:15 N:1 P (Redfield et al., 1963), it can be calculated that the footprints for nitrogen and phosphorus waste are 340 and 400 times larger than the cage area, respectively. When integrated with *Gracilaria*, the corresponding footprint is reduced to 150 and 25 times the cage area for nitrogen and phosphorus, respectively. This means that the carrying capacity of the area, in terms of nutrient absorption, is increased and that it should be possible to place salmon cages at somewhat higher densities before risking negative environmental effects from eutrophication.

Sustainable shrimp farming

Many suggestions have been put forward for 'sustainable' shrimp farming. Of particular interest is the idea that future development should, as a means of minimising land use (i.e. clearing of mangroves), focus on intensive rather than extensive systems (Flores-Nava, 1995; Menasveta, 1997). Another suggestion, and something that is already taking place, is to locate intensive shrimp systems on higher elevations inland of the mangroves (Ythoff, 1996; Menasveta, 1997). However, even if situated outside the mangroves, wastewater will usually be released to adjacent mangrove systems or coastal waters and thereby still constitute a threat to these environments. Such cultures would therefore preferably consist of enclosed systems, where water is being filtered for particles and dissolved nutrients before being released to the environment. This could be accomplished in different ways, either using high-technology cleaning solutions, re-using the water in enclosed systems or treating the wastewater before release to the environment using integrated farming techniques i.e. using filter feeders and seaweeds as biofilters (Enander & Hasselström, 1994; Flores-Nava, 1995). Due to difficulties and high costs involved in re-circulating systems i.e. high-technology solutions (Hopkins et al., 1995), it may be unrealistic to believe that such a development will take place. Instead, intensive systems combined with secondary biological treatment units should be promoted. In a study of an integrated farming system, consisting of oysters and *Gracilaria* cultivated in tanks in the shrimp farm effluent water, total nitrogen and total phosphorus could be reduced by 41 and 52%, respectively (over 90% removal of ammonium, nitrate, nitrite and phosphate) (A. Jones, Dept Botany, Univ. Queensland, pers. comm.). Enander and Hasselström (1994) also demonstrated effective uptake of nutrients by mussels and seaweeds cultivated in ponds receiving shrimp effluent waters. They were able to remove 81% ammonium and 19% nitrate from the dissolved waste. The possibility for cultivating seaweeds in shrimp ponds in Malaysian mangroves was studied by Phang et al. (1996), who compared the growth of *Gracilaria* in a shrimp pond with plants in an irrigation canal and in a natural mangrove habitat. A threefold increase in growth was found in the irrigation canal compared to the shrimp pond and the natural mangrove, the latter two having similar growth. This was explained by larger secchi depth and more frequent water exchange in the canal. The plants cultivated inside the shrimp pond were also heavily epiphytised and grazed by fish. Phang et al. (1996) concluded that the difficulties encountered in the shrimp pond could be solved through better pond management; for instance, flow around the plants could have been increased if the plants had been placed further towards the edge of the pond.

To increase the sustainability of shrimp farming in mangrove environments the benefits from aquasilviculture systems integrated with seaweeds needs to be investigated.

Conclusions

The renewed interest in adopting new farming techniques e.g. methods built on the principles of ecological engineering, have the potential to find solutions that could mitigate negative environmental impacts from intensive mariculture. However, systems like fish tank farming, fish cage farming and shrimp pond farming have different characteristics regarding how waste is being emitted to the surrounding ecosystems. Thus, a solution applied in one type of culture may not be feasible in another. The development of techniques where seaweeds are used as biofilters is just in its infancy and continued research on ecologically sound production systems is needed.

References

Ackefors H, Enell M (1994). The release of nutrients and organic matter from aquaculture systems in Nordic countries. J. appl. Ichth. 10: 225–241.

Berg H, Michélsen P, Troell M, Folke C, Kautsky N (1996) Managing aquaculture for sustainability in tropical Lake Kariba, Zimbabwe. Ecol. Econ. 18: 141–159.

Beveridge MCM, Lindsay LG, Kelly AL (1994) Aquaculture and biodiversity. Ambio 23: 497–502.

Black E, Carswell B (1987) Sechelt Inlet, Spring 1986, Impact of salmon farming on marine water quality. British Columbia Ministry of Agriculture and Fisheries. Commercial Fisheries Branch. Victoria, British Columbia, 43 pp.

Black E, Gowen R, Rosenthal H, Roth E, Stechy D, Taylor FJR (1997) The cost of eutrophication from salmon farming: Implication for policy- a comment. J. environ. Manag. 50: 105–109.

Boto, KG, Wellington JT (1983) Nitrogen and phosphorus nutritional status of a northern Australian mangrove forest. Mar. Ecol. Progr. Ser. 11: 63–69.

Brzeski V, Newkirk G (1997) Integrated coastal food production systems – a review of current literature. Ocean coast. Manag. 34: 66–71.

Briggs MRP, Funge-Smith SJ (1994) A nutrient budget of some intensive marine shrimp ponds in Thailand. Aquacult. fish. Manag. 25: 789–811.

Burkholder JM, Glasgow Jr HB (1997) *Pfiesteria piscicida* and other *Pfiesteria*-like dinoflagellates: behavior, impacts and environmental controls. Limnol. Oceanogr. 42: 1052–1075.

Buschmann AH, Cavilan M, Troell M, Kautsky N, Medina A, Guzman O (in press) Integrated tank cultivation of salmonids and seaweeds: a technical and economical analysis. Ecol. Econ.:

Buschmann, AH, Mora OA, Gómez P, Böttger M, Buitano S, Retamales C, Vergara PA, Gutierrez A (1994) *Gracilaria* tank cultivation in Chile: use of land based salmon culture effluents. Aquacult. Engineering 13: 283–300.

Buschmann AH, M. Troell M, Kautsky N, Kautsky L (1996) Integrated tank cultivation of salmonids and *Gracilaria chilensis* (Gracilariales, Rhodophyta). Hydrobiologia 326/327: 75–82.

Carlsson P, Granéli E, Olsson P (1990) Grazer elimination through poisoning: one of the mechanisms behind *Chrysochromulina polylepis* bloom. In Granéli E, Sundström B, Edler E, Andersson D (eds) Toxic Marine Phytoplankton. Elsevier Science Publishing Co, New York: 116–122.

Chandrkrachang S, Chinadit U, Chandayot P, Supasiri T, (1991) Profitable spin-offs from shrimp-seaweed polyculture. INFOFISH Int. 6/91: 26–28.

Claridge G (1996) Legal approaches to controlling the impacts of intensive shrimp aquaculture: adverse factors in the Thai situation. Paper presented to the International Law Institute workshop on the legal and regulatory aspects of aquaculture in India and Southeast Asia, Bangkok, Thailand, 11–13 March, 1996.

Cohen I, Neori A (1991) *Ulva lactuca* biofilters for marine fish pond effluent. I. Ammonia uptake kinetics and nitrogen content. Bot. mar. 34: 475–482.

Cripps SJ (1994) Minimizing outputs: treatment. J. appl. Ichtyol. 10: 284–294.

Daly HE, Cobb JB (1989) For the Common Good: Redirecting the Economy Toward Community, the Environment and a Sustainable Future. Beacon Press, Boston, USA, 482 pp.

Dierberg FE, Kiattisimkul W (1996) Issues, impacts and implications of shrimp aquaculture in Thailand. J. environ. Manag. 20: 649–666.

Edvardsen B, Moy F., Paasche E, (1990) Hemolytic activity in extracts of *Chrysochromulina polylepis* grown at different levels of selenite and phosphate. In Granéli E, Sundström B, Edler E, Andersson D (eds) Toxic Marine Phytoplankton. Elsevier Science Publishing Co, New York: 284–289.

Ekins P, Hillman M, Hutchison R (1994) The Gaia Atlas of Green Economics. Anchor Books, Doubleday, New York, 191 pp.

Enander M, Hasselström M (1994) An experimental wastewater treatment system for a shrimp farm. INFOFISH Int. 4/94: 56–61.

Flaherty M, Choomjet K (1995) Marine shrimp aquaculture and natural resource degradation in Thailand. J. environ. Manage. 19: 27–37.

Flores-Nava A (1995) Some considerations on sustainable shrimp mariculture. Jaina 6: 8.

FAO (1997) Aquaculture production statistics 1985–1995. FAO Fisheries Circular No. 815, Revision 9, FAO Rome.

Folke C, Kautsky N (1992) Aquaculture with its environment: prospects for sustainability. Ocean coast. Manag. 17: 5–24.

Folke C, Kautsky N, Troell M (1994) The costs of eutrophication from salmon farming: Implications for policy. J. environ. Manage. 40: 173–182.

Folke C, Kautsky N, Troell M, (1997) Salmon farming in context: response to Black et al. J. environ. Manage. 50: 95–103.

Gowen RJ, Weston DP, Ervik A (1991) Aquaculture and the benthic environment. In Cowey CB, Cho CY (eds), Nutritional Strategies and Aquaculture Waste, Proceedings of the first international symposium on nutritional strategies in management of aquaculture waste (NSMAW). Department of Nutritional Science, Univ. of Guelph, Guelph, Ontario, 1991: 187–205.

Gowen, RJ (1994) Managing eutrophication associated with aquaculture development. J. appl. Ichthyol. 10: 242–257.

Gowen RJ, Ezzi IA (1992) Assessment and prediction of the potential for hyper-nutrification and eutrophication associated with cage culture of salmonids in Scottish coastal waters. Dunstaffnage Marine Laboratory, Oban, Scotland, 136 pp.

Granéli E, Carlsson P, Olsson P, Sundström B, Granéli W, Lindahl O (1989) From anoxia to fish poisoning: The last ten years of phytoplankton blooms in Swedish marine waters. In Cosper EM, Bricelj VM, Carpenter EJ (eds) Novel Phytoplankton Blooms - Causes and Impacts of Recurrent Brown Tides and Other Unusual Blooms. Coast. estuar. Stud. 35: 407–428.

Haglund K, Pedersén M (1993) Outdoor pond cultivation of the subtropical marine red alga *Gracilaria tenuistipitata* in brackish water in Sweden. Growth, nutrient uptake, co-cultivation with rainbow trout and epiphyte control. J. appl. Phycol. 5: 271–284.

Haines KC (1975) Growth of the carrageenan-producing tropical red seaweed *Hypnea musciformis* in surface water, 870 m deep water effluent from a clam mariculture system, and in deep water enriched with artificial fertilizers or domestic sewage. In Persson G, Jaspers E (eds), Proc. 10th Eur. Symp. Mar. Biol.,

Ostend, Belgium, Sep. 17–23, 1975. Universa Press, Wetteren. Vol. 1: 207–220.

Hall, POJ, Holby O, Kollberg S, Samuelsson M-O, (1992) Chemical fluxes and mass balance in a marine fish cage farm. IV. Nitrogen. Mar. Ecol. Progr. Ser. 89: 81–91.

Harlin MM, Thorne-Miller B, Thursby GH (1979) Ammonium uptake by *Gracilaria* sp. (Florideophyceae) and *Ulva lactuca* (Chlorophyceae) in closed system fish culture. Proc. int. Seaweed Symp. 9: 258–292.

He X, Peng T, Liu S, Huang J, He Z, Xu Q, Huang L (1990) Studies on the elimination of stress factors in *Peneaus* diseases. Trop. Oceanol./Redai Haiyang 9: 61–67 (in Chinese with English Abstract).

Hirata H, Kohirata E (1993) Culture of sterile *Ulva* sp. in a marine fish farm. The Israeli J. Aquacult.- Bamidgeh 45: 164–168.

Hirata, H, Kohirata E, Guo F, Danakusumah E, Xu BT (1994a) Cage culture of the sterile *Ulva* sp. in a coastal fish farm. In Chou LM, Munro AD, Lam TJ, Chen TW, Cheong LKK, Ding JK, Hooi KK, Khoo HW, Phang VPE, Shim KF, Tan CH (eds) The Third Asian Fisheries Forum. Asian Fisheries Society, Manila, Philippines: 124–127.

Hirata H, Yamasaki S, Maenosono H, Nakazono T, Yamauchi T, Matsuda m (1994b) Relative budgets of $p\,O_2$ and $p\,CO_2$ in cage polycultured Red Sea bream, *Pagrus major* and sterile *Ulva* sp. Suisanzoshoku 42: 377–381.

Holby O, Hall POJ (1991) Chemical fluxes and mass balance in a marine fish cage farm. II. Phosphorus. Mar. Ecol. Progr. Ser. 70: 263–272.

Hopkins JS, Browdy CL, Hamilton II RD, Heffernan III JA (1995) The effect of low-rate sand filtration and modified feed management on effluent quality, pond water quality and production of intensive shrimp ponds. Estuaries 18: 116–123.

Håkansson L, Ervik A, Mäkinen T, Möller B (1988) Basic concepts concerning assessments of environmental effects of marine fish farms. Nordic council of ministers, Nord 1988, No. 90, 103 pp.

Iwama GK (1991) Interactions between aquaculture and the environment. Crit. Rev. envir. Contr. 21: 177–216.

Jiménez del Rio M, Ramazanov Z, García-Reina G (1994) Optimization of yield and biofiltering efficiencies of *Ulva rigida* C. Ag. cultivated with *Sparus aurata* L. waste waters. Sci. Mar. 58: 329–335.

Kaartvedt S, Johnsen TM, Aksnes DL, Lie U, Svendsen H (1991) Occurrence of the toxic phytoflagellate *Prymnesium parvum* and associated fish mortality in a Norwegian fjord system. Can. J. Fish. aquat. Sci. 48: 2316–2323.

Kautsky N, Folke C, Berg H, Jansson Å, Troell M (1997b) The ecological footprint concept for sustainable seafood production: a review. Ecol. Apps. 8: S63–S71.

Kautsky N, Troell M, Folke C (1997a) Ecological engineering for increased production and environmental improvement in open sea aquaculture. In Etnier C, Guterstam B (eds), Ecological engineering for wastewater treatment. Lewis Publisher, Chelsea, Michigan, 496 pp.

Krom MD, Ellner S, van Rijn J, Neori A (1995) Nitrogen and phosphorus cycling and transformations in a prototype 'non-polluting' integrated mariculture system, Eilat, Israel. Mar. Ecol. Prog. Ser. 118: 25–36.

Langton RW, Haines KC, Lyon RE (1977) Ammonia-nitrogen production by the bivalve mollusc *Tapes japonica* and its recovery by the red seaweed *Hypnea musciformis* in a tropical mariculture system. Helgoländer wiss. Meeresunters. 30: 217–229.

Lin KC, Raumthaveesub P, Wanuchsoontorn P (1993) Culture of the green mussel (*Perna viridis*) in waste water from an intensive shrimp pond: concept and practice. World Aquaculture 24: 68–73.

Lin CK (1995) Progression of intensive marine shrimp culture in Thailand. In Browdy CL, Hopkins SJ (eds), Swimming Through Troubled Water. Proceedings of the special session on shrimp farming, Aquaculture '95. World Aquaculture Society, Baton Rouge, Louisiana, USA: 13–23.

Lin CK, Ruamthaveesub P, Wanuchsoontorn P, Pokaphand C (1992) Integrated culture of green mussel (*Perna viridis*) and marine shrimps (*Penaeus monodon*). Book of abstracts and schedule, Aquaculture '92, Orlando, FL, 21–25 May.

Løland G (1993) Current forces on, and water flow through and around, floating fish farms. Aquacult. Int. 1: 72–89.

Menatsveta P (1997) Intensive and efficient shrimp culture system, the Thai way, can save mangroves. Aquacult. Asia 2: 38–44.

Neori A, Cohen I, Gordin H (1991) *Ulva lactuca* biofilters for marine fishpond effluent. II. Growth rate, yield and C:N ratio. Bot. mar. 34: 483–489.

Neori A, Krom MD, Ellner SP, Boyd CE, Popper D, Rabinovitch R, Davison PJ, Dvir O, Zuber D, Ucko M, Angel D, Gordin H (1996) Seaweed biofilters as regulators of water quality in integrated fish-seaweed culture units. Aquaculture 141: 183–199.

New MB, Wijkstrom UN (1990) Feed for thought: some observations on aquaculture feed production in Asia. World Aquaculture 21: 17–23.

Persson G (1988) Relationships between feed, productivity and pollution in the farming of large rainbow trout (*Salmo gairdneri*). Swedish Environmental Protection Agency, Stockholm, PM 3534, 48 pp.

Persson G (1991) Eutrophication resulting from salmonid fish culture in fresh and salt waters: Scandinavian experiences. In Cowey CB, Cho CY (eds), Nutritional Strategies and Aquaculture Waste: 163–185.

Petrell RJ, Mazhari Tabrizi K, Harrison PJ, Druehl LD (1993) Mathematical model of *Laminaria* production near a British Columbian salmon sea cage farm. J. appl. Phycol. 5: 1–14.

Phang S-M, Shaharuddin S, Noraishah H, Sasekumar A, (1996) Studies on *Gracilaria changii* (Gracilariales, Rhodophyta) from Malaysian mangroves. Hydrobiologia 326/327: 347–352.

Primavera JH (1993) A critical review of shrimp pond culture in the Philippines. Rev. fish. Sci. 1: 151–201.

Primavera JH (1994) Environmental and socioeconomic effects of shrimp farming: the Philippine experience. INFOFISH Int. 1/94: 44–49.

Robertson AI, Phillips MJ (1995) Mangroves as filters of shrimp pond effluents: predictions and biogeochemical research needs. Hydrobiologia 295: 311–321.

Roels OA, Haines KC, Sunderlin JB (1976) The potential yield of artificial upwelling mariculture. In Persson G, Jaspers E (eds), Proc. 10th Eur. Symp. Mar. Biol., Ostend, Belgium, Sept. 17–23, 1975. Universa Press, Wetteren. Vol. 1: 358–394.

Rönnberg O, Ådjers K, Roukolahti C, Bondestam M (1992) Effects of fish farming on growth, epiphytes and nutrient content of *Fucus vesiculosus* L. in the Åland archipelago, northern Baltic Sea. Aquat. Bot. 42: 109–120.

Rönnbäck P (in press) The ecological basis for economic value of Seafood production supported by mangrove ecosystems. Ecol. Econ.

Ruokolahti C (1988) Effects of fish farming on growth and chlorophyll a content of *Cladophora*. Mar. Poll. Bull. 19: 166–169.

Ryther JH, Goldman JC, Gifford JE, Huguenin JE, Wing AS, Clarner JP, Williams LD, Lapointe BE (1975) Physical models of integrated waste recycling-marine polyculture systems. Aquaculture 5: 163–177.

Tam NFY, Wong YS (1993) Retention of nutrients and metals in mangrove sediment receiving wastewater of different strengths. Envir. Technol. 14: 719–729.

Thongrak S, Prato T, Chiayvareesajja S, Kurtz W (1997) Economic and water quality evaluation of intensive shrimp production systems in Thailand. Agricultural Systems 53: 121–141.

Troell M (1996) Intensive fish cage farming: impacts, resource demands and increased sustainability through integration. PhD-thesis, Stockholm University, Stockholm, pp. 148.

Troell M, Berg H 1997. Cage farming in the tropical Lake Kariba, Zimbabwe: impact and biogeochemical changes in sediment. Aquaculture Res. 28: 527–544.

Troell M, Halling C, Nilsson A, Buschmann AH, Kautsky N, Kautsky L (1997) Integrated open sea cultivation of *Gracilaria chilensis* (Gracilariales, Rhodophyta) and salmonids for reduced environmental impact and increased economic output. Aquaculture 156: 45–61.

Troell M, Kautsky N, Folke C 1999 Comment: Applicability of integrated coastal aquaculture systems. Ocean coast. Manage. 42: 63–69.

Troell M, Norberg J (1998) Modelling output and retention of suspended solids in an integrated salmon- mussel culture. Ecological Modelling 110: 65–77.

Trott LA, Alongi DM, Olsen D (1996) Results from the monitoring program in mangrove creek ecosystems near Sea Ranch Pty Ltd. The Australian Prawn Farmers' Association 1996 Workshop and Annual General Meeting, 26–28 July, 1996, Cairns, Australia.

Uthoff D (1996) From traditional use to total destruction-forms and extent of economic utilization in the Southeast Asian mangroves. Natural resources and development, vol 43/44, Institute for Scientific Co-operation, Tübingen, Federal Republic of Germany: 58–94.

Vandermeulen H, Gordin H (1990) Ammonium uptake using *Ulva* (Clorophyceae) in intensive fishpond systems: mass culture and treatment of effluent. J. appl. Phycol. 2: 263–374.

Weber M (1996) So you say you want a blue revolution? The Amicus Journal, Fall 1996: 39–42.

Weston DP (1986) The environmental effects of floating cage mariculture in Puget Sound. School of Oceanography, College of Oceans and Fisheries Sci., University of Washington, Seattle, USA, 148 pp.

Wong YS, Lan CY, Chen GZ, Li SH, Chen XR, Liu ZP, Tam NFY (1995) Effects of wastewater discharge on nutrient contamination of mangrove soils and plants. Hydrobiologia 295: 243–254.

Wulff F, Stigebrandt A (1989) A time dependent budget model for nutrients in the Baltic Sea. Global biogeochem. Cycles 3: 63–78.

Occurrence of closely spaced genes in the nuclear genome of the agarophyte *Gracilaria gracilis*

Arturo O. Lluisma[1,2] & Mark A. Ragan[1,3,*]
[1]*Institute for Marine Biosciences, National Research Council of Canada, Halifax, Nova Scotia, B3H 3Z1 Canada*
[2]*Present address: Marine Science Institute, University of the Philippines, 1101 Quezon City, Philippines*
[3]*Canadian Institute for Advanced Research, Program in Evolutionary Biology*

(*Author for correspondence)

Received 1 April 1998; revised 13 November 1998; accepted 16 November 1998

Key words: genome structure, genome organisation, red algae, synteny

Abstract

Little is known about the structure and organisation of nuclear genomes in red algae. In particular, it is not known whether genes are densely or loosely packed, whether gene order is conserved, whether their genes tend to occur in one or multiple copies and whether their nuclear genes tend to be compact or interrupted by numerous introns. Sequencing of cloned genomic DNA from *Gracilaria gracilis* has begun to provide provisional answers to some of these questions. Four pairs of closely spaced genes have been found in *G. gracilis* upon sequencing genomic clones that contain genes for UDPglucose pyrophosphorylase, galactose-1-phosphate uridylyltransferase, the β subunit of tryptophan synthetase, and methionine sulphoxide reductase (a fifth pair of closely spaced genes, encoding polyubiquitin and aconitase, was reported earlier). An open reading frame with significant similarity to another known gene occurs close (<1.7 kbp) to each of these genes. In two pairs the intergenic region is less than 400 bp in length, and for these the location of the putative polyadenylation signals indicates that the gene transcripts, encoded on opposite strands, have overlapping (hence complementary) 3' regions. These somewhat unexpected findings begin to establish a basis for genome-level characterisation of red algae.

Abbreviations: EST – expressed sequence tag; GalT – galactose-1-phosphate uridylyltransferase; MSR – methionine sulphoxide reductase; NMIOR – NAD-dependent myo-inositol oxidoreductase; ORF – open reading frame; PCR – polymerase chain reaction; PTH – peptidyl tRNA hydrolase; SBE – starch branching enzyme; TSβS – tryptophan synthase-β subunit; UGPase – UDPglucose pyrophosphorylase

Introduction

Determining the structure and organisation of genomes is important for a number of reasons. As has been demonstrated for some groups of organisms (*e.g.*, grasses: Barakat et al., 1997; Bennetzen & Freeling, 1997), such characterisation is key not only to understanding how the genomes of extant members of a group are structurally and functionally related, but also to reconstructing the evolution of their genomes from an ancestral form. In addition, genomic characterisation is a prerequisite for comparative mapping, which facilitates the application of information from more readily manipulated 'model' species, to others that are typically of commercial importance (*e.g.*, cereals).

Current understanding of the structure, organisation and complexity of red algal genomes is based largely on studies that employed microspectrophotometry, DNA reassociation kinetics, and microscopy. Micro-spectrophotometry has been useful in quantifying the genomic DNA of a number of red algae; DNA reassociation kinetics in estimating the relative amounts of repetitive and unique DNA, and G+C content (molar fraction of guanine and cytosine in the DNA); and microscopy in estimating genome sizes and number of chromosomes. These techniques have been instrumental in establishing, for example, that in species of *Gracilaria* the chromosome number and nuclear genome size are highly conserved, but the G+C content and relative amounts of unique and repetitive DNA tend to vary (Kapraun et al., 1996a,b; Lopez-Bautista & Kapraun, 1995).

Beyond these basic features, however, most structural aspects of red algal genomes remain uninvestigated; in particular, essentially nothing is known about the order and distribution of genes along the chromosomes of red algae. Studies of other eukaryotic genomes have shown that genes may be somewhat clustered in some groups of organisms, *e.g.*, Gramineae, but relatively more scattered in others, *e.g.*, mammals (Bernardi, 1995). Groups that diverged relatively recently might be expected to share greater genomic co-linearity. Modern-day grasses, for example, have genomes that in general represent reorganisations of chromosomal segments of an ancestral grass genome (Moore et al., 1995; Bennetzen & Freeling, 1997). Similar studies have not been performed for red algae. Sequencing the entire genome of several representative red algae would be the ideal approach, but such an effort remains to be undertaken.

By sequencing the flanking regions of cloned nuclear genes, it should be possible to determine whether genes tend to be clustered. Several red algal genes have already been cloned, but their flanking regions have received little attention. Heretofore the only report of closely spaced nuclear genes in a red alga was that of Zhou and Ragan (1995), who found that the genes encoding mitochondrial aconitase and polyubiquitin are located about 1.5 kb apart in the genome of *G. gracilis*. The purpose of the present paper is to report that four more pairs of genes are closely spaced in the nuclear genome of *G. gracilis*.

Materials and methods

Isolation and sequencing of genomic clones

DNA was extracted from *G. gracilis* as described previously (Zhou & Ragan, 1993). The genomic library was constructed by partially digesting the DNA with *Sau*3AI, and ligating the genomic fragments to the Lambda-DASH vector (Stratagene, La Jolla, CA).

Table 1. Primers used for amplifying the terminal and intergenic regions from each pair of closely spaced genes

Gene pair	PCR round	F primer	R primer	Expected size of product
GalT-PtRH	1	galTsF-1	galTr-2	∼ 1.4 kb
	2	galTsF1	galTr-2	∼ 1.1 kb
UGPase-helicase	1	ugpsF-1	ugpr-6b	∼ 2.0 kb
	2	ugpsF3b	ugpr-6b	∼ 1.6 kb
NMIOR-TSβS	1	c313f-6	c313r3	∼ 2.0 kb
	2	c313f-5	c313r3	∼ 1.6 kb
MSR-mAT	1	c47f-1	mATr1b	∼ 2.5 kb
	2	c47sF1	mATr1b	∼ 2.0 kb

Clones were isolated from the genomic library by screening with probes for five putative genes from *G. gracilis*: GALT, MSR, SBE, TSβS, and UGPase. The probes used were generated either from PCR products (SBE: Lluisma & Ragan 1998a; UGPase: Lluisma & Ragan, unpublished data) or ESTs (GALT, MSR and TSβS: Lluisma & Ragan, 1997). The isolated clones were sequenced by primer walking from the region of previously known sequence (the region corresponding to the probe) out to the 5' and 3' flanking regions. Sequencing was carried out on an ABI 373 (Applied Biosystems) automated sequencer using ABI's Dye Terminator Cycle Sequencing protocol.

Sequence searches

The nucleotide sequences were used as query sequences to search the U.S. National Center for Biotechnology Information (NCBI) nr (non-redundant) peptide sequence database for similar sequences, using the algorithms BLASTX (Gish & Gates, 1993) and BEAUTY (Worley et al., 1995). Searches were carried out using the BCM Search Launcher facility, maintained by the Human Genome Center, Baylor College of Medicine (Smith et al., 1996).

PCR confirmation

Two rounds of PCR reactions were run to verify the close spacing of certain genes in the genome of *G. gracilis* (*i.e.*, to confirm that the observed proximity is not due to the artifactual formation of chimaeras during cloning). Primers specific for each gene in each pair of closely spaced genes were used in the first round, with genomic DNA from *G. gracilis* as template. In the second round, nested primers were used,

Table 2. Sequences of oligonucleotide primers (5′→3′)

galTsF-1	GAGAACCTCAAGGCGTATCATG
galTsF1	CACTTTCACATGCACTTCTATCC
galTr-2	GGGTGATGCGATGCTCAACTAC
ugpsF-1	AAGCTGGCTGTGCTCAAGCTC
ugpsF3b	AAGACACGTGCCGACATCAAG
ugpr-6b	AGGTACAATGCTGTGCTAGGAG
c313f-6	TCCAACAGGACGCCTGCTTTTG
c313F-5	CTGTGAGAAGCCCATTTCGAATG
c313r3	GGGGAGCGGCGCGTTTTGAAC
c47f-1	CTCAAGTCACCGGTCGACACAC
c47f1	GCACGTCGGTGGTTCCGATG
mATr1b	CACCATCTCAAGGGATATTCCG

with the product from the first round of PCR as template. The names and sequences of the primers are shown in Tables 1 and 2, respectively. The PCR reactions contain the following, per 100 μL reaction: 200 μM dNTP, 1X reaction buffer, 2.5 U Taq, 450 ng total DNA from *G. gracilis* (first-round PCR only), and 100 ng of each primer, except that the product of the first round (0.5 μL) was used as template for the second round. The PCR reaction buffer and *Taq* polymerase were obtained from Pharmacia. PCR reactions were performed using a Perkin Elmer 9600 thermal cycler, with the reaction parameters set as follows: 94 °C for 2 min; 30 cycles of denaturation (94 °C, 1 min), annealing (62 °C or 55 °C, 1 min) and extension (72 °C, 6 min); and final extension at 72 °C for 5 min.

Results

Proximity of genes in G. gracilis

We have isolated genomic clones containing putative homologs of five genes in *G. gracilis* (encoding GalT, MSR, SBE, TSβS, and UGPase), and sequenced at least 500 bp outward from the 5′ and 3′ ends of these protein-coding regions. Full sequences of the genes encoding GalT, SBE and UGPase, including their flanking regions, have been deposited in GenBank under accession numbers AF036247, AF042842 and AF100788 respectively. Partial sequences of putative genes encoding MSR and TSβS were reported previously (as ESTs: Lluisma & Ragan, 1997) and have been deposited in GenBank under accession numbers gi1140307 and gi1140279 respectively.

BLAST searches were used to compare translations of the flanking regions with sequences in the GenBank peptide database.

In the clone containing a gene (GalT) encoding galactose-1-phosphate uridylyltransferase (GALT) there occurs an ORF less than 1 kb downstream from GalT (Table 3). The BLAST results (score >100; Figure 1) suggest strongly that this downstream ORF encodes a peptidyl tRNA hydrolase (PTH). Sequence analysis, and the 3′ RACE data (Lluisma & Ragan 1998b), indicate that GALT and the PTH ORFs, which are encoded on opposite strands, may produce transcripts with overlapping and complementary 3′ ends; base-paring of the 3′ regions could theoretically interfere with the translation from the transcripts. Whether this occurs *in vivo,* and if so its functional significance, remain to be investigated.

We also found that a putative DNA helicase is adjacent to the gene for UDPglucose pyrophosphorylase (UGPase) in the nuclear genome of *G. gracilis* (Table 3). Although only a fragment of the DNA helicase gene has been sequenced (unpublished data), it nonetheless shows significant similarity (BLAST scores >100) with a number of DNA helicase entries in the GenBank. This putative helicase ORF has two interesting features. First, a potential 96-bp phase-0 GT/AG spliceosomal intron occurs near the 3′ end of this ORF; introns appear to be infrequent in red algal nuclear genes, and this is the first one localised in the 3′ end of a gene. Second, similar to the GalT-PTH pair, the putative UGPase and DNA helicase genes probably produce overlapping transcripts (unpublished data); again, possible physiological implications remain to be investigated.

A clone containing a potential gene for the β subunit of tryptophan synthase (TSβS) was isolated by hybridisation with the corresponding cDNA, reported earlier as part of our EST survey of *G. gracilis* (Lluisma & Ragan, 1997). Sequencing the region of the clone where TSβS is localised revealed that within 700 bp upstream of the gene can be found the 3′ end of an ORF whose conceptual translation is highly similar to sequences of hypothetical proteins in the GenBank peptide database that, in turn, are potential homologs of NAD-dependent myo-inositol oxidoreductase (NMIOR) of *B. subtilis* (Table 3). The *G. gracilis* ORF also matches *B. subtilis* NMIOR itself, although somewhat less strongly. Pairwise alignment reveals that the *G. gracilis* ORF is truncated in its 5′ region (data not shown).

Table 3. Results of BLAST searches revealing the potential identitites of ORFs found close to the putative GalT, MSR, TSβS, and UGPase genes in G. gracilis. ORF sequences have been deposited in GenBank with accession numbers AF036247, AF121271, AF121272 and AF100788. Relative orientation refers to the 5′→3′ direction of the coding strands; e.g., A → ← B means that sequences A and B are encoded on opposite strands and transcribed in opposite directions. Intergenic region refers to the number of nucleotides separating the putative protein-coding regions of the gene and the nearby ORF

Cloned gene	Relative orientation	Intergenic region	GenBank entry with highest similarity to ORF		Potential homolog of ORF
			Accession no.	BLAST score	
GalT	GalT→ ORF	179 bp	gi|586021	225	PTH[1]
MSR	MSR→ ORF	<1.4 kb*	gi|2739362[a]	201	Transporter[2]
TSβS	ORF→ TSβS→	<670 bp**	gi|2635242[b]	251	NMIOR(?)[3]
UGPase	UGP→ ORF	376 bp	gi|2495146 gi|2495145	205 205	DNA helicase[4]

* The intergenic region has not been fully sequenced; however, the tail-to-tail orientation of MSR and the ORF, and the alignment of the ORF with matching sequences, allows an estimate of a maximum length for the intergenic region.
** It is not possible to identify the start codon of TSβS, because 5'RACE experiments have not been done, and because we may expect a signal peptide not readily alignable with those of plant TSβS sequences to be encoded in this region. Thus at present we can determine only a maximum length for the intergenic region.
[1] Other PTH entries matching the ORF with BLAST scores >100: gi 2499989; gi 2507267; gi 2499987; gi 1346894; gi 1172722; gi 1346896; gi 2499986; gi 2499988; gi 1093598; gi 131550.
[2] Other entries matching the ORF with BLAST scores >100; mostly encoding membrane-bound transporter; some are sequences that encode mitochondrial ADP/ATP carrier protein: gi 2132987; gi 2393737; gi 1523933; gi 2132389; gi 2463664; gi 113465; gi 100424; gi 113457; gi 1729671; gi 2191150; gi 2804436; gi 2132884.
[3] Although the ORF was most similar to gi 2635242, the match between the ORF and NMIORs from B. subtilis and S. griseus also appear significant (BLAST score >100).
[4] Other DNA helicase entries matching the ORF with BLAST scores >100: gi 1709995; gi 101069; gi 5022; gi 2058510; gi 2134009; gi 119540; gi 296645; gi 131812; gi 2408082.
[a] This entry is indicated as sequence from *Arabidopsis thaliana* that is similar to peroxisomal calcium-dependent solute carrier.
[b] This entry was labelled 'similar to opine catabolism"; when used as a query sequence, it showed highest similarity with NMIORs from *Bacillus subtilis* (36% sequence identity) and *Streptomyces griseus* (30%).

Another pair of closely spaced genes from G. gracilis discovered via gene sequencing putatively encode methionine sulphoxide reductase (MSR) and a potential membrane-bound transporter, possibly a mitochondrial ATP/ADP transporter (Table 3). The MSR gene, like that for TSβS, was cloned following identification using a partial cDNA from the EST survey (Lluisma & Ragan, 1997) as a probe. The transporter-encoding ORF is located <1.4 kb downstream of the MSR coding region.

With only one of the five selected genes, that encoding starch branching enzyme (SBE; Lluisma & Ragan, 1998a), was no ORF detected in the proximal flanking regions. More than 1 kb (1483 bp) was sequenced upstream of the start codon, but only a short (318 bp) 3′ flanking region could be sequenced due to the proximity of the 3′ end of the gene to the insert-vector junction. BLAST searches with these regions yielded no significant matches.

We investigated the possibility that the close spacing of these four pairs of genes might be due to cloning artefacts. Two rounds of PCR reactions using primers designed for each gene within each pair were conducted, using genomic DNA from G. gracilis as template.

Figure 1 shows the results of the first and second round of PCR. In the first round, with genomic DNA from G. gracilis as template, products of the expected size (see Table 1) were obtained for each gene pair. These amplified fragments were used as template for the subsequent round, using nested primers (Table 2). The second-round PCR (Figure 1) shows amplification of smaller fragments of the expected size (Table 1), confirming that the PCR products of the first round

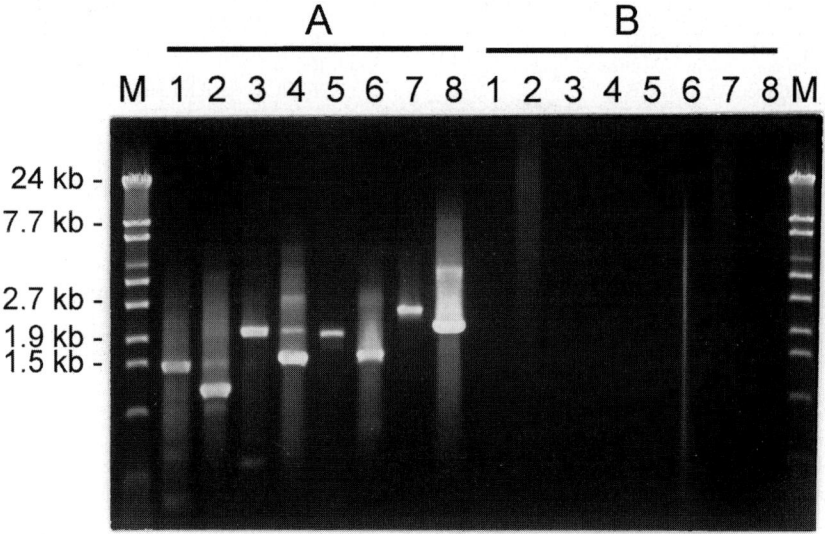

Figure 1. PCR amplification of intergenic regions from four pairs of closely spaced genes in *Gracilaria gracilis*, using the primers shown in Table 2. The PCR amplified products are from the pairs GalT-PTH (lanes 1 and 2), UGPase-DNA helicase (lanes 3 and 4), NMIOR-TSβS (lanes 5 and 6), and MSR-transporter (lanes 7 and 8). First round PCR: lanes 1, 3, 5, and 7. Second round PCR: lanes 2, 4, 6 and 8. Panel A, PCR reactions with genomic DNA template; Panel B, PCR reactions with no DNA template (negative controls).

were the desired genomic fragments. Negative controls (without DNA) yielded no amplification products (Table 1). These results confirm that the close spacing of genes indicated by sequencing of genomic clones reflects their actual proximity in the genome of *G. gracilis*.

Discussion

Characterization of red algal nuclear genomes has so far focused mainly on the most general properties, *e.g.*, genome size, number and size of chromosomes, G+C content, and relative sizes of kinetic components. Recombination studies have helped to determine linkage groups of loci whose alleles encode distinguishable phenotypic characteristics (for example, see van der Meer, 1990). Finer-scale resolution of genome structure in red algae has received little attention. One fundamental characteristic is the relative spacing of genes. Although extensive genomic sequencing would provide a global view, sequencing the flanking regions of cloned genes could be expected to provide at least a first indication of whether genes tend to be clustered or scattered in red algal genomes. The results we report herein confirm this expectation. An additional example, the 1.5-kb spacing between nuclear genes encoding polyubiquitin and mitochondrial aconitase in *G. gracilis* (as *G. verrucosa*; Zhou & Ragan, 1995), was reported previously. To the extent that this initial sample is representative, gene clustering may not be uncommon in the nuclear genome of *G. gracilis*.

The discovery of closely spaced genes in a red alga raises further questions, *e.g.*, whether the majority of genes in the *G. gracilis* genome tend to occur in clusters, and the extent to which corresponding genes in other red algae show the same (or similar) patterns. The results presented in this paper demonstrate that sequencing the flanking regions of cloned genes constitutes a valid and efficacious approach to surveying patterns of gene spacing in red algal nuclear DNA. Further characterisation of the structure of red algal genomes may be expected to yield practical benefits as well as fundamental insights into red algal biology.

Acknowledgements

We thank the staff of NRC-Institute for Marine Biosciences, Halifax, especially Ms Colleen Murphy, for excellent technical assistance. Issued as NRCC 42275.

References

Barakat A, Carels N, Bernardi G (1997) The distribution of genes in the genomes of Gramineae. Proc. natl Acad. Sci. U.S.A. 94: 6857–6861.

Bennetzen JL, Freeling M (1997) The unified grass genome: synergy in synteny. Genome Res. 7: 301–306.

Bernardi G (1995) The human genome: organization and evolutionary history. Ann. Rev. Genet. 29: 445–476.

Gish W, Gates DJ (1993) Identification of protein coding regions by database similarity search. Nature Genet. 3: 266–272.

Kapraun DF, Lopez-Bautista J, Bird KT (1996a) DNA base composition heterogeneity in some agarophytes (Gracilariales, Rhodophyta) from Mexico and the Philippines. J. appl. Phycol. 8: 229–237.

Kapraun DF, Lopez-Bautista J, Trono G, Bird KT (1996b) Quantification and characterization of nuclear genomes in commercial red seaweeds (Gracilariales) from the Philippines. J. appl. Phycol. 8: 125–130.

Lluisma AO, Ragan MA (1997) Expressed Sequence Tags (ESTs) from the marine red alga *Gracilaria gracilis*. J. appl. Phycol. 9: 287–293.

Lluisma AO, Ragan MA (1998a) Cloning and characterization of a nuclear gene encoding a starch-branching enzyme from the marine red alga *Gracilaria gracilis*. Curr. Genet. 34: 105–111.

Lluisma AO, Ragan MA (1998b) Characterization of galactose-1-phosphate uridylyltransferase gene from the marine red alga *Gracilaria gracilis*. Curr. Genet. 34: 112–119.

Lopez-Bautista J, Kapraun DF (1995) Agar analysis, nuclear genome quantification and characterization of four agarophytes (*Gracilaria*) from the Mexican Gulf Coast. J. appl. Phycol. 7: 351–357.

Moore G, Devos KM, Wang Z, Gale MD (1995) Grasses, line up and form a circle. Curr. Biol. 5: 737–739.

Smith RF, Wiese BA, Wojzynski MK, Davidson DB, Worley KC (1996) BCM Search Launcher – an integrated interface to molecular biology database research and analysis services available on the World Wide Web. Genome Res. 6: 454–462.

Worley KC, Wiese BA, Smith RF (1995) BEAUTY: an enhanced BLAST-based search tool that integrates multiple biological information resources into sequence similarity results. Genome Res. 5: 173–184.

van der Meer J (1990) Genetics. In Cole KM, Sheath RG (eds), Biology of the Red Algae. Cambridge University Press, New York: 103–121.

Zhou Y-H, Ragan MA (1993) cDNA cloning and characterization of the nuclear gene encoding chloroplast glyceraldehyde-3-phosphate dehydrogenease from the marine red alga *Gracilaria verrucosa*. Curr. Genet. 23: 483–489.

Zhou Y-H, Ragan MA (1995) Characterization of the nuclear gene encoding mitochondrial aconitase in the marine red algal *Gracilaria verrucosa*. Plant mol. Biol. 28: 635–646.

'Seed' production of *Porphyra* spp. by tissue culture

Masahiro Notoya
Laboratory of Applied Phycology, Tokyo University of Fisheries, Konan-4, Minato-ku, Tokyo, 108 Japan

(phone +81–3–5463-0532; fax +81–3–5463-0688)

Received 1 April 1998; revised 2 November 1998; accepted 29 November 1998

Key words: archeospore, polarity, *Porphyra* spp., regeneration, seed production, tissue culture

Abstract

Differentiation of archeospores was observed from excised tissue of young thalli of various monoecious *Porphyra* species (*P. tenera, P. yezoensis, P. suborbiculata, P. okamurae*) after 4–8 days in culture at temperatures of 20 and 25 °C. Excised tissue from adult thalli did not differentiate into archeospores, but rather regenerated directly into blades and rhizoids of foliose thalli. Tissues from young thalli of two dioecious *Porphyra* species (*P. dentata* and *P. pseudolinearis*) also regenerated into blades and rhizoids after manipulation of the culture conditions. In addition, 1–2 celled tissue pieces of both monoecious and dioecious species were also seen to develop directly into blades. Polarity of regeneration of blades and rhizoids was observed in these species. These results suggest that 'seed' can be obtained through tissue culture instead of using conventional conchocelis culture for commercial nori aquaculture in Japan.

Introduction

Although there have been many reports on seaweed tissue culture, especially in species of the Laminariales (Aguirre-Lipperhide et al., 1995; Kirihara et al., 1997; Notoya, 1997a), there have been relatively few studies on *Porphyra* species (Liu & Gordon, 1987; Liu & Kloareg, 1991; Notoya, 1995, 1996, 1997a,b; Notoya & Kim, 1996; Kim & Notoya, 1997). Using modern tissue culture techniques, it may be possible to apply this technology to the nori aquaculture industry in Japan. At present, *Porphyra* sea-farming must utilize the conchocelis phase to obtain seedlings for propagation of leafy-thalli to produce nori sheets. This procedure requires several months to grow the conchocelis filaments in sophisticated large indoor facilities to produce conchospores for seed production. *Porphyra* tissue culture might allow this procedure to be bypassed, leading to considerable savings in time and expense.

Our preliminary studies on *Porphyra* tissue culture have shown that excised pieces of foliose thallus tissue either differentiate into archeospores or regenerate into blades and rhizoids. The aim of the present study is to examine the effects of culture conditions, especially temperature, on excised tissue thallus of several cultivars and some wild species of *Porphyra* with the aim of future industrial application of tissue culture.

Materials and methods

Mature foliose thalli of *Porphyra dentata* and *P. suborbiculata* were collected at Enoshima, Kanagawa Prefecture on 27 January 1990, and 16 December 1990. Samples of *P. okamurae, P. pseudolinearis* and *P. tenera* were collected at Hakodate, Hokkaido, on 14 April 1995, at Fukaura, Aomori Prefecture, on 2 December 1992 and at Matsukawaura, Fukushima Prefecture, on 4 March 1995. Conchocelis of *P. yezoensis* was isolated from stock culture of a cultivar strain (Kisarazu) maintained in our laboratory.

Young and adult thalli were cultured in the laboratory at 15 °C and a photon irradiance of 40 μmol photon m^{-2}s^{-1}under a photoperiod of 10L:14D (Light:Dark). Prior to tissue isolation, the foliose thalli

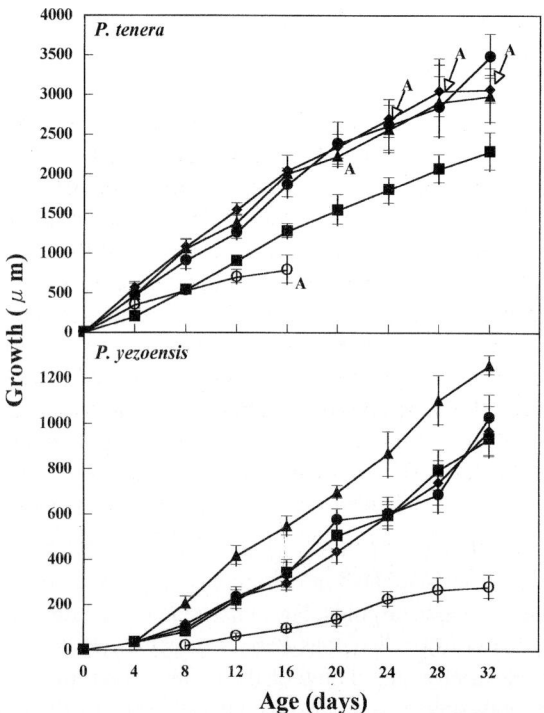

Figure 1. Growth and differentiation of the excised tissues from young thalli of *Porphyra tenera* and adult thalli of *P. yezoensis* at various temperatures with a photoperiod of 10L:14D under 40 μmol photon m^{-2}s^{-1}. Rectangles, 5 °C; closed circles, 10 °C; triangles, 15 °C; diamonds, 20 °C; open circles, 25 °C. 'A', indicates archeospore release. Data are mean of 10 pieces of tissue ± standard deviation.

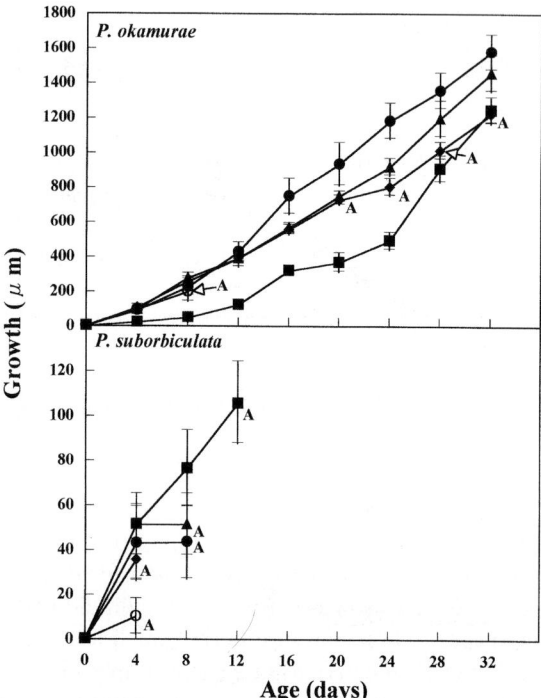

Figure 2. Growth and differentiation of the excised tissues from adult thallus of *Porphyra okamurae* and *P. suborbiculata* at various temperatures with a photoperiod of 10L:14D under 40 μmol photon m^{-2}s^{-1}. Rectangles, 5 °C; closed circles, 10 °C; triangles, 15 °C; diamonds, 20 °C; open circles, 25 °C. 'A', indicates archeospore release. Data are mean of 10 pieces of tissue ± standard deviation.

were kept overnight in an antibiotic solution (Polne-Fuller & Gibor, 1987) and then, the pieces of tissue were excised and placed in Petri dishes. Discs of excised tissues (about 1 mm diameter) or small tissue pieces containing one or two cells were excised from young (about 0.5–1 cm length) and adult thallus (> 5 cm length). These pieces were cultured at temperatures of 5, 10, 15, 20, 25 and 30 °C, and a photon irradiance of 40 μmol photon m^{-2}s^{-1} under a photoperiod of 10L:14D. A modified Grund medium (McLachlan, 1973) was used, and this was renewed every four days during the culture period.

Results

Influence of temperature

Tissues from young thalli of *P. tenera*, differentiated and liberated archeospores at 15, 20 and 25 °C (Figure 1). Although the isolated pieces of tissue did not grow at 25 °C, all vegetative cells of excised tissue differentiated into archeospores after 16 days in culture. Portions of excised pieces of tissue differentiated into archeospores at 15 and 20 °C after 24 and 20 days, respectively. At 10 and 5 °C, the excised tissue pieces regenerated only rhizoids with no differentiated reproductive cells, although growth of the tissue continued. Differentiation of archeospores was observed from excised tissue of young thalli of *P. yezoensis* after 4–8 days in culture, at 20 and 25 °C. Pieces of excised tissue from adult thalli of *P. yezoensis*, grew at all temperatures during the culture period. The excised tissue grew largest at 15 °C, and showed the least growth at 25 °C (Figure 1).

Growth of tissues and differentiation of archeospore from the excised tissue from young thalli of the wild species of *P. okamurae* and *P. suborbiculata* are shown in Figure 2. In *P. okamurae*, all pieces of tissue differentiated into archeospores at 25 °C within 8 days. After 20–32 days at 20 °C, only a few pieces of tissue showed differentiation into archeospores. At 5,

Regeneration of foliose thallus from one or two cells tissue

Foliose thalli developed from pieces of tissue consisting of one or two cells from adult thalli of the monoecious species *P. yezoensis* and from young thalli of the dioecious species *P. pseudolinearis*. Some small 1–2 celled tissue pieces of *P. yezoensis* differentiated into spermatangia, but the liberated cells did not germinate. Other small tissue pieces germinated into filamentous thallus and developed conchosporangia after several days in culture. Some 1–2 celled tissue pieces of both species regenerated directly into foliose thallus or callus-like clumps after producing rhizoids and bladelets (Figure 4A-C). Small 1–2 celled pieces of tissue, which were attached to synthetic strings, grew to adult thalli of about 14 cm after 8 weeks in culture at 15 °C (Figure 4D).

Polarity in regeneration of blades and rhizoids

Polarity in regeneration of blades and rhizoids was observed, especially in the excised tissue of *P. dentata*. *Porphyra dentata* developed denticulate cells arranged at the lower margin of foliose thalli. This feature could be used as a marker of blade direction. Denticulate cell arrangements were produced on the side margin of foliose thallus and bent slightly downwards to the basal portion (Figure 5A). Moreover, they were produced on the side cut edge of the pieces of tissue whereas bladelets and rhizoids were always produced on the upper and lower cut edges of the excised pieces of tissue, respectively (Figure 5B-C).

Figure 6 shows the production of bladelets and rhizoids from excised pieces of tissue on several parts of a young foliose thallus of *P. dentata*. In general, excised pieces of tissue from apical, central and side parts, with two or three cell layers (Figure 5I), produced bladelets and rhizoids on the upper and lower cut edges, respectively (Figure 6A, a-e). Discs removed from young foliose thalli, regenerated blades and rhizoids on the lower and upper cut edges, respectively (Figure 5D-E; Figure 6B, a), whereas excised discoid tissue regenerated bladelets and rhizoids on the upper and lower cut edges, respectively (Figure 5F-G, H; Figure 6B, b). Thus, the excised pieces of tissue had a polarity for regeneration of bladelets and rhizoids, with the bladelets always produced on the upward cut edge of the foliose thallus, and rhizoids always produced on the downward cut edge.

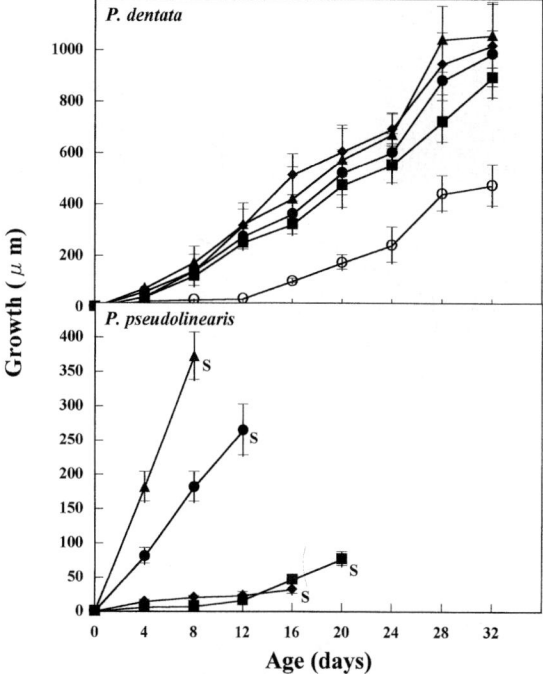

Figure 3. Growth and differentiation of the excised tissues from adult thalli of *Porphyra dentata* and *P. pseudolinearis* at various temperatures with a photoperiod of 10L:14D under 40 μmol photon m^{-2}s^{-1}. Rectangles, 5 °C; closed circles, 10 °C; triangles, 15 °C; diamonds, 20 °C; open circles, 25 °C. 'S', indicates spermatangium. Data are mean of 10 pieces of tissue ± standard deviation.

10 and 15 °C, the pieces of tissue grew continually without differentiation of archeospores. All tissues of *P. suborbiculata* differentiated into archeospores within 12 days at all temperatures from 5–25 °C; tissues differentiated earlier at higher temperatures.

Growth and differentiation of the excised tissue of the dioecious species, *P. dentata* and *P. pseudolinearis* are shown in Figure 3. Excised pieces of tissue from young thalli of *P. dentata* regenerated new bladelets and rhizoids, and grew at all temperatures from 5–25 °C, with no differentiation of spores or reproductive cells. At 25 °C, regenerated bladelets grew more slowly. Excised pieces of tissue from adult thalli of *P. pseudolinearis*, produced spermatangia within 12–24 days at all temperatures between 5 and 20 °C. Earlier differentiation of spermatangia was observed at higher temperatures. At 25 °C, the tissue did not survive. The excised tissue of neither dioecious species differentiated into archeospores.

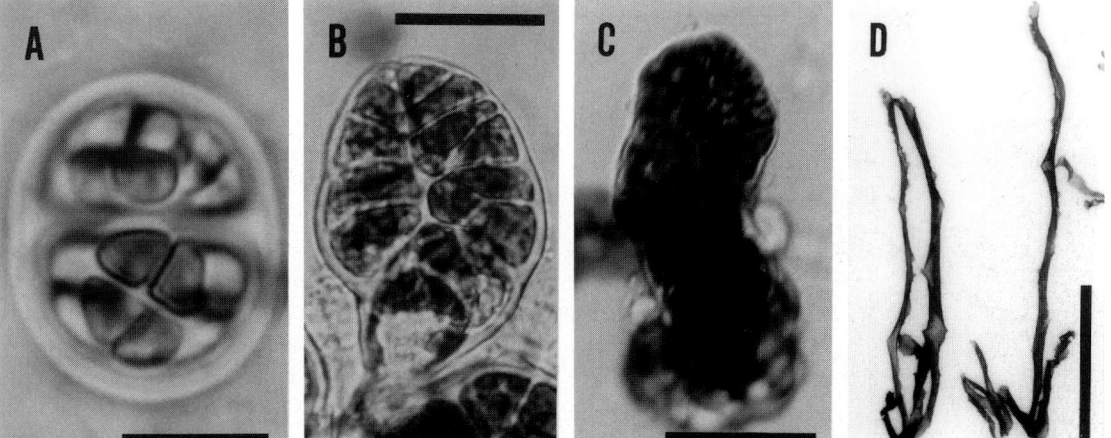

Figure 4. A, Division of a cell from the excised vegetative tissue of *Porphyra yezoensis* after 4 days at 15 °C and 40 μmol photon m^{-2}s^{-1} under 10L:14D. B, Development of foliose thallus from the excised tissue of *Porphyra yezoensis* after 8 days at 15 °C and 40 μmol photon m^{-2}s^{-1} under 10L:14D. C, Development of foliose thallus from the excised small tissue of *Porphyra pseudolinearis* after 12 days at 15 °C and 40 μmol photon m^{-2}s^{-1} under 10L:14D. D, Large thalli grown from excised 1–2 cells tissues of *Porphyra yezoensis* at temperature of 15 °C and 40 μmol photon m^{-2}s^{-1} under 10L:14D after 8 weeks in culture. Scale bars: A, 20 μm; B, 40 μm; C, 80 μ; D, 5 cm.

Discussion

Two different forms of differentiation of excised pieces of tissue from *Porphyra* spp. were observed, direct regeneration of the thallus and formation of reproductive cells. Direct regeneration of thallus was observed in the monoecious species *P. tenera* and *P. okamurae* from young thalli cultured at lower temperatures and from adult thalli of *P. yezoensis* and from young thalli of the dioecious species *P. dentata*. Tissue pieces consisting of 1–2 cells of vegetative tissue from *P. yezoensis* and *P. pseudolinearis* regenerated directly into new thalli. There appeared to be two methods of direct regeneration of bladelets: callus or embryo-like clumps, and normal regeneration or elongation of blade tissues. Callus or embryo-like clump differentiation was observed in 1–2 celled pieces of *P. yezoensis* in culture; blades were subsequently produced from these clumps. Callus clump differentiation has previously been reported in *P. umbilicalis* from experiments on the effects of plant hormones on callus induction (Liu & Kloareg, 1991). Polne-Fuller et al. (1984) and Polne-Fuller and Gibor (1987) reported that in *Porphyra* spp. callus clumps were produced from the protoplasts in agitated liquid medium and on solid medium. From these results, it appears that callus may be produced under abnormal culture conditions for growth of the foliose thallus.

Reproductive archeospores were produced from the tissue of young thalli of the monoecious species *P. tenera, P. okamurae* and *P. suborbiculata*, especially at higher temperatures. These archeospores germinated to form foliose thalli. The phenomenon of archeospore differentiation at higher temperatures resembles the development of early stages of foliose thallus in these species (Matsuo et al., 1994; Notoya et al., 1996a,b,c). Reproductive spermatangia were produced from the tissues of adult thalli of *P. pseudolinearis*. The differentiation of these cells may reveal the character of the original cells in the tissue.

Archeospores were not differentiated from the tissues of young or adult thalli of *P. dentata* or *P. pseudolinearis,* since the foliose thallus of these species do not produce archeospores.

Small pieces consisting of 1–2 cells of tissue from monoecious and dioecious species regenerated to form foliose thalli. The process of regeneration was similar to protoplast germination. Cell division occurred before germination, and later grew to produce thallus and/or rhizoids. After several weeks in culture, they grew into adult thalli.

In conclusion, these results suggest the possibility of using excised pieces of tissue or 1–2 celled tissues for obtaining archeospores from vegetative tissues in culture at higher temperatures. This 'seed production' could be used instead of conventional conchocelis culture for production of spores in commercial nori aquaculture. The polarity for growth of blades and rhizoids on excised tissues of *P. dentata* was similar to the results reported previously for *Schottera nicaeese* (as *Petroglossum nicaeense*) (Perrne & Felicini, 1972).

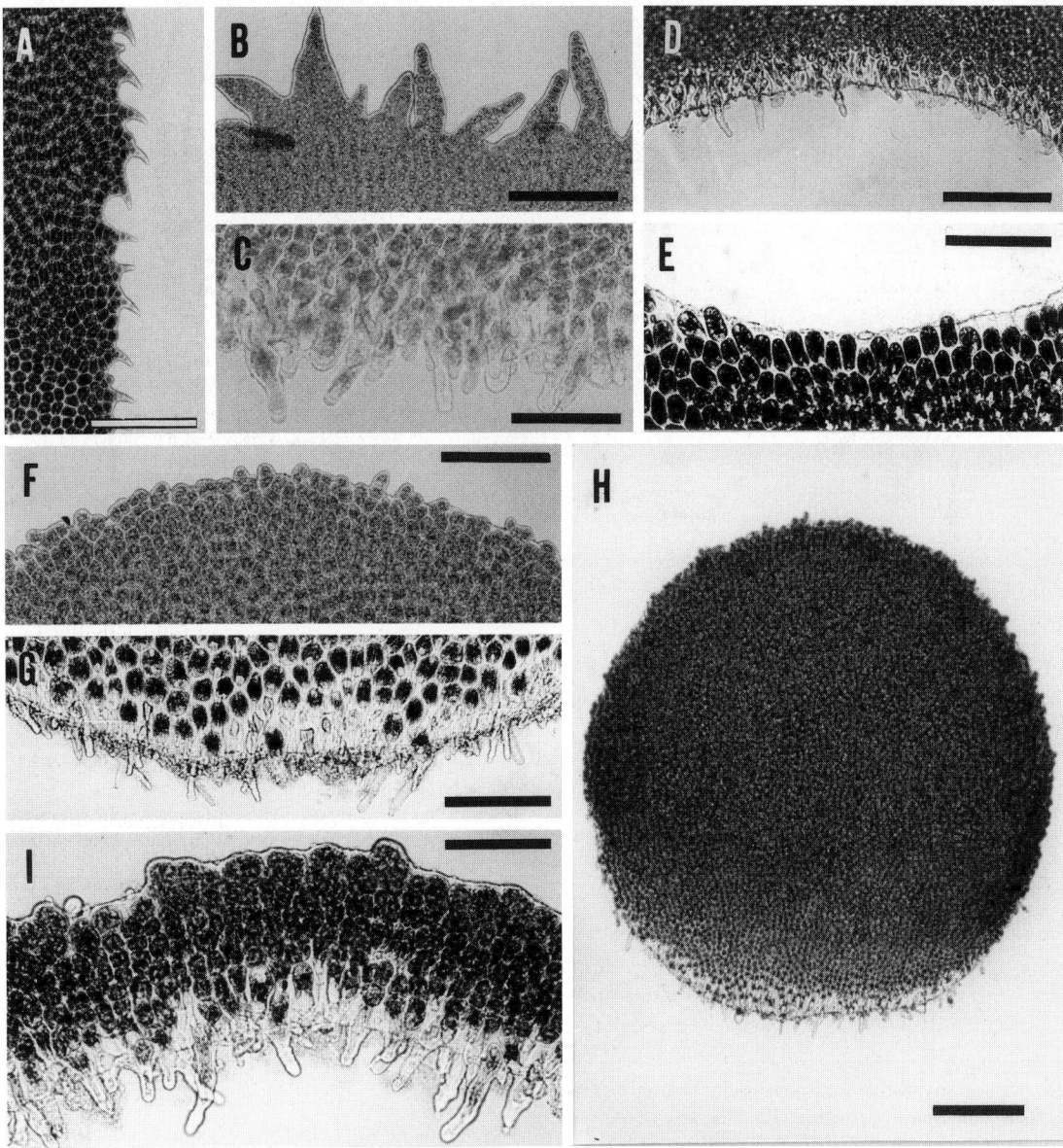

Figure 5. Regenerated bladelets and rhizoid from young thalli of *Porphyra dentata* tissue. A, Denticulate cell arrangement of *Porphyra dentata* at the side margin of the lower part of foliose thallus. B, Regenerated bladelets on the cut edge at the upper side of the tissue. C, Regenerated rhizoids on the cut edge at the lower side of the excised pieces of tissue. D, Regenerated rhizoids on the cut edge of the upper side of the punched out area in young foliose thalli. E, Regenerated blade cells on the cut edge of the lower side of the punched out area in young foliose thalli. F, Regenerated blade cells on the cut edge of upper side of the excised piece of tissue. G, Regenerated blade cells on the cut edge of lower side of the excised piece of tissue. H, Regenerated blade cells and rhizoid on the cut edge of the excised piece of tissue. I, Regenerated blade cells and rhizoid on the cut edge from the excised tissue which was cross-sectioned with 2–3-cell thickness. Scale bars: A-B and D-G, 150 μm; C, E and I, 80 μm; H, 200 μm.

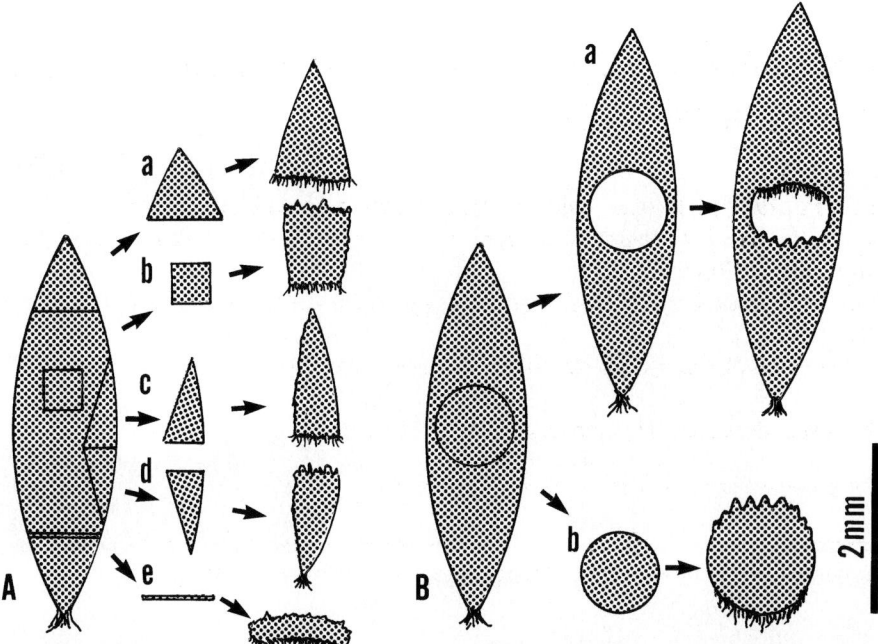

Figure 6. Schematic diagram of the excised tissue from young thalli of *Porphyra dentata* showing the regeneration of rhizoid and bladelets on the cut edges of the excised pieces of tissue. Scale bars: A, 20 μm; B, 40 μm; C, 80 μm; D, 5 cm.

However, further studies are required to determine the mechanisms responsible for the polarity in the pieces of tissue of *Porphyra* spp.

Acknowledgments

I thank Dr L.M.-C. Chen for his valuable comments and critical reading of the manuscript.

References

Aguirre-Lipperhide M, Estrada-Rodriguez FJ, Evans LV (1995) Facts, problems, and needs in seaweed tissue culture: an appraisal. J. Phycol. 31: 677–688.

Kim NG, Notoya, M (1997) Tissue culture of *Porphyra suborbiculata* and *P. pseudolinearis* (Bangiales, Rhodophyta). Abstracts of 4th International Marine Biotechnology Conference, 218 pp.

Kirihara S, Fujikawa Y, Notoya M (1997) Axenic tissue culture of *Sargassum confusum* C. Agardh (Phaeophyta) as a source of seeds for artificial marine forests. J. mar. Biotechnol. 5: 142–146.

Liu XW, Gordon ME (1987) Tissue and cell culture of New Zealand *Pterocladia* and *Porphyra* species. Hydrobiologia 151/152: 147–154.

Liu XW, Kloareg B (1991) Tissue culture of *Porphyra umbilicalis* (Bangiales, Rhodophyta). I. The effects of plant hormones on callus induction from tissue explants. Compt. r. Acad. Sci. Paris, Ser. 3: 517–522.

Matsuo M, Notoya M, Aruga Y (1994) Life history of *Porphyra suborbiculata* Kjellman (Bangiales, Rhodophyta) in culture. La Mer 32: 57–63. (in Japanese)

McLachlan J (1973) Growth media – marine. In Stein JR (ed.), Handbook of Phycological Methods. Cambridge Univ. Press, New York: 25–51.

Notoya M (1995) Concrete examples of tissue culture on *Porphyra* spp. and perspective for its application. Kaiyo Monthly 27: 639–642. (in Japanese)

Notoya M (1996) Tissue culture of *Porphyra* spp. and its application for cultivation. Abstracts of 2nd Japan-Korea Aquaculture Symposium: 25–26.

Notoya M (1997a) Chloroplast changes and differentiation of callus cells in *Eckloniopsis radicosa* (Kjellman) Okamura (Pheophyta, Lamianariales). J. appl. Phycol. 9: 175–178.

Notoya M (1997b) A new technique of seed production on *Porphyra* spp. by tissue culture. Abstracts of 4th International Marine Biotechnology Conference: 211.

Notoya M, Hara M, Kim NG, Yasui H (1996a) Field and culture studies on blade growth and morphology of *Porphyra okamurae* Ueda (Bangiales, Rhodophyta). Abstracts of First Asia-Pacific Phycological Forum: 51.

Notoya M, Kim HG (1996b) Excised pieces of the foliose thallus on *Porphyra suborbiculata* and *P. pseudolinearis* (Bangiales, Rhodophyta) in culture. Jap. J. Phycol. 44: 72. (in Japanese)

Notoya M, Kim NG, Suda M (1996c) Culture studies of *Porphyra tenera* from Matsukawaura, Fukushima Prefecture. Jap. J. Phycol. 44: 75. (in Japanese)

Perrone G, Felicini GP (1972) Sur le bourgeons adventifs de *Petroglossum nicaeese* (Duby) Schotter (Rhodophycées, Gigartinales) en culture. Phycologia 11: 87–95.

Polne-Fuller M, Biniaminov M, Gibor A (1984) Vegetative propagation of *Porphyra perforata*. Hydrobiologia 116/117: 308–313.

Polne-Fuller M, Gibor A (1987) Calluses and callus-like growth in seaweeds: Induction and culture. Hydrobiologia 151/152: 131–138.

Strain selection in *Kappaphycus alvarezii* var. *alvarezii* (Solieriaceae, Rhodophyta) using tetraspore progeny

Edison José de Paula[1,*], Ricardo Toledo Lima Pereira[2] & Masao Ohno[3]
[1] *Departamento de Botânica, Instituto de Biociências, Universidade de São Paulo, Caixa Postal 11.461, CEP 05422-970, São Paulo, SP, Brazil*
[2] *Instituto de Pesca da Coordenadoria de Pesquisa Agropecuária da Secretaria de Agricultura e Abastecimento do Estado de São Paulo, Caixa Postal 28, CEP 11.680-970, Ubatuba, SP, Brazil*
[3] *USA Marine Biological Institute, Kochi University, Usa-cho, Tosa, Kochi, 781-11, Japan*

(* Author for correspondence; e-mail: ejdpaula@usp.br)

Received 1 April 1998; revised 7 December 1998; accepted 8 December 1998

Key words: Kappaphycus, strain selection, tetraspore progeny, growth rate, carrageenan

Abstract

A brown strain of *Kappaphycus alvarezii* from the Philippines produced tetraspores in the summer and autumn (December 1995 to May 1996) in cultivation experiments in the sea at Ubatuba, São Paulo State, Brazil. *In vitro* tetraspore release and germination experiments showed a mass mortality two to four days after release. Only 20 plants derived from tetraspores were grown successfully for over a year in the laboratory. Large differences in morphology, colour, size and growth rates were observed amongst these plants. The individual plants differed from one another in one or more characteristic. Differences appeared in the early developmental stages and persisted through time. After ten months, the plants that grew best in laboratory culture were transferred into the sea, but the others remained very small (3 to 5 mm), even after two years. In the sea, the plants also showed individual differences in their ability to survive and grow. These results emphasise the potential of the tetraspore progeny for strain selection in *K. alvarezii*. The results also suggest that the tetrasporophyte used in these studies is of hybrid origin.

Introduction

Eucheuma denticulatum (Burman) Collins et Harvey and *Kappaphycus alvarezii* (Doty) Doty ex. Silva have been farmed as raw materials for carrageenan production in the Philippines since the 1970s and more recently in other tropical countries, mainly Indonesia and Tanzania (Adnan & Porse, 1987; Doty, 1987; Collén et al., 1995).

Vegetative propagation of these species is easily done by breaking off and planting individual large pieces of the thalli (Doty, 1987). This is the only farming method currently used (Trono, 1993, 1994; Collén et al., 1995; Azanza-Corrales et al., 1996). An impressive example of its efficiency is provided by Collén et al. (1995); they report that over 200 t fresh weight, which is now harvested each month in Zanzibar, started from approximately 4 kg of *E. denticulatum* brought from the Philippines in 1989. However, this labour-intensive method may bring a decrease in adaptability to environmental factors, particularly when cultivated plants originate from a single mother plant (Akatsuka, 1990; Collén et al., 1995). In fact, problems such as decreasing productivity and vigour of the stocks, seasonality in production, low quality of product, susceptibility of various strains of *Eucheuma* and *Kappaphycus* to 'ice-ice' and other diseases, which occur in the Philippines have been attributed to a variety of biological and economic factors (Trono & Ohno, 1989; Trono & Luisma, 1992; Trono, 1994; Azanza-Corrales et al., 1996). These problems are believed to result from the vegetative propagation of a single clone.

Several distinct phenotypes for both species have been commonly reported. However, the electrophoretic examination of isozyme variation in populations of *Eucheuma* in Florida (Cheney & Babbel, 1978) is the only formal genetic studies of these genera (Patwary & van der Meer, 1992). As pointed out by Kapraun & Lopez-Bautista (1997), strain improvement of these carragenophytes today relies on phenotypic selection of vigorous clones just as it did in the early 1970s (Doty, 1973; Doty & Alvarez, 1975; Trono & Luisma, 1992; Trono, 1994). Traditional hybridisation methods have been suggested as the logical solution to produce new strains, but *in vitro* work is limited due the large size of the thalli that requires large volumes of water and adequate facilities (Doty, 1987; Azanza-Corrales, 1990; Trono, 1994).

There has been considerable effort towards the production and selection of better strains of *E. denticulatum* and *K. alvarezii*. For example, the description of vegetative and reproductive structures and the occurrence of tetrasporic and gametophytic populations (Azanza-Corrales, 1990; Azanza-Corrales et al., 1992), *in vitro* branch, micropropagule and callus culture (Azanza-Corrales & Dawes, 1989; Azanza-Corrales et al., 1994; Dawes & Koch, 1991; Dawes et al., 1993, 1994; Polne-Fuller & Gibor, 1987), *in vitro* cystocarp formation, carpospore release, germination and development (Doty, 1987; Mairh et al., 1986; Azanza-Corrales & Aliaza, 1994), recruitment in the sea (Azanza-Corrales et al., 1996) and karyology (Kapraun & Lopez-Bautista, 1997). Sexual reproduction and the rearing of crops from spores have been suggested, but not as yet attempted (Doty,1987; Azanza-Corrales & Aliaza, 1994).

During a programme initiated in 1995 to evaluate the economic feasibility and the environmental risks of *K. alvarezii* mariculture in Brazil, plants introduced into the sea produced tetraspores in the summer and autumn months. Theoretically, higher genetic variability in tetraspore than carpospore progeny would be expected, since the former are produced after meiosis (putatively haploid) and the latter are produced after fertilisation followed by vegetative propagation in the carposporophyte (putatively diploid). The objective of the present study is to analyse the potential of tetraspore progeny for strain selection.

Materials and methods

Unialgal cultures were established from a 2.5 g brown branch of *K. alvarezii*, brought to São Paulo, Brazil, from 'USA' Marine Biological Institute, Kochi University Japan. After 10 months, 20 branches, with a mean range of 2.0 to 4.0 g wet weight, were produced and transferred to the sea at Ubatuba, SP, Brazil (23°26.9′ S, 45°0.3′ W).

Three excised fertile branches, about 10 cm in length and 50 g fresh weight, were selected each month from the sea during the reproductive period (December 1995 to May 1996). The branches were washed in running tap water followed by sterilised sea water, cleaned with paper towels, washed again in sterilised sea water and transported to the laboratory individually in Erlenmeyer flasks with 500 mL sterilised sea water. This material was maintained in a culture chamber at 25 ± 2 °C, under 60–100 μmol photon m^{-1} s^{-1} using 240-W fluorescent tubes with a 14L:10D (Light:Dark) photoperiod. After 1 or 2 days in these conditions, and when many tetraspores were being released, the branches were removed from the flasks. The seawater was first passed through a nylon net of 50 μm mesh to retain algal and other residues, followed by a 10 μm mesh filtration to retain the tetraspores.

The collected tetraspores were transferred to 50 or 100 mL beakers, decanted and washed with sterilised seawater several times until a clean suspension was obtained with minimum loss of spores. Healthy tetraspores were transferred to sterile 15 cm diameter Petri dishes, containing 100 mL of F/2 enriched seawater prepared as F/20 (Guillard & Ryther, 1962; Azanza-Corrales & Aliaza, 1994). After 1 to 2 days, the F/2 enriched seawater was siphoned and replaced by sterile seawater. The culture medium was changed twice a week with the period of nutrient enrichment (F/2 as F/20) limited to 1–2 d wk^{-1}. During the remaining 5-6 d wk^{-1} the tetra-sporelings were exposed to sterile seawater. Culture conditions for sporelings were similar except that the irradiance was adjusted to 170–200 μmol photons m^{-2} s^{-1} after the first month and media was changed once a week, alternating 7 days of nutrient enrichment (F/2 as F/5) with 7 days in sterile seawater. Salinity was 32–35 ‰ during the entire experimental period. Germanium dioxide (GeO$_2$) was added at 3 μg L^{-1} only after 1–2 months of culture to eliminate a minor contamination of diatoms.

Germlings were first grown fixed to the bottom of Petri dishes but, after one month, they were transferred to Erlenmeyer flasks of increasing size, from 50 to 500 mL, and were provided with aeration. During the first 4 months, all germlings were grown together in the same Erlenmeyer flasks. After that, they were transferred individually to separate flasks using a ratio of algal biomass: volume of culture medium below 10 g L^{-1}. From the eighth to tenth months of culture, each plant was weighed at weekly intervals. Growth rates were estimated from increases in fresh weight and are presented as percentage growth per day (Penniman et al., 1986):

G.R. = $[(Wt/Wo)^{1/t} - 1] \times 100$, where Wo = initial weight and Wt = weight after t days (7 days). After ten months, cuttings that grew best in laboratory culture were transferred to the sea. The raft method of cultivation was used, where the cuttings were attached to polypropylene long-lines positioned horizontally at a constant depth (30 cm) from the sea surface. The distance between long-lines was about 60 cm and the distance between cuttings varied from 10 to 25 cm. Growth rates were estimated at monthly intervals, based on fresh weight measurements. Growth rates were compared with clones of the parent grown under the same condition.

Semi-refined carrageenan content was analysed using the alkali treatment method (Ohno et al., 1996). Determinations were done in duplicate. About 50 g of material dried at 60 °C to a constant weight was incubated in 2 L of 6% KOH, followed by washing in gently running tap water overnight and then dried as before. Results are expressed as percentage of dry weight.

Results

Tetraspore release

Large numbers of tetraspores were obtained in laboratory experiments, from the branches collected in the sea between December 1995 and May 1996. However, in all 6 experiments carried out at monthly intervals a mass mortality of spores was observed 3–4 days after they had been released. Overall, very few (1 to 5) tetrasporelings germinated and survived 7 to 15 days after release, but in January 1996 21 tetra-sporelings survived and all but one survived in the laboratory for at least two years.

Laboratory experiments

Large differences in size, colour, morphology and growth were observed between the 20 survivors. These

Table 1. Colour, mean growth rates (G.R.) and final fresh weight (F.W.) of tetraspore progeny of *Kappaphycus alvarezii* in laboratory culture for 10 months

Plant	Colour	G.R. (% d^{-1})			F.W. (g)
		8th M	9th M	10th M	
Pl. 1	Green	–	–	–	0.01
Pl. 2	Green ↗ Black	2.8	0.8	0.7	0.53
Pl. 3	Green	2.6	1.1	1.3	0.59
Pl. 4	Red	0.9	1.6	1.3	0.12
Pl. 5	Red ↗ Black	1.6	0.7	–	0.68
Pl. 6	Red	3.1	2.6	1.9	3.97
Pl. 7	Red	3.7	1.9	1.2	2.50
Pl. 8	Red	2.3	4.3	3.5	11.66
Pl. 9	Pale brown	2.7	2.1	0.6	0.19
Pl. 10	Pale brown	2.1	2.3	2.5	1.26
Pl. 11	Pale brown	2.2	2.8	2.4	6.51
Pl. 12	Brown	0.9	0.4	0.2	2.78
Pl. 13	Brown	2.4	1.6	1.1	3.18
Pl. 14	Brown	1.9	2.8	1.6	4.43
Pl. 15	Brown ↗ Black	1.3	1.4	–	0.53
Pl. 16	Brown	2.2	2.3	1.7	17.19
Pl. 17	Brown	2.5	2.9	2.0	6.75
Pl. 18	Brown	2.5	4.5	2.9	11.08
Pl. 19	Brown	3.2	5.4	4.0	18.05
Pl. 20	Dark Brown	1.6	2.3	2.5	3.21

The notations: Green ↗ Black, Red ↗ Black and Brown ↗ Black were used to indicate that the plants changed their colour to black with time. Mean growth rates for 3 consecutive months (8th M = mean of 3 weeks, 9th M = mean of 4 weeks, 10th M = mean of 3 weeks).

differences appeared in the early stages of development and persisted through time. Each of the 20 plants differed from each other in one or more characteristic. By the end of 10 months the plants varied in length from about 10 to 100 mm and from unbranched to very densely branched (Figures 1, 2). Anastomosis occurred in the more branched plants. Colour of sporelings varied from green, red, to brown but typically remained constant with only small variations in tone. The exceptions were plants 2, 5 and 15 which became black with time; on changing colour the growth rates of these three plants decreased (Table 1). After 10 months, the plants differed in their growth rates and final fresh weight, ranging from 0.01 to 18.05 g. The best growing plants (Pl. 8, Pl. 11, Pl. 16, Pl. 17, Pl. 18, Pl. 19) were those that maintained relatively high and almost constant growth rates during the experiment (Table 1). Some plants (Pl. 1, Pl. 2, Pl. 4, Pl. 9) never attained sufficient size to be transferred to the sea even after 22 months in the laboratory.

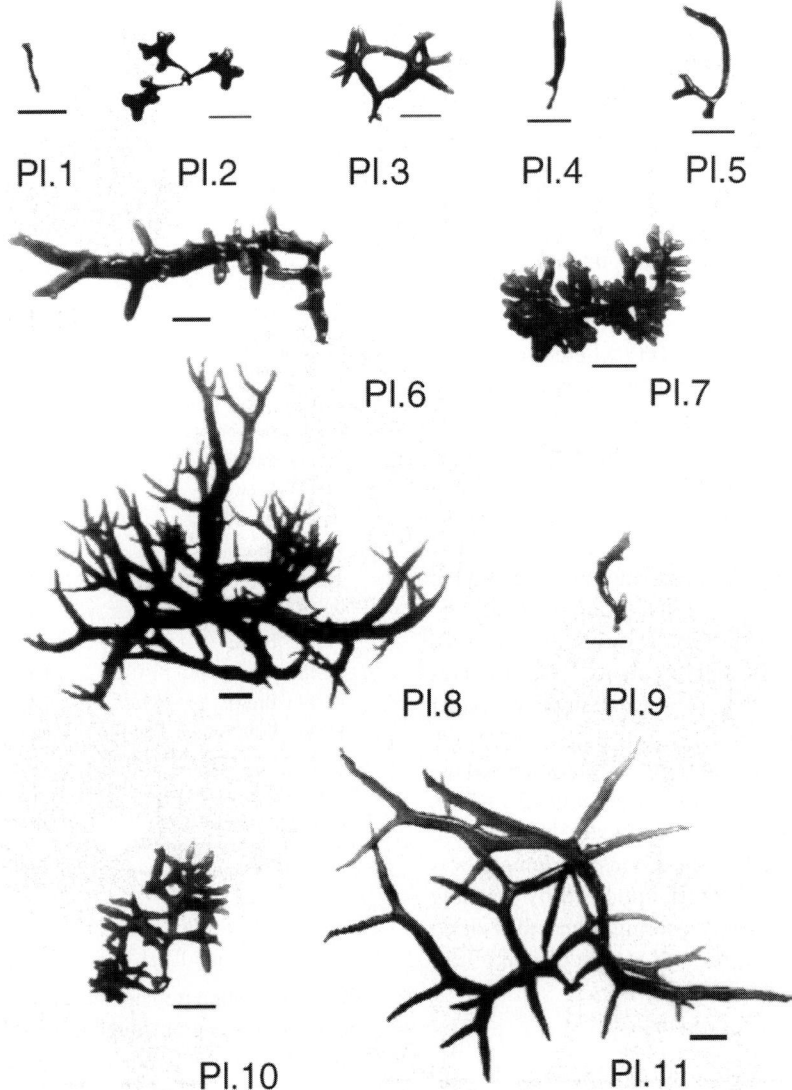

Figure 1. Morphology of tetraspore progeny of *Kappaphycus alvarezii* (Pl.1 to Pl. 11) after 10 months in laboratory culture. Scale bar = 1.0 cm (to be continued in Figure 2).

Sea experiments

Some of the cuttings produced from tetraspore progeny survived and were propagated in the sea at Ubatuba, SP, Brazil. Of these, some grew well by vegetative propagation (Pl. 8, Pl. 10, Pl. 11, Pl. 16, Pl. 17, Pl. 18, Pl. 19), while others were lost during the first few months in the sea due to very low growth rates associated with epibiosis and necrosis of the thallus.

The majority of the tetraspore progeny (9) attained a weight less than 10.0 g during the first 3 months in the sea but the remainder (6) attained fresh weights between about 200.0 and 800.0 g over the same period. Higher growth rates occurred from early spring to early autumn, with a decline occurring in May at the time of a drop in the seawater temperature. Large variations in growth rates were observed between different plants during the seven months in the sea (Table 2). Only two plants (Pl. 11 and Pl. 18) survived for the whole year. Cuttings from these plants had lower growth rates but higher semi-refined carrageenan content than the tetrasporophyte (Table 3).

The morphology and branch details of the best growing plants (Pl. 8, 11, 16, 17, 18 and 19) in the

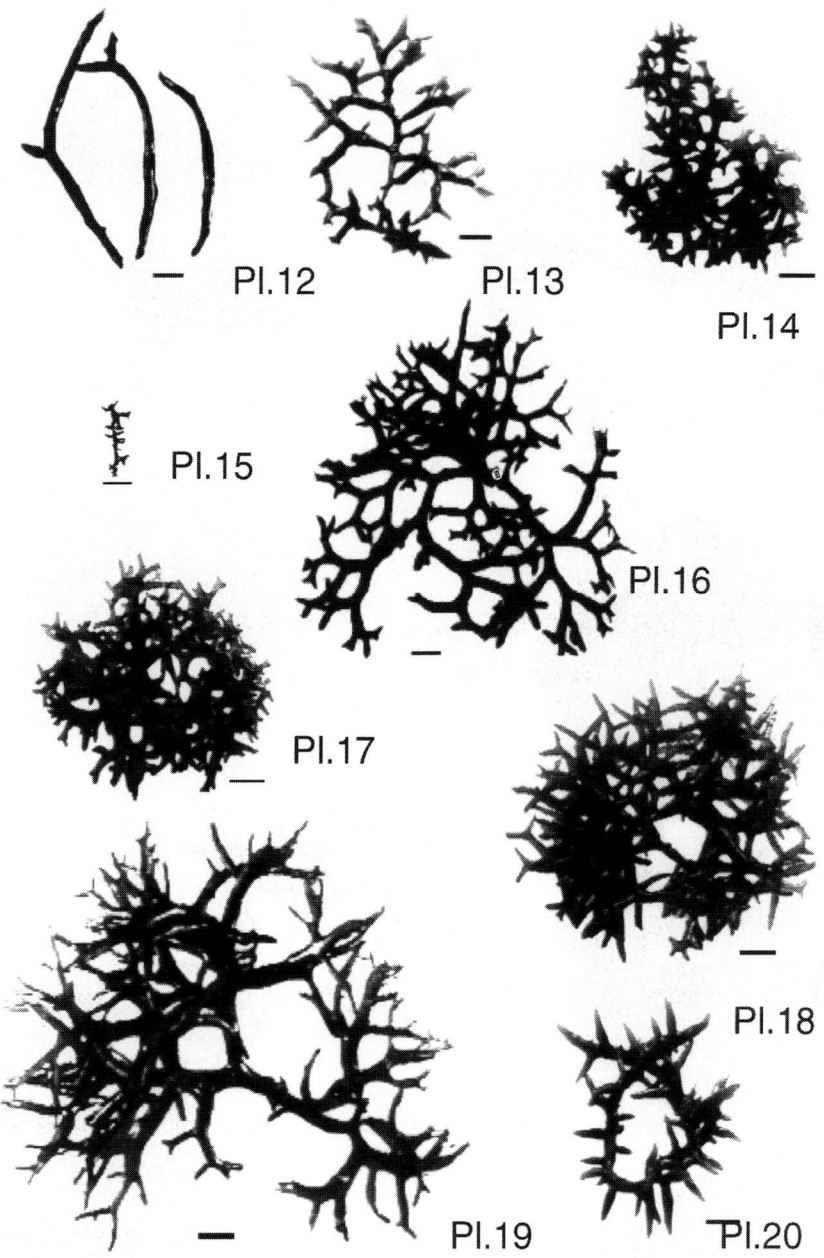

Figure 2. (Continued from Figure 1). Morphology of tetraspore progeny of *Kappaphycus alvarezii* (Pl.12 to Pl.20) after 10 months in laboratory culture. Scale bar = 1.0 cm.

sea were observed to differ from one another and from their parent plants grown under identical conditions (Figures 3, 4, 5). The observed differences remained almost constant during the entire experimental period (seven months). Some of the original characteristics observed in laboratory culture persisted in the sea including colour and branching pattern.

Plants Pl. 11 and Pl. 18 were the most robust and most similar to the parent plants; the former differed from the parent in details of branching with these being twisted and pale brown in colour (Figures 3–5). The main distinctive characteristics of Pl. 18 were the decumbent habit (Figure 3) and almost right angle branching (Figure 4). The other plants, Pl. 8, and par-

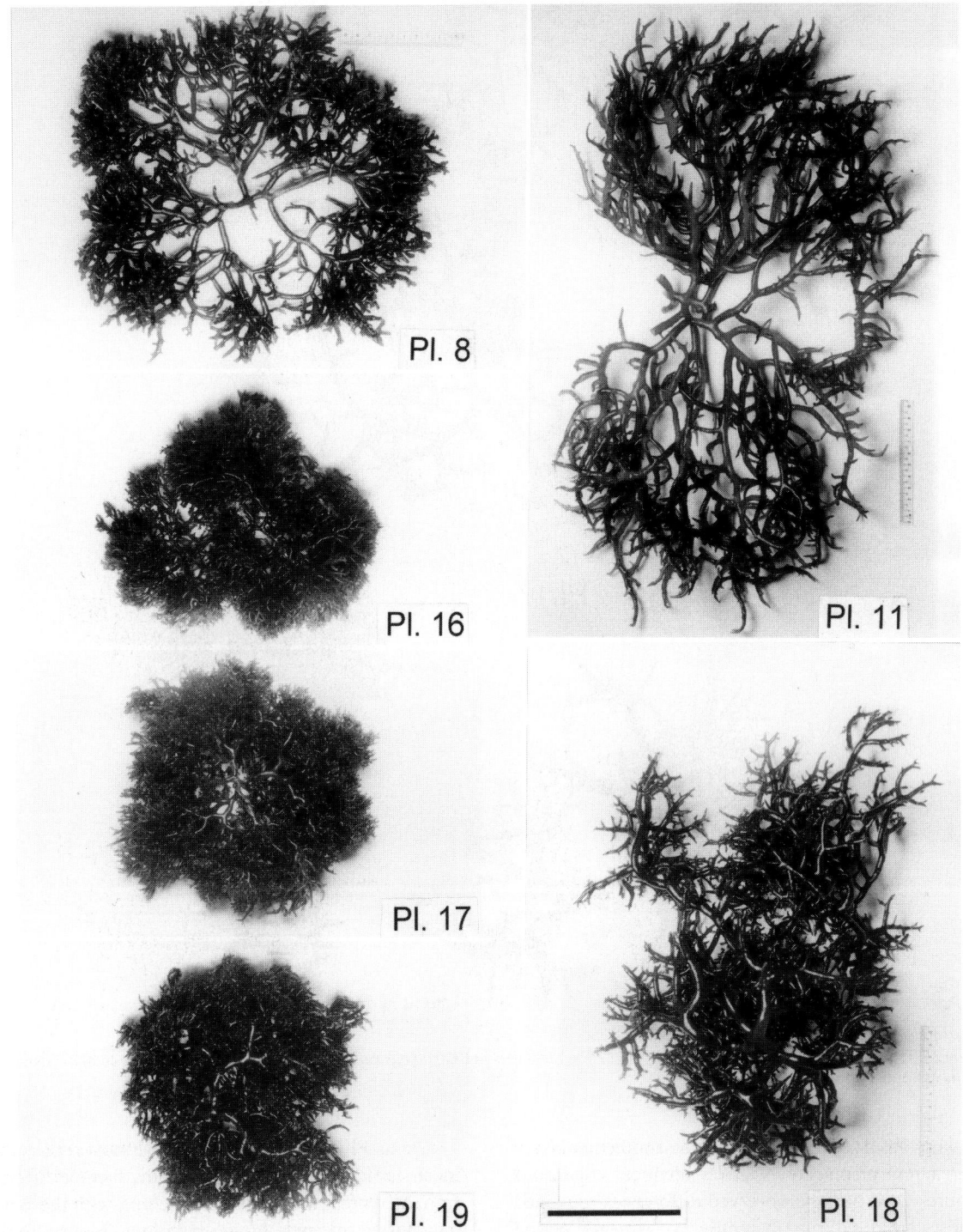

Figure 3. Morphology of the best growing plants from the tetraspore progeny of *Kappaphycus alvarezii* grown in the sea at Ubatuba, SP, Brazil. Scale bar = 10.0 cm.

Figure 4. Details of branches of the best growing plants from the tetraspore progeny of *Kappaphycus alvarezii* grown in the sea at Ubatuba, SP, Brazil. Scale bar = 5.0 cm.

ticularly Pl. 16, Pl. 17 and Pl. 19 were smaller and more densely branched with a bushy habit.

Discussion

The differences in morphology, colour, size and growth rates found amongst the various tetraspore progeny appeared early in their development and persisted through the period of cultivation in the laboratory and in the sea. Therefore, the variation described here for *K. alvarezii* as phenotypic may in fact be due to genotypic differences.

Collén et al. (1995) suggested that a potential threat to the cultivation of *Eucheuma* in Zanzibar is the

Figure 5. Morphology (top) and details (bottom) of branches of the tetrasporophyte of *Kappaphycus alvarezii* grown in the sea at Ubatuba, SP, Brazil. Scale bar, top = 10.0 cm and bottom = 5.0 cm.

narrow genetic base for the production, since cultivation is based on one or possibly two clones. According to the authors, through screening and strain selection it should be possible to find highly productive varieties which can grow rapidly and resist ice-ice and other potential diseases. In the present study, two of the best growing plants from tetraspore progeny had lower growth rates but higher semi-refined carrageenan content when compared with the parent tetrasporophyte. These findings emphasise the potential of the tetraspore progeny for strain selection in *K. alvarezii*.

Although higher genetic variability in tetraspore than carpospore progeny would be expected, the amount of individual variability of the tetraspore progeny, both in laboratory and in the sea, found in the present study is unexpected and apparently has no parallel for any algal species. It should be noted

Table 2. Growth rates of selected K*appaphycus alvarezii* plants (Pl.) derived from tetraspores grown in the sea at Ubatuba, SP, Brazil, for 7 months

Season	Spring	Summer		Autumn			
Month	96 Nov.	Dec.	97 Jan.	Feb.	Mar.	Apr.	May
Temp. (°C)	22.4	23.6	26.8	26.0	26.7	26.1	22.3
Plant							
Growth rate (% d^{-1})							
Pl. 8	6.5	3.8	2.7	3.8	5.2	5.3	1.3
Pl. 11	6.8	5.8	6.4	4.6	5.7	7.2	5.4
Pl. 16	3.5	3.1	3.6	4.4	3.8	4.0	−0.1
Pl. 17	4.8	4.5	5.1	4.7	4.6	4.7	1.4
Pl. 18	5.8	4.2	5.2	5.1	4.6	5.4	3.7
Pl. 19	3.8	2.5	3.1	2.6	5.3	2.5	0.6

Temp = Mean monthly seawater temperature.

Table 3. Growth rate and semi-refined carrageenan yield of *Kappaphycus alvarezii* parent plants and two plants derived from tetraspores (Pl. 11 and Pl. 18) grown in the sea at Ubatuba, SP, Brazil

Plant	Growth rate (% d^{-1})			Semi-refined carrageenan (%)		
	Parent	Pl. 11	Pl. 18	Parent	Pl. 11	Pl. 18
Month						
March 97	8.9	5.7	4.6	–	–	–
April	8.3	7.2	5.2	–	–	–
May	7.3	5.8	3.8	–	–	–
June	5.3	3.7	2.4	–	–	–
July	4.5	3.3	1.7	28.7	38.5	–
September	6.1	3.6	2.7	31.8	46.5	51.2
October	5.3	3.2	2.7	28.0	37.8	40.5
November	7.0	3.8	2.6	29.4	39.8	36.5
December	7.4	4.4	2.9	34.3	43.3	40.8
February 98	7.2	4.9	3.0	26.8	34.0	35.3

Number of plants analysed for growth rate, $n \geq 10$.

that previous work (Doty, 1987; Azanza-Corrales & Aliaza, 1994) made no mention of variability in carpospore progeny for either *K. alvarezii* or *Eucheuma* sp.

Besides the individual variability of the tetraspore progeny, a high mortality of tetraspores was also observed, suggesting that the strain studied is almost sterile. These two characteristics i.e. low fertility and variability of the progeny are classically attributed to some hybrids. *Kappaphycus alvarezii* is a vigorous species, possessing a third characteristic shared with some hybrids. Doty (1987) referred to *K. alvarezii* as 'a better variety' found early in the development of 'Eucheuma' farming, by a selection programme involving over 23 strains and species of wild 'cottonii'. Most of these early 'cottonii' were what would now be identified as *K. striatum* (Doty, 1973; Parker, 1974; Doty & Alvarez, 1975; Doty & Santos, 1978; Doty, 1985; Doty & Norris, 1985). It is noticeable that *K. striatum* is considered a highly variable species, and includes both dichotomous and decumbent forms (Doty, 1988; Trono, 1993).

The presence of two dichotomous and one decumbent plant among the tetraspore progeny obtained in the present study provides another argument supporting the hypothesis of a hybrid origin for *K. alvarezii*. The presence of these forms among the tetraspore progeny may represent the segregation of forms similar to the ancestral parent plants (putatively *K. striatum*).

Acknowledgements

The study was supported by Conselho Nacional de Desenvolvimento Científico e Tecnológico, CNPq and Fundação de Amparo à Pesquisa do Estado de São Paulo, FAPESP. Help provided by Luiz Carlos Silva Jr. and Marcelo A. da Silva, members of Instituto de Pesca, Base de Ubatuba, SP, during the field work, Ms Valéria Cress Gelli, head of the Instituto de Pesca for the use of local facilities, Dr Flavio A. de S. Berchez and Dra Estela M. Plastino for help in many ways, and Dr John A. West for critically reviewing the paper are all gratefully acknowledged.

References

Adnan H, Porse H (1987) Culture of *Eucheuma cottoni* and *Eucheuma spinosum* in Indonesia. Hydrobiologia 151/152: 355–358.

Akatsuka I (1990) Seaweed production in Japan. In Oliveira EC, Kautsky N (eds), Cultivation of seaweeds in Latin America. Workshop Univ. S. Paulo – Int. Foundation for Science, University of São Paulo, São Paulo, Brazil: 17–26.

Azanza-Corrales R (1990) The farmed *Eucheuma* species in Danajon Reef, Philippines: vegetative and reproductive structures. J. appl. Phycol. 2: 57–62.

Azanza-Corrales R, Aliaza TT (1994) *In vitro* carpospores release and germination of *Eucheuma* sp. from Tawi-Tawi, Philippines. Paper presented during the Second Asia – Pacific Conference on Algal Biotechnology held on 25–27 April, Singapore.

Azanza-Corrales R, Dawes CJ (1989) Wound healing in cultured *Eucheuma alvarezii* var. *tambalang* Doty. Bot. mar. 32: 229–234.

Azanza-Corrales R, Aliaza TT, Montaño NE (1996) Recruitment of *Eucheuma* and *Kappaphycus* on a farm in Tawi-Tawi, Philippines. Hydrobiologia 326/327: 235–244.

Azanza-Corrales R, Mamauag SS, Alfiler E, Orolfo MJ (1992) Reproduction in *Eucheuma denticulatum* (Burman) Collins and Hervey and *Kappaphycus alvarezii* (Doty) Doty farmed in Danajon Reef, Philippines. Aquaculture 103: 29–34.

Azanza-Corrales R, Mamauag S, Martin M, Sa-a P (1994) Laboratory branch culture of some Philippine farmed carrageenophytes. In Phang SM, Lee YK, Borowitzka MA, Whitton BA (eds), Algal Biotechnology in the Asia – Pacific Region. University of Malaya, Malaysia: 221–225.

Cheney DP, Babbel GR (1978) Biosystematic studies of the red algal genus *Eucheuma*. I. Electrophoretic variation among Florida populations. Mar. Biol. 47: 251–264.

Collén J, Mtolera M, Abrahamsson K, Semesi A, Pedersén M (1995) Farming and physiology of the red algae *Eucheuma*: growing commercial importance in East Africa. Ambio 24: 497–501.

Dawes CJ, Koch EW (1991) Branch micropropagule and tissue culture of the red algae *Eucheuma denticulatum* and *Kappaphycus alvarezii* farmed in the Philippines. J. appl. Phycol. 3: 247–257.

Dawes CJ, Lluisma AO, Trono GC, Jr (1994) Laboratory and field growth studies of commercial strains of *Eucheuma denticulatum* and *Kappaphycus alvarezii* in the Philippines. J. appl. Phycol. 6: 21–24.

Dawes CJ, Trono GC, Jr, Lluisma A (1993) Clonal propagation of *Eucheuma denticulatum* and *Kappaphycus alvarezii* for Philippine seaweed farms. Hydrobiologia 260/261: 379–383.

Doty MS (1973) Farming the red seaweed, *Eucheuma*, for carrageenans. Micronesica 9: 59–73.

Doty MS (1979) Status of marine agronomy, with special reference to the tropics. Proc. int. Seaweed Symp. 9: 35–58.

Doty MS (1985) *Eucheuma alvarezii*, sp. nov. (Gigartinales, Rhodophyta) from Malaysia. In Abbot IA, Norris JN (eds), Taxonomy of Economic Seaweeds: With Reference to Some Pacific and Caribbean Species. California Sea Grant College Program, Rep. T-CSGCP-011, La Jolla: 37–45.

Doty MS (1987) The production and use of *Eucheuma*. In Doty MS, Caddy JF, Santelices B (eds), Case Studies of Seven Commercial Seaweed Resources. FAO Fish. Tech. Pap. 281, Rome: 123–161.

Doty MS (1988) *Prodomus ad systematica Eucheumatoideorum*: A tribe of commercial seaweeds related to *Eucheuma* (Solieriaceae, Gigartinales). In Abbott IA (ed.), Taxonomy of Economic Seaweeds with Reference to Some Pacific and Caribbean Species. California Sea Grant College Program, Rep. T-CSGCP-018, La Jolla. Vol.2: 159–207.

Doty MS, Alvarez V (1975) Status, problems, advances and economics of *Eucheuma* farms. Mar. Technol. Soc. J. 9: 30–35.

Doty MS, Norris JN (1985) *Eucheuma* species (Solieriaceae, Rhodophyta) that are major sources of carrageenan. In Abbot IA, Norris JN (eds), Taxonomy of Economic Seaweeds: With Reference to Some Pacific and Caribbean Species. California Sea Grant College Program, Rep. T-CSGCP-011, La Jolla: 47–65.

Doty MS, Santos GA (1978) Carrageenans from tetrasporic and cystocarpic *Eucheuma* species. Aquat. Bot. 4: 143–149.

Guillard RRL, Ryther JH (1962) Studies of marine planktonic diatoms. I. *Cyclotella nana* Hustedt, and *Detonula confervacea* (Cleve) Gran. Can. J. Microbiol. 8: 229–239.

Kapraun DF, Lopez-Bautista J (1997) Karyology, nuclear genome quantification and characterization of carrageenophytes *Eucheuma* and *Kappaphycus* (Gigartinales). J. appl. Phycol. 8: 465–471.

Mairh OP, Soe-Htsun U, Ohno M (1986) Culture of *Eucheuma striatum* (Rhodophyta, Solieriaceae) in Sub-tropical Waters of Shikoku, Japan. Bot. mar. 29: 185–191.

Ohno M, Largo DB, Ikumoto T (1994) Growth rate, carrageenan yield and gel properties of cultured kappa-carrageenan producing red alga *Kappaphycus alvarezii* (Doty) Doty in the subtropical waters of Shikoku, Japan. J. appl. Phycol. 6: 1–5.

Ohno M, Nang HQ, Hirase S (1996) Cultivation and carragenan yield and quality of *Kappaphycus alvarezii* in the waters of Vietnam. J. appl. Phycol. 6: 431–437.

Parker HS (1974) The culture of the red algal genus *Eucheuma* in the Philippines. Aquaculture 3: 425–439.

Patwary MU, van der Meer JP (1992) Genetics and breeding of cultivated seaweeds. Kor. J. Phycol. 7: 281–318.

Penniman CA, Mathieson AC, Penniman CE (1986) Reproductive phenology and growth of *Gracilaria tikvahiae* McLachlan (Gigartinales, Rhodophyta) in the Great Bay Estuary, New Hampshire. Bot. mar. 29: 147–154.

Polne-Fuller M, Gibor A (1987) Calluses and callus-like growth in seaweeeds: Induction and culture. Proc. int. Seaweed Symp. 12: 131–138.

Trono GC, Jr (1993) *Eucheuma* and *Kappaphycus*: taxonomy and cultivation . In Ohno M, Critchley AT (eds), Seaweed Cultivation and Marine Ranching . Japan International Cooperation Agency, Kanagawa: 75–88.

Trono GC, Jr (1994) The mariculture of seaweeds in the tropical Asia-Pacific Region. In Phang SM, Lee YK, Borowitzka MA,

Whitton BA (eds), Algal Biotechnology in the Asia – Pacific Region. University of Malaya, Malaysia: 198–210.

Trono GC, Jr, Luisma AO (1982) Differences in biomass production and carrageenan yields among four strains of farmed carrageeenophytes in Northern Bohol, Philippines. Hydrobiologia 247: 223–227.

Trono GC, Jr, Ohno M (1989) Seasonality in the biomass production of the *Eucheuma* strains in Northern Bohol, Philippines. In Umezaki I (ed.), Scientific Survey of Marine Algae and their Resources in the Philippine Islands. Monbushio International Scientific Research Program, Japan: 71–80.

Efficient utilisation of high photon irradiance for mass production of photoautotrophic micro-organisms

Amos Richmond* & Ning Zou
Microalgal Biotechnology Laboratory, The Jacob Blaustein Institute for Desert Research Ben-Gurion University at Sede-Boker, Israel 84990

(*Author for correspondence; phone: 972–7–6596797; fax: 972–7–6596802; e-mail: AMOSR@bgumail.bgu.ac.il)

Received 1 April 1998; revised 10 November 1998; accepted 29 November 1998

Key words: mass cultures, light utilisation, light regime, optimal cell density, light path, mixing rates

Abstract

Basic issues involved in effective use of a high photon irradiance for mass production of microalgae are elucidated: efficient utilisation of high irradiance requires cultures of high cell density grown in reactors with a narrow light path. The smaller the light-path, the higher the growth rate and the volume output rate (g L^{-1}d^{-1}) of cell mass. Areal productivity (g m^{-2}d^{-1}) may be inversely related to the length of light-path (e.g. *Spirulina platensis*) or directly related to it, as is the case with *Nannochloropsis* sp., in which the areal output rate increased with the increase in the light-path and the areal volume (L m^{-2}). Inhibition of cell growth in *Nannochloropsis* became evident as cell concentration increased above a certain point. Response in cell growth to elevated irradiance was therefore possible only when the growth medium of ultrahigh cell density cultures was frequently changed. Inhibitory activity to culture growth may be directly involved in determining the optimal cell density (which results in the highest output of cell mass) and hence the optimal light-path. Under optimal growth conditions, cultures of high cell densities responded well to the rate of stirring, the relative beneficial effect of mixing increasing with the increase in cell density.

Introduction

The difficulty with designing efficient culture devices to mass-produce photoautotrophic microorganisms rests with the fact that light is attenuated exponentially upon penetrating into the culture. Unless the culture is optically very thin, light is available only in a narrow zone close to the culture surface. Therefore, stringent limits on cell density in photoautotrophic cultures and conventional reactors for mass production of microalgae used presently mandate very high areal volumes (e.g. 200 L m^{-2}) which contain relatively small cell concentrations, a costly and basically inefficient mode of production. Culturing photoautotrophic micro-organisms outdoors is based, conceptually, on effective utilisation of the high energy solar irradiance, which represents, as a rule, excessive radiation for the photosynthetic machinery of the individual cells. Efficient utilisation of high intensity solar irradiance (a condition on which outdoor production of photoautotrophy rests essentially), becomes therefore difficult to attain as the solar flux rises to mid-day. Reactors available today for industrial mass production fall short of facilitating the maximal photosynthetic potential, hence the challenge is to develop more effective, species specific modes for mass cultivation of photoautotrophic microorganisms.

Cultures of a very high cell density ('ultrahigh' cell densities - UHCD), arbitrarily defined as cultures with cell concentrations maintained above 10 g dry wt L^{-1} (or 150 mg chlorophyll L^{-1}) have been investigated recently (Hu et al. 1996b; Hu & Richmond, 1996). Ultra-high cell densities entail extreme mutual shading, due to which the cells are exposed to significant light intermittence. The limitation of light to growth must thus be defined in terms of the light regime, i.e.

the extent of cells' exposure to photon irradiance as modified by the characteristics of light intermittence. The latter is defined in terms of the overall duration of the light-dark cycles (related to the sizes of the lit and dark volumes in the reactor), the duration of the light and dark periods in an average cycle and, most importantly, the cycle frequency (Hu et al., 1998). This work describes some basic issues involved in effective use of irradiance for mass production of photoautotrophic algae.

Materials and methods

Reactor design, cooling and mixing by the use of compressed air as well as measurements of optimal cell density (OCD), photon irradiance reaching the reactor surfaces and the dry weight were all described by Hu et al. (1996b) and Hu & Richmond (1996). Hu et al. (1996a) described estimation of photoinhibition and Hu & Richmond (1994) described imposing shade on outdoor cultures.

Results and discussion

Optimal cell density

A meaningful difference between an open and an enclosed reactor concerns the optimal cell density (OCD). As in any biological system which entails productivity per unit area, there would be, at steady state under given conditions, an optimal population size (or cell concentration) that would result in the highest possible output rate of photoautotrophic cell mass per unit area.

The intensity of the light source in a light-limited system exerts a major influence on cell density as well as OCD. As irradiance increases, cell density rises to a maximal concentration which is characteristic of the species. The OCD for any given photon irradiance, as a rough rule, is in the range of one half the maximal cell concentration and at steady state, significant differences in output rate (productivity) between species are mainly due to differences in the specific growth rate at optimal cell mass (Richmond, 1996).

At a low photon irradiance (Figure 1A), OCD is attained at a relatively low cell concentration, resulting in a low output rate. At a high photon irradiance (Figure 1B), OCD is attained at a much higher cell concentration, yielding a much higher output rate. The

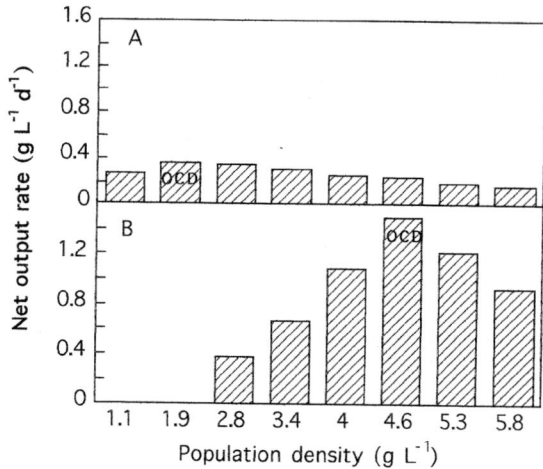

Figure 1. Interrelationships between incident photon flux density, optimal population density and net output rate of dry cell mass of *Isochrysis galbana*, grown outdoors in 2.6-cm diameter glass tubes (after Hu & Richmond, 1994).

importance of maintaining cell concentrations adjusted to the available irradiance is seen in Figure 1B; when cell density is too low in relation to irradiance, photodamage may cause culture collapse.

Damage by excess light depends to an extent on the state of photoacclimation of the cells (Falkowski et al., 1985; Falkowski & LaRoche, 1991). Cells acclimated to relatively low light (shade adapted) prior to exposure to high intensity solar radiation (as in Figure 1) will become photodamaged at a lower irradiation dose than cells which have been high light-acclimated. No matter what the physiological state of the cells is, taking care to maintain the optimal cell density in continuous cultures (albeit at the upper value of the OCD range) represents a practical approach by which to eliminate or greatly reduce photodamage induced by excess light (Hu et al., 1996b). In batch cultures grown outdoors, it is advisable to have the inoculum high light acclimated.

Length of light-path

The most significant factor, in addition to the intensity of the light source in affecting OCD and productivity, is the length of the light-path of the photobioreactor (Table 1). A reduction in the light-path will affect, in a strictly light-limited system, an increase in OCD as well as the specific growth rate, reflecting an improved surface to volume ratio that accompanies shortening of the light-path. The reduction in the light path, however, is associated with a proportional reduction in

Table 1. Comparison of the effect of the length of the light-path on optimal cell density (OCD) in *Spirulina platensis* (adapted from Hu et al, 1996) and *Nannochloropsis* sp.

Light path (mm)	Optimal Cell Density (OCD) g dry wt L^{-1}	
	Spirulina	*Nannochloropsis*
13	16	17
26	8	10
52	3	6
104	2	4
170	1	3

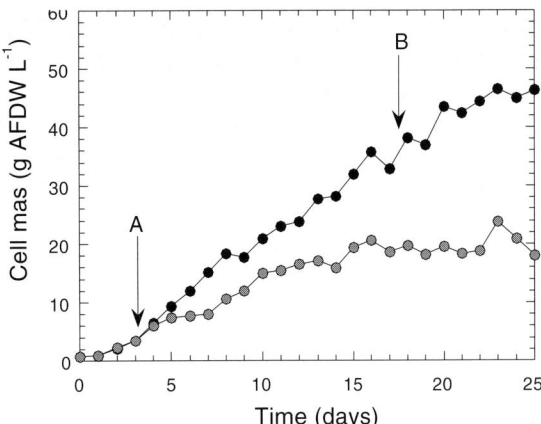

Figure 3. Growth curves of *Nannochloropsis*, as affected by irradiance and continuous removal of inhibitory activity. Arrow A: Beginning of the replacement for 'new medium culture'. Arrow B: Irradiance increased from 1000 to 2000 μmoles photon m^{-2}s^{-1}. New medium culture (black circles): culture centrifuged every 48 h, cell dispressed in new growth medium. Old medium culture (gray circles): as in new medium culture, except cells dispressed in old growth medium, to which nutrients were added every 72 h.

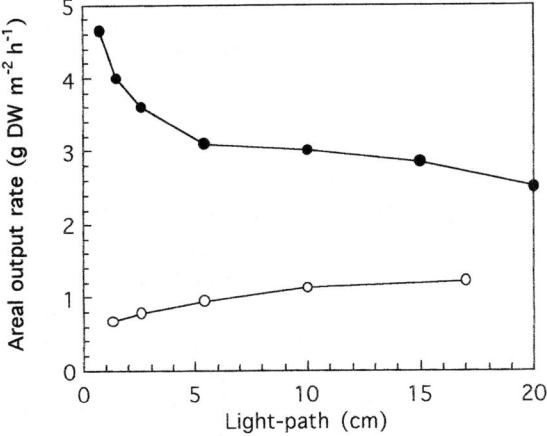

Figure 2. Comparison of the effect of the light-path on the areal output rate in *Spirulina platensis* (adapted from Hu et al., 1998) and outdoor cultures of *Nannochloropsis* sp. Full circles = *Spirulina*. Empty circles = *Nannochloropsis*.

the areal volume (L m^{-2}). Since productivity (P) of the culture at steady state is a function of the specific growth rate (μ), cell density (x) and culture volume (v) i.e. (P = $\mu \times$ v), a reduction in light path and culture volume could result in either increased or decreased areal productivity of cell mass (Figure 2), depending on the value of OCD and on the growth rate characteristic to the species at OCD. The former is clearly the case with *Spirulina*, in which the specific growth rate and areal productivity were accelerated as the light-path decreased (Hu et al., 1998). This is readily explained in that reduction in light-path increases the frequency of cells moving from the dark to the photic volume in the reactor, i.e. an increased frequency of the light-dark (L-D) cycle. The increased frequency affects a better utilisation of strong light and results in higher photosynthetic rates. This had been already proposed by Kok (1953) as well as Phillips & Mayers (1954) and was recently demonstrated by Grobbelaar et al. (1996) who used light emitting diodes to affect a range of L-D cycles.

The results obtained with *Nannochloropsis*, in contrast to *Spirulina*, are perhaps explainable by its intrinsic slow growth rate (doubling time of ca. 24 h compared with ca. 8 h of *Spirulina*). This prevents a large quantitative response to a sharp improvement in the light regime associated with a reduction in the light-path. It is worth noting that the OCD for *Spirulina* sp. and *Nannochloropsis* sp. is similar in the short light paths (i.e. 13 and 26 mm). However, with increasing light path, the difference in OCD between these species increases and the OCD in the 170-mm reactor is ca. 3 times higher in *Nannochloropsis* (Table 1). Indeed, as the light path increases, the productivity gap between these species decreases (Figure 2). Another possible explanation pertains to the observation that, cell growth inhibitory activity has been detected in association with ultrahigh cell densities (Javanmardian & Palsson, 1991), mandating continuous removal of growth-inhibitory materials, e.g. by frequent change of the growth medium. The cell concentration threshold at which the inhibitory activity begins to reduce the growth of *Nannochloropsis* is in the range of 6–7 g dry wt L^{-1}. This is indicated by comparing growth rates of two *Nannochloropsis* cultures (Figure 3); in one (black circles), the growth

Table 2. Interrelationships of the output rate, photosynthetic efficiency (PE) and the range of optimal cell density (OCD) in Spirulina platensis as affected by incident light intensity and the rate of mixing (adapted from Hu et al., 1996)

Photon Flux Density (μmol m^{-2} s^{-1})	Mixing rate[1] (L L^{-1} min^{-1})	OCD (g L^{-1})	Maximal output rate (mg L^{-1} h^{-1})	PE (%)	R[2] (%)
500	0.6	3.0	7.0	9.6	100
	2.1	5.0	98.0	13.4	133
	4.2	5.5	103.0	14.0	140
1800	0.6	6.0	200.0	7.6	100
	2.1	11.0	300.0	11.1	145
	4.2	16.0	400.0	15.3	200
	6.3	14.5	320.0	11.8	155

[1] litre of air per litre of culture suspension per minute.
[2] R = Relative increase in mixing-induced productivity. Mixing rate of 0.6 L L^{-1} min^{-1} serves as control (100%).

medium is completely replaced every 48 h by centrifuging the entire culture (5000 g for 15 min) and then resuspending the cells in fresh growth medium, in the other (gray circles), the culture is treated in the same manner, except that the cells are resuspended in the old medium to which the entire mineral nutrients comprising the growth medium are added every 72 h to prevent nutrient limitation. Up to a cell concentration of ca. 6–7 g dry wt L^{-1}, there is no difference in growth between treatments. Above this cell concentration, net growth in old medium is first curtailed, and eventually ceases (Figure 3). Clearly, if the sharp increase in population density expected with the reduction in light-path couldn't be fully expressed because of rising inhibitory activity, then maximal areal productivity would be obtained with a relatively long light path, associated with a larger areal volume and lower OCD, and thus no inhibitory activity. Maximal productivity of *Nannochloropsis* was obtained with long light-path (10 and 17 cm) reactors maintained at an OCD of 3 to 4 g L^{-1} (Figure 1). Clearly, the length of the light path should be carefully optimised for the cultured species. The relationships between the length of the light-path, the concentration of cell mass and the extent of inhibitory activity to cell growth are presently being investigated at the author's laboratory.

Mixing

The rate of mixing or the degree of turbulence in the culture may greatly affect the optimal population density, as shown in a culture of *Spirulina*, in which high turbulence is necessary to achieve the full potential for photosynthetic productivity obtainable by increased irradiance (Table 2). For a given irradiance there is, under given conditions, an optimal population density, which in turn is affected by the rate of mixing. As the population density increases (having been adjusted to the increase in irradiance), the relative effect of mixing rate on productivity increases. Mixing rates above a certain optimal, however, are harmful, resulting in decreased outputs. The rate of mixing in the culture therefore represents a factor, which facilitates a more efficient utilisation of light for photosynthesis by improving the light regime to which the average cell is exposed. Turbulent streaming is particularly effective in *Spirulina platensis* with long (100–200 μm) filaments, which would readily sink to the bottom when mixing is insufficient. This would be most probably the case with other large-size species, requiring a significant energy input to affect homogeneous dispersion of cell-mass in the growth medium. In such species, it would seem that the higher the cell concentration, the higher the relative effect (R) of optimal mixing rates (Table 2).

Conclusions

The conditions required to obtain highest photosynthetic productivity of photoautotrophic cell mass in continuous cultures exposed to high PFD have been identified as follows: a) the interaction between the intensity of the light source and the population density must be optimised. It determines a basic parameter in a light-limited continuous culture i.e. the light regime for the individual cells, b) a relatively narrow light path which ensures L-D cycles of the highest practical frequency is mandatory for maintaining ultra-high cell densities which result in high output rates. The high L-D frequency affects a higher OCD and may affect (e.g. in *Spirulina*, Hu et al., 1998) an exponential increase in output rate, reflecting efficient light utilisation for photosynthesis. However, this work illustrates that the length of light path must be carefully adjusted to the species. If cell-growth inhibitory activity associated with increased cell concentration cannot be practically or efficiently removed or, in species unable to respond quickly to a high light source (e.g., *Nannochloropsis*), a longer light-path with higher areal volume and lower cell concentrations will be optimal, c) a high (carefully optimised) rate of turbulent mixing must prevail to facilitate full expression of the photosynthetic potential. The higher the cell concentration, the higher the relative effect that the rate of mixing exerts on productivity.

These guidelines should be useful for designing efficient photobioreactors, which requires utilisation of high irradiance to produce high volume and areal output rates (Richmond, 1996). 'Reactor efficiency' may be estimated by relating Vm, the volumetric output rate (which as a rule would increase with a decrease in the light-path), to the overall irradiated photobioreactor area (A_I) required to produce a given amount of photoautotrophic cell mass i.e. Vm A_I^{-1}. Reactor efficiency pertains to an economic entity, the most effective being a reactor which yields high volume output rates (associated with relatively small areal volumes and high cell concentrations), yet requires the smallest irradiated area to produce a given quantity of cell mass or cell product.

References

Falkowski PG, Dubinsky Z, Wyman K (1985) Growth-irradiance relationships in phytoplankton. Limnol. Oceanogr. 30: 311—321.

Falkowski PG, LaRoche J (1991) Acclimation to spectral irradiance in algae. J. Phycol. 27: 8–14.

Grobbelaar JU, Nedbal L, Tichy V (1996) Influence of high frequency light/dark fluctuations on photosynthetic characteristics of microalgae photoacclimated to different light intensities and implication for mass algal cultivation. J. appl. Phycol. 8: 335–343.

Hu Q, Guterman H, Richmond A (1996a) A flat inclined modular photobioreactor (FIMP) for outdoor mass cultivation of photoautotrophs. Biotechnol. Bioengng 51: 51–60.

Hu Q, Guterman H, Richmond A (1996b) Physiological characteristics of *Spirulina platensis* cultured at ultrahigh cell densities. J. Phycol. 32: 1066–1073.

Hu Q, Richmond A (1994) Optimizing the population density in *Isochrysis galbana* grown outdoors in a glass column photobioreactor. J. appl. Phycol. 6: 391–396.

Hu Q, Richmond A (1996) Productivity and photosynthetic efficiency of *Spirulina platensis* affected by light intensity, cell density and rate of mixing in a flat plate photobioreactor. J. appl. Phycol. 8: 139–145.

Hu Q, Zarmi Y, Richmond A (1998) Combined effects of light intensity, light-path and culture density on output rate of *Spirulina platensis* (cyanobacteria). Eur. J. Phycol. 33: 165–171.

Javanmardian M, Palsson BO (1991) High-density photoautotrophic algal cultures: design, construction, and open operation of a novel photobioreactor system. Biotechnol. Bioengng 38: 1182–1189.

Kok B (1953) Experiments on photosynthesis by *Chlorella* in flashing light. In Burlew JS (ed.), Algal Culture from Laboratory to Pilot Plant. Publ. no. 600, Carnegie Institution, Washington DC., 63 pp.

Phillips JN, Meyers J (1954) Growth rate of *Chlorella* in flashing light. Plant Physiol. 29: 152–161.

Richmond A (1996) Efficient utilization of high irradiance for production of photoautotrophic cell mass: A survey. J. appl. Phycol. 8: 381–387.

Time-dependent attachment mechanism of bacterial pathogen during ice-ice infection in *Kappaphycus alvarezii* (Gigartinales, Rhodophyta)

Danilo B. Largo[1,2,*], Kimio Fukami[1] & Toshitaka Nishijima[1]
[1] *Laboratory of Aquatic Environmental Science, Department of Aquaculture, Faculty of Agriculture, Kochi University, Otsu 200 Monobe, Nankoku City, Kochi 783-0091, Japan*
[2] *Present address: Department of Biology, University of San Carlos, Cebu City 6000, Philippines*

(* Author for correspondence; e-mail biology@mangga.use.edu.ph)

Received 1 April 1998; revised and accepted 8 December 1998

Key words: Kappaphycus/Eucheuma, ice-ice disease, infection mechanism, seaweed-bacteria interaction, *Vibrio* sp. P11

Abstract

The mechanism of infection by *Vibrio* sp. P11 promoting the ice-ice disease in *Kappaphycus alvarezii* was investigated *in vitro*. Its intensity of infection differs from that of another ice-ice promoter (*Cytophaga* sp. P25) by promoting the disease much faster. However, when secondary infection by other bacteria starts, its ability to compete with these bacteria gradually diminishes, whereas, infection by P25, although not displaying such drastic effects as P11, shows consistent competitive ability against other bacteria. Time-series infection experiments with application of polyclonal antibodies to specifically detect *Vibrio* sp. P11 revealed that this bacterium has a high affinity for the seaweed especially when the latter is stressed. It promotes the disease after a rapid increase in cell density of up to 10^7 g^{-1} (wet wt.) in the first 24 h. This bacterial cell build-up may take only 1–2 h on stressed thalli, but takes about 24 h on non-stressed thalli. Build-up is not sustainable in non-stressed thalli as high density is usually followed by a sudden decline in cell number believed to result from an algal defence against potential pathogens. Inoculation of the bacterium on thalli incubated in continuous culture system extends the time of bacterial attachment due to laminar flow and, possibly, competition by existing bacteria on the seaweed surface and in ambient seawater medium. Motility-driven cell attachment by this bacterium is suggested as an important factor for infection.

List of abbreviations: DAPI-DC, Diamidinophenylindole – Direct Count; IFA, Indirect Fluorescence Antibody; Pabs, Polyclonal Antibodies.

Introduction

Bacterial disease in seaweeds is a rare phenomenon (Rheinheimer, 1992) compared to terrestrial plants. However, reviews by Andrews (1979) and, more recently, by Correa (1997) on seaweed diseases, suggest that algal pathology as a discipline is still in its infancy and much is yet to be learned on the existence of an algal pathosystem.

The recent phenomenon reported by Littler and Littler (1994, 1995) of a bacterial pathogen they called coralline lethal orange disease (CLOD), consuming a large population of reef-building coralline red algae in the South Pacific, underscores the potential of a bacterial pathogen to wreak havoc on coral reef ecosystems. Similar threats of this magnitude could possibly happen among cultivated seaweed species as farming practices of these algae mainly rely on clonal propagation methods (e.g. *Kappaphycus/Eucheuma* and *Gracilaria*), allowing the seaweed to become susceptible to a potential pathogen (Santelices, 1992).

The ice-ice disease of the red algae *Kappaphycus/Eucheuma* initially thought of as mainly a non-infectious disease and which could be triggered by unfavourable environmental conditions such as extremes of temperature, irradiance and salinity (Largo et al., 1995b), could also be attributed to some opportunistic bacterial pathogens (Largo et al., 1995a). So far, the mechanism of the seaweed-bacteria interaction during the process of infection has not been elucidated or demonstrated.

In an attempt to determine the existence of a seaweed-bacteria pathosystem for the ice-ice disease, the growth behaviour of ice-ice-promoting bacterial strains on *Kappaphycus alvarezii* (Doty) Doty was investigated under various culture conditions, based on the premise that the ability of a bacterial pathogen to colonise the alga will depend on the prevailing environmental conditions in which the alga is grown. In this paper, we report the time-dependent influence of water movement, temperature and conditions of the seaweed host on the ice-ice infection of *K. alvarezii* by the opportunistic bacterium *Vibrio* sp. P11; this bacterial strain was also compared with another ice-ice disease promoting strain *Cytophaga* sp. P25 on *K. alvarezii*.

Materials and methods

Source of experimental seaweed materials

Healthy, epiphyte-free parts of main branches of *Kappaphycus alvarezii* were selected from a seaweed stock cultured at Usa Marine Biological Institute, Kochi University (Japan). Prior to experiments, they were washed and brushed several times in clean seawater, followed by another washing with autoclaved seawater. To remove loosely-associated bacteria, the thalli were sonicated for 10 min at 38 kHz using a washing sonicator (Iuchi Co., Japan) with temperature adjusted to 25 °C. The branches were then treated with an antibiotic mixture as previously described (Largo et al., 1995a) to inhibit, if possible, growth of naturally existing bacteria. The branches were then cut into 5 cm long pieces and distributed individually into ten 100-mL flasks containing 100 mL autoclaved seawater prior to further treatment as described below.

Infection intensities of P11 and P25 in stressed and non-stressed seaweed

To compare the intensities of infection between *Vibrio* sp. P11 and *Cytophaga* sp. P25, bacteria cultured in FeTY medium (see Largo et al., 1995a) were inoculated separately in 10 replicate flasks containing *K. alvarezii* branches; half of these flasks contained stressed branches. Stress conditioning was conducted using a salinity of 20‰ for 3–5 days after which thalli were returned to seawater (autoclaved) at normal salinity (34‰) for 3 days prior to bacterial inoculation. Approximately 4.41×10^4 cells mL^{-1} of *Vibrio* sp. P11 and 1.5×10^3 cells mL^{-1} of *Cytophaga* sp. P25 were inoculated into the flasks containing seaweed. Since this experiment was performed before we could develop an immunofluorescent probe with polyclonal antibodies against *Vibrio* sp. P11 (Largo et al., 1998), the standard agar plate method was used to enumerate the bacteria (Largo et al., 1995a). Identification of strains P11 and P25 from infected thalli was based on the morphological characteristics of the re-isolated colony-forming units (CFUs) formed after spreading samples onto 1.5% agar as previously described (Largo et al., 1995a). The alga-bacterial consortium was incubated, without shaking, at the alga's optimum culture conditions of 25 °C, 12:12 L:D cycle and 50 μmol photons m^{-2} s^{-1}. Samples of both healthy and ice-ice diseased parts from the same and/or different thalli were obtained for bacterial enumeration.

Behaviour of *Vibrio* sp. P11 in stressed and non-stressed seaweed

Vibrio sp. P11 culture was inoculated into replicate flasks containing *K. alvarezii* to a final inoculum size of 3.19×10^5 cells mL^{-1}. The flasks were then incubated as described above. Sampling for bacterial counting involved the removal of small duplicate pieces of thalli, each less than 1 g (about 4–6 mm long), every hour for the first 24 h and every 1–2 days thereafter, from cultures incubated for between 14 and 21 days. In addition, parallel ambient seawater samples of 1 mL were obtained. Bacteria were counted by epifluorescence microscopy using DAPI staining for the total number (DAPI-DC) and a specific polyclonal antibody for *Vibrio* sp. P11 counting (IFA method; see Largo et al., 1998).

Behaviour of P11 on thalli under the influence of seawater movement

Vibrio sp. P11 was in this experiment allowed to interact, in moving water, with naturally occurring bacteria from seawater supplied in a continuous culture system (Figure 1). The effect of water turbulence compared

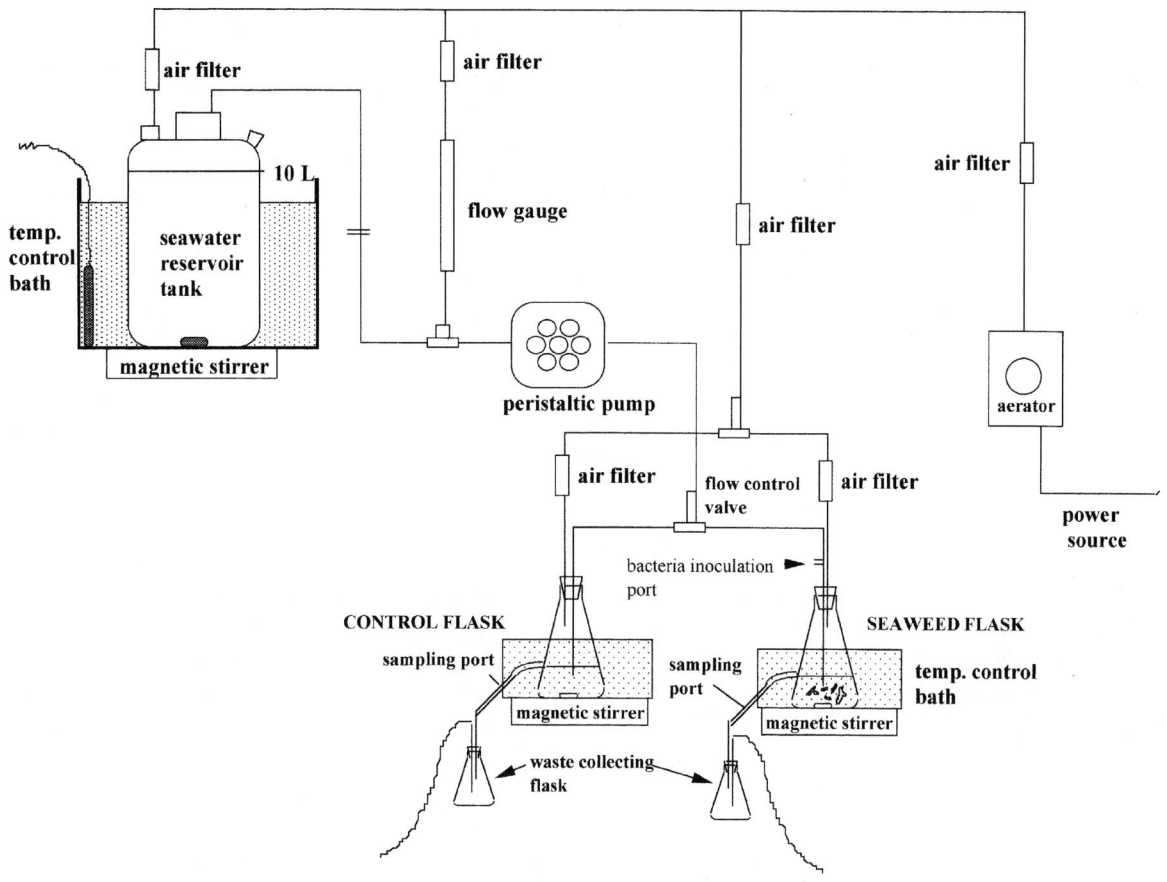

Figure 1. Diagram of the continuous culture system consisting of a seawater reservoir pulse-feeding, via a peristaltic pump, two 300-mL incubation flasks supplied with seawater at a culture volume exchange rate of 0.5–1.0 volume per day. Freshly collected seawater (from Uranouchi Inlet of Tosa Bay, southern Japan) was used in the experiment after passing it through a GF/C filter to remove grazers and large organic particles while keeping the natural bacteria intact. One of the incubation flasks contained seaweed branches cut into 5-cm pieces as described above while the other flask contained nothing except seawater, as control.

to laminar flow on bacterial attachment was determined by moving the seaweed with a magnetic stirrer at half the maximum speed of a Pasolina mini-stirrer (Iuchi Co., Japan). The results were compared to those from an identical experiment but without stirring. The desired temperature of 25 °C for the seaweed and control flasks were maintained using a heated water bath, and they were illuminated under the same conditions as described above. The seaweed was acclimatised for a 4 d period before strain P11 was inoculated. When depleted, the seawater in the reservoir was replenished with the same seawater stock as used from the start of each experiment in order to provide natural bacteria from the same population.

One of the experiments was performed to determine the effect of temperature on P11 attachment on non-stressed thalli within 24 h without stirring. Two flasks containing the seaweed were adjusted to the upper (30 °C) and lower (20 °C) temperatures of the optimal range required by the seaweed. Sampling and counting of bacteria were performed as above.

Results

Infection intensities of Vibrio sp. P11 and Cytophaga sp. P25 on K. alvarezii

The re-isolation of bacteria from both deteriorated and healthy branches of P11- and P25-inoculated cultivars is shown in Figure 2. P11-like bacteria, forming translucent and arboreous colonies, were recovered only from ice-ice diseased branches of both stressed and non-stressed thalli, and not from the apparently

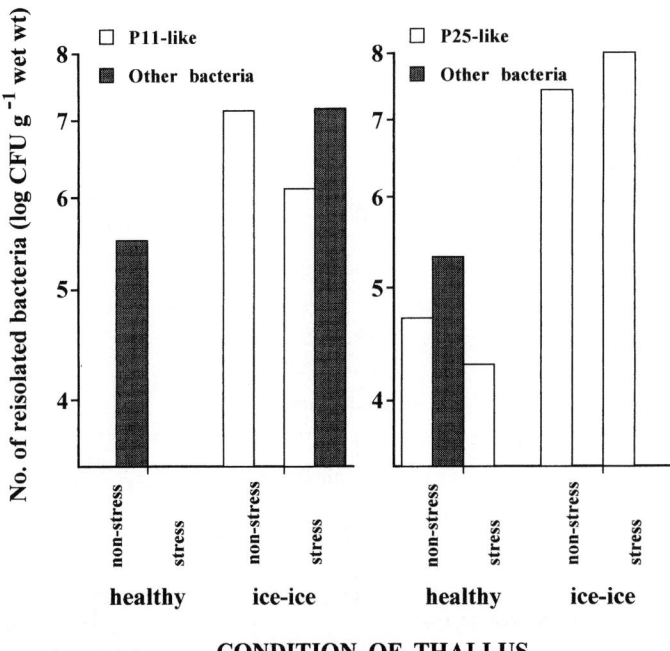

Figure 2. Re-isolated bacteria from P11- and P25-infected thalli of *K. alvarezii* in a batch culture experiment. Seaweed samples consisting of healthy and ice-ice diseased parts were obtained at the first sign of the disease.

healthy ones. Bacteria showing morphologically different CFUs from that of P11 were also found to increase in density, up to about one order of magnitude higher (1.64×10^7 CFUs g^{-1} wet wt.) than that of P11 (1.07×10^6 CFUs g^{-1} wet wt.) in stressed thalli. While other bacterial strains could be successfully suppressed by antibiotic treatment, P11-like CFUs in non-stressed thalli were high (1.34×10^7 CFUs g^{-1} wet wt.). Although P25-like CFUs were recovered from both healthy and diseased branches of both stressed and non-stressed thalli, diseased branches yielded higher numbers. These trends were similar to those obtained in a separate experiment (data not shown).

Behaviour of Vibrio *sp. P11 in stressed and non-stressed seaweed*

The fluctuation in strain P11 number counted by P11 polyclonal antibodies (PAbs) and that of total bacteria determined by DAPI-DC in batch cultures of stressed and non-stressed thalli of *K. alvarezii* are presented in Figure 3. *Vibrio* sp. P11 on both stressed and non-stressed thalli increased in cell density during the first hour of inoculation but always reached considerably higher numbers in stressed thalli than in non-stressed ones until day 12; at that time, the number of bacteria on the stressed thalli was lower than on the non-stressed thalli. P11 on stressed thalli reached cell densities of over 10^7 cells g^{-1} after 24 h but only less than 10^6 cells g^{-1} in non-stressed thalli over the same period. The early build-up of P11 cell density on stressed thalli was followed by an early onset of the ice-ice disease on the day 2 (Figure 3A, upper). However, the build-up of P11 cell density on non-stressed thalli was not immediately followed by an ice-ice condition as there was a sharp decline in its density after 3 d. Ice-ice started to develop in non-stressed thalli as the number of P11 started to increase again, reaching a high cell density on day 11; this time P11 underwent a stationary phase followed by a decline from day 14 onwards. A separate but identical experiment (data not shown) using P11 revealed a typical declining trend in the first few days following a rapid cell increase. This phenomenon was observed only in non-stressed thalli.

Compared to P11 cells, the total bacteria numbers on stressed and non-stressed thalli did not differ widely in the first 10 to 11 h (Figure 3A). Total bacteria growth followed that of P11 in the first 6 days, but then numbers tended to increase due to the appearance of other bacterial strains in the culture.

In the ambient seawater medium (Figure 3B), a consistent increase of P11 cell density was observed

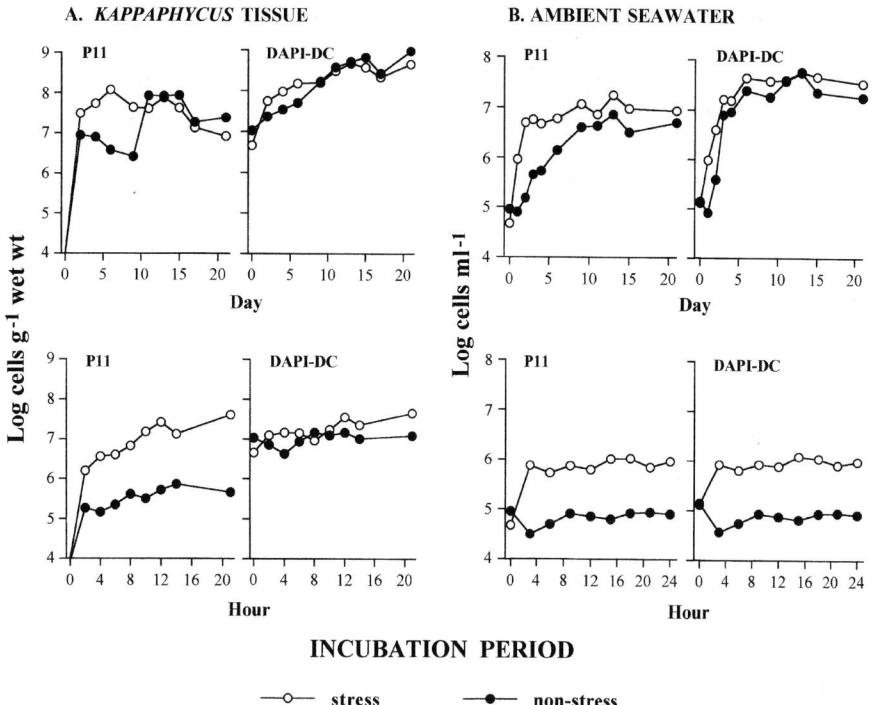

Figure 3. Behaviour of *Vibrio* sp. P11 under stressed and non-stressed condition in a batch culture system. Ice-ice developed (horizontal bars) after 2 days in stressed- and 8 days in non-stressed thalli. A: counts of bacteria from seaweed tissue. B: Counts of bacteria from ambient seawater. Each point represents the average count of duplicate subsamples. Legends apply to all panels.

in both stressed and non-stressed thalli, with stressed thalli giving consistently higher numbers. A similar pattern was observed for total bacteria.

Behaviour of P11 on seaweed thalli under the influence of seawater movement

Using a continuous culture system with stirring, strain P11 failed to attach to *K. alvarezii* tissue and a gradual disappearance of the bacterium in the ambient seawater was observed (Figure 4). When the same system was used in another experiment, but without stirring, strain P11 was able to colonise the seaweed tissue. When attachment was observed within the first 24 h, P11 cell density peaked at a value just over 10^6 cells g^{-1}. This was immediately followed by a slight decline for 2 d after which a generally increasing trend in cell density was observed up to about day 14 before equilibrium was reached. Ice-ice disease was observed to develop after about 5 d. The timing of ice-ice manifestation in the thalli was quite similar to our previous observations in batch culture, i.e. within a week (Largo et al., 1995b).

The density of P11 after inoculation in the ambient seawater of the continuous culture system with and without stirring dropped to about 50% from its initial density of $2 - 3 \times 10^5$ cells mL^{-1} after 24 h. In the stirred system, P11 gradually disappeared after 14 days but in the non-stirred system the decline occurred in the first 7 days after which it equilibrated (Figure 4B). In the system without stirring, total bacteria density from the *Kappaphycus* tissue only increased very slightly during the first 24 h but those from the culture medium declined dramatically, reflecting a more or less similar pattern to that of P11.

There seemed to be no distinct behavioural pattern of P11 in response to the two temperature regimes tested (20 and 30 °C; Figure 5). P11 appeared to be in a very unstable condition of attachment in the *Kappaphycus* tissue at both temperatures (Figure 5A) and this was accompanied by a gradual loss in the ambient seawater (Figure 5B). In the case of total bacteria, a slightly higher density was observed at 30 °C than at 20 °C in both seaweed tissue and ambient seawater, with no marked trend during the 24 h observation period.

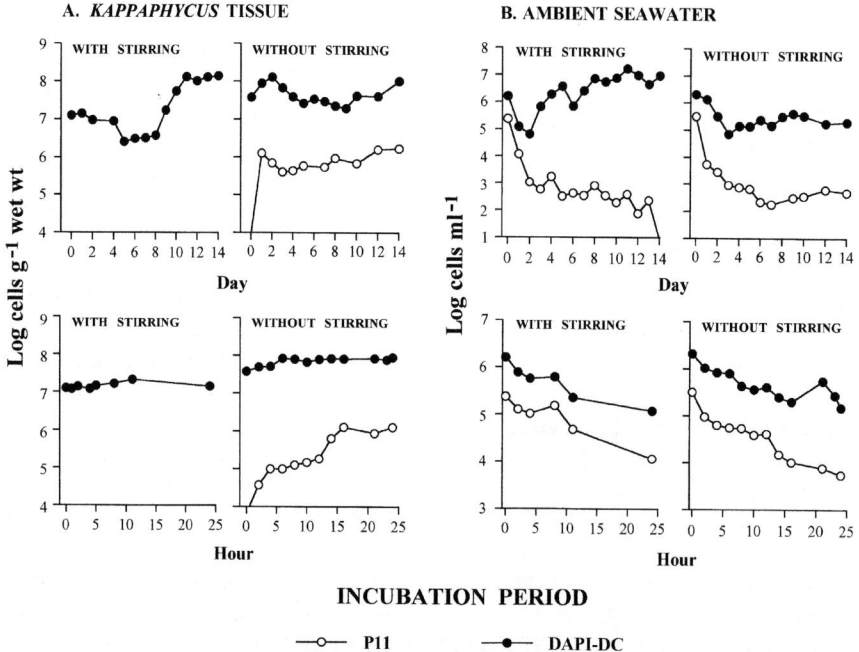

Figure 4. Behaviour of *Vibrio* sp. P11 in non-stressed *K. alvarezii* in a continuous culture system with an exchange rate of 1 volume d^{-1} with and without stirring. A: counts of bacteria from seaweed. B: counts of bacteria from ambient seawater medium. Note the absence of P11 in seaweed tissue from the stirred system. Each count represents the average of duplicate subsamples. Legends apply to all panels.

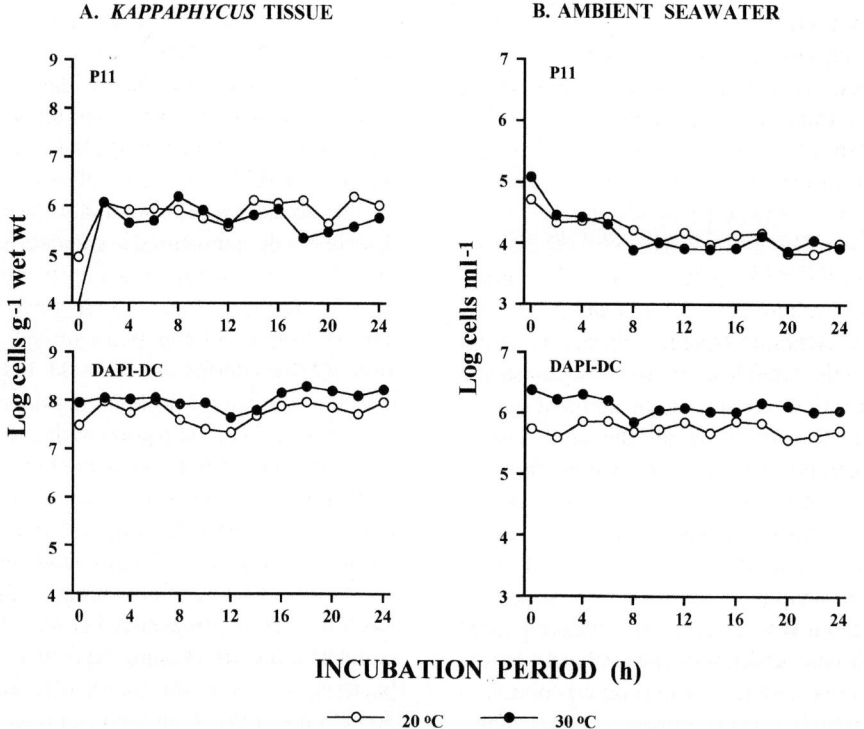

Figure 5. Effect of lower (20 °C) and upper (30 °C) range of temperature requirement of *K. alvarezii* on the behaviour of P11 on non-stressed thalli incubated using a continuous culture system without stirring. A) counts of bacteria from seaweed tissue. B) counts of bacteria from ambient seawater medium. Each point represents the average count of duplicate subsamples. Legends apply to all panels.

Discussion

The development of ice-ice disease in *Kappaphycus alvarezii* depends on several factors to which the seaweed is exposed. The combined effect of stress and biotic agents, such as opportunistic pathogenic bacteria, are primary factors of the ice-ice disease. The infection of the seaweed by these pathogens may depend, initially, on the bacteria's ability to establish themselves on the seaweed surface. This ability, however, could be influenced by the members of the bacterial community or, possibly, with other co-existing micro-organisms. Initial experiments using the agar plate method showed that *Vibrio* sp. P11 could be recovered only from ice-ice diseased thalli parts but its competitive ability against other bacteria was weak. In the absence of other bacteria causing ice-ice disease, non-stressed thalli created an environment in which P11 promoted the ice-ice disease. Our previous study showed that P11 induces the ice-ice disease within 7 days after infection (Largo et al., 1995a). On the other hand, the high cell density of *Cytophaga* sp. P25 recovered from both healthy and ice-ice diseased thalli parts suggests that this bacterium has more competitive ability against other bacteria and that it is part of the natural flora. Unfortunately, we were unable to produce an immunofluorescent probe for *Cytophaga* sp. P25 and thus, we could not study in detail the behaviour of this strain on the seaweed.

The attachment of *Vibrio* sp. P11 to the seaweed thalli is required for the bacterium to promote the ice-ice disease and it is enhanced by the flagellum of the micro-organism (Largo et al., unpublished data). Previous studies on the behaviour of motile and non-motile mutants of *Pseudomonas fluorescens* (Korber et al., 1994) and *Ps. syringae* (Haefele & Lindow, 1987) showed that motile strains colonised the substrate more successfully than non-motile ones, due to their active flagella. In the present study, the quick build-up of P11 cells on the thalli, reaching concentrations above 10^6 cells g^{-1}, is necessary in the ice-ice disease development and is one of the advantageous properties of P11 as an ice-ice causative bacterium. The development of ice-ice symptoms after a rapid increase in numbers of bacterial cells is typical of that found in higher plants. Alfano and Collmer (1996) described this phenomenon as a compatible response (disease-causing) of the plant host against an increased number of pathogenic bacteria. However, in non-stressed thalli of *K. alvarezii*, the observed decline in bacterial density, delaying for several days the development of the ice-ice symptoms, could result from a form of algal defence mechanism against invasive bacterial pathogen. The reversible attachment of P11 could be due to the presence of an inhibitory factor. *Kappaphycus alvarezii* was found to possess an antibacterial agent which selectively inhibits several strains of bacteria, depending on the season of seaweed collection and the thalli part where the antibacterial extract was obtained (Largo et al., unpublished results).

Another factor shown to affect the general attachment of P11 was water movement; continuous stirring clearly diminishes bacterial fixation. In the continuous culture system, when stirring was not used, some of the inoculated P11 cells were able to colonise the alga, but most of them were easily lost during the course of incubation. This has also been observed by Korber et al. (1994) for the motile strain of *Pseudomonas fluorescens* in a continuous culture system and may be due to physical factors such as hydrodynamic conditions and shear stress continually affecting the attaching cells. Even among already attached cells, after they reached an equilibrium, they remain dynamic as they are continually adhering and desorbing from the surface (Lawrence et al., 1995). Moreover, the number of P11 on the seaweed tissue, especially after a period of incubation, can also be affected by the increase in number of the pre-existing bacteria. Since these bacteria already occupied most of the seaweed surface prior to inoculation, despite the antibiotic treatment, this gave them a more competitive edge over that of the inoculated P11.

The attachment of P11 did not vary much in response to water temperatures within the lower or upper optimal range required for *K. alvarezii*. Both temperature regimes allowed attachment of P11 into the seaweed within the first 24 h although they were in a very unstable condition, suggesting reversibility in its attachment, at least during the early period.

Based on the findings presented in this study, it can be concluded that *Vibrio* sp. P11, by virtue of its motility, has the ability to attach quickly and colonise seaweed tissue as a first step of infection, provided that the prevailing conditions, both on the seaweed surface and in the ambient seawater, allow successful attachment of the bacterium. The subsequent build-up of the bacterial cells is a prerequisite to promoting the ice-ice disease. The second step of infection by *Vibrio* sp. P11 is based on its ability to utilise carrageenan in place of agar as a culture medium (Largo et al., 1995a), and therefore could be using this cellular polysaccharide as a carbon source which is abundant in

K. alvarezii cell wall matrix. The bacterium has been microscopically observed to penetrate the medullary layer of an P11-infected thalli (Largo et al., 1998), suggesting the possible involvement of hydrolytic enzymes such as carrageenase. Zablackis et al. (1993) showed that cellulase and carrageenase, which were originally isolated from bacteria, release epidermal as well as medullary cell protoplasts in *K. alvarezii*. This hydrolytic activity could be the reason for thalli whitening, symptomatic of ice-ice disease. Once P11 bacterial cells have colonised the thalli at high concentrations and start to utilise carrageenan (1–2 days), penetration into the medulla could cause the further weakening and eventual collapse of the thalli in the affected part. Water current in the cultivation site then causes the affected thalli to fragment prematurely.

From the standpoint of seaweed crop management, the disadvantageous factors affecting P11 attachment into *K. alvarezii* are also advantages when considering preventive measures of ice-ice disease. Areas with strong water exchange are normally ideal sites for seaweed farming. Thus, healthy thalli of *K. alvarezii*, as a result of efficient uptake of nutrients in well-flushed and growth-conducive areas, can themselves ward-off potential pathogens possibly by the production of an antimicrobial compound. This is an important consideration when selecting suitable sites for *Kappaphycus/Eucheuma* farming.

Acknowledgement

D.B. Largo is grateful to the Ministry of Education, Science and Culture, Government of Japan, for the fellowship grant under which this study was conducted as part of his PhD dissertation.

References

Alfano J, Collmer A (1996) Bacterial pathogens in plants: life against the wall. The Plant Cell 8: 1683–1698.

Andrews JH (1979) Pathology of seaweeds: Current status and future prospects. Experientia 35: 429–570.

Correa JA (1997) Infectious diseases of marine algae: current knowledge and approaches. In Round FE, Chapman DJ (eds), Progress in Phycological Research 12: 149–180.

Haefele DM, Lindow SE (1987) Flagellar motility confers epiphytic fitness advantages upon *Pseudomonas syringae*. Appl. envir. Microbiol. 53: 2528–2533.

Korber DR, Lawrence JR, Caldwell DE (1994) Effect of motility on surface colonization and reproductive success of *Pseudomonas fluorescens* in dual-dilution continuous culture and batch culture systems. Appl. envir. Microbiol. 60: 1421–1429.

Largo DB, Fukami K, Adachi M, Nishijima T (1998) Immunofluorescent detection of ice-ice disease-promoting bacterial strain *Vibrio* sp. P11 of the farmed macroalga, *Kappaphycus alvarezii* (Gigartinales, Rhodophyta). J. mar. Biotechnol. 6: 178–182.

Largo DB, Fukami K, Nishijima T (1995a) Occasional pathogenic bacteria promoting *ice-ice* disease in the carrageenan-producing red algae *Kappaphycus alvarezii* and *Eucheuma denticulatum* (Solieriaceae, Gigartinales, Rhodophyta). J. appl. Phycol. 7: 545–554.

Largo DB, Fukami K, Nishijima T, Ohno M (1995b) Laboratory-induced development of the *ice-ice* disease of the farmed red algae *Kappaphycus alvarezii* and *Eucheuma denticulatum* (Solieriaceae, Gigartinales, Rhodophyta). J. appl. Phycol. 7: 539–543.

Lawrence JR, Korber DR, Wolfaardt GM, Caldwell DE (1995) Behavioral strategies of surface-colonizing bacteria. In Jones JG (ed.), Advances in Microbial Ecology, Plenum Press, New York 14: 1–75.

Littler MM, Littler DS (1994) A pathogen of reef-building coralline algae discovered in the South Pacific. Coral Reefs 13: 202.

Littler MM, Littler DS (1995) Impact of CLOD pathogen on Pacific coral reefs. Science 267: 1356–1360.

Rheinheimer G (1992) Aquatic Microbiology. John Wiley and Sons, Chichester, U.K., 363 pp.

Santelices B (1992) Strain selection of clonal seaweeds. In Round FE, Chapman DJ (eds), Progress in Phycological Research 8: 85–116.

Zablackis E, Vreeland V, Kloareg B (1993) Isolation of protoplasts from *Kappaphycus alvarezii* var. *tambalang* (Rhodophyta) and secretion of *i*-carrageenan fragments by cultured cells. J. exp. Bot. 44: 1515–1522.

Subject index

Page numbers refer to the first page of a paper in which the entry is employed (from page 515 onwards this is the number at the bottom of the page)

abalone feed, 183, 495
abalone, 487
absorption spectra, 231
activation energy, 361
adelphoparasite, 401
adhesion, 241
Adriatic, 75
aeration, 329
AFRIKELP, 495
agarophytes, 1, 7, 173, 329, 443, 463, 487
agglutination, 563
agronomic diligence, 15
Alaska, 291
albumin, 241
algal blooms, 583, 603
alkali modification, 541
amphipods, 183
anatomy, 57
Ancud, 137, 427, 435
animal feed, 1
chromatography, 549
antheridia, 247
antibacterial agent, 1, 643
antibiotics, 329, 619, 469, 643
antifungal agents, 1
antiviral agents, 1
apical segments, 339
aquaculture, 291, 417, 487, 603
aquaria, 261, 443
archeospores, 115, 183, 299, 619
Argentina, 397
asexual cycle, 115
Atlantic Ocean, 25, 39
Atoll Seaweed Company, 477
atomic absorption analysis, 523, 593
attachment, 241
attenuation coefficient, 361
Australia, 65, 275
auxiliary cell, 25
auxins, 495
axenic explants, 339

bacteria, 7, 571, 643
Baja California, 501
Banda, 299
barren ground, 261
beach-cast, 487
beche de mer, 477
benthic diatoms, 253
biculture, 443
biodegradation, 515
biofilters, 603
biogeography, 25, 39, 65, 91
biolistic bombardment, 469

biomass, 137, 149, 159, 173, 183, 201, 275, 361, 509
biosorption, 593
biphasic, 121, 127
bipolar, 115
blades, 137, 191, 361
bladelets, 361
bleach, 253
bootstrap, 25, 39, 47
Brazil, 397, 625
Brittany, 515
bryozoans, 501

C/N ratio, 455
Calbuco, 137
California, 253
callus, 339, 619
canopy, 275
Cape Agulhas, 495
Caribbean Sea, 39
carpogonia, 25, 127
carposporangia, 57
carpospore shedding, 101, 285
carpospores, 149, 285, 427, 463
carposporophytes, 101
carrageenophyte, 1, 149, 159, 25, 417
cast material, 495
cell walls, 375
cell-cell recognition, 81
Chapaco, 217
checklist, 65
chemical composition, 167, 391
chemical structure, 385
chemical substitution, 391
chemical variation, 267
chemical-enzymatic degradation, 515
Chile, 115, 137, 149, 217, 375, 397, 427, 435, 571, 397, 469
chloroplasts, 375
chromatography, 549, 563
cicatrigenous branches, 91
clones, 15, 339, 473
colorimetry, 515
colour, 625
commercialization, 477, 509
community, 361
compensation irradiance, 361
conceptacle, 261
concession, 495, 121, 127, 291, 299
conchosporangia, 619
conchosporangial branches, 121, 127, 299
conchospores, 121
conductimetry, 515
confocal microscope, 81
consistency index, 25
continuous irradiance, 329

co-operative units, 477
copra, 477
coral cay, 65
coral islands, 477
Coral Sea Plateau, 65
cordate, 299
correspondence analysis, 191
cosmopolitan species, 75
cover, 173
cross-breeding, 473
crustose algae, 275
cultivars, 7, 15
cultivation, 1, 7, 15, 285, 291, 305, 315, 329, 397, 427, 435, 455, 473, 477, 487, 625
culture, 81, 91, 127
cyclic technologies, 7
cystocarpic thalli, 173, 463
cystocarps, 25, 57, 101, 137, 149, 159, 167, 285
cytoplasm, 375
cytoplasmic granules, 571

daylength, 91, 285
defense, 267
dehydration, 355
demography, 137
density, 137, 149, 183, 201, 275, 435
desiccation, 355
dialysis, 549
diatoms, 435
dichotomous branching, 625
differentiation, 619
dioecious species, 127, 619
discharge, 217
diurnal, 101
diversity, 571
Donnan model, 593
Droop equation, 305
dry matter, 167
drying, 7, 285

ecological engineering, 603
ecological footprint, 603
economics, 1, 7, 477
ecotoxicology, 217
effluent, 603
electrophoritic behaviour, 563
electrostatic attraction, 593
El-Nino, 477
emersion, 355
encrusting coralline algae, 261
endophytism, 247
endosymbiotic gametophytes, 247
Enoshima, 299
enthalpy, 361
entropy, 361
environmental changes, 75
environmental impact, 603
epiliths, 299
epiphytes, 115, 261, 299, 435, 443, 463, 501
epiphytic bacteria, 571
epithallial shedding, 261
erythrocytes, 563
ethoxyzolamide, 349
eulittoral zone, 183

eutrophication, 583, 603
evolution, 39, 47
excised tissue, 619
excitation probability, 361
exploited seaweeds, 7, 501
exports, 487
exposure, 275
expressed sequence tag, 613
extinction coefficient, 361
extra-plastidic lipid bodies, 529

farming, 1, 477
fatty acid profile, 529
fecundity, 149
feedstock, 487
fertilization, 81, 101
fertilizers, 1, 487
filamentous red algae, 247
fish farm waste, 455, 603
fish feed, 495
floristic composition, 75
fluorescence microscopy, 253
fluorescent labelling, 253
foliose thalli, 121, 127, 299, 619
food grade, 167
food, 1
footprint, 603
form, 191
fractionation, 555, 549
fragmentation, 417
frames, 435
fringing reef, 321
fronds, 137, 217, 375
FT-IR spectroscopy, 541
fused-silica column, 529

gamete binding, 81
gametophytes, 159, 247, 253, 339, 435, 469, 473
gas chromatography, 529, 555, 549
mass spectrometry, 549
gel filtration, 549, 563
gel properties, 541, 555
gel strength, 167, 173, 329, 391, 509, 541, 555
gelling properties, 549
gelling temperature, 167, 173, 391, 501
gene tree, 47
genes, 613
genetic component, 267
genetic transformation, 469
genome organization, 613
genomic clones, 613
germination, 241, 427
giant kelp, 253
Gilbert Islands, 477
Gondwana, 25
gonimoblasts, 25, 57, 101
granules, 571
grazers, 583
green algal epiphytes, 329
greenhouse, 435
growth promoting substances, 1, 495, 625
growth rate, 137, 191, 291, 305, 315, 329, 339, 417, 443, 455, 463
growth stages, 211, 411

growth, 121, 127, 137, 167, 253, 435, 91
Guam, 267
Gulf of California, 159, 509

haemocytometer, 529
hair cells, 81
halotolerant microalga, 529
harvesting, 7, 173, 183, 477, 487, 495, 501
hatchery, 435
Herald Cay, 65
herbivorous reef fishes, 267
herbivory, 571
heterogeneity, 275
holdfast communities, 217
holdfasts, 183, 375
Hus formula, 121, 127
hybrids, 473, 625

ice-ice disease, 643
immunofluorescent probe, 643
imports, 487
Indo-Pacific Ocean, 25, 39
infection, 643
infralittoral zone, 391
infra-red gas analyzer, 355
infra-red spectroscopy, 523, 541
inhibitor, 349
inoculation, 643
inorganic carbon uptake, 349
insolation, 361
intensive cage culture, 603
intercalary segments, 339
interfamilial divergence, 47
intergenic region, 613
intertidal assemblages, 329, 571
intertidal zone, 201, 211, 321
intrageneric divergence, 47
intra-plastidic lipid bodies, 529
ion exchange, 523, 563, 593
ionic strength, 593
Irish moss, 417
iron pollution, 217
irradiance, 91, 121, 127, 285, 291, 329, 339, 361, 427, 443, 529, 637
isopods, 183
isotherm model, 593
isotopes, 455
ITS, 39

Japan, 121, 299, 361, 411, 619

kelp gametophytes, 247
kelp systematics, 47
kelp, 191, 275, 411, 469, 495
Korea, 127
KwaZulu-Natal, 201

lagoons, 65, 477
lamellae, 375
lanceolate, 299
Langmuir model, 593
life history, 121, 127, 299
ligand, 241
light, 305, 361, 583, 637

light:dark cycle, 101
limpets, 183
Line Islands, 477
linkage, 385
littorinid snails, 183
long days, 291
lyophilization, 549

macroalgal assemblage, 275
macroinvertebrates, 217
makonbu, 411
management, 137, 183, 583
mangroves, 603
mariculture, 7, 285, 397, 463, 603
marine agronomy, 15
markets, 7, 495
Marmion Lagoon, 275
mass culture, 637
mass spectrometer, 455
mathematical model, 361
maximum likelihood tree, 47
Mediterranean Sea, 39, 75, 91
medullary cells, 57
melting temperature, 167, 173, 391, 501
meristoderm cells, 375
Mexico, 159, 285, 501, 509
Michilla, 217
microbial degradation, 7
microphotometry, 231
microscopic stages, 253
midrib, 191
milkfish, 443
mining activity, 217, 375, 571
mixing rates, 637
molecular systematics, 25
molecular weight, 329, 549, 563
monoculture, 443
monoecious species, 127, 619
monoline, 463
monophyly, 25, 39
moratorium, 487
Morocco, 391
morphology, 15, 57, 191, 625
morphotype, 417
mortality, 137, 361
most parsimonious tree, 39
multicellular rhizoids, 91
multiplicative scatter correction (MSC), 523
multivariate calibration, 523

Namibia, 397
necrosis, 625
neighbour joining tree, 39, 47
netlon, 455
New Zealand, 57, 191, 487
NMR spectroscopy, 385, 515, 541
nori, 183, 291
North Pacific Ocean, 47
nucleotide sequences, 613
nursery, 241
nutrients, 15, 305, 417, 455, 583, 625
nylon cord, 427, 435

obligate epiphyte, 121

ocean swell, 275
oogonium, 247
open reading frame, 613
optical rotation, 549
optimal cell density, 637
ordination, 275
orogenetic processes, 217
osmotic shock, 285
Otago Harbour, 191
outdoor tank, 427
outplanting, 417, 435, 463

Pacific Ocean, 39
paraphyly, 25, 39
parsimonious trees, 47
parsimony, 25, 39
parthenogenesis, 469
partial least squares (PLS), 523
patchy distribution, 183, 275
pathogens, 643
pellets, 495
perennation, 159
periaxial cells, 25
pericarp, 57, 101
pericentral cells, 91
periplasmalemmal space, 375
permanent quadrats, 183
permit, 487
perturbation, 217
pet fish, 477
pH, 349, 443, 593
phase balance, 137
phenology, 91, 149
phenotypic modulation, 191
phenotypic variation, 625
Philippines, 25, 173, 211, 321, 443, 463, 541, 625
photoacclimation, 637
photobioreactors, 637
photodamage, 637
photoperiod, 121, 299
photosynthesis, 321, 349, 355, 361
phycocolloids, 1, 391, 487
phylogeny, 25, 39, 47
pilot scale, 435
pinnae, 191
placenta, 25
plant growth regulators, 339
plantlets, 159
plasmalemma, 375
plastids, 231
polarimeter, 549
polarity, 619
pollution, 75, 217, 571
polyadenylation signals, 613
polyclonal antibodies, 643
polyfilament, 435
polymerase chain reaction, 613
polypropylene cord, 241
polysiphonous red algae, 247
ponds, 443
population studies, 137
preservation, 535
Prince Edward Island, 417
procarps, 25

processing, 7
production, 7, 361, 477
productivity, 15, 637
proliferations, 91
propagation, 427
propagules, 231, 241, 571
proton binding, 593
pruning, 463
pseudoperennial, 159
pulse frequency, 315

quadrant, 509
quadrats, 173, 211, 275

R/P index, 75
raft cultivation, 455, 625
rainfall, 173, 443
rbcL sequences, 25
re-attachment, 159
receptacles, 211
recruitment, 159, 241, 253, 361
reef flat, 321
reflectance, 523
regeneration, 159, 261, 427, 619
reproductive effort, 159
reproductive phenology, 137, 149, 201, 391
reproductive stages, 211
reserves, 305
resonance, 401
resource management, 7, 183
respiration, 321, 361
retention index, 25
reticulate evolution, 47
reverse light:dark cycle, 101
rheological properties, 391
rheology, 555
rhizoids, 91, 115, 619
rhythmicity, 101
RuBisCo spacer, 47
Russia, 555

Saldanha Bay, 455
salinity, 8, 173, 291, 321, 339
salmon farming, 603
salt preservation, 535
San Juan Islands, 247
sandy beaches, 509
satellite positioner, 509
scanner, 91
scanning spectrophotometer, 523
sea star, 261
sea vegetables, 487
seasonality, 137, 167, 173, 191, 201, 211, 391, 411, 501, 509
seaweed flora, 1
seaweed salads, 1
Secchi disc, 417
secondary metabolites, 267
sediments, 217
seeding, 253, 473, 619
senescence, 211
sequencing, 47, 613
set theory, 385
shear viscosity, 555
shell collectors, 159

shrimp farming, 603
signal assignments, 515
site fertility, 15
size classes, 137
skiophilous species, 75
sloughing, 261
socio-economics, 477
South Africa, 183, 201, 495
spatial variation, 267
spectral microphotometer, 231
spectrophotometer, 231
spermatangia, 121
spermatia, 81, 101, 115, 127, 299
spermatozoids, 247
spontaneous release, 285
spore shedding, 241, 285
sporelings, 473, 463
spores, 149, 231, 241, 435
sporophyll length, 191
sporophytes, 191, 473
sporulation, 435
stable isotopes, 455
standing stock, 201
starch branching enzyme, 613
starvation, 305
sterilization, 253
stipes, 191, 217, 361, 375, 523
stochastic spatial distribution, 183
storage, 7, 231, 535
strain selection, 625
stress, 643
structural linkages, 385
sublittoral zone, 91
submergence, 361
subsistence quota, 305
substitution, 385, 401
substratum, 253
subtidal zone, 75, 159, 211, 217, 253, 261, 275, 321
sugar composition, 515, 549
surface shedding, 261
survival, 137
survivorship, 427
suspension cultures, 305
sustainable development, 7
SV40, 469
symbiosis, 247
sympatric fauna, 183
syneresis, 541
synteny, 613

tagging, 417
target species, 15
taxonomy, 65
temperature, 91, 121, 127, 173, 201, 285, 291, 299, 321, 339, 361, 555, 583, 619
tent, 253
Tethys Sea, 25, 39
tetrasporangia, 149, 159
tetrasporangial nemathecia, 57
tetraspores, 149, 231, 427, 435, 625

tetrasporic plants, 321
tetrasporophytes, 159, 339
thallus length, 391
thallus weight, 167
thermal ecotypes, 91
thermal limit, 91
thermal stratification, 455
thin-layer chromatography, 529
Three Kings Islands, 57
thylakoids, 375, 571
tidepools, 321
tissue culture, 619
toxicity, 571
transects, 173, 211
transformation, 469
transgenic plants, 469
transition probability, 361
transmission electron microscopy, 375, 427
transplantation, 417, 267
transportation cost, 7, 477
Tremiti Islands, 75
trichoblasts, 91
trichogynes, 81
tropical reef, 65
trough vessels, 305
turbid water, 583
turbidimetric method, 549
turbulence, 637
turf, 91

ultrastructure, 375, 427
ulvan lyase degradation, 515
unattached rhodophytes, 487, 417
understorey, 275
unialgal culture, 339
urea, 315
utilization, 1, 487, 495

vacuolar system, 375
value-added products, 495
vegetative plants, 321
vegetative regeneration, 91
vicariance, 39
viscosity, 173, 211, 541, 555

wastewater, 593
water movement, 15, 643
water quality, 443
wave action, 201, 275
women farmers, 477
wound healing, 427

yield, 173, 211, 477
Yucatan peninsula, 285, 315

Zanzibar, 349
zoospores, 231, 473
zygotosporangia, 121
zygotospores, 121, 127, 299

Chemicals index

Page numbers refer to the first page of a paper in which the entry is employed (from page 515 onwards this is the number at the bottom of the page)

2,4-dichlorophenoxyacetic acid, 339
2-mercaptoethanol 563
4,4'-diisothiocyanatostilbene-2,2'-disulphonic acid, 349
6-benzylaminopurine, 339

acetazolamide, 349
aconitase, 613
agar(s), 1, 7, 57, 167, 173, 201, 241, 315, 321, 329, 339, 391, 397, 487, 501, 509, 541, 549, 603, 643
agarans, 397
agarobiose, 541
agarobiose, 329
agarocolloid(s), 329, 391, 541
agarose, 329
agglutinin, 81
albumin, 241
alginate(s), 1, 7, 201, 211, 217, 375, 411, 469, 495, 523, 535, 593
alginic acid, 201, 211, 217, 487
amino acids, 315
ammonium nitrate, 315
ammonium sulphate, 563
ammonium, 305, 315, 443, 603
ampicillin, 469
amylase, 541
antibiotic(s), 329, 469, 643
apakaochtodenes, 267
arsenic, 555
asialofetuin, 563
auxins, 339, 495

β-D-glucuronidase, 469, 515
bicarbonate, 349, 355
boron, 515, 555
bromide, 349

cadmium, 593
cadmium, 217, 375, 555, 593
calcium chloride, 535
calcium, 523
calcium acetate, 523
carbohydrate(s), 81, 315, 417, 563
carbon dioxide, 305, 349, 355
carbon, 305, 349, 361, 455, 603
carbonic acid, 349
carbonic anhydrase, 349, 355
carotene, β- 529
carotenoids, 1, 315, 529
carrabiose 2,4'-disulphate, 25, 401
carrabiose 2-sulphate, 25, 401
carrabiose 2-sulphate, 4',6'-O-(1-carboxyethylidene)-, 25, 401
carrabiose 4'-sulphate, 401
carrabiose, 401
carrageenan, alpha-, 25, 401
carrageenan, beta-, 25, 555

carrageenan, iota-, 25, 401, 549, 555
carrageenan, kappa-, 25, 159, 401, 417, 541, 549, 555
carrageenan, lambda-, 25, 385, 417, 549, 555
carrageenan, mu-, 417
carrageenan, xi-, 159,
carrageenan(s), 1, 7, 15, 25, 137, 149, 159, 173, 329, 401, 417, 427, 435, 487, 541, 549, 555, 625, 643
carrageenase, 643
cellulose, 593, 643
chloramphenicol acetyltransferase, 469
chloramphenicol, 469
chlorophyll, 231, 253, 315, 529, 603, 637
chondroitinase, 563
chromium, 217, 375
cobalt, 217
colloids, 397
concanavalin A, 81
copper, 217, 375, 555, 571, 593
cytokinin(s), 339, 495
cytosine, 613

diacylglycerols, 529
diamidinophenyindole, 643
DNA, 25, 39, 47, 339, 397, 417, 469, 613
dulcitol, 549

ethanol, 515
ethoxyzolamide, 349

fatty acid(s), 1, 529
fetuin, 563
floridean starch, 167, 329, 397, 541, 549, 555
floridoside, 57, 329
fluorescein isothiocyanate, 81
folic acid, 1
folinic acid, 1
formaldehyde, 253, 535
fucans, 411
fucoidan, 217, 593
fucose, 81, 411, 563
fulvic acid, 593

galactan(s), 25, 41, 385, 391, 541, 549, 563
galacto(pyrano)se, 3,6-anhydro-, 167, 173, 329, 391, 397, 401, 417, 509, 541, 549, 555
galacto(pyrano)se, 2-O-methyl-3,6-anhydro-, 329, 541, 549
galacto(pyrano)se, 4,6-O-(1-carboxyethylidene)-, 329, 401
galacto(pyrano)se, 6-O-methyl-, 167, 329, 391, 401, 541, 549
galacto(pyrano)se, 167, 329, 385, 397, 401, 411, 417, 515, 541, 549, 555, 563
galactopyranose 2,4-disulphate, 401
galactopyranose 2-sulphate, anhydro-, 401
galacto(pyrano)se 4-sulphate, 401, 555
galactose, GDP-, 329
galactose, UDP-, 329

galactose, 2-O-methyl, 549
galactose, 4-O-methyl-, 391
galactose 6-sulphate, 167, 329, 541, 549
galactose-1-phosphate uridyltransferase, 613
gelan, 1
gelatine, 241
germanium dioxide, 625
glucosamine, 563
glucose, 81, 167, 329, 391, 401, 515, 549, 563
glucuronan, 515
glucuronanase, 515
glucuronic acid, 411, 515
glutaraldehyde, 231
glycerine, 241
glycoproteins, 81, 563
gold, 217
guanine, 613
guluronic acid, 515, 593

heavy metals, 217, 375, 593
hemagglutinin, 563
herbicides, 469
hexitols, 549
humic acid, 593
hydrogen sulphide, 583
hygromycin, 469

iduronic acid, 515
indole-3-acetic acid , 339
inositol, 339
iodide, 349
iodine, 469
iron, 217, 375

kanamycin, 469

laminaran, 411
lead, 217, 375, 555
lectins, 81, 563
lincomycin, 469
lipid(s), 1, 375
lithium, 217

manganese, 217, 375, 555
mannan(s), 549
mannitol, 1, 411 , 469, 535
mannose, 81, 167, 411, 515, 549
mannuronic acid, 515, 593
mercury, 217
methionine sulphoxide reductase, 613
methyl galactose, 401
molybdenum, 217
monoterpenes, 267
myo-inositol oxidoreductase, 613

N-acetyl-galactosamine, 81
neoagarobiose, 391
neomycin, 469
nitrate, 173, 305, 315, 349, 443, 603
nitrite, 603
nitrogen, 267, 305, 315, 417, 443, 455, 501, 603, 529, 535, 583, 603

oleic acid, 529
oxygen, 361, 443, 535, 583, 603

peptidyl tRNA hydrolase, 613
periodate, 563
phenol, 1
phosphate, 173, 305, 315, 443, 563, 603
phosphatidylethanolamine, 529
phospholipids, 305
phosphoric acid, 253
phosphorus, 211, 267, 305, 315, 417, 443, 583, 603,
phycobiliproteins, 315
phycocolloid(s), 1, 7, 25, 57, 173, 315, 391, 417, 487, 541
phycocyanin, 315, 321
phycoerythrin, 305, 315, 321
polylysine, 241
polyphenol(s), 217, 523, 535
polysaccharide(s), 1, 7, 57, 329, 385, 391, 401, 411, 417, 487, 501, 515, 523, 541, 549, 555, 563
polyubiquitin, 613
potassium chloride, 549, 555
pronase, 563
propylene oxide, 593
protein(s), 1, 315, 349, 455, 469, 515, 549, 555, 563
proteoglycan(s), 563
pyruvate, 25, 167, 329, 385, 391, 401, 541, 549

resorcinol, 137, 159, 417
rhamnose 3-sulphate, 515
rhamnose, 391, 515
ribulose-1,5-biphosphate carboxylase/oxigenase, 47
RNA, 39

silver, 217
sodium chloride, 535, 549
sodium hydroxide, 549
sodium nitrate, 593
sodium, 593
steroid, 1
streptomycin, 469
sucrose, 339
sulfonamide, 349
sulphate, 167, 173, 329, 385, 391, 397, 401, 411, 417, 509, 515, 541, 549, 555, 563, 593

tannins, 1
terpene, 1
thyroglobulin, 563
tocopherol, 1
triacylglycerols, 529
triglyceride(s), 529
tryptophan synthase, 613

UDP-glucose pyrophosphorylase, 613
ulvan(s), 515
ulvan-lyase, 515
ulvanobiuronic acid, 515
urea, 315, 549
uronic acid(s), 391, 411, 549

vitamin, 1

xylogalactan(s), 549
xylomannan(s), 549
xylose, 391, 401, 515, 549, 563

zeocin, 469
zinc, 217, 375, 571

Taxonomic index

Page numbers refer to the first page of a paper in which the entry is employed (from page 515 onwards this is the number at the bottom of the page); names within ( ) denote synonyms and within [ ] denote basionyms

Acanthophora muscoides, 1
– *spicifera*, 1, 173
Acetabularia major, 1
Acmaea pallida, 261
Acrochaete, 247
Acrochaetiales, 25, 247
Acrosymphytaceae, 65
Acrosymphyton, 65
– *taylorii*, 65
Acrotylaceae, 25
Agardhiella, 25, 401
– *floridana*, 25
– *ramosissima*, 25
– *subulata*, 25, 339, 401, 417
(– *tenera*), 401
Agardhielleae, 25, 401
Agarum, 247
– *fimbriatum* 247
Aglaothamnion, 81, 101
– *neglectum*, 81
– *oosumiense*, 81
– *tenuissimum*, 75
Ahnfeltia, 427
– *durvillaei*, 401
Ahnfeltiopsis, 427
Alaria, 47
– *crassifolia*, 47
– *esculenta*, 47
– *marginata*, 47, 247
– *praelonga*, 47
– (*pinnatifida*), 191
Alariaceae, 47, 191
Amphiroa crassa, 65
– *anceps*, 275
Anadyomenaceae, 65
Anadyomene stellata, 65
Anatheca, 25, 401
– *furcata*, 25
– *montagnei*, 401
Anotrichium secundum, 65
– *tenue* 65
Antithamnion, 81, 91, 571
– *heterocladum*, 75
– *makroklonion*, 65
– *nipponicum* 81
– *sparsum* 81
Antithamnionella pacifica, 247
Aphelasterias japonica, 261
Aplysia parvula, 267
Arabidopsis thaliana, 613
Areschougia, 25, 401
– *congesta*, 25
– *ligulata*, 25
Areschougiaceae, 25, 401

Areshougieae, 25
Ascophyllum nodosum, 355, 487, 509, 523, 535
Asparagopsis, 159
– *armarta*, 487
– *taxiformis*, 1
Asterias amurensis, 261
Asterina pectinifera, 261
Asteromenia, 65
– *peltata* 65
Austroclonium, 25, 401
Avicennia, 603
Avrainvillea, 65
– *lacerata*, 65

Bacillus subtilis, 613
Balliella cladoderma, 75
Bangiaceae, 115, 121
Bangiales, 115, 121, 127, 183, 291, 299, 339, 619
Bangiophyceae, 91
Bangiophycidae, 47
(*Betaphycus*), 25
– (*gelatinum*), 25
– *philippinensis*, 25, 401
– *speciosum*, 25
Bifurcaria bifurcata, 411
Bonnemaisoniaceae, 1
Boodlea, 65
– *coacta*, 563
Bornetia secundiflora, 159
Bostrychia 81, 101, 549
– *bispora*, 159
– *montagnei*, 549
– *moritziana*, 101
– *radicans*, 101
Bostrychioideae, 549
Bryopsidales, 39, 65
Bryopsidophyceae, 39
Bryopsis plumosa, 191

Calliblepharis celatospora, 25, 401
– *ciliata* , 401
– *jubata*, 401
Callithamnion, 231
– *acutum*, 247
– *biseriatum*, 231
– *cordatum*, 101
Callophycus, 25, 401
– *africanus*, 25
– *oppositifolius*, 25
Callophyllis variegata, 137
Caloglossa, 101
– *apomeiotica*, 101
– *continua*, 101
– *leprieurii*, 101

– *ogasawaraensis*, 101
Carpoblepharis, 101
Carpopeltis, 1
– *flabellata*, 563
Catenella nipae, 401
Caulacanthaceae, 25
Caulacanthus, 25
– *ustulatus*, 25
Caulerpa, 65, 91, 267, 339
– *bartlettii*, 1
– *cupressoides*, 65
– *intricatum*, 1
– *lentillifera*, 1
– *peltata*, 1
– *prolifera*, 339
– *racemosa*, 1, 65
– *sertularioides*, 1
– *taxifolia*, 1, 7, 65
Caulerpaceae, 65
Caulerpales, 267
Centroceras clavulatum 65, 571
Ceramiaceae 65, 81, 101, 231
Ceramiales, 1, 65, 81, 91, 101, 275, 549
Ceramium, 1, 101, 583
– cf. *caudatum*, 65
– *flaccidum*, 65
– *marshallense*, 65
– *nodulosum*, 583
Ceriops, 603
Chaetomorpha, 375
– *brachygona*, 375
– *crassa*, 443
– *linum*, 315, 443, 571
– *melagonium*, 91
Chamaedoris, 65
– *peniculum*, 65
Champia caespitosa, 65
– *novae-zealandiae*, 385, 487
– *parvula*, 65
Champiaceae, 65
Chanos chanos, 443
Chauviniella coriifolia, 275
Chlamydomonas, 81
– *eugametos*, 81
– *monica*, 81
– *reinhardtii*, 349
Chlorella, 571, 637
Chlorophyceae, 1, 7, 65, 75, 91, 137, 339, 375, 435, 529, 603
Chlorophycophyta, 65
Chlorophyta, 1, 7, 39, 65, 91, 241, 267, 305, 315, 349, 355, 515, 571, 583
Chondracanthus canaliculatus, 509
– *pectinatus*, 159, 509
Chondria, 65
– *armata*, 1
– *macrocarpa*, 549
– *simpliciuscula*, 65
– *succulenta*, 65
Chondrus, 427
– *armatus*, 555
– *crispus*, 15, 101, 137, 159, 305, 315, 339, 349, 417, 427, 463, 555, 571
– *pinnulatus*, 555
– *yendoi*, 555

Chorda filum, 47, 121
– *tomentosa*, 247
Chordaceae, 47, 247
Chordaria flagelliformis, 137
Chordariaceae, 47
Chordariales, 247
Chrysochromulina polylepis, 603
Cladophora, 1, 603
– *feredayi* 191
Cladophorales 65, 91
Cladophoropsis, 65
– *membranacea*, 91
Codiaceae, 1, 65
Codiales, 7
Codium, 1, 65
– *edule*, 1
– *fragile*, 7, 191, 305
– *fragile* subsp. *tomentosoides*, 7, 305
– *ovale*, 65
Collisella, 241
Colpomenia, 509
– *peregrina*, 355
– *sinuosa*, 509
Constantinea subulifera, 137
Corallina, 65
– (annulata), 65
– *auriculata*, 65
– *delphinii*, 65
– *elongata*, 315
– *officinalis*, 549
– *orientalis*, 65
– [*peniculum*], 65
– *pilulifera*, 355
Corallinaceae, 1, 65, 261
Corallinales, 65, 261
Coscinodiscus, 443
Costaria, 191
– *costata*, 247
Crouania minutissima, 65
Cryptonemaciaceae, 385
Cryptonemiales, 1, 65, 261, 555
Cryptopleura, 501
Cubiculosporaceae, 25
Curdiea, 57
– *angustata*, 57
– *balthazar*, 57
– *codioides*, 57
– *coriacea*, 57, 541
– *crassa*, 57
– *flabellata*, 57
– *irwinii*, 57
– *obesa*, 57
– *racovitzae*, 57, 247
Cyanophyta, 1
Cyclotella nana, 81, 625
Cystocloniaceae, 25
Cystoclonium purpureum, 401, 563
Cystophora, 275
Cystoseira, 75
– *amentacea* var. *stricta*, 75
– *barbata*, 375
– *barbata* f. *repens*, 375
– *crinita*, 75
– *humilis*, 75

– *schiffneri*, 75
– *spicata* var. *elegans* 75
– *spinosa*, 75
Cytophaga, 643

Dasyaceae, 65
Dasycladaceae, 1
Dasycladales, 39, 65
Delesseriaceae, 65, 101
Delisea pulchra, 267
Desmarestia, 253
– *antarctica*, 247
– *ligulata*, 191, 253
Desmarestiaceae, 39
Desmarestiales, 47, 247
Detonula confervacea, 81, 625
Devalerea ramentacea, 91
Dicranemataceae, 25, 401
Dictymenia sonderi, 275
Dictyosphaeria, 65
– *cavernosa*, 1, 65
– *versluysii*, 65
Dictyota dichotoma, 65
Dictyotaceae, 65
Dictyotales, 65, 375
Digenea simplex, 1
Dinophyta, 231
Dipterosiphonia rigens, 91
Dolichos biflorus, 81
Dumontiaceae, 65
Dunaliella, 529
– *bardawil*, 529
– *salina*, 529
– *tertiolecta*, 529
Durvillaea, 115, 121, 487
– *potatorum*, 487

Echinometra mathaei, 275
Ecklonia, 47, 201, 487
– *cava*, 47, 361
– *maxima*, 183, 487, 495
– *radiata*, 217, 241, 275, 487, 509
– *stolonifera*, 47
Eckloniopsis, 47
– *radicosa*, 47, 619
Ectocarpales, 47, 247
Ectocarpus, 47, 81, 435
– *siliculosus*, 81
Egregia, 47
– *menziesii*, 47
Eiseina, 47
– *arborea*, 47, 247
– *bicyclis*, 47, 361
Endocladia muricata, 321, 555
Endophyton, 247
Enhalus acoroides, 173
Enteromorpha, 1, 65, 305, 435, 463, 583
– *clathrata*, 443
– *compressa*, 1, 191, 355, 571
– *flexuosa*, 231, 443
– *intestinalis*, 65, 159, 305, 349, 355
– *linza*, 355
Ereosmosphera viridis, 529
Erythroclonium, 25, 401

– *angustatum*, 25
– *muelleri*, 25
Escherichia coli, 563
Ethelia biradiata, 65
Eucheuma, 1, 7, 15, 25, 137, 173, 349, 417, 427, 435, 463, 469, 477, 625, 643
– [*alvarezii*], 25, 477
– *alvarezii* var. *tambalang*, 625
– *amakusaensis*, 25
– *arnoldii*, 1, 25
– (*cottonii*), 401, 417, 477, 555, 625
– *denticulatum*, 1, 7, 25, 339, 349, 401, 417, 427, 625, 643
– (*gelatinae*), 1, 25, 555
– *gelatinum*, 25
– *isiforme*, 25, 401
– sect. Anaxiferae, 25
– sect. Cottoniformia, 25
– sect. Eucheuma, 25
– sect. Gelatiformia, 25
– *serra*, 25
– *speciosa*, 555
– (*spinosum*), 401, 417, 625
– (*striatum*), 417
– *uncinatum*, 125, 159, 401, 417
Eupogodon iridescens, 65

Fauchea peltata, 65
Flabellariaceae, 1
Flabellarieae, 65
Flabellia, 39
Flahaultia, 25
Florideophyceae, 1, 7, 315, 417, 501, 603
Fucales, 65, 75, 183, 201, 211, 275, 339, 463
Fucophyceae, 75
Fucus, 81, 201, 253, 571, 583
– *distichus*, 201, 267, 355
– *gardneri*, 183
– *serratus*, 81, 339, 349, 463
– *vesiculosus*, 217, 253, 267, 375, 509, 563, 571, 583, 603
– *virsoides*, 375
Furcellaria fastigiata, 401
– *lumbricalis* 329, 401, 417, 463, 555
Furcellariaceae, 25

Galaxaura, 1
– *fastigiata*, 1
– *marginata*, 65
– *rugosa*, 65
Gammarus locusta, 583
Gardneriella, 25
Gelidiaceae, 65, 173, 321, 329, 501
Gelidiales, 7, 15, 25, 65, 167, 173, 285, 321, 329, 391, 501
Gelidiella, 7
– *acerosa*, 1, 7, 101, 321
– *lubrica*, 65
– *pannosa*, 65
– *womersleyana*, 65
Gelidium, 1, 7, 25, 154, 201, 321, 329, 391, 501, 541, 571
– *abbottiorum*, 201
– *canariensis*, 321, 391, 501
– *caulacantheum*, 487
– *coulteri*, 173, 321, 329
– *latifolium*, 15, 167, 173, 329, 391, 501, 571
– *pulchellum*, 329

– *rex*, 7, 501
– *robustum*, 101, 159, 173, 329, 501, 509
– *sesquipedale*, 167, 329, 391, 501
Gibsmithia, 65
– *hawaiiensis*, 65
Giffordia, 435, 443
Gigartina, 7, 15, 137, 149, 435, 487
– (*acicularis*), 91
– *alveata*, 487, 555
– *atropurpurea*, 401
– [*canaliculata*], 159, 401
– (*chamissoi*), 401
– *chapmanii*, 487
– [*circumcincta*], 487
– *clavifera*, 487
– *decipiens*, 385, 487, 555
– (*exasperata*), 435
– [*johnstonii*], 159
– *ochotensis*, 555
– *papillata*, 159
– *pectinata*, 159, 509
– *petrocelis*, 101
– *skottsbergii*, 137, 149, 427, 435, 549
– (*stellata*), 137
– *tenera*, 25, 401
Gigartinaceae, 25, 137, 149, 159, 183, 285, 385, 427, 435, 487, 555
Gigartinales, 1, 7, 25, 65, 91, 137, 149, 159, 167, 267, 285, 329, 339, 397, 401, 417, 427, 435, 443, 463, 509, 541, 555, 563, 619, 625, 643
Gloiopeltis cervicornis, 401
– *furcata*, 355
Gracilaria, 1, 7, 15, 47, 137, 167, 173, 201, 211, 285, 315, 321, 329, 339, 355, 391, 397, 417, 427, 435, 443, 463, 477, 487, 509, 541, 563, 583, 603, 613, 643
– *arcuata*, 1, 173, 443, 541
– *asiatica*, 463
– *blodgettii*, 173
– *bursa pastoris*, 7, 509, 541, 563
– *caudata*, 397
– cf. *conferta*, 329
– *changii*, 167, 463, 509, 603
– *chilensis*, 1, 7, 15, 57, 115, 285, 315, 329, 339, 355, 391, 397, 427, 435, 487, 603
– *chorda*, 167, 391
– *conferta*, 315, 443
– *cornea*, 285, 315
– *coronopifolia*, 15, 509
– *corticata*, 285, 329
– *crassissima*, 329
– *domingensis*, 167
– *dura*, 329
– *edulis*, 101, 285, 443
– *eucheumoides*, 1, 173, 541
– *firma*, 1
– *foliifera*, 137, 285, 305, 315, 391
– *foliifera* var. *angustissima*, 285, 315, 339
– *gracilis*, 7, 285, 315, 391, 397, 455, 613
– *heterocladia*, 1, 509, 167
– (*lemaneiformis*), 167, 339
– *mammillaris* 167, 541
– *manilaensis* 463, 509
– *multipartita*, 167
– *multipartita* var. *angustissima*, 167

– *pacifica*, 509
– *parvispora*, 285
– *pseudoverrucosa*, 167, 329
– *salicornia*, 173, 329
– *secundata*, 315, 355, 463
– *sjeostedtii*, 329, 463
– (*sordida*), 329, 339
– *tennuistipitata*, 1, 315, 339, 349, 463, 541, 603
– *tenuispitata* var. *liui*, 397, 443, 463
– *textorii* 285, 329, 339
– *tikvahiae* 167, 305, 315, 329, 339, 391, 397, 443, 463, 501, 509, 625
– *truncata*, 391
– *vermiculophylla*, 339
– *verrucosa*, 1, 47, 159, 167, 173, 285, 305, 329, 339, 391, 397, 435, 443, 455, 463, 509, 563, 613
Gracilariaceae, 57, 167, 329, 391, 397, 463, 509, 541
Gracilariales, 1, 15, 47, 57, 167, 285, 315, 339, 391, 397, 427, 455, 463, 509, 541, 603, 613
Gracilariopsis, 47, 397, 443, 509
– *bailinae* 443, 463, 509
– (*heteroclada*), 443, 463
– *lemaneiformis*, 159, 509
– *sjoestedtii*, 285
– *tenuifrons*, 339
Graminae, 613
Grateloupia, 115, 427
– *acuminata*, 427
– *dichotoma*, 339
– *filicina*, 1
– *filiformis*, 339
Griffithsia pacifica, 81, 247
Griffithsieae, 65
Gulsoniopsis, 101
Gymnogongrus, 1
– (*furcellatus*), 401

Halichrysis, 65
– cf. *micans*, 65
Halimeda, 1, 39, 65
– *copiosa*, 39
– *cryptica*, 39
– *cylindracea*, 39, 65
– *discoidea*, 39
– *fragilis*, 39, 65
– *goreauii*, 39
– *gracilis*, 39
– *incrassata*, 39
– *lacrimosa*, 39
– *lacunalis f. lata*, 39
– *macroloba*, 1, 39
– *monile*, 39
– *opuntia*, 39, 65
– sect. Crypticae, 39
– sect. Halimeda, 39
– sect. Micronesicae, 39
– sect. Opuntia, 39
– sect. Rhipsalis, 39
– *simulans*, 39
– *tuna*, 39, 65
Halimedaceae, 65, 267
Haliotis midae, 183, 495
– *roei*, 275
Halobacterium, 535

Halomonas, 535
Haloplegma duperreyi, 65
Halopteris hordacea , 571
Halosaccion americanum, 355
Halymenia, 65
– *durvilleai*, 1
Halymeniaceae, 65, 427, 487
Halymeniales, 65, 385
Hensonella cylindrica, 39
Heringia, 25
Herposiphonia, 91
– *arcuata*, 65
– *plumula*, 247
Hesperophycus harveyanus, 355
Heterodoxia denticulata, 275
Heterosiphonia crispella, 65
– *wurdemannii*, 65
Hildenbrandiales, 25
Himanthalia elongata, 201
– *lorea*, 411
Hormophysa cuneiformis, 1
Hyale, 183
Hydroclathrus, 1
– *clathratus*, 1
Hypnea, 1, 7, 555
– *japonica*, 563
– *musciformis*, 7, 329, 401, 603
Hypneaceae, 401
Hypneocolax stellaris, 401
Hypoglossum cf. *geminatum*, 65

Idotea, 583
Iridaea, 7, 435, 487
– *capensis*, 183
– (*ciliata*), 149, 285, 427, 435
– *cordata*, 137, 149, 159, 321, 435
– (*cornucopiae*), 555
– *cornucopiae* subsp.*japonica*, 555
– [*laminarioides*], 137, 241
– *splendens*, 149
– *undulosa*, 549, 555
Iridophycus, 137
Ishige okamurae, 299, 355
Isochrysis galbana, 637

Jania, 65
Jeannerettia pedicellata, 275

Kallymenia berggrenii, 487
Kappaphycus, 1, 25, 7, 173, 427, 435, 463, 477, 625, 643
– *alvarezii*, 1, 25, 7, 339, 401, 417, 427, 477, 625, 643
– *alvarezii* var. *alvarezii*, 625
– *alvarezii* var. *tambalang*, 643
– *cottonii*, 1, 25
– *procrusteanum*, 1
– *striatum*, 1, 25, 625
Kuckuckia, 47

Laminaria, 15, 47, 191, 201, 253, 435, 469, 473, 509, 603
– *angustata*, 247
– *digitata*, 411, 487, 535
– *farlowii*, 231
– (*groenlandica*), 247
– *hyperborea*, 247, 523

– *japonica*, 47, 121, 411, 469, 473
– *japonica* var. *ochotensis*, 247
– *longicruris*, 315
– *pallida*, 495
– *religiosa*, 247
– *saccharina*, 247, 253, 349, 535
Laminariaceae, 47
Laminariales, 47, 121, 159, 191, 217, 231, 247, 253, 361, 469, 487, 495, 523, 619
Laminariocolax, 247
Lates calcarifer, 443
Laurencia, 1, 501
– *nipponica*, 549
– *perforata*, 65
– *pinnatifida*, 541, 549
Lessonia, 217, 375, 487
– *nigrescens*, 159, 217, 241, 375, 571
– *trabeculata* 217, 375, 411
Lessoniaceae, 47
Liagora, 65
– *caenomyce*, 1
– *ceranoides*, 65
Liagoraceae, 65
Lithophylloideae, 261
Lithophyllum neoataleyense, 261
– *yessoense*, 261
Littorina, 201
– *littorea*, 583
– *peruviana*, 241
– *saxatilis*, 583
Lobophora variegata, 275
Lophocladia kipukaia, 65
– *minor*, 65
Lophosiphonia bermudensis, 91
– *scopulorum*, 91
Lyngbya majuscula, 443

Macrocystis, 15, 231, 487
– *pyrifera*, 15, 137, 217, 231, 247, 253, 275, 305, 361, 375, 487, 501, 509
Martensia australis, 65
Mastocarpus pacifica, 555
– *papillatus* 101, 321
Mastogloia, 443
Mastophoroideae, 261
Mazzaella, 7, 427, 435
– *cornucopiae*, 183, 201
– *laminarioides* 149, 241
– *membranacea* 115
– *splendens*, 137, 159
Mediothamnion lyallii, 191
Melanema dumosum, 25
Membraniphora tuberculata, 501
Meristiella, 25, 401
– *gelidium*, 25
Meristotheca, 25, 401
– *papulosa*, 25
– *senegalensis*, 401
Microcladia, 101
Microdictyon, 65
– *montagnei*, 65
– *setchellianum*, 65
– *umbilicatum*, 65
– *vanbosseae*, 65

Micropeuce setosus, 65
Murrayella periclados, 91, 101
Mychodeaceae, 25
Mychodeophyllaceae, 25
Myelophycus simplex, 299
Mytilus, 583
– *edulis*, 417, 583

Nannochloropsis, 529, 637
Navicula, 443
Nemaliales, 25, 65
Nemalionales, 1
Nemastomataceae, 65
(*Neoagardhiella*), 401
(– *baileyi*), 305, 315
– *gaudichaudii*, 25, 501
Neomeris annulata, 267
Neorhodomela larix, 267
Nereocystis luetkeana, 191, 231, 247
(*Nesea annulata*), 65
Nitophyllum punctatum, 101
– *tristromaticum*, 75
Nodilittorina africana, 183

Ochtodes secundiramea, 267
Oncorhynchus kisutch, 603
– *mykiss*, 603
Onikusa pristoides, 501
Opuntiella, 25
Ossiella pacifica, 65
Oxystele variegata, 183

Pachymenia, 385, 487
– *hymantophora*, 401
– *lusoria*, 385
Padina, 65
– *gymnospora*, 217, 375
– *pavonia*, 411
Pagrus major, 603
Palmaria, 81
Palmariales, 81, 355
Parisocladus, 183
Patella granularis, 183
– *vulgata*, 241
Patiria pectinifera, 261
Pelvetia canaliculata, 375, 411
– *fastigiata* f. *gracilis*, 355
Penaeus, 603
– *monodon*, 443, 603
Penicillus, 39
(– *annulatus*) , 65
Perna viridis, 603
Petrocelis, 401
(*Petroglossum nicaeense*), 619
Peyssonnelia capensis, 65
– *inamoena*, 65
– *neocaledonica*, 65
Peyssonneliaceae, 65
Pfiesteria, 603
– *piscicida*, 603
Phacelocarpus labillardieri, 57
Phaeophila, 247
Phaeophyceae, 1, 39, 47, 65, 81, 191, 201, 231, 247, 275, 375, 469, 523, 535

Phaeophyta, 1, 47, 91, 137, 159, 183, 191, 201, 211, 217, 241, 275, 339, 349, 355, 361, 463, 469, 487, 509, 535, 571, 619
Phycodrys, 47
– *quercifolia*, 191
– *rubens*, 91, 137
Phyllariaceae, 247
Phyllodictyon, 65
Phyllophora (*brodiei*), 401
Phyllophoraceae, 25
Pilayella, 583
– *littoralis*, 583
Pinnularia, 443
Placentophora, 25
Pleonosporium, 247
– *vancouverianum*, 247
Plocamiaceae, 385
Plocamiales, 385
Plocamium cartilagineum, 191, 385
– *costatum*, 385
Pneophyllum, 261
Polycavernosa, 285
Polyneura latissima, 247
Polysiphonia, 91, 435, 571, 583
– *akkeshiensis*, 91
– *boldii*, 91
– *denudata*, 91
– *ferulacea*, 91
– *fibrillosa*, 91
– *hemisphaerica*, 91
– *japonica*, 91
– *lanosa*, 549
– *setacea*, 91
– *sphaerocarpa*, 65
– *stricta*, 91
– *urceolata*, 91
– *violacea*, 91
Porphyra, 1, 7, 15, 47, 115, 121, 127, 183, 291, 299, 305, 355, 435, 469, 487, 619
– *abbottae*, 291
– *argentinensis*, 115
– (*augustinae*), 183
– *capensis*, 183
– *columbina*, 115, 183, 487
– *crispata*, 127
– *dentata*, 127, 619
– *fucicola*, 115
– *gardneri*, 115, 121
– *haitanensis*, 127
– *kinositae*, 121, 127
– *lacerata*, 121, 127, 299
– *lanceolata*, 127
– *leucosticta*, 305, 349
– *linearis*, 115, 127, 183
– (*maculosa*), 115
– *miniata*, 121, 469
– *moriensis*, 121
– *mumfordii*, 127
– *nereocystis*, 291
– *okamurae*, 619
– *perforata*, 115, 121, 183, 305, 619
– *pseudolanceolata*, 127
– *pseudolinearis*, 127, 291, 619
– *purpurea*, 127
– *rosengurtii*, 115

– *saldanhae*, 183
– (*sanjuanensis*), 115
– *suborbiculata*, 127, 619
– *subtumens* 115, 121
– *tanegashimensis* 121, 127
– *tenera*, 101, 121, 127, 305, 339, 619
– *tenuipedalis* 121, 127
– *torta*, 291
– *umbilicalis*, 15, 115, 305, 339, 619
– *yezoensis*, 15, 115, 127, 305, 355, 619
(*Porphyrella gardneri*), 115
Porphyridium, 7
Portieria hornemannii, 267
Postelsia palmaeformis, 47, 191, 253
Potamogeton perfoliatus, 583
Prasiales, 355
Prasiola stipitata, 355
Protoflorideae, 1 *Prymnesium parvum*, 603
Pseudochorda nagaii, 247
Pseudochordaceae, 247
Pseudomonas fluorescens, 643
– *syringae*, 643
Pterocladia, 1, 7, 321, 329, 487, 501, 541, 619
– *caerulescens*, 329
– *capillacea*, 7, 159, 501
– *lucida*, 275, 487
Pterosiphonia, 247
– *complanata*, 159
– *pennata*, 101
Pterygophora, 47, 231
– *californica*, 47, 231, 253, 275
Ptilosarcus, 247
– *gurneyi*, 247
Pyrocystis lunula, 231

Reinboldiella, 101
Rhabdonia, 25, 401
(– *africana*), 401
– *clavigera*, 25
– *coccinea*, 25
Rhabdoniaceae, 25, 401
Rhipilia, 65
– *orientalis*, 65
– *tenaculosa*, 65
Rhipiliella, 65
Rhipiliopsis, 65
– *gracilis*, 65
Rhipocephalus, 39
Rhizophora, 603
Rhizophyllidaceae, 267
Rhodoglossum affine, 137, 159, 321
– *hemisphaericum*, 555
Rhodomela larix, 275
Rhodomelaceae, 65, 91, 101, 159, 541, 549
Rhodophyceae, 1, 65, 75, 91, 101, 115, 127, 137, 167, 173, 315, 329, 339, 397, 401, 417, 509, 555
Rhodophycophyta, 285, 463
Rhodophyllidaceae, 25, 401
Rhodophyta, 1, 7, 15, 25, 47, 57, 65, 81, 91, 101, 115, 121, 127, 137, 149, 159, 167, 173, 183, 231, 241, 261, 267, 275, 285, 291, 299, 315, 321, 329, 339, 349, 355, 385, 391, 397, 401, 417, 427, 435, 443, 455, 463, 469, 487, 501, 509, 541, 549, 563, 571, 603, 613, 619, 625, 643
Rhodymenia foliifera, 191

– *leptophylla*, 191
– *pertusa*, 101
– *sonderi*, 275
Rhodymeniaceae, 65
Rhodymeniales, 1, 65, 385, 487
Rissoella verruculosa, 417
Rissoellaceae, 25
Rodriguezella pinnata, 75

Saccorhiza, 247
– *bulbosa* 247
– *dermatodea*, 247
– *polyschides*, 191, 523
Salmo gairdneri, 603
Sarcodiaceae, 401
Sarcodiotheca, 25, 401
– *furcata*, 25
Sarconema, 25
– *filiforme*, 25, 401
– *scinaioides*, 25
Sarcothalia, 15, 427, 435
– *crispata*, 137, 149, 435
Sargassaceae, 65, 201, 211, 241, 275
Sargassum, 1, 65, 75, 159, 201, 211, 275, 509, 593
– *acinarium*, 75
– *binderi*, 1
– *cinctum*, 1
– *confusum*, 619
– *crassifolium*, 1, 211
– *cristaefolium*, 1, 211
– *cymosum*, 201
– *elegans*, 201
– *feldmannii*, 1, 211
– *fluitans*, 593
– *furcatum*, 211
– *hemiphyllum*, 1
– *herporhizum*, 159, 509
– *heterophyllum*, 201
– *hornschuchii*, 75
– *ilicifolium*, 211
– *incisifolium*, 201
– *johnsonii*, 57, 159, 509
– *linifolium*, 411
– *mcclurei*, 201
– *muticum*, 201, 275, 349
– *oligocystum*, 1, 201, 211
– *paniculatum*, 1, 211
– *polycystum*, 1, 201, 211
– *pteropleuron*, 201
– *ringgoldianum*, 411
– *siliquosum*, 1, 201, 211
– *sinicola*, 159, 509
– *spinuligerum*, 241, 275
– *thunbergii*, 201
Scagelia, 231
– *corallina*, 65
– *pylaisaei*, 231, 247
Scenedesmus, 375, 571
Schottera nicaeese, 619
Scinaia hormoides, 1
(*Scopularia annulata*), 65
Scurria, 571
Scytothalia, 275
Sebdenia rodrigueziana, 75

Siganus argentus, 267
Siphonales, 7, 39, 65, 339
Siphonaria, 183
– *lessonii*, 241
Siphonocladaceae, 65
Siphonocladales, 65, 91
Solieria, 25, 401
– *anastomosa*, 25, 401
– *chordalis*, 401, 563
– *dichotoma*, 25, 401
– *dura*, 25, 401
– *filiformis*, 25, 159, 339, 401
– *jaasundii*, 25, 401
– *pacifica*, 25, 401
– *robusta*, 25, 401, 563
– [*tenera*], 401
– *tenuis*, 25
Solieriaceae, 25, 401, 417, 555, 625, 643
Solierieae, 25, 401
Sparus aurata, 603
Spatoglossum, 247
Sphacelariales, 91, 247
Spirogyra submargaritata, 375
Spirulina, 637
– *platensis*, 637
Spongites yendoi, 261
Sporochnales, 47, 247
Spyridia filamentosa, 101
Staphylococcus, 535
– *aureus*, 535
Stenogramme interrupta, 385
Stictosiphonia 101, 549
Stigeoclonium tenue, 375
Streblonema, 247
Streptomyces griseus, 613
Strongylocentrotus nudus, 261
Struvea, 65
Stypocaulon scoparium, 91
Synedra, 443

Tapes japonica, 603
Tegula atra, 241
Tetraselmis, 529
Tichocarpaceae, 25, 555
Tichocarpus crinitus, 555
Tikvahiella, 25
– *candida*, 25, 401
Titanoderma, 261
Titanophora, 65
– *weberae*, 65
Trichodesmium, 231
Turbinaria, 1, 65

– *conoides*, 1
– *ornata*, 1, 65
Turnerella, 25

Udotea, 39, 65
– *orientalis*, 65
Udoteaceae, 65
Udoteae, 1, 65
(*Ulopteryx pinnatifida*), 191
Ulotrichales, 375
Ulva, 1, 137, 305, 349, 435, 463, 487, 509, 515, 555, 583, 603
– *armoricana*, 515
– *fasciata*, 231
– *fenestrata*, 305
– *gigantea*, 515
– *lactuca*, 7, 305, 349, 455, 515, 603
– *olivascens*, 515
– *pertusa*, 1
– *rigida*, 191, 241, 349, 515, 583, 603
– *rotundata*, 515
– *scandinavica*, 515
Ulvaceae, 65
Ulvales, 39, 65, 305, 515
Undaria, 47, 191, 435
– *peterseniana*, 47
– *pinnatifida*, 47, 121, 191, 411, 469, 473, 487
– *pinnatifida* f. *distans*, 191
– *pinnatifida* f. *narutensis*, 191
– *pinnatifida* f. *typica*, 191
– *undarioides*, 47, 191

Valonia, 65
– *aegagropila*, 65
– *ventricosa*, 65
Valoniaceae, 65
Valoniopsis, 65
– *pachynema*, 65
Veleroa, 65
– *elongata*, 65
– *karulvalensis*, 65
– *subulata*, 65
Ventricaria, 65
– *ventricosa*, 65
Verosphacela, 247
Vibrio, 643

(*Womersleyella*), 91
(– *pacifica*), 91
(– *setacea*), 91
Wrangelia argus, 65

Zostera, 583

J.M. Kain (Jones), M.T. Brown & M. Lahaye (eds), Sixteenth International Seaweed Symposium.

Author index

From page 515 onwards the number is at the bottom of the page

Aedo, D., 241
Aguilar, A., 137
Alcantara, L. B., 443
Anderson, R. J., 183, 455
Anzai, M., 411
Araki, Y., 411
Arellano-Carbajal, F., 509
Aruga, Y., 355
Ask, E., 435
Avila, M., 149, 435
Azanza, R. V., 463

Bacic, A., 401
Belyea, E., 417
Benharbet, O., 391
Bolton, J. J., 183
Boo, S. M., 47
Bremner, G., 487
Brown, M. T., 191, 571
Buschmann, A. H., 427, 603

Calderón, M., 39
Calumpong, H. P., 173, 211, 443
Candia, A., 115, 149
Castilla, J. C., 571
Chee-Barragán, A., 509
Chiadmi, N., 167
Chiovitti, A., 401
Chopin, T., 417
Cimadevilla, E. A.-C., 515
Cinelli, F., 91
Coelho, S., 329
Cormaci, M., 75
Correa, J. A., 427, 571
Correa-Díaz, F., 509
Craik, D. J., 401
Critchley, A. T., 201

Dion, P., 515
Duarte, M. E. R., 549
Duggins, D., 247

Edwards, M. S., 253
El Gourji, A., 167

Falshaw, R., 57
Felga, A., 329
Fredericq, S., 25
Freile-Pelegrín, Y., 501
Freshwater, D. W., 25
Fujita, D., 261
Fukami, K., 643
Fukuoka, S., 563
Furnari, G., 75

Ganzon-Fortes, E. T., 321
Gao, K., 355
Garbary, D. J., 247
García R. G., 529
Garófalo, G., 397
Gillespie, R. D., 201
Ginsburg, D. W., 267
Givernaud, T., 167, 391
Graham, M. H., 231
Granbom, M., 349
Griffin, N. J., 183
Guiry, M. D., 91
Guzmán-Urióstegui, A., 285

Hafting, J. T., 305
Halling, C., 603
Hassani, L. A., 391
Hillis, L. W., 39
Hommersand, M. H., 25
Honda, M., 361
Honya, M., 411
Horn, S. J., 523
Hurd, C. L., 191, 487

Isakov, V. V., 555
Israel, A., 443

Jensen, A., 535
Ji, Y., 355
Jiang, P., 469
Jiménez del Río, M., 529
Jones, D., 417

Kakita, H., 339, 563
Kamishima, H., 563
Kapraun, D. F., 7
Kautsky, N., 603
Kendrick, G. A., 275
Kim, G. H., 81
Kim, K. Y., 247
Kim, N.-G., 127
Kim, S.-Ho, 81
Kim, Y. H., 555
Kitamura, T., 339
Klinger, T., 247
Knight, G. A., 57
Kooistra, W. H. C. F., 39
Kraft, G. T., 401
Kuhlenkamp, R., 515

Lagos, N., 571
Lahaye, M., 515
Largo, D. B., 643
Larsen, B., 535
Lavery, P. S., 275

Lemoine, Y., 167, 391
Leonardi, P. I., 375
Levitt, G. J., 455
Li, D., 473
Liao, M.-L., 401
Lin, R., 291
Lindstrom, S. C., 115, 291
Liu, H., 473
Lluisma, A. O., 613
Lognoné, V., 515
Luxton, D. M., 477
Luxton, P. M., 477

Macchiavello, J., 397
Magbanua, M., 173, 211
Martel, A., 529
Martinez-Goss, M. R., 443
Matlock, D. B., 267
Matsuhiro, B., 217
Maypa, A. P., 173, 211
McBride, D. L., 101
Mendoza, H., 529
Meñez, E. G., 443
Millar, A. J. K., 65
Miller, I. J., 385
Mitchell, B. G., 231
Miyashita, A., 121
Moen, E., 523, 535
Moenne, A., 571
Montaño, N. E., 541
Morales, J., 137
Mori, H., 411
Mouradi-Givernaud, A., 167, 391
Murano, E., 329

Nagaura, K., 299
Navarro-Angulo, L., 315
Nelson, W. A., 57
Nishijima, T., 643
Nisizawa, K., 411
Norambuena, R., 435
Noseda, M. D., 549
Notoya, M., 121, 299, 619
Nuñez, M., 149, 435

Obika, H., 339, 563
Ohno, M., 625
Oliveira, E. C., 397
Østgaard, K., 523, 535

Pacheco-Ruíz, I., 159, 509
Paul, V. J., 267
Paula, E. J. de, 625
Pedersén, M., 349
Pereira, R. T. L., 625
Pereira, R., 329
Phillips, J. C., 275

Qin, S., 469
Quemener, B., 515
Quintanilla, J., 137

Rabanal, S. F., 463
Ragan, M. A., 613
Ramírez, M., 571
Reyes, E., 115
Richmond, A., 637
Rindi, F., 91
Robledo, D., 285, 315, 501
Román, D., 571
Romero, J. B., 541
Romo, H., 149
Rönnbäck, P., 603
Rotmann, K. W. G., 495
Rudolph, B., 435

Saito, R., 397
Santelices, B., 15, 241
Schiewer, S., 593
Schramm, W., 583
Semple, R., 417
Serviere-Zaragoza, E., 501
Sharp, G., 417
Sigel, J., 137
Smit, A. J., 455
Solov'eva, T. F., 555
Sousa-Pinto, I., 329
Stekoll, M. S., 291
Stuart, M. D., 191
Suarez, P., 173
Sun, G.-Q., 469

Titlynov, E. A., 555
Troell, M., 603
Trono, G. C. Jr., 1
Tseng, C.-K., 469
Tulio, S., 549

Urzúa, C., 217

Varas, M., 571
Vasquez, J. A., 217, 375
Vega, J. M. A., 217
Vergara, S., 571
Villanueva, R. D., 541

West, J. A., 101
Westermeier, R., 137, 427
Wu, C., 473
Wu, Y., 469

Yermak, I. M., 555
Yokoya, N. S., 339
Yoon, H. Su, 47

Zemke-White, W. L., 487
Zertuche-González, J. A., 159, 509
Zhou, Z.-G., 473
Zou, L.-H., 469
Zou, N., 637